Good Microbes in Medicine, Food Production, Biotechnology, Bioremediation, and Agriculture

Good Microbes in Medicine, Food Production, Biotechnology, Bioremediation, and Agriculture

Edited by

Frans J. de Bruijn
INRAE/CNRS, LIPME, Castanet Tolosan Cedex, France

Hauke Smidt
Wageningen University and Research, Wageningen, Netherlands

Luca S. Cocolin
University of Torino, Torino, Italy

Michael Sauer
University of Natural Resources and Life Sciences, Vienna, Austria

David Dowling
Institute of Technology Carlow, Carlow, Ireland

Linda Thomashow
USDA-ARS, Washington State University, Pullman, Washington, USA

WILEY

This edition first published 2023

Registered Offices
John Wiley & Sons, Inc., 111 River Street, Hoboken, NJ 07030, USA
John Wiley & Sons Ltd, The Atrium, Southern Gate, Chichester, West Sussex, PO19 8SQ, UK

Editorial Office
The Atrium, Southern Gate, Chichester, West Sussex, PO19 8SQ, UK

For details of our global editorial offices, customer services, and more information about Wiley products visit us at www.wiley.com.

Wiley also publishes its books in a variety of electronic formats and by print-on-demand. Some content that appears in standard print versions of this book may not be available in other formats.

Library of Congress Cataloging-in-Publication Data
Names: de Bruijn, Frans J., editor. | Smidt, Hauke, editor. | Cocolin, Luca S., editor. | Sauer, Michael, editor. | Dowling, David, editor. | Thomashow, Linda, editor.
Title: Good microbes in medicine, food production, biotechnology, bioremediation and agriculture | edited by Frans J. de Bruijn, Hauke Smidt, Luca S. Cocolin, Michael Sauer, David Dowling, Linda Thomashow.
Description: Hoboken, NJ : John Wiley & Sons, 2023. | Includes bibliographical references and index.
Identifiers: LCCN 2021042854 (print) | LCCN 2021042855 (ebook) | ISBN 9781119762546 (hardback) | ISBN 9781119762379 (pdf) | ISBN 9781119762461 (epub) | ISBN 9781119762621 (obook)
Subjects: LCSH: Microbiology. | Microorganisms.
Classification: LCC QR41.2 .G66 2023 (print) | LCC QR41.2 (ebook) | DDC 579--dc23/eng/20211029
LC record available at https://lccn.loc.gov/2021042854
LC ebook record available at https://lccn.loc.gov/2021042855

Cover images: © Morsa Images/Getty Images; © David Papazian/Getty Images; © vitranc/Getty Images; © STEVE GSCHMEISSNER/Getty Images; © sturti/Getty Images; © Raycat/Getty Images; © Dmitry Kalinovsky/Shutterstock
Cover design by Wiley

Set in 10.5/13pt STIXTwoText by Integra Software Services Pvt. Ltd, Pondicherry, India

This Book is dedicated to my two daughters, Waverly Klaw and Vanessa Mina for their love, support and interest even from a distance, and to my grandchildren Isabella, Ethan, Cassidy and Elliot, and to Sandrine Faure for her keen interest in the Book and its relevance to sustainability.

Contents

Preface

The idea for a book on "Good Microbes" came out of a discussion among scientists of different backgrounds touching on microbiology when the question was raised "why do we always talk and write about microbes in a negative fashion, as pathogens especially in medicine, but also in the preparation of food and in agriculture?". Human bacteria are often associated with antibiotics, food bacteria with spoilage and sterilization, and agricultural bacteria with pesticides. So the challenge became to think of "good microbes" in different environments and their unique benefits to humanity. Out of these discussions an impressive list of beneficial microbes and their mode of action resulted and a book on "Good Microbes in Medicine, Food Production, Biotechnology, Bioremediation, and Agriculture" materialized. A book that does not aim to exhaustively cover all "Good Microbes" in the latter five disciplines but to introduce several examples of interest in each of the five areas. The book is written for non-specialist scientists as well as an informed public with some basic knowledge of microbiology. The idea is to have five co-editors, who specialize in Medical Microbiology, Food Microbiology, Biotechnical Microbiology, Bio(phyto)remediation, or Agricultural Microbiology. The five specialists have chosen up to ten topics of interest and have selected corresponding authors to contribute the chapters, resulting in a book of approximately 500 pages surveying the five fields. The resulting chapters were edited by the section editor and subsequently by the editor-in-chief, especially for the scientific content and format.

The book aims to highlight the positive aspects of microbes, therefore, the topics chosen clearly emphasize this assumption. The chosen topics are such that non-scientists can also relate to it. For example, in the medical section the reader is led to consider the use of probiotics, in the food section the reader is able to relate to the microbe-initiated fermentation of beer, dough, wine, and cheese, and in the biotechnology section to biofuels, biohydrogen, or bioplastics production or degradation. In the bioremediation section the reader will become familiar with various methods using microbes, sometimes in conjunction with plants in environmental clean-up and in the agriculture section with various approaches without pesticides to use good bacteria and fungi to enhance plant growth and defense against pathogens and under stressful environmental conditions. This is clearly reflected in the introduction of each chapter. It will be clear that often consortia of good microbes (microbiota) are involved and their analysis is complex but possible because of high power "omics" technologies, which can even be used to identify and characterize non-culturable microbes.

The science presented is of high quality and has been reviewed/edited as such by the co-editors and editor-in-chief. The book is written for non-specialist scientists, as well as well-educated citizens, physicians, teachers, and regulators.

Frans J. de Bruijn (Editor in Chief)

List of Contributors

Ahmed Abdelfattah
Leibniz Institute for Agricultural Engineering and Bioeconomy (ATB)
Potsdam, Germany

Altaf AlBaho
Queen's University Belfast
Belfast, Ireland

Hana Ameur
Free University of Bolzano
Bolzano, Italy

Tomás Aparicio
Universidad Autónoma de Madrid
Madrid, Spain

Sara Arbulu
University College Cork
Cork, Ireland

and

Teagasc Food Research Centre
Cork, Ireland

Anthoula A. Argyri
Hellenic Agricultural Organization DIMITRA
Athens, Greece

Kashika Arora
Free University of Bolzano
Bolzano, Italy

Özge Ata
University of Natural Resources and Life Sciences
1190 Vienna, Austria

Carmela Belloch
Instituto de Agroquímica y Tecnología de Alimentos
Valencia, Spain

Gemma Beltran
Universitat Rovira i Virgili
Tarragona, Spain

Gabriele Berg
Graz University of Technology
Graz, Austria

Lars M. Blank
RWTH Aachen University
Aachen, Germany

Varsha Bohra
Hong Kong Baptist University
Kowloon Tong, Hong Kong

Larissa Brumano
University of Sao Paulo
Federal University of ABC
Sao Paulo, Brazil

Chris Callewaert
Ghent University
Gent, Belgium

Tomislav Cernava
Graz University of Technology
Graz, Austria

and

Austrian Center for Industrial Biotechnology
Graz, Austria

John Cleary
Institute of Technology
Carlow, Ireland

Luca S.Cocolin
University of Turin
Torino, Italy

Robert Conlon
Institute of Technology Carlow
Carlow, Ireland

Paul D. Cotter
University College Cork
Cork, Ireland

and

Teagasc Food Research Centre
Cork, Ireland

Rune Daneels
Ghent University
Gent, Belgium

Frans J. de Bruijn
INRA/CNRS Laboratoire des Interactions Plant-Microorganismes et Environment
Castanet-Tolosan, France

Jorgen De Jonge
National Institute for Public Health and the Environment (RIVM)
The Netherlands

Victor de Lorenzo
Universidad Autónoma de Madrid
Madrid, Spain

Britta De Pessemier
Ghent University
Gent, Belgium

Jonas De Roos
Vrije Universiteit Brussel
Brussels, Belgium

Luc De Vuyst
Vrije Universiteit Brussel
Brussels, Belgium

Mirella Di Lorenzo
University of Bath
Bath, UK

Rory Doherty
Queen's University Belfast
Belfast, Ireland

Sharon L. Doty
University of Washington
Seattle, USA

David Dowling
Institute of Technology Carlow
Carlow, Ireland

Jakub Dziegielowski
University of Bath
Bath, UK

Michael Egermeier
BOKU-VIBT University of Natural Resources and Life Sciences
Vienna, Austria

Valeria Ellena
Austrian Centre of Industrial Biotechnology (ACIB GmbH)
Vienna, Austria

and

Institute of Chemical, Environmental and Bioscience Engineering
Vienna, Austria

Brian Federici
University of California
Riverside, USA

Leónides Fernández
Universidad Complutense de Madrid
Spain

Ilario Ferrocino
University of Turin
Torino, Italy

Valentina Fiorilli
University of Turin
Torino, Italy

Mónica Flores
Instituto de Agroquímica y Tecnología de Alimentos
Valencia, Spain

Irene Franciosa
University of Turin
Torino, Italy

Susana Fuentes
National Institute for Public Health and the Environment (RIVM)
The Netherlands

Amparo Gamero
Universitat de València
Valencia, Spain

Paolina Garbeva
Netherlands Institute of Ecology (NIOO-KNAW)
Wageningen, The Netherlands

Enriqueta Garcia-Gutierrez
University College Cork
Cork, Ireland

and

Teagasc Food Research Centre
Cork, Ireland

Daniel Garrido-Sanz
Universidad Autónoma de Madrid
Madrid, Spain

Thomas Gassler
University of Natural Resources and
Life Sciences
1190 Vienna, Austria

Kieran J. Germaine
Institute of Technology
Carlow, Ireland

Xuemei Liu Germaine
Institute of Technology
Carlow, Ireland

Marco Gobbetti
Free University of Bolzano
Bolzano, Italy

Beatriz Gómez-Sala
University College Cork
Cork, Ireland

and

Teagasc Food Research Centre
Cork, Ireland

Helena Ipe Pinheiro Guimaraes
National Institute for Public Health and the
Environment (RIVM)
The Netherlands

Z. H. Hassan
Indonesian Center for Agricultural Postharvest
Research and Development
Indonesian Agency for Agricultural
Research and Development (IAARD)
West Java, Indonesia

and

Wageningen University & Research
Wageningen, The Netherlands

Britta E. Heiss
University of California Davis
Davis, USA

Henric M.T. Hintzen
RWTH Aachen University
Aachen, Germany

F. Hugenholtz
Wageningen University & Research
The Netherlands

and

The Dutch Research Council (NWO)
The Hague, The Netherlands

Mariangela Hungria
EMBRAPA Soja
Paraná, Brazil

Nick M. Jensen
University of California Davis
Davis, USA

Maria Kazou
Agricultural University of Athens
Athens, Greece

Bongkyu Kim
University of Bath
Bath, UK

Eduardo Kleingesinds
University of Sao Paulo
Sao Paulo, Brazil

Petia Kovatcheva-Datchary
University of Würzburg
Würzburg, Germany

Rajat Kumar
Hong Kong Baptist University
Kowloon Tong, Hong Kong

Peter Kusstatscher
Graz University of Technology
Graz, Austria

Luisa Lanfranco
University of Turin
Torino, Italy

Ana Shein Lee Diaz
Netherlands Institute of Ecology (NIOO-KNAW)
Wageningen, The Netherlands

Melissa LeTourneau
Wheat Health, Genetics and Quality Research Unit
Pullman, USA

Agnes S. Y. Leung
The Chinese University of Hong Kong
Hong Kong

Wenyin Loh
Department of Paediatrics
Singapore

Matteo Lorito
University of Naples Federico II
Naples, Italy

and

Institute for Sustainable Plant Protection
National Research Council (CNR-ISPP)
Naples, Italy

Ben Lugtenberg
Leiden University
Leiden, The Netherlands

Tomas Macek
University of Chemistry and Technology, Prague
Prague, Czechia

Piyush Malaviya
University of Jammu
Jammu and Kashmir, India

Rajesh Mali
Institute of Technology Carlow
Carlow, Ireland

Manu Mk
Hong Kong Baptist University
Kowloon Tong, Hong Kong

Marta Martin
Universidad Autónoma de Madrid
Madrid, Spain

Esteban Martínez-García
Universidad Autónoma de Madrid
Madrid, Spain

Albert Mas
Universitat Rovira i Virgili
Tarragona, Spain

Diethard Mattanovich
University of Natural Resources
and Life Sciences
Vienna, Austria

David A. Mills
University of California Davis
Davis, USA

Jatziri Mota-Gutierrez
University of Turin
Torino, Italy

Marta Mozota
Universidad Complutense de Madrid
Spain

Loriane Murphy
Institute of Technology Carlow
Carlow, Ireland

Deepak Pant
Flemish Institute for Technological Research VITO
Mol, Belgium

Jakub Papik
University of Chemistry and Technology, Prague
Prague, Czechia

Letícia Parizotto
University of Sao Paulo
Sao Paulo, Brazil

Adalberto Pessoa Junior
University of Sao Paulo
Sao Paulo, Brazil

Andrea Polo
Free University of Bolzano
Bolzano, Italy

Kalliopi Rantsiou
University of Turin
Torino, Italy

Miguel Redondo-Nieto
Universidad Autónoma de Madrid
Madrid, Spain

Rafael Rivilla
Universidad Autónoma de Madrid
Madrid, Spain

Juan Miguel Rodríguez
Universidad Complutense de Madrid
Madrid, Spain

Lily Roney
Queen's University Belfast
Belfast, Ireland

Hannes Russmayer
BOKU-VIBT University of Natural Resources and Life Sciences
Vienna, Austria

Paula Sansegundo-Lobato
Universidad Autónoma de Madrid
Madrid, Spain

Michael Sauer
University of Natural Resources and Microbial Biotechnology
Vienna, Austria

Rozi Sharma
University of Jammu
Jammu and Kashmir, India

Smiley Sharma
University of Jammu
Jammu and Kashmir, India

Hauke Smidt
Wageningen University & Research
Wageningen, The Netherlands

Matthias Steiger
Austrian Centre of Industrial Biotechnology ACIB
Vienna, Austria

Michal Strejcek
University of Chemistry and Technology
Prague, Czechia

Jachym Suman
University of Chemistry and Technology
Prague, Czechia

Jyoti Prakash Tamang
Sikkim University
Sikkim, India

Mimi L. K. Tang
The Royal Children's Hospital
Melbourne, Australia

Chrysoula C. Tassou
Hellenic Agricultural Organization DIMITRA
Athens, Greece

Namrata Thapa
Sikkim University
Sikkim, India

Linda Thomashow
USDA ARS, Washington State University
Washington, USA

Till Tiso
RWTH Aachen University
Aachen, Germany

María Jesús Torija
Universitat Rovira i Virgili
Tarragona, Spain

Robert J. Tournay
University of Washington
Seattle, USA

Effie Tsakalidou
Agricultural University of Athens
Athens, Greece

Ondrej Uhlik
University of Chemistry and Technology, Prague
Prague, Czechia

Debbie Van Baarle
National Institute for Public Health and the Environment (RIVM)
The Netherlands

and

University Medical Center Groningen (UMCG)
The Netherlands

Tom Van De Wiele
Ghent University
Gent, Belgium

Mutian Wang
Institute of Technology
Carlow, Ireland

Birgit Wassermann
Graz University of Technology
Graz, Austria

and

Austrian Center for Industrial Biotechnology
Graz, Austria

David M. Weller
USDA ARS, Washington State University
Washington, USA

Gina Welsing
RWTH Aachen University
Aachen, Germany

Birger Wolter
RWTH Aachen University
Aachen, Germany

Jonathan W. C. Wong
Hong Kong Baptist University
Kowloon Tong, Hong Kong

Sheridan Lois Woo
University of Naples Federico II
Naples, Italy

Mingming Yang
Northwest A&F University
Yangling, P. R. China

E. G. Zoetendal
Wageningen University
Wageningen, The Netherlands

Acknowledgments

Frans J. de Bruijn would like to thank *INRAE/CNRS* and the Labex Tulip for their support of his editorial work. He would also like to thank Claudine Hendriksen for her relentless support during the edition of this volume. Linda Thomashow would like to thank the USDA for their support for her editorial work on the book.

Introduction

The majority of well educated lay people and even researchers in non-related fields, when asked "what is the impact of bacteria or microbes on medicine, food production, and agriculture?", are likely to respond that the microbes are nocive, pathogenic, must be controlled by antibiotics and sterilization and combatted by pesticides. This general negative notion is quite strong, ubiquitous and general throughout the population. Microbiology as a discipline is slated the same way and many microbiology texts are oriented in this direction. Here we would like to introduce an opposite point of view that bacteria and other microbes play important positive roles in medicine, food production, biotechnology, bioremediation and agriculture. In this book "Good Microbes in Medicine, Food, Biotechnology, Bioremediation and Agriculture" we would like to present a number of strong examples of the positive application of microbes in the above mentioned five fields.

SECTION 1: GOOD MICROBES IN MEDICINE, CO-EDITED BY HAUKE SMIDT

In Section 1 Chapter 1, we will focus on the role of good microbes in modern medicine; after that the focus will shift to their roles in a new field of medicine, namely probiotics or the use of microbes to positively influence the human microbiota, especially of the gastrointestinal tract, and thereby control diseases. The human gastrointestinal (GI) tract harbours a complex community of 3.9×10^{13} microbial cells, equalling the number of human cells that is 3.7×10^{13} (Chapter 8). This complex community, known as microbiota, comprises diverse microorganisms, including bacteria, archaea, and eukaryotes, of which bacteria is the predominant domain, that contributes to host metabolism and the immune system (Chapter 8). The composition and metabolic activity of the microbiota are considered to be among the critical factors in maintaining and improving host health. In turn, imbalances regarding microbial composition and metabolic activity have been associated with development of several intestinal and other diseases including, e.g., inflammatory bowel disease, obesity, diabetes, allergic diseases and psychiatric disorders such as cardiovascular disease and psychiatric disorders such as depression. Modulating the GI tract microbiota through dietary intervention, including administration of probiotics, is considered to be a promising strategy to restore and maintain the composition and metabolic activity of the GI tract microbiota (Chapter 8). Probiotics are preparations of microbes that have a beneficial effect on the intestinal flora. Several examples of their use to positively influence the intestinal microbiota of nursing babies, mitigate medical problems associated with vaccine responses, allergies, COVID-19 infections and bad body odor will be introduced below.

Good Microbes in Modern Medicine

Our survival relies on many good microorganisms. They can act in their living form directly on our bodies, and they are also widely used as microbioreactors to produce relevant therapeutic products for the treatment of diseases. Good microorganisms have played an essential role in the development of modern medicine, which started with the fight against infectious diseases that promoted the search for antipathogenic agents. The discovery of penicillin, the first therapeutic product isolated from microbial sources against pathogens, was the precursor of new biological medicines. Antibiotics revolutionized the treatment of infectious diseases. However, due to very resistant superbugs, we have recently been facing a threat to public health. Therefore, new antimicrobial agents are needed. In addition, we deal with new epidemics, a growing number of cancer cases, and other genetic conditions for which a cure seemed improbable. Fortunately, DNA recombinant technology has given new applications to good microorganisms in treating these diseases as well. They can be used as production platforms (hosts), tools for replication, plasmid stock or as a source of enzymes necessary for methodologies employed in genetic modification processes. Thus, new sophisticated biopharmaceuticals have been developed, such as vaccines, anticancer drugs, and gene therapy. Thanks to good microorganisms, vaccines can be produced more safely, economically, and quickly. Moreover, microorganisms contribute to producing anticancer drugs as natural producers of biomolecules, platforms for recombinant expression, or agents in immunotherapies. Recently, they have also been applied as carriers of nucleic acids in gene therapy, mainly to treat hereditary diseases. Therefore, good microorganisms have been supporting modern treatments that led to rapid responses to epidemic outbreaks, new therapies for preventable diseases and a chance of curing incurable diseases. In Chapter 1, some of the main applications of good microorganisms in modern medicine are presented.

Beneficial Infant-associated Bifibacterium

Bifidobacterium species are common residents of the human gastrointestinal tract. While they colonize people of all ages, they are most strikingly found in the infant gut, where they degrade human milk oligosaccharides (HMOs) found in breast milk. Individual *Bifidobacterium* strains encode unique sets of transport proteins and enzymes called glycoside hydrolases (GHs) to metabolize these diverse carbohydrates in milk (see Chapter 2). By fermenting HMOs, bifidobacteria produce bioactive end products such as lactate and acetate. These molecules reduce the pH of the gut, protect against invasion by harmful pathogens, and likely support healthy childhood growth and immune development. Although *Bifidobacterium* colonization promotes infant health, levels of these bacteria vary based on geography, breastfeeding rates and other factors. However, recent trials suggest that providing infants with probiotic *Bifidobacterium*, in combination with HMOs that sustain their nutrient niche, can support the persistent growth of these beneficial bacteria. These observations will be presented in Chapter 2.

Role of the Microbiome in Vaccine Response

Vaccination mediated protection is one of the best ways to control infectious diseases. However, inter-individual variation in vaccination responses, regardless of the type of vaccine, affects the effectiveness of vaccines. Demographics, comorbidities and aging are a few variables that may influence the differences in vaccination response and effectiveness. Novel strategies aimed at increasing immunogenicity of vaccines and thereby protecting all individuals is a major public health interest. The gastrointestinal tract is an ecosystem for various microorganisms and a hotspot for microbiota-derived molecules.

These compounds found in the gut can travel through the bloodstream and are able to enter the systemic circulation from the intestinal lumen. Thus, bacteria or bacteria-derived products may act as natural adjuvants and their interplay with immune cells could ultimately impact immune responses to infectious diseases and vaccination (see Chapter 3). Animal and human population studies indeed indicate that this hypothesis might play an important role in the shaping of a proper systemic immune response to vaccination. The effector mechanisms by which the gut microbiota can impact immune response and how these can be adjusted for a more efficient response to vaccinations will be discussed in Chapter 3.

Probiotics and Food Allergies

The search for effective preventive and treatment strategies for food allergy is becoming ever more important as the prevalence of food allergy increases. Since the hygiene hypothesis was first proposed, there has been growing evidence that dysbiosis plays an important role in the pathogenesis and development of food allergy. Chapter 4 presents the evidence for probiotics in the prevention and treatment of food allergy. Published studies show an inconsistent role for probiotics in prevention of disease, likely due to differences in probiotic strains used, as well as the dose and duration of therapy in the different studies. Application of probiotics for the treatment of food allergy appear more promising, with favourable results in both clinical and preclinical studies. It is noteworthy that probiotic effects appear to be species and strain-specific, with benefits primarily reported for *L. rhamnosus* GG and certain *Clostridia* strains (see Chapter 4).

COVID-19, Microbiota and Probiotics

SARS-CoV-2 is the agent of the current COVID-2 pandemic. This virus interacts with the mucosal surface lining the upper respiratory tract, which is associated with complex microbiota. Recent studies suggest an implication of the microbiota of the respiratory and gastrointestinal tracts in the modulation of the infection by this virus (see Chapter 5). In fact, the susceptibility to the infection and the severity of the disease seems higher when there is a depletion of butyrate-producing strict anaerobes in the gut microbiome. Few clinical trials involving the use of probiotics to prevent or minimize the impact of COVID-19 have been published so far, but they show a significant impact by shortening the duration of diarrheal episodes and decreasing the risk of respiratory failure and death. These observations will be discussed in Chapter 5.

Underarm Microbiome and Probiotics

Bromhidrosis or excessively foul-smelling body odor is a chronic pathologic condition that usually comes from axillary skin regions. This offensive body odor can have a significant impact on the quality of life and can cause professional, social, and emotional distress. There are two main reasons for underarm malodor: the presence of apocrine sweat, which is the precursor of body odor, and the underarm microbiome, which converts the precursors into malodorous volatiles. Secretions from either apocrine and eccrine glands are primarily odorless but become malodorous after bacterial catabolism. Factors that worsen the condition are poor hygiene or underlying complications that promote bacterial overgrowth, including obesity, intertrigo, diabetes, and erythrasma. Conventional therapies usually focus on antimicrobial treatments (e.g., deodorants containing antibacterial agents and antibiotics) or prevention of sweating (e.g., antiperspirants, medication, botox, surgery). However, alternative therapies promote the growth of beneficial and good-smelling bacteria and inhibit the growth of odor-causing bacteria. Promicrobiome therapies could rebalance the composition of the cutaneous microbiota and reduce smelly body odor in a sustainable manner. The latter will be discussed in Chapter 6.

Prevotella Copri: Beneficial or Detrimental

The gut microbiota plays an important role in human metabolism and health by interacting with host diet. An example of such a bacterium is *Prevotella copri*, the most abundant *Prevotella* species in the human gut that has been associated with diet, improved or impaired host metabolic health, and gut inflammation. Thus, there is a growing body of interest in research to further elucidate the drivers of the potentially beneficial or detrimental effect of *P. copri* on human health, which will be discussed in Chapter 7.

Future Perspectives of Probiotics and Prebiotics

The future perspectives of probiotics and prebiotics in foods and food supplements will be covered in Chapter 8. This chapter highlights the current scientific progress on the exploration, identification, and characterization of novel probiotic and prebiotic candidates as well as the latest development of probiotic/prebiotic-containing food products. Furthermore, future perspectives, especially on the regulatory and safety aspects, of prebiotics and probiotics and their application in foods and food supplements will be discussed (see Chapter 8).

SECTION 2: GOOD MICROBES IN FOOD PRODUCTION, CO-EDITED BY LUCA S. COCOLIN

Section 2 will address the use of good microbes in food spoilage control (bioprotection) and the production of a myriad of fermented (ethnic) foods and drinks using good microbes for their production.

Bioprotection for Food Safety

Bioprotection is a natural method for food preservation that uses microorganisms, mostly bacteria, or their by-products as food preservatives to enhance food safety and extend food shelf life. When microbes compete with each other, they produce antimicrobial molecules, such as bacteriocins, whose potential can be harnessed to inhibit the growth of spoilage and pathogen microorganisms in food. Bacteriocins and bacteriocin-producing microbes, especially lactic acid bacteria, stand out as an alternative or complement to traditional preservation methods. This role of good microbes is discussed in Chapter 9.

Aromatic Yeasts in Food Fermentations

Many different yeast species are involved in food fermentations. These microorganisms not only participate in the development of the fermentation process but also synthesize immense range of volatile compounds, contributing to the global aroma (see Chapter 10). Higher alcohols, esters, pyrazines, ketones, short chain fatty acids, nitrogen and sulfur compounds are, among others, different families of volatiles that can contribute to an immense variety of aromatic nuances in fermented products, such as fruity, sweet, toasted, green, acid, savory, spicy, sulfury, etc. In Chapter 10, the contribution of different yeast species to overall aroma of alcoholic beverages, fermented sausages, cheese, vegetables, and traditional fermented products is described.

Microbiota in Ethnic Fermented Foods

Ethnic fermented foods and beverages are community-specific as well as region-specific, traditionally based on the food preferences of the consumers. Diverse types of microorganisms appear from the local environment and raw/cooked bio-resources during natural fermentation to convert into edible products that are organoleptically and socially acceptable to the ethnic consumers. Some of these microorganisms are claimed to be beneficial and contribute several functional properties imparting health benefits in fermented foods and beverages. These are described in Chapter 11.

No Microbes, no Cheese!

The advent of dairying was a major innovation in prehistoric societies that enabled the supply of nutritious food without slaughtering precious livestock. In particular, the production of cheese was a momentous development; it allowed the preservation of milk in a transportable and non-perishable form, with cheese being also a more digestible food than milk for early prehistoric humans.

Iconographic and written evidence from the mid-third millennium BC faintly documents the history of cheese making, although its origins probably lie much earlier in prehistory. The finding of ample milk residues in pottery pots from seventh millennium BC sites in North Western Anatolia provided the earliest evidence of milk processing, although the technology could not be defined (see Chapter 12).

Cheese is a universal fermented dairy product, which appears in a great diversity of flavors, aromas, textures, and shapes around the globe. Most cheese varieties evolved unintentionally due to particular local conditions, such as species and breeds of dairy animals available, vegetation, and climate, factors that in turn shaped the milk chemical composition and microbiota. Additionally, human skills along with social, cultural and religious backgrounds played a significant role. One thing is clear: "no microbes, no cheese" (see Chapter 12).

Microbial Diversity and Function in Fermented Meat

Regarding meat-based products, it is well known that, since the ancient Egyptian age, fermentation has been a successful strategy for the preservation of meat. Countries in the Mediterranean area (Italy, Spain, Portugal, Turkey, and France) are the main producers of fermented meat (also referred to as fermented sausages or salami), followed by Germany and Hungary, and in recent decades by China, which is exponentially increasing the manufacturing of fermented sausages by the production of several typical products from different Chinese areas (see Chapter 13).

Due to the increase in demand for fermented sausages, thousands of companies spread around the world are interested in increasing the production of artisanal sausages (mostly without the use of starter culture) in order to confer on this product an additional value for consumers and to have the guarantee to produce an original product (see Chapter 13). The fermentation process is a complex succession of biochemical transformations that start from animal enzymes, but the main role is played by the microbiome (see Chapter 13).

The Sourdough Microbiota

The era of sourdough fermentation came into existence in ancient times, with continuous developments owed to the scientific community. The role of starter lactic acid bacteria and yeasts, constituting the sourdough microbiota, is undeniable in imparting characteristic nutritional and functional features to sourdough. This complex community is influenced by sourdough ingredients (flour, water and other additional ingredients) and the house microbiota leading to an optimized protocol for its preparation and propagation (see Chapter 14).

Nowadays, leavened bread is an essential component of the diet of people all around the world, and it is declared an intangible heritage of humanity by UNESCO. Its production will be described in Chapter 14.

Beneficial Role of Microorganisms in Olives

Table olives, like other plant origin products, are by nature functional foods due to their high content in nutraceuticals that promote several beneficial physiological effects and can also selectively stimulate the growth of the colon microflora and act as prebiotics. They are also a plentiful source of live microorganisms, which may contribute to human health in a manner similar to probiotics (see Chapter 15).

Thus, considerable research has been performed in the last decade concerning the isolation of multifunctional microorganisms from table olives to serve as fortifying agents in several foods or as a multifunctional starter or adjunct cultures in the fermentation of table olives (see Chapter 15).

Behind the Chocolate Scene

Lactic acid bacteria (LAB) are a heterogeneous group that inhabits several nutrient-rich niches in humans, animals, plants, and food. From the food perspective, *Lactiplantibacillus plantarum*, *Limosilactobacillus fermentum* and *Liquorilactobacillus cacaonum* have been identified and isolated during cocoa bean-pulp fermentation to exploit the full potential of these microbes in promoting green technologies that can also produce health benefits (see Chapter 16).

The growing interest in the field of beneficial microbes over the last decades has revolutionized the way we process our food. In contrast to the technological and scientific progress achieved, there is a historical consumer misinterpretation about microbes, which are associated with disease or disgust. In Chapter 16, a robust applied microbiological framework will be presented. This framework will highlight that microbes can be helpful to humans and that they can be found in the production of confectionery snack foods such as chocolate.

Microbial Control of Wines

Wine can be defined as an alcoholic beverage produced after the fermentation of grape juice. Behind this simple definition, there is an important world of microorganisms with a broad array of interactions that exhibit a clear succession during the wine-making process (see Chapter 17).

This chapter focusses on the primary transformation process, that is, alcoholic fermentation and the primary microbes involved in this process: yeasts. This review analyzes the methods used to monitor the yeast population during fermentation, differentiating consolidated knowledge gained through classical methods and the emerging new methodologies derived from advances in molecular biology (see Chapter 17).

Lambic Beer Microorganisms

With its complex consortium of spontaneously inoculated bacteria and yeasts, lambic beer takes up a unique spot in the global food and beer market (see Chapter 18). Regarded as one of the oldest commercial brewing styles, lambic beers possess a unique taste and aroma profile, due to the combined metabolic activities of specific yeasts, acetic acid bacteria, and lactic acid bacteria. In addition, these different microorganisms consume various substrates during the long-lasting fermentation and maturation process, which results in beneficial compositional changes and the production of possible health-promoting compounds (see Chapter 18).

SECTION 3: GOOD MICROBES IN BIOTECHNOLOGY, CO-EDITED BY MICHAEL SAUER

Bioeconomy and Microbiology

While microbes are tiny organisms, their impact on our planet is huge. The first chapter in Section 3 (Chapter 19) will shed some light on the fact that microorganisms are not only important in nature, but they have also many beneficial roles in economic terms. In the first part of this chapter, the author attempts to quantify their impact by looking at food production and the pharmaceutical industry – the two most prominent fields where microorganisms play a role. In the second part of the chapter, he will attempt to look into the future and outline how the importance of microorganisms will be increasing on our society's way to sustainability (see Chapter 19).

Role of Microorganisms in Environmental Remediation and Resource Recovery

In Chapter 20, the authors deal with the role of microorganisms in pollution management, CO_2 sequestration, production of single-cell proteins, biofertilizers, biogas, and biofuels. Also, various microbe-based technologies, including omics-based approaches, nanobioremediation, electrobioremediation, microbial electrosynthesis, microbial fuel cells, microbial electrolysis, consolidated bioprocessing, bioaugmentation, and biogranulation, are discussed to highlight the role of good microorganisms as efficient tactics for biotechnical remediation of harmful polluting agents (see Chapter 20; see also Section 4).

Microbial Carbon Dioxide Fixation

Around 2.4 to 2.0 billion years ago, the levels of atmospheric oxygen steadily increased and peaked in the so-called Great Oxidation Event, an environmental change that enabled great biological innovations and impacted all life on Earth as we know it today. A more recent environmental change is the ongoing increase of atmospheric CO_2 levels, which is primarily due to excess CO_2 emissions by human activities and which is the main driver of climate change (see Chapter 21).

We simply emit more CO_2 than can be captured by biological systems, which is an urgent call for change and for innovation. When we think of biological CO_2 fixation, we tend to focus on land plants and overlook microbial CO_2 fixation. Marine microbes bind almost as much CO_2 globally as land plants, so the sole focus in land plants is clearly not justified. Microbial CO_2 fixation is diverse and spread over all domains of life. The CO_2 fixation features of some prokaryotic and eukaryotic organisms have been exploited by humankind for a long time (see Chapter 21).

During the last decades the use of naturally occurring CO_2 fixing organisms in industrial applications has gained momentum. Further, more recent advances in genetic engineering have enabled the transfer of these CO_2 fixation pathways to industrial microorganisms. Beyond the natural fixation pathways, new-to-nature CO_2 fixation pathways have also been proposed, offering great opportunities by synthetic biology. Combined use of all three strategies will allow us to use CO_2 as the preferred carbon source of the future, which may in turn mitigate the greenhouse effect and help to limit climate change. If not stopped, the climate crisis will destroy our environments – but can CO_2 fixing microbes save us from this catastrophe? This question will be raised in Chapter 21.

The Biodiesel Biorefinery

Fossil oil based products and energy (such as fuel) are an integral part of our daily life. The increasing demand for fossil oil due to the growing world population and expanding economics will lead to a shortage of this feedstock in the future. In addition to this fact, burning of fossil fuels generates emissions of greenhouse gasses (CO_2), making it the primary contributor to global warming and climate change, which negatively impacts our environment (see Chapter 22).

In order to allow a life worth living for future generations, our current lifestyle must be changed to a more sustainable way of living based on renewable biomass. Several green alternatives to common fossil-based products already exist. One of the most prominent is biodiesel. Biodiesel constitutes the renewable alternative to common diesel and is produced by transesterification of plant derived fatty acids. Unfortunately, biodiesel production is so far not economically feasible. The application of an integrated biorefinery concept could solve this problem. Microorganisms have several important roles in almost all biorefinery concepts. They are used as cell factories for microbial chemical production from various feedstocks and waste streams. Furthermore, they can be used as the feedstock itself (biodiesel feedstock) or find application in the pretreatment of raw materials for production of chemicals from renewable biomass (see Chapter 22).

Applications of Good Fungi in Biotechnology

Fungi are an important group of microorganisms which are widely diverse and versatile. Decomposing organic matter and recycling nutrients, they significantly contribute to maintaining healthy and balanced natural ecosystems. Although exploited by humans since pre-recorded history, since the last century knowledge about fungi has increased substantially, marking the beginning of fungal biotechnology (see Chapter 23).

Nowadays fungi and their products are employed in a vast number of industrial applications. Replacing chemicals derived by fossil-fuels and chemical synthesis, fungal products provide environmentally-friendly alternatives for a sustainable future on earth. The scope of Chapter 23 is to give an overview on the beneficial contribution given by fungal biotechnology to humankind.

Good Microbes and Plastics

Plastic products have become a substantial part of our everyday lives, and production is expected to increase further in the future. Properties responsible for the rise in use of plastic products are, among others, durability, flexibility, and low production costs. However, some of these raise major environmental challenges in terms of end-of-life management of these products. Further, low recycling rates and increasing accumulation of plastics in the environment are major problems that require sustainable solutions (see Chapter 24).

For this reason, not only is the development of biodegradable plastic polymers becoming increasingly important in our society and economy, but also the establishment of alternative end-of-life technologies is urgently needed. In Chapter 24, the interplay of microbes and plastics as a potential solution is outlined. Microbial biotechnology offers various approaches to solve the sketched challenges.

Food Waste Bioconversion Using Microorganisms

Food waste is a global issue which affects all sectors of human society. As huge amounts of waste are generated on a daily basis, both living environments and natural habitats are under constant burden. Further, conventional treatment of this accumulating waste requires expenditure of important resources such as water, fertile land, labor and energy. This places a huge stress on the economy and climate. Alternatively, the nutrients trapped in food waste, i.e., carbohydrates, proteins, and lipids, can also be used as a commodity (see Chapter 25). To utilize these resources, good microorganisms can be used as an agent for bioconversion. The natural diversity in microbes is useful as different bacteria, yeast, and fungi work alone or in combination with other microbes to produce various kinds of environment-friendly byproducts (see Chapter 25). These important microbiomes of bacteria are dominated by members of *Bacteroidetes*, *Firmicutes*, *Proteobacteria*, and *Actinobacteria*, whereas fungal groups are enriched by *Trichocomaceae*, *Saccharomycetaceae*, and *Nectriaceae*. Derived value-added products include biofertilizer, bioenergy, biochemicals, biosurfactants, bioplastics, and biocatalysts. The chapter focuses on the inherent potential of microbial bioconversion, their interaction with food waste, different techniques employed, and various products obtained with different applications. Advances in these areas are described in Chapter 25.

SECTION 4: GOOD MICROBES IN BIO(PHYTO)REMEDIATION, CO-EDITED BY DAVID DOWLING

Bio(phyto)remediation is the use of beneficial bacteria, fungi and plants alone or in combinations to degrade or sequester pollutants in the environment.

Environmental Clean-up beyond the Tipping Point

After decades of realization of the environmental impact of chemical emissions stemming from industrial and urban activities, a large number of balances that have thus far sustained the functioning of the biosphere are close to reaching an irreversible tipping point (see Chapter 26). Besides badly needed socioeconomic changes, this state of affairs calls for developing technologies that mitigate and ultimately revert the effect of human action at a very large scale. Because of its mass, ubiquity, and catalytic activity, the global environmental microbiome appears as the key actor to this end. The onset of systems and synthetic biology enables not only the design of agents capable of remediating site-specific pollution problems, but also programming propagation of beneficial traits through complex microbial communities. Such qualities involve not just biodegradation capacities, but also endurance to environmental conditions, physical properties and niche-creating functions. This far-reaching application scenario imposes a shift of the earlier emphasis on pathway design and strain construction towards strategies for broad-host range, even inter-kingdom spread of purposely engineered DNA. This effort has to go hand in hand with sound models for dispersion of both strains and DNA deliberately introduced in the natural milieu based on the best contemporary ecological theory. For a detailed coverage of these concepts see Chapter 26.

Good Microbes for Enhancing Soil Bioremediation

Bioremediation can be used for many environmental pollutants, such as Polycyclic Aromatic Hydrocarbons (PAHs), Polychlorinated Biphenyls (PCBs), Petroleum Hydrocarbons, and Pesticides. There are many different forms of bioremediation, including biostimulation, bioaugmentation, biopiling, landfarming, and phytoremediation. The success of these clean-up technologies is largely dependent on a healthy population of highly active microbes (seen Chapter 27). Ecopiling is a modification of traditional passive biopiling in that, instead of enclosing the biopile in black plastic, the pile is planted with suitable phytoremediation plants. The ecopile process involves bio-stimulation of indigenous hydrocarbon-degrading microbes, bioaugmentation through the addition of a consortia of hydrocarbon-degrading bacteria and phytoremediation, through the effect of root growth and penetration throughout the soil and the resulting stimulation of microbial activity in the rhizosphere (rhizoremediation). In Chapter 27, some of the current bioremediation technologies are briefly covered and the process of ecopiling is looked at more fully. Some of the examples where ecopiling has been used to remediate large volumes of contaminated soil are presented and the good microbes whose metabolic processes are essential for the success of the process are discussed.

Plant-microbe Interactions in Biorestoration

Environmental pollution is widespread and poses serious human health risks. The levels of pollution range from extremely toxic albeit in small areas to low levels in expansive areas, with the full range causing harm. Conventional remediation strategies can be effective; however, they are often based on the movement of the contaminant from one area to another, and are prohibitively expensive. Phytoremediation, the use of trees to clean up pollution, is a low-cost and effective alternative for many sites (see Chapter 28). But the technology can be limited due to phytotoxic effects of the pollutant, presence of multiple classes of pollutants, and slow rates of removal. Microbial endophyte-assisted phytoremediation is a relatively new technology that harnesses the power of the plant microbiome, combining the benefits of phytoremediation with those of bioremediation. Increased awareness of this plant–microbe team approach could lead to treatment of many polluted sites currently abandoned due to the high cost of conventional remediation (see Chapter 28).

Endophytes for Pollutant Clean-up

Environmental pollution is widespread and poses serious human health risks. The levels of pollution range from extremely toxic albeit in small areas to low levels in expansive areas, with the full range causing harm. Conventional remediation strategies can be effective; however, they are often based on the movement of the contaminant from one area to another, and are prohibitively expensive.

Phytoremediation, the use of trees to clean up pollution, is a low-cost and effective alternative for many sites. But the technology can be limited due to phytotoxic effects of the pollutant, presence of multiple classes of pollutants, and slow rates of removal. Microbial endophyte-assisted phytoremediation is a relatively new technology that harnesses the power of the plant microbiome, combining the benefits of phytoremediation with those of bioremediation. Increased awareness of this plant-microbe team approach could lead to treatment of many polluted sites currently abandoned due to the high cost of conventional remediation, as discussed in Chapter 29.

Bacterial Consortia in Bioremediation and Metagenomics

Among bioremediation technologies, bioaugmentation with autochthonous bacterial consortia can result in the complete degradation of the target pollutant. The use of consortia isolated from environmental samples presents certain advantages over single inoculants or synthetic communities, but also some challenges. Metagenomics applied to bacterial consortia can serve to analyze their structure and composition and, more importantly, to elucidate the trophic interdependence and redundancy of pathways for the degradation of diverse compounds. Functional analysis could also serve to identify other key features important for the survival of the consortium or for a more effective bioremediation process. In Chapter 30, the advantages and challenges of isolating autochthonous bacterial consortia for bioremediation purposes, the current metagenomics analyses that are applied to study these consortia, and key examples of consortia for the degradation of organic pollutants are reviewed. Finally, the future of bacterial consortia analysis through combined omics to reveal the activity of the consortium members and the determination of carbon fluxes is presented (see Chapter 30).

Soil Microbial Fuel Cells

Soil microbial fuel cells (SMFC) are emerging as a new technology for the bioremediation of contaminated soils, which is cost-effective, environmentally friendly, and allows for concurrently harvesting energy while removing pollutants for self-powered operations. The electrochemical processes generated by SMFCs allow removal efficiency greater than traditional bioremediation approaches while reducing the treatment time. In the last decades, SMFCs have demonstrated bioremediation efficacy towards various contaminants in soil, such as petroleum hydrocarbons and pesticides. Starting from the working principles of SMFC technology, Chapter 31 provides an overview on the incidence that design solutions and microbial composition of the soil have on the bioremediation performance of the system. Future perspectives on the technology are also provided (see Chapter 31).

Biotechnology Strategies for Remediation of Microplastics

Environmental contamination due to plastics and microplastics has become a serious global issue. Due to careless use and inefficient waste management practices, enormous quantities of plastic waste enter the environment every year, and the environmental degradation of these materials leads to the formation of microplastics, small pieces of plastic < 5 mm in size. Microplastics have become almost ubiquitous in the environment and are associated with a range of negative impacts. They have been shown to have harmful effects on organisms and ecosystems and are potentially damaging to human health (see Chapter 32). Biotechnology offers possible solutions to the problem of plastic waste in two significant ways; by developing improved biodegradable bioplastics which can be fully recycled, and by enhancing the biodegradation of synthetic petroleum-based plastics by organisms and biomolecules. Both approaches can help to "close the loop" and develop a circular economy for plastic materials. In Chapter 32, the contributions that biotechnology can make to plastic waste management, as well as some of the current limitations and barriers to large-scale implementation of biodegradative technology, are highlighted. The potential of biotechnology to enhance the biodegradation of environmental microplastics is also examined: this poses challenges distinct from the management of larger plastic waste (see Chapter 32).

Bio-electrial Systems for Groundwater Bioremediation

Bio-electrochemical systems (BES) are an innovative method of breaking down organic contaminants, and could be a successful candidate as a sustainable remediation technology. A BES is a bioreactor device that consists of an anode where biodegradation of contaminants produces electrons that travel to a cathode via an external circuit. BES have been shown to work well at degrading hydrocarbons but have not yet been fully scaled up or adopted by the engineering community. In Chapter 33, modifications of standard contaminated land remediation approaches that could be modified into more efficient technologies by adopting a BES approach are proposed. Sustainable and eco-design approaches towards materials selection and the technology transfer of monitoring methods from geophysics are essential tools that biotechnologists must embrace to scale up the BESs that will be quickly adopted by the remediation industry (see Chapter 33).

SECTION 5: GOOD MICROBES IN AGRICULTURE, CO-EDITED BY LINDA THOMASHOW

This part of the book will focus on the discovery, understanding and application of microbes in agriculture. The importance of microbiology in sustainable agriculture will be emphasized, including the use of microbiomes as a source for combinations of beneficial microbes and the role of volatile organic compounds in communication between microbes and plants. Possible new commercial microbial disease control and plant growth promotion microbes will be presented. The life of microbes inside plants (endophytes) and their applications for protecting plants against the effects of climate change and improving the quality of grasses will be discussed. The well known processes of biological nitrogen fixation by microbes for plant growth and phosphate mobilization will be described.

Good Microbes for Agriculture

During the past several decades, intensive agricultural practices have led to severe biodiversity losses worldwide. Microbial diversity and function, which are associated with plant, human, and environmental health, are affected as well. To restore and conserve soil and plant microbial diversity, environmentally friendly and sustainable alternatives to chemicals for plant protection and fertilization are urgently needed. Understanding environmental microbiomes is a key approach to developing alternative solutions. Chapter 34 describes knowledge-based discovery and application strategies of beneficial bacteria for agriculture in the highly technology-driven field of microbiome research. New discoveries and the replacement of traditional chemical products are promising trends towards the transition to sustainable and healthy agriculture, food security and safety, ultimately contributing to planetary health (see Chapter 34).

Biological Control of Plant Diseases

Long appreciated for their beneficial influence on agriculture, plant-associated microbial communities are now seen as a plant's second genome, increasing the plant host's functional repertoire and helping it to grow and adapt to its environment. Technological advances in recent years have given insight into the plant's own mechanisms of defense and the contributions of the microbial community to protection against pests and pathogens. In Chapter 35, some of the historical challenges and limitations to our ability to implement plant disease control by beneficial microorganisms are considered and recent progress towards overcoming these obstacles is reviewed. Examples illustrating the interactions of microbes with each other, their plant hosts, plant pathogens, and insects pests are presented. Finally, the need for continued exploration of the physiological and molecular complexities underpinning beneficial microbe-plant relationships as well as increased practical implementation of the technology in response to the urgent need for greater agricultural sustainability will be assessed (see Chapter 35).

Disease-Suppressive Soils

Crop plants generally lack resistance to many soilborne fungal pathogens and parasitic nematodes, and therefore rely on natural microbial communities as the first line of defense against soilborne pathogens. Upon attack by pathogens, roots issue an "S.O.S." to recruit members of the soil and rhizosphere microbiome that can mitigate or deflect the attack. Disease-suppressive soils are the best examples of natural microbial-based defense of plant roots, and are defined as "soils in which, because of their microbial makeup and activity, a pathogen does not establish or persist, establishes but causes little or no disease, or establishes and causes disease at first but then the disease declines with successive cropping of a susceptible host even though the pathogen may still persist in the soil."(see Chapter 36). Suppressive soils have been described for diseases caused by a wide variety of soilborne pathogens including fungi, oomycota, bacteria and nematodes. "Long standing" suppressive soils have disease suppressiveness of unknown origin that is naturally associated with the soil and persists in the absence of the plant. In "induced" suppressive soils, the suppressiveness is initiated and sustained by continuously growing crops that are susceptible to the target pathogen or by adding inoculum of the target pathogen into the soil. Suppressiveness is transferrable in the field or under controlled conditions by adding small amounts (1–10% w/w) of suppressive soil to a conducive (non-suppressive) soil. Historically, suppressive soils have been thought to owe their activity to individual or select groups of microbes in the soil or rhizosphere. However, the application of new "omics" approaches, especially analysis of the microbiome by next generation sequencing, coupled with classical methods of characterizing suppressive soils, has revealed that the microbial basis of some suppressive soils is much more complex than previously thought (See Chapter 36).

Biological Nitrogen Fixation

Nitrogen (N) is arguably the most important nutrient required by plants, being an essential component of all amino acids and nucleic acids. However, the availability of N is limited in many soils and although the earth's atmosphere consists of 78.1% nitrogen gas (N_2), plants are unable to use this form of nitrogen. To compensate, modern agriculture has been highly reliant on industrial N fertilizers to achieve maximum crop productivity. However, a great deal of fossil fuel is required for the production and delivery of N fertilizer. Indeed, industrial nitrogen fixation alone accounts for 50% of fossil fuel used in agriculture and can be exceedingly expensive. Moreover, carbon dioxide (CO_2), which is released during fossil fuel combustion, contributes to the greenhouse effect, as does the decomposition of N fertilizer, which releases nitrous oxide (NOx), itself about 292 times more active as a greenhouse gas (GHG) than CO_2. In addition, applying chemical fertilizers is largely inefficient, as in general 30–50% of applied N fertilizer is lost by leaching, resulting in significant environmental problems, such as the eutrophication of waterways. Thus, there is a strong need to reduce our reliance on chemical N fertilizers and instead optimize alternative N inputs. Biological nitrogen fixation (BNF) is one alternative to N fertilizer (see Chapter 37). It is carried out by beneficial prokaryotes using an enzyme complex called nitrogenase and results in atmospheric N_2 being reduced into a form that both microorganisms and plants associated with them are able to use. It is this process and its major players which will be discussed in Chapter 37.

Efficacy and Safety of Bacillus Thuringiensis

The insecticidal bacterium *Bacillus thuringiensis* (Bt) was discovered early in the twentieth century, but effective commercial pest control products based on this species were not available until the 1960s. The first successful formulations used *B. t.* subspecies *kurstaki* (Btk), which had high efficacy against numerous pest caterpillar species. Within a few years these preparations were used widely in organic and integrated pest management (IPM) programs for agriculture and forestry, where caterpillars were the dominant pests. Btk's high specificity and environmental safety compared to broad-spectrum synthetic chemical insecticides led to its use worldwide (see Chapter 38). Subsequently, other Bt subspecies active against mosquito and black fly larvae and those of beetles were discovered in the 1970s and 1980s. One, *B. t.* subsp. *israelensis* (Bti), was rapidly commercialized during the 1980s, and again, owing to Bti's environmental safety compared to chemical insecticides, it quickly became used around the world for controlling nuisance mosquitoes and disease vectors. Advances in molecular biology and genetic engineering during the 1970s demonstrated that the active ingredients of Btk and Bti were primarily crystalline (Cry) proteins produced during sporulation, which, when eaten, selectively destroyed the stomachs of targeted insects. By the 1990s, Cry proteins had been genetically engineered into major commodity crops, yielding Bt cotton, Bt maize, and later, Bt soybeans, protecting these from caterpillar pests. Bt crops are grown in Australia, Argentina, Brazil, China, and the US, among others, and are now a multi-billion dollar industry. Bt insecticides remain widely used in organic and IPM programs, having a remarkable safety record of more than 50 years without a single verified human case of illness. Chapter 38 summarizes Bt's main features and uses, with a focus on safety over the last several years.

Beneficial Fungi Living in Plants

Fungi are notably diverse and proliferate in different environments (soil, air, water) but, remarkably, the highest numbers of fungi are found in the soil or associated with plants. Fungal endophytes, those living within plant tissues, are important members of the plant microbiota, but the nature of the relationship established with the host can be variable. The majority of endophytes do not affect the plant at all; some are pathogenic and some may even be beneficial, influencing plant growth and health by modulating different biological processes.

In Chapter 39, fungal endophytes with a recognized beneficial impact on plants, and a special group of endophytes, the mycorrhizal fungi, will be considered. The interaction at the cellular level will be presented, highlighting the functional significance of the associations and the impact of these fungal symbionts on the host plant, and illustrating potential applications.

The plant–endophyte alliance may increase plant phenotypic plasticity, which is perhaps an advantage in fluctuating environments such as those emerging from global climate change. Understanding the complex interactions between fungal endophytes and plants is, therefore, of relevance for the development of tailored microbial inocula to improve crop production and quality under varying environmental conditions. This appears to be a promising means of developing more environmentally friendly agricultural crop production systems and increasing global food security.

Volatile Mediated Plant–microbe Interactions

It has recently become clear that microorganisms release a wealth of microbial volatile organic compounds, unique substances that act over short and long distances. These compounds have both plant-growth promoting and disease suppressive activities, providing a novel and sustainable source of natural products (see Chapter 40). Most of the knowledge on plant beneficial volatiles originates from laboratory studies on single compounds or strains, but a few recent studies have shown that microbial volatiles are able to protect plants under greenhouse and field conditions as well. This chapter summarizes the current knowledge on the biological function of microbial volatiles important for plant growth and health, and their possible application.

Trichoderma in Agriculture

Trichoderma (teleomorph Hypocrea) are filamentous fungi, commonly named "green mold" due to their production of green spores, with many species able to produce plant beneficial effects that improve agricultural production. Some species are antagonists to plant disease agents and are registered as microbial biological control agents (BCA). But other than these direct effects controlling the agents that damage crops, these fungi are also able to indirectly protect the plant by inducing resistance, activating plant defense responses to many biotic factors, such as pathogens and pests, plus abiotic stress, including high temperatures, drought and salinity (see Chapter 41). These beneficial fungi, with their close association to the plant rhizosphere, may act as plant biostimulators, having plant growth promotion effects, increasing nutrient availability, and stimulating important physiological processes that improve crop productivity, yield, and quality. *Trichoderma* is a component in over 250 commercial biological products for agriculture, as biofungicides, biostimulants, biofertilizers, soil integrators, etc., where they are used as alternatives to chemical products in integrated pest management and organic farming systems, helping to reduce chemical use and risks to environmental and human health (see Chapter 41). *Trichoderma* can play an important role in the development of global sustainable agriculture, contributing to solutions for enhancing cultivation, crop protection and productivity, providing possibilities to increase food security. Some reflections are presented in Chapter 41 for how *Trichoderma* can be usefully applied in the agricultural systems of developing countries. In particular, the introduction of these microbes in existing agricultural systems may provide benefits to subsistence farmers, giving them the opportunity to farm in sub-optimal climatic and field conditions, plus cultivating marginal lands, aiding in the autonomous production not only for their own food supply, but also with the possibility for social and economic gains for the local community (see Chapter 41).

Selling Good Microbes for the Production of Healthy Plants

Good microbes are essential to produce healthy plants, seeds, fruits, and vegetables for a growing world population. Microbes are used to increase plant heath and robustness and to protect them against diseases.

In Chapter 42, the long and complicated road from the discovery of a potentially useful microbe until it is applied as a product is described. Players involved include scientists, entrepreneurs, product developers, consultants, regulatory authorities, policy makers, public opinion and, last but not least, the end user.

In this chapter several examples of microbes and players, and their activities, are described.

Frans J. de Bruijn and the authors of all the chapters.

SECTION 1

GOOD MICROBES IN MEDICINE

Co-Edited by Hauke Smidt and Frans J. de Bruijn

CHAPTER 1

Modern Medicine Relies on the Help of Microorganisms – From Vaccine Production to Cancer Medication

Letícia Parizotto[1], Larissa Brumano[1,2], Eduardo Kleingesinds[1], and Adalberto Pessoa Junior[1]

[1]Laboratory of Pharmaceutical Biotechnology, School of Pharmaceutical Sciences, University of Sao Paulo, 05508–000, Sao Paulo, Brazil

[2]Center for Natural and Human Sciences (CCNH), Federal University of ABC, Rua Oratório, 305 - Bangú - CEP 09280-550, Santo André, Sao Paulo, Brazil

1.1 INTRODUCTION: GOOD MICROORGANISMS AND OUR HEALTH

Since the advent of microbiology, people have been interested in the microbes found in our bodies (Sam et al., 2017). They are everywhere – and within us; most are harmless and many are beneficial (McLeod et al., 2019). Our survival depends on many microorganisms, as they help us digest nutrients, fight unwanted bacteria and fungi, in addition to stimulating the immune response and preventing pathogens from colonizing the skin (Franklin, 2013; Lin, n.d.; Staughton, 2019; Triangle Dentistry, 2018).

Without the billions of good microorganisms in our bodies, we would not be alive or healthy and, therefore, understanding the benefits of using microorganisms to our advantage means gaining quality of life (Staughton, 2019). Our microbiome is developed from the moment of birth to adulthood, when an individual may have more than 700 different types of bacteria in the mouth, more than 1000 species on the skin and hair follicles, and up to 100 trillion microbes in the intestinal microbiome.

Good microorganisms can act in their living form directly on our body, but they are also widely used as microbioreactors for the production of relevant therapeutic products for the treatment of infections, injuries, burns, lactose intolerance, dystonia, rheumatoid arthritis, multiple sclerosis, osteoporosis, diabetes mellitus, diabetic ulcer, anticoagulant, gout, hepatitis, leukemia, lymphoma, tumor lysis syndrome, and neutropenia, in addition to being used in the production of several vaccines (Jozala et al., 2016).

Among the drugs produced by microorganisms, penicillin is noteworthy as it represents the first therapeutic product isolated from microbial sources against disease-causing microbes and was the precursor to the development of all antibiotics produced to date. Penicillin is the drug with the greatest therapeutic impact of all time, not only due to its mechanism of action but also due to the technology developed for its production. These same biotechnologies have become a model for the discovery, development, and production of new bioproducts, mainly biopharmaceuticals (Kardos and Demain, 2011).

In this chapter, we present the application of good microorganisms in the development of antibiotics, vaccines, anticancer drugs, and in gene therapy. But first, let us understand how the interaction between humans and bad microorganisms has contributed to establishing new

Good Microbes in Medicine, Food Production, Biotechnology, Bioremediation, and Agriculture, First Edition. Edited by Frans J. de Bruijn, Hauke Smidt, Luca S. Cocolin, Michael Sauer, David Dowling, and Linda Thomashow.

technologies based on good microorganisms that are revolutionizing modern medicine.

1.2 BAD MICROORGANISMS: EPIDEMICS BOOSTED MODERN MEDICINE

Bad microorganisms plus war and hunger have been one of the major forces shaping human history over the decades. Commonly, people live in balance with microorganisms in their natural reservoirs (e.g., animals, soil, and water). However, with economic development, epidemics have become more frequent, as urbanization and population growth have resulted in the occupation of wild ecosystems providing contact between people and these natural reservoirs.

In addition the expansion of world trade and tourism increased intercontinental travel, which facilitated the spread of disease-causing microorganisms. The result was a series of new epidemic outbreaks, such as SARS (2003), avian influenza (2005), swine flu (2009), MERS (2012), polio (2014), Ebola (2014), Zika virus (2016), and COVID-19 (2019) even more impactful than bubonic plague, Spanish flu, or smallpox (Graham and Sullivan, 2018; Morens et al., 2004).

Diseases caused by microorganisms are called infectious diseases and are responsible for about 18% of annual global deaths (University of Washington, 2017), in addition to causing long-term economic and social damage.

One of the most emblematic cases was the COVID-19 pandemic that started in the province of Hubei, China, in 2019 and quickly spread to all continents and infected millions of people around the planet (Johns Hopkins University, 2020). This disease is caused by a virus of the coronavirus family, the same family of pathogens that caused two other epidemics of the 2000s, SARS and MERS. Within a few months, the outbreak became a pandemic and imposed drastic changes on our lifestyle. Also, it has resulted in enormous economic damage which, according to the World Economic Forum (Schwab, 2020), was at least 8 trillion dollars with losses in GDP and jobs, and increased deaths.

Infectious diseases are not only manifested in outbreaks, but also they can resist for years in stable percentages of infected people until a definitive cure is achieved. An example is the diseases transmitted by mosquitoes, such as dengue (virus), yellow fever (virus), Zika (virus), chikungunya (virus), and malaria (protozoan), which periodically affect tropical developing countries (Feachem et al., 2019; Song et al., 2017). Infectious diseases are also not only acute as they can become chronic. The most relevant example is AIDS, a disease transmitted by the HIV retrovirus, which became epidemic in the 1980s and 1990s (Broder, 2010).

The events cited are some examples of our constant battle against microorganisms that cause infectious diseases. The technologies developed to combat pathogenic microorganisms drove the development of modern medicine. This process started with the discovery of antibiotics that is presented in the next section.

1.3 ANTIMICROBIAL PEPTIDES: A NEW THERAPEUTIC ALTERNATIVE TO ANTIBIOTICS?

After the second half of the 19th century, when scientists such as Robert Koch identified bad microorganisms responsible for diseases, the search for chemical agents capable of inhibiting the growth or causing the death of fungi and bacteria grew. In 1910, Ehrlich developed the first synthetic antibiotic, Salvarsan, used against syphilis (Baquero and Levin, 2020).

Little progress was made in the following years in the development of antibiotics until the breakthrough in the treatment of bacterial infections occurred in 1929 with the discovery of penicillin by Alexander Fleming. The discovery that fungi produce substances capable of controlling bacterial proliferation has motivated a new area of research in the pursuit of antibiotics by prospecting in cultures of microorganisms, especially fungi and actinobacteria (Zhu et al., 2020). After the process of industrialization of penicillin, boosted by the Second World War, rapid growth in the discovery and development of new antibiotics took place.

The action of antimicrobials is not only directed against bacteria but also fungi, protozoa, viruses, and other organisms. Basically, there are two mechanisms of action: inhibition of microbial growth with reversible character; or death of the microorganism (Baquero and Levin, 2020). Antibiotics were initially defined as chemical substances synthesized by several species of microorganisms, plants, and animals, which hinder the growth of others. The advent of the pharmaceutical industry facilitated the production of antibiotics of semi-synthetic and synthetic origin. They differ from each other in terms of chemical properties, their spectra and mechanisms of action and are classified chemically as derivatives of amino acids, sugars, acetates, propionates, among others.

Several decades after the expansion of the use of antibiotics, bacterial infections remain a threat to public health worldwide, compromising the clinical, social, technological, and economic benefits achieved after their discovery (Wallace et al., 2020). In addition, the expansion and continued use of antibiotics contributed to the emergence of cases of bacterial multidrug resistance at the same time as there was a dramatic reduction in the identification of new potential antibiotics (Zhu et al., 2020). Antibiotic resistance is one of the most important scientific and public health challenges today (Nadeem et al., 2020). As outbreaks of resistance become increasingly widespread, the search for new strategies to control the phenomenon of antibiotic resistance becomes necessary.

This search motivated interest and studies into a group of antimicrobial peptides that have inhibitory properties against the growth of some pathogenic microorganisms (Mercer et al., 2020). Antimicrobial peptides are biomolecules naturally produced by a wide variety of living organisms, such as bacteria, fungi, plants, invertebrate animals, and mammals, including humans (León-Buitimea et al., 2020).

Their mechanisms of action involve the interaction of the peptide with components of the cell membrane of the pathogen through a combination of hydrophobic and electrostatic effects that interfere with the fluidity and lipid composition of the cell membrane and, consequently, the acquisition of resistance has become unlikely (León-Buitimea et al., 2020). Additionally, antimicrobial activity is not only restricted to a single type of pathogenic agent, such as gram-positive or gram-negative bacteria but also fungi, protozoa, and viruses (Mercer et al., 2020).

Among the microbial peptides studied are the bacteriocins. They are cationic peptides, produced by gram-positive and gram-negative bacteria, capable of creating pores in the target cells, resulting in interferences in the membrane potential. Nisin, the most well-known lactic acid bacteriocin, has been approved for use in food preservation in more than 40 countries for more than 50 years (see chapter 9). It is produced by the bacteria *Lactococcus lactis* and inhibits the growth of gram-positive bacteria, such as *Listeria monocytogenes*, *Clostridium botulinum*, *Bacillus cereus*, and *Staphylococcus aureus*. Despite the high concentrations required for some antimicrobial peptides to be effective, their mechanisms of action become interesting starting points for the development of new antimicrobial agents.

These antimicrobial agents demonstrate how good microorganisms are essential in combating harmful ones. Since the discovery of these agents, with the evolution of bioinformatics and microbiology tools, such as recombinant DNA technology, more sophisticated drugs have been formulated. Medications that not only treat and prevent infectious diseases but also other conditions that affect the population, such as cancer, diabetes, neurological disorders, and many others. Next, we show how good microorganisms were essential in these discoveries.

1.4 MICROORGANISMS AS TOOLS: RECOMBINANT DNA TECHNOLOGY (rDNAT)

Since the development of genetic engineering techniques also called recombinant DNA technology (rDNAT), several breakthroughs have been made in scientific areas, including drug production, in recent years. Basically, rDNAT allows genetic material to be recombined from different sources, and in this way, reprogramming some or all genetic information in target cells, allowing the production of molecules of interest (Khan et al., 2016). Microorganisms play an essential role as tools for this technology.

The technique has been successfully used for the development of biopharmaceuticals since the 1970s when the genes encoding human insulin and growth hormone were cloned and expressed in *Escherichia coli*. The first licensed drug produced with recombinant DNA technology was human insulin in 1982 (Baeshen et al., 2014). In the past, insulin applied to diabetics was extracted from the bovine or swine pancreas. In addition to causing adverse reactions, the treatment had an extremely high cost, about 250 000 pigs were needed to generate just 4.5 kilos of insulin. With the use of *E. coli* and rDNAT, it was possible to produce the human enzyme in microorganisms with relevant improvements both in the production and in the treatment of diabetic patients.

We can highlight the importance of microorganisms, either as a production platform (hosts), tools for replication, plasmid stock, or as a source of enzymes necessary for the steps and methodologies employed in genetic modification processes.

Figure 1.1 presents, as a model, the simplified scheme of recombinant production of the biopharmaceutical *L-asparaginase*, used for the treatment of acute lymphoblastic leukemia. Highlighted are the main steps that involve the use of microorganisms and/or their tools.

The basic science knowledge used today in complex processes, such as the production of recombinant drugs, was developed from simple microorganisms, such as the bacterium *E. coli*. This bacterium has a fast growth rate and it is easy to manipulate its genetic material using its plasmids, which are extrachromosomal DNA fragments, that confer various phenotypes to the bacterium (Chen et al., 2013).

Currently, gene manipulation, cloning, and expression techniques are routinely applied to a variety of eukaryotic models (Amann et al., 2019), such as the yeast *Saccharomyces cerevisiae*. Since, in addition to their well-known potential for genetic manipulation (Griffiths et al., 2004), yeasts are capable of making post-translational modifications (PTMs), such as N-glycosylation, which are important for the *in vivo* function of the biomolecule and its plasma half-life, in addition to preventing immunological reactions, essential requirements for biopharmaceuticals.

Today, most biopharmaceuticals based on glycosylated proteins require expression in mammalian cell lines (Amann et al., 2019), or in systems with modified glycosylation to approach humans (Wildt and Gerngross, 2005). Even when mammalian cells are used as hosts to produce biopharmaceuticals, microorganisms continue to be important in the process, especially in the steps that involve plasmid replication and storage.

As sources of specific enzymes used in rDNAT, we can highlight the contribution of microorganisms in two classic examples: Taq polymerase and, more recently, CRISPR-Cas9.

The polymerase is an essential enzyme for a PCR (polymerase chain reaction) step, a technique widely used for mass replication of portions of DNA, facilitating the analysis and search for genes of interest. The methodology became more convenient and feasible after the discovery of a specific polymerase, the Taq polymerase. It is extracted from the bacterium *Thermus aquaticus*, which was isolated from the seabed in regions of high temperature. Therefore, it has the

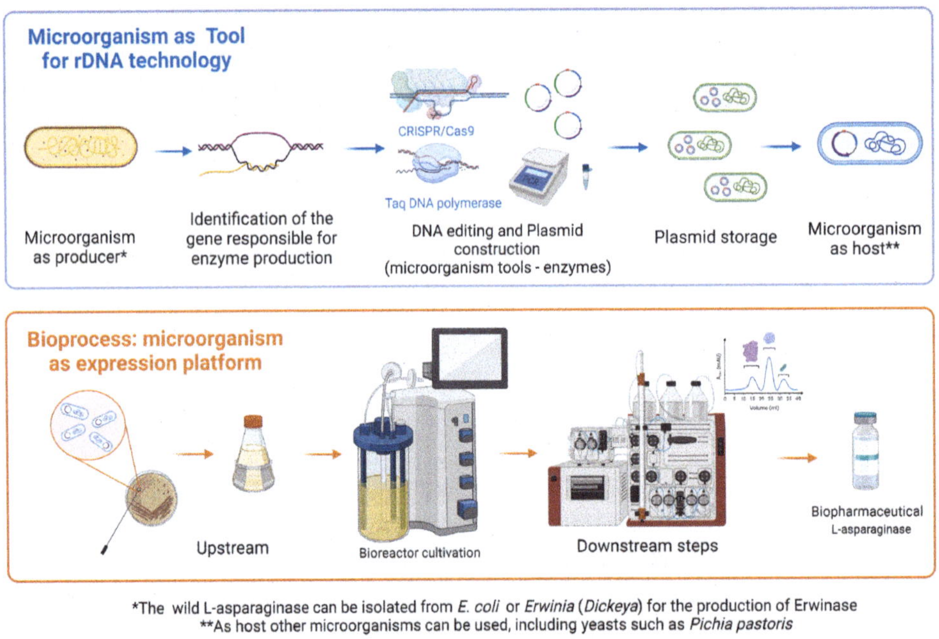

FIGURE 1.1 Steps and methodologies in which microorganisms are fundamental for recombinant biopharmaceutical production: the example of *L-asparaginase*. The wild *L-asparaginase* (ASNase) can be isolated from the producer bacteria *E. coli* or *Erwinia (Dickeya) chrysanthemi*. Later, the gene responsible for the ASNase expression can be selected and amplified by using enzymes provided by other good microorganisms, such as CRISPR-Cas9 and Taq polymerases, respectively. The gene is inserted in a plasmid that is introduced in a host microorganism, which can be bacteria or yeast, such as *Pichia pastoris*. Then the host microorganism is cultivated at different scales, from flasks to bioreactors, to obtain high-density population and the ASNase production is induced. This first phase of the process is called upstream. Later, the cells are harvested and the ASNase is separated initiating the downstream operations to obtain a purified ASNase ready for the biopharmaceutical formulation. (created with ©BioRender - biorender.com).

thermal stability essential for the operation of this technique.

The other microbial tool that is transforming genetic engineering is CRISPR-Cas9. The researchers responsible for the development of the CRISPR-Cas9 method for genome editing, Emmanuelle Charpentier and Jennifer A. Doudna, won the Nobel Prize in Chemistry in 2020.

Actually, CRISPR-Cas9 participates in a set of ancient and natural defense mechanisms found in several bacteria, however it has been used for DNA editing with a precision never achieved with efficiency and flexibility (Doudna and Charpentier, 2014). This tool allowed scientists to integrate or exclude genes from the target organisms which can be used to improve host cell production, increase product yields, and develop humanization of glycosylation pathways. It has great potential to revolutionize the production of biopharmaceuticals.

Next, we present recombinant biopharmaceuticals that were only feasible with the rDNAT.

1.5 VACCINES: THE USE OF MICROORGANISMS IN THE FRONTLINE AGAINST DISEASES

It is estimated that about 2.5 million deaths are prevented annually due to vaccination. Vaccination is the most effective way to prevent and eliminate infectious diseases (World Health Organization (WHO), 2013). The vaccine is a biological preparation that aims to stimulate the immune system by introducing an antigen, which can be a microorganism or part of it, to generate a protective response without causing the disease or its complications. In addition to the antigen, vaccines contain adjuvants (compounds that help the immune response, such as aluminum salts), preservatives and excipients (see also chapter 3).

The development of vaccines is deeply linked to microorganisms, whether using them as an active principle or as suppliers or transporters of antigens that cause the immune response. Thanks to good microorganisms, vaccines can be developed more safely, economically, and quickly. These characteristics allow rapid responses to epidemic outbreaks, new treatments for preventable diseases, and an opportunity of finding a cure for currently incurable diseases.

The first generation of vaccines was developed empirically during the 18th century by observing individuals who, once infected, did not contract the disease again. Jenner (1796) developed the technique that would give rise to vaccines by exposing people to the smallpox virus (*vacca*), a mild form of the disease, to obtain protection against human smallpox.

It was only after the discovery of microorganisms that the development of vaccines gained new impulse in the 19th century. Thus, vaccines containing whole microorganisms in attenuated or inactive forms were developed. Examples of attenuated vaccines are those which protect against mumps, measles, rubella, rotavirus, chickenpox, smallpox, and yellow fever; and inactive forms against hepatitis A, influenza, polio, and rabies (De Gregorio and Rappuoli, 2014).

During the 20th century, the second generation of vaccines, subunit vaccines (SUVs), was developed through rational design based on synthetic biology and bioinformatics tools, such as the rDNAT, aiming to identify the subunits of pathogens that cause immunity without causing infection. These subunits, like proteins, are reproduced in good microorganisms, such as the bacterium *E. coli* and the yeast *S. cerevisiae*, in the laboratory. Thus, safer and better quality antigens are quickly obtained than in the first generation (De Gregorio and Rappuoli, 2014).

Also, it was possible to improve the immune response by creating antigens that mimic the viral structure, virus-like particles (VLP), and that are likewise produced in good microorganisms. These technologies have allowed the creation of new vaccines, such as for HPV (human papillomavirus) and HBV (hepatitis B) and are being tested not only for more complex infectious diseases, such as AIDS and malaria but also for non-infectious diseases, such as Alzheimer's, diabetes, cancer, allergies, and many others (Kushnir et al., 2012).

Currently, vaccines can be developed from synthetic messenger RNA (mRNA) (produced in the laboratory) for the intracellular production of antigens. The mRNA works by translating the DNA information into proteins that, in the case of vaccines, cause the immune response. In this technique, microorganisms act as suppliers of the tools necessary for this synthetic production, such as the DNA template to be copied and the enzymes that make the transcription that, for example, can be produced in *E. coli* (Pardi et al., 2018).

Another method of safely delivering the genetic material of a pathogen is through viral vectors. Good viruses, such as Adenovirus, receive only the genes from the pathogenic virus that can produce a safe antigen. Therefore, when the viral vector is injected in the patients, it stimulates the immune response but does not cause the disease and its complications (Bolhassani and Zahedifard, 2012).

These techniques have increased security, allowed scalability, and abruptly reduced the time for vaccine development, which is why they are being used more and more, mainly in the control of epidemic outbreaks, as in the COVID-19 pandemic. Among the candidate vaccines for COVID-19 in clinical trials, the majority are based on mRNA, viral vectors, and SUVs (WHO, 2020).

Soon we will have increasingly sophisticated and safe vaccines that also work against non-infectious diseases, such as cancer. Cancer vaccines have different modes of action, but they have the same objective: to overcome the mechanisms of camouflage of the tumor and stimulate the response of the immune system.

In one approach, good microorganisms act by providing the tools to assemble an antigen that contains specific parts of the tumor and that, when injected into the patient, activates the protective cells against the antigen and, consequently, against the tumor. Based on this mechanism, mRNA vaccines are being developed for breast, prostate, ovarian, brain, and skin cancer, among others (Pardi et al., 2018).

In another approach, good microorganisms, such as viral, bacterial, and protozoan vectors, act as carriers of antigens for tumor cells, alerting the protective cells that begin to attack the tumor. About 8% of cancer vaccines in clinical trials are based on viral vectors (Bolhassani and Zahedifard, 2012). In addition to vaccines, good microorganisms have contributed to the formulation of other anticancer drugs that are described below.

1.6 ANTICANCER DRUGS: MANY WAYS TO FIGHT CANCER WITH GOOD MICROORGANISMS

Cancer remains one of the crucial health challenges of the 21st century. The growing number of cases is not accompanied by adequate progress in therapy. Therefore, it is fundamental to find new and more effective treatments (Łukasiewicz and Fol, 2018).

Microorganisms contribute to the production and development of anticancer drugs in many ways, such as natural producers of biomolecules with anticancer activity, platforms for recombinant expression of anticancer biopharmaceuticals (such as cytokines), or agents in immunotherapeutic treatments.

Among the molecules naturally produced by microorganisms used in anticancer treatment, we can highlight the enzyme L-asparaginase (ASNase). This enzyme revolutionized the treatment of cancers of the lymphatic system, mainly acute lymphoblastic leukemia (ALL) since, with its use, the survival rate of children and adolescents with ALL increased from 0.5% in 1960 to approximately 90% today (Egler et al., 2016). Its mechanism of action is based on the selective death of tumor cells (El-Nagga et al., 2014). Currently, on the market, it is possible to find ASNase produced by the bacteria *E. coli* and *E. chrysanthemi* (Erwinase) (Brumano et al., 2019).

Other microbial enzymes are also aids in antitumor treatment or prevention, such as T4 Endonuclease V (or dimericine), a bacterial enzyme isolated from *E. coli* infected with the bacteriophage T4 virus which helps in the initial prevention of skin cancer (Camp et al., 2011). Additionally, enzymes such as urate oxidase (or rasburicase), which is an enzyme extracted from the fungus *Aspergillus flavus* and expressed by genetic engineering in yeasts, act as an antidrip medicine and also assist in the anticancer treatment of leukemia, lymphoma, and solid tumors by reducing high levels of uric acid in the blood (hyperuricemia) that can occur during treatment (Bosly et al., 2003). Another example is the enzyme glucarpidase, produced naturally by *Pseudomonas sp.* and expressed by genetic engineering in. *E. coli*, that acts as an auxiliary biopharmaceutical in the treatment of cancer patients who have kidney failure and high levels of methotrexate (Ramsey et al., 2018).

With the advent of rDNAT (described in topic 4), other types of non-microbial proteins important for anticancer treatment could be produced in substantial amounts in microorganisms, such as cytokines, small signaling molecules of the immune system used in immunotherapies.

Immunotherapy is a type of biological treatment that aims to enhance the immune system so that it can fight infections and other diseases such as cancer. This stimulation can be performed by using cytokines or by using attenuated microorganisms or part of them (similar to vaccines), associated or not with chemotherapy (Łukasiewicz and Fol, 2018).

An example of a recombinant cytokine is INFα produced mainly in *E. coli*. The FDA has now approved the use of IFN against more than 14 types of cancer, including melanoma, renal carcinoma, AIDS-related Kaposi's sarcoma, and hairy cell leukemia. In addition to cancer treatment, cytokines are widely used in the treatment of autoimmune diseases, viral diseases, allergies, and transplants (Ferreira et al., 2019).

The other type of immunotherapy is based on the insertion of microorganisms in the body to stimulate the immune system in two ways: either as an antigen or as carriers of specific antineoplastic agents (for example, chemotherapy) to the tumor (Felfoul et al., 2016), similar to vaccination exemplified in topic 5.

Another technique is based on microorganisms that grow in anaerobic conditions, due to the lack of oxygenation inside tumors. Instead of spreading throughout the body, like chemotherapies, bacteria are directed only to the tumor site, where they have the ideal conditions for growth. The main bacteria used in this technique are: *Bacillus Calmette-Guérin* (BCG), the same as the antituberculosis vaccine; *Streptococcus pyogenes*, applied in the treatment of lymphangiomas in children; *Clostridium novyi*, capable of producing specific toxins that destroy cancer cells; and *Salmonella Typhimurium*, which also prevents the growth of cancer cells due to nutrient absorption (Broadway and Scharf, 2019; Łukasiewicz and Fol, 2018).

New approaches are being studied for anticancer treatment. In a recent study, a recombinant *E. coli* strain was developed as a live biotherapeutic for the treatment of cancer and which can activate an antitumor immune response. This work demonstrates the transformative potential of using recombinant microorganisms for the treatment of diseases when drug development criteria are incorporated into the design process of live medicines (Leventhal et al., 2020).

Furthermore, a recent treatment that has gained pace in modern medicine against cancer is gene therapy, which is presented in the next topic.

1.7 GENE THERAPY: THE FUTURE OF MODERN MEDICINE

Gene therapy is a field of medicine that focuses on the use of nucleic acids (DNA) to treat diseases. This technique is based on the insertion, replacement, extinction, or repair of genes in cells and tissues to treat conditions; mainly, hereditary diseases (Scherman, 2014). In most studies on gene therapy, a functional gene is inserted into the genome to replace an "anomalous" gene that causes the disease. The rDNAT and the use of bacteria and yeast as biological models were essential for the development of this treatment (Kresina, 2001).

Therapy can be performed in two ways, *ex vivo* or *in vivo*. In the *ex vivo* approach, specific patient cells are harvested, cultured in a laboratory, genetically modified, and reintroduced into the patient. In the *in vivo* treatment, the genetic material is injected directly into the patient, which can be directly into the affected tissue or organ or intramuscularly, intravenously, or even by nasal administration.

In both cases, a carrier, called a vector, is needed to deliver the therapeutic gene into the patient's target cells. Within the cells, the vector discharges its genetic material, containing the therapeutic gene that starts to produce functional proteins that restore the target cells, curing the disease transiently or permanently (Kresina, 2001; Panno, 2005; Scherman, 2014).

Currently, the most commonly used vectors in gene therapy are viruses that have been altered to carry the therapeutic gene. Viruses are good vectors for being able to efficiently deliver their genetic material into host cells. Ironically, the deadliest viruses are the most suitable transporters (Panno, 2005). However, for this, the virus genome is manipulated to remove the disease-causing genes and insert the therapeutic genes. Thus, after transfection, the host cells receive genetic information and produce additional copies of the virus, which can infect other cells. The most used viruses are adenoviruses (causing cold), retroviruses (as HIV), and adeno-associated viruses (Scherman, 2014).

Other widely used vectors are plasmids, generally produced in bacteria, such as *E. coli*. The therapeutic gene is introduced into these DNA molecules, which are inserted into bacteria. The bacteria multiply and, consequently, the number of plasmids increases. They are then disrupted releasing the plasmids that are separated and purified. Plasmids can be directly injected into the patient. However, they are less efficient carriers than viruses and do not integrate with the human genome; therefore, they only induce transient healing (Panno, 2005).

Commonly, this technique targets monogenic diseases, such as cystic fibrosis, sickle cell anemia, phenylketonuria, and some immunodeficiencies (Kresina, 2001). But also, therapies for multifactorial diseases involving a set of genes are being developed, such as: cancer; neurodegenerative diseases (Alzheimer's and Parkinson's); hematological, liver, cardiovascular, and immunological diseases; infectious diseases, such as AIDS and hepatitis (Kresina, 2001). Currently, 20 gene therapies have been approved, most of them in the last five years (Shahryari et al., 2019).

Especially regarding cancer treatment, some strategies have proved successful, as: replacement of abnormal tumor suppressor genes, restoring their functionality; delivery of intracellular therapeutic drugs, in which the therapeutic gene activates a non-toxic substance in a cytotoxin within malignant cells, such as bacterial nitroreductases, causing their death; modification of the antitumor immune response, helping protective cells to identify and fight cancer (like a vaccine); insertion of suicidal genes or other apoptosis inducers in malignant cells. Therefore, many therapies will soon be available even for cancers that are more resistant to chemotherapy (Kresina, 2001).

Although the technology is still at an early stage, it has shown potential. However, some challenges still need to be overcome, such as the specificity and efficiency of the vectors, and the more in-depth knowledge of the mechanisms of diseases (Kresina, 2001; Panno, 2005). In addition to technical questions, ethical questions also need to be answered, such as the use of therapy for the genetic engineering of a fetus or aesthetic application. Moreover, because it is difficult to scale the technique, in some cases cells from the patient themself are needed, it is still not very accessible. However, despite these issues, gene therapy has revolutionized modern medicine thanks to good microorganisms.

1.8 CONCLUDING REMARKS AND PERSPECTIVES

Our relationship with microorganisms starts at the moment of our birth when we acquire good microorganisms that allow our survival. However, this relationship is not always friendly; pathogenic microorganisms have caused social and economic damage throughout our history.

However, in defense of good microorganisms, medicine would not have advanced as much without them. In this chapter, we presented some of the applications of good microorganisms such as: an active antigen, vectors, natural producers of biomolecules, platforms for recombinant expression of biopharmaceuticals or as agents in immunotherapeutic treatments. Good microorganisms allowed the development of modern treatments that lead to rapid responses to epidemic outbreaks, new therapies for preventable diseases, and a chance of cure for currently incurable diseases.

The discovery of antibiotics revolutionized medicine and the technology for the manufacture of biopharmaceuticals. Despite the advance against pathogenic microorganisms, resistance to antibiotics is still one of the biggest public health problems today and solutions such as antimicrobial peptides are needed.

The development of more sophisticated biopharmaceuticals has gained force, mainly due to the new genetic engineering tools and the process should become even faster with the acquisition of the CRISPR-Cas9 system by the pharmaceutical industry.

The application of the recombinant DNA technique (rDNAT) resulted in safer and more efficient vaccines because the rational design and the use of good microorganisms as vectors (viral vectors) or as suppliers of proteins (SUV) or nucleic acids (mRNA) resulted in antigens capable of generating an immune response, but no infection. Consequently, vaccines have become suitable not only against infectious diseases but also to treat cancers.

About cancer, several approaches to the use of good microorganisms in the treatment were presented. Microorganisms can produce anticancer drugs naturally or synthetically. Additionally, they act as tumor signalers for the innate immune system. One of the main advantages of live drugs is their high specificity in contrast to traditional chemotherapies, which affect the patient's entire organism and, therefore, result in a better quality of life for the patient.

Modern medicine tends to customize treatments, with gene therapy being one of its main drivers. As presented, each individual may have their genetic code modified to correct disease-causing mutations or even for the still controversial genetic improvement.

The prospects for expanding the use of good microorganisms are promising since they can be genetically improved, cultivated, and used as microfactories for the production of innovative biopharmaceuticals to cure untreated diseases or for new treatment options that are cheap, easy to produce, and non-invasive. Multi- and interdisciplinary studies in genetic engineering, biotechnology, microbiology, biochemistry, pharmaceutical sciences, and chemical engineering have been essential to understanding the mechanisms of production of drugs by microorganisms in a safe, efficient, and large scale.

ACKNOWLEDGMENTS

The authors would like to thank the São Paulo Research Foundation for their support (FAPESP, grants No. 2017/25065–9, No. 2017/20384–9, No. 2017/21819–9, and No. 2019/23620–0) and the Brazilian National Council for Scientific and Technological Development (CNPq, grant No. 404791/2018–2).

REFERENCES

1. Amann T, Schmieder V, Faustrup Kildegaard H, Borth N, Andersen MR. 2019. Genetic engineering approaches to improve posttranslational modification of biopharmaceuticals in different production platforms. *Biotechnology and Bioengineering* 116: 2778–2796.
2. Baeshen NA, Baeshen MN, Sheikh A, Bora RS, Ahmed MMM, et al. 2014. Cell factories for insulin production. *Microbial Cell Factories* 13: 141.
3. Baquero F, Levin BR. 2020. Proximate and ultimate causes of the bactericidal action of antibiotics. *Nature Reviews Microbiology* 19: 123–132.
4. Bolhassani A, Zahedifard F. 2012. Therapeutic live vaccines as a potential anticancer strategy. *International Journal of Cancer* 131: 1733–1743.
5. Bosly A, Sonet A, Pinkerton CR, McCowage G, Bron D, et al. 2003. Rasburicase (recombinant urate oxidase) for the management of hyperuricemia in patients with cancer: Report of an international compassionate use study. *Cancer* 98: 1048–1054.
6. Broadway KM, Scharf BE. 2019. *Salmonella Typhimurium* as an Anticancer Therapy: Recent Advances and Perspectives. *Current Clinical Microbiology Reports* 6: 225–239.
7. Broder S. 2010. The development of antiretroviral therapy and its impact on the HIV 1/AIDS pandemic. *Antiviral Research* 85: 1–18.
8. Brumano LP, da Silva FVS, Costa-Silva TA, Apolinário AC, Santos JHPM, et al. 2019. Development of L-asparaginase biobetters: Current research status and review of the desirable quality profiles. *Frontiers in Bioengineering and Biotechnology* 6: 1–22.
9. Camp WL, Turnham JW, Athar M, Elmets CA. 2011. New agents for prevention of ultraviolet-induced nonmelanoma skin cancer. *Seminars in Cutaneous Medicine and Surgery* 30: 6–13.
10. Chen X, Zhou L, Tian K, Kumar A, Singh S, et al. 2013. Metabolic engineering of *Escherichia coli*: A sustainable industrial platform for bio-based chemical production. *Biotechnology Advances* 31: 1200–1223.
11. De Gregorio E, Rappuoli R. 2014. From empiricism to rational design: A personal perspective of the evolution of vaccine development. *Nature Reviews Immunology* 14: 505–514.
12. Doudna JA, Charpentier E. 2014. The new frontier of genome engineering with CRISPR-Cas9. *Science* 346: 1258096.
13. Egler R, Ahuja S, Matloub Y. 2016. L-asparaginase in the treatment of patients with acute lymphoblastic leukemia. *Journal of Pharmacology and Pharmacotherapeutics* 7: 62–71.
14. El-Nagga NEA, El-Ewasy SM, El-Shweihy NM. 2014. Microbial L-asparaginase as a potential therapeutic Agent for the Treatment of acute lymphoblastic leukemia: The pros and cons. *International Journal of Pharmacology* 10: 182–199.
15. Feachem RGA, Chen I, Akbari O, Bertozzi-Villa A, Bhatt S, et al. 2019. Malaria eradication within a generation: Ambitious, achievable, and necessary. *The Lancet* 394: 1056–1112.
16. Felfoul O, Mohammadi M, Taherkhani S, de Lanauze D, Zhong Xu Y, et al. 2016. Magneto-aerotactic bacteria deliver drug-containing nanoliposomes to tumour hypoxic regions. *Nature Nanotechnology* 11: 941–947.
17. Ferreira VL, Borba H, Bonetti A, Leonart L, Pontarolo R. 2019. Cytokines and interferons: Types and functions. In: *Autoantibodies and Cytokines*, IntechOpen, 74550.
18. Franklin D. 2013. Good Bacteria for Bad Breath. *Scientific American* 308: 30–32.
19. Graham BS, Sullivan NJ. 2018. Emerging viral diseases from a vaccinology perspective: Preparing for the next pandemic review-article. *Nature Immunology* 19: 20–28.
20. Griffiths A, Wessler S, Lewontin R, Gelbart W, Suzuki D. et al. 2004. *An Introduction to Genetic Analysis*. 8: 1–800.
21. Johns Hopkins University. 2020. Johns Hopkins Coronavirus Resource Center. Retrieved September 29, 2020, from https://coronavirus.jhu.edu
22. Jozala AF, Geraldes DC, Tundisi LL, Feitosa VA, Breyer CA, et al. 2016. Biopharmaceuticals from microorganisms: From production to purification. *Brazilian Journal of Microbiology* 47: 51–63.
23. Kardos N, Demain AL. 2011. Penicillin: The medicine with the greatest impact on therapeutic outcomes. *Applied Microbiology and Biotechnology* 92: 677–687.
24. Khan S, Ullah MW, Siddique R, Nabi G, Manan S, et al. 2016. Role of recombinant DNA technology to improve life. *International Journal of Genomics* 2016: 2405954.
25. Kresina TF 2001. *An Introduction to molecular medicine and gene therapy*: 9780471391883.
26. Kushnir N, Streatfield SJ, Yusibov V. 2012. Virus-like particles as a highly efficient vaccine platform: Diversity of targets and production systems and advances in clinical development. *Vaccine* 31: 58–83.

27. León-Buitimea A, Garza-Cárdenas CR, Garza-Cervantes JA, Lerma-Escalera JA, Morones-Ramírez JR. 2020. The demand for new antibiotics: Antimicrobial peptides, nanoparticles, and combinatorial therapies as future strategies in antibacterial Agent Design. *Frontiers in Microbiology* 11: 1669.
28. Leventhal DS, Sokolovska A, Li N, Plescia C, Kolodziej SA, et al. 2020. Immunotherapy with engineered bacteria by targeting the STING pathway for anti-tumor immunity. *Nature Communications* 11: 2739.
29. Lin S. n.d. *Good Bacteria in the Mouth and Gut Flora.* Retrieved September 28, 2020, from https://www.drstevenlin.com/good-bacteria-mouth-gut-flora.
30. Łukasiewicz K, Fol M. 2018. Microorganisms in the treatment of cancer: Advantages and limitations. *Journal of Immunology Research* 2018: 2397808.
31. McLeod C, Nerlich B, Jaspal R. 2019. Fecal microbiota transplants: Emerging social representations in the English-language print media. *New Genetics and Society* 38: 331–351.
32. Mercer DK, Torres MDT, Duay SS, Lovie E, Simpson L et al., 2020. Antimicrobial Susceptibility Testing of Antimicrobial Peptides to Better Predict Efficacy. *Frontiers in Cellular and Infection Microbiology* 10: 326.
33. Morens DM, Folkers GK, Fauci AS. 2004. The challenge of emerging and re-emerging infectious diseases. *Nature* 430: 242–249.
34. Nadeem SF, Gohar UF, Tahir SF, Mukhtar H, Pornpukdeewattana S, et al. 2020. Antimicrobial resistance: More than 70 years of war between humans and bacteria. *Critical Reviews in Microbiology* 46: 578–599.
35. Panno J. 2005. Gene Therapy - Treating Disease by Repairing Genes. *The New Biology* 1: 1–172.
36. Pardi N, Hogan MJ, Porter FW, Weissman D. 2018. mRNA vaccines — A new era in vaccinology. *Nature Reviews Drug Discovery* 17: 261–279.
37. Ramsey LB, Balis FM, O'Brien MM, Schmiegelow K, Pauley JL, et al. 2018. Consensus guideline for use of Glucarpidase in patients with high-dose methotrexate induced acute kidney injury and delayed methotrexate clearance. *The Oncologist* 23: 52–61.
38. Sam Q, Chang M, Chai L. 2017. The fungal Mycobiome and its interaction with gut bacteria in the host. *International Journal of Molecular Sciences* 18: 330.
39. Scherman D. 2014. *Advanced Textbook on Gene Transfer, Gene Therapy and Genetic Pharmacology: Principles, Delivery and Pharmacological and Biomedical Applications of Nucleotide-Based Therapies Edited. Imperial College Press.* 1: 1–544.
40. Schwab J. 2020. Fighting COVID-19 could cost 500 times as much as pandemic prevention measures. World Economic Forum Covid Action Platform. Retrieved September 28, 2020, from https://www.weforum.org/agenda/2020/08/pandemic-fight-costs-500x-more-than-preventing-one-futurity
41. Shahryari A, Jazi MS, Mohammadi S, Nikoo HR, Nazari Z, et al. 2019. Development and clinical translation of approved gene therapy products for genetic disorders. *Frontiers in Genetics* 10: 868.
42. Song B-H, Yun S-I, Woolley M, Lee Y-M. 2017. Zika virus: History, epidemiology, transmission, and clinical presentation. *Journal of Neuroimmunology* 308: 50–64.
43. Staughton J 2019. Good Bacteria Vs. Bad Bacteria: How Bacteria Can Be Healthy Too. Retrieved September 27, 2020, from https://www.scienceabc.com/humans/good-bad-gut-bacteria-human-body-probiotics-healthy.html
44. Triangle Dentistry. 2018. Balancing good and bad Bacteria in your mouth. Retrieved September 28, 2020, from https://triangledentistry.com/balancing-good-and-bad-bacteria-in-your-mouth
45. University of Washington. 2017. IHME - Institute for Health Metrics and Evaluation. Retrieved September 29, 2020, from http://www.healthdata.org
46. Wallace MJ, Fishbein SRS, Dantas G. 2020. Antimicrobial resistance in enteric bacteria: Current state and next-generation solutions. *Gut Microbes* 12: 1799654.
47. WHO. 2020. *Draft landscape of COVID-19 candidate vaccines. Covid-19.* Retrieved September 29, 2020, from https://www.who.int/publications/m/item/draft-landscape-of-covid-19-candidate-vaccines
48. Wildt S, Gerngross TU. 2005. The humanization of N-glycosylation pathways in yeast. *Nature Reviews Microbiology* 3: 119–128.
49. World Health Organization (WHO). 2013. *Global Vaccine Action Plan 2011–2020.*
50. Zhu J-W, Zhang S-J, Wang W-G, Jiang H. 2020. Strategies for discovering new antibiotics from Bacteria in the post-Genomic Era. *Current Microbiology* 77: 3213–3223.

CHAPTER 2

How Nursing Mothers Protect Their Babies with Bifidobacteria

Nick M. Jensen[1,2], Britta E. Heiss[1,2], and David A. Mills[1,2]
[1]Department of Food Science and Technology
[2]Foods for Health Institute, University of California Davis, One Shields Avenue, Davis, CA 95616, USA

2.1 BIFIDOBACTERIUM SPECIES AND DIVERSITY

Just as humans require air, water, food, and a place to live, microbes require a stable, moist habitat with an abundant store of energy. But for some bacteria, air can be deadly. Unlike most multicellular organisms, many microbes will die in the presence of oxygen, leading them to seek out environments that are reliably anaerobic. For many of these bacteria, animal guts offer not only a respite from dangerous oxygen, but also a reliable source of food in the form of undigested carbohydrates like dietary fiber that pass through the digestive tract (Deehan et al., 2016).

Bifidobacterium is one such genus of carbohydrate-degrading, oxygen-sensitive microbes. In the human gut, the majority of bacteria belong to the two large phyla Bacteroidetes and Firmicutes, with a few exceptions like the mucus-consuming *Akkermansia* (Gilbert et al., 2018). *Bifidobacterium* is the main representative from an entirely different phylum called *Actinobacteria*, which also includes the soil-dwelling *Streptomyces* that produces many medically-important antibiotics (Biavati and Mattarelli, 2018; see chapter 1). Like their environmental relatives, bifidobacteria excel at metabolizing carbohydrates, but they are too sensitive to oxygen to make a living on the forest floor. Despite this handicap, bifidobacteria can live inside a remarkable range of hosts. While many bacteria exhibit restricted associations, bifidobacteria have been isolated from animals as diverse as bees, chickens, cattle, lemurs, and humans (Mattarelli and Biavati, 2018). From humans alone, at least nine species have been discovered (Bottacini et al., 2014b).

However, these species are not evenly distributed among all humans. Indeed, their ability to colonize humans appears strikingly age dependent. The genus *Bifidobacterium* was originally discovered because of its specific association with babies. In 1899, Henri Tissier identified the first known *Bifidobacterium* when he discovered that the feces of a breastfed infant was dominated by distinctive Y-shaped bacteria (Tissier, 1900). We now know that some *Bifidobacterium* species are found almost exclusively in babies and toddlers, such as *B. breve*. Other species, like *B. catenulatum* and *B. adolescentis*, generally do not appear until after weaning. Some species are more cosmopolitan. *B. bifidum* and *B. longum*, for example, inhabit people of all ages (Kato et al., 2017). However, when considering all *Bifidobacterium* strains, their relative abundance appears to decline with the age of the human host (Kato et al., 2017).

Good Microbes in Medicine, Food Production, Biotechnology, Bioremediation, and Agriculture, First Edition. Edited by Frans J. de Bruijn, Hauke Smidt, Luca S. Cocolin, Michael Sauer, David Dowling, and Linda Thomashow.

2.2 HUMAN MILK OLIGOSACCHARIDES

While multiple factors may underlie age-dependent changes in *Bifidobacterium* levels, diet is likely the primary reason that bifidobacteria tend to be more common in infants. Bifidobacteria lack the enzymes needed to catabolize the large, intact fiber molecules that many other gut microbes can process. After these other bacteria, such as *Bacteroides*, break up complex fiber molecules into simpler carbohydrates called oligosaccharides, bifidobacteria latch onto these substrates as a more accessible energy source (Cockburn and Koropatkin, 2016). Since most undigested carbohydrates in our diet are bulky fibers, bifidobacteria generally rely on pre-degraded byproducts of other microbes to obtain their nutrition.

But there is one food in the human diet whose carbohydrates are already the perfect size for bifidobacteria: breast milk. After lactose and lipids, the third-largest component of breast milk is a wide array of human milk oligosaccharides (HMOs) (LoCascio et al., 2007). Although other mammals also produce milk oligosaccharides, human milk contains the most diverse, complex, and abundant oligosaccharides of any species (Tao et al., 2011). Their structures are elaborately decorated with rare monomeric sugars like sialic acid and fucose that are found in few plant foods (Bode and Jantscher-Krenn, 2012). Yet the enzymes required to digest them are not expressed in humans — adults or infants. On their own, HMOs may provide babies with some benefits, such as strengthening the integrity of the intestinal barrier (Chichlowski et al., 2012). But most of them pass untouched through the gut, where infant-specialized bifidobacteria like *B. longum* subsp. *infantis* and *B. breve* readily metabolize them. When HMOs are provided, infant-associated *Bifidobacterium* gain a metabolic advantage, and few other gut bacteria can outcompete them for this resource (Egan and van Sinderen, 2018). While they primarily consume free HMOs, infant-associated bifidobacteria may also metabolize other components of milk, such as milk glycans (the carbohydrate components of glycoproteins) (González-Morelo et al., 2020; Kirmiz et al., 2018).

2.3 BIFIDOBACTERIAL METABOLISM

In order to feed on HMOs, bifidobacteria rely on specialized classes of genes. *Bifidobacterium* species possess numerous genes encoding enzymes called glycoside hydrolases (GHs), which are necessary for metabolizing complex carbohydrates, as well as transport systems that import GH breakdown products (Egan and van Sinderen, 2018). Most bifidobacteria import oligosaccharides directly into the cell and metabolize the entire molecule, although some may only break off pieces outside the cell and leave leftover HMO portions for other microbes to consume (Figure 2.1) (Shani et al., 2018).

Either way, all *Bifidobacterium* species encode a wide repertoire of GHs and transporters that allow them to target specific oligosaccharides. *Bifidobacterium* strains exhibit significant heterogeneity in their GH content, with closely related strains often metabolizing distinct substrates. Most *B. longum* subsp. *longum* strains express a range of GHs that degrade plant-derived oligosaccharides; however, some strains encode a gene cluster that allows them to consume fucosylated HMOs (Garrido et al., 2016).

These metabolic differences afford each species and strain a degree of specialization that allows it to establish its own ecological niche in the intestine. Since HMOs are so diverse, many bifidobacteria preferentially consume certain subtypes, like fucosylated HMOs. Only one subspecies, *B. longum* subsp. *infantis*, has enough GHs and transporters to metabolize all known HMOs (Sela et al., 2008). Some of these transporters may import multiple HMOs, while others assimilate a restricted set, though many remain uncharacterized (Sakanaka et al., 2020). Transporters are often encoded in clusters with substrate-binding proteins (SBPs) that allow individual strains to sequester particular HMOs (Garrido et al., 2011). Operating in concert, these proteins equip each *Bifidobacterium* strain with unique metabolic capabilities that explain the wide range of growth phenotypes observed among related strains (Bottacini et al., 2014a; Milani et al., 2016).

These diverse growth phenotypes, whereby some strains preferentially consume a selective portion of the HMO pool while others degrade portions of that pool externally, leaving simple sugars behind, may foster resource sharing communities of *Bifidobacterium* strains. In fact, multiple *Bifidobacterium* species routinely co-occur in the microbiome of a single infant (Turroni et al., 2012). The specific composition of *Bifidobacterium* strains may vary widely between individuals in both infants and adults, suggesting that there is no single phenotype that consistently outperforms competitors over the entire human lifespan and range of diets (Turroni et al., 2009, 2018b). Instead, local conditions, such as HMO composition (Smilowitz et al., 2013), host genetics (Lewis et al., 2015), and background

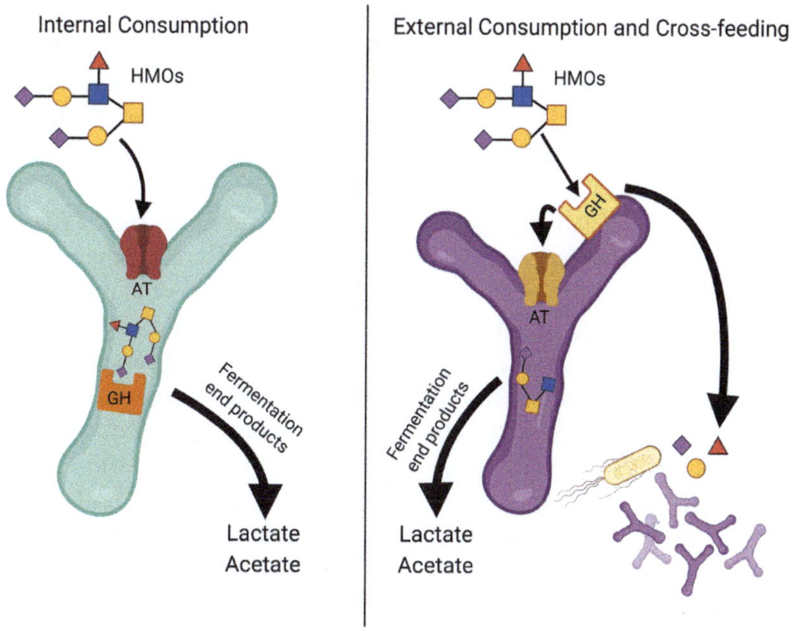

FIGURE 2.1 Modes of *Bifidobacterium* HMO consumption. Left: Many *Bifidobacterium* species, including *B. longum* subsp. *infantis*, use transporters (such as ABC transporters) to import intact human milk oligosaccharide (HMO) molecules. These bifidobacteria primarily encode internal glycoside hydrolases (GHs) to degrade the HMOs, allowing the bacteria to ferment these carbohydrates into the bioactive end products lactate and acetate. In this mode of consumption, entire HMO molecules are sequestered and metabolized. Right: Other *Bifidobacterium* species, such as *B. bifidum*, encode extracellular GHs that break up HMOs into smaller components which are then imported by transporters. These bifidobacteria also partially ferment HMOs into lactate and acetate. However, this mode of consumption releases small sugars into the gut environment, where they are metabolized by other microbes in a process termed "cross-feeding." These other microbes may include both bifidobacteria and potential pathogens such as *Enterobacteriaceae*. HMOs = human milk oligosaccharides. GH = glycoside hydrolase. AT = ABC transporter.

microbiota (Turroni et al., 2016), may lead to a unique partitioning of resources between colonizing strains.

2.4 BENEFITS OF *BIFIDOBACTERIUM*

As bifidobacteria thrive on the HMOs in breast milk, their growth also offers several benefits for their infant hosts. Bifidobacteria convert HMOs into metabolites that help babies grow and avoid illness. When bifidobacteria metabolize HMOs, they ferment them into bioactive end products. Using the enzyme fructose-6-phosphate phosphoketolase in a specialized pathway called the "bifid shunt," *Bifidobacterium* strains produce a characteristic 3:2 molar ratio of acetate to lactate (Sela et al., 2008). Both of these molecules are acidic, so they lower the pH of the gut. In a cohort of Bangladeshi children, an elevated fecal pH was linked to increased levels of childhood stunting, suggesting that the lactate and acetate may be protective (Hossain et al., 2019). Another study of Bangladeshi children recovering from malnutrition suggested that certain gut microbes, including *Bifidobacterium*, were associated with proteins that regulate childhood growth (Gehrig et al., 2019). Acetate is a short-chain fatty acid, a major class of microbial metabolites that has been intensely studied for its benefits to humans (Blaak et al., 2020; Postler and Ghosh, 2017). In mice, acetate produced by the microbiota can protect against infection by respiratory syncytial virus, a major cause of infant mortality, by signaling through the host receptor Gpr43 (Antunes et al., 2019). A similar mechanism may operate in human infants.

Bifidobacterium that ferment HMOs may also offer indirect benefits by preventing infection by pathogenic bacteria. Protection from harmful invaders, a phenomenon termed colonization resistance to pathogens, is another important service provided by gut microbes like *Bifidobacterium* (Romond et al., 1997). Low pH provides an obstacle to infection by dangerous bacteria like the foodborne pathogen *E. coli* O157:H7. In mice, acetate produced by *Bifidobacterium* limits the dissemination of Shiga toxin by strengthening the intestinal barrier (Fukuda et al., 2011). During the first month of life, fecal samples from a cohort of Japanese infants colonized by *Bifidobacterium* had lower pH and higher acetate concentrations (Matsuki et al., 2016). In addition, *Bifidobacterium* levels were negatively correlated with the relative abundance of *Enterobacteriaceae*, a family of bacteria that includes the most common human pathogens. In breastfed infants supplemented with *B. longum* subsp. *infantis*, fecal levels of proinflammatory cytokines were associated with *Enterobacteriaceae* and negatively correlated with *Bifidobacterium* relative abundance (Henrick et al., 2019). Recently, Casaburi et al. reported that a single strain of *Bifidobacterium longum* subsp. *infantis* provided to breastfed infants could reduce levels of fecal inflammatory markers and virulent endotoxin-producing bacteria and possibly protect intestinal barrier integrity by limiting mucus breakdown (Casaburi et al., 2019).

Blocking *Enterobacteriaceae* may also reduce the level of antibiotic resistance genes in the gut. When the gut microbiome of Bangladeshi infants was dominated by *Bifidobacterium*, it was found to contain fewer total antibiotic resistance genes and fewer classes of antibiotic resistance genes. This protective effect appeared to last until age two, possibly because the infants were colonized by a more diverse microbiota during weaning (Taft et al., 2018). *Bifidobacterium* may also assist in immune system development by boosting vaccine responses. At age two, mean *Bifidobacterium* abundance was associated with improved CD4 + T cell responses to essential vaccines against tetanus and hepatitis B (Huda et al., 2019).

2.5 GLOBAL DISTRIBUTION OF BIFIDOBACTERIUM

Given the many positive health outcomes associated with early, and predominant, colonization of infants by *Bifidobacterium*, supporting these microbes could benefit public health. However, fecal metagenome samples from around the world suggest that, during the first 6 months of life, even exclusively breastfed infants vary in *Bifidobacterium* species composition and prevalence, with some regions exhibiting high levels of *Bifidobacterium* colonization (often with > 50% relative abundance) (Bäckhed et al., 2015; Huda et al., 2019; Lawley et al., 2019; Taft et al., 2018; Vatanen et al., 2018), and others exhibiting a lower abundance (0–30%) (Ferretti et al., 2018; Lewis et al., 2015; Vatanen et al., 2016). For example, *Bifidobacterium* relative abundance appears higher in Russia than Finland and Estonia (Vatanen et al., 2016), and higher in Indonesia than New Zealand (Lawley et al., 2019).

Many variables may drive this variation in *Bifidobacterium* levels, including population-level breastfeeding rates, sanitation and hygiene practices, antibiotic usage, and birth delivery mode. Such factors may diminish *Bifidobacterium* levels or deplete specific HMO-catabolizing strains in the environment of the mother-infant dyad, thereby altering the level and time course of *Bifidobacterium* colonization in early life. Vatanen et al. found that infants were dominated by either *B. longum*, *B. bifidum*, or *B. breve*, with *B. longum* containing HMO-utilization genes being more common in breastfed infants (Vatanen et al., 2018). In a Swedish cohort, *Bifidobacterium* was common in infants at 4 months of age, but only in breastfed infants was it still dominant at 12 months (Bäckhed et al., 2015). Metagenomic samples from an Italian cohort suggested that individual *Bifidobacterium* strains may be transferred from mother to infant, so factors that affect a mother's gut, skin, and vaginal microbiota may indirectly impact the colonization of her baby (Ferretti et al., 2018). These studies reveal that the contemporary picture of the *Bifidobacterium*-infant association is far more complex than initial reports suggested. While it is not completely certain that *Bifidobacterium* levels have changed over time since their discovery in 1899, the average pH of infant feces has increased from 5.0 to 6.5 between 1926 and 2017 (Henrick et al., 2018). Such a dramatic increase in pH suggests that the composition of the infant microbiome has shifted over a relatively short time, possibly because of inconsistent breastfeeding rates.

2.6 SUPPORTING PERSISTENT BIFIDOBACTERIUM POPULATIONS

Since *Bifidobacterium* do not predominate in the microbiota of all infant cohorts, researchers hope to determine how to successfully seed and support the growth of these beneficial bacteria. Strains within the genus *Bifidobacterium* are often administered as probiotics for newborns and infants. Probiotics containing *Bifidobacterium* may help prevent or treat necrotizing enterocolitis, infantile colitis, acute diarrhea, obesity, allergies, and celiac disease (Di Gioia et al., 2014). Despite strong health associations, the mechanism of action for such probiotics is poorly understood, although it may involve known pathways such as acetate production. Regardless, these benefits depend on establishing persistent populations of *Bifidobacterium* within the host gut.

Breast milk may provide a nutrient niche that allows *Bifidobacterium* strains to persist if they have the appropriate metabolic genes. In a trial comparing two *Bifidobacterium* strains as probiotics for preterm infants, only *B. longum* subsp. *infantis* persisted in infants fed human milk. *Bifidobacterium animalis* subsp. *lactis*, which lacks the genes to metabolize HMOs, was unable to compete even when it was given as a supplement (Underwood et al., 2013). Recently, Frese et al. showed dramatic and persistent colonization of an HMO-catabolizing strain of *B. longum* subsp. *infantis* that was given as a supplement to breastfed infants (Frese et al., 2017). In infants with persistent *B. longum* subsp. *infantis*, fecal acetate and lactate were elevated, while pH was reduced. These infant cohorts demonstrate that the provision of an HMO-catabolizing *Bifidobacterium* can support stable engraftment across an infant population.

Bifidobacterium may be more likely to persist than other beneficial bacteria when HMOs are provided. Like *Bifidobacterium*, *Lactobacillus* species are often administered, both alone and in consortia, in infant probiotics designed to reduce necrotizing enterocolitis and infantile colic (Liu et al., 2020; Sung et al., 2018). In preterm infants supplemented with *Lactobacillus* as well as *Bifidobacterium*, only *Bifidobacterium* robustly persisted. Furthermore, *Bifidobacterium* levels correlated with the metabolism of HMOs, as measured by the production of fecal acetate. While *Lactobacillus* persistence was merely transient, the supplemented *Bifidobacterium* either persisted itself or improved the persistence of other *Bifidobacterium* species (Alcon-Giner et al., 2020). Clearly, the two microbes were not equivalent for this probiotic application.

Research on other microbes may explain the discrepancy by demonstrating the fundamental importance of a "metabolic niche." In both mice and humans recovering from antibiotics, a mixed-species probiotic failed to persist and also delayed the reconstitution of the microbiota compared to a fecal microbial transplant derived from a pre-antibiotic sample (Suez et al., 2018). Press coverage of this study prompted a wave of articles questioning the overall efficacy of probiotics. However, the *Lactobacillus* and *Bifidobacterium* species in this study's probiotic were not administered along with an appropriate carbohydrate source. "Microbiota-accessible carbohydrates" are critical for supporting populations of gut bacteria, and declining rates of dietary fiber consumption may cause the extinction of microbes bereft of a carbon source (Sonnenburg et al., 2016; Sonnenburg and Sonnenburg, 2014). Conversely, providing a unique metabolic niche can permit a single strain of a gut bacterium like *Bacteroides* to outcompete related microbes that lack the enzymatic machinery needed to degrade the target carbohydrate (Shepherd et al., 2018). *Bifidobacterium* species, not *Lactobacillus*, generally encode the enzymes and transporters needed to degrade the HMOs in the infant diet, explaining why they were able to persist. Probiotics may prove more effective if they are supported by an appropriate dietary modification such as supplementation of an accessible carbohydrate.

The metabolic niche established by HMO-consuming *Bifidobacterium* species may also depend on transport proteins. Just as a unique enzyme can allow a microbe to access a new carbohydrate, strains possessing specific transporters may enjoy a competitive advantage and more effectively colonize the human gut (Maldonado-Gómez et al., 2016). For example, an analysis of infant metagenomes revealed that HMO-specific transporters influence *Bifidobacterium* ability to colonize the infant gut. Sakanaka et al. found that the abundance of an HMO-specific transporter was correlated with *Bifidobacterium* relative abundance (Sakanaka et al., 2019). However, such an association may be geographically-dependent. When examining infant cohorts by country, a strong correlation between transporters and *Bifidobacterium* levels was detected for Malawian and Venezuelan infants, but not infants in the United States.

Fitness differences in the infant gut may also be explained by distinct mechanisms of HMO catabolism among *Bifidobacterium* species. Species such as *B. bifidum* often employ extracellular glycoside hydrolases instead of importing HMOs via transporters (Sakanaka et al., 2020). While these bifidobacteria are generally present at low levels in surveys of infant microbial communities, species that more commonly take in HMOs via transporters (such as *B. breve* and *B. longum* subsp. *infantis*) are often detected at a higher relative abundance (Lewis et al., 2015; Matsuki et al., 2016).

Such mechanistic differences underscore the importance of studying individual *Bifidobacterium* strains. While *Bifidobacterium* species and subspecies are traditionally characterized based on generalized estimates of metabolic capabilities, recognizing strain-specific differences in HMO catabolism is critical for selecting appropriate species to promote beneficial outcomes in the infant gut. A comparison of *Bifidobacterium* HMO utilization genes suggested that 14 *B. longum* subsp. *infantis* isolates could be classified into three distinct genomic variants. *B. longum* subsp. *infantis* genetically lacking an ABC-type transporter displayed a growth disadvantage in competitive assays against a strain possessing the transporter when grown on an isolated HMO. While these data were obtained *in vitro*, they indicate that genotypic variation can underlie differences in HMO growth ability (Duar et al., 2020).

These genotypic differences can result in fitness differences in the infant gut environment. In a retrospective cohort study of preterm, breastfed infants provided with a probiotic containing *B. longum* subsp. *infantis* and *L. acidophilus*, 75% of the infants still maintained a population of *B. longum* subsp. *infantis* four weeks after probiotic cessation. However, after a second round of probiotic administration, *B. longum* subsp. *longum* failed to persist, likely due to its competitive disadvantage in metabolizing HMOs. While we cannot be certain that the probiotic *B. longum* subsp. *infantis* was identical to the detected strain, as transient populations of *Bifidobacterium* circulate in NICUs (neonatal intensive care units), these data suggest that *Bifidobacterium* differ at the strain level in their ability to persist in the infant gut. Underwood et al. obtained similar results by administering HMO-catabolizing *B. longum* subsp. *infantis* and a non-HMO-catabolizing *B. animalis* subsp. *lactis* in a crossover trial of breastfed infants (Underwood et al., 2013). Regardless of the probiotic administered, HMO-catabolizing *Bifidobacterium* were the most highly represented within the infant microbiome, demonstrating that they enjoy an advantage in infant colonization.

2.7 SUMMARY

Many health benefits are clearly linked to the common colonization of the breastfed infant gut by *Bifidobacterium*. *Bifidobacterium* species produce lactate and acetate that may exclude pathogenic bacteria and support growth and immune development. These protective molecules are generated by fermenting complex HMOs, which support a specialized ecology of *Bifidobacterium* species. This natural enrichment demonstrates that appropriate dietary carbohydrates can nourish specific, and beneficial, gut microbial niches. Further research on the mechanisms underlying this enrichment will drive novel dietary and probiotic interventions that enable persistence of beneficial *Bifidobacterium* strains in vulnerable populations at any stage of life.

ACKNOWLEDGMENTS

This study was supported by National Institutes of Health awards 5T32AI060555 (BEH), RO1AT008759 (DAM) and the Peter J. Shields Endowed Chair in Dairy Foods (DAM). Figure 2.1 was created, in part, with BioRender.com

REFERENCES

1. Alcon-Giner C, Dalby MJ, Caim S, Ketskemety J, Shaw A, et al. 2020. Microbiota supplementation with *Bifidobacterium* and *Lactobacillus* modifies the preterm infant gut microbiota and metabolome: An observational study. *Cell Reports Medicine* 1: 100077.
2. Antunes KH, Luís Fachi J, de Paula R, Fraga da Silva E, Passariello Pral L, et al. 2019. Microbiota-derived acetate protects against respiratory syncytial virus infection through a GPR43-type 1 interferon response. *Nature Communications* 10: 1–17.
3. Bäckhed F, Roswall J, Peng Y, Feng Q, Jia H, et al. 2015. Dynamics and stabilization of the human gut microbiome during the first year of life. *Cell Host & Microbe* 17: 690–703.
4. Biavati B, Mattarelli P, 2018. Related genera within the family Bifidobacteriaceae. In: Mattarelli P, Biavati B, Holzaphel WH, Wood BJB (Eds.), *The Bifidobacteria and Related Organisms*, Academic Press - Elsevier, London UK. pp. 49–66.
5. Blaak EE, Canfora EE, Theis S̊, Frost G, Goren AK, et al. 2020. Short chain fatty acids in human gut and metabolic health. *Beneficial Microbes* 11: 411–455.
6. Bode L, Jantscher-Krenn E. 2012. The glycobiology of human milk oligosaccharides structure-function relationships of human milk oligosaccharides. *Advances in Nutrition* 3: 383–391.
7. Bottacini F, O'Connell Motherway MO, Kuczynski J, O'Connell KJ, Serafini F, et al. 2014a. Comparative genomics of the *Bifidobacterium breve* taxon. *BMC Genomics* 15: 170.

8. Bottacini F, Ventura M, van Sinderen D, Motherway MOC. 2014b. Diversity, ecology and intestinal function of bifidobacteria. *Microbial Cell Factories* 13: S4.

9. Casaburi G, Karav S, Frese S, Henrick B. 2019. Gut barrier function is improved in infants colonized by *Bifidobacterium longum* subsp. *Infantis EVC001 (FS04-05-19). Current Developments in Nutrition* 3: (Supplement_1) 932.

10. Chichlowski M, De Lartigue G, German JB, Raybould HE, Mills DA. 2012. Bifidobacteria isolated from infants and cultured on human milk oligosaccharides affect intestinal epithelial function. *Journal of Pediatric Gastroenterology and Nutrition* 55: 321–327.

11. Cockburn DW, Koropatkin NM. 2016. Polysaccharide degradation by the intestinal microbiota and its influence on human health and disease. *Journal of Molecular Biology* 428: 3230–3252.

12. Deehan EC, Walter J. 2016. The fiber gap and the disappearing gut microbiome: Implications for human nutrition. *Trends in Endocrinology & Metabolism* 27: 239–242.

13. Di Gioia D, Aloisio I, Mazzola G, Biavati B 2014. Bifidobacteria: Their impact on gut microbiota composition and their applications as probiotics in infants. *Applied Microbiology and Biotechnology* 98: 563–577.

14. Duar RM, Casaburi G, Mitchell RD, Scofield LNC, Ortega Ramírez CA. et al. 2020. Comparative genome analysis of *Bifidobacterium longum* subsp. *infantis* strains reveals variation in human milk oligosaccharide utilization genes among commercial probiotics. *Nutrients* 12: 3247.

15. Egan M, and van Sinderen D, 2018. Carbohydrate metabolism in bifidobacteria. In: Mattarelli P, Biavati B, Holzaphel WH, Wood BJB (Eds.), *The Bifidobacteria and Related Organisms*, Academic Press - Elsevier, London UK. pp. 145–164.

16. Ferretti P, Pasolli E, Tett A, Ascinar F, Gorfer V, et al. 2018. Mother-to-infant microbial transmission from different body sites shapes the developing infant gut microbiome. *Cell Host and Microbe* 24: 133–145.

17. Frese SA, Hutton AA, Contreras LN, Shaw CA, Palumbo MC, et al. 2017. Persistence of supplemented *Bifidobacterium longum* subsp. *infantis* EVC001 in breastfed infants. *mSphere* 2: e00501-17.

18. Fukuda S, Toh H, Hase K, Oshima K, Nakanishi Y, Yoshimura K, et al. 2011. Bifidobacteria can protect from enteropathogenic infection through production of acetate. *Nature* 469: 543–547.

19. Garrido D, Kim JH, German JB, Raybould HE, Mills DA, et al. 2011. Oligosaccharide binding proteins from *Bifidobacterium longum* subsp. *infantis* reveal a preference for host glycans. *PLoS One* 6: e17315.

20. Garrido D, Ruíz-Moyano S, Kirmiz N, Davis JC, Totten SM, et al. 2016. A novel gene cluster allows preferential utilization of fucosylated milk oligosaccharides in *Bifidobacterium longum* subsp. *longum* SC596. *Scientific Reports* 6: 35045.

21. Gehrig JL, Venkatesh S, Chang H-W, Hibberd MC, Kung VL, et al. 2019. Effects of microbiota-directed foods in gnotobiotic animals and undernourished children. *Science* 365: eaau4732.

22. Gilbert JA, Blaser MJ, Gregory Caporaso J, Jansson JK, Lynch SV. et al. 2018. Current understanding of the human microbiome. *Nature Medicine* 24: 392–400.

23. González-Morelo KJ, Vega-Sagardía M, Garrido D. 2020. Molecular insights into O-linked glycan utilization by gut microbes. *Frontiers in Microbiology* 11: 591568.

24. Henrick BM, Chew S, Casaburi G, Brown HK, Frese SA, et al. 2019. Colonization by *B. infantis* EVC001 modulates enteric inflammation in exclusively breastfed infants. *Pediatric Research* 86: 749–757.

25. Henrick BM, Hutton AA, Palumbo MC, Casaburi G, Mitchell RD, et al. 2018. Elevated fecal pH indicates a profound change in the breastfed infant gut microbiome due to reduction of *Bifidobacterium* over the past century. *mSphere* 3: e00041-18.

26. Hossain MS, Das S, Gazi MA, Alam MA, Haque NMS, et al. 2019. Association of faecal pH with childhood stunting: Results from a cross-sectional study. *BMJ Paediatrics Open* 3: 1–6.

27. Huda MN, Ahmad SM, Alam MJ, Khanam A, Kalanetra KM, et al. 2019. *Bifidobacterium* abundance in early infancy and vaccine response at 2 years of age. *Pediatrics* 143: e20181489.

28. Kato K, Odamaki T, Mitsuyama E, Sugahara H, Xiao JZ, et al. 2017. Age-related Changes in the Composition of Gut *Bifidobacterium* Species. *Current Microbiology* 74: 987–995.

29. Kirmiz N, Robinson RC, Shah IM, Barile D, Mills DA, 2018. Milk glycans and their interaction with the infant gut microbiota. *Annual Reviews in Food Science and Technology* 9: 429–450.

30. Lawley B, Otal A, Moloney-Geany K, Diana A, Houghton L, et al. 2019. Fecal microbiotas of Indonesian and New Zealand children differ in complexity and

bifidobacterial taxa during the first year of life. *Applied and Environmental Microbiology* 85: 1–14.

31. Lewis ZT, Totten SM, Smilowitz JT, Popovic M, Parker E, et al. 2015. Maternal fucosyltransferase 2 status affects the gut bifidobacterial communities of breastfed infants. *Microbiome* 3: 13.
32. Liu D, Shao L, Zhang Y, Kang W, 2020. Safety and efficacy of *Lactobacillus* for preventing necrotizing enterocolitis in preterm infants. *International Journal of Surgery* 76: 79–87.
33. LoCascio RG, Ninonuevo MR, Freeman SL, Sela DA, Grimm R, et al. 2007. Glycoprofiling of bifidobacterial consumption of human milk oligosaccharides demonstrates strain specific, preferential consumption of small chain glycans secreted in early human lactation. *Journal of Agricultural and Food Chemistry* 55: 8914–8919.
34. Maldonado-Gómez MX, Martínez I, Bottacini F, O'Callaghan A, Ventura M, et al. 2016. Stable engraftment of *Bifidobacterium longum* AH1206 in the human gut depends on individualized features of the resident microbiome. *Cell Host & Microbe* 20: 515–526.
35. Matsuki T, Yahagi K, Mori H, Matsumoto H, Hara T, et al. 2016. A key genetic factor for fucosyllactose utilization affects infant gut microbiota development. *Nature Communications* 7: 11939.
36. Mattarelli P, and Biavati B. 2018. Species in the genus *Bifidobacterium*. In: Mattarelli P, Biavati B, Holzaphel WH, Wood BJB (Eds.), *The Bifidobacteria and Related Organisms*, Academic Press - Elsevier, London UK. pp. 9–48.
37. Milani C, Turroni F, Duranti S, Lugli GA, Mancabelli L, et al. 2016. Genomics of the genus *Bifidobacterium* reveals species-specific adaptation to the glycan-rich gut environment. *Applied and Environmental Microbiology* 82: 980–991.
38. Postler TS, Ghosh S. 2017. Understanding the holobiont: How microbial metabolites affect human health and shape the immune system. *Cell Metabolism* 26: 110–130.
39. Romond MB, Haddou Z, Mielcareck C, Romond C. 1997. Bifidobacteria and human health: Regulatory effect of indigenous bifidobacteria on *Escherichia coli* intestinal colonization. *Anaerobe* 3: 131–136.
40. Sakanaka M, Gotoh A, Yoshida K, Odamaki T, Koghuchi H, et al. 2020. Varied pathways of infant gut-associated *Bifidobacterium* to assimilate human milk oligosaccharides: Prevalence of the gene set and its correlation with bifidobacteria-rich microbiota formation. *Nutrients* 12: 1–21.
41. Sakanaka M, Hansen ME, Gotoh A, Katoh T, Yoshida K, et al. 2019. Evolutionary adaptation in fucosyllactose uptake systems supports bifidobacteria-infant symbiosis. *Science Advances* 5: 1–16.
42. Sela DA, Chapman J, Adeuya A, Kim JH, Chen F, et al. 2008. The genome sequence of *Bifidobacterium longum* subsp. *infantis* reveals adaptations for milk utilization within the infant microbiome. *Proceedings of the National Academy of Sciences*. 105:18964–18969.
43. Shani GI, Lewis ZT, Robinson AM, Mills DA. 2018. Interactions between bifidobacteria, milk oligosaccharides, and neonate hosts. In: 2018. Mattarelli P, Biavati B, Holzaphel WH, Wood BJB (Eds.), *The Bifidobacteria and Related Organisms*, Academic Press - Elsevier, London UK. pp. 165–175.
44. Shepherd ES, Deloache WC, Pruss KM, Whitaker WR, Sonnenburg JL. 2018. An exclusive metabolic niche enables strain engraftment in the gut microbiota. *Nature* 557: 434–438.
45. Smilowitz JT, O'Sullivan A, Barile D, German JB, Lönnerdal B, et al. 2013. The human milk metabolome reveals diverse oligosaccharide profiles. *Journal of Nutrition* 143: 1709–1718.
46. Sonnenburg ED, Smits SA, Tikhonov M, Higginbottom SK, Wingreen NS, et al. 2016. Diet-induced extinctions in the gut microbiota compound over generations. *Nature* 529: 212–215.
47. Sonnenburg ED, Sonnenburg JL. 2014. Starving our microbial self: The deleterious consequences of a diet deficient in microbiota-accessible carbohydrates. *Cell Metabolism* 20: 779–786.
48. Suez J, Zmora N, Zilberman-Schapira G, Mor U, Dori-Bachash M, et al. 2018. Post-antibiotic gut mucosal microbiome reconstitution is impaired by probiotics and improved by autologous FMT. *Cell* 174: 1406–1423.
49. Sung V, D'Amico F, Cabana MD, Chau K, Koren G, et al. 2018. *Lactobacillus reuteri* to treat infant colic: A meta-analysis. *Pediatrics* 141: e20171811.
50. Taft DH, Liu J, Maldonado-Gómez MX, Akre S, Huda MN, et al. 2018. Bifidobacterial dominance of the gut in early life and acquisition of antimicrobial resistance. *mSphere* 3: 1–24.
51. Tissier H. 1900. Recherchés sur la flore intestinale des nourrissons (État normal et pathologique). *Thèse de médecine de l'Université de Paris*. - Paris France.

52. Tao N, Wu S, Kim J, AnHJ, HindeK, et al. 2011. Evolutionary glycomics: Characterization of milk oligosaccharides in primates. *J Proteome Res.* 101548–1557.

53. Turroni F, Marchesi JR, Foroni E, Gueimonde M, Shanahan F, et al. 2009. Microbiomic analysis of the bifidobacterial population in the human distal gut. *ISME Journal* 3: 745–751.

54. Turroni F, Milani C, Duranti S, Ferrario C, Lugli GA, et al. 2018a. Bifidobacteria and the infant gut: An example of co-evolution and natural selection. *Cell and Molecular Life Sciences* 75: 103–118.

55. Turroni F, Milani C, Duranti S, Mancabelli L, Mangifesta M, et al. 2016. Deciphering bifidobacterial-mediated metabolic interactions and their impact on gut microbiota by a multi-omics approach. *ISME Journal* 10: 1656–1668.

56. Turroni F, Milani C, van Sinderen D, Ventura M. 2018b. Bifidobacteria: Ecology and coevolution with the host. In: Mattarelli P, Biavati B, Holzaphel WH, Wood BJB (Eds.), *The Bifidobacteria and Related Organisms*, Academic Press - Elsevier, London UK. pp. 213–220.

57. Turroni F, Peano C, Pass DA, Foroni E, Severgnini M, et al. 2012. Diversity of bifidobacteria within the infant gut microbiota. *PLoS One* 7: 20–24.

58. Underwood MA, Kalanetra KM, Bokulich NA, Lewis ZT, Mirmiran M, et al. 2013. A comparison of two probiotic strains of bifidobacteria in premature infants. *Journal of Pediatrics* 163: 1585–1591.

59. Vatanen T, Franzosa EA, Schwager R, Tripathi S, Arthur TD, et al. 2018. The human gut microbiome in early-onset type 1 diabetes from the TEDDY study. *Nature* 562: 589–594.

60. Vatanen T, Kostic AD, d'Hennezel E, Siljander H, Franzosa EA, et al. 2016. Variation in microbiome LPS immunogenicity contributes to autoimmunity in humans. *Cell* 165: 842–853.

CHAPTER 3

Gut Microbiome and the Immune System

Role in Vaccine Response

Helena Ipe Pinheiro Guimaraes[1], Jorgen De Jonge[1], Debbie Van Baarle[1,2], and Susana Fuentes[1]

[1]Center for Infectious Disease Control, National Institute for Public Health and the Environment (RIVM), The Netherlands

[2]Dept. Medical Microbiology and Infection prevention, Virology and Immunology research Group, University Medical Center Groningen (UMCG), The Netherlands

3.1 IMMUNOLOGY OF VACCINES

3.1.1 Induction of Protective Immunity by Vaccination

The development of vaccines is one of science's main breakthroughs that has been saving millions of lives since the 18th century, from smallpox disease (Plotkin et al., 2011) to the H1N1 Influenza pandemic in 2009 (Borse et al., 2013), and more recently the SARS-CoV2 pandemic (Chodick et al., 2021). A vaccine works by allowing the immune system to "practice" on a weakened or killed version of the pathogen or components thereof, including surface proteins, polysaccharides, or toxins. This leads to a primary immune response, consisting of activation of the innate immune system, which in turn will stimulate adaptive associated immunity for a more specific response to eliminate and memorize the pathogen or antigen. Innate immunity is usually nonspecific in the recognition of a microorganism, acting via pathogen recognition receptors (PRRs) and the processing of microbial derived particles. This process will result in the upregulation of membrane bound molecules and cytokines that will contribute to the activation of the adaptive immune system, after specific antigen recognition by B and T cells. Innate signals will stimulate T cell proliferation and differentiation as well as B cell maturation into antibody-secreting plasma cells, leading to the elimination of the pathogen and the establishment of long-term effective memory against it (Kang et al., 2009) (Figure 3.1). While the innate arm of the immune system is the first to react, both innate and adaptive immune responses must work together for the effective control of microbial infections and the generation of immune memory after infection and vaccination.

The effector mechanisms by which the adaptive immune system can lead to clearance of an infection are mainly via (i) neutralization of the pathogen by antibodies that can bind to the antigen or induce antibody-mediated cytotoxicity (ADCC), antibody-dependent phagocytosis (ADCP) or complement dependent cytotoxicity (CDC) (further reviewed on (Van Erp et al., 2019)); (ii) killing of infected cells by CD8 + (cytotoxic) T cells; and (iii) cytokine release and B cell activation/maturation by CD4 + (helper) T cells, to reduce, control, and clear pathogens (Siegrist, 2018). Immunological memory eventually

Good Microbes in Medicine, Food Production, Biotechnology, Bioremediation, and Agriculture, First Edition. Edited by Frans J. de Bruijn, Hauke Smidt, Luca S. Cocolin, Michael Sauer, David Dowling, and Linda Thomashow.

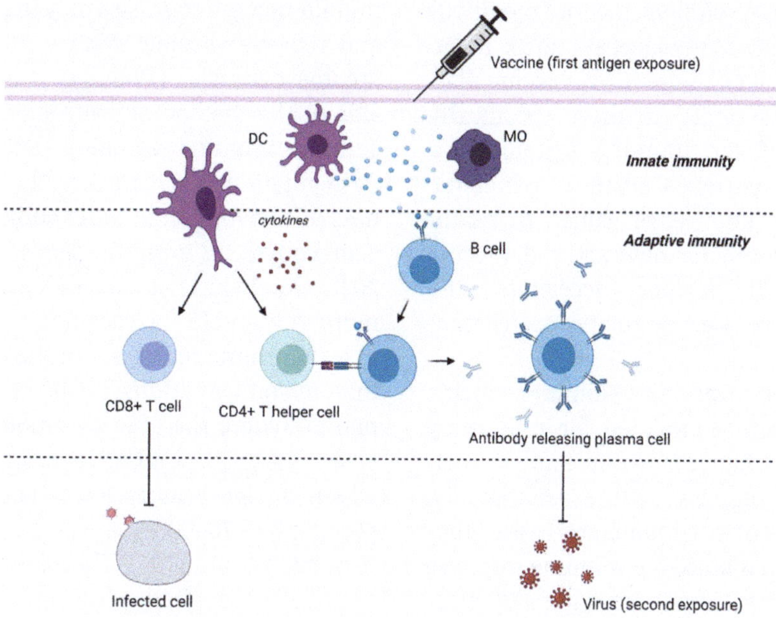

FIGURE 3.1 Figure scheme of vaccine response. The vaccine contains an antigen that is responsible for a first exposure of a pathogenic microorganism in a way that is unable to cause disease but can still confer immunity to the host. Macrophages (MO) and dendritic cells (DC) are the first line of response of the innate arm of the immune system and can process the vaccine or the pathogen itself resulting in antigen presentation. The antigen presenting cell (APC), in here illustrated as a dendritic cell, is able to migrate to the draining lymph nodes to activate T and B cells (adaptive immunity). The receptor stimulation of naive B and T cells leads to cell proliferation and maturation. CD4+ T cells will be responsible to further activate and mature B cells leading to the efficient differentiation of B cells and subsequently antibody-producing plasma cells that can prevent infection or induce effector mechanisms that can kill the pathogen (e.g., ADCC). CD8+ T cells do not prevent infection but are able to control infection by clearance of infected cells, thereby reducing disease. (Figure created by BioRender.)

guarantees long-term individual protection through the persistence of vaccine-induced antibodies above protective thresholds. In the event of a novel encounter with the vaccinated pathogen, immune memory will lead to the expansion of antibody-producing plasma cells in a faster rate, along with increased induction of T cell responses (Dutton et al., 1998; Siegrist, 2018). Besides inducing individual protective immunity, vaccines reduce the number of susceptible individuals in a population by conferring a larger number of immunized people and lowering the pathogen incidence, an indirect protection called "herd immunity" (Metcalf et al., 2015).

3.1.2 Evolution of Vaccines

One of the first techniques used to induce protective immunity was called "variolation" and it was performed by exposing the skin of the healthy individual to powdered viral scabs of a patient with a light case of smallpox, with the idea that this would result in a mild infection but protective immunity (Leung, 2011). The first official vaccine was a cowpox virus (*Variolae vaccinae*), applied by Edward Jenner, who observed that milkmaids, frequently exposed to cowpox, were less susceptible to the disease (Baxby, 2011). Three centuries later and the idea behind variolation is still used, but with a greater refinement. The key developments include methods to attenuate the pathogen via serial cultivation – a technique first used on the tuberculosis vaccine (Ravenel, 1928). By the end of the 19th century, it was discovered that some bacteria could remain immunogenic after being killed, and the complete inactivation of pathogens was used as a vaccine against the second plague pandemics (Haffkine, 1897) and as the first Influenza vaccine (Francis et al., 1937). The use of purified proteins or polysaccharides, rather than the pathogen itself, came along later in the 20th century to reduce side effects and increase safety. This approach was further narrowed down to genetically engineered antigens, allowing structure-based vaccine design, quick adaptation, and use of alternative expression systems. Finally, the human body itself is considered the antigen production platform by administering the code of the antigen via DNA, viral vectors or (modified) mRNA (Folgori et al., 2012; Mascola et al., 2020; Rajão et al., 2018; see also chapter 1). These novel vaccine platforms can be rationally designed to more specifically induce a certain arm of the immune system or a combination thereof. The current COVID-19 pandemic greatly benefits from this knowledge gained in recent years, and is one big test-case for new and old vaccine technologies which will further accelerate the development of innovative vaccine platforms (Krammer, 2020).

To further improve vaccine immunogenicity, administration strategies can be modified, for instance by a prime/boost strategy and change of the time interval between immunizations (Zhang et al., 2015). Additionally, vaccines can be formulated in combination with adjuvants, substances that can enhance the immune response e.g. through activation of the innate immune system or through providing a depot effect. For example, the bacterial polysaccharides and other constituents of the gram negative cell wall have been described for their antiviral protection via toll-like receptor signaling (Ichinohe et al., 2011; Stefan et al., 2020). Toll-like receptors (TLRs) are a specific type of PRRs expressed on innate immune cells, and its activation leads to the induction of transcription factors, the production of proinflammatory cytokines, and the upregulation of antigen-presenting cells (APCs) upon agonist stimulation. This subsequently leads to a better induction of the adaptive immune response (both B and T cells) (Tomai et al., 2012). Thus, molecules of bacterial origin allow for steering the innate response for boosting and finetuning the eventual adaptive response.

3.1.3 Vaccine Limitations

There are several hurdles and concerns to overcome in order to achieve a safe and successful immunization among different individuals. Adequate access to vaccines, high development and production costs and antigenic drift are a few of them (Oyston et al., 2012). The latter requires constant adaptation of vaccine formulation to better adapt to the seasonal strain (Valkenburg et al., 2018).

Furthermore, despite the wide protection that vaccines induce at the population level, distinct differences are observed in protection at the individual level. Vaccine efficacy can greatly vary between individuals, independently of the type of vaccine. For example, in infants, it has been shown that environmental differences, parasitic infections, and malnutrition associated with low- and middle-income countries may impair the immune system and thereby vaccine efficacy when compared to high-income countries (Praharaj et al., 2015). On the other side of the life course, with aging, both the innate and adaptive immune responses deteriorate due to immune dysregulation that can directly influence the effectiveness of vaccination in older subjects. This phenomenon, also known as immune senescence (Franceschi et al., 2020; Goodwin et al., 2006), leaves the aging population more vulnerable for vaccine preventable infectious diseases. Immune responses in the elderly are less vigorous than in younger adults, as shown for the influenza vaccination, where the efficacy among older adults was between 17% to 53% for both seroconversion and protection, which is significantly lower compared to the younger population (70–90%), as reviewed by Goodwin in 2006 (Goodwin et al., 2006). Therefore, there is a current focus to understand whether vaccination offers sufficient protection for these individuals (Murasko et al., 2002).

While age is one very important driver of vaccine efficacy, other factors should be considered, including demographics, comorbidities, previous vaccinations, and medication use, but also the intestinal microbiome. Several interventions that target the intestinal microbiome for the improvement of vaccine efficacy have been reported (Boge et al., 2009; Harris et al., 2018; Kostinov et al., 2018; Praharaj et al., 2015; Vitetta et al., 2017). Innovative strategies aiming at protecting all individuals with a high burden of morbidity and mortality against vaccine preventable diseases is of major interest for public health. In this chapter we aim to describe how the gut microbiome has both local and systemic impacts on the immune response, and how these can be used to improve vaccine efficacy.

3.2 GUT MICROBIOME AND THE IMMUNE SYSTEM

3.2.1 Microbiome Development in Life

The human gastrointestinal (GI) microbiota refers to all living microorganisms (including bacteria, archaea, protozoa, and fungi) that inhabit the human body. While plasmids, viruses and mobile genetic elements are generally excluded when referring to the microbiota, they are part of the more global term known as the microbiome. The microbiome refers to not only the microorganisms involved, but it also encompasses their ecological niches and integrative dynamics with the environment (Berg et al., 2020). The human organism is therefore the result of stable and coevolutionary associations with the microbiome, that interacts with many organs within the host, such as the skin, respiratory tract, genitourinary tract, brain, liver, and, the most studied, the gastrointestinal tract (Adolph et al., 2018; Barnard et al., 2017; Dang et al., 2019; Kundu et al., 2017; Morais et al., 2020; Neuman et al., 2015; Santoro et al., 2020).

The GI tract is a semipermeable multilayer ecosystem that harbors very diverse niches for bacterial colonization, with varying levels of colonization by different microorganisms, and allows for a multitude of host–microbe interactions (Baquero et al., 2012; Kundu et al., 2017). The gut microbiome evolves throughout different periods in life, and it is considered a potent physiological contributor to the development and homeostasis of the immune system (Moles et al., 2013; Moore et al., 2019). It is debated whether intestinal colonization starts inside the uterus, with first colonizers entering via amniotic fluid, and progressively developing during the very first weeks of life (Dominguez-Bello et al., 2010; Reyman et al., 2019). Microbiome development in early life is influenced by several factors, such as mode of delivery, breastfeeding or formula feeding, introduction to solid food, antibiotic usage, and environmental exposures (Reyman et al., 2019). It has been shown that delivery mode was associated with different fecal microbiota profiles from caesarean and vaginal delivered infants (Dominguez-Bello et al., 2010; Shao et al., 2019), which in turn had an impact on health (Depner et al., 2020). The gradual microbiome transition from a highly variable infant profile to a more stable adult-like microbiome coincides with the maturation of the immune system (Moles et al., 2013). The first days of life are thought to be critical for the training of host immunity by commensal microbiota, during the so-called "window of opportunity" (Barnard and Li, 2017; Zheng et al., 2020).

In adulthood, a healthy microbiome comprises bacteria mainly of the Firmicutes and Bacteroidetes phyla, and is characterized by high diversity and functional redundancy which can cope with environmental changes (Medina et al., 2019). In contrast, the gut microbiota of infants and older adults is characterized by lower diversity, and is more vulnerable to environmental factors (Derrien et al., 2019; Santoro et al., 2020). With the advancement of age, besides the loss in bacterial diversity, the microbiota is also characterized by high heterogeneity, with the progressive decrease in anti-inflammatory genera and increase in the proportion of potentially pathogenic bacteria (Claesson et al., 2012; Lakshminarayanan et al., 2014). As previously shown (Fuentes et al., 2021), microbiota heterogeneity in an older adult population was increased during influenza-like illness (ILI), as well as levels of pro-inflammatory taxa. This indicates that microbiota stability and composition in the aging population also plays a role in disease onset and development.

3.2.2 Host–microbe Interactions: Impact on Health

Studies associating differences in the gut microbiota composition and functionality from a healthy to a disease-state are abundant, and vary from metabolic (San-Cristobal et al., 2020) or cardiovascular diseases (Wang et al., 2018) to neurodevelopment disorders (Kang et al., 2019). The link between early exposure to microorganisms and allergy and autoimmune disorders has been explored since 1989, when David Strachan introduced the hygiene hypothesis (Strachan, 1989). Interestingly, chronic lung conditions, such as chronic obstructive pulmonary disease (COPD) and asthma are often associated with GI diseases such as irritable bowel syndrome (IBS) (Arrieta et al., 2015; Roussos et al., 2003; Yazar et al., 2001). Therefore, it is evident that the gut microbiome has an impact on health through the interaction with the host, both on a local and a global scale.

The balance of the microbiota composition in the intestinal ecosystem is controlled at different levels. Intestinal epithelial cells (IECs) are part of the host's immune defense, acting as physical barriers to foreign pathogens, producing mucin and antimicrobial peptides, also able to respond to stimuli and recruit other cells for immune activation (Peterson et al., 2014). IECs mediate local interactions of the microbiota with other immune cells (B, T, and APCs), able to detect microbial-derived products and their metabolites (De Jong et al., 2020). Here, we also find Dectin-1, a receptor associated with fungal infections through ß-glucan (a component of the fungal cell wall) recognition (Brown et al., 2005), which triggers a downstream signaling cascade resulting in the production of antimicrobial peptides directly affecting the microbiota (Iliev, 2015). Furthermore, the presence of bacteriophages and other viruses in the gut can also have an impact on the overall bacterial composition and therefore the downstream immune response. Shifts in the enteric bacteriophages present in the host and the predator–prey bacteria–bacteriophages relationship (Lim et al., 2015), or by the cross-talk between commensal bacteria and systemic viral infections (Abt et al., 2012) can affect microbiota composition. In the adaptive arm of immunity in the gut, plasma B cells release antibodies that can coat and inactivate its targets, controlling the microbial community. The repertoire of IgA antibodies is constantly adjusting to changes in the microbiome community and are able to tag and establish a controlled microbiota and restrain inflammatory processes, enhancing mutualism between the microbes and the host (Kawamoto et al., 2014; Lindner et al., 2015). Indeed cells in the lymphoid tissue known as Peyer's patches and in the lamina propria that express IgA are reduced in germ-free animals (Honda et al., 2016), and heavily coated IgA bacteria have often been associated with an exacerbated inflammation in the gut (Kau et al., 2015).

Bacteria are able to bind to IECs in the small intestine and induce the proliferation and differentiation of Th17 CD4 T cells (Atarashi et al., 2015). These are usually associated to neutrophil recruitment and activation of cells for clearance of extracellular pathogens in concert with IgA secreting cells and innate lymphoid cells (Weaver et al., 2013). It is also often associated with fungi infection, such as *Candida albicans* (Hernández-Santos et al., 2012), and with autoimmune diseases (Bedoya et al., 2013). In contrast to the pro-inflammatory profile of Th17 CD4 T cells, CD4 regulatory T cells (Treg) are crucial for controlling inflammation and immune tolerance to the microbiota in the gut (Bilate et al., 2012), and its expansion and differentiation occurs primarily in the colon, and not in the thymus, in the presence of commensal bacteria (Ohnmacht et al., 2015; Sefik et al., 2015). The colonization of mice with a consortia of clostridia species was shown to increase the number of Treg cells and the production of IgA in the mouse colon (Geuking et al., 2011). The mechanism by which a determined group of bacteria can stimulate induction of these cells and establish immune regulation is still under investigation. One possible hypothesis is that they act through the production of short-chain fatty acids (SCFAs), which are found in high concentrations in the intestinal tract as a product of bacterial fermentation (Corrêa-Oliveira et al., 2016; Furusawa et al., 2013). These molecules have been extensively studied (Den Besten et al., 2013; Peterson and Artis, 2014; Tominaga et al., 2021) and are commonly acetic, propionic, and butyric acids that signal through the activation of G protein receptors (GPCRs), inhibition of histone deacetylases (HDACs), and other pathways that are able to modify cellular responses (Figure 3.2) (Corrêa-Oliveira et al., 2016). Although these molecules are primarily a

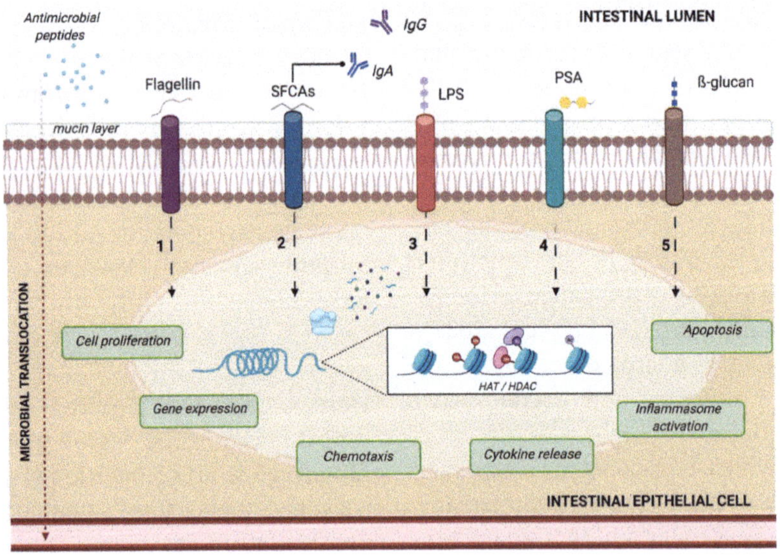

FIGURE 3.2 Microbial translocation and microbial associated molecular patterns. The intestinal lumen is filled with microbial associated molecular patterns (MAMPs) derived from the microbiota. These molecules can enter the blood stream and interact with various tissues of the host via PRR recognition. The interface of the epithelial cell layer is covered with mucin as well as antimicrobial peptides and immunoglobulins (IgA) to help control the microbial growth. When these microbial fragments encounter the receptors on the cell membrane (e.g., TLRs and dectin-1), they start the internal signaling cascade which leads to triggering of an immune response such as chemotaxis, cytokine release, and inflammasome activation or other physiologic responses of upregulating gene expression or apoptosis. The MAMPs are also able to enter the IECs and directly impact gene expression by histone acetylase (HAT) or histone deacetylase (HDAC) enzyme activation or enter the blood stream through microbial translocation and affect other tissues. (Figure created by BioRender).

waste product for the bacteria, IECs are usually able to take up SCFAs and use them as a source of energy for cell proliferation. The rapid increase of butyrate acts as an internal regulation mechanism of inhibition of HDACs, leading to a pro-apoptotic effect (Donohoe et al., 2012). Additionally, SCFAs are able to activate neutrophils and induce chemotaxis and recruitment of cells during intestinal inflammation through stimulation of G-couple membrane proteins, which are cells most involved in mounting an initial response against infectious agents (Sina et al., 2009). SCFAs also have the potential to influence B cell responses, as shown by the work of Kim et al., where levels of IgA and IgG were elevated in blood after SCFA administration (Kim et al., 2016). Indeed, luminal microbial associated molecular patterns (MAMPs) can be absorbed into the blood circulation and influence other non-GI distal tissues. Microbiota-derived molecules generated in the gut were found in various tissues after two hours of ^{13}C tagged bacteria had been administrated to germ-free mice (Uchimura et al., 2018) in a phenomenon called microbial translocation, i.e. when microbial elements are able to enter the systemic circulation from the intestinal lumen in the absence of a bacteremia (Sandler et al., 2012) (Figure 3.2). It is therefore evident that the role of the gut microbiota has both local and systemic impacts on the host, mostly achieved by the interplay of immune cells and bacteria or bacteria-derived products, which will ultimately impact the host response to infections and potentially vaccination.

3.3 MICROBIOME AND VACCINE RESPONSE

3.3.1 Mechanistic Studies in Animal Models

Animal models are usually used so that direct causality can be inferred between the composition of the microbiota to a given treatment and/or infection. In this regard, in 2011 Ichinohe et al. showed that antibiotic treated mice had an elevated viral titer for a longer period than wild-type mice upon influenza infection (Ichinohe et al., 2011). The restoration of antibody and T cell responses against influenza infection in antibiotic treated mice following use of LPS stimuli indicates that signals coming from bacterial products might be sufficient to support immune priming in the lungs via TLR-4 activation, which could be a primary signal for the inflammasome cascade. The study of Oh et al. (Oh et al., 2015) in 2015 shed light on the molecular mechanisms of the use of bacterial flagellin from the gut as a potential adjuvant for the trivalent inactivated influenza vaccine (TIV). Analysis of the transcript responses on the blood of influenza vaccinated adults revealed a strong correlation between the TLR-5 coding gene expression, an immune receptor that senses bacterial flagellin, and antibody responses against TIV. The influence of the microbiota in regard to this receptor when tested in murine models was found to be important for the antibody responses against the TIV non-adjuvanted vaccine, but not for adjuvanted or live-attenuated vaccines (Oh et al., 2015). These results indicate a microbiota-dependent mechanism for immunity against non-adjuvanted vaccines or to those poor in immunogenicity. Furthermore, TLR agonists are known to directly stimulate APCs into antigen uptake, processing, and cytokine release (Ichinohe et al., 2011; Reiser et al., 2017). Along this line, the TLR-5 receptor and flagellin have been extensively studied, since TLR-5 is usually expressed on dendritic cells in intestinal lamina propria and has been shown to participate in the production of IgA and Th17 cell differentiation (Uematsu et al., 2008). Therefore, the use of TLR agonists as adjuvants is an interesting strategy to increase vaccine efficacy.

For mucosal adjuvanted vaccines, Aoshi et al. in 2017 described two modes of action for adjuvants to work; (i) through PRR agonist and direct recognition of the immune system or (ii) via damage-associated molecular pattern (DAMP) and innate immune response activation through inflammation (Aoshi, 2017). Usually, microbial compounds can work as adjuvants through the first mode of action. Besides LPS and flagellin, a wide range of bacterial agonists are able to signal through different TLRs (further reviewed in (Kumar et al., 2019)). Upon its induction, intracellular pathways such as NF-kB can be activated and upregulate the expression of essential factors for mature B cell differentiation (Bekeredjian-Ding et al., 2009).

Not only have external receptors been described to work as a mode of action for natural adjuvants, but also intracellular cytosolic PRRs, like the Nod-like receptor family have been known to play a role in MAMP recognition and downstream signaling of inflammasomes (Sellin et al., 2018). The work of Kim et al. in 2016 explores the use of cholera toxin (CT) as a possible vaccine adjuvant that works through the Nod-2 intracellular receptor and promotes an increased immunogenicity in the mucosa of mice (Kim et al.,

2016). The IgG production to a model antigen (human serum albumin) was shown to be dependent on CT administration. Antibiotic-treated animals followed by intranasal immunization of human serum albumin (HSA) and CT had an impaired antibody response compared to untreated mice. The Nod 2-/- also had an impaired IgG and IgA production to the antigen. This work showed that pre-treatment of DCs with CT enhanced cytokine production in the cells when compared to a Nod-2 agonist, indicating that the adjuvanted activity of CT requires the Nod-2 intracellular receptor for proper generation of T helper and plasma cells, as well as for activation of DCs (Kim et al., 2016). Together, these results show a mechanistic model of how the microbiota can impact immune responses induced by immunization as adjuvants (Table 3.1). However, questions are raised when it comes to the use of the microbiota as a therapeutic intervention for humans. Randomized control-trials have been trying to observe how this knowledge can be applied to improve immunogenicity and a proper vaccine response through different interventions targeting the GI microbiota.

3.4 ROLE OF THE MICROBIOME IN VACCINE RESPONSE IN HUMAN STUDIES

The latest research in the human microbiome field has made evident that the gut microbiome can impact vaccine responses (Ciabattini et al., 2019; De Jong et al., 2020; Hagan et al., 2019; Zimmermann et al., 2018), from vaccine efficacy to its effect on extending immune memory against a specific pathogen (Ferreira et al., 2010; McCoy et al., 2019). To study the influence of the microbiota on the host immune response, antibiotic treatment is usually the standard methodology for both animal and human studies. It was found that the impact of antibiotic treatment is in general associated with a worsened immune response in unadjuvanted vaccines such as TIV and polio, in contrast to the alum-adjuvanted yellow fever vaccine, which remained unaffected (Oh et al., 2015; Pabst et al., 2014).

Bifidobacterial abundance was found positively associated with vaccine responses against tetanus toxoid (TT), when the vaccine-specific T cell stimulation index was measured fifteen weeks and two years after

TABLE 3.1 Studies that explore the effect of the gut microbiota as adjuvants in vaccination. The study outcome and design for each type of vaccine and host is under the references stated in the table

Study Design and Outcome	Type of Vaccine	Host	Reference
Germ-free and antibiotic-treated animals. Antibody responses in germ-free or antibiotic-treated animals were impaired but restored with a flagellated strain of *E. coli*	Trivalent inactivated influenza	Mice	(Oh et al., 2015)
Vaccine-specific T cell measurement after bifidobacteria administration in infants. *Bifidobacterium* abundance was positively correlated with CD4-Tcell responses to different vaccines in early life.	Tetanus toxoid (TT) and hepatitis B virus (HBV)	Human (2 years of age)	(Huda et al., 2019)
Lactobacillus-enriched cereal and vaccine-antibody measurement in infants from 4 to 13 months of age; infants were immunized with DTaP, polio, and Hib-conjugated vaccines. There was an interaction of the intervention and colonization during vaccination. Although it didn't prevent infections, immune responses to protein antigens were increased (especially in breastfed infants).	Diphtheria/tetanus toxoid/acellular pertussis, polio, and *Haemophilus influenza* type B (Hib)-conjugated vaccines	Human (from 4 to 13 months of age)	(West et al., 2008)
Review on responders and non-responders. Besides heterogeneity between studies, higher relative abundance of Actinobacteria and Firmicutes and higher vaccine responses was a consistent result. As well as the association of a higher abundance of Proteobacteria and Bacteroidetes with low responses.	Rotavirus, poliovirus, and *Salmonella typhi* oral vaccines	Human (adults and infants)	(Ciabattini et al., 2019)
Analysis of the cholera toxin as a natural adjuvant against model antigen. Antibiotic and germ-free mice had reduced IgG response against HSA. Commensal microbiota and Nod-2 receptor are required for adjuvant activity of cholera toxin.	Humoral responses against human serum albumin (HSA)	Mice	(Kim et al., 2016)

administration, and against the hepatitis B virus (HBV), measured fifteen weeks after administration (Huda et al., 2019), indicating that intestinal colonization with these bacteria is beneficial for proper vaccine responses and sustainable systemic vaccine-specific memory. Furthermore, several differences at the genus level of the overall bacterial composition in the infant gut microbiome were reported between rotavirus responders and non-responders (Harris et al., 2017). In a double-blind, placebo-controlled randomized trial, four to thirteen month old infants were fed *Lactobacillus* F19 (LF19) enriched cereal at least once a day, and antibody responses against diphtheria/tetanus toxoid/acellular pertussis, polio, and *Haemophilus influenza* type B (Hib)-conjugated vaccines were measured (West et al., 2007). Infants carrying LF19 had higher anti-diphtheria antibodies, and exclusively breastfed infants (often associated with more abundant Bifidobacteria) produced higher antibodies concentrations in response to Hib vaccine than formula-fed infants. In 2018, a systematic review (Zimmermann and Curtis, 2018) compared different vaccine delivery systems and microbiota composition among responders and non-responders in three infant and one adult study. Overall, this review stated that Firmicutes and Actinobacteria were associated with improved vaccine responses in adults and infants, whereas Bacteroidetes and Proteobacteria (mainly Enterobacteriaceae) were correlated with worsening of the immune response.

Although the healthy adult-like profile of the human intestinal microbiota is mainly stable over time (Lozupone et al., 2012), with aging, microbiota disturbances associated with an impaired immune response have been described, (Pangrazzi et al., 2020) usually accompanied by a decrease in vaccine efficacy (Goodwin et al., 2006). A study involving elderly adults consuming a fermented dairy drink with the probiotic strain *Lactobacillus paracasei* combined with *Streptococcus thermophilus* and *Lactobacillus bulgaricus* after being vaccinated for influenza, showed that the antibody titer against each strain of influenza (H1N1, H3N2, and B) was higher in the probiotic group when compared to the control group, and remained higher up to nine weeks after vaccination, corresponding to the period of consumption of the product (Boge et al., 2009). This would indicate not only the possible beneficial use of probiotics in vaccination response in the aging population, but also its contribution to the maintenance of higher antibody levels over time.

3.5 CONCLUSIONS AND FUTURE PERSPECTIVES

Since high morbidity and mortality due to respiratory infections are often due to the lack of vaccines or limited vaccine efficacy, it is important to improve response to infection and vaccine effectiveness (VE), especially in at-risk populations. As shown in this review, the effect of bacterial derived molecules originating in the gut is not limited to this ecosystem, as these molecules are able to travel through the bloodstream and stimulate immune cells at distal sites, and can therefore potentially act as natural adjuvants for vaccine responses. When used as added adjuvants into vaccines, they have been proven not only to increase immunogenicity and vaccine efficacy, but also to broaden protection against genetically close variants of the vaccine antigen (Ansaldi et al., 2010; Lefebvre et al., 2013).

The host–microbe crosstalk at the gut interphase is a fine balance between immunological responses that control microbiome homeostasis, and the ability of microbes to cause disease. This balance is established through the training of the immune system by the early life microbiome and is susceptible to different exposures and changes throughout life (Figure 3.3). One possible way to modulate the microbiome is through diet. For example, by promoting the growth of commensal bacteria that are able to produce molecules such as SCFAs through a high fiber-content diet, in turn leading to changes on the microbiota with positive effects on allergic inflammation (Trompette et al., 2014). Next to changes to diet, supplementation with probiotics (viable non-pathogenic microbial species) and prebiotics (food supplements for selectively stimulating bacterial groups related to a healthy profile) are also able to alter the immune response by modulating the intestinal microbiota (Chattha et al., 2013; Lakshminarayanan et al., 2014; Shida et al., 2008; Vitetta et al., 2017). These could be used for promoting positive changes in the microbiota that correlates with better vaccine efficacy, by diminishing the inter-individual differences between immune responses. When considering the microbiota in vaccination strategies, it is important to include the reverse scenario, where vaccines can impact the microbiota. Few studies have explored the bi-directionality of the gut-vaccine axis in a vaccination scenario, investigating both the effect of the infection and the effect of vaccination on the gut microbiota (Groves et al., 2018). In this study, although the attenuated influenza virus

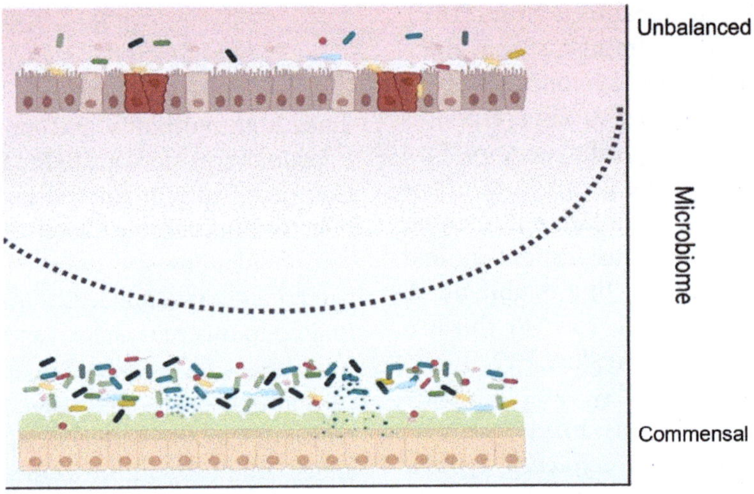

FIGURE 3.3 Microbial latency adapted from Pirofski and Casadevall 2020. Microbial latency is described as the microbial state in a disease context, where it characterizes the state of asymptotic host and non-perturbed immune homeostasis. Clinical manifestation of the "reactivation" of these microbial consortia in an unbalanced rate is usually associated with conditions that impair the immune response and it happens in a weak immune response scenario (risk-group and immune compromised). However, strong immune responses result in a dysregulation of inflammation and can affect the host–microbe interactions, leading to reoccurrence of disease (e.g., through excessive inflammation). (Figure created by BioRender).

vaccine showed to be indifferent, live viral infection was able to modulate gut microbiota composition in animals.

The microbiome is a complex variable of great relevance in vaccine studies. As discussed here, vaccine efficacy could be boosted directly, by the addition of microbiome-based adjuvants, or indirectly, via modulation of the microbiome toward a specific composition or functionality. Systematic studies in humans that investigate the GI microbiota prior to vaccination and how this relates to the induced immune response could help to identify specific groups or biomarkers for enhanced vaccine responses. Furthermore, mechanistic studies are needed to understand how these can be modulated to improve immune responses and long-lasting memory to vaccination. Other factors need to be addressed as well, such as the proper administration to achieve this: at what age, how long, use of bacteria or their metabolites, and so on. There is still much to be explored in the host immune-microbiota axis, and the important role that the gut environment plays at a systemic level and on vaccination.

REFERENCES

1. Abt MC, Osborne LC, Monticelli LA, Doering TA, Alenghat T, et al. 2012. Commensal bacteria calibrate the activation threshold of innate antiviral immunity. *Immunity* 37: 158–170.
2. Adolph TE, Grander C, Moschen AR, Tilg H. 2018. Liver–Microbiome axis in health and disease. *Trends in Immunology* 39: 712–723.
3. Ansaldi F, Zancolli M, Durando P, Montomoli E, Sticchi L, et al. 2010. Antibody response against heterogeneous circulating influenza virus strains elicited by MF59- and non-adjuvanted vaccines during seasons with good or partial matching between vaccine strain and clinical isolates. *Vaccine* 28: 4123–4129.
4. Aoshi T. 2017. Modes of action for mucosal vaccine adjuvants. *Viral Immunology* 30: 463–470.
5. Arrieta MC, Stiemsma LT, Dimitriu PA, Thorson L, Russell S, et al. 2015. Early infancy microbial and metabolic alterations affect risk of childhood asthma. *Science Translational Medicine* 7: 307RA152.
6. Atarashi K, Tanoue T, Ando M, Kamada N, Nagano Y, et al. 2015. Th17 cell induction by adhesion of microbes to intestinal epithelial cells. *Cell* 163: 367–380.
7. Baquero F, Nombela C. 2012. The microbiome as a human organ. *Clinical Microbiology and Infection* 18: 2–4.
8. Barnard E, Li H. 2017. Shaping of cutaneous function by encounters with commensals. *The Journal of Physiology* 595: 437–450.
9. Baxby D. 2011. Edward Jenner's role in the introduction of smallpox vaccine. 13–19.

10. Bedoya SK, Lam B, Lau K, Larkin J. 2013. Th17 cells in immunity and autoimmunity. *Clinical and Developmental Immunology* 2013: 16.
11. Bekeredjian-Ding I, Jego G. 2009. Toll-like receptors – Sentries in the B-cell response. *Immunology* 128: 311–323.
12. Berg G, Rybakova D, Fischer D, Cernava T, Vergès MCC, et al. 2020. Microbiome definition re-visited: Old concepts and new challenges. *Microbiome* 8: 1–22.
13. Bilate AM, Lafaille JJ. 2012. Induced CD4+Foxp3+ regulatory T cells in immune tolerance. *Annual Review of Immunology* 30: 733–758.
14. Boge T, Rémigy M, Vaudaine S, Tanguy J, Bourdet-Sicard R, et al. 2009. A probiotic fermented dairy drink improves antibody response to influenza vaccination in the elderly in two randomised controlled trials. *Vaccine* 27: 5677–5684.
15. Borse RH, Shrestha SS, Fiore AE, Atkins CY, Singleton JA, et al. 2013. Effects of vaccine program against pandemic influenza A(H1N1) virus, United States, 2009-2010. *Emerging Infectious Diseases* 19: 439–448.
16. Brown GD, Gordon S. 2005. Immune recognition of fungal β-glucans. *Cellular Microbiology* 7: 471–479.
17. Chattha KS, Vlasova AN, Kandasamy S, Rajashekara G, Saif LJ. 2013. Divergent immunomodulating effects of probiotics on T cell responses to oral attenuated human rotavirus vaccine and virulent human rotavirus infection in a neonatal gnotobiotic piglet disease model. *Journal of Immunology* 191: 2446–2456.
18. Chodick G, Tene L, Patalon T, Gazit S, Tov BA, et al. 2021. The effectiveness of the first dose of BNT162b2 vaccine in reducing SARS-CoV-2 infection 13-24 days after immunization: Real-world evidence. *medRxiv*.
19. Ciabattini A, Olivieri R, Lazzeri E, Medaglini D. 2019. Role of the microbiota in the modulation of vaccine immune responses. *Frontiers in Microbiology*.
20. Claesson MJ, Jeffery IB, Conde S, Power SE, O'Connor EM, et al. 2012. Gut microbiota composition correlates with diet and health in the elderly. *Nature* 488: 178–184.
21. Corrêa-Oliveira R, Fachi JL, Vieira A, Sato FT, Vinolo MAR. 2016. Regulation of immune cell function by short-chain fatty acids. *Clinical and Translational Immunology* 5: 1–8.
22. Dang AT, Marsland BJ. 2019. Microbes, metabolites, and the gut–lung axis. *Mucosal Immunology* 12: 843–850.
23. De Jong SE, Olin A, Pulendran B. 2020. The Impact of the Microbiome on Immunity to Vaccination in Humans. *Cell Host and Microbe* 28: 169–179.
24. Den Besten G, Van Eunen K, Groen AK, Venema K, Reijngoud DJ, et al. 2013. The role of short-chain fatty acids in the interplay between diet, gut microbiota, and host energy metabolism. *Journal of Lipid Research* 54: 2325–2340.
25. Depner M, Taft DH, Kirjavainen VP, Kalanetra KM, Karvonen AM, et al. 2020. Maturation of the gut microbiome during the first year of life contributes to the protective farm effect on childhood asthma. *Nature Medicine* 26: 1766–1775.
26. Derrien M, Alvarez A-S, De Vos WM. 2019. The Gut Microbiota in the First Decade of Life. *Trends in Microbiology* 27: 997–1010.
27. Dominguez-Bello MG, Costello EK, Contreras M, Magris M, Hidalgo G, et al. 2010. Delivery mode shapes the acquisition and structure of the initial microbiota across multiple body habitats in newborns. *Proceedings of the National Academy of Sciences* 107: 11971 LP - 11975.
28. Donohoe DR, Collins LB, Wali A, Bigler R, Sun W, et al. 2012. The Warburg effect dictates the mechanism of butyrate-mediated histone acetylation and cell proliferation. *Molecular Cell* 48: 612–626.
29. Dutton RW, Bradley LM, Swain SL. 1998. T Cell memory. *Annual Review of Immunology* 16: 201–223.
30. Ferreira RBR, Antunes LCM, Brett Finlay B. 2010. Should the human microbiome be considered when developing vaccines?. *PLoS Pathogens* 6: 1–2.
31. Folgori A, Capone S. 2012. Engineered Viruses as Vaccine Platforms. In: *Innovation in Vaccinology: From Design, through to Delivery and Testing* (Selene Baschieri), Nature Publishing Group, pp. 65–86.
32. Franceschi C, Santoro A, Capri M. 2020. The complex relationship between Immunosenescence and Inflammaging: Special issue on the New Biomedical Perspectives. *Seminars in Immunopathology* 42: 517–520.
33. Francis T, Magill TP. 1937. The antibody response of human subjects vaccinated with the virus of human influenza. *The Journal of Experimental Medicine* 65: 251–259.
34. Fuentes S, Den Hartog G, Nanlohy NM, Wijnands L, Ferreira JA, et al. 2021. Associations of faecal microbiota with influenza-like illness in participants aged 60 years or older: An observational study. *The Lancet Healthy Longevity* 2: 13–23.

35. Furusawa Y, Obata Y, Fukuda S, Endo TA, Nakato G, et al. 2013. Commensal microbe-derived butyrate induces the differentiation of colonic regulatory T cells. *Nature* 504: 446–450.
36. Geuking MB, Cahenzli J, Lawson MAE, Ng DCK, Slack E, et al. 2011. Intestinal bacterial colonization induces mutualistic regulatory T cell responses. *Immunity* 34: 794–806.
37. Goodwin K, Viboud C, Simonsen L. 2006. Antibody response to influenza vaccination in the elderly: A quantitative review. *Vaccine* 24: 1159–1169.
38. Groves HT, Cuthbertson L, James P, Moffatt MF, Cox MJ, et al. 2018. Respiratory disease following viral lung infection alters the murine gut microbiota. *Frontiers in Immunology* 9: 182.
39. Haffkine WM. 1897. Remarks on the plague prophylactic fluid. *British Medical Journal* 1: 1461–1462.
40. Hagan T, Cortese M, Rouphael N, Boudreau C, Linde C, et al. 2019. Antibiotics-Driven gut microbiome perturbation alters immunity to vaccines in humans. *Cell* 178: 1313–1328.
41. Harris VC, Armah G, Fuentes S, Korpela KE, Parashar U, et al. 2017. Significant correlation between the infant gut microbiome and rotavirus vaccine response in rural Ghana. *The Journal of Infectious Diseases* 215: 34–41.
42. Harris VC, Haak BW, Handley SA, Jiang B, Velasquez DE, et al. 2018. Effect of antibiotic-mediated microbiome modulation on rotavirus vaccine immunogenicity: A human, randomized-control proof-of-concept trial. *Cell Host and Microbe* 24: 197–207.
43. Hayashi A, Mikami Y, Miyamoto K, Kamada N, Sato T, et al. 2017. Intestinal dysbiosis and biotin deprivation induce alopecia through overgrowth of *Lactobacillus murinus* in mice. *Cell Reports* 20: 1513–1524.
44. Hernández-Santos N, Gaffen SL. 2012. Th17 cells in immunity to *Candida albicans*. *Cell Host and Microbe* 11: 425–435.
45. Honda K, Littman DR. 2016. The microbiota in adaptive immune homeostasis and disease. *Nature* 535: 75–84.
46. Huda MN, Ahmad SM, Alam MJ, Khanam A, Kalanetra KM, Taft DH, Raqib R, Underwood MA, Mills DA 2019. *Bifidobacterium* abundance in early infancy and vaccine response at 2 years of age. *Pediatrics* 143(2): e20181489. Feb. doi: 10.1542/peds.2018-1489. PMID: 30674610; PMCID: PMC6361348.
47. Ichinohe T, Pang IK, Kumamoto Y, Peaper DR, Ho JH, et al. 2011. Microbiota regulates immune defense against respiratory tract influenza a virus infection. *Proceedings of the National Academy of Sciences* 108: 5354–5359.
48. Iliev ID. 2015. Dectin-1 exerts dual control in the gut. *Cell Host and Microbe* 18: 139–141.
49. Kang D-W, Adams JB, Coleman DM, Pollard EL, Maldonado J, et al. 2019. Long-term benefit of Microbiota Transfer Therapy on autism symptoms and gut microbiota. *Scientific Reports* 9: 5821.
50. Kang S-M, Compans RW. 2009. Host Responses from innate to adaptive iImmunity after vaccination: Molecular and cellular events. *Molecules and Cells* 27: 5–14.
51. Kau AL, Planer JD, Liu J, Rao S, Yatsunenko T, et al. 2015. Functional characterization of IgA-targeted bacterial taxa from undernourished Malawian children that produce diet-dependent enteropathy. *Science Translational Medicine* 7: 276RA224.
52. Kawamoto S, Maruya M, Kato Lucia M, Suda W, Atarashi K, et al. 2014. Foxp3+ T cells regulate immunoglobulin A selection and facilitate diversification of bacterial species responsible for immune homeostasis. *Immunity* 41: 152–165.
53. Kim D, Kim Y-G, Seo S-U, Kim D-J, Kamada N, et al. 2016. Nod2-mediated recognition of the microbiota is critical for mucosal adjuvant activity of cholera toxin. *Nature Medicine* 22: 524–530.
54. Kostinov MP, Akhmatova NK, Khromova EA, Skhodova SA, Stolpnikova VN, et al. 2018. The impact of adjuvanted and non-adjuvanted Influenza vaccines on the innate and adaptive immunity effector. In: Saxena, Shailendra K. (Edited by), *Influenza - Therapeutics and Challenges*, IntechOpen, pp. 38.
55. Krammer F. 2020. SARS-CoV-2 vaccines in development. *Nature* 586: 516–527.
56. Kumar S, Sunagar R, Gosselin E. 2019. Bacterial protein Toll-like-receptor agonists: A novel perspective on vaccine adjuvants. *Frontiers in Immunology* 10: 1144.
57. Kundu P, Blacher E, Elinav E, Pettersson S. 2017. Our Gut Microbiome: The Evolving Inner Self. *Cell* 171: 1481–1493.
58. Lakshminarayanan B, Stanton C, O'Toole PW, Ross RP. 2014. Compositional dynamics of the human intestinal microbiota with aging: Implications for health. *The Journal of Nutrition, Health and Aging* 18: 773–786.
59. Lefebvre JS, Haynes L. 2013. Vaccine strategies to enhance immune responses in the aged. *Current Opinion in Immunology* 25: 523–528.

60. Leung AKC. 2011. "Variolation" and vaccination in late imperial China, ca 1570–1911. In: Plotkin S. (eds), *History of Vaccine Development*, Springer, New York, NY, pp. 5–12.

61. Lim ES, Zhou Y, Zhao G, Bauer IK, Droit L, et al. 2015. Early life dynamics of the human gut virome and bacterial microbiome in infants. *Nature Medicine* 21: 1228–1234.

62. Lindner C, Thomsen I, Wahl B, Ugur M, Sethi MK, et al. 2015. Diversification of memory B cells drives the continuous adaptation of secretory antibodies to gut microbiota. *Nature Immunology* 16: 880–888.

63. Lozupone CA, Stombaugh JI, Gordon JI, Jansson JK, Knight R. 2012. Diversity, stability and resilience of the human gut microbiota. *Nature* 489: 220–230.

64. Mascola JR, Fauci AS. 2020. Novel vaccine technologies for the 21st century. *Nature Reviews Immunology* 20: 87–88.

65. McCoy KD, Burkhard R, Geuking MB. 2019. The microbiome and immune memory formation. *Immunology and Cell Biology* 97: 625–635.

66. Medina DA, Li T, Thomson P, Artacho A, Pérez-Brocal V, et al. 2019. Cross-regional view of functional and taxonomic microbiota composition in obesity and post-obesity treatment shows country specific microbial contribution. *Frontiers in Microbiology* 10: 2346.

67. Metcalf CJE, Ferrari M, Graham AL, Grenfell BT. 2015. Understanding herd immunity. *Trends in Immunology* 36: 753–755.

68. Moles L, Gómez M, Heilig H, Bustos G, Fuentes S, et al. 2013. Bacterial diversity in meconium of preterm neonates and evolution of their fecal microbiota during the first month of life. *PLoS ONE* 8.

69. Moore RE, Townsend SD. 2019. Temporal development of the infant gut microbiome. *Open Biology* 9.

70. Morais LH, Schreiber HL, Mazmanian SK. 2020. The gut microbiota–brain axis in behaviour and brain disorders. *Nature Reviews Microbiology* 19: 241–255.

71. Murasko DM, Bernstein ED, Gardner EM, Gross P, Munk G, et al. 2002. Role of humoral and cell-mediated immunity in protection from influenza disease after immunization of healthy elderly. *Experimental Gerontology* 37: 427–439.

72. Neuman H, Debelius JW, Knight R, Koren O. 2015. Microbial endocrinology: The interplay between the microbiota and the endocrine system. *FEMS Microbiology Reviews* 39: 509–521.

73. Oh JZ, Ravindran R, Chassaing B, Carvalho FA, Maddur S, et al. 2015. TLR5-Mediated sensing of gut microbiota is necessary for antibody responses to seasonal Influenza vaccination. *Immunity* 41: 478–492.

74. Ohnmacht C, Park J-H, Cording S, Wing JB, Atarashi K, et al. 2015. The microbiota regulates type 2 immunity through RORγt+ T cells. *Science* 349: 989 LP - 993.

75. Oyston P, Robinson K. 2012. The current challenges for vaccine development. *Journal of Medical Microbiology* 61: 889–894.

76. Pabst O, Hornef M. 2014. Gut microbiota: A natural adjuvant for vaccination. *Immunity* 41: 349–351.

77. Pangrazzi L, Weinberger B. 2020. T cells, aging and senescence. *Experimental Gerontology* 134: 110887.

78. Peterson LW, Artis D. 2014. Intestinal epithelial cells: Regulators of barrier function and immune homeostasis. *Nature Reviews Immunology* 14: 141–153.

79. Plotkin SA, Plotkin SL. 2011. The development of vaccines: How the past led to the future. *Nature Reviews Microbiology* 9: 889–893.

80. Praharaj I, John SM, Bandyopadhyay R, Kang G. 2015. Probiotics, antibiotics and the immune responses to vaccines. *Philosophical Transactions of the Royal Society B: Biological Sciences* 370.

81. Rajão DS, Pérez DR. 2018. Universal vaccines and vaccine platforms to protect against Influenza viruses in humans and agriculture. *Frontiers in Microbiology* 9: 123.

82. Ravenel MP. 1928. La vaccination préventive contre la Tuberculose par le "BCG." *American Journal of Public Health and the Nations Health* 18: 1075.

83. Reiser ML, Mosaheb MM, Lisk C, Platt A, Wetzler LM. 2017. The TLR2 binding neisserial porin PorB enhances antigen presenting cell trafficking and cross-presentation. *Scientific Reports* 7: 736.

84. Reyman M, Van Houten MA, Van Baarle D, Bosch AATM, Man WH, et al. 2019. Impact of delivery mode-associated gut microbiota dynamics on health in the first year of life. *Nature Communications* 10: 4997

85. Roussos A, Koursarakos P, Patsopoulos D, Gerogianni I, Philippou N. 2003. Increased prevalence of irritable bowel syndrome in patients with bronchial asthma. *Respiratory Medicine* 97: 75–79.

86. San-Cristobal R, Navas-Carretero S, Martínez-González MÁ, Ordovas JM, Martínez JA. 2020. Contribution of macronutrients to obesity: Implications for precision nutrition. *Nature Reviews Endocrinology* 16: 305–320.

87. Sandler NG, Douek DC. 2012. Microbial translocation in HIV infection: Causes, consequences and treatment opportunities. *Nature Reviews Microbiology* 10: 655–666.

88. Santoro A, Zhao J, Wu L, Carru C, Biagi E, et al. 2020. Microbiomes other than the gut: Inflammaging and age-related diseases. *Seminars in Immunopathology* 42: 589–605.

89. Sefik E, Geva-Zatorsky N, Oh S, Konnikova L, Zemmour D, et al. 2015. Individual intestinal symbionts induce a distinct population of RORγ regulatory T cells. *Science* 349: 993 LP - 997.

90. Sellin ME, Müller AA, Hardt WD. 2018. Consequences of epithelial inflammasome activation by bacterial pathogens. *Journal of Molecular Biology* 430: 193–206.

91. Shao Y, Forster SC, Tsaliki E, Vervier K, Strang A, et al. 2019. Stunted microbiota and opportunistic pathogen colonization in caesarean-section birth. *Nature* 574: 117–121.

92. Shida K, Nanno M. 2008. Probiotics and immunology: Separating the wheat from the chaff. *Trends in Immunology* 29: 565–573.

93. Siegrist C-A. 2018. Vaccine Immunology.

94. Sina C, Gavrilova O, Förster M, Till A, Derer S, et al. 2009. G Protein-coupled receptor 43 is essential for neutrophil recruitment during intestinal inflammation. *The Journal of Immunology* 183: 7514–7522.

95. Stefan KL, Kim VM, Iwasaki A, Kasper DL. 2020. Commensal microbiota modulation of natural resistance to virus infection. *Cell* 183: 1–13.

96. Strachan DP. 1989. Hay fever, hygiene, and household size. *British Medical Journal* 20(299): 1260.

97. Tomai M, Vasilakos J. 2012. TLR agonists as vaccine adjuvants. *Innovation in Vaccinology: From Design, through to Delivery and Testing* 205: 205–228.

98. Tominaga K, Tsuchiya A, Mizusawa T, Matsumoto A, Minemura A, et al. 2021. Evaluation of intestinal microbiota, short-chain fatty acids, and immunoglobulin a in diversion colitis. *Biochemistry and Biophysics Reports* 25: 100892.

99. Trompette A, Gollwitzer ES, Yadava K, Sichelstiel AK, Sprenger N, et al. 2014. Gut microbiota metabolism of dietary fiber influences allergic airway disease and hematopoiesis. *Nature Medicine* 20: 159–166.

100. Uchimura Y, Fuhrer T, Li H, Lawson MA, Zimmermann M, et al. 2018. Antibodies set boundaries limiting microbial metabolite penetration and the resultant mammalian host response. *Immunity* 49: 545–559.

101. Uematsu S, Fujimoto K, Jang MH, Yang B-G, Jung Y-J, et al. 2008. Regulation of humoral and cellular gut immunity by lamina propria dendritic cells expressing Toll-like receptor 5. *Nature Immunology* 9: 769–776.

102. Valkenburg SA, Leung NHL, Bull MB, Yan L-M, Li APY, et al. 2018. The hurdles from bench to bedside in the realization and implementation of a universal Influenza vaccine. *Frontiers in Immunology* 9: 1479.

103. Van Erp EA, Luytjes W, Ferwerda G, Van Kasteren PB. 2019. Fc-mediated antibody effector functions during Respiratory Syncytial Virus infection and disease. *Frontiers in Immunology* 10: 548.

104. Vitetta L, Saltzman ET, Thomsen M, Nikov T, Hall S. 2017. Adjuvant probiotics and the intestinal microbiome: Enhancing vaccines and immunotherapy outcomes. *Vaccines* 5: 1–17.

105. Wang Z, Zhao Y. 2018. Gut microbiota derived metabolites in cardiovascular health and disease. *Protein and Cell* 9: 416–431.

106. Weaver CT, Elson CO, Fouser LA, Kolls JK. 2013. The Th17 pathway and inflammatory diseases of the intestines, lungs, and skin. *Annual Review of Pathology: Mechanisms of Disease* 8: 477–512.

107. West C, Gothefors L, Granström M, Käyhty H, Hammarström M-L, et al. 2007. Effects of feeding probiotics during weaning on infections and antibody responses to diphtheria, tetanus and Hib vaccines. *Pediatric Allergy and Immunology* 19: 53–60.

108. West CE, Gothefors L, Granström M, Käyhty H, Hammarström ML, Hernell O. 2008, Feb. Effects of feeding probiotics during weaning on infections and antibody responses to diphtheria, tetanus and Hib vaccines. *Pediatr Allergy Immunol* 19(1): 53–60. doi: 10.1111/j.1399-3038.2007.00583.x. Epub 2007 Dec 11. PMID: 18086218.

109. Yazar A, Atis S, Konca K, Pata C, Akbay E, et al. 2001. Respiratory symptoms and pulmonary functional changes in patients with irritable bowel syndrome. *The American Journal of Gastroenterology* 96: 1511–1516.

110. Zhang L, Wang W, Wang S. 2015. Effect of vaccine administration modality on immunogenicity and efficacy. *Expert Review of Vaccines* 14: 1509–1523.

111. Zheng D, Liwinski T, Elinav E. 2020. Interaction between microbiota and immunity in health and disease. *Cell Research* 30: 492–506.

112. Zimmermann P, Curtis N. 2018. The influence of the intestinal microbiome on vaccine responses. *Vaccine* 36: 4433–4439.

CHAPTER 4

Probiotics for Prevention or Treatment of Food Allergies

Agnes S. Y. Leung[1]*, Wenyin Loh[2], and Mimi L. K. Tang[3,4,5]

[1]Department of Paediatrics, Prince of Wales Hospital, The Chinese University of Hong Kong, Hong Kong
[2]Allergy Service, Department of Paediatrics, KK Women's and Children's Hospital, Singapore 229899, Singapore
[3]Allergy Immunology, Murdoch Children's Research Institute, Melbourne 3052, Australia
[4]Department of Allergy and Immunology, The Royal Children's Hospital, Melbourne 3052, Australia
[5]Department of Paediatrics, University of Melbourne, Melbourne 3010, Australia
*Agnes Leung is an Honorary Research Fellow at Murdoch Children's Research Institute, Melbourne, Australia

4.1 INTRODUCTION

Food allergy prevalence has increased exponentially in recent decades across all regions of the world (Loh and Tang, 2018). There are currently few effective prevention strategies and prevalence continues to increase, with recent studies showing that food allergy now affects 1 in 10 babies and 1 in 20 children. Even more dire is the lack of effective treatments that offer a long-term solution for patients with food allergies. Current standard of care remains focused on strict allergen avoidance and management of allergic reactions when avoidance fails. The constant vigilance required to avoid allergen(s) and the unpredictability of reactions due to accidental exposure cause significant psychological distress and lifestyle restrictions, which together result in severely reduced quality of life for both the patient and their family. The recently approved desensitizing peanut oral immunotherapy, Palforzia, can reduce the risk of allergic reaction following accidental peanut exposure; however, patients must continue indefinite daily dosing and reactions to treatment are frequent. Moreover, patients taking Palforzia must continue to strictly avoid peanut, which means that lifestyle restrictions associated with avoidance are not removed. Based on the above, some experts have suggested that desensitization may not be an optimal outcome for patients with a food allergy. There remains a significant unmet need for a treatment that can provide longer-lasting protection through inducing sustained unresponsiveness (also referred to as remission (Burks et al., 2018)) and has reduced side effects.

The rapid rise in food allergy is attributed to changes in the environment, particularly those associated with the western lifestyle. One dominant theory is the hygiene hypothesis which proposes that the lack of early childhood exposure to microbes predisposes individuals to allergic diseases by disrupting the healthy development of the immune system (Strachan, 1989; see also chapter 3). There is now growing evidence that dysbiosis, as a result of the modern living environment, diet, and lifestyle, has an important role in the pathogenesis and development of food allergy (Bunyavanich and Berin, 2019). Furthermore, studies have shown that the gut microbiota differ in populations with and without food allergy. Infants with allergies have been shown to have significantly lower counts of some fecal species, such as Bifidobacteria, compared to healthy infants (Björkstén, 2001).

If the gut microbiota composition can indeed influence the risk of developing food allergy and modu

Good Microbes in Medicine, Food Production, Biotechnology, Bioremediation, and Agriculture, First Edition. Edited by Frans J. de Bruijn, Hauke Smidt, Luca S. Cocolin, Michael Sauer, David Dowling, and Linda Thomashow.

late immune mechanisms driving the allergic response, it seems logical to investigate the use of microbial therapeutics for the prevention and/or treatment of food allergy. Probiotic bacteria, defined as live bacteria that provide a beneficial effect when administered in adequate amounts, may offer a means of altering the gut microbiota toward a healthy composition. Probiotics have a long record of safe oral consumption and are generally regarded as safe (GRAS) by the FDA. The effects of probiotics appear to be primarily mediated via the innate immune system while selected *Lactobacillus* species have been shown to support tolerogenic immune responses including promoting Th1 cytokine responses with increased production of IFN-γ, T regulatory responses with increased production of IL-10 (Mohamadzadeh et al., 2005; Pessi et al., 2000), and allergen-specific IgA responses (Isolauri et al., 1995; Rautava et al., 2006).

This chapter summarizes the evidence for probiotics in the prevention and treatment of food allergy.

4.2 PREVENTION OF FOOD ALLERGY

Strategies for food allergy prevention include primary prevention which aims to prevent the onset of IgE sensitization, and secondary prevention which aims to interrupt the development of food allergy in IgE-sensitized children. Probiotics used for prevention of food allergies can either be given directly to the infant in a milk formula or as an oral supplement, or indirectly through the mother during breastfeeding or pregnancy.

Early studies suggested that the patterns of gut microbial colonization in the first weeks of life may influence immune development, including development of allergies (Björkstén, 2004). One of the earliest randomized controlled trials assessing the efficacy of probiotics in the primary prevention of atopic disease recruited 159 pregnant mothers who had at least one first-degree relative (or partner) with atopic eczema, allergic rhinitis, or asthma. Subjects were randomized to receive placebo or 1×10^{10} colony-forming units of *Lactobacillus rhamnosus* strain GG (ATCC 53103; Valio Ltd, Helsinki, Finland) daily for 4 weeks prior to expected delivery and continued postnatally administered to their infants for 6 months. While the frequency of atopic eczema in the probiotic group was half that of the placebo group at 2-year follow up (23 vs 46%; $p = 0.008$), there was no significant difference in rates of food sensitization (Kalliomaki et al., 2001). The same group subsequently reported that the preventive effect of *Lactobacillus* GG on atopic eczema extended to 7 years of age (Kalliomaki et al., 2003, 2007). Interestingly, increased rates of allergic rhinitis (22 vs 10%,; RR, 2.30; 95% CI, 0.93–5.70) and asthma (17% vs 5%; RR, 3.44; 95% CI, 0.98–12.1) were reported in the *Lactobacillus* GG group compared to the placebo group at 7 years of age. Another primary prevention trial from Sweden randomized 188 pregnant mothers whose fetuses were at high risk for allergy at 35 weeks of gestation to receive probiotics or placebo (Abrahamsson et al., 2007). Pregnant mothers in the probiotic group received *Lactobacillus reuteri* ATCC 55730 (1×10^8 colony forming units) daily from gestational week 36 until delivery, and their babies then continued the same *L. reuteri* supplementation from birth until 12 months of age. At 2 years, the *L. reuteri* group had a lower incidence of atopic dermatitis and allergen sensitization, but not wheeze, compared with the placebo group. Another trial of similar study design randomized high risk mothers to receive either a probiotic mixture including *Lactobacillus* GG (ATCC; 5×10^9 colony-forming units [cfu]), *L. rhamnosus* LC705 (DSM 7061; 5×10^9 cfu), *Bifidobacterium breve* Bb99 (DSM 13692; 2×10^8 cfu), and *Propionibacterium freundii* spp. Sharmonii JS (DSM 7076; 2×10^9 cfu) or placebo twice daily starting from 36 weeks of gestation (Kuitunen et al., 2009). Their infants then received the same probiotic combination mixed with galacto-oligosaccharides once daily from birth until 6 months. Results showed that the risk of allergen sensitization and eczema was only significantly reduced in the caesarean-delivered probiotic group compared with the caesarean-delivered placebo group at 5 years, but no allergy preventive effect was evident in the vaginally delivered children. The authors concluded that probiotics may have restored the deficient microbiota of children delivered via caesarean section.

Conversely, a double-blind, placebo-controlled randomized trial involving 253 Asian infants at risk of allergic disease and supplemented with *Bifidobacterium longum* 1×10^7 colony-forming units and *Lactobacillus rhamnosus* 2×10^7 colony-forming units in cow's milk formula daily for the first 6 months showed no effect on prevention of eczema or allergen sensitization in the first year of life (Soh et al., 2009). In a randomized controlled trial of 178 infants born to atopic mothers in Australia, either probiotic supplementation with *Lactobacillus acidophilus* (LAVRI-A1) or placebo was administered daily for the first 6 months of life. At 6 and 12 months, not only was the rate of atopic dermatitis and food allergy not significantly different between groups, but the rate of allergen sensitization was found to have paradoxically increased in the probiotic group (Taylor et al., 2007). At 5 years, the proportion of children with allergic diseases was not significantly different between groups while a trend toward development of eczema and recurrent wheezing was noted in the probiotic group (Jensen et al., 2012). Another Australian study randomised 250 pregnant women carrying infants at high risk of allergic disease to receive *L rhamnosus* GG (ATCC 53103; 1.8×10^{10} colony-forming units once daily) from 36 weeks gestation until delivery. Prenatal supplementation did not reduce the risk of infant eczema at 12 months of age (Boyle et al., 2011). A German study randomized 105 pregnant women carrying high risk infants to receive either the probiotic *Lactobacillus rhamnosus* GG (ATCC 53103; 5×10^{9} colony-forming units twice daily) or placebo from 4 to 6 weeks before expected delivery and continued postnatally for 6 months (taken by breastfeeding mothers for 3 months followed by direct oral intake by infants for another 3 months) (Kopp et al., 2008). At 2 years, the incidence and severity of atopic dermatitis were not different between the groups, instead an increased rate of recurrent episodes of wheezing bronchitis was found in the probiotic group.

These varied and inconsistent results may be due to differences in the probiotic strains, dose, timing and duration of therapy used in the different studies. Furthermore, most of these randomized trials assessed the role of probiotics in preventing eczema; however, there are few high-quality randomized trials that specifically address the effects of probiotics on food allergy. A systematic review of 29 randomized trials assessing the effects of any probiotic administered to pregnant women, breastfeeding mothers, and/or infants, found that probiotics administered prenatally and/or postnatally reduced the risk of eczema but had little effect on other allergic conditions, including food allergy (Cuello-Garcia et al., 2015). Based on the results of the systemic review, the World Allergy Organisation published the GLAD-p guideline recommending probiotic supplementation to mothers during pregnancy and infants in the first year of life with the open acknowledgment that this was supported by very low-quality evidence and that the likely net benefit was on the prevention of eczema (Fiocchi et al., 2015). Nevertheless, contrary to the recommendations of the WAO, most expert guidelines including in the US (Thomas and Greer, 2010), EU (Muraro et al., 2014), UK (Turner et al., 2018), and Australia do not recommend the use of probiotics for prevention of food allergy or other allergic diseases. Further research looking directly at the effect of probiotics on food allergy is clearly warranted.

4.3 TREATMENT OF FOOD ALLERGY

The search for effective therapies for food allergy has become extremely important as disease prevalence continues to rise (Gupta et al., 2019; Osborne et al., 2011). Initial studies of subcutaneous immunotherapy (SCIT) for the treatment of peanut allergy showed that this approach was associated with high rates of systemic reactions and was subsequently abandoned (Oppenheimer et al., 1992). Current research has instead focused on oral, sublingual, and epicutaneous food immunotherapy or SCIT using modified allergens that do not trigger the immediate allergic response.

There are several outcomes that can be achieved with a food immunotherapy treatment. Desensitization (DS) – a temporary increase in the threshold for allergen reactivity (Burks et al., 2018) – provides protection from accidental exposure to the allergen, however, requires continued daily dosing to maintain protection and patients must continue strict allergen avoidance as the underlying allergic response has not been addressed. Two recent studies of peanut OIT showed that discontinuation of regular peanut intake/OIT after initial therapy to induce desensitization resulted in a reduced level of protection and increased reactions (Chinthrajah et al., 2019; Vickery et al., 2020), highlighting the importance of remaining on daily dosing indefinitely. Sustained unresponsiveness (SU), also referred to as remission, is defined as a lack of clinical reaction to a food allergen after active therapy has been discontinued for a period of time. SU is a preferable outcome compared to DS as it allows patients to cease treatment and introduce the allergen into their diet if they wish, freeing them from the burden of allergen avoidance. Acquisition of SU which allowed free intake of peanut in the diet was shown to result in a sustained beneficial effect on quality of life (Dunn Galvin et al., 2018).

Randomized trials of peanut, cow's milk, and egg oral immunotherapy (OIT) have consistently demonstrated efficacy at inducing desensitization. A review of randomized controlled, open label, and uncontrolled trials reported a desensitization rate with peanut OIT of between 49% and 87%, with the level of desensitization in the various studies ranging from 300 mg to 10 g peanut protein (Tang et al., 2020). Desensitization is however associated with increased reactions and a recent systematic review and a meta-analysis (Chu et al., 2019) reported that although current peanut OIT regimens were effective in achieving desensitization, the risks of anaphylaxis (risk ratio, 3.12), anaphylaxis frequency (incidence rate ratio, 2.72), and use of epinephrine (RR, 2.21) were significantly higher compared to peanut avoidance. It is therefore uncertain whether desensitization offers an optimal outcome for patients. Few OIT studies have examined for SU and those few studies which have evaluated for this outcome have reported limited efficacy with around 30%–35% of treated patients successfully achieving sustained unresponsiveness (Chinthrajah et al., 2019; Syed et al., 2014). Novel approaches to improve the efficacy and safety of OIT are warranted.

4.3.1 Clinical Use of Probiotics in Food Immunotherapy

Clinical studies have investigated the efficacy of selected probiotic strains for treatment of food allergy. Tang et al. (Tang et al., 2015) evaluated the coadministration of a bacterial adjuvant with peanut OIT in young children aged 1–10 years. This phase 2 proof-of-concept double-blind, placebo-controlled trial randomized 62 children with peanut allergy 1:1 to receive a combination of probiotic *Lactobacillus rhamnosus* CGMCC 1.3724 and peanut OIT (PPOIT) or placebo (placebo probiotic and placebo OIT) for 18 months. Sustained unresponsiveness was the primary outcome, determined by DBPCFC conducted 2 to 6 weeks after discontinuation of treatment. The PPOIT group achieved a significantly higher rate of SU (82.1%) compared with the placebo group (3.6%, $P < .001$). PPOIT was also highly effective in desensitizing peanut-allergic individuals (90%), while maintaining a favorable safety profile. At the end of treatment, subjects who achieved SU were advised to introduce peanut freely into the diet without any specific instructions regarding frequency or amount, while all other subjects (including those who were desensitized without achieving SU and those who were not desensitized) were advised to continue with strict peanut avoidance. Long-term follow-up of trial participants 4 years after treatment completion showed long-term SU (assessed by DBPCFC after 8 weeks of peanut elimination at 4 years post-treatment) in 58.3% PPOIT-treatment participants compared to 6.7% placebo participants ($P = .012$) (Hsiao et al., 2017). Among initial treatment responders, 70% had persistence of SU at 4 years post-treatment. The PPOIT-treated participants had smaller peanut skin prick test wheal size and higher peanut sIgG4:sIgE ratio compared to the placebo participants, both at end-of-treatment and at 4 years post-treatment. Sixty-seven percent of PPOIT-treated participants had continuing peanut intake at 4 years post-treatment, confirming that the majority of children who achieved SU had continued to include peanut in their diet as recommended by the study team. A phase 2b trial directly comparing the combination probiotic peanut OIT with OIT alone and placebo is underway to clarify whether addition of the probiotic leads to improved efficacy or safety of OIT (or both).

The effect of *L. rhamnosus* GG probiotic on tolerance acquisition in cow's milk allergic patients was examined by Berni Canani et al. (Berni Canani et al., 2013) in an open-controlled study which randomly allocated 260 infants into three groups: an intervention group that received extensively hydrolyzed casein formula (EHCF) ($n = 55$), a group that received EHCF containing the probiotic strain *Lactobacillus rhamnosus* ATCC 53103 (at least 1.4×10^7 colony-forming units/100 mL) ($n = 71$), and control groups that received either hydrolyzed rice formula ($n = 46$), soy formula ($n = 55$), or amino acid-based formula ($n = 33$). Use of EHCF could be considered a quasi OIT – the cow's milk allergen was modified (hydrolyzed) to remove IgE-reactive epitopes, but residual casein peptides may continue to support immune-modulatory effects (Gill et al., 2000; Sandré et al., 2001). The control groups on hydrolyzed rice, soy, and amino acid-based formula were considered to have maintained complete allergen avoidance. These challenge-proven cow's milk allergic infants underwent additional cow's milk challenges after 6 and 12 months of treatment. The rate of tolerance after 12 months was significantly higher in the groups receiving EHCF with *L. rhamnosus* GG (78.9%) compared to EHCF (43.6%) only and significantly higher than the control groups (hydrolyzed rice formula [32.6%], soy formula [23.6%], amino acid-based formula [18.2%]), supporting the notion that supplementation of modified hypoallergenic cow's milk formula with a probiotic, specifically *L. rhamnosus* GG, could enhance tolerance acquisition to whole cow's milk protein compared with intake of hypoallergenic formula alone. Moreover, intake of some modified cow's milk allergens appeared to support a higher rate of tolerance than strict cow's milk allergen avoidance. The same author observed that the *L. rhamnosus* GG supplemented EHCF-treated cow's milk allergic children who acquired oral tolerance possessed a forkhead box P3 (FOXP3) gene DNA methylation pattern and T_H1/T_H2 cytokine profile that were more closely related to that observed in healthy control subjects whereas this was not observed in the other treatment groups (Berni Canani et al., 2017). In contrast to the findings of Berni-Canani et al., another study by

Hol et al. (Hol et al., 2008), showed no benefit from supplementation with *Lactobacillus casei* CRL431 and *Bifidobacterium lactis* Bb-12 for 12 months on acquisition of cow's milk tolerance compared to no probiotics in infants with cow's milk allergy. In this study, both groups of infants were taking hypoallergenic extensively hydrolyzed cow's milk formula (EHCF).

A meta-analysis of studies evaluating probiotics for the treatment of cow's milk allergy (Tan-Lim and Esteban-Ipac, 2018) found that probiotic-induced symptom reduction and tolerogenic effects were strain-specific, with subgroup analysis of strains showing beneficial effects with *Lactobacillus rhamnosus* GG but not with other species or strains. The *L. rhamnosus* GG has been shown to express strain-specific proteins that comprise mucus-binding pili on its surface, which enable binding and persistence of this strain in the human intestinal tract (Kankainen et al., 2009). It was postulated that pili from this gram-positive bacterial strain have immunostimulatory effects (Barocchi et al., 2006) including induction of IL10 secretion that are strain specific and not observed in three other commonly consumed lactobacilli species (Cai et al., 2010).

4.3.2 Preclinical Studies of the Effects of Probiotics for Treatment of Food Allergy

Conceptually, the balance between atopy and tolerance is considered to reflect a balance between allergen-specific T_H2 cells (atopy) and allergen specific T_{reg} cells (tolerance) (Yu et al., 2016). Bacterial adjuvants in food immunotherapy have been postulated to mediate beneficial effects by redirecting the immune response from a T_H2-predominant to a T_{reg}-predominant state.

Stefka et al. (2014) demonstrated in a gnotobiotic mouse model that Clostridia-group bacteria were able to reduce sensitization to food allergens and severity of anaphylaxis. This protective effect was mediated by early induction of IL22 production, which in turn led to enhanced barrier protective responses, proliferation of goblet cells, and increased mucus production. More delayed adaptive immune responses were also noted with an expansion of IL-22+CD4+T cells and colonic Foxp3+Treg cells, and increased IgA secretion into the intestinal lumen. Rachid and Chatila et al. (Abdel-Gadir et al., 2019) using a similar mouse model, showed that oral administration of a defined consortium of six human-origin Clostridiales species to hyper-allergic mice significantly inhibited food allergic reactions, increased expression of ROR-γt in Treg cells and suppressed Th2 responses. In a "humanised mouse model" (Atarashi et al., 2013), germ-free mice colonized with bacteria from healthy, but not cow's milk allergic infants, were protected against anaphylactic responses to cow's milk allergen, with this effect mediated by an increase in Treg cells in the colonic lamina propria. Correlation of the ileal bacteria with genes upregulated in the ileum of healthy or CMA colonized mice identified a clostridial species that utilizes lactate and acetate and produces butyrate, *Anaerostipes caccae*, as a key bacteria mediating the protective effect.

Taken together, data from mouse models suggest that interventions to modulate bacterial communities may have a therapeutic role in food allergy. To date, these data have been translated into three human clinical trials evaluating probiotics for the treatment of peanut allergy, including a small trial of oral encapsulated fecal microbiota transplant (NCT02960074), a trial evaluating OIT in combination with a probiotic capsule that contains a mix of Clostridia strains selected for their immune tolerogenic effects (NCT03936998), and the earlier described PPOIT 001 trial evaluating the efficacy of *L. rhamnosus* CGMCC 1.3724 peanut OIT.

4.4 CONCLUSION

Food allergy has been the subject of active research in the past decade due to its rising global prevalence. Recent research advances have pointed to the potential application of probiotics for the prevention or treatment of food allergy. Currently there is insufficient evidence to recommend probiotics for the prevention of food allergy. Probiotic interventions hold greater promise for the treatment of established food allergy, with preclinical and clinical studies suggesting beneficial effects in reducing the severity of reactions as well as inducing sustained unresponsiveness. Probiotic effects appear to be species and strain-specific, with benefits reported for *L. rhamnosus* GG and certain Clostridia strains. Additional well designed randomized trials with clearly defined outcomes are warranted to further evaluate the role of probiotics in prevention and treatment of food allergy.

REFERENCES

1. Abdel-Gadir A, Stephen-Victor E, Gerber GK, Noval Rivas M, Wang S, et al. 2019. Microbiota therapy acts via a regulatory T cell MyD88/RORγt pathway to suppress food allergy. *Nature Medicine* 25: 1164–1174.

2. Abrahamsson TR, Jakobsson T, Bottcher MF, Fredrikson M, Jenmalm MC, et al. 2007. Probiotics in prevention of IgE-associated eczema: A double-blind, randomized, placebo-controlled trial. *Journal of Allergy and Clinical Immunology* 119: 1174–1180.
3. ASCIA Guidelines - Infant Feeding and Allergy Prevention [Internet]. Australasian Society of Clinical Immunology and Allergy (ASCIA). 2021 [cited 15 March 2021]. Available from: https://www.allergy.org.au/hp/papers/infant-feeding-and-allergy-prevention.
4. Atarashi K, Tanoue T, Oshima K, Suda W, Nagano Y, et al. 2013. Treg induction by a rationally selected mixture of Clostridia strains from the human microbiota. *Nature* 500: 232–236.
5. Barocchi MA, Ries J, Zogaj X, Hemsley C, Albiger B, et al. 2006. A pneumococcal pilus influences virulence and host inflammatory responses. *Proceedings of the National Academy of Sciences of the United States of America* 103: 2857–2862.
6. Berni Canani R, Di Costanzo M, Bedogni G, Amoroso A, Cosenza L, et al. 2017. Extensively hydrolyzed casein formula containing *Lactobacillus rhamnosus* GG reduces the occurrence of other allergic manifestations in children with cow's milk allergy: 3-year randomized controlled trial. *Journal of Allergy and Clinical Immunology* 139: 1906–1913.e4.
7. Berni Canani R, Nocerino R, Terrin G, Frediani T, Lucarelli S, et al. 2013. Formula selection for management of children with cow's milk allergy influences the rate of acquisition of tolerance: A prospective multicenter study. *Journal of Pediatrics* 163: 771–777.e1.
8. Björkstén B. 2001. The epidemiology of food allergy. *Current Opinion in Allergy and Clinical Immunology* 1: 225–227.
9. Björkstén B. 2004. Effects of intestinal microflora and the environment on the development of asthma and allergy. *Seminars in Immunopathology* 25: 257–270.
10. Boyle RJ, Ismal IH, Kivivuori S, Licciardi PV, Robins-Browne RM, et al. 2011. *Lactobacillus* GG treatment during pregnancy for the prevention of eczema: a randomized controlled trial. *Allergy* 66: 509–516.
11. Bunyavanich S, Berin MC. 2019. Food allergy and the microbiome: Current understandings and future directions. *Journal of Allergy and Clinical Immunology* 144: 1468–1477.
12. Burks AW, Sampson HA, Plaut M, Lack G, Akdis CA. 2018. Treatment for food allergy. *Journal of Allergy and Clinical Immunology* 141: 1–9.
13. Cai S, Bay BH, Lee YK, Lu J, Mahendran R. 2010. Live and lyophilized *Lactobacillus* species elicit differential immunomodulatory effects on immune cells. *FEMS Microbiology Letters* 302: 189–196.
14. Chinthrajah RS, Purington N, Andorf S, Long A, O'Laughlin KL, et al. 2019. Sustained outcomes in oral immunotherapy for peanut allergy (POISED study): A large, randomised, double-blind, placebo-controlled, phase 2 study. *Lancet* 394: 1437–1449.
15. Chu DK, Wood RA, French S, Fiocchi A, Jordana M, et al. 2019. Oral immunotherapy for peanut allergy (PACE): A systematic review and meta-analysis of efficacy and safety. *Lancet* 393: 2222–2232.
16. Cuello-Garcia CA, Brożek JL, Fiocchi A, Pawankar R, Yepes-Nuñez JJ, et al. 2015. Probiotics for the prevention of allergy: A systematic review and meta-analysis of randomized controlled trials. *Journal of Allergy and Clinical Immunology* 136: 952–961.
17. Dunn Galvin A, McMahon S, Ponsonby AL, Hsiao KC, Tang MLK. 2018. The longitudinal impact of probiotic and peanut oral immunotherapy on health-related quality of life. *Allergy* 73: 560–568.
18. Fiocchi A, Pawankar R, Cuello-Garcia C, Ahn K, Al-Hammadi S, et al. 2015. World Allergy Organization-McMaster University Guidelines for Allergic Disease Prevention (GLAD-P): Probiotics. *World Allergy Organization Journal* 8: 4.
19. Gill HS, Doull F, Rutherfurd KJ, Cross ML. 2000. Immunoregulatory peptides in bovine milk. *British Journal of Nutrition* 84: S111–117.
20. Gupta RS, Warren CM, Smith BM, Jiang J, Blumenstock JA, et al. 2019. Prevalence and severity of food allergies among US adults. *JAMA Network Open* 2: e185630–e185630.
21. Hol J, Van Leer EH, Elink Schuurman BE, De Ruiter LF, Samsom JN, et al. 2008. The acquisition of tolerance toward cow's milk through probiotic supplementation: A randomized, controlled trial. *Journal of Allergy and Clinical Immunology* 121: 1448–1454.
22. Hsiao KC, Ponsonby AL, Axelrad C, Pitkin S, Tang MLK. 2017. Long-term clinical and immunological effects of probiotic and peanut oral immunotherapy after treatment cessation: 4-year follow-up of a randomised, double-blind, placebo-controlled trial. *The Lancet Child and Adolescent Health* 1: 97–105.
23. Isolauri E, Joensuu J, Suomalainen H, Luomala M, Vesikari T. 1995. Improved immunogenicity of oral D x RRV reassortant rotavirus vaccine by *Lactobacillus casei* GG. *Vaccine* 13: 310–312.

24. Jensen MP, Meldrum S, Taylor AL, Dunstan JA, Prescott SL. 2012. Early probiotic supplementation for allergy prevention: Long-term outcomes. *Journal of Allergy and Clinical Immunology* 130: 1209-1211.e5.

25. Kalliomaki M, Salminen S, Arvilommi H, Kero P, Koskinen IE. 2001. Probiotics in primary prevention of atopic disease: A randomised placebo-controlled trial. *Lancet* 357: 1076–1079.

26. Kalliomaki M, Salminen S, Poussa T, Arvilommi H, Isolauri E. 2003. Probiotics and prevention of atopic disease: 4-year follow-up of a randomised placebo-controlled trial. *Lancet* 361: 1869–1871.

27. Kalliomaki M, Salminen S, Poussa T, Isolauri E. 2007. Probiotics during the first 7 years of life: A cumulative risk reduction of eczema in a randomized, placebo-controlled trial. *Journal of Allergy and Clinical Immunology* 119: 1019–1021.

28. Kankainen M, Paulin L, Tynkkynen S, Von Ossowski I, Reunanen J, et al. 2009. Comparative genomic analysis of *Lactobacillus rhamnosus* GG reveals pili containing a human- mucus binding protein. *Proceedings of the National Academy of Sciences of the United States of America* 106: 17193–17198.

29. Kopp MV, Hennemuth I, Heinzmann A, Urbanek R. 2008. Randomized, double-blind, placebo-controlled trial of probiotics for primary prevention: No clinical effects of *Lactobacillus* GG supplementation. *Pediatrics* 121: e850–856.

30. Kuitunen M, Kukkonen K, Juntunen-Backman K, Korpela R, Poussa T, et al. 2009. Probiotics prevent IgE-associated allergy until age 5 years in cesarean-delivered children but not in the total cohort. *Journal of Allergy and Clinical Immunology* 123: 335–341.

31. Loh W, Tang MLK. 2018. The Epidemiology of Food Allergy in the Global Context. *International Journal of Environmental Research and Public Health* 15.

32. Mohamadzadeh M, Olson S, Kalina WV, Ruthel G, Demmin GL, et al. 2005. Lactobacilli activate human dendritic cells that skew T cells toward T helper 1 polarization. *Proceedings of the National Academy of Sciences of the United States of America* 102: 2880.

33. Muraro A, Halken S, Arshad SH, Beyer K, Dubois AE, et al. 2014. EAACI food allergy and anaphylaxis guidelines. Primary prevention of food allergy. *Allergy* 69: 590–601.

34. Oppenheimer JJ, Nelson HS, Bock SA, Christensen F, Leung DY. 1992. Treatment of peanut allergy with rush immunotherapy. *Journal of Allergy and Clinical Immunology* 90: 256–262.

35. Osborne NJ, Koplin JJ, Martin PE, Gurrin LC, Lowe AJ, et al. 2011. Prevalence of challenge-proven IgE-mediated food allergy using population-based sampling and predetermined challenge criteria in infants. *Journal of Allergy and Clinical Immunology* 127: 668-676.e1-2.

36. Pessi T, Sutas Y, Hurme M, Isolauri E. 2000. Interleukin-10 generation in atopic children following oral *Lactobacillus rhamnosus* GG. *Clinical and Experimental Allergy* 30: 1804–1808.

37. Probiotics in food - Health and nutritional properties and guidelines for evaluation [Internet]. Fao.org. 2021 [cited 15 March 2021]. Available from: http://www.fao.org/3/a-a0512e.pdf.

38. Rautava S, Arvilommi H, Isolauri E. 2006. Specific probiotics in enhancing maturation of IgA responses in formula-fed infants. *Pediatric Research* 60: 221–224.

39. Sandré C, Gleizes A, Forestier F, Gorges-Kergot R, Chilmonczyk S, et al. 2001. A peptide derived from bovine beta-casein modulates functional properties of bone marrow-derived macrophages from germfree and human flora-associated mice. *Journal of Nutrition* 131: 2936–2942.

40. Soh SE, Aw M, Gerez I, Chong YS, Rauff M, Ng YP, et al. 2009. Probiotic supplementation in the first 6 months of life in at risk Asian infants–effects on eczema and atopic sensitization at the age of 1 year. *Clinical and Experimental Allergy* 39: 571–578.

41. Stefka AT, Feehley T, Tripathi P, Qiu J, McCoy K, et al. 2014. Commensal bacteria protect against food allergen sensitization. *Proceedings of the National Academy of Sciences of the United States of America* 111: 13145–13150.

42. Strachan DP. 1989. Hay fever, hygiene, and household size. *BMJ Clinical Research Ed* 299: 1259–1260.

43. Syed A, Garcia MA, Lyu SC, Bucayu R, Kohli A, et al. 2014. Peanut oral immunotherapy results in increased antigen-induced regulatory T-cell function and hypomethylation of forkhead box protein 3 (FOXP3). *Journal of Allergy and Clinical Immunology* 133: 500–510.

44. Tang MLK, Lozinsky AC, Loke P. 2020. Peanut oral immunotherapy: State of the art. *Immunology and Allergy Clinics of North America* 40: 97–110.

45. Tang MLK, Ponsonby AL, Orsini F, Tey D, Robinson M, et al. 2015. Administration of a probiotic with peanut oral immunotherapy: A randomized trial. *Journal of Allergy and Clinical Immunology* 135: 737–744.e8.

46. Tan-Lim CSC, Esteban-Ipac NAR. 2018. Probiotics as treatment for food allergies among pediatric patients: A meta-analysis. *World Allergy Organization Journal* 11: 25.

47. Taylor AL, Dunstan JA, Prescott SL. 2007. Probiotic supplementation for the first 6 months of life fails to reduce the risk of atopic dermatitis and increases the risk of allergen sensitization in high-risk children: A randomized controlled trial. *Journal of Allergy and Clinical Immunology* 119: 184–191.

48. Thomas DW, Greer FR. 2010. Probiotics and prebiotics in pediatrics. *Pediatrics* 126: 1217–1231.

49. Turner PJ, Feeney M, Meyer R, Perkin MR, Fox AT. 2018. Implementing primary prevention of food allergy in infants: New BSACI guidance published. *Clinical and Experimental Allergy* 48: 912–915.

50. Vickery BP, Vereda A, Nilsson C, Du Toit G, Shreffler WG, et al. 2020. Continuous and daily oral immunotherapy for peanut allergy: Results from a 2-year open-label follow-on study. *The Journal of Allergy and Clinical Immunology: In Practice.*

51. Yu W, Freeland DMH, Nadeau KC. 2016. Food allergy: Immune mechanisms, diagnosis and immunotherapy. *Nature Reviews Immunolology* 16: 751–765.

CHAPTER 5

COVID-19, Microbiota, and Probiotics

Marta Mozota[1], Leónides Fernández[2], and Juan Miguel Rodríguez[1]
[1]Dpto. Nutrición y Ciencia de los Alimentos, Universidad Complutense de Madrid, Spain
[2]Dpto. Farmacia Galénica y Tecnología Alimentaria, Universidad Complutense de Madrid, Spain

5.1 INTRODUCTION

Coronavirus 2, universally known as SARS-CoV-2 (severe acute respiratory syndrome coronavirus 2), was identified only a year ago as the causative agent of a new acute respiratory disease: the "coronavirus disease 2019" (COVID-19). This disease began in the province of Hubei (China) but its rapid expansion meant that on January 30, 2020, it was declared a health emergency of international concern by the World Health Organization. Soon after, on March 11, it was recognized as a pandemic by that same international organization. Up to March 2022, more than 450 million cases of the disease have been confirmed, with more than six million official deaths, although the real numbers, both of cases and deaths, are probably much higher. People aged ≥ 60 years, with underlying health conditions, such as hypertension, diabetes, obesity, cardiovascular disease, chronic respiratory disease, and weakened immune systems, and/or those living in long-term care facilities, are more likely to have serious conditions. However, many of those infected remain asymptomatic, a fact that complicates the control of the disease.

The pathognomonic symptoms of COVID-19 are cough, fever, and dyspnea. The infection can be divided into three phases. In the first stage, mild symptoms are usually observed (general malaise, low fever, and dry cough). In a second stage, some patients (~20%) develop atypical pneumonia, with high fever, cough, and dyspnea, often accompanied by hypoxia. Finally, a minority of patients progress to a third stage, characterized by the now famous "cytokine storm" that translates into a systemic hyperinflammatory state that can lead to acute respiratory distress syndrome, multiple organ failure, sepsis, and death. However, the disease can present a very wide spectrum of clinical manifestations that affect various organs and systems, including gastrointestinal symptoms, such as nausea, vomiting, abdominal pain, and, especially, diarrhea (Beattie et al., 2020; Cha et al., 2020; Liang et al., 2020). Such gastrointestinal symptoms are present in more than half of the patients (Han et al., 2020) and, on occasion, they can be the first manifestation of the infection (Buscaini et al., 2020).

A recent meta-analysis that included 17 studies indicated that, on average, SARS-CoV-2 RNA is present in the stools of 44% of COVID-19 patients, being significantly more frequent among individuals with gastrointestinal symptoms (77%) than among those who did not present them (57%), and among individuals with severe disease (68%) than among those with mild or moderate symptoms (34%) (Wong et al., 2020). In fact, the presence of the viral RNA in stools can persist long after its disappearance from the respiratory system (Wu et al., 2020), which is why it has been suggested that it would be convenient to perform a stool detection test before hospital discharge. The presence of the virus in feces indicates the possibility of fecal–oral transmission and the spread of the virus through sewage (Kitajima et al., 2020), although a recent study has questioned the real importance of this route in detecting the RNA of the virus but

Good Microbes in Medicine, Food Production, Biotechnology, Bioremediation, and Agriculture, First Edition. Edited by Frans J. de Bruijn, Hauke Smidt, Luca S. Cocolin, Michael Sauer, David Dowling, and Linda Thomashow.

not the virus as such in stool samples from patients with COVID-19 (Wolfel et al., 2020).

Although vaccines are now available, specific etiological treatments for preventing or treating the disease are not available yet, and the in-practice treatments include respiratory support and the empirical use of drugs aimed at avoiding or reducing the state of (hyper) inflammation, antivirals and, also, antibiotics and antifungals intended for the prevention and/or treatment of sepsis and secondary infections.

5.2 RELATIONSHIP BETWEEN COVID-19 AND THE MICROBIOTA

The fact that SARS-CoV-2 initially interacts with mucosal epithelia associated with a complex microbiota (naso-oro-pharyngeal cavity and gastrointestinal tract), and that its mechanism of infection is related to the angiotensin-converting enzyme receptor 2 (ACE2), whose activity is influenced and, in turn, influences the microbiota, suggests an implication of the microbiota in the modulation of infection by this virus. The modulatory role of the microbiota has already been observed with other respiratory viruses, including other coronaviruses and the respiratory syncytial virus (De Steenhuijsen Piters et al., 2016; Hasegawa et al., 2016; McGillivary et al., 2009).

The renin-angiotensin system is essential to understand the pathogenesis of SARS-CoV-2 infection (Figure 1.5.1). This system catalyzes the transformation of angiotensinogen into angiotensin I (AngI) through the action of renin and the subsequent conversion of AngI to angiotensin II (AngII) by the angiotensin-converting enzyme (ACE). The accumulation of AngII elicits vasoconstrictive and inflammatory responses. To counteract these effects of AngII, angiotensin-converting enzyme 2 (ACE2) transforms part of AngII into angiotensin 1–7 (Ang1–7), which is transported into cells through the Mas receptor. Angiotensin II (AngII) exerts vasodilator and anti-inflammatory effects. Any situation involving a decrease in ACE2 and Ang1–7 or an increase in AngII usually leads to negative effects (inflammation, clotting) on various essential organs, including the heart, lungs, kidneys, brain, liver, pancreas, or bone marrow (Figure 1.5.1).

Precisely, ACE2 acts as a receptor for the SARS-CoV-2 protein S in the cells of the respiratory epithelium, an essential interaction in infection by this virus that affects the functioning of the renin–angiotensin system. This fact poses a dilemma: on the one hand, the virus has more binding sites available as the number of

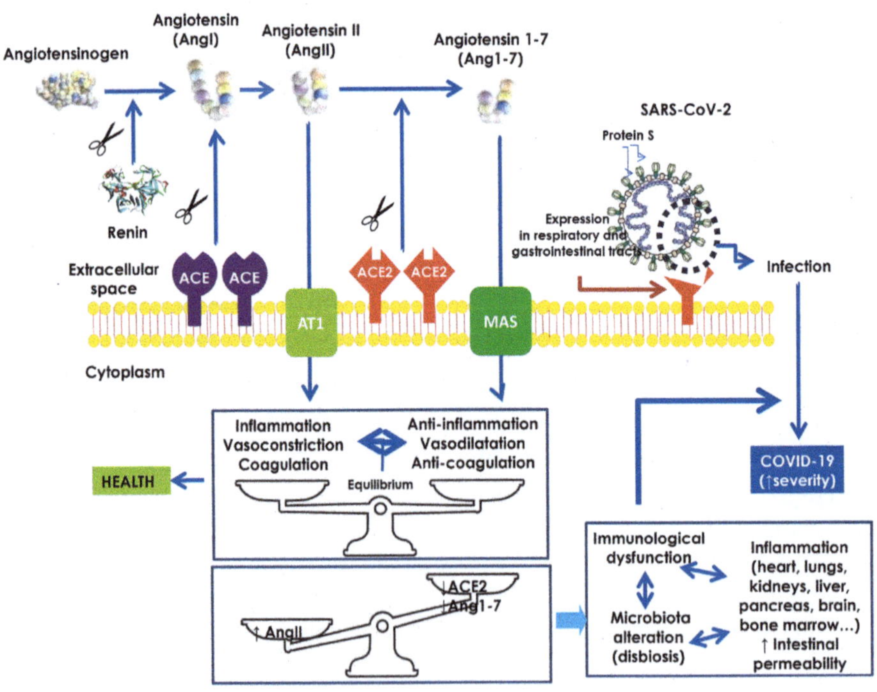

FIGURE 5.1 Interactions between the renin–angiotensin system, SARS-CoV-2, and the microbiota that can exert an influence on the degree of severity of this infection.

ACE2 molecules increases; but, on the other hand, if the number of ACE2 receptors decreases, the negative effect of AngII on the organs mentioned above can be exacerbated. This dilemma limits the treatments aimed to inactivate ACE2. ACE2 expression is elevated in respiratory epithelial cells but also in gastrointestinal epithelial cells (Xiao et al., 2020), and both mucosal tracts seem to be key sites for the acquisition, action, and spread of SARS-CoV-2 (Figure 1.5.1).

In murine models, exposure of alveolar epithelial cells (AT1/AT2) to bacterial lipopolysaccharide (LPS) causes a marked reduction in the expression of ACE2 and an increase in the expression of renin, AngII, and the ACE and AT1 receptors. In addition, the reduction of ACE2 expression is associated with an alteration in intestinal permeability associated with alterations in the microbiota (Viana et al., 2020). These facts suggest that the composition of the microbiota of the upper respiratory and gastrointestinal tracts may possibly play a relevant role in individual susceptibility to COVID-19 (Dhar et al., 2020; He et al., 2020a; Khatiwada et al., 2020) (Figure 1.5.1).

In this sense, it has been postulated that those with "eubiotic" respiratory and gastrointestinal microbiotas would be more likely to have an asymptomatic or mild SARS-CoV-2 infection given that they are associated with appropriate mucosal immune responses. In contrast, people with respiratory or intestinal dysbiosis would not be able to develop correct responses and would be more prone to more severe conditions, secondary infections, and sepsis. It has also been postulated that the coronavirus infection itself can cause alterations in the microbiota that favor the development of inflammatory conditions.

5.3 RESPIRATORY MICROBIOTA IN PATIENTS WITH COVID-19

To date, only two studies have been published to assess whether there are differences between the respiratory microbiota of healthy people and those with COVID-19, and in both cases the sample size was small. The first of them analyzed the composition of the microbiota present in fluid samples from bronchoalveolar lavage from eight patients with pneumonia associated with COVID-19, 25 patients with pneumonia not associated with COVID-19, and 20 healthy controls using metatranscriptomic techniques (Shen et al. al., 2020). The results showed that the microbial diversity was significantly lower in the samples of the patients with pneumonia (regardless of the cause) than in the healthy controls, reflecting a situation of dysbiosis associated with pneumonia. However, no specific patterns were found associated with each type of pneumonia, a fact that could be due to the high number of antibiotic doses that the two types of patients had received. In any case, the microbial profiles could be classified into three groups. The first group was characterized by the high frequency and abundance of pathobionts and included six of the eight patients with COVID-19. The second group contained a very low diversity while the third group (which included most of the healthy controls) was characterized by a much wider microbial diversity than in the other two groups.

The second study performed a metataxonomic analysis from nasopharyngeal samples obtained from 18 patients with COVID-19 (all of them with moderate symptoms) and from 18 controls who did not suffer from the infection (De Maio et al., 2020). In contrast to the previous study, the results indicated that there were no significant differences between the two groups in terms of bacterial diversity or frequency of detection or abundance of any phylum or genus.

5.4 GUT MICROBIOTA IN PATIENTS WITH COVID-19

One of the first indications that the intestinal microbiota could be a relevant actor in the pathogenesis of COVID-19 was obtained in gnotobiotic rats, in which it was observed that the microbiota regulated the colonic expression of ACE2 (Yang et al., 2020). For some years now, the importance of ACE2 in the regulation of blood pressure has been known and, more recently, hypertension has been associated with a state of intestinal dysbiosis characterized by a reduction in ACE2 (Galla et al., 2018; Mell et al., 2015; Yang et al., 2016).

As with other organs, the intestine is highly influenced by the protective effect of ACE2, which tries to prevent inflammatory processes in a microbiota-dependent manner. In this sense, there is a close link between ACE2, the transport of certain amino acids (tryptophan), and the mammalian target of rapamycin (mTOR) pathway. These interactions, in which various elements of the immune system are also relevant, are responsible for the intestinal cells producing antimicrobial peptides that contribute to the maintenance of a diverse and balanced microbiota. Consequently, the composition of the intestinal microbiota, characterized by high interindividual variability, may be a protective or susceptibility factor against COVID-19.

A study that analyzed the fecal microbiota of 15 patients with COVID-19 and 15 healthy controls revealed the existence of significant differences between both populations (Zuo et al., 2020a). More specifically, the fecal microbiome of healthy people was characterized by a high frequency and abundance of some families (*Lachnospiraceae*) and genera (*Faecalibacterium*, *Eubacterium*, *Roseburia*), considered as commensals or mutualists and that stand out for their great capacity to produce short chain fatty acids from (SCFAs), such as butyrate, which exert an immunomodulatory and anti-inflammatory effect. In contrast, the microbiome of COVID-19 patients reflected a notable reduction in these same bacterial groups and a relative increase in sequences corresponding to opportunistic pathogens.

Interestingly, Zuo et al. (2020a) also found a negative association between some species of the *Bacteroides* genus (*B. dorei*, *B. tethaiotamicron*, *B. massilensis*, *B. ovatus*) and the presence of SARS-CoV-2. Apart from SCFA production, these *Bacteroides* species could play beneficial roles in the context of COVID-19 by, at least, two other complementary mechanisms. First, through the production of sphingolipids, such as ceramide, which play roles similar to those of SCFAs, including the stimulation of the differentiation of regulatory T cells (Tregs) and the maturation of dendritic cells, which, ultimately, increases the antiviral responses associated with T lymphocytes, reduces inflammatory responses, and improves the function of the mucosal barrier (Peng et al., 2020). Second, through an extremely high capacity to produce glycosidases that participate in the catabolism of heparan sulfate, a substance that seems essential for the binding of the protein S of SARS-CoV-2 to ACE2 (Martino et al., 2020). As mentioned above, the presence of the aforementioned *Bacteroides* species is notably lower in patients with COVID-19, which translates into a reduction in the presence of their glycosidases in bronchoalveolar samples (Martino et al., 2020). Precisely, the presence of these *Bacteroides* tends to decrease with aging, a fact that could help explain why the elderly constitute a particularly vulnerable population for this disease.

Another study analyzed the fecal microbiota of 30 patients with COVID-19, 24 patients with influenza A (H1N1), and 30 controls (Gu et al., 2020). The results indicated the presence of a dysbiosis situation between the two groups of patients, regardless of their disease, characterized by a much lower bacterial diversity than that existing in the control samples. In addition, there were significant differences between the group of patients with COVID-19 and the healthy controls in terms of the presence of various bacterial groups. Thus, for example, while sequences of the genera *Faecalibacterium* and *Eubacterium* and *torquens* group *Ruminococcus* were more abundant among controls, those of the genera *Streptococcus*, *Rothia*, *Veilonella*, and *Actinomyces* were more numerous among people affected with COVID-19. A few days after the publication of this work, the same journal published a letter in which it was stated that the depletion observed in bacterial groups that are large producers of butyrate, such as *Faecalibacterium*, *Eubacterium*, or *Ruminococcus*, both among patients with COVID-19 as in those with influenza A, could be simply due to the presence of diarrhea (Klann et al., 2020). Diarrhea, regardless of its origin, is associated with a notable loss of members of the families *Ruminococcaceae* and *Lachnospiraceae* and a reduction in alpha diversity (Duvallet et al., 2017). The authors of the letter suggested that such findings could be associated with a decrease in transit time through the intestine and not have a causal relationship with these viral diseases.

5.5 PROBIOTICS AND COVID-19

Given that the composition of the microbiota is relevant to our health and that COVID-19 does not seem to be an exception, various articles have postulated the application of strategies that allow modulating the respiratory and/or digestive microbiota (reducing the presence of pathobionts), that do not generate resistance and that are respectful or that stimulate the presence of those members of the microbiota with beneficial functions for our health. Among these strategies are prebiotics, paraprobiotics, postbiotics, and, especially, probiotics, which are the ones that have generated the most expectations to date in the context of COVID-19.

Although it has been pointed out that there are no precedents regarding the possible efficacy of the use of probiotics in patients with COVID-19 (Mak et al., 2020), the same situation is applicable to any of the drugs (antivirals, anti-inflammatories, etc.) that have been empirically tested to treat a previously unknown disease, including some that have shown relevant adverse effects (Giannoni et al., 2020). In this sense, it should be remembered that to date there is no known effective etiological treatment against the SARS-CoV-2 infection and that although vaccines are now available, they may not be the panacea against this disease in the short term.

Practically all the systematic reviews and meta-analyses carried out to date on the use of probiotics for the treatment of acute respiratory infections, both in children and adults, have been favorable toward the use of probiotics despite the heterogeneity of strains, dosage, and objectives among the clinical trials included in them (Hao et al., 2015; King et al., 2014; Wang et al., 2016). Overall, they all show a significant reduction in incidence, duration of the episodes, use of antibiotics, and absenteeism from work or school.

A high percentage of people hospitalized for COVID-19 require the use of respirators, which may be essential but also carry certain side effects. In this sense, it has been observed that the prophylactic administration of probiotics in patients who need a respirator significantly reduces cases of ventilator-associated pneumonia, cases of *Clostridioides difficile*-associated diarrhea, and days of antibiotic therapy (Morrow et al., 2010; Su et al., 2020; van Ruissenet al., 2019), showing no adverse effects even among those patients in a critical situation (Yang et al., 2016).

Regarding the use of probiotics in patients with COVID-19, an article related to the management of initial cases from China contained a section entitled "Maintaining microecological balance and reducing bacterial translocation and infection," recommending the performing of fecal microbiota analysis and the use of probiotics and prebiotics given that a depletion of lactobacilli and bifidobacteria populations had been observed in many patients (Xu et al., 2020). However, this article did not provide experimental data to support such recommendations. Since then, several articles have appeared on probiotics and COVID-19, but nearly all of them simply discuss the general benefits of probiotics for health or speculate or hypothesize about the possible effects that probiotics or some of their metabolites could have on patients who suffer from this infection (Antunes et al., 2020; Anwar et al. al., 2020; Baud et al., 2020; Bottari et al., 2020; Conte et al., 2020; Di Pierro, 2020; He et al., 2020b; Infusino et al., 2020; Mahooti et al., 2020; Morais et al., 2020; Pagnini et al., 2020; Rozga et al., 2020; Sundadaraman et al., 2020).

Some people, on their own initiative or on the advice of family, friends, or healthcare personnel, have consumed probiotics trying to prevent or alleviate the effect of COVID-19. In some hospitals, the empirical use of some probiotics has also been used, especially when diarrhea appeared, which was sometimes associated with infection but which, in many others, was related to the intensive antibiotic therapy that patients received for the prevention of secondary infections and sepsis. To date, there are only reports of two cases of adverse effects (sepsis that resolved favorably) linked to this empirical administration of probiotics and both associated to the use of a strain of *Saccharomyces cerevisiae* (Ventoulis et al., 2020). It should be remembered that this species is included in the EFSA list of microorganisms with a qualified presumption of safety (QPS) status, but it is the only species on the list whose use is contraindicated in people with fragile health and in patients with catheters, due to the risk of fungal contamination (EFSA (EFSA Panel on Biological Hazards), 2016).

However, data from most clinical trials specifically designed for the prevention or as an aid in the treatment of COVID-19 through the use of probiotics are not yet available. This fact is logical considering the short time that has elapsed since the onset of the pandemic and the time required for clinical trials to obtain rigorous results. The first published trial included 70 positive hospitalized patients, all of whom had fever and pulmonary involvement greater than 50% but did not require invasive oxygen therapy. Among these patients, 42 received hydroxychloroquine, antibiotics, and tocilizumab, alone or in combination, while the remaining 28 subjects received similar therapies but supplemented with a multi-strain probiotic formulation that was administered orally (d'Ettorre et al., 2020). Treatment with probiotics had a significant impact by shortening the duration of diarrheal episodes and decreasing the risk of respiratory failure. It should be noted that some of the drugs that patients received in that trial are currently contraindicated for the treatment of COVID-19. Later, the same group published another trial in which the same probiotic product was tested in a population of 200 COVID-19 patients with severe pneumonia. All patients received similar treatments, which included various combinations of heparin, hydroxychloroquine, antibiotics, antivirals, and/or tocilizumab. Of the 200 patients, 88 additionally received the probiotic product. Its administration was associated with a significant decrease in the risk of death (Ceccarelli et al., 2021). There are other trials in progress, but time is required to know and compare their results and, therefore, to know which strains may be useful as an aid in the treatment or prevention of this disease.

In any case, COVID-19 contains numerous aspects that constitute clear targets for the application of probiotics (Angurana and Bansal, 2020) (Figure 5.2): (1) the presence of diarrhea and other digestive symptoms in a high percentage of cases (Villapol, 2020); (2) modulation

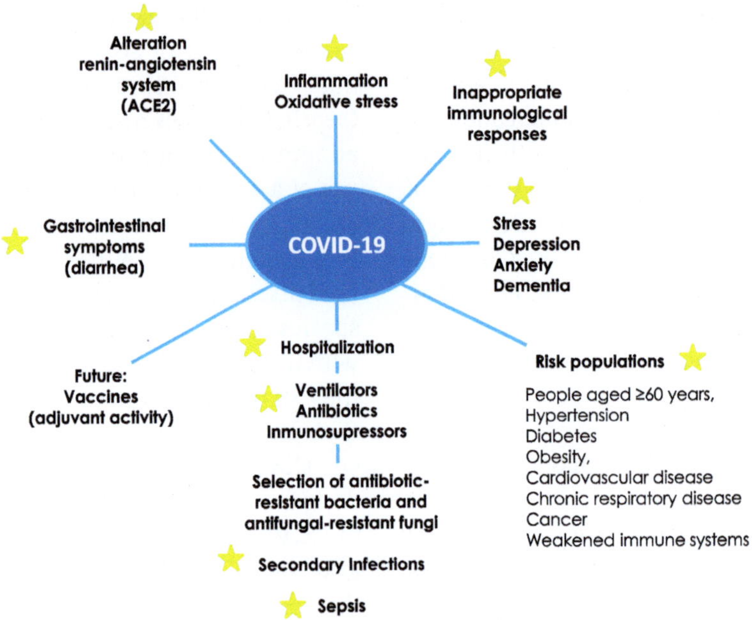

FIGURE 5.2 Frequent situations in the course of COVID-19 that are associated or lead to a situation of dysbiosis in the respiratory and intestinal microbiota and that, consequently, constitute targets for probiotics and other strategies for modulating the microbiota. Yellow star: conditions or factors usually associated with dysbiosis in the respiratory and/or gastrointestinal tracts.

of the immune and inflammatory response to the virus; (3) prevention of bacterial or fungal co-infections (Chen et al., 2020; Zuo et al., 2020); (4) prevention or treatment of microbiota alterations associated to the use of respirators, corticosteroids, antibiotics, and antifungals and which, in the case of antimicrobials, adds to the growing concern about the selection of (multi)resistant pathogens; (5) the pre-existing situation of dysbiosis that usually characterizes the population most vulnerable to COVID-19 (elderly, diabetic, immunosuppressed, etc.), and that can also be a consequence of the treatments received; or (6) the rise in cases of anxiety, depression, or stress associated with the pandemic in the case of the so-called psychobiotics (Cryan et al., 2019; Follmer, 2020).

Even now when vaccines against SARS-CoV-2 have become widely available, probiotics may be very useful as adjuvants that improve antibody titers and the functions of some immune cells subpopulations after vaccination.

REFERENCES

1. Angurana SK, Bansal A. 2020. Probiotics and COVID-19: Think about the link. *British Journal of Nutrition* Sep 14: 1–26.
2. Antunes AEC, Vinderola G, Xavier-Santos D, Sivieri K. 2020. Potential contribution of beneficial microbes to face the COVID-19 pandemic. *Food Research International* 136: 109577.
3. Anwar F, Altayb HN, Al-Abbasi FA, Al-Malki AL, Kamal MA, Kumar V. 2020. Antiviral effects of probiotic metabolites on COVID-19. *Journal Biomolecular Structure Dynamics* June 9: 1–10.
4. Baud D, Dimopoulou Agri V, Gibson GR, Reid G, Giannoni E. 2020. Using probiotics to flatten the curve of Coronavirus disease COVID-2019 pandemic. *Frontiers Public Health* 8: 186.
5. Beattie RM, Ashton JJ, Penman ID. 2020. COVID-19 and the gastrointestinal tract: Emerging clinical data. *Frontline Gastroenterology* 11: 290–292.
6. Bottari B, Castellone V, Neviani E. 2020. Probiotics and Covid-19. *International Journal of Food Science Nutrition* 72: 293–299.
7. Buscarini E, Manfredi G, Brambilla G, Menozzi F, Londoni C, Alicante S, et al. 2020. GI symptoms as early signs of COVID-19 in hospitalised Italian patients. *Gut* 69: 1547–1548.
8. Ceccarelli G, Borrazzo C, Pinacchio C, Santinelli L, Innocenti GP, et al. 2021. Oral bacteriotherapy in patients with COVID-19: A retrospective cohort study. *Frontiers in Nutrition* 7: 613928.
9. Cha MH, Regueiro M, Sandhu DS. 2020. Gastrointestinal and hepatic manifestations of COVID-19: A compre-

hensive review. *World Journal Gastroenterology* 26: 2323–2332.

10. Chen X, Liao B, Cheng L, Peng X, Xu X. 2020. The microbial coinfection in COVID-19. *Applied Microbiology Biotechnology* 104: 7777–7785.
11. Conte L, Toraldo DM. 2020. Targeting the gut-lung microbiota axis by means of a high-fibre diet and probiotics may have anti-inflammatory effects in COVID-19 infection. *Therapy Advances Respiratory Disease* 14: 1753466620937170.
12. Cryan JF, O'Riordan KJ, Cowan CSM, Sandhu KV, Bastiaanssen TFS. 2019. The microbiota-gut-brain axis. *Physiology Reviews* 99: 1877–2013.
13. d'Ettorre G, Ceccarelli G, Marazzato M, Campagna G, Pinacchio C, et al. 2020. Challenges in the management of SARS-CoV2 Infection: The role of oral bacteriotherapy as complementary therapeutic strategy to avoid the progression of COVID-19. *Frontiers in Medicine* 7: 389.
14. De Maio F, Posteraro B, Ponziani FR, Cattani P, Gasbarrini A, Sanguinetti M. 2020. Nasopharyngeal microbiota profiling of SARS-CoV-2 infected patients. *Biology Proceedings Online* 18.
15. De Steenhuijsen Piters WA, Heinonen S, Hasrat R, Bunsow E, Smith B, et al. 2016. Nasopharyngeal microbiota, host transcriptome, and disease severity in children with respiratory syncytial virus infection. *American Journal of Respiratory and Critical Care Medicine* 194: 1104–1115.
16. Dhar D, Mohanty A. 2020. Gut microbiota and Covid-19-possible link and implications. *Virus Research* 285: 198018.
17. Di Pierro F. 2020. A possible probiotic (*S. salivarius* K12) approach to improve oral and lung microbio tas and raise defenses against SAR S-CoV-2. *Minerva Medicine* 111: 281–283.
18. Duvallet C, Gibbons SM, Gurry T, Irizarry RA, Alm EJ. 2017. Meta-analysis of gut microbiome studies identifies disease-specific and shared responses. *Nature Communications* 8: 1784.
19. EFSA (EFSA Panel on Biological Hazards). 2016. Statement on the update of the list of QPS-recommended biological agents intentionally added to food or feed as notified to EFSA 4: Suitabilityof taxonomic units notified to EFSA until March 2016. *EFSA Journal* 14: 4522.
20. Follmer C. 2020. Gut microbiome imbalance and neuroinflammation: Impact of COVID-19 on Parkinson's Disease. *Movement Disorders* 35: 1495–1496.
21. Galla S, Chakraborty S, Cheng X, Yeo J, Mell B, et al. 2018. Disparate effects of antibiotics on hypertension. *Physiological Genomics* 50: 837–845.
22. Giannoni E, Baud D, Agri VD, Gibson GR, Reid G. 2020. Probiotics and COVID-19. *Lancet Gastroenterology Hepatology* 5: 720–721.
23. Gu S, Chen Y, Wu Z, Chen Y, Gao H, et al. 2020. Alterations of the gut microbiota in patients with COVID-19 or H1N1 influenza. *Clinical Infectious Diseases* 71: 2669–2678.
24. Han C, Duan C, Zhang S, Spiegel B, Shi H, et al. 2020. Digestive symptoms in COVID-19 patients with mild disease severity: Clinical presentation, stool viral RNA testing, and outcomes. *American Journal of Gastroenterology* 115: 916–923.
25. Hao Q, Dong BR, Wu T. 2015. Probiotics for preventing acute upper respiratory tract infections. *Cochrane Database Systems Reviews* 2: CD006895.
26. Hasegawa K, Mansbach JM, Ajami NJ, Espinola JA, Henke DM, et al. 2016. The MARC-35 Investigators. Association of nasopharyngeal microbiota profiles with bronchiolitis severity in infants hospitalised for bronchiolitis. *European Respiration Journal* 48: 1329–1339.
27. He LH, Ren LF, Li JF, Wu YN, Li X, Zhang L. 2020a. Intestinal flora as a potential strategy to fight SARS-CoV-2 infection. *Frontiers in Microbiology* 11: 1388.
28. He Y, Wang J, Li F, Shi Y. 2020b. Main clinical features of COVID-19 and potential prognostic and therapeutic value of the microbiota in SARS-CoV-2 infections. *Frontiers in Microbiology* 11: 1302.
29. Infusino F, Marazzato M, Mancone M, Fedele F, Mastroianni CM, et al. 2020. Diet supplementation, probiotics, and nutraceuticals in SARS-CoV-2 infection: A scoping review. *Nutrients* 12: 1718.
30. Khatiwada S, Subedi A. 2020. Lung microbiome and coronavirus disease 2019 (COVID-19): Possible link and implications. *Human Microbiology Journal* 17: 100073.
31. King S, Glanville J, Sanders ME, Fitzgerald A, Varley D. 2014. Effectiveness of probiotics on the duration of illness in healthy children and adults who develop common acute respiratory infectious conditions: A systematic review and meta-analysis. *British Journal of Nutrition* 112: 41–54.
32. Kitajima M, Ahmed W, Bibby K, Carducci A, Gerba CP, et al. 2020. SARS-CoV-2 in wastewater: State of the knowledge and research needs. *Science of Total Environment* 739: 139076.

33. Klann E, Rich S, Mai V. 2020. Gut microbiota and COVID-19: A superfluous diagnostic biomarker or therapeutic target? *Clinical and Infectious Diseases* ciaa1191.

34. Liang W, Feng Z, Rao S, Xiao C, Xue X, et al. 2020. Diarrhoea may be underestimated: A missing link in 2019 novel coronavirus. *Gut* 69: 1141–1143.

35. Mahooti M, Miri SM, Abdolalipour E, Ghaemi A. 2020. The immunomodulatory effects of probiotics on respiratory viral infections: A hint for COVID-19 treatment? *Microbial Pathogenesis* 48: 104452.

36. Mak JWY, Chan FKL, Ng SC. 2020. Probiotics and COVID-19: One size does not fit all. *Lancet Gastroenterology and Hepatology* 5: 644–645.

37. Martino C, Kellman BP, Sandoval DR, Clausen TM, Marotz CA, et al. 2020. Bacterial modification of the host glycosaminoglycan heparan sulfate modulates SARS-CoV-2 infectivity. *bioR* xiv doi: 10.1101/2020.08.17.238444.

38. McGillivary G, Mason KM, Jurcisek JA, Peeples ME, Bakaletz LO. 2009. Respiratory syncytial virus-induced dysregulation of expression of a mucosal beta-defensin augments colonization of the upper airway by non-typeable. *Haemophilus Influenzae. Cellular Microbiology* 11: 1399–1408.

39. Mell B, Jala VR, Mathew AV, Byun J, Waghulde H, et al. 2015. Evidence for a link between gut microbiota and hypertension in the Dahl rat. *Physiological Genomics* 47: 187–197.

40. Morais AHA, Passos TS, Maciel BLL, Da Silva-maia JK. 2020. Can probiotics and diet promote beneficial immune modulation and purine control in coronavirus infection? *Nutrients* 12: 1737.

41. Morrow LE, Kollef MH, Casale TB. 2010. Probiotic prophylaxis of ventilator-associated pneumonia: A blinded, randomized, controlled trial. *American Journal of Respiratory and Critical Care Medicine* 182: 1058–1064.

42. Pagnini C, Di Paolo MC, D'alba L, Urgesi R, Pallotta L, et al. 2020. Probiotics in COVID-19 pandemic: When evidence lacks, rationale can still ease the way. *Minerva Gastroenterology Dietology* doi: 10.23736/S1121-421X.20.02760-9.

43. Peng Y, Zhao J, Tun HM. 2020. The New Foe and old friends: Are we ready for microbiota-based therapeutics in treating COVID-19 patients? *Gastroenterology* 160: 2192–2193.

44. Rozga M, Cheng FW, Handu D. 2020. Effects of probiotics in conditions or infections similar to COVID-19 on health outcomes: An evidence analysis center scoping review. *Journal Academy Nutrition and Diet* S2212–2672(20): 30930–30938.

45. Shen Z, Xiao Y, Kang L, Ma W, Shi L, et al. 2020. Genomic diversity of severe acute respiratory syndrome-coronavirus 2 in patients with Coronavirus Disease 2019. *Clinical Infectious Diseases* 71: 713–720.

46. Su M, Jia Y, Li Y, Zhou D, Jia J. 2020. Probiotics for the prevention of ventilator-associated pneumonia: A meta-analysis of randomized controlled trials. *Respiratory Care* 65: 673–685.

47. Sundararaman A, Ray M, Ravindra PV, Halami PM. 2020. Role of probiotics to combat viral infections with emphasis on COVID-19. *Applied Microbiology Biotechnology* 104: 8089–8104.

48. Van Ruissen MCE, Bos LD, Dickson RP, Dondorp AM, Schultsz C, Schultz MJ. 2019. Manipulation of the microbiome in critical illness-probiotics as a preventive measure against ventilator-associated pneumonia. *Intensive Care Medicine Experiments* 7: (Suppl 1) 37.

49. Ventoulis I, Sarmourli T, Amoiridou P, Mantzana P, Exindari M, et al. 2020. Bloodstream infection by *Saccharomyces cerevisiae* in two COVID-19 patients after receiving supplementation of *Saccharomyces* in the ICU. *Journal of Fung* 6: E98.

50. Viana SD, Nunes S, Reis F. 2020. ACE2 imbalance as a key player for the poor outcomes in COVID-19 patients with age-related comorbidities - Role of gut microbiota dysbiosis. *Ageing Research Reviews* 62: 101123.

51. Villapol S. 2020. Gastrointestinal symptoms associated with COVID-19: Impact on the gut microbiome. *Translational Research* 226: 57–69.

52. Wang Y, Li X, Ge T, Xiao Y, Liao Y, et al. 2016. Probiotics for prevention and treatment of respiratory tract infections in children: A systematic review and meta-analysis of randomized controlled trials. *Medicine* 95: e4509.

53. Wölfel R, Corman VM, Guggemos W, Seilmaier M, Zange S, et al. 2020. Virological assessment of hospitalized patients with COVID-2019. *Nature* 581: 465–469.

54. Wong MC, Huang J, Lai C, Ng R, Chan FKL, Chan PKS. 2020. Detection of SARS-CoV-2 RNA in fecal specimens of patients with confirmed COVID-19: A meta-analysis. *Journal of Infection* 81: e31–e38.

55. Wu Y, Guo C, Tang L, Hong Z, Zhou J, et al. 2020. Prolonged presence of SARS-CoV-2 viral RNA in faecal samples. *Lancet Gastroenterology Hepatology* 5: 434–435.

56. Xiao F, Tang M, Zheng X, Liu Y, Li X, Shan H. 2020. Evidence for gastrointestinal infection of SARS-CoV-2. *Gastroenterology* 158: 1831–1833.e3.

57. Xu K, Cai H, Shen Y, Ni Q, Chen Y, et al. 2020. [Management of COVID-19: The Zhejiang experience]. *Zhejiang Da Xue Xue Bao Yi Xue Ban* 49: 147–157.

58. Yang T, Chakraborty S, Saha P, Mell B, Cheng X, et al. 2020. Gnotobiotic rats reveal that gut microbiota regulates colonic mRNA of *Ace2*, the receptor for SARS-CoV-2 infectivity. *Hypertension* 76: e1–e3.

59. Yang T, Santisteban MM, Rodriguez V, Li E, Ahmari N, et al. 2016. Effect of probiotics on the incidence of ventilator-associated pneumonia in critically ill patients: A randomized controlled multicenter trial. *Intensive Care Medicine* 42: 1018–1028.

60. Zuo T, Zhan H, Zhang F, Liu Q, Tso EYK, et al. 2020. Alterations in fecal fungal microbiome of patients with COVID-19 during time of hospitalization until discharge. *Gastroenterology* 159: 1302–1310.

61. Zuo T, Zhan H, Zhang F, Liu Q, Tso EYK, et al. 2020a. Alterations in gut microbiota of patients with COVID-19 during time of hospitalization. *Gastroenterology* 159: 944–955.e8.

CHAPTER 6

Underarm Body Odor, the Microbiome, and Probiotic Treatment

Britta De Pessemier, Rune Daneels, Tom Van De Wiele, and Chris Callewaert
Ghent University, Coupure Links 653, 9000 Gent, Belgium

6.1 SKIN STRUCTURE AND FUNCTION

The skin is the body's largest and most exposed organ that comes in contact with the environment. The main job of the skin is to form a physical barrier to protect the body from unwanted external influences. The epidermis and the dermis are the two main layers of the skin (SanMiguel and Grice, 2015). The epidermis is the upper layer and is the one that will contribute most to the task of being a physical barrier by resisting the penetration of microorganisms. Furthermore, it also keeps nutrients and moisture inside the body. The epidermis can be divided into four distinct layers: the *stratum basale*, *stratum spinosum*, *stratum granulosum*, *stratum corneum*. With the *stratum basale* being the bottom layer, it is responsible for renewing the cells of the epidermis (Nestle et al., 2009).

Besides, keratinocytes or squames are produced in this layer and pushed upward toward the stratum corneum, the top layer, where they occur as maturing keratinocytes. They are built of keratin fibrils, cornified envelopes embedded in lipid bilayers, forming the "bricks and mortar" of the epidermis (Grice and Segre, 2011). Other specialized cells of the epidermis are the Langerhans cells, which are the only skin-resident immune-cells and the melanocytes, which produce the pigment melanin. Additionally, T-cells are present in the stratum basale and the stratum spinosum. Under the epidermis, a much thicker layer can be found, the dermis. This is a structural framework made of connective tissue elements, which can also contain hair follicles, sebaceous glands, sweat glands, blood vessels, and organs for the sensation of touch, pressure, pain, and temperature. The hair follicle with its associated glands can be seen as the most important structure of the dermis when it comes to microbial colonization because it is directly open to the external environment (Figure 6.1) (Van De Wiele, 2011).

Migrating cells, such as nutrients, can travel across the dermis through lymphatic and vascular conduits to supply the epidermis (Nestle et al., 2009). Further, the dermis is important for the skin's flexibility, its mechanical resistance, and it is also involved in the regulation of the body's temperature (Van De Wiele, 2011). Underneath these strata, there is another layer called the subcutis, consisting of fat, which is crucial for the protection of deeper tissues and bones (SanMiguel and Grice, 2015).

6.2 SWEAT

The principal function of sweat is to cool down the body, which is essential for survival. If the temperature of the body rose above 40°C, it could lead to protein denaturation and cell death which could cause organ failure.

Good Microbes in Medicine, Food Production, Biotechnology, Bioremediation, and Agriculture, First Edition. Edited by Frans J. de Bruijn, Hauke Smidt, Luca S. Cocolin, Michael Sauer, David Dowling, and Linda Thomashow.

The body could reach a state of hyperthermia. Therefore, the body temperature lowers with the release of thermal energy by evaporation of water from the surface of the skin. Perspiration occurs under conditions of high environmental temperatures and/or physiological stress (Wilke et al., 2007).

6.2.1 Sweat Glands

Among the different types of appendages, there are two main sweat glands located in the dermis, the eccrine and apocrine sweat glands with the eccrine being the most abundant and the apocrine being present in limited areas. The eccrine glands are so called cutaneous glands, because the secretion takes place via a duct to the surface of the skin (Wilke et al., 2007). Both forms are distinguished based on their mode of secretion, morphology, and function (Grice and Segre, 2011).

6.2.1.1 Eccrine Glands

The eccrine sweat glands are mostly involved with the thermoregulatory sweating during exposure to a hot environment or physical exercise, but they also respond to emotional and gustatory stimuli. The thermoregulation is controlled by the hypothalamus which is a part of the central nervous system. The glands are present on all skin surfaces except for two locations, the lips and the glans penis, and they are the only glands with direct access to the surface of the skin (Figure 6.1). The sweat that is produced is odorless and contains mainly sodium and chloride causing the epidermis to be constantly bathed in a water- and salt-based sweat solution (SanMiguel and Grice, 2015). Other compounds are potassium, bicarbonate, lactate, urea, amino acids, and calcium (Groscurth, 2002). The glands consist of a single tubular epithelium ranging from 4 to 8 mm, and are composed of a highly secretory coil and a duct coil (Wilke et al., 2007). Besides thermoregulation, acidification of the skin is another important function of the eccrine sweat glands to prevent the colonization and growth of microorganisms (Grice and Segre, 2011; Wilke et al., 2007).

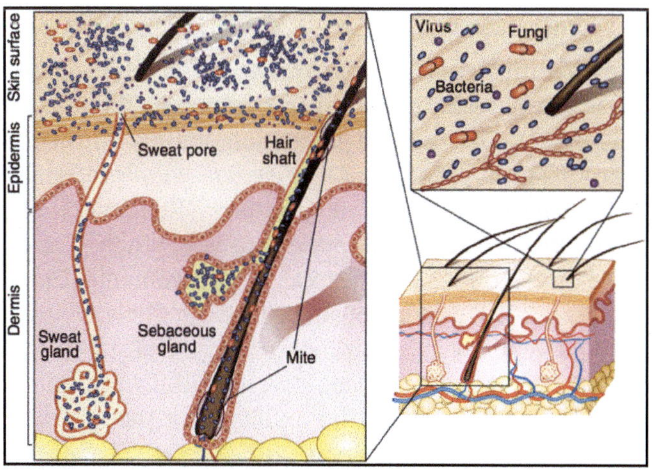

FIGURE 6.1 Structure of the skin viewed in cross-section with skin appendages, microorganisms, and an indication of the main skin layers. The schematic figure of Kong and Segre (2012).

6.2.1.2 Apocrine Glands

The apocrine sweat glands depend on sex hormones and become active with puberty. They are restricted to the hairy body areas because they open to the hair canal (Figure 6.1). Therefore, they can only be found in the axilla, mammary, perineal, and genital region (Wilke et al., 2007). Apocrine glands respond to adrenaline, producing a secretion that is a thick, sterile, milky, and odorless fluid containing proteins, lipids, and steroids. The secretion is released into the upper hair follicles (SanMiguel and Grice, 2015). Potent odorous compounds can be generated when the secretion undergoes a bacterial decomposition leading to a characteristic smell (Groscurth, 2002). Therefore, the apocrine glands can be called the scent glands. The glands consist of a basal secretory and a short straight duct and can be found close to the hair follicle (Groscurth, 2002; Wilke et al., 2007). Further, men have larger apocrine glands than women, and they may be more active in order to supply nutrients for bacterial growth, but women have 75% more apocrine glands in their armpits (Troccaz et al., 2015).

6.2.1.3 Apoeccrine Glands

Besides the eccrine and apocrine sweat glands, another type of sweat gland exists. The so-called apoeccrine sweat glands, known as the mixed sweat glands, are also found in the hairy body areas. The glands secrete eccrine-like watery fluid. The composition of this secretion could resemble that of the eccrine sweat because of the similarities these glands share (e.g. morphology and mode of secretion). However, this is hard to determine as it is yet not possible to discriminate eccrine and apoeccrine sweat during sweat sampling of the axilla (Kong and Segre, 2012; Wilke et al., 2007).

6.2.1.4 Sebaceous Glands

Finally, the body also contains sebaceous glands, which are connected to the hair follicle (Figure 6.1). Through this connection they will form pilosebaceous units concentrated on the face and upper body. These glands are relatively anoxic, but the pilosebaceous follicles support an array of microorganisms that can thrive under these circumstances. A common skin bacterium known to grow in this environment is for example the facultative anaerobe *Cutibacterium acnes* (Grice and Segre, 2011). The sebaceous glands secrete an oily, lipid-rich substance called sebum that protects and emolliates the skin and hair. It also provides skin flexibility and waterproofing. By metabolizing the lipids present in the sebum, free fatty acids are formed, which contribute to the acidic pH of the skin (San-Miguel and Grice, 2015).

6.3 SKIN AND UNDERARM MICROBIOME

The skin is home to a large variety of microorganisms and they can be divided in two groups: (1) resident microbes that are routinely found on the skin and that are often commensal bacteria. Upon perturbation they are able to re-establish themselves. (2) Transient microbes arise from the environment and persist for hours or days and are therefore not permanently present on the skin surface (Kong and Segre, 2012). Further, which bacteria are more abundant depends greatly on the location of the skin. Specific bacteria are associated with moist, sebaceous, and dry microenvironments. The dry areas contain the least bacterial load and are the most diverse as there is a mix of phyla such as Actinobacteria, Proteobacteria, Firmicutes, and Bacteroidetes. Some examples of dry areas are the forearm, the buttock, and various parts of the hand (Grice and Segre, 2011; Kong and Segre, 2012). *Cutibacterium* spp. are known to be more dominant in sebaceous areas like the forehead, and *Staphylococcus* and *Corynebacterium* spp. are more likely to be present in moist areas, such as the axillary vault (Figure 6.2) (Chen and Tsao, 2013). Diversity seems to be the lowest when it comes to sebaceous sites, suggesting that there is a specific selection of microorganisms that can tolerate these conditions (Grice and Segre, 2011). Cutibacteria also appear to contribute to a more efficient immunological response to general infections and prefer a slightly acidic pH, whereas other bacteria favor more neutral and alkaline pHs. The number of microbial species present on the skin is limited by a variety of physical and biochemical factors. The flow of secretions, the integrity of barrier function, skin pH, number and size of follicles and glands, and the gland function are examples of physical factors that are mainly defined by the host. As for the biochemical factors, chemical compounds, biochemical molecules influence the colonization (Krutmann, 2009).

The resident microbiota has many advantages and its preservation is thought to be beneficial to the host, but may become dangerous when skin integrity is disturbed (Krutmann, 2009). An unbalanced ecosystem could result in skin disorders or infections such as atopic dermatitis, chimera, psoriasis, and seborrheic dermatitis. For example, the presence of *Cutibacterium acnes* is strongly associated with acne, a common condition, mostly among teenagers whereas infections with *Staphylococcus aureus* are linked with atopic dermatitis (AD), an itchy chronic skin disease. *Staphylococcus epidermidis* is a skin commensal, which may protect humans from pathogenic bacteria, but can be an opportunistic pathogen in immunocompromised individuals (Kong, 2011). Besides being involved in disorders and infections, the commensal microbes have also evolved together with human beings to provide mutual benefit. *Cutibacterium acnes* for example, hydrolyses triglycerides present in sebum, releases free fatty acids that promote bacterial adherence and then colonizes sebaceous units (Kong and Segre, 2012).

6.4 AXILLARY MICROBIOME

The axilla supports a large permanent population of Gram-positive bacteria that thrive on secretions from the eccrine, apocrine, and sebaceous glands (Austin and Ellis, 2003). The resident microbiota mainly consist of Gram-positive bacteria of the genera *Staphylococcus, Corynebacterium*, and *Cutibacterium* (Rudden et al., 2020). Besides these abundant species, other species make up less than 1% of the total skin microbiome (Chen and Tsao, 2013; Rudden et al., 2020).

The key player is *Staphylococcus*, a Gram-positive bacterium that is highly adaptive and likely present at occluded areas such as the axilla. Even though they can be easily found in areas with a relatively higher moisture content, pH and temperature, they are also present on dry, exposed areas. Besides, staphylococci can also survive in the low oxygen environment of the

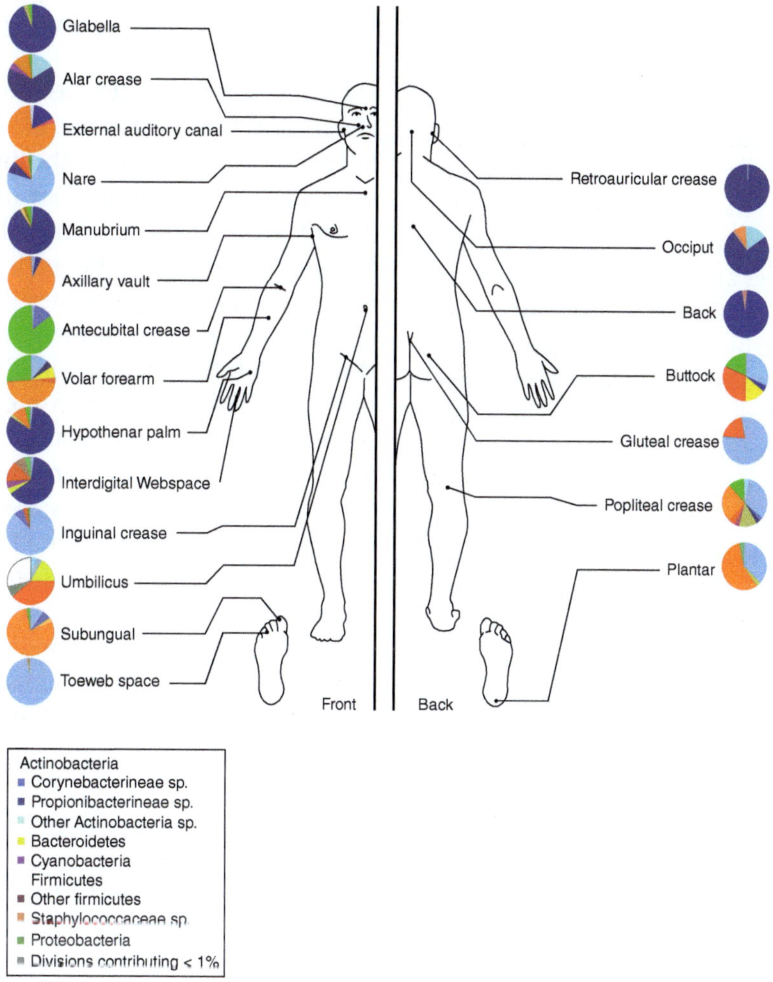

FIGURE 6.2 Topographical distribution of bacteria on the skin site: microbial composition differs among habitats. Four major phyla are shown: *Actinobacteria, Firmicutes, Proteobacteria, Bacteriodetes* as well as the most abundant families: Corynebacteriaceae, Staphylococcaceae, and Propionibacteriaceae (formerly *Propionibacterium acnes* renamed to *Cutibacterium acnes*). The schematic figure of Kong and Segre (2012).

hair follicle as they are both aerobes and facultative anaerobes. Another important trait of the staphylococci is their halotolerance, which means that they can live in high salt conditions and therefore they can even use the urea present in sweat as a source of nitrogen (Scharschmidt and Fischbach, 2013). When it comes to body odor, staphylococci are rather associated with faint acid, non-apocrine odor quality (James et al., 2013). The most prevalent skin species is *S. epidermidis*, as well as other *Staphylococcus* species such as *S. hominis, S. capitis*, and *S. saprophyticus*. *S. aureus*, on the other hand, is usually considered as a pathogen, but the methicillin-resistant strain is also found on the skin of healthy people (Scharschmidt and Fischbach, 2013).

Another important group when it comes to the axilla are the corynebacteria, which are Gram-positive aerobes or facultative anaerobes. Similar to the staphylococci, corynebacteria are also halotolerant. Therefore, they grow on sites rich in eccrine glands. Besides, they are lipophilic, which means they obtain lipids from the environment as they cannot produce them (Scharschmidt and Fischbach, 2013). Corynebacteria are associated with pungent, apocrine odor quality while staphylococci produce a more acid odor (James et al., 2013). Some of the most abundant species are *C. accolens*, *C. jeikeium*, and *C. urealyticum* (Scharschmidt and Fischbach, 2013).

Finally, cutibacteria is another group of Gram-positive bacteria that can be found in the axilla whereby *C. acnes* is the most well-known member. Cutibacteria can be cultured from the skin surface but also grow where oxygen is low, next to the hair follicles (Scharschmidt and Fischbach, 2013).

6.5 BROMHIDROSIS PATHOPHYSIOLOGY

Body odor is a characteristic human trait that is generated by the bacterial degradation of odorless natural secretions into volatile odorous molecules (Austin and Ellis, 2003). Therefore, malodor formation highly depends on the interplay between commensal bacteria with their catalytic enzymes and the unique sweat secretions of the human host (Callewaert et al., 2014). It is mainly the sweat produced by the apocrine glands that is responsible for malodor and is composed of a mixture of volatile organic compounds with volatile fatty acids (VFAs) and thioalcohols being the primary components (James et al., 2013). Under emotional stress, strong body odor formation is often observed instantly, and hormones can trigger molecular processes in secretory cells of apocrine sweat glands (Barzantny et al., 2012). Two main genera dominate the axillary region, *Staphylococcus* and *Corynebacterium*, whereby staphylococci only reveal low levels of body odor, while corynebacteria are considered to be responsible for a strong axillary malodor or bromhidrosis (Barzantny and Brunel, 2012; Callewaert et al., 2014). Short- and medium-chain VFAs, 16-androstene steroids, and thioalcohols are different molecular classes that form a potential pathway implicated in bromhidrosis. Corynebacteria are involved in three out of four processes leading to malodor (Figure 6.3) (Barzantny and Brunel, 2012; Callewaert et al., 2014; James et al., 2013).

6.5.1 Steroid-based Malodor

Odorous steroids, such as 16-androstenes, 5-androstenol, and 5-androstenone, are known to be present in apocrine sweat and are partly implicated in axillary malodor. *Corynebacterium* spp. are linked with odor strength as they are identified as being the key steroid biotransforming organism on the axilla, indicating the possession of a number of enzymes that can modify steroids (Figure 6.3). Biotransformation is only efficient when there is an interplay between different bacterial species in the human axilla, which means that bacterial communities collectively operate to generate odorous steroids. These bacteria convert odorless precursor steroids into 16-androstenes. The stereochemistry of the precursor determines the extent of malodor as for example the α-epimers of different 3-sterols were more readily transformed into odorous 3-oxo-steroids by aerobic corynebacteria than the corresponding β-epimers (Barzantny and Brunel, 2012). Nevertheless, the contribution of steroids is rather minimal in comparison to other pathways (Austin and Ellis, 2003).

6.5.2 Long-chain Fatty Acids (LCFAs)

The bioconversion of LCFAs into short-chain VFAs also leads to body odor. A wide range of lipids, originating from sebaceous and apocrine glands, are

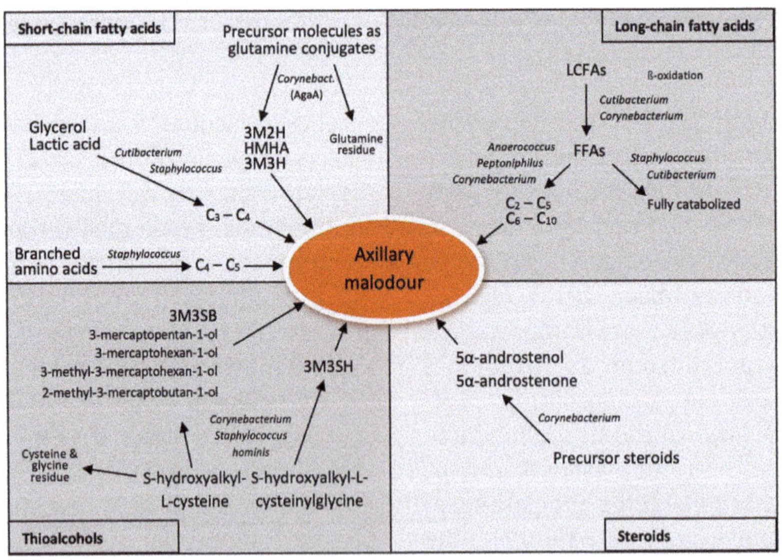

FIGURE 6.3 Microbiological and biochemical origins of axillary malodor; LCFA = long-chain fatty acid; FFA = free fatty acid; 3M2H = 3-methyl-2-hexenoic acid; HMHA = 3-methyl-3-hydroxyhexanoic acid; 3M3SH = 3-methyl-3-sulfanylhexanol; 2M3SB = 2-methyl-3-sulfanylbutan-1-ol, 3-hydroxy-3-methylhexanoic acid (3M3H) (Barzantny and Brunel, 2012; Bawdon et al., 2015; Gower et al., 1994; Hasegawa et al., 2004; James et al., 2013, 2004; Natsch et al., 2006; Rudden et al., 2020; Zeng et al., 1991).

distributed on the human skin surface. First, the skin lipids are catalyzed by secreted lipases to generate free fatty acids. Especially cutibacteria and aerobic coryneforms exhibit a strong lipase activity. Subsequently, the LCFAs are partially degraded by corynebacteria to VFAs through a β-oxidation, which are a prominent cause of body odor development. Thus, a strong malodor is associated with a dense population of corynebacteria (Figure 6.3). Bacteria like *Micrococcus*, on the other hand, have the ability to fully catabolize the LCFAs which does not result in malodor (Barzantny and Brunel, 2012).

6.5.3 VFA-based Malodor

It is widely accepted that short-chain (C_2–C_5) and medium-chain (C_6–C_{10}) VFAs contribute to axillary malodor. Glycerol and lactic acid are fermented to acetic and propionic acid by cutibacteria and staphylococci. Moreover, staphylococci convert branched aliphatic amino acids to highly odorous short-chain (C_4–C_5) methyl-branched VFAs, like isovaleric acid which is associated with the acidic smell of axillary malodor. Bacterial proteases can break down proteins coming from the keratinized epidermis and the apocrine secretion to form amino acids or these amino acids can also be present in eccrine sweat. Yet, corynebacteria metabolize fatty acids at a higher rate than staphylococci metabolize branched aliphatic amino acids (James et al., 2013). Another important component causing body odor are the medium-chain VFAs, such as 3-methyl-2-hexenoic acid (3M2H), which is the most dominant medium-chain VFA. 3M2H is present in apocrine sweat and covalently-bound to l-glutamine residues and released on the skin surface by a corynebacteria enzyme, N-acyl-glutamine aminoacylase (N-AGA) (Natsch et al., 2006).

6.5.4 Thioalcohol-based Malodor

The last group of components that causes underarm malodor are the thioalcohols which, despite their presence in trace amounts, are the most pungent odor volatiles (Rudden et al., 2020). Four different thioalcohols involved with underarm odor were identified in the axilla secretions consisting of two sets of isomers (James et al., 2013). In each case, one isomer has a less objectionable and offensive, occasionally fruity odor, being the 3-mercaptohexan-1-ol, while the other one possesses a meaty, onion-like smell, being the 2-methyl-3-mercaptobutan-1-ol (2M3SB). This is the characteristic axillary odor. The most abundant thioalcohol is 3-methyl-3-sulfanhexan-1-ol (3M3SH), which is generated from the odorless precursor Cys-Gly-3M3SH secreted onto the surface of the skin by apocrine glands (Rudden et al., 2020).

6.6 METHODS TO TREAT BODY ODOR

6.6.1 Conventional Methods

The overproduction of sweat and body odor can be very unpleasant for social life and therefore, different products are designed to combat these undesirable features. Especially when it comes to the underarm site, antiperspirants and deodorants are used daily by a large share of the adult population. Whereas antiperspirants act like a deodorant, deodorants do not necessarily function as an antiperspirant. However, controversies exist about the effect these products have on the epidermis, as they affect the bacterial community when applied to the axillary region. The overall bacterial density decreases which opens up space for new species. Hence, the overproduction of these underarm cosmetics could lead to a selective colonization of malodor-causing microbiota (Callewaert et al., 2017). A more natural way to reduce body odor is by simply keeping your armpit clean and therefore hygiene is the most important factor limiting the bacterial load and deposits of apocrine sweat compounds in the axilla. Shaving the hairy armpit will already reduce the release of odorous volatile compounds (Piérard et al., 2003).

6.6.1.1 Deodorants

Deodorants are classified as cosmetics, intended to make the skin smell better by deodorizing and masking the body odor. Deodorants have two main functions: (1) to reduce the bacterial counts and (2) to mask (or deodorize) body odor. Deodorants can contain antibacterial agents such as ethanol, triclosan, quaternary ammonium compounds, fatty acid esters of glycerol, octoxyglycerol, usnic acid, propylene glycol, benzalkonium chloride, and metal salts to decrease the density of the skin bacterial community (Benohanian, 2001). These molecules cause a bacterial inhibition or prevent bacterial adhesion to the stratum corneum. Unfortunately, these molecules can also cause an unbalanced microbiome. Besides the antimicrobial compounds, deodorants often contain odor-masking fragrances. Examples are cinnamaldehyde, eugenol, and geraniol that also possess antimicrobial properties (Ali et al., 2005; Johny et al., 2010). The addition of perfume is thus an important aspect. In addition, sodium bicarbonate and zinc carbonate are sometimes added as chemical neutralizers of the odorous VFAs. Finally, decreasing the skin pH could also improve the efficacy of some deodorants (Saba M. Ali and Yosipovitch, 2013).

6.6.1.2 Antiperspirants

Where deodorants are more about masking the body odor, antiperspirants are a subgroup of deodorants that affect odor as well as prevent sweating by blocking the sweat glands (Callewaert et al., 2014). Their method of action focuses on (1) the application of perfumes to mask malodors; (2) the usage of antimicrobial compounds to reduce the microbial counts in the axillae; and (3) the usage of aluminum chloride hexahydrate to block the sweat pores (Benohanian, 2001). Antiperspirants are classified as an over-the-counter drug by the US Food and Drug Administration and must undergo rigorous testing. While most antiperspirants function as a deodorant, deodorants do not necessarily act as an antiperspirant (United States Office of the Federal Register, 2005). According to the European legislation, both deodorants and antiperspirants are categorized as cosmetics (EC 1976).

The use of antiperspirants to reduce sweating is generally regarded as the first line of therapy for most patients (Lillis and Coleman, 1990). There are also prescription antiperspirants (or medical antiperspirants) with a higher concentration of aluminum salts. Most antiperspirants will interfere directly or indirectly with bacterial growth by using chemicals whereby metal salt solutions are most frequently used. The pH of the skin surface and sweat is an important parameter to determine the right metal salt solution to obtain the most effectiveness (Callewaert et al., 2014; Piérard et al., 2003). Unfortunately, the use of these compounds also leads to a side effect especially when it comes to antiperspirants as they seem to open the skin niche, leading toward a higher diversity (Callewaert et al., 2014).

6.6.1.3 Antibiotics

Topical antibiotics, applied in the axillary region, are occasionally prescribed to patients in order to eliminate the malodor-causing microbiota (Soriano et al., 1995). The reduction of the microbial density in the axillae has proven to diminish underarm odor (Leyden et al., 1981). Nonetheless, overuse of antimicrobial compounds, and especially antibiotics, can lead to bacterial mutation and resistant microbiota.

6.6.1.4 Medication

Oral medication, such as propantheline bromide, is another possibility to reduce sweating. In the case of olfactory reference syndrome, antidepressants, antipsychotics, or a combination of both are used and are regarded as helpful (Begum and McKenna, 2011).

6.6.1.5 Botox

Botulinum toxin injections are used to block the release of acetylcholine neurotransmitter, paralyzing the muscles which steer the sweat secretion and consequently reduce sweating (Heckmann et al., 2001). The botulinum injection results in a temporary sweat reduction of two to six months.

6.6.1.6 Surgery

Sympathectomy, on the other hand, is a permanent way to stop heavy sweating. A surgeon cuts or clamps the nerves and thus prevents nerve signals from passing through (Drott et al., 1995). Axillary liposuction is often performed to permanently remove or disrupt the eccrine and apocrine sweat glands (Tung and Wei, 1997). Other treatments include laser- or microwave-based ablation of sweat glands (Kim et al., 1999; Lee et al., 2013). These procedures are used as a last resort if other steps, such as antiperspirants or medications, did not work.

6.6.2 Alternative Methods

The conventional methods mainly focus on the limitation of (eccrine) sweat production or use broad-spectrum antimicrobial ingredients. It is important to focus on promoting the growth of beneficial and good-smelling bacteria and inhibiting the growth of odor-causing bacteria when trying to find a better solution to improve body odor (Callewaert et al., 2014; Krutmann and Humbert, 2010). Instead of using deodorants and antiperspirants, promicrobiome therapies could rebalance the composition of the cutaneous microbiota and reduce body odor (Krutmann and Humbert, 2010).

6.6.2.1 Pre-, Pro-, and Postbiotics

When it comes to prebiotics, bioactives or nutrients are added to inhibit the growth of the pathogenic species and support the growth of the beneficial bacteria. In this strategy the growth of the commensal *S. epidermidis* should be preserved by prebiotics, but the growth of *S. aureus* should be inhibited. This method is clearly superior to antibacterial cosmetics that do not selectively reduce all bacterial growth (Figure 6.4) (Krutmann, 2009).

Probiotics are defined by the WHO and FAO as "live organisms which, when administered in adequate amounts, confer a health benefit to the host." The term is often considered in a broader sense as not only viable microorganisms are used, but also non-viable microbial preparations such as supernatants, extracts, and cell lysates (Krutmann, 2009; Krutmann and Humbert, 2010). Probiotics are beneficial for the user as they improve the characteristics of the intestinal microbiome. The skin effect induced by probiotic skin microbiome is thought to be represented by changes in systemic immune responses and could also be beneficial not only for diseased skin but also for healthy skin. Another advantage of probiotics could be improving the skin barrier function. Similar to the improvements obtained by the use of oral probiotics, the same beneficial effects can be reached by the topical application of probiotic bacteria on the skin (Figure 6.4) (Krutmann, 2009). The topical or oral application of living bacteria can form a valuable and durable alternative for existing treatments like deodorants and antiperspirants (Callewaert et al., 2017).

Postbiotics are molecules secreted by living bacteria, or are released after bacterial lysis, providing a physiological benefit to the host, and/or can selectively inhibit the growth of pathogens. They are also known as bacterial metabolites, metabiotics, or biogenics. Increasing evidence shows that particular probiotic-derived components, such as bacteriocins, lipoteichoic acids, surface layer protein, and secreted proteins, can also have a protective role for the host, in a similar fashion as probiotics (Figure 6.4) (Gao et al., 2019). The usage of postbiotics in skin cosmetics is generally preferred, as the consumer is often not ready to apply

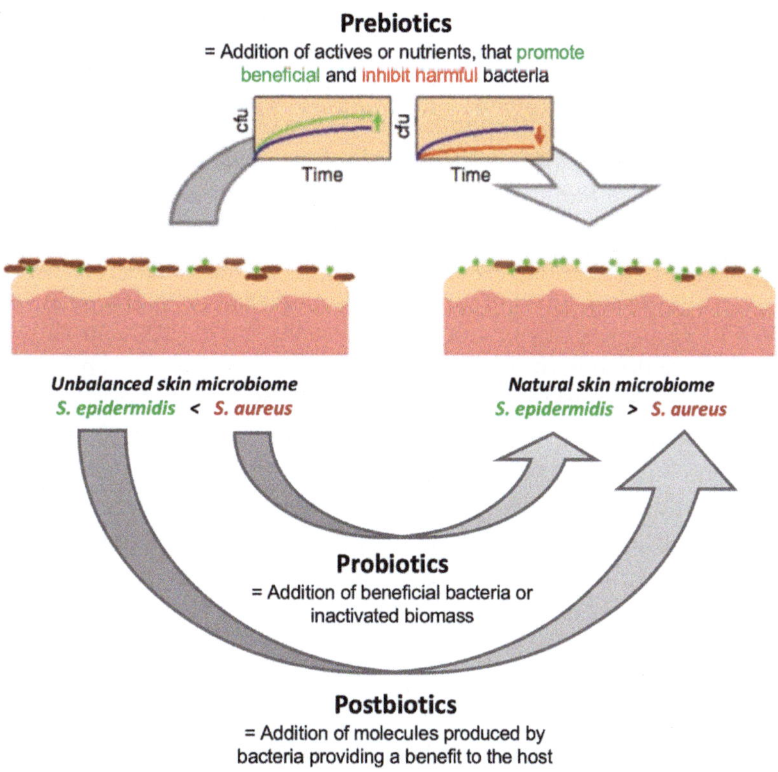

FIGURE 6.4 Schematic illustration of pre-, pro-, and postbiotic cosmetics, redrafted after the image of Krutmann (2009). Topical application of a prebiotic will promote the growth of beneficial (*Staphylococcus epidermidis*) and inhibit harmful bacteria (*Staphylococcus aureus*). Topical application of a postbiotic will use the molecules produced by *S. epidermidis* to outcompete *S. aureus*.

live bacteria on their skin, or as legislation concerning live biotherapeutics is lagging behind.

6.6.2.2 Armpit Bacterial Transplant

The good microbes, causing no or neutral body odor, and bad microbes, associated with malodor development, are known. The microbiome plays an important role in malodor production, and thus microbiome manipulation is a plausible alternative. We proposed a new treatment by means of bacterial transplantation in the axillae. Transplanting the microbiome is an emerging field of medical research and practice. Fecal transplantations were successful in treating intestinal *Clostridium difficile* infections (Borody et al., 2004). More recently, skin bacterial transplantations have been carried out in the treatment of atopic dermatitis, and skin neoplasia (Myles et al., 2016; Nakatsuji et al., 2017; Nakatsuji et al., 2018). A recent review summarized all the skin microbiome manipulation therapies tried to date and their preliminary results (Callewaert et al., 2021). According to the principle of preemptive colonization, the novel microbiota occupy space and nutrition and thus prevent the growth of other bacteria colonizing the niche.

Through armpit bacterial transplantation, the autochthonous malodor-causing microbiome is removed and subsequently replaced with a non-odor-causing microbiome. We are replacing autochthonous axillary microbiomes dominated by *Corynebacterium* spp., by a microbiome dominated with *Staphylococcus* spp., from a healthy subject without body odor. A total of 17 subjects were treated with microbial transplants with the objective of improving underarm odor in a durable manner. This was done in a series of iterations over time. Transplants were done between non-related individuals, who did not experience immediate success, and between related individuals, where the transplants worked best and were also easier to accept. Appropriate donors were evaluated and selected in order to obtain a successful bacterial transmission. In some cases, an initial sterilization was performed using topical antibiotics and antibacterial washes, in order to remove the autochthonous microbiome. This also helped in the successful bacterial transmission and colonization. Subsequently, a high bacterial load from the donor was transferred to the acceptor's clean axillary skin. The results are promising and we were able to shift the microbiome towards a better equilibrium and a better underarm odor (Callewaert et al., 2017).

6.6.2.3 Bacteriotherapy

Frequently used treatments mainly focus on either sweat reduction, odor masking, or reducing the bacterial load. This last example includes usage of deodorants, antiperspirants, and the application of topical antibiotics which minimizes the growth of all bacterial species present on the skin surface. Afterward, the bacteria will recolonize on the surface of the epidermis as well as beneath it. As *Staphylococcus* and *Cutibacterium* spp. are correlated with a better underarm odor, it is preferable that they occupy the axillary skin regions rather than malodor-causing bacteria such as *Corynebacterium* species. To create a shift in the skin microbiome towards more non-odor-causing microbiota, a topical application and probiotic use of existing skin bacteria can be used. The application of isolated and cultured non-odor-causing skin residents can antagonize the autochthonous malodor-causing microbiota in a durable way (Callewaert et al., 2017).

A probiotic spray was designed containing liquefied *Staphylococcus*. The sprays were tested on 79 subjects suffering from very intense body odor and led to positive results. Upon use of the bacteria, the underarm microbiome was shifted from a high abundance of *Actinobacteria* towards a high abundance of *Firmicutes*. The dominance of *Staphylococcus* also caused the eradication of one or multiple malodor-associated bacteria in the underarm. This led to a significantly improved underarm odor, as determined by a trained odor panel. The subjects themselves also noticed an improved odor. Analysis by GC/MS also showed a clear difference in VOC pattern, with much lower malodorous molecules upon spray application of the bacteria. We are still developing the spray to obtain a more durable result. Nonetheless, the results show a clear influence of the underarm microbiome on odor development. This proves that the "right bugs" can improve a skin condition and opens perspectives in other areas.

6.7 CONCLUSIONS

Sterile sweat that is freshly secreted on the skin is odorless, and does not smell bad. The underarm microbiome is the main influencer that converts sweat into odor molecules. Conventional deodorants and antiperspirants focus on the inhibition of the entire underarm microbiome. Once the antimicrobial molecules are sweated out, the bacteria can grow back. These bacteria are the most adapted to the underarm conditions, as well as the presence of antimicrobial agents. As a result, stress-resistant bacteria also develop, and are often linked to malodor development. The conventional underarm cosmetics are therefore not a durable solution in the long run. New therapies are looking into bacterial transplants of complete underarm microbiomes, as well as bacteriotherapy, using targeted microbes that are linked to a better underarm odor. The preliminary results are promising and show that a promicrobiome approach may be the best solution for heavy bromhidrosis. This opens new perspectives and introduces a completely new way of looking at the conventional methods. The future of underarm cosmetics is promicrobial.

ACKNOWLEDGMENTS

This study was supported by Research Foundation – Flanders (grant FWO19/PDS/084) and Ghent University.

REFERENCES

1. Ali SM, Khan AA, Ahmed I, Musaddiq M, Ahmed KS, et al. 2005. Antimicrobial activities of Eugenol and Cinnamaldehyde against the human gastric pathogen. *Helicobacter Pylori. Annals of Clinical Microbiology and Antimicrobials* 4: 20.
2. Ali SM, Yosipovitch G. 2013. Skin pH: From basic science to basic skin care. *Acta Dermato-Venereologica* 93: 261–267.
3. Austin C, Ellis J. 2003. Microbial pathways leading to steroidal malodour in the axilla. *The Journal of Steroid Biochemistry and Molecular Biology* 87: 105–110.
4. Barzantny H, Brunel TA. 2012. Molecular basis of human body odour formation: Insights deduced from corynebacterial genome sequences. *International Journal of Cosmetic Science* 341: 2–11.
5. Bawdon D, Cox DS, Ashford D, James AG, Thomas GH. 2015. Identification of axillary *Staphylococcus* sp. involved in the production of the malodorous thioalcohol 3-methyl-3-sufanylhexan-1-ol. *FEMS Microbiology Letters* 362: fnv111.
6. Begum M, McKenna PJ. 2011. Olfactory reference syndrome: A systematic review of the world literature. *Psychological Medicine* 41: 453–461.
7. Benohanian A. 2001. Antiperspirants and deodorants. *Clinics in Dermatology* 19: 398–405.
8. Borody TJ, Warren EF, Leis SM, Surace R, Ashman O, Siarakas S. 2004. Bacteriotherapy using fecal flora: Toying with human motions. *Journal of Clinical Gastroenterology* 38: 475–483.
9. Callewaert C, Hutapea P, Van De Wiele T, Boon N. 2014. Deodorants and antiperspirants affect the axillary bacterial community. *Archives of Dermatological Research* 306: 701–710.
10. Callewaert C, Knödlseder N, Karoglan A, Güell M, Paetzold B. 2021. Skin microbiome transplantation and manipulation: Current state of the art. *Computational and Structural Biotechnology Journal* 19: 624–631.
11. Callewaert C, Lambert J, Van De Wiele T. 2017. Towards a bacterial treatment for armpit malodour. *Experimental Dermatology* 26: 388–391.
12. Chen YE, Tsao H. 2013. The skin microbiome: Current perspectives and future challenges. *J Am Acad Dermatol.* 69(1): 143–155e.3. Jul. https://www.ncbi.nlm.nih.gov/pmc/articles/PMC3686918.
13. Drott C, Göthberg G, Claes G. 1995. Endoscopic transthoracic sympathectomy: An efficient and safe method for the treatment of hyperhidrosis. *Journal of the American Academy of Dermatology* 33: 78–81.
14. Gao J, Li Y, Wan Y, Hu T, Liu L, et al. 2019. A novel postbiotic from *Lactobacillus rhamnosus* GG with a beneficial effect on intestinal barrier function. *Frontiers in Microbiology* 10: 3389.
15. Gower DB, Holland KT, Mallet AI, Rennie PJ, Watkins WJ. 1994. Comparison of 16-Androstene steroid concentrations in sterile apocrine sweat and axillary secretions: Interconversions of 16-Androstenes by the axillary microflora—a mechanism for axillary odour production in man? *The Journal of Steroid Biochemistry and Molecular Biology* 48: 409–418.
16. Grice EA, Segre JA. 2011. The skin microbiome. *Nature Reviews. Microbiology* 9: 244–253.
17. Groscurth P. 2002. Anatomy of sweat glands. *Current Problems in Dermatology* 30: 1–9.
18. Hasegawa Y, Yabuki M, Matsukane M. 2004. Identification of new odoriferous compounds in human axillary sweat. *Chemistry & Biodiversity* 1: 2042–2050.

19. Heckmann M, Ceballos-Baumann AO, Plewig G. 2001. Botulinum Toxin A for axillary hyyperhidrosis (excessive sweating). *New England Journal of Medicine* 344: 488–493.
20. James AG, Austin CJ, Cox DS, Taylor D, Calvert R. 2013. Microbiological and biochemical origins of human axillary odour. *FEMS Microbiology Ecology* 83: 527–540.
21. James AG, Hyliands D, Johnston H. 2004. Generation of volatile fatty acids by axillary bacterial1. *International Journal of Cosmetic Science* 26: 149–156.
22. Kim IH, Seo SL, Oh CH. 1999. Minimally invasive surgery for axillary osmidrosis: Combined operation with CO2 laser and subcutaneous tissue remover. *Dermatologic Surgery* 25: 875–879.
23. Kollanoor-Johny A, Darre MJ, Donoghue AM, Donoghue DJ, Venkitanarayanan K. 2010. Antibacterial effect of trans-cinnamaldehyde, eugenol, carvacrol, and thymol on *Salmonella enteritidis* and *Campylobacter jejuni* in chicken cecal contents in vitro. *Journal of Applied Poultry Research* 19: 237–244.
24. Kong HH. 2011. Skin microbiome: Genomics-based insights into the diversity and role of skin microbes. *Trends in Molecular Medicine* 17: 320–328.
25. Kong HH, Segre JA. 2012. Skin microbiome: Looking back to move forward. *The Journal of Investigative Dermatology* 132: 933–939.
26. Krutmann J. 2009. Pre- and probiotics for human skin. *Journal of Dermatological Science* 54: 1–5.
27. Krutmann J, Humbert P. (Eds.). 2010. *Nutrition for Healthy Skin: Strategies for Clinical and Cosmetic Practice* [PDF]. Berlin Heidelberg: Springer-Verlag. https://www.springer.com/gp/book/9783642122637.
28. Lee SJ, Chang KY, Suh DH, Song KY, Ryu HJ. 2013. The efficacy of a microwave device for treating axillary hyperhidrosis and osmidrosis in Asians: A preliminary study. *Journal of Cosmetic and Laser Therapy: Official Publication of the European Society for Laser Dermatology* 15: 255–259.
29. Leyden JJ, McGinley KJ, Hölzle E, Labows JN, Kligman AM. 1981. The microbiology of the human axilla and its relationship to axillary odor. *Journal of Investigative Dermatology* 77: 413–416.
30. Lillis PJ, Coleman WP. 1990. Liposuction for treatment of axillary hyperhidrosis. *Dermatologic Clinics* 8: 479–482.
31. Myles IA, Williams KW, Reckhow JD, Jammeh ML, Pincus NB, et al. 2016. Transplantation of human skin microbiota in models of atopic dermatitis. *JCI Insight* 1(10): e86955. Jul 7. https://www.ncbi.nlm.nih.gov/pmc/articles/PMC4963067.
32. Nakatsuji T, Chen TH, Butcher AM, Trzoss LL, Nam S, et al. 2018. A commensal strain of protects against skin neoplasia. *Science Advances* 4(2): eaao4502. Feb 28. https://pubmed.ncbi.nlm.nih.gov/29507878.
33. Nakatsuji T, Chen TH, Narala S, Chun KA, Two AM, et al. 2017. Antimicrobials from human skin commensal bacteria protect against and are deficient in atopic dermatitis. *Science Translational Medicine* 9: 378.
34. Natsch A, Derrer S, Flachsmann F, Schmid J. 2006. A broad diversity of volatile carboxylic acids, released by a bacterial aminoacylase from axilla secretions, as candidate molecules for the determination of human-body odor type. *Chemistry & Biodiversity* 3: 1–20.
35. Nestle FO, Di Meglio P, Qin JZ, Nickoloff BJ. 2009. Skin immune sentinels in health and disease. *Nature Reviews Immunology* 9: 679–691.
36. Piérard GE, Elsner P, Marks R, Masson P, Paye M, EEMCO Group. 2003. EEMCO guidance for the efficacy assessment of antiperspirants and deodorants. *Skin Pharmacology and Applied Skin Physiology* 16: 324–342.
37. Rudden M, Herman R, Rose M, Bawdon D, Cox DS, et al. 2020. The molecular basis of thioalcohol production in human body odour. *Scientific Reports* 10: 12500.
38. SanMiguel A, Grice EA. 2015. Interactions between host factors and the skin microbiome. *Cellular and Molecular Life Sciences* 72: 1499–1515.
39. Scharschmidt TC, Fischbach MA. 2013. What lives on our skin: Ecology, genomics and therapeutic opportunities of the skin microbiome. *Drug Discovery Today. Disease Mechanisms* 10: e83–e89.
40. Soriano F, Zapardiel J, Nieto E. 1995. Antimicrobial susceptibilities of *Corynebacterium* species and other non- spore-forming gram-positive bacilli to 18 antimicrobial agents. *Antimicrobial Agents and Chemotherapy* 39: 208–214.
41. Troccaz M, Gaïa N, Beccucci S, Schrenzel J, Cayeux I, et al. 2015. Mapping axillary microbiota responsible for body odours using a culture-independent approach. *Microbiome* 3: 3.
42. Tung TC, Wei FC. 1997. Excision of subcutaneous tissue for the treatment of axillary osmidrosis. *British Journal of Plastic Surgery* 50: 61–66.

43. United States. Office of the Federal Register. 2005. Code of Federal Regulations: Title 21: Food and Drugs. Office of the Federal Register.

44. Van De Wiele T. 2011. Interphase processes of host associated microorganisms. *Cursus, Universiteit Gent, Faculteit Bio-Ingenieurswetenschappen*.

45. Wilke K, Martin A, Terstegen L, Biel SS. 2007. A short history of sweat gland biology. *International Journal of Cosmetic Science* 29: 169–179.

46. Zeng XN, Leyden JJ, Lawley HJ, Sawano K, Nohara I, Preti G. 1991. Analysis of characteristic odors from human male axillae. *Journal of Chemical Ecology* 17: 1469–1492.

CHAPTER 7

The Enigma of *Prevotella copri*

Petia Kovatcheva-Datchary
Institute of Molecular Infection Biology, University of Würzburg, 97070, Würzburg, Germany

7.1 INTRODUCTION

The human microbiome is a complex collection of microbes that inhabit various sites of the human body, including the skin, the oral cavity, the respiratory tract, the gastrointestinal tract, the urinary tract, and the reproductive tract (Sender et al., 2016; Shreiner et al., 2015). This assembly of different microorganisms includes Eukaryotes, Archaea, Bacteria, and viruses, and can influence most of the systems in our body (Kovatcheva-Datchary et al., 2013). The human gut microbiota is a highly diverse and dynamic ecosystem estimated to be composed of 39 trillion microbial cells (Sender et al., 2016) and with more than 1000-fold more genes (Tierney et al., 2019). The gut microbiota is strongly influenced by the host's diet (David et al., 2014; Wu et al., 2011) and changes in gut ecology affect the inflammatory and metabolic properties of the microbes and thereby the host's metabolism (Tremaroli and Backhed, 2012). Epidemiological studies have associated specific gut bacteria to improved or impaired host metabolic health (Karlsson et al., 2013; Kovatcheva-Datchary et al., 2015; Pedersen et al., 2016; Qin et al., 2012). *Prevotella copri*, the most abundant *Prevotella* species in the human gut (Consortium, 2012) also associated with habitual diet (De Filippis et al., 2019), represents one of the clearest cases of dissimilar associations with either health or disease (Cani, 2018).

Representatives of the genus *Prevotella* predominantly colonize the human oral cavity, where they are considered as part of the oral microbiome but also as clinically relevant pathogens in periodontal infections (Szafrański et al., 2015). In the human gut *P. copri* is by far the most abundant and relevant *Prevotella* species (De Filippis et al., 2019; Hayashi et al., 2007). To date apart from metagenomic studies relatively little is known about this bacterium.

This chapter summarizes the recent knowledge about *P. copri* and the potential of this bacterium as a valuable member of the healthy human microbiome.

7.2 *PREVOTELLA COPRI* PHYSIOLOGY, GROWTH, AND METABOLISM

The genus *Prevotella* belongs to the Bacteroidetes phylum and consists of Gram-negative, obligately anaerobic, and immobile rods (Hayashi et al., 2007). *Prevotella copri* DSM 18205 is the only commercially available type strain that has been the gold standard for all *P. copri*-based research.

Data from growth experiments and from analytic profile index (API) tests, showed that *P. copri* can absorb and metabolize a large number of different sugars (Hayashi et al., 2007). In contrast to other *Prevotella* species that occupy the human oral cavity and mainly metabolize proteins, peptides, amino acids, and amino sugar compounds (Takahashi and Yamada, 2000), *P. copri* hardly ferments any amino acids and peptides (Hayashi et al., 2007). In contrast, *P. copri* has a large

Good Microbes in Medicine, Food Production, Biotechnology, Bioremediation, and Agriculture, First Edition. Edited by Frans J. de Bruijn, Hauke Smidt, Luca S. Cocolin, Michael Sauer, David Dowling, and Linda Thomashow.

repertoire of enzymes to break down plant fibers such as starch, xylan, and pectin, which makes it comparable to other species of *Prevotella* that inhabit the rumen of ruminants or intestinal *Bacteroides* species (Accetto and Avguštin, 2019; Flint et al., 2008; Sonnenburg et al., 2010). In addition, *P. copri* is characterized by extremely competitive anaerobic growth, which potentially allows the bacterium to outcompete a vast number of other inhabitants in the gut. To this end, under laboratory conditions the organism is able to achieve considerable cell densities with relatively short doubling times of 40 min and a maximum optimum density of $OD_{600} = 2.6$ in PYG (peptone yeast extract glucose) medium and doubling times of 2.5 h and maximum optimum density of $OD_{600} = 1.5$ in DMMG (defined minimal medium glucose) (Franke and Deppenmeier, 2018).

As some other members of the gut including *Bacteroides* species, *P. copri* has restricted capacity of heme and quinone synthesis. Therefore, hemin and vitamin K_1 are supplemented to the growth media (Fenn et al., 2017; Franke and Deppenmeier, 2018; Macy et al., 1975; Sperry et al., 1977). An important characteristic of *P. copri* is the pronounced requirement of CO_2 or bicarbonate for growth, which makes this bacterium different than *Bacteroides* species. This unique characteristic of *P. copri* might be of high importance in the context of microbial competition in the gut. Another major difference between *P. copri* and *Bacteroides* species is the different type of end metabolites the bacteria produces from carbohydrate fermentation. Whereas *Bacteroides* species are known to release mostly propionate (Macy et al., 1975), *P. copri* generates succinate, acetate, and formate (Hayashi et al., 2007). Microbially produced metabolites, in particular short-chain fatty acids (SCFAs) and organic acids as succinate, lactate, and formate, are essential not only for balancing the microbe–microbe interactions in the gut, but are also an important factor in host health and disease (Koh et al., 2016, Koh and Bäckhed, 2020). Differences in the end metabolite pool will create a different environment in the gut, promoting the growth of different microbes and different effects on the host.

7.3 PREVOTELLA COPRI, AN IMPORTANT MEMBER OF THE HUMAN GUT MICROBIOTA

Gut microbiota composition is known to vary greatly between healthy individuals. The human gut microbiota of westernized populations is dominated by two bacterial phyla, the Firmicutes and the Bacteroidetes, the latter being mostly dominated by the two genera *Bacteroides* and *Prevotella* (Kovatcheva-Datchary et al., 2013).

Metagenome analysis revealed three discrete enterotypes that represent specific configurations of the microbial community in the healthy human gut (Arumugam et al., 2011; Bäckhed et al., 2015; David et al., 2014). These enterotypes were defined by their predominating taxa, which are either *Bacteroides*, *Ruminococcus*, or *Prevotella*. The *Prevotella*-dominated microbiomes have been shown to occur in populations consuming agrarian fiber-rich diets, such as indigenous populations in Africa and South America (De Filippo et al., 2010; Yatsunenko et al., 2012; Rampelli et al., 2015).

Although the concept of enterotypes is currently a matter of debate (Knights et al., 2014), it is well recognized that *P. copri* is a frequent colonizer of the human gut (Consortium, 2012; Qin et al., 2010) that occurs abundantly (Arumugam et al., 2011) in some individuals/populations (Koren et al., 2013). The high presence of this bacterium is reported in the gut of humans on health-promoting diets such as fiber-rich and agrarian plant-based diets, and is associated with vegetarians and indigenous people, that supports the notion that *P. copri* is highly relevant for human health (Clemente et al., 2012; Kovatcheva-Datchary et al., 2015; Ley, 2016; Tett et al., 2019; Yatsunenko et al., 2012). Nevertheless, whereas in some studies, *P. copri* has been associated with improved glucose metabolism (Kovatcheva-Datchary et al., 2015), in other studies increased abundance of *Prevotella* was measured in patients suffering from hypertension (Li et al., 2017) and with a higher risk for the development of rheumatoid arthritis (Scher et al., 2013). However, these seemingly contradictory roles of *P. copri* in human physiology are likely due to a poor understanding of *Prevotella* species diversity and in particular of *P. copri* strains.

7.4 THE UNEXPLORED DIVERSITY OF PREVOTELLA COPRI

The human gut is populated by microbial communities that exhibit a remarkable degree of strain-level diversity (Consortium, 2012). This diversity gives each species, or even each strain, distinct functional capacities that can explain the different responses reported for certain bacteria when physically interacting with and utilizing various components of the diets consumed by their hosts.

In the context of *P. copri*, recent research revealed high strain-level diversity and interindividual strain specificity between *P. copri* populations in the gut (Truong et al., 2017). Analysis of the pangenome of *P. copri* recovered exclusively from Italian subjects showed strain-specific diet associations (De Filippis et al., 2019). Indeed, subjects with habitual diets rich in fibers had *P. copri* strains with enhanced potential to convert carbohydrates. In contrast, *P. copri* strains associated with omnivore diets had a higher functional capacity for branched-chain amino acid (BCAA) biosynthesis (De Filippis et al., 2019). The latter has been identified as a risk factor for glucose intolerance and type 2 diabetes (Pedersen et al., 2016). A large-scale cross-continent meta-analysis of reconstructed *P. copri* genomes recovered about 1000 *P. copri* high-quality genomes for population-based analysis (Tett et al., 2019). A major finding was that *P. copri* is not a monotypic species, but encompasses four distinct clades that are highly associated with non-westernized populations and is widely distributed. In westernized populations *P. copri* was mostly represented by a single clade, with a very small fraction of the westernized individuals (4.6%) for whom microbiomes contained the four *P. copri* clades. Previously, migration from non-western countries to the USA was associated with loss of microbiome diversity and function, also related to *P. copri* strains (Vangay et al., 2018). In the microbiome of US individuals only a single strain of *P. copri* was detected, while the gut microbiome of non-westernized individuals had up to four strains of *P. copri* (Vangay et al., 2018). Further, analysis of ancient stool microbiomes featured a similar pattern of *P. copri* presence consistent with modern non-westernized populations, which allows the time of clade delineation to predate human migration waves out of Africa (Tett et al., 2019) to be estimated. In summary, this reveals that *P. copri* is far more diverse than previously appreciated.

A consequent study on 83 new *P. copri* isolates from 11 donors, showed that most of the *P. copri* isolates fell within one of the four clades, identified as clade A by (Tett et al., 2019), whereas clades B and D had only one representing clone each (Fehlner-Peach et al., 2019). Further functional and genomic analysis of the new *P. copri* isolates confirmed a high genomic diversity within the *P. copri* complex that was illustrated by a relatively modest core genome of 1750 genes, whereas the accessory genome ranged from 1250 to up to 2250 genes in some strains.

Analysis of the genomic diversity of well-characterized *P. copri* genes encoding polysaccharide catabolism enzymes that are commonly grouped in polysaccharide utilization loci (PULs), revealed that some of the *P. copri* polysaccharide utilizing abilities were clade specific. For example, all *P. copri* isolates except those in clade B, were able to grow on arabinoxylan and arabinan. In contrast, only the clade B isolate was able to degrade heparin. In addition, previous analysis of the enzymatic repertoire of *P. copri* to metabolize carbohydrates, showed clade-specific distribution of carbohydrate active enzymes (CAZymes), essential for the utilization of plant-derived carbohydrates (Tett et al., 2019). All this further justifies that *P. copri* strains from different clades that might co-inhabit the same host, own different functionalities with respect to conversions of complex plant-derived polysaccharides, given that they likely use complementary enzymatic activities. Strain-level diversity of *P. copri* in the human gut microbiome is affected by diet and should be considered when examining host–microbe relationships.

In summary, the gut microbiota awards us with microbes the functionality of which can contribute to host metabolic benefits. We still lack essential knowledge on how microbes in the gut ecosystem interact and affect the host. This book chapter summarizes recent findings on *Prevotella copri*, one of the gut microbes the functionality of which strongly depends on the host's diet, emphasizing the relevance of this bacterium for host nutrition and health. *Prevotella copri* possesses the ability to break down dietary polysaccharides, a function critical for human survival. A strong body of evidence points toward the loss of this functionality in *P. copri* as a result of the westernization of our diet. To this end, in order to achieve modulation of *P. copri* functionality for host benefits, personalized strategies have to be considered, either through manipulating the host's diet or editing the microbial functionality by targeting microbial enzymes with specific antisense oligonucleotides.

REFERENCES

1. Accetto T, Avguštin G. 2019. The diverse and extensive plant polysaccharide degradative apparatuses of the rumen and hindgut *Prevotella* species: A factor in their ubiquity? *Systematic Applied Microbiology* 42: 107–116.
2. Arumugam M, Raes J, Pelletier E, Le Paslier D, Yamada T, et al. 2011. Enterotypes of the human gut microbiome. *Nature* 473: 174–180.
3. Bäckhed F, Roswall J, Peng Y, Feng Q, Jia H, et al. 2015. Dynamics and stabilization of the human gut micro-

biome during the first year of life. *Cell Host & Microbe* 17: 690–703.

4. Cani PD. 2018. Human gut microbiome: Hopes, threats and promises. *Gut* 67: 1716–1725.
5. Clemente JC, Ursell LK, Parfrey LW, Knight R. 2012. The impact of the gut microbiota on human health: An integrative view. *Cell* 148: 1258–1270.
6. Consortium HMP. 2012. Structure, function and diversity of the healthy human microbiome. *Nature* 486: 207–214.
7. David LA, Maurice CF, Carmody RN, Gootenberg DB, Button JE, et al. 2014. Diet rapidly and reproducibly alters the human gut microbiome. *Nature* 505: 559–563.
8. De Filippis F, Pasolli E, Tett A, Tarallo S, Naccarati A, et al. 2019. Distinct genetic and functional traits of human intestinal *Prevotella copri* strains are associated with different habitual diets. *Cell Host Microbe* 25(444-453): e443.
9. De Filippo C, Cavalieri D, Di Paola M, Ramazzotti M, Poullet JB, et al. 2010. Impact of diet in shaping gut microbiota revealed by a comparative study in children from Europe and rural Africa. *Proceedings National. Academy Science. U.S.A* 107: 14691–14696.
10. Fehlner-Peach H, Magnabosco C, Raghavan V, Scher JU, Tett A, et al. 2019. Distinct polysaccharide utilization profiles of human intestinal *Prevotella copri* isolates. *Cell Host Microbe* 26(680-690): e685.
11. Fenn K, Strandwitz P, Stewart EJ, Dimise E, Rubin S, et al. 2017. Quinones are growth factors for the human gut microbiota. *Microbiome* 5: 161.
12. Flint HJ, Bayer EA, Rincon MT, Lamed R, White BA. 2008. Polysaccharide utilization by gut bacteria: Potential for new insights from genomic analysis. *Nature Reviews in Microbiology* 6: 121–131.
13. Franke T, Deppenmeier U. 2018. Physiology and central carbon metabolism of the gut bacterium *Prevotella copri*. *Molecular Microbiology* 109: 528–540.
14. Hayashi H, Shibata K, Sakamoto M, Tomita S, Benno Y. 2007. *Prevotella copri* sp. nov. and *Prevotella stercorea* sp. nov., isolated from human faeces. *Internatial Journal Systematic. Evolutionary Microbiology* 57: 941–946.
15. Karlsson FH, Tremaroli V, Nookaew I, Bergstrom G, Behre CJ, et al. 2013. Gut metagenome in European women with normal, impaired and diabetic glucose control. *Nature* 498: 99–103.
16. Knights D, Ward TL, McKinlay CE, Miller H, Gonzalez A, et al. 2014. Rethinking "enterotypes." *Cell Host Microbe* 16: 433–437.
17. Koh A, Bäckhed F. 2020. From association to causality: The Role of the gut microbiota and its functional products on host metabolism. *Molecular Cell* 78: 584–596.
18. Koh A, De Vadder F, Kovatcheva-Datchary P, Bäckhed F. 2016. From dietary fiber to host physiology: Short-chain fatty acids as key bacterial metabolites. *Cell* 165: 1332–1345.
19. Koren O, Knights D, Gonzalez A, Waldron L, Segata N, et al. 2013. A guide to enterotypes across the human body: Meta-analysis of microbial community structures in human microbiome datasets. *PLoS Computational Biology* 9: e1002863.
20. Kovatcheva-Datchary P, Nilsson A, Akrami R, Lee Ying S, De Vadder F, et al. 2015. Dietary fiber-induced improvement in glucose metabolism Is associated with increased abundance of *Prevotella*. *Cellular Metabolism* 22: 971–982.
21. Kovatcheva-Datchary P, Tremaroli V, Bäckhed F. 2013. Gut Microbiota. In: Eugene Rosenberg et al. (eds). *The Prokaryotes - Human Microbiology*. *Springer-Verlag*, p. 3.
22. Ley RE. 2016. Gut microbiota in 2015: *Prevotella* in the gut: Choose carefully. *Nature Reviews Gastroenterology Hepatology* 13: 69–70.
23. Li J, Zhao F, Wang Y, Chen J, Tao J, et al. 2017. Gut microbiota dysbiosis contributes to the development of hypertension. *Microbiome* 5: 14.
24. Macy J, Probst I, Gottschalk G. 1975. Evidence for cytochrome involvement in fumarate reduction and adenosine 5′-triphosphate synthesis by *Bacteroides fragilis* grown in the presence of hemin. *Journal of Bacteriology* 123: 436–442.
25. Pedersen HK, Gudmundsdottir V, Nielsen HB, Hyotylainen T, Nielsen T, et al. 2016. Human gut microbes impact host serum metabolome and insulin sensitivity. *Nature* 535: 376–381.
26. Qin J, Li R, Raes J, Arumugam M, Burgdorf KS, et al. 2010. A human gut microbial gene catalog established by metagenomic sequencing. *Nature* 464: 59–65.
27. Qin J, Li Y, Cai Z, Li S, Zhu J, et al. 2012. A metagenome-wide association study of gut microbiota in type 2 diabetes. *Nature* 490: 55–60.
28. Rampelli S, Schnorr SL, Consolandi C, Turroni S, Severgnini M, et al. 2015. Metagenome Sequencing of the Hadza Hunter-Gatherer Gut Microbiota. *Current Biology* 25: 1682–1693.
29. Scher JU, Sczesnak A, Longman RS, Segata N, Ubeda C, et al. 2013. Expansion of intestinal *Prevotella copri* correlates with enhanced susceptibility to arthritis. *Elife* 2: e01202.

30. Sender R, Fuchs S, Milo R. 2016. Are we really vastly outnumbered? Revisiting the ratio of bacterial to host cells in humans. *Cell* 164: 337–340.
31. Shreiner AB, Kao JY, Young VB. 2015. The gut microbiome in health and in disease. *Current Opinion in Gastroenterology* 31: 69–75.
32. Sonnenburg ED, Zheng H, Joglekar P, Higginbottom SK, Firbank SJ, et al. 2010. Specificity of polysaccharide use in intestinal bacteroides species determines diet-induced microbiota alterations. *Cell* 141: 1241–1252.
33. Sperry JF, Appleman MD, Wilkins TD. 1977. Requirement of heme for growth of *Bacteroides fragilis*. *Applied Environmental Microbiology* 34: 386–390.
34. Szafrański SP, Deng ZL, Tomasch J, Jarek M, Bhuju S, et al. 2015. Functional biomarkers for chronic periodontitis and insights into the roles of *Prevotella nigrescens* and *Fusobacterium nucleatum*; a metatranscriptome analysis. *NPJ Biofilms Microbiomes* 1: 15017.
35. Takahashi N, Yamada T. 2000. Pathways for amino acid metabolism by *Prevotella intermedia* and *Prevotella nigrescens*. *Oral Microbiology Immunology* 15: 96–102.
36. Tett A, Huang KD, Asnicar F, Fehlner-Peach H, Pasolli E, et al. 2019. The *Prevotella copri* complex comprises four distinct clades underrepresented in westernized populations. *Cell Host Microbe* 26(666-679): e667.
37. Tierney BT, Yang Z, Luber JM, Beaudin M, Wibowo MC, et al. 2019. The landscape of genetic content in the gut and oral human microbiome. *Cell Host Microbe* 26(283-295): e288.
38. Tremaroli V, Backhed F. 2012. Functional interactions between the gut microbiota and host metabolism. *Nature* 489: 242–249.
39. Truong DT, Tett A, Pasolli E, Huttenhower C, Segata N. 2017. Microbial strain-level population structure and genetic diversity from metagenomes. *Genome Research* 27: 626–638.
40. Vangay P, Johnson AJ, Ward TL, Al-Ghalith GA, Shields-Cutler RR, et al. 2018. US immigration westernizes the human gut microbiome. *Cell* 175: 962–972.
41. Wu GD, Chen J, Hoffmann C, Bittinger K, Chen -Y-Y, et al. 2011. Linking long-term dietary patterns with gut microbial enterotypes. *Science* 334: 105–108.
42. Yatsunenko T, Rey FE, Manary MJ, Trehan I, Dominguez-Bello MG, et al. 2012. Human gut microbiome viewed across age and geography. *Nature* 486: 222–227.

CHAPTER 8

Future Perspectives of Probiotics and Prebiotics in Foods and Food Supplements

Z. H. Hassan[1,2], F. Hugenholtz[2,3], E. G. Zoetendal[2], and Hauke Smidt[2]

[1]Indonesian Center for Agricultural Postharvest Research and Development, Indonesian Agency for Agricultural Research and Development (IAARD), Jl. Tentara Pelajar No. 12 Bogor 16114 West Java, Indonesia

[2]Laboratory of Microbiology, Wageningen University & Research, Stippeneng 4, 6708 WE Wageningen, The Netherlands

[3]Current address: The Dutch Research Council (NWO), Laan van Nieuw Oost-Indië 300, 2593 CE The Hague, The Netherlands

8.1 INTRODUCTION

The human gastrointestinal (GI) tract harbours a complex community of 3.9×10^{13} microbial cells (Sender et al., 2016), equalling the number of human cells that is 3.7×10^{13} (Bianconi et al., 2013). This complex community, known as microbiota, comprises diverse microorganisms, including bacteria, archaea, and eukaryotes, of which bacteria is the predominant domain (Rajilic-Stojanovic and de Vos, 2014), as well as viruses and phages.

Most (>99%) of the bacterial members of the GI tract microbiota are obligate anaerobes (Geva-Zatorsky et al., 2015). *Bacteroidetes* and *Firmicutes* are the most predominant phyla, followed by *Actinobacteria* and *Proteobacteria* (Rajilic-Stojanovic and de Vos, 2014). Other phyla, including *Verrucomicrobia*, *Lentisphaerae*, *Synergistetes*, *Planctomycetes*, *Tenericutes*, and *Deinococcus-Thermus* have also been detected as members of the bacterial microbiota, but at lesser proportions (Rajilic-Stojanovic and de Vos, 2014). The *Bacteroidetes* include mainly the genera *Bacteroides* and *Prevotella*, while the phylum *Firmicutes* encompasses diverse genera ranging from fastidious facultative anaerobes such as *Lactobacillus*, *Streptococcus*, and *Mycoplasma* to obligate anaerobes, such as *Eubacterium*, *Faecalibacterium*, *Roseburia*, *Clostridium*, and *Ruminococcus* (Figure 8.1) (Rajilic-Stojanovic and de Vos, 2014). Furthermore, the microbes residing in the GI tract add ~ 150 times more genes to our total gene pool than our own human genome (Qin et al., 2010). With these facts, the GI tract, specifically the colon, is found to harbour the largest and most diverse proportion and composition of microbiota compared to other human body regions such as the skin, the oral cavity, the respiratory tract, and the genitourinary tract.

Although the structure and function of the human GI tract microbiota have been studied intensively, it remains a challenge to define a normal human GI tract microbiota. Nonetheless, it has been established that the

Good Microbes in Medicine, Food Production, Biotechnology, Bioremediation, and Agriculture, First Edition. Edited by Frans J. de Bruijn, Hauke Smidt, Luca S. Cocolin, Michael Sauer, David Dowling, and Linda Thomashow.

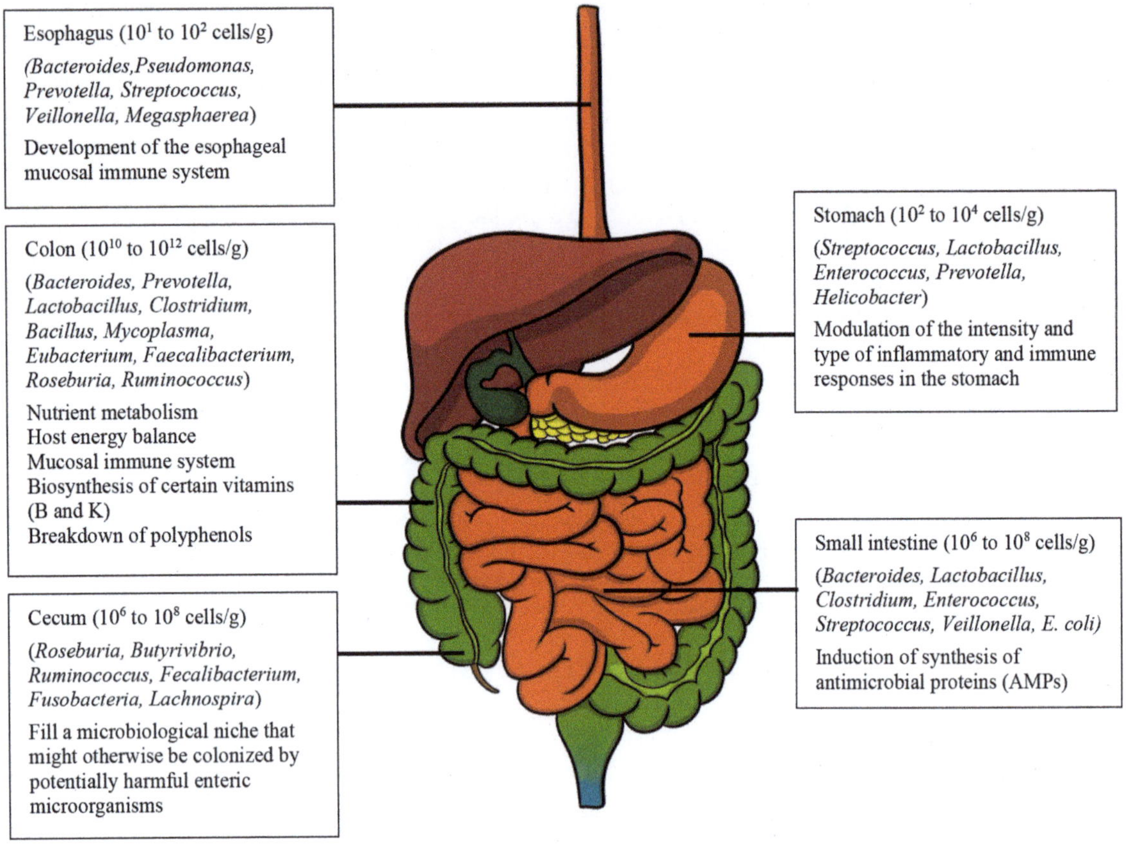

FIGURE 8.1 Schematic representation of the numbers and predominant genera of the microbiota at different locations and their respective functions. Modified from (Jandhyala et al., 2015).

GI tract of healthy adults is characterized by a highly resilient and stable (homeostatic) microbiota (Fassarella et al., 2021), which is mutualistic to the host through the defence against colonization by opportunistic pathogens. Aberrations in composition and function of GI tract microbiota may occur, which is affected by a number of divergent factors, including, among others, antibiotics, non-steroidal anti-inflammatory drugs, and diet (Figure 8.2) (David et al., 2014). The resulting microbial aberrations may subsequently lead to developing degenerative diseases in the GI tract, such as inflammatory bowel disease (Crohn's disease, ulcerative colitis), irritable bowel syndrome, as well as extra-intestinal diseases such as obesity, diabetes, allergic diseases, or cardiovascular disease. In addition, microbial aberrations have also been associated with a number of other systemic disorders including neurological pathologies, such as autism, anxiety, and depression (Maiuolo et al., 2021).

The GI tract microbiota is established during early life. However, throughout life, its composition and activity can be altered by various factors. Dietary components have been suggested to be a major factor in the development of GI tract microbiota (Figure 8.2). While it has been shown that a number of dietary components can promote the growth of opportunistic pathogens, it is also known from many studies that some diets have the potential to positively modulate the composition of the GI tract microbiota by promoting the growth of commensal microbes which have beneficial effects on intestinal functioning and systemic host health and thus protect against the development of certain microbiota-associated diseases.

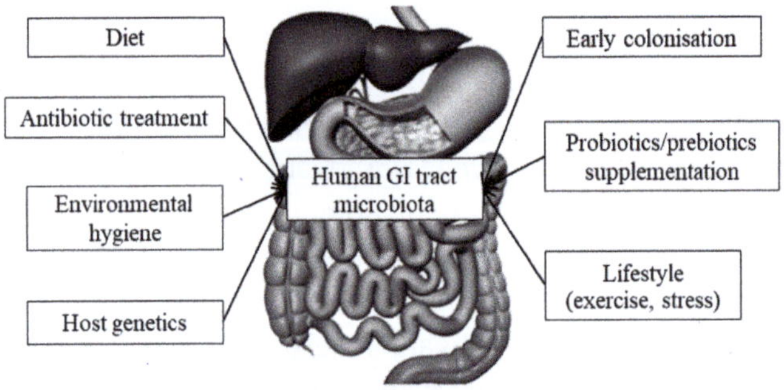

FIGURE 8.2 Factors influencing human GI tract microbiota composition and function.

There are a number of strategies to modulate the GI tract microbiota through dietary intervention. These include, among others, the use of probiotics, prebiotics, or synbiotic combinations of both. Probiotics and prebiotics target the GI tract microbiota by distinct, yet complementary mechanisms of actions. A better understanding of their mode of action is essential in order to define which strategy to use so they can effectively deliver their beneficial effects on the host.

8.2 FUNCTION OF THE GI TRACT MICROBIOTA

The microbiota in the GI tract has been reported to play numerous essential roles related to the health of the host. This involves a number of physiological and metabolic functions in the human body, which cannot be performed by the host itself. Primarily, the major roles of the GI tract microbiota include regulation of energy metabolism (Musso et al., 2010), nutrient acquisition (Allen and Stabler, 2008) as well as immunomodulation (Rajilić-Stojanović, 2013) (Figure 8.1).

With regard to its function as an energy metabolism regulator, the microbiota in the GI tract influences the energy-balance during its metabolism, including the conservation, storage, and release of energy obtained from the diet. Short chain fatty acids (SCFAs), primarily acetate, propionate, and butyrate, produced by colonic microbiota upon the fermentation of fibres are absorbed and used as energy sources by host cells. Acetate is the main energy source for brain, heart, and muscle. Butyrate is the main energy source for colonocytes, whereas propionate is mainly taken up by the liver. Therefore, the development of metabolic disorders is at least, in part, caused by alterations in the composition and function of the GI tract microbiota. Although the exact mechanisms by which the GI tract microbiota affects host energy metabolism are still uncertain, several potential avenues have been proposed. These include: increased energy harvest from the diet (Rosenbaum et al., 2015), regulation of GI tract-derived peptides involved in energy homeostasis (Holzer et al., 2012), regulation of the activation of free fatty acid receptors (FFA2 and FFA3) by SCFAs (Byrne et al., 2015), modulation of intestinal barrier by glucagon-like peptide-2 secretion (Tolhurst et al., 2012), activation of innate immunity and hepatic fibrogenesis through the lipopolysaccharide (LPS)-toll-like receptor-4 axis (Carotti et al., 2015), and the regulation of fat storage (Musso et al., 2010). Nevertheless, the impact of the microbiota on the metabolism of the host has mainly been reported in controlled mouse studies, whereas recent human intervention studies failed to reproduce those drastic impacts (Blaak et al., 2020; Reijnders et al., 2016).

From the perspective of nutrient acquisition, the GI tract microbiota is involved in the biosynthesis of certain vitamins which cannot be produced by humans or for which their availability in the food is insufficient, particularly the B-group vitamins such as biotin, cobalamin, folates, nicotinic acid, pantothenic acid, pyridoxine, riboflavin, and thiamine, as well as vitamin K (LeBlanc et al., 2013). These microbially produced vitamins are mostly absorbed in the colon.

Next to their contribution to host metabolic capacity, the GI tract microbiota provides the first line of defence against other (pathogenic) microbes, and in this sense it supports our immune system (Rajilić-Stojanović, 2013). In addition, microorganisms are constantly interacting with each other, for example, by secretion of metabolites such as anti-microbial compounds and bacteriocins. On top of that, SCFAs produced as end products of microbial fermentation of indigestible dietary components have been suggested to have a major contribution in regulating the host immune system. Butyrate, which has anti-carcinogenic and anti-inflammatory properties was shown to promote the intestinal immune system by modulating the function of intestinal macrophages, propionate was found to be involved in immune system regulation by suppressing the proliferation of cancer cells in the liver, and acetate has been shown to induce the proliferation and accumulation of T_{reg} cells in the colon. Furthermore, GI tract microbes not only interact with each other, but also with their host, as the GI tract contains most of the body's immune cells, hormone producing cells, and the body's nervous system is embedded throughout the lining of the GI tract (Mann and Li, 2014).

8.3 MODULATING THE GI TRACT MICROBIOTA TO IMPROVE HEALTH

Modulating the composition and/or functionality of the GI tract microbiota to a diverse yet well-balanced intestinal ecosystem is considered to be a valuable approach in improving and re-establishing bacterial homeostasis which might have been altered. One way is to stimulate the proliferation and/or activity of beneficial microbial populations already residing in the GI tract and preventing an increased level of undesirable and harmful microbial populations and their metabolic activity. Thus, understanding the difference between a healthy and an abnormal microbiota will provide valuable information for the design of potential interventions to positively modulate the composition and functionality of the GI tract microbiota.

Diets, including the type, composition, and proportion of constituents, as well as the main dietary macronutrients can affect the composition and function of GI tract microbiota, and subsequently affect health. A number of dietary intervention studies have already shown that altering diets to target the microbiome can improve host health, such as for example in children suffering from malnutrition (Chen et al., 2021) or in the frail elderly (Ghosh et al., 2020).

8.3.1 Modulating the GI Tract Microbiota with Probiotics

Probiotics have had a long history in food production and consumption. First introduced by Elie Metchnikoff in the 1900s, the word probiotics was derived from the Greek which means "for life". It was used to indicate the so called "friendly" or "good" bacteria. In 1908 Metchnikoff discovered that certain inhabitants of the Bulgarian population living in the Caucasus Mountains drinking a fermented yoghurt drink on a daily basis lived much longer than others. Later he found that the fermented yoghurt drink contained a high number of lactic acid bacteria, which when present in sufficient numbers within the GI tract, could prolong life by preventing deterioration related to old age (Metchnikoff, 1908). Based on the observations of Stamen Girgorov (1878–1945), he recognized the possibility of modifying the microbial populations in the human GI tract by replacing the harmful microbes with those that are more useful, which later became the basis for the term probiotics.

The definition of probiotics has evolved over time, but the term is basically used to refer to specific microbes that have beneficial effects on host health. Fuller (Fuller, 1989) defined probiotics as a live microbial food supplement, which beneficially affects the host by improving its intestinal microbial balance. Based on the internationally endorsed definition set up by FAO/WHO, probiotics are defined as live microorganisms which when administered in adequate amounts confer a health benefit on the host, especially by improving intestinal microbial balance (FAO/WHO, 2001). During the International Scientific Association for Probiotics and Prebiotics (ISAPP) consensus meeting in 2014, this definition was updated to clarify the scope and the appropriate use of the probiotic term. The panel suggested keeping the FAO/WHO definition of probiotics with a recommendation that the term probiotic be used only for products that deliver live microorganisms with a suitable viable count of well-defined strains with a reasonable expectation of delivering benefits for the wellbeing of the host (Hill et al., 2014).

8.3.2 Criteria for a Microorganism to Be Classified as Probiotic

It has been proposed that an ideal probiotic strain must have the following properties to be effective and commercially attractive (Fontana et al., 2013):

- the ability to survive the passage through the GI tract (tolerant to acids, bile, and digestive enzymes), until they reach the part of the GI tract where they exert their intended effect;
- a demonstrated beneficial effect on the host;
- safe for (human) consumption (non-pathogenic, non-toxic and free of significant adverse side effects);
- compatible with product matrix, processing, and storage conditions to maintain the desired properties (amenable to cultivation on an industrial scale, high viability, stable on storage, and oxygen tolerant).

Although understanding the molecular mechanisms underlying the efficacy of probiotics has been lagging behind, it has been shown that the host duodenal epithelium responds differently to different probiotics strains (van Baarlen et al., 2011). Remarkably, even consumption of the same strain harvested at different growth phases showed differential responses (van Baarlen et al., 2009).

Most of the probiotic bacterial strains currently available on the market are notably selected based on specific properties as mentioned above, such as resistance to acids and bile salts, high viability or being on the list of microbes safe for human consumption (EFSA, 2013). Moreover, some strains are selected for their functional properties, such as stimulation of immune cells (Geíza Alves de et al., 2010), adherence to epithelial cells (Serrano-Nino et al., 2016), their ability to inhibit growth of potential pathogens (Bratz et al., 2015), and their ability to reduce cholesterol levels (Jones et al., 2012).

TABLE 8.1 Selected examples of probiotic bacterial strains established on market

Probiotics	Sources	References
Lacticaseibacillus rhamnosus GG *Bifidobacterium longum* BB536 *Lactobacillus delbrueckii* subsp. *bulgaricus* 2038 *Streptococcus salivarius* subsp. *thermophilus* 1131 *Lacticaseibacillus casei* Shirota *Bifidobacterium breve* Yakult *Bifidobacterium lactis* FK120 *Bifidobacterium lactis* LKM512 *Lactobacillus acidophilus* CK92 *Lactobacillus helveticus* CK60 *Lacticaseibacillus casei* SBR1202 *Lactobacillus gasseri* SP *Bifidobacterium* SP, *L. casei* NY1301 *Lactobacillus* LC1 *Bifidobacterium lactis* Bb-12	Fermented (dairy) products	(Grzeskowiak et al., 2011)
Lacticaseibacillus rhamnosus strain GG (ATCC 53103) *Lacticaseibacillus casei* strain Shirota YIT9029 *Lactobacillus acidophilus* strain LB *Lactobacillus johnsonii* NCC 533 (CNCM I-1225) *Lacticaseibacillus casei* DN-114 001 (CNCM I-1518) *Limosilactobacillus reuteri* DSM 17938 (ATCC 55730 *Lactobacillus acidophilus* strain NCFM (ATCC 700396) *Lactiplantibacillus plantarum* 299v (DSM 6595) *Limosilactobacillus fermentum* ME-3 (DSM 14241) *Escherichia coli* Nissle 1917	Human GI tract	(Rosander et al., 2008)
Bacillus substilis 3 *Bacillus licheniformis* 31	Soil	(Sorokulova, 1996)

Most probiotic bacteria currently used belong to the genera *Bifidobacterium* and *Lactobacillus*, whereas a few belong to other genera, such as *Enterococcus*, *Escherichia*, *Bacillus*, or *Streptococcus*. In addition to isolating potential probiotics from fermented food products, probiotic bacterial strains were also isolated from the human GI tract (Table 8.1).

In the early period of the development of probiotics, one of the most important properties of effective probiotic strains was that they were of human origin. It was assumed that microbes that are part of the complex microbial communities residing in the GI tract would be safer and more effective to be used as probiotics and thus could contribute most positively to the health of the host. Among these are also commensal butyrate-producing bacteria (as will be discussed later), which have been suggested to have probiotic potential for patients suffering from inflammatory bowel disease (IBD).

Furthermore, the fact that different strains may exert different effects based on specific capabilities and enzymatic activities, even within one species, was brought to the idea of using multi-strain, rather than single-strain probiotics. One aspect to be taken into consideration when developing multi-strain or even multi-species probiotics is the interactions between the strains. In addition to the individual effects, the strains used in multi-strain probiotics should be compatible and have synergistic activity. Currently, an increasing number of multi-strain and multi-species probiotic products are on the market.

8.4 MODULATING THE GI TRACT MICROBIOTA WITH PREBIOTICS

The concept of prebiotics was introduced by Gibson and Roberfroid in 1995, who first identified and defined prebiotics as a non-digestible food ingredient that specifically affects the host by selectively stimulating growth and/or activity of one or a limited number of bacteria in

TABLE 8.2 Selected examples of established prebiotics

Prebiotics	Sources	Type of study	Effect on microbiota	References
Fructo-oligosaccharides (FOS)	Onion, chicory, garlic, artichoke, leek, banana	Animal study	*Bifidobacterium* spp.	(Howard et al., 1995)
		Animal study	*Lactobacillus* spp.	(Saminathan et al., 2011)
		Human study	*Bifidobacterium* spp.	(Bouhnik et al., 2006)
		Human study	*Bifidobacterium* spp., *Lactobacillus* spp.	(Ten Bruggencate et al., 2006)
Galacto-oligosaccharides (GOS)	Human milk, mammal milk	Human study	*Bifidobacterium* spp.	(Ito et al., 1993)
		In vitro	*Bifidobacterium* spp.	(Garrido et al., 2013)
Inulin	Bananas, garlic, and wheat, chicory root, artichokes, agave, asparagus, leeks, wheat, onions, wild yams	Human study	*Bifidobacterium* spp., *Faecalibacterium prausnitzii*	(Ramirez-Farias et al., 2009)
		In vitro	*Bifidobacterium* spp.	(Chung et al., 2016)
		Human study	*Bifidobacterium* spp., *Lactobacillus* spp.	(Costabile et al., 2010)
Lactulose	Synthetic polymer of lactose	*In vitro*	*Lactobacillus* spp.	(Venema et al., 2003)
		In vitro	*Bifidobacterium* spp., *Lactobacillus* spp.	(Mao et al., 2014)
		Human study	*Bifidobacterium* spp., *Lactobacillus* spp.	(Tuohy et al., 2002)

the colon, thereby improving health. During the International Scientific Association for Probiotics and Prebiotics (ISAPP) consensus meeting in 2016, this definition was modified to "a substrate that is selectively utilized by host microorganisms conferring a health benefit" (Gibson et al., 2017).

Early studies on prebiotics have suggested dietary fibres as the most useful potential substrates. Such fibres are even commonly considered as prebiotics due to their resistance to gastric acidity and digestive enzymes. However, in fact dietary fibres are different from prebiotics in terms of the selectivity of the fermentation. Although many prebiotics currently available are of fibre origin, not all fibres or other indigestible substrates that are left undigested or unhydrolyzed by host enzymes and reach the colon can be classified as prebiotics, since further than these criteria, they also must comply with another requirement, i.e., only be fermented by one or a few colonic bacteria, and not by all bacteria (Slavin, 2013). Accepted examples of established prebiotics are fructo-oligosaccharides (FOS), galacto-oligosaccharides (GOS), inulin and lactulose (Table 8.2).

Furthermore, in addition to the above requirements, potential prebiotics also require good evidence and confirmation in human studies, next to *in vitro* and animal studies, before they can be fully established as prebiotic.

Many studies on the relationships between the consumption of prebiotics and health benefits have demonstrated that, in addition to their capacity to selectively stimulate the growth of certain gut microbes, consuming prebiotics is associated with the reduction of symptoms of intestinal discomfort such as abdominal pain, flatulence, and bloating; prevention of gastroenteritis, suppression of diarrhoea; lowering the atherosclerotic cardiovascular disease associated with dyslipidemia, insulin resistance, obesity, and possibly type 2 diabetes; and reduction of risks of osteoporosis (Slavin, 2013). However, as reported in a recent study (Wang et al., 2021), in addition to their ability to alter the composition of gut microbiota, prebiotics did not demonstrate significant benefits on glucose metabolism, lipid metabolism, and body composition on pre-diabetes.

8.5 MODULATING THE GI TRACT MICROBIOTA WITH SYNBIOTICS

Next to probiotics and prebiotics, another important strategy in modulating the GI tract microbiota to improve host health is the use of synbiotics. Synbiotics are a mixture comprising live microorganisms and substrate(s) selectively utilized by host microorganisms that confer a health benefit on the host (Swanson et al., 2020). Recent studies indicated that the use of synbiotics yields higher efficacy and has more extensive positive effects in improving host health compared to either probiotics or prebiotics alone.

The combination of prebiotics and probiotics is considered to have synergistic or complementary effects since in addition to the supplementation of newly added probiotic microorganisms, its prebiotic component also enhances the survival and growth of existing beneficial microorganisms in the GI tract. In other words, prebiotics are the food for probiotic and beneficial commensal microorganisms. Furthermore, combining prebiotics and probiotics also allows effects simultaneously both in the small and large intestine. Another advantage is that the combination of probiotics and prebiotics enables the probiotics to already adapt to the prebiotics, which might lead to better synergy *in vivo* ((Martinez et al., 2015). Hence, studying the effectiveness of a combination of a prebiotic compound and certain probiotic bacteria as a synbiotic is becoming a major focus area in the development of new synbiotic products.

During the last decade, there has been an increase in the number of studies, both in models (*in vitro* or *in vivo*) as well as in humans, that addressed the efficacy of synbiotics in enhancing the survival and growth of probiotic bacteria as well as their health-promoting effects as outlined below.

An *in vitro* study using the Simulator of Human Intestinal Microbial Ecosystem (SHIME®) model by Gmeiner *et al.* showed that supplementation of *Lactobacillus acidophilus* strain 74-2 with fructo-oligosaccharides induced an increase in the number of *Bifidobacterium* spp. and decrease in the number of *E. coli* (Gmeiner et al., 2000). Similar effects have been also seen with a combination of lactitol and *Lactobacillus acidophilus* NCFM in a study performed with an *in vitro* semi-continuous colon fermentation simulation. The administration of lactitol in conjunction with *Lactobacillus acidophilus* was found to exhibit synergistic beneficial effects on colon microbial composition (increased numbers of bifidobacteria) and activity (increased butyrate production) (Makivuokko et al., 2010). Another *in vitro* study using batch culture demonstrated a combination of one of four probiotic bacteria (*Limosilactobacillus fermentum* ME-3, *Lactiplantibacillus plantarum* WCFS1, *Lacticaseibacillus paracasei* 8700:2 or *Bifidobacterium longum*) with either fructo-oligosaccharides or short-chain fructo-oligosaccharides considerably increased the number of *Bifidobacterium* spp. and *Eubacterium rectale–Clostridium coccoides*-group and lowered the number of *E. coli* compared to the use of individual components (Saulnier et al., 2008). Interestingly, most of these studies observed an increase in the abundance of *Bifidobacterium* when a combination of *Lactobacillus* with a certain type of prebiotic was used. This was suggested to be the result of the metabolic interaction between two or more microbial species. It is well known that members of the GI tract microbiota degrade complex carbohydrates to low molecular weight oligosaccharides, which can subsequently be degraded into monosaccharides and further fermented into SCFAs. This is also what was observed in the cases mentioned above, as degradation of certain oligosaccharides by *Bifidobacterium* or *Lactobacillus* spp. led to the production of lactate and acetate. Lactate and acetate could be subsequently converted into propionate and butyrate by lactate-fermenting bacteria such as *Veillonella* spp. or butyrate-producing bacteria such as *Anaerobutyricum hallii* and *Anaerostipes caccae*, which ultimately, in consequence of their butyrogenic effect, stimulate the growth of *Bifidobacterium*. This is also often referred to as metabolic cross-feeding.

Next to *in vitro* studies, a large number of studies using animal models have also been conducted to assess the effectiveness of the combination of selective prebiotics to certain probiotic bacteria as synbiotics. An experiment using a rat model revealed that administration of raffinose supplemented with encapsulated *Bifidobacterium breve* JCM 1192 altered the bacterial composition in the cecum of rats, including an increase of the relative abundance of *B. breve* to about 6.3% of total bacteria (Dinoto et al., 2006). Similar results were observed in another *in vivo* rat study. An increase in the total number of anaerobic bacteria was observed in both adult and elderly rats fed with a diet containing a mixture of *Lacticaseibacillus rhamnosus* GG and *Bifidobacterium animalis* subsp. *lactis* Bb-12 with the prebiotic inulin. Interestingly, the changes were age dependent. The increase in the abundance of lactobacilli was significant in adult rats, whereas the increase in bifidobacteria was significant in elderly rats (Lesniewska et al., 2006).

There are, however, only few studies regarding the functionality of synbiotics in humans. Ouwehand et al. found that consumption of lactitol combined with *Lactobacillus acidophilus* NCFM by healthy elderly subjects resulted in a significant increase in faecal levels of *L. acidophilus* NCFM levels and *Bifidobacterium* compared to the control group (Ouwehand et al., 2009). In another human volunteer study, intake of inulin in combination with *Bifidobacterium animalis* subsp. *lactis* Bb-12 increased the population of bifidobacteria and decreased the population of clostridia (Palaria et al., 2012). Okazaki et al. found that ingestion of a synbiotic containing GOS, *Bifidobacterium breve* and *Lacticaseibacillus casei* by elderly patients undergoing gastroenterological surgery significantly improved the counts of bifidobacteria and lactobacilli and decreased the number of potentially pathogenic *Enterobacteriaceae, Staphylococcus*, and *Pseudomonas* (Okazaki et al., 2013).

Apparently, the application of synbiotics may be more advantageous than the application of probiotics or prebiotics alone in terms of modulating the microbial composition. It was also reported that the application of synbiotics was more effective than probiotics alone in blood glycemic control, improved FPG, fasting insulin levels, HbA1c, insulin resistance and insulin sensitivity, as well as inflammation (Wang et al., 2021). However, only few studies have indicated advantages concerning alterations of the functional capacity or metabolic activity. Furthermore, one of the shortcomings in the use of synbiotics is that it is difficult to predict the selectivity and specificity of each component and what the resulting mechanism of action will be (Wang et al., 2021). Therefore, further studies are still required to confirm these results. For this, an integrated approach should be employed in order to fully understand the mechanisms underlying the success of either probiotics, prebiotics or synbiotics.

8.6 FUTURE PERSPECTIVES

One obstacle in studying the effect of probiotics, prebiotics, and synbiotics on the human GI tract, including composition and activity of the residing microbiota, is the requirement to perform invasive sampling from several GI locations in large numbers of human volunteers. Since this is intractable due to practical and ethical reasons, *in vitro* fermentation models simulating different parts of the human GI tract have been frequently used in studies of the ecological interactions between the microbes as well as the dynamics of host–microbe interactions. There is also an increasing trend towards the preferred use of synthetic microbial cultures as inoculum. As compared to single-strain cultures, mixed synthetic microbial cultures provide a more representative human *in vivo*-like system, while compared to complex natural microbial cultures, they are preferred due to their reduced complexity and increased controllability (Elzinga et al., 2019; Jia et al., 2016).

Furthermore, several approaches have been used to study the mechanisms of probiotics, prebiotics, and synbiotics in modulating GI tract microbial composition and function. Among these are various "omics" approaches such as transcriptomics, proteomics, and metabolomics. However, instead of using an integrated omics approach, most of these studies employed only a single approach, and thus did not provide a holistic view on the complex microbial system. In the future, ecological or mechanistic studies using integrated omics approaches will offer a better understanding in elucidating the mechanisms underlying the effects of probiotics, prebiotics, and synbiotics (Edwards et al., 2020; Kiousi et al., 2021; Rebollar et al., 2016).

There is a considerable interest in the application of faecal microbial transplantation (FMT), i.e., the transfer of microbiota from one individual (donor) to another individual (recipient). FMT has been considered to be a promising and effective therapy to restore the homeostasis of GI tract microbiota composition. Although many studies have reported the success of this intervention, especially for patients of recurrent *Clostridioides difficile* infection (CDI), there is no clear explanation yet on the exact mechanisms of this treatment. The mechanism explaining the high success rate in CDI is the reinstallation of a highly diverse microbiota which enables the level of the opportunistic pathogen *Clostridioides difficile* levels to be pushed back to normal. However, FMT as a treatment for other diseases often shows a lower success rate with large individual differences with respect to their responsiveness to the treatment. Identifying why not all individuals respond positively to FMT remains a challenge. Hence, further studies are needed to understand the basic principles that underlie the mechanisms by which FMT shows therapeutic efficacy, and which donor and recipient factors are crucial for success (Danne et al., 2021; Rohlke and Stollman, 2012). This could lead to the identification of signature microbes which are responsible for the efficacy and hence, follow-up interventions could include their stimulation by specific prebiotics or isolating them as candidate next generation probiotics.

The volume of research on (potential) probiotics and prebiotics is continuously growing. Significant progress has been achieved, however, after decades of research in this field, not many new players have made it into the arena. Only a limited number of microbes have been documented as new probiotic. Similarly, only a few prebiotics are currently established. Many issues remain to be established, especially concerning the aspects of the legislation on food safety and regulation on health claims.

For every microbe, an unambiguous safety status is a prerequisite to be eligible as a probiotic strain. Several of the currently used probiotic strains have a GRAS status for human consumption, because they have been found in fermented foods that we have been consuming for years, or even centuries. Many of the "traditional" lactic acid bacteria used in foods and dietary supplements have a Qualified Presumption of Safety (QPS) status, which is used by the European Food Safety Authority (EFSA) to assess the safety of probiotic strains (EFSA, 2013). However, for the use of "non-traditional" species for probiotic applications, safety is a concern. In addition, the more recent use of intestinal isolates that are suggested to be consumed in high cell numbers by individuals with potentially compromised health, has raised the question of safety. How to approach the safety evaluation of probiotic strains is still debated (Sanders et al., 2010), however, as stated in recent papers that reviewed the safety of probiotics "there is no such thing as zero risk". Nevertheless, the development of protocols that combine the use of propidium monoazide (PMA) with metagenomics or polyphasic approaches including PMA real time PCR or flow cytometry for viability assessment, and whole genome sequence analysis for identification and typing of the probiotic strain, enables producers and regulators to control the purity of the commercialized products (Fusco et al., 2021).

For next generation probiotic strains it is still difficult to get guidance as to which safety studies have to be performed to ensure safe application in humans. Considerations that have to be taken into account related to safe consumption of live microbes include their potential to produce toxins and to cause infections. At a genomic level the presence of transferrable antibiotic-resistance genes and virulence-related genes will immediately raise concerns about using such a strain as probiotic. However, in particular the presence of possible virulence factors is difficult to assess, because the properties that are virulence factors for some pathogens, are colonization factors for other commensal microbes, and thus might actually contribute to their probiotic features. Nevertheless, many human studies have been conducted regarding potential next generation probiotics. Recent studies with live and pasteurized *Akkermansia muciniphila* for subjects with excess body weight, demonstrated that administration of live or pasteurized *A. muciniphila* is safe in humans (Depommier et al., 2019; Plovier et al., 2017).

The other current key issue for both probiotics and prebiotics concerns the regulation on health claims. As for prebiotics, it is still difficult to provide scientific evidence on the "measurable" beneficial physiological effects of novel candidate compounds. This is due to the lack of truly validated biomarkers for GI health. Currently, the efficacy of prebiotics is mostly still expressed as the relative increase in numbers of beneficial bacteria, of which bifidobacteria are likely to be the most commonly used target bacteria (known as bifidogenic factors or bifidus factors), and changes in levels of individual bacterial metabolites. Although a lot of research has been undertaken to develop new techniques for measuring the efficacy of prebiotics in terms of GI health, only a limited number of biomarkers are currently identified and accepted to use. Yet, they still need to be validated before they are well-established as GI health biomarkers. A scientific committee working on GI health issues has defined five major criteria for a healthy GI system, i.e., an effective digestion and absorption of food, an absence of GI complaints/illness, a normal and stable intestinal microbiota, an effective immune status, and a status of well-being (Bischoff, 2011). Other criteria are glucose tolerance and glycaemic index which have been used as biomarker of diabetes; C-reactive protein levels, apolipoprotein levels, and blood pressure, all of which are biomarkers for heart disease; plasma nitrate level which is used as biomarker for pediatric gastroenteritis; faecal calprotectin, lactoferrin, and faecal serine protease which have been considered as reliable biomarkers for irritable bowel syndrome.

Next to the lack of real validated biomarkers for GI health, another reason related to the difficulty of classifying candidate prebiotics as established prebiotics is the challenge to translate *in vitro* or animal study findings into significant clinical effectiveness in humans. Many factors influence this process as the environment controls the mechanisms and the expression of certain functionalities in the human body. Hence, additional evidence involving nutritional or clinical intervention studies in human subjects is still required to substantiate the potential benefits of these novel candidate prebiotics to obtain approval and bring them to the market.

TABLE 8.3 Selected examples of probiotics candidates

Probiotics	Sources	References
Faecalibacterium prausnitzii	Human GI tract	(Miquel et al., 2013; Sokol et al., 2009; Van Immerseel et al., 2010; Walker et al., 2011)
Roseburia hominis	Human GI tract	(Machiels et al., 2014; Tilg and Danese, 2014)
Butyricicoccus pullicaecorum	Human GI tract	(Eeckhaut et al., 2013)
Intestinimonas butyriciproducens AF 211	Human GI tract	(Bui et al., 2015)
Akkermansia muciniphila	Human GI tract	(Belzer and de Vos, 2012; Depommier et al., 2019; Druart et al., 2021; Everard et al., 2013; Plovier et al., 2017)
Methanogenic archaea	Human GI tract	(Fadhlaoui et al., 2020)
Commensal *E. coli* K12	Human GI tract	(Behnsen et al., 2013; Croxen et al., 2013)
Non-pathogenic strains of *Clostridioides difficile*	Human GI tract	(Nagaro et al., 2013)
Non-pathogenic strains of *Clostridium butyricum*	Human GI tract	(Hayashi et al., 2013)
Synthetic communities or minimal microbiomes	Human GI tract	(Rode et al., 2021; Shetty et al., 2019; Tvede et al., 2015)

8.6.1 Next Generation Probiotics

Recently a number of bacterial strains have been proposed as potential next generation probiotic candidates (Table 8.3).

One of these is the butyrate-producing bacterium *Faecalibacterium prausnitzii*, an abundant species in the human GI tract (Walker et al., 2011). Although this species is found in the top 57 of most common species found in the human GI tract in ≥ 90% of the studied population, the range in relative abundance can be over a thousand-fold (Qin et al., 2010). Since the first report on the association between low abundance of *F. prausnitzii* in the GI tract and remission time of Crohn's disease (Sokol et al., 2009), numerous studies have shown a correlation between the relative abundance of *F. prausnitzii* and a number of other diseases. In a study using a mouse model for IBD, the supplementation of *F. prausnitzii*, and notably its spent medium, resulted in a marked reduction of colitis and a tendency to counterbalance the colitis-associated aberrations (Van Immerseel et al., 2010). In these studies, the characterization of compositional aberrations has led to the identification of possible commensal bacterial species that could be used to restore microbial imbalance and subsequently improve health. Based on these findings, it has been suggested that the relative abundance of *F. prausnitzii* can serve as a biomarker for GI health (Miquel et al., 2013).

Another proposed butyrate-producing probiotic candidate is *Roseburia hominis*. Decreased numbers of *R. hominis* (and *F. prausnitzii*) were observed in ulcerative colitis patients (Machiels et al., 2014). Based on these findings, a probiotic potential for *R. hominis* as butyrate-producing commensal has been suggested as well (Tilg and Danese, 2014). Another butyrate-producer with probiotic potential is *Butyricicoccus pullicaecorum*. A decreased abundance of the genus *Butyricicoccus* has been found in patients suffering from IBD (Eeckhaut et al., 2013). Furthermore, supplementation of *Bu. pullicaecorum* resulted in an attenuation of disease in a rat colitis model (Eeckhaut et al., 2013). A recent study reported a novel butyrate-producing bacterium isolated from the human GI tract, *Intestinimonas butyriciproducens* strain AF 211, which has the capacity to convert lysine and the Amadori product fructoselysine into butyrate, thus helping to promote and maintain colonic health (Bui et al., 2015).

Next to butyrate-producing bacteria, another intestinal commensal that has been suggested to be a biomarker for GI health is *Akkermansia muciniphila*, a mucus-degrading bacterium that belongs to the phylum *Verrucomicrobia* (Belzer and de Vos, 2012). Recently, the administration of *A. muciniphila* to high-fat diet-induced obese mice was shown to restore GI tract barrier function and increase the thickness of the mucus layer (Everard et al., 2013). Based on these observations a probiotic potential for *A. muciniphila* has been proposed for the prevention or treatment of obesity and associated metabolic disorders. Importantly, more detailed mechanistic insight in the mode of action of *A. muciniphila* has recently been obtained. To this end, pasteurization of *A. muciniphila* surprisingly increased its activity towards reduction of fat mass development, insulin resistance and dyslipidemia in mice. Furthermore, an isolated outer membrane protein of *A. muciniphila*, Amuc_1100, could be shown to be at least partly responsible for the effects observed for live or pasteurized whole cells (Plovier et al., 2017). A first proof-of-concept study in humans has recently been conducted and showed improvements on some metabolic markers (Depommier et al., 2019) and that it is safe for human consumption (Druart et al., 2021).

Last but not least, the use of methanogenic archaea as next generation probiotics is also investigated. Methanogenic archaea can use trimethylamine (TMA), an important molecule for increased risk of developing cardiovascular pathologies, as electron acceptor in their respiration and could potentially reduce TMA circulation in the human host (Fadhlaoui et al., 2020).

Nevertheless, it is still debatable whether the most effective human probiotic strains should originate from the human GI tract as it only refers to from what source the particular strain was isolated, and it does not necessarily imply adaptation to this environment and its effectiveness. The limited number of bacterial genera that contain currently acknowledged probiotic strains does not cover the complete diversity of the commensal GI microbiota. Therefore it is possible, maybe even likely, that there are other genera that include strains with higher probiotic effectiveness than "traditional" probiotics, which are generally selected based on their GRAS status. Furthermore, considering that health benefits are more important than only viable cell counts, focus of current probiotic research has now slightly changed from finding "good" bacterial strains to finding effective strains. The fact that some strains within a species are pathogenic does not mean that it is impossible for other strains of that same species to have beneficial properties. There is growing evidence that these non-pathogenic strains can be beneficial by similar mechanisms as "traditional" probiotics. This notion is exemplified by *E. coli* strain Nissle 1917. *E.coli* is well-known for its pathogenic capacity to cause diarrhoea and extra-intestinal diseases, whereas the non-pathogenic strain Nissle 1917 is currently available on the market as a probiotic strain for the prevention of infectious diarrhoea and the management of IBD (Behnsen et al., 2013).

The same applies to the genus *Clostridium*, which in general has a notorious reputation, and members of this genus have been considered to be "bad", if not "deadly". Since the nomenclature of *Clostridium* was traditionally based on being an anaerobic spore-forming bacterium, the phylogenetic diversity of the genus was enormous and hence, many of the traditional *Clostridium* isolates have been renamed. Besides, also within this genus, examples of potentially beneficial counterparts of pathogenic strains can be found. An example comes from *Clostridioides difficile*, previously known as *Clostridium difficile*, a well-known toxigenic pathogen. Colonization by non-pathogenic strains of *C. difficile* has been demonstrated to protect against the colonization by toxigenic strains (Nagaro et al., 2013). Another example is *Clostridium butyricum*. Some strains of this species have pathogenic properties due to their ability to produce toxins. However, not all strains of *C. butyricum* have the capacity to produce toxins, and some atoxigenic *C. butyricum* strains have even shown to have probiotic potential, mainly attributed to their ability to produce butyrate and their interaction with the host immune system (Hayashi et al., 2013). In Japan, a particular strain of *C. butyricum* (MIYAIRI) has been used for years for the prevention of antibiotic-associated diarrhoea.

In line with the use of mixtures of classical probiotics, combinations of intestinal microbes, often termed as synthetic communities or minimal microbiomes, have also received therapeutic interest. Use of synthetic microbiomes can aid in understanding the key trophic interactions between intestinal microbes and thereby improve our understanding of the ecosystem (Shetty et al., 2019). In addition, in the field of FMT minimal microbiomes are considered as interesting alternative bacteriotherapy as these might introduce the necessary combined microbial activities required for successful therapy without introducing unknown risk factors associated with faecal samples. The first observation of rectal bacteriotherapy with CDI showed its potential although its efficacy did not reach that of FMT (Rode et al., 2021; Tvede et al., 2015).

Although many of the strains/species described in the aforementioned studies have not been tested in humans or are not available yet on the probiotic market, they do show the potential of using microbiological insights to pinpoint specific commensal microbes that may have a beneficial effect on host health. Hence, several important steps are still required to identify, characterize, and confirm specific groups of commensal bacteria in the GI tract that may potentially be exploited for their probiotic properties and developed as new probiotics. The use of modern molecular methods for species and strain identification has revolutionized microbiological research and led to another strategy for the identification and selection of novel probiotic strains. Culture-independent studies have delivered and will continue to deliver a wealth of information on microbial composition and diversity, and more recently also on functionality, in health and disease. This information can be used to identify specific microbial species or strains that are under-represented in specific diseases and may potentially be supplemented in order to improve health as next-generation probiotics.

8.6.2 Next Generation Prebiotics

During the past years, many studies have been done towards developing novel candidate prebiotics. A range of non-digestible food carbohydrates such as arabinoxylan (Neyrinck et al., 2011; Riviere et al., 2014), resistant starch (Fuentes-Zaragoza et al., 2011; Zaman and Sarbini, 2016), polydextrose (Stowell, 2009), xylo-oligosaccharides (Jain et al., 2015; Samanta et al., 2015) and isomalt (Gostner et al., 2006) have been considered as promising novel candidate prebiotics as they demonstrated similar properties as established prebiotics with respect to the stimulation of the growth and/or activity of certain bacteria (Table 8.4).

Next to these food-derived carbohydrates, human milk oligosaccharides (HMOs) are also considered as promising novel candidate prebiotics (Barile and Rastall, 2013). An *in vitro* study found that HMOs might affect the composition of the GI microbiota, particularly in the infant GI tract, by selectively stimulating the growth of beneficial bacteria such as *Bifidobacterium* spp. Furthermore, fucosylated and sialylated HMOs were reported as anti-adherence factors for a range of bacteria and viruses which consequently prevent the adhesion of pathogenic bacteria to the epithelial surface in the GI tract (Kulinich and Liu, 2016). A recent intervention study with 2′-fucosyllactose demonstrated that consumption of this HMO was able to increase faecal bifidobacterial levels in IBS patients (Iribarren et al., 2020).

TABLE 8.4 Selected examples of prebiotics candidates

Prebiotics	Sources	Type of study	Effect on microbiota	References
Arabinoxylan	Cereal grains, wheat bran	*In vitro*	*Bifidobacterium* spp.	(Riviere et al., 2014)
		Animal study	*Bifidobacterium* spp., *Roseburia intestinalis*, *Bacteroides-Prevotella* spp.	(Neyrinck et al., 2011)
		Animal study	*Bifidobacterium* spp.	(Van Craeyveld et al., 2008)
		Animal study	*Roseburia intestinalis*, *Eubacterium rectale*, *Anaerostipes caccae*, *Bifidobacterium longum*	(Van den Abbeele et al., 2011)
		Animal study	*Faecalibacterium prausnitzii*, *Roseburia intestinalis*, *Blautia coccoides–Eubacterium rectale*, *Bifidobacterium* spp. and *Lactobacillus* spp.	(Nielsen et al., 2014)
		Human study	*Bifidobacterium* *Lactobacillus* *Bifidobacterium*	(Francois et al., 2012) (Walton et al., 2012)
Resistant starch	Grains, cereals, vegetables (especially potatoes), tuber, fruits, legumes, seeds, and some nuts	Animal study	*Lactobacillus* spp. *Bifidobacterium*	(Silvi et al., 1999)
		Animal study	*Faecalibacterium prausnitzii*	(Haenen et al., 2013)
		Animal study	*Bifidobacterium*	(Regmi et al., 2011)
		Human study	*Bifidobacterium adolescentis*, *Parabacteroides distasonis*, *Ruminococcus bromii*, *Eubacterium rectale*	(Martinez et al., 2010)
		Human study	*Ruminococcus bromii*	(Abell et al., 2008)
Isomalt	Beet sugar	Human study	*Bifidobacterium* spp.	(Gostner et al., 2006)

(Continued)

TABLE 8.4 (Continued)

Prebiotics	Sources	Type of study	Effect on microbiota	References
Polydextrose	Synthetic polymer of (corn) glucose	*In vitro*	*Bifidobacterium* spp.	(Probert et al., 2004)
		Human study	*C. histolyticum, R. intestinalis, C. leptum, Lactobacillus, Enterococcus*	(Costabile et al., 2012)
		Human study	*Parabacteroides, Eubacterium, Ruminococcus, Roseburia, Dorea*	(Vester Boler et al., 2011)
		Human study	*Faecalibacterium, Phascolarctobacterium*, and *Dialister*	(Hooda et al., 2012)
		Human study	*Lactobacillus and Bifidobacterium, B. fragilis, B. vulgatus, and B. intermedius*	(Jie et al., 2000)
Xylo-oligosaccharides (XOS)	Cereal grains, grasses, green coconut husk, sugarcane bagasse, tobacco stalks, sunflower stalks	*In vitro*	*Bifidobacterium adolescentis, Bifidobacterium bifidum, Levilactobacillus brevis, Lactiplantibacillus plantarum*	(Manisseri and Gudipati, 2012)
		In vitro	*B. lactis*	(Makelainen et al., 2010)
		Animal study	*Lactobacillus* spp.	(Pourabedin et al., 2015)
		Animal study	*Lactobacillus crispatus, Anaerostipes butyraticus*	(De Maesschalck et al., 2015)
		Human study	*Bifidobacterium*	(Finegold et al., 2014)
		Human study	*Lactobacillus* spp., *Bifidobacterium* spp., *Clostridium perfringens*	(Lin et al., 2016)
Human milk oligosaccharides	Breast milk	*In vitro*	*Bifidobacterium infantis* and *Bifidobacterium bifidum*	(Ward et al., 2007)
		Human study	*Bifidobacterium*	(De Leoz et al., 2015; Elison et al., 2016; Gostner et al., 2006; Iribarren et al., 2020; Steenhout et al., 2016)
Polyphenols	Vegetables, herbs, fruits and seeds	*In vitro*	*Lactobacillus* and *Bifidobacterium*	(Ma and Chen, 2020)
Polyunsaturated fatty acids and conjugated fatty acid	Plant and animal-based foods such as meat, fish, vegetable and fish oils or seeds	Human study	*Bifidobacterium, Roseburia*, and *Lactobacillus*	(Watson et al., 2018)
Proteins, hydrolysates and peptides	cheese whey, human milk, casein, soybean, or peptides resulting from the hydrolysis of poultry bones and feathers	*In vitro*	*Lactobacillus* and *Bifidobacterium*	(Cardoso et al., 2021; Zhang et al., 2020)

The modification on the definition of a prebiotic which was released during the ISAPP consensus meeting in 2016 (Gibson et al., 2017) has opened the opportunity for the development of non-carbohydrate-based prebiotics such as polyphenols (Ma and Chen, 2020), polyunsaturated fatty acids converted to respective conjugated fatty acids (Watson et al., 2018), proteins, hydrolysates, and peptides (Cardoso et al., 2021; Zhang et al., 2020). Although many studies have been performed, either *in vitro*, in animal models or human trials, to examine their potential health benefit to the human host in order to fulfil their prebiotic status, not many of them are currently approved and available on the market.

Last but not least, during the last decades, most of the studies on prebiotic potential were focused on western-origin plant materials, such as asparagus, burdock, chicory, dandelion root, Jerusalem artichoke, leeks, and onions. However, as these plants are not always available around the world, the access to these prebiotic products is often limited. As a result, the development of prebiotic-containing products in some other countries might be not as extensive. Many studies have been conducted to explore and assess the possibility to develop local-based novel prebiotics. Dwivedi et al. has listed a number of food crops, including some local plants, which contain prebiotic carbohydrates with significant prebiotic properties and have a potential as novel sources of prebiotics (Dwivedi et al., 2014). Judprasong et al. evaluated the presence of inulin-type fructans, fructo-oligosaccharides (1-kestose (1-kestotriose; GF_2), nystose (1,1-kestotetraose; GF_3), and 1 F-β-fructofuranosylnystose (1,1,1-kestopentaose; GF_4)) and sum of sugars (fructose, glucose, and sucrose) potential as prebiotics in various foods consumed in Thailand (Judprasong et al., 2011). Hence, exploring local based-novel prebiotics could be one option to cope with this challenge.

ACKNOWLEDGMENTS

This work has been funded by Indonesian Agency for Agricultural Research and Development (IAARD), Indonesia. FH was supported by the Gravitation grant SIAM (Soehngen Institute for Anaerobic Microbiology) Grant number 024-002-002. The authors thank Coline Gerritsen for valuable and inspiring discussions.

REFERENCES

1. Abell GCJ, Cooke CM, Bennett CN, Conlon MA, McOrist AL. 2008. Phylotypes related to *Ruminococcus bromii* are abundant in the large bowel of humans and increase in response to a diet high in resistant starch. *FEMS Microbiology Ecology* 66: 505–515.
2. Allen RH, Stabler SP. 2008. Identification and quantitation of cobalamin and cobalamin analogues in human feces. *The American Journal of Clinical Nutrition* 87: 1324–1335.
3. Barile D, Rastall RA. 2013. Human milk and related oligosaccharides as prebiotics. *Current Opinion in Biotechnology* 24: 214–219.
4. Behnsen J, Deriu E, Sassone-Corsi M, Raffatellu M. 2013. Probiotics: Properties, examples, and specific applications. *Cold Spring Harbor Perspectives in Medicine* 3: 15.
5. Belzer C, de Vos WM. 2012. Microbes inside–from diversity to function: The case of *Akkermansia*. *The ISME Journal* 6: 1449–1458.
6. Bianconi E, Piovesan A, Facchin F, Beraudi A, Casadei R, et al. 2013. An estimation of the number of cells in the human body. *Annals of Human Biology* 40: 463–471.
7. Bischoff SC. 2011. 'Gut health': A new objective in medicine? *BMC Medicine* 9: 24.
8. Blaak EE, Canfora EE, Theis S, Frost G, Groen AK, et al. 2020. Short chain fatty acids in human gut and metabolic health. *Beneficial Microbes* 11: 411–455.
9. Bouhnik Y, Raskine L, Simoneau G, Paineau D, Bornet F. 2006. The capacity of short-chain fructo-oligosaccharides to stimulate faecal bifidobacteria: A dose-response relationship study in healthy humans. *Nutrition Journal* 5: 8.
10. Bratz K, Golz G, Janczyk P, Nockler K, Alter T. 2015. Analysis of *in vitro* and *in vivo* effects of probiotics against *Campylobacter* spp. *Berliner Und Munchener Tierarztliche Wochenschrift* 128: 155–162.
11. Bui TP, Ritari J, Boeren S, de Waard P, Plugge CM, et al. 2015. Production of butyrate from lysine and the Amadori product fructoselysine by a human gut commensal. *Nature Communications* 6: 10062.
12. Byrne CS, Chambers ES, Morrison DJ, Frost G. 2015. The role of short chain fatty acids in appetite regulation and energy homeostasis. *International Journal of Obesity* 39: 1331–1338.
13. Cardoso BB, Amorim C, Silvério SC, Rodrigues LR. 2021. Novel and emerging prebiotics: Advances and

opportunities. *Advances in Food and Nutrition Research* 95: 41–95.

14. Carotti S, Guarino MPL, Vespasiani-Gentilucci U, Morini S. 2015. Starring role of toll-like receptor-4 activation in the gut-liver axis. *World Journal of Gastrointestinal Pathophysiology* 6: 99–109.
15. Chapman CMC, Gibson GR, Rowland I. 2011. Health benefits of probiotics: Are mixtures more effective than single strains? *European Journal of Nutrition* 50: 1–17.
16. Chen RY, Mostafa I, Hibberd MC, Das S, Mahfuz M, et al. 2021. A microbiota-directed food intervention for undernourished children. *New England Journal of Medicine* 384: 1517–1528.
17. Chung WSF, Walker AW, Louis P, Parkhill J, Vermeiren J, et al. 2016. Modulation of the human gut microbiota by dietary fibres occurs at the species level. *BMC Biology* 14: 1–13.
18. Costabile A, Fava F, Roytio H, Forssten SD, Olli K, et al. 2012. Impact of polydextrose on the faecal microbiota: A double-blind, crossover, placebo-controlled feeding study in healthy human subjects. *British Journal of Nutrition* 108: 471–481.
19. Costabile A, Kolida S, Klinder A, Gietl E, Bäuerlein M, et al. 2010. A double-blind, placebo-controlled, cross-over study to establish the bifidogenic effect of a very-long-chain inulin extracted from globe artichoke (*Cynara scolymus*) in healthy human subjects. *British Journal of Nutrition* 104: 1007–1017.
20. Croxen MA, Law RJ, Scholz R, Keeney KM, Wlodarska M, et al. 2013. Recent advances in understanding enteric pathogenic Escherichia *coli*. *Clinical Microbiology Reviews* 26: 822–880.
21. Danne C, Rolhion N, Sokol H. 2021. Recipient factors in faecal microbiota transplantation: One stool does not fit all. *Nature Reviews Gastroenterology and Hepatolology*.
22. David LA, Maurice CF, Carmody RN, Gootenberg DB, Button JE, et al. 2014. Diet rapidly and reproducibly alters the human gut microbiome. *Nature* 505.
23. De Leoz ML, Kalanetra KM, Bokulich NA, Strum JS, Underwood MA, et al. 2015. Human milk glycomics and gut microbial genomics in infant feces show a correlation between human milk oligosaccharides and gut microbiota: A proof-of-concept study. *Journal of Proteome Research* 14: 491–502.
24. De Maesschalck C, Eeckhaut V, Maertens L, De Lange L, Marchal L, et al. 2015. Effects of xylo-oligosaccharides on broiler chicken performance and microbiota. *Applied and Environmental Microbiology* 81: 5880–5888.
25. de Simone C. 2019. The unregulated probiotic market. *Clinical Gastroenterology and Hepatology* 17: 809–817.
26. Depommier C, Everard A, Druart C, Plovier H, Van Hul M, et al. 2019. Supplementation with *Akkermansia muciniphila* in overweight and obese human volunteers: A proof-of-concept exploratory study. *Nature Medicine* 25: 1096–1103.
27. Dinoto A, Suksomcheep A, Ishizuka S, Kimura H, Hanada S, et al. 2006. Modulation of rat cecal microbiota by administration of raffinose and encapsulated *Bifidobacterium breve*. *Applied and Environmental Microbiology* 72: 784–792.
28. Druart C, Plovier H, Van Hul M, Brient A, Phipps KR, et al. 2021. Toxicological safety evaluation of pasteurized *Akkermansia muciniphila*. *Journal of Applied Toxicology* 41: 276–290.
29. Dwivedi S, Sahrawat K, Puppala N, Ortiz R. 2014. Plant prebiotics and human health: Biotechnology to breed prebiotic-rich nutritious food crops. *Electronic Journal of Biotechnology* 17: 238–245.
30. Edwards PT, Kashyap PC, Preidis GA. 2020. Microbiota on biotics: Probiotics, prebiotics, and synbiotics to optimize growth and metabolism. *American Journal of Physiology-Gastrointestinal and Liver Physiology* 319: G382–g390.
31. Eeckhaut V, Machiels K, Perrier C, Romero C, Maes S, et al. 2013. *Butyricicoccus pullicaecorum* in inflammatory bowel disease. *Gut* 62: 1745–1752.
32. EFSA 2013. Scientific Opinion on the maintenance of the list of QPS biological agents intentionally added to food and feed (2013 update). *The EFSA Journal* 11: 3349.
33. Elison E, Vigsnaes LK, Rindom Krogsgaard L, Rasmussen J, Sørensen N, et al. 2016. Oral supplementation of healthy adults with 2'-O-fucosyllactose and lacto-N-neotetraose is well tolerated and shifts the intestinal microbiota. *British Journal of Nutrition* 116: 1356–1368.
34. Elzinga J, van der Oost J, de Vos WM, Smidt H. 2019. The use of defined microbial communities to model host-microbe interactions in the human gut. *Microbiology and Molecular Biology Reviews* 83.
35. Everard A, Belzer C, Geurts L, Ouwerkerk JP, Druart C, et al. 2013. Cross-talk between *Akkermansia muciniphila* and intestinal epithelium controls diet-induced obesity. *Proceedings of the National Academy of Sciences of the United States of America* 110: 9066–9071.
36. Fadhlaoui K, Arnal M-E, Martineau M, Camponova P, Ollivier B, et al. 2020. Archaea, specific genetic traits,

and development of improved bacterial live biotherapeutic products: Another face of next-generation probiotics. *Applied Microbiology and Biotechnology* 104: 4705–4716.

37. FAO/WHO 2001. Food and Agricultural Organization of the United Nations and World Health Organization. *Health and Nutritional Properties of Probiotics in Food Including Powder Milk with Live Lactic Acid Bacteria.* Cordoba, Argentina.

38. Fassarella M, Blaak EE, Penders J, Nauta A, Smidt H, et al. 2021. Gut microbiome stability and resilience: Elucidating the response to perturbations in order to modulate gut health. *Gut* 70: 595.

39. Fijan S. 2014. Microorganisms with claimed probiotic properties: An overview of recent literature. *International Journal of Environmental Research and Public Health* 11: 4745–4767.

40. Finegold SM, Li Z, Summanen PH, Downes J, Thames G, et al. 2014. Xylooligosaccharide increases bifidobacteria but not lactobacilli in human gut microbiota. *Food & Function* 5: 436–445.

41. Fontana L, Bermudez-Brito M, Plaza-Diaz J, Munoz-Quezada S, Gil A 2013. Sources, isolation, characterisation and evaluation of probiotics. *The British Journal of Nutrition* 109 (Suppl 2): S35–50.

42. Francois IE, Lescroart O, Veraverbeke WS, Marzorati M, Possemiers S, et al. 2012. Effects of a wheat bran extract containing arabinoxylan oligosaccharides on gastrointestinal health parameters in healthy adult human volunteers: A double-blind, randomised, placebo-controlled, cross-over trial. *British Journal of Nutrition* 108: 2229–2242.

43. Fuentes-Zaragoza E, Sánchez-Zapata E, Sendra E, Sayas E, Navarro C, et al. 2011. Resistant starch as prebiotic: A review. *Starch-Stärke* 63: 406–415.

44. Fuller R. 1989. Probiotics in man and animals. *Journal of Applied Bacteriology* 66: 365–378.

45. Fusco V, Fanelli F, Chieffi D. 2021. Authenticity of probiotic foods and dietary supplements: A pivotal issue to address. *Critical Reviews in Food Science and Nutrition*: 1–18.

46. Garrido D, Ruiz-Moyano S, Jimenez-Espinoza R, Eom H-J, Block DE, et al. 2013. Utilization of galactooligosaccharides by *Bifidobacterium longum* subsp. *infantis* isolates. *Food Microbiology* 33: 262–270.

47. Geíza Alves de A, Tânia Lúcia Montenegro S, Evandro Leite de S, Flávio Fonseca V, Edvaldo Rodrigues de A, et al. 2010. *In vivo* assessment of possible probiotic properties of *Zymomonas mobilis* in a Wistar rat model. *Electronic Journal of Biotechnology* 13: 7.

48. Geva-Zatorsky N, Alvarez D, Hudak JE, Reading NC, Erturk-Hasdemir D, et al. 2015. *In vivo* imaging and tracking of host-microbiota interactions via metabolic labeling of gut anaerobic bacteria. *Nature Medicine* 21: 1091–1100.

49. Ghosh TS, Rampelli S, Jeffery IB, Santoro A, Neto M, et al. 2020. Mediterranean diet intervention alters the gut microbiome in older people reducing frailty and improving health status: The NU-AGE 1-year dietary intervention across five European countries. *Gut* 69: 1218–1228.

50. Gibson GR, Hutkins R, Sanders ME, Prescott SL, Reimer RA, et al. 2017. Expert consensus document: The International Scientific Association for Probiotics and Prebiotics (ISAPP) consensus statement on the definition and scope of prebiotics. *Nature Reviews Gastroenterology & Hepatology.*

51. Gibson GR, Scott KP, Rastall RA, Tuohy KM, Hotchkiss A, et al. 2010. Dietary prebiotics: Current status and new definition. *Food Science & Technology Bulletin Functional Foods* 1–19.

52. Gmeiner M, Kneifel W, Kulbe KD, Wouters R, De Boever P, et al. 2000. Influence of a synbiotic mixture consisting of *Lactobacillus acidophilus* 74-2 and a fructooligosaccharide preparation on the microbial ecology sustained in a simulation of the human intestinal microbial ecosystem (SHIME reactor). *Applied Microbiology and Biotechnology* 53: 219–223.

53. Gostner A, Blaut M, Schaffer V, Kozianowski G, Theis S et al. 2006. Effect of isomalt consumption on faecal microflora and colonic metabolism in healthy volunteers. *British Journal of Nutrition* 95: 40–50.

54. Grossenbacher F, Gashi A, Vaart IB-vd. 2016. Use of the multispecies probiotic Winclove 500/BactoSan pro FOS leads to less gastrointestinal complaints in adults - An observational *in vivo* pilot study. *Advances in Microbiology* 06: 12.

55. Grzeskowiak L, Isolauri E, Salminen S, Gueimonde M. 2011. Manufacturing process influences properties of probiotic bacteria. *British Journal of Nutrition* 105: 887–894.

56. Haenen D, Zhang J, Souza da Silva C, Bosch G, van der Meer IM, et al. 2013. A diet high in resistant starch modulates microbiota composition, SCFA concentrations, and gene expression in pig intestine. *Journal of Nutrition* 143: 274–283.

57. Hayashi A, Sato T, Kamada N, Mikami Y, Matsuoka K, et al. 2013. A single strain of *Clostridium butyricum* induces intestinal IL-10-producing macrophages to suppress acute experimental colitis in mice. *Cell Host Microbe* 13: 711–722.
58. Hill C, Guarner F, Reid G, Gibson GR, Merenstein DJ, et al. 2014. Expert consensus document. The International Scientific Association for Probiotics and Prebiotics consensus statement on the scope and appropriate use of the term probiotic. *Nature Clinical Practice Gastroenterology & Hepatology* 11: 506–514.
59. Holzer P, Reichmann F, Farzi A. 2012. Neuropeptide Y, peptide YY and pancreatic polypeptide in the gut–brain axis. *Neuropeptides* 46: 261–274.
60. Hooda S, Boler BM, Serao MC, Brulc JM, Staeger MA, et al. 2012. 454 pyrosequencing reveals a shift in fecal microbiota of healthy adult men consuming polydextrose or soluble corn fiber. *The Journal of Nutrition* 142: 1259–1265.
61. Howard MD, Gordon DT, Pace LW, Garleb KA, Kerley MS. 1995. Effects of dietary supplementation with fructooligosaccharides on colonic microbiota populations and epithelial cell proliferation in neonatal pigs. *Journal of Pediatric Gastroenterology and Nutrition* 21: 297–303.
62. Iribarren C, Törnblom H, Aziz I, Magnusson MK, Sundin J, et al. 2020. Human milk oligosaccharide supplementation in irritable bowel syndrome patients: A parallel, randomized, double-blind, placebo-controlled study. *Neurogastroenterology and Motility* 32: e13920.
63. Ito M, Deguchi Y, Matsumoto K, Kimura M, Onodera N, et al. 1993. Influence of galactooligosaccharides on the human fecal microflora. *Journal of Nutritional Sciences and Vitaminology (Tokyo)* 39: 635–640.
64. Jain I, Kumar V, Satyanarayana T. 2015. Xylooligosaccharides: An economical prebiotic from agroresidues and their health benefits. *Indian Journal of Experimental Biology* 53: 131–142.
65. Jandhyala SM, Talukdar R, Subramanyam C, Vuyyuru H, Sasikala M, et al. 2015. Role of the normal gut microbiota. *World Journal of Gastroenterology* 21: 8787–8803.
66. Jia X, Liu C, Song H, Ding M, Du J et al. 2016. Design, analysis and application of synthetic microbial consortia. *Synthetic and Systems Biotechnology* 1: 109–117.
67. Jie Z, Bang-yao L, Ming-jie X, Hai-wei L, Zu-kang Z, et al. 2000. Studies on the effects of polydextrose intake on physiologic functions in Chinese people. *The American Journal of Clinical Nutrition* 72: 1503–1509.
68. Jones ML, Martoni CJ, Prakash S 2012. Cholesterol lowering and inhibition of sterol absorption by *Lactobacillus reuteri* NCIMB 30242: A randomized controlled trial. *European Journal of Clinical Nutrition* 66: 1234–1241.
69. Judprasong K, Tanjor S, Puwastien P, Sungpuag P. 2011. Investigation of Thai plants for potential sources of inulin-type fructans. *Journal of Food Composition and Analysis* 24: 642–649.
70. Kiousi DE, Rathosi M, Tsifintaris M, Chondrou P, Galanis A. 2021. Pro-biomics: Omics technologies to unravel the role of probiotics in health and disease. *Advances in Nutrition.*
71. Kulinich A, Liu L. 2016. Human milk oligosaccharides: The role in the fine-tuning of innate immune responses. *Carbohydrate Research* 432: 62–70.
72. Lang FC. 2010. Use of a multi-species probiotic for the prevention of antibiotic associated diarrhea. *Nutrafoods* 9: 27–31.
73. LeBlanc JG, Milani C, de Giori GS, Sesma F, van Sinderen D, et al. 2013. Bacteria as vitamin suppliers to their host: A gut microbiota perspective. *Current Opinion in Biotechnology* 24: 160–168.
74. Lesniewska V, Rowland I, Cani PD, Neyrinck AM, Delzenne NM, et al. 2006. Effect on components of the intestinal microflora and plasma neuropeptide levels of feeding *Lactobacillus delbrueckii, Bifidobacterium lactis*, and inulin to adult and elderly rats. *Applied Microbiology and Biotechnology* 72: 6533–6538.
75. Lin S-H, Chou L-M, Chien Y-W, Chang J-S, Lin C-I. 2016. Prebiotic effects of xylooligosaccharides on the improvement of microbiota balance in human subjects. *Gastroenterology Research and Practice* 2016: 6.
76. Ma G, Chen Y. 2020. Polyphenol supplementation benefits human health via gut microbiota: A systematic review via meta-analysis. *Journal of Functional Foods* 66: 103829.
77. Machiels K, Joossens M, Sabino J, De Preter V, Arijs I, et al. 2014. A decrease of the butyrate-producing species *Roseburia hominis* and *Faecalibacterium prausnitzii* defines dysbiosis in patients with ulcerative colitis. *Gut* 63: 1275–1283.
78. Maiuolo J, Gliozzi M, Musolino V, Carresi C, Scarano F, et al. 2021. The contribution of gut microbiota–brain axis in the development of brain disorders. *Frontiers in Neuroscience* 15.
79. Makelainen H, Forssten S, Saarinen M, Stowell J, Rautonen N, et al. 2010. Xylo-oligosaccharides enhance

the growth of bifidobacteria and *Bifidobacterium lactis* in a simulated colon model. *Beneficial Microbes* 1: 81–91.

80. Makivuokko H, Forssten S, Saarinen M, Ouwehand A, Rautonen N. 2010. Synbiotic effects of lactitol and *Lactobacillus acidophilus* NCFM in a semi-continuous colon fermentation model. *Beneficial Microbes* 1: 131–137.
81. Manisseri C, Gudipati M. 2012. Prebiotic activity of purified xylobiose obtained from ragi (*Eleusine coracana*, Indaf-15) bran. *Indian Journal of Microbiology* 52: 251–257.
82. Mann ER, Li X. 2014. Intestinal antigen-presenting cells in mucosal immune homeostasis: Crosstalk between dendritic cells, macrophages and B-cells. *World Journal of Gastroenterology* 20: 9653–9664.
83. Mao B, Li D, Zhao J, Liu X, Gu Z, et al. 2014. *In vitro* fermentation of lactulose by human gut bacteria. *Journal of Agricultural and Food Chemistry* 62: 10970–10977.
84. Martinez I, Kim J, Duffy PR, Schlegel VL, Walter J. 2010. Resistant starches types 2 and 4 have differential effects on the composition of the fecal microbiota in human subjects. *PloS One* 5: e15046.
85. Martinez RC, Bedani R, Saad SM. 2015. Scientific evidence for health effects attributed to the consumption of probiotics and prebiotics: An update for current perspectives and future challenges. *British Journal of Nutrition* 114: 1993–2015.
86. Metchnikoff É 1908. *The Prolongation of Life: Optimistic Studies*. New York & London: G.P. Putnam's Sons.
87. Miquel S, Martín R, Rossi O, Bermúdez-Humarán LG, Chatel JM, et al. 2013. Faecalibacterium prausnitzii and human intestinal health. *Current Opinion in Microbiology* 16: 255–261.
88. Musso G, Gambino R, Cassader M. 2010. Gut microbiota as a regulator of energy homeostasis and ectopic fat deposition: Mechanisms and implications for metabolic disorders. *Current Opinion in Lipidology* 21: 76–83.
89. Nagaro KJ, Phillips ST, Cheknis AK, Sambol SP, Zukowski WE, et al. 2013. Nontoxigenic *Clostridium difficile* protects hamsters against challenge with historic and epidemic strains of toxigenic BI/NAP1/027 *C. difficile*. *Antimicrobial Agents and Chemotherapy* 57: 5266–5270.
90. Neyrinck AM, Possemiers S, Druart C, Van de Wiele T, De Backer F, et al. 2011. Prebiotic effects of wheat arabinoxylan related to the increase in bifidobacteria, *Roseburia* and *Bacteroides/Prevotella* in diet-induced obese mice. *PLoS One* 6: e20944.
91. Nielsen TS, Lærke HN, Theil PK, Sørensen JF, Saarinen M, et al. 2014. Diets high in resistant starch and arabinoxylan modulate digestion processes and SCFA pool size in the large intestine and faecal microbial composition in pigs. *British Journal of Nutrition* 112: 1837–1849.
92. Okazaki M, Matsukuma S, Suto R, Miyazaki K, Hidaka M, et al. 2013. Perioperative synbiotic therapy in elderly patients undergoing gastroenterological surgery: A prospective, randomized control trial. *Nutrition* 29: 1224–1230.
93. Ouwehand AC, Tiihonen K, Saarinen M, Putaala H, Rautonen N. 2009. Influence of a combination of *Lactobacillus acidophilus* NCFM and lactitol on healthy elderly: Intestinal and immune parameters. *British Journal of Nutrition* 101: 367–375.
94. Palaria A, Johnson-Kanda I, O'Sullivan DJ. 2012. Effect of a synbiotic yogurt on levels of fecal bifidobacteria, clostridia, and enterobacteria. *Applied Microbiology and Biotechnology* 78: 933–940.
95. Plovier H, Everard A, Druart C, Depommier C, Van Hul M, et al. 2017. A purified membrane protein from *Akkermansia muciniphila* or the pasteurized bacterium improves metabolism in obese and diabetic mice. *Nature Medicine* 23: 107–113.
96. Pourabedin M, Guan L, Zhao X. 2015. Xylo-oligosaccharides and virginiamycin differentially modulate gut microbial composition in chickens. *Microbiome* 3: 15.
97. Probert HM, Apajalahti JHA, Rautonen N, Stowell J, Gibson GR. 2004. Polydextrose, lactitol, and fructo-oligosaccharide fermentation by colonic bacteria in a three-stage continuous culture system. *Applied and Environmental Microbiology* 70: 4505–4511.
98. Qin J, Li R, Raes J, Arumugam M, Burgdorf KS, et al. 2010. A human gut microbial gene catalogue established by metagenomic sequencing. *Nature* 464: 59–65.
99. Rajilić-Stojanović M. 2013. Function of the microbiota. *Best Practice & Research Clinical Gastroenterology* 27: 5–16.
100. Rajilic-Stojanovic M, de Vos WM. 2014. The first 1000 cultured species of the human gastrointestinal microbiota. *FEMS Microbiology Reviews* 38: 996–1047.
101. Ramirez-Farias C, Slezak K, Fuller Z, Duncan A, Holtrop G, et al. 2009. Effect of inulin on the human gut microbiota: Stimulation of *Bifidobacterium adolescentis* and *Faecalibacterium prausnitzii*. *British Journal of Nutrition* 101: 541–550.
102. Rebollar EA, Antwis RE, Becker MH, Belden LK, Bletz MC, et al. 2016. Using "omics" and integrated multi-

omics approaches to guide probiotic selection to mitigate Chytridiomycosis and other emerging infectious diseases. *Frontiers in Microbiology* 7: 68.

103. Regmi PR, Metzler-Zebeli BU, Gänzle MG, van Kempen TATG, Zijlstra RT. 2011. Starch with high amylose content and low *in vitro* digestibility increases intestinal nutrient flow and microbial fermentation and selectively promotes bifidobacteria in pigs. *The Journal of Nutrition* 141: 1273–1280.

104. Reijnders D, Goossens Gijs H, Hermes Gerben DA, Neis Evelien PJG, van der Beek Christina M, et al. 2016. Effects of gut microbiota manipulation by antibiotics on host metabolism in obese humans: A randomized double-blind placebo-controlled trial. *Cell Metabolism* 24: 63–74.

105. Riviere A, Moens F, Selak M, Maes D, Weckx S, et al. 2014. The ability of bifidobacteria to degrade arabinoxylan oligosaccharide constituents and derived oligosaccharides is strain dependent. *Applied and Environmental Microbiology* 80: 204–217.

106. Rode AA, Chehri M, Krogsgaard LR, Heno KK, Svendsen AT, et al. 2021. Randomised clinical trial: A 12-strain bacterial mixture versus faecal microbiota transplantation versus vancomycin for recurrent *Clostridioides difficile* infections. *Alimentary Pharmacology & Therapeutics* 53: 999–1009.

107. Rohlke F, Stollman N. 2012. Fecal microbiota transplantation in relapsing *Clostridium difficile* infection. *Therapeutic Advances in Gastroenterology* 5: 403–420.

108. Rosander A, Connolly E, Roos S. 2008. Removal of antibiotic resistance gene-carrying plasmids from *Lactobacillus reuteri* ATCC 55730 and characterization of the resulting daughter strain, *L. reuteri* DSM 17938. *Applied and Environmental Microbiology* 74: 6032–6040.

109. Rosenbaum M, Knight R, Leibel RL. 2015. The gut microbiota in human energy homeostasis and obesity. *Trends in Endocrinology and Metabolism* 26: 493–501.

110. Samanta AK, Jayapal N, Jayaram C, Roy S, Kolte AP, et al. 2015. Xylooligosaccharides as prebiotics from agricultural by-products: Production and applications. *Bioactive Carbohydrates and Dietary Fibre* 5: 62–71.

111. Saminathan M, Sieo C, Kalavathy R, Abdullah N, Ho Y. 2011. Effect of prebiotic oligosaccharides on growth of *Lactobacillus* strains used as a probiotic for chickens. *African Journal of Microbiology Research* 5: 57–64.

112. Sanders ME, Akkermans LM, Haller D, Hammerman C, Heimbach J, et al. 2010. Safety assessment of probiotics for human use. *Gut Microbes* 1: 164–185.

113. Saulnier DM, Gibson GR, Kolida S. 2008. In vitro effects of selected synbiotics on the human faecal microbiota composition. *FEMS Microbiology Ecology* 66: 516–527.

114. Sender R, Fuchs S, Milo R. 2016. *Revised Estimates for the Number of Human and Bacteria Cells in the Body*. bioRxiv.

115. Serrano-Nino JC, Solis-Pacheco JR, Gutierrez-Padilla JA, Cobian-Garcia A, Cavazos-Garduno A, et al. 2016. Isolation and identification of lactic acid bacteria from human milk with potential probiotic role. *Journal of Food and Nutrition Research* 4: 170–177.

116. Shetty SA, Smidt H, de Vos WM. 2019. Reconstructing functional networks in the human intestinal tract using synthetic microbiomes. *Current Opinion in Biotechnology* 58: 146–154.

117. Silvi S, Rumney CJ, Cresci A, Rowland IR. 1999. Resistant starch modifies gut microflora and microbial metabolism in human flora-associated rats inoculated with faeces from Italian and UK donors. *Journal of Applied Microbiology* 86: 521–530.

118. Slavin J. 2013. Fiber and prebiotics: Mechanisms and health benefits. *Nutrients* 5: 1417–1435.

119. Sokol H, Seksik P, Furet JP, Firmesse O, Nion-Larmurier I, et al. 2009. Low counts of *Faecalibacterium prausnitzii* in colitis microbiota. *Inflammatory Bowel Diseases* 15: 1183–1189.

120. Sorokulova IB. 1996. Outlook for using bacteria of the genus *Bacillus* for the design of new biopreparations. *Antibiot Khimioter* 41: 13–15.

121. Steenhout P, Sperisen P, Martin F-P, Sprenger N, Wernimont S, et al. 2016. Term infant formula supplemented with human milk oligosaccharides (2'fucosyllactose and lacto-N-neotetraose) shifts stool microbiota and metabolic signatures closer to that of breastfed infants. *The FASEB Journal* 30: 275–277.

122. Stowell JD. 2009. Prebiotic potential of polydextrose. In: *Prebiotics and Probiotics Science and Technology*. Springer, pp. 337–352.

123. Swanson KS, Gibson GR, Hutkins R, Reimer RA, Reid G, et al. 2020. The International Scientific Association for Probiotics and Prebiotics (ISAPP) consensus statement on the definition and scope of synbiotics. *Nature Reviews Gastroenterology & Hepatology* 17: 687–701.

124. Ten Bruggencate SJ, Bovee-Oudenhoven IM, Lettink-Wissink ML, Katan MB, van der Meer R. 2006. Dietary fructooligosaccharides affect intestinal barrier function in healthy men. *The Journal of Nutrition* 136: 70–74.

125. Tilg H, Danese S. 2014. *Roseburia hominis*: A novel guilty player in ulcerative colitis pathogenesis? *Gut* 63: 1204–1205.

126. Tolhurst G, Heffron H, Lam YS, Parker HE, Habib AM, et al. 2012. Short-chain fatty acids stimulate glucagon-like peptide-1 secretion via the G-protein–coupled receptor FFAR2. *Diabetes* 61: 364–371.

127. Tvede M, Tinggaard M, Helms M. 2015. Rectal bacteriotherapy for recurrent Clostridium difficile-associated diarrhoea: Results from a case series of 55 patients in Denmark 2000–2012. *Clinical Microbiology and Infection* 21: 48–53.

128. van Baarlen P, Troost F, van der Meer C, Hooiveld G, Boekschoten M, et al. 2011. Human mucosal *in vivo* transcriptome responses to three lactobacilli indicate how probiotics may modulate human cellular pathways. *Proceedings of the National Academy of Sciences of the United States of America* 108 (Suppl 1): 4562–4569.
129. van Baarlen P, Troost FJ, van Hemert S, van der Meer C, de Vos WM. et al. 2009. Differential NF-κB pathways induction by *Lactobacillus plantarum* in the duodenum of healthy humans correlating with immune tolerance. *Proceedings of the National Academy of Sciences of the United States of America* 106: 2371–2376.
130. Van Craeyveld V, Swennen K, Dornez E, Van de Wiele T, Marzorati M, et al. 2008. Structurally different wheat-derived arabinoxylooligosaccharides have different prebiotic and fermentation properties in rats. *Journal of Nutrition* 138: 2348–2355.
131. Van den Abbeele P, Gérard P, Rabot S, Bruneau A, El Aidy S, et al. 2011. Arabinoxylans and inulin differentially modulate the mucosal and luminal gut microbiota and mucin-degradation in humanized rats. *Environmental Microbiology* 13: 2667–2680.
132. Van Immerseel F, Ducatelle R, De Vos M, Boon N, Van De Wiele T, et al. 2010. Butyric acid-producing anaerobic bacteria as a novel probiotic treatment approach for inflammatory bowel disease. *Journal of Medical Microbiology* 59: 141–143.
133. Venema K, van Nuenen MHMC, van den Heuvel EG, Pool W, van der Vossen JMBM. 2003. The effect of lactulose on the composition of the intestinal microbiota and short-chain fatty acid production in human volunteers and a computercontrolled model of the proximal large intestine. *Microbial Ecology in Health and Disease* 15: 12.
134. Vester Boler BM, Rossoni Serao MC, Bauer LL, Staeger MA, Boileau TW, et al. 2011. Digestive physiological outcomes related to polydextrose and soluble maize fibre consumption by healthy adult men. *British Journal of Nutrition* 106: 1864–1871.
135. Walker AW, Ince J, Duncan SH, Webster LM, Holtrop G, et al. 2011. Dominant and diet-responsive groups of bacteria within the human colonic microbiota. *The ISME Journal* 5: 220–230.
136. Walton GE, Lu C, Trogh I, Arnaut F, Gibson GR, 2012. A randomised, double-blind, placebo controlled cross-over study to determine the gastrointestinal effects of consumption of arabinoxylan-oligosaccharides enriched bread in healthy volunteers. *Nutrition Journal* 11: 36-36.
137. Wang X, Yang J, Qiu X, Wen Q, Liu M, et al. 2021. Probiotics, pre-biotics and synbiotics in the treatment of pre-diabetes: A systematic review of randomized controlled trials. In *Frontiers in Public Health* 9.
138. Ward RE, Ninonuevo M, Mills DA, Lebrilla CB, German JB. 2007. *In vitro* fermentability of human milk oligosaccharides by several strains of bifidobacteria. *Molecular Nutrition & Food Research* 51: 1398–1405.
139. Watson H, Mitra S, Croden FC, Taylor M, Wood HM. et al. 2018. A randomised trial of the effect of omega-3 polyunsaturated fatty acid supplements on the human intestinal microbiota. *Gut* 67: 1974–1983.
140. Zaman SA, Sarbini SR. 2016. The potential of resistant starch as a prebiotic. *Critical Reviews in Biotechnology* 36: 578–584.
141. Zhang C, Zhang Y, Li H, Liu X. 2020. The potential of proteins, hydrolysates and peptides as growth factors for *Lactobacillus* and *Bifidobacterium*: Current research and future perspectives. *Food & Function* 11: 1946–1957.

SECTION 2

GOOD MICROBES IN FOOD PRODUCTION

Co-Edited by Luca S. Cocolin and Frans J. de Bruijn

CHAPTER 9

Bioprotective Cultures and Bacteriocins for Food

Sara Arbulu[1,2], Beatriz Gómez-Sala[1,2], Enriqueta Garcia-Gutierrez[1,2], and Paul D. Cotter[1,2]
[1]APC Microbiome Ireland, University College Cork, T12 YN60, Co. Cork, Ireland
[2]Teagasc Food Research Centre, Moorepark, Fermoy P61 C996, Co. Cork, Ireland

9.1 INTRODUCTION

Food preservation is a challenge that humankind has faced since food excesses first occurred after the agriculture and cattle revolution. Through history, humans have developed different strategies to extend the shelf life of foods and to avoid or delay food spoilage and the presence of food-borne pathogenic microorganisms that might cause food poisoning. According to the Food and Agriculture Organization (FAO), 600 million people suffer from food-borne related illnesses annually worldwide (FAO, 2019). Unsafe food not only threatens human health, but it also results in huge economic losses and food insecurity. Food safety is key to accomplishing the Sustainable Development Goals (SDGs) set by the United Nations (UN, 2015) and it is linked to SDG 3 (Good health and wellbeing), SDG1 (No poverty), SDG 6 (Clean water and sanitation), and SDG 2 (Zero hunger), among other goals (Grace, 2017).

Food preservation has evolved over time from the traditional fermented, brined, and smoked foods to a food industry that aims for strategies that maintain organoleptic properties for longer periods, while ensuring food safety. Moreover, modern food and beverage production industries are not only focused on the food itself but also on the production, packaging, and distribution processes (Cotter et al., 2005; Woraprayote et al., 2018). The industry must ensure the rigorous fulfillment of the regulations established by national and supranational organizations, such as the European Food Safety Authority (EFSA) in Europe or the Food and Drug Administration (FDA) in the United States of America. These organizations are responsible for the implementation and compliance of the regulatory guidelines that safeguard consumers' health.

9.1.1 Food Safety Hazards

Four major hazards can threaten food safety, including biological, chemical, physical, and allergenic hazards (Ricci et al., 2017).

Biological hazards refer to bacteria, yeasts, moulds, viruses, and parasites that spoil food and cause food-borne infections when the food is consumed. Food intrinsic factors specific to the food itself (pH, water content, redox potential, physical structure of the food and presence of antimicrobials), and extrinsic factors associated with the surrounding environment (temperature, humidity, oxygen, and carbon dioxide atmosphere) determine the microbiological hazards of each food.

Microorganisms that are part of foods can be collectively referred to as the "food microbiota", which is specific to each food. Some microorganisms can spoil the food during refrigeration, ripening or, if left exposed, exploiting the rich source of nutrients available. In addition, foods can be home to pathogens, which can also be acquired at different points through the food chain.

Good Microbes in Medicine, Food Production, Biotechnology, Bioremediation, and Agriculture, First Edition. Edited by Frans J. de Bruijn, Hauke Smidt, Luca S. Cocolin, Michael Sauer, David Dowling, and Linda Thomashow.

9.1.2 Bioprotection: Fermentation, Protective Cultures, and Bacteriocins

Currently, the use of chemical additives with antagonistic and antimicrobial properties against the undesired microorganisms continues to be used extensively as a means of food and beverage preservation. However, consumers and the industry seek more environmentally friendly and sustainable approaches that ensure food safety and quality while keeping the highest organoleptic properties. Bioprotection, or biopreservation, technology refers to the use of microorganisms and/or their metabolites (Stiles, 1996) to protect foods from biological hazards. Microorganisms take part in microbiological competition and produce antimicrobial substances such as acids or bacteriocins to control food spoilage and pathogen microorganisms (Cotter et al., 2005).

9.1.3 Fermented Foods

Fermentation is a traditional form of food preservation in which microbes typically break down carbohydrates in fresh food substrates to produce alcohols or acids and, ultimately, a new food product. The fermentation process can be spontaneous, with the microorganisms naturally present in food, or controlled, with selected microorganisms, usually isolated from the food in the first place. The latter is the preferred option by the industry because it enables a more homogeneous and safer fermentation. Microorganisms that trigger the fermentation process are called "starter cultures".

Fermentation can change the nutritional values of foods while also introducing new functionalities, flavours or textures, and, importantly, preserve foods for longer periods. Until the development of modern microbiology, fermented products have traditionally been prepared on the basis of empirical knowledge, without understanding the microbes involved in the process. Their stability and quality not only rely on the characteristics of the raw material, but they are also attributed to the combination of a number of hurdles such as pH, salt concentration, and the action of a series of antimicrobial compounds produced by microorganisms during fermentation. These compounds include many organic acids, ethanol, hydrogen peroxide, and bacteriocins, which inhibit the growth of or inactivate spoilage and pathogenic microorganisms (El Bassi et al., 2009; Skåra et al., 2015; Zang et al., 2020).

9.1.4 Protective Cultures

Microbes can be used as bioprotective cultures when they have inhibitory abilities against spoilage and pathogenic microorganisms. This is normally due to the production of antimicrobial compounds, as mentioned earlier. Most lactic acid bacteria (LAB) species have a long history of safe human use, and therefore have been assigned a qualified presumption of safety (QPS) status by the EFSA in Europe (EFSA, 2017, 2017; EFSA, 2007, 2005) for use in food and feed.

9.1.5 Bacteriocins

Bacteriocins are bacterial by-products that have attracted a lot of interest as food preservatives, especially those produced by LAB. They are antimicrobial peptides produced by bacteria that exert activity against other bacteria. The production of these antimicrobials provides advantages to the bacterial strains that produce them, generally against other bacteria that share the same ecological niches. They show a low toxicity profile, with a narrow or broader spectrum of action, and they are active at small concentrations. As ribosomally synthesized molecules, they can be bioengineered to enhance their production and activity using genetically modified microorganisms (GMMs).

Interestingly, bacteriocins can be produced *in situ* by bacteria used as bioprotective cultures or bacteria involved in fermentation (Silva et al., 2018, Settanni and Moschetti, 2010). A comprehensive list of bacteriocins isolated from food matrixes can be seen in Garcia-Gutierrez (2019). Bacteriocins are generally considered as safe because they can be easily degraded by proteolytic enzymes at the mammalian gastrointestinal tract. The majority of those investigated with a view to their use in food are produced by LAB. Applications include the incorporation of purified or semi-purified bacteriocins and bacteriocin-producing LAB either directly onto the food or by immersion, on bioactive biofilm and coatings for packaging purposes, and are to be used on food surfaces (Cotter et al., 2005).

9.1.6 Bacteriocin Classification

Bacteriocins are a very heterogeneous group of peptides. Here, we will refer to bacteriocin classification according to their structure (Cotter et al., 2013). Class I bacteriocins are small heat-stable peptides of less than 10 kDa that have modified amino acids. Class I bacteriocins include different subgroups depending on their characteristics and include lantibiotics, lasso peptides, microcins, sactibiotics, linear-azole peptides, linaridins, proteusins, cyanobactins, bottromycins, and thiopeptides. Class II bacteriocins are the largest group. They are small heat-stable peptides of less than 10 kDa too, but they are unmodified. They are classified in four subgroups: IIa or pediocin-like peptides; IIb, formed by two peptides acting synergistically; IIc, circular peptides and IId, linear non-pediocin like. Bacteriolysins, formerly known as Class III, are large heat labile proteins bigger than 30 kDa.

Over the rest of this chapter, we will provide an overview of the current and potential use of bacteriocin-producing bacteria and their bacteriocins in different groups of foods, including milk and dairy products, meat, fish, and fruit and vegetables, with special focus on fermented foods.

9.2 BIOPROTECTION OF MILK AND DAIRY PRODUCTS

9.2.1 Milk Products and Their Importance in Society

Milk is a staple food in many cultures. Millions of litres of milk from different origins (cow, sheep, camel, donkey, etc.) are consumed worldwide daily (Faye and Konuspayeva, 2012). Milk contains proteins, carbohydrates (mainly lactose), lipids, and minerals. In addition to the milk itself, there are a number of milk-derived dairy products, most of them fermented, such as yogurt, cheese or kefir, which are widely consumed. These fermented foods can be a source of beneficial microbes that remain in the human gut and provide us with health benefits (Mathur et al., 2020, Pasolli et al., 2020). Therefore, milk and dairy products have a paramount importance from an industry, diet, and health perspective.

9.2.2 Spoilage and Food-borne Pathogenic Bacteria in Milk and Dairy Products

Several bacterial species can spoil milk and dairy products. Other bacteria can be pathogenic for humans and cause food poisoning.

Refrigerated raw milk can develop *Pseudomonas*, which metabolize proteins to change flavours, producing lipases and heat-resistant proteinases. Coliforms produce lactic, acetic and formic acids, carbon dioxide, and hydrogen with the result that the milk gets foamy and is not suitable for consumption. *Alcaligenes* can also be present forming viscous polysaccharides. If raw milk is not properly refrigerated, there is a different set of microorganisms that can develop, such as *Lactococcus*, *Lactobacillus*, *Bacillus*, *Micrococcus*, *Enterococcus*, or *Clostridium* (Huck et al., 2008). Pasteurized milk can be contaminated by the growth of thermoresistant bacteria that can survive the pasteurization process (such as spores from *Bacillus* and *Clostridium*) or by post-pasteurization contaminants (*Pseudomonas*, *Alcaligenes*, and *Flavobacterium*). On the other hand, ultra-high temperature (UHT) treated milk is much closer to being sterile and it is not likely to be spoiled over a short period.

Milk by-products, like butter, can be spoiled by *Pseudomonas*, *Candida*, and *Geotrichum* on the surface. Yogurts can be affected by spoilage bacteria that can alter their organoleptic characteristics and make them too acidic, due to the growth of yeast and other fungi on the surface. Cheese spoilage depends on the acidity and humidity of different cheese types, with bacteria from the genera *Alcaligenes*, *Pseudomonas*, *Leuconostoc*, *Clostridium*, *Lactococcus*, and *Enterococcus* having the potential to be problematic (Odeyemi et al., 2020). However, pathogenic bacteria with serious effects on human health, such as *Listeria monocytogenes*, *Staphylococcus aureus*, *Escherichia coli*, and *Salmonella* spp. can develop (De Buyser et al. 2001). Table 9.1 summarizes some of the key spoilage and pathogenic bacteria of relevance to milk and dairy products.

9.3 FERMENTED DAIRY PRODUCTS

Milk can undergo a variety of different processes in order to produce different fermented dairy products (Figure 9.1).

Briefly, cheeses are produced by coagulation (curdling) or by a soft pasteurization of milk followed by the action of LAB fermenters, added or naturally present in the milk, which produce a coagulant, called curd. After that, the serum (whey) is extracted, either by heat or by mechanical action and left for maturation for a determined period depending on the cheese type (see also chapter 12).

The bacteria that are most frequently used in cheese production are *Lactococcus lactis* subsp. *lactis*, *Lc. lactis* subsp. *cremoris*, *Streptococcus thermophilus*, and *Lactobacillus bulgaricus*. Emmental cheese also uses *Propionibacterium freudenreichii*, while cheeses such as Roquefort, Cabrales, and other blue cheeses contain

TABLE 9.1 Spoilage and pathogenic bacteria in milk and dairy products

Milk/dairy product	Spoilage and pathogenic bacteria
Raw milk (refrigerated)	Pseudomonas, coliforms, Alcaligenes
Raw milk (non-refrigated)	Lactococcus, Lactobacillus, Bacillus, Micrococcus, Enterococcus or Clostridium
Pasteurized milk	Micrococcus, Enterococcus, Streptococcus, Corynebacterium, spores from B. cereus and Clostridium, Pseudomonas, Alcaligenes, and Flavobacterium
UHT milk	Thermophilic spores
Butter	Pseudomonas, Candida, and Geotrichum
Cheese	Alcaligenes, Pseudomonas, Leuconostoc, Clostridium, Lactococcus, Enterococcus, Listeria monocytogenes, Staphylococcus aureus, Escherichia coli, and Salmonella spp.

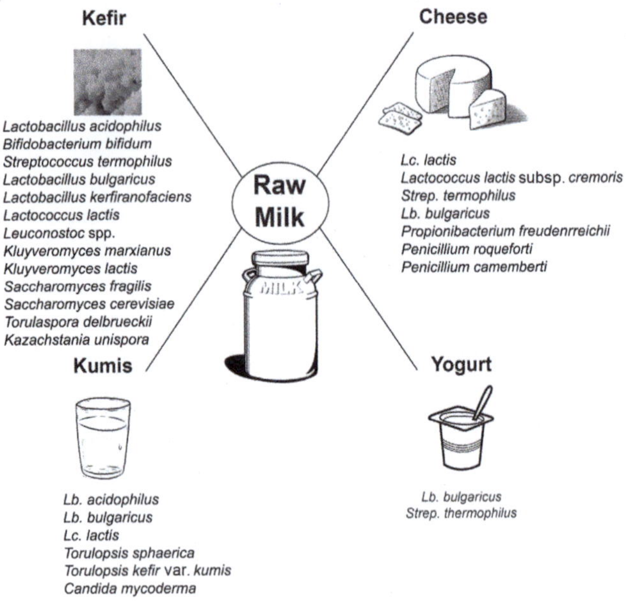

FIGURE 9.1 Examples of beneficial bacteria used in milk-derived fermented foods.

Penicillium roqueforti. Soft cheeses such as brie or camembert contain *Penicillium camemberti*.

The production of yogurts typically relies on the addition of *Lb. bulgaricus* and *Strep. thermophilus* to the pasteurized milk and incubation for 3–4 hours at 40–43°C before refrigeration for a determined time. The process is called cooperative fermentation, in that *Lb. bulgaricus* produces peptides and free amino acids that *Strep. thermophilus* transforms into formic and pyruvic acids and carbon dioxide.

There are also very popular fermented drinks derived from milk, such as *kefir* or *kumis*. Both are considered mixed fermented milks, since they require LAB and lactose-fermenting yeasts. They are acidic, can contain low levels of alcohol and have been associated with a variety of health benefits (Bourrie et al., 2018). *Kefir* is produced using consortia of microorganisms contained within physical structures known as *kefir* grains. These *kefir* grains can include bacteria such as *Lactobacillus acidophilus*, *Lb. bulgaricus*, *Lactobacillus helveticus*, *Lactobacillus kefiranofaciens*, *Lc. lactis*, and *Leuconostoc* sp. The yeast representatives include both species such as *Kluyveromyces marxianus*, *Kluyveromyces lactis*, *Saccharomyces fragilis*, *Saccharomyces cerevisiae*, *Torulaspora delbrueckii*, or *Kazachstania unispora* (Bourrie et al., 2016). *Kumis* is produced using a liquid starter culture that includes different combinations of *Lb. acidophilus*, *Lb. bulgaricus*, *Lc. lactis*, *Torulopsis sphaerica*, *Torulopsis kumis*, and *Candida mycoderma* (Kozhahmetova and Kasenova, 2013).

A number of the above mentioned bacteria involved in the fermentation of milk and dairy products are able to produce bacteriocins that can control spoilage and pathogenic bacteria in milk-related products.

9.4 APPLICATION OF BACTERIOCINS AND THEIR PROTECTIVE CULTURES IN MILK AND DAIRY PRODUCTS

The most extensively used and studied bacteriocins are nisin A and Z, both frequently produced by LAB isolated from milk and dairy products. They are particularly effective against bacteria from these niches. Indeed, the cheese industry has used them to prevent the proliferation of endospore formers, such as gas-producing clostridia and the pathogen *Clostridium botulinum*. Another important dairy pathogen is *L. monocytogenes*, which is also likely to contaminate after the production process (Ibarra-Sánchez et al., 2020). Bacteriocins such as lacticins 3147 and 481 or thermophilin, also isolated from

LAB, show inhibition of *L. monocytogenes* and *Clostridium tyrobutyricum* in cheese production (Gálvez et al., 2014), while bacteriocin AS-48, produced by *Enterococcus faecalis* A-48-32 in cheese, inhibits *Bacillus cereus* growth (Muñoz et al., 2004). Another example of beneficial bacteria that could control *L. monocytogenes* and *Staph. aureus*, this time in yogurt, is the bacteriocin producer *Streptococcus salivarius* subsp. *thermophilus* B (Benkerroum et al., 2002). These are a few examples of bacteriocins of relevance in the dairy industry. Further reading on the topic can be found in (Silva et al., 2018; Arqués et al., 2015, and Ibarra-Sánchez et al., 2020).

9.5 BIOPROTECTION OF MEAT AND MEAT PRODUCTS

9.5.1 Meat and Meat Products and Their Importance in Society

Meat is a heterogeneous food with a composition that varies according to the origin of the muscle (ruminants, pigs, and poultry), the type of muscles, and the processes used to prepare different dishes (curing, drying, fermentation, etc.). Meat and meat products contribute significantly to the intake of energy, protein, and important micronutrients, at least when consumed frequently (De Smet and Contreras, 2005; Santé-Lhoutellier, 2014). At a global level, both the average per capita consumption of meat and the total amount of meat consumed are rising, driven by increasing average individual incomes and by population growth (Godfray et al., 2018; Henchion et al., 2014).

9.5.2 Spoilage and Food-borne Pathogenic Bacteria in Meat and Meat Products

The nature of the microbial communities present on meat is very diverse as it depends on the animal and on the nature of the subsequent processing steps (Zagorec and Champomier-Vergès, 2017). Fresh raw meat constitutes a highly perishable food, due to factors associated to the product (intrinsic parameters) and the direct exposure of the carcass to the environment (extrinsic parameters), especially once the natural anatomical barrier of the skin/hide is removed. The high water content ($a_W > 0.99$), the pH (5.5–6.5), which corresponds to the optimal range for microbial growth, and the availability of nutrients (sugars, amino acids, vitamins, minerals, etc.), account for meat being a foodstuff with a short shelf-life (Pothakos et al., 2015; Sofos, 2014a, 2014b; Zagorec and Champomier-Vergès, 2017).

The spoilage microbiota is dominated by psychrotrophic Gram-negative rod-shaped aerobic, facultative, and obligated anaerobic bacteria, mainly belonging to a few genera, including *Pseudomonas, Shewanella, Acinetobacter, Moraxella, Psychrobacter, Alcaligenes,* Enterobacteriaceae, *Flavobacterium*, and several Gram-positive bacteria such as *Staphylococcus, Micrococcus, Clostridium, Brochothrix, Corynebacterium*, and LAB, that dominate under different conditions. However, the growth and the predominance of the different strains is variable because of packaging, temperature, and time of storage (Comi, 2017; Pothakos et al., 2015; Sofos, 2014a, 2014b; Zagorec and Champomier-Vergès, 2017). LAB are frequently associated with fresh meat and cooked meat products but their impact can vary, with some species and strains contributing to the generation of offensive metabolites and the subsequent organoleptic downgrading of meat, and others serving as bioprotective agents by reducing spoilage capacities and inhibiting spoiling microbiota (Pothakos et al., 2015).

The levels of food-borne pathogens in meat processed under sanitary and hygienic conditions are generally very low relative to the levels of spoilage microorganisms present. The most important food-borne pathogens associated with meat are *Salmonella* spp., *Staph. aureus, E. coli, Campylobacter jejuni, L. monocytogenes, Cl. perfringes, Yersinia enterocolitica*, and *Aeromonas hydrophila*. Among them, *Salmonella* species, *Campy. jejuni, L. monocytogenes*, and verocytotoxin producing *E. coli* O157 are a major public health problem. All these may be harbored in the gastrointestinal tract of food producing animals. The most frequent chain of events leading to meat-borne illness involves food animals, which are healthy carriers of the pathogens that are subsequently transferred to humans through production, handling, and consumption of meat and meat products (Nørrung and Buncic, 2008; Sofos, 2014a, 2014b).

9.6 FERMENTED MEAT PRODUCTS

The origin of fermented meat products can be traced back to Mediterranean countries in Roman times. Production was then spread throughout America, Africa, and Australia by immigrants from European countries. There are hundreds of different formulations for fermented meat products all over the world as a consequence of variations in the raw materials, formulations, and manufacturing processes, which come from the habits and customs of the different countries. European countries are major producers of fermented meat products, and fermented sausages constitute a large part of that market

TABLE 9.2 Microorganisms isolated from fermented meat products from different countries

Product	Description	Origin	Microorganisms	Reference
Chorizo	Dry-cured sausage	Spain	*Lactobacillus* spp., *Staphylococcus* spp.	(Aymerich et al., 2003)
Fuet	Dry-cured sausage	Spain	*Lactobacillus* spp., *Staphylococcus* spp.	(Aymerich et al., 2003)
Salchichón	Dry-cured sausage	Spain	*Lactobacillus* spp., *Macrococcus* spp., Enterobacteriaceae	(Fernández-López et al., 2008)
Alheira	Traditional fermented sausage	Portugal	*Lactobacillus* spp., *Enterococcus* spp., *Leuconostoc* spp. *Pediococcus* spp.	(Albano et al., 2009)
Sucuk	Dry sausage	Turkey	*Lactobacillus* spp., *Macrococcus* spp., Enterobacteriaceae	(Gençcelep et al., 2008)
Salame friulano	Dry sausage	Italy	*Lactobacillus* spp., *Staphylococcus* spp., *Macrococcus* spp., *Candida* spp., *Debaryomyces* spp.	(Cocolin et al., 2001; Rantsiou et al., 2005, Rantsiou et al., 2004)
Urutan	Dry-fermented sausage	Bali	*Lactobacillus* spp., *P. acidilactici*	(Endo et al., 2014)

(Anagnostopoulos and Tsaltas, 2019; Kumar et al., 2017; Leroy and De Vuyst, 2016; Talon and Leroy, 2014).

Fermented sausage is generally made from a mixture of meat (often pork), pork fat, and salt, eventually sugar, nitrate, and/or nitrite. Meat to fat ratios and total fat content are variable depending on the product. Then, the mixture is seasoned with various spices depending on the recipe. Traditional products are frequently not inoculated and, thus, fermentation relies on the indigenous microbiota, whereas industrial products are inoculated by starter cultures (Kanninen and Puolanne, 2008; Kumar et al., 2017; Lebert et al., 2007; see also chapter 13). The major microorganisms involved in meat fermentation are species of LAB and gram-positive coagulase-negative cocci (CNC). However, yeasts and enterococci are also present in some meat products (Lebert et al., 2007; Rantsiou and Cocolin, 2008). The LAB species most commonly identified in traditional fermented sausages are *Lactobacillus sakei*, *Lactobacillus curvatus*, and *Lactobacillus plantarum*, with *Lb. sakei* generally dominating. Regarding CNC, *Staphylococcus* spp. and *Kocuria* spp. are used as starter cultures for manufacturing industrial dry sausages in Europe. In traditional products at the end of ripening, *Staphylococcus xylosus* is often the most common species and is identified in all the sausages. Moulds and yeasts traditionally play an important role by bringing about a characteristic surface appearance and flavours due to their well-known proteolytic and lipolytic activity. In some fermented sausages, particularly those produced in Spain and Italy, the sensory properties are influenced by the development of moulds and yeast in the surface of the product (Fontana et al., 2005; Rantsiou and Cocolin, 2008). Well-known examples of fermented sausages include different types of *salami*, *chorizo*, and *alheira* (Table 9.2) as well as some fermented meats, which are semidry or even moist, such as German Mettwurst, and have distinctive texture properties (Leroy and De Vuyst, 2016; see also chapter 13).

9.7 APPLICATION OF PROTECTIVE CULTURES AND THEIR BACTERIOCINS IN MEAT AND MEAT PRODUCTS

Fresh meat and fermented meat products provide an excellent environment for the growth of pathogenic and spoilage organisms. The application of antimicrobial interventions to improve the microbiological quality of meat during meat conditioning will assure safe raw fresh meat in compliance with regulatory requirements. Properly processed meat should have a low pH and minimum numbers of contaminating bacteria, and any contamination that occurs is likely reduced by chilling and other methods used to restrict microbial growth. Still, contamination of carcasses at slaughter can occur and different combinations of antimicrobials have been tested for the decontamination of the animal carcasses, such as (i) acids antimicrobials (acetic, citric, and lactic acids) (Arthur et al., 2008; Carlson et al., 2008) and their salts (Bosilevac et al., 2006), (ii) bacteriocins (de Barboza

Martínez et al., 2002), (iii) combination of bacteriocin and acid (nisin in combination with lactic acid) (de Barboza Martínez et al., 2002), and (iv) bacteriocin-producing bioprotective cultures (Castellano and Vignolo, 2006; Dortu et al., 2008; Maragkoudakis et al., 2009) have been used in beef and poultry carcasses and/or raw meat with good results. The bioprotective cultures *Enterococcus faecium* PCD71 and *Lactobacillus fermentum* ACA-DC179 were applied in chicken meat and were effective against *L. monocytogenes* and *Salmonella enterica* serovar Enteritidis, respectively (Maragkoudakis et al., 2009). The sakacin P producer *Lb. curvatus* CWBI-B28 and sakacin G producer *Lb. sakei* CWBI-B1365 successfully controlled *L. monocytogenes* on raw beef and poultry meat (Dortu et al., 2008). Similarly, in refrigerated vacuum-packaged bovine meat the use of bacteriocin producing *Lb. curvatus* CRL705 inhibited *Listeria innocua* and *Brochothrix thermosphacta* (Castellano and Vignolo, 2006). For a review on this topic see Castellano et al. (2017).

In fermented meat products, such as sausages, the use of starters such as LAB can increase the safety of the fermented products by means of rapid matrix acidification and/or due to the production of antimicrobial substances such as bacteriocins. In this way, the selection of bacteriocin-producing starter cultures or their use as co-cultures jointly with a commercial starter culture would be a useful approach for protection against food-borne pathogenic bacteria (Laranjo et al., 2019). Several LAB strains, including the commonly meat starter cultures *Lb. curvatus*, *Lb. sakei*, and *Lb. plantarum*, have been reported as bacteriocin producers and have been used as bioprotective starter cultures, and their activity against *L. monocytogenes* has been proved in meat products (Diaz-Ruiz et al., 2012; Dicks et al., 2004; Hugas et al., 1995; Jacobsen et al., 2003; Todorov et al., 2007). Nonetheless, the results of using these bioprotective starters depends on different factors such as sausage recipes, initial inoculation levels, and/or antilisteria starters applied. For a review in this topic see Castellano et al. (2017), Laranjo et al. (2019) and chapter 13.

Purified and semi-purified bacteriocins such as nisin, pediocin, and different combinations of them have also been tested for preservation of meat products (Hartmann et al., 2011; Kalschne et al., 2014; Tu and Mustapha, 2002). However, there are some limitations to the use of purified or semi-purified bacteriocins in meat products such as the possibility of enzymatic destruction by proteases, binding to food components such as fat particles or protein surfaces, inactivation by some additives used in meat fermentations, low solubility of the bacteriocin, pH effects on bacteriocin stability and activity and/or uneven distribution in the food matrix (Da Costa et al., 2019). For this reason, to ensure the effectiveness of the bacteriocin activity in a meat environment, it is essential to carry out a case-by-case study against specific target and non-target bacteria on the food matrix in which the bacteriocin wants to be applied (Castellano et al., 2017).

9.8 BIOPROTECTION OF FRESH FISH AND FISH PRODUCTS

9.8.1 Fish and Fish Products and Their Importance in Society

Fish and fish products are generally recognized as having an excellent nutritional value, providing high-quality protein, long-chain polyunsaturated fatty acids, and a wide variety of vitamins and minerals. The fisheries sector is crucial to food and nutrition security, and its importance is growing. This is confirmed by the fact that total production, trade, and consumption of fish and fish products reached an all-time record in 2018 (FAO, 2020; Zang et al., 2020).

Fish are also one of the most internationally traded foods with the added challenge of being a highly perishable product with a short shelf life (Wiernasz et al., 2017). Throughout the world, post-harvest fish losses are a major concern and occur in most fish distribution chains, with an estimated 35% of global harvest of fisheries and aquaculture production being either lost or wasted every year (FAO, 2020).

9.8.2 Spoilage and Food-borne Pathogenic Bacteria in Fish and Fish Products

Soon after a fish dies, endogenous bacteria begin to invade the tissues through the gills, along blood vessels and directly through the skin and the lining of the belly cavity. The composition of fish provides sufficient nutrients for the development of spoilage and food-borne pathogenic microorganisms (Figure 9.2). In addition, the changes brought about by catching, handling, and processing are also a source of contamination (Françoise, 2010; Ghanbari et al., 2013; Gram and Huss, 1996).

Spoilage of fresh fish and fish products is caused by a complex process in which physical, chemical and microbiological factors are involved (Huss, H.H., 1995). Among the microorganisms present in fish, only a fraction of the initial microbiota, known as specific spoilage organisms, are responsible for fish spoilage (Gram and Dalgaard,

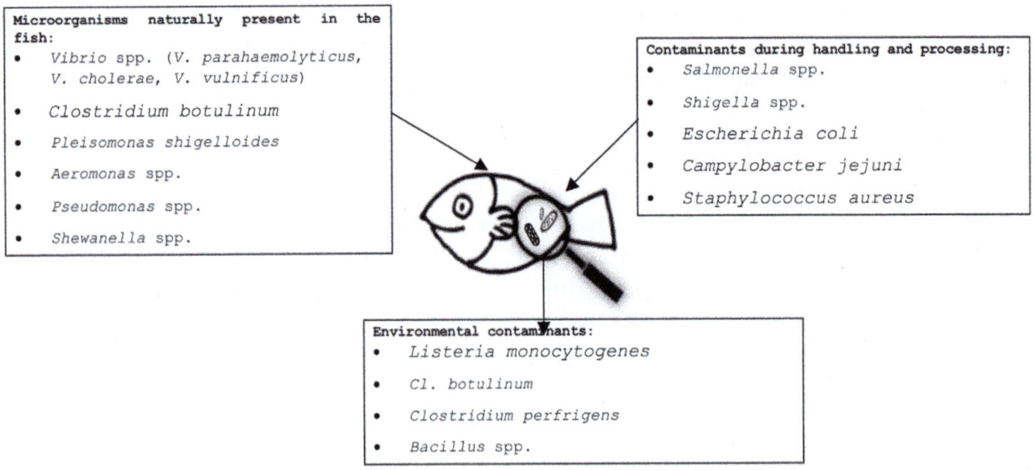

FIGURE 9.2 Spoilage and food-borne pathogenic bacteria that can be found in fish and fish products.

2002). The high *post mortem* pH (>6), combined with the nutrient availability in fresh fish allow the quick growth of Gram-negative psychotropic bacteria naturally present in fish like *Pseudomonas* spp. and *Shewanella* spp. At ambient temperature, fish is dominated by mesophilic *Vibrionaceae* and, if the fish is caught in polluted waters, mesophilic Enterobacteriaceae (Françoise, 2010; Gram and Dalgaard, 2002; Gram and Huss, 1996) can grow. The use of vacuum or modified atmosphere packaging selects for the spoilage bacteria *Photobacterium phosphoreum* and also for LAB. In chilled seafood, Gram-positive pathogens are normally of lesser concern than the Gram-negative spoilage bacteria present, such as *Pseudomonas* spp., *Shewanella* spp., and Enterobacteriaceae, which grow more rapidly (Huss et al., 2000).

Regarding food-borne pathogens, the most important are *Vibrio* spp. and *Cl. botulinum* type E, which are natural constituents of aquatic environments, and *L. monocytogenes*, *Salmonella* sp. and *Staphylococcus* as resulting from post-contamination during manufacturing processes (Françoise, 2010; Ghanbari et al., 2013; Huss et al., 2000). Some human pathogens present in fish such as *Shigella* spp. and *E.coli* can be spread by aquatic birds, mainly in aquaculture environments, but normal cooking procedures eliminate the risk of these pathogens, so the safety concerns are therefore primarily related to consumption of raw or insufficiently cooked fish (Figure 9.2) (Huss et al., 2000).

9.9 FERMENTED FISH PRODUCTS

Fermented fish products are produced and consumed in different parts of the world and are an integral part of many food cultures. They are especially popular in Asia, Africa, and Europe. Many of these countries have their own unique types of fermented products based on differences in raw materials, environmental conditions, microorganisms, and dietary traditions (Skåra et al., 2015; Xu et al., 2020; Zang et al., 2020). In European cuisines, fermented fish is an old staple food, for instance, the ancient Greeks and Romans made a famous sauce from fermented fish called *garum*. However, nowadays only a few traditional fermented fish products are still produced in northern Europe, such as *hákarl* (Iceland), *surströmming* (Sweden) and *rakfisk* (Norway) (Skåra et al., 2015). Artisanal fish processing, including fermentation of fish, remains the predominant and most important method of fish preservation in Africa. Fermented fishery products in Africa are usually whole or in cut pieces such as *lanhouin* (Benin), *momone* (Ghana) and *feseek* (Egypt) (Sheikha, 2014; Zang et al., 2020). In Southeast Asia, fermented fish products have a long history and they are of great nutritional importance. Various different types of fermented fish products including whole fish or fish pieces such as *plaa-som* (Thailand), sauces such as *bakasang* (Indonesia) and pastes such as *prahok* (Cambodia) can be found in Asian markets (Ohshima and Giri, 2014; Xu et al., 2020, 2020).

The overall processing steps of fish fermentation are similar throughout the world. In general, the enzymes are most significant with respect to changing texture as well as producing some of the flavor, and the microorganisms help in the development of aroma and flavor (Zang et al., 2020). There are different types of fermented fish products and they can be divided in different groups according to different criteria (Table 9.3).

TABLE 9.3 Types and examples of traditional fermented fish around the world

Types		Product	Raw material	Microorganisms	Country	Reference
According to the final product appearance	Fermented whole or in pieces	Hákarl	Greenland shark	*Moraxella*, *Acinetobacter*, *Lactobacillus* sp.	Iceland	(Skåra et al., 2015)
		Plaa-som	Snakehead fish	*P. pentosaceus*, *Lactobacillus alimentarius*, *Weissella*, *Lactobacillus planetarium*, *Lactococcus garvieae*, *Zygosaccharomyces rouxii*	Thailand	(Kopermsub and Yunchalard, 2010)
	Fermented fish pastes	Prahok	Striped snakehead	*Staphylococcus*, *Tetragenococcus*, *Clostridium*	Cambodia	(Chuon et al., 2014)
	Fermented fish sauces	Nam-pla	Anchovy/ Mrigal carp and salt	Extremely halophilic bacterial strains	Thailand	(Tanasupawat et al., 2006; Zang et al., 2020)
According to the processing method	Spontaneous fermentation	Garum	Atlantic bluefin tuna/ Atlantic mackerel	Not reported	Italy (Rome) and Greece	(Skåra et al., 2015; Zang et al., 2020)
	Fermentation using starter cultures	Som-fug	Bigeye snapper	*Lactobacillus plantarum*, *Pediococcus acidilactici*, *P. pentosaceus*	Thailand	(Riebroy et al., 2008; Zang et al., 2020)
According to the types of substrates	Products using fish and salt	Lanhouin	Cassava fish/ Spanish mackerel and salt	*Bacillus subtilis*, *Bacillus licheniformis*, *Bacillus megaterium*, *Bacillus mycoides*, *Bacillus cereus*, *Staph. lentus*, *Staph. xylosus*	Benin	(Anihouvi et al., 2007; Sheikha, 2014)
	Products using fish and salt and carbohydrates	Suan yu	Common carp, roasted rice, salt and sugar	*Lb. plantarum*, *P. pentosaceus*, *Leuconostoc*, *Paralimentarius*, *Saccharomyces cerevisiae*, *Hansenula anomala*	China	(Zang et al., 2020; Zeng et al., 2013)
		Momoni	African jack mackerel/ Snakehead fish, salt, palm sirup, roasted rice	*Micrococcus*, *Staphylococcus aureus*, *Staphylococcus* sp., *Bacillus* sp., *Lactobacillus* sp., *Pseudomonas*, *Pediococcus*, *Klebsiella*, *Debaryomyces*, *Hansenula*, *Aspergillus* spp.	Ghana	(Sheikha, 2014; Zang et al., 2020)
According to the proportion of salt added	High salt products (>20% of the total weight)	Feseekh	Silversides/ tigerfish/ flathead gray mullet and salt	*Staphylococcus equorum*, *B. subtilis*, *Lactobacillus* sp., *Teratogenococcus halophilus*, *Clostridium bifermentans*, *Clostridium butyricum*, and *Cl. cochlearium*	Egypt	(Sheikha, 2014; Zang et al., 2020)
	Low salt products (3-8%)	Rakfish	Salmonid, salt	*Lactobacillus*	Norway	(Skåra et al., 2015)
	No salt products	Ngari	Pool barb	*Lactococcus lactis* subsp. *cremoris*, *Lb. plantarum*, *B. subtilis*, *Candida*, *Saccharomycopsis*	India	(Thapa et al., 2004; Zang et al., 2020)

Adapted from (Thapa et al., 2004; Zang et al., 2020)

9.10 APPLICATION OF PROTECTIVE CULTURES AND THEIR BACTERIOCINS IN FISH AND FISH PRODUCTS

In fish and fish products, the development of biopreservation strategies should consider the employment of suitable LAB strains isolated from marine species, since the strains already acclimatized to a seafood habitat would be advantageous in terms of biopreservation (El Bassi et al., 2009; Gómez-Sala et al., 2016, 2015; Tahiri et al., 2004). While several bacteriocins have been identified from bacteria isolates from fish, very few commercial applications have been developed in fish and fish products. One of the reasons is that the selection of potential protective bacteria for fish and fish products is still a challenge as they must be able to: (i) survive the steps of food processing, (ii) adapt and develop in a specific media such as fish, with low carbohydrate concentration, and (iii) produce active antagonistic metabolites but their metabolic activities should not affect the organoleptic characteristics of the product (Françoise, 2010; Ghanbari et al., 2013). Many of the studies regarding seafood biopreservation have been mainly conducted in lightly preserved fish products such as cold-smoked salmon (CSS) because they are considered risky products of foodstuffs but also because the implementation of LAB strains in fresh fish is more difficult (Françoise, 2010; Ghanbari et al., 2013) (Table 9.4).

The bioprotection potential of bacteriocin-producing strains belonging to the genera *Lactobacillus* has been tested in fresh fish (Gómez-Sala et al., 2016), minced fish (El Bassi et al., 2009) and fish fillets (Kannappan et al., 2017). In all cases, the use of bioprotective *Lactobacillus* strains improved the final quality of the bioprotected fish or fish product by reducing the microbial counts (Gómez-Sala et al., 2016), limiting the development of total coliform and slightly reducing the production of trimethylamine (TMA) and total volatile basic nitrogen contents (TVB-N) (El Bassi et al. (2009) and regulating the growth of specific and native spoilage bacteria and reducing the production of TMA-N and TVB-N (Kannappan et al., 2017). For a more extensive review on this topic see Gómez-Sala et al. (2019) and Ghanbari and Jami (2013).

It should be also noted that bacteriocins are not always indispensable for the desirable bioprotective effect of the strains as several studies have shown that the use of bacteriocin-producing and non-producing strains can have a similar effect in the reduction of *L. monocytogenes* in CSS (Katla et al., 2001; Nilsson et al., 2004).

Similarly, the biopreservative potential of bacteriocin PSY2 produced by *Lc. lactis* PSY2 was evaluated using fillets of reef cod (Sarika et al., 2012). The bacteriocin dissolved in sterile distilled water was sprayed into the fish fillets and the results showed that the biopreservation approach applied extended the shelf-life of the fillets, reduced the total count of spoilage bacteria and the sensory analysis was better in bacteriocin-treated samples than in the control without bacteriocin. Katla et al. observed a delay in the growth of *L. monocytogenes* when adding sakacin P or nisin in CSS (Katla et al., 2001).

9.11 BIOPROTECTION OF FRUITS AND VEGETABLES

9.11.1 Fruit and Vegetables and Their Importance in Society

Fruits and vegetables are promoted as healthy and recommended as part of a healthy diet worldwide. They are a rich source of vitamins and minerals, dietary fiber, phytosterols, flavonoids, and other antioxidants. However, the increasing demand and availability of fruit and vegetables through global trade, together with poor manipulation practices has resulted in an increase in the number of human infections and outbreaks related to these foods.

Fruit and vegetables are perishable products marketed raw or minimally processed. This includes ready to eat products, such as peeled or chopped fruits and salads; frozen, preserved (canned, dehydrated, juices, purees, etc.), and fermented foods.

After the harvesting of fruit and vegetables, they are stored under different conditions for variable periods. Then, they are transported directly to markets or to processing plants, where preservation methods are applied for their transformation into a variety of food products. Each of the stages in the food supply chain can raise a food safety issue and must be considered before the product reaches the consumer. Here, we summarize the microbiological hazards associated with fruit and vegetables, and discuss how bacteria and their bacteriocins can provide bioprotection to enhance food safety.

TABLE 9.4 Bioprotection of fish and fish products using LAB and/or bacteriocins

LAB strain	Bacteriocinogenic (Y/N) and bacteriocin(s) produced[a]	Isolation source	Product applied	Reported effects	Reference(s)
Lactobacillus curvatus BCS35	Sakacins P and X	Dry-salted cod	Young hake and megrim	Improvement of the organoleptic characteristics, inhibition of *Listeria **monocytogenes*** and a higher value in the market	(Gómez-Sala et al., 2016)
Lb. curvatus ET06	Y, ND	CSS	CSS	Inhibition of *Listeria innocua*	(Tomé et al., 2008)
Lactobacillus rhamnosus ATCC 53103	Y, ND	ND	Sea bass fillets	Inhibition of spoilage bacteria	(Kannappan et al., 2017)
Lactobacillus sakei Lb790	Sakacin P	ND	CSS	Inhibition of *L. monocytogenes*	(Katla, T et al., 2001)
Lactobacillus delbrueckii	ND	Sardine	Minced sardine with salt and sugar	Inhibition of coliforms, *Salmonella*, *Staphylococcus*, and *Clostridium*	(Ndaw et al., 2008)
Lb. delbrueckii ET32	Y, ND	CSS	CSS	Inhibition of *L. innocua*	(Tomé et al., 2008)
Lactobacillus plantarum 3	Y, ND	Sea products	Refrigerated and VP minced *Dicentrarchus labrax*	Suppression of TVB-N and TMA production	(El Bassi et al., 2009)
Lactobacillus pentosus 7	N	Sea products	Refrigerated and VP minced *D. labrax*	Suppression of TVB-N and TMA production	(El Bassi et al., 2009)
Lactobacillus casei T3	Y, ND	Dairy	CSS	Inhibition of *L. innocua*	(Vescovo et al., 2006)
Enterococcus faecium BNM58	Enterocin L50	Albacore	Young hake and megrim	Improvement of the organoleptic characteristics, inhibition of *L. monocytogenes* and a higher value in the market	(Gómez-Sala et al., 2016)
En. faecium ET05	Y, ND	CSS	CSS	Inhibition of *L. innocua*	(Tomé et al., 2008)
Enterocococcus faecalis L04	Bacteriocin EFL4	Sea bass	RTE fresh salmon fillets	Inhibition of spoilage and food-borne pathogenic bacteria	(Mei et al., 2020)
Lactococcus lactis PSY2	Bacteriocin PSY2	Marine perch	Reef cod fillets	Extension of shelf-life and better protection against spoilage bacteria	(Sarika et al., 2012)

(*Continued*)

TABLE 9.4 (Continued)

LAB strain	Bacteriocinogenic (Y/N) and bacteriocin(s) produced[a]	Isolation source	Product applied	Reported effects	Reference(s)
Lactococcus piscium EU2241	N	Seafood	VP shrimp	Reduction of the number of *Listeria monocytogenes* and *Staphylococcus aureus*	(Matamoros et al., 2009)
Carnobacterium piscicola A9b	Carnobacteriocin B2	VP CSS	CSS	Inhibition of *L. monocytogenes*	(Nilsson et al., 2004)
Ca. piscicola V1	Piscicocin V1a and V1b	Trout intestine	CSS	Inhibition of *L. monocytogenes*	(Brillet et al., 2004)
Carnobacterium divergens V41	Divercin V41	Salmon intestine	CSS	Inhibition of *L. monocytogenes*	(Brillet et al., 2004)
Ca. divergens M35	Divergicin M35	Frozen smoked mussels	CSS	Inactivation of *L. monocytogenes*	(Tahiri et al., 2004)
Carnobacterium maltaromaticum CS526	Piscicocin CS526	Surimi	CSS	Decrease the population of *L. monocytogenes*	(Yamazaki et al., 2003)
Lactococcus gelidum EU2247	Y, ND	Seafood	VP shrimp	Reduction of the number of *L. monocytogenes* and *Staphylococcus aureus*	(Matamoros et al., 2009)
Pediococcus acidilactici ET34	Y, ND	CSS	CSS	Inhibition of *Listeria innocua*	(Tomé et al., 2008)

Abbreviations: ND, not determined; VP, vacuum packaged; CSS, cold-smoked salmon; RTE, ready-to-eat.

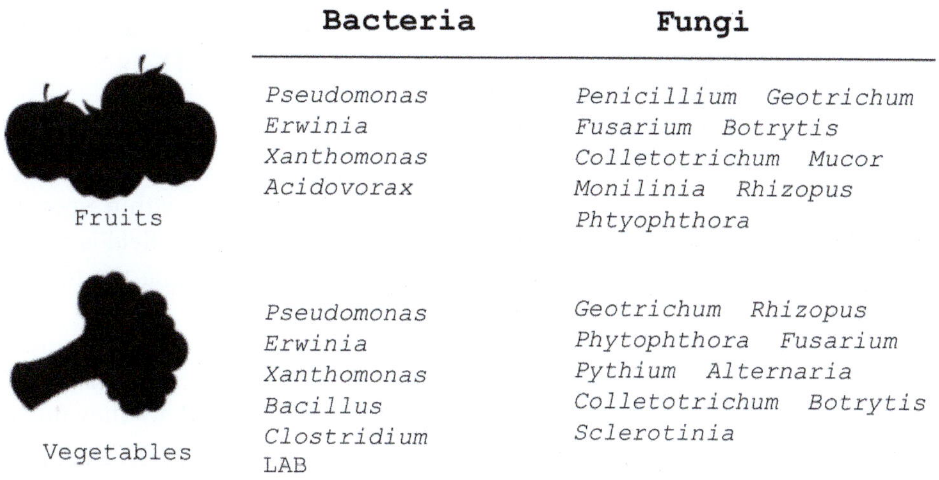

FIGURE 9.3 Main spoilage microorganisms of fruit and vegetables.

9.11.2 Spoilage and Pathogenic Bacteria in Fruit and Vegetables

Microorganisms are part of the ecological niche where fruits and vegetables grow. Bacterial groups from the phyla Actinobacteria, Bacteroidetes, Firmicutes, and Proteobacteria, and fungi such as Glomeromycota, Ascomycota, Basidiomycota, and Zygomycota are part of their complex microbiota. Spoilage microorganisms of fruit and vegetables are summarized in Figure 9.3.

Pathogens most commonly associated with fruit and vegetables are the enterohaemorrhagic bacteria *E. coli* O157:H7 and *S. enterica*, *B. cereus*, *Campylobacter*, *Cl. botulinum*, *L. monocytogenes*, *Yersinia pseudotuberculosis*, and *Y. enterocolitica* and *Shigella*. Among the viruses, hepatitis A virus and noroviruses, and fungi such as *Cyclospora cayatenensis*, *Cryptosporidium* are present.

Microbiological hazards on fruits and vegetables are divided into three risk categories according to FAO: level 1 (leafy green vegetables), level 2 (berries, green onions, melons, sprouted seeds, and tomatoes) and level 3 (carrots, cucumbers, almonds, baby corn, sesame seeds, onions and garlic, mango, *paw paw*, celery, and *maimai*), with level 1 being the most susceptible group.

9.12 FERMENTED FRUITS AND VEGETABLES PRODUCTS

Microbial groups that have been associated with fruit and vegetables fermentation include yeasts (*Saccharomyces*, *Pichia*, *Candida*, and *Torulaspora* genera), fungi (*Rhizopus*), and bacteria such as *Bacillus*, *Acetobacter*, and LAB. The latter are usually responsible for the spontaneous fermentation of raw vegetables and fruits, even if they are not the most abundant microbial group. Among them, *Leuconostoc*, *Lactobacillus* (frequently *Lb. plantarum*), *Weissella*, *Enterococcus* and *Pediococcus* are the most common LAB in fermentation.

Almost every culture around the globe has their own fermented fruit and vegetable products. Here, we highlight some of the most popular ones. *Kombucha* is a tea beverage produced by the spontaneous fermentation of black tea and white sugar, although many variants exist nowadays (Jakubczyk et al., 2020). A combination of bacteria and yeasts, known as the symbiotic culture of bacteria and yeast (SCOBY) form a jelly-type mass during fermentation that can be reused for several batches. SCOBY microorganisms transform the sugar into different compounds producing a sweet tea that contains little alcohol and carbon dioxide as compared to wine or beer. The microorganism profile is variable depending on the fermentation conditions and matrix, but in general acetic acid bacteria (*Acetobacter*, *Gluconobacter*), LAB (*Lactobacillus*, *Lactococcus*) and yeasts (*Candida*, *Saccharomyces*, *Zygosaccharomyces*) have been detected (Coton et al., 2017; Marsh et al., 2014). *Lb. plantarum* has been shown to produce the bacteriocin SLG10 which may play a role in *kombucha* preservation (Pei et al., 2020).

Sauerkraut is a preserved cabbage prepared with shredded cabbage and salt originally from Asian countries, but now spread all over the world. LAB such as *Leuconostoc*, *Lactobacillus*, and *Pediococcus* start the spontaneous fermentation and transform and preserve the cabbage. *Lactobacillus* and *Leuconostoc* isolates from *sauerkraut* have demonstrated probiotic potential and antimicrobial activity against the foodborne pathogen *L. monocytogenes* (Beganović et al., 2014; Touret et al., 2018; Yu et al., 2013), while the bacteriocin paracin 1.7, produced by *Lactobacillus paracasei* shows broad-spectrum antimicrobial activity that may play a role in *sauerkraut* preservation.

Kimchi is a salty combination of fermented cabbage and radish with different seasonings (garlic, spring onions, pepper, ginger, chili, etc.) originated in Korea. Its fermentation is led by the LAB *Leuconostoc*. A study with *Leuconostoc citreum* GJ7, a bacteriocin producer isolated from Kimchi was shown to decrease the level of pathogen microorganisms as compared to natural fermentation (Chang and Chang, 2011).

Other popular fermented vegetables include table-olives with *Lactobacillus pentosus* leading the fermentation (Lucena-Padrós et al., 2014), or pickling, an ancient fermentation method in which cucumbers or peppers are immersed in salty vinegar. Again the *Lactobacillus* group dominates this fermentation (Pérez-Díaz et al., 2020). Soybean LAB fermented by-products are also widely consumed (Rollán et al., 2019).

Fermentation can also use waste food, which would otherwise be discarded, as a substrate. For instance, pineapple peels are used in Latin America to produce vinegar. *Sac. cerevisiae* and *Acetobacter pasteurianus* are involved in this process (Tanamool et al., 2020). Fresh coconut leftovers can be fermented by *B. subtilis* to produce semayi, which is widely consumed in Indonesia (Deshpande et al., 2000).

9.13 APPLICATION OF PROTECTIVE CULTURES AND THEIR BACTERIOCINS IN FRUIT, VEGETABLES, AND BY-PRODUCTS

Raw fruits and vegetables can have lifespans from days to months if no treatment is applied. Factors such as the high water content, the ripening and senescence processes, environmental conditions or the harvesting procedure itself can favor the microbial growth and potential contamination and poison of fruit and vegetables.

Bioprotective LAB cultures and bacteriocins have shown great potential to extend the life span of fruit and vegetables, including through the fermentation of these foods (Tables 9.5 and 9.6). Some non-LAB cultures including *Pseudomonas* or *Candida* have also been suggested for bioprotection purposes (Table 9.5).

9.14 REGULATORY ISSUES IN BIOPROTECTION

The use of microbial food cultures and their bacteriocins in food relies on the different regulatory bodies across the world and within specific countries. The number of patents in this field is increasing, and slowly new products are reaching the market (López-Cuellar et al., 2016).

Microbial cultures are considered as food ingredients under the General Food Law, EU Regulation No. 178/2002, which says that "they must be safe for their intended use and the sole responsibility for this lies with the distributor." The safety of the selected strains is one of the key aspects to consider. In this sense, the qualified presumption of safety (QPS) approach developed by the EU can be considered a risk assessment tool for any microorganism which is aimed to enter to the market (EFSA, 2017, 2018). Similarly, in the US, the generally recognized as safe (GRAS) status exists for food and substances used in food (FDA, 2020). In both cases, regularly updated lists of QPS recommended biological agents or GRAS food substances are provided by the corresponding regulatory bodies (EFSA and FDA, respectively). The first QPS list made in 2007 (EFSA, 2007) consisted of 72 species of microorganisms. The last published update from December 2017 included 91 species (EFSA, 2017, 2018). It should be noted that microorganisms not included on the QPS list are not necessarily considered to be unsafe. It can be simply that a certain species has not been evaluated by EFSA. Currently, the QPS list is updated with the new taxonomic units two times a year, and it is reasonably conceivable that new bioprotective strains would obtain the QPS status if the necessary documentation is notified to EFSA (Françoise, 2010; Lau and Dunn, 2018). On the other hand, all GMMs and their by-products to be used in the food industry must comply with EU directive No 2001/18, EU regulation No. 1830/2003, and Commission regulation No. 641/2004. Table 9.7 shows examples of commercial bioprotective cultures.

With regard to bacteriocins, they are considered as food additives and must be approved under Regulation (EC) 1333/2008. Only nisin as E234 (Nisaplin®, Danisco) and natamycin, also called piramicin, as E235 (Natamax®, Danisco) are approved as food preservatives in Europe. The latter is produced by *Streptomyces* and it is active against yeasts and moulds. However, introduction

TABLE 9.5 Microorganisms used as bioprotective cultures in fruit and vegetables in research studies

Bioprotective culture	Target pathogen	Food	Reference
Lactobacillus plantarum	*Listeria monocytogenes*	Sliced apples Lamb's lettuce	(Siroli et al., 2015)
Leuconostoc *Lb. plantarum* *Lactococcus lactis* *Weisella*	*Salmonella typhimurium* *Escherichia coli* *L. monocytogenes*	Wounded Golden Delicious apples and Iceberg lettuce leaf cuts	(Trias et al., 2008)
Lb. plantarum *Lb. fermentum*	*L. monocytogenes E. coli* O157:H7	Pineapple pieces	(Russo et al., 2014)
Candida oleophila	*E. coli* O157:H7	Fuji apple wounds	(Wang et al., 2012)
Pseudomonas graminis	*L. monocytogenes* *S. enterica* subsp. *Enterica*	Golden Delicious apple pieces	(Alegre et al., 2013)
Lb. plantarum	*Staphylococcus aureus*	Fresh vegetable juice of carrots, endive, iceberg lettuce, garden rocke	(Scolari et al., 2004)
Ps. Fluorescens Ps. viridiflava	*L. monocytogenes*	Potato tuber slices	(Liao and Sapers, 1999)
Ps. fluorescens	*Salmonella* Chester and *L. monocytogenes*	Green pepper disks	(Liao and Fett, 2001)

TABLE 9.6 Bacteriocins used for pathogen control in fruit and vegetables in research studies

Bacteriocin	Target pathogen	Food
Enterocin AS-48	*Staphylococcus aureus* *Listeria monocytogenes* *Bacillus cereus*	Lettuce juice
Combination nisin–phytic acid and nisin–pediocin	*L. monocytogenes*	Cabbage and broccoli
Nisin	*L. monocytogenes*	Fresh-cut honeydew melon Fresh-cut red delicious apples
Bovicin HC5 Nisin	*Clostridium tyrobutyricum*	Mango pulp
Nisin	*Clostridium sporogenes* and *Cl. tyrobutyricum* spores *B. cereus* and *B. subtilis*	Mashed potato
Nisin	*Alicyclobacillus kacidoterrestris*	Orange and mixed commercial fruit drinks
Nisin	*A. acidoterrestris*	Orange, grapefruit and apple commercial juices
Enterocin CCM 4231	*L. monocytogenes* *Staph. aureus*	Soya milk
Nisin	*Bacillus stearothermophilus*	Soya milk

TABLE 9.7 Commercial bioprotective cultures

Product	Bioprotective bacteria	Target microorganism	Application
Lyoflora FP-18 (Sacco System)	*Carnobacterium*	*Listeria monocytogenes*	Fish products
F-LC SafePro® F-Lc (CHR Hansen)	*Pediococcus acidilactici, Lactobacillus curvatus*, and *Staphylococcus xylos*	*L. monocytogenes*	Sausages
Bioprox (Brioprox)	*Lactococcus rhamnosus*	Yeasts and moulds	Dairy products
Holdbac® YM-B (Dupont)	*Propionibacterium freudenreichii* subsp. *shermanii*	Yeasts and moulds	Dairy products

of a bacteriocin as a secondary metabolite, through *in situ* production by a strain or addition of a fermentate not designed specifically for antimicrobial application, is not formally regulated as a bacteriocin.

Pediocin (ALTA®2351, Kerry Bioscience) and a combination of the pediocin-sakacin producers *P. acidilactici, Lb. curvatus*, and *Staph. xylosus* (Bactoferm FLC®,Chr. Hansen) are commercialized for meat fermentation to improve organoleptic properties but can also contribute to protection against *L. monocytogenes* and. On the other hand, micocin, produced by *Carnobacterium maltaromaticum* and effective against *L. monocytogenes* is approved in Canada and commercialized as Microcin ® by Griffith Laboratories.

9.15 CONCLUSIONS

Food safety relies on good quality raw materials, good manipulation practises and preservation strategies to extend the shelf-life of food products. Bioprotection methods by bacteria and their by-products arise as a natural preservation and cost effective alternative with plenty of possibilities. LAB stand out as microbes that have been used in food making and consumed as part of fermented foods since ancient times. Because of this, they hold the GRAS and QPS status from FDA and EFSA, respectively. LAB relevant to food bioprotection include *Lactobacillus, Lactococcus, Leuconostoc, Pediococcus*, and *Streptococcus*. However, non-LAB microorganisms have also shown great potential and could be approved if they meet the required criteria. The use of bacteria isolated from the same food where they will be reintroduced as a preservative is preferred, as these bacteria are better adapted to the food ecological niche and would not disturb the microbial diversity. Their protective effect is due to their ability to produce different compounds that hamper the growth of other microorganisms. Bacteriocins are particularly interesting as killer peptides that could be used alone or in combination with other preservation methods. Many studies have aimed at the characterization of bacteria and bacteriocins on different foods and applied different strategies for food bioprotection at a research level but their broad use is still a chimera due to regulatory issues and high production costs.

ACKNOWLEDGMENTS

This publication has emanated from research conducted with the financial support of Science Foundation Ireland (SFI) under grant number SFI/12/RC/2273. SA has received funding from the European Union's Horizon 2020 research and innovation programme under the Marie Skłodowska-Curie grant agreement number 754535. EG-G has received funding from Enterprise Ireland and the European Union's Horizon 2020 research and innovation programme under the Marie Skłodowska-Curie grant agreement number 847402.

REFERENCES

1. Albano H, Van Reenen CA, Todorov SD, Cruz D, Fraga L, et al. 2009. Phenotypic and genetic heterogeneity of lactic acid bacteria isolated from "alheira," a traditional fermented sausage produced in Portugal. Meat Science 82: 389–398.
2. Alegre I, Viñas I, Usall J, Anguera M, Altisent R, et al. 2013. Antagonistic effect of Pseudomonas graminis CPA-7 against foodborne pathogens in fresh-cut apples under simulated commercial conditions. Food Microbiology 33: 139–148.
3. Anagnostopoulos DA, Tsaltas D. 2019. Chapter 10 - fermented foods and beverages. In: Galanakis CM (Ed.), *Innovations in Traditional Foods*. Woodhead Publishing, pp. 257–291.

4. Anihouvi VB, Sakyi-Dawson E, Ayernor GS, Hounhouigan JD. 2007. Microbiological changes in naturally fermented cassava fish (Pseudotolithus sp.) for lanhouin production. *International Journal of Food Microbiology* 116: 287–291.
5. Arqués JL, Rodríguez E, Langa S, Landete JM, Medina M. 2015. Antimicrobial activity of lactic acid bacteria in dairy products and gut: effect on pathogens. *Biomed Research International* 2015: 584183.
6. Arthur TM, Kalchayanand N, Bosilevac JM, Brichta-Harhay DM, Shackelford SD, Bono JL, Wheeler TL, Koohmaraie M. 2008. Comparison of effects of antimicrobial interventions on multidrug-resistant *Salmonella*, susceptible *Salmonella*, and *Escherichia coli* O157:H7. *Journal of Food Protection* 71(11): 2177–81.
7. Aymerich T, Martín B, Garriga M, Hugas M. 2003. Microbial quality and direct PCR identification of lactic acid bacteria and nonpathogenic Staphylococci from artisanal low-acid sausages. *Applied and Environmental Microbiology* 69: 4583–4594.
8. de Barboza Martinez Y, Ferrer K, Salas EM. 2002. Combined effects of lactic acid and nisin solution in reducing levels of microbiological contamination in red meat carcasses. *Journal of Food Protection* 65(11): 1780–1783.
9. Bari ML, Ukuku DO, Kawasaki T, Inatsu Y, Isshiki K, et al. 2005. Combined efficacy of nisin and pediocin with sodium lactate, citric acid, phytic acid, and potassium sorbate and edta in reducing the listeria monocytogenes population of inoculated fresh-cut produce. *Journal of Food Protection* 68: 1381–1387.
10. Beganović J, Kos B, Leboš Pavunc A, Uroić K, Jokić M, et al. 2014. Traditionally produced sauerkraut as source of autochthonous functional starter cultures. *Microbiological Research* 169: 623–632.
11. Benkerroum N, Oubel H, Mimoun LB. 2002. Behavior of *Listeria monocytogenes* and *Staphylococcus aureus* in yogurt fermented with a bacteriocin-producing thermophilic starter. *Journal of Food Protection* 65(5): 799–805.
12. Bosilevac JM, Nou X, Barkocy-Gallagher GA, Arthur TM, Koohmaraie M. 2006. Treatments using hot water instead of lactic acid reduce levels of aerobic bacteria and Enterobacteriaceae and reduce the prevalence of *Escherichia coli* O157:H7 on preevisceration beef carcasses. *Journal of Food Protection* 69(8): 1808–13.
13. Bourrie BC, Willing BP, Cotter PD. 2016. The Microbiota and Health Promoting Characteristics of the Fermented Beverage Kefir. *Frontiers in Microbiology* 7: 647.
14. Bourrie, BCT, Cotter PD, Willinga, BP. 2018. Traditional kefir reduces weight gain and improves plasma and liver lipid profiles more successfully than a commercial equivalent in a mouse model of obesity. *Journal of Functional Foods* 46: 29–37.
15. Brillet A, Pilet M-F, Prevost H, Bouttefroy A, Leroi F. 2004. Biodiversity of *Listeria monocytogenes* sensitivity to bacteriocin-producing *Carnobacterium* strains and application in sterile cold-smoked salmon. *Journal of Applied Microbiology* 97: 1029–1037.
16. De Buyser ML, Dufour B, Maire M, Lafarge V. 2001. Implication of milk and milk products in food-borne diseases in France and in different industrialised countries. *International Journal of Food Microbiology* 67(1-2): 1–17.
17. Carvalho AATD, Mantovani HC, Vanetti MCD. 2007. Bactericidal effect of bovicin HC5 and nisin against *Clostridium tyrobutyricum*isolated from spoiled mango pulp. *Letters in Applied Microbiology* 45: 68–74.
18. Carlson BA, Geornaras I, Yoon Y, Scanga JA, Sofos JN, Smith GC, Belk KE. 2008. Studies to evaluate chemicals and conditions with low-pressure applications for reducing microbial counts on cattle hides. *Journal of Food Protection* 71(7): 1343–8.
19. Castellano P, Pérez Ibarreche M, Blanco Massani M, Fontana C, Vignolo GM. 2017. Strategies for pathogen biocontrol using lactic acid bacteria and their metabolites: A focus on meat ecosystems and industrial environments. *Microorganisms* 5: 38.
20. Castellano P, Vignolo G. 2006. Inhibition of *Listeria innocua* and *Brochothrix thermosphacta* in vacuum-packaged meat by addition of bacteriocinogenic *Lactobacillus curvatus* CRL705 and its bacteriocins. *Letters in Applied Microbiology* 43: 194–199.
21. Chang JY, Chang HC. 2011. Growth inhibition of food-borne pathogens by kimchi prepared with bacteriocin-producing starter culture. *Journal of Food Science* 76: M72–78.
22. Chuon MR, Shiomoto M, Koyanagi T, Sasaki T, Michihata T, et al. 2014. Microbial and chemical properties of Cambodian traditional fermented fish products. *Journal of Science of Food and Agriculture* 94: 1124–1131.
23. Cocolin L, Manzano M, Cantoni C, Comi G. 2001. Denaturing gradient gel electrophoresis analysis of the 16S rRNA gene V1 region to monitor dynamic changes in the bacterial population during fermentation of Italian sausages. *Applied and Environmental Microbiology* 67: 5113–5121.
24. Comi G. 2017. Chapter 8 - spoilage of meat and fish. In: Bevilacqua A, Corbo MR, Sinigaglia M (Eds.), *The Microbiological Quality of Food*. Woodhead Publishing, pp. 179–210.
25. Coton M, Pawtowski A, Taminiau B, Burgaud G, Deniel F, et al. 2017. Unraveling microbial ecology of industrial-scale Kombucha fermentations by metabarcoding and

culture-based methods. *FEMS Microbiology Ecology* 93: fix048.

26. Cotter PD, Hill C, Ross RP. 2005. Bacteriocins: Developing innate immunity for food. *Nature Reviews Microbiology* 3: 777–788.
27. Cotter PD, Ross RP, Hill C. 2013. Bacteriocins - a viable alternative to antibiotics? *Nature Reviews Microbiology* 11(2):95–105.
28. Da Costa RJ, Voloski FLS, Mondadori RG, Duval EH, Âm F. 2019. Preservation of meat products with bacteriocins produced by lactic acid bacteria isolated from meat. *Journal of Food Quality* article ID 4756210.
29. De Smet K, Contreras R. 2005. Human antimicrobial peptides: Defensins, cathelicidins and histatins. *Biotechnology Letters* 27: 1337–1347.
30. Deshpande SS, Salunkle DK, Oyewole OB, Azam-Ali S, Battcock M, et al. 2000. *Fermented Grain Legumes, Seeds and Nuts: A Global Perspective.* Food and Agriculture Organisation of the United Nations.
31. Diaz-Ruiz G, Omar NB, Abriouel H, Martinez Canamero M, Galvez A. 2012. Inhibition of *Listeria monocytogenes* and *Escherichia coli* by bacteriocin-producing *Lactobacillus plantarum* EC52 in a meat sausage model system. *African Journal of Microbiology Research* 6: 1103–1108.
32. Dicks LMT, Mellett FD, Hoffman LC. 2004. Use of bacteriocin-producing starter cultures of *Lactobacillus plantarum* and *Lactobacillus curvatus* in production of ostrich meat salami. *Meat Science* 66: 703–708.
33. Dortu C, Huch M, Holzapfel WH, Franz CMP, Thonart P. 2008. Anti-listerial activity of bacteriocin-producing *Lactobacillus curvatus* CWBI-B28 and *Lactobacillus sakei* CWBI-B1365 on raw beef and poultry meat. *Letters in Applied Microbiology* 47: 581–586.
34. EFSA, 2005. QPS-Qualified Presumption of Safety micro-organisms in food and feed. EFSA Scientific Colloquium, Summary Report.
35. EFSA. 2007. Introduction of a Qualified Presumption of Safety (QPS) approach for assessment of selected microorganisms referred to EFSA. *EFSA Journal* 587: 1–16.
36. EFSA. 2017. Scientific Opinion on the update of the list of QPS-recommended biological agents intentionally added to food or feed as notified to EFSA. *EFSA Journal* 15: 4664.
37. EFSA BIOHAZ Panel (EFSA Panel on Biological Hazards). 2018. Statement on the update of the list of QPS-recommended biological agents intentionally added to food or feed as notified to EFSA 7: suitability of taxonomic units notified to EFSA until September 2017. *EFSA Journal* 18: 5131.
38. El Bassi L, Hassouna M, Shinzato N, Matsui T. 2009. Biopreservation of refrigerated and vacuum-packed *Dicentrarchus labrax* by lactic acid bacteria. *Journal of Food Science* 74: M335–339.
39. Endo A, Irisawa T, Dicks L, Tanasupawat S. 2014. Fermented foods | Fermentations of east and southeast Asia. In: Batt CA, Tortorello ML (Eds.), *Encyclopedia of Food Microbiology*, Second Edition. Academic Press, Oxford, pp. 846–851.
40. FAO, 2019. Food safety is everyone's business. http://www.fao.org/news/story/en/item/1196961/icode (accessed 7.9.20).
41. FAO, 2020. The state of world fisheries and aquaculture 2020. Contributing to food security and nutrition for all. Sustainability in action.
42. Faye B, Konuspayeva G. 2012. The sustainability challenge to the dairy sector – The growing importance of non-cattle milk production worldwide. *International Dairy journal* 24(2):50–56.
43. FDA. 2020. SCOGS (Select Committee on GRAS Substances). Available from: https://www.cfsanappsexternal.fda.gov/scripts/fdcc/?set=SCOGS.
44. Fernández-López J, Sendra E, Sayas-Barberá E, Navarro C, Pérez-Alvarez JA. 2008. Physico-chemical and microbiological profiles of "salchichón" (Spanish dry-fermented sausage) enriched with orange fiber. *Meat Science* 80: 410–417.
45. Fontana C, Cocconcelli PS, Vignolo G. 2005. Monitoring the bacterial population dynamics during fermentation of artisanal Argentinean sausages. *International Journal of Food Microbiology* 103: 131–142.
46. Françoise L. 2010. Occurrence and role of lactic acid bacteria in seafood products. *Food Microbiology* (6):698–709.
47. Gálvez A, López RL, Pulido RP, Burgos MJG. 2014. Biopreservation of milk and dairy products. In: *Food Biopreservation. Springer Briefs in Food, Health, and Nutrition*. Springer, New York, NY.
48. Garcia-Gutierrez E. 2019. *New Antimicrobials to Target Gut and Food Pathogens*. University of East Anglia.
49. Gençcelep H, Kaban G, Aksu Mİ, Öz F, Kaya M. 2008. Determination of biogenic amines in sucuk. *Food Control* 19: 868–872.
50. Ghanbari M, Jami M. 2013. *Lactic Acid Bacteria and Their Bacteriocins: A Promising Approach to Seafood Biopreservation*. Lactic Acid Bacteria - R & D for Food, Health and Livestock Purposes.

51. Ghanbari M, Jami M, Domig KJ, Kneifel W. 2013. Seafood biopreservation by lactic acid bacteria – A review. *LWT - Food Science and Technology* 54: 315–324.
52. Godfray HCJ, Aveyard P, Garnett T, Hall JW, Key TJ, et al. 2018. Meat consumption, health, and the environment. *Science* 361: eaam5324.
53. Gómez-Sala B, Hermida JF, Cruza PEH, Izarra LMC, Hermida JF, et al. 2019. Lactic acid bacteria in aquatic environments and their applications. In: Vinderola G, Ouwehand AC, Salminen S, Von Wright A (Eds.), *Lactic Acid Bacteria. Microbiological and Functional Aspects.* CRC Press, pp. 555–577.
54. Gómez-Sala B, Herranz C, Diaz-Freitas B, Hernandez PE, Sala A, et al. 2016. Strategies to increase the hygienic and economic value of fresh fish: Biopreservation using lactic acid bacteria of marine origin. *Internationa Journal of Food Microbiology* 223: 41–49.
55. Gómez-Sala B, Muñoz-Atienza E, Sánchez J, Basanta A, Herranz C, et al. 2015. Bacteriocin production by lactic acid bacteria isolated from fish, seafood and fish products. *European Food Research and Technology* 241: 341–356.
56. Grace D. 2017. *Food Safety and the Sustainable Development Goals.* International Livestoc Research Institute.
57. Gram L, Dalgaard P. 2002. Fish spoilage bacteria–problems and solutions. *Current Opinion in Biotechnology* 13: 262–266.
58. Gram L, Huss HH. 1996. Microbiological spoilage of fish and fish products. *Internationa Journal of Food Microbiology* 33: 121–137.
59. Grande MJ, Lucas R, Valdivia E, Abriouel H, Maqueda M, et al. 2005. Stability of enterocin as-48 in fruit and vegetable juices. *Journal of Food Protection* 68: 2085–2094.
60. Hartmann HA, Wilke T, Erdmann R. 2011. Efficacy of bacteriocin-containing cell-free culture supernatants from lactic acid bacteria to control *Listeria monocytogenes* in food. *Internationa Journal of Food Microbiology* 146: 192–199.
61. Henchion M, McCarthy M, Resconi VC, Troy D. 2014. Meat consumption: Trends and quality matters. *Meat Science* 98: 561–568.
62. Huck JR, Sonnen M, Boor KJ. 2008. Tracking heat-resistant, cold-thriving fluid milk spoilage bacteria from farm to packaged product. *Journal of Dairy Science* 91(3):1218–28.
63. Hugas M, Garriga M, Aymerich MT, Monfort JM. 1995. Inhibition of *Listeria* in dry fermented sausages by the bacteriocinogenic *Lactobacillus sake* CTC494. *Journal of Applied Microbiology* 79: 322–330.
64. Huss HH, 1995. Quality and quality changes in fresh fish. FAO Fisheries Technical Paper 348. Food and Agriculture Organization of the United Nations.
65. Huss HH, Reilly A, Karim Ben Embarek P. 2000. Prevention and control of hazards in seafood. *Food Control* 11: 149–156.
66. Ibarra-Sanchez LA, El-Haddad N, Mahamoud D, Miller MJ, Karam L. 2020. Invited review: Advances in nisin use for preservation of dairy products. *Journal or Daity Science* 103(3): 2041–2052.
67. Jacobsen T, Budde BB, Koch AG. 2003. Application of *Leuconostoc carnosum* for biopreservation of cooked meat products. *Journal of Applied Microbiology* 95: 242–249.
68. Jakubczyk K, Kałduńska J, Kochman J, Janda K. 2020. Chemical profile and antioxidant activity of the kombucha beverage derived from white, green, black and red tea. *Antioxidants (Basel)* 9: 447.
69. Kalschne DL, Geitenes S, Veit MR, Sarmento CMP, Colla E. 2014. Growth inhibition of lactic acid bacteria in ham by nisin: A model approach. *Meat Science* 98: 744–752.
70. Kannappan S, Sivakumar K, Sivagnanam S. 2017. Effect of *Lactobacillus rhamnosus* cells against specific and native fish spoilage bacteria and their spoilage indices on Asian seabass fish chunks. *Journal of Environmental Biology* 38: 841–847.
71. Kanninen EP, Puolanne E, 2008. Principles of meat fermentation. In: Toldrà F, Hui YH, Astiasaran I, Sebranek JG, Talon R (Eds.), *Handbook of Fermented Meat and Poultry.* John Wiley & Sons, Ltd, pp. 31–36.
72. Katla T, Moretro T, Aasen IM, Holck A, Axelsson L, et al. 2001. Inhibition of *Listeria monocytogenes* in cold smoked salmon by addition of sakacin P and/or live *Lactobacillus sakei* cultures. *Food Microbiology* 18: 431–439.
73. Komitopoulou E, Boziaris IS, Davies EA, Delves-Broughton J, Adams MR. 1999. *Alicyclobacillus acidoterrestris* in fruit juices and its control by nisin. *International Journal of Food Science* 34: 81–85.
74. Kopermsub P, Yunchalard S. 2010. Identification of lactic acid bacteria associated with the production of plaa-som, a traditional fermented fish product of Thailand. *Internationa Journal of Food Microbiology* 138: 200–204.
75. Kumar P, Chatli MK, Mehta N, Singh P, Malav OP, Verma AK. 2017. Meat analogues: Health promising sustainable meat substitutes. *Critital Reviews in Food Science and Nutrition* 57: 923–932.

76. Laranjo M, Potes ME, Elias M. 2019. Role of starter cultures on the safety of fermented meat products. *Frontiers in Microbiology* 10: 853.

77. Lau JL, Dunn MK. 2018. Therapeutic peptides: Historical perspectives, current development trends, and future directions. *Bioorganic & Medical Chemistry* 26: 2700–2707.

78. Lauková A, Czikková S. 1999. The use of enterocin CCM 4231 in soy milk to control the growth of *Listeria monocytogenes* and *Staphylococcus aureus*. *Journal of Applied Microbiology* 87: 182–186.

79. Lebert I, Leroy S, Giammarinaro P, Lebert A, Chacornac JP, et al. 2007. Diversity of microorganisms in the environment and dry fermented sausages of small traditional French processing units. *Meat Science* 76: 112–122.

80. Leroy F, De Vuyst L. 2016. Advances in production and simplified methods for recovery and quantification of exopolysaccharides for applications in food and health. *Internationa Journal of Dairy Science* 99: 3229–3238.

81. Leverentz B, Conway WS, Camp MJ, Janisiewicz WJ, Abuladze T, et al. 2003. Biocontrol of *Listeria monocytogenes* on fresh-cut produce by treatment with lytic bacteriophages and a bacteriocin. *Applied and Environmental Microbiology* 69: 4519–4526.

82. Liao CH, Fett WF. 2001. Analysis of native microflora and selection of strains antagonistic to human pathogens on fresh produce. *Journal of Food Protection* 64: 1110–1115.

83. Liao CH, Sapers GM. 1999. Influence of soft rot bacteria on growth of *Listeria monocytogenes* on potato tuber slices. *Journal of Food Protection* 62: 343–348.

84. López-Cuellar MDR, Rodríguez-Hernández A-I, Chavarría-Hernández N. 2016. LAB bacteriocin applications in the last decade. *Biotechnology and Biotechnological Equipment* 30: 1039–1050.

85. Lucena-Padrós H, Caballero-Guerrero B, Maldonado-Barragán A, Ruiz-Barba JL. 2014. Microbial diversity and dynamics of Spanish-style green table-olive fermentations in large manufacturing companies through culture-dependent techniques. *Food Microbiology* 42: 154–165.

86. Maragkoudakis PA, Mountzouris KC, Psyrras D, Cremonese S, Fischer J, Cantor MD, Tsakalidou E. 2009. Functional properties of novel protective lactic acid bacteria and application in raw chicken meat against *Listeria monocytogenes* and *Salmonella enteritidis*. *Internationa Journal of Food Microbiology* 130: 219–226.

87. Marsh AJ, O'Sullivan O, Hill C, Ross RP, Cotter PD. 2014. Sequence-based analysis of the bacterial and fungal compositions of multiple kombucha (tea fungus) samples. *Food Microbiology* 38: 171–178.

88. Matamoros S, Leroi F, Cardinal M, Gigout F, Kasbi Chadli F, et al. 2009. Psychrotrophic lactic acid bacteria used to improve the safety and quality of vacuum-packaged cooked and peeled tropical shrimp and cold-smoked salmon. *Journal of Food Protection* 72: 365–374.

89. Mathur H, Beresford TP, Cotter PD. 2020. Health Benefits of Lactic Acid Bacteria (LAB) Fermentates. *Nutrients* 12(6):1679.

90. Mei J, Shen Y, Liu W, Lan W, Li N, et al. 2020. Effectiveness of sodium alginate active coatings containing bacteriocin Efl4 for the quality improvement of ready-to-eat fresh salmon fillets during cold storage. *Coatings* 10: 506.

91. Muñoz A, Maqueda M, Gálvez A, Martínez-Bueno M, Rodríguez A, Valdivia E. 2004. Biocontrol of psychrotrophic enterotoxigenic *Bacillus cereus* in a nonfat hard cheese by an enterococcal strain-producing enterocin AS-48. *Journal of Food Protection.*

92. Ndaw A, Zinedine A, Faid M, Bouseta A. 2008. Effect of controlled lactic acid bacterial fermentation on the microbiological and chemical qualities of Moroccan sardines (*Sardina pilchardus*). *Acta Microbiologica Et Immunologica Hungarica* 55: 295–310.

93. Nilsson L, Ng YY, Christiansen JN, Jørgensen BL, Grótinum D, Gram L. 2004. The contribution of bacteriocin to inhibition of *Listeria monocytogenes* by *Carnobacterium piscicola* strains in cold-smoked salmon systems. *Journal of Applied Microbiology* 96: 133–143.

94. Nørrung B, Buncic S. 2008. Microbial safety of meat in the European Union. *Meat Science* 78: 14–24.

95. Odeyemi OA, Alegbeleye OO, Strateva M, Stratev D. 2020. Understanding spoilage microbial community and spoilage mechanisms in foods of animal origin. *Comprehensive Reviews in Food Science and Food Safety* 19(2):311–331.

96. Ohshima T, Giri A. 2014. Fermented foods | Traditional fish fermentation technology and recent developments. In: Batt CA, Tortorello ML (Eds.), *Encyclopedia of Food Microbiology (Second Edition)*. Academic Press, Oxford, pp. 852–869.

97. Ozhahmetova, Z, Kasenova G. 2013. Selection of lactic acid bacteria and yeast for koumiss starter and its impact on quality of koumiss. *Applied Technologies and Innovations* 9(4): 138.

98. Pasolli E, De Filippis F, Mauriello IE, Cumbo F, Walsh AM, Leech J, Cotter PD, Segata N, Ercolini D. 2020. Large-scale genome-wide analysis links lactic acid bacteria from food with the gut microbiome. *Nature Communications* 11(1):2610.

99. Pei J, Jin W, Abd El-Aty AM, Baranenko DA, Gou X, et al. 2020. Isolation, purification, and structural identification of a new bacteriocin made by *Lactobacillus plantarum* found in conventional kombucha. *Food Control* 110: 106923.
100. Pérez-Díaz IM, Dickey AN, Fitria R, Ravishankar N, Hayes J, et al. 2020. Modulation of the bacterial population in commercial cucumber fermentations by brining salt type. *Journal of Applied Microbiology* 128: 1678–1693.
101. Pothakos V, Devlieghere F, Villani F, Björkroth J, Ercolini D. 2015. Lactic acid bacteria and their controversial role in fresh meat spoilage. *Meat Science* 109: 66–74.
102. Rantsiou K, Cocolin L. 2008. Fermented meat products. In: Cocolin L, Ercolini D (Eds.), *Molecular Techniques in the Microbial Ecology of Fermented Foods*. Springer New York, New York, pp. 91–118.
103. Rantsiou K, Comi G, Cocolin L. 2004. The *rpoB* gene as a target for PCR-DGGE analysis to follow lactic acid bacterial population dynamics during food fermentations. *Food Microbiology* 21: 481–487.
104. Rantsiou K, Urso R, Iacumin L, Cantoni C, Cattaneo P, et al. 2005. Culture-dependent and -independent methods to investigate the microbial ecology of Italian fermented sausages. *Applied and Environmental Microbiology* 71: 1977–1986.
105. Ricci A, Chemaly M, Davies R, Psf E, Girones R, et al. 2017. Hazard analysis approaches for certain small retail establishments in view of the application of their food safety management systems. *EFSA Journal* 15: e04697.
106. Riebroy S, Benjakul S, Visessanguan W. 2008. Properties and acceptability of Som-fug, a Thai fermented fish mince, inoculated with lactic acid bacteria starters. *LWT - Food Science and Technology* 41: 569–580.
107. Rollán GC, Gerez CL, LeBlanc JG. 2019. Lactic fermentation as a strategy to improve the nutritional and functional values of pseudocereals. *Frontiers in Nutrition* 6: 98.
108. Russo P, Mlv DC, Vernile A, Amodio ML, Arena MP, et al. 2014. Fresh-cut pineapple as a new carrier of probiotic lactic acid bacteria. *Biomed Research Internatinal* 2014: 309183.
109. Santé-Lhoutellier V. 2014. Human nutrition | Meat and human diet: Facts and myths. In: Dikeman M, Devine C (Eds.), *Encyclopedia of Meat Sciences*, Second Edition. Academic Press, Oxford, pp. 118–123.
110. Sarika AR, Lipton AP, Aishwarya MS, Dhivya RS. 2012. Isolation of a bacteriocin-producing *Lactococcus lactis* and application of its bacteriocin to manage spoilage bacteria in high-value marine fish under different storage temperatures. *Applied Biochemistry and Biotechnology* 167: 1280–1289.
111. Scolari G, Vescovo M, Zacconi C, Bonadé A. 2004. Influence of *Lactobacillus plantarum* on *Staphylococcus aureus* growth in a fresh vegetable model system. *European Food Research Technology* 218: 274–277.
112. Settanni L, Moschetti G. 2010. Non-starter lactic acid bacteria used to improve cheese quality and provide health benefits. *Food Microbiology* 27(6): 691–7.
113. Sheikha E. 2014. African fermented fish products in scope of risks. *International Food Research Journal* 21: 425.
114. Silva C, Silva S, Ribeiro S. C. 2018. Application of bacteriocins and protective cultures in dairy food preservation. *Frontiers in microbiology* 9, 594.
115. Siroli L, Patrignani F, Serrazanetti DI, Tabanelli G, Montanari C, et al. 2015. Lactic acid bacteria and natural antimicrobials to improve the safety and shelf-life of minimally processed sliced apples and lamb's lettuce. *Food Microbiology* 47: 74–84.
116. Skåra T, Axelsson L, Stefánsson G, Ekstrand B, Hagen H. 2015. Fermented and ripened fish products in the northern European countries. *Journal of Ethnic Food* 2: 18–24.
117. Sofos JN. 2014a. Safety of food and beverages: Meat and meat products. In: Motarjemi Y (Ed.), *Encyclopedia of Food Safety*. Academic Press, Waltham, pp. 268–279.
118. Sofos JN. 2014b. Chapter 6 - Meat and Meat Products. In: Motarjemi Y, Lelieveld H (Eds.), *Food Safety Management*. Academic Press, San Diego, pp. 119–162.
119. Stiles ME. 1996. Biopreservation by lactic acid bacteria. *Antonie Van Leeuwenhoek* 70: 331–345.
120. Tahiri I, Desbiens M, Benech R, Kheadr E, Lacroix C, et al. 2004. Purification, characterization and amino acid sequencing of divergicin M35: A novel class IIa bacteriocin produced by *Carnobacterium divergens* M35. *International Journal of Food Microbiology* 97: 123–136.
121. Talon R, Leroy S. 2014. Fermented foods | Fermented meat products and the role of starter cultures. In: Batt CA, Tortorello ML (Eds.), *Encyclopedia of Food Microbiology (Second Edition)*. Academic Press, Oxford, pp. 870–874.
122. Tanamool V, Chantarangsee M, Soemphol W. 2020. Simultaneous vinegar fermentation from a pineapple by-product using the co-inoculation of yeast and thermotolerant acetic acid bacteria and their physiochemical properties. *3 Biotechnology* 10: 115.
123. Tanasupawat S, Pakdeeto A, Namwong S, Thawai C, Kudo T, et al. 2006. Lentibacillus halophilus sp. nov., from fish sauce in Thailand. *International Journal of Systematic and Evolutionary Microbiology* 56: 1859–1863.
124. Thapa N, Pal J, Tamang JP. 2004. Microbial Diversity in Ngari, Hentak and Tungtap, Fermented Fish Products of North-East India. *World Journal of Microbiology and Biotechnology* 20: 599.

125. Thomas LV, Ingram RE, Bevis HE, Davies EA, Milne CF, et al. 2002. Effective use of nisin to control *Bacillus* and *Clostridium* spoilage of a pasteurized mashed potato product. *Journal of Food Protection* 65: 1580–1585.
126. Todorov SD, Koep KSC, Van Reenen CA, Hoffman LC, Slinde E, et al. 2007. Production of salami from beef, horse, mutton, Blesbok (*Damaliscus dorcas phillipsi*) and Springbok (*Antidorcas marsupialis*) with bacteriocinogenic strains of *Lactobacillus plantarum* and *Lactobacillus curvatus*. *Meat Science* 77: 405–412.
127. Tomé E, Gibbs PA, Teixeira PC. 2008. Growth control of *Listeria innocua* 2030c on vacuum-packaged cold-smoked salmon by lactic acid bacteria. *Internationa Journal of Food Microbiology* 121: 285–294.
128. Touret T, Oliveira M, Semedo-Lemsaddek T. 2018. Putative probiotic lactic acid bacteria isolated from sauerkraut fermentations. *Plos ONE* 13: e0203501.
129. Trias R, Bañeras L, Badosa E, Montesinos E. 2008. Bioprotection of Golden Delicious apples and Iceberg lettuce against foodborne bacterial pathogens by lactic acid bacteria. *Internationa Journal of Food Microbiology* 123: 50–60.
130. Tu L, Mustapha A. 2002. Reduction of *Brochothrix thermosphacta* and *Salmonella* serotype Typhimurium on vacuum-packaged fresh beef treated with nisin and nisin combined with EDTA. *Journal of Food Science* 67: 302–306.
131. United Nations, 2015. Envision 2030: 17 goals to transform the world for persons with disabilities. United Nations Enable https://www.un.org/development/desa/disabilities/envision2030.html (accessed 7.10.2020).
132. Vescovo M, Scolari G, Zacconi C. 2006. Inhibition of *Listeria innocua* growth by antimicrobial-producing lactic acid cultures in vacuum-packed cold-smoked salmon. *Food Microbiology* 23: 689–693.
133. Wang Y, Wei A, Li H. 2012. Using *Candida oleophila* as a biocontriol agens to prevent foodborne *Escherichia coli* O157 EHEC infections. *Springerplus* 1: 82.
134. Wiernasz N, Cornet J, Cardinal M, Pilet M-F, Passerini D, Leroi F. 2017. Lactic acid bacteria selection for biopreservation as a part of a hurdle technology approach applied on seafood. *Frontiers in Marine Sciences* 4: 119.
135. Xu Y, Zang J, Regenstein JM, Xia W. 2020. Technological roles of microorganisms in fish fermentation: A review. *Critical Reviews in Food Science and Nutrition*. doi: 10.1080/10408398.2020.1750342.
136. Yamazaki K, Murakami M, Kawai Y, Inoue N, Matsuda T. 2000. Use of nisin for inhibition of *Alicyclobacillus acidoterrestris* in acidic drinks. *Food Microbiology* 17: 315–320.
137. Yamazaki K, Suzuki M, Kawai Y, Inoue N, Montville TJ. 2003. Inhibition of *Listeria monocytogenes* in cold-smoked salmon by *Carnobacterium piscicola* CS526 isolated from frozen surimi. *Journal of Food Protection* 66: 1420–1425.
138. Yu Z, Zhang X, Li S, Li C, Li D, et al. 2013. Evaluation of probiotic properties of *Lactobacillus plantarum* strains isolated from Chinese sauerkraut. *World Journal of Microbiology and Biotechnology* 29: 489–498.
139. Zagorec M, Champomier-Vergès MC. 2017. *Lactobacillus sakei*: A starter for sausage fermentation, a protective culture for meat products. *Microorganisms* 5: 56.
140. Woraprayote W, Pumpuang L, Tosukhowong A, Zendo T, Sonomoto K, Benjakul S, Visessanguan W. 2018. Antimicrobial biodegradable food packaging impregnated with bacteriocin 7293 for control of pathogenic bacteria in pangasius fish fillets. *LWT-Food Science and Technology* 89: 427–433.
141. Zang J, Xu Y, Xia W, Regenstein JM. 2020. Quality, functionality, and microbiology of fermented fish: A review. *Critical Reviews in Food Science and Nutrition* 60: 1228–1242.
142. Zeng X, Xia W, Yang F, Jiang Q. 2013. Changes of biogenic amines in Chinese low-salt fermented fish pieces (Suan yu) inoculated with mixed starter cultures. *International Journal of Food Science and Technology* 48: 685–692.

CHAPTER 10

Aromatic Yeasts

Revealing Their Flavor Potential in Food Fermentations

Amparo Gamero[1], Mónica Flores[2], and Carmela Belloch[2]
[1]Dep. Preventive Medicine and Public Health, Food Science, Toxicology and Forensic Medicine, Faculty of Pharmacy, Universitat de València, Avda. Vicent Andrés Estellés s/n, 46100 Burjassot (Valencia), Spain
[2]Instituto de Agroquímica y Tecnología de Alimentos (CSIC), Avenue Agustín Escardino 7, 46980 Paterna (Valencia), Spain

10.1 INTRODUCTION

The flavor of fermented foods may differ depending on many factors, especially the raw material used for fermentation (vegetal and animal raw materials), the ingredients and additives used in the formulation, the processing conditions, and finally, the microorganisms used in the fermentation. The main components of the raw materials used in terms of protein, lipid, and carbohydrates are essential for flavor formation. These components are hydrolyzed during the processing and generate aroma precursors such as free fatty acids, free amino acids and sugars. All of these are substrates of chemical and microbial reactions producing volatile aroma compounds (Flores, 2009).

The main players in the generation of fermentative aromas are microorganisms. In addition to lactic acid bacteria, yeasts are key in the production of fermented food (Terefe and Augustin, 2020). Yeasts play the main role in the fermentation of fruits and vegetables for production of alcoholic beverages. In fermented foods from animal sources, such as meat and dairy products, different yeast species contribute to the generation of different aromas. In Asian countries many fermented food products are based on yeast and mold fermentation processes. In summary, the fungal biota in fermented food plays an important role in flavor development (Copetti, 2019).

Flavor includes the impression of taste, odor, tactile, and other impressions; however, this chapter will focus on the volatile compounds producing aroma sensations in fermented foods and the contribution of yeast to the aroma sensations.

10.2 YEAST AROMA IN ALCOHOLIC BEVERAGES

The most important fermented beverages worldwide are wine and beer. Wine is the resulting product of fermenting crushed grapes whereas beer is produced from malted barley (see also chapters 17 and 18). Yeasts are crucial for the development of these products and their acceptance by the consumers. This group of microorganisms not only carries out the conversion of fermentable sugars to ethanol and carbon dioxide, but also enables the release of volatiles from odorless precursors present in the raw materials as well as *de novo* synthesize a high diversity of aromatic

Good Microbes in Medicine, Food Production, Biotechnology, Bioremediation, and Agriculture, First Edition. Edited by Frans J. de Bruijn, Hauke Smidt, Luca S. Cocolin, Michael Sauer, David Dowling, and Linda Thomashow.

compounds. In this way, the contribution of yeast to global aroma of alcoholic beverages is fundamental and determines in a great deal the overall quality and acceptability of these products. Figure 10.1 depicts the effect in aroma profile of wine must fermented by *S. cerevisiae.*

10.2.1 Yeast: Saccharomyces and Non-Saccharomyces

The main yeast genus involved in the fermentation of alcoholic beverages is *Saccharomyces*, especially the species *S. cerevisiae* in wine and ale beer and the hybrid *S. pastorianus* (*S. cerevisiae* × *S. eubayanus*) in lager beer (Capece et al., 2018; Cubillos et al., 2019; Holt et al., 2019). Recent publications have demonstrated the relevance and potential in these processes of other *Saccharomyces* hybrids (Gamero et al., 2013; Su et al., 2019) as well as non-*Saccharomyces* yeasts. The latter frequently lack good fermentative performance, but they display excellent skills for aroma enhancement when combined with *Saccharomyces* species in either sequential or mixed fermentations. Some outstanding examples of these non-conventional yeasts are *Torulaspora delbrueckii*, *Lachancea thermotolerans*, or *Metschnikowia pulcherrima*

FIGURE 10.1 Modification of the aroma profile of wine must (A) after alcoholic fermentation by *S. cerevisiae* (B) (adapted from Gamero et al., 2011a).

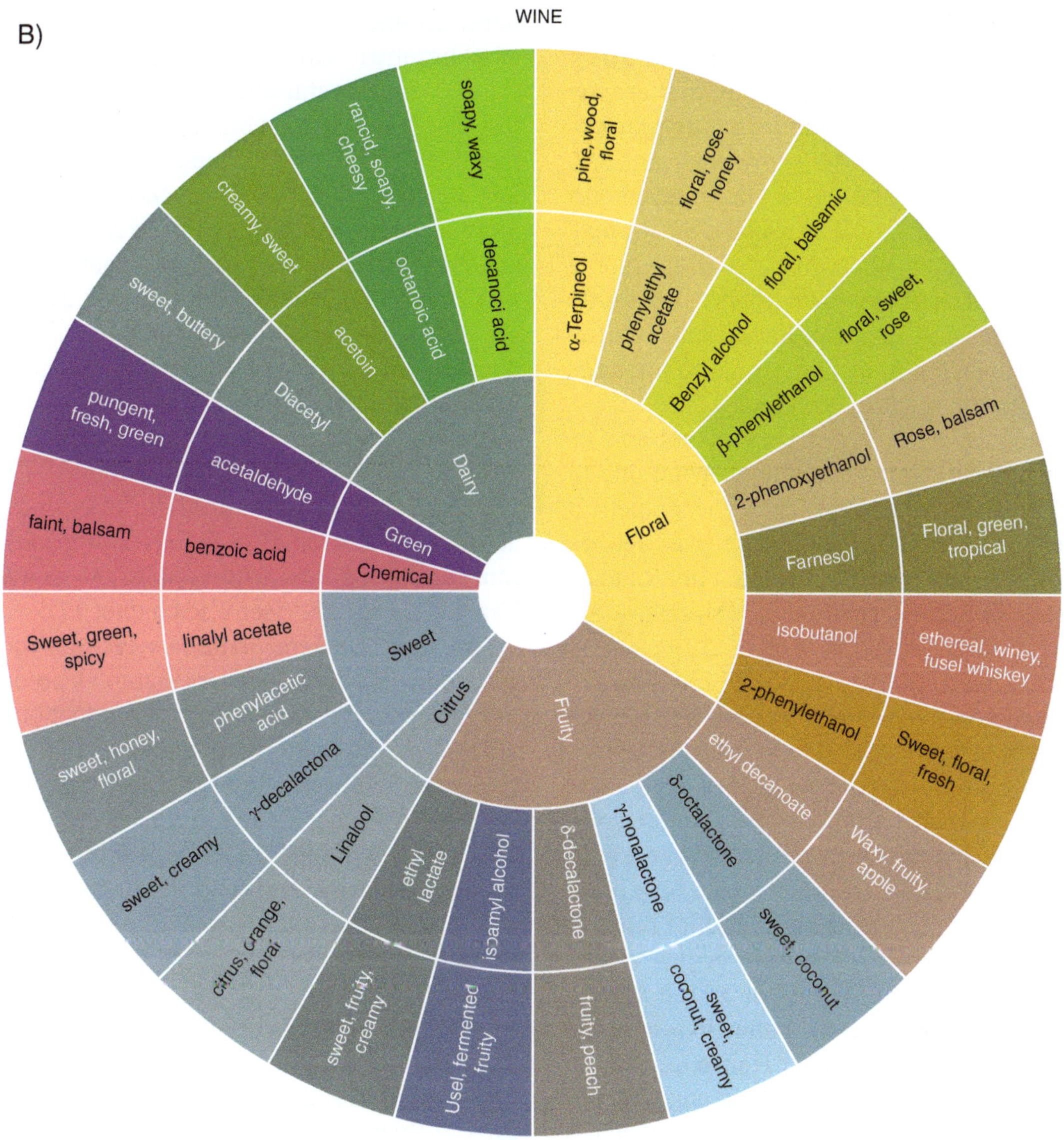

FIGURE 10.1 (Continued)

among others (Capece et al., 2018; Cubillos et al., 2019; Morata et al., 2020; Tempere et al., 2018; Van Wyk et al., 2019). Some of these yeasts are nowadays commercially available as dry or frozen products to be used as starters in alcoholic fermentations and their presence in the market is rapidly growing (Morata et al., 2020).

10.2.2 Aromatic Precursors

In wines, varietal aroma is the aroma coming from the grapes. However, grapes also contain aroma precursors in form of glycoconjugated terpenes and glutathionated or cysteinated thiols, among others. These compounds are not volatile and lack aroma, but some yeasts are able to either detach sugars from their link to the aromatic part called aglycone or eliminate glutathionation or cysteinylation, respectively. In this way, terpenes and thiols become volatile and are able to contribute to increase varietal aroma (Liu et al., 2017; Van Wyk et al., 2019). This action is done through the action of several enzymes naturally occurring in some yeasts, such as ß-glucosidases or carbon sulfur-lyases (Morata et al., 2020; Van Wyk et al., 2019).

Released terpenes from glycosides through the action of yeast enzymes results in a significant contribution to the fruity and floral notes, especially in certain wine must varieties, such as Muscat and Gewürztraminer (Parker et al., 2018). Some examples of released terpenes in wines are linalool, geraniol, nerol, citronellol, and α-terpineol (Liu et al., 2017). Among yeasts, ß-glucosidases of the non-*Saccharomyces* yeasts *Pichia anomala, Candida molischiana, Candida wickerhamii, Hanseniaspora uvarum*, and *M. pulcherrima* have been shown to be relevant in increasing wine varietal aroma through this strategy (Liu et al., 2017; Morata et al., 2020). In addition, some *Saccharomyces* hybrids have shown potential not only in releasing terpenes but also in either *de novo* synthesization or bioconversion into other aroma compounds (Gamero et al., 2011a, b; see also chapter 17).

Volatile thiols, such as 3-mercaptohexanol and its acetate, are very desirable compounds in wine due to their strong and pleasant passion fruit and grapefruit aromatic notes (Morata et al., 2020). The yeast species *Pichia Kluyveri* and *T. delbrueckii* have been proved to be efficient in releasing these compounds (Morata et al., 2020; Tempère et al., 2018; Van Wyk et al., 2019).

On the other hand, odorless glycoside and glutathionated or cysteinated compounds can also occur in the hops employed in brewing as well as in the wood of barrels used for maturation (Capece et al., 2018; Holt et al., 2019). The hydrolysis of hop glycosides is most commonly observed in spontaneously fermented beers with fruit maceration, such as Lambic and Gueuze base beers with cherries (Kriek), or in sour beer varieties (Holt et al., 2019; see also chapter 18). The yeast genus responsible is *Dekkera* (*Brettanomyces*), such as the species *Dekkera bruxellensis*, which has the ability to synthesize ß-glucosidases (Capece et al., 2018; Holt et al., 2019). Regarding the release of thiols, there is a commercialized patented yeast of the species *P. kluyveri*, which is able to enhance thiol levels in joint beer fermentation with *S. pastorianus* (Holt et al., 2019).

10.2.3 Fermentative Aroma Compounds

In addition to increasing varietal aroma, yeasts are able to synthesize *de novo* several aromatic compounds. The majority of these are higher alcohols, acetate esters and ethyl esters (Capece et al., 2018; Morata et al., 2020). Higher alcohols are mainly formed from amino acid catabolism by the Ehrlich pathway and can afterwards be esterified with acetic acid to produce low threshold aromatic esters, such as 2-phenylethyl acetate, isoamyl acetate, and isobutyl acetate, which confer fruity and floral aromatic notes. Ethyl esters are formed by acylation with acyl-CoA of fatty acids and confer sweet and floral aromas. Some examples are ethyl butyrate and ethyl hexanoate (Morata et al., 2020).

In wine, several attempts have been carried out in the last years employing non-conventional yeasts to increase fruity aroma through their high ester synthesis. For instance, *Hanseniaspora vineae* and *L. thermotolerans* excelled in the production of 2-phenylethyl acetate and ethyl lactate, giving rose, strawberry, and toffee aromatic notes to the wine (Morata et al., 2020; Van Wyk et al., 2019). *Wickerhamomyces anomalus* not only produces high quantity of 2-phenylethyl acetate but also isoamyl acetate (banana) and ethyl acetate (fruity), whereas *M. pulcherrima* is a good producer of the higher alcohol 2-phenylethanol (rose aroma). *T. delbrueckii* improves global aroma through the synthesis of linalyl acetate (citric aroma), 3-ethoxy-1-propanol (blackcurrant) and ethyl hexanoate (apple) (Morata et al., 2020; Van Wyk et al., 2019; see also chapter 17).

In beer, *Dekkera/Brettanomyces*, the most relevant species involved in sour beers (Lambic), produces a peculiar flavor profile due to its high production of phenolic aromas in combination with high amounts of ethyl esters, presenting fruity and floral characteristics (Capece et al., 2018; see also chapter 18). *T. delbrueckii* yields low-alcohol beers presenting high levels of amyl alcohol and fruity flavor, often associated with wheat beer. *L. thermotolerans* appears as an alternative to the addition of lactic acid bacteria in sour beers (Capece et al., 2018; Cubillos et al., 2019).

Yeasts diversity is not limited to the few species isolated from food. There are many yeast species poorly investigated but few attempts have been carried out to explore their potential for bioflavoring in alcoholic fermentations (Dashko et al., 2015; Gamero et al., 2016, 2020).

10.3 YEAST AROMA IN FOODS FROM ANIMAL SOURCES

The most popular foods from animal sources are meat and dairy products, and among these the best representative of yeast fermentation are fermented sausages (see Chapter 13) and cheese (see Chapter 12).

Dry and semidry sausages are produced using lean meat and pork back fat, with several ingredients (salt, sugar, spices, etc.) and starter cultures, followed by stuffing in casings, a fermentation stage, and the final ripening–drying stage (Flores and Olivares, 2015; see also chapter 13). The combination of starter cultures like lactic acid bacteria, and yeast and molds together with different processing parameters (temperature, relative humidity, etc.) produce dry and semidry sausages with completely different sensory characteristics (Toldrá and Flores, 2014). In fermented sausage flavor, the hydrolysis of proteins and lipids produce free fatty acids and amino acids that act as substrates of chemical and microbial reactions (Flores and Olivares, 2015).

Dairy products like cheese are produced by the fermentation of milk followed by different types of ripening stages (see also chapter 12). Coagulation of milk can be achieved by acidification using starter cultures of lactic acid bacteria, addition of enzymes such as rennet or lyophilized *Cynara cardunculus* or other vegetable coagulants (Roseiro et al., 2003). The rate and extent of fermentation and acid development are essential steps to develop the desired sensory characteristics of cheese (Johnson, 2017).

Flavor development occurs mainly during the ripening process. As described for fermented sausages, the ripening of cheese is carried out with the participation of cheese microbiota which plays an important role in the formation of flavor compounds. Cheese microbiota participates in the fermentation of lactose, the consumption of lactate, and the proteolysis and lipolysis reactions providing flavor precursors (Smit et al., 2005).

10.3.1 Yeast: Debaryomyces and Kluyveromyces

The fermentation step for the production of meat and dairy products involves mixed populations of microorganisms, namely lactic acid bacteria, yeasts, and fungi. Generally, lactic acid bacteria play the main role in the fermentation of foods from animal sources, although other bacteria such as staphylococci or *Brevibacterium* contribute to the production of meat products and cheese, respectively (Cocconcelli and Fontana, 2008; Cogan, 2011). Many yeast species have been frequently isolated from meat and dairy products, but very few species have been found relevant for their contribution to food aroma. The yeast species primarily investigated for its beneficial contribution to meat and dairy production is *Debaryomyces hansenii*, although *Kluyveromyes marxianus* and *K. lactis* have relevance in dairy products due to their ability to ferment lactose (Belloch et al., 2011).

The importance of yeasts in the manufacture of sausages and cheese, among other foods from animal source, has been recognized since the 1970s with the use of *D. hansenii*, *Geotrichum candidum*, and *K. lactis* as starter cultures (Fröhlich-Wyder et al., 2018; Hammes and Kauf, 1994).

The contribution of these yeasts to food aroma is based on their proteolytic and lipolytic potential, as well as their ability to transform sugars and amino acids into aroma compounds (Bertuzzi et al., 2017; Centeno et al., 2017; Flores et al., 2017; Padilla et al., 2014; Perea-Sanz et al., 2019).

10.3.2 Fermentation Aroma Compounds

Aroma formation in food from animal source begins with the enzymatic hydrolysis of protein and fat during the fermentation and ripening/drying stages. The free amino acids and free fatty acids generated act as substrates for further chemical and microbial reactions generating aroma compounds. The chemical fat auto-oxidation that affects primarily the unsaturated free fatty acids contributes to a desirable flavor being modulated by the fatty acid profile. Also, fermentation of carbohydrates and the presence of compounds derived from spices contributes to animal fermented flavor (Flores and Olivares, 2015; Johansen, 2018).

The microbial metabolism is responsible for the generation of many volatile compounds. Different reactions are involved but essentially are: fermentation of carbohydrates, microbial amino acids degradation, lipid beta-oxidation, and staphylococci esterase activity (Flores, 2018; Ravyts et al., 2012). The generation of the aroma compounds is highly dependent on the substrates present in the raw food components and the microbial strains from *Lactobacillus*, *Staphylococcus*, and yeast present or inoculated in the food (Flores et al., 2015). In meat products, the fermentation of carbohydrates is performed by lactic acid bacteria in anaerobiosis producing lactic acid and other volatile compounds with strong aroma properties (diacetyl, acetaldehyde, ethanol, acetic, propionic, and butanoic acids, etc.) (Ravyts et al., 2012).

Microbial amino acid degradation is mainly carried out by coagulase negative staphylococci and yeasts (*D. hansenii*) generating many aroma compounds through the degradation of amino acids (Flores et al., 2015). Generally, these reactions are performed through a transamination followed by a decarboxylation of the amino acid, producing aromatic aldehydes. The lipid beta-oxidation is an oxidation reaction performed by the microbial metabolism (mainly *Staphylococcus* and yeasts), generating short chain fatty acids and beta-ketoacids that are

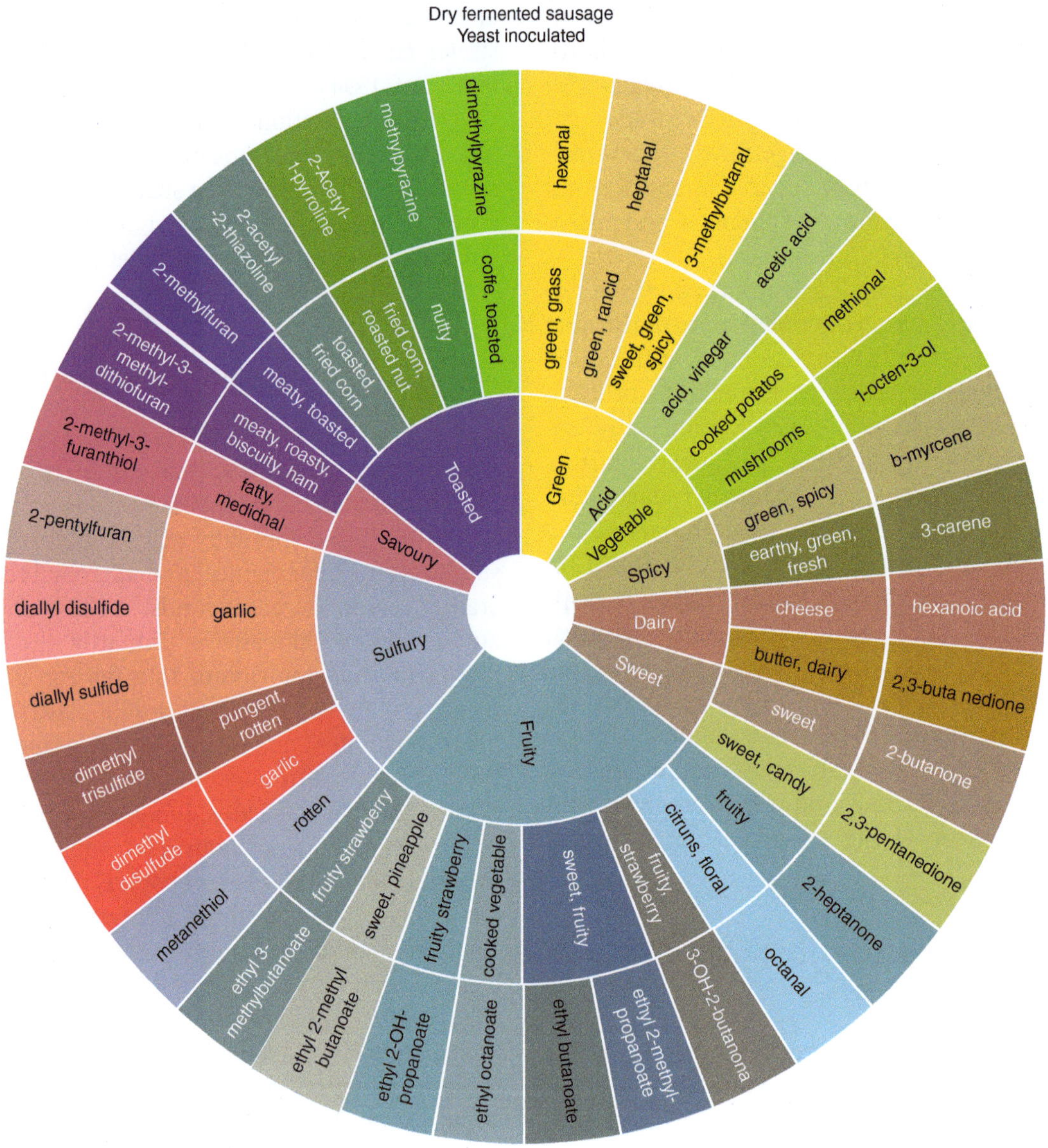

FIGURE 10.2 Aroma profile of dry fermented sausage (adapted from Perea-Sanz et al., 2018, 2020).

further decarboxylated producing methyl ketones that can be reduced to secondary alcohols (Montel et al., 1998). Another important microbial activity is the *Staphylococcus* esterase, which produces ester compounds (mainly ethyl esters) through the reaction of an acid and alcohol (mainly ethanol) (Ravyts et al., 2012).

The aroma of fermented sausages is mainly produced by the combination of compounds that produce different aroma notes (Figure 10.2). Among the aroma compounds, those derived from oxidation reactions may produce a detrimental effect on the fermented aroma. In those cases, the control of oxidation in fermented animal products, can be achieved by using yeast such *D. hansenii* and molds that growth on the surface. These molds protect lipids against light and oxidation by consuming oxygen, together with the physical protection against access of oxygen to the fat (Cano-García et al., 2014; Corral et al., 2015).

The aroma profile in dry fermented sausages is composed of many aroma notes like savory (meaty), sulfur, toasted, green, acid, vegetable, spicy, dairy, sweet, and fruity (Flores, 2018). These aroma notes are produced individually by each volatile compound. However, their presence in the meat product produces interactions among compounds including the possibility of synergistic, antagonistic or additive effects. Furthermore, the interactions of these volatile compounds with the meat matrix results in the final fermented aroma. Many different chemical classes of volatile compounds produced the aroma notes as indicated in Figures 10.2 and 10.4 for fermented meat and cheese, respectively. The contribution of aldehydes to green fresh odors is characteristic, while short chain fatty acids produced dairy cheesy notes. At the same time, ketone produced sweet aroma odors and fruity notes are produced by ester compounds. The potent meat and toasted notes are produced by sulfur and nitrogen compounds as well as heterocyclic compounds that have been detected as potent odorants in fermented products.

In recent years, yeasts have been used to enhance the aroma of fermented sausages (see also chapter 13). In addition to the ability of *D. hansenii* to inhibit lipid oxidation in fermented sausages, by reducing the formation of aldehydes, one strain produced ethyl acetate, ethyl 2-methylpropanoate and 3-methylbutyl acetate contributing to fruity odor notes in fermented sausages (Cano-García et al., 2014). Other *D. hansenii* strains produce dimethyl disulfide and dimethyl trisulfide, contributing to vegetable and cured odors. Moreover, D. *hansenii* has been used in reformulated fermented sausages where fat and salt were reduced. *D. hansenii* contributed to fruity odor notes by the production of ethyl ester compounds and decrease the green odors produced by linear aldehydes derived from oxidation reactions (Corral et al., 2015, 2017). The yeast effect was observed especially in the generation of compounds derived from amino acid degradation (3-methylbutanoic acid, methional, and benzothiazole) and by their ester activity (ethyl 2-methylpropanoate, ethyl 2-methylbutanoate and ethyl 3-methylbutanoate). These results are seen in Figure 10.3 that represents a heat map with the aroma produced in different inoculated and non-inoculated fermented sausages with reduced content of fat and salt. Based on the aroma profile, two groups can be clearly distinguished: non-inoculated and inoculated sausages. The main differences are in the highest presence of ester compounds, branched-chain acids and sulfur compounds in the yeast inoculated sausages. These compounds have been detected as important contributors to the aroma as seen in Figure 10.2.

D. hansenii yeast has also been used to mask flavors in fermented sausages such as the boar taint odor by producing ester compounds and fruity odors (Corral et al., 2018) and enhance the aroma of nitrite reduced fermented sausages (Perea-Sanz et al., 2020). In recent years, the potential of *D. hansenii* isolated from fermented sausages to produce volatiles with meaty odor notes, like sulfur compounds, has been shown by the degradation of methionine and the production of the thioester methyl thioacetate (Perea-Sanz et al., 2019). However, the impact of this sulfur compound on fermented sausage aroma is not clear yet. In summary, yeast contributed to fruity notes in fermented sausages but at the same time modulated the sausage aroma profile by reducing oxidized aromas and increased the conversion of amino acids into aroma compounds (Figure 10.3).

The volatile profile of cheese is also very complex and is composed of hundreds of volatile compounds (Figure 10.4; see also chapter 12). The aroma is mainly produced by the combination of compounds that impact different aroma notes. As can be seen, many of volatile compounds present in fermented meat are also important in the aroma profile of cheese. Nevertheless, it can be seen that cheese aroma is highly characterized by the presence of short chain fatty acids contributing to cheesy notes while longer chain acids produce fatty aroma odors.

However, other volatile compounds are also relevant to the overall cheese aroma. For instance, the compounds derived from amino acids like 3-methylbutanal and 3-methylbutanol, sulfur compounds (methanethiol, dimethyl sidulfide, and trimethyltrisulfide) are important in Gouda aroma and besides acids (butyric acid) (Smit et al., 2005). Other compounds are essential for aroma in other types of cheese. Dimethyl trisulfide has been indicated as an impact compound as other sulfur compounds such as methionol (methylthiopropanol) are characteristic in blue cheese aroma. Also, nitrogen compounds like pyrazines, such as trimethyl pyrazine, tetramethylpyrazine, and diethylmethyl pyrazines, have aroma impact in hard grating cheeses (Frank et al., 2004; Figure 10.4).

In addition to *D. hansenii*, *K. marxianus* and *G. candidum* are frequently used as starter cultures in cheese. The main contribution of *D. hansenii* to cheese aroma is the production of branched-chain aldehydes and alcohols (Padilla et al., 2014). Additionally, Bertuzzi et al. (2018) demonstrated the contribution of *D. hansenii* to the generation of carboxylic acids and alcohols. The ability of *K. marxianus* to ferment lactose plays an important role in the potential of this yeast to produce aromas in cheese. *K. marxianus* has been found responsible for generation of carboxylic acids from lactose, ethanol, esters and acetaldehyde producing acidic, cidery, alcoholic fermented, and fruity flavors (Belloch et al., 2011; Binetti et al., 2013; Cholet et al., 2007; Padilla et al.,

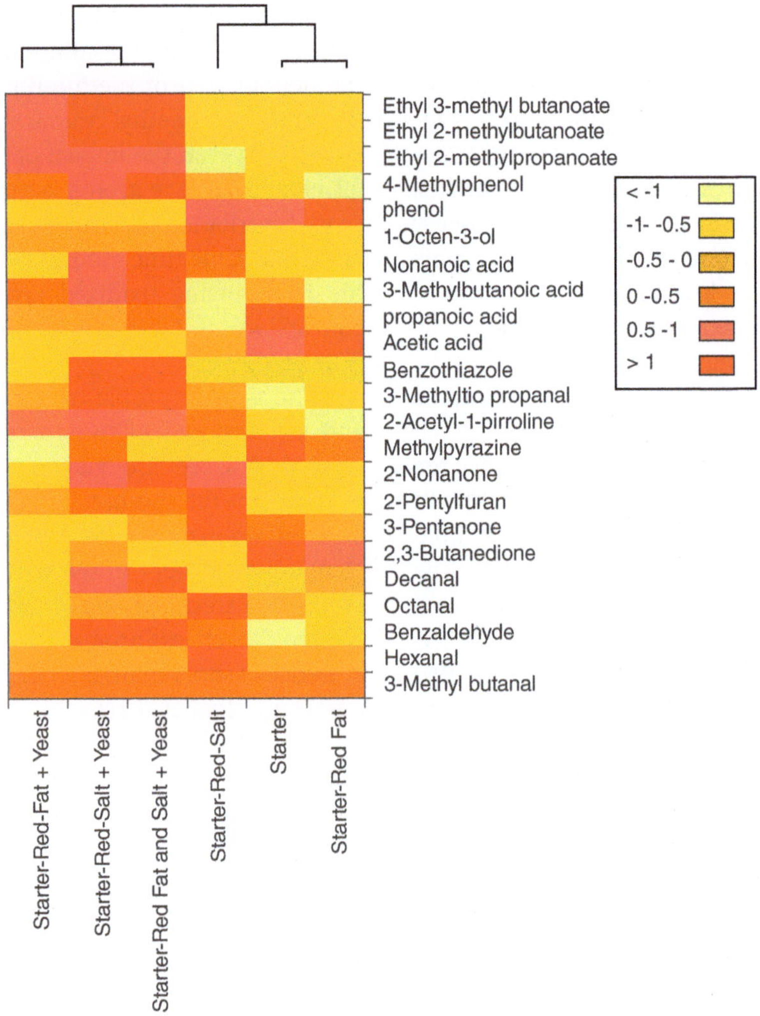

FIGURE 10.3 *D. hansenii* effect on the aroma profile of fermented sausages (adapted from Corral et al., 2015a). The colors represent the following: red indicates relatively high abundance whereas yellow indicates relatively low abundance of each aroma compound and orange indicates no difference.

2014). In contrast, *Geotrichum candidum* exhibits an important proteolytic and lipolytic activity and contributes to the generation of secondary alcohols, phenyl compounds, lactones, esters, sulfur compounds (methanethiol, dimethyl disulphide), and ketones (Arfi et al., 2006; Bertuzzi et al., 2018; Eliskases-Lechner et al., 2011). These compounds have been associated with cheese, moldy, yeasty, fermented, cidery, and fruity aromas.

10.4 YEAST AROMA IN OTHER FERMENTATIONS

Vegetables are the main staple for the production of many fermented foods. Among the most well-known foods from vegetable sources are soy sauce, cocoa, and coffee because of their worldwide popularity. However, many other foods from vegetable sources are traditionally produced around the world (Hittinger et al., 2018; Steinkraus, 1997). Among these, fermentation of cabbages, olives, peppers, legumes, and cereals, are some of the most representative. Vegetable fermentations are usually carried out by complex microbial communities composed mainly by lactic acid bacteria and yeasts (Chang et al., 2008; Devanthi and Gkatzionis, 2019; Elhalis et al., 2020; Gandul-Rojas and Gallardo-Guerrero, 2020; Ho et al., 2014; Kim et al., 2019; Lara-Hidalgo et al., 2020; Lee et al., 2020; Li et al., 2017; Palla et al., 2020). The yeast species involved in vegetable fermentations are diverse and, generally, more than one species actively contributes to the biochemical changes

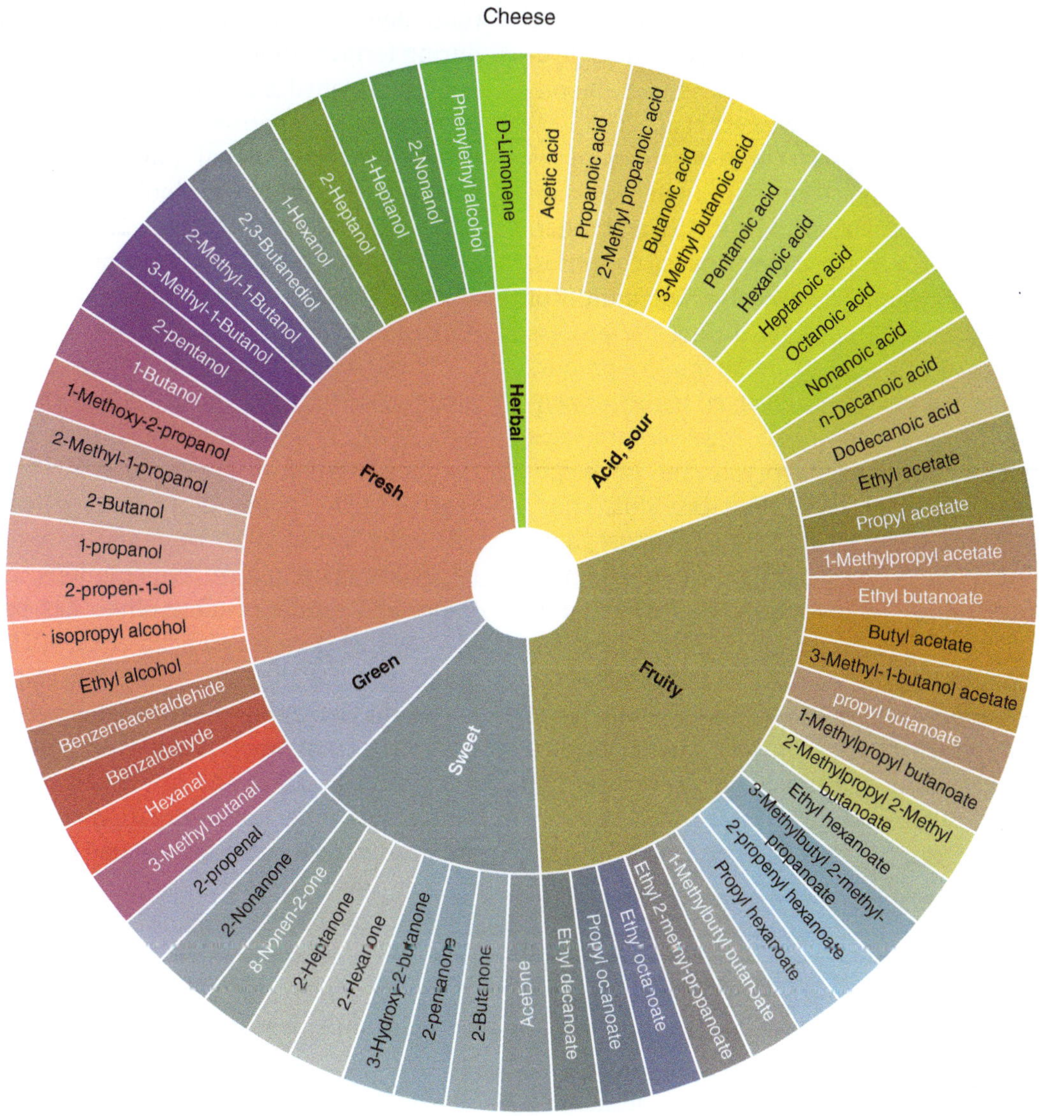

FIGURE 10.4 Volatile profile of ewe's and goat cheeses (adapted from Padilla et al., 2014).

involved in the generation of aroma compounds during fermentation. The fermentation of vegetables with autochthonous yeast starters produces many changes in the sensory characteristics (Terefe and Augustin, 2020). However, the production of volatile compounds depends on fermentation time and inoculated yeast species (Benitez-Cabello et al., 2019; Lara-Hidalgo et al., 2020).

10.4.1 Vegetables

The aroma of vegetable fermentations is complex and strongly influenced by the composition of pulp sugars and the breakdown of the plant proteins.

In the case of cocoa, glucose and fructose fermentation by *H. guilliermondii*, *P. kudriavzevii*, and *K. marxianus* produces higher alcohols, esters, aldehydes, and ketones, which have been found key contributors to chocolate aroma and flavor (see also chapter 16). Yeasts have proven to be essential for cocoa bean fermentation, not only by the generation of the fermentative aromas. Moreover, they also contribute to the generation of precursors of pyrazines and other key compounds for the aromas produced during bean roasting (Ho et al., 2014).

The wet fermentation of coffee beans involves the crucial participation of *H. uvarum* and *P. kudriavzevii*. The presence of these yeasts during the fermentation step increased the proportion of isoamyl alcohol, ethanol, acetaldehyde, and ethyl acetate in the coffee beans, and the generation of compounds such as glycerol, a key component of the sweet taste and smooth mouthfeel (Elhalis et al., 2020; Swiegers et al., 2005).

Soy sauce is greatly appreciated in Asian countries for its intense umami taste. Four main ingredients are used in its preparation, soybeans, wheat, salt, and water. The fermentation of soy sauce varies between countries, although in all cases a complex microbial community of fungi, yeast, and bacteria develops during the process (Devanthi and Gkatzionis, 2019). In the modern production of soy sauce, the key yeasts involved in the fermentative step of its preparation are *Zygosaccharomyces rouxii* and several *Candida*, *Meyerozyma*, and *Trichosporon* species (Tanaka et al., 2012; Van Der Sluis et al., 2001; Wei et al., 2013; Yang et al., 2017). The most abundant aroma compounds found during the initial fermentative step are alcohols, followed by esters in the middle stage and alkanes and heterocyclic compounds at the end of fermentation (Wei et al., 2013). In addition of the fermentative aromas, nearly 300 aroma compounds are detected in soy sauce such as phenols, furanones, pyrazines, pyridines, and sulfur-containing compounds (Nunomura et al., 1980).

10.4.2 Traditional Fermentations

A large number of traditional foods are produced at household scale, as a way to preserve overabundant food staples from spoilage. Some are typical products of geographical areas, and others are only produced for family celebrations or local festivities. These fermented foods provide not only nutrients but also a diversity of flavors, aromas and textures enriching the human diet (Steinkraus, 1997; see also chapter 11).

Korean kimchi is produced by the fermentation of kimchi cabbage, radishes, leeks and chives (Lee et al., 2020). Kimchi microbial communities have been described as composed by lactic acid bacteria and yeast from the genera *Pichia*, *Kazachstania*, and *Rhodotorula* (Moon et al., 2014), as well as, *Lodderomyces*, *Trichosporon*, *Saccharomyces*, *Candida*, and *Kluyveromyces* (Chang et al., 2008). Analysis of Kimchi detected 77 aroma active compounds (Cha et al., 1998). Most of them were sulfur-containing compounds, but also fermentative aromas such as alcohols, aldehydes, and ketones were detected. The strong odor of cooked cabbage, garlic, and green onion represented by some sulfur compounds (dimethyl trisulfide and dimethyl tetrasulfide), decreased during fermentation which impacted positively Kimchi quality. Aldehydes generated during fermentation were reported as major aroma compounds in Kimchi. Among these, 3-methyl-butanal (dark chocolate aroma) has been reported from microbiological origin (Collin et al., 1993). Ketones, particularly 2,3-butanedione (buttery and cheese aroma) and alcohols including terpene alcohols (fruity and floral aroma) also increased during fermentation.

Table olives are an essential component of the Mediterranean diet (Gandul-Rojas and Gallardo-Guerrero, 2020). The main microbial groups participating in the fermentation of olives are lactic acid bacteria and yeasts (Arroyo-Lopez et al., 2012; see also chapter 15). The yeast species *Wicherhamomyces anomalus* seems to be predominant in olive fermentations, and its impact in the aroma profile of table olives has been thoroughly investigated (Benítez-Cabello et al., 2019; De Angelis et al., 2015; Tufariello et al., 2015). Inoculation of *W. anomalus* in the fermentation of olives increased the formation of volatile compounds, such as alcohols (1-butanol and 2-methyl-1-butanol) and esters (2-phenylethyl acetate). Moreover, sequential inoculation of lactic acid bacteria and yeasts produced differentiated high quality products with improved aromatic profile.

Chili peppers (*Capsicum annum sp.*) are an integral part of the Mexican diet. They are consumed fresh, in sauces, powdered and pickled (Laborde and Pozo, 1982). Guajillo chili pepper sauce is usually the product of spontaneous fermentation. The microbiota participating in this fermentation is usually composed of lactic acid bacteria and yeasts (Watts et al., 2018). The inoculation of different yeast strains (*Hanseniaspora opuntiae*, *P. kudriavzevii*, and *W. anomalus*) in a minced chili pepper for the manufacture of a chili sauce, was key for modification of the aroma profile (Lara-Hidalgo et al., 2020). The yeast fermentation process produced a change in the aroma profile as seen in Figure 10.5. An increase in green, fruity, and herbal notes were observed after fermentation changing the aroma profile and producing the fermented chili sauce aroma. After fermentation, the increase of impact aroma compounds ethyl 2-methylbutanoate (fruity), and 6-methyl-5-hepten-2-one (strong, citrus) was important. However, aroma compounds production depended on the type of inoculated yeast. Among the studied yeast, two produced a similar aroma profile in Chili Guajillo sauce, *P.*

kudriavzevii and *W. anomalus*, with fruity, green-herbal, and cheesy odor notes based on ester compounds, alcohol, and branched-chain acids production. However, *W. anomalus* inoculation increased the cheesy character by the production of propanoic acid. In contrast, *H. opuntiae* inoculation produced a different aroma profile characterized by herbal and green notes, based on aldehydes, ketones, and acetic acid production, and a light fruity character.

Cereals and legumes are major staple crops and a primary source of nutrients for millions of people all over the world (Adebo et al., 2020). In addition to the production of alcoholic beverages, cereals, and legumes are used for the production of sourdough breads, pastes, porridges, and soups all over the world (Adebo et al., 2020; Feng et al., 2006; Li et al., 2017; Özdemir et al., 2018; Palla et al., 2020; Wakil et al., 2008). The microbial communities found in these foods are

FIGURE 10.5 Aroma profile of *Capsicum annum* (A) (adapted from Zimmerman and Schieberle, 2000) and yeast fermented *Capsicum annum* (B) (adapted from Lara-Hidalgo et al., 2020).

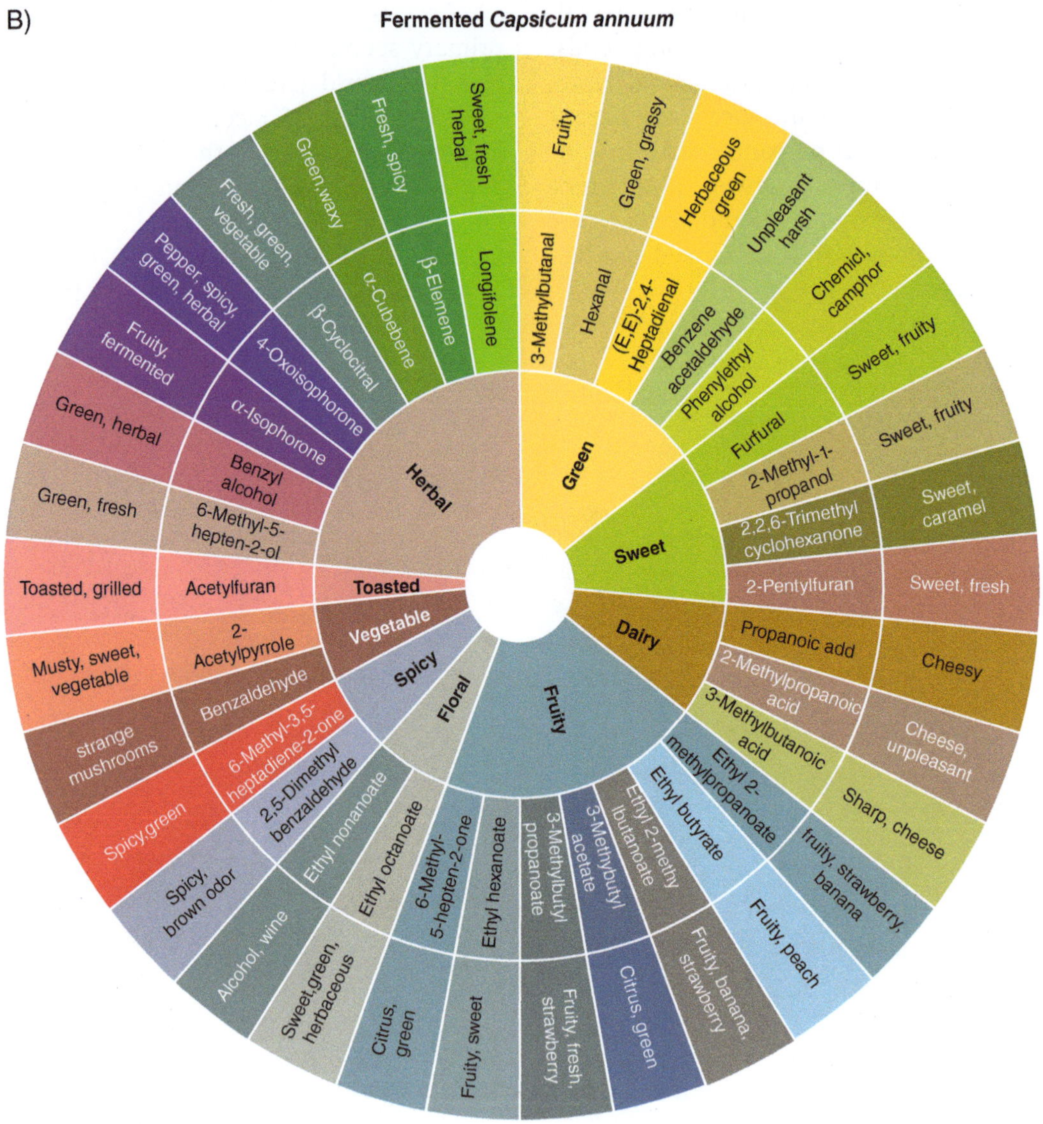

FIGURE 10.5 (Continued)

often complex, consisting primarily of lactic acid bacteria and yeasts which are characteristic of each cereal and legume fermented. The flavor of these fermented products depends on the substrates and also on the microbial community involved in the fermentative process. The hydrolysis of starch and other carbohydrates and proteolytic activities carried out by the lactic acid bacteria and yeasts determine the composition and fraction of precursors used for the generation of aroma compounds.

Sourdoughs can be prepared with different cereals and using different process parameters (Palla et al., 2020). The yeast primarily used in the fermentation of sourdough is *S. cerevisiae*, followed by *Kazachstania humilis* and *W. anomalus* (Gänzle and Ripari, 2016); although, many other yeast species have been found contributing to the process (Gobbetti, 1998). Yeasts consume glucose and/or maltose competing with acidifying lactic acid bacteria. The survival of *S. cerevisiae* during the fermentation process is usually low due to its low tolerance to acids, and usually other yeasts are used in co-fermentation (Damiani et al., 1996). Besides carbon dioxide, yeasts produce ethanol and secondary metabolites, which influence the quality of fermented products

(Palla et al., 2020). Moreover, yeasts have a significant impact on the flavor of breads producing different aroma compounds such as esters, aldehydes, and ketones (Damiani et al., 1996). Among these, ethyl alcohol, propanol, isobutanol, diacetyl, 2,3-butanediol, and 2-heptanone are produced by yeasts during the fermentation of sourdough (Plessas et al., 2008).

Other examples of food products from the fermentation of cereals and beans are Pixian broad bean paste (Li et al., 2017) and tarhana (Özdemir et al., 2018). The dominant yeasts are *Issatchenkia orientalis* and *M. pulcherrima* in Pixian bean paste, while in similar fermentations such as miso and doengang, the main yeasts are *Z. rouxii* and *D. hansenii*, respectively (Li et al., 2017). The traditionally fermented Pixian bean paste has a complex aromatic profile. Among these, ethyl caprylate, hexadecanoic acid methyl ester, 2-methyl-pyrazine, 2,6-dimethyl-pyrazine, 2-pentyl furan, 2-acetyl furan, and hexadecanoic acid are considered aroma active compounds. However, most of them were not found in industrial preparations of the food. A traditional cereal fermentation is tarhana which is consumed as a soup. The main yeasts involved in tarhana fermentation are *P. kudriavzevii* and *Candida humilis* (Özdemir et al., 2018), but other species have been isolated and used in the fermentation process (Özel et al., 2015). The fermentative yeasts contribute to the synthesis of alcohols and esters, as well as ketones followed by terpenes and aldehydes. Moreover, different combinations of yeasts and lactic acid bacteria have been found to produce different aromatic profiles (Özdemir et al., 2018).

10.5 FINAL REMARKS

The ability of yeasts to develop and enhance the aroma of fermented food products is immense. The potential for aroma generation of many yeasts species has been demonstrated in a huge variety of fermented foods. The flavor of these fermented products depends on the raw materials and also on the microbial community involved in the fermentative process. The ability of yeast to survive in the fermented environment and interact with acidifying lactic acid bacteria diversifies the aroma profile of fermented foods. However, the potential of many other yeast species has been poorly studied and constitutes an untapped source of new aroma compounds and new food products that are waiting to be discovered.

ACKNOWLEDGMENTS

Financial support from RTI2018-098074-B-I00 founded by MCIU/AEI/10.13039/501100011033 by "ERDF A way of making Europe" and by GV/2021/065 Conselleria de Innovación, Universidades, Ciencia y Sociedad Digital Generalitat Valenciana (Spain) are fully acknowledged.

REFERENCES

1. Adebo OA, Oyayinka SA, Adebiyi JA, Feng X, Wilkin JD, et al. 2021. Application of gas chromatography-mass spectrometry (GC-MS)-based metabolomics for the study of fermented cereal and legume foods: A review. *International Journal of Food Science and Technology* 56: 1514–1534.
2. Arfi K, Landaud S, Bonnarme P. 2006. Evidence for distinct L-methionine catabolic pathways in the yeast *Geotrichum candidum* and the bacterium. *Brevibacterium Linens. Applied and Environmental Microbiology* 72: 2155–2162.
3. Arroyo-López FN, Bautista-Gallego J, Domínguez-Manzano J, Romero-Gil V, Rodríguez-Gómez F, et al. 2012. Formation of lactic acid bacteria-yeasts communities on the olive surface during Spanish-style Manzanilla fermentations. *Food Microbiology* 32: 295–301.
4. Belloch C, Querol A, Barrio E. 2011. *Kluyveromyces* spp. In: Fuquay JW (Ed.), *Encyclopedia of Dairy Sciences*, 2nd edition. Elsevier Ltd, Oxford, pp. 754–764.
5. Benitez-Cabello A, Rodríguez-Gómez F, Morales ML, Garrido-Fernandez A, Jimenez-Díaz R, et al. 2019. Lactic acid bacteria and yeast inocula modulate the volatile profile of Spanish-style green table olives. *Foods* 8: 280.
6. Bertuzzi AS, Kilcawley KN, Sheehan JJ, O'Sullivan MG, Kennedy D, et al. 2017. Use of smear bacteria and yeasts to modify flavour and appearance of Cheddar cheese. *International Dairy Journal* 72: 44–54.
7. Bertuzzi AS, Walsh AM, Sheehan JJ, Cotter PD, Crispie F, et al. 2018. Omics-based insights into flavor development and microbial succession within surface-ripened cheese. *MSystems* 3(1): e00211–e00217.
8. Binetti A, Carrasco M, Reinheimer J, Suarez V. 2013. Yeasts from autochthonal cheese starters: Technological and functional properties. *Journal of Applied Microbiology* 115: 434–444.
9. Cano-García L, Belloch C, Flores M. 2014. Impact of *Debaryomyces hansenii* strains inoculation on the quality of slow dry-cured fermented sausages. *Meat Science* 96: 1469–1477.

10. Capece A, Romaniello R, Siesto G, Romano P. 2018. conventional and non-conventional yeasts in beer production. *Fermentation* 4: 38.
11. Centeno JA, Garabal JI, Docampo F, Lorenzo JM, Carballo J. 2017. Recovering traditional raw-milk Tetilla cheese flavour and sensory attributes by using *Kocuria varians* and *Yarrowia lipolytica* adjunct cultures. *International Journal of Food Microbiology* 251: 33–40.
12. Cha YJ, Kim H, Cadwallader KR. 1998. Aroma-active compounds in kimchi during fermentation. *Journal Agricultural Food Chemistry* 46: 1944–1953.
13. Chang HW, Kim KH, Nam YD, Roh SW, Kim MS, et al. 2008. Analysis of yeasts and archaeal population dynamics in Kimchi using denaturing gradient gel electrophoresis. *International Journal of Food Microbiology* 126: 159–166.
14. Cholet O, Henaut A, Casaregola S, Bonnarme P. 2007. Gene expression and biochemical analysis of cheese-ripening yeasts: Focus on catabolism of L-methionine, lactate, and lactose. *Applied and Environmental Microbiology* 73: 2561–2570.
15. Cocconcelli PS, Fontana C. 2008. Characteristics and applications of microbial starters in meat fermentations. Chapter 5. In: Toldrá F (Ed.), *Meat Biotechnology, Springer Verlag*, New York, pp. 129–148.
16. Cogan TM. 2011. *Brevibacterium linens, Brevibacterium aurantiacum* and other smear microorganisms. In: Fuquay JW (Ed.), *Encyclopedia of Dairy Sciences*, 2nd edition. Elsevier Ltd., Oxford, pp. 395–400.
17. Collin S, Osman K, Delcambre S, El-Zayat AI, Dufour JP. 1993. Investigation of volatile flavor compounds in fresh and ripened domiati cheeses. *Journal of Agricultural and Food Chemistry* 41: 1659–1663.
18. Copetti MV. 2019. Yeasts and molds in fermented food production: An ancient bioprocess. *Current Opinion in Food Science* 25: 57–61.
19. Corral S, Belloch C, López-Díez JJ, Flores M. 2018. Lipolysis and aroma generation as mechanisms involved in masking boar taint in sodium reduced fermented sausages inoculated with *Debaryomyces hansenii* yeast. *Journal of the Science of Food and Agriculture* 98: 2121–2130.
20. Corral S, Belloch C, López-Díez JJ, Salvador A, Flores M. 2017. Yeast inoculation as a strategy to improve the physico-chemical and sensory properties of reduced salt fermented sausages produced with entire male fat. *Meat Science* 123: 1–7.
21. Corral S, Salvador A, Belloch C, Flores M. 2015. Improvement the aroma of reduced fat and salt fermented sausages by *Debaromyces hansenii* inoculation. *Food Control* 47: 526–535.
22. Cubillos FA, Gibson B, Grijalva-Vallejos N, Krogerus K, Nikulin J. 2019. Bioprospecting for brewers: Exploiting natural diversity for naturally diverse beers. *Yeast* 36: 383–398.
23. Damiani P, Gobbetti M, Cossignani L, Corsetti A, Simonetti MS, et al. 1996. The sourdough microflora. Characterization of hetero- and homofermentative lactic acid bacteria, yeasts and their interactions on the basis of the volatile compounds produced. *Lebensmittel Wissenschaft Und Technologie* 29: 63–70.
24. Dashko S, Zhou N, Tinta T, Sivilotti P, Lemut MS, et al. 2015. Use of non-conventional yeast improves the wine aroma profile of RibollaGialla. *Journal of Industrial Microbiology and Biotechnology* 42(7): 997–1010.
25. De Angelis M, Campanella D, Cosmai L, Summo C, Rizzello CG, et al. 2015. Microbiota and metabolome of un-started and started Greek-type fermentation of Bella di Cerignola table olives. *Food Microbiology* 52: 18–30.
26. Devanthi PVP, Gkatzionis K. 2019. Soy sauce fermentation: Microorganisms, aroma formation and process modification. *Food Research International* 120: 364–374.
27. Elhalis H, Cox J, Frank D, Zhao J. 2020. The crucial role of yeasts in the wet fermentation of coffee beans and quality. *International Journal of Food Microbiology* 333: 108796.
28. Eliskases-Lechner F, Guéguen M, Panoff JM. 2011. Geotrichum candidum. In: Fuquay JW, Fox PF, McSweeney PLH (Eds.), *Encyclopedia of Dairy Sciences*, Vol. 4. Elsevier Academic Press, London, pp. 765–777.
29. Feng XM, Passoth V, Eklund-Jonsson C, Alminger ML, Schnürer J. 2006. *Rhizopus oligosporus* and yeast co-cultivation during barley tempeh fermentation-nutritional impact and real-time PCR quantification of fungal growth dynamics. *Food Microbiology* 24: 393–402.
30. Flores M. 2009. Flavor of Meat Products (Chap. 19). In: Nollet MLL, Toldrá F (Eds.), *Handbook of Processed Meats and Poultry Analysis*, CRC Press, USA, pp. 385–398.
31. Flores M. 2018. Understanding the implications of current health trends on the aroma of wet and dry cured meat products. *Meat Science* 144: 53–61.
32. Flores M, Corral S, Cano-García L, Salvador A, Belloch C. 2015. Yeast strains as potential aroma enhancers in

dry fermented sausages. *International Journal of Food Microbiology* 212: 16–24.

33. Flores M, Moncunill D, Montero R, López-Díez JJ, Belloch C. 2017. Screening of *Debaryomyces hansenii* strains for flavour production under reduced concentration of nitrifying preservatives used in meat products. *Journal of Agricultural and Food Chemistry* 65(19): 3900–3909.

34. Flores M, Olivares A. 2015. Flavor (Chap. 25). In: Toldrá F (Ed.), *Handbook of Fermented Meat and Poultry*, 2nd Edition. John Wiley & Sons Ltd, USA, pp. 217–225.

35. Frank DC, Owen CM, Patterson J. 2004. Solid phase microextraction (SPME) combined with gas-chromatography and olfactometry-mass spectrometry for characterization of cheese aroma compounds. *Lebensmittel-Wissenschaft & Technologie* 37: 139–154.

36. Fröhlich-Wyder MT, Arias-Roth E, Jakob E. 2018. Cheese yeasts. *Yeast* 36: 129–141.

37. Gamero A, Dijkstra A, Smit B, De Jong C. 2020. Aromatic potential of diverse non-conventional yeast species for winemaking and brewing. *Fermentation* 6(50): 50.

38. Gamero A, Hernández-Orte P, Querol A, Ferreira V. 2011a. Effect of aromatic precursor addition to wine fermentations carried out with different *Saccharomyces* species and their hybrids. *International Journal of Food Microbiology* 147(1): 33–44.

39. Gamero A, Manzanares P, Querol A, Belloch C. 2011b. Monoterpene alcohols release and bioconversion by *Saccharomyces* species and hybrids. *International Journal of Food Microbiology* 145(1): 92–97.

40. Gamero A, Quintilla R, Groenewald M, Alkema W, Boekhout T, et al. 2016. High-throughput screening of a large collection of non-conventional yeasts reveals their potential for aroma formation in food fermentation. *Food Microbiology* 60: 147–159.

41. Gamero A, Tronchoni J, Belloch C, Querol A. 2013. Production of aroma compounds by cryotolerant *Saccharomyces* species and hybrids at low and moderate fermentation temperatures. *Journal of Applied Microbiology* 114: 1405–1414.

42. Gandul-Rojas B, Gallardo-Guerrero L. 2020. Characterization and processing of table olives: A special issue. *Foods* 9: 1469.

43. Gänzle MG, Ripari V. 2016. Composition and function of sourdough microbiota: From ecological theory to bread quality. *International Journal Food Microbiology* 239: 19–25.

44. Gobbetti M. 1998. The sourdough microflora: Interactions of lactic acid bacteria and yeasts. *Trends in Food Science and Technology* 9: 267–274.

45. Hammes WP, Knauf HJ. 1994. Starters in the processing of meat products. *Meat Science* 36: 155–168.

46. Hittinger CT, Steele JL, Ryder DS. 2018. Diverse yeasts for diverse fermented beverages and foods. *Current Opinion in Biotechnology* 49: 199–206.

47. Ho VTT, Zhao J, Fleet G. 2014. Yeasts are essential for cocoa bean fermentation. *International Journal of Food Microbiology* 174: 72–87.

48. Holt S, Miks MH, Trindade De Carvalho B, Foulquié-Moreno MR, Thevelein JM. 2019. The molecular biology of fruity and floral aromas in beer and other alcoholic beverages. *FEMS Microbiology Reviews* 43: 193–222.

49. Johansen E. 2018. Use of natural selection and evolution to develop new starter cultures for fermented foods. *Annual Reviews in Food Science and Technology* 9: 411–428.

50. Johnson ME 2017. A 100-Year Review: Cheese production and quality. *Journal of Dairy Science* 100: 9952–9965.

51. Kim JY, Kim J, Cha IT, Jung MY, Song HS, et al. 2019. Community structures and genomic features of undesirable white colony-forming yeasts on fermented vegetables. *Journal of Microbiology* 57: 30–37.

52. Laborde J, Pozo O. 1982. *Presente Y Pasado Del Chile En México. Secretaria De Agricultura Y Recursos Hidraúlicos.* México City: Secretaria de Agricultura y Recursos Hidraulicos (SARH), D.F. Mexico.

53. Lara-Hidalgo C, Belloch C, Dorantes-Alvarez L, Flores M. 2020. Evaluation of aroma profile generated by autochthonous yeasts with probiotic potential in the fermentation of Guajillo pepper sauce. *Journal of the Science of Food and Agriculture* 100: 4940–4949.

54. Lee SH, Whon TW, Roh SW, Jeon CO. 2020. Unraveling microbial fermentation features in kimchi: From classical to meta-omics approaches. *Applied Microbiology and Biotechnology* 104: 7731–7744.

55. Li X, Zhao C, Zheng C, Liu J, Vu VH, et al. 2017. Characteristics of microbial community and aroma compounds in traditional fermentation of Pixian broad bean paste as compared to industrial fermentation. *International Journal of Food Properties* 20: S2520–S2531.

56. Liu J, Zhu XL, Ullah N, Tao YS. 2017. Aroma Glycosides in Grapes and Wine. *Journal of Food Science* 82(2): 248–259.

57. Montel MC, Masson F, Talon R. 1998. Bacterial role in flavor development. *Meat Science* 49: 111–123.

58. Moon SH, Chang M, Kim HY, Chang HC. 2014. *Pichia kudriavzevii* is the major yeast involved in film-formation, off-odor production, and texture-softening in over-ripened Kimchi. *Food Science and Biotechnology* 23(2): 489–497.

59. Morata A, Escott C, Bañuelos MA, Loira I, Del Fresno JM, et al. 2020. Contribution of non-*Saccharomyces* yeasts to wine freshness. A review. *Biomolecules* 10: 34.

60. Nunomura N, Sasaki M, Yokotsuka T. 1980. *Shoyu* (Soy Sauce) flavor components: Acidic fractions and the characteristic flavor component. *Agricultural and Biological Chemistry* 44: 339–351.

61. Özdemir N, Yazici G, Şimşek Ö, Özkal SG, Çon AH. 2018. The effect of lactic acid bacteria and yeasts usage on aroma development during tarhana fermentation. *Food Bioscience* 26: 30–37.

62. Özel S, Sabanoğlu S, Çon AH, Şimşek Ö. 2015. Diversity and stability of yeast species during the fermentation of tarhana. *Food Biotechnology* 29(1): 117–129.

63. Padilla B, Belloch C, López-Díez JJ, Flores M, Manzanares P. 2014. Potential impact of dairy yeasts on the typical flavour of traditional ewes' and goats' cheeses. *International Dairy Journal* 35: 122–129.

64. Palla M, Blandino M, Grassi A, Giordano D, Sgherri C, et al. 2020. Characterization and selection of functional yeast strains during sourdough fermentation of different cereal wholegrain flours. *Scientific Reports* 10: 12856.

65. Parker M, Capone DL, Francis IL, Herderich MJ. 2018. Aroma precursors in grapes and wine: Flavor release during wine production and consumption. *Journal of Agricultural and Food Chemistry* 66: 2281–2286.

66. Perea-Sanz L, López-Díez JJ, Belloch C, Flores M. 2020. Counteracting the effect of reducing nitrate/nitrite levels on dry fermented sausage aroma by *Debaryomyces hansenii* inoculation. *Meat Science* 164: 108103.

67. Perea-Sanz L, Montero R, Belloch C, Flores M. 2018. Nitrate reduction in the fermentation process of salt reduced dry sausages: Impact on microbial and physicochemical parameters and aroma profile. *International Journal of Food Microbiology* 282: 84–91.

68. Perea-Sanz L, Peris D, Belloch C, Flores M. 2019. *Debaryomyces hansenii* metabolism of sulfur amino acids as precursors of volatile sulfur compounds of interest in meat products. *Journal of Agricultural and Food Chemistry* 67: 9335–9343.

69. Plessas S, Bakatorou A, Gallanagh J, Koutinas AA, Psarianos C. 2008. Evolution of aroma volatiles during storage of sourdough breads made by mixed cultures of *Kluyveromyces marxianus* and *Lactobacillus delbrueckii* ssp. *bulgaricus* or. *Lactobacillus Helveticus. Food Chemistry* 107: 883–889.

70. Ravyts F, De Vuyst L, Leroy F. 2012. Bacterial diversity and functionalities in food fermentations. *Engineering in Life Sciences* 12(4): 356–367.

71. Roseiro LB, Barbosa M, Ames JM, Wilbey RA. 2003. Cheesemaking with vegetable coagulants, the use of *Cynara L.* for the production of ovine milk cheeses. *International Journal of Dairy Technology* 56: 76–83.

72. Smit G, Smit BA, Engels WJM. 2005. Flavour formation by lactic acid bacteria and biochemical flavour profiling of cheese products. *FEMS Microbiology Reviews* 29(3): 591–610.

73. Steinkraus KH. 1997. Classification of fermented foods: Worldwide review of household fermentation techniques. *Food Control* 8: 311–317.

74. Su Y, Gamero A, Rodríguez ME, Lopes CA, Querol A, et al. 2019. Interspecific hybridisation among diverse *Saccharomyces* species: A combined biotechnological solution for low-temperature and nitrogen-limited wine fermentations. *International Journal of Food Microbiology* 310: 108331.

75. Swiegers JH, Bartowsky EJ, Henschke PA, Pretorius IS. 2005. Yeast and bacterial modulation of wine aroma and flavour. *Australian Journal of Grape Wine Research* 11: 139–173.

76. Tanaka Y, Watanabe J, Mogi Y. 2012. Monitoring of the microbial communities involved in the soy sauce manufacturing process by PCR-denaturing gradient gel electrophoresis. *Food Microbiology* 31: 100–106.

77. Tempère S, Marchal A, Barbe JC, Bely M, Masneuf-Pomarede I, et al. 2018. The complexity of wine: Clarifying the role of microorganisms. *Applied Microbiology and Biotechnology* 102: 3995–4007.

78. Terefe NS, Augustin MA. 2020. Fermentation for tailoring the technological and health related functionality of food products. *Critical Reviews in Food Science and Nutrition* 60(17): 2887–2913.

79. Toldrá F, Flores M. 2014. Sausages, types of: Dry and semidry. In: Devine C, Dikeman M (Eds.), *Encyclopedia of Meat Sciences*, 2nd edition, Vol. 3. Elsevier, Oxford, pp. 248–255.

80. Tufariello M, Durante M, Ramires F, Grieco F, Tommasi L, et al. 2015. New process for production of fermented black table olives using selected autochthonous microbial resources. *Frontiers Microbiology* 6: 1007.

81. Van Der Sluis C, Tramper J, Wijffels RH. 2001. Enhancing and accelerating flavour formation by salt-tolerant yeasts in Japanese soy-sauce processes. *Trends in Food Science and Technology* 12(9): 322–327.

82. Van Wyk N, Grossmann M, Wendland J, Von Wallbrunn C, Pretorius IS. 2019. The whiff of wine yeast innovation: Strategies for enhancing aroma production by yeast during wine fermentation. *Journal of Agricultural and Food Chemistry* 67: 13496–13505.

83. Wakil SM, Onilude AA, Adetutu EM, Ball AS. 2008. PCR-DGGE fingerprints of microbial successional changes during fermentation of cereal-legume weaning foods. *African Journal of Biotechnology* 7: 4643–4652.

84. Watts EG, Janes ME, Prinyawiwatkul W, Shen Y, Xu Z, et al. 2018. Microbiological changes and their impact on quality characteristics of red hot chilli pepper mash during natural fermentation. *International Journal of Food Science and Technology* 53: 1816–1823.

85. Wei CL, Chao SH, Tsai WB, Lee PS, Tsau NH, et al. 2013. Analysis of bacterial diversity during the fermentation of *inyu*, a high-temperature fermented soy sauce, using nested PCR-denaturing gradient gel electrophoresis and the plate count method. *Food Microbiology* 33(2): 252–261.

86. Yang Y, Deng Y, Jin Y, Liu Y, Xia B, et al. 2017. Dynamics of microbial community during the extremely long-term fermentation process of a traditional soy sauce. *Journal of the Science of Food and Agriculture* 97(10): 3220–3227.

87. Zimmermann M, Schieberle P. 2000. Important odorants of sweet bell pepper powder(*Capsicum annuum cv. annuum*): Differences between samples of Hungarian and Morrocan origin. *European Food Research and Technology* 211: 175–180.

CHAPTER 11

Beneficial Microbiota in Ethnic Fermented Foods and Beverages

Jyoti Prakash Tamang and Namrata Thapa
DAICENTER and Bioinformatics Centre, Department of Microbiology, School of Life Sciences, Sikkim University Tadong, Gangtok, 737102, Sikkim, India

11.1 INTRODUCTION

The classical and conventional text books in microbiology/biology/life sciences discuss pathogenic and disease-causing food poisoning microorganisms, giving the impression of "bad bugs" in to young students, rarely discussing "good bugs" or beneficial microorganisms in microbiology text books. However, there are several beneficial microorganisms which have been used or "cultured" by ethnic people since the beginning of human civilization for the production of foods, use in agriculture, medicine, etc., without the insights of scientific mechanisms. Ethnic fermented foods and beverages are a biological hub for many beneficial live microorganisms (Cocolin et al., 2016; Hutkins, 2018; Marco et al., 2021; Rezac et al., 2018; Stiemsma et al., 2020; Tamang et al., 2020).

11.2 ETHNIC FERMENTED FOODS

Every ethnic community on Earth has unsurpassed ethnic foods including fermented foods and alcoholic drinks depending upon the locally available domesticated raw bio-resources of plant and animal origins. Food preference, religious taboo, geographical location, dietary habits, traditional gastronomy and cuisine have shaped the distinct food culture of the people. Ethnic fermented foods and beverages are the community-specific fermented foods and beverages prepared by diverse groups of ethnic people in each and every region of the world using their indigenous knowledge of food fermentation and native culinary skills to produce various artisan foods and edible foods. Some ethnic fermented foods are artisan (Cocolin et al., 2016), exotic (Tamang, 2020a), and even endangered due to rapid urbanization, modernization and the shifting of traditional food habits to fast foods (Bricas, 2019; Tamang et al., 2020), which may result in the disappearance of native microorganisms from food ecosystems. Ethnic fermented foods and alcoholic beverages are traditionally prepared by natural or spontaneous fermentation (Cocolin and Ercolini, 2008) or sometimes by using traditionally prepared starter cultures or "back-slopping" methods (Whittington et al., 2019). Approximately 5000 varieties of common, uncommon, region-specific, community-specific, exotic, and endangered unlisted ethnic fermented foods and alcoholic beverages represent the gastronomy of every major or minor ethnic community in all continents of the world (Tables 11.1–11.6), which are consumed as a staple, side dish, soup, curry, condiment, seasoning, taste-maker, pickle, confectionery, savory, colorant, alcoholic, and non-alcoholic beverages (Benkerroum, 2013; Cocolin and Ercolini, 2008; Franz et al., 2014; Hesseltine, 1983; Hutkins, 2018; Steinkraus, 1996; Tamang, 2010, 2015a, 2016, 2020a; Tamang and

Good Microbes in Medicine, Food Production, Biotechnology, Bioremediation, and Agriculture, First Edition. Edited by Frans J. de Bruijn, Hauke Smidt, Luca S. Cocolin, Michael Sauer, David Dowling, and Linda Thomashow.

Kailasapathy, 2010). Daily per capita consumption of fermented foods and alcoholic beverages are estimated as approximately 50–400 g globally (Tamang et al., 2016a). Since preparation and preferences of ethnic fermented foods depend on the availability of locally grown agricultural products and domesticated animal resources and also geographical locations, most of the ethnic fermented foods are substrate-specific. For example, soybeans are cultivated in South East Asian countries including North East India, eastern Nepal and Bhutan, and are commonly consumed both as non-fermented and fermented soya-products such as *kinema, natto, cheonggkukjang, douchi, miso, tempe, furu/rufu*, and *shoyu*. (Kadar et al., 2020; Kharnaior and Tamang, 2021; Kubo et al., 2011; Kwon et al., 2019b; Shin and Jeong, 2015; Tamang, 2015b; Xie et al., 2019). However, such soybean-based foods are traditionally unknown in other parts of the world, despite the major cultivation of soybeans in countries such as USA, Brazil, and Argentina (Shea et al., 2020), soybean as an ethnic food has never been used. Interestingly, ethnic fermented soybean products with a characteristic stickiness, due to viscous materials, presumably γ-polyglutamic acid (PGA), D-glutamic acid polymer produced by *Bacillus subtilis* (Hsueh et al., 2017), are produced and eaten in a few South East Asian countries within the "*natto* triangle" (Nakao, 1972) or "*Kinema-Natto-Thua* (KNT) triangle" (Tamang, 2010) including North East India (*kinema, hawijar, bekang, peruyaan, peron naming, axone, tyrang-bai*), eastern Nepal and southern Bhutan (*kinema*), Myanmar (*pe poke*), Northern Thailand (*thua nao*), Laos (*sieng*) Yunnan province of China (*douchi*), Korea (*cheonggukjang*), and Japan (*natto*) (Tamang et al., 2020). The ethnic fermented rice–legume mixture, locally called *idli, dosa*, and *appam*, of South India and Sri Lanka (Antony et al., 2020), are unique artisan food products produced by South Indian people with balanced protein-rich and calorie-rich cultural diets. Agave (*Agave americana*) is a popular ornamental succulent plant in many tropical countries, it is traditionally fermented into a mildly alcoholic beverage called *pulque* in Latin America (mostly in Mexico), which is a traditional drink in Latin America (Álvarez-Ríos et al., 2020; Escalante et al., 2016). Some major and minor region-specific foods and beverages of Africa and Asia are listed in Tables 11.1 and 11.2.

Varieties of community-specific and region-specific ethnic cheese products, prepared from the milk of cows/sheep, are a traditional delicacy of the people of Australia (Table 11.3), Europe (Table 11.4), North America (Table 11.5) and South America (Table 11.6) (Johnson, 2017; Tamang et al., 2020).

The flavor, taste, texture, and other sensory properties are determined by some beneficial microorganisms present in community-specific cheese products (Choi et al., 2020; Irlinger et al., 2015). Similarly,

TABLE 11.1 Some ethnic fermented foods and beverages of Africa (Benkerroum, 2013; Dabire et al., 2018; Diaz et al., 2019; Franz et al., 2014; Johansen et al., 2019)

Fermented Products	Local Names of Foods
Fermented milk	Amabere, Amaruranu, Amasi, Ergo, Fènè, Gariss, Kefir, Kule Naoto, Leben, Lben, Mabisi, Mafi, Masai, Mursik, Mutandabota, Nunu, Omashikwa, Pendidam, Nyarmie, Sethemi, Suusac, Zabady
Fermented cereals	Busa, Ben-saalga, Enjera/Injera, Gowé, Hussuwa, Kenkey, Kunu-zaki, Kisra, koko, Mawè, Mbege, Ogi, Pito, Poto poto, Togwa, Uji
Fermented meat	Basterma, Basturma Gueddid, Khlii, Khlia, Msrana, Merguez, Naqaneq, Pastirma, Pastrami, Sujuk, Soudjouk
Fermented fish	Feseekh, Momone
Fermented legumes	Bikalga, Dawadawa, Iru, Kawal, Kinda, Ogiri, Ogili, Okpehe, Soumbala, Ugba
Fermented soybeans	No record yet
Fermented vegetables	Fermented olives
Fermented roots (cassava)	Chikwangue, Cingwada, Fufu, Gari, Lafun, Konkonte
Alcoholic beverages	Bantu beer, Bouza, Bussa, Bushera, Gowé, Kachasu, Mangisi, Mbege, Merrisa, Palm wine, Sifanu, Tchoukoutou
Miscellaneous	Cocoa

NB: *The authors apologize for any incorrect phonetics and/or spellings of vernacular names of some ethnic foods.*

TABLE 11.2 Some of the ethnic fermented foods and beverages of Asia (Nout and Aidoo, 2002; Tamang, 2010, 2016; Tamang et al., 2020, 2020; Wach et al., 2018)

Fermented Products	Local Names of Foods
Fermented milk	Airag, Chhu, Chhurpi, Churkam, Dahi, Dadih, Kalari, Lassi, Mar, Misti dahi/Lal dahi/Payodhi, Mohi, Phiu, Shrikhand, Somar, Sua Chua, Tarag
Fermented cereals	Ang-kak, Appam, Dosa, Idli, Jalabi, Khamak (Kao-mak), Lao-chao, Nan, Puto, Rabadi, Selroti, Tape Ketan
Fermented meat	Arjia, Chartayshya, Kargyong, Khyopeh, Nham (Musom), Nem-chua, Pastirma, Sai-krok-prieo, Sa-um, Satchu, Suka ko masu, Tocin
Fermented fish	Balao-Balao (Burong Hipon Tagbilao), Bordia, Belacan (Blacan), Bakasang, Burong Bangus, Budu, Gnuchi, Gulbi, Hentak, Hoi-Malaeng, Ika-Shiokara, Jeotkal, Karati, Kusaya, Lashim, Myulchijeot, Ngari, Narezushi, Nam Pla (Nampla-Dee, Nampla-Sod), Nuoc Mam, Patis, Pla-Paeng-Daeng, Pla-Som (Pla-Khao-Sug), Pu-Dong, Saeoo Jeot, Sheedal, Sidra, Suka ko Maacha, Sukuti, Shottsuru, Sikhae, Surströmming, Tungtap
Fermented legumes	Bhallae, Dhokla, Maseura, Oncom Hitam (Black Oncom), Oncom Merah (Orange Oncom), Papad, Wari
Fermented soybeans	Axone/Aakhoni, Bekang, Cheonggukjang, Doenjang, Douchi, Furu, Grep Chhurpi, Gochujang, Hawaijar, Kinema, Kanjang, Kecap, Ketjap, Natto, Meitauza, Meju, Meju, Miso, Pe poke, Peruyaan, Peron Naming, Pheha Shoyu, Sieng, Sufu, Tauco, Thua Nao, Tempe, Tungrymbai, Yandou
Fermented vegetables and bamboo shoots	Burong mustala, Dha muoi, Ekung, Eup, Fu-tsai, Gundruk, Goyang, Hom-dong, Hiring, khalpi, Jiang-gua, Jiang-sun, Kimchi, Naw-mai-dong, Mesu, Oiji, Pak-gard-dong, Pak-sian-dong, Pao cai, Soidon, Soibum, Sayur asin, Sinki, Sunki, Suan-cai, Suan-tsai Takuanzuke, Tuaithur
Fermented roots	Tape
Alcoholic beverages	Aarak, Aitanga, Angoori, Apong, Basi, Brem, Bhaati jaanr, Baijiu, Chulli, Chyang, Daru, Darassun, Duizou, Ennog, Ewhaju, Feni, Handia, Kanji, Khao maak, Kodo ko jaanr, Lugri, Madhu, Mingri, Makgeolli, Nam khao, Nchiangne, Oh, Ou, Poko, Pona, Ruou de, Raksi, Ruou nep, Saké, Sato, Sura, Soju, Shochu, Toddy Takju, Tapuy, Tapai pulut, Tapai ubi, Tapé-kekan, Tien-chiu-niang, Yakju, Yu, Zu, Zutho
Amylolytic dry starters	Bakhar, Benh-men, Bubod, Chiu yueh, Chowan, Dawdim, Daqu, Dombea/medombae, Hamei, Humao, Khekhrii, Koji, Krachae, Loogpang, Mana, Marcha, Men, Manapu, Nuruk, Phab, Phut, Paa, Pee, Ragi, Thiat, Torani, Xiaoqu
Miscellaneous	Kombucha, miang, vinegar

NB: *The authors apologize for any incorrect phonetics and/or spellings of vernacular names of some ethnic foods.* LAB, lactic acid bacteria; CNS, coagulase-negative staphylococci.

some beneficial lactic acid bacteria and coagulase-negative staphylococci contribute a typical organoleptic property of hundreds of lesser-known and common ethnic fermented sausages, ham, and salami (Alfaia et al., 2018; Laranjo et al., 2019) which are part of the traditional food culture of Europeans, Americans, and Australians (Holck et al., 2017; Van Reckem et al., 2019).

Every region in Europe has specific wines, traditionally and naturally fermented, with typical taste and flavor preferred by the ethnic communities (Pretorius, 2020; Walker, 2014). The quality and sensory attributes including the flavor of the wine are determined by species/strains of *Saccharomyces* and non-*Saccharomyces* (Hu et al., 2020; Pulvirent et al., 2020).

11.3 DIVERSITY OF BENEFICIAL MICROORGANISMS IN ETHNIC FERMENTED FOODS

Microorganisms are associated with edible plant and animal sources and have contributed to human survival since the evolution of humans. Accidently or unknowingly, people have invented traditional techniques for the utilization of unseen beneficial microorganisms present in and around their environments for the preservation and fermentation of perishable plants or animal substrates to obtain desirable and culturally acceptable ethnic fermented foods and alcoholic beverages (Ashaolu, 2020; Capozzi et al., 2020; Tamang, 2010; Tamang et al., 2016a). Some of these microorganisms

TABLE 11.3 Some of the ethnic fermented foods and beverages of Australia (Gubag et al., 1996; Hutkins, 2018; Tamang et al., 2020, 2016a; Tamang and Kailasapathy, 2010; Wade et al., 2019)

Fermented Products	Local Names of Foods
Fermented milk	Cheese, yogurt
Fermented cereals	Sourdough, loaves
Fermented meat	Fermented sausages, ham, salami,
Fermented fish	Smoked fish
Fermented legumes	No record yet
Fermented soybeans	No record yet
Fermented vegetables	*Sapal* (fermented *Colocasia esculenta* and coconut) in Papua New Guinea
Fermented roots	No record yet
Alcoholic beverages	Beer, wine

NB: *The authors apologize for any incorrect phonetics and/or spellings of vernacular names of some ethnic foods.* LAB, lactic acid bacteria; CNS, coagulase-negative staphylococci.

are beneficial and grow in/on raw, cooked, water-soaked, or smoked bio-resource during natural fermentation to convert the food into edible products that are organoleptically and socially acceptable to the ethnic consumers (Rezac et al., 2018; Tamang et al., 2016b; Xiang et al., 2019). Naturally fermented foods and beverages harbor different groups of classified and unclassified microorganisms including bacteria, yeasts, filamentous molds, viruses which are mostly bacteriophage and archaea (De Filippis et al., 2017; Leech et al., 2020).

11.3.1 Lactic Acid Bacteria

Among 111 bacterial phyla (Parks et al., 2018), *Firmicutes* is the most abundant phylum in several ethnic fermented foods and beverages of the world, followed by *Proteobacteria* and *Actinobacteria* (Dimidi et al., 2019; Bhutia et al., 2021a, b). Lactic acid bacteria (LAB) belonging to families *Lactobacillaceae*, *Streptococcaceae*, *Streptococcaceae*, and *Enterococcaceae* are generally found to be dominant in acidic fermented foods (Tamang et al., 2000, 2005, 2008; Holzapfel and Wood, 2014; Rezac

TABLE 11.4 Some of the ethnic fermented foods and beverages of Europe (Baschali et al., 2017; Johnson, 2017; Ropars et al., 2020; Van Reckem et al., 2019)

Fermented Products	Local Names of Foods
Fermented milk	Hundreds of European traditional Cheeses with different vernacular names (*Camembert, Cheddar, Brunost, Dubliner, Kefir, Koumiss, Manchego, Serra da Estrela, Skyr, Västerbottensost*); *Viili, Tarhana*, yogurt
Fermented cereals	Sourdough, loaves
Fermented meat	Hundreds of lesser-known and common traditional fermented sausages such as *Alheira, Androlla, Chorizo, Morcilla, Pastirma, Peperoni, Salchichon/saucisson, Salsiccia, Soppressata, Sucuk*; hams,
Fermented fish	*Surströmming, rakfisk, hákarl*
Fermented legumes	No record yet
Fermented soybeans	No record yet
Fermented vegetables	Sauerkraut
Fermented roots	No record yet
Alcoholic beverages	*Bagni*, Beer, *Boza*, Champagne, Cider, *Kyass*, Sparkling Wine, *Tarahana*, Vodka, Wine

NB: *The authors apologize for any incorrect phonetics and/or spellings of vernacular names of some ethnic foods.* LAB, lactic acid bacteria; CNS, coagulase-negative staphylococci.

TABLE 11.5 Some of the ethnic fermented foods and beverages of North America (Hutkins, 2018; Palla et al., 2020; Tamang et al., 2020, 2016a; Tamang and Kailasapathy, 2010)

Fermented Products	Local Names of Foods
Fermented milk	Cheese, yogurt
Fermented cereals	Sourdough, loaves,
Fermented meat	Fermented sausages, ham, jerky
Fermented fish	Smoked fish
Fermented legumes	No report yet
Fermented soybeans	No report yet
Fermented vegetables	Sauerkraut
Fermented roots	No report yet
Alcoholic beverages	Beer, Champagne, cider, sparkling wine, wine,

NB: *The authors apologize for any incorrect phonetics and/or spellings of vernacular names of some ethnic foods.* LAB, lactic acid bacteria; CNS, coagulase-negative staphylococci.

TABLE 11.6 Some of the ethnic fermented foods and beverages of South America (Álvarez-Ríos et al., 2020; Mayorga et al., 2020; Penna et al., 2016; Romero-Luna et al., 2017)

Fermented Products	Local Names of Foods
Fermented milk	Cheese: *Coalho, Corrientes, Minas, Pategrás, Reggianito Argentino Serrano*; yogurt
Fermented cereals	*Pozol*, Sourdough
Fermented meat	Sausages
Fermented fish	Smoked fish
Fermented legumes	No report yet
Fermented soybeans	No report yet
Fermented vegetables	Fermented olives
Fermented roots (cassava)	*Beiju, Calugi, Cauim, Caxiri, Puba, Tarubá, Tucupi, Yakupa, Y Parakari*
Alcoholic beverages	Beer, *Cachaça Chicha, Champu, Masato Pulque,* Tequila

NB: *The authors apologize for any incorrect phonetics and/or spellings of vernacular names of some ethnic foods.* LAB, lactic acid bacteria; CNS, coagulase-negative staphylococci.

et al., 2018; Ashaolu and Reale, 2020). Major genera of LAB isolated from ethnic fermented foods and beverages are *Lactobacillus, Lacticaseibacillus, Lentilactobacillus, Levilactobacillus, Loigolactobacillus, Lactococcus, Leuconostoc, Pediococcus, Enterococcus, Streptococcus, Weissella, Tetragenococcus, Oenococcus, Alkalibacterium, Carnobacterium, Fructobacillus, Vagococcus* (Ashaolu and Reale, 2020; Holzapfel and Wood, 2014; Mathur et al., 2020; Pradhan and Tamang, 2019; Zheng et al., 2020). Most of the LAB present in the fermented foods are listed in the microbial food cultures (MFC) safety inventory (Bourdichon et al., 2018, 2019). Among LAB, *Lactiplantibacillus plantarum* (previously *Lactobacillus plantarum*) is abundantly present in many ethnic fermented foods and beverages (Bachtarzi et al., 2020; Chen et al., 2020; Goel et al., 2020), which is widely used as beneficial bacterium in food safety (Behera et al., 2018; Hu et al., 2019), probiotics (Angelescu et al., 2019), safe starter culture in dairy products (Dan et al., 2019; Li et al., 2017), fermented vegetable (Jung et al., 2019a), and fermented sausage (Ba et al., 2018). Other dominant LAB genera viz. *Lactococcus, Leuconostoc, Pediococcus, Enterococcus, Streptococcus, Weissella* have also been reported from many ethnic fermented foods and alcoholic beverages (Azam et al., 2017; Li et al., 2020; Nami et al., 2019; Wade et al., 2019). Among *Lactobacillus* isolated from some ethnic fermented foods, *Lactiplantibacillus (Lactobacillus) kimchii*, a bacteriocin-producing lactic acid bacterium in *kimchi*, an ethnic fermented vegetable product of Korea, has been named after *kimchi* from where it was originally isolated (Yoon et al., 2000). Nomenclature of *Lactiplantibacillus kefiri* and *Lactiplantibacillus kefiranofaciens* have derived from *kefir*, an ethnic fermented milk and kefir grains drink from Russia (Bengoa et al., 2019).

11.3.2 Non-Lactic Acid Bacteria

Non-lactic acid bacteria of the family *Bacillaceae* of *Firmicutes* are also abundant in the alkaline fermented foods of Asia and Africa represented by *Bacillus* (Chettri et al., 2016; Dabire et al., 2018; Elshaghabee et al., 2017; Kharnaior and Tamang, 2021; Kumar et al., 2019). Some species of *Bacillus* reported in fermented soybeans, non-soybean legumes, and other fermented foods are *Bacillus subtilis, B. subtilis* variety *natto, B. amyloliquefaciens, B. circulans, B. coagulans, B. firmus, B. licheniformis, B. megaterium, B. pumilus, B. cereus, B. thuringiensis, B. sphaericus, B. thermoamylovorans,* and *B. sonorensis* (Chettri et al., 2016; Chettri and Tamang, 2015; Jeyaram et al., 2008, 2011; Kamada et al., 2015; Kharnaior and Tamang, 2021; Kiers et al., 2000; Kimura and Yokoyama, 2019; Kubo et al., 2011; Kumar et al., 2019; Lee and

Chang, 2017; Nwagu et al., 2020; Tamang and Nikkuni, 1996, 1998; Urushibata et al., 2002). *Bacillus subtilis* is the main fermenting organism in sticky, non-salted fermented soybean foods such as *kinema, bekang, hawaijar, tungrymbai, peha, peron-namsing, peruyaan*, and *aakhone/axone* of the north east states of India, *pepok* of Myanmar, *douchi* of the Yunnan province of China, *thua-nao* of Thailand, *sieng* of Laos, *cheonggukjang* of Korea and *natto* of Japan (Nam et al., 2012; Chettri and Tamang, 2015a; Tamang, 2015; Tamang et al., 2016a; Liu et al., 2020).

Some beneficial non-lactis such as species of *Bifidobacterium, Brevibacterium*, and *Propionibacterium* are associated with some ethnic fermented milk and meat products (Linares et al., 2017; Redondo-Useros et al., 2019; Song et al., 2018). *Bdellovibrio bacteriovorus* (phylum *Proteobacteria*) has been reported in exotic dried meat products of Sikkim (Bhutia et al., 2021a), which is a predatory bacterium (Bonfiglio et al., 2020) and may act as a bio-control agent for Gram-negative pathogenic bacteria (Negus et al., 2017). *Rummeliibacillus* (phylum: *Firmicutes*) has been reported in *sidra*, a dried fish product of India (Bhutia et al., 2021b) which is used as a probiotic for fish (Tan et al., 2019).

Some coagulase-negative staphylococci such as *Staphylococcus carnosus, S. saprophyticus, S. equorum, S. succinus*, and *S. xylosus* are reported from many fermented meats (Sánchez Mainar et al., 2017) and fermented soybean foods (Jeong et al., 2019) and some of them are used as starter cultures (Heo et al., 2020).

11.3.3 Yeasts

Of nine classified fungal phyla (Naranjo-Ortiz and Gabaldón, 2019), yeasts belonging to phyla *Ascomycota, Mucoromycota*, and *Basidiomycota* are reported in ethnic fermented foods and beverages (Aidoo and Nout, 2010; Hesseltine and Kurtzman, 1990; Ji et al., 2018). Genera of yeasts generally present in ethnic fermented foods and beverages are *Brettanomyces, Candida, Cryptococcus, Debaryomyces, Dekkera, Galactomyces, Geotrichum, Hansenula, Hanseniaspora, Hyphopichia, Issatchenkia, Kazachstania, Kluyveromyces, Metschnikowia, Meyerozyma, Pichia, Rhodotorula, Rhodosporidium, Saccharomyces, Saccharomycodes, Saccharomycopsis, Schizosaccharomyces, Sporobolomyces, Torulaspora, Torulopsis, Trichosporon, Wickerhamomyces, Yarrowia*, and *Zygosaccharomyces* (Hittinger et al., 2018; Johansen et al., 2019; Sha et al., 2018, 2019; Tamang and Fleet, 2009). *Saccharomyces cerevisiae* is the most abundant yeast species present in traditionally prepared dry starters to ferment alcoholic beverages in Asia (Chen et al., 2014; Sha et al., 2018, 2019; Tsuyoshi et al., 2005), alcoholic beverages (Walker and Stewart, 2016) including wine (Bagheri et al., 2017; Pulvirent et al., 2020; Walker, 2014) and palm alcoholic drink (Olowonibi, 2017; Onwumah et al., 2019; Das and Tamang, 2021), sourdough (Palla et al., 2020), ethnic fermented dairy products (de Lima et al., 2017). *Saccharomycopsis fibuligera* is a unique genus to many Asian fermented foods mostly present in dry starters and alcoholic beverages (Lee et al., 2018; Sha et al., 2019; Su et al., 2020)

11.3.4 Filamentous Molds

Mucoromycota and *Ascomycota* are the dominant fungal phyla in fermented foods and beverages, followed by *Basidiomycota* (Hesseltine, 1991; Londoño Hernández et al., 2017; Takeshita, 2020). Species of *Actinomucor, Amylomyces, Aspergillus, Monascus, Mucor, Neurospora, Parcilomyces, Penicillium, Rhizopus*, and *Ustilago* are reported from many fermented foods, non-food amylolytic starters, and alcoholic beverages (Anupma and Tamang, 2020; Chen et al., 2015, 2014; Hartanti et al., 2015; Ji et al., 2018; Nout and Aidoo, 2002; Yasui et al., 2020). Filamentous molds are mostly considered as beneficial fungi by Asian people and some genera are traditionally cultured for production of many ethnic fermented foods such as *tempe*, a fermented soybean food of Indonesia (Kadar et al., 2020), *koji*, a starter culture for production of *shoyu* and *miso* fermented soybean foods and *saké*, a traditional alcoholic drink of Japan (Takeshita, 2020), *doenjang-meju*, a fermented soybean paste of Korea (Lee et al., 2020), *furu, sufu*, and *da-jiang*, fermented soybean foods of China (Xie et al., 2019; Zhang et al., 2018) and traditionally prepared dry amylolytic and alcohol-producing starters of Asia such as *marcha, bakhar, hamei, thiat, humao, chowan, dawdim, phut* of India, *ragi* of Indonesia, *bubod* of the Philippines, *daqu/xiaoqu* of China, *nuruk* of Korea, *benh or men* of Vietnam, *loogpang* of Thailand, and *dombea* or *medombae* of Cambodia (Anupma and Tamang, 2020; Jung et al., 2012; Fronteras and Bullo, 2017; Ly et al., 2018; Roslan et al., 2018; Pradhan and Tamang, 2019; Tamang, 2016a; Tang et al., 2019). Filamentous molds are generally considered as spoilage and toxin-producing organisms in other parts of the world, or are rarely used in blue-cheese fermentation such as *Penicillium roqueforti* in Roquefort or Gorgonzola cheeses (Ropars et al., 2017), and *Penicillium camemberti* in Camembert or Brie cheeses (Ropars et al., 2020).

Among mycelial molds, *Rhizopus oligosporous* has been used as a starter for *tempe*, an ethnic mold-fermented soybean food, its fermentation has been practiced in Central or East Java in Indonesia prior to 1800 (Shurtleff and Aoyagi, 2007), and contributes several health promoting benefits as well as nutraceutical values in the product (Kameda et al., 2018; Polanowska et al., 2020). *Aspergillus oryzae*, non-toxin producer (Frisvad et al., 2018), has been cultivated as a beneficial fungus in the form of solid-state cultivation (rice grain, soybean, and wheat bran) known as *koji* (Soccol et al., 2017) by the ethnic Chinese for 3000 years. This was later introduced to Japan during the Yayoi period around 10 BCE-3 CE (Shurtleff and Aoyagi, 2012) for the preparation of fermented soybean foods such as *miso, shoyu, doenjang, da-jiang*, and alcoholic beverages such as *saké, shochu, soju*, and *baijiu*. (Hong and Kim, 2020; Sakandar et al., 2020; Takeshita, 2020). *Angkak* pigment is an example of a natural colorant produced by *Monascus purpureus* (Wu et al., 2019), which is used as a food colorant in China, Japan, Thailand, Philippines, and Indonesia (Danuri, 2008; Manan et al., 2017) and has several functional properties (Zhu et al., 2019).

11.3.5 Probiotic Strains from Ethnic Fermented Foods

Commercial probiotic microorganisms generally originate from the human gastrointestinal tract (Zommiti et al., 2020); however, several probiotic microorganisms have also been isolated from fermented foods (Angelescu et al., 2019; Zhao et al., 2019; Zielińska and Kołożyn-Krajewska, 2018). Several species of LAB have proved to be potential probiotic candidates in many ethnic fermented foods and beverages (AlKalbani et al., 2019; Colombo et al., 2018; García-Ruiz et al., 2014; Ghosh et al., 2019; Pradhan and Tamang, 2021; Sohn et al., 2020; Somashekaraiah et al., 2019). Some of the promising probiotic bacterial strains viz. *Lactobacillus plantarum* ACA-DC 2640 and ACA-DC4039 strains, and *Steptococcus thermophilus* ACA-DC 26 and ACA-DC 170 are isolated from Greek ethnic fermented dairy products (Zoumpopoulou et al., 2017), *Lb. kefiranofaciens* XL10 isolated from Tibetan kefir (Xing et al., 2017), *Lb. plantarum* subsp. *plantarum* SKI19 isolated from *naem*, an ethnic fermented pork sausage of Thailand (Botthoulath et al., 2018), *Lactococcus lactis* KC24 isolated from Korean *kimchi* (Son et al., 2018), *Lb. plantarum* L4 and *Leuconostoc mesenteroides* LMG 7954 isolated from *sauerkraut*, an ethnic fermented vegetable product of Germany (Beganović et al., 2011), *Leuconostoc mesenteroides* strain P45 isolated from Mexican *pulque* (Giles-Gómez et al., 2016), *Pediococcus pentosaceus* SMB13-1, *P. acidilactici* AKB3 *Enterococcus faecium* BPB11 and *E. durans* DMB13 isolated from a traditionally prepared dry starter to ferment an alcoholic beverage of the Eastern Himalayan regions of India, Nepal, and Bhutan (Pradhan and Tamang, 2021). Some yeasts including *Saccharomyces cerevisiae* and *Kluyveromyces marxianus* have shown probiotic properties in many African fermented foods (Cissé et al., 2019; Franz et al., 2014). Recently several probiotic yeasts have been reported from some ethnic fermented foods and beverages such as *Saccharomyces cerevisiae, S. boulardii, Debaryomyces, Pichia, Torulaspora, Kluyveromyces, Hanseniaspora, Rhodotorula, Wickerhamomyces*, and *Candida* (Maïworé et al., 2019; Ogunremi et al., 2015).

11.3.6 Functional Profiles of Beneficial Microorganisms

Some beneficial microorganisms present in ethnic fermented foods and beverages have claimed to have functional properties and therapeutic uses, and also to combat some human diseases (Fernández et al., 2015; Kok and Hutkins, 2018; Marco et al., 2021; Melini et al., 2019; Rezac et al., 2018; Tamang et al., 2016b). The functional properties of beneficial microorganisms exhibited in food fermentations are mainly the production of antimicrobial compounds for the biological preservation of perishable foods (Da Costa et al., 2019; Tamang et al., 2009), food safety (Coban, 2020), bioenrichment of nutrients (Melini et al., 2019), enhance desirable sensory quality of the food (Xiang et al., 2019), degradation of toxic components, and antinutritive factors (Adeyemo and Onilude, 2013; Sokrab et al., 2014), biosynthesis of vitamins (Capozzi et al., 2012; Hati et al., 2019; Pradhan and Tamang, 2021), probiotic properties (Ghosh et al., 2019; Marco et al., 2021), antioxidant activities (Chettri and Tamang, 2014; Menezes et al., 2020; Verni et al., 2019), disease combating effects (Antunes et al., 2020; Lee et al., 2017a; Mota De Carvalho et al., 2018; Tamang, 2015b), and fortification with nutritional value (García-Burgos et al., 2020; Nkhata et al., 2018; Taylor et al., 2020) and production of health-promoting bioactive compounds (Ferraz Da Costa et al., 2020; Şanlier et al., 2019; Verardo et al., 2020).

Some native fungi isolated from the ethnic fermented foods of Asia produce various enzymes for industrial applications. *Rhizopus oryzae, Aspergillus orzuyae*, and *A. sojae* syntheses enzymes such as xxx-amylase, amyloglucosidase, maltase, invertase, pectinase, ß-galactosidase, cellulase, hemi-cellulase; acid, and alkaline proteases; and lipases (Aidoo and Nout, 2010; Ichishima, 2016; Nout and Aidoo, 2002) and essential secondary metabolites in fermented foods and alcoholic beverages (Cairns et al., 2019; Stewart, 2017) including untargeted metabolites (Kwon et al., 2019b; Mu et al., 2020). Many species of beneficial LAB isolated from ethnic fermented foods and beverages show antimicrobial activities against pathogenic and spoilage organisms (Bungenstock et al., 2020; Jiang et al., 2012; Mokoena, 2017; Tamang et al., 2009), due to production of antimicrobial compounds such as bacteriocin and nicin (Goel et al., 2020; Khan et al., 2010; O'Connor et al., 2020; Rasheed et al., 2020).

There are several unparalleled and community-specific ethnic fermented foods and beverages across the world where beneficial native microorganisms have been isolated and validated for their various nutraceutical and health benefits (Tamang et al., 2020; Xiang et al., 2019). Some good examples of utilization and validation of native beneficial microorganisms originally isolated from ethnic fermented foods and beverages for health benefits to consumers are cited below.

Natto, an ethnic fermented soybean food of Japan, is now produced using a monoculture of *Bacillus subtilis* variety *natto* strain (Kubo et al., 2011). There are many strains of *B. subtilis natto* which are commercially available (Kobayashi et al., 2016), however few strains have been completely sequenced such *B. subtilis natto* BEST 195 (Kamada et al., 2015; Nishito et al., 2010; Parks et al., 2019) and *Bacillus subtilis* natto VK161. *B. subtilis natto* has proven to produce several functional properties, health benefits, and therapeutics uses (Chen et al., 2018; Dabbagh et al., 2014; Ju et al., 2019; Parks et al., 2019). More than 500 small and large-scale commercial *natto* production factories are operated in Japan according to the Federation of Japan Natto Manufactures Cooperative Society. *Bacillus* spp. have also been reported to show several health-promoting benefits in other fermented soybean foods of Asia (*kinema* and *hawijar* of India, *cheonggukjang* of Korea, *thua nao* of Thailand, *douchi* of China) such as fibrinolytic activity (Kotb, 2012; Meerak et al., 2007; Singh et al., 2014; Weng et al., 2017), bio-synthesis of poly-glutamic acid (Chettri et al., 2016; Ratha and Jhon, 2018), probiotics, and immunomodulating activities (Elshaghabee et al., 2017), antioxidant activities (Chettri and Tamang, 2014; Keishing et al., 2015) and also in various community-specific fermented ethnic locust beans of Africa (Diaz et al., 2019; Franz et al., 2014; Iheke et al., 2017; Nwagu et al., 2020).

Historically, *yogurt* is an ancient fermented milk product of Turkey in Eastern Europe (Fisberg and Machado, 2015), which is similar to *dahi* of India originated around 8000 years ago (Tamang, 2020b). Indigenous beneficial lactic acid bacteria viz. *Lactobacillus delbrueckii* subsp. *bulgaricus* and *Streptococcus thermophilus* were isolated from yogurt (Soomro and Masud, 2008) and are now successfully used as starter cultures for production of commercial probiotic yogurt (Rul, 2017).

Kimchi, an ethnic fermented vegetable product of Korea probably originated in Korea about 4000 years ago (Jang et al., 2015), and since then has been consumed as one of the major side dishes in the Korean diet (Kim et al., 2016). Genomic analysis has revealed the health benefits of *kimchi* fermented by beneficial indigenous bacteria such as *Leuconostoc kimchi* (Jung et al., 2019b), *Lactiplantibacillus curvatus* (Lee et al., 2017a), and *Lactiplantibacillus plantarum* (Beck et al., 2019). Due to the healthy nature of *kimchi* (Han et al., 2015; Park et al., 2014) it has been recognized as one of the most beneficial foods in the world (Chang, 2018).

Ethnic people of Indonesia historically invented the "ethnic-microbiology" to use the golden-mold *Rhizopus oryzae* (Londoño Hernández et al., 2017) to produce *tempe*, a fermented soybean food (Allia et al., 2015) with several functional properties, health benefits, and therapeutic uses (Astuti, 2015; Athaillah et al., 2019; Denardi-Souza et al., 2018; Huang et al., 2018; Polanowska et al., 2020). The Dutch community has adopted and included *tempe* in their cultural food (Stahel, 1946).

Lactiplantibacillus kefir and *Lactiplantibacillus kefiranofaciens* isolated from *kefir*, an ethnic fermented milk and kefir grain beverage from Russia (Tamang et al., 2020) have several functional properties and therapeutic uses (Bengoa et al., 2019; Federici et al., 2018; Slattery et al., 2019).

11.4 CONCLUSION

Health promoting related properties of beneficial microorganisms isolated from many ethnic fermented foods and beverages of the world have been studied and some of them have been validated, are marketed globally as functional, therapeutic, and nutraceuticals food and drinks. Some examples are probiotic yogurt, wine, *shoyu*, *natto*, *kimchi*, cheese, *tempe*, *pulque*, *kombucha*, etc. However, 90% of community-specific, naturally fermented foods, and alcoholic beverages in the world are still prepared traditionally at household level. It has also been observed that consumption of some lesser known, and endangered ethnic fermented foods is declining due to modernization and urbanization of villages in many countries in terms of development, changes in lifestyle, shifting from cultural food consumption to commercial foodstuffs and fast foods, and also changes due to climate change in some places, with severe effects on traditional culinary practices. Academic programs on the importance of beneficial microorganisms in ethnic fermented foods and beverages may be incorporated in curricula at the school level. Training on molecular identification in next generation sequencing techniques, application of machine learning to understand metataxonomic and metabolomics for functional profiles of microorganisms in ethnic fermented foods and beverages may be taught to young researchers.

REFERENCES

1. Adeyemo SM, Onilude AA. 2013. Enzymatic reduction of anti-nutritional factors in fermenting soybeans by *Lactobacillus plantarum* isolates from fermenting cereals. *Nigerian Food Journal* 31(2): 84–90.
2. Aidoo KE, Nout MJR. 2010. Functional yeasts and molds in fermented foods and beverages. In: Tamang JP, Kailasapathy K (Eds.), *Fermented Foods and Beverages of the World*. CRC Press, Taylor & Francis Group, New York, pp. 127–148.
3. Alfaia C, Gouveia I, Fernandes M, Fernandes M, Semedo-Lemsaddek, et al. 2018. Assessment of coagulase-negative staphylococci and lactic acid bacteria isolated from Portuguese dry fermented sausages as potential starters based on their biogenic amine profile: Biogenic amines on staphylococci. *Journal of Food Science* 83. doi. 10.1111/1750-3841.14298.
4. AlKalbani NS, Turner MS, Ayyash MM. 2019. Isolation, identification, and potential probiotic characterization of isolated lactic acid bacteria and in vitro investigation of the cytotoxicity, antioxidant, and antidiabetic activities in fermented sausage. *Microbial Cell Factories* 18: 188. doi.org/10.1186/s12934-019-1239-1.
5. Allia N, Lilis N, Suliantari AS. 2015. Microbial growth dynamics during tempe fermentation in two different home industries. *International Food Research Journal* 22: 1668–1674.
6. Álvarez-Ríos GD, Figueredo-Urbina CJ, Casas A. 2020. Physical, chemical, and microbiological characteristics of pulque: Management of a fermented beverage in Michoacán, Mexico. *Foods* 9(3): 361. doi: 10.3390/foods9030361.
7. Angelescu IR, Zamfir M, Stancu MM, Grosu-Tudor SS. 2019. Identification and probiotic properties of lactobacilli isolated from two different fermented beverages. *Annals of Microbiology* 69: 1557–1565.
8. Antony U, Ilango S, Chelliah R, Ramakrishnan SR, Ravichandran K. 2020. Ethnic Fermented Foods and Beverages of Tamil Nadu. In: Tamang JP (Ed.), *Ethnic Fermented Foods and Alcoholic Beverages of India: Science History and Culture*. Springer Nature, Singapore, pp. 539–560. ISBN: 978-981-15-1486-9.
9. Antunes A, Vinderol G, Xavier-Santos D, Sivieri K. 2020. Potential contribution of beneficial microbes to face the COVID-19 pandemic. *Food Research International (Ottawa, Ont.)* 136: 109577. https://doi.org/10.1016/j.foodres.2020.109577.
10. Anupma A, Tamang JP. 2020. Diversity of filamentous fungi isolated from some amylase and alcohol-producing starters of India. *Frontiers in Microbiology* 11: 905. doi: 10.3389/fmicb.2020.00905.
11. Ashaolu TJ. 2020. Safety and quality of bacterially fermented functional foods and beverages: A mini review. *Food Quality and Safety* 4(3): 123–127.
12. Ashaolu TJ, Reale A. 2020. A holistic review on Euro-Asian lactic acid bacteria fermented cereals and vegetables. *Microorganisms* 8(8): 1176. doi.org/10.3390/microorganisms8081176.
13. Astuti M. 2015. Health benefits of *Tempe*. In: Tamang JP (Ed.), *Health Benefits of Fermented Foods*. CRC Press, Taylor & Francis Group, New York, pp. 371–394.
14. Athaillah ZA, Muzdalifah D, Lestari A, Devi AF, Udin LZ, et al. 2019. Phenolic compound profile functionality of aqueous overripe tempe extracts. *Current Research in Nutrition and Food Science* 7(2). doi.org/10.12944/CRNFSJ.7.2.08.
15. Azam M, Mohsin M, Ijaz H, Tulain UR, Ashraf MA, Fayyaz A, Abadeen Z, Kamran Q. 2017. Review - Lactic acid bacteria in traditional fermented Asian foods. *Pakistan Journal of Pharmaceutical Science* 30(5): 1803–1814.
16. Ba HV, Seo HW, Seong PN, Kang SM, Kim YS, et al. 2018. *Lactobacillus plantarum* (KACC 92189) as a potential probiotic starter culture for quality improvement of fermented sausages. *Korean Journal for Food Science of Animal Resources* 38(1): 189–202.

17. Bachtarzi N, Speciale I, Kharroub K, De Castro C, Ruiz L, et al. 2020. Selection of exopolysaccharide-producing *Lactobacillus plantarum* (*Lactiplantibacillus plantarum*) isolated from Algerian fermented foods for the manufacture of skim-milk fermented products. *Microorganisms* 23;8(8): 1101. doi: 10.3390/microorganisms8081101.

18. Bagheri B, Bauer FF, Setati ME. 2017. The Impact of *Saccharomyces cerevisiae* on a wine yeast consortium in natural and inoculated fermentations. *Frontiers in Microbiology* 8: 1988. https://doi.org/10.3389/fmicb.2017.01988.

19. Baschali A, Tsakalidou E, Kyriacou A, Karavasiloglou N, Matalas AL. 2017. Traditional low-alcoholic and non-alcoholic fermented beverages consumed in European countries: A neglected food group. *Nutrition Research Reviews* 30(1): 1–24.

20. Beck BR, Park G-S, Lee YH, Im S, Jeong DY, Kang J. 2019. Whole genome analysis of *Lactobacillus plantarum* strains isolated from kimchi and determination of probiotic properties to treat mucosal infections by *Candida albicans* and *Gardnerella vaginalis*. *Frontiers in Microbiology* 10: 433. doi: 10.3389/fmicb.2019.00433.

21. Beganović J, Pavunc AL, Gjuračić K, Špoljarec M, Šušković J, et al. 2011. Improved sauerkraut production with probiotic strain *Lactobacillus plantarum* L4 and *Leuconostoc mesenteroides* LMG 7954. *Journal of Food Science* 76: M124–M129.

22. Behera SS, Ray RC, Zdolec N. 2018. *Lactobacillus plantarum* with functional properties: An approach to increase safety and shelf-life of fermented foods. *BioMed Research International* 9361614. https://doi.org/10.1155/2018/9361614.

23. Bengoa A, Iraporda C, Garrote G, Abraham A. 2019. Kefir micro-organisms: Their role in grain assembly and health properties of fermented milk. *Journal of Applied Microbiology* 126: 686–700.

24. Benkerroum N. 2013. Traditional fermented foods of North African countries: Technology and food safety challenges with regard to microbiological risks. *Comprehensive Reviews in Food Science and Food Safety* 12: 54–89.

25. Bhutia MO, Thapa N, Shangpliang HNK, Tamang JP. 2021a. Metataxonomic profiling of bacterial communities and their predictive functional profiles in traditionally preserved meat products of Sikkim state in India. *Food Research International* 140: 110002. doi.org/10.1016/j.foodres.2020.110002.

26. Bhutia MO, Thapa N, Shangpliang HNK, Tamang JP. 2021b. High-throughput sequence analysis of bacterial communities and their predictive functionalities in traditionally preserved fish products of Sikkim, India. *Food Research International* 109885. doi.org/10.1016/j.foodres.2020.109885.

27. Bonfiglio G, Neroni B, Radocchia G, Marazzato M, Pantanella F, et al. 2020. Insight into the possible use of the predator *Bdellovibrio bacteriovorus* as a probiotic. *Nutrients* 12, 8(2252). https://doi.org/10.3390/nu12082252.

28. Botthoulath V, Upaichit A, Thumarat U. 2018. Identification and in vitro assessment of potential probiotic characteristics and antibacterial effects of *Lactobacillus plantarum* subsp. *plantarum* SKI19, a bacteriocinogenic strain isolated from Thai fermented pork sausage. *Journal of Food Science and Technology* 55(7): 2774–2785.

29. Bourdichon F, Alper I, Bibiloni R, Dubois A, Laulund S, et al. 2018. Inventory of microbial food cultures with safety demonstration in fermented food products. *Bulletin. International Dairy Federation* 495: 2018.

30. Bourdichon F, Laulund S, Tenning P. 2019. Inventory of microbial species with a rationale: A comparison of the IDF/EFFCA inventory of microbial food cultures with the EFSA Biohazard Panel qualified presumption of safety. *FEMS Microbiology. Letters* 366: 1–6. https://doi.org/10.1093/femsle/fnz048.

31. Bricas N. 2019. Urbanization issues affecting food system sustainability. In: Brand C, et al. (Ed.), *Designing Urban Food Policies. Urban Agriculture*. Springer, Cham. https://doi.org/10.1007/978-3-030-13958-2_1.

32. Bungenstock L, Abdulmawjood A, Reich F. 2020. Evaluation of antibacterial properties of lactic acid bacteria from traditionally and industrially produced fermented sausages from Germany. *PLOS ONE* 15(3): e0230345. https://doi.org/10.1371/journal.pone.0230345.

33. Cairns TC, Zheng X, Zheng P, Sun J, Meyer V. 2019. Moulding the mould: Understanding and reprogramming filamentous fungal growth and morphogenesis for next generation cell factories. *Biotechnology for Biofuels* 12: 77. https://doi.org/10.1186/s13068-019-1400-4.

34. Capozzi V, Fragasso M, Russo P. 2020. Microbiological safety and the management of microbial resources in artisanal foods and beverages: The need for a transdisciplinary assessment to conciliate actual trends and risks avoidance. *Microorganisms* 8: 306. doi: 10.3390/microorganisms8020306.

35. Capozzi V, Russo P, Dueñas MT, López P, Spano G. 2012. Lactic acid bacteria producing B-group vitamins: A

great potential for functional cereals products. *Applied Microbiology and Biotechnology* 96(6): 1383–1394.

36. Chang HC. 2018. Healthy safe Korean traditional fermented foods: Kimchi and cheonggukjang. *Journal of Ethnic Foods* 5(3): 161–166.
37. Chen B, Wu Q, Xu Y. 2014. Filamentous fungal diversity and community structure associated with the solid state fermentation of Chinese Maotai-flavor liquor. *International Journal of Food Microbiology* 179: 80–84.
38. Chen H, McGowan EM, Ren N, Lal S, Nassif N, et al. 2018. Nattokinase: A promising alternative in prevention and treatment of cardiovascular diseases. *Biomarker Insights* 13: 1177271918785130. https://doi.org/10.1177/1177271918785130.
39. Chen W, He Y, Zhou Y, Shao Y, Feng Y, et al. 2015. Edible filamentous fungi from the species *Monascus*: Early traditional fermentations, modern molecular biology, and future genomics. *Comprehensive Reviews in Food Science and Food Safety* 14: 555–567.
40. Chen Z, Kang J, Zhang Y, Yi X, Pang X, et al. 2020. Differences in the bacterial profiles and physicochemical between natural and inoculated fermentation of vegetables from Shanxi Province. *Annals of Microbiology* 70: 66. https://doi.org/10.1186/s13213-020-01605-5.
41. Chettri R, Bhutia M, Tamang JP. 2016. Poly-γ-glutamic acid (PGA)-producing *Bacillus* species isolated from *Kinema*, Indian fermented soybean food. *Frontiers in Microbiology* 7: 971. https://doi.org/10.3389/fmicb.2016.00971.
42. Chettri R, Tamang JP. 2014. Functional properties of *tungrymbai* and *bekang*, naturally fermented soybean foods of North East India. *International Journal of Fermented Foods* 3: 87–103.
43. Chettri R, Tamang JP. 2015. *Bacillus* species isolated from tungrymbai and bekang, naturally fermented soybean foods of India. *International Journal of Food Microbiology* 197: 72–76.
44. Choi J, Lee SI, Rackerby B, Goddik L, Frojen R, et al. 2020. Microbial communities of a variety of cheeses and comparison between core and rind region of cheeses. *Journal of Dairy Science* 103(5): 4026–4042.
45. Cissé H, Kagambèga B, Sawadogo A, Tankoano A, Sangaré G, et al. 2019. Molecular characterization of *Bacillus*, lactic acid bacteria and yeast as potential probiotic isolated from fermented food. *Scientific African* 6: e00175. doi.org/10.1016/j.sciaf.2019.e00175.
46. Coban HB. 2020. Organic acids as antimicrobial food agents: Applications and microbial productions. *Bioprocess and Biosystems Engineering* 43: 569–591.
47. Cocolin L, Ercolini D. 2008. *Molecular Techniques in the Microbial Ecology of Fermented Foods*. Springer. ISBN-10:1441925600.
48. Cocolin L, Gobbetti M, Neviani E, Daffonchio D. 2016. Ensuring safety in artisanal food microbiology. *Nature Microbiology* 1: 16171. doi.org/10.1038/nmicrobiol.2016.171.
49. Colombo M, Castilho NPA, Todorov SD, Nero LA. 2018. Beneficial properties of lactic acid bacteria naturally present in dairy production. *BMC Microbiology* 18: 219. https://doi.org/10.1186/s12866-018-1356-8.
50. Da Costa RJ, Flávia LS, Voloski RG, Mondadori EH, Duval ÂMF. 2019. Preservation of meat products with bacteriocins produced by lactic acid bacteria isolated from meat. *Journal of Food Quality* Article ID 4726510. doi.org/10.1155/2019/4726510.
51. Dabbagh F, Negahdaripour M, Berenjian A, Behfar A, Mohammdi F, et al. 2014. Nattokinase: Production and application. *Applied Microbiology and Biotechnology* 98: 9199–9206.
52. Dabire Y, Somda MK, Ugwuanyi J, Ezeogu LI, Traore AS. 2018. Diversity of *Bacillus* species and their antimicrobial compounds involved in alkaline-fermentation of indigenous food condiments used in Africa. *International Journal of Advanced Research* 6: 331–355.
53. Dan T, Chen H, Li T, Tian J, Ren W, et al. 2019. Influence of *Lactobacillus plantarum* P-8 on fermented milk flavor and storage stability. *Frontiers in Microbiology* 9: 3133. https://doi.org/10.3389/fmicb.2018.03133.
54. Danuri H. 2008. Optimizing angkak pigments and lovastatin production by *Monascus purpureus*. *HAYATI Journal of Biosciences* 15(2): 61–66.
55. De Filippis F, Parente E, Ercolini D. 2017. Metagenomics insights into food fermentations. *Microbial Biotechnology* 10(1): 91–102.
56. Denardi-Souza T, Massarolo KC, Tralamazza SM, Badiale-Furlong E. 2018. Monitoring of fungal biomass changed by *Rhizopus oryzae* in relation to amino acid and essential fatty acids profile in soybean meal, wheat and rice. *CyTA - Journal of Food* 16(1): 156–164.
57. Diaz M, Kellingray L, Akinyemi N, Adefranye OO, Olaonipekun AB, et al. 2019. Comparison of the microbial composition of African fermented foods using amplicon sequencing. *Scientific Reports* 9(2019): 13863. https://doi.org/10.1038/s41598-019-50190-4.
58. Dimidi E, Co SR, Rossi M, Whelan K. 2019. Fermented foods: Definitions and characteristics, impact on the gut microbiota and effects on gastrointestinal health and

disease. *Nutrients* 11(8): 1806. https://doi.org/10.3390/nu11081806.

59. Elshaghabee F, Rokana N, Gulhane RD, Sharma C, Panwar H. 2017. *Bacillus* As potential probiotics: Status, concerns, and future perspectives. *Frontiers in Microbiology* 8: 1490. https://doi.org/10.3389/fmicb.2017.01490.
60. Escalante A, López Soto DR, Velázquez Gutiérrez JE, Giles-Gómez M, Bolívar F, et al. 2016. *Pulque*, a traditional Mexican alcoholic fermented beverage: Historical, microbiological, and technical aspects. *Frontiers in Microbiology* 7: 1026. doi: 10.3389/fmicb.2016.01026.
61. Federici F, Manna L, Rizzi E, Galantini E, Marini U. 2018. Draft genome sequence of *Lactobacillus kefiri* SGL 13, a potential probiotic strain isolated from kefir grains. *Microbiology Resource Announcements* 7(4): e00937-18. doi: 10.1128/MRA.00937-18.
62. Fernández M, Hudson JA, Korpela R, De Los Reyes-gavilán CG. 2015. Impact on human health of microorganisms present in fermented dairy products: An overview. *BioMed Research International* ID 412714, 13. https://doi.org/10.1155/2015/412714.
63. Ferraz Da Costa DC, Pereira Rangel L, Quarti J, Santos RA, Silva JL, et al. 2020. Bioactive compounds and metabolites from grapes and red wine in breast cancer chemoprevention and therapy. *Molecules* (Basel, Switzerland) 25(15): 3531. https://doi.org/10.3390/molecules25153531.
64. Fisberg M, Machado R. 2015. History of yogurt and current patterns of consumption. *Nutrition Reviews* 73(Suppl. 1): 4–7.
65. Franz CM, Huch M, Mathara JM, Abriouel H, Benomar N, Reid G, et al. 2014. African fermented foods and probiotics. *International Journal of Food Microbiology* 190: 84–96.
66. Frisvad JC, Llh M, Larsen TO, Kumar R, Arnau J. 2018. Safety of the fungal workhorses of industrial biotechnology: Update on the mycotoxin and secondary metabolite potential of *Aspergillus niger, Aspergillus oryzae*, and *Trichoderma reesei*. *Applied Microbiology and Biotechnology* 102: 9481–9515.
67. Fronteras JP, Bullo LLR. 2017. Raw starch-digesting amylase from *Saccharomycopsis fibuligera* 2074 isolated from bubod starter. *Philippine Journal of Science* 146(1): 27–35.
68. García-Burgos M, Moreno-Fernández J, Mjm A, Díaz-Castro J, López-Aliaga I. 2020. New perspectives in fermented dairy products and their health relevance. *Journal of Functional Foods* 72: 104059. doi.org/10.1016/j.jff.2020.104059.
69. García-Ruiz A, González De Llano D, Esteban-Fernández A, Requena T, Bartolomé B, Moreno-Arribas MV. 2014. Assessment of probiotics properties in lactic acid bacteria isolated from wine. *Food Microbiology* 44: 220–225.
70. Ghosh T, Beniwal A, Semwal A, Navani NK. 2019. Mechanistic insights into probiotic properties of lactic acid bacteria associated with ethnic fermented dairy products. *Frontiers in Microbiology* 10: 502. doi: 10.3389/fmicb.2019.00502.
71. Giles-Gómez M, García JGS, Matus V, Quintana IC, Bolívar F, et al. 2016. In vitro and in vivo probiotic assessment of *Leuconostoc mesenteroides* P45 isolated from pulque, a Mexican traditional alcoholic beverage. *Springer Plus* 5(1).
72. Goel A, Halami PM, Tamang JP. 2020. Genome analysis of *Lactobacillus plantarum* isolated from some Indian fermented foods for bacteriocin production and probiotic marker genes. *Frontiers in Microbiology* 11: 40. doi: 10.3389/fmicb.2020.00040.
73. Gubag R, Omoloso DA, Owens JD. 1996. Sapal: A traditional fermented taro [*Colocasia esculenta* (L.) Schott] corm and coconut cream mixture from Papua New Guinea. *International Journal of Food Microbiology* 28(3): 361–367.
74. Han ES, Kim HJ, Choi HK. 2015. Health benefits of *Kimchi*. In: Tamang JP (Ed.), *Health Benefits of Fermented Foods*. CRC Press, Taylor & Francis Group, New York, pp. 343–370.
75. Hartanti AT, Rahayu G, Hidayat I. 2015. *Rhizopus* species from fresh tempeh collected from several regions in Indonesia. *HAYATI Journal of Biosciences* 22(3): 136–142.
76. Hati S, Patel M, Mishra BK, Das S. 2019. Short-chain fatty acid and vitamin production potentials of *Lactobacillus* isolated from fermented foods of Khasi Tribes, Meghalaya, India. *Annals of Microbiology* 69: 1191–1199.
77. Heo S, Lee JH, Jeong DW. 2020. Food-derived coagulase-negative *Staphylococcus* as starter cultures for fermented foods. *Food Science and Biotechnology* 29(8): 1023–1035.
78. Hesseltine CW. 1983. Microbiology of oriental fermented foods. *Annual Review of Microbiology* 37: 575–601.
79. Hesseltine CW. 1991. Zygomycetes in food fermentations. *Mycologist* 5: 162–169.
80. Hesseltine CW, Kurtzman CP. 1990. Yeasts in amylolytic food starters. *Anales del instituto de biologia de la universidad nacional autonoma de Mexico. Serie Botanica* 60: 1–7.
81. Hittinger CT, Steele JL, Ryder DS. 2018. Diverse yeasts for diverse fermented beverages and foods. *Current Opinion in Biotechnology* 49: 199–206.

82. Holck A, Axelsson L, McLeod A, Rode TM, Heir E. 2017. Health and safety considerations of fermented sausages. *Journal of Food Quality* Article ID 9753894. doi.org/10.1155/2017/9753894.
83. Holzapfel WH, Wood BJB. 2014. *Lactic Acid Bacteria: Biodiversity and Taxonomy*. Wiley-Blackwell. ISBN-10: 9781444333831.
84. Hong HH, Kim MK. 2020. Physiochemical quality and sensory characteristics of *koji* made with soybean, rice, and wheat for commercial *doenjang* production. *Foods* 9, 8: 975. doi.org/10.3390/foods9080975.
85. Hsueh YH, Huang KY, Kunene S, Lee TY. 2017. Poly-γ-glutamic acid synthesis, gene regulation, phylogenetic relationships, and role in fermentation. *International Journal of Molecular Sciences* 18(12): 2644. https://doi.org/10.3390/ijms18122644.
86. Hu CH, Ren LQ, Zhou Y, Ye C. 2019. Characterization of antimicrobial activity of three *Lactobacillus plantarum* strains isolated from Chinese traditional dairy food. *Food Science and Nutrition* 7: 1997–2005.
87. Hu L, Liu R, Wang X, Zhang X. 2020. The sensory quality improvement of citrus wine through co-fermentations with selected non-*Saccharomyces* yeast strains and *Saccharomyces cerevisiae*. *Microorganisms* 3: 323. doi.org/10.3390/microorganisms8030323.
88. Huang YC, Wu BH, Chu YL, Chang WC, Wu MC. 2018. Effects of tempeh fermentation with *Lactobacillus plantarum* and *Rhizopus oligosporus* on Streptozotocin-Induced Type II diabetes mellitus in rats. *Nutrients* 10(9): 1143. https://doi.org/10.3390/nu10091143.
89. Hutkins RW. 2018. *Microbiology and Technology of Fermented Foods*, 2nd edition. Wiley, Hoboken, NJ.
90. Ichishima E. 2016. Development of enzyme technology for *Aspergillus oryzae, A. sojae* and *A. luchuensis*, the national microorganisms of Japan. *Bioscience, Biotechnology and Biochemistry* 80(9): 1681–1692.
91. Iheke E, Oshodi A, Omoboye A, Ogunlalu O. 2017. Effect of fermentation on the physicochemical properties and nutritionally valuable minerals of locust bean (*Parkia biglobosa*). *American Journal of Food Technology* 12: 379–384.
92. Irlinger F, Layec S, Hélinck S, Dugat-Bony E. 2015. Cheese rind microbial communities: Diversity, composition and origin. *FEMS Microbiology Letters* 362(2): 1–11.
93. Jang DJ, Chung KR, Yang HJ, Kim KS, Kwon DY. 2015. Discussion on the origin of *kimchi*, representative of Korean unique fermented vegetables. *Journal of Ethnic Foods* 2(3): 126–136.
94. Jeong DW, Le H, Jeong K, Kim CT, Shim ST, et al. 2019. Effects of starter candidates and NaCl on the production of volatile compounds during soybean fermentation. *Journal of Microbiology and Biotechnology* 29: 191–199.
95. Jeyaram K, Mohendro Singh W, Premarani T, Devi AR, Chanu KS, et al. 2008. Molecular identification of dominant microflora associated with "Hawaijar" - A traditional fermented soybean [*Glycine max* (L.)] food of Manipur, India. *International Journal of Food Microbiology* 122: 259–268.
96. Jeyaram K, Romi W, Singh TA, Adewumi GA, Basanti K, et al. 2011. Distinct differentiation of closely related species of *Bacillus subtilis* group with industrial importance. *Journal of Microbiological Methods* 87. https://doi.org/10.1016/j.mimet.2011.08.011.
97. Ji Z, Jin J, Yu G, Mou R, Mao J, et al. 2018. Characteristic of filamentous fungal diversity and dynamics associated with wheat Qu and the traditional fermentation of Chinese rice wine. *International Journal of Food Science and Technology* 53: 1611–1621.
98. Jiang J, Shi B, Zhu D, Cai Q, Chen Y, et al. 2012. Characterization of a novel bacteriocin produced by *Lactobacillus sakei* LSJ618 isolated from traditional Chinese fermented radish. *Food Control* 23: 338–344.
99. Johansen PG, Owusu-Kwarteng J, Parkouda C, Padonou SW, Jespersen L. 2019. Occurrence and importance of yeasts in indigenous fermented food and beverages produced in sub-Saharan Africa. *Frontier in Microbiology* 10: 1789. doi: 10.3389/fmicb.2019.01789.
100. Johnson ME. 2017. A 100-year review: Cheese production and quality. *Journal of Dairy Science* 100(12): 9952–9965.
101. Ju S, Cao Z, Wong C, Liu Y, Foda MF, et al. 2019. Isolation and optimal fermentation condition of *Bacillus subtilis* subsp. *natto* strain WTC016 for nattokinase production. *Fermentation* 5, 4: 92. doi: 10.3390/fermentation5040092.
102. Jung JH, Kim SJ, Lee JY, Yoon SR, You SY, et al. 2019a. Multifunctional properties of *Lactobacillus plantarum* strains WiKim83 and WiKim87 as a starter culture for fermented food. *Food Science & Nutrition* 7(8): 2505–2516.
103. Jung JY, Jeong JW, Lee SY, Jin HM, Choi HW, et al. 2019b. Complete genome sequence of *Leuconostoc kimchii* Strain NKJ218, isolated from homemade kimchi. *Microbiology Resource Announcements* 8(27): e00367–19. doi: 10.1128/MRA.00367-19.
104. Jung MJ, Nam YD, Roh SW, Bae JW. 2012. Unexpected convergence of fungal and bacterial communities during fermentation of traditional Korean alcoholic beverages inoculated with various natural starters. *Food Microbiology* 30(1): 112–123.

105. Kadar AD, Astawan M, Putri SP, Fukusaki E. 2020. Metabolomics-based study of the effect of raw materials to the end product of tempe—an Indonesian fermented soybean. *Metabolites* 10: 367. doi: 10.3390/metabo10090367.
106. Kamada M, Hase S, Fujii K, Miyake M, Sato K, et al. 2015. Whole-genome sequencing and comparative genome analysis of *Bacillus subtilis* strains isolated from non-salted fermented soybean foods. *PLoS ONE* 10(10): e0141369. https://doi.org/10.1371/journal.pone.0141369.
107. Kameda T, Aoki H, Yang Y, Nirmagustina DE, Iwamoto A, et al. 2018. Beneficial effects of dietary tempeh prepared with *Rhizopus stolonifer* on liver function in rats fed with a high-fat diet. *Journal of Nutritional Science and Vitaminology* (Tokyo) 64(5): 379–383.
108. Keishing S, Banu T, Umadevi M. 2015. Effect of fermentation on the nutrient content, antioxidant and antidiabetic activities of Hawaijar, an indigeneous fermented soya of Manipur, India. *Journal of Human Nutrition & Food Science* 3(3): 1066.
109. Khan H, Flint S, Yu PL. 2010. Enterocins in food preservation. *International Journal of Food Microbiology* 141: 1–10.
110. Kharnaior P, Tamang JP. 2021. Bacterial and fungal communities and their predictive functional profiles in *kinema*, a naturally fermented soybean food of India, Nepal and Bhutan. *Food Research International* 110055. doi.org/10.1016/j.foodres.2020.110055.
111. Kiers JL, Van Laeken AEA, Rombouts FM, Nout MJR. 2000. In vitro digestibility of *Bacillus* fermented soya bean. *International Journal of Food Microbiology* 60: 163–169.
112. Kim EK, Ha AW, Choi EO, Ju SY. 2016. Analysis of *kimchi*, vegetable and fruit consumption trends among Korean adults: Data from the Korea National Health and Nutrition Examination Survey (1998-2012). *Nutrition Research and Practice* 10(2): 188–197.
113. Kimura K, Yokoyama S. 2019. Trends in the application of *Bacillus* in fermented foods. *Current Opinion in Biotechnology* 56: 36–42.
114. Kobayashi K, Shimojo S, Watanabe S. 2016. Contribution of a fermentation process using *Bacillus subtilis* (natto) to high polyamine contents of natto, a traditional Japanese fermented soy food. *Food Science and Technology Research* 22(1): 153–157.
115. Kok CR, Hutkins R. 2018. Yogurt and other fermented foods as sources of health-promoting bacteria. *Nutrition Reviews* 76(Supplement 1): 4–15.
116. Kotb E. 2012. Fibrinolytic bacterial enzymes with thrombolytic activity. *Springer Briefs Microbiology*. doi: 10.1007/978-3-642-24980-8_1.
117. Kubo Y, Rooney AP, Tsukakoshi Y, Nakagawa R, Hasegawa H, Kimura K. 2011. Phylogenetic analysis of *Bacillus subtilis* strains applicable to natto (fermented soybean) production. *Applied and Environmental Microbiology* 77(18): 6463–6469.
118. Kumar J, Sharma N, Kaushal G, Samurailatpam S, Sahoo D, Rai AK, Singh SP. 2019. Metagenomic insights into the taxonomic and functional features of kinema, a traditional fermented soybean product of Sikkim Himalaya. *Frontiers in Microbiology* 10: 1744. https://doi.org/10.3389/fmicb.2019.01744.
119. Kwon DY, Chung KR, Jang DJ. 2019a. The history and science of Cheonggukjang, a Korean fermented soybean product. *Journal of Ethnic Food* 6: 5. doi.org/10.1186/s42779-019-0004-8.
120. Kwon YS, Lee S, Lee SH, Kim HJ, Lee CH. 2019b. Comparative evaluation of six traditional fermented soybean products in East Asia: A metabolomics approach. *Metabolites* 9(9): 183. https://doi.org/10.3390/metabo9090183.
121. Laranjo M, Potes ME, Elias M. 2019. Role of starter cultures on the safety of fermented meat products. *Frontiers in Microbiology* 10: 853. https://doi.org/10.3389/fmicb.2019.00853.
122. Lee JH, Paek SH, Shin HW, Lee SY, Moon BS, et al. 2017a. Effect of fermented soybean products intake on the overall immune safety and function in mice. *Journal of Veterinary Science* 18(1): 25–32.
123. Lee JM, Heo S, Kim YS, Lee JH, Jeong DW. 2020. Culture-dependent and -independent investigations of bacterial migration into doenjang from its components meju and solar salt. *PLoS ONE* 15(10): e0239971. https://doi.org/10.1371/journal.pone.0239971.
124. Lee SG, Chang HC. 2017. Assessment of *Bacillus subtilis* SN7 as a starter culture for *Cheonggukjang*, a Korean traditional fermented soybean food, and its capability to control *Bacillus cereus* in *Cheonggukjang*. *Food Control* 73: 946–953.
125. Lee SH, Jung MY, Song JH, Lee M, Chang JY. 2017. Complete genome sequence of *Lactobacillus curvatus* strain WiKim38 isolated from kimchi. *Genome Announcements* 5(18): e00273–17. https://doi.org/10.1128/genomeA.00273-17.
126. Lee SM, Jung JH, Seo JA, Kim YS. 2018. Bioformation of volatile and nonvolatile metabolites by *Saccharomycopsis fibuligera* KJJ81 cultivated under different conditions—carbon sources and cultivation Times. *Molecules* 23: 2762. doi.org/10.3390/molecules23112762.
127. Leech J, Cabrera-Rubio R, Walsh AM, Macori G, Walsh CJ, et al. 2020. Fermented-food metagenomics reveals substrate-associated differences in taxonomy and health-associated and antibiotic resistance determinants. *mSystems* 5(6): e00522–20. doi: 10.1128/mSystems.00522-20.
128. Li C, Song J, Kwok L, Wang J, Dong Y, et al. 2017. Influence of *Lactobacillus plantarum* on yogurt fermentation properties and subsequent changes during post-

fermentation storage. *Journal of Dairy Science* 100(4): 2512–2525.

129. Li W, Ren M, Duo L, Li J, Wang S, et al. 2020. Fermentation characteristics of *Lactococcus lactis* subsp. *lactis* isolated from naturally fermented dairy products and screening of potential starter Isolates. *Frontiers in Microbiology* 11: 1794. doi: 10.3389/fmicb.2020.01794.
130. Linares DM, Gómez C, Renes E, Fresno JM, Tornadijo ME, et al. 2017. Lactic acid bacteria and Bifidobacteria with potential to design natural biofunctional health-promoting dairy foods. *Frontiers in Microbiology* 8: 846. doi: 10.3389/fmicb.2017.00846.
131. Liu XF, Liu CJ, Zeng XQ, Zhang HY, Luo YY, et al. 2020. Metagenomic and metatranscriptomic analysis of the microbial community structure and metabolic potential of fermented soybean in Yunnan Province. *Food Science and Technology* 40(1): 18–25.
132. Londoño Hernández L, Toro C, Ruiz H, Ascacio-Valdés J, Aguilar M, et al. 2017. *Rhizopus oryzae* – Ancient microbial resource with importance in modern food industry. *International Journal of Food Microbiology* 257. doi: 10.1016/j.ijfoodmicro.2017.06.012.
133. Ly S, Mith H, Tarayre C, Taminiau B, Daube G, et al. 2018. Impact of microbial composition of Cambodian traditional dried starters (Dombea) on flavor compounds of rice wine: Combining amplicon sequencing with HP-SPME-GCMS. *Frontiers in Microbiology* 9: 894. doi: 10.3389/fmicb.2018.00894.
134. Maïworé J, Ngoune LT, Piro-Metayer I, Montet D. 2019. Identification of yeasts present in artisanal yoghurt and traditionally fermented milks consumed in the northern part of Cameroon. *Scientific African* 6: e00159. doi.org/10.1016/j.sciaf.2019.e00159.
135. Manan M, Mohamad R, Ariff A. 2017. *Monascus* spp.: A source of natural microbial color through fungal biofermentation. *Journal of Microbiology and Experimentation* 5: 19. doi:10.15406/jmen.2017.05.00148.
136. Marco ML, Sanders ME, Gänzle M, Arrieta MC, Cotter PD, et al. 2021. The International Scientific Association for Probiotics and Prebiotics (ISAPP) consensus statement on fermented foods. *Nature Reviews Gastroenterology and Hepatology*. doi.org/10.1038/s41575-020-00390-5.
137. Mathur H, Beresford TP, Cotter PD. 2020. Health benefits of lactic acid bacteria (LAB) fermentates. *Nutrients* 12: 1679. doi: 10.3390/nu12061679.
138. Mayorga GAC, Palma GBA, Sandoval-Cañas GJ, Ordoñez-Araque RH. 2020. Ancestral fermented indigenous beverages from South America made from cassava (*Manihot esculenta*). *Food Science and Technology* 19. doi.org/10.1590/fst.15220.
139. Meerak J, Lida H, Watanabe Y, Miyashita M, Sato H, Nakagawa Y, Tahara Y. 2007. Phylogeny of poly-γ-glutamic acid-producing *Bacillus* strains isolated from fermented soybean foods manufactured in Asian countries. *Journal of General and Applied Microbiology* 53: 315–323.
140. Melini F, Melini V, Luziatelli F, Ficca AG, Ruzzi M. 2019. Health-promoting components in fermented foods: An up-to-date systematic review. *Nutrients* 11(5): 1189. https://doi.org/10.3390/nu11051189.
141. Menezes AGT, Ramos CL, Cenzi G, Melo DS, Dias DR, et al. 2020. Probiotic potential, antioxidant activity, and phytase production of indigenous yeasts isolated from indigenous fermented foods. *Probiotics and Antimicrobial Protection* 12: 280–288.
142. Mokoena MP. 2017. Lactic acid bacteria and their bacteriocins: Classification, biosynthesis and applications against uropathogens: A mini-review. *Molecules (Basel, Switzerland)* 22(8): 1255. doi.org/10.3390/molecules22081255.
143. Mota De Carvalho N, Costa EM, Silva S, Pimentel L, Fernandes TH, et al. 2018. Fermented foods and beverages in human diet and their influence on gut microbiota and health. *Fermentation* 4(no. 4): 90. doi: 10.3390/fermentation4040090.
144. Msf DL, Kms DS, Albuquerque WWC, Teixeira JAC, Cavalcanti MTH, et al. 2017. *Saccharomyces cerevisiae* from Brazilian kefir-fermented milk: An in vitro evaluation of probiotic properties. *Microbial Pathogenesis* 110: 670–677.
145. Mu Y, Su W, Mu Y, Jiang L. 2020. Combined application of high-throughput sequencing and metabolomics reveals metabolically active microorganisms during Panxian ham processing. *Frontiers in Microbiology* 10: 3012. doi: 10.3389/fmicb.2019.03012.
146. Nakao S. 1972. *Ryori No Kigen* (in Japanese). NHK Press, Tokyo, Tokyo, p. 121.
147. Nam YD, Yi SH, Lim SI. 2012. Bacterial diversity of cheonggukjang, a traditional Korean fermented food, analyzed by barcoded pyrosequencing. *Food Control* 28(1): 135–142.
148. Nami Y, Vaseghi BR, Mohammadzadeh JH, Lotfi H, Eslami S, Hejazi MA. 2019. Probiotic Properties of *Enterococcus* Isolated from artisanal dairy products. *Frontiers in Microbiology* 10: 300. doi: 10.3389/fmicb.2019.00300.
149. Naranjo-Ortiz MA, Gabaldón T. 2019. Fungal evolution: Diversity, taxonomy and phylogeny of the Fungi. *Biological Reviews* 94: 2101–2137.
150. Negus D, Moore C, Baker M, Raghunathan D, Tyson J, Sockett RE. 2017. Predator versus pathogen: How does predatory *Bdellovibrio bacteriovorus* interface with the challenges of killing Gram-negative pathogens in a host setting? *Annual Review of Microbiology* 71: 441–457.
151. Nishito Y, Osana Y, Hachiya T, Popendorf K, Toyoda A, et al. 2010. Whole genome assembly of a natto produc-

tion strain *Bacillus subtilis natto* from very short read data. *BMC Genomics* 11: 243–255.
152. Nkhata SG, Ayua E, Kamau EH, Shingiro JB. 2018. Fermentation and germination improve nutritional value of cereals and legumes through activation of endogenous enzymes. *Food Science & Nutrition* 6(8): 2446–2458.
153. Nout MJR, Aidoo KE. 2002. Asian fungal fermented food. In: Osiewacz HD (Ed.), *The Mycota*. Springer-Verlag, New York, pp. 23–47.
154. Nwagu TN, Ugwuodo CJ, Onwosi CO, Inyima O, Uchendu OC, et al. 2020. Evaluation of the probiotic attributes of *Bacillus* strains isolated from traditional fermented African locust bean seeds (*Parkia biglobosa*), "daddawa." *Annals of Microbiology* 70: 20. doi.org/10.1186/s13213-020-01564-x.
155. O'Connor PM, Kuniyoshi TM, Oliveira RPS, Hill C, Ross RP. 2020. Antimicrobials for food and feed; a bacteriocin perspective. *Current Opinion in Biotechnology* 61: 160–167.
156. Ogunremi OR, Sanni AI, Agrawal R. 2015. Probiotic potentials of yeasts isolated from some cereal-based Nigerian traditional fermented food products. *Journal of Applied Microbiology* 119(3): 797–808.
157. Olowonibi OO. 2017. Isolation and characterization of palm wine strains of *Saccharomyces cerevisiae* potentially useful as bakery yeasts. *European Journal of Experimental Biology* 7: 11. doi: 10.21767/2248-9215.100011.
158. Onwumah M, Okoronkwo CP, Effiong E. 2019. Molecular characterization of yeast isolated from palm wine in Alakahia, Rivers State, Nigeria. *World Science News* 130: 297–304.
159. Palla M, Blandino M, Grassi A, Giordano D, Sgherri C, et al. 2020. Characterization and selection of functional yeast strains during sourdough fermentation of different cereal wholegrain flours. *Scientific Reporters* 10: 12856. doi.org/10.1038/s41598-020-69774-6.
160. Park KY, Jeong JK, Lee YE, Daily J. 2014. Health benefits of kimchi (Korean fermented vegetables) as a probiotic food. *Journal of Medicinal Food* 17(1): 6–20.
161. Parks D, Chung IH, Lee IK, Kim EJ, Niraula S. 2019. Genome sequence of *Bacillus subtilis* natto VK161, a novel strain that produces vitamin K_2. *Microbiology Resource Announcements* 8(35): e00444–19. doi: 10.1128/MRA.00444-19.
162. Parks DH, Chuvochina M, Waite DW, Rinke C, Skarshewski A, et al. 2018. A standardized bacterial taxonomy based on genome phylogeny substantially revises the tree of life. *Nature Biotechnology* 36: 996. https://doi.org/10.1038/nbt.4229.
163. Penna ALB, Nero LA, Todorov SD. 2016. *Fermented Foods of Latin America from Traditional Knowledge to Innovative Applications*. Tylor and Francis, Boca Raton.
164. Polanowska K, Grygier A, Kuligowski M, Rudzińska M, Nowak J. 2020. Effect of tempe fermentation by three different strains of *Rhizopus oligosporus* on nutritional characteristics of faba beans. *LWT-Food Science and Technology* 122: 109024. https://doi.org/10.1016/j.lwt.2020.109024.
165. Pradhan P, Tamang JP. 2019. Phenotypic and genotypic identification of bacteria isolated from traditionally prepared dry starters of the Eastern Himalayas. *Frontiers in Microbiology* 10: 2526. doi: 10.3389/fmicb.2019.02526.
166. Pradhan P, Tamang JP. 2021. Probiotic properties of lactic acid bacteria isolated from traditionally prepared dry starters of the Eastern Himalayas. *World Journal of Microbiology and Biotechnology* 37: 7. doi.10.1007/s11274-020-02975-3.
167. Pretorius IS. 2020. Tasting the *terroir* of wine yeast innovation. *FEMS Yeast Research* 20(1): foz084. https://doi.org/10.1093/femsyr/foz084.
168. Pulvirent A, De Vero L, Blaiotta G, Sidari R, Iosca G, et al. 2020. Selection of wine *Saccharomyces cerevisiae* strains and their screening for the adsorption activity of pigments, phenolics and ochratoxin A. *Fermentation* 6: 80. doi: 10.3390/fermentation6030080.
169. Rasheed HA, Tuoheti T, Zhang Y, Azi F, Tekliye M, et al. 2020. Purification and partial characterization of a novel bacteriocin produced by bacteriocinogenic *Lactobacillus fermentum* BZ532 isolated from Chinese fermented cereal beverage (Bozai). *LWT-Food Science and Technology* 124: 109113. doi.org/10.1016/j.lwt.2020.109113.
170. Ratha P, Jhon DY. 2018. Factors increasing poly-γ-glutamic acid content of cheongguk-jang fermented by *Bacillus subtilis* 168. *Food Science and Biotechnology* 28(1): 103–110.
171. Redondo-Useros N, Gheorghe A, Díaz-Prieto LE, Villavisencio B, Marcos A, et al. 2019. Associations of probiotic fermented milk (PFM) and yogurt consumption with *Bifidobacterium* and *Lactobacillus* components of the gut microbiota in healthy adults. *Nutrients* 11: 651. doi: 10.3390/nu11030651.
172. Rezac S, Kok CR, Heermann M, Hutkins R. 2018. Fermented foods as a dietary source of live organisms. *Frontiers in Microbiology* 9: 1785. doi: 10.3389/fmicb.2018.01785.
173. Romero-Luna HE, Hernández-Sánchez H, Dávila-Ortiz G. 2017. Traditional fermented beverages from Mexico as a potential probiotic source. *Annals of Microbiology* 67: 577–586.
174. Ropars J, Didiot E, Rodríguez De La Vega RC, Bennetot B, Coton M, et al. 2020. Domestication of the emblematic white cheese-making fungus *Penicillium camemberti* and its diversification into two varieties. *Current Biology* 30(22): 4441–4453.

175. Ropars J, López-Villavicencio M, Snirc A, Lacoste S, Giraud T. 2017. Blue cheese-making has shaped the population genetic structure of the mould *Penicillium roqueforti*. *PLoS ONE* 12(3): e0171387. doi.org/10.1371/journal.pone.0171387.
176. Roslan R, Rehan MM, Kamarudin KR, Noor HM, Huda-Faujan N, et al. 2018. Isolation and identification of amylolytic bacteria from Ragi. *Malaysian Applied Biology Journal* 47(2): 83–88.
177. Rul F. 2017. Yogurt: Microbiology, organoleptic properties and probiotic potential. In: *Fermented Foods, Part II: Technological Interventions*, CRC Press, Food Biology Series, 9781138637849. hal–01579303.
178. Sakandar HA, Hussain R, Khan QF, Zhang H. 2020. Functional microbiota in Chinese traditional Baijiu and Mijiu Qu (starters): A review. *Food Research International* 138: 109830. doi.org/10.1016/j.foodres.2020.109830.
179. Sánchez Mainar M, Stavropoulou DA, Leroy F. 2017. Exploring the metabolic heterogeneity of coagulase-negative staphylococci to improve the quality and safety of fermented meats: A review. *International Journal of Food Microbiology* 17(247): 24–37.
180. Şanlier N, Gökcen BB, Sezgin AC. 2019. Health benefits of fermented foods. *Critical Review of Food Science and Nutrition* 59(3): 506–527.
181. Sha SP, Suryavanshi MS, Jani K, Sharma A, Shouche Y, et al. 2018. Diversity of yeasts and molds by culture-dependent and culture-independent methods for mycobiome surveillance of traditionally prepared dried starters for the production of Indian alcoholic beverages. *Frontiers in Microbiology* 9: 2237. doi: 10.3389/fmicb.2018.02237.
182. Sha SP, Suryavanshi MS, Tamang JP. 2019. Mycobiome diversity in traditionally prepared starters for alcoholic beverages in India by high-throughput sequencing method. *Frontiers in Microbiology* 10: 348. doi: 10.3389/fmicb.2019.003482237.
183. Shea Z, Singer WM, Zhang B. 2020. *Soybean Production, Versatility, and Improvement, Legume Crops - Prospects, Production and Uses. Mirza Hasanuzzaman*. IntechOpen. doi: 10.5772/intechopen.91778.
184. Shin DH, Jeong D. 2015. Korean traditional fermented soybean products: Jang. *Journal of Ethnic Foods* 2(1): 2–7.
185. Shurtleff W, Aoyagi A. 2007. *History of Tempeh*. Soyinfo Center, Lafayette, California.
186. Shurtleff W, Aoyagi A. 2012. *History of Koji - Grains And/or Soybeans Enrobed with a Mold Culture (300 BCE to 2012)*. Soyinfo Center, Lafayette, California.
187. Singh TA, Devi KR, Ahmed G, Jeyaram K. 2014. Microbial and endogenous origin of fibrinolytic activity in traditional fermented foods of Northeast India. *Food Research International* 55: 356–362.
188. Slattery C, Cotter PD, O'Toole PW. 2019. Analysis of health benefits conferred by *Lactobacillus* species from kefir. *Nutrients* 11(6): 1252. doi.org/10.3390/nu11061252.
189. Soccol CR, Ferreira Da Costa ES, Letti LAJ, Karp SG, Woiciechowski AL, et al. 2017. Recent developments and innovations in solid state fermentation. *Biotechnology Research and Innovation* 1(1): 52–71.
190. Sohn H, Chang YH, Yune JH, Jeong CH, Shin DM, et al. 2020. Probiotic properties of *Lactiplantibacillus plantarum* LB5 Isolated from kimchi based on nitrate reducing capability. *Foods* 9: 1777. doi.org/10.3390/foods9121777.
191. Sokrab AM, Mohamed Ahmed IA, Babiker EE. 2014. Effect of fermentation on antinutrients, and total and extractable minerals of high and low phytate corn genotypes. *Journal of Food Science and Technology* 51(10): 2608–2615.
192. Somashekaraiah R, Shruthi B, Deepthi BV, Sreenivasa MY. 2019. Probiotic properties of Lactic Acid Bacteria isolated from neera: A naturally fermenting coconut palm nectar. *Frontiers in Microbiology* 10: 1382. doi: 10.3389/fmicb.2019.01382.
193. Son SH, Jeon HL, Yang SJ, Sim MH, Kim YJ, et al. 2018. Probiotic lactic acid bacteria isolated from traditional Korean fermented foods based on β-glucosidase activity. *Food Science and Biotechnology* 27(1): 123–129.
194. Song MY, Van-Ba H, Park WS, Yoo JY, Kang HB, et al. 2018. Quality Characteristics of functional fermented sausages added with encapsulated probiotic *Bifidobacterium longum* KACC 91563. *Korean Journal for Food Science of Animal Resources* 38(5): 981–994.
195. Soomro AH, Masud T. 2008. Selection of yoghurt starter culture from indigenous isolates of *Streptococcus thermophilus* and *Lactobacillus delbrueckii* subsp. *bulgaricus* on the basis of technological properties. *Annals of Microbiology* 58: 67–71.
196. Stahel G. 1946. Foods from fermented soybeans; as prepared in the Netherlands Indies; *tempe*, a tropical staple. *Memoirs of the New York Botanical Garden* 47(564): 285–296.
197. Steinkraus KH. 1996. *Handbook of Indigenous Fermented Food*, 2nd edition. Marcel Dekker, Inc, New York.
198. Stewart GG. 2017. The production of secondary metabolites with flavour potential during brewing and distilling wort fermentations. *Fermentation* 3(4): 63. doi.org/10.3390/fermentation3040063.
199. Stiemsma LT, Nakamura RE, Nguyen JG, Michels KB. 2020. Does consumption of fermented foods modify the human gut microbiota? *The Journal of Nutrition* 150(7): 1680–1692.
200. Su C, Zhang KZ, Cao XZ, Yang JG. 2020. Effects of *Saccharomycopsis fibuligera* and *Saccharomyces*

cerevisiae inoculation on small fermentation starters in Sichuan-style Xiaoqu liquor. *Food Research International* 137: 109425. doi.org/10.1016/j.foodres.2020.109425.
201. Takeshita N. 2020. Fungal research in Japan: Tradition and future. *Fungal Biology and Biotechnology* 7: 14. https://doi.org/10.1186/s40694-020-00104-1.
202. Tamang B, Tamang JP, Schillinger U, Franz CMAP, Gores M, et al. 2008. Phenotypic and genotypic identification of lactic acid bacteria isolated from ethnic fermented tender bamboo shoots of North East India. *International Journal of Food Microbiology* 121: 35–40.
203. Tamang JP. 2010. *Himalayan Fermented Foods: Microbiology, Nutrition, and Ethnic Values*. CRC Press, Taylor & Francis Group, New York.
204. Tamang JP. 2015a. Naturally fermented ethnic soybean foods of India. *Journal of Ethnic Foods* 2: 8–17.
205. Tamang JP. 2015b. *Health Benefits of Fermented Foods and Beverages*. CRC Press, Taylor & Francis Group, New York, pages 636. ISBN: 978–1–4665–88097.
206. Tamang JP. 2016. *Ethnic Fermented Foods and Alcoholic Beverages of Asia*. Springer, New Delhi, pages 409. ISBN: 978–81–322–2798–4.
207. Tamang JP. 2020a. *Ethnic Fermented Foods and Alcoholic Beverages of India: Science History and Culture*. Springer Nature, Singapore, pages 685. ISBN: 978–981–15–1486–9.
208. Tamang JP. 2020b. History and Culture of Indian Ethnic Fermented Foods and Beverages. In: Tamang JP (Ed.), *Ethnic Fermented Foods and Alcoholic Beverages of India: Science History and Culture*. Springer Nature, Singapore, pp. 1–40. ISBN: 978–981–15–1486–9.
209. Tamang JP, Cotter P, Endo A, Han NS, Kort R, et al. 2020. Fermented foods in a global age: East meets west. *Comprehensive Reviews in Food Science and Food Safety* 19: 184–217. doi: 10.1111/1541-4337.12520.
210. Tamang JP, Dewan S, Thapa S, Olasupo NA, Schillinger U, et al. 2000. Identification and enzymatic profiles of predominant lactic acid bacteria isolated from soft-variety *chhurpi*, a traditional cheese typical of the Sikkim Himalayas. *Food Biotechnology* 14: 99–112.
211. Tamang JP, Fleet GH. 2009. Yeasts diversity in fermented foods and beverages. In: Satyanarayana T, Kunze G (Eds.), *Yeasts Biotechnology: Diversity and Applications*. Springer, New York, pp. 169–198.
212. Tamang JP, Holzapfel WH, Watanabe K. 2016a. Diversity of microorganisms in global fermented foods and beverages. *Frontiers in Microbiology* 7: 377. doi: 10.3389/fmicb.2016.00377.
213. Tamang JP, Kailasapathy K. 2010. *Fermented Foods and Beverages of the World*. CRC Press, Taylor & Francis Group, New York, pages 448. ISBN: 9781420094954.
214. Tamang JP, Nikkuni S. 1996. Selection of starter culture for production of kinema, fermented soybean food of the Himalaya. *World Journal of Microbiology and Biotechnology* 12(6): 629–635.
215. Tamang JP, Nikkuni S. 1998. Effect of temperatures during pure culture fermentation of Kinema. *World Journal of Microbiology and Biotechnology* 14(6): 847–850.
216. Tamang JP, Shin DH, Jung SJ, Chae SW. 2016b. Functional properties of microorganisms in fermented foods. *Frontiers in Microbiology* 7: 578. doi: 10.3389/fmicb.2016.00578.
217. Tamang JP, Tamang B, Schillinger U, Franz CMAP, Gores M, et al. 2005. Identification of predominant lactic acid bacteria isolated from traditional fermented vegetable products of the Eastern Himalayas. *International Journal of Food Microbiology* 105(3): 347–356.
218. Tamang JP, Tamang B, Schillinger U, Guigas C, Holzapfel WH. 2009. Functional properties of lactic acid bacteria isolated from ethnic fermented vegetables of the Himalayas. *International Journal of Food Microbiology* 135(1): 28–33.
219. Tan HY, Chen SW, Hu SY. 2019. Improvements in the growth performance, immunity, disease resistance, and gut microbiota by the probiotic *Rummeliibacillus stabekisii* in Nile tilapia (*Oreochromis niloticus*). *Fish Shellfish Immunology* 92: 265–275.
220. Tang Q, He G, Huang J, Wu C, Jin Y, et al. 2019. Characterizing relationship of microbial diversity and metabolite in Sichuan *xiaoqu*. *Frontiers in Microbiology* 10: 696. doi.org/10.3389/fmicb.2019.00696.
221. Taylor BC, Lejzerowicz F, Poirel M, Shaffer JP, Jiang L, et al. 2020. Consumption of fermented foods is associated with systematic differences in the gut microbiome and metabolome. *mSystems* 5(2): e00901–19. doi: 10.1128/mSystems.00901-19.
222. Tsuyoshi N, Fudou R, Yamanaka S, Kozaki M, Tamang N, et al. 2005. Identification of yeast strains isolated from marcha in Sikkim, a microbial starter for amylolytic fermentation. *International Journal of Food Microbiology* 99(2): 135–146.
223. Urushibata Y, Tokuyama S, Tahara Y. 2002. Characterization of the *Bacillus subtilis ywsC* gene, involved in –polyglutamic acid production. *Journal of Bacteriology* 184(2): 337–343.
224. Van Reckem E, Geeraerts W, Charmpi C, Van Der Veken D, De Vuyst L, et al. 2019. Exploring the link between the geographical origin of European fermented foods and the diversity of their bacterial communities: The case of fermented meats. *Frontiers in Microbiology* 10: 2302. https://doi.org/10.3389/fmicb.2019.02302.
225. Verardo V, Gómez-Caravaca AM, Tabanelli G. 2020. Bioactive components in fermented foods and food by-products. *Foods* 9(2): 153. doi: 10.3390/foods9020153.
226. Verni M, Verardo V, Rizzello CG. 2019. How fermentation affects the antioxidant properties of cereals and

legumes. *Foods (Basel, Switzerland)* 8(9): 362. doi.org/10.3390/foods8090362.

227. Wach Y, Do TL, Do TBH, Do TY, Haure M, et al. 2018. Prospects for food fermentation in South-East Asia, topics from the tropical fermentation and biotechnology network at the end of the AsiFood Erasmus+Project. *Frontiers in Microbiology* 9: 2278. doi: 10.3389/fmicb.2018.02278.
228. Wade M, Strickland M, Osborne J, Edwards C. 2019. Role of *Pediococcus* in winemaking. *Australian Journal of Grape and Wine Research* 25: 7–24.
229. Walker GM. 2014. Microbiology of Winemaking. In: Batt C, Tortorello MA (Eds.), *Encyclopaedia of Food Microbiology*, 2nd edition. Elsevier Ltd, Oxford, pp. 787–792.
230. Walker GM, Stewart GG. 2016. *Saccharomyces cerevisiae* in the production of fermented beverages. *Beverages* 2: 30.
231. Weng Y, Yao J, Sparks S, Wang KY. 2017. Nattokinase: An oral antithrombotic agent for the prevention of cardiovascular disease. *International Journal of Molecular Sciences* 18(3): 523. https://doi.org/10.3390/ijms18030523.
232. Whittington HD, Dagher SF, Bruno-Bárcena JM. 2019. Production and conservation of starter cultures: From "backslopping" to controlled fermentations. In: Azcarate-Peril M, Arnold R, Bruno-Bárcena J (Eds.), *How Fermented Foods Feed a Healthy Gut Microbiota*. Springer, Cham. doi.org/10.1007/978-3-030-28737-5_5.
233. Wu HC, Cheng MJ, Wu MD, Chen JJ, Chen YL, et al. 2019. Secondary metabolites from the fermented rice of the fungus *Monascus purpureus* and their bioactivities. *Natural Product Research* 33(24): 3541–3550.
234. Xiang H, Sun-Waterhouse D, Geoffrey IN, Waterhouse CC, Zheng RZ. 2019. Fermentation-enabled wellness foods: A fresh perspective. *Food Science and Human Wellness* 8(3): 203–243.
235. Xie C, Zeng H, Li J, Qin L. 2019. Comprehensive explorations of nutritional, functional and potential tasty components of various types of Sufu, a Chinese fermented soybean appetizer. *Food Science and Technology* 39(Suppl. 1): 105–114.
236. Xing Z, Tang W, Geng W, Zheng Y, Wang Y. 2017. In vitro and in vivo evaluation of the probiotic attributes of *Lactobacillus kefiranofaciens* XL10 isolated from Tibetan kefir grain. *Applied Microbiology and Biotechnology* 101(6): 2467–2477.
237. Yasui M, Oda K, Masuo S, Hosoda S, Takuya Katayama T, et al. 2020. Invasive growth of *Aspergillus oryzae* in rice *koji* and increase of nuclear number. *Fungal Biology and Biotechnology* 7(8). doi.org/10.1186/s40694-020-00099-9.
238. Yoon JW, Kang S, Mheen T, Ahn JS, Lee HJ, et al. 2000. *Lactobacillus kimchii* sp. nov., a new species from Kimchi. *International Journal of Systematic and Evolutionary Microbiology* 50(5): 1789–1795.
239. Zhang P, Zhang P, Xie M, An F, Qiu B, Wu R. 2018. Metaproteomics of microbiota in naturally fermented soybean paste, *da-jiang*. *Journal of Food Science* 83: 1342–1349.
240. Zhao W, Liu Y, Latta M, Ma W, Wu Z, et al. 2019. Probiotics database: A potential source of fermented foods. *International Journal of Food Properties* 22(1): 198–217.
241. Zheng J, Wittouck S, Salvett E, Franz CMAP, Harris MB, et al. 2020. A taxonomic note on the genus *Lactobacillus*: Description of 23 novel genera, emended description of the genus *Lactobacillus beijerinck* 1901, and union of Lactobacillaceae and Leuconostocaceae. *International Journal of Systematic and Evolutionary Microbiology* 70(4): 2782–2858.
242. Zhu B, Qi F, Wu J, Yin G, Hua J, et al. 2019. Red yeast rice: A systematic review of the traditional uses, chemistry, pharmacology, and quality control of an important Chinese folk medicine. *Frontiers in Pharmacology* 10: 1449. https://doi.org/10.3389/fphar.2019.01449.
243. Zielińska D, Kołożyn-Krajewska D. 2018. Food-origin lactic acid bacteria may exhibit probiotic properties: Review. *BioMed Research International* ID 5063185. doi.org/10.1155/2018/5063185.
244. Zommiti M, Feuilloley MGJ, Nathalie C. 2020. Update of probiotics in human world: A nonstop source of benefactions till the end of time. *Microorganisms* 8(12): 1907. doi: 10.3390/microorganisms8121907.
245. Zoumpopoulou G, Tzouvanou A, Mavrogonatou E, Alexandraki V, Georgalaki M, et al. 2017. Probiotic features of lactic acid bacteria isolated from a diverse pool of traditional Greek dairy products regarding specific strain–host interactions. *Probiotics and Antimicrobial Proteins* 10(2): 313–322.

CHAPTER 12

No Microbes, No Cheese

Maria Kazou and Effie Tsakalidou

Laboratory of Dairy Research, Department of Food Science and Human Nutrition, Agricultural University of Athens, Athens 11855, Greece

12.1 CHEESE FOR LIFE: THE HISTORY

The advent of dairying was a major innovation in prehistoric societies that enabled a supply of nutritious food without the need to slaughter precious livestock. In particular, the production of cheese was a momentous development; it allowed the preservation of milk in a transportable and non-perishable form, with cheese being also a more digestible food than milk for early prehistoric humans. Iconographic and written evidence from the mid-third millennium BC faintly documents the history of cheese making, although its origins probably lie much earlier in prehistory. The finding of ample milk residues in clay pots from seventh millennium BC sites in North Western Anatolia provided the earliest evidence of milk processing, although the technology could not be defined. The discovery of pottery vessels with small holes appears at early Neolithic sites in Europe in the sixth millennium BC and these have been typologically interpreted as cheese sieves. Organic residues preserved in pottery containers have provided direct evidence for early milk use during the Neolithic period in the Near East and South Eastern Europe, Northern Africa, Denmark, and the British Isles (Salque et al., 2013).

The dissemination of the cheese making craft went hand in hand with the spread of civilization through the Near East, Greece, and Rome, as indicated by references to cheese in the Old Testament, and ancient Greek and Roman literature. After the fall of the Western Roman Empire and during the Middle Ages in Europe (fifth–fifteenth century AD), cheese making technology was significantly improved and diversified, mainly in monasteries and feudal estates, which were self-contained communities that developed many other agriculture and food commodities as well. The colonization by Europeans of the Americas in fifteenth century AD and later of Oceania introduced cheese making to these regions, and cheese became an important component of the culinary tradition in North and South America, Australia, and New Zealand. Europeans and later Americans introduced cheese to Asia. Despite the fact that cheese is still not a regular staple food in the diet of most Asian countries, the Tibetans and Mongolians have a long history in dairy food production and might have played a role in the development of cheese making in China (Kindstedt, 2012).

Cheese making remained a craft until quite recently. Cheese manufacture was a home or farm activity until the mid-nineteenth century AD. There were thousands of small-scale cheese makers and there must have been big variations within a given cheese type. Interestingly, despite the current scientific and technological advances, and the fact that nowadays cheese is mostly produced on an industrial scale, either in big dairy companies or in small and medium dairy enterprises, there are significant inter- and intra-factory variations regarding the quality and safety of even well-defined cheese

Good Microbes in Medicine, Food Production, Biotechnology, Bioremediation, and Agriculture, First Edition. Edited by Frans J. de Bruijn, Hauke Smidt, Luca S. Cocolin, Michael Sauer, David Dowling, and Linda Thomashow.

varieties. It should be highlighted though, that the cheese curds of many famous cheese varieties, e.g. Parmigiano Reggiano, Grana Padano, Emmental, and Roquefort, are produced in many farm-level dairies under the supervision of the respective producers' consortia and the cheeses then age in and are marketed by the central consortium facilities (Cheese problems solved 2007; Cheese and Microbes, 2014; Encyclopedia of Dairy Sciences, 2022).

12.2 THE TECHNOLOGY

Cheese is a universal fermented dairy product, which appears in a great diversity of flavors, aromas, textures, and shapes around the globe. Most cheese varieties evolved unintentionally due to particular local conditions, such as species and breeds of dairy animals available, vegetation, and climate, factors that in turn shaped the chemical composition and microbiota of the milk. Additionally, human skills along with social, cultural, and religious backgrounds played a significant role.

In the early days of cheese making history, "mishaps" during manufacture and storage, e.g. growth of molds, could not be always avoided. Most probably, those "mishaps" that led to well appreciated quality changes were subsequently incorporated in the manufacturing procedure, which, over the years, underwent modifications and improvements, leading to widely diversified cheese varieties. Behind every traditional type of cheese there is a fascinating story. By exploring the role of the cheese makers throughout world history and by understanding a few basic principles of cheese science and technology, we can see how different cheeses have been shaped by and tailored to their surrounding environment, as well as have been defined by social and cultural context. Since the beginning of research on cheese microbiology, chemistry, and technology at the end of the nineteenth century AD, cheese making is quite well understood. However, there are still gaps in our knowledge, and, thus, research has to be continued and intensified (Kindstedt, 2012).

The production of cheese is technically a rather simple fermentation process, which involves four major actors, i.e. milk, rennet, microorganisms and salt. It starts with the coagulation of milk proteins, which is mostly achieved enzymatically, i.e. rennet, and concurrent acidification, i.e. via the organic acids produced by microorganisms, namely lactic acid bacteria (LAB). The coagulation yields a semi-solid curd, which contains caseins, the main milk proteins, and the major part of milk fat. The liquid phase, called whey, contains the water-soluble lactose and the so-called whey proteins, and it is expelled. The straining process is commonly achieved using a coarse textile, called cheesecloth, or plastic and metal strainers. The curd is further processed in various ways, e.g. by either salting, cooking, pressing, immersing in brine, and it is either consumed fresh or left to ripen under varying conditions, depending on the cheese type (Cheese problems solved, 2007; Cheese and Microbes, 2014; Encyclopedia of Dairy Sciences, 2022).

Fresh cheeses, such as cottage cheese, cream cheese, and quark, are mainly produced via acidification only, and in some countries these are the principal cheese varieties, which are used widely in other food products as well, e.g. cheese-cakes and cheese-based dips and sauces. On the other hand, when milk coagulation is achieved by both enzymes, i.e. rennet, and milk acidification, curds are left to ripen; most cheese varieties are produced by this dual mechanism. Finally, thermal treatment of the whey results in the denaturation, and, thus, coagulation of whey proteins, leading to the production of the so-called whey cheeses, such as Ricotta (Italy), Brunost (Norway) and Myzithra (Greece) (Cheese and Microbes, 2014; Encyclopedia of Dairy Sciences, 2022).

The technology of coagulating milk using animal rennet, i.e. the tissue material deriving from the abomasum, which is the fourth and final stomach compartment in ruminants, e.g. lambs, kids, and calves, or a brine extract thereof, may have been appreciated fairly soon after milk collection commenced in the Neolithic period (Kindstedt, 2012). In addition, plant origin coagulating enzymes, e.g. from *Cynara cardunculus* L., are used in the production of several artisanal cheeses, mainly in Southern Europe, e.g. Portugal (Serra da Estrela cheese) and Spain (Flor de Guía cheese), and in West Africa as well (Bivar Roseiro et al., 2003). It should be stressed, however, that the high cheese production of modern times has led to the shortage of such milk clotting enzymes, demanding the introduction of safe substitutes, namely the biotechnological rennet produced by microorganisms, such as *Aspergillus oryzae*, *Rhizomucor miehei*, *Rhizomucor pusillus*, *Endothia parasitica*, and *Irpex lactis*, which are widely used to produce rennet at large industrial scale (Ben Amira et al., 2017).

From the very first moment of cheese making until the end of cheese ripening, numerous microbes are present and metabolically active. Their enzymes, along with the indigenous milk enzymes and the added rennet, act on the milk components, i.e. lactose, proteins, fat, and citrate, resulting in changes in the flavor, aroma, texture, and nutritional value of cheese (see below).

12.3 THE MARKET

Nowadays, more than 1000 cheese varieties are produced around the globe, mainly from cow, sheep, goat, and buffalo milk. Interestingly, a number of European cheeses have been granted Protected Geographical Status under EU law through (a) the Protected Designation of Origin (PDO) regime, which identifies products that are produced, processed, and prepared in a specific geographical area, using the recognized know-how of local producers and ingredients from the region concerned, and (b) the Protected Geographical Indication (PGI), which identifies products, whose quality or reputation is linked to the place or region where they are produced, processed or prepared, although the ingredients used need not necessarily come from that geographical area. The legislation has been designed to protect regional and traditional European foods, among them cheeses, and came into force in 1992. Italy, France, Spain, and Greece are the leading countries as far as the number of cheeses under the PDO regime is concerned (https://www.pdopgi.eu/european-certification-system).

According to recent Statista data, the global market of cheese has been valued at approximately 70 billion $US in 2019 and is forecast to reach to 113 billion $US by 2025 (https://www.statista.com/statistics/602542/cheese-market-value-worldwide). The EU, the US, New Zealand, Belarus, and Australia account for 75% of the total world trade (FAO, 2018). The EU remains the largest global exporter of cheese with about 25% of its exports shipped to the US and Japan. The recent imposition of US tariffs on cheese (and butter) from the EU are expected to reduce exports but the impact is expected to be rather limited since demand for imported high quality cheeses is assumed to be relatively inelastic (USDA, 2019)

12.4 MICROBES, MILK, AND CHEESE: A LONG LASTING THREESOME LOVE AFFAIR

A wide variety of fermented foods and beverages, such as cheese, yogurt, kefir, koumiss, bread, table olives, wine, beer, kimchi, sauerkraut, tea, chocolate etc., are significant components of the human diet around the world for millennia. We now know that during fermentation, microorganisms convert the raw material into new foods and beverages that are even more complex and rich in flavor and aroma, and in most cases they are of higher nutritional value as well.

Louis Pasteur, the French chemist and biologist, popularly known as the "Father of Microbiology," was the first to reveal, in the mid-nineteenth century AD, the existence of microorganisms and their role in food fermentations. His research also showed that the growth of many of these microbes can spoil liquids, such as milk, and that these microbes can be inactivated by heating, a process called after him pasteurization, i.e. exposure of a food to an elevated temperature for a period of time sufficient to destroy certain microorganisms, such as those that can cause spoilage or disease, without drastically changing the food chemical composition.

Milk is sterile at secretion in the udder; however, it is contaminated by microbes even before it leaves the udder. Except in the case of mastisis, which is an inflammatory reaction of the udder tissue due to physical trauma or microbial infections, the microbes at this point are harmless and in low numbers. Further contamination of the milk by microorganisms takes place during milking, handling, storage, and other processing actions. The indigenous milk microbiota, which is either fully (in the case of raw milk) or partially (in the case of pasteurized milk) transferred to cheese, can be roughly divided into three groups (Olajide and LaPointe, 2022):

(a) The so-called pro-technological microorganisms, which mostly comprise LAB. When milk is transformed to cheese, some LAB [e.g. species of the genera *Lactococcus*, *Streptococcus*, and some species of the genus *Lactobacillus* before its taxonomic revision (Zheng et al. 2020)] rapidly convert milk lactose to lactic acid, which decreases the cheese pH, and in this way the cheese safety is ensured; these are the so-called starter LAB (SLAB). On the other hand, other LAB [e.g. species of the genera *Lactobacillus* before its taxonomic revision (Zheng et al. 2020), *Pediococcus*, and *Leuconostoc*] are mostly responsible for shaping the sensory profile of cheese; these are called non-starter LAB (NSLAB) (Table 12.1).

(b) Spoilage microorganisms, which are responsible for the deterioration of cheese flavor, aroma, texture, or color, and make it unattractive or unsuitable for consumption. The main milk spoilage microorganisms are psychrothrophic (cold-loving)

TABLE 12.1 Main microbes involved in cheese making

Microbial Group/Species	Cheese
Starter Cultures	
Starter lactic acid bacteria (SLAB)	
Lactococcus lactis subsp. *lactis*	Cheddar, Dutch-type cheeses (e.g. Gouda and Edam), Pasta-filata cheeses (e.g. Mozzarella, Scamorza, Caciocavallo, Ragusano, Kashkaval, Provolone), Swiss-type cheeses (e.g. Emmental and Gruyère), surface-ripened cheeses (e.g., Stilton, Roquefort, Camembert, Limburger, Brick), Cottage cheese
Lactococcus lactis subsp. *cremoris*	Cheddar, Dutch-type cheeses (e.g. Gouda and Edam), Pasta-filata cheeses (e.g. Mozzarella, Scamorza, Caciocavallo, Ragusano, Kashkaval, Provolone), Swiss-type cheeses (e.g. Emmental and Gruyère), surface-ripened cheeses (e.g., Stilton, Roquefort, Camembert, Limburger, Brick)
Streptococcus thermophilus	Feta, Pecorino Romano, Pasta-filata cheeses (e.g. Mozzarella, Scamorza, Caciocavallo, Caciocavallo Pugliese, Ragusano, Kashkaval, Provolone), Swiss-type cheeses (e.g. Emmental and Gruyère)
Lactobacillus delbrueckii subsp. *bulgaricus*	Feta, Grana-type cheeses (e.g. Parmigiano Reggiano and Grana Padano), Pasta-filata cheeses (e.g. Mozzarella, Scamorza, Caciocavallo, Caciocavallo Pugliese, Ragusano, Kashkaval, Provolone), Swiss-type cheeses (e.g. Emmental and Gruyère)
Lactobacillus delbrueckii subsp. *lactis*	Pecorino Romano, Grana-type cheeses (e.g. Parmigiano Reggiano and Grana Padano), Caciocavallo Pugliese
Limosilactobacillus fermentum	Grana-type cheeses (e.g. Parmigiano Reggiano and Grana Padano), Caciocavallo Pugliese
Lactobacillus helveticus	Pecorino Romano, Grana-type cheeses (e.g. Parmigiano Reggiano and Grana Padano), Pasta-filata cheeses (e.g. Mozzarella, Scamorza, Caciocavallo, Caciocavallo Pugliese, Ragusano, Kashkaval, Provolone), Swiss-type cheeses (e.g. Emmental and Gruyère)
Leuconostoc mesenteroides	Dutch-type cheeses (e.g. Gouda and Edam)
Non-starter lactic acid bacteria (NSLAB)	
Levilactobacillus brevis *Lacticaseibacillus casei* *Latilactobacillus curvatus* *Lacticaseibacillus paracasei* *Lactiplantibacillus plantarum* *Lacticaseibacillus rhamnosus* *Pediococcus* spp. *Leuconostoc* spp.	most cheeses
Propionic acid bacteria (PAB)	
Propionibacterium freudenreichii subsp. *shermanii*	Swiss-type cheeses (e.g. Emmental and Gruyère), Maasdam, Graviera
Molds	
Penicillium candidum	Brie
Penicillium camemberti	Camembert, Brie
Penicillium roqueforti	Roquefort, Gorgonzola, Stilton, Danablu

(Continued)

TABLE 12.1 (Continued)

Microbial Group/Species	Cheese
Smear bacteria and yeasts	
Arthrobacter spp. *Brevibacterium linens* *Candida* spp. *Corynebacterium casei* *Debaryomyces hansenii* *Geotrichum candidum* *Kluyveromyces* spp. *Microbacterium gubbeenense* *Micrococcus caseolyticus* *Micrococcus freudenreichii* *Rhodotorula* spp. *Staphylococcus equorum* *Trichosporon* spp. *Yarrowia lipolytica*	Smear-ripened cheeses (e.g. Camembert, Limburger, Munster, Tilsit, Brick, Appenzeller)

and are destroyed by pasteurization; however, some species, such as *Pseudomonas fluorescens* and *Pseudomonas fragi* produce proteolytic and lipolytic extracellular enzymes, which are heat stable and cause spoilage. Moreover, some species or strains of the genera *Bacillus* and *Clostridium*, can survive pasteurization and grow at refrigeration temperatures, causing spoilage problems as well.

(c) Pathogenic microorganisms, which are responsible for milk-borne diseases. Diseases, such as tuberculosis, brucellosis and typhoid fever, result from the ingestion of raw milk or dairy products made with milk that was not properly pasteurized or was poorly handled causing post-processing contamination. Bacterial pathogens, such as *Bacillus cereus, Listeria monocytogenes, Yersinia enterocolitica, Salmonella* spp., *Escherichia coli* O157:H7 and *Campylobacter jejuni*, are still of concern today in raw milk and other dairy products. It should also be noted that molds, mainly species of the genera *Aspergillus, Fusarium*, and *Penicillium*, can grow in milk and dairy products, and if the conditions are favorable, they may produce health hazardous mycotoxins. On the other hand, though, molds like *Penicillium camemberti* and *Penicillium roqueforti* are significant members of the pro-technological microbiota in Camembert and Roquefort, respectively.

Microbes associated with cheese production are deriving not only from milk but they are also introduced either unintentionally via, e.g. animal teats, animal feed, bedding material, cheese makers, equipment and utensils, cheese production and ripening facilities (in fact, these facilities develop their own microbial ecosystem symbiotically with the cheeses produced in those), or deliberately, i.e. added as starter and/or non-starter cultures (see section below).

It should be stressed that the successful transformation of milk to cheese is not an aseptic, sterile process, and, thus, it demands good hygiene practices. Depending on the individual characteristics of the contaminating microbes, these can either have a beneficial impact on cheese, or can lead to the development of off-flavors, bitterness, rancidity, poor texture, and unpleasant appearance (spoilage), or even pose health risks (pathogens). As a general rule to prevent cheese making contamination by spoilage and pathogenic microbes, (a) these undesirable microbes must be kept out (via sanitation), (b) they should be eliminated (via heat or antimicrobial treatment), and (c) their growth should be inhibited (via pH, temperature, salt content, moisture). Moreover, extensive monitoring of cheese making along with compositional testing analysis, using either classical and/or advanced physicochemical and microbiological methodologies, should be in place to control contamination.

Given the complexity of the cheese microbiota as well as its plasticity, i.e. the dynamic succession of microbial groups during cheese making and ripening, the major challenge is to reliably assess the route and fate of each microbial group or even certain microbial strains during cheese making, and to understand their role and contribution in the quality and safety of the final product. Indeed, it has been shown that microbial communities and sensory profiles for the same cheese variety significantly fluctuate at the intra- and inter-factory level (Cheese problems solved 2007; Cheese and Microbes, 2014; Encyclopedia of Dairy Sciences, 2022).

The above challenges surpass not only the limits of the established classical microbiological but also chemical analysis. The response to this so-called biogeography challenge, i.e. the microbiota changes over time and space as well as the microbiota functionality fluctuations, can be nowadays encountered by the employment of the so-called omics approaches. These include Next Generation Sequencing (NGS) based methods, such as metagenomics (i.e. the study of the metagenome, which is the collective genome of all microorganisms in an environmental sample, so as to provide information on the microbial diversity and ecology of this specific environment) and metatranscriptomics (i.e. the sequencing of the complete metatranscriptome of a microbial community, which unravels the genes that are expressed by the community as a whole at a given time point). Additionally, metaproteomics (i.e. the large-scale characterization of the entire protein complement of the microbiota in an environmental sample at a given time point, which unravels the mechanistic details of microbial interactions and the functional dynamics of the microbiome), and metabolomics (i.e. the systematic identification and quantification of the small metabolic products, called the metabolome, in a biological community) are of high significance as well (De Filippis et al., 2018; Ferrocino and Cocolin, 2017).

12.5 RAW MILK CHEESE VERSUS PASTEURIZED MILK CHEESE: A THOUGHTFUL DEBATE ABOUT CHEESE QUALITY AND SAFETY

For centuries, traditional cheese making has relied on the use of raw milk as well as the microbes that naturally occur in milk and the cheese making milieu. More than 400 LAB species, Gram- and catalase-positive bacteria, Gram-negative bacteria, yeasts, and molds are present in raw milk. During cheese making, this biodiversity decreases in the cheese core, where a small number of LAB species are numerically dominant, but persists on the cheese surface, which harbors numerous species of bacteria, yeasts and molds. Flavor is more intense and rich in raw milk cheeses than in those prepared from pasteurized milk, and the abundant native milk microbiota is the focal player. It should be stressed, however, that safe cheese production from raw milk requires milk of high microbial quality and employment of good hygiene practices throughout the entire cheese making process and during storage. The risk of microbial contamination can be further reduced by proper cheese ripening procedures. Moreover, LAB, as significant members of the cheese microbiota, produce antimicrobial compounds, such as organic acids (lactic and acetic acid), bacteriocins (i.e. ribosomally synthesized antimicrobial peptides/proteins), hydrogen peroxide, diacetyl etc., which inhibit cheese contamination by spoilage and pathogenic microorganisms. Nevertheless, outbreaks of food-borne diseases have been linked to the consumption of raw milk cheeses, and, thus, thorough microbiological monitoring during cheese making and ripening is highly recommended. Interestingly, in recent years, consumer preference for raw milk cheeses has been continually growing. France is a country with a long tradition in the production of high quality and safe raw milk cheeses, such as Camembert, Comte, Reblochon, Roquefort etc., which make up almost 70% of the French PDO cheeses, while, in contrast, raw milk cheeses are practically banned in the US (Montel et al., 2014; Quigley et al., 2011; Yoon et al., 2016).

On the other hand, nowadays, the major proportion of cheeses is produced at the industrial level, which requires standardized processing and ripening conditions, as well as the use of pasteurized milk of customary physicochemical and microbiological features, well defined rennet type and quality, and, last but not least, a properly designed cocktail of starter and non-starter cultures (Table 12.1; see also section below), depending on the type of cheese. Starter and non-starter cultures are classified as either mesophilic, which means mild temperature-loving, such as *Lactococcus lactis* subsp. *lactis* and *Lactococcus lactis* subsp. *cremoris*, or thermophilic, meaning heat-loving, such as *Streptococcus thermophilus*, *Lactobacillus helveticus*, and *Lactobacillus delbreuckii* spp., based on their respective optimal growth temperatures. These microorganisms, in turn, dictate the temperatures used in the cheese making process (Gobbetti et al., 2018).

Overall, there is a great variety of starter and non-starter cultures employed to facilitate the making of a wide range of cheeses. Both types of cultures are industrially developed and produced, and dedicated companies supply not only big dairy industries but also small-scale cheese makers. Reliance on industrial starters/non-starters results not only in a better controlled and more reproducible cheese making procedure, but also in final products with standard and typical quality and safety. Not surprisingly, almost all starter and non-starter cultures used in cheese manufacture are microorganisms that were originally isolated from milk or naturally (i.e. by fermentation relying on the indigenous microbiota of milk and the environment) fermented dairy products. These "wild" microbes are initially screened for their technological performance in the milk and dairy products milieu, and the best actors are cautiously developed to well-targeted starter and non-starter cultures cocktails for specific cheese varieties.

Regardless of whether raw or pasteurized milk is used for cheese making, the initial indigenous microbiota of milk along with the starters/non-starters used (pasteurized milk) or not used (raw milk), there are vibrant competition and cooperation interactions among microbes, also known as microbial ecology. This dynamic symbiosis plays a significant role in the conversion of milk to cheese. Successions of various microbial groups/species/strains, a kind of relay race, take place until the formation of the final product. The cheese microbiota is initially dominated by the starter cultures (the sprinters), but during the course of ripening, the secondary microbiota (the marathon runners) take over. In some of the more complex cheese varieties, the starter microbiota is subsequently dominated by yeasts, and then surface mold/bacteria microbiota take over to facilitate cheese ripening. Yeasts are known for their neutralizing activity and commonly prevail on the outside of washed rind cheeses to neutralize the surface pH, enabling bacteria, such as *Brevibacterium linens*, to grow there. Indeed, the identity of most cheese types is based on the microbial communities that make the cheese (Cheese problems solved, 2007; Cheese and Microbes, 2014; Encyclopedia of Dairy Sciences, 2022).

In both raw and pasteurized milk cheese, the key to quality cheese manufacture is maintaining control over the process regardless of the production scale. Parameters, such as initial milk composition, temperature, moisture level, lactic acid content and pH, salt addition, the size of curd particles at cutting, and stirring and draining rates, all influence the cheese making process and the quality of the resulting cheese. These parameters define the survival and succession of the different microbial groups, which in turn, through their metabolic activities, shape the organoleptic and safety traits of the final product. Hence, large-scale cheese makers must employ adequate testing throughout the cheese manufacturing process, while small-scale makers must pay close attention to consistency with their particular cheese making procedures. Additionally, dairy bacteriophages, i.e. viruses that infect only bacteria in milk and dairy products but not humans, can also contaminate the microorganisms involved in cheese making and inhibit the initial fermentation step, thus leading to failed manufacturing processes (Cheese problems solved, 2007; Cheese and Microbes, 2014; Encyclopedia of Dairy Sciences, 2022).

12.6 STARTER CULTURES VERSUS NON-STARTER CULTURES, ALIAS, SPRINTERS VERSUS MARATHON RUNNERS

The principal role of starter cultures, the sprinters, being mainly SLAB (Table 12.1), is the rapid conversion of milk lactose to lactic acid, which leads to the initial pH drop; ideally, milk pH should drop from 6.6 to 5.3 within the first 6 h of cheese making. The accumulation of lactic acid and the pH drop at the initial fermentation steps guarantee that the cheese will be protected against spoilage and pathogenic microorganisms assuming, of course, that the following steps are performed under good hygiene practices. As mentioned above, cheese making is not performed under sterile/aseptic conditions, and, thus, for the safe transformation of milk to cheese basic hygiene rules are required all the way to the final product. Besides their principal role, starter cultures also contribute by shaping the sensorial features of cheese, i.e. flavor, aroma, and texture, through the production of enzymes, which act on the main milk components, i.e. proteins, fat, and citrate, and produce a wide range of primary and secondary metabolites (see section below) (Gobbetti et al., 2015, 2018).

On the other hand, non-starter cultures, the marathon runners (Table 12.1), are not good performers in acidifying milk; instead, they practically dominate the cheese microbial ecosystem during the cheese ripening period and play the most significant role in shaping the sensory profile of each individual cheese variety. Non-starter cultures are added as adjunct cultures in pasteurized milk cheeses or appear as secondary microbiota in raw milk cheeses. Depending on the cheese variety, this category of microorganisms comprises four major microbial groups: (a) the NSLAB (that are present in almost all cheese varieties), (b) the propionic acid bacteria (PAB;

such as *Propionibacterium freudenreichii* subsp. *shermanii* that are involved in the production of Swiss type cheeses, e.g. Emmental and Gruyere), (c) molds (such as *P. camemberti* and *P. roqueforti* with major contribution to cheeses like Camembert, Brie, Roquefort, Gorgonzola, Stilton, and Danish Blue) and, finally, (d) the smear microbiota (comprising various coryneform bacteria, e.g. *Corynebacterium, Brevibacterium*, and *Arthrobacter*, as well as micrococci and staphylococci, in the so-called surface-ripened/smear cheeses like Limburger, Munster, and Tilsit) (Gobbetti et al., 2015, 2018).

Both types of cultures, either intentionally added (in pasteurized milk cheeses) or as autochthonous microbiota of milk and the cheese making environment (in raw milk cheeses) change the cheese microenvironment, affecting a variety of factors, including pH, redox potential, levels of organic acids, such as lactate, acetate, and propionate, and production of flavor and aroma compounds. Although some of the starter cultures begin to die during ripening, their cells release enzymes that further break down milk components, mainly caseins, while, at the same time, the dead starter culture cells and their debris become a food source for the non-starter microbes.

12.7 CHEESE MICROBIAL COMMUNITIES THRIVE WHILE CHEESE IS AGING AND MAKE A FORTUNE IN AROMA, FLAVOR, TEXTURE, AND COLOR

The milk chemical composition, the cheese making technology and the microbial groups involved in cheese making, all together dictate the flavor, aroma, texture and color of cheese. The microbes are particularly essential in this process, since, through their metabolic activities degrade the milk components and produce hundreds of compounds that affect the cheese sensory profile.

The breakdown of milk lactose, but mostly of caseins and fat by the proteolytic and lipolytic enzymes of the cheese microbiota, respectively, and the secondary degradation of free amino acids and fatty acids, along with the milk citrate catabolism, result in the production of numerous compounds, such as carboxylic acids, alcohols, aldehydes, ketones, esters, lactones, *S*-compounds, *N*-compounds, pyrazines, and furans, which are responsible for key cheese flavors and aromas, such as buttery, fruity, malty, nutty, chocolate-like, bitter almond, banana, sweet cherry, pungent, sulfur, etc., which can be either light or bold and intense, depending on the cheese variety (Table 12.2).

Most literature data have been generated on *L. lactis*, which is the predominant organism among the starter cultures used for cheese making, but also on *Lactobacillus* before its taxonomic revision (Zheng et al. 2020), *Streptococcus, Propionibacterium* and species used for surface-ripened/smear cheeses (Marilley et al., 2004; Smit et al., 2005).

The cheese texture is formed upon the initial acidification rate, which is defined by the starter cultures, the moisture content, which is defined by the curd heating/cooking steps and the cheese aging time, as well as the proteolytic degradation of the milk proteins, initially by rennet, most of which is expelled during curd drainage or inactivated during curd heating/cooking though, and subsequently by the proteolytic enzymes of the cheese microbiota. For example, Camembert is soft and creamy because of the extensive protein degradation by the mold involved, namely *P. camemberti*. On the other hand, Feta is a soft cheese as well, not due to extensive casein degradation though, but because of the high moisture content. Parmigiano Reggiano is very hard and brittle, mainly because of the low moisture content due to the curd heating/cooking steps at the initial cheese making stages and the long ripening time. Cheese texture can also be affected by crystal formation. The ratio of salt to moisture influences the final pH that can cause calcium lactate crystals formation on the surface of and/or inside the cheese. Crunchy crystals can be formed in aged, hard cheeses. In some cheeses, such as Parmigiano Reggiano, this trait can be regarded as desirable; however, gritty crystals are considered a costly defect in cheeses like Cheddar (Cheese problems solved, 2007; Cheese and Microbes, 2014; Encyclopedia of Dairy Sciences, 2022).

Cheese color can range from white to ivory, yellow, golden, orange, grey and blue-streaked. The color may be due to natural milk pigments, such as carotenes, which come from the grasses in the animal feed. Among ruminants, only cows and buffalos accumulate high concentrations of carotenoids, mainly beta-carotene, possibly due to low vitamin A synthesis efficiency in enterocytes (Noziere et al., 2006). Additionally, the microbes involved in cheese making can affect cheese color by producing pigments themselves. In blue cheeses, such as Roquefort, the blue veins are due to the growth of the mold *P. roqueforti*, which is used as adjunct culture, while the orange-tinted bacterium *B. linens* gives the color to the rinds of surface-ripened/smear cheeses, such Limburger (Cheese problems solved, 2007; Cheese and Microbes, 2014; Encyclopedia of Dairy Sciences, 2022).

The puzzle, however, whether all microbes in cheese are metabolically active throughout cheese making and

TABLE 12.2 Representative significant cheese flavor compounds, their source compounds, and their description (adapted from Marilley and Casey, 2004; Smit et al., 2005)

Flavor Compound	Source Compound	Description of Flavor	Cheese
Diacetyl	Lactose, citrate	Buttery, strong	Cheddar, Gouda, Swiss-type cheeses (e.g. Emmental and Gruyère), Maasdam
Acetaldehyde	Lactose, citrate	Yogurt, green, nutty, pungent	Cheddar
Propionic acid	Lactose, lactate	Pungent, sour milk, cheese	Swiss-type cheeses (e.g. Emmental and Gruyère), Maasdam
Butyric acid	Fat	Sweaty, butter, cheese, strong, acid	Cheddar, Gouda, Camembert
Ethyl-butyrate	Fat	Fruity, buttery, ripe fruit	Cheddar
2-methyl-propanal	Val	Banana, malty, chocolate-like	Cheddar
3-methyl-butanal	Ile, Leu	Malty, powerful, cheese	Cheddar, Gouda, Camembert, Swiss-type cheeses (e.g. Emmental and Gruyère), Maasdam
3-methyl-butanol	Ile, Leu	Fresh cheese, breathtaking, alcoholic	Cheddar, Gouda
3-methyl-butyrate	Leu	Rancid, sweat, cheese, putrid	Camembert
Methional	Met	Cooked potato, meat like, sulphur	Cheddar, Camembert, Swiss-type cheeses (e.g. Emmental and Gruyère), Maasdam
Methane-thiol	Met	Rotting cabbage, cheese, vegetative, sulphur	Cheddar, Gouda, Camembert
Benzaldehyde	Phe	Bitter almond, sweet cherry	Camembert
Phenyl-acetate	Phe	Rough, lily-jasmine with metallic note	Cheddar

Val: Valine; Ile: Isoleucine; Leu: Leucine; Met: Methionine; Phe: Phenylalanine (amino acids)

in which degree (even though some of these microorganisms cannot be cultivated under conventional laboratory conditions) remains. However, the novel omics approaches mentioned above are expected to give sufficient and reliable clues in response, i.e. which are the protagonists, which are the best performers and what is the key role each of them plays in shaping the sensory profile of cheese. This may further facilitate diversifying the shepherding of microbes so that new cheese varieties with novel organoleptic properties can be developed.

12.8 CHEESE MICROBIOTA AND HUMAN HEALTH: MYTH OR REALITY?

Cheese is a rich source of essential nutrients, namely proteins, fat, vitamins and minerals. Milk proteins, in particular whey proteins, are of high nutritional value on the basis of the essential amino acids content. Moreover, the casein breakdown during cheese making, by the rennet and the proteolytic system of the cheese microbes, produces the so-called bioactive peptides, which are credited with health-promoting properties, such as anti-hypertensive, anti-thrombotic, anti-oxidant and hypo-cholesterolemic peptides (related to the cardiovascular system), anti-obesity and satiety peptides (related to the gastrointestinal system), antimicrobial and immune-modulatory peptides (related to the immune system), and opioid peptides (related to the nervous system). Despite the presence of notable amounts of saturated fatty acids, so far, there is no clinical evidence that clearly correlates the consumption of cheese to any disease, in particular cardiovascular diseases, while conjugated linoleic acid (CLA) present in several cheeses may reduce total and low-density lipoprotein (LDL) cholesterol or even improve blood pressure. The high concentration of calcium in cheese contributes to the formation and maintenance of strong bones and teeth. Finally, ripened cheeses are free of lactose and therefore suitable for the nutrition of lactose-intolerant individuals (Walther et al., 2008).

As well as its chemical components, cheese contains billions of microbes ranging from a million up to a billion microorganisms per gram of cheese. Practically, humans have been consuming cheese microbes for millennia, which are generally considered neither particularly healthful nor harmful, apart from the spoilage and pathogenic contaminants that make cheese non-safe and non-consumable. It should be stressed that in recent years in order to better regulate the safety of the microorganisms used in fermented foods, among others new starter and non-starter cultures in cheese production, both the US (FDA, Food and Drug Administration) and the EU (EFSA, European Food Safety Authority) have introduced the GRAS (Generally Recognized As Safe) and the QPS (Qualified Presumption of Safety) regimes, respectively. In both systems, safety status is based either on a history of safe use in cheese prior to 1958 (GRAS) and 1997 (QPS), or on the employment of scientific protocols, which require the same quantity and quality of evidence as for any food additive (Bourdichon et al., 2012).

Nevertheless, as with any scientific inquiry, the more we learn about cheese, the more questions arise. Are there members of the cheese microbiota with specific beneficial impact on human health? Can we incorporate healthful microbes into cheese making, e.g. probiotics, as starters or non-starters, which can survive the cheese manufacturing conditions and at the same time interact with the indigenous microbes of the human gut so as to assist in maintaining a healthy gut microbiota, or even render additional health benefits to humans? So far, the dairy products, which are generally considered as the best vehicles for administering probiotic microorganisms to humans, are yogurt and fermented milks, which mostly comprise rather simple microbiomes and have short production time. In contrast, cheese as a proper food matrix for probiotic administration has not yet been sufficiently studied. Evidence of probiotic action is deriving mainly from in vitro and ex vivo experiments, and animal trials. The few human trials that have been performed so far imply beneficial impact, such as stimulation of the innate immune response (Gouda cheese), and improved lipid profile or reduction of oral candidiasis risk (Minas Frescal cheese) (Gomes Da Cruz et al., 2009; Rolim et al., 2020). Interestingly, in recent years, there is growing evidence that dairy PAB, such as *P. freudenreichii*, which is a significant player in Swiss type cheeses, may have important potential as a probiotic culture, by serving as a modulator of the colon microbiota and preventing colon cancer (Rabah et al., 2017). For sure, on the way to the general application of probiotics in cheese, there are some major hurdles to overcome. For any probiotic microbial culture with specific and well defined health beneficial features to be successfully included in cheese making, essential perquisites refer to the endurance of the culture over all cheese manufacture and storage conditions of the specific cheese variety, as well as its stamina to deal with the competition it faces alongside the continuous succession of and interplay among microbial groups that take place during cheese making. Once these hurdles are overcome, then cheese probiotic microbes can reach the human gastrointestinal tract and start exhibiting their specific health beneficial impact; this, however, should be proved by well-designed and properly performed human trials (i.e. number of subjects, double-blinded, inclusion of control cheese without probiotics etc.), as this is the case for all probiotics no matter which food is used as vehicle.

12.9 CONCLUSIONS

Cheese making goes back to the seventh millennium BC and remained a craft until quite recently. Not surprisingly, microorganisms and their role in cheese making were unknown, until Pasteur, in the late nineteenth century AD, talked for the first time about the involvement of microbes in food fermentations. Since then, cheese microbiology began to blossom. Despite the progress made so far in cheese science and technology, as well as the numerous and excellent cheese varieties produced around the world, there are still gaps in our knowledge and new challenges that need to be addressed. No microbes–no cheese is a fact. However, the more we learn about cheese, the more questions arise. Recent advanced technologies, such as the omics ones, are expected to help us understand the interplay dynamics among the members of the cheese microbial ecosystem, as well as the role of each of them in shaping cheese quality, safety and nutritional value. Moreover, taking into account the consumers' trend toward foods with additional beneficial impact on nutrition and health, our attention should also focus on how and in what degree cheese microbes can be involved and contribute to the development of cheeses with human health related functionalities.

REFERENCES

1. Ben Amira A, Besbes S, Attia H, Blecker C. 2017. Milk-clotting properties of plant rennets and their enzymatic, rheological, and sensory role in cheese making: A review. *International Journal of Food Properties* 20: Sup1 S76–S93.
2. Bivar Roseiro L, Barbosa M, Ames JM, Wilbey RA. 2003. Cheese making with vegetable coagulants—the use of *Cynara* L. for the production of ovine milk cheeses. *International Journal of Dairy Technology* 56: 76–85.
3. Bourdichon F, Casaregola S, Farrokh C, Frisvad JC, Gerds ML, et al. 2012. Food fermentations: Microorganisms with technological beneficial use. *International Journal of Food Microbiology* 154: 87–97.
4. De Filippis F, Parente E, Ercolini D. 2018. Recent past, present, and future of the food Microbiome. *Annual Review of Food Science and Technology* 9: 589–608.
5. Ferrocino I, Cocolin L. 2017. Current perspectives in food-based studies exploiting multi-omics approaches. *Current Opinion in Food Science* 13: 10–15.
6. Gobbetti M, De Angelis M, Di Cagno R, Mancini L, Fox PF. 2015. Pros and cons for using non-starter lactic acid bacteria (NSLAB) as secondary/adjunct starters for cheese ripening. *Trends in Food Science & Technology* 45: 167–178.
7. Gobbetti M, Di Cagno R, Calasso M, Neviani E, Fox PF, et al. 2018. Drivers that establish and assembly the lactic acid bacteria biota in cheeses. *Trends in Food Science & Technology* 78: 244–254.
8. Gomes da Cruz A, Alonso Buriti FC, Hoch Batista de Souza C, Fonseca Faria JA, Isay Saad SM. 2009. Probiotic cheese: Health benefits, technological and stability aspects. *Trends in Food Science & Technology* 20: 344–354.
9. Marilley L, Casey MG. 2004. Flavours of cheese products: Metabolic pathways, analytical tools and identification of producing strains. *International Journal of Food Microbiology* 90: 139–159.
10. Montel MC, Buchin S, Mallet A, Delbes-Paus C, Vuitton DA, et al. 2014. Traditional cheeses: Rich and diverse microbiota with associated benefits. *International Journal of Food Microbiology* 177: 136–154.
11. Noziere P, Graulet B, Lucas A, Martin B, Grolier P, et al. 2006. Carotenoids for ruminants: From forages to dairy products. *Animal Feed Science and Technology* 131: 418–450.
12. Olajide AM, LaPointe G. 2022. Microorganisms associated with raw milk. In: McSweeney PLH, McNamara JP (Eds.), *Encyclopedia of Dairy Sciences*, pp. 319–328 3rd Ed. Elsevier, ISBN: 978-0-12-818767-8, doi: 10.1016/B978-0-12-818766-1.00023-4.
13. Quigley L, O'Sullivan O, Beresford TP, Ross P, Fitzgerald GF, et al. 2011. Molecular approaches to analysing the microbial composition of raw milk and raw milk cheese. *International Journal of Food Microbiology* 150: 81–94.
14. Rabah H, do Carmo FLR, Jan G. 2017. Dairy Propionibacteria: Versatile Probiotics. *Microorganisms* 5: 24.
15. Rolim FRL, Freitas Neto OC, Oliveira MEG, Oliveira CJB, Queiroga RCRE. 2020. Cheeses as food matrixes for probiotics: *In vitro* and *in vivo* tests. *Trends in Food Science & Technology* 100: 138–154.
16. Salque M, Bogucki PI, Pyzel J, Sobkowiak-Tabaka I, Grygiel R, et al. 2013. Earliest evidence for cheese making in the sixth millennium BC in Northern Europe. *Nature* 493: 522–525.
17. Smit G, Smit BA, Engels WJM. 2005. Flavour formation by lactic acid bacteria and biochemical flavour profiling of cheese products. *FEMS Microbiology Reviews* 29: 591–610.
18. Walther B, Schmidt A, Sieber R, Wehrmüller K. 2008. Cheese in nutrition and health. *Dairy Science & Technology* 88: 389–405.
19. Yoon Y, Lee S, Choi KH. 2016. Microbial benefits and risks of raw milk cheese. *Food Control* 63: 201–215.
20. Zheng J, Wittouck S, Salvetti E, Franz CMAP, Harris HMB, et al. 2020. A taxonomic note on the genus *Lactobacillus*: Description of 23 novel genera, emended description of the genus *Lactobacillus* Beijerinck 1901, and union of *Lactobacillaceae* and *Leuconostocaceae*. *International Journal of Systematic and Evolutionary Microbiology* 70: 2782–2858.

FURTHER READING

1. Cheese problems solved. 2007. McSweeney, PLH (Ed.), Woodhead Publishing Limited, ISBN 978-1-84569-060-1.
2. Encyclopedia of Dairy Sciences. 2022. McSweeney PLH, McNamara JP (Eds.), 3rd Ed. Elsevier, ISBN: 978-0-12-818767-8.
3. Kindstedt PS. (Ed.). 2012. *Cheese and culture - A history of cheese and its place in Western civilization*. Chelsea Green Publishing, White River Junction, Vermont, ISBN 978-1-60358-411-1.
4. Cheese and Microbes. 2014. Connelly CW (Ed.), ASM Press, ISBN 978-1-55581-586-8.
5. FAO. 2018. Dairy market review - Overview of global dairy market developments in 2018.
6. USDA. 2019. Dairy: World markets and trade.

CHAPTER 13

The Microbiome of Fermented Sausages

Ilario Ferrocino, Irene Franciosa, Kalliopi Rantsiou, and Luca S. Cocolin
DISAFA – Microbiology and Food Technology Sector, University of Turin, Grugliasco, Torino, Italy

13.1 INTRODUCTION

Improving our knowledge of microbiota/microbiomes has become a popular topic over the past two decades not only in the scientific community and among the general public, but also in the food industry because of the growing demand to use the microbiome as a tool to improve quality and safety in the food chain. With the advent of new technologies among which are -omics technologies, rapid detection tools and blockchain technology, we are able to generate billions of data. As a consequence, an in depth characterization of the microbial complexity in food systems is proceeding very fast and we are now able to track and map shifts in microbial communities, discover new molecules, new metabolic pathways and also new strains.

This flow of data and results also helped by scientific dissemination campaigns on social channels has made the food industry considerably closer to the microbiome theme, increasingly pushing the possibility of using it as a tool to improve the quality of products along the entire supply chain. One of the main targets of the use of the microbiome is the fermented foods sector that includes milk and dairy products, vegetables, and meat-based products. Regarding meat-based products it is well know that since the Egyptian age fermentation has been a successful strategy for the preservation of meat. Countries in the Mediterranean area (Italy, Spain, Portugal, Turkey, and France) are the main producers of fermented meat (also referred to as fermented sausages or salami) followed by Germany and Hungary, and also by China that in the last decades has been exponentially increasing the manufacture of fermented sausages by the production of several typical products from different Chinese areas.

Due to the increase in demand of fermented sausages thousands of companies spread around the world are interested in increasing the production of artisanal sausages (mostly without the use of starter culture) in order to confer to this product an additional value for the consumers and to have the guarantee of producing a unique product. Compared to industrial products, with artisanal products it is very difficult to guarantee the same organoleptic characteristics to every production, since the indigenous microbes that are present on the carcass or in the production environment are subject to a high degree of variation. This has a huge impact on the final characteristics of the product, not to mention that fermentation could also fail.

The fermentation process is a complex succession of biochemical transformations that start from animal enzymes, but the main role is played by the microbiome.

As recently reported, the definition of the microbiome can be summarized as follows: a characteristic microbial community (composed by bacteria, archaea, fungi, algae, and small protists) in a well-defined habitat

Good Microbes in Medicine, Food Production, Biotechnology, Bioremediation, and Agriculture, First Edition. Edited by Frans J. de Bruijn, Hauke Smidt, Luca S. Cocolin, Michael Sauer, David Dowling, and Linda Thomashow.

which has distinct physio-chemical properties where microbes interact with each other, with existing microbial networks and with the host, also encompassing their theatre of activity, which results in the formation of specific ecological niches (Berg et al., 2020).

With the massive application of -omics techniques we are able to obtain, in various ways and by different powerful instruments, a huge amount of data that require computing tools for an in-depth analysis in a reasonable timeframe. With all this information we are able to integrate data and to decipher what happens in a complex ecosystem like fermented sausages.

The first challenge in monitoring or modulating the sausage's microbiome is the control of pathogens, but pathogens represent only a minor fraction of the microbial population in fermented sausages. The main fraction is involved in other metabolic functions where the main targets are the macronutrients of the meat (carbohydrates, amino acids, and lipids). In this light it is evident that the use or the modulation of the microbiome in this food system can help to improve the final quality of the products.

In the case of industrial production, microbes as a starter culture are deliberately added with the aim of controlling the fermentation process and guaranteeing reproducibility. By definition microbial food starter cultures are the live bacteria, yeasts or molds used in food production that with their metabolic function confer the desired effects in the final product (Bourdichon et al., 2012; Hammes and Hertel, 1998). Starter standardization is a challenge since a multidisciplinary approach is needed to ensure in primis safety and secondarily to satisfy consumer expectations and to preserve the traditional aspects of the typical fermented meat products. However, it is also possible to modulate the microbiome by acting on the process parameters or by acting on the indigenous microbiota of the environment.

13.2 THE MICROBIOTA OF FERMENTED SAUSAGES

Due to their composition which is rich in proteins, carbohydrates, lipids, vitamins, and salts, meat and meat products are a perfect growth substrate for microbes. Several factors influence the occurrence of microbes in meat starting from the slaughtering process. Muscle meats can be contaminated by indigenous microorganisms from the first step in the abattoir directly from animal skin or from the gastrointestinal tract of the animal itself, or from water, air, soil, workers, and equipment involved in the processing. In the later steps of the meat chain, factors such as temperature, gaseous atmosphere, pH, and NaCl levels select specific microbes (Stellato et al., 2016). In sausage production an important factor are the ingredients used (especially spices) since they can be a source of microbes mostly causing spoilage (Figure 13.1). The production process of fermented sausages is usually driven by the metabolic actions of lactic acid bacteria (LAB) and coagulase-negative staphylococci (CNS).

The main role of the lactic acid bacteria is to lower the pH through the production of organic acids and as a consequence the production of many volatile compounds, while coagulase-negative cocci (CNC) are often responsible for proteolysis, lipolysis, and the decomposition of free amino acids and peroxides (Cardinali et al., 2018) (Figure 13.1). Among LAB, hetero and homo-fermentative genera belonging to *Leuconostoc*, *Lactobacillus*, *Weissella*, *Enterococcus*, and *Pediococcus* were variously identified. However, among them *Latilactobacillus sakei*, *Lactiplantibacillus plantarum* as well as *Latilactobacillus curvatus* are considered the main microbes involved in meat fermentation (Figure 13.2). LAB in fermented meat are often derived from the animal intestine and have been found to be present in natural sausage casings (Quijada et al., 2018).

L. sakei is a ubiquitous psychotrophic heterofermentative bacterium with a versatile redox metabolism and shows great tolerance to both high salt concentration (up to 9% sodium chloride) and low temperatures (4°C). It produces lactic acid products and bacteriocins and its ability is mainly recognized to metabolize meat components such as proteins and this reflects its adaptability to the meat environment (Xu et al., 2015). *L. sakei* contains genes encoding several amino acids and peptide transporters which contribute to aroma formation from amino acid substrates and are involved in stress responses in protein-rich environments (Xu et al., 2015). In this light *L. sakei* can be considered as one of the main characters involved in the flavor development during fermentation. Fermented meat is considered an ecological niche for *L. sakei* since many ecological studies report its strong predominance in the microbiota since the first days of ripening and assess it to be never lower than 40% of the total microbial population (Aquilanti et al., 2016). This capability is also obtained through its highest competitive traits and high growth rates, its

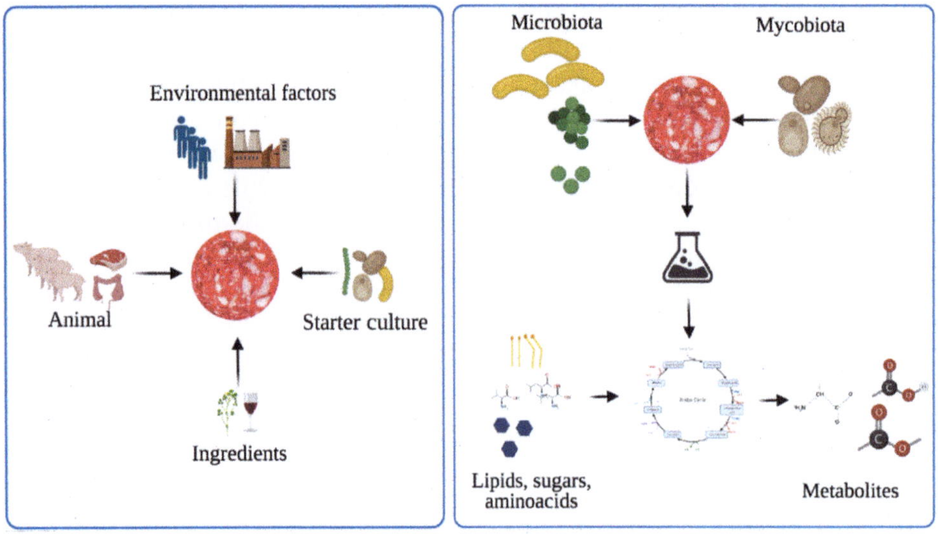

FIGURE 13.1 Representation of the main factors that influence the occurrence of microbes in fermented meat and their function during the fermentation. Meats can be contaminated by indigenous microorganisms from the first step in the abattoir directly from animal skin or from the gastrointestinal tract of the animal itself, or from environment, workers, and equipment involved in the processing. Microbes can be also added as a starter culture in order to drive the fermentation process. Microbiota and mycobiota are often responsible for proteolysis, lipolysis, and the decomposition of free amino acids and peroxides through the production of several metabolites that improves the overall texture and consistency of the final products. Created with BioRender.com.

oxidative metabolism, production of bacteriocins and small peptides and amino acids from the hydrolysis of muscle sarcoplasmic and myofibrillar proteins (Cardinali et al., 2018; Janßen et al., 2020). The importance of *L. sakei* during meat fermentation is also connected to its ability to improve safety by bacteriocin production which results in the suppression of competing microbes, like pathogens such as *Listeria monocytogenes* and *Staph. aureus* or other spoilage microbes (Castellano et al., 2012; Janßen et al., 2020).

Glucose is the preferred carbon source for *L. sakei* and this can be the primary way to ensure its ability to colonize meat, but only in the first stage of the fermentation process because this compound is rapidly exhausted in meat (Chaillou et al., 2005). Glucose metabolism of *L. sakei* is an important factor since by this pathway it is able to rapidly lower the pH through the formation of acetic acid, one of the main precursors of volatile compounds (Ferrocino et al., 2018). The acidification produced by *L. sakei* in sausages also favors their dehydration and improves the overall texture and consistency of the final products. In addition, *L. sakei* possesses a heme-dependent catalase activity that hydrolyses hydrogen peroxide and prevents rancidity and discoloration of meat (Quijada et al., 2018). After glucose declines, free amino acids are the preferred source for microbial growth on meat, in particular the degradation of arginine plays an important role in the survival of *L. sakei*. Arginine, inosine, and adenosine can be used as energy sources by all *L. sakei* strains and these energy sources offer a competitive advantage for this species in a meat environment (Rimaux et al., 2011). It is also recognized that *L. sakei* has strain-dependent properties and its large diversity (or distinct ecotypes, likely linked or specialized to different environmental reservoirs) represents a valuable and exploitable asset in the development of a variety of industrial applications (Chaillou et al., 2013). In fact, one major challenge for improving and controlling industrial fermentation processes is the uncovering of adaptation and assertiveness mechanisms at strain level (Janßen et al., 2020).

Among LABs, *L. curvatus* is a facultative heterofermentative lactic acid bacterium, commonly associated with fermented and vacuum-packaged refrigerated meat and fish products and also isolated from several other food systems, but also identified in the feces or gut of many animal species (Terán et al., 2018). The *L. sakei* and *L. curvatus* species are clearly separated at the genomic level (40–50% identity) but the typical ecological niches harboring *L. sakei* do not differ significantly with respect to the requirements needed for the growth of *L. curvatus*. In more detail, despite the genetic diversity a similarity was observed in the volatilomes related to the two species that show a certain qualitative uniformity (Freiding et al., 2011). In fermented meat systems the predominance of *L. sakei* or *L. curvatus* depends essentially on the interspecies competition mediated by the production of bacteriocins. In particular, strains of *L. curvatus* and *L. sakei* have been reported to produce different bacteriocins, the most known of which are referred as curvacins and sakacins. These possess strong antilisterial activity and are classified both in class IIa (single-peptide bacteriocins) and class IIb (two-peptide bacteriocins) (Casaburi et al., 2016). The ability to produce bacteriocins is strain dependent, therefore when the natural *L. sakei* population is not able to produce bacteriocins, *L. curvatus* is the dominant species and vice versa (Janßen et al., 2020).

Among LAB one of the main microbes often isolated is *Pediococcus pentosaceous*, which plays a leading role in many traditional fermented sausages and exhibits a higher proteolytic activity than other strains. In particular it is able to generate flavor precursors such as free amino acids, soluble peptides, and smaller molecule compounds that contribute to the formation of specific flavors (Sun et al., 2019). In addition, strains of *P. pentosaceous* possess an antioxidant ability and the ability to inhibit LPS (oxidative stress)-induced THP-1 monocyte–endothelial cell adhesion mainly due to the increase of p-coumaric acid (Kim et al., 2019). *P. pentosaceus* prevents an excessive lipid oxidation that can be the cause of quality deterioration in meat products (off-flavor) and shows a high acidification capability which can be exploited to quickly lower pH. This capability could be a positive survival strategy in fermented food and a positive quality for the use of this species as a starter culture in fermented meat for consumers that prefer high acidity products (Chen et al., 2015a, 2015b).

The microbiota of fermented meat is also the niche of CNC (Figure 13.2), in particular *Staphyloccous xylosus* that dominates meat fermentation especially at the beginning, decreasing through the ripening process in favor of the development of LAB (in primis *L. sakei*) (Połka et al., 2015). Even if *Staphylococcus* appears less than LAB, it plays an important role in desirable sensory qualities of sausages because of its ability to produce flavor compounds, by using isoleucine and methionine as precursors to produce sulfur compounds like 2-methyl-1-butanol, and some ethyl esters (Bertuzzi et al., 2018). The use of *S. xylosus* has been mentioned to yield a rounder and less acidic taste compared to *S. equorum* (Quijada et al., 2018). Compared to *S. simulans* or *S. carnosum* it appears that *S. xylosus* strains possess the ability to convert metmyoglobin into nitrosomyoglobin and thus contribute to color formation in nitrate- and nitrite-free sausages (Cardinali et al., 2018), promote lipid hydrolysis and the release of FFAs via lipase secretion and possess highest esterase activity (Xiao et al., 2020; Figure 13.2). Overall it is not clear yet which species are best for flavor enhancement during

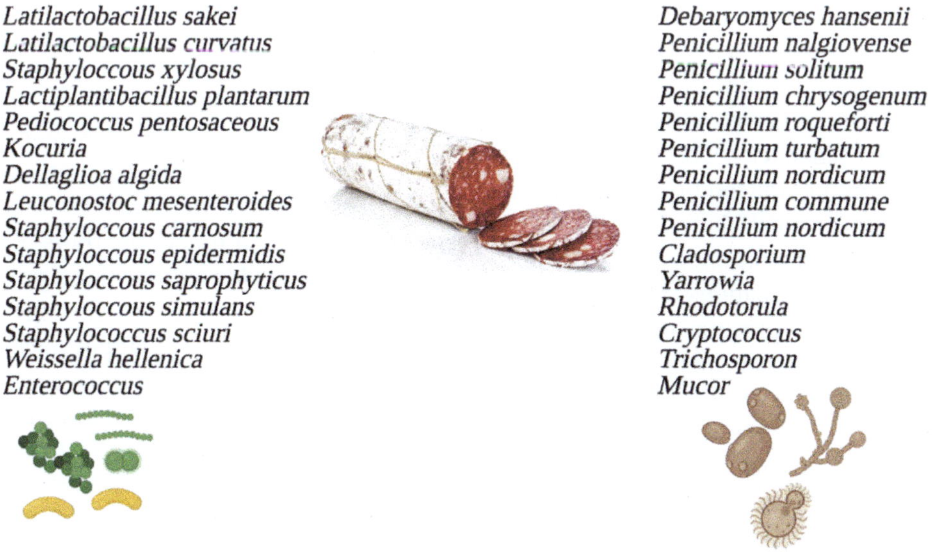

FIGURE 13.2 Overview of the main bacteria and fungi as reported in the literature involved during the sausage's fermentation. Created with BioRender.com.

fermented sausage production and consequently only some species belonging to CNS are often proposed as starter cultures, such as *Staphylococcus xylosus* or less frequently *Staphylococcus carnosus*.

The presence of indigenous strains naturally colonizing meat or the environment on the one hand is considered as positive because it confers the typicity to the products, but on the other hand can often confer negative proprieties to the fermentation. For example, opportunistic potential pathogens such as *S. epidermidis* and *S. saprophyticus* are frequently found in spontaneously fermented meats (Lopez et al., 2020). A not well controlled fermentation can result in products with not controlled lipolytic or proteolytic activity that can cause flaking of the slice when cut, early rancidity of the fat part, off-flavor or meat discoloration. Spontaneously fermented sausages are commonly considered safe for consumption, because the acidification by natural lactic acid flora is one of the main preserving factors. However, outbreaks of serious sausage-borne gastrointestinal infections mediated by verocytotoxic *Escherichia coli* (VTEC/STEC), *Salmonella*, and *Listeria monocytogenes* can occur (Zaiko et al., 2020). In this context an appropriate choice of a combination of strains in a starter formulation is fundamental to obtain products of the expected quality.

A correct selection and mixture of autochthonous strains from same genus or from different genera is fundamental to better control the fermentation process. It should also be noted that an in depth study of strains should be performed, since strains belonging to the same genus can exert a different impact on the system. In most cases one single strain is not able to confer specific characteristics to the product, but often a mixture of different genetic repertoires gives a more important contribution.

Understanding the interactions regulating microbial development in complex ecosystems like fermented food products is the new frontier of the food-omics approach. By using this approach, it is possible to investigate the metabolic potential of the autochthonous microbes and to show how external perturbations of the system (presence of contaminants, process parameters, or presence of starter culture) impact ripening, safety, and flavor development (Ferrocino et al., 2018). Metagenomic sequencing can help in the characterization of microbiota but should necessarily be coupled with the phenotypic analysis of the strains. Deeply understanding changes in genomic features between strains belonging to the same species will give us the ability to easily modulate the microbiome.

13.3 THE IMPORTANCE OF THE SAUSAGE'S MYCOBIOTA

In food ecology yeasts and molds, also grouped under the term mycobiota, are often still understudied, probably because of difficulties in the specific identification related to the use of alternative marker genes (unlike 16S rRNA for bacteria) or because taxonomic information available in current databases is relatively incomplete if compared to bacteria. Mycobiota of fermented sausage products is particularly important especially in the casing since a layer of mold has an antioxidant effect, prevents rancidity, preserves color, gives to the sausage its typical appearance and allows the development of positive conditions that prevents a sticky or slimy surface (Franciosa et al., 2018).

Sausages are the perfect environment for the development of filamentous fungi since the water activity (*aw*) and the salt concentration provide an optimum substrate. It should be noted that *aw* of the products is one of the main factors influencing mold growth together with the ripening temperature, since different values can affect the mycobiota development on the surfaces of dry-cured meat products (Zadravec et al., 2020). For example the presence of *Mucor* in some sausages is directly associated with low temperatures and is often isolated only from winter sausages (Zadravec et al., 2020).

Penicillium is frequently found and in this genus *P. solitum*, *P. nalgiovense*, and *P. chrysogenum*, *P. roqueforti*, *P. turbatum*, *P. nordicum*, P. *commune*, and *P. nordicum* are the species often identified in fermented meat (Figure 13.2) (Álvarez et al., 2020; Belleggia et al., 2020b; Bourdichon et al., 2012; Iacumin et al., 2020; Zadravec et al., 2020). Most of them are considered technological molds because they penetrate deeply with their mycelia, contributing to flavor formation mediated by the production of enzymes such as amylase, protease, and lipase. They are also important for the improvement of the nutritional and functional traits of fermented foods because they provide a protection against oxidation and detrimental microorganisms (Belleggia et al., 2020b; Osimani et al., 2019). The importance of *Penicillium* is also assessed by the recently sequenced genomes that show innate adaptations in the fermented foods environment, with a genomic repertoire suggesting functions involved in antagonism with other microorganisms found in foods (Papadimitriou et al., 2015).

The presence of fungi is often associated with the environment, especially the ones transported from air (like *Aspergillus*, *Fusarium*, or xerophilic ones) or the ones coming from the spices used as a ingredients for sausage manufacturing. Among the indigenous ones some are considered good in sausages, like *Cladosporium cladosporioides* recently found in South American sausages (Parussolo et al., 2019) and in Portuguese sausages (Belleggia et al., 2020a). Members of the genus *Cladosporium* are also considered Generally Regarded as Safe (GRAS) due to their ability to delay the germination of toxigenic fungi, which leads to their use as biological control agents (BCA) (Parussolo et al., 2019).

Together with filamentous fungi, yeasts are also important members of the sausage mycobiota. Compared with filamentous fungi, yeasts seem to change during processing, although variations depend on meat origin and factory environment. Among them several genera such as *Rhodotorula*, *Cryptococcus*, and *Trichosporon* (at the initial ripening stage) and *Candida Debaryomyces*, *Yarrowia*, *Pichia* (at later stages) are often found. Their main action is to protect against the detrimental effect of oxygen and to facilitate the drying process by protecting the sausage against fluctuations in humidity (Flores et al., 2015).

Debaryomyces hansenii (*Candida famata* in anamorph stage) is often recognized as a good producer of desirable aromas (such as aldehydes, ketones, alcohols, esters, and sulphur compounds) in dry fermented sausages due to its esterase activity (often associated with the ones from *Staphylococcus*) mediated by its lipolytic and proteolytic activities (Cano-García et al., 2014; Flores et al., 2015). Conversely *D. hansenii*, by its ability to degrade peroxides, can contribute to the stabilization of the reddening effect. Moreover, it can be considered as a BCA able to control the development of ochratoxigenic molds, as a biocontrol agent against fungi, and as a consequence mycotoxin contamination (Murgia et al., 2019).

13.4 USE OF THE AUTOCHTHONOUS MICROBIOME TO IMPROVE THE QUALITY AND SAFETY OF FERMENTED SAUSAGES

There is a huge variety of artisanal fermented sausages that differ in the ingredients used and in the process parameters, each one having its own peculiar characteristics that are deeply rooted in tradition and linked to the territory (Franciosa et al., 2018). The increased consumer interest in typical products pushes the industry to find a good compromise to guarantee safety and most importantly the same organoleptic characteristics of artisanal products. One action is to use autochthonous microbes as a starter culture. It is widely reported that strains from the same genus possess different abilities (Casaburi et al., 2011; Ercolini et al., 2010) and it should be pointed out that no single strain is able to reproduce the natural microbiome of a sausage, not even if an autochthonous one is used.

The interaction among bacteria and fungi is also of great interest because several bacteria are not able to grow in the absence of certain fungi or are inhibited by their presence (Wolfe et al., 2014). Besides guaranteeing a standard of the product's organoleptic characteristics (taste, texture, and aromatic notes) the application of an autochthonous microbiome can cover several safety issues. Among them it is worth mentioning the biocontrol of pathogens and the reduction of nitrites/nitrates, antimicrobial resistance genes (ARGs), mycotoxins, and biogenic amines. It is widely demonstrated that an accurate selection of lactic acid bacteria with the ability to produce bacteriocins as antimicrobial agents (Casaburi et al., 2016; Chaillou et al., 2005; Janßen et al., 2020) and fungi to produce molecules against other fungi (Murgia et al., 2019) can prevent the development of undesired microorganisms.

The use of an autochthonous microbiome is also important in respect to the accumulation of mycotoxins in fermented meat due to the presence of indigenous fungi, in particular ochratoxin A (OTA) that has very negative effects including nephrotoxicity, immunotoxicity, and neurotoxicity (Álvarez et al., 2020). The meat industry normally uses synthetic antifungal compounds, but the same degree of protection can be attained with the use of an autochthonous microbiome. Through microbial fermentation this can produce natural microbial metabolites, like enzymes (β-1,3 glucanases, lytic proteases, and chitinases) able to hydrolyze microorganism cell wall components (Cence et al., 2019). The use of an autochthonous microbiome can also allow the use of nitrites and nitrates to be reduced. These are used as curing agents in dry fermented sausages because of their antioxidant and antimicrobial properties as well as their positive effect on aroma and can be partially replaced by the action of *D. hansenii* which performs similar actions (Perea-Sanz et al., 2020).

The process chain of fermented sausages can be considered a major source of the spread of antibiotic resistance genes (ARGs) from farm to the final product. The intensification of livestock farming with the extensive use of veterinary drugs and high levels of antibiotics has led to an increased acquisition of resistance in microorganisms (Zaiko et al., 2020). ARGs can be transferred from indigenous microorganisms in fermented sausages to opportunistic or pathogenic bacteria present in humans by horizontal gene transfer, which represents a worldwide public health concern. In this particular food system the main microbes involved in the spread of ARGs belong to *Staphylococcus* (Lopez et al., 2020). Risks often linked with CNS (*Staphylococcus*) and LAB (*Lactobacillus, Carnobacterium*, and *Enterococcus*) are the production of decarboxylases that can cause biogenic amine production typically linked with fermented foods. In fermented meats, the most frequently found biogenic amines are tyramine, putrescine, and cadaverine, while histamine is randomly detected (Van der Veken et al., 2020). In this light the transfer of ARGs from fermented sausages can be limited by the use of specific autochthonous microbiome cultures lacking in AR genes, never compromising on safety through an assessment of their capability to produce biogenic amines that remains an important aspect when designing food processes.

13.5 CONCLUSION

Several tools, detection methods and bioinformatics pipelines are available to understand what happens during meat fermentation, particularly concerning both bacteria and fungi for which fermented sausage is a preferred substrate. With the aim of preserving traditional product characteristics while standardizing industrial production processes and ensuring the highest quality and safety, an intense effort should be made in isolating and characterizing both groups of microorganisms present in traditional food environments. Studying the interaction, functions, and diversity of each of the microbial species harbored in this complex system is a key factor toward selecting autochthonous microbiomes that include bacteria, fungi, and their metabolites to easily manipulate a food system like fermented sausages increasing the quality and safety of the products.

REFERENCES

1. Álvarez M, Rodríguez A, Núñez F, Silva A, Andrade MJ. 2020. In vitro antifungal effects of spices on ochratoxin A production and related gene expression in *Penicillium nordicum* on a dry-cured fermented sausage medium. *Food Control* 114: 107222.
2. Aquilanti L, Garofalo C, Osimani A, Clementi F. 2016. Ecology of lactic acid bacteria and coagulase negative cocci in fermented dry sausages manufactured in Italy and other Mediterranean countries: An overview. *International Food Research Journal* 23: 429–445.
3. Belleggia L, Ferrocino I, Reale A, Boscaino F, Di T, et al. 2020a. Portuguese cacholeira blood sausage: A first taste of its microbiota and volatile organic compounds. *Food Research International* 136: 109567.
4. Belleggia L, Milanović V, Ferrocino I, Cocolin L, Haouet MN et al. 2020b. Is there any still undisclosed biodiversity in Ciauscolo salami? A new glance into the microbiota of an artisan production as revealed by high-throughput sequencing. *Meat Science* 165: 108128.
5. Berg G, Rybakova D, Fischer D, Cernava T, Vergès MCC, et al. 2020. Microbiome definition re-visited: Old concepts and new challenges. *Microbiome* 8: 103.
6. Bertuzzi AS, Walsh AM, Sheehan JJ, Cotte PD, Crispie F, et al. 2018. Omics-based insights into flavor development and microbial succession within surface-ripened cheese. *mSystems* 3: e00211–17.
7. Bourdichon F, Casaregola S, Farrokh C, Frisva JC, Gerds ML, et al. 2012. Food fermentations: Microorganisms with technological beneficial use. *Internation Journal of Food Microbiology* 154: 87–97.
8. Cano-García L, Rivera-Jiménez S, Belloch C, Flores M. 2014. Generation of aroma compounds in a fermented sausage meat model system by *Debaryomyces hansenii* strains. *Food Chemistry* 151: 364–373.
9. Cardinali F, Milanovic V, Osimani A, Aquilanti L, Taccari M. et al. 2018. Microbial dynamics of model Fabriano-like fermented sausages as affected by starter cultures, nitrates and nitrites. *Internation Journal of Food Microbiology* 278: 61–72.
10. Casaburi A, Di Martino V, Ferranti P, Picariello L, Villani F. 2016. Technological properties and bacteriocins production by *Lactobacillus curvatus* 54M16 and its use as starter culture for fermented sausage manufacture. *Food Control* 59: 31–45.
11. Casaburi A, Nasi A, Ferrocino I, Di Monaco R, Mauriello G. et al. 2011. Spoilage-related activity of Carnobacterium maltaromaticum strains in air-stored and vacuum-packed meat. *Applied and Environmental Microbiology* 77: 7382–7393.
12. Castellano P, Aristoy MC, Sentandreu MA, Vignol G, Toldrá F. 2012. *Lactobacillus sakei* CRL1862 improves

safety and protein hydrolysis in meat systems. *Journal of Applied Microbiology* 113: 1407–1416.

13. Cence K, Santos P, Garcia MV, Copetti MV, Valduga E, et al. 2019. Enzymatic biocontrol of spoilage fungi from salami. *LWT - Food Science and Technology* 115: 108457.

14. Chaillou S, Champomier-Vergès MC, Cornet M, Crutz-Le C, et al. 2005. The complete genome sequence of the meat-borne lactic acid bacterium *Lactobacillus sakei* 23K. *Nature Biotechnology* 23: 1527–1533.

15. Chaillou S, Lucquin I, Najjari A, Zagorec M, Champomier-Vergès MC. 2013. Population genetics of *Lactobacillus sakei* reveals three lineages with distinct evolutionary histories. *PLoS One* 8: e73253.

16. Chen Q, Kong B, Sun Q, Dong F, Liu Q. 2015a. Antioxidant potential of a unique LAB culture isolated from Harbin dry sausage: In vitro and in a sausage model. *Meat Science* 110: 180–188.

17. Chen Q, Liu Q, Sun Q, Kong B, Xiong Y, 2015b. Flavour formation from hydrolysis of pork sarcoplasmic protein extract by a unique LAB culture isolated from Harbin dry sausage. *Meat Science* 100: 110–117.

18. Ercolini D, Casaburi A, Nasi A, Ferrocino I, Di Monaco R, et al. 2010. Different molecular types of *Pseudomonas fragi* have the same overall behaviour as meat spoilers. *Internation Journal of Food Microbiology* 142: 120–131.

19. Ferrocino I, Bellio A, Giordano M, Macori G, Romano A, et al. 2018. Shotgun metagenomics and volatilome profile of the microbiota of fermented sausages. *Applied and Environmental Microbiology* 84: 1–15.

20. Flores M, Corral S, Cano-García L, Salvador A, Belloch C. 2015. Yeast strains as potential aroma enhancers in dry fermented sausages. *Internation Journal of Food Microbiology* 212: 16–24. https://doi.org/10.1016/j.ijfoodmicro.2015.02.028.

21. Franciosa I, Alessandria V, Dolc P, Rantsiou K, Cocolin L. 2018. Sausage fermentation and starter cultures in the era of molecular biology methods. *Internation Journal of Food Microbiology* 279: 26–32.

22. Freiding S, Gutsche KA, Ehrman MA, Vogel RF 2011. Genetic screening of *Lactobacillus sakei* and *Lactobacillus curvatus* strains for their peptidolytic system and amino acid metabolism, and comparison of their volatilomes in a model system. *Systematic and Applied Microbiology* 34: 311–320.

23. Hamme WP, Hertel C. 1998. New developments in meat starter cultures. *Meat Science* 49: S125–S138.

24. Iacumin L, Osualdini M, Bovolenta S, Boscolo D, Chiesa L, et al. 2020. Microbial, chemico-physical and volatile aromatic compounds characterization of Pitina PGI, a peculiar sausage-like product of North East Italy. *Meat Science* 163: 108081.

25. Janßen D, Dworschak L, Ludwig C, Ehrmann MA, Vogel RF. 2020. Interspecies assertiveness of *Lactobacillus curvatus* and *Lactobacillus sakei* in sausage fermentations. *Internation Journal of Food Microbiology* 331: 108689.

26. Kim JS, Kim JH, Arunachalam PS, Lee CC, You JW, et al. 2019. Yak-Kong soybean (Glycine max) fermented by a novel *Pediococcus pentosaceus* inhibits the oxidative stress-induced monocyte–endothelial cell adhesion. *Nutrients* 11: 1380.

27. Lopez CM, Callegari ML, Patrone V, Rebecchi A 2020. Assessment of antibiotic resistance in staphylococci involved in fermented meat product processing. *Current Opinion in Food Science* 31: 17–23.

28. Murgia MA, Marongiu A, Aponte M, Blaiotta G, Deiana P, et al. 2019. Impact of a selected *Debaryomyces hansenii* strain's inoculation on the quality of Sardinian fermented sausages. *Food Research International* 121: 144–150.

29. Osimani A, Ferrocino I, Agnolucci M, Cocolin L, Giovannetti M, et al. 2019. Unveiling hákarl: A study of the microbiota of the traditional Icelandic fermented fish. *Food Microbiology* 82: 560–572.

30. Papadimitriou K, Pot B, Tsakalidou E. 2015. How microbes adapt to a diversity of food niches. *Currunt Opinion in Food Science* 2: 29–35.

31. Parussolo G, Bernardi AO, Garcia MV, Stefanello A, Silva T, et al. 2019. Fungi in air, raw materials and surface of dry fermented sausage produced in Brazil. *LWT - Food Science and Technology* 108: 190–198.

32. Perea-Sanz L, López Díez JJ, Belloch C, Flores M. 2020. Counteracting the effect of reducing nitrate/nitrite levels on dry fermented sausage aroma by *Debaryomyces hansenii* inoculation. *Meat Science* 164: 108103.

33. Połka J, Rebecchi A, Pisacane V, Morelli L, Puglisi E. 2015. Bacterial diversity in typical Italian salami at different ripening stages as revealed by high-throughput sequencing of 16S rRNA amplicons. *Food Microbiology* 46: 342–356.

34. Quijada NM, De Filippis F, Javier Sanz J, Del Camino García-Fernandez M, Rodríguez Lazaro D, et al. 2018. Different *Lactobacillus* populations dominate in Chorizo de Leon; manufacturing performed in different production plants. *Food Microbiology* 70: 94–102.

35. Rimaux T, Vrancken G, Vuylsteke B, De Vuyst L, Leroy F. 2011. The pentose moiety of adenosine and inosine

is an important energy source for the fermented-meat starter culture *Lactobacillus sakei* CTC 494. *Applied and Environmental Microbiology* 77: 6539–6550.

36. Stellato G, La Storia A, De Filippis F, Borriello G, Villani F, et al. 2016. Overlap of spoilage-associated microbiota between meat and the meat processing environment in small-scale and large-scale retail. *Applied and Environmental Microbiology* 82: 4045–4054.
37. Sun F, Hu Y, Chen Q, Kong B, Liu Q. 2019. Purification and biochemical characteristics of the extracellular protease from *Pediococcus pentosaceus* isolated from Harbin dry sausages. *Meat Science* 156: 156–165.
38. Terán LC, Coeuret G, Raya R, Zagorec M, Champomier-Vergès MC, et al. 2018. Phylogenomic analysis of *Lactobacillus curvatus* reveals two lineages distinguished by genes for fermenting plant-derived carbohydrates. *Genome Biology and Evolution* 10: 1516–1525.
39. Van der Veken D, Benhachemi R, Charmpi C, Ockerman L, Poortmans M, et al. 2020. Exploring the ambiguous status of coagulase-negative staphylococci in the biosafety of fermented meats: The case of anti-bacterial activity versus biogenic amine formation. *Microorganisms* 8: 167.
40. Wolfe BE, Button JE, Santarelli M, Dutton RJ. 2014. cheese rind communities provide tractable systems for in situ and in vitro studies of microbial diversity. *Cell* 158: 422–433.
41. Xiao Y, Liu Y, Chen C, Xie T, Li P. 2020. Effect of *Lactobacillus plantarum* and *Staphylococcus xylosus* on flavour development and bacterial communities in Chinese dry fermented sausages. *Food Research International* 135: 109247.
42. Xu H, Gao L, Jiang Y, Tian Y, Peng J, et al. 2015. Transcriptome response of *Lactobacillus sakei* to meat protein environment. *Journal of Basic Microbiology* 55: 490–499.
43. Zadravec M, Vahčić N, Brnić D, Markov K, Frece J, et al. 2020. A study of surface moulds and mycotoxins in Croatian traditional dry-cured meat products. *Internation Journal of Food Microbiology* 317: 108459.
44. Zaiko EV, Bataeva DS, Yushina YK, Makhova AA, Minaev MY. 2020. Antibiotic resistant of microorganisms in fermented sausages. *Earth and Environmental Science* 421: 052013.

CHAPTER 14

The Sourdough Microbiota and Its Sensory and Nutritional Performances

Hana Ameur, Kashika Arora, Andrea Polo, and Marco Gobbetti

Department of Food Engineering and Biotechnology, Faculty of Science and Technology, Free University of Bolzano, Italy, 39100

14.1 INTRODUCTION

As stated by historical sources, sourdough represents one of the first natural starters used for food processing by humans. In the first century AC, Pliny the Elder first mentioned it in the encyclopedia *Naturalis Historia* (1972). Ancient Egypt is considered the most probable motherland of sourdough bread (Guidotti, 2005). The importance of bread in Egyptian culture is highlighted by the language that used the same term to define both ferment and bread, and the term means "life." The meaning of leavened bread remained crucial for human spirituality and for several civilizations also in subsequent ages. It represented the "food of eternal life" for the Catholic culture and it was used in many religious rituals. The nutritional properties of bread due to fermentation were already claimed in the Middle Ages, when it was considered useful for human wellness. Nowadays, leavened bread is an essential component of diet for people all around the word, and it is declared an intangible heritage of humanity by UNESCO.

Until the late 1900s, baker's yeast was almost the only starter used in the industrial and artisanal manufacture of baked goods. This practice did not encourage a systematic research investigation into sourdough bread; that has only developed since 1900, much later than other fermented matrices. However, since around 1990, sourdough fermentation has been studied more and more with the final goal being to clarify the properties of sourdough fermentation and (re)exploit this old biotechnology. Originally, the interest of scientists was mainly attracted by the technological effects of sourdough fermentation in terms of flavor, rheology and shelf life, and on the ecological relationship among the microorganisms involved. Soon, other topics like the sourdough assembly and composition were also investigated, aiming at elucidating the microbial diversity, relative abundances of lactic acid bacteria, and the evolution of the ecosystem during fermentation (both qualitatively and in terms of relative abundances). However, around 35 years ago a number of research groups consistently focused their attention on sourdough fermentation, and the first description of a key lactic acid bacterium playing a major role in sourdough fermentation was published only in the 1970s when *Fructilactobacillus sanfranciscensis* (formerly *Lactobacillus brevis* subsp. *Lindneri*) was described, which represented a milestone for this field of study. Nowadays, the scientific community is united to define the sourdough as a mixture of flour and water, spontaneously fermented by lactic acid bacteria and yeasts, and having acidification and leavening capacities (Gobbetti, 1998). These studies opened the door to more recent research activities that

Good Microbes in Medicine, Food Production, Biotechnology, Bioremediation, and Agriculture, First Edition. Edited by Frans J. de Bruijn, Hauke Smidt, Luca S. Cocolin, Michael Sauer, David Dowling, and Linda Thomashow.

elucidated the source of sourdough microbiota (i.e. flours, house microbiota, additional ingredients and tap water) and/or drivers to establish the potential of this natural starter (Minervini et al., 2019). In recent years, research has addressed the multiple nutritional advantages supplied by sourdough fermentation with respect to other leavening agents. In the last 20 years, several reviews have been published summarizing the state of the art of different research topics on sourdough. In 2021, the first systematic review was also undertaken with the purpose of highlighting the findings and knowledge derived from the last thirty years of research on the microbiological, biochemical, technological, and nutritional potential of sourdough fermentation (Arora et al., 2021).

Based on the most recent scientific literature, this chapter aims to summarize the current knowledge about the mechanisms driving the sourdough microbiota assembly, its uses and characteristics, its potential to exploit non-conventional flours, and its sensory and nutritional performances in baked goods products. Finally, the main conclusions and still open questions will be highlighted, as well as potential future purposes and perspectives.

14.2 HOW THE SOURDOUGH MICROBIOTA IS ASSEMBLED

The microbial composition of sourdough is influenced by a number of factors, primarily, the house microbiota, type of flour, water, and other added ingredients. Despite several thorough investigations, the researchers have not been successful in associating species diversity with geographical location. The probable origin(s) of sourdough microbiota are discussed in the following sub-sections.

14.2.1 House Microbiota

Scientists began to investigate all the probable sources contributing to the species diversity. Several artisanal bakeries participated for the sampling of bacterial isolates from the sourdoughs, bakery environment and equipment as well as the hands of the bakers. The molecular techniques, such as amplified fragment length polymorphism (AFLP) fingerprinting, identified *F. sanfranciscensis* and *Lactiplantibacillus plantarum* to be genetically persistent in the sourdough and bakery ecosystems (Scheirlinck et al., 2008). It was demonstrated that several strains of lactic acid bacteria were associated with the bakery sourdoughs and surroundings. With the advancement in technology, the consistency of lactic acid bacteria strains was further verified by metagenomic studies. In particular, the RNA extracted from the flour, sourdough, dough mixer and storage box of four bakeries identified *F. sanfranciscensis* strains from all the sources. However, a higher relative abundance of some species (e.g., *Weissella cibaria*) was observed only in sourdoughs indicating their main origin as flour contaminants (Minervini et al., 2015a). Experiments were conducted to differentiate the house microbiota of sourdoughs propagated at artisan bakery and laboratory levels (Minervini et al., 2012a). The back-slopping of seven mature type I sourdoughs from Southern Italy was performed for 80 days mimicking the technology parameters at artisan bakery and laboratory scale. The microbial identification was elaborated by culture-dependent and -independent (Denaturing Gradient Gel Electrophoresis (DGGE)) techniques demonstrating a significant variation in the same type of sourdough. Although some strains belonging to *F. sanfranciscensis* persisted in some sourdoughs propagated at different locations (bakery or laboratory), many others were identified only at either artisan bakeries (*L. plantarum*) or the laboratory (e.g., *Leuconostoc citreum*) levels. Hence, it can be assumed that the species diversity in sourdoughs can be largely attributed to the environment of propagation. In fact, some researchers highlighted the shift of sourdough microbiota in terms of yeast density and the differences observed in the persistence of *Saccharomyces cerevisiae* when sourdoughs were propagated at the bakery or laboratory level (Minervini et al., 2012a). This shift in microbiota was also shown to be influenced by the switch from firm to liquid sourdough fermentation. Di Cagno et al. (2014) collected typical firm sourdoughs from traditional bakeries in Southern Italy and propagated them for one week in laboratory under bakery conditions. Then, these sourdoughs were subjected to comparative propagation for 28 days under firm and liquid conditions to study the probable shift in house microbiota. Consequently, it was reported that the liquid sourdoughs harbored a lower microbial diversity as compared to firm sourdoughs. Out of the few persistent strains, *L. plantarum* dominated in firm conditions over time, while other lactic acid bacteria, such as *F. sanfranciscensis*, *Leuconostoc lactis* and *Levilactobacillus brevis*, dominated only in some firm sourdoughs. Besides, *L. citreum* was the only species always identified in liquid sourdoughs or flanked by *Leuconostoc mesenteroides* in others. Scheirlinck et al. (2009) suggested that the house microbiota of sourdough can be built by continuous propagation of type I sourdough, that will eventually act as an inoculum for the next fermentation. The addition of flour plays a crucial role in stimulating the sourdough fermentation, as detailed in the next section.

14.2.2 Flour

Apart from the bakery environment, the microbial diversity of sourdough is widely attributed to the type of flour as well. Figure 14.1 lists the *Lactobacillus* species isolated and identified from different flours (cereals, pseudo-cereals, and legumes) and/or related sourdough. De Vuyst et al. (2009) isolated and identified the lactic acid bacteria directly from the flour. The microbial population and their relative abundance on the cereal grains is influenced by several parameters, which includes the climatic conditions (temperature and humidity), physical rupture by insects or fungi attacks as well as the use of insecticides and fungicides.

The microbes usually colonize the outer layers of kernels and the bran-rich fractions during milling. Although the flour should contain a lower bacterial density after milling as compared to caryopses, the increase in microbial content of flour can be attributed to caryopsis conditioning. Corsetti et al. (2007) investigated the differences in the microbial population of grains, bran and flour and mature sourdoughs through culture-dependent methods. They reported the occurrence of a sub-dominant population of lactic acid bacteria which includes *Enterococcus faecium*, *Enterococcus mundtii*, and *Lactobacillus graminis*. Furthermore, Alfonzo et al. (2013) characterized the lactic acid bacteria isolated from wheat flours used in traditional bread recipes in Sicily. The identification by partial sequencing of 16S rRNA gene grouped the strains into six genera with the most prevalent species as *L. plantarum*, *W. cibaria*, *Leuconostoc pseudomesenteroides* and *L. citreum*. Based on the similar approach, 19 Italian sourdoughs were investigated for their house microbiota by Minervini et al. (2012b). Consequently, the researchers were able to find a correlation between the type of flour and the microbial community in sourdough. Moreover, high levels of glucose, fructose, maltose and total free amino acids were characterized in *Triticum turgidum* ssp. durum flours. These higher levels can be correlated to the presence of heterofermentative lactic acid bacteria and a lower cell density of yeasts in the mature sourdoughs. The abundance of bran in the flour can also influence the sourdough microbiota due to enriched content of dietary fibers and bioactive compounds.

It is important to note that the autochthonous lactic acid bacteria strains characterized in flours are also members of the plant endophytic microbiota. This was proven by the isolation and identification of epiphytic lactic acid bacteria from forage crops and the

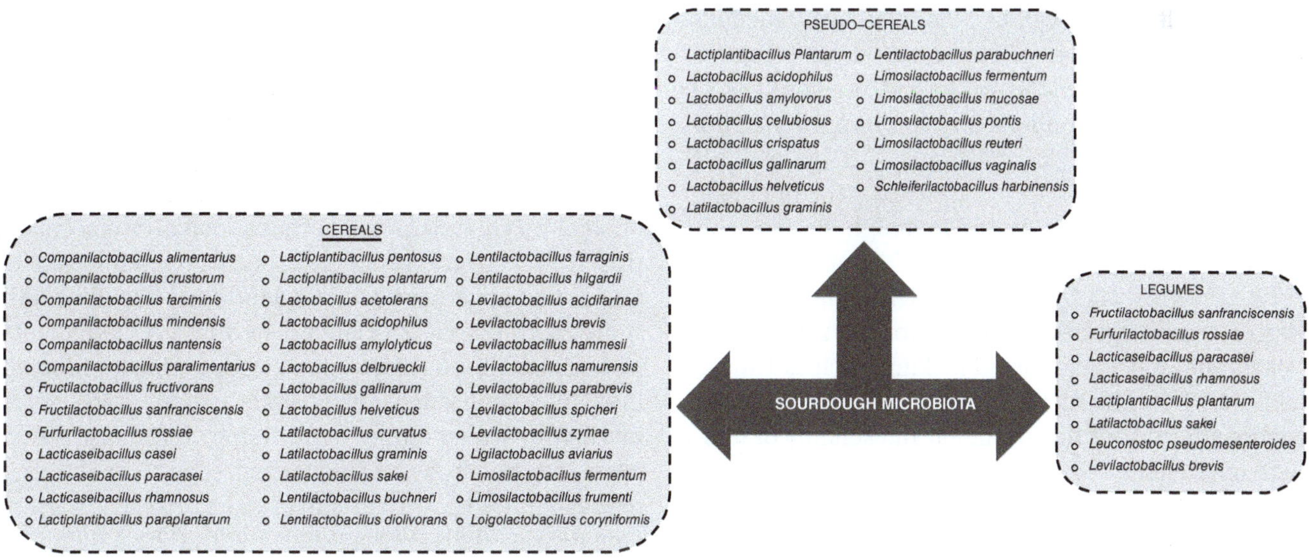

FIGURE 14.1 *Lactobacillus* species isolated and identified from different flours and/or related sourdough (cereals, pseudo-cereals, and legumes) constituting the sourdough microbiota.

inoculation of these lactobacilli into *Lolium perenne* (ryegrass) was able to cause colonization of the roots at the endophytic level. Similarly, the agricultural lands for two durum wheat cultivars were elaborated in terms of kinetics of the Firmicutes population during the phenological stages of wheat (Minervini et al., 2015a). The endophytic and epiphytic microbial populations were analyzed using metagenomic studies. The early stages of wheat growth revealed *Lactobacillus, Lactococcus, Streptococcus,* and *Enterococcus* as the main epiphytic and endophytic genera of lactic acid bacteria. In addition, the core microbiota was also comprised of *Paenibacillus, Bacillus, Exiguobacterium,* and *Staphylococcus* genera. Overall, the relative abundance of these genera was impacted by the cultivars, phenological stage, other Firmicutes genera, atmospheric temperature, and water activity of the plants. However, the genotypic typing of strains belonging to *L. plantarum*, a prominent sourdough species, showed its presence throughout the phenological stages of all the cultivars, as well as milled flour and bran. Conclusively, lactic acid bacteria can be considered as endophytic microbiota of the wheat plant, and consequently of the flour used for sourdough fermentation and propagation.

14.2.3 Water

Being an integral part of any sourdough preparation, water is also considered to be an important ingredient influencing the house microbiota. The addition of water can affect the viscoelastic properties of the gluten network in the dough as well as other chemical reactions that occur during mixing, fermentation, and baking. Moreover, the pH of water can impact the hardness of dough, and hence the mineral content (or nutrients) utilized by pro-technological microbes. An optimal pH value in the range of 5.0–6.0 is critical for the microbial and enzymatic activities in the sourdough. Besides, this pH range also increases the volume of steamed bread by favoring the growth of yeast during fermentation. Some artisan bakers suggested guidelines for the use of water in traditional sourdough bread making, such as Pane di Altamura PDO. The metal ions and trace elements are also thought to affect the formation and stability of sourdough microbiota.

In this context, Minervini et al. (2019) investigated the effect of tap water, collected from ten different Italian regions differing in cultural heritage for leavened baked goods, on type I sourdough microbiota. The hardness of all the tap waters was shown to be very high (120–300 mg of $CaCO_3$/L) as opposed to the standard range (50 to 200 mg of $CaCO_3$/L) suggested by other researchers. It is interesting to note that regardless of the origin of the tap water, differences were observed at the level of sourdough use, i.e., preparation or propagation. The researchers observed that the propagation of a traditional mature Apulian sourdough for nine days did not change the cell density of lactic acid bacteria. Although the composition of LAB strains in mature sourdoughs was also shown to be 60–80% identical, some strains of *L. plantarum* and *Pediococcus pentosaceus* were identified only in the sourdoughs after propagation. This could be attributed to the dominant nature of these strains over sub-dominant ones (Gobbetti et al., 2016). The type of tap water moderately affected the LAB population and acidifying capacity during propagation under laboratory conditions, whereas the yeast cell density was decreased due to less contamination with baker's yeast (as expected in bakeries) (Minervini et al., 2012b). Although the textural properties of sourdough breads, in terms of gumminess, chewiness and hardness, vary greatly due to the type of tap water, the sourdough microbiota was minimally affected by the water utilized during the preparation, propagation and baking conditions.

14.2.4 Other Ingredients

Besides the crucial ingredients mentioned in the previous sections, there are other ingredients which can affect the sourdough microbiota depending on the type of baked goods. For instance, the additional ingredients, like sugar, salt, fruits, and grape must, may induce the pro-technological lactic acid bacteria and yeasts, consequently, influencing the intrinsic parameters of sourdough (in terms of e.g., water activity, concentration of monosaccharides and di-saccharides) (Minervini et al., 2014). The sourdough used for the manufacture of Pane di Matera PGI and Coppia Ferrarese contains macerated ripe fruit and grape must, respectively. While apple, grape, and sugarcane are frequently used in Brazilian sourdough recipes. The impact of different ingredients on sourdough microbiota was elaborated by producing individual sourdoughs using baker's yeast, grape must, macerated pears, honey, and water from macerated pears, separately as additional ingredients in the first propagation step (Minervini et al., 2016). Each sourdough was daily propagated by the back-slopping procedure until stability was achieved in terms of dough acidification and leavening power. The findings from Gordún et al. (2015), also suggested that the diversity of lactic acid bacteria in sourdough varied depending upon the type of additional ingredients, such as apple, grape or yoghurt. In particular, *P. pentosaceus* was identified only in sourdoughs prepared with yoghurt, whereas the sourdoughs prepared with apple or grape were inhabited by *L. plantarum*, *L. brevis*, *Latilactobacillus sakei*, and *W. cibaria* (only in grape sourdough). The metagenomic analysis using the Illumina MiSeq platform detected *Lactobacillus* as the dominant species in baker's yeast as well as in macerated pear and its water. Other subdominant genera included *Leuconostoc* (macerated pear and related water), *Streptococcus* (honey), and *Lactococcus* (baker's yeast). In comparison to the typical sourdough (only flour and water as ingredients), the alpha-diversity indices were found to be lower throughout the fermentation and propagation of sourdoughs prepared with additional ingredients. Nevertheless, *Lactobacillus* sp. was found to be dominant in the mature sourdoughs, irrespective of the type of ingredients added to the sourdough preparation. However, differences were observed at the strain level of *L. plantarum* depending on the additional ingredient. Other species isolated from sourdoughs with additional ingredients included *L. brevis*, *Latilactobacillus curvatus*, and *Limosilactobacillus fermentum* (from baker's yeast), *L. mesenteroides* (from grape must), and *Lactiplantibacillus pentosus* (from macerated pears). Furthermore, a higher cell density of lactic acid bacteria was observed for the sourdoughs with additional ingredients as compared to the control sourdough after the first fermentation. In fact, the sourdough prepared with macerated pear or its water dropped the pH to approx. 4.0 after the second backslopping. Also, these additional ingredients also impacted the concentration of free amino acids (e.g., glutamine, isoleucine, leucine, arginine, and phenyl alanine) and the related metabolites (ornithine and γ-aminobutyric acid) in the mature sourdoughs (Minervini et al., 2016).

14.3 WHERE AND HOW TO USE THE SOURDOUGH

Sourdough is used worldwide for making fermented baked goods. This natural starter improves the performance of salty and sweet products (e.g., bread rolls, pasta, burger buns, pizza, biscuits, cakes, brioches) compared to baker's yeast (Figure 14.2). Sourdough fermentations necessitate a thorough understanding of the effects of process conditions, raw materials, and microorganisms to achieve a consistent, reproducible sourdough, and baked goods quality.

14.3.1 Baked Goods and Flours

Currently, the literature is particularly rich in results, which shows the uniqueness of the microbial composition and functionality of each sourdough baked good. A wide variety of sourdough baked goods have been characterized in almost 50 countries around the world. They are mainly salty products (breads and substitutes) and sweet baked goods such as traditional Italian salty and sweet leavened baked goods, the industrial sourdough breads, rolls, crackers, and cookies in the United States, and the Egyptian and Chinese typical sourdough breads (Arora et al., 2021). Sourdough fermentation has been demonstrated as a powerful tool to ferment many flours and agro-food by-products. The use of sourdough has been the most suitable and sustainable processes to increase the nutritional, functional, rheology, sensory and shelf-life attributes of baked goods made with flours from 23 species/varieties of cereals, sprouted seeds, 10 pseudo-cereals, 19 varieties of legumes, and 25 miscellaneous vegetables. In detail, the nutritional value, and health-promoting compounds, such as free amino acids (FAA), γ-amino butyric acid (GABA), polyphenols, dietary fibers (DF), bioavailable minerals, antioxidant activities, and *in vitro* protein digestibility were increased, whereas the glycemic index (GI) and the anti-nutritional factors were decreased in fermented legume flours (Gabriele et al., 2019). Furthermore, the rheology, sensory, shelf life, and nutritional attributes of gluten-free formulations and baked goods fortified with bran or cereal germ has been improved with sourdough fermentation (Pontonio et al., 2020).

14.3.2 Conditions of Use

Based on the technology applied, sourdoughs have been grouped into three types. Type I sourdough or traditional sourdough; which is restarted using previously made dough (mother dough) to keep the microorganisms metabolically active; type II sourdough or accelerated sourdough is generally propagated at relatively high temperature (> 30°C) and long fermentation time (up to five days); and type III sourdough, or dried sourdough, is generally used as a flavoring agent (De Vuyst and Neysens, 2005). The use of mature sourdough depends on the type of process and traditional products. Traditional sourdough is the most studied; nevertheless, the other

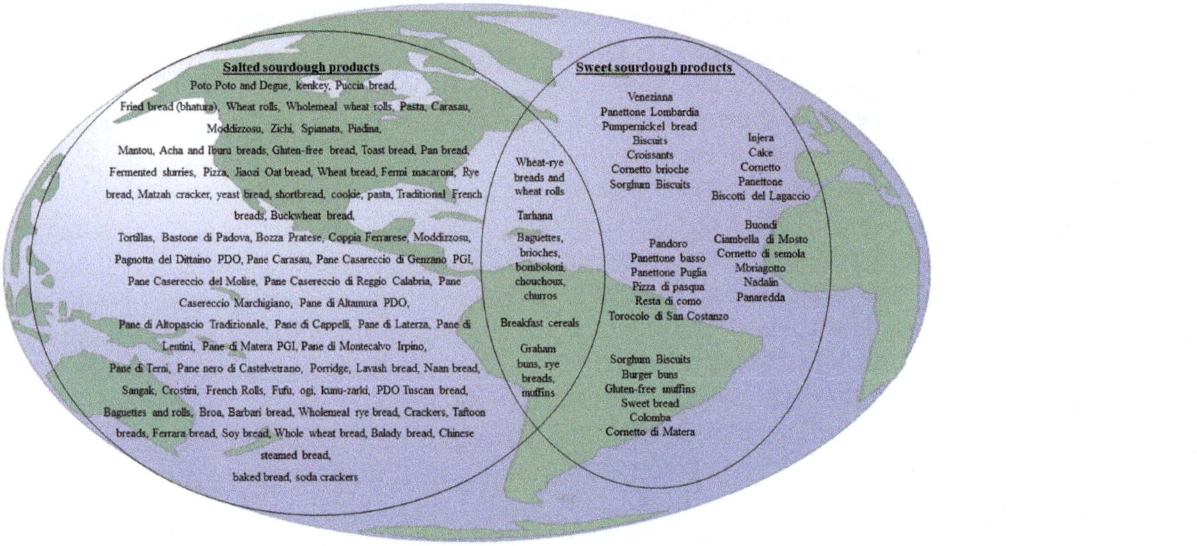

FIGURE 14.2 Worldwide map listing sourdough products (salty and sweet) adapted from Arora et al. (2021).

two types have gained increasing interest recently especially on the industrial level. Overall, a one-step process, 20% sourdough inoculum, 4 h of fermentation at 30°C are the most common practice before baking. Of course, these conditions can vary especially when some unconventional matrices are used (e.g., wheat germ, brewer's spent grains) (Rizzello et al., 2010). The number of metabolically active lactic acid bacteria should be at 10^8–10^9 cfu g^{-1} and yeasts at 10^6–10^7 cfu g^{-1} for the proper process of acidification and leavening. During fermentation for optimal activities, the ratio between lactic acid bacteria and yeast should be 100:1, for several sourdoughs, it narrows to 10:1. When using commercial dried sourdoughs or when adding starter cultures to the dough, the above-mentioned cell density for lactic acid bacteria can exceptionally go below 10^3 cfu g^{-1} or above 10^9 cfu g^{-1}. Regarding the yeasts, we cannot obtain the desired leavening power with a cell density inferior to 10^5 cfu g^{-1}. Another important technological parameter that can greatly influence the sourdough fermentation is the dough yield (DY). Most firm sourdoughs have a DY between 150 and 160 g, while semi-liquid sourdough has a range between 200 and 300 g. An extremely high DY of 500 has been used to make bread with durum wheat semolina rendered gluten-free by sourdough lactic acid bacteria and fungal proteases (Rizzello et al., 2007). Spreading the use of sourdough on an industrial and artisanal level requires optimizing the use conditions toward shorter times and lower temperatures of fermentation.

14.3.3 Microbiological and Biochemical Characteristics

The sourdough characterization can be performed by culture-dependent, culture-independent methods or the use of both approaches. Furthermore, some biochemical parameters such as pH, total titratable acidity, the concentration of organic acids, and total free amino acids allow the sourdough performance to be described. Sourdough is considered to be a unique food ecosystem because of its complex microbial composition. Microbiological studies have revealed that almost 60 bacterial genera occur in sourdoughs: 10 belonging to lactic acid bacteria. Most of the non-lactic acid bacteria, i.e. the other bacteria in the sourdough, act within the community as satellite members. More than 80 species of mainly obligate and facultative hetero-fermentative Lactic acid bacteria belonging to the genus *Lactobacillus* have been found. *L. plantarum*, *L. brevis*, *F. sanfranciscensis*, and *L. fermentum* are the most common isolated bacterial species from sourdoughs worldwide. Eighty species of yeasts were identifiable from sourdoughs; *Saccharomyces*, *Candida*, *Kazachstania*, *Torulopsis*, *Yarrowia*, and *Pichia* were the main genera. Although *Kazachstania exigua* and *Kazachstania humilis* were found in stable associations with lactic acid bacteria to produce sweet baked goods, the species most identified is *S. cerevisiae*. Whether *S. cerevisiae* exists as a wild strain or as a result of environmental cross-contamination from the excessive use of baker's yeast is still a debate. The dominant microbiota identified in sourdoughs is mainly due to the resilience and resistance of several *Lactobacillus* species. For instance, rye flour autochthonous *F. sanfranciscensis* and *L. fermentum*, became dominant during sourdough fermentation. *E. faecium* and *P. pentosaceus* present in durum wheat flour rapidly acidified the dough, making the ecosystem suitable for sourdough maturation by other *Lactobacillus* species (Corsetti and Settanni, 2007). The microbial dynamics of many spelt sourdoughs mirrored the autochthonous microbiota of flour, which was influenced by flour composition, cultivation methods, and storage conditions. *E. faecium, Enterococcus durans, L. brevis, L. pentosus, and Lacticaseibacillus paracasei* showed the interesting capacity to overcome the stressful conditions during the harvest and storage of wheat grains (Gaglio et al., 2020). The focus of attention of further fundamental research should be on deepening the sourdough microbiota assembly and functionality using meta-genome and metatranscriptome approaches.

Acidity is an important characteristic of sourdough. Lactic acid produced by LAB starters during fermentation is responsible for lowering the pH and increasing the total titratable acidity in the final dough. The median value for pH is 4.1, with differences usually related to the protocol and the type of flours used. Values of pH below 3.0 were detectable using particular ingredients and very long fermentation times. The median value of total titratable acidity (TTA) is 11.0 ml of 0.1 M NaOH/10 g of dough. Extremely high values of TTA (40.1 ml of 0.1 M NaOH/10 g of dough) were found during seven days of hemp sourdough fermentation (Nionelli et al., 2018). The determination of organic acids showed that the median concentration of lactic acids found in sourdough is 75 mM, while for acetic acid, it is 20 mM Usually, the high acetic acid content is associated with a sourdough with a better flavor. The most recommended quotient of fermentation (QF, the molar ratio between lactic and acetic acids) is below 5.0. The concentration of free amino formed during fermentation through proteolysis is another crucial biochemical indicator of the performance of sourdough. Flour type, fermentation conditions, and strains used can greatly influence FAA concentration after sourdough fermentation; besides, an increase compared to the initial flour concentration would guarantee the accumulation of enough FAA acting as flavor precursors. In general, FAA have a median value of 1360 mg/kg.

14.4 SOURDOUGH TO EXPLOIT THE POTENTIAL OF NON-CONVENTIONAL FLOURS

Despite the inevitable benefits of sourdough over baker's yeast bread, the choice of more nutritional and functional flours is the current requirement for a healthy lifestyle. In this context, the non-conventional flours such as pseudo-cereals, cereal bran, and germ as well as legumes have gained much attention in the past decade. However, these non-conventional flours contain the beneficial compounds in non-bioavailable form and the anti-nutritional factors (e.g., phytase, raffinose, condensed tannins, and trypsin inhibitory activity) which can negatively impact the digestibility of baked products (Bartkiene et al., 2016). Furthermore, many researchers have demonstrated adverse textural and sensorial properties with the use of legumes and pseudo-cereals due to the absence of gluten (Scarnato et al., 2016). The lack of gluten was substituted by additives (e.g., starches or hydrocolloids) or preprocessing by technological procedures (e.g., extrusion cooking or high-pressure treatments) and enzymatic treatments. In fact, the microbial transglutaminase was shown to improve the textural and sensorial features of gluten-free matrices by altering the microstructure of dough with the catalysis of protein cross-links (Scarnato et al., 2016). Hence, sourdough fermentation with lactic acid bacteria can be regarded as an alternative to improve the nutritional and functional value of non-conventional matrices. This section highlights most commonly used non-conventional flours for sourdough fermentation.

14.4.1 Legumes

Legumes are the most abundant food crops after cereals throughout the world with *Leguminosae* as one of the largest families of flowering plants. Legumes have gained interest for use in baked goods due to their high nutritional value in terms of essential amino acids, dietary fibers, vitamins, minerals, carbohydrates, phenolic compounds, and other functional features. The fortification of legumes has been suggested by lactic acid fermentation for the improvement of protein digestibility and the availability of dietary fibers and phenolics. An overview on the benefits of legume fortification in sourdough biotechnology has been detailed in a recent review (Gobbetti et al., 2020). Table 14.1 summarizes the

TABLE 14.1 Main findings from the application of sourdough biotechnology to legume flours and legume-fortified wheat bread. The table summarizes data collected by Gobbetti et al. (2020)

Starter microorganism	Legume	Overall observed effects
Lactiplantibacillus plantarum	Bean, chickpea, grass pea, lentil, pea (local cultivars), fava bean, mixture of chickpea and pseudo-cereals, kidney beans, soybean, cowpea, mottled cowpea, speckled kidney bean, small rice bean.	Increase in phytase and antioxidant activity, protein digestibility, peptides, free amino acids, GABA, soluble fibers and total phenols concentrations; decrease in ANF (raffinose, phytic acid, vicine, convicine and condensed tannins) concentrations; decrease in trypsin inhibitory activity, starch availability and starch hydrolysis index; increased palatability and overall acceptability of bread; conversion of isoflavones into daidzein and genistein; release of lunasin-like polypeptides and bioactive peptides; inhibition of the proliferation of human adenocarcinoma Caco-2 cells.
Levilactobacillus brevis	Bean, chickpea, grass pea, lentil, pea (local cultivars).	Increase in phytase and antioxidant activity; increase in free amino acids, GABA, soluble fibers, and total phenols concentrations; decrease in raffinose and condensed tannins concentrations; release of lunasin-like polypeptides; inhibition of the proliferation of human adenocarcinoma Caco-2 cells.
Lactococcus lactis	Chickpea, mixture of chickpea and pseudo-cereals, adzuki bean.	Increase in free amino acids and GABA concentrations, and antioxidant activity; improved sensorial features (palatability and overall acceptability) of bread; decreased starch hydrolysis index.
Pediococcus acidilactici	Lupin.	Decrease in the trypsin inhibitor activity and hydrolysis index; increase in protein digestibility and bioavailability, and antioxidant activity.
Lacticaseibacillus paracasei	Soybean, cowpea, mottled cowpea, speckled kidney bean, small rice bean.	Conversion of isoflavones into daidzein and genistein; increase in antioxidant activity.
Lacticaseibacillus rhamnosus	Adzuki bean.	Increase in GABA concentration.
Latilactobacillus sakei	Lupin.	Increase in protein bioavailability and digestibility.
Pediococcus pentosaceus	Fava bean, lupin.	Increase in protein bioavailability, digestibility, and protein biological indices; increase in volume and hardness of bread; decrease in glycemic index.
Weissella confusa	Fava bean.	Increase in protein concentration, viscoelastic behavior, and specific volume of bread; decrease in crumb hardness of bread.
Leuconostoc pseudomesenteroides	Fava bean.	Increase in protein concentration, viscoelastic behavior and specific volume of bread; decrease in crumb hardness of bread.
Furfurilactobacillus rossiae	Chickpea, lentil.	Increase in the concentrations of peptides, free amino acids and GABA; increase in protein digestibility; decrease in starch availability, phytic acid, condensed tannins and raffinose concentrations, and trypsin inhibitory activity.
Fructilactobacillus sanfranciscensis	Chickpea, lentil.	Increase in the concentrations of peptides, free amino acids and GABA; increase in protein digestibility; decrease in starch availability, phytic acid, condensed tannins and raffinose concentrations, and trypsin inhibitory activity.
Spontaneous fermentation	Soybean, cowpea, mottled cowpea, kidney bean, small rice bean, African breadfruit.	Improved nutritional quality and sensory properties; increase in antioxidant activity, GABA, lysine and essential amino acids concentration, protein digestibility; decrease in α-galactosides, phytic acid, trypsin inhibitors and condensed tannins concentrations.
Type I sourdough	Chickpea/lentil/bean (mixture).	Increase in free amino acid concentration, and antioxidant and phytase activities.

main findings due to the application of sourdough biotechnology to legume flours and legume-fortified wheat bread. Traditional sourdough starters, such as selected strains of *L. plantarum* and *L. brevis*, were exploited for the fermentation of 19 Italian legume flours (Curiel et al., 2015). As a result, increased concentrations of free amino acids, total phenolics and fibers were quantified. In addition, many research studies have reported a marked increase in the levels of GABA due to sourdough fermentation (Coda et al., 2010), a crucial non-protein amino acid used extensively in pharmaceutical preparations due to its health benefits. Until today, many researchers have exploited the potential of legumes in the form of sprouts, flours and hydrolysates by the use of lactic acid fermentation (Montemurro et al., 2019). Fermentation of cowpea (*Vigna unguiculata*), mottled cowpea, small rice bean, and speckled kidney bean flours with lactic acid bacteria also increased the antioxidant activity as a result of biotransformation between soluble phenols and the release of bound phenolic compounds assisted by microbial enzymes. Glycosyl hydrolase, phenolic acid decarboxylase and reductase, and esterase are some of the enzymes involved in the conversion of phenolic compounds into biologically active metabolites (Filannino et al., 2018). Many legume flours (e.g., chickpea, lentil, beans, lupin, and fava beans) have also demonstrated their ability to improve protein digestibility *in vitro* (IVPD), and lower glycemic and starch hydrolysis index when fortified in baked goods (Coda et al., 2017; Rizzello et al., 2017). Simultaneously, the potential of lactic acid bacteria enzymes (e.g., α-galactosidase, β-glucosidase, phytases, tannases) to decrease the content of anti-nutritional factors present in legumes added beneficial value (Montemurro et al., 2019).

14.4.2 Pseudo-cereals

In the current proactive world, pseudo-cereals meet the needs of the producers and the consumers due to their adaptive agro-ecological nature and/or crop yields, and their nutritional potential, respectively. The most common pseudo-cereals utilized in food industries are quinoa (*Chenopodium quinoa* Willd.), amaranth (*Amaranthus hypochondriacus*), and buckwheat (*Fagopyrum esculentum*), while recently hemp (*Cannabis sativa*) has also received attention. Numerous health-beneficial attributes have been linked to these pseudo-cereals in the past decade and research is still ongoing to exploit their potential in baked goods. Sourdough fermentation with autochthonous lactic acid bacteria isolated from quinoa (*L. plantarum* and *Furfurilactobacillus rossiae* (Rizzello et al., 2016)), amaranth (*L. plantarum*, *L. sakei* and *P. pentosaceus*), buckwheat (*L. paracasei* and *P. pentosaceus* (Rocchetti et al., 2019)), and hemp (*L. plantarum*, *Pediococcus acidilactici*, and *Leuc. mesenteroides* (Nionelli et al., 2018)) has been studied extensively for nutritional/functional, rheological and sensorial properties. These non-conventional cereals are of special interest due to their use in gluten-free preparations increasing the choice of food for gluten-intolerant patients. Positive feedback has been received for the composite (wheat blend) or gluten-free breads prepared with these pseudo-cereals (Polo et al., 2020).

14.4.3 Milling By-products

The milling of cereal kernels is fractionated into flour, germ, and bran, where germ and bran are considered as milling by-products. These milling by-products were earlier discarded as waste or livestock feed. The use of milling by-products is intended to support the sustainable agriculture and ensure food security. In fact, the dietary benefits of germ and bran encouraged the use of these by-products in baked good formulas. Bran is rich in dietary fibers, vitamins, minerals, and bioactive compounds (e.g., phenols, tocopherols, proteins, and folates) while the germ fraction is enriched with vitamins E and B, dietary fibers, essential amino acids, flavonoids, and sterols. However, the supplementation with bran has been shown to weaken the dough structure and gluten network as well as to cause a bitter taste to the final product. Nevertheless, sourdough fermentation can be regarded as a promising alternative to overcome these limitations (Coda et al., 2014). The bran was subjected to pre-fermentation or micronization and air fractionation before its supplementation to dough formulas undergoing fermentation with lactic acid bacteria (*L. brevis*, *F. sanfranciscensis*, and *L. plantarum*). The lactic acid fermentation improves the textural and sensorial features as well as the digestibility of insoluble dietary fibers facilitated by microbial enzymes (Coda et al., 2014). On the other hand, the germ fraction contains high levels of unsaturated fatty acids and oxidative enzymes leading to high rancidity during storage. Sourdough fermentation with *L. plantarum* and *F. rossiae* has proven to inactivate the germ endogenous enzymes (lipases) decreasing the pH and therefore, reducing the rancidity (Coda et al., 2010). Sourdough fermented wheat germ increased the content of free amino acids by approximately 50%, catalyzed the conversion of glutamate into GABA, and decreased the raffinose content and elevated the phenolic compounds and antioxidant and phytase activities (Rizzello et al., 2010). In fact, when compared to control wheat germ, sourdough fermented wheat germ showed anti-proliferative activity *ex vivo* toward cancerous cell lines (Rizzello et al., 2013).

Overall, it can be concluded that non-conventional cereals or flours can be selected as an ingredient of choice for enhancing their nutritional significance with a sourdough fermentation approach.

14.5 THE SENSORY PERFORMANCES OF SOURDOUGH BAKED GOODS

Abundant literature has dealt with the sensory features of sourdough fermentation. It showed how sourdough behaves with respect to baker's yeast using mainly descriptive panel analysis or through the determination of volatile components (VOC). Among the 27 sensory attributes retrieved, we can list sourness, acidic flavor, intense aftertaste and aroma, attractiveness, pronounced crumb and crust color, crispness, freshness, fruitiness, and high porosity, which clearly distinguish sourdough bread (Figure 14.3). Almost 90 volatile components belonging to numerous chemical classes: alcohols, carbonyls, esters, acids, and others were identified in sourdough bread using mass spectrometry techniques. Researchers highlighted how *Lactobacillus* species contribute to synthesizing a large and broad spectrum of VOC in sourdough bread. The liberation of FAA during fermentation act as precursors of VOC which directly affect the taste and flavor of sourdough baked goods (Zhao and Gänzle, 2016).

14.6 THE NUTRITIONAL PERFORMANCES OF SOURDOUGH BAKED GOODS

Recently, scientific research has moved toward the nutritional features of sourdough fermentation. Sourdough fermentation is undoubtedly efficient in increasing mineral, protein and dietary fibers bioavailability, lowering glycemic index, and degrading anti-nutritional factors.

14.6.1 Mineral Bioavailability

Cereals, pseudo-cereals, and legumes are an important source of minerals such as K, P, Mg, or Zn, but unfortunately mineral utilization is limited by the presence of phytic acid. Phytic acid (myo-inositol hexaphosphate) is highly charged with six phosphate groups, and it forms insoluble complexes with dietary cations, thus impairing mineral bioavailability. Phytate-degrading enzymes (myo-inositol hexakisphosphate phosphohydrolase; EC

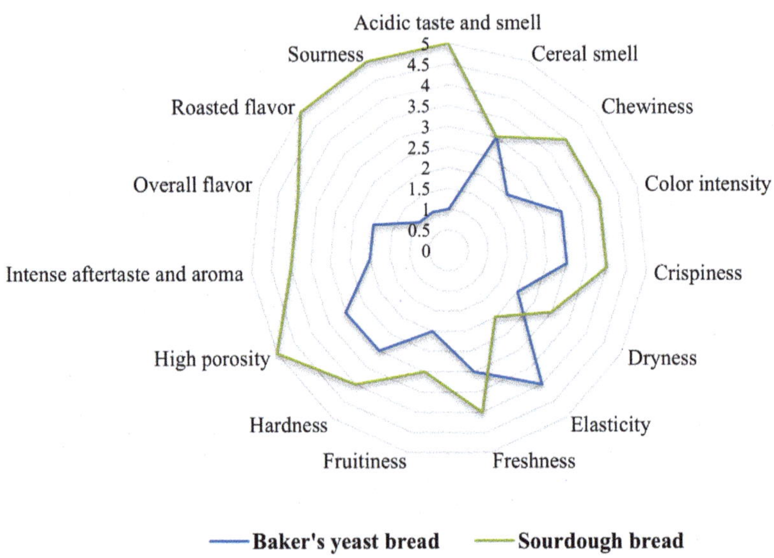

FIGURE 14.3 Summarized characteristics and respective average scores based on descriptive sensory analyses of sourdough versus baker's yeast breads as assessed by trained panelists. Adapted from Arora et al. (2021).

3.1.3.8/EC 3.1.3.26), release soluble inorganic phosphate, low size inositol phosphate and myo-inositol. The activity of endogenous grain phytase is stimulated under acidic conditions, which are promoted through sourdough fermentation. In general, the optimum pH value is in the range of 4.3–4.6 to observe a phytic acid breakdown of more than 70%. Sourdough bread contained higher concentrations of several minerals, including calcium, sodium, magnesium, iron, and zinc (Di Cagno et al., 2008). According to the literature, 30 species of lactic acid bacteria mainly belonging to the *Lactobacillus* genus and five species of yeasts have potential phytase activities. Overall, sourdough biotechnology is a reliable tool for improving the mineral bioavailability of baked goods.

14.6.2 Dietary Fibers

The recognized health benefits of DF have been demonstrated in many studies. In modern societies, diets contain less than the recommended daily intake, which is 25 g/day according to the World Health Organization (WHO). Thus, there is a growing interest in increasing the fiber content of various foods to compensate for this deficiency. Despite the most common sources of DF (e.g., hemicellulose, resistant starch, β-glucans, arabinoxylans) are cereals and pseudo-cereals (Williams et al., 2019), the fortification of baked goods with bran, wheat germ, milling by-products, and legume flours is becoming more widespread. However, additions of these matrices, especially in high amounts, can affect the sensory and rheology features of baked goods. Based on the scientific evidence, the use of sourdough allows the technological drawbacks of high-fiber flours to be overcome, making their use possible in baked goods. For instance, fermentation allowed the increase of DF in various gluten-free products (Di Cagno et al., 2008), a bran fortification up to 20%, and the exploitation of matrices naturally rich in DF (e.g., fava bean and hemp) (Wang et al., 2019) without compromising the quality of bread.

14.6.3 Glycemic Index

The glycemic index (GI) is a measurement of a food's ability to increase glucose levels in the blood after a meal. Foods are classified into high (≥70), moderate (between 69 and 55), and low (≤55) GI, according to the Harvard Medical School. Both *in vitro* studies and *in vivo* challenges were conducted to demonstrate how only sourdough fermentation has the capability to shift the bread GI from high to moderate. In fact, when sourdough fermentation combines with the addition of DF (5–10%), the GI decreases to values lower than 55, which rank these baked goods as low GI foods, recommendable for all dietary habits (Wolter et al., 2014). The lowering of GI can be attributed to biological acidification, increased resistant starch, the liberation of peptides, FAA, polyphenols and water-soluble DF, fast gastric emptying, stimulation of satiety hormones, and use of fermentable cereals and legumes (Rizzello et al., 2019). Nevertheless, further studies should investigate the relation between GI after bread ingestion and gut microbiome composition to solidify the findings mentioned above.

14.6.4 Protein Digestibility

The ability of sourdough fermentation to increase protein digestibility has been proven through empirical and *in vitro* scientific evidence, and also two recent *in vivo* challenges confirmed it. Almost all *in vitro* studies concluded that sourdough fermentation improved IVPD, which is expressed as the percentage of the total protein, which was solubilized after enzyme hydrolysis. Furthermore, other indices such as the amount of protein required to provide the minimal essential amino acid pattern (chemical score, CS); the protein nutritional quality based on the amino acid profile after hydrolysis (protein efficiency ratio, PER); and the nutritional index (NI), which normalizes the qualitative and quantitative variations of the protein compared to its nutritional status were also enhanced under long-time fermentation. In a recent study conducted by (Rizzello et al., 2019), 36 healthy volunteers underwent an *in vivo* challenge in response to bread consumption. Sourdough bread with moderate acidification stimulated more appetite and induced lower satiety. The sourdough bread with the most intense acidic taste triggered the fast and high fullness perception. Gall bladder response did not vary among breads, while gastric emptying was more rapid with sourdough versus baker's yeast breads. Orocecal transit was prolonged for baker's yeast bread and faster for sourdough breads, especially when made with long-time fermentation whose transit lasted ca. 20 min less than baker's yeast bread. Differences in carbohydrate digestibility and absorption determined different post-prandial glycemic responses. Sourdough breads showed the lowest values. After ingesting sourdough breads, which had the highest total FAA content, the levels of FAA in blood plasma constantly maintained at high levels for an extended time. While the increase in digestibility is apparent, the processes that promote it need to be further investigated. Any possible hypotheses include biological acidification per se or via the indirect activation of flour endogenous proteases, secondary proteolysis by a portfolio of lactic acid bacteria peptidases, and, more broadly, modulation of the gluten network, which becomes more sensitive to digestive enzymes.

In recent years, there has been significant progress in the research on the role of sourdoughs in enabling individuals affected by gluten sensitivity or various gastrointestinal disorders to consume cereal-based products (Polo et al., 2020). The key evidence concerned the removal of residues of gluten, which avoided cross-contamination of gluten-free products, selective hydrolysis of gluten, which increased digestibility, and complete gluten digestion for making gluten-free baked goods produced with soft or durum wheat flour. Under semi-liquid sourdough fermentation lasting approximately 24 hours, selected strains of lactic acid bacteria and food-grade fungal proteases were capable of reducing the residual gluten content to less than 10 ppm. Based on immunological and serological tests and biopsy specimens from celiac patients, clinical trials (Di Cagno et al., 2010; Mandile et al., 2017) proved the safety of baked goods made with completely hydrolyzed flours.

14.6.5 Degradation of Anti-nutritional Factors

Tailored sourdough fermentations are the most suitable and sustainable option to limit or eliminate anti-nutritional factors in cereals, pseudo-cereals, and legume flours. The presence of anti-nutritional factors (ANF), namely phytic acid (already discussed in Section 14.6.1.), raffinose, condensed tannins, vicine and convicine, saponins, and trypsin inhibitors decrease the digestibility and reduce the bioavailability of other nutrients, and, exceptionally, are responsible for diseases (Figure 14.4). Raffinose is not digestible by pancreatic enzymes, but it is fermentable by gas-producing bacteria in the large intestine, causing digestive problems. Condensed tannins and trypsin antagonists block digestive enzymes, resulting in low protein and food digestibility. Saponins, vicine, and convicine are biologically active glycosides that induce hemolysis of red blood cells and form complexes with nutrients, blocking their absorption. Vicine and convicine, in particular, are precursors of the aglycones divicine and isouramil, which are the primary causes of fauvism. This hereditary disorder causes severe hemolysis after consuming fava beans. Although heat treatments fully inactivate trypsin inhibitors, ANF is heat tolerant. Dehulling, soaking, germination, air classification, and extrusion are only partially successful in reducing ANF content; thus, other options, including sourdough fermentation, were investigated. Several sourdough lactic acid bacteria enzymes such as α-galactosidase, β-glucosidase, and tannases, can mitigate the presence of multiple ANF. Indeed, *L. plantarum* thoroughly degraded vicine and convicine within 48 h of fermentation, with aglycone derivatives, not detectable (Rizzello et al., 2016). *Ex-vivo* tests on human blood indicated that sourdough fermented fava bean is not toxic. Sourdough fermentation of whole wheat, barley, chickpea, lentils, and quinoa flours, as well as yellow and red lentil, white and black bean, chickpea, and pea flours, reduced the concentrations of raffinose (62–80%), condensed tannins (23%),

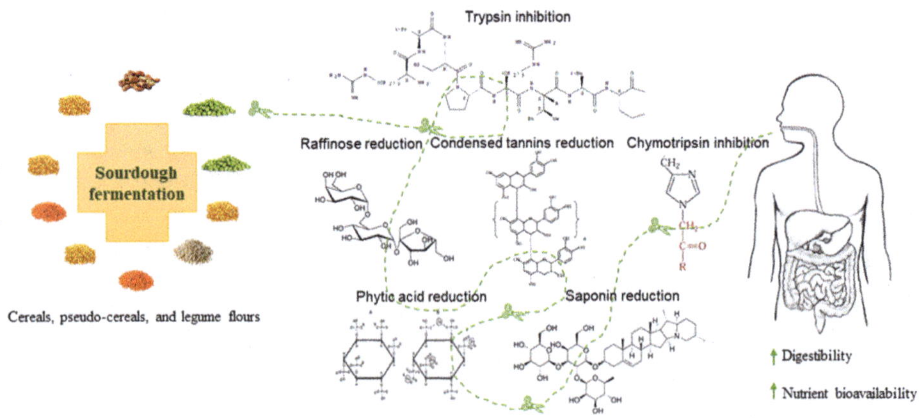

FIGURE 14.4 Schematic representation of the effect of sourdough fermentation on the reduction of anti-nutritional factors (ANF) in flours (cereals, pseudo-cereals, and legumes).

trypsin inhibitors (23–44%), and saponins (68%) (Montemurro et al., 2019). Combining, gelatinization and sourdough fermentation further decreased condensed tannins (62%) and trypsin inhibitors (70%) residual concentrations (De Pasquale et al., 2020).

14.7 CONCLUSIONS

Sourdough technology has been extensively studied, elaborated, and investigated in the past decades to emphasize its beneficial aspects. The fermentation of sourdough is driven by the house microbiota derived from the flour, water, endophytic plant materials, and/or other additional ingredients. The sourdough microbiota is rich in lactic acid bacteria and yeasts functioning as autochthonous (type I sourdough) or allochthonous (type II and III sourdough) starter cultures, influencing the nutritional, and functional characteristics of sourdough. In this context, the use of sourdough has been defined with the inoculum percentage, time, and temperature of fermentation being the key parameters. Moreover, several conventional and non-conventional cereals have been exploited for sourdough fermentation to meet the needs of the consumers for improved nutritional, rheological, and sensorial characteristics. In fact, the use of milling by-products in the baking industry is being promoted to drive a sustainable agriculture and food security worldwide. Sourdough has been known since ancient times for imparting health benefits with improved digestibility. In this regard, exhaustive research has been done, and is still in progress, to examine the nutritional attributes of sourdough (along with the corresponding baked goods), in terms of mineral bioavailability, dietary fibers, starch hydrolysis, glycemic index, protein digestibility etc. Finally, an increased knowledge of the assembly, organization, management, and functions of the sourdough complex meta-communities is fundamental to building up very robust sourdoughs, which are resistant, resilient to perturbations, and more stable for long-time use.

REFERENCES

1. Alfonzo A, Ventimiglia G, Corona O, Di Gerlando R, Gaglio R, et al. 2013. Diversity and technological potential of lactic acid bacteria of wheat flours. *Food Microbiology* 36(2): 343–354. https://doi.org/10.1016/j.fm.2013.07.003.
2. Arora K, Ameur H, Polo A, Di Cagno R, Rizzello CG, et al. 2021. Thirty years of knowledge on sourdough fermentation: A systematic review. *Trends in Food Science & Technology* 108: 71–83. https://doi.org/10.1016/j.tifs.2020.12.008.
3. Bartkiene E, Schleining G, Krungleviciute V, Zadeike D, Zavistanaviciute P, et al. 2016. Development and quality evaluation of lacto-fermented product based on hulled and not hulled hempseed (*Cannabis sativa* L.). *LWT - Food Science and Technology* 72: 544–551. https://doi.org/10.1016/j.lwt.2016.05.027.
4. Coda R, Rizzello CG, Curiel JA, Poutanen K, Katina K. 2014. Effect of bioprocessing and particle size on the nutritional properties of wheat bran fractions. *Innovative Food Science & Emerging Technologies* 25: 19–27. https://doi.org/10.1016/j.ifset.2013.11.012.
5. Coda R, Rizzello CG, Gobbetti M. 2010. Use of sourdough fermentation and pseudo-cereals and leguminous flours for the making of a functional bread enriched of γ-aminobutyric acid (GABA). *International Journal of Food Microbiology* 137(2–3): 236–245. https://doi.org/10.1016/j.ijfoodmicro.2009.12.010.
6. Coda R, Varis J, Verni M, Rizzello CG, Katina K. 2017. Improvement of the protein quality of wheat bread through faba bean sourdough addition. *LWT - Food Science and Technology* 82: 296–302. https://doi.org/10.1016/j.lwt.2017.04.062.
7. Corsetti A, Settanni L. 2007. Lactobacilli in sourdough fermentation. *Food Research International* 40(5): 539–558. https://doi.org/10.1016/j.foodres.2006.11.001.
8. Corsetti A, Settanni L, López CC, Felis GE, Mastrangelo M, et al. 2007. A taxonomic survey of lactic acid bacteria isolated from wheat (*Triticum durum*) kernels and non-conventional flours. *Systematic and Applied Microbiology* 30(7): 561–571. https://doi.org/10.1016/j.syapm.2007.07.001.
9. Curiel JA, Coda R, Centomani I, Summo C, Gobbetti M, et al. 2015. Exploitation of the nutritional and functional characteristics of traditional Italian legumes: The potential of sourdough fermentation. *International Journal of Food Microbiology* 196: 51–61. https://doi.org/10.1016/j.ijfoodmicro.2014.11.032.
10. De Pasquale I, Pontonio E, Gobbetti M, Rizzello CG. 2020. Nutritional and functional effects of the lactic acid bacteria fermentation on gelatinized legume flours. *International Journal of Food Microbiology* 316: 108426. https://doi.org/10.1016/j.ijfoodmicro.2019.108426.
11. De Vuyst L, Vrancken G, Ravyts F, Rimaux T, Weckx S. 2009. Biodiversity, ecological determinants, and

metabolic exploitation of sourdough microbiota. *Food Microbiology* 26(7): 666–675. https://doi.org/10.1016/j.fm.2009.07.012.

12. De Vuyst L, Neysens P. 2005. The sourdough microflora: Biodiversity and metabolic interactions. *Trends in Food Science & Technology* 16(1–3): 43–56. https://doi.org/10.1016/j.tifs.2004.02.012.
13. Di Cagno R, Barbato M, Di Camillo C, Rizzello CG, De Angelis M, et al. 2010. Gluten-free sourdough wheat baked goods appear safe for young celiac patients: A pilot study. *Journal of Pediatric Gastroenterology & Nutrition* 51(6): 777–783. https://doi.org/10.1097/MPG.0b013e3181f22ba4.
14. Di Cagno R, Pontonio E, Buchin S, De Angelis M, Lattanzi A, et al. 2014. Diversity of the lactic acid bacterium and yeast microbiota in the switch from firm- to liquid-sourdough fermentation. *Applied and Environmental Microbiology* 80(10): 3161–3172. https://doi.org/10.1128/AEM.00309-14.
15. Di Cagno R, Rizzello CG, De Angelis M, Cassone A, Giuliani G, et al. 2008. Use of selected sourdough strains of *Lactobacillus* for removing gluten and enhancing the nutritional properties of gluten-free bread. *Journal of Food Protection* 71(7): 1491–1495. https://doi.org/10.4315/0362-028X-71.7.1491.
16. Filannino P, Di Cagno R, Gobbetti M. 2018. Metabolic and functional paths of lactic acid bacteria in plant foods: Get out of the labyrinth. *Current Opinion in Biotechnology* 49: 64–72. https://doi.org/10.1016/j.copbio.2017.07.016.
17. Gabriele M, Sparvoli F, Bollini R, Lubrano V, Longo V, et al. 2019. The Impact of Sourdough Fermentation on Non-nutritive Compounds and Antioxidant Activities of Flours from Different Phaseolus Vulgaris L. genotypes. *Journal of Food Science* 84(7): 1929–1936. https://doi.org/10.1111/1750-3841.14672.
18. Gaglio R, Cirlincione F, Di Miceli G, Franciosi E, Di Gerlando R, et al. 2020. Microbial dynamics in durum wheat kernels during aging. *International Journal of Food Microbiology* 324: 108631. https://doi.org/10.1016/j.ijfoodmicro.2020.108631.
19. Gobbetti M. 1998. The sourdough microflora: Interactions of lactic acid bacteria and yeasts. *Trends in Food Science & Technology* 9(7): 267–274. https://doi.org/10.1016/S0924-2244(98)00053-3.
20. Gobbetti M, De Angelis M, Di Cagno R, Polo A, Rizzello CG. 2020. The sourdough fermentation is the powerful process to exploit the potential of legumes, pseudo-cereals and milling by-products in baking industry. *Critical Reviews in Food Science and Nutrition* 60(13): 2158–2173. https://doi.org/10.1080/10408398.2019.1631753.
21. Gobbetti M, Minervini F, Pontonio E, Di Cagno R, De Angelis M. 2016. Drivers for the establishment and composition of the sourdough lactic acid bacteria biota. *International Journal of Food Microbiology* 239: 3–18. https://doi.org/10.1016/j.ijfoodmicro.2016.05.022.
22. Gordún E, del Valle LJ, Ginovart M, Carbó R. 2015. Comparison of the microbial dynamics and biochemistry of laboratory sourdoughs prepared with grape, apple and yogurt. *Food Science and Technology International* 21(6): 428–439. https://doi.org/10.1177/1082013214543033.
23. Guidotti MC. 2005. L'alimentazione nell'antico Egitto. Cibi e Sapori Nel Mondo Antico, 18–24.
24. Mandile R, Picascia S, Parrella C, Camarca A, Gobbetti M, et al. 2017. Lack of immunogenicity of hydrolysed wheat flour in patients with coeliac disease after a short-term oral challenge. *Alimentary Pharmacology & Therapeutics* 46(4): 440–446. https://doi.org/10.1111/apt.14175.
25. Minervini F, Celano G, Lattanzi A, De Angelis M, Gobbetti M. 2016. Added ingredients affect the microbiota and biochemical characteristics of durum wheat type-I sourdough. *Food Microbiology* 60: 112–123. https://doi.org/10.1016/j.fm.2016.05.016.
26. Minervini F, Celano G, Lattanzi A, Tedone L, De Mastro G, et al. 2015a. Lactic acid bacteria in durum wheat flour are endophytic components of the plant during its entire life cycle. *Applied and Environmental Microbiology* 81(19): 6736–6748. https://doi.org/10.1128/AEM.01852-15.
27. Minervini F, De Angelis M, Di Cagno R, Gobbetti M. 2014. Ecological parameters influencing microbial diversity and stability of traditional sourdough. *International Journal of Food Microbiology* 171: 136–146. https://doi.org/10.1016/j.ijfoodmicro.2013.11.021.
28. Minervini F, Di Cagno R, Lattanzi A, De Angelis M, Antonielli L, et al. 2012a. Lactic acid bacterium and yeast microbiotas of 19 sourdoughs used for traditional/typical Italian breads: Interactions between ingredients and microbial species diversity. *Applied and Environmental Microbiology* 78(4): 1251–1264. https://doi.org/10.1128/AEM.07721-11.
29. Minervini F, Dinardo FR, De Angelis M, Gobbetti M. 2019. Tap water is one of the drivers that establish and assembly the lactic acid bacterium biota during sourdough preparation. *Scientific Reports* 9(1): 570. https://doi.org/10.1038/s41598-018-36786-2.

30. Minervini F, Lattanzi A, De Angelis M, Di Cagno R, Gobbetti M. 2012b. Influence of artisan bakery- or laboratory-propagated sourdoughs on the diversity of lactic acid bacterium and yeast microbiotas. *Applied and Environmental Microbiology* 78(15): 5328–5340. https://doi.org/10.1128/AEM.00572-12.

31. Montemurro M, Pontonio E, Gobbetti M, Rizzello CG. 2019. Investigation of the nutritional, functional and technological effects of the sourdough fermentation of sprouted flours. *International Journal of Food Microbiology* 302: 47–58. https://doi.org/10.1016/j.ijfoodmicro.2018.08.005.

32. Nionelli L, Montemurro M, Pontonio E, Verni M, Gobbetti M, et al. 2018. Pro-technological and functional characterization of lactic acid bacteria to be used as starters for hemp (*Cannabis sativa* L.) sourdough fermentation and wheat bread fortification. *International Journal of Food Microbiology* 279: 14–25. https://doi.org/10.1016/j.ijfoodmicro.2018.04.036.

33. Polo A, Arora K, Ameur H, Di Cagno R, De Angelis M, et al. 2020. Gluten-free diet and gut microbiome. *Journal of Cereal Science* 95: 103058. https://doi.org/10.1016/j.jcs.2020.103058.

34. Pontonio E, Dingeo C, Di Cagno R, Blandino M, Gobbetti M, et al. 2020. Brans from hull-less barley, emmer and pigmented wheat varieties: From by-products to bread nutritional improvers using selected lactic acid bacteria and xylanase. *International Journal of Food Microbiology* 313: 108384. https://doi.org/10.1016/j.ijfoodmicro.2019.108384.

35. Rizzello CG, De Angelis M, Di Cagno R, Camarca A, Silano M, et al. 2007. Highly efficient gluten degradation by lactobacilli and fungal proteases during food processing: New perspectives for celiac disease. *Applied and Environmental Microbiology* 73(14): 4499–4507. https://doi.org/10.1128/AEM.00260-07.

36. Rizzello CG, Verni M, Koivula H, Montemurro M, Seppa L, et al. 2017. Influence of fermented faba bean flour on the nutritional, technological and sensory quality of fortified pasta. *Food & Function* 8(2): 860–871. https://doi.org/10.1039/C6FO01808D.

37. Rizzello CG, Lorusso A, Montemurro M, Gobbetti M. 2016. Use of sourdough made with quinoa (*Chenopodium quinoa*) flour and autochthonous selected lactic acid bacteria for enhancing the nutritional, textural and sensory features of white bread. *Food Microbiology* 56: 1–13. https://doi.org/10.1016/j.fm.2015.11.018.

38. Rizzello CG, Mueller T, Coda R, Reipsch F, Nionelli L, et al. 2013. Synthesis of 2-methoxy benzoquinone and 2,6-dimethoxybenzoquinone by selected lactic acid bacteria during sourdough fermentation of wheat germ. *Microbial Cell Factories* 12(1): 105. https://doi.org/10.1186/1475-2859-12-105.

39. Rizzello CG, Nionelli L, Coda R, De Angelis M, Gobbetti M. 2010. Effect of sourdough fermentation on stabilisation, and chemical and nutritional characteristics of wheat germ. *Food Chemistry* 119(3): 1079–1089. https://doi.org/10.1016/j.foodchem.2009.08.016.

40. Rizzello CG, Portincasa P, Montemurro M, Di Palo DM, Lorusso MP, et al. 2019. Sourdough fermented breads are more digestible than those started with baker's yeast alone: An *in vivo* challenge dissecting distinct gastrointestinal responses. *Nutrients* 11(12): 2954. https://doi.org/10.3390/nu11122954.

41. Rocchetti G, Miragoli F, Zacconi C, Lucini L, Rebecchi A. 2019. Impact of cooking and fermentation by lactic acid bacteria on phenolic profile of quinoa and buckwheat seeds. *Food Research International* 119: 886–894. https://doi.org/10.1016/j.foodres.2018.10.073.

42. Scarnato L, Serrazanetti DI, Aloisi I, Montanari C, Del Duca S, et al. 2016. Combination of transglutaminase and sourdough on gluten-free flours to improve dough structure. *Amino Acids* 48(10): 2453–2465. https://doi.org/10.1007/s00726-016-2258-4.

43. Scheirlinck I, Van der Meulen R, De Vuyst L, Vandamme P, Huys G. 2009. Molecular source tracking of predominant lactic acid bacteria in traditional Belgian sourdoughs and their production environments. *Journal of Applied Microbiology* 106(4): 1081–1092. https://doi.org/10.1111/j.1365-2672.2008.04094.x.

44. Scheirlinck I, Van der Meulen R, Van Schoor A, Vancanneyt M, De Vuyst L, et al. 2008. Taxonomic structure and stability of the bacterial community in Belgian sourdough ecosystems as assessed by culture and population fingerprinting. *Applied and Environmental Microbiology* 74(8): 2414–2423. https://doi.org/10.1128/AEM.02771-07.

45. Wang Y, Compaoré-Sérémé D, Sawadogo-Lingani H, Coda R, Katina K, et al. 2019. Influence of dextran synthesized in situ on the rheological, technological and nutritional properties of whole grain pearl millet bread. *Food Chemistry* 285: 221–230. https://doi.org/10.1016/j.foodchem.2019.01.126.

46. Williams BA, Mikkelsen D, Flanagan BM, Gidley MJ. 2019. "Dietary fibre": Moving beyond the "soluble/

insoluble" classification for monogastric nutrition, with an emphasis on humans and pigs. *Journal of Animal Science and Biotechnology* 10(1): 45. https://doi.org/10.1186/s40104-019-0350-9.

47. Wolter A, Hager AS, Zannini E, Arendt EK. 2014. Influence of sourdough on *in vitro* starch digestibility and predicted glycemic indices of gluten-free breads. *Food & Function* 5(3): 564. https://doi.org/10.1039/c3fo60505a.

48. Zhao CJ, Gänzle MG. 2016. Synthesis of taste-active γ-glutamyl dipeptides during sourdough fermentation by *Lactobacillus reuteri*. *Journal of Agricultural and Food Chemistry* 64(40): 7561–7568. https://doi.org/10.1021/acs.jafc.6b02298.

CHAPTER 15

Beneficial Role of Microorganisms in Olives

Anthoula A. Argyri and Chrysoula C. Tassou
Institute of Technology of Agricultural Products, Hellenic Agricultural Organization - DIMITRA, Attica, Athens, 14123, Greece

15.1 TABLE OLIVES AS FERMENTED FOOD

Table olives are considered to be the most widespread fermented food in the Mediterranean area, with 30% of world production originating from EU member countries such as Spain, Italy, and Greece and their consumption is expanding all over the world (IOC, 2020). The main benefits of table olive fermentation are the control of microbial growth, the improvement of sensory characteristics, the enhancement of the safety and prolonged shelf life of the final product (Argyri et al., 2013a; Nout and Rombouts, 1992; Sánchez Gómez et al., 2006; Tassou et al., 2010). Spanish style green olives, naturally black olives (Greek style), and black ripe olives (California style) are the most applied and well described trade preparations in the international market (Garrido-Fernández et al., 1997; Kailis and Harris, 2007). Both green and black olives are fermented "naturally" in brine by diverse microbial groups, with lactic acid bacteria (LAB) and yeasts being the dominant microorganisms involved in olive fermentation (Arroyo-López et al., 2008; Hurtado et al., 2012). LAB influence fermentation in a variety of ways, mainly due to the transformation of the substrate to lactic acid and the consequent pH decrease and acidity increase ensures a higher microbiological stability for extended storage periods. On the other hand, an unexpected growth of yeast population or other contaminating microorganisms may influence the fermentation process and result in a less acceptable final product with a limited shelf life (Abriouel et al., 2012; Bevilacqua et al., 2009; Hurtado et al., 2009; Panagou et al., 2003, 2008). In this respect, the use of appropriate starter cultures can offer a more predictable fermentation process and eliminate the possibility of spoilage due to the growth of unwanted microbiota (Panagou and Tassou, 2006).

15.1.1 Microbiota of Fermented Olives

The microbial ecology of table olive fermentation in both spontaneous and inoculated processes has been studied through the application of molecular methods based on both culture-dependent and culture-independent approaches (Abriouel et al., 2011; Arroyo-López et al., 2016; Botta and Cocolin, 2012; Cocolin et al., 2013; Lucena-Padrós et al. 2014; 2015; Nychas et al., 2002; Randazzo et al., 2012; 2017). The dominance of LAB usually characterizes the Spanish-style fermentation process and the Greek-style fermentations with low salt levels (Blana et al., 2014; Cocolin et al., 2013; Panagou et al., 2008; Tassou et al., 2002). Greek-style fermentations with high salt levels and natural fermentations in general, favor the yeast growth, with LAB and yeasts usually being detected in similar levels (Arroyo-López et al., 2012; Bautista-Gallego et al., 2015; Benitez-Cabello et al., 2015; Nisiotou et al., 2010; Porru et al., 2018). The growth of *Enterobacteriaceae, Pseudomonas, Staphylococcus*, and *Clostridium* is usually observed at the onset of the natural fermentation of black and green olives (Grounta and Panagou, 2017; Heperkan, 2013; Panagou, 2002; Tassou et al., 2010).

Recently, the rapid spread of omics technologies (Vaccalluzzo et al., 2020) has represented an important

Good Microbes in Medicine, Food Production, Biotechnology, Bioremediation, and Agriculture, First Edition. Edited by Frans J. de Bruijn, Hauke Smidt, Luca S. Cocolin, Michael Sauer, David Dowling, and Linda Thomashow.

challenge to better understand the function, adaptation and exploitation of microbial diversity in different complex ecosystems, such as table olives. An "omics" approach used to evaluate the bacterial biodiversity changes during the fermentation process of natural green cracked Aloreña de Málaga table olives (Medina et al., 2016), showed the presence of 131 bacterial genera, that were not detected by the conventional approach. The presence of *Enterobacteriaceae* and *Lactobacillaceae* species was scarce during the four months of study. In contrast, the most important genus at the end of fermentation was *Celerinatantimonas* in both brine (95.3% of frequency) and fruit (89.4%) samples, while the presence of well-known spoilage microorganisms (*Pseudomonas* and *Propionibacterium*) and halophilic bacteria (*Modestobacter, Rhodovibrio, Salinibacter*) were also common during the course of fermentation.

The biogeography of the microbial communities at species and strain levels associated with Spanish-style green olive fermentations was studied by Lucena-Padros and Ruis-Barba (2019). A total of 376 distinct genotypes were identified, belonging to 41 bacterial and 16 yeast species, with 16 bacterial species not being described before in table olives. The *Lb. pentosus* species showed a ubiquitous presence in all 30 fermenters and *Pediococcus parvulus, Lb. collinoides/paracollinoides, Lb. coryniformis, Lb. plantarum, P. manshurica*, and *C. thaimueangensis* were found in every sample.

Recently, Kazou et al. (2020) identified bacterial and yeast/fungal microbiota of both olives and brines obtained from cv. Kalamata olive samples produced in different geographical regions of Greece. The dominant bacterial family identified in olive and brine samples from both regions was *Lactobacillaceae*, presenting, however, higher average abundances in the samples from Aitoloakarnania compared to Messinia/Lakonia. At the genus level, *Lactobacillus, Celerinatantimonas, Propionibacterium*, and *Pseudomonas* were the most abundant. In addition, the yeasts/fungal communities were less diverse compared to those of bacteria, with *Pichiaceae* being the dominant family and *Pichia, Ogataea*, and *Saccharomyces* being the most abundant genera.

The bacterial and yeast diversity, the microbial biogeography and association between certain taxa and geographical area of fermented olives of two main Greek varieties from different regions were assessed by Argyri et al. (2020). *Lactobacillaceae* (*Lb. acidipiscis, Lb. coryniformis, Lb. paracollinoides, Lb. parafarraginis, Lb. harbinensis, Lb. kisonensis, Pediococcus parvulus*, and *P. ethanolidurans*) was the most abundant family in all samples, while *Leuconostocaceae, Nostocaceae, Shewanellaceae, Propionibacteriaceae* and *Gloeobacteraceae, Nostocaceae, Enterobacteriaceae, Gloeobacteraceae, Bacillaceae, Phormidiaceae, Colwelliaceae*, Gloeobacteraceae, and *Vibrionaceae*, were also detected in different samples. In relation to yeast diversity, *Phaffomycetaceae* was the most abundant yeast family in Konservolia olives from the Magnesia region, while *Pichiaceae* dominated the yeast microbiota in Konservolia olives from Fthiotida and in Halkidiki olives from both regions. Other families detected in lower proportions were *Debaryomycetaceae* and *Aureobasidiaceae*, while *P. manshurica, Brettanomyces custersianus, P. membranifaciens, Aureobasidium pullulans, Schwanniomyces etchelsii*, and *W. anomalus* were characterized at species level. Further analysis of the data employing multivariate analysis allowed for the first time the discrimination of cv. Konservolia and cv. Halkidiki table olives according to their geographical origin.

15.1.2 Microbial Starters in Olive Fermentation

Inoculation of the brine with an appropriate starter culture reduces the probability of spoilage and helps to achieve an improved fermentation process (Panagou and Tassou, 2006). The benefits of starter culture applications in table olive processing, has been repeatedly reported (Chorianopoulos et al., 2005; De Bellis et al., 2010; de Castro et al.; 2002; Hurtado et al., 2010; Panagou et al., 2008; Panagou and Tassou, 2006; Panagou et al., 2003; Ruiz-Barba et al., 1994; Spyropoulou et al., 2001). Many LAB strains can be used as starter cultures with the most frequent ones being *Lb. plantarum, Lb. pentosus* or both, with some strains being commercially available (Corsetti et al., 2012; Hurtado et al., 2010; Panagou et al., 2003). However, yeasts have also been reported to be beneficial to the fermentation process, usually in combination with LAB (Arroyo-López et al., 2008; Ciafardini and Zullo, 2019; D'Antuono et al., 2018; Hernández et al., 2007; Psani and Kotzekidou, 2006).

15.2 TABLE OLIVES AS FUNCTIONAL/PROBIOTIC FOOD

Table olives like other plant origin products are by nature functional foods due to their high content in nutraceuticals, such as dietary fibers, polyunsaturated fatty acids, proteins, peptides, amino acids, keto acids, minerals, antioxidative vitamins, and other antioxidants. Some of the neutraceuticals promote several beneficial physiological effects and can also selectively stimulate the growth of the colon microflora and act as prebiotics (Gupta and Abu-Ghannam, 2012).

Although limited clinical studies have been performed on fermented foods, these foods may deliver additional health benefits as a source of live microorganisms, which may contribute to human health in a manner similar to probiotics (Marco et al., 2017; Peres et al., 2012; Rezac et al., 2018; Xiang et al., 2019). According to FAO/WHO (2002) probiotic foods are in general fermented foods containing an amount of viable and active microorganisms large enough to reach the intestine and exert an equilibrating action on the intestinal microflora.

Probiotic bacteria are mainly LAB beneficial for human health, with an increasing research interest in the recent years regarding the probiotic LAB isolated from fermented foods of plant origin (Peres et al., 2012). However, the beneficial effects of yeasts for human health have also been reported (Arroyo-López et al., 2012).

The basic requirements for probiotics include that they should be generally recognized as safe (GRAS), tolerant to acid, pepsin, and bile, show antibacterial activity against pathogenic bacteria and good adherence to the intestinal mucus (FAO/WHO, 2006). The ability to adhere to the mucosal surfaces of the intestine and the subsequent long- or short-term colonization as well as to inhibit pathogen adhesion to intestinal cell lines, has been one of the most commonly encountered criteria for the probiotic strain selection (Collado et al., 2009; Hudault et al., 1997; Lebeer et al., 2008; Lievin-Le Moal et al., 2002; Marco et al., 2006). On the other hand, it is increasingly supported that probiotics can exert positive effects only through the passage of GI tract (without colonization) (Ohland and MacNaughton, 2010). Safety aspects such as hemolytic activity and antibiotic susceptibility for the selection of probiotic strains were also proposed in guidelines (FAO/WHO, 2002).

Other benefits include alleviation of intolerances (such as lactose intolerance), enhancement of nutrient bioavailability, and prevention or reduction of the prevalence of allergies in susceptible individuals (Chiang and Pan, 2012; Isolauri 2001; Jahreis et al., 2002). Probiotics may also have antimutagenic, anticarcinogenic, hypocholesterolemic, antihypertensive, anti-osteoporosis, and immunomodulatory effects (Chiang and Pan, 2012). They relieve the symptoms of inflammatory bowel diseases, irritable bowel syndrome, colitis, alcoholic liver disease, constipation and reduce the risk for colon, liver, and breast cancers (Prado et al., 2008).

15.2.1 Probiotic Microorganisms of Olives

A significant number of studies in recent years have demonstrated the probiotic properties of olive isolates (Table 15.1), while Argyri et al. (2016) have extensively reviewed studies on probiotics from olive microbiota.

Different strains of *Lactobacillus* spp. isolated from different sources were screened *in vitro* for their tolerance to low pH (2.5), bile salts, and simulated digestion (pH 2 and 7), antagonistic activity against *Yersinia enterocolitica*, and adhesion ability to Caco-2 cells (Lavermicocca et al., 2008). Among the 4 strains isolated from olives, *Lb. plantarum* ITM5BG showed good survival at low pH and in bile salts and strong adhesion to Caco-2 cells. The other 3 olive strains (Table 15.1), showed tolerance to low pH and bile salts but did not survive after the simulated digestion (Table 15.1).

The technological and probiotic properties of 19 LAB (18 *Lb. plantarum* and 1 *Enterococcus faecalis*) isolated from Bella di Cerignola Italian table olives were assessed by Bevilacqua et al. (2010). All bacteria were able to grow at a pH 4.0 to 10.0, in media supplemented with 2.5 to 7.5% NaCl and 0.3% bile salts, to survive in acidified (pH 2.5) MRS broth, to inhibit *E. coli* O157:H7 and compete with *Staph. aureus*. Two (2) isolates of *Lb. plantarum* (C16 and C19) showed higher adhesion values (up to 16%) to IPEC-J2 cells and the authors proposed at least 3 *Lb. plantarum* isolates as potential multifunctional starters for Bella di Cerignola olives (Table 15.1).

In another work, 144 LAB (including *Lb. pentosus*, *Leuc. pseudomesenteroides*, and *Pediococcus parvulus*) isolated from Manzanilla Aloreña naturally fermented green table olives were studied for their properties (Abriouel et al., 2012). Some strains exhibited antimicrobial activity against pathogens, utilization of raffinose and stachylose, bile salt hydrolase, phytase and haeme-dependent catalase activities, growth at 10°C and at 6.5% NaCl, good acidifying capacity, and resistance to freezing. Thus, 16 strains with potential traits were selected (15 *Lb. pentosus*, 1 *Leuc. mesenteroides*) and tested for their survival at low pH and tolerance to bile salts (Table 15.1). These were shown to survive under acidic conditions (pH 2 for 30 min) and 2% of bile salts, with 43.75% of them surviving at pH 1.5, with 37.5% of them tolerating at 3% bile salts (Table 15.1).

TABLE 15.1 Probiotic microorganisms isolated from fermented olives

Strains	Reference
Lb. plantarum (strains ITM5BG, ITM4TG, ITM2TP, ITM4TP1)	Lavermicocca et al., 2008
Lb. plantarum (strains 16, 19)	Bevilacqua et al., 2010
Lb. pentosus (strains CF1-19, CF1-20, CF1-23, CF1-33, CF1-37, CF1-43, CF2-3, CF2-10, Lp-1, Lp-5, Lp-7, Lp-8, MP-10, AP2-15, AP2-16),	
Lc. pseudomesenteroides CF2-33	Abriouel et al., 2012
Lb. pentosus (strains B281, E97, E104, E108) *Lb. plantarum* (strains B282, E10, E69), *Lb. paracasei* subsp. *paracasei* (strains E93, E94)	Argyri et al., 2013b
Lb. pentosus (strains GG2S-T2-168, GM2F-T5-327)	Bautista Gallego et al., 2015
Lb. pentosus S3T60C	
Lb. plantarum (strains S11T3E, S1T10A, S2T10D, O2T60C)	Botta et al., 2014
Lb. plantarum (strains 17.2 b, B95, 69B, 607, P, FF28, 33)	
Lb. paraplantarum (strains B13, K, O1)	Peres et al., 2014
P. guilliermondii Y16, *P. kluyveri* Y17, *C. silvae* Y19, *M. pulcherrima* Y20, *S. cerevisiae* Y22, *P. manshurica* Y37, *R. mucilaginosa* Y38	Bonatsou et al., 2015
Lb. pentosus (strains AP2-15N, CF1-39, CF2-10N, CF2-12, MP-10, AP2-16N, CF1-6, CF2-5, 5C2, CF2-10N, CF1-6, AP2-16N)	Montoro et al., 2016
Lb. pentosus D303.36, *Lb. coryniformis* H307.6	Guantario et al., 2018
S. cerevisiae 15A, *C. tropicalis* 1A, *C. norvegica* 7A,	
P. guilliermondii 25A	Oliveira et al., 2017
S. cerevisiae Sc24, *C. boidinii* Cb60	Porru et al., 2018
S. cerevisiae Y34	Bonatsou et al., 2018
Lb. pentosus (strains B356, B362, B363, B364, B366, B368, B370, B371, B377, B378, B385, B399), *Lb. plantarum* (strains B372, B373, B375, B377, B380, B384)	Pavli et al., 2019
Lb. pentosus (strains Lp13, LPG1)	Benitez Cabello et al., 2019

The probiotic potential of 71 LAB (17 *Leuconostoc mesenteroides*, 1 *Ln. pseudomesenteroides*, 13 *Lb. plantarum*, 37 *Lb. pentosus*, 1 *Lb. paraplantarum*, and 2 *Lb. paracasei* subsp. *paracasei*) isolated from naturally fermented olives was also evaluated (Argyri et al., 2013b). Three (3) *Lb. pentosus*, 4 *Lb. plantarum*, and 2 *Lb. paracasei* strains demonstrated high survival (>8 log cfu/ml) after exposure at low pH (pH 2.5 for 3 h). The majority of the tested strains were resistant to bile salts, while 5 *Lb. plantarum* and 7 *Lb. pentosus* strains exhibited partial bile salt hydrolase activity. Variability was observed to Caco-2 cells adherence ratios as well as to the strains' susceptibility toward different antibiotics. None of the strains exhibited antimicrobial or β-haemolytic activity. Concluding, 4 *Lb. pentosus*, 3 *Lb. plantarum*, and 2 *Lb. paracasei* strains were found to possess desirable *in vitro* probiotic properties similar to or even better than the reference *Lb. casei* Shirota and *Lb. rhamnosus* GG strains and constitute potential probiotic starters for table olive fermentation (Table 15.1).

In continuance, two of the aforementioned isolated strains (Argyri et al., 2013b), namely *Lb. pentosus* B281 and *L. plantarum* B282 were further evaluated for their properties by Saxami et al.(2016). Both strains were found to exhibit a significant higher adherence to Caco-2 cells in comparison to the reference strain *L. casei* ATCC 393. Quantitative analysis and confocal microscopy showed that treatment with conditioned media (CM) of the two strains caused significant reduction of cell proliferation, as demonstrated by Sulforhodamine B (SRB) and clonogenic assays. Moreover, the CM of the two strains induced a G1-phase arrest and down-regulation of specific cyclin genes, as indicated by flow cytometry and real-time PCR analysis. The anti-proliferative activity of heat-treated CM was similar to the non-heated CM in a time- and dose-dependent manner, indicating the presence of thermostable bioactive compounds (Table 15.1).

Bautista Gallego et al. (2013) have evaluated the probiotic potential of 109 LAB isolates (107 *Lb. pentosus*, 1 *Lb. plantarum*, and 1 *Lb. paraplantarum*) from fermented table olives. The majority of the isolates exhibited high autoaggregation and low hydrophobicity properties, lower survival to gastric than to pancreatic digestion, while none of the isolates showed bacteriocin, haemolytic or bile salt hydrolase activities. Some *Lb. pentosus* (*Lb. pentosus* GG2S-T2-168, *Lb. pentosus* GM2F-T5-327) isolates were found with promising potential probiotic characteristics, better than probiotic reference strains (*Lb. rhamnosus* GG and *Lb. casei* Shirota) (Table 15.1).

Botta et al. (2014) screened 238 LAB (*Lb. plantarum, Lb. pentosus,* and *Ln. mesenteroides* species), isolated from Nocellara Etnea table olives. Seventeen (17) promising probiotics (15 *Lb. plantarum*, 1 *Lb. pentosus*, 2 *Ln.* mesenteroides) were selected due to transit tolerance in the upper human gastrointestinal tract combined with autoaggregation and hydrophobicity. None of the 17 selected strains showed resistance to a broad spectrum of antibiotics and they were confirmed to be not cytotoxic when in contact with H4-1 human epithelial cells. A high adhesion ratio was observed for *Lb. plantarum* O2T60C (over 9%) when tested in the 3D functional model (simulation of real intestinal conditions). The *Lb. plantarum* S11T3E strain showed high resistance to simulated gastric digestion, increased transepithelial resistance of polarized H4-1 cells and significant reduction in *L. monocytogenes* invasion in undifferentiated gut model cells, exhibiting thus the best probiotic performance (Table 15.1). In continuance, the aforementioned strains were evaluated for their ability to produce short chain fatty acids (SCFA) which display beneficial effects on human health (Pessione et al., 2015), while the extracellular proteomes of the two most interesting strains were investigated. All the strains were found to produce significant amounts of acetic and lactic acids, while adhesive ability was attributed for *Lb. plantarum* S11T3E to adhesive proteins and for *Lb. pentosus S3T60C* to non-proteinaceous molecules.

The potential probiotic features of 156 LAB strains isolated from cv. Galega fermented olives have been screened and characterized by Peres et al. (2014). According to the results, 10 strains (7 *Lb. plantarum* and 3 *Lb. paraplantarum*) were acid – and bile salt tolerant and exhibited survival rates up to 48% under simulated digestion (Table 15.1). All strains exhibited auto- (4–12%) and co-aggregation features (≥30%), hydrophobicity (5–20%) and exopolysaccharide-producing abilities, while no strain possessed haemolytic capacity or ability to hydrolyze mucin. Antibiotic resistance, oleuropein degradation, proteolytic activity, and antimicrobial activity were strain-dependent features.

A collection of 31 *Lb. pentosus* strains isolated from naturally fermented Aloreña green table olives were screened for their probiotic potential (Montoro et al., 2016). Several strains could be considered promising probiotic candidates, since they showed good growth capacity and survival under simulated gastro-intestinal conditions, good auto-aggregation ability and co-aggregation with pathogenic bacteria and adherence to intestinal and vaginal cell lines. They presented antagonistic activity against pathogenic bacteria, ability to ferment several prebiotics and capacity to ferment lactose (Table 15.1). *Lb. pentosus* CF2-10 N was selected to have the best probiotic profile being of great interest in further studies.

In another study, 19 *Lb. pentosus* and 7 *Lb. coryniformis* strains from Nocellara del Belice table olives fermented with the Spanish or Castelvetrano methods, were evaluated for their probiotic potential (Guantario et al., 2018). The majority showed high survival rates under *in vitro* simulated gastro-intestinal conditions and positive antimicrobial activity against pathogens. Three (3) *Lb. coryniformis* strains were very good performers in the initial *in vitro* testing screens, they were antibiotic susceptible, as well as capable of inhibiting pathogen growth. One (1) *Lb. pentosus* and 1 *Lb. coryniformis* strain significantly induced prolongevity effects and protection from pathogen-mediated infection when fed to *Caenorhabditis elegans*. Both strains displayed adhesion to Caco-2 cells and were able to outcompete foodborne pathogens for cell adhesion. Overall, these results render these strains as promising starter candidates (Table 15.1).

Recently, 33 strains LAB previously isolated from natural black Conservolea olives were assessed for their probiotic potential *in vitro*, as well as for their autoinducer-2 (AI-2) activity under standard growth conditions and the production of γ-aminobutyric acid (GABA) (Pavli et al., 2019). The results indicated that 17 strains were able to survive in low pH and in the presence of bile, with 15 of them also exhibiting partial bile salt hydrolase activity (Table 15.1). None of the strains exhibited haemolytic activity or inhibited the growth of any of the examined pathogens. Moreover, the strains displayed generally low AI-2 activity under the growth conditions tested, regardless of the species and none of the isolates were found to produce GABA after 48 h of growth.

Benítez-Cabello et al. (2019) have characterized and evaluated a total of 554 LAB isolates obtained from the olive surface of Manzanilla, Gordal, and Aloreña cultivars processed as green Spanish-style or directly brined (natural) olives for their technological and probiotic characteristics. *Lb. pentosus* Lp13 or LPG1 would be the appropriate multifunctional starters that showed both technological and probiotic characteristics (genotype more frequently found, high titratable acidity production (homofermentative metabolism), possibility of inoculation at high pH levels, esterase (production of aromas and degradation of bitter compounds), β-galactosidase (important in lactose assimilation), and phytase activity (necessary for assimilation of phosphate and other minerals) (Table 15.1).

In recent years there has been significant research effort to elucidate the role of yeasts in the fermentation process and select yeasts species with appropriate technological and probiotic potential in order to develop multipurpose starter cultures for table olive fermentation. The beneficial effects for human health that justify the use of yeasts in table olive processing include: probiotic activity, biodegradation of phytate complexes, biofortification with folates and vitamins, reduction of cholesterol levels, biodegradation or bioabsorption of mycotoxins, and production of bioactive compounds and natural antioxidants (Arroyo-López et al., 2012).

In this direction, 12 yeast strains originally isolated from Greek natural black table olive fermentations assessed for their potential technological (resistance to salt, lipase, esterase and β-glucosidase activities), probiotic (survival to gastric -pH 2.0- and pancreatic digestions), and prebiotic (phytase activity) features (Bonatsou et al., 2015). It was shown that *W. anomalus* Y18 showed interesting technological properties, *Pichia guilliermondii* Y16 exhibited probiotic potential (showed overall the highest resistance to salt and simulated digestions) and *Metschnikowia pulcherrima* Y20 was the strain with the highest phytase activity levels. Thus, the selected yeasts could be evaluated as multifunctional starters, alone or in combination with LAB, in table olive fermentations (Table 15.1).

Oliveira et al. (2017) evaluated the probiotic potential of 16 yeast strains previously isolated during the natural fermentation of table olives *Negrinha de Freixo* cultivar. *Saccharomyces cerevisiae* (strains 15A and 15B) exhibited the highest antioxidant activity (similar to the reference strain *S. boulardii*). *Candida norvegica* 7A and *Galactomyces reessii* 34A showed antifungal ability against to *Cryptococcus neoformans*. *S. cerevisiae* 15A, *C. tropicalis* 1A, and *C. norvegica* 7A showed > 80% autoaggregation capacity after 24 h. *P. guilliermondii* 25A and *C. norvegica* 7A were shown to have the highest resistant under simulated digestion, similar to the reference strain *S. boulardii* (Table 15.1).

Porru et al. (2018) evaluated a total of 72 yeast isolates previously obtained from Bosana black olive brines to be applied as multifunctional starters. The dominant species were *Wickerhamomyces anomalus* and *Nakazawaea molendini-olei*, albeit *C. diddensiae, C. boidinii, Zygotorulaspora mrakii*, and *Saccharomyces cerevisiae* were also present in lower proportions. According to the results multivariate analysis of the technological (esterase, lipase, and β-glucosidase activities, growth in presence of oleuropein, resistance, and susceptibility to NaCl) and probiotic (cholesterol removal, gastric and pancreatic digestions, biofilms assays) features of the different yeast biotypes, *W. anomalus* Wa1 showed the best technological characteristics, while *S. cerevisiae* Sc24 and *C. boidinii* Cb60 showed probiotic potential (Table 15.1).

In another study, 49 yeast strains previously isolated from cv. Kalamata table olives were assessed for their probiotic potential and technological characteristics (Bonatsou et al., 2018). The obtained results indicated that 42 strains presented overall survival rate higher than 50%, while 24 strains showed a survival percentage higher than 70% at the end of the simulated gastric and pancreatic digestions. The assayed strains presented hydrophobicity > 75%, while the autoaggregation ability ranged between 72 and 91%. None of the strains showed haemolytic activity. The majority of the strains presented high salt tolerance (some strains being tolerant to salt > 200 g/L). Concerning the enzymatic activity, 45 strains presented valine and cystine arylamidase activity, 27 and 14 strains showed β- and α-glucosidase activity, respectively and 11 strains presented α-galactosidase and alkaline phosphatase activity. In conclusion, the strain *S. cerevisiae* Y34 presented positive results in the majority of both probiotic and technological assays and thus it could be considered a potential starter either as a single or a combined culture with LAB in the fermentation of Greek-style natural black olives (Table 15.1).

15.2.2 Probiotic Microorganisms as Starters in Olive Fermentation

The surface microarchitecture of the plants, due to the presence of ridges and natural prebiotic compounds (such as oligosaccharides), protect probiotic microorganisms from the passage through the GI tract being in parallel a source of nutrients (Martins et al., 2013; Ranadheera et al., 2010). Thus, considerable research efforts have been made to enhance the already nutritional value of table olives with numbers of beneficial bacteria, offering thus a multifunctional food to the consumer.

The ability of LAB to colonize the surface of olives and form biofilms during fermentation provided new perspectives for the use of table olives as biological carrier of probiotic microorganisms. Nychas et al. (2002) provided the first evidence (through scanning electron microscopy) that on the surface of the fermented olive epidermis, areas with a coating of epicuticular wax provided a site of attachment for both yeasts and bacteria.

Subsequently, Lavermicocca et al. (2005) exploited fermented table olives (final products) as a vehicle for incorporating probiotic bacterial species of human origin. According to the survival performance of *Lb. rhamnosus* GG on different olive samples, brines (4% NaCl) of Spanish black pitted olives, Hojiblanca variety, were inoculated with *Lb. rhamnosus* IMPC11 and IMPC19, *Lb. paracasei* IMPC2.1 and IMPC4.1, *B. bifidum* ATCC 15696, and *B. longum* ATCC 15708 (Table 15.1). The results demonstrated that bifidobacteria and *Lb. rhamnosus* GG showed a good survival rate (10^6 CFU/ g after 30 days)(Boylston et al., 2004; Oliveira et al., 2001). *Lb. paracasei* IMPC2.1 exhibited high survival rates (>10^7 CFU/g during the 3-month storage period) and was evaluated *in vivo* through the passage from the human GI tract. It was recovered from fecal samples in four out of five volunteers fed 10 to 15 olives per day carrying about 10^9 to 10^{10} viable cells for 10 days.

In recent years, a series of experiments have been performed to evaluate the imposition and survival of LAB (of human, dairy or olive origin) applied as adjunct, starter or fortification cultures in table olives and the research results are given in detail below.

15.2.2.1 Non-olive Origin Probiotic Starters

Several works have reported on the inoculation of table olives with probiotic LAB originating from other sources than the olive environment. Initially, the performance of the probiotics *Lb. plantarum* ACA-DC 146 and *Lb. paracasei* ACA-DC 4037 of dairy origin as starter cultures in the Spanish-style fermentation of green table olives was evaluated by Saravanos et al. (2008). It was shown that the two strains accelerated the fermentation process of green olives (compared to the control) and were shown to survive during the whole fermentation period, while they were the only species present in the brine according to pulsed field gel electrophoresis (PFGE) profiles of the LAB isolates.

Further, the dynamics of microbial populations adhering on the surface of debittered green olives cv. Bella di Cerignola in inoculated fermentations with the human probiotic strain *Lb. paracasei* IMPC2.1 was studied in different brining conditions (4% and 8% NaCl) at refrigerated (4°C) and room temperature (RT) (De Bellis et al., 2010). *Lb. paracasei* IMPC2.1 strain was found to accelerate the fermentation process and reduce the survival period of *Enterobacteriaceae* and yeasts. Repetitive extragenic palindromic-PCR (REP-PCR) was applied to monitor the presence of *Lb. paracasei* IMPC2.1 strain during the fermentation process, while multiplex PCR assay differentiated *Lb. paracasei* from *Lb. casei* strains. The corresponding survival rates of the probiotic were 31.1% and 41.9% at the end of the fermentation process in 4% and 8% NaCl brines respectively at RT but 0% at 4%NaCl/4°C.

Randazzo et al. (2014, evaluated the effects of selected starters (*Lb. plantarum* GC3 and *Lb. paracasei* BS21) and probiotic strains (*Lb. rhamnosus* H25 (previously isolated from cheese) and commercial probiotic *Lb. rhamnosus* GG) alone or in combinations on the evolution of microbiota, the trend of volatile compounds, and the final sensorial traits of green table olives. The survival of the probiotic strains at the end of the fermentation ranged between ca. 21–30% in the brines depending on the different cases.

15.2.2.2 Olive Origin Probiotic Starters

Autochthonous functional starter cultures contribute to microbial safety or offer one or more organoleptic, technological, nutritional, or health advantages. Wild-type strains that dominate spontaneous fermentations tend to have higher metabolic dynamics (compared to industrial starter cultures), features that may improve the quality and safety characteristics of table olives. Natural selection is likely to have forced such strains to be more competitive by endowing them with ecological advantages (Ayad et al., 2000; Maldonado et al., 2002)).

In this respect, Rodríguez-Gómez et al. (2013) evaluated four potential probiotic *Lb. pentosus* strains (TOMC LAB2, TOMC LAB3, TOMC LAB4 and TOMC LAB5) as starters in green table olive Spanish-style fermentations. Faster acidification of the brines and a decrease in *Enterobacteriaceae* levels was observed in the inoculated cases (compared to the control). However, RAPD-PCR analysis at the end of fermentation, revealed that the inoculated strains were not always found in the cover brines of the small-scale fermentation vessels. In addition, two of the aforementioned strains (TOMC LAB2 and TOMC LAB4) were investigated as multifunctional starters in Spanish-style fermentation of Manzanilla olives (Rodríguez-Gómez et al., 2014a). Both inoculated strains were recovered at high rates (60% and 53% for *Lb. pentosus* TOMC-LAB2 and TOMC-LAB4, respectively) at the end of the fermentation process. In continuance, Rodríguez-Gómez et al. (2014b) have studied the survival of probiotic strain *Lb. pentosus* TOMC-LAB2 when inoculated in fresh brines of previously fermented green Spanish-style olives with or without the putative probiotic strain. The presence of *Lb. pentosus* TOMC-LAB2 on the olive surface as confirmed by REP-PCR, that at the end of the shelf life (200 days), the recovery rates of 52.6%, 57.9%, and 100.0% for olives that were previously spontaneously fermented, fermented using *Lb. pentosus* TOMC-LAB2 as starter and spontaneously fermented and then thermally treated, respectively.

The performance of the potential probiotics *Lb. pentosus* B281 and *Lb. plantarum* B282 as starters in Spanish-style fermentation of heat shocked green olives cv. Halkidiki at two initial salt levels (10% and 8% NaCl) was assessed by Argyri et al. (2014). LAB reached their highest populations earlier in the inoculated samples (compared to the control) while their final population was maintained above 10^6 CFU/g in olive flesh. Both the strains dominated over indigenous LAB, with strain B281 exhibiting higher recovery rates (100% or 94.7% for B281 and 58.8% or 55.0% for B282 in 10% or 8% NaCl, respectively) in olives. *Lb.* pentosus B281 also dominated over *Lb. plantarum* B282, when the two strains were co-inoculated in olive fermentations. The sensory assessment showed higher preference for inoculated fermentations of *Lb. pentosus* and *Lb. plantarum* separately in 8% NaCl, followed by the *Lb. plantarum* in 10% NaCl. Further, the aforementioned fermented olives were packed under modified atmospheres (70% N_2–30% CO_2) and shelf life of the samples was monitored after storage at 4°C and 20°C for 12 months (Argyri et al., 2015). Both strains were recovered in high percentages (64.7% for B281 and 94.1% for B282) after 6 months of storage and in lower survival rates after one year (35.0% for B281 and 13.3% for B282) at 4°C, with lower rates were also observed during storage at 20°C. Taking into account the strains' survival rates and the sensory analysis results, a shelf life of 6 months at refrigerating temperature was proposed to deliver a promising probiotic food to the consumer.

In agreement with the previous study, the performance of the potential probiotic *Lb. pentosus* B281 and *Lb. plantarum* B282 as starter cultures in Spanish-style fermentation of cv. Halkidiki green olives without previous heat-treatment (Blana et al., 2014). PFGE analysis revealed that *Lb. pentosus* B281 presented higher colonization in both salt levels (8% and 10% NaCl) at the end of fermentation (81.2% and 93.3% in 8% and 10% NaCl brines, respectively). *Lb. plantarum* B282 presented a high survival rate (83.3%) in 8% NaCl brines, but in 10% NaCl the strain could not colonize the surface of olives. As in the previous study (Argyri et al., 2014), *Lb. pentosus* B281 also dominated over *Lb. plantarum* B282 in inoculated fermentations when the two strains were used as combined culture. Further, the aforementioned fermented olives were packed in polyethylene pouches, covered with freshly prepared brine (9% NaCl, 2‰ citric and 1.5‰ ascorbic acid) and stored at 4°C and 20°C for 357 days. PFGE analysis revealed that *Lb. pentosus* B281 and *Lb. plantarum* B282 showed a high survival rate (100% and 96%, respectively) at 4°C, and low survival rate (< 20% for both strains) at 20°C (Blana et al., 2016). The latter finding comes in agreement with the previous reported study of Argyri et al. (2015), that lower survival rates were observed after storage at 20°C.

Recently, combinations of *C. boidinii* TOMC-Y13 strain with different *Lb. pentosus* strains were evaluated as starters in green table olive fermentation in an attempt to modulate the antiadhesive properties of brine exopolysaccharides (EPS) against enterotoxigenic *Escherichia coli* (ETEC) (Zhu et al., 2019). *Lb. pentosus* 119-14MT (high EPS variant strain), improved highly the antiadhesive properties of brine EPS and increased its ability to block the ETEC K88 fimbria. These results illustrate that the antiadhesive properties of the EPSs produced during the traditional fermentation of olives could be modulated by the use of defined starters, aiming to produce functional table olives for ETEC diarrhea prevention.

Finally, *Lb. pentosus* LPG1, *Lb. pentosus* Lp13, *Lb. plantarum* Lpl15, and *Wickerhanomyces anomalous* Y12, (isolated from fermented table olive biofilms) alone or in combinations were studied as multifunctional starters for Manzanilla Spanish-style green table olive fermentations (Benítez-Cabello et al., 2020). According to rep-PCR analysis, LPG1 strain and, particularly Lp13, were excellent biofilm formers and showed the major imposition on the fruit epidermis. Production of AI-2 was lower in the treatment inoculated with yeast Y12 but had the highest presence in the sequential yeast-LAB inoculum, with its maximum concentration and maximum LAB population on fruits (19 days) strongly related. The authors proposed that co-inoculation of Spanish-style table olive fermentations with yeast and LAB could be an advisable practice.

15.3 CONCLUSIONS

Fermented olives contain a complex microbial consortium and include beneficial microorganisms that may contribute to the well-being of consumers. Multifunctional cultures with probiotic properties isolated from table olives may serve as food fortifying agents or as a multifunctional starter or adjunct cultures in the fermentation of table olives. In this respect, one of the oldest known fermented food products, table olives, with an already high nutritional value, may become a multifunctional food product that combines prebiotic, probiotic, and nutraceutical properties.

REFERENCES

1. Abriouel H, Benomar N, Lucas R, Gálvez A. 2011. Culture-independent study of the diversity of microbial populations in brines during fermentation of naturally-fermented Aloreña green table olives. *International Journal of Food Microbiology* 144: 487–496.
2. Abriouel H, Benomar N, Cobo A, Caballero N, Fernández Fuentes MA, et al. 2012. Characterization of lactic acid bacteria from naturally-fermented Manzanilla Aloreña green table olives. Food Microbiology 32: 308–316.
3. Argyri AA, Lyra E, Nisiotou A, Panagou EZ, Tassou C. 2013a. Fate of Escherichia coli O157: H7,Salmonella Enteritidis and Listeria monocytogenes during storage of fermented green table olives in brine. Food Microbiology 36: 1–6.
4. Argyri AA, Zoumpopoulou G, Karatzas K-AG, Tsakalidou E, Nychas G-JE, Panagou EZ, Tassou CC. 2013b. Selection of potential probiotic lactic acid bacteria from fermented olives by in vitro tests. Food Microbiology 33: 282–291.
5. Argyri AA, Nisiotou A, Mallouchos A, Panagou EZ, Tassou C. 2014. Performance of two potential probiotic Lactobacillus strains from the olive microbiota as starters in the fermentation of heat shocked green olives. International Journal of Food Microbiology 171: 68–76.
6. Argyri AA, Nisiotou A, Pramateutaki P, Doulgeraki AI, Panagou EZ, Tassou CC. 2015. Shelf life of green table olives fermented with lactic acid bacteria with probiotic potential in modified atmosphere packaging at 4 and 20°C. LWT- Food Research and Technology 62(1): 783–790.
7. Argyri AA, Panagou EZ, Tassou CC. 2016. Probiotics from the olive microbiota. In: Watson R, Preedy V Eds., Bioactive Foods in Promoting Health: Probiotics, Prebiotics, and Synbiotics, 2nd Edition. Elsevier Chapter 25: 371–389.
8. Argyri K, Doulgeraki A, Manthou E, Grounta A, Argyri A, et al. 2020. Microbial Diversity of Fermented Greek Table Olives of Halkidiki and Konservolia Varieties from Different Regions as Revealed by Metagenomic Analysis. Microorganisms 8(8): 1241. https://doi.org/10.3390/microorganisms8081241.
9. Arroyo-López FN, Querol A, Bautista-Gallego J, Garrido-Fernández A. 2008. Role of yeasts in table olive production. *International Journal of Food Microbiology* 128: 189–196.
10. Arroyo-López FN, Romero-Gil V, Bautista-Gallego J, Jiménez-Díaz R, García-García P, et al. 2012. Yeasts in table olive processing: Desirable or spoilage microorganisms? *International Journal of Food Microbiology* 160: 42–49.
11. Arroyo-López FN, Medina E, Ruiz-Bellido MA, Romero-Gil V, Montes-Borrego M, Landa BB. 2016. Enhancement of the knowledge on fungal communities in directly brined Aloreña de Málaga green olive fermentations by metabarcoding analysis. *PLoS ONE* 11: e0163135.
12. Ayad EHE, Verheul A, Wouters JTM, Smit G. 2000. Application of wild starter cultures for flavour development in pilot plant cheese making. *International Dairy Journal* 10: 169–179.
13. Bautista-Gallego J, Arroyo-López FN, Rantsiou K, Jiménez-Díaz R, Garrido-Fernández A, Cocolin L 2013. Screening of lactic acid bacteria isolated from fermented table olives with probiotic potential. *Food Research International* 50: 135–142.
14. Bautista Gallego J, Arroyo-Lopez FN, Romero-Gil V, Rodriguez-Gomez F, Garcia-Garcia P, Garrido-Fernandez A. 2015. Fermentation profile of green

Spanish-style Manzanilla olives according to NaCl content in brine. *Food Microbiology* 49: 56–64.

15. Benítez-Cabello A, Calero-Delgado B, Rodríguez-Gómez F, Garrido-Fernández A, et al. 2019. Biodiversity and multifunctional features of Lactic Acid Bacteria Isolated From Table Olive Biofilms. *Frontiers in Microbiology* 17:: https://doi.org/10.3389/fmicb.2019.00836.
16. Benítez-Cabello A, Calero-Delgado B, Rodríguez-Gómez F, Bautista-Gallego J, Garrido-Fernández A, et al. 2020. The use of multifunctional yeast-lactobacilli starter cultures improves fermentation performance of Spanish-style green table olives. *Food Microbiology* 91: 103497.
17. Benítez-Cabello A, Romero-Gil V, Rodríguez-Gómez F, Garrido-Fernández A, Jiménez-Díaz R, Arroyo-López FN. 2015. Evaluation and identification of polymicrobial biofilms on natural green Gordal table olives. *Antonie Van Leeuwenhoek* 108: 597–610.
18. Bevilacqua A, Perricone M, Cannarsi M, Corbo MR, Sinigaglia M. 2009. Technological and spoiling characteristics of the yeast microflora isolated from Bella di Cerignola table olives. *International Journal of Food Science and Technology* 44: 2198–2207.
19. Bevilacqua A, Altieri C, Corbo MR, Sinigaglia M, Ouoba LI. 2010. Characterization of lactic acid bacteria isolated from Italian Bella di Cerignola Table Olives: Selection of potential multifunctional starter cultures. *Journal of Food Science* 75: 536–544.
20. Blana VA, Grounta A, Tassou CC, Nychas G-JE, Panagou EZ. 2014. Inoculated fermentation of green olives with potential probiotic *Lactobacillus pentosus* and *Lactobacillus plantarum* starter cultures isolated from industrially fermented olives. *Food Microbiology* 38: 208–218.
21. Blana VA, Polymeneas N, Tassou C, Panagou EZ. 2016. Survival of potential probiotic lactic acid bacteria on fermented green table olives during packaging in polyethylene pouches at 4 and 20 °C. *Food Microbiology* 53: 71–75.
22. Bonatsou S, Benítez A, Rodríguez-Gomez F, Panagou EZ, Arroyo-Lopez FN. 2015. Selection of yeasts with multifunctional features for application as starters in natural black table olive processing. *Food Microbiology* 46: 66–73.
23. Bonatsou S, Karamouza M, Zoumpopoulou G, Mavrogonatou E, Kletsas D, et al. 2018. Evaluating the probiotic potential and technological characteristics of yeasts implicated in cv. Kalamata natural black olive fermentation. *International Journal of Food Microbiology* 271: 48–59.
24. Botta C, Cocolin L. 2012. Microbial dynamics and biodiversity in table olive fermentation: Culture-dependent and -independent approaches. *Frontiers in Microbiology* 3: 245.
25. Botta C, Langerholc T, Cenčič A, Cocolin L. 2014. *In vitro* selection and characterization of new probiotic candidates from table olive microbiota. *PLOS ONE* 9(4): e94457.
26. Boylston TD, Vinderola CG, Ghoddusi HB, Reinheimer JA. 2004. Incorporation of bifidobacteria into cheeses: Challenges and rewards. *International Dairy Journal* 14: 375–387.
27. Chiang SS, Pan TM. 2012. Beneficial effects of *Lactobacillus paracasei* subsp. *paracasei* NTU 101 and its fermented products. *Applied Microbiology and Biotechnology* 93: 903–916.
28. Chorianopoulos NG, Boziaris IS, Stamatiou A, Nychas G-JE. 2005. Microbial association and acidity development of unheated and pasteurised green table olives fermented using glucose or sucrose supplements at various levels. *Food Microbiology* 22: 117–124.
29. Ciafardini G, Zullo BA. 2019. Use of selected yeast starter cultures in industrial-scale processing of brined Taggiasca black table olives. *Food Microbiology* 84: 103250.
30. Cocolin L, Alessandria V, Botta C, Gorra R, De Fillipis F, et al. 2013. NaOH-debittering induces changes in bacterial ecology during table olives fermentation. *PLoS ONE* 8: e69074.
31. Collado MC, Isolauri E, Salminen S, Sanz Y. 2009. The impact of probiotic on gut health. Development and application of an in vitro methodology to determine the transit tolerance of potentially probiotic *Lactobacillus* and *Bifidobacterium* species in the upper human gastrointestinal tract. *Current Drug Metabolism* 10: 68–78.
32. Corsetti A, Perpetuini G, Schirone M, Tofalo R, Suzzi G. 2012. Application of starter cultures to table olive fermentation: An overview on the experimental studies. *Frontiers in Microbiology* 3: 1–6.
33. D' Antuono I, Bruno A, Linsalat V, Minervini F, Garbetta A, et al. 2018. Fermented Apulian table olives: Effect of selected microbial starters on polyphenols composition, antioxidant activities and bioaccessibility. *Food Chemistry* 248: 137–145.

34. De Bellis P, Valerio F, Sisto A, Lonigro SL, Lavermicocca P. 2010. Probiotic table olives: Microbial populations adhering on olive surface in fermentation sets inoculated with the probiotic strain *Lactobacillus paracasei* IMPC2.1 in an industrial plant. *International Journal of Food Microbiology* 140: 6–13.

35. De Castro A, Montaño A, Casado F-J, Sánchez H, Rejano L. 2002. Utilization of *Enterococcus casselinflavus* and *Lactobacillus pentosus* as starter cultures for Spanish style green olive fermentation. *Food Microbiology* 19: 637–644.

36. FAO/WHO. 2002. Joint FAO/WHO Working Group Report on Drafting Guidelines for the Evaluation of Probiotics in Food London. Ontario, Canada, April 30 and May 1, 2002.

37. FAO/WHO. 2006. Probiotic in foods. Health and nutritional properties and guidelines for evaluation. In FAO Food and Nutrition, 2006 pp 85. ISBN 92-5-105513-0.

38. Guantario B, Zinno P, Schifano E, Roselli M, Perozzi G, et al. 2018. *In vitro* and *in vivo* selection of potentially probiotic Lactobacilli From Nocellara del Belice table olives. *Frontiers in Microbiology* 9: 595.

39. Garrido-Fernández A, Fernández Díez MJ, Adams MR. 1997. *Table Olives: Production and Processing*.Chapman & Hall, London.

40. Gupta S, Abu-Ghannam N. 2012. Probiotic fermentation of plant based products: Possibilities and opportunities. *Critical Reviews in Food Science and Nutrition* 52: 183–199.

41. Grounta A, Panagou EZ. 2017. Olives fermentation. In: Paramithiotis S (Ed.), *Lactic Acid Fermentation of Fruits and Vegetables*, (1st ed.). CRC Press, 157–180.

42. Hernández A, Martin A, Aranda E, Pérez-Nevado F, Córdoba MG. 2007. Identification and characterization of yeast isolated from the elaboration of seasoned green table olives. *Food Microbiology* 24: 346–351.

43. Heperkan D. 2013. Microbiota of table olive fermentations and criteria of selection for their use as starters. *Frontiers in Microbiology* 4: 143.

44. Hudault S, Liévin V, Bernet-Camard MF, Servin AL. 1997. Antagonistic activity *in vitro* and *in vivo* exerted by *Lactobacillus casei* (strain GG) against *Salmonella typhimurium* infection. *Applied and Environmental Microbiology* 63: 513–518.

45. Hurtado A, Reguant C, Bordons A, Rozès N. 2009. Influence of fruit ripeness and salt concentration on the microbial processing of Arbequina table olives. *Food Microbiology* 26, 827–833.

46. Hurtado A, Reguant C, Bordons A, Rozès N. 2010. Evaluation of a single and combined inoculation of a *Lactobacillus pentosus* starter for processing cv. Arbequina natural green olives. *Food Microbiology* 27: 731–740.

47. Hurtado A, Reguant C, Bordons A, Rozès N. 2012. Lactic acid bacteria from fermented table olives. *Food Microbiology* 31: 1–8.

48. International Olive Council (IOC). 2020. Trade standard applying to table olives. November 2020 COI/OT/NC no. 1. Madrid. https://www.internationaloliveoil.org/wp-content/uploads/2020/12/HO-W901-23-11-2020-P.pdf

49. International Olive Council (IOC). 2014. World table olive figures. http://www.internationaloliveoil.org/estaticos/view/132-world-table-olive-figures.

50. International Olive Council (IOC). 2017. World table olive figures. http://old.internationaloliveoil.org/estaticos/view/132-world-table-olive-figures

51. Isolauri E. 2001. Probiotics in the prevention and treatment of allergic disease. *Pediatric Allergy and Immunology* 12: 56–59.

52. Jahreis G, Vogelsang H, Kiessling G, Schubert R, Hammes WP. 2002. Influence of probiotic sausage (*Lactobacillus paracasei*) on blood lipids and immunological parameters of healthy volunteers. *Food Research International* 35: 133–138.

53. Kailis S, Harris D. 2007. Quality and safety. In: Kailis S, Harris D (Eds.), *Producing Table Olives*. Landlinks Press, Collingwood, Australia, pp. 243–294.

54. Kazou M, Tzamourani A, Panagou EZ, Tsakalidou E. 2020. Unraveling the Microbiota of Natural Black cv. Kalamata fermented olives through 16S and ITS metataxonomic analysis. *Microorganisms* 8: 672. https://doi.org/10.3390/microorganisms8050672.

55. Lavermicocca P, Valerio F, Lonigro SL, De Angelis M, Morelli L, et al. 2005. Study of adhesion and survival of lactobacilli and bifidobacteria on table olives with the aim of formulating a new probiotic food. *Applied and Environmental Microbiology* 71: 4233–4240.

56. Lavermicocca P, Valerio F, Lonigro SL, Di Leo A, Visconti A. 2008. Antagonistic activity of potential probiotic lactobacilli against the ureolytic pathogen. *Yersinia Enterocolitica. Current Microbiology* 56: 175–181.

57. Lievin-Le Moal V, Amsellem R, Servin AL, Coconnier MH. 2002. *Lactobacillus acidophilus* (strain LB) from the resident adult human gastrointestinal microflora exerts activity against brush border damage promoted

by a diarrhoeagenic *Escherichia coli* in human enterocyte-like cells. *Gut* 50: 803–811.

58. Lebeer S, Vanderleyden J, De Keersmaecker SC. 2008. Genes and molecules of lactobacilli supporting probiotic action. *Microbiology and Molecular Biology Reviews* 72: 728–764.
59. Lucena-Padrós H, Caballero-Guerrero B, Maldonado-Barragán A, Ruiz-Barba, JL. 2014. Genetic diversity and dynamics of bacterial and yeast strains associated to Spanish-style green table olive fermentations in large manufacturing companies. *International Journal of Food Microbiology* 190, 72–78.
60. Lucena-Padrós H, Jiménez E, Maldonado-Barragán A, Rodríguez JM, Ruiz-Barba JL. 2015. PCR-DGGE assessment of the bacterial diversity in Spanish-style green table olive fermentations. *International Journal of Food Microbiology* 205: 47–53.
61. Lucena-Padrós H, Ruiz-Barba JL. 2019. Microbial biogeography of Spanish-style green olive fermentations in the province of Seville, Spain. *Food Microbiology* 82: 259–268.
62. Maldonado A, Ruiz-Barba JL, Floriano B, Jiménez-Díaz R. 2002. The locus responsible for production of plantaricin S, a class IIb bacteriocin produced by *Lactobacillus plantarum* LPCO10, is widely distributed among wild-type *Lact. plantarum* strains isolated from olive fermentations. *International Journal of Food Microbiology* 77: 117–124.
63. Martins EMF, Ramos AM, Vanzela ESL, Stringheta PC, de Oliveira Pinto CL, Martins JM. 2013. Products of vegetable origin: A new alternative for the consumption of probiotic bacteria. *Food Research International* 51: 764–770.
64. Marco ML, Pavan S, Kleerebezem M. 2006. Towards understanding molecular modes of probiotic action. *Current Opinion in Biotechnology* 17: 204–210.
65. Marco ML, Heeney D, Binda S, Cifelli CJ, Cotter PD, et al. 2017. Health benefits of fermented foods: microbiota and beyond. *Current opinion in biotechnology* 44: 94–102.
66. Medina E, Ruiz-Bellido MA, Romero-Gil V, Rodríguez-Gómez F, Montes-Borrego M, et al. 2016. Assessment of the bacterial community in directly brined Aloreña de Málaga table olive fermentations by metagenetic analysis. *International Journal of Food Microbiology* 236: 47–55.
67. Montoro BP, Benomar N, Lerma LL, Gutiérrez SC, Gálvez A, Abriouel H. 2016. Fermented aloreña table olives as a source of potential probiotic *Lactobacillus pentosus* strains. *Frontiers in Microbiology* 7: 1583.
68. Nisiotou AA, Chorianopoulos N, Nychas G-JN, Panagou EZ. 2010. Yeast heterogeneity during spontaneous fermentation of black Conservolea olives in different brine solutions. *Journal of Applied Microbiology* 108: 396–405.
69. Nychas G-JN, Panagou EZ, Parker ML, Waldron KW, Tassou CC. 2002. Microbial colonization of naturally black olives during fermentation and associated biochemical activities in the cover brine. *Letters in Applied Microbiology* 34: 173–177.
70. Nout MJR, Rombouts FM. 1992. Fermentative preservation of plant foods. *Journal of Applied Bacteriology* 73: 136–147.
71. Oliveira MN, Sodini I, Remeuf F, Corrieu G. 2001. Effect of milk supplementation on acidification, textural properties and microbiological stability of fermented milks containing probiotic bacteria. *International Dairy Journal* 11: 935–942.
72. Oliveira T, Ramalhosa E, Nunes L, Pereira JA, Colla E, Pereira EL. 2017. Probiotic potential of indigenous yeasts isolated during the fermentation of table olives from Northeast of Portugal. *Innovative Food Science and Emerging Technologies* 44: 167–172.
73. Ohland CL, MacNaughton WK. 2010. Probiotic bacteria and intestinal epithelial barrier function. *American Journal of Physiology-Gastrointestinal and Liver Physiology* 298: 807–819.
74. Panagou EZ. 2002. Fermentation, Preservation and Microbial Ecology of Fermented Table Olives. Ph.D. thesis. Agricultural University of Athens.
75. Panagou EZ, Tassou CC, Katsaboxakis CZ. 2003. Induced lactic acid fermentation of untreated green olives of the Conservolea cultivar by. *Lactobacillus Pentosus. Journal of the Science of Food and Agriculture* 83: 667–674.
76. Panagou EZ, Tassou CC. 2006. Changes in volatile compounds and related biochemical profile during controlled fermentation of cv. Conservolea green olives. *Food Microbiology* 23: 738–746.
77. Panagou EZ, Schillinger U, Franz C, Nychas G-JE. 2008. Microbiological and biochemical profile of cv. Conservolea naturally black olives during controlled fermentation with selected strains of lactic acid bacteria. *Food Microbiology* 25: 348–358.
78. Pavli F, Gkana E, Adebambo O, Karatzas K-A, Panagou E, Nychas G-JE. 2019. *In Vitro* Screening of

γ-Aminobutyric acid and autoinducer-2 signalling in lactic acid bacteria exhibiting probiotic potential isolated from natural black Conservolea olives. *Foods* 8: 640. https://doi.org/10.3390/foods8120640.

79. Peres CM, Peres C, Hernandez -Mendoza A, Xavier Malcata F. 2012. Review on fermented plant materials as carriers and sources of potentially probiotic lactic acid bacteria - With an emphasis on table olives. *Trends in Food Science and Technology* 26: 31–42.

80. Peres CM, Alves M, Hernandez-Mendoza A, Moreira L, Silva S, et al. 2014. Novel isolates of lactobacilli from fermented Portuguese olive as potential probiotics. *LWT - Food Science and Technology* 59: 234–246.

81. Pessione A, Lo Bianco G, Mangiapane E, Cirrincione S, Pessione E. 2015. Characterization of potentially probiotic lactic acid bacteria isolated from olives: Evaluation of short chain fatty acids production and analysis of the extracellular proteome. *Food Research International* 67: 247–254.

82. Porru C, Rodríguez-Gómez F, Benítez-Cabello A, Jiménez-Díaz R, Zara G, et al. 2018. Genotyping, identification and multifunctional features of yeasts associated to Bosana naturally black table olive fermentations. *Food Microbiology* 69: 33–42.

83. Prado FC, Parada JL, Pandey A, Soccol CR. 2008. Trends in non-dairy probiotic beverages. *Food Research International* 41: 111 123.

84. Psani M, Kotzekidou P. 2006. Technological charac teristics of yeast strains and their potential as starter adjuncts in Greek-style black olive fermentation. *World Journal of Microbiology and Biotechnology* 22: 1329–1336.

85. Ranadheera RDCS, Baines SK, Adams MC. 2010. Importance of food in probiotic efficacy. *Food Research International* 43: 1–7.

86. Randazzo CL, Ribbera A, Pitino I, Romeo FV, Caggia C. 2012. Diversity of bacterial population of table olives assessed by PCR-DGGE analysis. *Food Microbiology* 32: 87–96.

87. Randazzo CL, Todaro A, Pino A, Pitino I, Corona O, Mazzaglia A, Caggia C. 2014. Giarraffa and Grossa di Spagna naturally fermented table olives: Effect of starter and probiotic cultures on chemical, microbiological and sensory traits. *Food Research International* 62: 1154–1164.

88. Randazzo CL, Todaro A, Pino A, Pitino I, Corona O, Caggia C. 2017. Microbiota and metabolome during controlled and spontaneous fermentation of Nocellara Etnea table olives. *Food Microbiology* 65: 136–148.

89. Rezac S, Kok CR, Heermann M, Hutkins R. 2018. Fermented foods as a dietary source of live microorganisms. *Frontiers in Microbiology* 9: 1785.

90. Rodríguez-Gómez F, Bautista-Gallego J, Arroyo-López FN, Romero-Gil V, Jiménez-Díaz R, et al. 2013. Table olive fermentation with multifunctional *Lactobacillus pentosus* strains. *Food Control* 34: 96–105.

91. Rodríguez-Gómez F, Romero-Gil V, Bautista-Gallego J, García-García P, Garrido-Fernandez A, Arroyo-Lopez FN. 2014a. Production of potential probiotic Spanish-style green table olives at pilot plant scale using multifunctional starters. *Food Microbiology* 44: 278–287.

92. Rodríguez-Gómez F, Romero-Gil V, García-García P, Garrido-Fernandez A, Arroyo-Lopez FN. 2014b. Fortification of table olive packing with the potential probiotic bacteria *Lactobacillus pentosus* TOMC-LAB2. *Frontiers in Microbiology* 5: 1–9.

93. Ruiz-Barba JL, Cathcart DP, Warner PJ, Jiménez-Díaz R. 1994. Use of *Lactobacillus plantarum* LPCO10, a bacteriocin producer, as a starter culture in Spanish style green olive fermentations. *Applied and Environmental Microbiology* 60: 2059–2064.

94. Saxami G, Karapetsas A, Lamprianidou E, Kotsianidis I, Chlichlia A, et al. 2016. Two potential probiotic lactobacillus strains isolated from olive microbiota exhibit adhesion and anti-proliferative effects in cancer cell lines. *Journal of functional foods* 24: 461–471.

95. Saravanos E, Kagli D, Zoumpopoulou G, Panagou EZ, Tassou CC. 2008. Use of probiotic lactic acid bacteria as starter cultures in Spanish style green olive fermentation and determination of their survival using PFGE. *Proceedings of the 21st International ICFMH Symposium*, 1–4 September, Aberdeen, UK, p. 188

96. Spyropoulou KE, Chorianopoulos NG, Skandamis PN, Nychas GJE. 2001. Control of *Escherichia coli* O157: H7 during the fermentation of Spanish-style green table olives (Conservolea variety) supplemented with different carbon sources. *International Journal of Food Microbiology* 66: 3–11.

97. Tassou CC, Panagou EZ, Katsaboxakis KZ. 2002. Microbiological and physicochemical changes of naturally black olives fermented at different temperatures and NaCl levels in the brines. *Food Microbiology* 19: 605–615.

98. Tassou CC, Panagou EZ, Nychas G-JE. 2010. Microbial colonization of naturallyfermented olives. In: Preedy VR,

Ross Watson R (Eds.), *Olives and Olive Oil in Health and Disease Prevention* Academic Press, Oxford, pp. 397–406.

99. Vaccalluzzo A, Pino A, Russo N, De Angelis M, Caggia C, Luci C. 2020. FoodOmics as a new frontier to reveal microbial community and metabolic processes occurring on table olives fermentation. *Food Microbiology* 92: 103606.

100. Xiang H, Sun-Waterhouse D, Waterhouse GIN, Cui C, Ruan Z. 2019. Fermentation-enabled wellness foods: A fresh perspective. *Food Science and Human Wellness* 8: 203–243.

101. Zhu Y, González-Ortiz G, Benítez-Cabello A, Calero-Delgado B, Jiménez-Díaz R, Martín-Orúe SM. 2019. The use of starter cultures in the table olive fermentation can modulate the antiadhesive properties of brine exopolysaccharides against enterotoxigenic. *Escherichia coli*. *Food and Function* 10(6): 3738–3747.

CHAPTER 16

The Functional and Nutritional Aspects of Cocobiota: Lactobacilli

Jatziri Mota-Gutierrez and Luca S. Cocolin
Department of Agricultural, Forest and Food Sciences, University of Turin, Largo Braccini 2, 10095 Grugliasco, Turin, Italy

16.1 INTRODUCTION

Good microbes are involved in the production of characteristic cocoa aromas such as malty, buttery, fruity, and nutty during fermentation (Mota-Gutierrez et al., 2019). The cocoa pulp-bean fermentation process includes the establishment of microbial populations that interact indirectly as they compete for common resources (Figure 16.1). Sugars are first metabolized into ethanol and lactic acid and other volatile compounds (VOCs), such as higher alcohols and esters by yeast and lactic acid bacteria (LAB) (Mota-Gutierrez et al., 2018). The production of lactic acid from yeast or LAB includes the hydrolysis of glucose, fructose, and sucrose present in the cocoa pulp-bean mass. LAB can also produce other compounds such as acetic acid, mannitol, succinic acid, and VOCs (diacetyl, acetoin, and 2,3-butanediol, which are characterized with buttery and creamy odors) during the heterofermentative cocoa pulp-bean fermentation (Adler et al., 2013; De Vuyst et al., 2010; Lefeber et al., 2010, 2011). Once there is oxygen availability and an excess production of ethanol and lactic acid, acetic acid bacteria (ABB) transform these compounds into acetic acid and acetoin (Adler et al., 2014; Moens et al., 2014). Finally, the rise in temperature, the increase in pH, and the overproduction of acetic and lactic acid that penetrate the cocoa beans, kill the cocoa embryo, and stop the germination (Schwan and Wheals, 2004). Allowing microorganism to inhabit the cocoa pulp-beans not only prevents seed germination but also reduces the astringency and bitterness of the beans and produces desirable aromatic notes.

What are these microbes that naturally inhabit the cocoa beans? The core of the cocobiota includes mainly the presence of yeasts (*Saccharomyces cerevisiae*), AAB (*Acetobacter pasteurianus*), LAB (*Lactiplantibacillus plantarum*), and in some cases spore-forming bacteria (*Bacillus*) (Agyirifo et al., 2019; Arana-Sánchez et al., 2015; Ho et al., 2015; Illeghems et al., 2012; Jespersen et al., 2005; Lefeber et al., 2012; Mota-Gutierrez et al., 2018; Papalexandratou et al., 2019; Serra et al., 2019; Visintin et al., 2016). The presence of *Limosilactobacillus fermentum* during cocoa pulp-bean fermentation remains controversial. This bacterium has been identified in spontaneous cocoa pulp-bean fermentation in Nigeria (Kostinek et al., 2008), Ghana (Agyirifo et al., 2019; Garcia-Armisen et al., 2010), Brazil (Da Veiga Moreira et al., 2013; Garcia-Armisen et al., 2010; Papalexandratou et al., 2011a), Malaysia (Papalexandratou et al., 2013), Ecuador (Papalexandratou et al., 2011a), Nicaragua (Papalexandratou et al., 2019), and Australia (Ho et al., 2015), but it has not been identified in Mexico, while *Liquorilactobacillus cacaonum* has been only detected in Latin American and Ghanaian cocoa

Good Microbes in Medicine, Food Production, Biotechnology, Bioremediation, and Agriculture, First Edition. Edited by Frans J. de Bruijn, Hauke Smidt, Luca S. Cocolin, Michael Sauer, David Dowling, and Linda Thomashow.

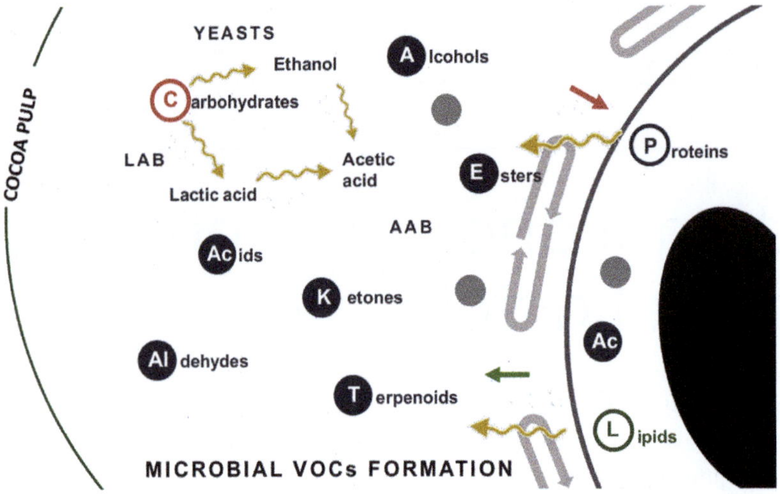

FIGURE 16.1 Microbial reactions occurring within the fermented cocoa ecosystem.

pulp-bean fermentation (De Bruyne et al., 2009; Papalexandratou et al., 2011a; Romanens et al., 2018; Serra et al., 2019).

The absence of one lactobacilli species from different cocoa fermentations around the world highlights the importance of the environmental conditions to determine the core cocobiota (Serra et al., 2019). The genetics of the cocoa variety, characteristics of the plant surfaces, ecological factors, nutrient availability, and composition, and harvesting and post-harvesting procedures influence also the microbial ecology during cocoa pulp-bean fermentation (Da Veiga Moreira et al., 2013; De Melo Pereira et al., 2013; Fernández et al., 2016; Hamdouche et al., 2019; Mota-Gutierrez et al., 2018). This high species richness created by external factors promotes functional diversity, which is relevant for food and health applications. In this chapter, we will describe the nutritional value, functional and technological properties of *Liq. cacaonum, Lim. fermentum*, and *Lacp. plantarum* isolated from fermented cocoa pulp-bean and explore the current perspectives on lactobacilli food regulatory landscapes.

16.2 CHARACTERISTICS OF LIQUORILACTOBACILLUS CACAONUM, LIMOSILACTOBACILLUS FERMENTUM, AND LACTIPLANTIBACILLUS PLANTARUM

Lactobacilli are rod-shaped, Gram-positive, catalase-negative, non-spore-forming bacteria that frequently grow better in microaerophilic conditions (Goldstein et al., 2015). The taxonomy of LAB has undergone significant revisions over the past decade, leading to the establishment of 26 phylogenetic groups and 261 species (Zheng et al., 2020). In both the human and animal gastrointestinal tracts several lactobacilli occur naturally, including *Lacp. plantarum, Lim. fermentum, Lacticaseibacillus casei*, and *Lacticaseibacillus rhamnosus* (Goldstein et al., 2015). Nonetheless lactobacilli are not only commonly isolated from the gut, but also food environments, such as cocoa pulp-bean fermentation. Whether bacteria can be used within food production or health applications, an up-to-date list of microorganisms used during food processing and its origin are not accessible to consumers, when it comes to product information. Figure 16.2 maps the most up-to-date information regarding the origin (14 cocoa-producing countries) and distribution of *Liq. cacaonum, Lim. fermentum*, and *Lacp. plantarum* isolated from fermented cocoa pulp-bean to embrace transparency on microbial research achievement with all actors involved in the food system (from food producers to consumers). In detail, a total of nine *Liq. cacaocum*, 66 *Lacp. plantarum*, and 70 *Lim. fermentum* have been identified from 36 different published papers (Figure 16.2).

16.2.1 Nutrition and Growth

Lactobacilli species diversity also generates headlines of different nutritional, genetic potential and metabolic function capacities between species and strains. In 2009, *Liq. cacaonum* was described as a facultative anaerobic with non-motile cells. Colonies are beige, opaque, and smooth. At 37°C growth is observed in De Man, Rogosa,

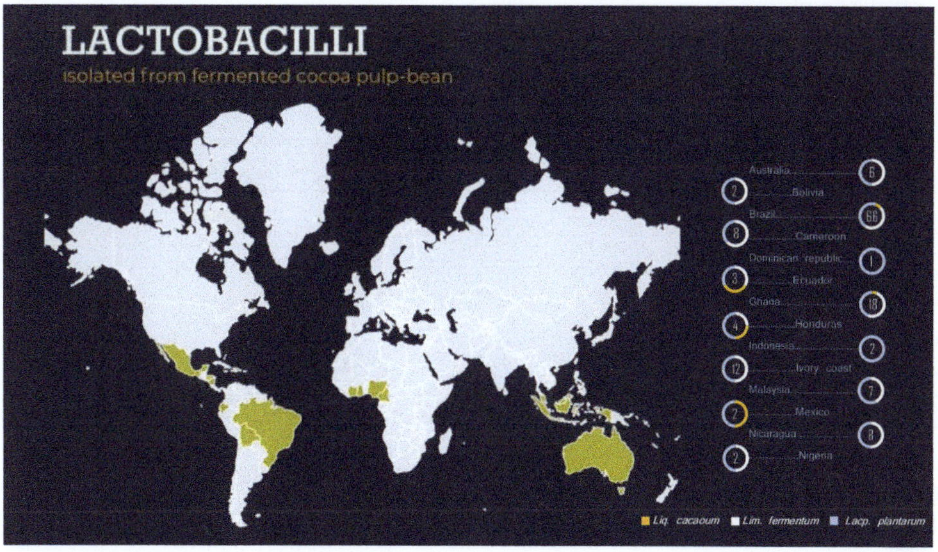

FIGURE 16.2 Incidence of *Liq. cacaonum* (yellow), *Lim. fermentum* (white), and *Lacp. plantarum* (light blue) in fermented cocoa beans expressed as a percentage of total microbial isolates from the cocoa-producing country (total number of isolates are described inside the donut chart).

and Sharpe broth supplemented with NaCl (pH 3.9). *Liq. cacaonum* produces lactic acid and acetic acid as an end product of glucose metabolism. It mostly produces L-lactic acid isomers, and very few D-lactic isomers (90:10). This species is capable to deaminate arginine, and acid is produced from glucose, fructose, mannose, aesculin, salicin, cellobiose, and maltose (Table 8.1). The guanine plus cytosine (G + C) content of the DNA is 34.5 mol% (De Bruyne et al., 2009).

In 1901, *Lim. fermentum* was described as a heterofermentative LAB. This type of LAB is able to ferment fructose, galactose, gluconate, lactose, maltose, mannose, melibiose, raffinose, ribose, and sucrose (Table 16.1). However, some strains may also produce acid from amygdalin, arabinose, cellobiose, aesculin, trehalose, and xylose. *Lim. fermentum* mainly produces L- or DL- lactic acid and NH_3 from arginine. At 15°C or 45°C growth is observed. The genome G + C range content is 52–54 mol% (Dellaglio et al., 2004).

Lacp. plantarum strains are highly heterogeneous (Dellaglio et al., 1975). This species can hardly be distinguished from the phenotypes of its closely related lactobacilli, *Lactiplantibacillus pentosus* and *Lactiplantibacillus paraplantarum* (Curk et al., 1996; Zanoni et al., 1987). These three species contain strains with similar biochemical characteristics, while not many similarities are found at the DNA level. Noteworthy, this intraspecific differentiation between these lactobacilli at the DNA level cannot be distinguished by 16S rRNA sequence analysis (Dellaglio et al., 1975).

Lacp. plantarum is a facultative anaerobic bacterium. At 15°C but not at 45°C growth is observed. This type of LAB is capable of fermenting amygdalin, cellobiose, aesculin, D-fructose, galactose, D-glucose, lactose, maltose, mannitol, and D-mannose (Table 16.1). In contrast to *Liq. cacaonum* and *Lim. fermentum* this bacterium does not deaminate arginine. It produces both lactic acid isomers. The G + C content of the DNA in the range 44–46 mol% (Bringel et al., 2005).

16.2.2 Genetics

Whole-genome sequencing of several *Lim. fermentum* and *Lacp. plantarum* strains originating from different environmental sources has revealed specific functionalities, such as probiotic properties (Grover et al., 2013a, 2013b; Jayashree et al., 2013; Jiménez et al., 2010; Karlyshev and Abramov, 2014; Qureshi et al., 2014), prophage occurrence (Crowley et al., 2013; Zhang et al., 2009), bacteriocin production (Laursen et al., 2015), metabolic properties (D'Auria et al., 2014; Jang et al., 2014; Kim et al., 2013; Li et al., 2013; Naz et al., 2013), and niche adaptations (Karlyshev et al., 2013; Kleerebezem et al., 2003; Ladero et al., 2013). However, to the best of the author's knowledge whole-genome sequencing of *Liq. cacaonum* strains has not been performed yet. With the rise of novel omics technologies, further insights into the diversity and functionality of the microorganisms involved in the fermented cocoa ecosystem are certainly

TABLE 16.1 Phenotypic features of *Liquorilactobacillus cacaonum*, *Limosilactobacillus fermentum*, and *Lactiplantibacillus plantarum*

Characteristics	Liq. cacaonum	Lim. fermentum	Lacp. plantarum	
			ATCC 14197	DKO 22
Growth at:				
10–30° C	+			
37° C	+	+	+	+
Ammonia from arginine	+			
Aesculin	+	-		
Amygdalin	-	-	+	+
Arabinose	-	V	+	+
Cellobiose	+	V		
Galactose	-			
Maltose	+			
Melibiose	-	+	+	
Mannitol	-			
Rhamnose	-		+	
Raffinose	-		+	
Salicin	+	-		
Sorbitol	-		+	+
Sucrose	-	+	+	+
Trehalose	-	+		
Lactic acid isomer	DL	L	+	+

V, Variable. **(Sources:** De Bruyne et al., 2009; Dellaglio et al., 2004)

altering our understanding of the role of the cocobiota during this process and its potential applications (Illeghems et al., 2015). Studies of the genome sequencing of *Lim. fermentum* and *Lacp. plantarum* (222 and 80, respectively) isolated from Ghanaian cocoa pulp-bean fermentation had generated headlines of the strain-dependent genetic potential, core genome, and pangenome of these species (Illeghems et al., 2015).

Overall, the genome sequence of *Lim. fermentum* 222 and *Lacp. plantarum* 80 contains 1864 and 3123 predicted protein-encoding genes. Neither lactobacilli contain virulence factors or antibiotic resistance genes, which indicates a safe use of these microorganisms in the food and health industry. During cocoa pulp-bean fermentation, microorganisms are constantly exposed to stress; as a consequence genes involved in stress adaptation were found in both genome sequences (Illeghems et al., 2015). In agreement with this, *Lacp. plantarum* strains retrieved KEGG genes encoding cluster regularly interspaced short palindromic repeat (CRISP) and phage genes, which confirms the adaptation to stress of this bacteria during cocoa fermentation (Mota-Gutierrez et al., 2021).

Based on a comparative genomic analysis of *Lim. fermentum* 222 with 11 complete and draft genomes of the same species has revealed that different *Lim. fermentum* strains isolated from different genetic material were not associated with any other strain, which might indicate specific adaptation to the fermented cocoa pulp-bean mass ecosystem. In addition, 182 and 223 strain-specific genes were found in *Lim. fermentum* 222 and *Lacp. plantarum* 80 without a homolog to currently available genome sequences respectively, confirming niche adaptation (Illeghems et al., 2015). The strain variability of *Lacp. plantarum* during cocoa pulp-bean fermentation has been recently highlighted, confirming the environmental niche adaptation of this species during this process (Mota-Gutierrez et al., 2021).

16.2.3 Metabolic Properties

Understanding the functions underlying the adaptive behavior of *Liq. cacaonum*, *Lim. fermentum*, and *Lacp. plantarum*, during cocoa pulp-bean fermentation will help us to select appropriate starter cultures and design new protocols that guarantee food quality and safety. In terms of the reconstruction of metabolic pathways of *Lim. fermentum* 222 and *Lacp. plantarum* 80, both bacterial strains contain the genetic capability to metabolize carbohydrate and amino acid, in particular external electron acceptors, citrate transporters, several enzymes involved in the amino acid conversion (transaminases, endopeptidases, and aminopeptidases) and extended pyruvate pathways (Illeghems et al., 2015). Genes encoding a complete arginine deiminase and heterolactic fermentation pathways were attributed to *Lim. fermentum* 222, while genes involved in the homo and heterolatic fermentation and the metabolism of fructose and/or sorbose were retrieved from *Lacp. plantarum* 80 (Illeghems et al., 2015). Arginine deiminase is an important enzyme that has been shown to play a key role in determining the dominance of one bacterial strain during food fermentation.

Whereas the absence of genes involved in the metabolism of glycerol or production of biogenic amines in both strains was reported (Illeghems et al., 2015). Enzymes such as invertase, specifically sucrose invertase (Almeida et al., 2020), aminotransferase (Agyirifo et al., 2019), aldehyde-alcohol dehydrogenase, transaldolase, phosphonoacetate hydrolase, succinate-semialdehyde dehydrogenase, transketolase, and shikimate dehydrogenase have been also retrieved from *Lacp. plantarum* metagenome-assembled genomes (MAGS) from fermented cocoa bean (Mota-Gutierrez et al., 2021). Interestingly, a recent study demonstrates that the associations between reconstructed *Lacp. plantarum* and pangenes were dependent on the type of cocoa variety used to ferment, which highlights the strain-specific functional adaptation (Mota-Gutierrez et al., 2021). It was noted by the authors that a lack of available information regarding the metabolic potential of *Liq. cacaonum* requires care, since finding analogies could indicate similar metabolic pathways used for *Lacp. plantarum*.

16.2.4 Potential Food Application of Lactobacilli from Fermented Cocoa Pulp-bean Mass

For decades, the use of lactobacilli has been exploited as a part of food processing operations and use as therapeutic delivery. The biochemical and physiological attributes of *Liq. cacaonum, Lim. fermentum*, and *Lacp. plantarum* isolated from fermented cocoa beans presumed to be natural biological agents that enhance food quality by improving the appearance, color, flavor and texture for overall acceptability and/or prolong the shelf-life (Figure 16.3). The use of lactobacilli isolated from fermented cocoa bean has been either used directly or indirectly during food manufacture to improve food quality and safety. In this section, we will discuss the challenges and perspectives of using *Liq. cacaonum, Lim. fermentum*, and *Lacp. plantarum* isolated from fermented cocoa pulp-bean mass as starter cultures during cocoa fermentation and other foodstuff and the potential of these microorganisms to be used as biocontrol agents. The ability of lactobacilli to produce antimicrobial compounds, organic acids, aromatic compounds, bacteriocins, exopolysaccharides, vitamins, and enzymes are also discussed in this section to exploit the potential of engineering this lactobacilli metabolites for the production of food additives with specific functions such as, acidulants, antioxidants, antimicrobial agents and discover its ability to produce vitamins that can be used to fortify food for health benefits.

16.2.5 Starter Cultures

From the chocolate production point of view, lactobacilli occur naturally in the raw material during the fermentation process or they can be introduced during this process to boost lactic acid production (Schwan and Wheals, 2004). Spontaneous cocoa pulp-bean fermentation does not only prolong the shelf-life and ensure microbiological safety of the cocoa beans but also improves texture and contributes to the pleasant and expected sensory characteristics of the end-products, cocoa powder, and chocolate. The ability of lactobacilli to produce antimicrobial and aromatic compounds, bacteriocins, exopolysaccharides, vitamins, enzymes, and its capability to survive in high acidic environments are suitable characteristics of starter culture for fermented food such as dairy products, kimchi, mustard, sourdough, meat, fish, vegetable items, and chocolate (Goldstein et al., 2015; Leroy and De Vuyst, 2004).

The growing demand for high-quality chocolate has been increasing in recent years and, as a result, much research has contributed to optimizing the production process, leading to the exploration of starter cultures during cocoa pulp-bean fermentation. In 1998, a defined microbial cocktail inoculum (*S. cerevisiae* var. *chevalieri, Lactobacillus delbrueckii* subsp. *lactis, Lacp. plantarum, Acetobacter aceti*, and *Gluconobacter oxydans*) was used for the first time to conduct cocoa pulp-bean fermentation with the main goal of enhancing chocolate quality (Schwan, 1998). Following this, to ensure and hence increase the quality of cocoa beans and chocolate, different microbial cocktail inocula have been exploited over the last decade (Figueroa-Hernández et al., 2019; Mota-Gutierrez et al., 2019). Four *Lim. fermentum* strains (222, L18, FJ462686.1 and UFLA CHBB 8.12) and two

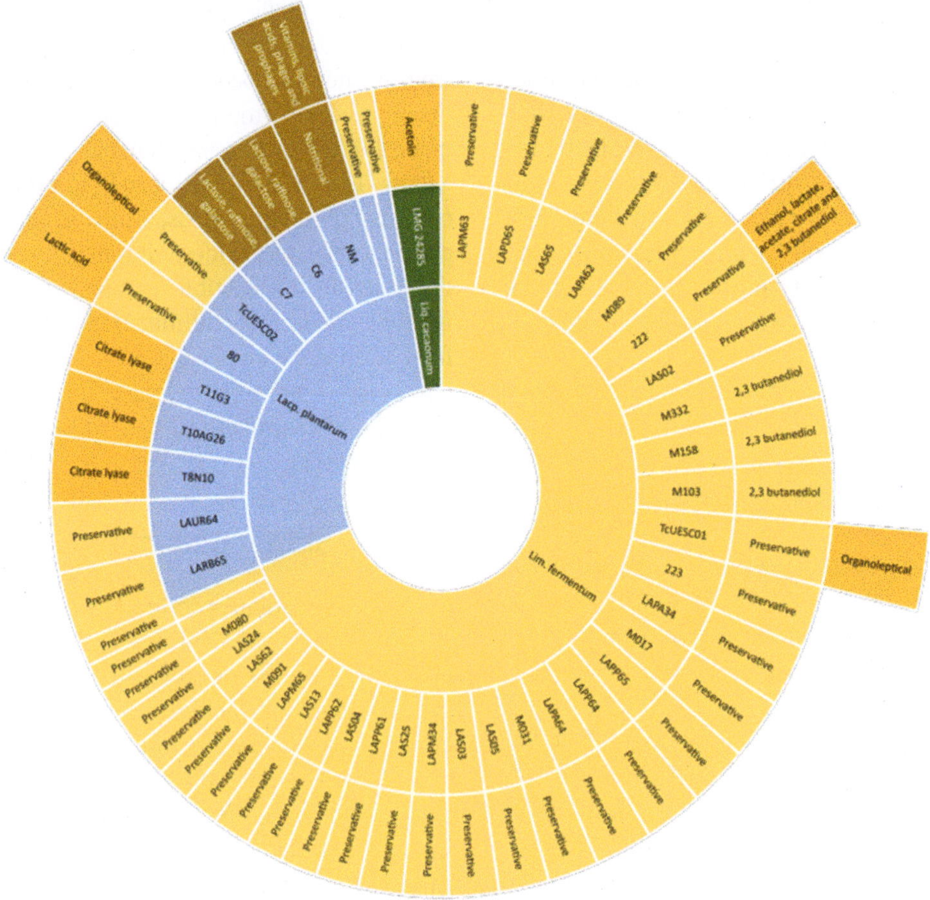

FIGURE 16.3 Lactobacilli strains isolated from fermented cocoa pulp-beans mass with potential food applications. *Lim. fermentum* strains are colored light yellow, *Lacp. plantarum* strains are colored blue, and *Liq. cacaonum* strains are colored green. Organoleptical properties of specific LAB strains are colored dark orange, nutritional properties are colored brown, and preservative properties are colored light yellow. (**Source**: Ruggirello et al., 2019; E Romanens et al., 2019; T Lefeber et al., 2011; Ouattara et al., 2017; Agyirifo et al., 2019; Illeghems et al. 2015).

Lacp. plantarum strains (80, HQ117897.1) isolated from fermented cocoa pulp-beans mass have been used as part of a mixed starter culture (2014; Crafack et al., 2013; Ho et al., 2018; Lefeber et al., 2010, 2012; Miguel et al., 2017; Moens et al., 2014). Overall, the use of starter cultures during cocoa fermentation influenced the volatilome profile of fermented cocoa bean, while contrasting findings on the improvement of the sensorial attributes in the end product are not conclusive. Consumer perceptions are a major factor determining the success of the inclusion of starter cultures in chocolate. Therefore, studies investigating the consumer acceptance of chocolate previously inoculated with microbial cultures urgently need to be assessed to guarantee food acceptability and increase consumer transparency.

Studies on the effect of the use of only LAB isolated from fermented cocoa beans are scarce in large scale production. Research into microbial ecology together with the automation of the fermentation process is thus recommended. Starter cultures suitable for large-scale fermentation production should have the capability, among others, to grow in sucrose and glucose, tolerate heat, oxidize lactic acid, and survive in alcoholic and acidic environments and to produce a high rate of aromatic compounds. The fermentation of single cocoa varieties for large- or small-scale production is highly advisable. Recent literature demonstrated that the degree of fermentation is influenced by the variety of cocoa used and highlights the importance of generating a method or marker to monitor the fermentation degree (Mota-Gutierrez et al., 2021). Reducing the fermentation time using starter cultures or implementing new fermentation management could optimize the production cost of chocolate and strengthen the taste and aroma of this confectionary snack food.

16.2.6 Food Preservation Applications

During cocoa fermentation, spore-forming bacteria and filamentous fungi can also grow, mainly in the most-aerated and colder outer layers of the fermentation mass. Its occurrence has been associated with the production of off-flavors and mycotoxins and causing defects of the shell fracture, germination-damage and insect attack (Schwan and Fleet, 2014). As a strategy to mitigate the growth of undesirable microorganisms during the cocoa fermentation or storage processes, LAB and yeast species isolated from fermented cocoa bean have been used to control mold growth in cocoa beans (Romanens et al., 2019). A total of 30 *Lim. fermentum* isolates and four *Lacp. plantarum* isolates from fermented cocoa have shown a strong antifungal potential against *Aspergillus* and *Penicillium* (Abbasiliasi et al., 2012; Romanens et al., 2019; Ruggirello et al., 2019), while nine *Lim. fermentum* strains showed inhibition growth of *Gibberella moniliformis* after 10–11 days, which reveals a broadband antifungal activity (Romanens et al., 2019).

The inhibition of filamentous fungal growth by *Lim. fermentum* M107 was confirmed by *in vivo* antifungal assays and the genomes of *Lim. fermentum* M017 and 223 showed no virulence genes and chloramphenicol resistance gene catB10 from *Pseudomonas aeruginosa* (Romanens et al., 2019). Overall, the antifungal assessments of different LAB species evaluated showed a variable strength and intraspecific variation, highlighting the importance of selecting appropriate strains as biopreservatives. Noteworthy, the antifungal mechanism of the action of LAB isolated from fermented cocoa environments remains unclear, while the antibacterial mode of action from other LAB strains isolated from different food environments has been associated with bacteriocin production (Leroy and De Vuyst, 2004).

Lacp. plantarum and *Lim. fermentum* isolated from fermented cocoa also improve the hygiene of the fermentation process through inhibition or elimination of food pathogens. *Lim. fermentum* TcUESC01 and *Lacp. plantarum* TcUESC02 isolated from Brazilian cocoa fermentation had shown an inhibition growth of opportunistic human pathogens such as *Staphylococcus aureus, Salmonella typhimurium, Escherichia coli, Enterococcis faecalis, Listeria monocytogenes*, and *Shigella flexneri* (Melo et al., 2016; Oliveira et al., 2018). Noteworthy, *Lim. fermentum* TCUESC01 reduced significantly the thickness of the biofilm and regulated the expression of *icaA* and *icaR* genes, which modulated the production of polysaccharides adhesion and oligomers (Melo et al., 2016). From the genomic point of view, *Lacp. plantarum* 80 has retrieved genes involved in the production of bacteriocin (plantaricin) and, *Lim. fermentum* 222 had retrieved genes encoding a prophage-related genomic region (Illeghems et al., 2015). Bacteriophage and bacteriocin are used in the food industry mostly in the primary production or storage to combat pathogenic bacteria and guarantee food quality (Loc Carrillo et al., 2005). Apart from the potential application as a biological control of *Lim. fermentum* and *Lacp. plantarum* strains isolated from fermented cocoa, it has been stipulated that the settlement of LAB in the food environment could have a positive correlation with fermentative yeast (Lima et al., 2011; Mota-Gutierrez et al., 2018).

16.2.7 Organoleptic Applications

Microorganisms are necessary to turn the astringent and bitter flavors from the cotyledon into precursors of characteristic chocolate aroma and flavor (De Vuyst and Weckx, 2016). The high concentration of fructose and citric acid in the cocoa pulp is an ideal substrate for the growth of *Lim. fermentum* and *Liq. cacaonum* (Lefeber et al., 2011). Recent literature demonstrated the capability of *Lim. fermentum* strains (222, M103, M158, 332), *Liq. cacaonum* (LMG 24285), and *Lacp. plantarum* to produce butter-like odors (2,3-butanediol and acetoin, Figure 16.3) (Lefeber et al., 2011; Ouattara et al., 2017). The production of several aroma compounds suggests the potential of the above strains to produce the desirable aromas of chocolate.

The capability of *Lacp. plantarum* strains to metabolize citrate is controversial. Three *Lacp. plantarum* strain (T11G3, T10AG26, and T8N10) isolated from fermented cocoa from the Ivory Coast showed different levels of citrate lyase in media containing glucose and citrate (Ouattara et al., 2017). However, *Lacp. plantarum* 80 seems not to be a very ideal strain to be used in cocoa fermentation due to its low capability to ferment citrate (Lefeber et al., 2011). Citrate-degrading enzyme is often used to provide flavor and texture to several fermented products by breaking down citrate into oxaloacetate and acetate.

Lacp. plantarum strains identified on fermented Mexican cocoa had been associated with the presence of alcohol-aldehyde dehydrogenase, phosphonoacetate hydrolase, and shikimate dehydrogenase enzymes (Mota-Gutierrez et al., 2021). These pangenes are involved in ethanol metabolism leading to the generation of acetic acid and aromatic compounds at the beginning of cocoa pulp-bean fermentation. In addition, *Lacp. plantarum* strains were also associated with the presence of novel enzymatic activities such as D-alanyl-D-alanine carboxypeptidase, leucine, -isoleucine, -valine, -threonine, and -alanine binding protein precursors, pyruvate oxidase, succinate semialdehyde dehydrogenase, which may lead to the production of interesting metabolites such as succinate and amino acids (L-alanine) (Mota-Gutierrez et al., 2021). These two compounds are widely used for multiple applications in the food industry as a flavor enhancer.

LAB has been also used for the food industry to produce natural texture-improving sugar polymers (exopolysaccharides) for dairy products (yogurt, sour cream and whipped toppings, ice cream, and mozzarella) (Leroy and De Vuyst, 2004). In this regard, genome sequencing analysis has revealed the potential of *Lim. fermentum* 222 to synthesis exopolysaccharide, which highlights the potential technological implementation of this strain for industrial applications (Illeghems et al., 2015).

The autochthonous *Lim. fermentum* and *Lacp. plantarum* isolated from fermented cocoa beans had shown promising future for potential industrial applications for the development of new functional food. Recently, *Lim. fermentum* TcUESC01 and *Lacp. plantarum* TcUESC02 showed great potential to be used as starter culture to produce soy yogurt due to desirable functional properties (decrease in pH, consequently increasing the titratable activity and cell growth). From the consumer perspective, acceptable hedonic scores in the sensory evaluation of soy yogurt inoculated with *Lim. fermentum* TcUESC01 and *Lacp. plantarum* TcUESC02 were reported (Saito et al., 2014). In contrast, *Lim. fermentum* 222 was not considered a good starter strain used to prepare sourdoughs based on its low fermentation quotient (molar ratio of lactic acid and acetic acid) if compared with other *Lim. fermentum* strains (Ravyts and De Vuyst, 2011).

16.2.8 Nutritional Applications

Functional food ingredients such as food additives, nutraceuticals, and dietary supplements are either intentionally added to food and claim a specific technological or health-related function or are used to enhance the intake of nutrients in a normal diet (Augustin and Sanguansri, 2012). The functionalities of LAB isolated from fermented cocoa bean are of high interest since they can be used for several applications, including nutritional (Figure 16.3). In this regard, recently the metagenome sequence of *Lacp. plantarum* strains retrieved sequences encoding biotin, riboflavin, lipoic acid biosynthesis, bacteriophages, and prophages, indicating the great potential of this bacteria to produce vitamins, lipoic acids, and beneficial virus which can be used to fortify food to provide health benefits (Agyirifo et al., 2019; Illeghems et al., 2015). *Lacp. plantarum* strains (C7 and C9) isolated from fermented cocoa may also lead to the removal of toxic or antinutritive factors, such as lactose, raffinose, and galactose (Abbasiliasi et al., 2012). These characteristics of *Lacp. plantarum* could be used for application in fermented milk and soy to prevent lactose intolerance, flatulence, and intestinal cramp and accumulation of galactose (Leroy and De Vuyst, 2004). On the other hand, *Lim. fermentum* 222 has shown great potential in producing lactic acid that can be a source of dietary mineral supply and is widely used as a food preservative and flavoring agent (Illeghems et al., 2015).

Microbial peptidases are of great acceptance as industrial enzymes with several applications in the food industry. Aminopeptidases are widely used in the food industry as a debittering agent, removing the bitterness of peptides and synthesizing biopeptides and amino acids, while endopeptidases are mainly used to hydrolase meat proteins into smaller peptides and oligopeptides (Nandan and Nampoothiri, 2020). In that sense, the metagenome sequence of *Lim. fermentum* 222 and *Lacp. plantarum* 80 isolated from fermented cocoa has led to the retrieval of genes encoding peptidase, such as endopeptidases and aminopeptidases, indicating a high potential to be used in the food industry (Illeghems et al., 2015).

The interest in seeking more stable, functional, and sustainable food additives that can be incorporated into a wide range of food products during processing to improve food quality is in a constant momentum (Augustin and Sanguansri, 2012). Microbial metabolites are used as food additives to improve technological functions and organoleptic aspects. Genes involved in the production of fumarate and succinate have been retrieved in LAB isolated from fermented cocoa (*Lim. fermentum* 222 and *Lacp. plantarum* 80), indicating the potential of these strains to control pH of foods (Illeghems et al., 2015). In addition, none of *Lacp. plantarum* and *Lim. fermentum* strains isolated from fermented cocoa are able to produce biogenic amines due to the lack of enzymes involved in decarboxylation of histidine and aromatic amino acids, indicating no safety concern of using these strains in the food industry (Illeghems et al., 2015). Overall, the industrial exploitation of LAB isolated from fermented cocoa are still seen more at the research and development stage.

16.3 EUROPEAN REGULATION OF FOOD CULTURES

The European Food and Feed Cultures Association (EFFCA) in 2018, introduces the term "food cultures" (FC) defined as safe live bacteria, yeasts, or molds used during food production. This denomination includes the terms of starter cultures, microbial food cultures, probiotics, and starters use in the food industry (ripening, sourdough, dairy, sausage, wine, yeast starters) (European Food and Feed Cultures Association, 2018). FC are commonly used as food ingredients at different stages of food processing to develop a unique property of food products. Thus, it seems obvious that if FC is used in the manufacture or preparation of foodstuff should be legally required to be listed on the ingredients label; at the moment no regulation requires a mandatory declaration of the use and general description of the FC used during food manufacturing. Reflective and conscious consumers with an internet connection could have access to scientific platforms such as EFFCA that briefly describes the use of FC in different food production industries, such as dairy, wine, and meat to find more information beyond the label (European Food and Feed Cultures Association (EFFCA), 2018). Noteworthy, the authors note the lack of the description of the use of FC in the chocolate, brewery, bakery, fruit and vegetable industries, and a full description of microorganisms used, including the environmental source on the EFFCA platform.

Consumer beliefs regarding the use of microorganisms during food processing or as a source of protein is an argument that deserves consideration. Although there is a positive trend around fermented foods and edible microbial biomass a key factor that will influence adoption rates of these new food trends is the cost of these foodstuffs, and consumer aversion to microorganisms raising mostly safety concerns of this type of commodity. A strategy that might motivate the acceptability of the use of microbes during food processing or as a microbial sources of food and feed is to provide recognition of the importance of food cultures in accomplishing a new food production system that is more sustainable and more resilient (Mota-Gutierrez and Cocolin 2021). However, we should bear in mind that this strategy may not change the consumer's negative attitude and perception toward microbial food.

16.3.1 Food Safety Assessment

In terms of food safety assessment, in Europe, several entities are involved in the evaluation of the performance of single strains used in different food manufacturing processes (including shelf-life) (Laulund et al., 2017). As a result of the assessment of a broad range of biological agents the European Food Safety Authority (EFSA) has published a Qualified Presumption of Safety (QPS) status which includes a list of bacteria, yeasts, and molds that are safe to use as sources of food and feed additives (Ricci et al., 2017). *Lim. fermentum* and *Lacp. plantarum* are granted with a QPS status and are not only safe to use in food production but also for human consumption (European Food Safety Authority (EFSA), 2007). In 2019, the QPS list was updated and it now comprises a safety assessment of several strains submitted to the EFSA (European Food Safety Authority (EFSA), 2020). It should be stressed that the absence of *Liq. cacaonum* in the QPS list does not imply a food safety concern of its use but it should be considered as a microorganism to be subjected to a full safety assessment due to lack of existing knowledge of the taxonomic description. More research is needed concerning FC management at industrial or small-scale production of microbial strains granted with a QPS status and recommendations of what kind of information should accompany the label of products that use FC during food processing or use FC as food ingredients. Industrial guidelines can be used as a foundation for further discussion to generate appropriate food regulations.

16.4 CONCLUSIONS

The fermentation process of cocoa pulp-beans not only favors the generation of aromatic compounds and reduce the bitterness of the cocoa beans, it also represents a source of microbes, such as *Liq. cacaonum*, *Lim. fermentum*, and *Lacp. plantarum* that can be used in the food industry as antimicrobial and antifungal agents, organoleptic enhancer, food ingredient, food additive, and as food culture. It should be noted that the unique functional properties imparting food quality and safety of *Lim. fermentum* and *Lacp. plantarum* isolated from fermented cocoa pulp-beans are strain dependent. Therefore, care should be taken when selecting a LAB strain as a food ingredient or starter culture. Properties such as acidification, enzymatic activity, production of antimicrobial compounds, and sensory analysis are key elements under investigation to understand the role of LAB as a starter culture in the food production. The functional properties of *Liq. cacaonum*, requires further assessment, as finding could indicate similar features revealed from *Lacp. plantarum*. Research determining a microbial marker for monitoring the quality and degree of fermentation and food safety as well as the optimization of the cocoa fermentation could form the basis for the development of coherent guidelines and regulations on the use of starter cultures as an enhancer of food quality and reduce production costs. From the perspective of consumer acceptance, chocolate inoculated with microorganisms during fermentation may appear on the food market in the present or the future. This type of product should thus be labeled with a clear description of the origin and strain-description of the FC to increase the consumer's awareness and transparency.

REFERENCES

1. Abbasiliasi S, Tan JS, Ibrahim TAT, Ramanan RN, Vakhshiteh F, et al. 2012. Isolation of *Pediococcus Acidilactici* Kp10 with ability to secrete bacteriocin-like inhibitory substance from milk products for applications in Food Industry. *BMC Microbiology* 12: 1–12.
2. Adler P, Bolten CJ, Dohnt K, Hansen CE, Wittmann C. 2013. Core fluxome and metafluxome of lactic acid bacteria under simulated cocoa pulp fermentation conditions. *Applied and Environmental Microbiology* 79: 5670–5681.
3. Adler P, Frey LJ, Berger A, Bolten CJ, Hansen CE, et al. 2014. The key to acetate: Metabolic fluxes of acetic acid bacteria under cocoa pulp fermentation-simulating conditions. *Applied and Environmental Microbiology* 80: 4702–4716.
4. Agyirifo DS, Wamalwa M, Otwe EP, Galyuon I, Runo S, et al. 2019. Metagenomics analysis of cocoa bean fermentation microbiome identifying species diversity and putative functional capabilities. *Heliyon* 5: e02170.
5. Almeida OGG, Pinto UM, Matos CB, Frazilio DA, Braga VF, et al. 2020. Does quorum sensing play a role in microbial shifts along spontaneous fermentation of cocoa beans? An *in silico* perspective. *Food Research International* 131: 109034.
6. Arana-Sánchez A, Segura-García LE, Kirchmayr M, Orozco-Ávila I, Lugo-Cervantes E, et al. 2015. Identification of predominant yeasts associated with artisan Mexican cocoa fermentations using culture-dependent and culture-independent approaches. *World Journal of Microbiology and Biotechnology* 31: 359–369.
7. Augustin MA, Sanguansri L. 2012. Challenges in developing delivery systems for food additives, nutraceuticals and dietary supplements. In: Nissim Garti and D. Julian McClements (Edited by), *Encapsulation Technologies and Delivery Systems for Food Ingredients and Nutraceuticals*, 19–48. Elsevier Ltd.
8. Bringel F, Castioni A, Olukoya DK, Felis GE, Torriani S, et al. 2005. *Lactobacillus plantarum* subsp. *argentoratensis* nov., isolated from vegetable matrices. *International Journal of Systematic and Evolutionary Microbiology* 55: 1629–1634.
9. Crafack M, Keul H, Eskildsen CE, Petersen MA, Saerens S, et al. 2014. Impact of starter cultures and fermentation techniques on the volatile aroma and sensory profile of chocolate. *Food Research International* 63: 306–316.
10. Crafack M, Mikkelsen MB, Saerens S, Knudsen M, Blennow A, et al. 2013. Influencing cocoa flavour using *Pichia kluyveri* and *Kluyveromyces marxianus* in a defined mixed starter culture for cocoa fermentation. *International Journal of Food Microbiology* 167: 103–116.
11. Crowley S, Bottacini F, Mahony J, Van Sinderen D. 2013. Complete genome sequence of *Lactobacillus plantarum* strain 16, a broad-spectrum antifungal-producing lactic acid bacterium. *Genome Announcements* 1. doi:10.1128/genomeA.00533-13.
12. Curk MC, Hubert JC, Bringel F. 1996. *Lactobacillus paraplantarum* sp. nov., a new species related to *Lactobacillus plantarum*. *International Journal of Systematic Bacteriology* 46: 595–598.
13. D'Auria G, Džunková M, Moya A, Tomáška M, Kološta M, et al. 2014. Genome sequence of *Lactobacillus plantarum* 19L3, a strain proposed as a starter culture for Slovenská bryndza ovine cheese. *Genome Announcements* 2. doi:10.1128/genomeA.00292-14.
14. Da Veiga Moreira IM, Gabriela M, Miguel P, Ferreira W, Ribeiro D, et al. 2013. Microbial succession and the dynamics of metabolites and sugars during the fermentation of three different cocoa (*Theobroma cacao* L.) hybrids. *Food Research International* 54: 9–17.

15. De Bruyne K, Camu N, De Vuyst L, Vandamme P. 2009. *Lactobacillus fabifermentans* sp. nov. and *Lactobacillus cacaonum* sp. nov., isolated from Ghanaian cocoa fermentations. *International Journal of Systematic and Evolutionary Microbiology* 59: 7–12.
16. De Melo Pereira GV, Magalhães KT, De Almeida EG, Da Silva Coelho I, Schwan RF. 2013. Spontaneous cocoa bean fermentation carried out in a novel-design stainless steel tank: Influence on the dynamics of microbial populations and physical-chemical properties. *International Journal of Food Microbiology* 161: 121–133.
17. De Vuyst L, Lefeber T, Papalexandratou Z, Camu N. 2010. The functional role of lactic acid bacteria in cocoa bean fermentation. *Biotechnology of Lactic Acid Bacteria: Novel Applications* 301: 325.
18. De Vuyst L, Weckx S. 2016. The cocoa bean fermentation process : From ecosystem analysis to starter culture development. *Journal of Applied Microbiology* 121: 5–17.
19. Dellaglio F, Bottazzi V, Vescovo M. 1975. Deoxyribonucleic acid homology among *Lactobacillus* species of the subgenus Streptobacterium orla jensen. *International Journal of Systematic Bacteriology* 25: 160–172.
20. Dellaglio F, Torriani S, Felis GE. 2004. Reclassification of *Lactobacillus cellobiosus* as a later synonym of *Lactobacillus fermentum* Beijerinck 1901. *International Journal of Systematic and Evolutionary Microbiology* 54: 809–812.
21. European Food and Feed Cultures Association (EFFCA). 2018. Definition of food cultures. 2018. https://effca.org/_library/_files/EFFCA_Definition_Food_Cultures_final_updated.pdf.Access date November 2020.
22. European Food Safety Authority (EFSA). 2007. Opinion of the scientific committee on the introduction of a Qualified Presumption of Safety (QPS) approach for assessment of selected microorganisms referred to EFSA. *EFSA Journal* 5: 1–16.
23. European Food Safety Authority (EFSA). 2020. The 2019 updated list of QPS status recommended biological agents in support of EFSA risk assessments. *EFSA Journal* 18: 5966.
24. Fernández Y, Balzarini T, Clapé P, Evrard P, De Vuyst L, et al. 2016. The environmental and intrinsic yeast diversity of Cuban cocoa bean heap fermentations. *International Journal of Food Microbiology* 233: 34–43.
25. Figueroa-Hernández C, Mota-Gutierrez J, Ferrocino I, Hernández-Estrada ZJ, González-Ríos O, et al. 2019. The challenges and perspectives of the selection of starter cultures for fermented cocoa beans. *International Journal of Food Microbiology* 301: 41–50.
26. Garcia-Armisen T, Papalexandratou Z, Hendryckx H, Camu N, Vrancken G, et al. 2010. Diversity of the total bacterial community associated with Ghanaian and Brazilian cocoa bean fermentation samples as revealed by a 16 S rRNA gene clone library. *Applied Microbiology and Biotechnology* 87: 2281–2292.
27. Goldstein EJC, Tyrrell KL, Citron DM. 2015. *Lactobacillus* species: Taxonomic complexity and controversial susceptibilities. *Clinical Infectious Diseases* 60: S98–107.
28. Grover S, Sharma VK, Mallapa RH, Batish VK. 2013a. Draft genome sequence of *Lactobacillus fermentum* Lf1, an Indian isolate of human gut origin. *Genome Announcements* 1. doi:10.1128/genomeA.00883-13.
29. Grover S, Sharma VK, Mallapa RH, Batish VK. 2013b. Draft genome sequence of *Lactobacillus plantarum* strain Lp91, a promising Indian probiotic isolate of human gut origin. *Genome Announcements* 1. doi:10.1128/genomeA.00976-13.
30. Hamdouche Y, Meile JC, Lebrun M, Guehi T, Boulanger R, et al. 2019. Impact of turning, pod storage and fermentation time on microbial ecology and volatile composition of cocoa beans. *Food Research International* 119: 477–491.
31. Ho VTT, Fleet GH, Zhao J. 2018. Unravelling the contribution of lactic acid bacteria and acetic acid bacteria to cocoa fermentation using inoculated organisms. *International Journal of Food Microbiology* 279: 43–56.
32. Ho VTT, Zhao J, Fleet G. 2015. The effect of lactic acid bacteria on cocoa bean fermentation. *International Journal of Food Microbiology* 205: 54–67.
33. Illeghems K, De Vuyst L, Papalexandratou Z, Weckx S. 2012. Phylogenetic analysis of a spontaneous cocoa bean fermentation metagenome reveals new insights into its bacterial and fungal community diversity. *PLoS ONE* 7: e38040.
34. Illeghems K, De Vuyst L, Weckx S. 2015. Comparative genome analysis of the candidate functional starter culture strains *Lactobacillus fermentum* 222 and *Lactobacillus plantarum* 80 for controlled cocoa bean fermentation processes. *BMC Genomics* 16: 1–13.
35. Jang JY, Lim HI, Park HW, Choi HJ, Kim TW, et al. 2014. Draft genome sequence of *Lactobacillus plan-*

tarum wikim18, isolated from Korean kimchi. *Genome Announcements* 2. doi:10.1128/genomeA.00467-14.

36. Jayashree S, Pooja S, Pushpanathan M, Vishnu U, Sankarasubramanian J, et al. 2013. Genome sequence of *Lactobacillus fermentum* strain MTCC 8711, a probiotic bacterium isolated from yogurt. *Genome Announcements* 1. doi:10.1128/genomeA.e00770-13.
37. Jespersen L, Nielsen DS, Hønholt S, Jakobsen M. 2005. Occurrence and diversity of yeasts involved in fermentation of West African cocoa beans. *FEMS Yeast Research* 5: 441–453.
38. Jiménez E, Langa S, Martín V, Arroyo R, Martín R, et al. 2010. Complete genome sequence of *Lactobacillus fermentum* CECT 5716, a probiotic strain isolated from human milk. *Journal of Bacteriology* 192: 4800.
39. Karlyshev AV, Abramov VM. 2014. Draft genome sequence of *Lactobacillus plantarum* 2165. *Genome Announcements* 2. doi:10.1128/genomeA.e01179-13.
40. Karlyshev AV, Raju K, Abramov VM. 2013. Draft genome sequence of *Lactobacillus fermentum* strain 3872. *Genome Announcements* 1. doi:10.1128/genomeA.01006-13.
41. Kim EK, Park YM, Lee OY, Lee WJ. 2013. Draft genome sequence of *Lactobacillus plantarum* strain WJL, a *Drosophila* gut symbiont. *Genome Announcements* 1. doi:10.1128/genomeA.00937-13.
42. Kleerebezem M, Boekhorst J, Van Kranenburg R, Molenaar D, Kuipers OP, et al. 2003. Complete genome sequence of *Lactobacillus plantarum* WCFS1. *Proceedings of the National Academy of Sciences of the United States of America* 100: 1990–1995.
43. Kostinek M, Ban-Koffi L, Ottah-Atikpo M, Teniola D, Schillinger U, et al. 2008. Diversity of predominant lactic acid bacteria associated with cocoa fermentation in Nigeria. *Current Microbiology* 56: 306–314.
44. Ladero V, Alvarez-Sieiro P, Redruello B, Del Rio B, Linares DM, et al. 2013. Draft genome sequence of *Lactobacillus plantarum* strain IPLA 88. *Genome Announcements* 1. doi:10.1128/genomeA.00524-13.
45. Laulund S, Wind A, Derkx P, Zuliani V. 2017. Regulatory and safety requirements for food cultures. *Microorganisms* 5: 28.
46. Laursen MF, Bahl MI, Licht TR, Gram L, Knudsen GM. 2015. A single exposure to a sublethal pediocin concentration initiates a resistance-associated temporal cell envelope and general stress response in. *Listeria Monocytogenes. Environmental Microbiology* 17: 1134–1151.
47. Lefeber T, Janssens M, Camu N, De Vuyst L. 2010. Kinetic analysis of strains of lactic acid bacteria and acetic acid bacteria in cocoa pulp simulation media toward development of a starter culture for cocoa bean fermentation. *Applied and Environmental Microbiology* 76: 7708–7716.
48. Lefeber T, Janssens M, Moens F, Gobert W, De Vuyst L. 2011. Interesting starter culture strains for controlled cocoa bean fermentation revealed by simulated cocoa pulp fermentations of cocoa-specific lactic acid bacteria. *Applied and Environmental Microbiology* 77: 6694–6698.
49. Lefeber T, Papalexandratou Z, Gobert W, Camu N, De Vuyst L. 2012. On-farm implementation of a starter culture for improved cocoa bean fermentation and its influence on the flavour of chocolates produced thereof. *Food Microbiology* 30: 379–392.
50. Leroy F, De Vuyst L. 2004. Lactic acid bacteria as functional starter cultures for the food fermentation industry. *Trends in Food Science and Technology* 15: 67–78.
51. Li XR, Gong FM, Zheng HJ, Zhang ZH, Luo YY, et al. 2013. Draft genome sequence of *Lactobacillus plantarum* strain AY01, isolated from the raw material of fermented goat milk cheese. *Genome Announcements* 1. doi:10.1128/genomeA.e00737-13.
52. Lima JR, Almeida MH, Rob Nout MJ, Zwietering MH. 2011. *Theobroma cacao* L., "the food of the gods": Quality determinants of commercial cocoa beans, with particular reference to the impact of fermentation. *Critical Reviews in Food Science and Nutrition* 51: 731–761.
53. Loc Carrillo C, Atterbury RJ, Dillon E, Scott A, Connerton IF, et al. 2005. Bacteriophage therapy to reduce *Campylobacter jejuni* colonization of broiler chickens. *Applied and Environmental Microbiology* 71: 6554–6563.
54. Melo TA, Dos Santos TF, De Almeida ME, Junior LAGF, Andrade EF, et al. 2016. Inhibition of *Staphylococcus aureus* biofilm by *Lactobacillus* isolated from fine cocoa. *BMC Microbiology* 16: 1–9.
55. Miguel MGDCP, Reis LVDC, Efraim P, Santos C, Lima N, et al. 2017. Cocoa fermentation: Microbial identification by MALDI-TOF MS, and sensory evaluation of produced chocolate. *LWT - Food Science and Technology* 77: 362–369.
56. Moens F, Lefeber T, De Vuyst L. 2014. Oxidation of metabolites highlights the microbial interactions and role of *Acetobacter pasteurianus* during cocoa bean

fermentation. *Applied and Environmental Microbiology* 80: 1848–1857.

57. Mota-Gutierrez J, and Cocolin L. 2021. Current trends and applications of plant origin lactobacilli in the promotion of sustainable food systems. *Trends in Food Science & Technology* 114: 198–211.
58. Mota-Gutierrez J, Barbosa-Pereira L, Ferrocino I, Cocolin L. 2019. Traceability of functional volatile compounds generated on inoculated cocoa fermentation and its potential health benefits. *Nutrients* 11: 884.
59. Mota-Gutierrez J, Botta C, Ferrocino I, Giordano M, Bertolino M, et al. 2018. Dynamics and biodiversity of bacterial and yeast communities during the fermentation of cocoa beans. *Applied and Environmental Microbiology* 84: e01164–18.
60. Mota-Gutierrez J, Ferrocino I, Giordano M, Suarez-Quiroz ML, Gonzalez-Ríos O, Cocolin L. 2021. Jun 25. Influence of taxonomic and functional content of microbial communities on the quality of fermented cocoa pulp-bean mass. *Applied and Environmental Microbiology* 87(14): e0042521. doi:10.1128/AEM.00425-21. Epub PMID: 33990301; PMCID: PMC8231445.
61. Nandan A, Nampoothiri KM. 2020. Therapeutic and biotechnological applications of substrate specific microbial aminopeptidases. *Applied Microbiology and Biotechnology* 104: 5243–5257.
62. Naz S, Tareb R, Bernardeau M, Vaisse M, Lucchetti Miganeh C, et al. 2013. Genome sequence of *Lactobacillus plantarum* strain UCMA 3037. *Genome Announcements* 1. doi:10.1128/genomeA.00251-13.
63. Oliveira JS, Costa K, Acurcio LB, Sandes SHC, Cassali GD, et al. 2018. *In vitro* and *in vivo* evaluation of two potential probiotic lactobacilli isolated from cocoa fermentation (*Theobroma cacao* L.). *Journal of Functional Foods* 47: 184–191.
64. Ouattara HD, Ouattara HG, Droux M, Reverchon S, Nasser W, et al. 2017. Lactic acid bacteria involved in cocoa beans fermentation from Ivory Coast: Species diversity and citrate lyase production. *International Journal of Food Microbiology* 256: 11–19.
65. Papalexandratou Z, Falony G, Romanens E, Jimenez JC, Amores F, et al. 2011a. Species diversity, community dynamics, and metabolite kinetics of the microbiota associated with traditional Ecuadorian spontaneous cocoa bean fermentations. *Applied and Environmental Microbiology* 77: 7698–7714.
66. Papalexandratou Z, Kaasik K, Kauffmann LV, Skorstengaard A, Bouillon G, et al. 2019. Linking cocoa varietals and microbial diversity of Nicaraguan fine cocoa bean fermentations and their impact on final cocoa quality appreciation. *International Journal of Food Microbiology* 304: 106–118.
67. Papalexandratou Z, Lefeber T, Bahrim B, Seng O, Daniel H, et al. 2013. *Acetobacter pasteurianus* predominate during well-performed Malaysian cocoa bean box fermentations, underlining the importance of these microbial species for a successful cocoa bean fermentation process. *Food Microbiology* 35: 73–85.
68. Qureshi A, Itankar Y, Ojha R, Mandal M, Khardenavis A, et al. 2014. Genome sequence of *Lactobacillus plantarum* EGD-AQ4, isolated from fermented product of Northeast India. *Genome Announcements* 2. doi:10.1128/genomeA.e01122-13.
69. Ravyts F, De Vuyst L. 2011. Prevalence and impact of single-strain starter cultures of lactic acid bacteria on metabolite formation in sourdough. *Food Microbiology* 28: 1129–1139.
70. Ricci A, Allende A, Bolton D, Chemaly M, Davies R, et al. 2017. Scientific opinion on the update of the list of QPS-recommended biological agents intentionally added to food or feed as notified to EFSA. *EFSA Journal* 15.(3):4664.
71. Romanens E, Freimüller Leischtfeld S, Volland A, Stevens M, Krähenmann U, et al. 2019. Screening of lactic acid bacteria and yeast strains to select adapted anti-fungal co-cultures for cocoa bean fermentation. *International Journal of Food Microbiology* 290: 262–272.
72. Romanens E, Näf R, Lobmaier T, Pedan V, Leischtfeld SF, et al. 2018. A lab-scale model system for cocoa bean fermentation. *Applied Microbiology and Biotechnology* 102: 3349–3362.
73. Ruggirello M, Nucera D, Cannoni M, Peraino A, Rosso F, et al. 2019. Antifungal activity of yeasts and lactic acid bacteria isolated from cocoa bean fermentations. *Food Research International* 115: 519–525.
74. Saito VST, Dos Santos TF, Vinderola CG, Romano C, Nicoli JR, et al. 2014. Viability and resistance of lactobacilli isolated from cocoa fermentation to simulated gastrointestinal digestive steps in soy yogurt. *Journal of Food Science* 79: M208–13.
75. Schwan RF. 1998. Cocoa fermentations conducted with a defined microbial cocktail inoculum. *Applied and Environmental Microbiology* 64: 1477–1483.
76. Schwan RF, Fleet GH. 2014. *Cocoa and Coffee Fermentations*. CRC Press.
77. Schwan RF, Wheals AE. 2004. The microbiology of cocoa fermentation and its role in chocolate qual-

ity. *Critical Reviews in Food Science and Nutrition* 44: 205–221.

78. Serra JL, Moura FG, GVdM P, Soccol CR, Rogez H, et al. 2019. Determination of the microbial community in Amazonian cocoa bean fermentation by Illumina-based metagenomic sequencing. *Lwt* 106: 229–239.

79. Visintin S, Alessandria V, Valente A, Dolci P, Cocolin L. 2016. Molecular identification and physiological characterization of yeasts, lactic acid bacteria and acetic acid bacteria isolated from heap and box cocoa bean fermentations in West Africa. *International Journal of Food Microbiology* 216: 69–78.

80. Zanoni P, Farrow JAE, Phillips BA, Collins MD. 1987. Lactobacillus pentosus. *International Journal of Systematic and Evolutionary Microbiology* 37: 339–341.

81. Zhang ZY, Liu C, Zhu YZ, Zhong Y, Zhu YQ, et al. 2009. Complete genome sequence of *Lactobacillus plantarum* JDM1. *Journal of Bacteriology* 191: 5020–5021.

82. Zheng J, Wittouck S, Salvetti E, Franz CMAP, Harris HMB, et al. 2020. A taxonomic note on the genus *Lactobacillus*: Description of 23 novel genera, emended description of the genus *Lactobacillus* Beijerinck 1901, and union of *Lactobacillaceae* and *Leuconostocaceae*. *International Journal of Systematic and Evolutionary Microbiology* 70: 2782–2858. doi:10.1128/genomeA.00533-13.

CHAPTER 17

Microbiological Control as a Tool to Improve Wine Aroma and Quality

Albert Mas, Gemma Beltran, and María Jesús Torija

Universitat Rovira i Virgili, Dept. Bioquímica I Biotecnologia, Facultat d'Enologia. Marcel·li Domingo, 1. 43007-Tarragona, Spain

17.1 INTRODUCTION

The fermentation of grape juice and its transformation into wine is one of the oldest processes used by humans. However, humanity was exposed to some alcoholic products earlier, as many fruits can spontaneously ferment when ripe to yield some alcoholic products. Those fruits or juices that spontaneously fermented generated a preserving product (ethanol) that could enhance their nutritive value and retain essential flavors.

The primary microorganisms involved in this transformation appeared on earth much earlier than humans. The primary task of microorganisms in nature is to recycle organic matter, including sugars. Among these microorganisms, yeasts have a high capacity to ferment sugars to alcohol whenever fruits are ripe. Additionally, ethanol can be oxidized to acetic acid and even further to carbon dioxide and water (Ribéreau-Gayon et al., 2006a).

The reactions involved in alcoholic fermentation were already understood by the end of the eighteenth century, but knowledge of the microorganisms involved in this process required another century to be elucidated and then another century to before the process could be efficiently controlled. Beyond the pioneering studies of Cagniard de La Tour and Pasteur, the tools used for the control of alcoholic fermentation appeared at the end of the nineteenth century, when Hansen started developing the selected starters for beer production (Mas, 2019; see chapter 18). The use of these starters in wine cellars took a long time to be implemented, mostly due to the link between harvest and a given season of the year, which prevented the continuous use of the starters and the need for laboratories to maintain the cultures (Fleet, 1993). It was during the second half of the twentieth century when active dry yeast preparations became available in a form considered "cellar-friendly," as large quantities of the product can be produced in the short time that the linked harvesting and winemaking processes take place. Although this is a very effective way of controlling alcoholic fermentation, currently, many winemakers are looking for alternatives to this method, as it is often considered to be a "uniformizing" or "artificial" means of producing wine (Mas et al., 2016).

17.2 METHODS OF ANALYSIS: CLASSICAL AND MOLECULAR METHODS

Any type of control of a microbiological process is heavily dependent on the corresponding analytical methodology. During the fermentation process, a succession of microorganisms occurs, and the methods available for analyzing microbiological dynamics can be classified as "classical" methods, molecular methods, and next-generation sequencing methods (or high-throughput methods due to the number of microorganisms

Good Microbes in Medicine, Food Production, Biotechnology, Bioremediation, and Agriculture, First Edition. Edited by Frans J. de Bruijn, Hauke Smidt, Luca S. Cocolin, Michael Sauer, David Dowling, and Linda Thomashow.

identified). To gain a true appreciation and precise control over the fermentation process, at least two methods are often necessary due to their complementarity (Mas et al., 2021).

Classical microbiology research has led to the development of tools used to detect and quantify the microorganisms involved in wine production. Thus, plating and traditional observation under a microscope have yielded the basic and general knowledge that we currently have. These methods combine morphological tests (shape, size, etc. of the microorganisms under the microscope as well as the colony morphology on different solid media) with several physiological tests (differential growth based on different nutritional needs) that can be assessed on either agar-solidified plates or in liquid cultures. The description of the shape, size, and arrangement of microorganisms under a phase-contrast microscope is an initial step. However, the results can be misleading, as these characteristics can change according to the age of the cells and the culture medium that has been used (Mas et al., 2021).

The identification of microorganisms using plating and differential growth is dependent on previously performed colony isolation, which has been used, for instance, by Barnett et al. (2000) to identify and classify most yeast species. However, numerous and time-consuming tests may be needed to identify and isolate at the species level, and interpretation of the results requires a certain degree of expertise.

Nevertheless, these methods allow the enumeration of microorganisms present in grapes, must or fermentation samples. Under the microscope, the total number of cells can be observed, although if agglomerated they can be difficult to count. In contrast, colonies are counted on plates, which are abbreviated as cfu (colony forming units). The cfu may not exactly represent the original number of cells present but is generally considered to indicate the number of viable cells, although they represent the number of cells able to grow in that given medium (Mas et al., 2021).

The basic initial approach of a winemaker to the control of fermentation is performed by monitoring the density of the grape must based on the transformation of sugars into ethanol and CO_2, which is lost as gas and which produces the density decrease. The estimation of the total microbial population can be performed by microscope-based counting (with the limitation of a minimal population) or by plating on a rich medium (YPD medium, the most commonly used yeast, comprises yeast extract, peptone, and glucose). The combined use of these three methodologies (density monitoring, cell counting under the microscope, and plating) can provide a good idea as to the development of fermentation and should be a highly advisable practice in cellars. The culture media can incorporate antimicrobial compounds specifically to enumerate different wine-related microorganisms. In Table 17.1, we summarize the most common media and the purposes for their use.

The development of molecular techniques based on DNA analysis for the identification of microorganisms has been an important step in advancing the study of the wine-making process. The prior knowledge obtained using classical methods has been greatly improved with the identification of microorganisms at the molecular level, a practice that is generally combined with their isolation after plating (König et al., 2017). The analysis of polymorphisms (variations in DNA sequences) in ribosomal RNA tandem repeats has allowed the identification

TABLE 17.1 Summary of the most common media used for wine microorganisms. LAB: Lactic Acid Bacteria, AAB Acetic Acid Bacteria. YPD: Yeast Extract, Peptone, Dextrose; MRS: Man, Rogosa, Sharpe medium; GY: Glucose and Yeast Extract. +: growth, - absence of growth

Medium	Yeast	LAB	AAB
YPD	+	+	+
YPD + Chloramphenicol	+	-	-
MRS	+	+	+
MRS + Nystatin	-	+	+
MRS + Nystatin + Sodium azide	-	+	-
GY	+	+	+
GY + Pimaricin/Nystatin	-	+	+
GY + Pimaricin+ penicillin	-	-	+

FIGURE 17.1 Structure of the ribosomal RNA coding region in yeasts.

of different species (Kurtzman and Robnett, 1998). The genes that encode different RNAs are present in the genome in specific arrangements, and as they encode elements that are crucial for life, they are extremely well conserved between species. However, genes harbor a few regions that show enough polymorphism to allow for identification at the species level. In yeasts, these polymorphic sequences are the non-coding internal transcribed spacers (ITS) and the D1–D2 region within the 26S coding region, the analysis of which has allowed for the identification of most wine-related yeast species (Montrocher et al., 1998) (Figure 17.1). The application of methods based on rDNA polymorphisms in routine studies to analyze wine ecology was pioneered by Guillamón et al. (1998) followed by Torija et al. (2001) and Beltran et al. (2002). These methodologies have been improved with the increased availability of sequencing protocols that are cheaper and faster.

One important aspect of molecular techniques is that some of them can be used without the need to culture strains on solid media. These culture-independent techniques have allowed for new insights to be made, especially after the observation of "viable but not culturable" (VBNC) yeast cells during wine production (Millet and Lonvaud-Funel, 2000). This cell status is likely to be very common during wine production due to several processes that induce it, including the common use of sulfur dioxide, which induces this VBNC state, the response to adverse conditions (especially increased ethanol levels and the absence of nutrients) during the wine-making process and the resulting biological warfare among the different wine microorganisms (see microbial interactions). The application of these culture-independent molecular techniques to wine making has been pioneered by Cocolin and Mills (2003), denaturing gradient gel electrophoresis, DGGE-PCR, Hierro et al. (2006), quantitative PCR, qPCR, and Andorrà et al. (2010a), DNA binding dye coupled qPCR, EMA, or PMA-qPCR.

An additional step has been the appearance of methodologies based on high-throughput sequencing or next-generation sequencing (NGS). A major aspect of this technique is the number of sequences (equivalent to microorganisms) that can be analyzed from a given sample. Previous culture-dependent techniques could not evaluate more than 30–100 microorganisms at a given time point, and culture-independent techniques could only estimate the primary (DGGE-PCR) or target microorganisms (qPCR), whereas NGS-based methods can identify as many as several million microorganisms if needed. Thus, the ability to analyze samples in depth, in terms of numbers, is very powerful. However, in terms of identification at the species level, the quality of the data is more questionable. When applied to ecological studies, the ability of this method to differentiate between taxa stops at the genus level or with some groups of species. Nevertheless, this technique has already started being used for microbial analyses in winemaking. However, despite the number of authors that have already published articles using this technique, it is still far from being a suitable and user-friendly technique. The first references of NGS studies in winemaking were performed by Setati et al. (2012) and Bokulich et al. (2013).

One of the challenges that still remains to be solved is the identification of microorganisms below the species level (at the strain level, for instance), especially for non-*Saccharomyces* species. This ability is crucial in wine making, as some wine characteristics are dependent on the strain used (Fleet, 1993; Ribéreau-Gayon et al., 2006a). To date, the identification of microorganisms at this level requires previous isolation in culture media. Furthermore, monitoring the strain during fermentation can only be performed using tagged systems (labeling the strain with a fluorescent tag or designing specific primers for use in qPCR).

17.3 GRAPE MICROBIOME

Grape surfaces are able to support a wide array of epiphytes that can grow on the substrates present in grape exudates. These exudates are typically rich in saccharides, as the grape berry is rich in sugars. As sugar levels increase with ripening and the berry cuticle becomes softer (Barata et al., 2012), the population of microorganisms that can be harvested increases at the end of ripening. The microbial population on intact, healthy grapes can range from 10^2 cfu/berry on green grapes to 10^5 cfu/berry at the end of ripening (Renouf et al., 2005). Grape berries can be damaged by several factors (insects, birds, bursting, etc.), which allows many microorganisms to easily proliferate on the surface, as the availability of nutrients will be higher, reaching values from 10^6 to 10^8 cfu/berry (Barata et al., 2012). Grape must derived from healthy grapes can contain between 10^4 and 10^6 cfu/ml of yeasts (Beltran et al., 2002; Padilla et al., 2016).

Yeasts present on the grape surfaces are generally ascomycetous yeasts, although some yeast-like ascomycetous molds and basidiomycetous fungi are also present (Ribéreau-Gayon et al., 2006a). *Aerobasidium pullulans* is the most frequent yeast-like ascomycetous mold present on grapes. The genera *Cryptococcus*, *Rhodotorula*, and *Sporodiobolus* are the most common basidiomycetous yeasts. However, ascomycetous yeasts are by far the greatest colonizers of the grape surface, with the genera *Hanseniaspora* (anaform *Kloeckera*), *Candida* (some species have been renamed *Starmerella*), *Metschnikowia*, *Debaryomyces*, *Issatchenkia*, and *Pichia* being the ones that are more universal and commonly found (Barata et al., 2012; Fleet, 1993; Ribéreau-Gayon et al., 2006a). However, the distribution between these genera and the species present in different grape growing areas around the world can be very diverse and may exhibit some specificity. The presence and distribution of the different yeast species can change with a variety of environmental aspects, such as vineyard treatments, climatic conditions, and geographic location. Some biotic and grapevine-related factors (grape variety, age, vintage, etc.) are also believed to alter the distribution of species on grapes (Barata et al., 2012; Kioroglou et al., 2019). Traditionally, some species have been detected almost everywhere in the world, such as *Hanseniaspora uvarum*, *Metschnikowia pulcherrima*, *Issatchenkia terricola*, and *Issatchenkia orientalis*. However, although NGS technologies are extending this list, it should be considered that with previous classical methodologies, only major genera and species were detected. This is the case for *Saccharomyces cerevisiae*, the primary wine yeast, which was barely detected on grapes due to its limited abundance in this environment. However, in extremely healthy grapes, it was detected using classical methods due to the low population of other yeasts (Beltran et al., 2002). Currently, in NGS studies, *S. cerevisiae* is easily detected, although its percentage is typically extremely low, in agreement with previous results obtained using classical methodologies (Bokulich et al., 2014; Kioroglou et al., 2019).

Furthermore, NGS studies have confirmed some previous theories regarding the specificity of the microbial population linked to a given area or variety in what could be considered a "microbial terroir" (Bokulich et al., 2014; Gilbert et al., 2014; Setati et al., 2015; reviewed by Belda et al., 2017). However, these authors note that this concept is strongly affected by other characteristics, such as the type of soil and the grape variety, which are stable (cannot be changed easily). Furthermore, these studies also evaluated the effects of other characteristics that are more variable, such as climate or crop handling. Thus, this microbial terroir should not be considered stable over time but rather continues to change as the climate and vineyard handling varies from year to year (Pretorius et al., 1999).

17.4 SUCCESSION OF MICROORGANISMS DURING ALCOHOLIC FERMENTATION

The yeasts present on the grape surface have a great opportunity to proliferate in grape must, a complex medium in which all the primary compounds necessary to support microbial growth are present. Grape must contains large amounts of easily metabolizable sugars (mostly glucose and fructose, typically in equimolar concentrations and variable from 180 to 260 g/l in normal musts), nitrogen sources (such as ammonium salts, amino acids, and small peptides), and organic acids (tartaric, citric, and malic acids) as well as variable amounts of vitamins, minerals, polyphenols, etc. Importantly, grape must also have diverse secondary metabolites that are either volatiles or can be transformed into volatiles and contribute to the aroma (Fleet, 1993). The volatiles present in the grape must and which remain in the wine are considered varietal aromas (characteristic of the grape variety), while the others that are developed by the yeasts are fermentative aromas (can be directly related to the yeast performing alcoholic fermentation) (Fleet, 1993). Although grape must could be considered an excellent culture medium, in reality, it is a highly selective medium in which only a few microorganisms are able to proliferate. For instance, the high sugar concentration in grape must results in a high osmotic pressure and reduced water activity, limiting the proliferation of many molds. Furthermore, the high organic acid content of grape must results in a highly acidic environment, with pH values typically between 3 and 4, which is not permissive for the growth of many microorganisms. Finally, there is a clear imbalance between carbon and nitrogen sources in grape must. Nitrogen becomes a limiting nutrient, as it is required for biomass build-up, and a low nitrogen content may result in a low yeast biomass and stuck or sluggish fermentations (Mas et al., 2014).

Under these conditions, yeasts and some bacteria are able to develop and grow. The differences among microorganisms result in a succession of different strains and species, resulting in a dramatic transformation of the culture medium. The most effective yeasts are those able to grow the fastest and take up as many nutrients as possible in a short time. *Saccharomyces*, specifically *S. cerevisiae*, is one such organism that has adapted its metabolic machinery to be able to grow fast in this changing medium and is the primary player in this transformation. This adaptation, which has taken several millennia, is considered a "domestication" process due to the human-driven vinification process (Fay and Benavides, 2005). Compared to these *Saccharomyces* yeasts, other yeasts are generically referred to as "non-*Saccharomyces*," a classification without any taxonomic significance but with high practical relevance (Fleet, 1993). Thus, non-*Saccharomyces* yeasts generally have a poor fermentation capacity (compared to *Saccharomyces*) and proceed with slower fermentation rates, whereas *Saccharomyces* exhibits a fast and effective fermentation rate that allows for wine production in a short time. However, as these *Saccharomyces* yeasts are not the most abundant on the grapes, they must overcome challenges to dominate the fermentation by outgrowing the non-*Saccharomyces* yeasts (Fleet, 1993).

The first step involved in this increase in *Saccharomyces* is the entrance to the cellar. Cellar machinery (presses, pumps, and fermentation vats) and the cellar itself (walls, surfaces, and floors) are typically contaminated by different yeast, but the presence of *Saccharomyces* yeast in this environment is much higher than observed in the natural habitat, although many other microorganisms unrelated to wine production (molds, yeast-like fungi, yeast with oxidative metabolism, etc.) are also present (Bokulich et al., 2013). The existence of this cellar resident microbiota increases as the operations in the cellar increase, both over time (years) and during the same harvest season. During the harvest season, cross-contamination among fermentation vats is extremely common due to the equipment being shared between fermentations (pumps and hoses, for instance) and to other biotic factors (such as insects and winery personnel) that can act as vectors of the microorganisms. From harvest to harvest, some microorganisms are also able to survive on cellar surfaces and equipment, allowing them to become prominent the following season (Beltran et al., 2002).

Although this concept is currently changing, winemakers have generally considered *Saccharomyces* yeasts are appropriate for making wine and that non-*Saccharomyces* yeasts are a source of problems (stuck and sluggish fermentations, wine spoilage, etc.). To prevent the growth of non-*Saccharomyces* yeasts, the use of different forms of sulphur dioxide (either as gas or as solubilized sulfites) is a common practice in cellars. This practice is intended to prevent the oxidation of musts as well as the growth of other undesirable microorganisms (Ribéreau-Gayon et al., 2006a). A step further has been the development of appropriate starter cultures. The use of starter cultures is a typical practice in fermentation technology to control the process (Fleet, 1993). The use of starter cultures can be the simple practice of using an already fermenting material to start a new fermentation or involve a more complex system with a selected microorganism that will provide a distinctive input to the fermented product. In winemaking, these selected starter cultures have been developed with different species of *Saccharomyces*, mostly *S. cerevisiae*. A big step for the use of starter cultures in wine production has been the development of active dry wine yeast (ADWY), which is a ready-to-use starter culture (Fleet, 1993). Dehydrated yeasts are rehydrated in warm water, sometimes with a small proportion of must, and in 30 minutes are ready to be added to new must and quickly and steadily initiate the fermentation process (Figure 17.2). Yeasts used as ADWY have been selected after the results of different tests and the application of different criteria for their selection. More than 200 different commercial ADWY are currently available and have the primary characteristic of initiating good fermentation and the capacity to finish the fermentation process. Furthermore, several other criteria for their selection can include resistance to higher concentrations of sugars, fermentation potential at different temperatures, and good aromatic expression (Fleet, 1993).

A rapid initiation of fermentation also generates an anaerobic environment, as fermentation transforms glucose or fructose into ethanol and carbon dioxide. The production of carbon dioxide makes the fermentation "tumultuous" and displaces all of the dissolved oxygen in the must. Thus, although most fermentations are performed in the open air, the process is essentially anaerobic. The production of ethanol is also an additional challenge for many microorganisms, as it is toxic, and only a few microorganisms can withstand and grow at the concentrations of ethanol that develop during the fermentation process (Ribéreau-Gayon et al., 2006a).

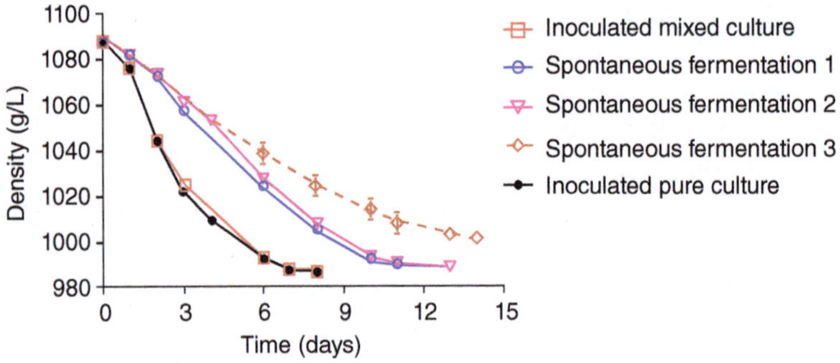

FIGURE 17.2 Monitoring of alcoholic fermentation by density. Comparison between inoculated and spontaneous fermentations.

Thus, the fermentation process can be seen as an obstacle course race in which the less fit microorganisms are continuously eliminated from competition. The process itself has selected microorganisms that are the most fit, and at the end of the race (the fermentation process), the yeast population will have been essentially reduced to a single species, *Saccharomyces cerevisiae* (Fleet, 1993). This succession of microorganisms starts with an overwhelming abundance of non-*Saccharomyces* that are progressively substituted by *Saccharomyces*. However, the succession of microorganisms does not stop here, as there are still some compounds of interest that can be transformed by different microorganisms. For instance, lactic acid bacteria can proceed with malolactic fermentation. This transformation is sometimes a positive process to reduce the acidity of wines and occurs due to the conversion of malic acid, a dicarboxylic acid, into lactic acid, a monocarboxylic acid. Another interesting characteristic is the transformation of ethanol into acetic acid, or wine into vinegar, which is performed by acetic acid bacteria (Ribéreau-Gayon et al., 2006a). If the transformation is only partial, it has to be considered spoilage, as it gives a vinegary taint to the wine. However, vinegar is a wine byproduct that is of great interest in the food industry as a natural preservative, as well as in gastronomy, as it can provide a distinctive sour smell and taste to many dishes. Thus, in this succession of microorganisms, wine is not necessarily the end of the race but an intermediate goal for some microorganisms.

17.5 MICROBIAL INTERACTIONS DURING ALCOHOLIC FERMENTATION

Importantly, the abovementioned hurdles that appear during fermentation are sometimes not enough to ensure the appropriate dominance of the desired species. It is generally agreed that good alcoholic fermentation requires the dominance of *Saccharomyces*. This yeast has unique characteristics, such as a high capacity to ferment sugars quickly and efficiently and the ability to produce and endure large concentrations of ethanol, which increases its performance in the wine-making environment (Ribéreau-Gayon et al., 2006a).

Although the domestication of this yeast has resulted in its adaptation to perform and succeed during alcoholic fermentation, under some conditions, fermentation dynamics allow for the survival of other non-*Saccharomyces* species. Some studies have reported the survival of *Hanseniaspora* sp. at low-temperature fermentations (Andorrà et al., 2010b), of *Zygosaccharomyces* sp. in botrytized and rotten grapes (Lleixá et al., 2018; Nisiotou et al., 2007) and of the codominance of *Starmerella* (previously classified as *Candida*) with *S. cerevisiae* during the late stages or end of some fermentations (Clemente-Jiménez et al., 2004; Llauradó et al., 2002). It should be emphasized that most of these observations on the tolerance and survival of non-*Saccharomyces* yeast during alcoholic fermentation have been performed with traditional culture-dependent methodologies.

The use of newly available methodologies has revealed that the competition between yeasts in the wine environment is more complex than the survival of some species during changes in the medium (mostly an increase in ethanol concentration). Other interactions, including cell-to-cell contact (Nissen and Arneborg, 2003), nutrient limitation (Wang et al., 2016) and the secretion of antimicrobial peptides that are different from the well-known killer peptides (Albergaria et al., 2010) have been recently described. These interactions will likely induce the VBNC state in non-*Saccharomyces* yeasts, which will prevent them from growing, including on plates (Branco et al., 2015; Wang et al., 2016). It is necessary to remember that the VBNC state is also induced by SO_2, and the response could be similar

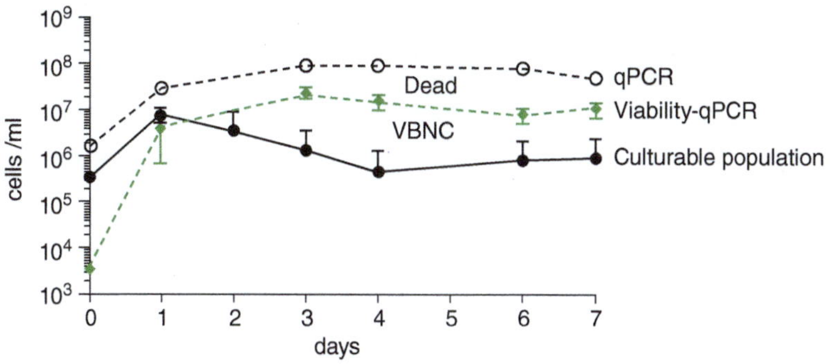

FIGURE 17.3 Monitoring of microbial populations during alcoholic fermentation. Analysis of the same microbial populations using three methods: plate culture (fully viable and culturable microbial population), viability qPCR (measurement of fully viable and culturable plus VBNC microbes), and whole population (dead and viable microbial population, qPCR).

(Millet and Lonvaud-Funel, 2000). However, it has been shown that in the absence of SO_2, *S. cerevisiae* can also enter the VBNC state in the presence of other non-*Saccharomyces* yeasts (Navarro et al., 2020) in what could be considered a generalized biological warfare among competing yeasts. It must be emphasized that because yeasts in this VBNC state are metabolically active their effect on wines can still be noticeable. Indeed, between fully viable and dead cells, there are a range of cells in different transient states. These transient states may include old cells that still retain the capacity to grow under optimal conditions, old cells that have completely lost the capacity to grow but which are still fully functional, and cells that are entering the autolytic process. All these types of cells are typically present during winemaking, which is a crucial aspect of why the use of culture-independent methods may be of great interest in addition to traditional protocols for microbiological control in cellars. This issue is especially important for wines that will undergo aging, when these undesirable yeasts, although with low metabolic activity, have time to spoil the final wine (Mas et al., 2021). Thus, the determination of cells in different states (viable, VBNC, and dead cells) is crucial and can be determined by their combined quantification on culture plates and some culture-independent methods such as qPCR, alone or combined with DNA-binding dyes (viability qPCR) (Figure 17.3).

17.6 PRODUCTION OF AROMAS AND WINE QUALITY

Microbiological control is necessary to produce wines with properties that define the quality and reproducibility of the wines. More than 200 compounds have been detected in wines, including three primary components (water, ethanol, and glycerol) that are present in all wines and account for 98–99% of the wine composition. The remaining compounds produce the distinctiveness of a specific wine. Among these characteristic components, some are varietal components (from the specific grape variety), while others are fermentative components (produced during fermentation, either de novo or by the modification of precursors from the grapes) (Morata et al., 2019) (Figure 17.4).

Different wine components can be identified using analytical technologies, with the most common being different types of chromatography, including liquid or gas chromatography, which are often coupled with very sensitive methods of detection. These chromatographic methods separate the different molecules in a given matrix (such as wine or must), whereas the detection methods allow for the identification of the specific molecule that may be associated with a given characteristic (Figure 17.5). The association of specific characteristics with molecules has been determined using pure compounds. Wine quality is mostly defined by its aromatic composition, which is essentially determined by specific compounds that we can be associated with known and pleasant products (fruits, flowers, spices, etc.) (Ribéreau-Gayon et al., 2006b).

The production of fermentative aromas is species- and strain-dependent and is primarily attributable to the production of esters, either acetate esters or fatty acid ethyl esters. The usual organoleptic descriptors for these compounds are fruity aromas, many of which are easily associated with specific fruits: banana (isoamyl acetate), pear (isobutyl acetate), pineapple (ethyl butyrate), apple (ethyl hexanoate), etc. Some additional esters are more associated with flowery descriptors, such as rose (2-phenyl acetate) or jasmine (benzyl acetate) (Morata et al., 2019). Thus, through the use of different

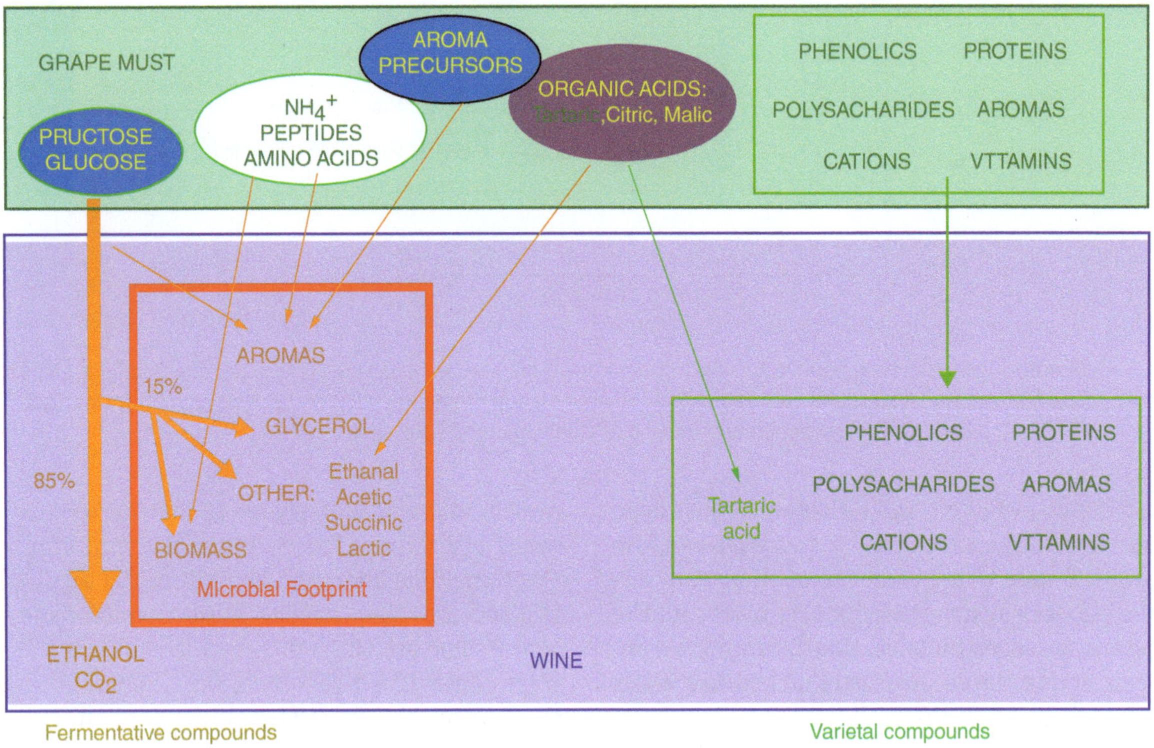

FIGURE 17.4 Changes in chemical composition during alcoholic fermentation.

descriptors (Figure 17.6), we can consider that microorganisms leave their distinct "microbial footprint."

When the native grape microbiota is allowed to freely proliferate, spontaneous fermentation occurs in which non-*Saccharomyces* yeasts will become dominant during a large part of the fermentation. The result of such fermentation is unpredictable, as there is a wide variety of yeasts, and the dynamics will proceed in an uncontrolled manner (Fleet, 1993). However, not all the characteristics provided by non-*Saccharomyces* yeasts are negative, and there is a growing interest in these yeasts to incorporate them into the wine-making process (Jolly et al., 2014).

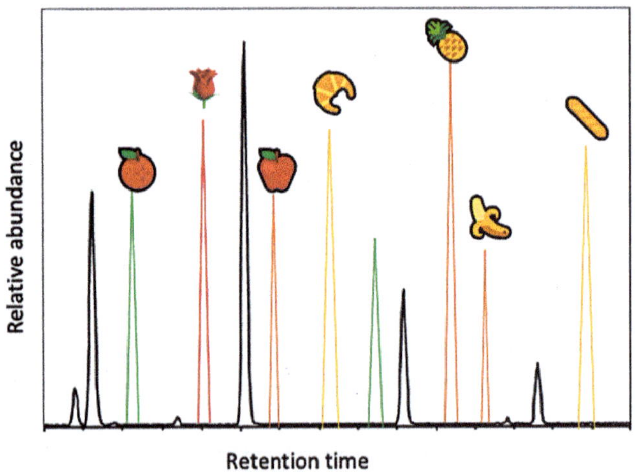

FIGURE 17.5 Analysis of volatile compounds by gas chromatography coupled with mass spectrometry. Chromatographic peaks are identified with an illustration of the associated aroma.

The selection of non-*Saccharomyces* strains is based on their effects on wine quality, which can be understood as the production of pleasant aromas or the removal of compounds that will reduce the quality of the wine. One of the most popular commercial non-*Saccharomyces* species is *Torulaspora delbrueckii*, which can reduce volatile acidity (Renault et al., 2009) but has also been used for wine production from botrytized grapes (Bely et al., 2008). Another yeast species available for commercial use is *Metschnikowia pulcherrima*, which in white wines increases the levels of volatile compounds, especially those derived from thiols and terpenes (González-Royo et al., 2015). *Pichia kluyveri* is also available and is mostly recommended for the generation of thiol-containing aromas. However, not all non-*Saccharomyces* are recommended to increase the aromatic profile of wines. For instance, *Lachancea thermotolerans* is associated with the production of glycerol and lactic acid (Gobbi et al., 2013), which can increase the mouthfeel characteristics of wines.

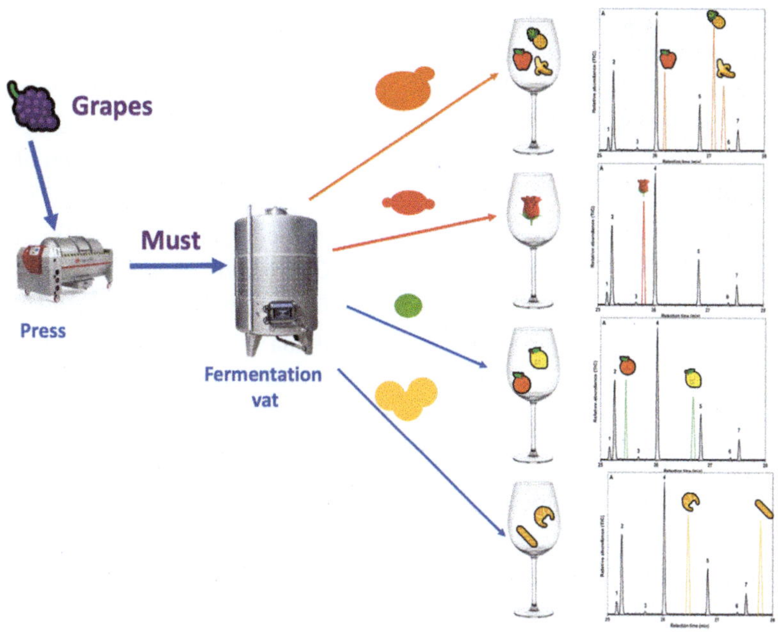

FIGURE 17.6 Development of alcoholic fermentation with different yeast species. The most relevant characteristics developed in each final wine are indicated in the chromatograms.

Due to the growing acceptance of these non-*Saccharomyces* yeast species, many are currently being tested for their use in winemaking, including some species of *Starmerellla* (previously included in the *Candida* genus) and *Hanseniaspora*. The primary characteristics sought by using *Starmerella* species are a reduction in volatile acidity, fructophilic characteristics (which would facilitate the end of fermentation) or the production of glycerol (Englezos et al., 2015). Among *Hanseniaspora* species, the most promising is *H. vineae*, which has been used to produce commercial wines in Uruguay (Martín et al., 2016) and Spain (Lleixà et al., 2016b). Some strains of this species are able to ferment up to an 11% or 12% alcohol content, having a notable aromatic impact. Moreover, this species has been described as producing large quantities of 2-phenyl acetate (rose aroma) and benzyl acetate (jasmine aroma), resulting in highly aromatic and flowery wines (Giorello et al., 2019).

However, the oenological capacity of most of these non-*Saccharomyces* species is very limited, as they have low fermentation rates and performances and are sensitive to SO_2 (Ciani et al., 2010), leading to their use in combination with *S. cerevisiae*. Due to the high fermentative takeover of *S. cerevisiae*, coinoculation (simultaneously) with non-*Saccharomyces* yeasts is not recommended, as the non-*Saccharomyces* "character" would not be developed. Thus, sequential inoculation is highly advisable, first with non-*Saccharomyces* and later with *S. cerevisiae*. Furthermore, to develop the desired characteristics, biomass growth is needed, which would be jeopardized by high concentrations of SO_2 or high loads of other yeast populations (which may happen in harvests with compromised health status). The inoculation of *S. cerevisiae* has to be performed 24–48 h after the inoculation of the non-*Saccharomyces* yeasts, as further delays may compromise the development of *S. cerevisiae* and lead to sluggish or stuck fermentations (Lleixà et al., 2016a).

There are two additional factors that are relevant to the use of non-*Saccharomyces* yeast in wine making. On the one hand, the extensive practice of inoculation with *S. cerevisiae* has brought some uniformity to the secondary characteristics of wine. Although there are many commercial ADWY products available for this species, there are some "best sellers" that have resulted in a level of uniformity (Mas et al., 2016). Although the selection of some "local" *S. cerevisiae* strains has been a means of combating this uniformity, the appearance of non-*Saccharomyces* yeasts has increased the biodiversity and complexity of wines (Jolly et al., 2014). On the other hand, the lower ethanol yields and, in some cases, the respiratory characteristics of some non-*Saccharomyces* (which reduces the formation of ethanol from sugars due to their respiratory conversion into CO_2 and water) has been proposed as a system to reduce the ethanol content in wines (Gonzalez et al., 2013). This aspect is highly relevant with respect to the present climatic change scenario, as global warming has the effect of increasing the sugar content in fruits, which will lead to higher ethanol concentrations. Several different strategies have been tested to date to decrease the ethanol content of wines, including the use of different non-*Saccharomyces* species, strains of *S. cerevisiae* or combinations of both approaches (Contreras et al., 2014; Gonzalez et al., 2013; Zhu et al., 2020).

17.7 CONCLUSIONS

Wine production has been performed during most of the recent history of humanity. Yeasts are the primary players involved in the transformation of grape must into wine, and the need for proper control of fermentation requires good microbiological knowledge and appropriate tools to monitor the process. Classical microbiology has provided tools to develop consolidated knowledge regarding wine microbiology that are currently available, but DNA-based molecular methods, involving both dependent- and independent-culture techniques, have improved the control and monitoring of fermentation. Moreover, the recent development of new technologies, especially those derived from NGS, will provide additional and powerful methods to increase the knowledge and capacity for microbial control. A huge step in controlling alcoholic fermentation and improving wine quality has been the development of cellar-friendly *Saccharomyces cerevisiae* starter cultures. However, the development of new starter cultures based on non-*Saccharomyces* yeasts seeks to ensure an enrichment of wine characteristics and increase wine quality.

REFERENCES

1. Albergaria H, Francisco D, Gori K, Arneborg N, Gírio F. 2010. *Saccharomyces cerevisiae* CCMI 885 secretes peptides that inhibit the growth of some non-*Saccharomyces* wine-related strains. *Applied Microbiology and Biotechnology* 86: 965–972.
2. Andorrà I, Esteve-Zarzoso B, Guillamón JM, Mas A. 2010a. Determination of viable wine yeast using DNA binding dyes and quantitative PCR. *International Journal of Food Microbiology* 144: 257–262.
3. Andorrà I, Landi S, Mas A, Esteve-Zarzoso B, Guillamón JM. 2010b. Effect of fermentation temperature on microbial population evolution using culture-independent and dependent techniques. *Food Research International* 43: 773–779.
4. Barata A, Malfeito-Ferreira M, Loureiro V. 2012. Changes in sour rotten grape berry microbiota during ripening and wine fermentation. *International Journal of Food Microbiology* 154: 152–161.
5. Barnett JA, Payne RW, Yarrow D. 2000. *Yeasts: Characteristics and Identification* (3rd ed.). Cambridge University Press Cambridge.
6. Belda I, Zarraonaindia I, Perisin M, Palacios A, Acedo A. 2017. From Vineyard soil to wine fermentation: Microbiome approximations to explain the "terroir" concept. *Frontiers in Microbiology* 8: 821.
7. Beltran G, Torija MJ, Novo M, Ferrer N, Poblet M, et al. 2002. Analysis of yeast populations during alcoholic fermentation: A six year follow-up study. *Systematic and Applied Microbiology* 25: 287–293.
8. Bely M, Stoeckle P, Masneuf-Pomarède I, Dubourdieu D. 2008. Impact of mixed *Torulaspora delbrueckii-Saccharomyces cerevisiae* culture on high-sugar fermentation. *International Journal of Food Microbiology* 122: 312–320.
9. Bokulich NA, Ohta M, Richardson PM, Mills DA. 2013. Monitoring seasonal changes in winery-resident microbiota. *PLoS ONE* 8: e66437.
10. Bokulich NA, Thorngate JA, Richardson PM, Mills DA. 2014. Microbial biogeography of wine grapes in conditioned by cultivar, vintage, and climate. *Proceedings of the National Academy of Sciences* 111: 139–148.
11. Branco P, Viana T, Albergaria H, Arneborg N. 2015. Antimicrobial peptides (AMPs) produced by *Saccharomyces cerevisiae* induce alterations in the intracellular pH, membrane permeability and culturability of *Hanseniaspora guilliermondii* cells. *International Journal of Food Microbiology* 205: 112–118.
12. Ciani M, Comitini F, Mannazzu I, Domizio P. 2010. Controlled mixed culture fermentation: A new perspective on the use of non-*Saccharomyces* yeasts in winemaking. *FEMS Yeast Research* 10: 123–133.
13. Clemente-Jiménez J, Mingorance-Cazorla L, Martínez-Rodríguez S, Heras-Vázquez FJL, Rodríguez-Vico E. 2004. Molecular characterization and oenological properties of wine yeasts isolated during spontaneous fermentation of six varieties of grape must. *Food Microbiology* 21: 149–155.
14. Cocolin L, Mills DA. 2003. Wine yeast inhibition by sulfur dioxide: A comparison of culture-dependent and independent methods. *American Journal of Enology and Viticulture* 54: 125–130.

15. Contreras A, Hidalgo C, Henschke PA, Chambers PJ, Curtin C, Varela C. 2014. Evaluation of non *Saccharomyces* yeasts for the reduction of alcohol content in wine. *Applied and Environmental Microbiology* 80: 1670–1678.

16. Englezos V, Rantsiou K, Torchio F, Rolle L, Gerbi V, Cocolin L. 2015. Exploitation of the non-*Saccharomyces* yeast *Starmerella bacillaris* (synonym *Candida zemplinina*) in wine fermentation: Physiological and molecular characterizations. *International Journal of Food Microbiology* 199: 33–40.

17. Fay JC, Benavides JA. 2005. Evidence for Domesticated and Wild populations of. *Saccharomyces Cerevisiae. PLoS Genetics* 1: e5.

18. Fleet GH. 1993. *Wine Microbiology and Biotechnology*. Harwood Academic Publishers, Chur, Switzerland.

19. Gilbert JA, van der Lelie D, Zarraonaindia I. 2014. Microbial *terroir* for wine grapes. *Proceedings of the National Academy of Sciences* 111: 5–6.

20. Giorello F, Valera MJ, Martin V, Parada A, Salzman V et al. 2019. Genomic and transcriptomic basis of *Hanseniaspora vineae*'s impact on flavor diversity and wine quality. *Applied and Environmental Microbiology* 85: e01959-18.

21. Gobbi M, Comitini F, Domizio P, Romani C, Lencioni L, Mannazzu I, Ciani M. 2013. *Lachancea thermotolerans* and *Saccharomyces cerevisiae* in simultaneous and sequential co-fermentation: A strategy to enhance acidity and improve the overall quality of wine. *Food Microbiology* 33: 271–281.

22. Gonzalez R, Quiros M, Morales P. 2013. Yeast respiration of sugars by non-*Saccharomyces* yeast species: A promising and barely explored approach to lowering alcohol content of wines. *Trends in Food Science and Technology* 29: 55–61.

23. González-Royo E, Pascual O, Kontoudakis N, Esteruelas M, Esteve-Zarzoso B, et al. 2015. Oenological consequences of sequential inoculation with non-*Saccharomyces* yeasts (*Torulaspora delbrueckii* or *Metschnikowia pulcherrima*) and *Saccharomyces cerevisiae* in base wine for sparkling wine production. *European Food Research and Technology* 240: 999–1012.

24. Guillamón JM, Sabate J, Barrio E, Cano J, Querol A. 1998. Rapid identification of wine yeast species based on RFLP analysis of the ribosomal internal transcribed spacer (ITS) region. *Archives of Microbiology* 169: 387–392.

25. Hierro N, Esteve-Zarzoso B, González A, Mas A, Guillamón JM. 2006. Real-time quantitative PCR (QPCR) and reverse transcription-QPCR (RT- QPCR) for the detection and enumeration of total yeasts in wine. *Applied and Environmental Microbiology* 72: 7148–7155.

26. Jolly NP, Varela C, Pretorius IS. 2014. Not your ordinary yeast: Non-*Saccharomyces* yeasts in wine production uncovered. *FEMS Yeast Research* 14: 215–237.

27. Kioroglou D, Kraeva-Deloire E, Schmidtke L, Mas A, Portillo M. 2019. Geographical origin has a greater impact on grape berry fungal community than grape variety and maturation state. *Microorganisms* 7: 669.

28. König H, Unden G, Fröhlich J. 2017. *Biology of Microorganisms on Grapes, in Must and in Wine*. Springer-Verlag, Berlin Heidelberg.

29. Kurtzman CP, Robnett CJ. 1998. Identification and phylogeny of ascomycetous yeast from analysis of nuclear large subunit 26S ribosomal DNA partial sequences. *Antonie Van Leeuwenhoek* 73: 331–371.

30. Llauradó J, Rozés N, Bobet R, Mas A, Constantí M. 2002. Low temperature alcoholic fermentations in high sugar concentration grapemusts. *Journal of Food Science* 67: 268–273.

31. Lleixá J, Kioroglou D, Mas A, Portillo MC. 2018. Microbiome dynamics during spontaneous fermentations of sound grapes in comparison with sour rot and *Botrytis* infected grapes. *International Journal of Food Microbiology* 281: 36–46.

32. Lleixà J, Manzano M, Mas A, Portillo MC. 2016a. *Saccharomyces* and non-*Saccharomyces* Competition during Microvinification under different sugar and Nitrogen Conditions. *Frontiers in Microbiology* 7: 1959.

33. Lleixà J, Martín V, Portillo MC, Carrau F, Beltran G, Mas A. 2016b. Comparison of fermentation and wines produced by inoculation of *Hanseniaspora vineae* and. *Saccharomyces cerevisiae. Frontiers in Microbiology* 7: 338.

34. Martín V, Mas A, Carrau F, Dellacasa E, Boido E. 2016. Effect of yeast assimilable nitrogen on the synthesis of phenolic aroma compounds by *Hanseniaspora vineae* strains. *Yeast* 33: 323–328.

35. Mas A. 2019. Wine. In: Schmidt M Eds., *Encyclopedia of Microbiology*, 4th ed. Academic Press, pp. 598–603.

36. Mas A, Guillamón JM, Torija MJ, Beltran G, Cerezo AB et al. 2014. Bioactive compounds derived from yeast metabolism of aromatic amino acids during alcoholic fermentation. BioMed Research International Article ID898045.

37. Mas A, Padilla B, Esteve-Zarzoso B, Beltran G, Reguant C, Bordons A. 2016. Taking advantage of natural bio-

diversity for Wine making: The WILDWINE project. *Agriculture and Agricultural Science Procedia* 8: 4–9.

38. Mas A, Torija MJ, Beltran G, Sengun I. 2021. Winemaking: Microbiology. Chapter 9. In: *Wine Making: Basics and Applied Aspects*, Joshi VK, Ray RC (Eds.), CRC Press, Boca Raton. CA, US.
39. Millet V, Lonvaud-Funel A. 2000. The viable but non-culturable state of wine micro-organisms during storage. *Letters in Applied Microbiology* 30: 136–141.
40. Montrocher R, Verner MC, Briolay J, Gautier C, Marmeisse R. 1998. Phylogenetic analysis of the *Saccharomyces cerevisiae* group based on polymorphisms of rDNA spacer sequences. *International Journal of Systematic and Evolutionary Microbiology* 48: 295–303.
41. Morata A, Escott C, Bañuelos MA, Loira I, Fresno J, González C, Suárez-Lepe JA. 2019. Contribution of Non-*Saccharomyces* yeasts to wine freshness. A review. *Biomolecules* 10: 34.
42. Navarro Y, Torija MJ, Mas A, Beltran G. 2020. Viability-PCR allows monitoring yeast population dynamics in mixed fermentations including viable but non-culturable Yeast. *Foods* 9: 1373.
43. Nisiotou AA, Spiropoulos AE, Nychas GE. 2007. Yeast community structures and dynamics in healthy and *Botrytis*-affected grape must fermentations. *Applied and Environmental Microbiology* 73: 6705–6713.
44. Nissen P, Arneborg N. 2003. Characterization of early deaths of non- *Saccharomyces* yeasts in mixed cultures with *Saccharomyces cerevisiae*. *Archives of Microbiology* 180: 257–263.
45. Padilla B, García-Fernández D, González B, Izidoro-Pacheco I, Esteve-Zarzoso B, et al. 2016. Yeast biodiversity from DOQ Priorat uninoculated fermentations. *Frontiers in Microbiology* 7: 930.
46. Pretorius IS, van der Westhuizen TJ, Augustyn OPH. 1999. Yeast biodiversity in vineyards and wineries and its importance to the South African wine industry. *South African Journal of Enology and Viticulture* 20: 61–70.
47. Renault P, Miot-Sertier C, Marullo P, Hernandez-Orte P, Lagarrigue L, Lonvaud-Funel A, Bely M. 2009. Genetic characterization and phenotypic variability in *Torulaspora delbrueckii* species: Potential applications in the wine industry. *International Journal of Food Microbiology* 134: 201–210.
48. Renouf V, Claisse O, Lonvaud-Funel A. 2005. Understanding the microbial ecosystem on the grape berry surface through numeration and identification of yeast and bacteria. *Australian Journal of Grape and Wine Research* 11: 316–327.
49. Ribéreau-Gayon P, Dubourdieu D, Doneche B, Lonvaud A. 2006a. *Handbook of Enology vol. 1. The Microbiology of Wine and Vinifications* (2nd ed.). John Wiley & Sons Ltd, West Sussex, England.
50. Ribéreau-Gayon P, Glories Y, Maujean A, Dubourdieu D. 2006b. *Handbook of Enology vol. 2. The Chemistry of Wine Stabilization and Treatments* (2nd ed.). John Wiley & Sons Ltd, West Sussex, England.
51. Setati ME, Jacobson D, Andong UC, Bauer F. 2012. The Vineyard Yeast Microbiome, a mixed model Microbial map. *PLoS ONE* 7: e52609.
52. Setati ME, Jacobson D, Bauer F. 2015. Sequence-based analysis of the *Vitis vinifera* L. cv Cabernet Sauvignon grape must mycobiome in three south African vineyards employing distinct agronomic systems. *Frontiers in Microbiology* 6: 1358.
53. Torija MJ, Rozès N, Poblet M, Guillamón JM, Mas A. 2001. Yeast population dynamics in spontaneous fermentations: Comparison between two different wine-producing areas over a period of three years. *Antonie van Leeuwenhoek* 79: 345–352.
54. Wang C, Mas A, Esteve-Zarzoso B. 2016. The Interaction between *Saccharomyces cerevisiae* and non-*Saccharomyces* yeast during alcoholic fermentation is species and strain specific. *Frontiers in Microbiology* 7: 502.
55. Zhu X, Navarro Y, Mas A, Torija MJ, Beltran G. 2020. A rapid method for selecting non-*Saccharomyces* strains with low ethanol yield. *Microorganisms* 8: 658.

CHAPTER 18

Lambic Beer, A Unique Blend of Tradition and Good Microorganisms

Jonas De Roos and Luc De Vuyst

Research Group of Industrial Microbiology and Food Biotechnology, Faculty of Sciences and Bioengineering Sciences, Vrije Universiteit Brussel, B-1050 Brussels, Belgium

18.1 INTRODUCTION

Beers are considered to be one of the oldest known fermentation processes, as written sources of beer-like beverages date back to 2800 BCE, and historians even suspect an earlier consumption far back to 9000 BCE (Anderson, 2003, 2006; Pires and Branyik, 2015). Historically, beer played a pivotal role in ensuring a safe water supply for rural populations. Since then, brewing has evolved into an enormous industry and still remains the most popular alcoholic beverage to this day (Briggs et al., 2004; Howard, 2014). The basic ingredients of beer are water, malt (germinated and kilned barley grains), and hops (*Humulus lupulus* L.), which has been stipulated in the Bavarian *Reinheitsgebot* from 1516. As modern beer production is made with a steered fermentation step, the *Reinheitsgebot* was expanded with the addition of brewers' yeast in the German *Vorläufiges Biergesetz* of 1993. Also, many cereals other than barley, whether or not malted, are used to produce today's beers. Malting enables controlled grain germination, thereby activating several endogenous starch- and protein-degrading enzymes, ensuring a better extraction yield of nutrients that can be used during fermentation (Palmer, 2006). Generally, during fermentation, carbohydrates (glucose, maltose, and maltotriose) are converted by yeasts into mainly ethanol and carbon dioxide (Pavsler and Buiatti, 2009). Based on the type of primary fermentation process applied, a general classification can be made for the production of four different categories of beer (Figure 18.1; Briggs et al., 2004; Bokulich and Bamforth, 2013; De Roos and De Vuyst, 2019). The first two types make use of the addition of a specific yeast strain (sometimes multiple strains) to perform the fermentation step and are commonly produced throughout the world. Bottom fermentation makes use of *Saccharomyces bayanus* or *Saccharomyces pastorianus* (synonym (syn.) *Saccharomyces carlsbergensis*) to produce lager beers at low temperature in stainless steel tanks (Bokulich and Bamforth, 2013). Lager beers are widely commercially available and make up the biggest part of the global beer market. Top fermentation makes use of *Saccharomyces cerevisiae* to produce ales. Ales are typically fermented at higher temperatures than lager beers, and generally result in a beer with more fruity undertones. They are fermented in stainless steel tanks and often subsequently subjected to bottle refermentation to enhance carbonation and alcohol content (Bokulich and Bamforth, 2013).

With Belgium sharing a border with France and Germany, brewers were influenced by both countries, leading to the massive diversity of Belgian beers known today. Moreover, in Belgium, beer brewing has always been a major part of the cultural heritage. This is reflected in both the conservation of traditional brewing styles as well as the continuous strive for developments leading to new beers. Some of these traditional brewing

Good Microbes in Medicine, Food Production, Biotechnology, Bioremediation, and Agriculture, First Edition. Edited by Frans J. de Bruijn, Hauke Smidt, Luca S. Cocolin, Michael Sauer, David Dowling, and Linda Thomashow.

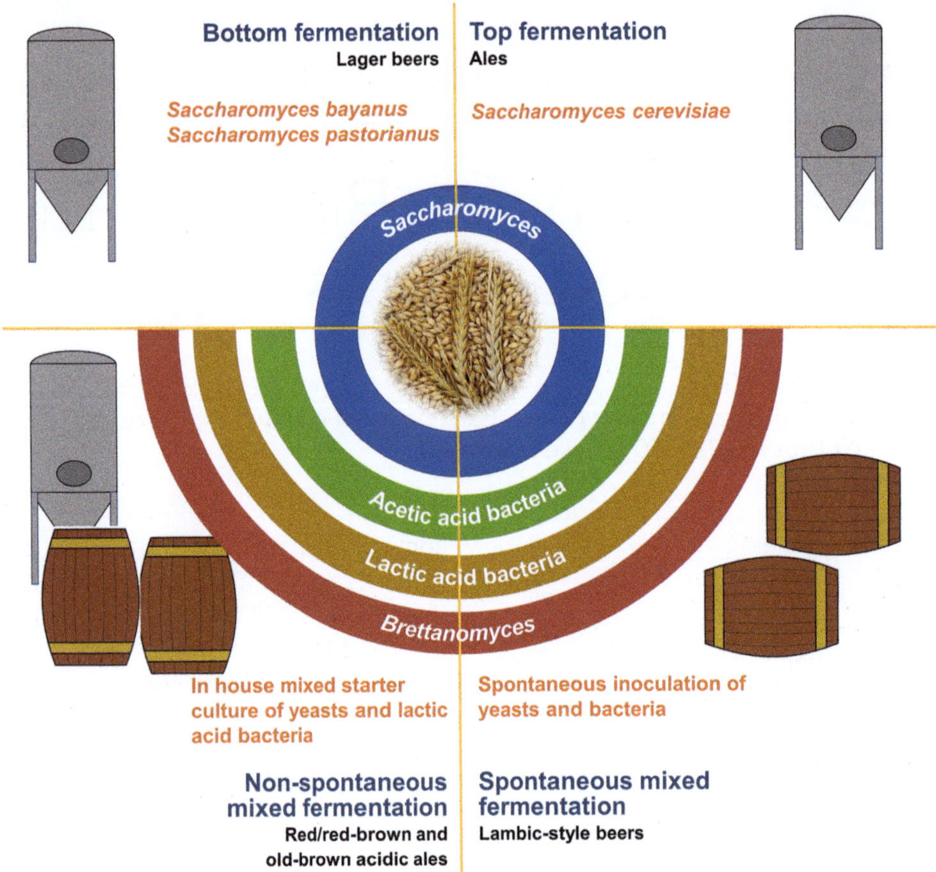

FIGURE 18.1 General classification of different beer styles based on their fermentation type. Four styles can be distinguished, namely bottom-fermented (lager) beer, top-fermented (ale) beer, non-spontaneous mixed-fermented (acidic ale) beer, and spontaneous mixed-fermented (lambic-style acidic ale) beer.

styles make up the last two beer categories based on fermentation type; they are less common but have continuously been applied in Belgium (De Roos and De Vuyst, 2019; Spitaels et al., 2017). Non-spontaneous, mixed fermentation traditionally makes use of an in-house starter culture that consists of both yeasts and lactic acid bacteria (LAB) to produce, for example, red/red–brown acidic ales (Flanders red ales; fermentation in stainless steel tanks, followed by maturation in oak barrels), and old-brown acidic ales (Flanders brown ales; fermentation and maturation in stainless steel tanks). Spontaneous, mixed fermentation makes use of bacteria and yeasts coming from the surrounding environment to produce lambic-type beers. Both mixed fermentation processes lead to acidic beers, which are becoming increasingly popular nowadays. In fact, their increasing popularity led to a global resurgence of acidic beer styles visible in emerging productions worldwide. These productions often reflect alternative fermentation styles, such as traditional sour-malted and sour-mashed productions, as well as kettle-soured, sour-fermented, and sour-matured ones (Bossaert et al., 2020; Dysvik et al., 2020). In this chapter we will focus on the microorganisms involved in one of the most traditional and peculiar acidic beers that is still brewed today, namely lambic beer.

18.2 LAMBIC BEER, A LONG-LASTING BREW

Lambic beer is a traditionally produced, spontaneously fermented beer that harbors a complex consortium of bacteria and yeasts. This combination of different microorganisms results in an acidic beer with a unique and complex flavor profile. Belgian lambic beer is regarded as the oldest commercial brewing style that is still in use today in much of the same way as performed many centuries ago (De Roos and De Vuyst, 2019; Pothakos et al., 2016; Spitaels et al., 2017). Once, all brews were most likely spontaneously fermented, and it is suspected that one of the rulers of the region that is now called Belgium introduced an ancient brewing process that later evolved into the lambic beer brewing process that flourished around Brussels (De Keersmaecker, 1996). Lambic beer, known as yellow beer at that time, soon became an undisputed part of social life, perfectly captured by many sixteenth century paintings by Pieter Brueghel the Elder and fellow Flemish painters. At the end of the nineteenth century, most of the breweries around the world, including Belgian breweries, switched to the use of pure, isolated yeast cultures of the genus *Saccharomyces*, thanks to its purification by Hansen at the Carlsberg laboratories in 1888, to diminish batch-to-batch variations, making the fermentation process more standardized (De Roos and De Vuyst, 2019). Despite the use of pure yeast cultures, and other developments aimed at increasing predictability of the produce outcome, lambic beer continued to survive and remained popular in the region around Brussels, even marking a remarkable resurgence in popularity over the last two decades.

Traditionally, lambic beers are produced through the spontaneous fermentation of water, barley malt, unmalted wheat, and aged dry hops in horizontal oak or chestnut barrels, a production method that is legally protected as a guaranteed traditional specialty (EEG 2301/97). The general scheme of the lambic beer production process is similar to that of lager or ale production and, hence, consists of the same four basic steps, namely malting, mashing (followed by lautering), fermentation, and post-fermentation (De Roos and De Vuyst, 2019; Spitaels et al., 2017). However, there are some important distinctions that cement the unique position of lambic beer in the ever expanding beer landscape. First, during mashing, as regulated by law, malted barley is mixed with at least 30% of unmalted wheat (Verachtert and Iserentant, 1995). The addition of unmalted wheat in lambic beer production leads to high concentrations of maltooligosaccharides (a glucose polymer containing a small number of glucose units) in the wort (the liquid extracted from the cereals during the mashing process), which plays a role in the longevity of the fermentation and maturation process of lambic beer (De Roos and De Vuyst, 2019; De Roos et al., 2018a, 2020). These relatively long saccharide chains help in enabling the growth and survival of *Brettanomyces* yeast species, which are commonly associated with lambic beer maturation and essential for their unique taste and aroma profile. Besides the use of unmalted wheat, the applied mashing method (turbid mashing) also adds to the high concentrations of maltooligosaccharides in the wort (De Keersmaecker, 1996; De Roos and De Vuyst, 2019; Spitaels et al., 2017; Verachtert and Derdelinckx, 2014). Mashing is the application of a temperature program to extract fermentable carbohydrates and other yeast nutrients from the mixture of malt, wheat, and water, resulting in wort. After mashing, the wort is boiled with the addition of typically old hops, which contain a less bitter taste than fresh hops. The use of these old hops avoids interference with the tart character of the beer, without hampering the inhibitory effect of hops toward various harmful bacteria (De Roos and De Vuyst, 2019; Verachtert and Iserentant, 1995). As a consequence of the high content of oxidized resins in old hops, which can give rise to off-flavors, a long boil of 3–5 h is unavoidable for lambic beer production (De Keersmaecker, 1996; De Roos and De Vuyst, 2019; Spitaels et al., 2017). After this unusually long boil, the most mythical and distinctive event in lambic beer production takes place. The finished wort is transferred to an open metal vessel (coolship), where it is left to cool down overnight, open to the surrounding air. This air-exposed cooling allows the inoculation of various yeasts and bacteria coming from the environment. In the case of traditional breweries, the natural cooling in a large open vessel limits the brewing period to the cold winter months, as the wort needs to cool down to 20° C within one night to avoid the growth of pathogenic bacteria, such as certain enterobacteria (De Keersmaecker, 1996; De Roos and De Vuyst, 2019; Verachtert and Derdelinckx, 2014; Verachtert and Iserentant, 1995). Some larger-scale breweries use enhanced cooling, enabling a continuous production of lambic beer throughout the year via the pre-chilling of the wort before its transfer to the open coolship (Spitaels et al., 2015b). After cooling, the spontaneously inoculated wort is transferred to wooden barrels wherein it ferments and matures for a minimum of one year to produce young lambic beer (De Keersmaecker, 1996; De Roos and De Vuyst, 2019; Verachtert and Derdelinckx, 2014; Verachtert and Iserentant, 1995). For the production of old lambic beer, fermentation and maturation proceeds for up to three years. This long fermentation and maturation process is characterized by a microbial succession of different yeast and bacterial species (Spitaels et al., 2014c, 2015b; De Roos et al., 2018a,b, 2020).

Nowadays, lambic-style beers can be found around the world but they were originally produced by specialized breweries located in close proximity of the Senne river valley in the Payottenland in and around Brussels. By public and popular belief, this region harbors the essential microbial cocktail for a successful lambic beer fermentation and maturation process. However, over the years, the scientific consensus suggests that under similar environmental conditions these beers can be produced in many regions spanning the globe. This is further strengthened by the existence of lambic breweries located in south-west Flanders and the existence of American coolship ales (ACAs) that have similar fermentation and maturation profiles harboring a very similar microbiota (Bokulich et al., 2012; De Roos and De Vuyst, 2019). Moreover, research has stipulated that the spontaneous inoculation can not only be ascribed to the environmental microorganisms entering the wort in the open coolship during overnight cooling. Also the wooden barrels and brewery equipment (house microbiota) contribute to the establishment of a typical and stable lambic beer microbial ecosystem (De Roos et al., 2019).

Lambic beer *strictu sensu* is the resulting beer of the production process described above. However, lambic beer is often not sold as such but rather blended into lambic-based beers, such as gueuze (bottle refermentation of a mixture of young and old lambic beers), faro (bottle refermentation of a mixture of light [originally lambic beer based on the last run-off fraction during lautering, now often young lambic beer] and old lambic beers supplemented with rock sugar), and fruit lambics (refermentation of young lambic beer with whole fruits, such as sour [Schaerbeek] cherries and raspberries, in tanks or casks) (De Keersmaecker, 1996; Verachtert and Derdelinckx, 2014). Those blended beers are not only produced by lambic breweries but also by so-called gueuze blenders, who purchase wort from various lambic breweries for further wooden-barrel maturation and blending (De Roos and De Vuyst, 2019).

18.3 A UNIQUE BLEND OF MICROORGANISMS

The spontaneous fermentation and maturation process of lambic beer comprises a very complex tapestry of different microbial groups (De Roos and De Vuyst, 2019; Spitaels et al., 2017). Despite the spontaneous nature of lambic beer, it presents no danger to human health due to the combination of a low pH, the presence of hop antimicrobial compounds, the nutrient depletion, the low oxygen concentration, and the high alcohol content of the final beer. This is known as the hurdle effect, where relatively low-intensity hurdles combined together pose a substantial antimicrobial effect, and which allows only very few and specifically adapted microbial species to survive (Dysvik et al., 2020).

Lambic beer fermentation and maturation takes place in an overlapping succession of typically four microbial phases (Figure 18.2; Spitaels et al., 2017; De Roos and De Vuyst, 2019).

The first phase, lasting for up to several weeks, is characterized by the presence of enterobacteria, (oxidative) yeasts, and a limited growth of acetic acid bacteria (AAB) (De Roos et al., 2018a,b; Spitaels et al., 2014c, 2015b). Enterobacteria generally grow during the initial stages of traditional lambic beer fermentations. By producing a combination of organic acids and specific long-chain fatty acids, among other compounds, they also can contribute to the taste and aroma of lambic beer (Spaepen et al., 1978). Although enterobacteria do appear in most lambic beer productions, multiple studies have hypothesized that they should not be considered necessary for lambic beer fermentation and could possibly be disregarded as a part of the lambic beer core microbiota, for two main reasons (De Roos and De Vuyst, 2019; De Roos et al., 2018b; Spitaels et al., 2015b). First, enterobacteria are a major group of intestinal microorganisms, with some species well-known as gastrointestinal pathogens. As a consequence, they are generally regarded as indicators of unclean production processes and potential fecal contamination (Baylis et al., 2011; Brenner and Farmer, 2005). Second, they can contribute to the formation of biogenic amines in fermented foods and beverages (De Roos et al., 2018a, 2020; Marino et al., 2000). Despite this negative connotation they are often present in fermented foods and beverages. Therefore, a common practice in today's lambic breweries is to acidify the wort manually before the start of the fermentation to minimize their growth (Spitaels et al., 2015b; De Roos et al., 2018a,b, 2020). However, even without this manual acidification step enterobacteria disappear quickly during fermentation, due to increasing ethanol concentrations, decreasing pH values, and monosaccharide depletion (De Keersmaecker, 1996; De Roos and De Vuyst, 2019; Martens et al., 1991; Priest and Stewart, 2006; Spitaels et al., 2014c; Verachtert and Iserentant, 1995). As a consequence of these increasing hurdles during fermentation enterobacteria are rarely found in the resulting food fermentation end-products, posing little threat to healthy human individuals (Wouters

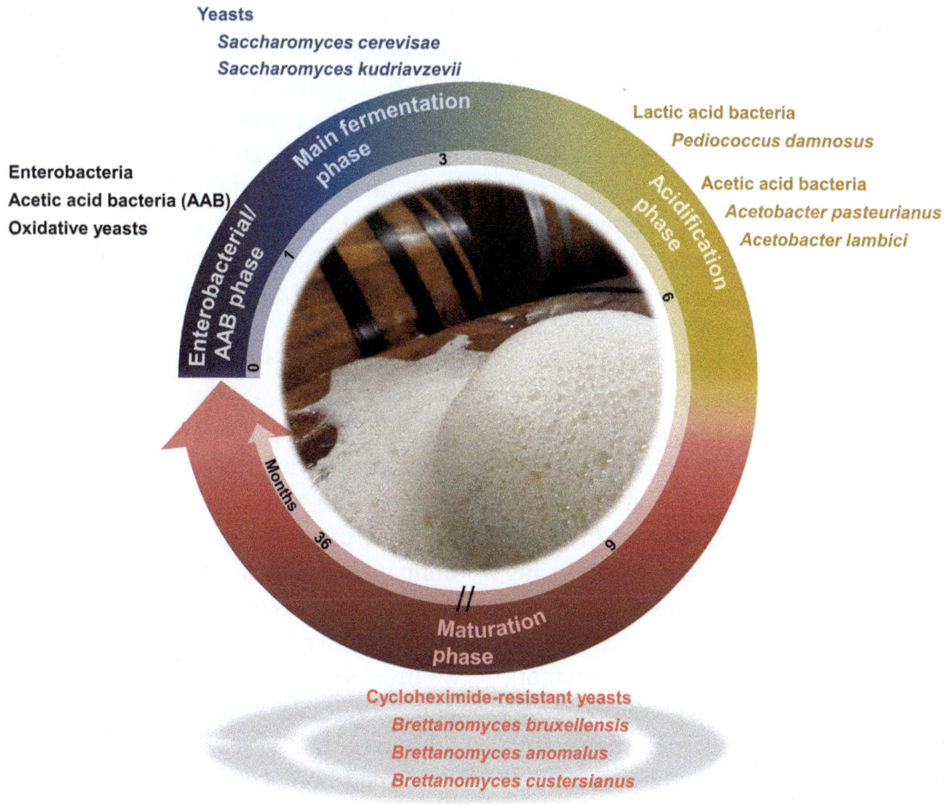

FIGURE 18.2 Overview of the general microbial succession during lambic beer production. Four phases can be distinguished, namely an enterobacterial and oxidative yeast phase (with a limited growth of acetic acid bacteria), an alcoholic or main fermentation phase performed by *Saccharomyces* yeasts, an acidification phase due to the growth of lactic acid bacteria and acetic acid bacteria, and a maturation phase with the growth of *Brettanomyces* yeasts and lactic acid bacteria.

et al., 2013). Together with the growth of enterobacteria, growth of maltose non-fermenting yeasts occurs during the early stages of lambic beer fermentation. These yeasts quickly disappear, due to the same increasing antimicrobial hurdles mentioned above, and they are also under debate whether they are a part of the lambic beer core microbiota or not (De Roos et al., 2018a, 2020; Spitaels et al., 2014c, 2015b).

The second phase is called the main fermentation phase and can last for several weeks to several months. It is characterized by the prevalence of yeast species from the *Saccharomyces* genus. These *Saccharomyces* species are widely found and used in other fermented food products (e.g. bread, see Chapter 14, cocoa, see Chapter 16, and wine, see Chapter 17) and are responsible for most of the ethanol production that will be present in the finished lambic beer. The third phase, lasting for several months, is called the acidification phase and is characterized by the presence of LAB and AAB. These bacteria are mainly responsible for the beer's refreshing taste by the production of lactic acid and acetic acid. Acetic acid also contributes to the aroma of the beer. The final phase is the maturation phase, which can last for several years and is characterized by the presence of yeast species from the *Brettanomyces* genus. These yeasts play an important role in the formation of the typical Brett aroma of lambic beers, characterized mainly by the production of 4-ethylguaiacol and 4-ethylphenol (De Roos and De Vuyst, 2019; Lentz and Harris, 2015; Steensels et al., 2015).

18.4 HOW BEER-SPOILING BACTERIA CAN BE WANTED

Generally, the antimicrobial hurdle effect makes beer a very safe beverage in view of microbial growth and stability. Specifically adapted microorganisms that can survive and grow in a beer matrix do exist and could potentially cause spoilage with loss of colloidal stability, appearance of ropiness, and occurrence of aroma and taste defects (Esmaeili et al., 2015). Major groups among these beer-spoiling microorganisms are LAB and AAB

(De Roos and De Vuyst, 2019; Dysvik et al., 2020; Garofalo et al., 2015; Van Vuuren et al., 1979). AAB and LAB are generally considered as safe and non-pathogenic bacteria and can be found in many fermented foods and beverages, such as various fermented dairy products (LAB, see Chapter 12), kombucha (AAB, see Chapter 11), and cocoa (both LAB and AAB; see chapter 16). Furthermore, evidence is increasing that AAB and LAB in fermented foods can exert host-specific health benefits and contribute to the microbial balance inside the human body (Kok and Hutkins, 2018; Marco et al., 2017, 2021; Veiga et al., 2014). In alcoholic fermentations, such as beer production, AAB and LAB are generally unwanted, as they convert ethanol and fermentable saccharides into acetic acid and lactic acid respectively, rendering an undesirable acidic flavor (De Roos and De Vuyst, 2018, 2019). Moreover, LAB can alter the viscosity and turbidity of beers, by producing exopolysaccharides (long saccharide chains). In contrast to ale or lager productions, this does not affect the salability of lambic beers, as this so-called slime formation, which is a common phenomenon during lambic beer production, usually disappears during the long maturation phase (De Roos et al., 2020). In general, these microbial groups have to be viewed in a different light in the context of lambic beer production, as their involvement is essential for the final taste and aroma profiles as highlighted below (De Roos et al., 2018a,b).

LAB and AAB account for the biggest contribution to the acidity of lambic beers, making them an essential part of the lambic beer ecosystem, as acidity not only adds to the microbial safety of these beers but is also an indispensable part of their taste profile. The most commonly isolated LAB species of the lambic beer production process is *Pediococcus damnosus*, which is a homofermentative LAB species and, hence, produces lactate only (Van Oevelen et al., 1977; Bokulich et al., 2012; Spitaels et al., 2014c, 2015b; Snauwaert et al., 2015; De Roos et al., 2018a,b, 2020). *Pediococcus damnosus* is a commonly found beer spoilage microorganism, well adapted to growth in beer, thanks to its resistance to the hop antimicrobial compounds that inhibit the growth of many Gram-positive bacteria (De Roos et al., 2020; Snauwaert et al., 2015; Vriesekoop et al., 2012). Besides being acidifiers, they also impact the aroma profile of lambic beer via the indirect formation of the ester ethyl lactate by providing lactic acid, which renders a fruity aroma (De Roos et al., 2018a). Moreover, they also perform a malolactic fermentation, converting the tart L-malic acid into softer-tasting L-lactic acid (De Roos et al., 2018a, 2020). Malolactic fermentation is performed in the commercial production of wine and generally occurs in other spontaneously fermented alcoholic beverages such as cider (Fugelsang and Edwards, 2007; González-Arenzana et al., 2013a,b, 2012; Salih et al., 1988; Swaffield and Scott, 1995). LAB are also part of the core microbiota in other spontaneously fermented acidic beers, such as ACAs and red–brown acidic ales (Bokulich et al., 2012; Martens et al., 1997; Snauwaert et al., 2016, 2015).

Although the acidic taste of lambic beers is often ascribed to a combination of acetic acid and lactic acid, the role and importance of AAB was always considered to be limited during lambic beer fermentation and maturation due to the microaerobic character of the production process (Spitaels et al., 2014c, 2015b). Being obligate aerobic bacteria, AAB species have only been sporadically recovered from lambic beer productions during many decades of research (Martens et al., 1991; Shanta Kumara and Verachtert, 1991; Van Oevelen et al., 1977; Verachtert and Dawoud, 1984; Verachtert and Iserentant, 1995). Yet, two AAB species have been described that seem to be specific for lambic beer production, namely *Acetobacter lambici* (Spitaels et al., 2014a) and *Gluconobacter cerevisiae* (Spitaels et al., 2014b). Moreover, by applying spatial sampling, it has been shown that the obligate aerobic AAB are concentrated at the wort–air interface of the barrels and, hence, may have always been a chronically underestimated microbial group when investigating lambic beer production (De Roos et al., 2018b). Indeed, various *Acetobacter* species have been shown to be significantly present during different phases of the lambic beer production process (De Roos et al., 2018b, 2020). AAB improve the shelf-life of lambic beer by the production of acetic acid. Acetic acid also leads to a satiety response in humans, thereby regulating food intake. Also other health benefits seem to be ascribed to acetic acid, such as protection against inflammatory bowel diseases and colon cancer (D'Argenio and Mazzacca, 1999; Wong et al., 2006). Besides through the production of acetic acid, AAB also impact the aroma profile of lambic beer via either direct or indirect formation of acetoin and ethyl acetate, which renders ethereal and fruity aromas. The use of well-sealed wooden casks that enable microaerobic conditions, the formation of a yeast pellicle at the air–liquid interface of the maturing beer, and certain technical practices applied by lambic brewers help to control AAB growth and avoids the production of excessive concentrations of acetic acid and acetoin, which may create an undesirable flavor profile in the finished product (De Roos et al., 2018b). As for LAB, also AAB have been isolated from other sour ale productions, such as Belgian red–brown acidic ales (Snauwaert et al., 2016). Their occurrence in these beers are more likely, as maturation proceeds in vertical wooden barrels as opposed to horizontal wooden barrels for lambic beers, thereby increasing headspace volume and possible oxygen influx.

18.5 YEASTS, MORE THAN A ONE-TRICK PONY

Although food-associated yeasts are predominantly known for their ability to convert fermentable saccharides into ethanol and carbon dioxide, they are also important for flavor production as well as for spoilage prevention. In many food fermentation processes yeasts from the *Saccharomyces* genus play a pivotal role and lambic beer is no exception to this. Generally, *Saccharomyces* yeasts are able to prevail during the main stage of food and beverage fermentation processes because of their quick conversion of the available carbohydrates into ethanol through alcoholic fermentation. This is known as the make-accumulate phase of their make-accumulate-consume metabolic strategy (Dashko et al., 2014; De Vuyst and Weckx, 2016; Hagman et al., 2013; Rozpędowska et al., 2011). The fast consumption of the available substrates, together with the production of high ethanol and carbon dioxide concentrations, impairs the growth of competing microorganisms, further stamping their authority during the beginning of the fermentation process. After exhaustion of the fermentable carbohydrates, these yeasts can remain viable and prevalent for a prolonged period by metabolizing the accumulated ethanol through aerobic respiration during the consume phase (De Keersmaecker, 1996; De Roos et al., 2018a, 2020; Piškur et al., 2006; Rozpędowska et al., 2011; Spitaels et al., 2015b).

Due to the spontaneous nature of the lambic beer production process, different *Saccharomyces* species appear during the main fermentation phase. Generally, the most prevalent species are *Saccharomyces cerevisiae*, *Saccharomyces kudriavzevii*, and *Saccharomyces pastorianus*, with the latter two becoming increasingly prevalent toward the end of the main fermentation phase (De Roos et al., 2018a, 2020; Spitaels et al., 2014c). Besides adding to the microbial safety due to the production of ethanol and carbon dioxide, they also impact the aroma profile via, for example, the biosynthesis of higher alcohols and esters (De Roos et al., 2018a, 2020; Procopio et al., 2011; Van Oevelen et al., 1976; Verstrepen et al., 2003).

Whereas *Saccharomyces* species are often the sole species present in beer fermentations, lambic beers typically possess an additional group of yeasts that appear late in the production process and were first isolated from non-lambic, English stock ales in Denmark in 1904, hence referred to as *Brettanomyces*, linking it with Great Britain (Claussen, 1904; Steensels et al., 2015). When many breweries in the early 1900s started to produce their beers using single-strain starter cultures instead of back-slopping practices, English brewers struggled to maintain the typical stout aroma with the application of these pure cultures (Claussen, 1904). Later, it was found that *Brettanomyces* species, which were first taxonomically named and described during a Belgian study of lambic beer, were responsible for the formation of this aroma (officially described and classified as *Dekkera bruxellensis* by Kufferath and Van Laer, 1921; Oelofse et al., 2008). Species belonging to the *Brettanomyces* genus have currently been isolated from not only other alcoholic fermentation processes, such as wine, cider, water kefir, tequila, etc., but also fermented dairy products, kombucha (see chapter 11), sourdough (see Chapter 14), and olives (see Chapter 15) (Steensels et al., 2015). Also, specific strains of *Brettanomyces bruxellensis* are used in the production of Orval, a Belgian trappist beer (De Roos and De Vuyst, 2019).

Brettanomyces species make use of the same make-accumulate-consume metabolic strategy as *Saccharomyces* species, but thanks to their nutritionally efficient metabolism, they are capable of surviving the harsh conditions during the long-lasting lambic beer production process, whereas *Saccharomyces* species cannot (Castro-Martinez et al., 2005; De Deken, 1966; Rozpędowska et al., 2011; Shanta Kumara and Verachtert, 1991; Spaepen and Verachtert, 1982; Spitaels et al., 2017; Steensels et al., 2015). Their higher tolerance against a number of inhibitors, combined with a higher metabolic efficiency, makes them competitive with *Saccharomyces* species under the nutrient-limiting conditions that appear toward the end of the lambic beer fermentation and maturation process (Blomqvist et al., 2010; Steensels et al., 2015). Moreover, *Brettanomyces* species possess both intra- and extracellular α-glucosidase activities, allowing the degradation of maltooligosaccharides into fermentable carbohydrates (De Roos et al., 2020; Shanta Kumara et al., 1993). On the other hand, *Saccharomyces* species are generally limited to the use of monosaccharides, maltose, and maltotriose. However, recently it has been found that *S. kudriavzevii*, which appears during the main fermentation phase of certain lambic beer productions, has the genetic capability to degrade maltooligosaccharides as well (De Roos et al., 2020).

The most isolated *Brettanomyces* species that prevail during lambic beer maturation are *Brettanomyces bruxellensis*, *Brettanomyces anomalus*, and *Brettanomyces custersianus* (De Keersmaecker, 1996; De Roos et al., 2018a, 2020; Smith et al., 1990; Spitaels et al., 2014c, 2015b; Van Oevelen et al., 1977; Verachtert and Iserentant, 1995). *Brettanomyces* species are mainly responsible for the typical Brett aroma of lambic beers, mostly through the production of the volatile phenolic compounds 4-ethylguaiacol and 4-ethylphenol (Steensels et al., 2015). They also produce several other volatile phenolic compounds and heavily impact the "fruity" part of the aroma profile by either forming or degrading many volatile esters (De Roos and De Vuyst, 2019; Spaepen and Verachtert, 1982). These combined effects lead to a volatile ester profile that is very distinct of other beers, and together with the Brett aroma, form the major constituents of the typically described lambic beer aroma (De Roos and De Vuyst, 2019; Van Oevelen et al., 1976).

Besides contributing to the flavor profile, *Brettanomyces* species together with other yeasts are responsible for pellicle or biofilm formation at the air–liquid interface of the maturing beer (De Keersmaecker, 1996; De Roos et al., 2018a, 2020; Van Oevelen et al., 1977). Pellicle formation is a typical characteristic during lambic beer maturation, which prevents excessive oxygen influx, thereby inhibiting beer oxidation and excessive growth of AAB (De Roos and De Vuyst, 2019). Another unique characteristic found in some *Brettanomyces* species that highlights the complex balance of the lambic beer ecosystem is their capability to degrade the exopolysaccharides produced by LAB with a specific β-glucosidase enzyme. This possibly explains why so-called slime formation occurs in some lambic beer production processes and not in others and why it usually disappears later on in the production process (De Roos et al., 2020).

Spontaneously inoculated yeasts also play an important role during the bottle conditioning of gueuze production by further refermenting the mixture of young and old lambic beers, adding more complexity to the already complex flavor profile. Extensive bottle conditioning of gueuze beers is not uncommon in traditional lambic beer breweries, where the aging of certain blends and bottles can take up to ten years and even longer (Spitaels et al., 2015a). Refermentation assures a well-carbonated end-product and enhances the sensory stability of the beer by preventing the formation of off-flavors due to oxidation (Štulíková et al., 2020). Generally, in beer production, bottle conditioning also contributes to preventing the growth of aerobic spoilage microorganisms. In gueuze specifically, it additionally prevents the excessive growth of AAB, which can render the beer into undrinkable vinegar (De Roos and De Vuyst, 2019; Spitaels et al., 2015b).

18.6 CONCLUSIONS

Various microorganisms are required for the production of lambic beer, resulting in a beer with a unique flavor profile, due to the combined metabolic activities of specific yeasts, AAB, and LAB. Lambic beer and gueuze are consumed unfiltered and, therefore, contain living microorganisms at the time of consumption. Due to methodological limitations, the role of these food- or fermentation-derived microorganisms in promoting human health has always been underreported. However, recent evidence of the role of fermentation-derived microorganisms in host-specific health benefits is increasing. Nonetheless, even without the presence of living microorganisms at the time of consumption, vitamins and bioactive molecules (such as antioxidants) produced by the microorganisms active during fermentation and maturation are thought to remain present. In addition, the complex consortium of lambic beer microorganisms consume various substrates during fermentation and maturation, which results in compositional changes such as a low carbohydrate content and the production of health-promoting compounds such as acetic acid.

REFERENCES

1. Anderson RG. 2003. Beer. In: Blocker J, Tyrrell I, Fahey D (Eds.), *Alcohol and Temperance in Modern History: An International Encyclopedia*, Volume 1, ABC-Clio, Santa Barbara, California, USA, pp. 92–100.
2. Anderson RG. 2006. History of industrial brewing. In: Priest FG, Stewart GG (Eds.), *Handbook of Brewing*, Second ed. CRC Press, Taylor & Francis Group, Boca Raton, Florida, USA, pp. 1–33.
3. Baylis C, Uyttendaele M, Joosten H, Davies A. 2011. *The Enterobacteriaceae and their Significance to the Food Industry. ILSI Europe Report Series.* International Life Sciences Institute, Brussels, Belgium, 48.
4. Blomqvist J, Eberhard T, Schnürer J, Passoth V. 2010. Fermentation characteristics of *Dekkera bruxellensis* strains. *Applied Microbiology and Biotechnology* 87: 1487–1497.

5. Bokulich NA, Bamforth CW. 2013. The microbiology of malting and brewing. *Microbiology and Molecular Biology Reviews* 77: 157–172.

6. Bokulich NA, Bamforth CW, Mills DA. 2012. Brewhouse-resident microbiota are responsible for multi-stage fermentation of American coolship ale. *PLoS One* 7: e35507.

7. Bossaert S, Crauwels S, De Rouck G, Lievens B. 2020. The power of sour – A review: Old traditions, new opportunities. *Brewing Science* 72: 78–88.

8. Brenner DJ, Farmer JJ. 2005. Enterobacteriaceae. In: Brenner DJ, Krieg NR, Staley JT (Eds.), *Bergey's Manual of Systematic Bacteriology: Volume Two: The Proteobacteria*, Springer, New York, New York, USA, pp. 587–849.

9. Briggs DE, Boulton C, Brookes P, Stevens R. 2004. *Brewing Science and Practice*. Woodhead Publishing, Cambridge, UK, 900.

10. Castro-Martinez C, Escudero-Abarca BI, Gomez Rodríguez J, Hayward-Jones PM, Aguilar-Uscanga MG. 2005. Effect of physical factors on acetic acid production in *Brettanomyces* strains. *Journal of Food Process Engineering* 28: 133–143.

11. Claussen NH. 1904. On a method for the application of Hansen's pure yeast system in the manufacturing of well-conditioned English stock beers. *Journal of the Institute of Brewing* 10: 308–331.

12. D'Argenio G, Mazzacca G. 1999. Short-chain fatty acid in the human colon. Relation to inflammatory bowel diseases and colon cancer. *Advances in Experimental Medicine and Biology* 472: 149–158.

13. Dashko S, Zhou N, Compagno C, Piškur J. 2014. Why, when, and how did yeast evolve alcoholic fermentation? *FEMS Yeast Research* 14: 826–832.

14. De Deken RH. 1966. The Crabtree effect: A regulatory system in yeast. *Journal of General Microbiology* 44: 149–156.

15. De Keersmaecker J. 1996. The mystery of lambic beer. *Scientific American* 275: 375–385.

16. De Vuyst L, Weckx S. 2016. The cocoa bean fermentation process: From ecosystem analysis to starter culture development. *Journal of Applied Microbiology* 121: 5–17.

17. De Roos J, De Vuyst L. 2018. Acetic acid bacteria in fermented foods and beverages. *Current Opinion in Biotechnology* 49: 115–119.

18. De Roos J, Vandamme P, De Vuyst L. 2018a. Wort substrate consumption and metabolite production during lambic beer fermentation and maturation explain the successive growth of specific bacterial and yeast species. *Frontiers in Microbiology* 18: 2763.

19. De Roos J, Verce M, Aerts M, Vandamme P, De Vuyst L. 2018b. Temporal and spatial distribution of the acetic acid bacterium communities throughout the wooden casks used for the fermentation and maturation of lambic beer underlines their functional role. *Applied and Environmental Microbiology* 84: e02846–17.

20. De Roos J, De Vuyst L. 2019. Microbial acidification, alcoholization, and aroma production during spontaneous lambic beer production. *Journal of the Science of Food and Agriculture* 99: 25–38.

21. De Roos J, Van Der Veken D, De Vuyst L. 2019. The interior surfaces of wooden barrels are an additional microbial inoculation source for lambic beer production. *Applied and Environmental Microbiology* 85: e02226–18.

22. De Roos J, Verce M, Weckx S, De Vuyst L. 2020. Temporal shotgun metagenomics revealed the potential metabolic capabilities of specific microorganisms during lambic beer production. *Frontiers in Microbiology* 11: 1692.

23. Dysvik A, Leanti La Rosa S, De Rouck G, Rukke E-O, Westereng B, Wicklund T. 2020. Microbial dynamics in traditional and modern sour beer production. *Applied and Environmental Microbiology* 86: e00566–20.

24. Esmaeili S, Mogharrabi M, Safi F, Sohrabvandi S, Mortazavian AM, Bagheripoor-Fallah N. 2015. The common spoilage microorganisms of beer: Occurrence, defects, and determination – a review. *Journal of Food Science and Technology* 7: 68–73.

25. Fugelsang KC, Edwards GC. 2007. *Wine Microbiology: Practical Applications and Procedures*. Second ed. Springer, Boston, Massachusetts, USA, 394.

26. Garofalo C, Osimani A, Milanovic V, Taccari M, Aquilanti L, Clementi F. 2015. The occurrence of beer spoilage lactic acid bacteria in craft beer production. *Journal of Food Science* 80: M2845–M2852.

27. González-Arenzana L, López R, Santamaría P, López-Alfaro I. 2013a. Dynamics of lactic acid bacteria populations in Rioja wines by PCR-DGGE, comparison with culture-dependent methods. *Applied Microbiology and Biotechnology* 97: 6931–6941.

28. González-Arenzana L, López R, Santamaría P, Tenorio C, López-Alfaro I. 2012. Dynamics of indigenous lactic acid bacteria populations in wine fermentations from La Rioja (Spain) during three vintages. *Microbial Ecology* 63: 12–19.

29. González-Arenzana L, Santamaría P, López R, López-Alfaro I. 2013b. Indigenous lactic acid bacteria communities in alcoholic and malolactic fermentations of Tempranillo wines elaborated in ten wineries of La Rioja (Spain). *Food Research International* 50: 438–445.

30. Hagman A, Säll T, Compagno C, Piškur J. 2013. Yeast "make-accumulate-consume" life strategy evolved as a multi-step process that predates the whole genome duplication. *PLoS ONE* 8: e68734.
31. Howard PH. 2014. Too big to ale? Globalization and consolidation in the beer industry. In: Patterson M, Pullen N (Eds.), *The Geography of Beer: Regions, Environment, and Society*, Springer, New York, New York, USA, pp. 155–165.
32. Kok CR, Hutkins R. 2018. Yogurt and other fermented foods as sources of health-promoting bacteria. *Nutrition Reviews* 76: 4–15.
33. Kufferath H, Van Laer M. 1921. Etudes sur les levures de lambic. Leur action chimique sur les milieux de culture. *Bulletin des Sociétés Chimiques Belges* 30: 270–276.
34. Lentz M, Harris C. 2015. Analysis of growth inhibition and metabolism of hydroxycinnamic acids by brewing and spoilage strains of *Brettanomyces* yeast. *Foods* 4: 581–593.
35. Marco ML, Heeney D, Binda S, Cifelli CJ, Cotter PD, et al. 2017. Health benefits of fermented foods: Microbiota and beyond. *Current Opinion in Biotechnology* 44: 94–102.
36. Marco ML, Sanders ME, Gänzle M, Arrieta MC, Cotter PD, et al. 2021. The International Scientific Assocation for Probiotics and Prebiotics (ISAPP) consensus statement on fermented foods. *Nature Reviews Gastroenterology & Hepatology* 18: 196–208.
37. Marino M, Maifreni M, Moret S, Rondinini G. 2000. The capacity of *Enterobacteriaceae* species to produce biogenic amines in cheese. *Letters in Applied Microbiology* 31: 169–173.
38. Martens H, Dawoud E, Verachtert H. 1991. Wort enterobacteria and other microbial populations involved during the first month of lambic fermentation. *Journal of the Institute of Brewing* 97: 435–439.
39. Martens H, Iserentant D, Verachtert H. 1997. Microbiological aspects of a mixed yeast-bacterial fermentation in the production of a special Belgian acidic ale. *Journal of the Institute of Brewing* 103: 85–91.
40. Oelofse A, Pretorius IS, du Toit M. 2008. Significance of *Brettanomyces* and *Dekkera* during winemaking: A synoptic review. *South African Journal of Enology and Viticulture* 29: 128–144.
41. Palmer GH. 2006. Barley and malt. In: Priest FG, Stewart GG (Eds.), *Handbook of Brewing*, Second ed. CRC Press, Taylor & Francis Group, Boca Raton, Florida, USA, pp. 139–160.
42. Pavsler A, Buiatti S. 2009. Non-lager beer & lager beer. In: Preedy VR (Ed.), *Beer in Health and Disease Prevention*, Academic Press, Elsevier, London, UK, pp. 17–43.
43. Pires E, Branyik T. 2015. *Biochemistry of Beer Fermentation*. Springer International Publishing AG, Cham, Switzerland, 80.
44. Piškur J, Rozpędowska E, Polakova S, Merico A, Compagno C. 2006. How did *Saccharomyces* evolve to become a good brewer? *Trends in Genetics* 22: 183–186.
45. Pothakos V, Illeghems K, Laureys D, Spitaels F, Vandamme P, De Vuyst L. 2016. Acetic acid bacteria in fermented food and beverage ecosystems. In: Matsushita K, Toyama H, Tonouchi N, Okamoto-Kainuma A (Eds.), *Acetic Acid Bacteria: Ecology and Physiology*, Springer, Tokyo, Japan, pp. 73–100.
46. Priest FG, Stewart GG. 2006. Microbiology and microbiological control in the brewery. In: Priest FG, Stewart GG (Eds.)., *Handbook of Brewing*, Second ed. CRC Press, Taylor & Francis Group, Boca Raton, Florida, USA, pp. 607–629.
47. Procopio S, Qian F, Becker T. 2011. Function and regulation of yeast genes involved in higher alcohol and ester metabolism during beverage fermentation. *European Food Research and Technology* 233: 721–729.
48. Rozpędowska E, Hellborg L, Ishchuk OP, Orhan F, Galafassi S et al. 2011. Parallel evolution of the make-accumulate-consume strategy in *Saccharomyces* and *Dekkera* yeasts. *Nature Communications* 2: 302.
49. Salih A, Drilleau J, Cavin F, Divies C, Bourgeois C. 1988. A survey of microbiological aspects of cider making. *Journal of the Institute of Brewing* 94: 5–8.
50. Shanta Kumara HMC, De Cort S, Verachtert H. 1993. Localization and characterization of α-glucosidase activity in *Brettanomyces lambicus*. *Applied and Environmental Microbiology* 59: 2352–2358.
51. Shanta Kumara HMC, Verachtert H. 1991. Identification of lambic superattenuating microorganisms by the use of selective antibiotics. *Journal of the Institute of Brewing* 97: 181–185.
52. Smith MT, Yamazaki M, Poot G. 1990. *Dekkera, Brettanomyces* and *Eeniella*: Electrophoretic comparison of enzymes and DNA-DNA homology. *Yeast* 6: 299–310.
53. Snauwaert I, Roels SP, Van Nieuwerburg F, Van Landschoot A, De Vuyst L, Vandamme P. 2016. Microbial diversity and metabolite composition of Belgian red-brown acidic ales. *International Journal of Food Microbiology* 221: 1–11.
54. Snauwaert I, Stragier P, De Vuyst L, Vandamme P. 2015. Comparative genome analysis of *Pediococcus damnosus*

LMG 28219, a strain well-adapted to the beer environment. *BMC Genomics* 16: 267.

55. Spaepen M, Van Oevelen D, Verachtert H. 1978. Fatty acids and esters produced during the spontaneous fermentation of lambic and gueuze. *Journal of the Institute of Brewing* 84: 278–282.
56. Spaepen M, Verachtert H. 1982. Esterase activity in the genus. *Brettanomyces. Journal of the Institute of Brewing* 88: 11–17.
57. Spitaels F, Li L, Wieme A, Balzarini T, Cleenwerck I, et al. 2014a. *Acetobacter lambici* sp. nov., isolated from fermenting lambic beer. *International Journal of Systematic and Evolutionary Microbiology* 64: 1083–1089.
58. Spitaels F, Van Kerrebroeck S, Wieme AD, Snauwaert I, Aerts M, et al. 2015a. Microbiota and metabolites of aged bottled gueuze beers converge to the same composition. *Food Microbiology* 47: 1–11.
59. Spitaels F, Wieme AD, Balzarini T, Cleenwerck I, Van Landschoot A, et al. 2014b. *Gluconobacter cerevisiae* sp. nov., isolated from the brewery environment. *International Journal of Systematic and Evolutionary Microbiology* 64: 1134–1141.
60. Spitaels F, Wieme AD, Janssens M, Aerts M, Daniel H-M et al. 2014c. The microbial diversity of traditional spontaneously fermented lambic beer. *PLoS ONE* 9: e95384.
61. Spitaels F, Wieme AD, Janssens M, Aerts M, Van Landschoot A, De Vuyst L, Vandamme P. 2015b. The microbial diversity of an industrially produced lambic beer shares members of a traditionally produced one and reveals a core microbiota for lambic beer fermentation. *Food Microbiology* 49: 23–32.
62. Spitaels F, Wieme AD, Snauwaert I, De Vuyst L, Vandamme P. 2017. Microbial ecology of traditional beer fermentations. In: Bokulich N, Bamforth C (Eds.), *Brewing Microbiology: Current Research, Omics and Microbial Ecology*, Caister Academic Press, Poole, UK, pp. 179–196.
63. Steensels J, Daenen L, Malcorps P, Derdelinckx G, Verachtert H, Verstrepen KJ. 2015. *Brettanomyces* yeasts – From spoilage organisms to valuable contributors to industrial fermentations. *International Journal of Food Microbiology* 206: 24–38.
64. Štulíková K, Novák J, Vlček J, Šavel J, Košin P, Dostálek P. 2020. Bottle conditioning: Technology and mechanisms applied in refermented beers. *Beverages* 6: 56.
65. Swaffield CH, Scott JA. 1995. Existence and development of natural microbial populations in wooden storage vats used for alcoholic cider maturation. *Journal of the American Society of Brewing Chemists* 53: 117–120.
66. Van Oevelen D, De L'Escaille F, Verachtert H. 1976. Synthesis of aroma compounds during the spontaneous fermentation of lambic and gueuze. *Journal of the Institute of Brewing* 82: 322–326.
67. Van Oevelen D, Spaepen M, Timmermans P, Verachtert H. 1977. Microbiological aspects of spontaneous wort fermentation in the production of lambic and gueuze. *Journal of the Institute of Brewing* 83: 356–360.
68. Van Vuuren HJJ, Loos MA, Louw HA, Meisel R. 1979. Distribution of bacterial contaminants in a South African lager brewery. *Journal of Applied Bacteriology* 47: 421–424.
69. Veiga P, Pons N, Agrawal A, Oozeer R, Guyonnet D et al. 2014. Changes of the human gut microbiome induced by a fermented milk product. *Scientific Reports* 4: 6328.
70. Verachtert H, Dawoud E. 1984. Microbiology of lambic-type beers. *Journal of Applied Bacteriology* 57: R11–R12.
71. Verachtert H, Derdelinckx G. 2014. Belgian acidic beers: Daily reminiscences of the past. *Cerevisia - Belgian Journal of Brewing and Biotechnology* 38: 121–128.
72. Verachtert H, Iserentant D. 1995. Properties of Belgian acidic beers and their microflora. I. The production of gueuze and related refreshing acid beers. *Cerevisia - Belgian Journal of Brewing and Biotechnology* 20: 37–41.
73. Verstrepen KJ, Derdelinckx G, Dufour JP, Winderickx J, Pretorius IS, et al. 2003. The *Saccharomyces cerevisiae* alcohol acetyl transferase gene *ATF1* is a target of the cAMP/PKA and FGM nutrient-signaling pathways. *FEMS Yeast Research* 4: 285–296.
74. Vriesekoop F, Krahl M, Hucker B, Menz G. 2012. 125th Anniversary review: Bacteria in brewing: The good, the bad and the ugly. *Journal of the Institute of Brewing* 118: 335–345.
75. Wong JM, de Souza R, Kendall CW, Emam A, Jenkins DJ. 2006. Colonic health: Fermentation and short chain fatty acids. *Journal of Clinical Gastroenterology* 40: 235–243.
76. Wouters D, Grosu-Tudor S, Zamfir M, De Vuyst L. 2013. Bacterial community dynamics, lactic acid bacteria species diversity and metabolite kinetics of traditional Romanian vegetable fermentations. *Journal of the Science of Food and Agriculture* 93: 749–760.

SECTION 3

GOOD MICROBES IN BIOTECHNOLOGY

Co-Edited by Michael Sauer and Frans J. de Bruijn

CHAPTER 19

Microbiology and Bio-economy – Sustainability by Nature

Michael Sauer
Institute of Microbiology and Microbial Biotechnology, Department of Biotechnology, BOKU – University of Natural Resources and Life Sciences, 1190 Vienna, Austria

19.1 INTRODUCTION

Microorganisms are tiny organisms, individually invisible to our eyes. Nevertheless, they are extremely powerful and their effects are visible everywhere around us. This book is dedicated to many of the beneficial roles microorganisms play in our world and our lives. This chapter sheds some light on the impact, which microorganisms have on our economy and our global societal development.

The economic impact of microorganisms – to start with one topic – is a factor which is very easily overlooked. The modern buzzword in the context of economy and biology is of course "bio-economy." However, one has to understand, that microorganisms already play key roles in many aspects related to our economy, even though the concept of "bio-" is not really embraced yet.

Some of the already existing roles of microorganisms in our economy are obvious. For example, they are heavily involved into our food production or pharmaceutical production. It appears useful, however, to set this into an economic perspective, to get an idea of the real importance of our little invisible helpers. To this end, it is good to take a look at some numbers. Estimations of the value of the world market of any product are generally very difficult to obtain. No-one knows the amounts of worldwide traded goods exactly. World-wide statistics about actual employment are also not available. However, often the order of magnitude can be estimated and generally gives sufficient insight. Indeed, there are many agencies and companies dedicated to assessing such data. Government agencies or the United Nations publish those data for free, but many other companies make their living from the provision of statistics (which are therefore not easily available for public use). In this chapter I try to set some of the (freely available) numbers into context, to highlight the impact of microorganisms on our real economy and employment. However, the reader should kindly note, that the numbers reflect orders of magnitude and are gross estimations – not more – but also not less. As such, they give a good overview and a feeling for the impact we want to look at, but for sure, none of those numbers is precise and correct, looked at individually.

19.2 ECONOMY, EMPLOYMENT, AND MICROBES – SOME NUMBERS

The entire economy of our global human community has been estimated to have amounted to about 87.5 trillion USD in the year 2019 (Worldbank, 2021). This number includes goods which have been traded independent of place and manner. It does not include the stock market, which amazingly reaches a similar number – a fact which should make us think, by the way.

Good Microbes in Medicine, Food Production, Biotechnology, Bioremediation, and Agriculture, First Edition. Edited by Frans J. de Bruijn, Hauke Smidt, Luca S. Cocolin, Michael Sauer, David Dowling, and Linda Thomashow.

This number is huge, but amazingly the fraction of products, which are microbially produced, is significant and visible, even when drawn in a circle diagram as shown in Figure 19.1. Here, only three examples of product groups, where microorganisms play a role, have been considered: food, pharmaceuticals, and chemicals.

The fermented food and ingredients market is projected to be worth 875.21 billion USD by 2027 (Emergen Research, 2021). This is roughly between 12% and 15% of the total world-wide food market, which is quite impressive. So while we enjoy the drinks and food produced with the help of microorganisms, this share is also of major importance from an economic point of view (see Section 2 of this volume).

Roughly a quarter of the huge pharmaceutical market is produced by microorganisms. The global pharmaceutical market amounted to about 1.25 trillion USD in 2019 (Mikulic, 2020), out of which 50 billion USD have been spent for antibiotics and 250 billion USD relate to recombinant proteins (insulin for treatment of diabetes or a number of vaccines for example). The final product group which should be mentioned, is the chemical market. This is also a huge market with about

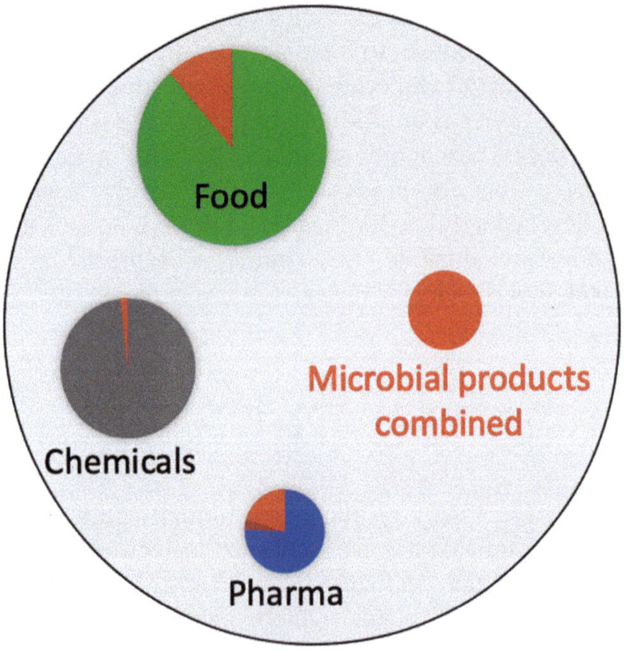

FIGURE 19.1 Size comparison of the global economic market (calculated in USD) with the global food, chemical and pharmaceutical markets. The shares of microbial products of the respective markets are shown in red. Pharma dark red: antibiotics, light red: recombinant proteins. The share of these three microbial product markets is also depicted combined for comparison. Clearly, there are more microbial products already on the market. However, the market share is already impressive and the size is likely to increase in future.

3.8 trillion USD (Statista, 2021). Of course, the major fraction of all chemicals is derived from petroleum and other fossil resources and mainly chemical processes are employed by the respective industries to obtain the desired products (from plastics and fibres to ingredients for cosmetics, just to name a few). Clearly, petroleum as a carbon source is a major driving factor for climate change and unsustainability. Therefore, major efforts are directed to tap other resources, which are more sustainable. The use of agricultural or forestry products or waste for the production of value added chemicals offers great possibilities in this regard. Interestingly, with the change from petroleum to biobased resources, often also a change to microbiological processes comes along. Right now only about 1.7% of the chemicals on the market (by value) is produced with microorganisms (Grandview, 2021). However, the potential is high in this respect – as shall be outlined below.

The numbers about market sizes indicate that microorganisms are also relevant for many jobs. To get a better grasp of this importance we can have a look to the European Union, which publishes useful statistics on a regular basis. About 17.5 million people have been employed in the bioeconomy 2017 – which corresponds to about 9% of all employees (Ronzon et al., 2020). About half of them work in agriculture to place this number into a context. Roughly 420000 people work in the manufacture of bio-based chemicals, pharmaceuticals, plastics, and rubber or the manufacture of liquid biofuels in the European Union. Most of those processes will have one or more relationship with microbiology. Of course this is only a fraction of the employees in microbiology related fields. However, it clearly indicates that over 400000 people's lives and the lives of their families in the European Union directly benefit from the actions of our tiny and invisible helpers and this number will increase decisively in future.

19.3 OUTLOOK INTO A SUSTAINABLE FUTURE – MICROBIAL CHEMICAL PRODUCTION AS AN EXAMPLE

We have seen that the impact of microorganisms is already visible within our economies and employment. However, their importance is susceptible to increase. Climate change is the major challenge for our societies on a global level. The major source for greenhouse gases is of course our excessive use of fossil resources. Also the setup of economies in a linear way (take–make–use–dispose)

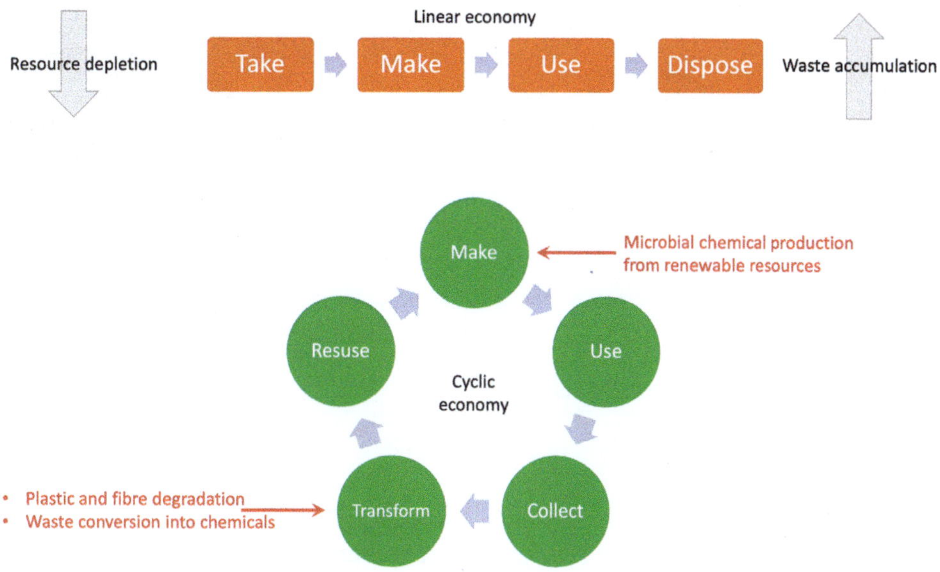

FIGURE 19.2 Our current economic model is linear (upper part of the figure), which leads to resource depletion and waste accumulation. A sustainable model of economy must be cyclic, thereby reusing all resources and avoiding waste accumulation. Industrial microbiology is one technology aiding this development.

is an obvious reason for unsustainability of our global human society (Figure 19.2, upper part). Nature gives us the prime example, how an incredible use of resources can be sustainable for millennia. The basis for nature's sustainability is, that nature is constructed in cycles. Carbon dioxide is fixed from the atmosphere in plants or microalgae, the plants (fixed carbon) are taken up as food and finally microorganisms degrade all organic material in the end again – closing the cycle by releasing the carbon as carbon dioxide into the atmosphere (see chapter 21). All other elements (nitrogen, sulphate, and the like) flow in similar cycles – and always the microbial population of our planet plays as decisive role, because many degradation reactions can only be performed by microbial life forms. Also many of the required reactions for degradation tasks need specialized extreme conditions to run, which only microbes can sustain.

Our economic models must be adapted to become cyclic too – like the cyles in nature – in order to let our societies become sustainable in the long run (Figure 19.2, lower part). A major part of this transition is the development of a viable bioeconomy, as it is pursued by the European Union and many other world regions.

As in nature, microbiology has the option to play a key-role in many aspects in this transition. Microorganisms can serve as cell factories for the production of chemicals from renewable resources. However, they can also take over tasks degrading materials, which would otherwise be waste. Currently, microbial communities are for example used to produce biogas from organic waste materials. This at least allows the energy stored in that waste to be obtained and made use of. Much better would be a further use of the material itself. For example, processes are being developed to degrade waste (including fibres and plastic) into monomeric substances, which can subsequently be used again to produce plastics and fibres (see chapter 24).

19.4 WHAT MAKES MICROORGANISMS USEFUL FOR THE CHEMICAL INDUSTRY?

Living microbes carry out a quite complex network of chemical reactions. Not unlike us they take up substrate (eat food), metabolize it to obtain energy and grow and excrete waste products (go to the toilet). However, what is a waste product from the microorganisms perspective, might be a valuable chemical from a human perspective.

One such example, which has been used for about 100 years now, is *Aspergillus niger* (see Figure 19.3). This fungus is a black mold, which might very well have been found by the reader already on rotting fruit. In 1917 it was discovered that this fungus produces large amounts of citric acid when cultured under the appropriate conditions (Currie, 1917). Citric acid was at that time already a sought after ingredient widely used by

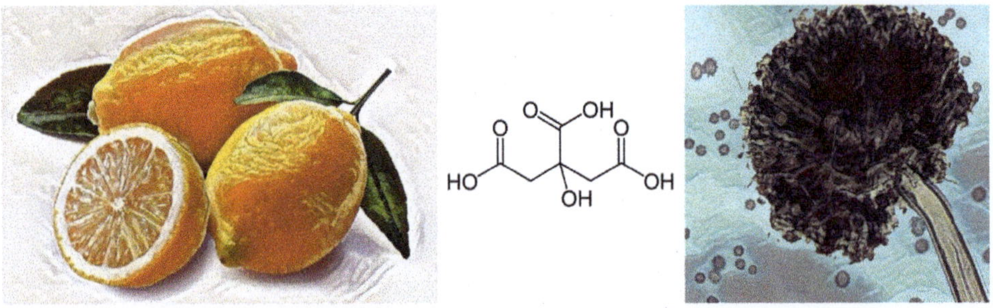

FIGURE 19.3 Lemons (left) have been the basis for citric acid production for a long time. Nowadays the filamentous fungus *Aspergillus niger* (right) is used to produce the acid, which is depicted in the center. Citric acid is frequently used in the food and beverage industry as acidulant, which has a very pleasant taste.

the food and beverage industry. However, citric acid had until then been exclusively produced from lemons, which was quite an expensive process. With the discovery of the ability of the fungus to produce this ingredient from sugar, industrial production changed to microbiology (Sauer et al., 2008). Nowadays, sugar is converted into citric acid in huge (up to 1000 m^3) steel tanks. About 2 million tons of this acid are used every year and the fungus obtained as a side product can be used as animal feed, underlining the sustainability of this approach (see chapter 23).

However, not only food ingredients can be produced by microbes, but also fuels, solvents, and other chemicals. Indeed, the major biotech product (by volume) currently on the market is ethanol as fuel (around 100 billion litres or 80 million tons per year). Like many microbial products ethanol is produced from sugar, which might be derived from sugar cane or sugar beet, or from cereals, such as corn, or wheat. Ethanol is a natural product from baker's or brewer's yeast. The fermentation capability of this fungus is used in the beverage industry for alcoholic drinks, such as beer and wine (see Chapters 17 and 18), but – as indicated – also industrially for ethanol as fuel. Biofuel production touches several important points, which must be considered on our way to a microbially powered sustainable future (see chapter 23). First of all the capacity and capability for microbial products at very large scale and at prices low enough to be sold and used as fuel clearly exist. However, when considering production at such a scale, the raw material provision becomes crucial. Ethanol production is currently based on food grade agricultural products, which raises ethical concerns, considering that we are still not able to provide all humans with the minimum amount of food. In any case, biofuel production is based on agriculture or forestry. Therefore, arable land area must be used for fuel production. This in turn raises concerns about land use change. Rain forest might disappear to free land for sugar cane, but also land, now used for food production, might be used for fuel production, indicating that it is not the use of the vegetable product itself, which should raise ethical concernd, but the fact that land which could be an area of unspoilt nature or land which could be used for food production are turned into areas for fuel or chemical production. This concern does not mean that such endeavors cannot be sustainable, but they clearly show that our transition to a biobased economy must be accompanied by well-reasoned policies, addressing factors such as land use change, otherwise a fantastic approach can go utterly wrong. Data quantifying the potential of our globally available arable land clearly show that we have enough space for food and chemical production, satisfying the world market for both (Vennestrøm et al., 2011). However, in contrast to the amount of chemicals for material use, the amount of energy fuels, which are currently derived from petroleum cannot easily be met by biomass. This seems also not really to be required, as other energy sources (solar energy, wind and water, to name just a few) are available.

The number of chemicals which are currently produced from renewable resources by microorganisms is still limited (as mentioned about 1.7% of the market). Figure 19.4 gives an illustration for chosen examples of chemicals, which can actually be produced by microbial processes. They range from fuels, base chemicals for plastic and fiber production, to food ingredients and pharmaceuticals. Some of them are produced by naturally occurring microbes, but some others can only be produced by microbial cell factories, because they have been designed appropriately.

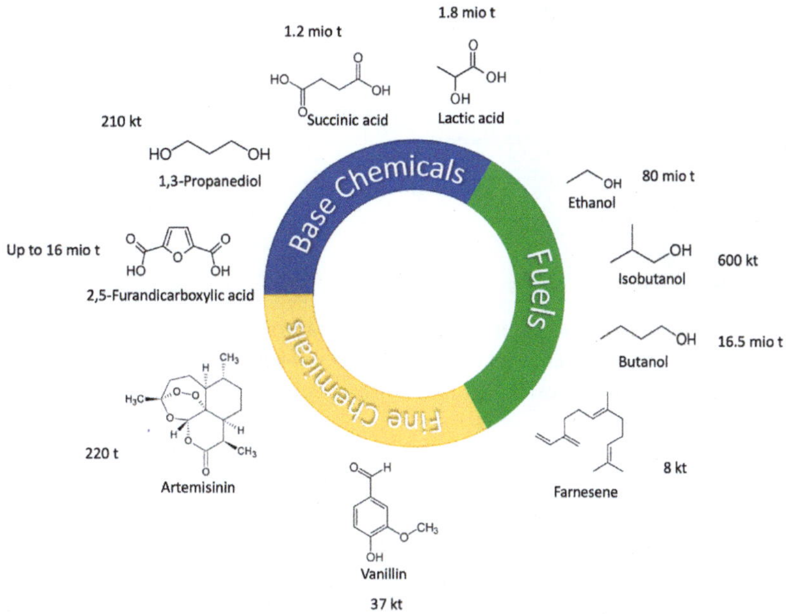

FIGURE 19.4 Examples for microbially produced chemicals. The chemicals are grouped into base chemicals and fine chemicals according to their market size and application. Base chemicals, as defined here, are used for fiber production and other wide ranges of applications. Fine chemicals have limited markets for special applications. The given market sizes are to give an order of magnitude only. Currently, some of the chemicals are only partly produced by microbial processes and partly by chemical processes from petroleum. Also the market sizes are apt to increase significantly, when the processes can be optimized and the prizes are falling. 2,5-furan dicarboxylic acid is such an example. The current production is very low (less than 10 kt). However, it has the potential to exchange terephtalic acid for plastic production (PET) and is, therefore, potentially a huge market. The given number of 16 million tons is the target market of terephtlaic acid – not the current market of 2,5-furan dicarboxylic acid. Similarly, succinic acid with a current market of up to 1.2 mio t has the potential to replace maleic acid with a current market exceeding 3 mio. t.

19.5 METABOLIC ENGINEERING ALLOWS THE DESIGN OF MICROBIAL CELL FACTORIES

While citric acid and ethanol are mass products formed by natural microorganisms, the full potential of the microbial world for the chemical industry only becomes available when we can apply our engineering principles not only to the production plant made of steel, but also to the microorganisms, which are at the center of the process. We call these microorganisms our microbial cell factories. As mentioned before, microbial cells depend on a more or less complicated set of chemical reactions fulfilling all tasks to sustain life. These chemical reactions form the metabolism of the cell. When genetic engineering became possible – starting in the 1970s – it became possible to tinker also with the metabolic setup of cells (see also chapter 1). Enzymes are catalysts, which enable chemical reactions in the cells. When a cell produces a certain enzyme it can foster a given reaction, when a cell ceases to produce this enzyme, the reaction will no longer proceed. We can use such knowledge and add or remove enzymes, thereby changing the metabolism of microbial cells at will (Choi et al., 2019). This allows us to enable microbes to produce chemicals, which they would naturally not produce. Although many of them are chemicals which are naturally produced in other living organisms. Isoprenoids for rubber production are one such example. Rubber is traditionally produced from tree saps. Microbial rubber production has the potential to ease our dependence on tropical tree plantations (Li et al., 2020). The malaria drug artemisinin is another example.

19.6 FROM PLANT TO MICROBE – PRODUCTION OF THE MALARIA MEDICATION ARTEMISININ

Traditional Chinese medicine has a long history of using the plant sweet wormwood (*Artemisia annua*) for the treatment of malaria (Figure 19.5). The active ingredient Artemisinin was identified in the 1970ies by Youyou Tu, who received the Nobel prize in physiology or medicine

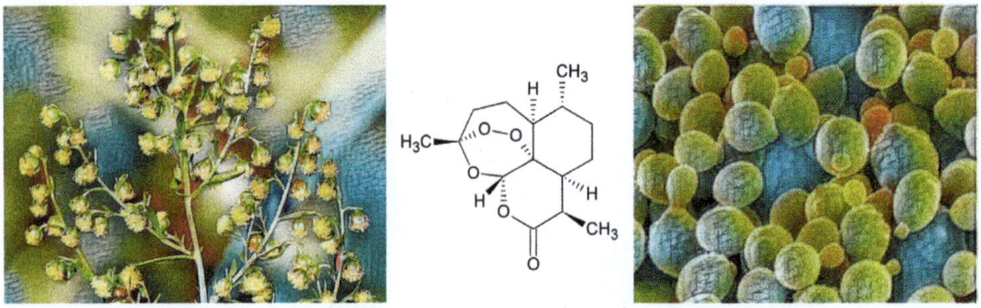

FIGURE 19.5 The malaria drug Artemisinin (structure in the center) is a natural product of the sweet wormwood (*Artemisia annua*) which is depicted to the left. Metabolic engineering allowed construction of a baker's yeast strain (right), which produces a precursor for this drug very efficiently. So nowadays, this drug can be microbially produced.

2015 for her achievements (NobelPrize.org, 2015). The WHO designated Artemisinin as a first-line antimalarial drug. However, providing patients world-wide with this substance remained difficult because it depends on extraction of the drug from plants, which do not contain large amounts of the substance and whose cultivation is weather dependent. Therefore, prices and availability fluctuated. Chemically, the substance is a complicated molecule, which is difficult to synthesize – so a purely chemical production was not an economically feasible option. Starting with a grant from the Bill and Melinda Gates Foundation scientists set out to develop a process for semi-synthetic synthesis of the drug. The strategy was to produce a precursor (artemisinic acid) for the drug with microorganisms and then to convert this precursor into the active ingredient artemisinin by synthetic organic chemistry (Paddon and Keasling, 2014). The task was to enable a microorganism of choice to produce the precursor artemisinic acid – which is a substance only produced by plants in nature. That plants produce it, proves that all enzymes exist to catalyze all the chemical reactions required for its production. The first task was therefore to identify all of the genes and enzymes involved. The next step was to transfer the required genes by gene technological interventions into a microorganism of choice. The metabolism of the microorganism had to be engineered to enable this microbial cell factory to produce a plant metabolite (Paddon and Keasling, 2014).

First approaches relied on *Escherichia coli* as a microbial cell factory. This bacterium is a natural inhabitant of the human gut and it is a model organism for research and production for decades. By recombinant DNA transfer of the genes coding for enzymes from baker's yeast and *A. annua* into this organism, it was possible to obtain the desired plant chemical from this gut bacterium. However, it turned out that the enzyme for the final reaction step forming artemisinic acid was not well produced by the bacterium, so the cell factory was changed. The next approach was using baker's yeast, into which the plant enzymes were transferred. This approach was quite successful and nowadays it is possible to obtain a constant supply of the desired drug at a predictable price – from a microbial process – not being dependent on plant extraction anymore.

19.7 OPENING UP THE CHEMICAL SPACE WITH THE TOOLS OF SYNTHETIC BIOLOGY

In the past the development of microbial cell factories has widely been a field of trial and error. Classical metabolic engineering relies on the cyclic procedure of design, build, and test and then start over with the next design build and test cycle. The required a change of host organism, described for artemisinic acid production can be seen as a typical step in this cycle. The *E. coli* strain has been designed and successfully built, but thorough analysis of the obtained strain revealed some technical hurdles which could barely be overcome. A change of host was the suggested solution and led to success. However, what is described in these few lines and sounds very simple was of course a task for many researchers for many years. Synthetic biology attempts to ease this process and shorten the times, by applying common rules for engineering to biological systems (Brooks and Alper, 2021). The development and use of standardized parts would help to increase the predictability of the reactions and allow robust cellular systems with novel functionalities that do not exist in nature to be constructed. Systems biology is a complementary attempt to gain understanding of how the cells work by looking at large data sets and using these data to construct mathematical models describing the cell in its entirety (Yu and Nielsen, 2019). Such large scale data sets comprise, for example, data about all genes present in a cell. Which genes are active and to which degree? Therefore, we gain understanding of our microbial cell factories by systems biology, while synthetic biology provides tools, which work in a predictable manner and allows easy construction of a desired metabolism. Such tools help, for example, to be able to combine metabolic pathways in different cell factories in a short time with limited effort. Finally, this allows us to produce chemicals of choice, creating a sustainable starting point for green chemistry with the potential to substitute the petroleum refinery as basis for our materials. Most of the processes to produce such unnatural substances are still under development. However, the potential in terms of economic success and improved sustainability is huge. One prominent example for such a chemical based on renewable resources with the potential to substitute a petroleum based counterpart is 2,5-furandicarboxylic acid (Wierckx et al., 2015). This chemical can be microbially produced from lignocellulosic biomass (from agricultural wastes) and can replace terephtalic acid for plastic production. Terephtalic acid is currently used for PET (polyethylene terephthalate) production – one of the most widely used plastics (see chapter 24). Various microbial cell factories are being developed for this process. These include bacteria and yeasts. Which of the strains will finally go large scale in industry is not decided yet.

Another promising example is adipic acid. This chemical is already a petroleum based mass product. About 3 million tons are used every year to produce fibres (GrandView, 2018). Nylon is the most widely used fibre produced from adipic acid. Production of this chemical is quite a polluting procedure, responsible for significant amounts of greenhouse gases released into the atmosphere (Kruyer and Peralta-Yahya, 2017). While adipic acid can be found in very small amounts in certain organisms (even in the liver of humans) it seems to be a degradation product of larger molecules, but essentially no natural metabolic pathway for production of this chemical has been found to date. Many groups worldwide are currently seeking to construct a microbial cell factory for providing this fibre precursor. We are not there yet, but when we arrive, a major environmental problem can potentially be solved and an 8 billion USD market satisfied. Many similar approaches are under development.

However, the economic success of such chemicals is currently often limited, because petroleum is very cheap at the moment and the price is the major market driver. Environmental impact is only a minor decision factor for industry and consumers at the time being. However, this will change.

19.8 CONCLUSIONS

Microorganisms already play a significant role for our economies. With the transition to a more sustainable bioeconomy this importance will increase. Particularly, the chemical industry will make use of the versatile capabilities of our tiny helpers. Synthetic biology and systems biology allow to design microbial cell factories to our needs, proving the chemical industry with unprecedented options to become green.

REFERENCES

1. Brooks SM, Alper HS. 2021. Applications, challenges, and needs for employing synthetic biology beyond the lab. *Nature Communication* 12(1): 1390. 10.1038/s41467-021-21740-0.
2. Choi KR, Jang WD, Yang D, Cho JS, Park D, Lee SY. 2019. Systems metabolic engineering strategies: Integrating systems and synthetic biology with metabolic engineering. *Trends in Biotechnology* 37(8): 817–837. https://doi.org/10.1016/j.tibtech.2019.01.003.
3. Currie JN. 1917. The citric acid fermentation of *Aspergillus niger*. *Journal of Biological Chemistry* 31: 15–37. https://doi.org/10.1016/S0021-9258(18)86708-4.
4. Emergen Research. 2021. https://www.globenewswire.com/news-release/2020/12/03/2138811/0/en/Fermented-Food-and-Ingredients-Market-To-Be-Worth-USD-875-21-Billion-by-2027-Emergen-Research.html (accessed, 4th March 2021).
5. Grandview. 2021. https://www.grandviewresearch.com/industry-analysis/fermentation-chemicals-market (accessed March 2021)
6. GrandView. 2018. https://www.grandviewresearch.com/press-release/global-adipic-acid-market (accessed March 2021)
7. Kruyer NS, Peralta-Yahya P. 2017. Metabolic engineering strategies to bio-adipic acid production. *Current Opinion in Biotechnology* 45: 136–143. https://doi.org/10.1016/j.copbio.2017.03.006.
8. Li M, Hou F, Wu T, Jiang X, Li F, et al. 2020. Recent advances of metabolic engineering strategies in

natural isoprenoid production using cell factories. *Natural Production Reports* 37(1): 80–99. https://doi.org/10.1039/c9np00016j.

9. MikulicM.2020.https://www.statista.com/topics/1764/global-pharmaceutical-industry/#dossierSummary__chapter2 (accessed March 2021)
10. NobelPrize.org. 2015. https://www.nobelprize.org/prizes/medicine/2015/tu/facts (accessed March 2021).
11. Paddon CJ, Keasling JD. 2014. Semi-synthetic artemisinin: A model for the use of synthetic biology in pharmaceutical development. *Nature Reviews in Microbiology* 12(5): 355–367. https://doi.org/10.1038/nrmicro3240.
12. Ronzon T, Piotrowski S, Tamosiunas S, Dammer L, Carus M M'barek. 2020. Developments of economic growth and employment in bioeconomy Sectors across the EU. *Sustainability*. 12(11): 4507. https://doi.org/10.3390/su12114507.
13. Sauer M, Porro D, Mattanovich D, Branduardi P. 2008. Microbial production of organic acids: Expanding the markets. *Trends in Biotechnology* 26: 100–108. https://doi.org/10.1016/j.tibtech.2007.11.006.
14. Statista. 2021. https://www.statista.com/statistics/302081/revenue-of-global-chemical-industry (accessed March 2021).
15. Vennestrøm PN, Osmundsen CM, Christensen CH, Taarning E. 2011. Beyond petrochemicals: The renewable chemicals industry. *Angewandte Chemie International. Ed Engl* 50(45): 10502–10509. https://doi.org/10.1002/anie.201102117.
16. Wierckx N, Schuurman TDE, Blank LM, Rui-jssenaars HJ. 2015. Whole-cell biocatalytic production of 2, 5-furandicarboxylic acid. In: Kamm B, Steinbüchel A (Eds.), *Micorbiology Micrographs volume 26: Microorganisms in Biorefineries*, Springer Verlag Berlin Heidelberg, Germany. pp. 207–223. DOI:10.1007/978-3-662-45209-7.
17. Worldbank. 2021. https://data.worldbank.org/indicator/NY.GDP.MKTP.CD?most_recent_value_desc=true (accessed March 29, 2021).
18. Yu R, Nielsen J. 2019. Big data in yeast systems biology. *FEMS Yeast Research* 19(7): foz070. https://doi.org/10.1093/femsyr/foz070.

CHAPTER 20

Role of Microorganisms in Environmental Remediation and Resource Recovery through Microbe-Based Technologies Having Major Potentials

Piyush Malaviya[1], Rozi Sharma[1], Smiley Sharma[1], and Deepak Pant[2]

[1]Department of Environmental Sciences, University of Jammu, Jammu, Jammu and Kashmir, India

[2]Separation & Conversion Technology, Flemish Institute for Technological Research (VITO), Mol 2400, Belgium

20.1 INTRODUCTION

Our environment has become an issue of immediate concern in present times because of the rapid industrialization, modernization, urbanization, and continuous population progression. Available natural resources are rapidly degrading and environmental pollution (including air, water, noise, and soil pollution) is rising with each passing day. This pollution is directly threatening the health and survival of all life forms, and disturbing the ecological balance of the planet (Sharma et al., 2021). Therefore, abatement of environmental pollution is urgently needed, for which many physical, chemical, and biological treatment methods are available (Mishra et al., 2020). Physico-chemical methods have many disadvantages associated with them that include formation and disposal of sludge, high treatment cost, low efficiency, and inapplicability to a wide variety of metals (Bilal and Iqbal, 2020). Thus, biological methods are preferred over physico-chemical methods owing to their lower cost, higher effectiveness, and eco-friendly nature. Of all the biological technologies, bioremediation has emerged as the most proficient and anticipated approach for wastewater treatment (Malaviya et al., 2019; Pant and Adholeya, 2009a).

Bioremediation utilizes microorganisms (bacteria, fungi, yeast, and algae) and even higher plants (in some applications), to break down the organic compounds and clean up the polluted sites (Kumar et al., 2019; Pant and Adholeya, 2009b; see section 4). Broken organic compounds serve as food for microbes. The ability of bioremediation to detoxify hazardous materials and transfer them from one environment medium to another makes it more significant over other remediation techniques (Rashid et al., 2019). One practical aspect of microbial ecology is its contribution to improvement in wastewater treatment and in situ bioremediation (Garcia-Garcia et al., 2016).

Since the environment-energy-food-climate change nexus seems set to become the most prominent in the next decades, therefore microbiological processes need to be promoted for global applications. Microbes have a major potential for these environmental needs (Singh et al., 2021). For instance, fungal biomass can treat metal-contaminated effluents, organic pollutants, nutrients, and dyes from wastewaters with great efficiency. Fungi can also produce valuable products like enzymes

Good Microbes in Medicine, Food Production, Biotechnology, Bioremediation, and Agriculture, First Edition. Edited by Frans J. de Bruijn, Hauke Smidt, Luca S. Cocolin, Michael Sauer, David Dowling, and Linda Thomashow.

(amylase, laccase, LiP, MnP, etc.) and single-celled proteins (Lacina et al., 2003; Pant and Adholeya, 2007). Besides, different microorganisms can also produce value-added products like nutritional foods, organic acids, biopesticides, biofertilizers, biofuels, biogas, bioenergy, biopolymers, reduce the greenhouse effect, bioplastics, etc., that are major environmental needs of the next generation (Singh et al., 2021). Thus, it is crucial to find innovative and potential microbe-based technologies in the framework of climate change, by-products valorization, and environmental sustainability.

20.2 MICROORGANISMS AS IMPORTANT BIOLOGICAL ENTITIES IN THE ENVIRONMENT

Microorganisms include a diversity of bacteria, fungi, viruses, yeast, algae, cyanobacteria, actinomycetes. Some of them are deleterious (pathogens) while many others are beneficial. Therefore, specific protocols must be adopted to identify beneficial microbes and different strategies should be developed to increase their population. Multiple groups of microorganisms with diverse physiological and metabolic abilities and phylogenetic affiliations are present in the environment (Lin et al., 2007). There are many microbe types like mesophilic and thermophilic fermentative microorganisms, hyperthermophiles, syntrophic, chemolithoautotrophic, autotrophic, and heterotrophic microorganisms. Microorganisms can produce a number of added-value products (acids, solvents, enzymes, gases, biosurfactants, biopolymers, biofuels, bioenergy, biopolymers, agro-based materials, and emulsifiers) and have the potential to degrade and transform a wide range of xenobiotics (organic chemicals, heavy metals, PAHs, crude-oil, phenol, etc.) present in aquatic ecosystems, soils, and air (Mishra et al., 2019). Microorganisms specific for biodegradation characteristics are isolated from soils, biofilms, or sludges that are polluted with particular xenobiotic compounds. Microorganisms can even facilitate the growth of plants in soils containing high concentrations of salts. Beneficial microorganisms especially plant-growth-promoting bacteria and mycorrhizae enable the plants to proliferate even in inhibitory levels of salts. They also account for human health and food safety (Berendonk et al., 2015).

Microorganisms are, therefore, regarded as efficient tools for tackling the severe environmental challenges of the next decades. They engage in high energy-efficient processes and show high pollutant removal rates (Read et al., 2011). They not only host a pivotal position in the wastewater treatment approach but also serve as the provider of renewable energy, nutrients, and important organics (Batstone and Virdis, 2014; Colombo et al., 2017). As microbial growth occurs exponentially, large amounts of cost-effective valuable products can be produced rapidly from renewable resources.

20.2.1 Role of Microorganisms in Urgent Environmental Needs

To abate pollution, microbial technologies have been developed and optimized since the beginning of the twentieth century (Ardern and Lockett, 1914), and such technologies have led to a number of applications in this framework. Innovations in microbial technologies have uplifted the bioremediation process from a mere wastewater treatment technology to a provider of value-added products as well (De Vrieze et al., 2016). This combination of wastewater management and single-cell protein production technologies influences the water-food-climate-energy nexus and many other related issues of present times in a very effective manner. However, in this chapter, we have focused only on issues related to climate change, environmental pollution, and sustainable development, as well as the role of microorganisms to tackle such urgent environmental issues.

20.2.1.1 Pollution Control

Alteration in the environment due to the continuous addition of xenobiotics and several other recalcitrant compounds is a serious threat to life on the planet. To eliminate such pollutants from the environment, bioremediation has emerged as an efficient pollution management key (Jaiswal and Shukla, 2020; see section 4). Microorganisms have degradation capabilities and their enzyme systems also have the potential to break, detoxify and transform the hazardous compounds present in the environment (Liu et al., 2019a). Most of the treatment processes use the natural self-purification capacity of aquatic environments which is due to the action of the microbial population present in them. Therefore, the design and development of different microbial bioremediation techniques are to be sought in accordance with the pollutants existing in the environment. Microbes with engineered catalytic potential are cultivated. High densities and activities of such microorganisms should be maintained in treatment plants to satisfy the various purification demands. Cloning, isolating strains, and enriching pollutant-degrading mixed microbe cultures can successfully enrich specific as well as non-dominant consortium members capable of degrading even intermediates of the pollutants (Dejonghe et al., 2003).

Microorganism species belonging to the genera *Acinetobacter, Arthrobacter, Acinetobacter, Enterobacter, Photobacterium, Pseudomonas* are the potential agents for pesticide degradation (Monica et al., 2016; Ravi et al., 2015). *Pseudomonas fluorescens, Pseudomonas aeruginosa, Lysinibacillus sphaericus, Microbacterium profundi, Aspergillus versicolor, Aspergillus fumigatus, Paecilomyces* sp., *Trichoderma* sp., *Microsporum* sp., *Cladosporium* sp. *Geobacter* sp. are the microorganisms that serve for removing heavy metals from the environment (Mirlahiji and Eisazadeh, 2014; Pena-Montenegro et al., 2015; Wu et al., 2020). Phenolic compounds are also one of the major pollutants present in wastewater and there are many phenol-digesting microbes like *Comamonas testosteroni* that can degrade these toxic compounds (Whitley and Bailey, 2000). Polycyclic aromatic hydrocarbons (PAHs) are the severely toxic pollutants present in the environment and are priority pollutants owing to their mutagenic and carcinogenic nature (Malaviya et al., 2019). A diverse group of microorganisms including bacteria and fungi is capable of degrading, oxidizing, mineralizing, and transforming a variety of recalcitrant PAHs into detoxified products (see Chapter 27). Major microbial strains capable of degrading PAHs include *Alcaligenes* sp., *Achromobacter* sp., *Burkholderia cepacia, Coprinellus radians, Mycobacterium* and *Microbacterium* sp., *Novosphingobium* sp., *Ralstonia* sp., *Rhodococcus* sp., *Sphingomonas paucimobilis, Stenotrophomonas maltophilia* (Aranda et al., 2010; Simarro et al., 2013; Wongwongsee et al., 2013; Sani et al., 2015). Several microorganisms have remediation potential for hydrocarbons present in oil waste, for example, *Sphingomonas, Sphingobium* strains (Zhao et al., 2017), *Aeromonas hydrophila, Yarrowia lipolytica* (Mishra et al., 2020), *Xanthomonas maltophila*, and *Candida maltosa* that can degrade up to 90% of diesel (Kaczorek et al., 2013).

20.2.1.2 Carbon Sequestration

The rise in CO_2 level increases global temperature and results in a "greenhouse effect" and makes the earth unsuitable for habitation (Kweku et al., 2017). To reduce CO_2 emissions, CO_2 sequestration is necessary. Carbon sequestration is the removal of CO_2 from the atmosphere by natural or deliberate processes (see Chapter 21). By sequestering about 4200 Mt of CO_2 per year, approximately 80% of emissions can be reduced (Mac Dowell et al., 2017). Broadly there are two sequestration strategies: biological and non-biological. The removal of CO_2 from the atmosphere by biological entities (plants, soil, and microbes) is known as bio-sequestration (Kondaveeti et al., 2020). Chemical reactions (like mineral carbonation), non-living means (like the storage of atmospheric carbon in deep oceans), or carbon storage in underground geological formations (like unmineable coal seams, saline aquifers, dissolved, porous sediments) (Mac Dowell et al., 2017; Nogia et al., 2016) contribute a lot to CO_2 reduction, however, alternative options like microbial processes for CO_2 sequestration also need further exploration. Microorganisms (including eukaryotes, prokaryotes, and microalgae) act as biofactories for CO_2 sequestration even from very low atmospheric CO_2 concentrations (Menon et al., 2007). Their CO_2 fixation efficiency is 10–50 times higher than terrestrial plants. Microbial CO_2 sequestration provides additional advantages like procurement of by-products like bioplastics, biofuels, and biomolecules like proteins, lipids, and carbohydrates (Mistry et al., 2019). Microbe-rich soil with little manipulation can reach a high degree of microbial homeostasis and contribute to high CO_2 sequestration (Cleveland and Liptzin, 2007). Since, carbon capture, storage, and utilization are the main focus during carbon reductions, therefore, microbial engineering holds a vast potential to maximize CO_2 sequestration (see Chapter 21).

20.2.1.3 Biofuel Production

Biofuels derived from microorganisms are third-generation fuels and considered sustainable fuel alternatives (Lopes da Silva et al., 2014). Biofuel includes biodiesel (see Chapter 22), biohydrogen, biomethane, and bioethanol (Rodionova et al., 2017). Fuels derived from biological sources or derived by the action of microorganisms are considered to be a most valuable and potential resource due to attractive features like renewability, degradability, environment-friendliness, high lipid content (50–70%), the high caloric value of 39–41 MJ/kg, and fewer greenhouse gas emission (Aghbashlo and Demirbas, 2016; Satputaley et al., 2017). These are the only fuels that can be produced plentifully in a renewable way and with minimal ecological harm. However, biofuels are required to be produced by using advanced microbial technologies because it is not possible to produce them abundantly from crops only as these could ultimately serve as food sources also. This leads to their high production cost. Some microbes can even produce added-value products like carbohydrates, fatty acids, proteins and are used in biorefinery integrated processes in industries like food, medicine, and cosmetics. In this way, the fuel production process can become economically feasible as recovered products can sustain biofuel production (Lopes da Silva et al., 2014). *Saccharomyces cerevisiae, Chlamydomonas perigranulata, Chlamydomonas reinhardtii, Nannochloropsis oculata, Chlorella protothecoides, Candida, Dunaliella, Scenedesmus, Spirulina, Desmodesmus* sp. are important microorganisms used for biofuel generation (Bhatia et al., 2017).

20.2.1.4 Biogas Production

Biogas is methane-rich gas that is produced by a consortium of microbes through hydrolysis and fermentation of organic waste under anaerobic conditions. This pollution-free and renewable energy source gas can be employed for electricity generation, heat production (Szarka et al., 2013), microbial feed (Strong et al., 2015), etc. Biogas can be produced through several engineered microbial technologies, and simpler anaerobic microbial ecosystems as well (like sewers, landfills, wetlands, manure pits, septic tanks) (Perez-Barberia, 2017). The process is carried out in four stages viz. hydrolysis, acidogenesis, acetogenesis, and methanogenesis catalyzed by specific microorganisms (Maus et al., 2014). *Clostridium bornimense, Clostridium ultunense, Herbinix hemicellulosilytica, Methanoculleus bourgensis, Peptoniphilus* sp., *Ruminiclostridium cellulosi* are some of the microorganisms known to have potential for biogas production. Although there are at present many experimental efforts focused on biogas production through microbial technologies, the contributing microorganisms and genes are still not known extensively (Heyer et al., 2016). So, there is a huge potential for microbial technologies in biogas production to deal with methane emissions, and hence climate change (Aryal et al., 2021).

20.2.1.5 Biofertilizer Production

Dependency on chemical fertilizers to increase the crop yield destroys environmental ecology, soil health as well as human health. Therefore, it is necessary to replace them with biofertilizers that are capable of restoring soil nutrients (White and Broadley, 2009). These eco-friendly bio-based fertilizers enhance plant growth, crop yield, and soil fertility, and promote sustainable agriculture. Epiphytic, endophytic, and rhizospheric microorganisms colonize the plant systems and help in nutrient uptake via certain mechanisms like biological nitrogen fixation (see Chapter 37), production of growth-promoting hormones, siderophores, hydrolytic enzymes, and solubilization of potassium, zinc, and phosphorus (Kour et al., 2020; see section 5).

20.2.1.6 Production of Single-cell Proteins

Microbes also help in the production of single-cell proteins (SCPs) that possess much higher nitrogen efficiencies, and that can be directly used as food sources (Matassa et al., 2016; Pikaar et al., 2017). The main advantage of SCPs as food and fodder lies in their capability to recover nitrogen from diverse sources, prevent excessive buildup of reactive nitrogen in the biosphere, and avoid additional costs for conversion of nitrogen into nitrogen gas (Luther et al., 2015).

20.3 DIFFERENT MICROBIAL TECHNOLOGIES WITH HIGH POTENTIAL FOR ENVIRONMENTAL EXIGENCIES

20.3.1 Omics Technologies

Omics is a newly emerging molecular technique that holds a significant position in the medical field, biology research, and the worlds of science and sustainability (Salehi Jouzani and Sharaf, 2018). These technologies are helpful in biosynthesis and/or degradation of metabolic precursors, intermediates, or end products, and also to regulate the metabolic flux and genetic manipulations (Mishra et al., 2019). Different omics technologies are available for different cellular molecules, including genes, RNAs, proteins, and metabolites leading to developments of technologies like genomics, transcriptomics, proteomics, and metabolomics, respectively. Genomics uses the combination of rDNA, DNA sequencing, and bioinformatics to identify genes and analyze their function under stress as well as their role in the recovery of value-added products (Stolze et al., 2015; see chapter 30). Transcriptomics studies the expression and regulation of all types of RNAs which help in finding active cellular processes (Maus et al., 2016). Proteomics is the identification of proteins and their physiological role in toxicity prediction and waste management (Heyer et al., 2016). Metabolomics studies the metabolites produced in cellular processes to understand the biology of organisms, their role in the environment, response to stress and utilization in drug formation, and many other ecological functions (Sasaki et al., 2014). All these "omics" approaches provide excellent opportunities for documentation of uncultivable microbes, unidentified genes, and unexplored pathways involved in the production of biogas (Salehi Jouzani and Sharaf, 2018), pharmaceuticals, nutraceuticals, cosmeceuticals (Mishra et al., 2019), wastewater treatment (Wilks and Scholes, 2018), nutrient removal (Perera et al., 2019), metal removal (Zhang et al., 2015), etc.

20.3.2 Nanobioremediation Technology

Nanobioremediation is a new microbial technology that uses nanoparticles and bioremediation in an integrated manner for the removal of pollutants in a sustainable manner (Cecchin et al., 2017). This technology removes pollutants at a higher rate (Koul and Taak, 2018). However, in certain cases, nanoparticles may harm the remediating microorganisms (Cecchin et al., 2017). Bokare et al. (2012) used nanoparticles of Pd/Fe and enzyme produced by *Trametes versicolor* to remove triclosan from contaminated solutions and reported total dechlorination of the chemical just 20 minutes after application. Le et al. (2015) have obtained 89% degradation in a PCB (arocolor 1248), and 90% degradation in biphenyls using Zero Valent Iron (nanoparticle) and *Burkholderia xenovorans*. Up to 99% efficiency removing lindane from contaminated soil by nanoremediation techniques using Carboxy-methyl-cellulose-Pd/Fe nanoparticles and *Sphingomonas* sp. have also been reported (Bokare et al., 2012).

20.3.3 Electrobioremediation

Electrobioremediation is a microbial technology that cleans organic as well as inorganic contaminants from polluted environmental matrices using a combination of remediation and electrokinetics techniques (Rabbani et al., 2021). In this technology, direct current is applied within media which induces electrokinetics and accelerates the transport of contaminants and microorganisms toward an electrode for remediation (see chapter 33). Electrokinetics involves several processes like diffusion, osmosis, electrophoresis, migration, and electrolysis under weak electric currents (0.2–2 V/cm) (Maszenan et al., 2011). The efficiency of electrobioremediation is influenced by electrolyte type, nature of contaminants, and appropriate electrode material (that should be inexpensive, durable, non-toxic to microbes, chemically stable, catalytically active, and selective for target pollutants) (Daghio et al., 2017). Design and operating parameter optimization are necessary to lessen the internal resistance in the device and increase the overall efficiency of the process. There are many studies available on this technology, for example, phenanthrene degradation by *Shewanella oneidensis* and *Pseudomonas aeruginosa* in a microbial fuel cell (Adelaja et al., 2014), degradation of PAHs by *Chloroflexi* sp. and *Nitrospira* sp. (Yan et al., 2012), remediation of tannery wastewater by *Bacillus* in electrooxidation reactor (Kanagasabi et al., 2013), etc.

20.3.4 Microbial Electrosynthesis for CO_2 Sequestration

Rabaey and Rozendal (2010) coined the term "microbial electrosynthesis" and defined it as an electrochemical process that uses microorganisms as biocatalysts to drive microbial metabolism and produce industrially important organic products as well as produce renewable energy in electrochemical cells. In this process, microbial cells capture electrons from the cathode and fix carbon (Marshall et al., 2017). Among different anodic and cathodic routes, the process of converting CO_2 into bio-products is the most significant, attractive, and interesting. Bio-based products obtained after CO_2 reduction have the potential to replace fossil-based commodities, and microbial technology has scientific novelty, and societal as well as ecological impact (Prevoteau et al., 2020).

Microbial electrosynthesis from CO_2 ensues through electron transfer mechanisms and the Wood–Ljungdahl pathway (Ragsdale and Pierce, 2008). The same has been presented in Figure 20.1 (modified from Prevoteau et al., 2020). By providing or removing electrons, certain biochemical reactions can be activated, for example, the reduction of CO_2 into acetic acid (Prevoteau et al., 2020). Electron transfer may take place by direct electron transfer or with the help of redox mediators. The Wood–Ljungdahl pathway occurs in the presence of homoacetogenic bacteria and avoids nearly all ATP-consuming reactions by combining endergonic reactions with other non-ATP-consuming exergonic reactions (Bar-Even et al., 2012).

Carbon (from CO_2 or other organic feedstocks) is fixed in the form of biotechnologically valuable products like acetic acid, butyric acid, ethanol, butanol, and many others (Arends et al., 2017; Gildemyn et al., 2015; Lovley, 2011; Srikanth et al., 2018). When these chemicals undergo microbial chain elongation, they can be

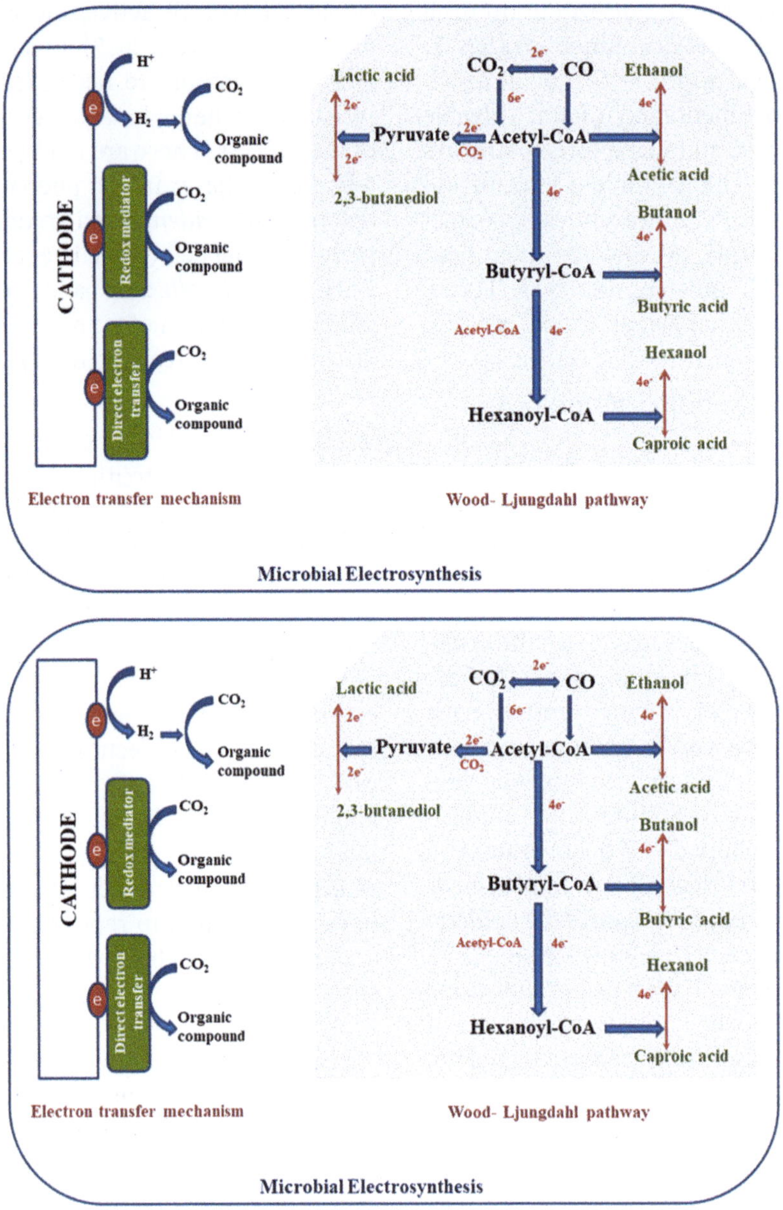

FIGURE 20.1 Two stage microbial electrosynthesis process converting CO_2 into essential organic compounds and certain other valuable acids and alcohols (modified from Prevoteau et al., 2020).

transformed into high-quality chemicals containing medium-chain fatty acids (such as caproate and caprylate). These modified chemicals possess higher economic value and work as eco-friendly building blocks for the chemical industry (Angenent et al., 2016; Gildemyn et al., 2015; Spirito et al., 2014). Nevin et al. (2010) reported that extracellular multicarbon compounds can be produced from CO_2 by providing electrons to the *Sporomusa ovate* grown at the cathode of a bioelectrochemical system. Marshall et al. (2017) used *Acetobacterium* (primary carbon fixer), *Sulfurospirillum*, and *Desulfovibrio* in reactors, at a small voltage of 590 mV against a standard hydrogen electrode, and operated the system for over 100 days. They obtained approximately 65% of acetate, 34% hydrogen, 0.4% formate, 0.3% propionate, and 0.2% of butyrate from CO_2 after 100 days.

20.3.5 Microbial Fuel Cells (MFCs) for Electricity Generation

Microbial fuel cells exploit the catabolic activities of microbes to generate electricity from a variety of complex organic waste and renewable biomass (Cario et al., 2019; Tang et al., 2020; see also chapter 31). MFCs are one of the best solutions for sustainable wastewater treatment (Gandu et al., 2020; Yaqoob et al., 2020). The main advantage of MFCs over other chemical fuel cells is that they do not require purified fuel (like hydrogen) to work (Chaturvedi and Verma, 2016). Moreover, this technology also helps in water reclamation, self-generation of microorganisms, produces a low carbon footprint, and generates less sludge (Rabbani et al., 2021). In the process of electricity generation in MFCs, the first step is the removal of non-utilizable fractions in settling tanks or debris basins to avoid clogging in succeeding steps of treatment. Thereafter, according to the properties of the wastewater, the appropriate path from the following three different pathways is chosen: (i) direct feeding into MFC system, if wastewater contains low organic content; (ii) pre-fermentation of wastewater in the anaerobic reactor before treatment in MFC if it contains a high level of organic wastes, and conversion of biogas thus produced into bioelectricity through the combination of heat and power systems; and (iii) struvite precipitation or pre-fermentation of phosphate-rich wastewater to generate favorable wastewater composition in MFC system and for nutrient recovery (Li et al., 2014). Figure 20.2 explains the process flow of the MFC system for wastewater treatment and electricity generation. Nevertheless, MFCs are not sufficient for removing all the nutrients from wastewater, especially those that are nutrient-rich. Therefore, the remaining organics in the MFC effluent are further treated with algae or undergo membrane filtration to improve effluent quality, recover a wide range of valuable products, and lift the energy output (Cai and Zheng, 2013). MFCs contribute to environmental sensing, production of biohydrogen, wastewater desalination, and CO_2 sequestration as well (Singh et al., 2020) which makes this technology economically beneficial too. Moreover, sludge generated after primary treatment and bioreactor processes can also be used as a source of energy. There are many research reports available on wastewater treatment and electricity generation in MFCs (Hiegemann et al., 2019; Jadhav et al., 2020).

Commonly used electron-producing microbes belong to bacteria (phyla – Actinobacteria, Firmicutes, Proteobacteria), fungi, algae, and yeast. *Aeromonas hydrophila, Clostridium butyricum, Escherichia coli, Geobacteraceae sulferreducens, Shewanella putrefaciens, Rhodoferax ferrireducens* are common microorganisms that are introduced into MFCs to transfer electrons to electrodes and generate electricity (Almatouq et al., 2020; Kaur et al., 2020).

Anodes are the sites where microorganisms attach, grow, form, and develop biofilms. Therefore, the morphology of electrodes affects the rate of redox reactions and the biocompatibility of microbes (Kaur et al., 2020; Tang et al., 2020; see chapter 31). Advanced anodes with enhanced electrical properties are thus used to enhance the transfer of electrons between electroactive microbes and electrode surfaces, and hence more bioelectricity production in MFCs (Cai et al., 2020). For example, certain nanomaterials as well as their composites can be used as anodes (sometimes also as cathodes) (Kaur et al., 2020). To improve the efficiency of MFCs, many engineered MFC reactors are also used. In engineered MFC reactors, the overall reactor size is increased although the distance between anode and cathode is narrowed to ensure efficient ion transfer and decrease in internal resistance (Fornero et al., 2010). The amount of active biomass is increased with high electrode-packing density in MFC which improves the system performance by reducing the pressure of footprint demand, limiting electron transfer and mass diffusion, preventing electrode clogging, and increasing efficiency

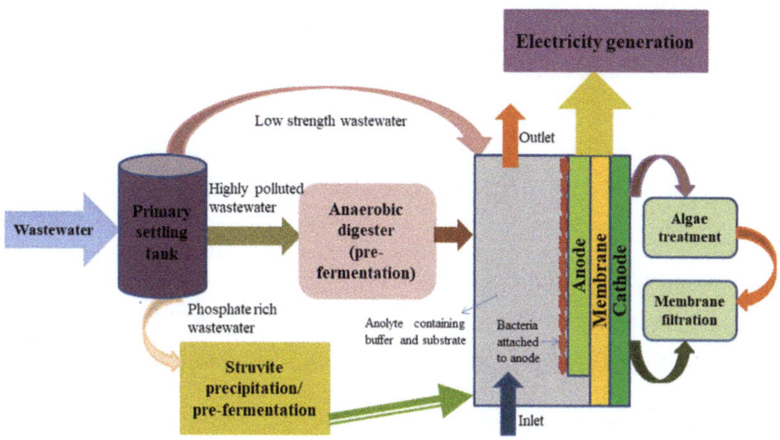

FIGURE 20.2 MFC system for wastewater treatment and electricity generation.

of wastewater treatment and energy production (Logan and Rabaey, 2012). Another feasible way to improve the efficiency of MFCs is their scaling-up, i.e. to use many moderately-scaled MFC units as a stack (Fornero et al., 2010). A series of MFC stacks can significantly increase the overall voltage output, offer high flexibility in their installation and modularized operation (Cheng and Logan, 2011).

Metal recovery has also been explored in MFCs. The pH of the solution strongly affects the Cr (VI) removal efficiency of MFC (Wang et al., 2008). At pH 2, 100% removal of the metal has been observed by Singhvi and Chhabra (2013). In an MFC with the graphitic rod as electrodes and *Shewanella oneidensis* MR-1, 67% removal of Cr (VI) and 32.5 mA/m^2 energy production was achieved after 8 days of operation, at natural pH range, and initial concentration of metal as 200 mg/L (Xafenias et al., 2013). Biochar addition can also be used to enhance the adsorption of dyes, for instance, 90% and 93% adsorption of methylene blue has been reported by using biochar prepared by citric acid modification of water hyacinth based-biochar, and coconut shell-biochar, respectively (Xu et al., 2016; Ayyappan et al., 2018). During degradation of xenobiotic compounds, certain compounds are supplemented as a substrate or co-substrate in the anodic chamber, while some are reductively cleaved or oxidized at the cathode during the bio-electro-Fenton process or hydroxide free radical oxidation mechanism of hydrogen peroxide production (Chakraborty et al., 2020). When comparing graphene oxide, platinum carbon, and reduced graphene oxide as cathode catalysts in MFCs aimed to remove Cu^{2+} and produce electricity, reduced graphene oxide and platinum carbon showed better performance. The dominant microbial community at anodes was found to be *Geobacter* whereas *Nitratireductor, Ochrobactrum*, and *Serratia* were found in cathodic biofilms of MFC containing platinum carbon as cathode catalysts, and *Azoarcus* were important contributors in the case of reduced graphene oxide-based MFCs (Wu et al., 2020).

20.3.6 Microbial Electrolysis for Hydrogen Production

Microbial electrolysis is carried out in microbial electrolytic cells (MEC) which consist of electrodes (an anode and a cathode), a membrane, electrochemically active microbes, a substrate, and a power supply (Bajracharya et al., 2019; Jiang et al., 2020). The electrochemically active microorganisms are referred to differently in different publications as exo-electrogens (Logan and Regan, 2006), electrogens (Lovley, 2006), anode-respiring bacteria (Torres et al., 2008). In an anodic chamber, organic matter breaks down through microbial action, and electrons and protons are released (Hosseinzadeh et al., 2020). Under the influence of electrode potential, the electrons travel through the wire and combine with protons to form hydrogen at the cathode (Rousseau et al., 2020). The electrode reactions are as follows:

Anodic reactions: $CH_3COO^- + 4H_2O = 2\,HCO_3^- + 8e^-$
Cathodic reactions: $2H^+ + 2e^- = H_2$

Different groups of bacteria that possess electron transfer activity to anodes include *Acinetobacter, Aeromonas, Clostridium, Desulfuromonas, Desulfomicrobium, Desulfobulbus, Desulfovibrio, Enterobacter, Geobacter, Geothrix, Geopsychrobacter, Pseudomonas, Rhodoferax, Sporumusa thiomonas* (Bond and Lovley, 2005; Chaudhuri and Lovley, 2003; Holmes et al., 2004). These active microbes can be extracted from ocean sediments (Reimers et al., 2001), domestic wastewater (Liu and Tay, 2004), or even anaerobic sewage sludge (Kim et al., 2005). Mostly, the mixed microbial cultures have been used in MECs (Hu et al., 2008; Rozendahl et al., 2007; Tartakovsky et al., 2009).

MEC is widely used nowadays; for instance, Aryal et al. (2017) used graphene paper (cathode) and *Sporumusa ovata*, Vassilev et al. (2018) used 3D Graphene Ni-foam (cathode) and mixed culture, and Kracke et al. (2019) used cathode prepared by NiMo deposition on doped Si wafer, and *S. ovate* and *A. woodii* as a microbial culture for electrosynthesis from CO_2 in MEC. Jayabalan et al. (2019) used nickel oxide-reduced graphene oxide and cobalt oxide-reduced graphene oxide cathodes for production of hydrogen in MEC from sugar industry wastewater and their results reported that at 1 V voltage, nickel oxide-reduced graphene oxide nanocomposite exhibited an H_2 production rate of ~ 4.38 mmol/L/d, coloumbic efficiency of 65.6%, and cathodic H_2 recovery of 20.8%. Co-culture of *Enterobacter aerogenes* and *Clostridium acetobutylicum* were used by Hasibar et al. (2020) for H_2 production through microbial electrolysis in electrolytic cells. A voltage of 0.8 V was applied in MEC and cellobiose was used as a carbon source for increased production of H_2 and maximum HER of 0.93 mmol/L/h. A MEC consisting of anode and cathode made from phenolic precursor-based carbon film dispersed with cerium oxide and reduced graphene oxide nanoparticles, respectively has the potential to generate

93% H_2 from wastewater and reduce its COD by 95% at a small voltage of 1 V under visible light irradiation (Pophali et al., 2020). MEC can also be applied for removing sulfate from wastewater and recovering elemental sulfur. Using sulfate-oxidizing bacteria (*Acinetobacter* and *Thiomonas*) and sulfate-reducing bacteria (*Desulfovibrio* and *Desulfomicrobium*) in novel phytoelectricity MEC, operating at 1.2 V voltage, Luo et al. (2020) observed sulfate reduction of ~ 200 mg/L/d and 46.3% (on average) conversion of reduced sulfate into sulfur. With an increase in voltage to 1.5 V, an increase in sulfate reduction rates was observed whereas the efficiency of elemental sulfur recovery decreased owing to high DO level diffusing from the anode side.

20.3.7 Consolidated Bioprocessing for Bioethanol Production

Consolidated bioprocessing is the method that involves the same microorganism (having saccharolytic and fermentative abilities) to produce hydrolytic enzymes and ferment sugars into ethanol (Cripwell et al., 2019; Loaces et al., 2017). The method combines enzyme production, enzymatic saccharification of cellulose, and sugar fermentation for efficient conversion and valorization of lignocellulosic biomass into ethanol (Olson et al., 2012). The process is carried out in the presence of microbes that contain several enzymes with specific properties, capable of expressing cellulosome at their surface. They are organized through several protein-protein interactions via different domains (including carbohydrate-binding domains) which enable them to bind with cellulose. The method also employs recombinant microorganisms as whole-cell biocatalysts ascribed to the higher selectivity and catalytic efficiency of such biocatalysts and as the process can be carried out even at milder operation conditions (Lin and Tao, 2017). Figure 20.3 explains how the process works on the degradation of pollutants and their conversion into valuable products.

Metabolic engineering provides naturally occurring fermenters that are capable of degrading cellulose or hemicellulose, and used for biofuel production such as *Clostridium cellulolyticum, Clostridium phytofermentans, Clostridium thermocellum, Thermoanaerobacter* sp., *Thermanaerobacterium* sp. (Olson et al., 2012). In one approach, strains capable of expressing and secreting heterologous soluble cellulases are created (Lynd et al., 2002); however, sometimes small quantities of heterologously expressed cellulases are also secreted owing to energy limitations under manufacturing conditions. Another approach is to express varied types of cellulolytic enzymes that are capable of producing artificial minicellulosomes on the cell surface. This feature helps the bacteria to tap into the synergistic effect of the physical organization of degradative enzymes on the catalytic efficiency of cellulose hydrolysis. For instance, three cellulases, i.e. endoglucanase, exoglucanase, and β-glucosidase from *C. thermocellum* and *C. cellulolyticum* accumulate as functional minicellulosomes on the cell surface of *Saccharomyces cerevisiae*, resulting in 2.6 times higher ethanol titre than that produced by an equal quantity of added purified cellulases (Tsai et al., 2010).

Consolidated bioprocessing of *Parthenium hysterophorus* biomass by an enzyme from *Aspergillus*

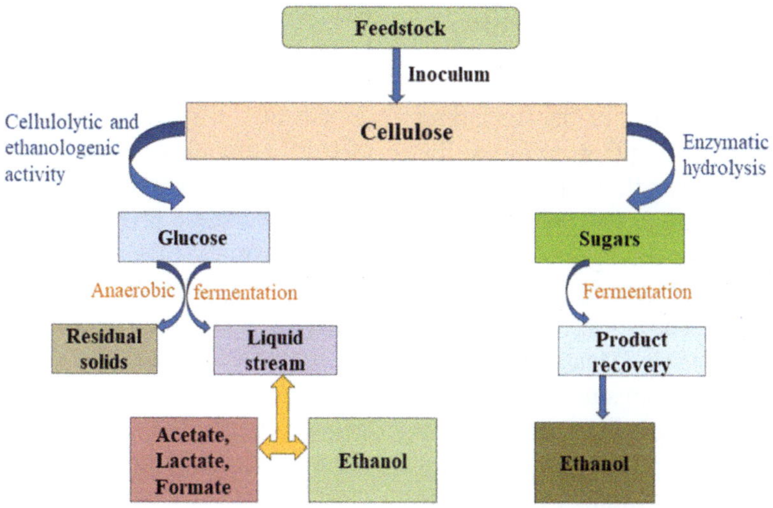

FIGURE 20.3 Consolidated bioprocessing showing conversion of cellulosic biomass into ethanol and other valuable products.

aculeatus PN14 can yield 40.1% sugar under optimized conditions of biomass loading, time, and temperature (Nargotra et al., 2019). Up to 7 g/L, isopropanol titer can be obtained from untreated corn cob through xylan utilization by *Clostridium beijerinckii* EMSD5 (Liu et al., 2019a). *Trametes hirsuta* (Bm-2) in co-culture with *Brosimum alicastrum* could yield 61% starch, and residual biomass after fermentation could produce 22.5% of protein for animal feeding (Olguin-Maciel et al., 2019). *S. cerevisiae* is a thermotolerant strain, resistant to inhibitors, hosts hemicellulolytic enzymes, can assimilate xylose, and develop whole-cell biocatalysts as well. Owing to its potential, this strain is extensively used for consolidated bioprocessing. *S. cerevisiae* helps in 11.1 g/L productions of bioethanol by hydrothermal pretreatment of corn cob and yield of 0.328 g/g of xylose and glucose and that too without the involvement of any external hydrolytic catalysts (Cunha et al., 2020). Co-culture of microbes is usually more effective than monoculture in ethanol production during consolidated processing. For instance, co-culture of *C. thermocellum, C. stercorarium*, and *Thermoanaerobacter thermohydrosulfuricus* when compared with monoculture *C. thermocellum* showed an increase in the average lactate:acetate: ethanol ratio from 6:45:49 in monoculture to 7:26:66 in tri-culture (Froese et al., 2020). Butanol has been also produced from cellulosic material using co-culture of *Clostridium saccharoperbutylacetonicum* and white-rot fungus *Phlebia* sp. MG-60-P2 (Tri and Kamei, 2020). In their study, the pyruvate decarboxylase gene was also knocked out from fungus to produce transformant line KO77 that inhibited ethanol fermentation and enhanced the butanol production further (2.5 in co-culture to 3.2 g/L in transformant).

20.3.8 Microbial Technologies for Biogas Production

Hydrolysis and fermentation of organic matter (including agricultural, industrial, municipal, plant residues) by microorganisms under anaerobic conditions in bioreactors leads to the production of biogas. Biogas is generally produced in four steps viz. hydrolysis–acidogenesis–cetogenesis–methanogenesis and the process is catalyzed by lignocellulose-degrading, acidogenic, acetogenic, and methanogenic microbes, respectively (Maus et al., 2014). In different stages, different microorganisms work together by forming a food chain in which products of one group forms the substrate for another group. The by-product produced at end of the process can be utilized as fertilizer (Ward et al., 2008). Feedstock characteristics and microbial activity determine the performance of anaerobic digestion (Batstone et al., 2002). New designs for bioreactors have been developed to increase the efficiency of biogas production like high rate up-flow anaerobic sludge blanket reactor and membrane bioreactors.

In a high rate up-flow anaerobic sludge blanket reactor, the process is carried out in two phases: hydrolysis and acid formation steps are separated from the methanogenesis step (Aslanzadeh, 2014). This modified reactor configuration has attractive features like low retention times for anaerobic treatment, high loading rates, compactness, low operational cost, capability to treat high strength wastewater, high degree of pathogen removal, minimized sludge production, dense, granular, flocculated, and well-settleable methanogenic sludge formation, minimal biofouling, and high biogas production rates. In membrane bioreactors, encapsulated methane-producing bacteria and membrane forming a selective barrier is used. Encased bacteria allow the penetration of waste into the cells and membrane allows the selective components to pass and retains the inhibitors, resulting in faster production of biogas that can easily escape from capsules (Youngsukkasem et al., 2013).

20.3.9 Bioaugmentation

Bioaugmentation is a microbial technology that involves the addition of pollutant degrading microbes to augment the degradation potential of microbial populations indigenously present in the contaminated sites. Figure 20.4 depicts how the addition of microorganisms speeds up the removal of contaminants. Added microorganisms may be natural, exotic, or engineered. Naturally occurring species are not sufficiently proficient to degrade specific compounds, so it is important to enhance their degradation capacity as well as population growth through engineering. Microorganisms are collected from a contaminated site, cultured separately, modified genetically, and then put again in the site. The efficiency of this technology depends on the capability of introduced strains to survive well in the remediation site and be retained there for a longer time. So far, only a few cases of small-scale bioaugmentation are known for wastewater treatment.

Jiang et al. (2020) investigated that bioaugmentation can improve the anaerobic digestion of food-based waste and produce more biogas. They reported that the introduction of 0.25 g/L/d of bioaugmentation seed after every three days in a digester enhanced (> 80%) the population of *Methanothrix* which increased the biogas production 12-fold when compared to non-bioaugmented control. In another study, a biofilm reactor augmented with

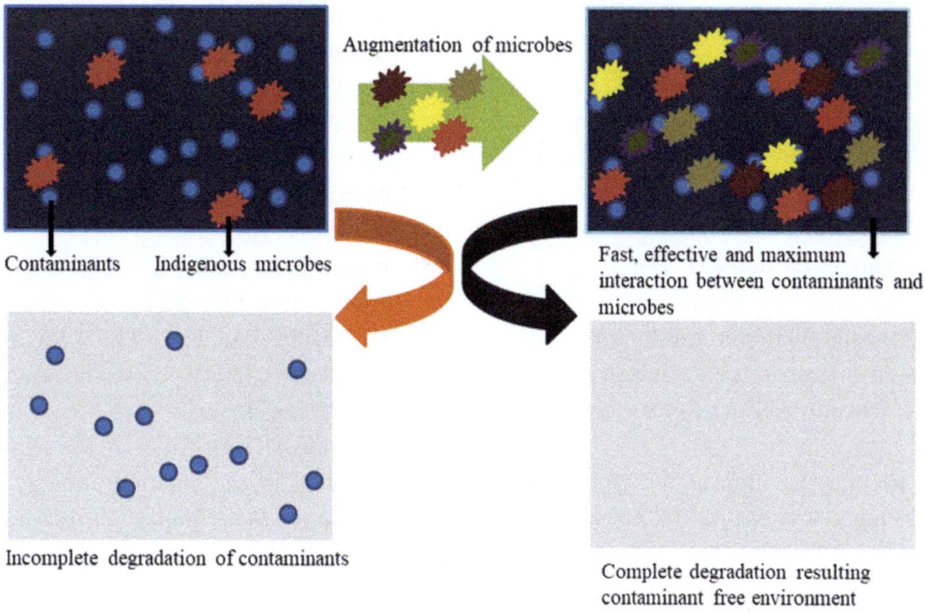

FIGURE 20.4 Enhanced remediation after augmentation of microorganisms in contaminated environment.

Pseudomonas mendocina IHB602 showed more rapid biofilm-formation than control (reactor without strain). The effluent removal for NH_4^+ (95%), NO_3^- (91%), and TN (88%) were higher than those in the control (NH_4^+: 91%, NO_3^-: 88%, TN: 82%) (Hong et al., 2020). Shia et al. (2020) performed the bioaugmentation of petroleum-pollution seawater by adding petroleum-degrading and biosurfactant-producing bacteria and indicated that the degradation rate of diesel by bioaugmentation exceeded by 38% that in the absence of bioaugmentation. In an anoxic filter-fluidized bed reactor system treating coking wastewater and containing *Comamonas* sp. ZF-3, COD, and NH_4^+ were found to be reduced by 93% and 97%, respectively (Yuan et al., 2020).

20.3.10 Biogranulation

Biogranulation is another wastewater treatment technology. Self-immobilization of microorganisms form biogranules which are the dense consortia of different bacterial species and contain many millions of organisms packed per gram of biomass. These packed biogranules are highly capable of degrading complex industrial wastes (Liu and Tay, 2004). This technology enhances the settling property and high biomass retention, and hence improves the treatment efficiency of the remediation process. There are generally two kinds of granulation, aerobic and anaerobic. Anaerobic biogranulation is carried in an up-flow anaerobic sludge blanket reactor while the aerobic type is performed in sequencing batch reactors. However, nowadays, aerobic biogranulation is also carried in pilot-scale aerobic up-flow sludge blanket reactors. Anaerobic biogranulation is used for the treatment of high-strength wastewater, and aerobic biogranulation is used for a medium or low-strength wastewater remediation.

20.4 CONCLUSION

Bioremediation tools are based on the coupling of mechanical engineering with biological diversity and functionality. The construction of genetic microbial tools to degrade pollutants is important from an environmental engineering perspective. Rational integration of multiple technologies including microbial technologies is needed to achieve maximized recovery of energy, wastewater treatment, added-value resources from wastewater, and to attain sustainable development. Although microbial technologies are gaining a lot of popularity these days due to their efficiencies, there is still the requirement for more field trials to engineer different such technologies to optimize degradation of pollutants, make them more cost-feasible, and increase their applicability widely. Modified reactor configurations, optimum process conditions, and a method of process coupling are also needed. Overall, a good exercise of the basics of the microbial communities in key remediation processes is needed to tackle diverse significant environmental issues in the coming decades.

REFERENCES

1. Adelaja O, Keshavarz T, Kyazze G. 2014. Enhanced biodegradation of phenanthrene using different inoculum types in a microbial fuel cell. *Engineeering in Life Sciences* 14(2): 218–228.
2. Aghbashlo M, Demirbas A. 2016. Biodiesel: Hopes and dreads. *Biofuel Research Journal* 3(2): 379.
3. Almatouq A, Babatunde AO, Khajah M, Webster G, Alfodari M. 2020. Microbial community structure of anode electrodes in microbial fuel cells and microbial electrolysis cells. *Journal of Water Process Engineering* 34: 101140.
4. Angenent LT, Richter H, Buckel W, Spirito CM, Steinbusch KJJ, Plugge CM et al. 2016. Chain elongation with reactor microbiomes: Open-culture biotechnology to produce biochemicals. *Environmental Science and Technology* 50: 2796–2810.
5. Aranda E, Ullrich R, Hofrichter M. 2010. Conversion of polycyclic aromatic hydrocarbons, methyl naphthalenes and dibenzofuran by two fungal peroxygenases. *Biodegradation* 21: 267–281.
6. Ardern E, Lockett WT. 1914. Experiments on the oxidation of sewage without the aid of filters. *Journal of the Society of Chemical Industry* 33: 523–539.
7. Arends JB, Patil SA, Roume H, Rabaey K. 2017. Continuous long-term electricity-driven bioproduction of carboxylates and isopropanol from CO_2 with a mixed microbial community. *Journal of CO2 Utilization* 20: 141–149.
8. Aryal N, Halder A, Zhang MW, Whelan PR, Tremblay PL, Chi QJ, Zhang T. 2017. Freestanding and flexible graphene papers as bioelectrochemical cathode for selective and efficient CO_2 conversion. *Scientific Reports* 7: 9107.
9. Aryal N, Ottosen LDM, Kofoed MVW, Pant D. (eds). 2021. *Emerging Technologies and Biological Systems for Biogas Upgrading*. Elsevier, Academic Press, London.
10. Aslanzadeh S. 2014. *Pretreatement of Cellulosic Waste and High Rate Biogas Production*. University of Boras, Boras, Sweden.
11. Ayyappan SC, Bhalambaal VM, Kumar S. 2018. Effect of biochar on bio-electrochemical dye degradation and energy production. *Bioresource Technology* 251: 165–170.
12. Bajracharya S, Aryal N, De Wever H, Pant D. 2019. Bioelectrochemical syntheses. In: Aresta M, Karimi I, Kawi S (Eds.)., *An Economy Based on Carbon Dioxide and Water: Potential of Large Scale Carbon Dioxide Utilization*, Springer International Publishing, Cham, pp. 327–358.
13. Bar-Even A, Flamholz A, Noor E, Milo R. 2012. Thermodynamic constraints shape the structure of carbon fixation pathways. *Biochimica Et Biophysica Acta* 1817: 1646–1659.
14. Batstone DJ, Keller J, Angelidaki I, Kalyuzhnyi SV, Pavlostathis SG, Rozzi A, Sanders WTM, Siegrist H, Vavilin VA. 2002. The IWA anaerobic digestion model no 1 (ADM 1). *Water Science and Technology* 45(10): 65–73.
15. Batstone DJ, Virdis B. 2014. The role of anaerobic digestion in the emerging energy economy. *Current Opinion in Biotechnology* 27: 142–149.
16. Berendonk TU, Manaia CM, Merlin C, Fatta-Kassinos D, Cytryn E, Walsh F et al. 2015. Tackling antibiotic resistance: The environmental framework. *Nature Reviews Microbiology* 13: 310–317.
17. Bhatia SK, Kim SH, Yoon JJ, Yang YH. 2017. Current status and strategies for second generation biofuel production using microbial systems. *Energy Conversion and Management* 148: 1142–1156.
18. Bilal M, Iqbal MNH. 2020. Microbial bioremediation as a robust process to mitigate pollutants of environmental concern. *Case Studies in Chemical and Environmental Engineering* 2: 100011.
19. Bokare V, Murugesan K, Kim JH, Kim EJ, Chang YS. 2012. Integrated hybrid treatment for the remediation of 2,3,7,8-tetrachlorodibenzo-pdioxin. *Science of the Total Environment* 435: 563566.
20. Bond DR, Lovley DR. 2005. Evidence for involvement of an electron shuttle in electricity generation by. *Geothrix Fermentans. Applied and Environmental Microbiology* 71(4): 2186–2189.
21. Cai J, Zheng P. 2013. Simultaneous anaerobic sulfide and nitrate removal in microbial fuel cell. *Bioresource Technology* 128: 760–764.
22. Cai T, Meng L, Chen G, Xi Y, Jiang N et al. 2020. Application of advanced anodes in microbial fuel cells for power generation: A review. *Chemosphere* 248: 125985.
23. Cario BP, Rossi R, Kim KY, Logan BE. 2019. Applying the electrode potential slope method as a tool to quantitatively evaluate the performance of individual microbial electrolysis cell components. *Bioresource Technology* 287: 121418.
24. Cecchin I, Krishna R, Thom A, Tessaro EF, Schnaid F. 2017. Nanobioremediation: Integration of nano-

particles and bioremediation for sustainable remediation of chlorinated organic contaminants in soils. *International Biodeterioration and Biodegradation* 119: 419–428.

25. Chakraborty I, Sathe SM, Khuman CN, Ghangrekar MM. 2020. Bioelectrochemically powered remediation of xenobiotic compounds and heavy metal toxicity using microbial fuel cell and microbial electrolysis cell. *Materials Science for Energy Technologies* 3: 104–115.
26. Chaturvedi V, Verma P. 2016. Microbial fuel cell: A green approach for the utilization of waste for the generation of bioelectricity. *Bioresources and Bioprocessing* 3(38).
27. Chaudhuri SK, Lovley DR. 2003. Electricity generation by direct oxidation of glucose in mediatorless-microbial fuel cells. *Nature Biotechnology* 21(10): 1229–1232.
28. Cheng S, Logan BE. 2011. High hydrogen production rate of microbial electrolysis cell (MEC) with reduced electrode spacing. *Bioresource Technology* 102: 3571–3574.
29. Cleveland CC, Liptzin D. 2007. C:N:P stoichiometry in soil: Is there a “Redfield ratio” for the microbial biomass? *Biogeochemistry* 85: 235–252.
30. Colombo A, Marzorati S, Lucchini G, Cristiani P, Pant D, Schievano A. 2017. Assisting cultivation of photosynthetic microorganisms by microbial fuel cells to enhance nutrients recovery from wastewater. *Bioresource Technology* 237: 240–248.
31. Cripwell RA, Rose SH, Favaro L, Van Zyl WH. 2019. Construction of industrial *Saccharomyces cerevisiae* strains for the efficient consolidated bioprocessing of raw starch. *Biotechnology for Biofuels* 12: 201.
32. Cunha JT, Romani A, Inokuma K, Johansson B, Hasunuma T et al. 2020. Consolidated bioprocessing of corn cob-derived hemicellulose: Engineered industrial *Saccharomyces cerevisiae* as efficient whole cell biocatalysts. *Biotechnology for Biofuels* 13: 138.
33. Daghio M, Aulenta F, Vaiopoulou E, Franzetti A, Arends JB, Sherry A et al. 2017. Electrobioremediation of oil spills. *Water Research* 114: 351–370.
34. De Vrieze J, Smet D, Klok J, Colsen J, Angenent LT, Vlaeminck SE. 2016. Thermophilic sludge digestion improves energy balance and nutrient recovery potential in full-scale municipal wastewater treatment plants. *Bioresource Technology* 218: 1237–1245.
35. Dejonghe W, Berteloot E, Goris J, Boon N, Crul K et al. 2003. Synergistic degradation of Linuron by a bacterial consortium and isolation of a single linuron-degrading *Varovorax* strain. *Applied and Environmental Microbiology* 69(3): 1532–1541.
36. Fornero JJ, Rosenbaum M, Angenent LT. 2010. Electric Power Generation from Municipal, Food, and Animal Wastewaters Using Microbial Fuel Cells. *Electroanalysis* 22: 832–843.
37. Froese AG, Nguyen TN, Ayele BT, Sparling R. 2020. Digestibility of wheat and cattail biomass using a co-culture of thermophilic Anaerobes for consolidated bioprocessing. *Bioenergy Research* 13: 325–333.
38. Gandu B, Rozenfeld S, Ouaknin Hirsch L, Schechter A, Cahan R. 2020. Immobilization of bacterial cells on carbon-cloth anode using alginate for hydrogen generation in a microbial electrolysis cell. *Journal of Power Sources* 455: 227986.
39. Garcia-Garcia JD, Sanchez-Thomas R, Moreno-Sanchez R. 2016. Bio-recovery of nonessential heavy metals by intra-and extracellular mechanisms in free-living microorganisms. *Biotechnology Advances* 34(5): 859–873.
40. Gildemyn S, Verbeeck K, Slabbinck R, Andersen SJ, Prevoteau A, Rabaey K. 2015. Integrated production, extraction, and concentration of acetic acid from CO_2 through microbial electrosynthesis. *Environmental Science and Technology Letters* 2(11): 325–328.
41. Hasibar B, Ergal I, Dias SA, Bochmann G, Simon KMR, Fuchs W. 2020. Increasing biohydrogen production with the use of a co-culture inside a microbial electrolysis cell. *Biochemical Engineering* 164: 107802.
42. Heyer R, Benndorf D, Kohrs F, De Vrieze J, Boon N, Hoffmann M. 2016. Proteotyping of biogas plant microbiomes separates biogas plants according to process temperature and reactor type. *Biotechnology for Biofuels* 9(1): 155.
43. Hiegemann H, Littfinski T, Krimmler S, Lubken M, Klein D et al. 2019. Performance and inorganic fouling of a submergible 255 L prototype microbial fuel cell module during continuous long-term operation with real municipal wastewater under practical conditions. *Bioresource Technology* 294: 122227.
44. Holmes DE, Bond DR, Lovley DR. 2004. Electron transfer by Desulfobulbus propionicus to Fe(III) and graphite electrodes. *Appllied and Environmental Microbiology* 70(2): 12–34.
45. Hong P, Wu X, Shu Y, Wang C, Tian C, Wu H, Xiao B. 2020. Bioaugmentation treatment of nitrogen-rich wastewater with a denitrifier with biofilm-formation and nitrogen-removal capacities in a sequencing batch biofilm reactor. *Bioresource Technology* 303: 122905.

46. Hosseinzadeh A, Zhou JL, Altaee A, Baziar M, Li D. 2020. Effective modelling of hydrogen and energy recovery in microbial electrolysis cell by artificial neural network and adaptive network-based fuzzy inference system. *Bioresource Technology* 316: 123967.

47. Hu H, Fan Y, Liu H. 2008. Hydrogen production using single-chamber membrane- microbial electrolysis cells. *Water Research* 42(15): 4172–4178.

48. Jadhav DA, Das I, Ghangrekar MM, Pant D. 2020. Moving towards practical applications of microbial fuel cells for sanitation and resource recovery. *Journal of Water Process Engineering* 38: 101566.

49. Jaiswal S, Shukla P. 2020. Alternative strategies for microbial remediation of pollutants via synthetic biology. *Frontiers in Microbiology* 11: 808.

50. Jayabalan T, Matheswaran M, Preethi V, Naina Mohamed S. 2019. Enhancing biohydrogen production from sugar industry wastewater using metal oxide/graphene nanocomposite catalysts in microbial electrolysis cell. *International Journal of Hydrogen Energy* 45(13).

51. Jiang Y, Chu N, Qian DK, Zeng RJ. 2020. Microbial electrochemical stimulation of caproate production from ethanol and carbon dioxide. *Bioresource Technology* 295: 122266.

52. Kaczorek E, Cieslak K, Bielicka-Daszkiewicz K, Olszanowski A. 2013. The influence of Rhamnolipids aliphatic fractions of diesel oil biodegradation by microorganisms. *Indian Journal of Microbiology* 53(1): 84–91.

53. Kanagasabi S, Kang YL, Manickam M, Ibrahim S, Pichiah S. 2013. Intimate coupling of electro and biooxidation of tannery wastewater. *Desalination and Water Treatment* 51(3436): 6617–6623.

54. Kaur R, Marwaha A, Chhabra VA, Kim KH, Tripathi SK. 2020. Recent developments on functional nanomaterial-based electrodes for microbial fuel cells. *Renewable and Sustainable Energy Reviews* 119: 109551.

55. Kim JR, Min B, Logan BE. 2005. Evaluation of procedures to acclimate a microbial fuel cell for electricity production. *Applied Microbiology and Biotechnology* 68(1): 23–30.

56. Kondaveeti S, Abu-Reesh IM, Mohanakrishna G, Bulut M, Pant D. 2020. Advanced routes of biological and bio-electrocatalytic carbon dioxide (CO_2) mitigation toward carbon neutrality. *Frontiers in Energy Research* 8: 94.

57. Koul B, Taak P. 2018. *Nanobioremediation. Biotechnological Strategies for Effective Remediation of Polluted Soils*. Springer, Singapore, pp. 197–220.

58. Kour D, Rana KL, Yadav AN, Yadav N, Kumar M et al. 2020. Microbial biofertilizers: Bioresources and eco-Consolidated bioprocessing performance of technologies for agricultural and environmental sustainability. *Biocatalysis and Agricultural Biotechnology* 23: 101487.

59. Kracke F, Wong AB, Maegaard K, Deutzmann JS, Hubert MA, Hahn C, Jaramillo TF, Spormann AM. 2019. Robust and biocompatible catalysts for efficient hydrogen-driven microbial electrosynthesis. *Chemosphere* 2(1): 45.

60. Kumar M, Jaiswal S, Sodhi KK, Shree P, Singh DK, Agrawal PK et al. 2019. Antibiotics bioremediation: Perspectives on its ecotoxicity and resistance. *Environment International* 124: 448–461.

61. Kweku DW, Bismark O, Maxwell A, Desmond KA, Danso KB et al. 2017. Greenhouse effect: Greenhouse gases and their impact on global warming. *Journal of Scientific Research & Reports* 17(6): 1–9.

62. Lacina C, Germain G, Spiros AN. 2003. Utilization of fungi for biotreatment of raw wastewaters. *African Journal of Biotechnology* 2(12): 620–630.

63. Le TT, Nguyen KH, Jeon JR, Francis AJ, Chang YS. 2015. Nano/bio treatment Xof polychlorinated biphenyls with evaluation of comparative toxicity. *Journal of Hazardous Materials* 287: 335–341.

64. Li WW, Yu HQ, Zhen HZ. 2014. Towards sustainable wastewater treatment by using microbial fuel cells-centered technologies. *Energy and Environmental Science* 7(3): 911–924.

65. Lin B, Tao Y. 2017. Whole-cell biocatalysts by design. *Microbial Cell Factories* 16: 106.

66. Lin CN, Wu SY, Lee KS et al. 2007. Integration of fermentative hydrogen process and fuel cell for online electricity generation. *International Journal of Hydrogen Energy* 32: 802–808.

67. Liu L, Yang J, Yang Y, Luo L, Wang R, Zhang Y, Yuan H. 2019. Consolidated bioprocessing performance of bacterial consortium EMSD5 on hemicellulose for isopropanol production. *Bioresource Technology* 292: 121965.

68. Liu Y, Tay JH. 2004. State of the art of biogranulation technology for wastewater treatment. *Biotechnology Advances* 22(7): 533–463.

69. Loaces I, Schein S, Noya F. 2017. Ethanol production by *Escherichia coli* from *Arundo donax* biomass under SSF, SHF or CBP process confgurations and in situ production of a multifunctional glucanase and xylanase. *Bioresource Technology* 224: 307–313.

70. Logan BE, Rabaey K. 2012. Conversion of wastes into bioelectricity and chemicals by using microbial electrochemical technologies. *Science* 337: 686–690.

71. Logan BE, Regan JM. 2006. Microbial fuel cells – Challenges and applications. *Environmental Science and Technology* 40: 5172–5180.

72. Lopes da Silva T, Luisa Gouveia L, Reis A. 2014. Integrated microbial processes for biofuels and high value-added products: The way to improve the cost effectiveness of biofuel production. *Applied Microbiology and Biotechnology* 98: 1043–1053.

73. Lovley DR. 2006. Bug Juice: Harvesting electricity with microorganisms. *Nature Reviews Microbiology* 4(7): 497–508.

74. Lovley DR. 2011. Powering microbes with electricity: Direct electron transfer from electrodes to microbes. *Environmental Microbiology Reports* 3: 27–35.

75. Luo H, Bai J, He J, Liu G, Lu Y, Zhang R, Zeng C. 2020. Sulfate reduction and elemental sulfur recovery using photoelectric microbial electrolysis cell. *Science of the Total Environment* 728: 138685.

76. Luther AK, Desloover J, Fennell DE, Rabaey K. 2015. Electrochemically driven extraction and recovery of ammonia from human urine. *Water Research* 87: 367–377.

77. Lynd LR, Weimer PJ, van Zyl WH, Pretorius IS. 2002. Microbial cellulose utilization: Fundamentals and biotechnology. *Microbiology and Molecular Biology Reviews* 66: 506–577.

78. Mac Dowell N, Fennell PS, Shah N, Maitland GC. 2017. The role of CO_2 capture and utilization in mitigating climate change. *Nature Climate Change* 7: 243–249.

79. Malaviya P, Sharma R, Sharma PK. 2019. Rain Gardens as stormwater management tool. In: Shah S *et al* (Ed.)., *Sustainable Green Technologies for Environmental Management*, Springer Nature Singapore Pte Ltd, Singapore, pp. 141–166.

80. Marshall CW, Ross DE, Handley KM, Weisenhorn PB, Edirisinghe JN et al. 2017. Metabolic reconstruction and modeling microbial electrosynthesis. *Scientific Reports* 7: 83–91.

81. Maszenan AM, Liu Y, Ng WJ. 2011. Bioremediation of wastewaters with recalcitrant organic compounds and metals by aerobic granules. *Biotechnology Advances* 29(1): 111–123.

82. Matassa S, Boon N, Pikaar I, Verstraete W. 2016. Microbial protein: future sustainable food supply route with low environmental footprint. *Microbial Biotechnology* 9(5): 568–S75.

83. Maus I, Stantscheff R, Wibberg D, Stolze Y, Winkler A, Puhler A, Konig H, Schluter A. 2014. Complete genome sequence of the methanogenic neotype strain *Methanobacterium formicicum* MFT. *Journal of Biotechnology* 192: 40–41.

84. Maus I, Wibberg D, Winkler A, Puhler A, Schnurer A. 2016. Complete genome sequence of the methanogen *Methanoculleus bourgensis* BA1 isolated from a biogas reactor. *Genome Announcements* 4(3): 00568.

85. Menon S, Denman KL, Brasseur G, Chidthaisong A, Ciais P, Cox PM et al. 2007. *Couplings between Changes in the Climate System and Biogeochemistry*. ED: Ernest Orlando, Lawrence Berkeley National Laboratory, Berkeley, CA, Medium.

86. Mirlahiji SG, Eisazadeh K. 2014. Bioremediation of Uranium by *Geobacter* spp. *Journal of Research and Development* 1: 52–58.

87. Mishra A, Medhi K, Malaviya P, Thakur IS. 2019. Omics approaches for microalgal applications: Prospects and challenges. *Bioresource Technology* 291: 121890.

88. Mishra B, Varjani S, Kumar G, Awasthi MK, Awasthi SK, Sindhu R, Binod P, Rene ER, Zhang Z. 2020. Microbial approaches for remediation of pollutants: Innovations, future outlook, and challenges. *Energy and Environment* 1–30.

89. Mistry AN, Ganta U, Chakrabarty J, Dutta S. 2019. A review on biological systems for CO_2 sequestration: Organisms and their pathways. *Environmental Progress & Sustainable Energy* 38(1): 127–136.

90. Monica P, Darwin RO, Manjunatha B, Zuniga JJ, Diego R et al. 2016. Evaluation of various pesticides-degrading pure bacterial cultures isolated from pesticide-contaminated soils in Ecuador. *African Journalof Biotechnology* 15: 2224–2233.

91. Nargotra P, Sharma V, Bajaj BK. 2019. Consolidated bioprocessing of surfactant-assisted ionic liquid-pretreated *Parthenium hysterophorus* L. biomass for bioethanol production. *Bioresource Technology* 289: 121611.

92. Nevin KP, Woodard TL, Franks AE, Summers ZM, Lovley DR. 2010. Microbial electrosynthesis: Feeding microbes electricity to convert carbon dioxide and water to multicarbon extracellular organic compounds. *mBio* 1: 103–110.

93. Nogia P, Sidhu GK, Mehrotra R, Mehrotra S. 2016. Capturing atmospheric carbon: Biological and nonbiological methods. *International Journal of Low-Carbon Technologies* 11: 266–274.

94. Olguin-Maciel E, Larque-Saavedra A, Lappe-Oliveras PE, Barahona-Perez LF. 2019. Consolidated bioprocess

for bioethanol production from raw flour of *Brosimum alicastrum* seeds using the native strain of *Trametes hirsuta* Bm-2. *Microorganisms* 7(11): 483.

95. Olson DG, McBride JE, Shaw AJ, Lynd LR. 2012. Recent progress in consolidated bioprocessing. *Current Opinion in Biotechnology* 23: 396–405.

96. Pant D, Adholeya A. 2007. Identification, ligninolytic enzyme activity and decolorization potential of two fungi isolated from a distillery effluent contaminated site. *Water, Air, and Soil Pollution* 183(1): 165–176.

97. Pant D, Adholeya A. 2009a. Nitrogen removal from biomethanated spentwash using hydroponic treatment followed by fungal decolorization. *Environmental Engineering Science* 26(3): 559–565.

98. Pant D, Adholeya A. 2009b. Concentration of fungal ligninolytic enzymes by ultrafiltration and their use in distillery effluent decolorization. *World Journal of Microbiology and Biotechnology* 25(10): 1793–1800.

99. Pena-Montenegro TD, Lozano L, Dussan J. 2015. Genome sequence and description of the mosquitocidal and heavy metal tolerant strain *Lysinibacillus sphaericus* CBAM5. *Standards in Genomic Sciences* 10: 1–10.

100. Perera IA, Abinandan S, Subashchandrabose SR, Venkateswarlu K, Naidu R, Megharaj M. 2019. Advances in the technologies for studying consortia of bacteria and cyanobacteria/microalgae in wastewaters. *Critical Reviews in Biotechnology* 39: 709–731.

101. Perez-Barberia FJ. 2017. Scaling methane emissions in ruminants and global estimates in wild populations. *Science of the Total Environment* 579: 1572–1580.

102. Pikaar I, Matassa S, Rabaey K, Bodirsky BL, Popp A, Herrero M, Verstraete W. 2017. Microbes and the Next Nitrogen Revolution. *Environmental Science and Technology* 51(13): 7297–7303.

103. Pophali A, Singh S, Verma N. 2020. Simultaneous hydrogen generation and COD reduction in a photoanode-based microbial electrolysis cell. *Journal of Hydrogen Energy* 45(48): 25985–25995.

104. Prevoteau A, Carvajal-Arroyo JM, Ganigue R, Rabaey K. 2020. Microbial electrosynthesis from CO_2: Forever a promise? *Current Opinion in Biotechnology* 62: 48–57.

105. Rabaey K, Rozendal RA. 2010. Microbial electrosynthesis d revisiting the electrical route for microbial production. *Nature Reviews Microbiology* 8: 706–716.

106. Rabbani A, Zainith S, Deb VK, Das P, Bharti P, Rawat DS, Kumar N, Saxena G. 2021. Microbial technologies for environmental remediation: Potential issues, challenges, and future prospects. In: Kumar A, Singh VK, Singh P, Mishra VK (Eds.), *Microbe Mediated Remediation of Environmental Contaminants*, Woodhead Publishing, Elsevier. pp. 271–286.

107. Ragsdale SW, Pierce E. 2008. Acetogenesis and the wood- ljungdahl pathway of CO_2 fixation. *Biochimica Et Biophysica Acta* 1784: 1873–1898.

108. Rashid US, Simsek S, Kanel SR, Bezbaruah AN. 2019. Modified tapioca starch for iron nanoparticle dispersion in aqueous media: Potential uses for environmental remediation. *SN Applied Sciences* 1(11): 1–14.

109. Ravi RK, Pathak B, Fulekar MH. 2015. Bioremediation of persistent pesticides in rice field soil environment using surface soil treatment reactor. *International Journal of Current Microbiology and Applied Sciences* 4: 359–369.

110. Read S, Marzorati M, Guimaraes B, Boon N. 2011. Microbial resource management revisited: Successful parameters and new concepts. *Applied Microbiology and Biotechnology* 90: 861–871.

111. Reimers CE, Leonard MT, Fertig S, Wang W. 2001. Harvesting energy from the marine sediment - Water interface. *Environmental Science and Technology* 35(1): 192–195.

112. Rodionova MV, Poudyal RS, Tiwari I, Voloshin RA, Zharmukhamedov SK et al. 2017.Biofuel production: Challenges and opportunities. *International Journal of Hydrogen Energy* 42(12):8450–8461.

113. Rousseau R, Etcheverry L, Roubaud E, Basseguy R, Delia ML, Bergel A. 2020. Microbial electrolysis cell (MEC): Strengths, weaknesses and research needs from electrochemical engineering standpoint. *Applied Energy* 257: 113938.

114. Rozendahl RA, Hamelers HVM, Molenkamp RJ, Buisman CJN. 2007. Performance of single chamber biocatalyzed electrolysis with different types of ion exchange membranes. *Water Research* 41(9): 1984–1994.

115. Salehi Jouzani G, Sharaf R.2018. New "omics" technologies and biogas production. In: Tabatabaei M, Ghanavati H (Eds.)., *Biogas, Biofuel and Biorefinery Technologies*, Springer International Publishing AG, part of Springer Nature.

116. Sani I, Safiyanu I, Rita SM. 2015. Review on bioremediation of oil spills using microbial approach. *International Journal of Educational Science and Research* 3: 41–45.

117. Sasaki D, Sasaki K, Tsuge Y, Morita M, Kondo A. 2014. Comparison of metabolomic profiles of microbial

communities between stable and deteriorated methanogenic processes. *Bioresource Technology* 172: 83–90.

118. Satputaley SS, Zodpe DB, Deshpande NV. 2017. Performance, combination and emission study on CI engine using microalgae oil and microalgae oil methyl esters. *Journal of the Energy Institute* 90: 513–521.
119. Sharma R, Vymazal J, Malaviya P. 2021. Application of floating treatment wetlands for stormwater runoff: A critical review of the recent developments with emphasis on heavy metals and nutrient removal. *Science of the Total Environment* 777: 146044.
120. Shia K, Zhang Q, Xue J, Chena X, Chen Y et al. 2020. Study on the degradation performance and bacterial community of bioaugmentation in petroleum-pollution seawater. *Journal of Environmental Chemical Engineering* 8: 103900.
121. Simarro R, Gonzalez N, Bautista LF, Molina MC. 2013. Assessment of the efficiency of in situ bioremediation techniques in a creosote polluted soil: Change in bacterial community. *Journal of Hazardous Materials* 262: 158–167.
122. Singh A, Sharma R, Pant D, Malaviya P. 2021. Engineered algal biochar for contaminant remediation and electrochemical applications. *Science of the Total Environment* 774: 145676.
123. Singh A, Srivastava S, Rathore D, Pant D. 2020. *Environmental Microbiology and Biotechnology: Volume 2: Bioenergy and Environmental Health.* Springer Nature, Singapore, .DOI: https://doi.org/10.1007/978-981-15-7493-1
124. Singhvi P, Chhabra M. 2013. Simultaneous chromium removal and power generation using algal biomass in a dual chambered salt bridge microbial fuel cell. *Journal of Bioremediation and Biodegradation* 4(5): 1000190.
125. Spirito CM, Richter H, Rabaey K, Stams AJM, Angenent LT. 2014. Chain elongation in anaerobic reactor microbiomes to recover resources from waste. *Current Opinion in Biotechnology* 27: 115–122.
126. Srikanth S, Singh D, Vanbroekhoven K, Pant D, Kumar M, Puri SK, Ramakumar SV. 2018. Electro-biocatalytic conversion of carbon dioxide to alcohols using gas diffusion electrode. *Bioresource Technology* 265: 45–51.
127. Stolze Y, Zakrzewski M, Maus I, Eikmeyer F, Jaenicke S, Rottmann N et al. 2015. Comparative metagenomics of biogas-producing microbial communities from production-scale biogas plants operating under wet or dry fermentation conditions. *Biotechnology for Biofuels* 8(1): 14.
128. Strong PJ, Xie S, Clarke WP. 2015. Methane as a resource: Can the methanotrophs add value? *Environmental Science and Technology* 49: 4001–4018.
129. Szarka N, Scholwin F, Trommler M, Jacobi HF, Eichhorn M, Ortwein A, Thran D. 2013. A novel role for bioenergy: A flexible, demand-oriented power supply. *Energy* 61: 18–26.
130. Tang RCO, Jang JH, Lan TH, Wu JC, Yan WM et al. 2020. Review on design factors of microbial fuel cells using Buckingham's Pi Theorem. *Renewable and Sustainable Energy Reviews* 130: 109878.
131. Tartakovsky B, Manuel MF, Wang H, Guiot SR. 2009. High rate membrane-less microbial electrolysis cell for continuous hydrogen production. *International Journal of Hydrogen Energy* 34(2): 672–677.
132. Torres CI, Marcus AK, Rittmann BE. 2008. Proton transport inside the biofilm limits electrical current generation by anode-respiring bacteria. *Biotechnology and Bioengineering* 100: 872–881.
133. Tri CL, Kamei I. 2020. Butanol production from cellulosic material by anaerobic co-culture of white-rot fungus *Phlebia* and bacterium *Clostridium* in consolidated bioprocessing. *Bioresource Technology* 305: 123065.
134. Tsai SL, Goyal G, Chen W. 2010. Surface display of a functional minicellulosome by intracellular complementation using a synthetic yeast consortium and its application to cellulose hydrolysis and ethanol production. *Applied and Environmental Microbiology* 76: 7514–7520.
135. Vassilev I, Hernandez PA, Batlle-Vilanova P, Freguia S, Kromer JO et al. 2018. Microbial electrosynthesis of isobutyric, butyric, caproic acids, and corresponding alcohols from carbon dioxide. *ACS Sustainable Chemistry and Engineering* 6(7): 8485–8493.
136. Wang G, Huang L, Zhang Y. 2008. Cathodic reduction of hexavalent chromium [Cr (VI)] coupled with electricity generation in microbial fuel cells. *Biotechnology Letters* 30: 19–59.
137. Ward AJ, Hobbs PJ, Holliman PJ, Jones DL. 2008. Optimisation of the anaerobic digestion of agricultural resources. *Bioresource Technology* 99(17): 7928–7940.
138. White PJ, Broadley MR. 2009. Biofortification of crops with seven mineral elements often lacking in human diets–iron, zinc, copper, calcium, magnesium, selenium and iodine. *New Phytologist* 182: 49–84.
139. Whitley AS, Bailey MJ. 2000. Bacterial community structure and physiological state within an industrial

phenol bioremediation system. *Applied Environmental Microbiology* 66(6): 2400–2407.

140. Wilks R, Scholes E 2018. DNA sequencing of pulp and paper wastewater treatment systems to inform process analysis. Appita J.: J. Tech. Assoc. Australian New Zealand Pulp Paper Ind. 71: 157.

141. Wongwongsee W, Chareanpat P, Pinyakong O. 2013. Abilities and genes for PAH biodegradation of bacteria isolated from mangrove sediments from the central of Thailand. *Marine Pollution Bulletin* 74: 95–104.

142. Wu Y, Wanga L, Jina M, Zhanga K. 2020. Simultaneous copper removal and electricity production and microbial community in microbial fuel cells with different cathode catalysts. *Bioresource Technology* 305: 123166.

143. Xafenias N, Zhang Y, Banks CJ. 2013. Enhanced performance of hexavalent chromium reducing cathodes in the presence of *Shewanella oneidensis* MR-1 and lactate. *Environmental Science and Technology* 47: 4512–4520.

144. Xu Y, Liu Y, Liu S, Tan X, Zeng G et al. 2016. Enhanced adsorption of methylene blue by citric acid modification of biochar derived from water hyacinth (*Eichornia crassipes*). *Environmental Science and Pollution Research* 23: 23606–23618.

145. Yan Z, Song N, Cai H, Tay JH, Jiang H. 2012. Enhanced degradation of phenanthrene and pyrene in freshwater sediments by combined employment of sediment microbial fuel cell and amorphous ferric hydroxide. *Journal of Hazardous Materials* 199: 217–225.

146. Yaqoob AA, Khatoon A, Setapar SHM, Umar K, Parveen T, Ibrahim MNM, Ahmad A, Rafatullah M. 2020. Outlook on the role of microbial fuel cells in remediation of environmental pollutants with electricity generation. *Catalysts* 10: 819.

147. Youngsukkasem S, Akinbomi J, Rakshit SK, Taherzadeh MJ. 2013. Biogas production by encased bacteria in synthetic membranes: Protective effects in toxic media and high loading rates. *Environmental Technology* 34(13-16): 2077–2084.

148. Yuan K, Li S, Zhong F. 2020. Treatment of coking wastewater in biofilm-based bioaugmentation process: Biofilm formation and microbial community analysis. *Journal of Hazardous Materials* 400: 123117.

149. Zhang W, Tan NG, Fu B, Li SF. 2015. Metallomics and NMR-based metabolomics of *Chlorella* sp. reveal the synergistic role of copper and cadmium in multi-metal toxicity and oxidative stress. *Metallomics* 7: 426–438.

150. Zhao Q, Yue S, Bilal M, Hu H, Wang W, Zhang X. 2017. Comparative genomic analysis of 26 *Sphingomonas* and *Sphingobium* strains: Dissemination of bioremediation capabilities, biodegradation potential and horizontal gene transfer. *Science of the Total Environment* 609: 1238–1247.

CHAPTER 21

Microbes Saving the World? How Microbial Carbon Dioxide Fixation Contributes to Storing Carbon in Goods of Our Daily Life

Diethard Mattanovich, Özge Ata, and Thomas Gassler

Institute of Microbiology and Microbial Biotechnology, Department of Biotechnology, University of Natural Resources and Life Sciences, 1190 Vienna, Austria

21.1 INTRODUCTION

The conditions leading to the development of life on Earth would be considered extremely hostile today: an atmosphere of carbon dioxide and water vapor with some carbon monoxide, ammonia, and methane, a runaway greenhouse effect, and no breathable oxygen around. It is supposed that the first cells lived on organic compounds that were formed chemically, by volcanic activity in a carbon-rich atmosphere. We can assume that this food source was not quite abundant and evolution found benefits for cells that developed other nutritional sources: so microbes developed autotrophy, the ability to reduce carbon dioxide and incorporate the carbon into their growing biomass. Carbon dioxide was abundant but it required a lot of energy to rip oxygen off the molecule to make it amenable for the chemical reactions to build life. The chemical energy equivalent is electrons: cells need electrons to reduce CO_2, and they had to find ways to make electrons available. One hugely abundant electron source was, and still is, the oxidation of minerals, like iron or sulfur (Amenabar and Boyd, 2019). Microbes take electrons from these reduced minerals, channel them through so-called electron-transfer chains in their membranes whereby energy is stored, and finally transfer the electrons to so-called electron acceptors which are reduced in this process. Oxidation of iron (from Fe^{2+} to Fe^{3+}) was probably invented first, as iron is one of the most abundant elements on Earth. However, the energy gain of the reaction is small so that early iron bacteria had to dig through huge amounts of minerals to feed themselves.

Another virtually unlimited energy source is sunlight, and evolution found ways to harvest light energy, turning it into reducing power to feed on CO_2 again. This invention came with an enormous threat, namely the production of oxygen from water. Until then free oxygen was scarce and life was not used to its high reac tivity. Initially, free oxygen was rapidly captured by minerals, mainly iron, and it enabled a leap in microbial energy gain through mineral oxidation as it serves as a very efficient electron acceptor. While it is obvious that also chemoautotrophic synthesis flourished with the onset of photosynthesis, there is an ongoing debate to which extent microbial or chemical oxidation of iron was responsible for capturing the majority of early oxygen release. In any case, chemical or biological, oxidation changes the color of iron minerals from green

Good Microbes in Medicine, Food Production, Biotechnology, Bioremediation, and Agriculture, First Edition. Edited by Frans J. de Bruijn, Hauke Smidt, Luca S. Cocolin, Michael Sauer, David Dowling, and Linda Thomashow.

to red which can still be seen in banded iron formations, billions of years old rock formations with red-to-black oxidized iron minerals. A recent hypothesis claims that the same may have happened on Mars about 4 billion years ago which would mean that the Red Planet owes its attribute to a yet unidentified source of oxygen (Liu et al., 2021). Some researchers speculate that this may be a signature of early microbial life on Mars.

By developing protective enzymes and chemicals new microbial species learned to survive the Great Oxygen Event, and they learned to tame the dangerous molecule by using oxygen as electron acceptor for their energy production: respiration was born, a hugely successful metabolic pathway due to its large energy yield. Since then, life on Earth found a balance of photosynthetic CO_2 consumption and O_2 production, and respiratory O_2 consumption and CO_2 emission, or, in other words, between reducing and oxidizing reactions catalyzed by living organisms.

The largest part of carbon on Earth is stored in various forms in the Earth's crust and the oceans, and as fossil remnants of prehistoric biomass, so that the perpetual turnover by autotrophic and heterotrophic organisms makes just a fraction of total carbon. Next to land plants, microbes are at par with the annual turnover of CO_2 into biomass (Woodward, 2007; Figure 21.1). More disturbing is the impact of human activity on the carbon cycle by the release of huge amounts of carbon from fossil deposits into the atmosphere which has led to an increase of CO_2 levels from 0.26% prior to industrialization to 0.42% today. Biotechnology is increasingly utilizing autotrophic microorganisms for

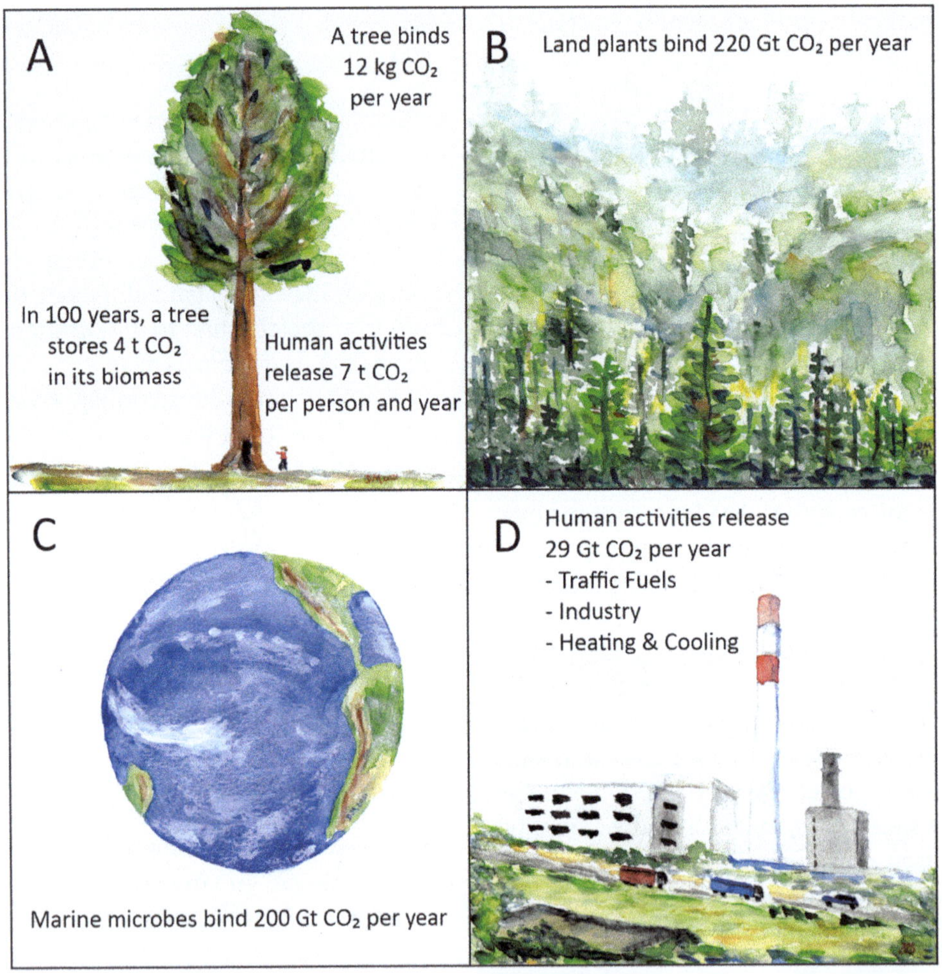

FIGURE 21.1 The global biological carbon cycle is driven equally by land plants and marine CO_2 assimilating microbes. (A) A single tree stores less CO_2 in a lifetime than human activities release per capita in a year. (B) All land plants together bind about half of the annually assimilated CO_2. (C) Marine microbes (microalgae, cyanobacteria and other autotrophic bacteria and archaea) are responsible for the other half of biological carbon capture. (D) While human activities, driven by industrialization, release an order of magnitude less CO_2 than autotrophic organisms bind annually this release triggers a dangerous increase of atmospheric CO_2 concentration.

CO_2 neutral or even CO_2 negative production, and synthetic biology is paving new ways to carbon capture and utilization.

21.2 PHOTOAUTROPHIC MICROORGANISMS

Photoautrophic organisms use photons from sunlight as an energy source to fix carbon from gaseous CO_2. The fixation of CO_2 energized by sunlight happens either by the Calvin–Benson–Bassham (CBB) cycle, the 3-hydroxypropionate bicycle or the reductive tricarboxylic acid (TCA) cycle. The most abundant pathway among them is the CBB cycle, both in nature and in biotech applications (Erb and Zarzycki, 2018). All phototrophs use energy from sunlight to fuel electron transport chains or direct electron pumps forcing adenosine triphosphate (ATP) synthase to make ATP, the energy currency of the cell. If they use the produced ATP solely for anabolic purposes to turn CO_2 into organic molecules, we speak of photoautrophs and photosynthesis. If they also use the ATP produced with the help of sunlight for utilizing other organic substrates, they are called photoheterotrophs. The phototrophic lifestyle relies on complex chemical pigments. Plants, algae, and cyanobacteria use chlorophylls, while purple and green bacteria predominantly use a wider class of bacteriochlorophylls to carry out photosynthesis. Chlorophylls absorb blue and red light and transmit green light, making terrestrial vegetation appear green. The light spectrum of the sun peaks in the green wavelengths at around 550 nm, a bit aside from the absorption maxima of chlorophyll with 465 nm and 665 nm. Bacteriochlorophylls generally have absorption maxima at higher wavelengths, which allows organisms containing them to peacefully occupy the same habitats as their green neighbors. Up to now we do not fully understand why chlorophyll has had such a great evolutionary success, although performing at "suboptimal" absorption maximum. A compelling argument is that the green light is avoided to protect the cells from extra energy. This hypothesis is supported by several microbial strategies. Several microalgae species accumulate astaxanthin to protect them from UV radiation in alpine and polar regions ("snow algae"), making them appear red. Some oceanic cyanobacteria and algae grow near to the surface where just enough light can enter, but where the full energy potential of the sunlight cannot harm them.

Cyanobacteria, formerly known as blue–green algae, are gram-negative bacteria able to carry out oxygenic photosynthesis. They represent one of the most ancient microbial lineages and have strongly enabled many biological innovations. They were so eager to spread that their oxygenic lifestyle literally led to a rusting of the Earth. With their rapid invasion of the ecosystems, they flooded the atmosphere with oxygen and erased many other species that were not able to cope with oxidizing conditions. Therefore, you might also consider them as great killers. This "microbial genocide" by cyanobacteria was facilitated by having a chemical companion. Oxygen produced by these early cyanobacteria was probably captured by natural oxygen sinks such as iron – Fe and Fe^{2+} – that formed insoluble compounds after oxidation to Fe^{3+} which we still observe today as banded iron formations. Once these natural oxygen sinks were exhausted, the atmospheric oxygen levels vastly increased due to the oxygenic photosynthesis of cyanobacteria leading to the Great Oxygenation Event (GOE).

Cyanobacteria are everywhere. They inhabit nearly all ecosystems including some of the harshest environments from limnic and marine environments to hot springs and salty water systems, where no other photosynthetic organism can survive. Most of the cyanobacteria live exclusively as photoautotrophs, but some are also known to survive long periods without sunlight. The morphologies of cyanobacteria are just as versatile as their natural habitats. They may appear unicellular, filamentous (as so called trichomes), or as colonies (Figure 21.2A). In addition to using CO_2 as a carbon source, some cyanobacteria can also fix elemental nitrogen (see chapter 37). The responsible enzyme nitrogenase is sensitive to oxygen and some cyanobacteria form distinct cellular structures, which are called heterocysts. This allows spatial separation of nitrogen fixation and hence protection from oxygenic reactions and further expands their morphological diversity.

The ability to use CO_2 as a carbon source in the presence of oxygen is a hallmark of evolution. In ecological niches where no organic carbon is present, higher organisms often enter into symbiotic relations with cyanobacteria to feed on carbon from CO_2 or to use molecular nitrogen. For example, in lichens, generally describing a multi-species consortia with several fungi, cyanobacteria may provide carbon for exchange of other nutrients. The degree of this mutualistic exchange varies from facultative to obligate, meaning that the symbiotic partners have lost the ability to survive without the other. In endosymbiosis, the most intimate form of symbiosis, the symbiont resides within the host cell. It is broadly accepted that photosynthesis in eukaryotes originates from primary endosymbiotic events between cyanobacteria and early protistan hosts. Currently it is assumed, that the oldest of these events resulted in the Archaeplastida lineage, from which all higher eukaryotic photoautotrophs (including algae and all land plants) either originate directly or indirectly, by entering a second endosymbiotic event on the evolutionary trajectory. These events have led to plastids and in the following to the evolution of chloroplasts, the remnants of the cyanobacterial endosymbionts (Rockwell et al., 2014).

The success of eukaryotic photosynthesis is beyond doubt. As primary producers, the representatives now dominate most terrestrial and marine environments and fix the majority of CO_2 globally (see Figure 21.1). As in the case for cyanobacteria, not all species summarized under the term of microalgae exclusively depend on CO_2, but also use alternative carbon sources. Microalgae are very diverse and morphologically varied. The prominent genus of *Clorella* comprises 13 species (including *C. vulgaris*) that appear spherical and unicellular. Others, such as members of the genus *Chlamydomonas* (example *C. reinhardtii*) or *Dunianella* (e.g. *D. salina*) have flagellae as locomotive organs (Figure 21.2B). Microalgae primarily use chlorophyll to harvest light energy, but also accumulate a plethora of other pigments for various reasons from nutritional needs to biological defense. For example, *D. salina* can accumulate β-carotene reaching up to 14% of its dry mass, which is thought to protect the cells from osmotic pressure in saline habitats – the environment where this organism is preferentially found.

21.2.1 Cultivation and Applications of Cyanobacteria and Microalgae

Cooperation is key for innovation. This dogma holds true for intellectual as well as biological innovations. Cyanobacteria and other organisms have cooperated to trade carbon and other nutrients which has led to the evolution of all eukaryotic photoautotrophs. In a world with emerging levels of CO_2 and oxygen, "higher" organisms have acquired photosynthesis and have learned to thrive in these new conditions. Changing environments not only drive metabolic but also "behavioral" innovations. The use of photosynthetic microbes, such as *Nostoc* or *Arthrospira* species, by humankind goes back for thousands of years. *Spirulina* (*Arthrospira sp.*) for instance was harvested as food sources by several cultures in South America and in Africa (e.g. by the Aztec from Lake Texcoco in Mexico). However, the awareness of the anthropogenic cause of the climate crisis and the need for sustainable production systems have pushed the development of applications with cyanobacteria and microalgae since the second half of the twentieth century.

Culturing of cyanobacteria and microalgae can be done in open systems or in closed photobioreactors. The oldest systems are open ponds which are obviously inspired by the natural habitats (Figure 21.2C). The setup of these open ponds is cost-efficient and they are easy to operate. Open pond cultivation also includes implementation in limnic freshwater and artificial circular systems in so-called raceway ponds. However, the scale up of open cultivation requires a large amount of land, and the maximum reachable cell concentration is severely limited. Further, the operation of these systems is prone to contamination, which can further lead to prolonged down-times of production. Cultivation in raceway ponds, in which the biomass is recirculated to prevent settlement and to ensure sufficient mixing, is commonly used by industry. One of the biggest raceway ponds was operated by the company Sapphire Energy and could yield 260 metric tons of dried algal biomass per year. A more modern approach to cultivate photosynthetic organisms is the cultivation in closed photobioreactors under controlled conditions. Compared to open systems, higher space-time yields lead to more efficient use of land. The scale-up of photobioreactors faces several challenges which are all primarily owed to the fact that light is the sole source of energy and that it has to penetrate the cell suspension. Several different geometries were developed to cope with the limited access of light. Flat-plate photobioreactors consist of rectangular transparent cultivation compartments in which the culture is mixed by recirculating airlift systems. Creative designs make the flat-plate based types the most efficient photobioreactors in terms of illumination and surface area. The second type are vertical column reactors which are mixed and supplied with CO_2 by sparging, which allows them to operate at high gas–liquid mass transfer rates. Due to the cylindrical shapes, the amount of light that can penetrate the cell suspension is low which is still limiting large-scale applications (reactors of this type typically have working volumes below 100 L). The third geometry type is tubular photobioreactors, which are widely used by the microalgae industry. The reactors are made of differently shaped transparent tubing systems and can be arranged in a fixed or moving manner to ensure maximum capture of light, letting them appear as "hanging gardens." The cells are circulated by air lifting and/or mechanical pumping, which

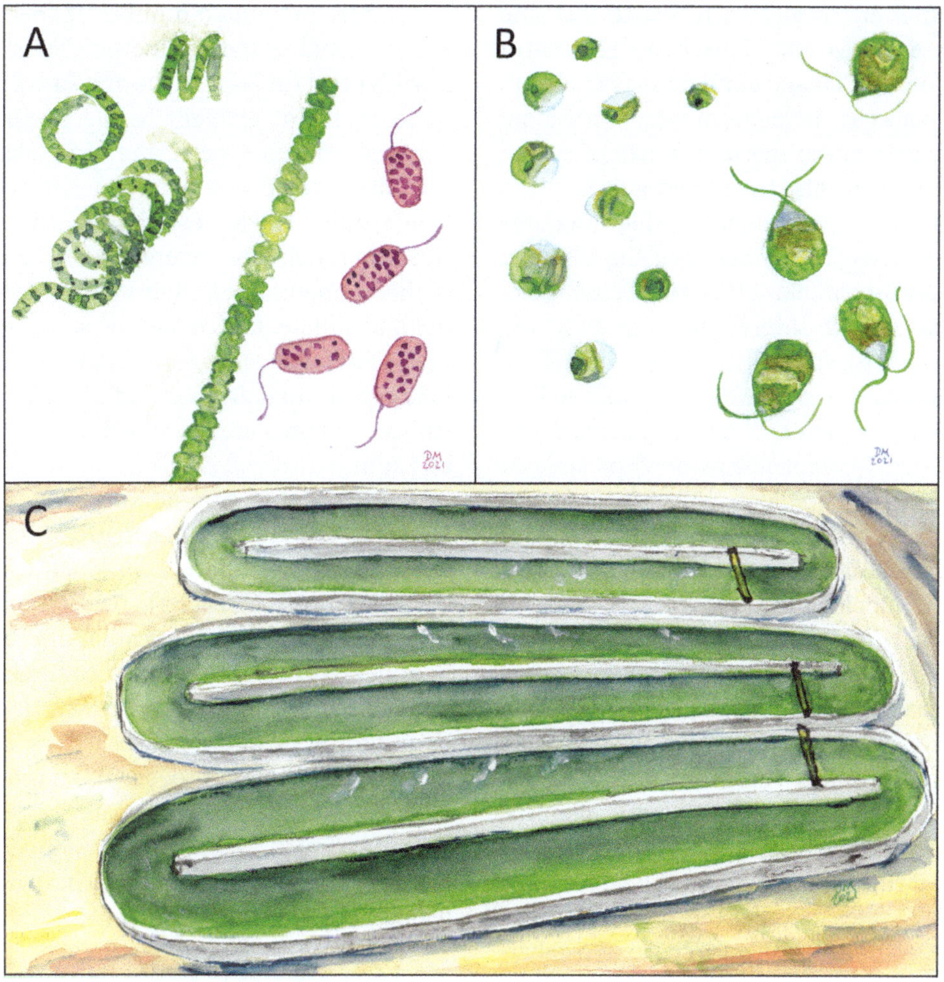

FIGURE 21.2 Photoautotrophic microorganisms assimilate CO_2 by harvesting light energy. They live mainly in water, both in marine and freshwater habitats. (A) Phototrophic prokaryotes are mainly chlorophyll containing bacteria, the so-called cyanobacteria (left: *Arthrospira sp.* – often called *Spirulina*; middle: *Nostoc sp.*). Some bacteria and archaea use red pigments to harvest light energy, the so-called purple bacteria (right: *Chromatium sp.*, sulfur purple bacteria which oxidize sulfide to sulfur that is deposited in the cell in form of granules). (B) Microalgae are mostly single celled eukaryotic algae (left: *Chlorella vulgaris*, right: *Dunaliella salina*). (C) Open pond cultivation is the simplest way of growing microalgae or cyanobacteria to produce biomass or some of their ingredients.

typically leads to drawbacks in terms of mass transfer and nutrient distribution. However, several reactors of this type are being used at industrial scale today (Daneshvar et al., 2021; Tan et al., 2020).

Although humankind has used microalgae as food for a long time, their industrial cultivation has only started during the last decades. The needs of a steadily growing world population have to be met in a sustainable way. The use of photoautrophs in industrial applications offers the opportunity to capture CO_2 and to turn it into value added products. Of special interest here is the favorable macromolecular composition of microalgae and cyanobacteria with protein contents typically over 50%, a carbohydrate content of around 20% and a high lipid content, sometimes exceeding 20%. The quality of the protein in terms of amino acid spectrum is comparable to vegetable or meat-based protein. The use of microalgae and cyanobacteria as farm animal and aqua culture feed has a long tradition. It is known that microalgae are not only a promising alternative protein source for animal feed, but their supplementation can also boost growth performance and quality in several animals. For example, supplementation of *Arthrospira platensis* can increase the body weight in lambs, and beta-carotene from *Dunianella salina* supplementation can protect against white spot syndrome virus in shrimp production. There are numerous applications in animal feed, but the technology of the Swedish company Volta Greentech is especially noteworthy. They report that supplementation of red algae can reduce methane production by cows up to 80%. This is an illustrative example how the use of CO_2 and sunlight may enable to reduce another climate gas, methane, in meat production. The nutritional value and the special features of cyanobacteria and microalgae are also largely recognized by the food industry. Oil from microalgae contains high amounts of triacylglycerides (TAGs), also including the essential fatty acids eicosapentaenoic acid (EPA) and docosahexaenoic acid (DHA) along with several other omega-3 fatty acids. Several *Chlorella* based products are on the market that claim to stimulate the immune system and reduce blood lipids due to high contents of beta-1,3-glucan. *Spirulina* is offered by many companies as dried or tableted food supplement or as an ingredient in other foods (e.g. in cereal and protein bars, cookies, dairy products, pasta, or desserts) or in juices. Besides potential health and nutritional benefits, the supplements are added to improve sensoric characteristics and texture of the processed foods (Dineshbabu et al., 2019). Besides the use in feed and food production, cyanobacteria and microalgae offer application potential in a variety of other sectors, including the production of various other lipids and fatty acids, conversion of microalgae and derivatives into biofuels, production of bioplastics (e.g. polyhydroxyalkanoate (PHA) production by the cyanobacterium *Calothrix scytonemicola*; see also chapter 24), and the production of a variety of bioactive compounds that can be used as pharmaceuticals and nutraceuticals.

21.3 CHEMOAUTOTROPHIC BACTERIA

Photosynthesis has been a great success story of evolution to enable life to flourish, essentially by harvesting energy from sunlight and carbon from atmospheric CO_2. Microbial life is, however, much more diverse and developed biochemical machineries to utilize chemical energy as well. Generally said, microbial oxidation processes may generate energy as they release electrons which the cells channel through membrane bound proteins called the electron transport chains. The electrons are finally transferred to an electron acceptor (oxygen, nitrate, sulfate, to name a few), which propels ATP synthases to produce the cellular energy carrier, ATP. Additionally, the electrons are needed to reduce CO_2 for autotrophic growth. For this purpose, a cellular redox carrier, mostly $NADP^+$, receives the electrons and is reduced. So, if it is not sunlight serving as energy source, but chemical oxidation, we are speaking of chemoautotrophic growth.

Nature provides a number of chemicals that can be oxidized, among them minerals that are abundant in the Earth's crust (at least at specific sites). Iron is the fourth most abundant element in the Earth's crust and it is believed to have been the first element that early microbes (called iron bacteria) oxidized to gain energy. While sulfur is way less abundant than iron, microbes can gain much more energy by oxidizing reduced forms of this element, like sulfide and elemental sulfur, to sulfite and sulfate. Consequently, sulfur oxidizing bacteria like *Acidithiobacillus ferrooxidans* are abundant in soil, lakes, rivers, and drainages of coal and ore mines, where sulfide minerals are found (Figure 21.3A). Besides iron and sulfur, nitrogen is a third abundant element that can be used as an electron donor for microbial autotrophy. Nitrogen is mainly utilized in the form of ammonia, which is a degradation product of biomass. These so-called nitrifying bacteria oxidize ammonia via nitrite to nitrate to generate energy that they use for CO_2 assimilation. Hydrogen is the most abundant element in the Universe. On the Earth's surface it exists mostly in bound form, in water, hydrocarbons, and other organic matter. Molecular hydrogen (H_2) is present as a trace gas in air, and dissolved in water. Hydrogen oxidizing bacteria use it as an energy source as its oxidation releases a high energy level. They are found in fresh water, soils, and the oceans.

Next to their energy source chemoautotrophic bacteria need a CO_2 assimilation pathway to form the chemical bonds between carbon atoms for building their biomass. The best known of these is the Calvin–Benson–Bassham (CBB) cycle that also microalgae and land plants use for their CO_2 harvest. Evolution of metabolism is inventive, however, and has developed six other known CO_2 assimilation pathways which are found in different types of bacteria and archaea. All these pathways have in common that they utilize enzymes called carboxylases which add an inorganic CO_2 molecule to an organic acceptor molecule and thus incorporate CO_2 into the global biomass pool. Then cyclic pathways transform the products of the carboxylation to regenerate the acceptor molecules and to release a fraction of the molecules to build biomass. The type of carboxylase, and the cyclic regeneration pathway defines the respective CO_2 assimilation pathway. Nature developed a plentiful of combinations of these assimilation pathways and the previously described oxidative energy harvesting processes. Unexpectedly, a seventh CO_2 assimilation pathway was discovered in 2020 that works without a regeneration cycle of the acceptor molecule: in the reductive glycine

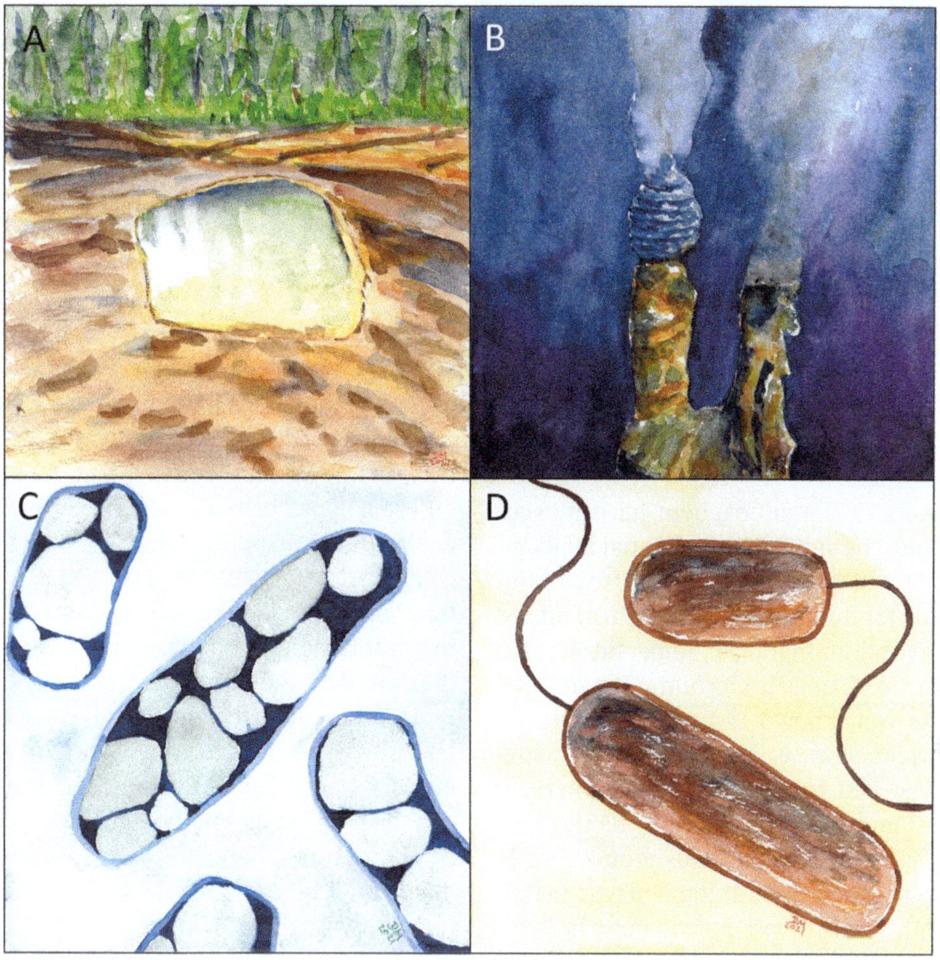

FIGURE 21.3 Chemoautotrophic microorganisms harvest energy by oxidizing minerals. (A) Iron bacteria live on the oxidation of iron and color soils and water red. They are often found in wet habitats and leave ochre behind (iron oxide and hydroxides) that was one of the first natural dyes humans have used, like from the Rocky Mountain Paint Pots. (B) Deep sea volcanic sites create hydrothermal vents where heat and gas release containing CO_2, H_2, and iron and sulfur elements providing energy and nutrients for chemoautotrophic microbes that feed a rich local fauna. (C) *Cupriavidus necator* live on hydrogen and CO_2 and store granules of polyhydroxyalcanoate which is produced as a biobased plastic material. (D) *Acidithiobacillus ferrooxidans* is used in bioleaching to extract metals from their ore by oxidation.

pathway of *Desulfovibrio desulfuricans*, CO_2 is directly reduced to formic acid and then coupled with another CO_2 molecule to form the amino acid glycine that serves as the starting point for biomass formation.

We tend to focus on terrestrial and freshwater habitats when thinking of microorganisms, but we should not overlook that 70% of our planet is covered by oceans, inhabited by microbes of all kinds. Autotrophic oceanic microbes play an immense role in feeding the oceans by generating new organic matter, and they are responsible for about half of Nature's CO_2 capture capacity. Near the surface, where sunlight can reach, phototrophic microorganisms dominate, but most of the oceans are dark, in sediments close to the coasts, in open deep-sea water, and at the bottom of the oceans. In these dark regions chemoautotrophs gain energy by oxidizing hydrogen, sulfide, or ammonia, and capture CO_2 that is dissolved in the water to grow new biomass (Hügler and Sievert, 2011). It is estimated that chemoautotrophs make up to 1.5% of the oceans' CO_2 capture capacity.

Arguably the most exotic habitats on our planet are deep sea hydrothermal vents, underwater volcanic sites where mineral and CO_2 rich hot water is discharged from the seafloor (Figure 21.3B). Independent from sunlight, chemoautotrophic bacteria make a living from oxidizing minerals like sulfides, capturing CO_2 to grow, and they feed a surprisingly rich fauna of giant tube worms, clams, crabs, and fish.

A third microbial strategy to gain energy and to reduce CO_2 for assimilation is called chemoorganoautotrophy where an organic molecule is oxidized to serve as energy source. Among those, methane oxidation as an energy source is standing out. A prominent member of these methylotrophic CO_2 fixating bacteria is called *Methylacidiphilum infernosum*. The name stands for itself: a methane and acid lover from hell, as it is found in geothermal sites where methane and CO_2 are released into hot acidic ponds.

21.3.1 Biotech Applications of Chemoautotrophs

The rising CO_2 levels in the atmosphere have moved chemoautotrophy into the spotlight of potential applications to develop microbial processes that capture carbon and make it available for human use in the form of materials as well as food and animal feed. Some autotrophs like the hydrogen oxidizing bacterium *Cupriavidus necator* has become well known to produce a natural polymer as a nutrient storage, polyhydroxyalkanoate (PHA) which is deposited inside the cells as spherical "plastic" granules (Figure 21.3C). PHA is a biodegradable thermoplastic biopolymer that holds promise for a mineral oil free replacement of traditional plastic materials (see chapter 24). While *C. necator* can grow heterotrophically on sugars, it is an autotroph as well, feeding on hydrogen (H_2) as energy source and fixating CO_2 through the CBB cycle. Thus, it is able to produce bioplastic from CO_2 and H_2, derived by electrolysis of water. With some genetic engineering, *C. necator* has also been enabled to produce biofuels like isobutanol from H_2 and CO_2, so that renewable electricity could be used to turn CO_2 into a carbon neutral transportation fuel (Yu, 2018). Back in the 1970s, *C. necator* has been explored as a potential renewable food source, recycling CO_2 to human nutrition in crewed space missions. This idea has been taken up recently by the start-up companies Solar Foods (Finland) and Air Protein (USA) to produce edible protein from electricity and CO_2.

Bioleaching is a process to extract metals from their ores with the aid of microbial oxidation by chemoautotrophs such as *Acidithiobacillus* species (Figure 21.3D). Its main environmental benefit is to make the process cleaner by avoiding the use of toxic cyanide. While the microbes grow on CO_2 also in this process, it is not their autotrophy that stands in the foreground here, but their pathways for inorganic energy production. The same principle can also be used for the desulfurization of coal and for the recycling of rare metals from electronic scrap, and thus bacteria may help us to conserve precious rare elements and reduce the exploitation of natural deposits (Dunbar, 2017).

21.4 SYNTHETIC BIOLOGY: NEW-TO-NATURE CO_2 FIXATION PATHWAYS

The natural biodiversity is tremendous. Thanks to this diversity, we are surrounded by microorganisms with different features. With their aid we are producing a very broad range of goods like enzymes, medicines, bioplastics, fuel, food supplements, and we are using them in different types of biotechnological processes including environmental applications.

About 440 Gigatons of CO_2 are fixed by plants and microorganisms every year (Figure 21.1). This number might sound huge, but it is not enough to compensate for the emissions that originate from human activity (Figure 21.1D). To reduce the global CO_2 emissions for a sustainable future, we need technologies that are CO_2-neutral or even CO_2-negative. In the fight against the climate crisis, can the CO_2 eating microorganisms offer a solution?

In nature, there are seven different pathways in plants and microbes to fix CO_2. But, natural carbon fixing pathways, like the most popular CBB cycle, are rate limited: although higher efficiencies can be reached theoretically, in reality, the photosynthetic efficiency in plants is typically around 1% whereas cyanobacterial photosynthetic efficiency is also limited to 1–2%. Therefore, we need to improve the carbon capturing efficiency first to achieve sustainable CO_2-neutral or even CO_2-consuming applications. Second, we should harvest the energy required for operating the CO_2 fixing reactions in an efficient, cheap, and environment-friendly way.

Here is the point where synthetic biology joins our story. Using synthetic biology, we are able to redesign the natural pathways or create novel ones so that the organisms have new abilities. We also have powerful tools for genetic engineering like CRISPR-Cas9 for easy and precise gene editing. These tools are mainly available for well-known microorganisms, which makes these microbes tempting to use as microbial cell factories. So, thanks to the developments in synthetic biology and metabolic engineering, we are not limited to the native autotrophic microbes for CO_2 capturing.

We can start with addressing the first question: how can we improve the CO_2 fixing efficiency? One of the obvious ideas is to implement the CBB or a similar CO_2 fixing cycle to heterotrophs and rewire the metabolism, thereby generating an efficient carbon capturing cycle. But this is not as easy as it sounds: with an attempt to reconfigure the metabolism of *Escherichia coli*, even though the full CO_2 capturing pathway can be successfully integrated to the metabolic network, the cells were not able to grow and produce the entire biomass precursors solely from CO_2. *E. coli* was not the only host to be engineered for growing on CO_2 as the only carbon source. A methylotrophic bacterium, *Methylobacterium extorquens*, was engineered by blocking its methanol assimilation and expressing key enzymes of CBB cycle. But this strain was not able to grow on CO_2 as the sole C source continuously, either.

Let us talk a bit more about methylotrophy here. Methylotrophic microorganisms do not grow on CO_2 but they can utilize other single carbon compounds like methanol or methane, which are found quite plentiful in nature. Methylotrophy is not limited to bacteria. A few yeast genera, such as *Ogataea* and *Komagataella*, can also metabolize single C compounds and utilize methanol as the sole carbon and energy source. The methanol assimilation pathway of these microbes can resemble a blueprint to implement a carbon fixing cycle into these heterotrophs. *Komagataella phaffii*, formerly known as *Pichia pastoris*, makes a good example here. Its methanol assimilation pathway, the so called the xylulose monophosphate pathway (XuMP), shows a high resemblance and shares many enzymes with the CBB cycle. Thereby, using synthetic biology tools, researchers made a metabolic design so that the heterotrophic *K. phaffii* could be converted into an autotroph that can grow on CO_2 as its sole carbon source while using methanol to harvest energy (Gassler et al., 2020). The slow growth rates of this synthetic autotrophic *K. phaffii* remain as a challenge, but through adaptive laboratory evolution and metabolic engineering, growth rates can be increased. *K. phaffii* is not the only microorganism that is rewired to grow on CO_2. Another research group could construct a synthetic *E. coli* strain that can grow on CO_2 as sole carbon source and harvest energy from formic acid (Gleizer et al., 2019). *E. coli* and *K. phaffii* are among the major workhorse microorganisms for biotechnological applications. With further attempts, these synthetic autotrophs might serve as microbial cell factories for value added chemicals (like organic acids) with a net CO_2-negative production rate, which can help mitigating atmospheric CO_2 to fight against the climate crisis.

As mentioned above, the CBB cycle – although being the most famous one – is not the only carbon fixing pathway. Again, with the help of synthetic biology, the central metabolism of *E. coli* was reconfigured so that it can grow on formic acid (or methanol) and CO_2 via a reductive glycine pathway. This synthetic pathway was designed from scratch based on a natural reductive acetyl-CoA pathway (rAcCoAP), but then it turned out that it naturally occurs in a chemolithoautotrophic microbe, *Desulfovibrio desulfuricans*. This is a very nice example of how evolution and rational metabolic engineering design found routes independently toward a common CO_2 fixing pathway.

But – keeping the enormous potential of biodiversity and evolution in mind – are we really limited to the natural pathways? The example above shows that, even without a natural blueprint, we might design a functional synthetic CO_2 fixing pathway and implement it into a heterologous host. If we can make a more efficient pathway to capture CO_2, eventually the resources can be allocated to other pathways, like producing chemicals or biomass in general for biotechnological purposes (Bar-Even et al., 2010). A good example is the crotonyl-CoA/ethylmalonyl-CoA/hydroxybutyryl-CoA (CETCH) cycle which includes 17 enzymes across all three domains of life and nine different organisms. Optimized by enzyme and metabolic engineering, the CO_2 capturing enzyme, enoyl-CoA carboxylase-reductase, has a 37-fold higher activity than Rubisco and has a lower energy requirement than other natural CO_2 fixing pathways. This cycle works *in vitro*, but it is quite challenging to integrate such a large, artificial, functional pathway into a cell. It requires many optimization processes in selecting the appropriate host organism, identifying and opening the bottlenecks in the host metabolism, and fine-tuning gene expression levels to achieve the highest flux. *In vitro* application of the CETCH cycle is also not feasible as the addition of ATP and NADH, the energy source for a functional CETCH cycle, would be very expensive for an economical CO_2 fixing bioprocess. Thus, the use of synthetic cycles like CETCH in biotechnological applications is still far from reality.

This brings us to another key challenge in CO_2 fixing processes: harvesting energy. So far, we discussed creating efficient CO_2 fixing pathways which is one key aspect in CO_2 capture for future biotechnological applications. The other key aspect is harvesting the energy (ATP) and reducing power (NADH/NADPH) to operate these cycles. Plants and other photoautotrophs use sunlight which is readily available (Figure 21.4A). Sunlight

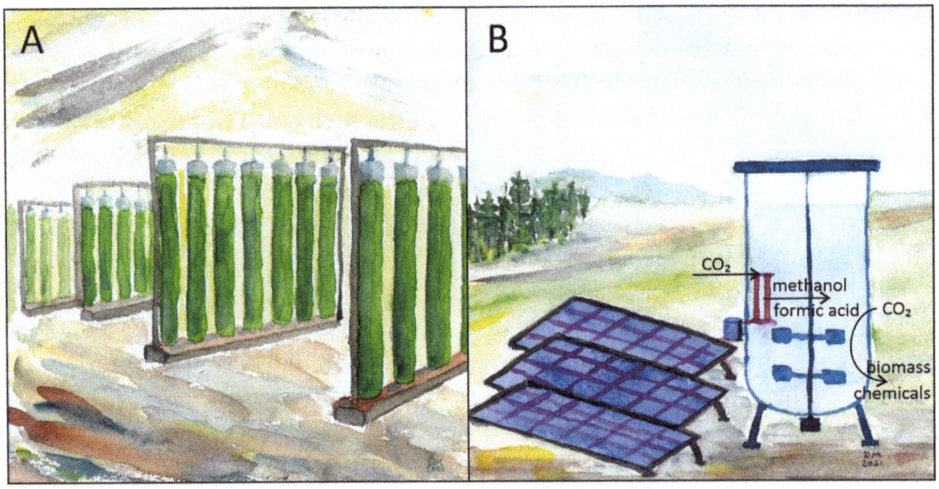

FIGURE 21.4 The next generation of biological carbon capture technology. (A) Vertical tube photobioreactors use sunlight as an energy source to grow microalgae. As a carbon source CO_2 is blown in at the bottom, which drives also vertical movement and mixing of the liquid (airlift reactors). (B) Synthetic biology allows to create new microorganisms that co-utilize different one-carbon molecules like methanol, formic acid, and CO_2. Electrochemistry allows to produce methanol and formic acid from CO_2 with high efficiency. With the use of renewable (e.g. photovoltaic) electricity, a concept is envisaged that captures CO_2 both electrochemically and biologically into microbial biomass and valuable chemicals.

is a cheap energy source but not very economically feasible as the photosynthetic efficiency is very low and vast areas are required to capture sunlight. According to FAO statistics, about one-third of the agricultural areas (38% of the world's land area) are used. With the increasing population, the demand for agricultural land is increasing too. In addition, the conversion of arable areas into non-agricultural lands, extensive use of chemicals, shorter fallow periods, and climate crisis result in a decrease in the lands suitable for agriculture. So, are there other energy sources that we can use for CO_2 fixation?

Electricity is another readily available and a cheaper option. We can use electricity to capture the CO_2 from air, electrolyze water into H_2 and O_2, and produce single carbon compounds like methanol or formic acid that can be utilized as an energy source. With the integrated use of CO_2 capture by electricity to produce methanol or formic acid and methylotrophic CO_2 fixing microbes, we can establish CO_2-negative, value-added chemical production systems (Figure 21.4B). But what if we step further and use electricity directly as the energy source? Microbial CO_2 conversion by using the electricity as a direct energy source is called microbial electrosynthesis, which can be done by a few acetogenic and methanogenic microbial species. Although proof-of-concept studies already showed its feasibility, industrial microbial electrosynthesis is still in its infancy. Constructing more robust and efficient microorganisms and design of 3D electrodes which enable a thick inner and outer biofilm to reach high current densities coupled with reactor configurations for an efficient mass and ion transfer remain as challenges to realize economically feasible microbial electrosynthesis applications (Jourdin and Burdyny, 2021).

We, as humans, are consuming and polluting the natural resources of our world and destroying the habitats of other living beings. In addition to the other environmental problems, the climate crisis has never been so close. Atmospheric level of CO_2 is increasing steadily and doubled its concentration since the industrial revolution. Leaving the world to explore new planets for human settlements is not a fantasy anymore. But is there nothing else left to save the world? Can't we use the high technology we developed for the benefit of our planet? It is very tempting to imagine a future with "good microbes" as our travel companions on space journeys, converting the CO_2 we produce to food or chemicals for the spare parts in our spaceship. But we have the tools and the knowledge for the paradigm shift to rebuild our future here on planet Earth, too (Mohan et al., 2016).

REFERENCES

1. Amenabar MJ, Boyd ES. 2019. A review of the mechanisms of mineral-based metabolism in early Earth analog rock-hosted hydrothermal ecosystems. *World Journal of Microbiology and Biotechnology* 35: 29. https://doi.org/10.1007/s11274-019-2604-2.

2. Bar-Even A, Noor E, Lewis NE, Milo R. 2010. Design and analysis of synthetic carbon fixation pathways. *Proceedings of the National Academy of Sciences of the U.S.A* 107: 8889–8894. https://doi.org/10.1073/pnas.0907176107.
3. Daneshvar E, Ok YS, Tavakoli S, Sarkar B, Shaheen SM, et al. 2021. Insights into upstream processing of microalgae: A review. *Bioresource Technology* 329: 124870. https://doi.org/10.1016/j.biortech.2021.124870.
4. Dineshbabu G, Goswami G, Kumar R, Sinha A, Das D. 2019. Microalgae–nutritious, sustainable aqua- and animal feed source. *Journal of Functional Foods* 62: 103545. https://doi.org/10.1016/j.jff.2019.103545.
5. Dunbar WS. 2017. Biotechnology and the mine of tomorrow. *Trends in Biotechnology* 35: 79–89. https://doi.org/10.1016/j.tibtech.2016.07.004.
6. Erb TJ, Zarzycki J. 2018. A short history of RuBisCO: The rise and fall (?) of nature's predominant CO_2 fixing enzyme. *Current Opinion in Biotechnology* 49: 100–107. https://doi.org/10.1016/J.COPBIO.2017.07.017.
7. Gassler T, Sauer M, Gasser B, Egermeier M, Troyer C, et al. 2020. The industrial yeast *Pichia pastoris* is converted from a heterotroph into an autotroph capable of growth on CO_2. *Nature Biotechnology* 38: 210–216. https://doi.org/10.1038/s41587-019-0363-0.
8. Gleizer S, Ben-Nissan R, Bar-On YM, Antonovsky N, Noor E, et al. 2019. Conversion of *Escherichia coli* to generate all biomass carbon from CO_2. *Cell* 179: 1255–1263. https://doi.org/10.1016/j.cell.2019.11.009
9. Hügler M, Sievert SM. 2011. Beyond the Calvin cycle: Autotrophic carbon fixation in the ocean. *Annual Review of Marine Science* 3: 261–289. https://doi.org/10.1146/annurev-marine-120709-142712.
10. Jourdin L, Burdyny T. 2021. Microbial electrosynthesis: Where do we go from here? *Trends in Biotechnology* 39: 359–369. https://doi.org/10.1016/j.tibtech.2020.10.014.
11. Liu J, Michalski JR, Tan W, He H, Ye B, Xiao L. 2021. Anoxic chemical weathering under a reducing greenhouse on early Mars. *Nature Astronomy*. https://doi.org/10.1038/s41550-021-01303-5.
12. Mohan SV, Modestra JA, Amulya K, Butti SK, Velvizhi G. 2016. A circular bioeconomy with biobased products from CO_2 sequestration. *Trends in Biotechnology* 34: 506–519. https://doi.org/10.1016/j.tibtech.2016.02.012.
13. Rockwell NC, Lagarias JC, Bhattacharya D. 2014. Primary endosymbiosis and the evolution of light and oxygen sensing in photosynthetic eukaryotes. *Frontiers in Ecology and Evolution* 2: 66. https://doi.org/10.3389/fevo.2014.00066.
14. Tan JS, Lee SY, Chew KW, Lam MK, Lim JW, et al. 2020. A review on microalgae cultivation and harvesting, and their biomass extraction processing using ionic liquids. *Bioengineered* 11: 116–129. https://doi.org/10.1080/21655979.2020.1711626.
15. Woodward FI. 2007. Global primary production. *Current Biology* 17: R269–R273. https://doi.org/10.1016/j.cub.2007.01.054.
16. Yu J. 2018. Fixation of carbon dioxide by a hydrogen-oxidizing bacterium for value-added products. *World Journal of Microbiology Biotechnology* 34: 89. https://doi.org/10.1007/s11274-018-2473-0.

CHAPTER 22

The Biodiesel Biorefinery

Opportunities and Challenges for Microbial Production of Fuels and Chemicals

Hannes Russmayer[1,2] and Michael Egermeier[2]
[1]CD Laboratory for Biotechnology of Glycerol, 1190 Vienna, Austria
[2]Department of Biotechnology, Institute of Microbiology and Microbial Biotechnology, BOKU-VIBT University of Natural Resources and Life Sciences, 1190 Vienna, Austria

22.1 THE CONCEPT OF A BIOREFINERY

The non-renewable character of fossil oil in combination with a rising demand will lead to an increasing scarcity of this feedstock. Therefore, the quest for renewable alternatives is on. A very promising feedstock for chemical and biofuel production in the future is renewable plant derived biomass.

The conversion of biomass in marketable products follows the concept of a biorefinery. This concept is similar to the one of a fossil oil based refinery, where crude fossil oil is separated into fuels and platform chemicals leading to a broad product portfolio. Each single product stream is efficiently used and adds value to the process chain. This basic idea is also true for a biorefinery, but with the important difference of using renewable biomass as a feedstock.

Therefore, a biorefinery can be defined as the efficient processing of renewable biomass into products of higher value (Cherubini, 2010). Additionally, CO_2 emissions can be reduced by capturing CO_2 during the growth of renewable biomass (plant biomass) (Figure 22.1; see chapter 21).

Biorefineries are assigned to different classes mainly by the starting biomass used and the obtained products (de Jong and Jungmeier, 2015). Examples for different classes are oleochemical biorefineries or biorefineries using lignocellulosic starting material. However, biorefining basically involves two steps, which are the decomposition of the biomass into its components (carbohydrates, proteins, lipids) and the further efficient use of these components for bio-based fuel and chemical production (Figure 22.1). The upgrading of biomass to marketable products involves thermo/chemical and microbial processes. Especially microbial processes, using microorganisms as efficient cell factories, are a key technology found in almost all biorefinery concepts. The range of application for microbial processes in biorefineries is wide and they can be applied in many different biorefining steps (Figure 22.1). For example, enzymes secreted by microorganisms are used for effective decomposition of biomass (enzymatic hydrolysis) or microbial fermentation of bioethanol from lignocellulosic material (Chen et al., 2017). However, biomass is a rather complex mixture of chemically diverse molecules and decomposition leads to various available substrates. In order to allow economic viability of a biorefinery all substrates should be used efficiently. Here nature provides the solution by the high biodiversity of the microbial community (Russmayer et al., 2019). The high diversity within the microbial community creates a certain

Good Microbes in Medicine, Food Production, Biotechnology, Bioremediation, and Agriculture, First Edition. Edited by Frans J. de Bruijn, Hauke Smidt, Luca S. Cocolin, Michael Sauer, David Dowling, and Linda Thomashow.

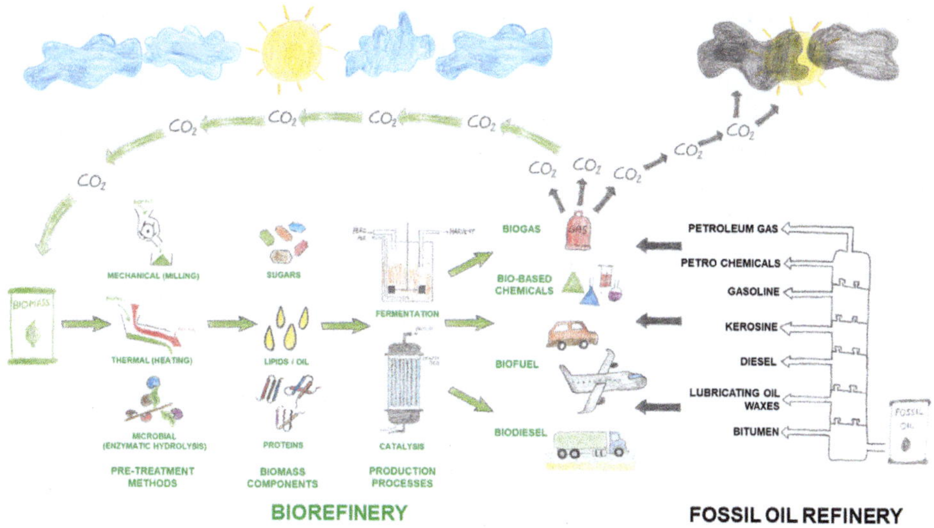

FIGURE 22.1 The general concept of a biomass biorefinery for sustainable production of industrial bulk chemicals.

metabolic flexibility in utilization of substrates with different chemical properties. Additionally, many microorganisms can use these complex substrates for the production of industrially relevant chemicals.

22.1.1 Biorefinery Concept for Biodiesel Production

A first step toward a sustainable future is the implementation of biodiesel as a renewable alternative to common fossil oil-based transport fuels. Biodiesel is mainly produced from oily plant biomass and is therefore considered as renewable. The production process is a simple transesterification process where the obtained plant oil reacts with methanol to form a mixture of fatty acid methyl esters, which is later named biodiesel. Besides its renewable character biodiesel can be directly applied to any motor engine in pure or blended form. Additionally, the net CO_2 emission is close to zero if biodiesel is used as a transport fuel, which helps to reduce the effects of climate change. In the last decade biodiesel production grew steadily and is predicted to further grow (Monteiro et al., 2018). Despite the increasing demand, biodiesel production is still not economically feasible and not competitive with common fossil oil-based refineries, which is mainly due to the higher production costs for biodiesel, which are highly dependent on the price of the available feedstock. A challenge is the implementation of technologies for the efficient use of almost all feedstock components and efficient recycling of accumulating waste streams from biodiesel production to allow an economically feasible process. Here, especially glycerol, the major by-product of the transesterification process, accumulates in vast amounts and constitutes a big problem for most biodiesel production plants. The following section will shed more light on the important role of microorganisms within an integrated biorefinery concept of a biodiesel plant, be it as a feedstock for biodiesel production or as a cell factory for the upgrading of waste streams.

22.1.2 Microorganisms as Feedstocks for Biodiesel Production

The choice of the feedstock is an important detail for the economically viable production of biodiesel. It is estimated that the used feedstocks account for around 70% of the overall production costs of biodiesel (Ma et al., 2018). Biodiesel production relies on plant oil derived fatty acids, which are bound as triglycerides. Basically, triglycerides are built by free fatty acids and a backbone molecule called glycerol. Depending on the plant or animal different fatty acid profiles can be found in triglycerides. A wide variety of oils can be used for production, but depending on their fatty acid profile and chemical properties they lead to biodiesel with different qualities (Hoekman et al., 2012). The source of the oil for biodiesel production is mainly determined by the location of the production site. In Europe often rape seed oil is used, whereas the biodiesel in the United States of America is produced from soybean oil and in Southeast Asia mainly palm oil is used. The production of biodiesel nowadays focuses mainly on the use of first generation feedstocks, which have the advantage of being easily processable to obtain the needed oil. On the downside they are edible and therefore interfere directly with the human food chain. A solution for this problem is found by using next generation feedstocks. Second generation feedstocks are a large group containing non-edible plant oil (such as jatropha oil) and unused animal fats (Singh et al., 2020). Despite being non-edible, such raw materials still interfere with the human food chain as farmland is occupied, which could be used for growing food crops. Furthermore, scaling of the feedstock production to meet the needs of industrial scale production of biodiesel is difficult, because farmland is a limited resource in some countries. Therefore, alternative feedstocks are needed and should have a fast and easy scalable production at low cost with no effect on the food supply chain. In this sense, oleaginous microorganisms have gained a lot of interest as potential feedstock. Oleaginous microorganisms, such as yeast, fungi, bacteria, and microalgae accumulate lipids up to 60–70% of their cell dry mass and with a similar fatty acid profile as plant oil (Thiru et al., 2011). High biomass productivity within a short time interval and no use of potential farmland for production make them very attractive. Furthermore, formation of lipid rich biomass is possible from cheap substrates, such as crude glycerol, waste oils, and CO_2.

Among this group microalgae and yeasts are the best studied organisms for microbial oil production. Microalgae constitute a rather diverse group of phototropic prokaryotic and eukaryotic organisms. Such organisms grow by using sunlight for energy production and fix atmospheric CO_2 as carbon source. Typically, microalgae (such as *Chlorella*, *Dunaliella*, *Nannochloris*) have a lipid content of 50–70%. The species *Chlorella vulgaris* and *Spirulina platensis* are of great interest, due to easier cultivation than other algae (Mata et al., 2010). Microalgae are favored as biodiesel feedstock for example because of faster biomass production and higher lipid content on simple and cheap media (waste water) than most food crops. Furthermore, the oil yield by hectare is much higher for microalgae with less land occupied for cultivation (Mata et al., 2010). The additional benefit is the reduction of greenhouse gas emissions (GHG) by fixation of atmospheric CO_2 and removal of pollutants from waste water by simple algae biomass production. Regarding the CO_2 fixation potential of algae biomass, it is estimated that 1 kg of algae biomass can roughly fix around 1.83 kg CO_2 (see chapter 21). On the one hand, this has a positive effect on atmospheric GHG concentrations, on the other hand, CO_2 is low cost and a carbon source, which is available in vast amounts (Khan et al., 2018). The cultivation basically can be carried out either in open culture systems (such as ponds) or closed photobioreactors. However, thinking of industrial scale production the production costs for microbial oil from microalgae are still too high and biodiesel production relying on such oil is not economically feasible. The higher production cost of biodiesel from microalgae oil is attributed to inefficient downstream methods for oil extraction and higher investment costs in the specific equipment needed for cultivation (photobioreactor) (Chisti, 2008).

Besides microalgae certain yeast species showed to have a high potential as a future feedstock for biodiesel production. Oleaginous yeast species accumulated between 35% and 74% (w/w) lipids and are, for example, *Yarrowia lipolytica*, *Rhodotorula glutinis*, *Rhodotorula graminis*, *Rhodosporidium toruloides*, *Cryptococcus curvatus*, and *Lipomyces starkeyi* (Russmayer et al., 2019).

The most promising and best studied organism in the sense of lipid accumulation is *Y. lipolytica*. It unites several aspects needed for industrial scale production as, for example, the high biomass production rate on cheap media with low cultivation costs. Furthermore, the availability of the genome sequence and genetic engineering tools allow the genetic modification of the yeast to pave the way for tailor-made optimization of naturally occurring product formation processes, but also production of modified oils better suited for biodiesel production (Bredeweg et al., 2017). *Y. lipolytica* is a very distant relative to baker's yeast and can utilize a wide range of substrates ranging from sugars (glucose, fructose), carboxylic acids (acetic, citric, pyruvic, lactic acid) to sweet alcohols (glycerol, erythritol) and – as the name of the yeast already implies – hydrophobic substrates (fatty acids and alkanes) (Liu et al., 2015). In contrast to other yeast species *Y. lipolytica* prefers glycerol over glucose as the primary carbon source, even when both substrates are available. (Workman et al., 2013). In addition, the ability to utilize industrial side-streams like waste water from olive mills (Papanikolaou et al., 2008) and especially crude glycerol from biodiesel production (André et al., 2010; Rakicka et al., 2015; Rywińska et al., 2009) underline the importance of this yeast for a circular economy. Depending on the strain and chosen cultivation conditions *Y. lipolytica* accumulates lipids in the range of 35–50 g per 100 g biomass. With additional genetic engineering and process engineering the lipid content is further increased (Blazeck et al., 2014).

Genetic engineering of *Y. lipolytica* with regard to lipid production mainly focuses on abolishing competing pathways (beta-oxidation or degradation pathways), overexpression of genes involved in fatty acid synthesis and engineering of the redox metabolism (Russmayer et al., 2019).

22.1.3 Microbial Upgrading of Waste Streams from Biodiesel Production

The valorization of biodiesel waste streams is of high importance not only because of economic considerations, but also because efficient carbon recycling ensures future sustainability of the whole biodiesel production. Additionally, a potential waste problem is solved. The major waste stream of biodiesel production is crude glycerol. Around 1 kg crude glycerol is produced per 10 kg of biodiesel. The accumulation of this waste stream constitutes a major bottleneck for a further increase of biodiesel production in the future. Hence, research focused on developing processes for crude glycerol upgrading. A straightforward way for valorization is refining of crude glycerol. Refined glycerol can then be sold for a higher price and has about 1500 possible applications (such as in cosmetics, pharma glycerol) (Monteiro et al., 2018; Russmayer et al., 2019).

A more sustainable way of upgrading is the microbial conversion of crude glycerol to industrial building blocks with a higher value (such as 1,3-propanediol) (Sauer et al., 2017). Glycerol is a versatile feedstock with several advantages over common sugar-based feedstocks, especially for microbial chemical production. Microbial production of highly reduced chemicals, such as alcohols, is more favorable with glycerol as a carbon source than with other sources (glucose) due to the higher degree of reduction for glycerol.

Another waste stream from biodiesel production is lignocellulosic biomass, which is most of the time the solid remainder after oil extraction from first and second generation biodiesel feedstocks. Such lignocellulosic biomass contains cellulose and hemicellulose, which are built mainly by glucose and xylose. To a lesser extent mannose and galactose can also be found. An example of these are lignocellulosic waste streams which are empty fruit bunches (EFB) obtained during the production of palm oil. Today, EFBs cause a massive waste problem and are mainly used as fertilizers or are simply burned, which is a rather inefficient and an environmentally questionable use of this biomass (Chin et al., 2019). An alternative approach for a sustainable way of EFB utilization is to use it as a feedstock for microbial chemical production. However, such waste streams need extensive pretreatment in order to obtain the typical cellulosic and hemicellulosic sugars. Pretreatment methods span mechanical, thermal, chemical, and biological methods and mostly a combination of different methods is necessary to allow proper pretreatment (Kumar et al., 2020).

However, microbial conversion of biodiesel production waste streams mainly focuses on production of value-added chemicals from crude glycerol, as this substance is by far the major waste stream.

The economic feasibility of microbial upgrading of crude glycerol (or general of waste streams) on an industrial scale relies on achieving high final product concentration and high productivity (= units of product per time) of the desired value-added product. The product purification steps after microbial fermentation of glycerol necessitate high final product concentration, due to product loss during the purification processes. Higher final concentrations massively reduce costs of downstream purification processes. Unfortunately, both parameters sometimes decrease due to harsh fermentation conditions. A major problem here is the accumulation of inhibitory or toxic compounds in the waste stream during the original production processes. For crude glycerol often the accumulation of alcohols (methanol) from the transesterification process or high salt concentrations causes severe problems for microbial growth and efficiency of microbial conversion processes (Chatzifragkou et al., 2011; Samul et al., 2014).

Luckily, nature provides the solution for this problem by a high biodiversity within the microbial community. Over millions of years of evolution nature developed microorganisms tolerating harsh conditions (Papanikolaou and Aggelis, 2009; Sauer et al., 2017). Such microorganisms are perfectly suited to withstand industrial fermentation processes with harsh production conditions. Furthermore, the biodiversity within the microbial community gives many microbial options to produce a broad portfolio of chemicals and biofuels with higher value (Lee et al., 2011).

Crude glycerol can be used by a wide variety of organisms, such as bacteria, yeast, and fungus, either to use it for growth and energy production or as a terminal electron acceptor under anaerobic conditions (Clomburg and Gonzalez, 2013). The features of the microbial hosts and the products of microbial glycerol metabolism with relevance for industry will be discussed in the following sections (Figure 22.2).

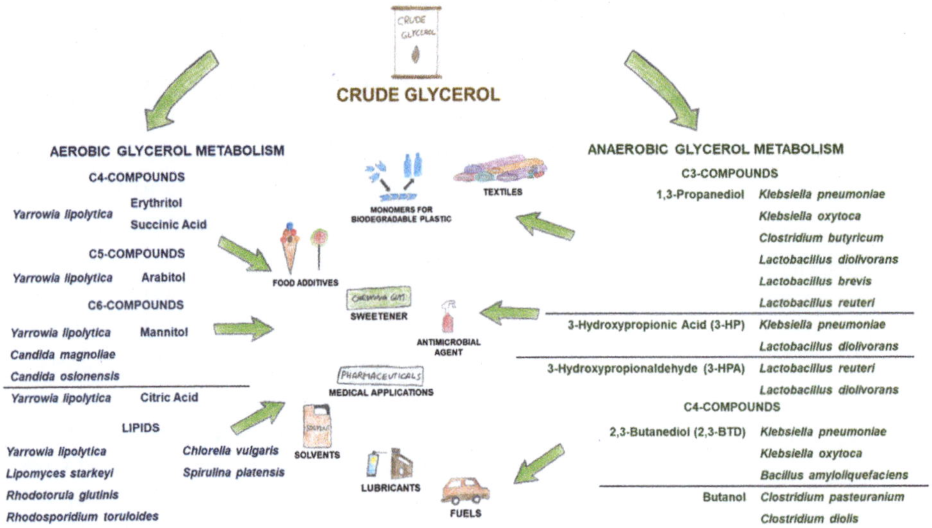

FIGURE 22.2 Microbial glycerol biorefining: Overview of industrial relevant chemicals and their final application in daily life.

22.2 HIGHER VALUE CHEMICALS FROM AEROBIC GLYCEROL METABOLISM

A wide variety of microorganisms, such as bacteria, yeast, and fungus, are able to oxidize glycerol for carbon and energy production (Figure 22.2). A very promising host for the production of value-added chemicals through glycerol oxidation is *Y. lipolytica*.

Under aerobic and non-limiting growth conditions *Y. lipolytica* utilizes glycerol as carbon and energy source and biomass is formed. However, if certain media components are limited (e.g. nitrogen) glycerol flux is diverted from biomass production toward the formation of metabolites. Typical metabolites are carboxylic acids (citric acid, isocitric acid), polyols (arabitol, erythritol, mannitol) to various fatty acids. The metabolic flexibility of this yeast is high and the product pattern is highly dependent on the origin of isolation of a specific yeast strain and the cultivation conditions (Egermeier et al., 2017). *Y. lipolytica* readily consumes crude glycerol and shows little disturbance by impurities in the substrate, while efficiently secreting organic acids and polyols into the supernatant of a fermentation broth (Morgunov et al., 2013; Rywińska et al., 2010). The limitation of nitrogen is the main trigger for metabolite production with this yeast. By changing the carbon to nitrogen ratio (C/N) the product pattern is easily switched from lipid production to polyol and organic acids. The potential of *Y. lipolytica* for lipid production has already been discussed in Section 22.1.3, therefore the focus now will be on organic acid and polyol production. Additional to nitrogen limitation also the pH has an important role on the product pattern. For example, a high initial pH with a high C/N ratio favors the production of citric acid, whereas a lower pH with the same C/N ratio leads to polyol formation (Egermeier et al., 2017).

Citric acid is the major organic acid produced by microbial fermentation processes and the filamentous fungi *Aspergillus niger* is the unchallenged frontrunner in terms of industrial scale production. However, *Y. lipolytica* is able to efficiently convert glycerol to citric acid with the advantage of higher fermentation rates and higher tolerance to metal ions, which allows the use of unrefined substrates (Levinson et al., 2007). Succinic acid, another top value chemical, can be efficiently produced by genetic engineering *Y. lipolytica*. The final titer of almost 200 g/L currently represents the highest value achieved by microbial conversion of glycerol (Li et al., 2017). Further industrially relevant chemicals produced by this yeast are pyruvic acid and α-ketoglutaric acid (Cybulski et al., 2018). A lot of effort has also been put into efficient production of erythritol and mannitol, the main type of polyols secreted by *Y. lipolytica* and the environmental triggers (nitrogen limitation, low pH conditions, high osmotic pressure) are known, but not completely understood (Rzechonek et al., 2018). Mannitol is a naturally occurring sweetener in fruits and vegetables, whereas erythritol is widely used as a non-caloric sweetener (60–70% sweetness of sucrose) in food and medical applications. While the proposed biosynthetic pathway for mannitol production in this yeast is still controversial, formation of erythritol is well studied and occurs via the non-oxidative branch of the pentose phosphate pathway. Genetic engineering of *Y. lipolytica* by transcription factor engineering and enzyme targeting enhanced titers and productivities (Cheng et al., 2018; Janek et al., 2017; Qiu et al., 2020). Besides *Y. lipolytica*, other yeasts classified in the genus *Yarrowia* have a high potential for polyol production. In particular, *Candida oslonenis* and *Candida magnoliae* are of interest industrial level mannitol production from glycerol (Khan et al., 2009).

22.2.1 Anaerobic Glycerol Metabolism for Industrial Chemical Production

22.2.1.1 Dehydration of Glycerol to Industrial Relevant Building Blocks

The metabolic pathway for the dehydration of glycerol is a short and simple two step pathway and is used by many different bacteria under anaerobic cultivation conditions. The first step is a simple dehydration of glycerol to 3-hydroxypropionaldehyde (3-HPA) catalyzed by the glycerol dehydratase, the key enzyme of the pathway. In a second step the built 3-HPA is in most cases reduced to 1,3-propanediol (1,3-PDO). However, under certain conditions 3-HPA is also oxidized to 3-hydroxypropionic acid (3-HP). The three mentioned metabolites are of considerable industrial interest (Figure 22.2). Nature provides a few organisms which are able to synthesize all three metabolites from glycerol. They belong to the bacterial genera of *Klebsiella*, *Clostridia*, *Lactobacillus*, and *Citrobacter* (Russmayer et al., 2019). 1,3-PDO is used for the synthesis of polytrimethylene terephthalate (PTT). PTT can be used for the production of fibers (Sorona, Du Pont), carpets (Corterra, Shell), monofilaments, or fibrous material (Almeida et al., 2012). About 100 years ago the human pathogen *Klebsiella pneumonia* was the first organism described to produce 1,3-PDO from glycerol (Freund, 1881). From the genus *Klebsiella*, also *Klebsiella oxytoca* is known to be a good producer of 1,3-PDO. The outstanding features of this group are a high productivity for 1,3-PDO and the tolerance to high concentration of glycerol. Additionally, *Klebsiella spp.* use glycerol for growth and 1,3-PDO production. The availability of various genetic engineering tools allows the production of high 1,3-PDO concentrations (Chen et al., 2015; Zhou et al., 2015). Genetic engineering strategies mainly focused on the elimination of competing pathways (lactate, ethanol formation), overexpression of genes directly involved in 1,3-PDO synthesis (Rhie et al., 2019; Yang et al., 2007). A disadvantage of this genus is the human pathogenic character of all strains, which raises concerns about industrial scale production.

Efficient natural 1,3-PDO producers from glycerol within the genus *Clostridia* are strictly anaerobic bacteria *Clostridium butyricum* and *Clostridium diols*. The main trigger for 1,3-PDO production in *Clostridia* is the initial glycerol concentration. High initial concentration favors 1,3-PDO, whereas lower concentration leads to the production of butyric acid and lactic acid. Both strains have the capability of high level 1,3-PDO production with industrial potential. However, final concentrations are higher dependent on the used cultivation conditions and initial glycerol concentrations (Kubiak et al., 2012). Genetic engineering approaches are scarce, due to the lack of engineering tools (Abbad-Andaloussi et al., 1995; Otte et al., 2009). The main focus for 1,3-PDO production with *Clostridia spp.* is on media and process optimization (Kubiak et al., 2012). The advantage of this bacterial genus is the high potential for 1,3-PDO production and non-pathogenic character. On the downside impurities of crude glycerol negatively impact the 1,3-PDO productivity of *Clostridia spp.* (Chatzifragkou et al., 2011). Furthermore, *Clostridia spp.* need anaerobic conditions to allow proper growth and 1,3-PDO production. The maintenance of anaerobic conditions on an industrial scale is often problematic and cost intensive.

The third bacterial genus containing natural producers of 1,3-PDO from glycerol are *Lactobacilli*. *Lactobacilli* have already been known for centuries because of their outstanding capabilities in food fermentation (bread, beer; see section 2). However, they are also very interesting cell factories for microbial chemical production. *Lactobacilli* have several advantages over *Klebsiella spp.* and *Clostridia spp.*, making them perhaps the most promising group for industrial production of 1,3-PDO from glycerol. They are generally classified as genetically safe organisms (GRAS), tolerate oxygen, and efficiently produce 1,3-PDO from glycerol. The difference to the former groups of natural producers is that *Lactobacilli* produce 1,3-PDO from glycerol only during glucose/glycerol co-fermentation, because glycerol is only used as a terminal electron acceptor. Hence, a second carbon source for growth and energy production must be supplemented (Sauer et al., 2017). *Lactobacilli* evolved in rich niches, therefore cultivation of this organism necessitates rather rich media, which is a disadvantage due to higher process costs.

The most promising members of the group *Lactobacilli* with an outstanding capability of 1,3-PDO production are *Lactobacillus diolivorans*, *Lactobacillus reuteri*, and *Lactobacillus brevis* (Pflügl et al., 2012; Vivek et al., 2017). Genetic engineering approaches for *L. diolivorans* focus on glycerol transport in the cell and the overexpression of the terminal 1,3-PDO oxidoreductase. Both approaches led to an increase in 1,3-PDO titers in batch cultivations (Lindlbauer et al., 2017a; Pflügl et al., 2013).

Also process engineering led to higher 1,3-PDO concentrations. For *L. diolivorans* simple glucose pulsing during fed batch cultivations led to an increase of the 1,3-PDO titer to 92 g/L (Lindlbauer et al., 2017b). Interestingly, *L. diolivorans* is able to use sugars derived from lignocellulosic material (Pflügl et al., 2014). This is also of potential interest for biodiesel biorefinery because it allows the efficient recycling of lignocellulosic waste streams accumulated during oil production. For *L. brevis* process engineering concentrated on cell immobilization during batch cultivations (Vivek et al., 2017).

The second metabolite of interest formed by glycerol dehydration pathway is 3-HPA. 3-HPA is industrially relevant as an antimicrobial agent and can also be used for the production of acrolein and acrylate. During normal growth conditions cells do not accumulate 3-HPA, due to its high toxicity. The production of 3-HPA is usually done by resting cells. Resting cells are viable and metabolically active, but growth and energy metabolism are decoupled leading to the accumulation of 3-HPA. The best studied producer of 3-HPA is *L. reuteri* (Lüthi-Peng et al., 2002). Another interesting host for 3-HPA production is *L. diolivorans* (Lindlbauer et al., 2017a). 3-HPA concentrations were further increased by engineering glycerol uptake of *L. diolivorans* (Lindlbauer et al., 2017a). The limiting factor for further increased titers is the toxicity of 3-HPA. Detoxification of 3-HPA was used as a process engineering strategy to increase 3-HPA concentration in *L. reuteri* (Krauter et al., 2012).

3-HP as the last but not least interesting metabolite is produced via oxidation of 3-HPA under anaerobic to microaerophilic conditions. It can be used as a monomer for polymerization to biodegradable polymers. The benchmark for 3-HP production is *K. pneumoniae*. However, 3-HP titers produced by wild type strains are rather low (by very high titers of 1,3-PDO). Only via intensive genetic engineering, by overexpressing the terminal aldehyde dehydrogenase and elimination of competing pathways (lactate, acetic acid formation) high level production of 3-HP was possible (Jers et al., 2019). A further promising cell factory 3-HP production is *L. diolivorans*. No further genetic engineering is needed to produce higher concentrations of 3-HP (Lindlbauer et al., 2017b).

22.2.1.2 Microbial Glycerol Reduction for Chemical Production

Industrially relevant metabolites originating from this branch of glycerol metabolism are butanol and 2,3-butanediol (2,3-BTD) isomers (Figure 22.2). Both metabolites are side products of glycerol metabolism during anaerobic (butanol) or microaerophilic (2,3-BTD) microbial growth. They have a significant industrial relevance as future biofuels or solvents. Butanol is predicted to be the biofuel of the future. The limitation for microbial butanol production again lies in its high toxicity and therefore only a few microorganisms are capable of butanol production. The most efficient butanol producers are found in the genus *Clostridia*, where butanol formation was reported for *Clostridium pasteurianum*, *Clostridium beijerinckii*, *C. diolis*, and *C. butyricum*. The benchmark for butanol production is *C. pasteurianum* (Biebl, 2001). Genetic engineering approaches led to increased titers and mainly focused on elimination of 1,3-PDO, which is also a product of anaerobic glycerol metabolism in *Clostridia spp.* (Pyne et al., 2016). Another key to more efficient butanol production is media optimization and process engineering. A strategy of media optimization was the supplementation of butyric acid, which caused increased butanol titers (Regestein et al., 2015). Process engineering mainly focused on detoxification of butanol through in situ removal using liquid/liquid extraction or membrane filtration (Lin et al., 2015).

For 2,3-BTD production *K. pneumonia* constitutes the benchmark. During microaerophilic growth conditions *K. pneumoniae* is able to produce industrial relevant 2,3-BTD concentrations (Petrov and Petrova, 2009). Genetic engineering proved to be a valuable tool for the improvement of 2,3-BTD by eliminating competing pathways (lactate formation). Additionally, process engineering positively influenced 2,3-BTD productivity (Petrov and Petrova, 2010). A non-pathogenic production host for 2,3-BTD is *Bacillus amyloliquefaciens*. An advantage of this organism is the possibility of 2,3-BTD production on simple and cheap media (Yang et al., 2013).

22.3 CONCLUDING REMARKS

The increasing scarcity of fossil oil in combination with growing awareness of environmental problems pushes the development of alternative sources for the production of chemicals and fuel used in daily life. In recent years renewable biomass comes to the fore as a future feedstock for chemical and fuel production. Especially in the sector of engine fuel, green alternatives (such as biodiesel) gained a lot of interest. Biodiesel is a renewable alternative to common diesel and is produced from various oily plant feedstocks through the process of transesterification. Unfortunately, biodiesel production costs are high, and the production is not economically feasible. Additionally, the accumulation of waste streams, especially crude glycerol cause tremendous problems. A solution for both problems can be the application of a biorefinery concept. A key within this concept is the application of microorganisms as cell factories or also as the feedstock itself. Microorganisms will help as a perfect alternative to already applied feedstocks. Furthermore, they will allow a sustainable biodiesel production process by efficient carbon recycling from waste streams to higher value daily life products through microbial fermentation processes. Due to the high metabolic flexibility of microorganism's various substrates (crude glycerol) can be converted into a broad portfolio of higher value chemicals. Biorefining of glycerol is possible by a wide variety of organisms spanning from yeasts (e.g., *Y. lipolytica*) to bacteria (genus *Klebsiella, Clostridia, Lactobacillus*). The mentioned organisms are potential future cell factories for biorefining because glycerol is efficiently converted at high rates and high product yields, which is a prerequisite for industrial scale production.

ACKNOWLEDGMENTS

The financial support of the Austrian Federal Ministry for Digital and Economic Affairs and the National Foundation for Research, Technology and Development through the Christian Doppler research Association is gratefully acknowledged.

REFERENCES

1. Abbad-Andaloussi S, Manginot-Durr C, Amine J, Petitdemange E, Petitdemange H. 1995. Isolation and characterization of *Clostridium butyricum* DSM 5431 mutants with increased resistance to 1,3-Propanediol and altered production of acids. *Applied and Environmental Microbiology* 61: 4413–4417.
2. Almeida JRM, Fávaro LCL, Quirino BF. 2012. Biodiesel biorefinery: Opportunities and challenges for microbial production of fuels and chemicals from glycerol waste. *Biotechnology for Biofuels* 5: 1–16.
3. André A, Diamantopoulou P, Philippoussis A, Sarris D, Komaitis M, Papanikolaou S. 2010. Biotechnological conversions of bio-diesel derived waste glycerol into added-value compounds by higher fungi: Production of biomass, single cell oil and oxalic acid. *Industrial Crops and Products* 31: 407–416.
4. Biebl H. 2001. Fermentation of glycerol by *Clostridium pasteurianum*–batch and continuous culture studies. *Journal of Industrial Microbiology and Biotechnology* 27: 18–26.
5. Blazeck J, Hill A, Liu L, Knight R, Miller J, Pan A, Otoupal P, Alper HS. 2014. Harnessing *Yarrowia lipolytica* lipogenesis to create a platform for lipid and biofuel production. *Nature Communications* 5: 3131.
6. Bredeweg EL, Pomraning KR, Dai Z, Nielsen J, Kerkhoven EJ, Baker SE. 2017. A molecular genetic toolbox for *Yarrowia lipolytica*. *Biotechnology for Biofuels* 10: 2.
7. Chatzifragkou A, Nychas G-JE, Doulgeraki AI, Zeng A-P, Papanikolaou S, Dietz D. 2011. Production of 1,3-propanediol by *Clostridium butyricum* growing on biodiesel-derived crude glycerol through a non-sterilized fermentation process. *Applied Microbiology and Biotechnology* 91: 101–112.
8. Chen H, Liu J, Chang X, Chen D, Xue Y, Liu P, Lin H, Han S. 2017. A review on the pretreatment of lignocellulose for high value chemicals. *Fuel Processing Technology* 160: 196–206.
9. Chen Z, Wu Y, Huang J, Liu D. 2015. Metabolic engineering of *Klebsiella pneumoniae* for the de novo production of 2-butanol as a potential biofuel. *Bioresource Technology* 197: 260–265.
10. Cheng H, Wang S, Bilal M, Ge X, Zhang C, Fickers P, Cheng H. 2018. Identification, characterization of two NADPH-dependent erythrose reductases in the yeast *Yarrowia lipolytica* and improvement of erythritol productivity using metabolic engineering. *Microbial Cell Factories* 17: 133.
11. Cherubini F. 2010. The biorefinery concept: Using biomass instead of oil for producing energy and chemicals. *Energy Conversion and Management* 51: 1412–1421.
12. Chin DW-K, Lim S, Pang Y-L, Leong L-K, Lim C-H. 2019. Investigation of organosolv pretreatment to natural microbial-degraded empty fruit bunch for sugar based substrate recovery. *Energy Procedia* 158: 1065–1071.
13. Chisti Y. 2008. Biodiesel from microalgae beats bioethanol. *Trends in Biotechnology* 26: 126–131.

14. Clomburg JM, Gonzalez R. 2013. Anaerobic fermentation of glycerol: A platform for renewable fuels and chemicals. *Trends in Biotechnology* 31: 20–28.

15. Cybulski K, Tomaszewska-Hetman L, Rymowicz W, Rakicka M, Rywińska A. 2018. *Yarrowia lipolytica* application as a prospective approach for biosynthesis of pyruvic acid from glycerol. *Chemical Papers* 72: 3077–3083.

16. de Jong E, Jungmeier G. 2015. Biorefinery concepts in comparison to petrochemical refineries. In: *Industrial Biorefineries and White Biotechnology*. Elsevier, pp. 3–33.

17. Egermeier M, Russmayer H, Sauer M, Marx H. 2017. Metabolic flexibility of *Yarrowia lipolytica* growing on glycerol. *Frontiers in Microbiology* 8: 1–9.

18. Freund A. 1881. Über die Bildung und Darstellung von Trimethylenalkohol aus Glycerin. *Monatshefte Für Chemie* 2: 636–641.

19. Hoekman SK, Broch A, Robbins C, Ceniceros E, Natarajan M. 2012. Review of biodiesel composition, properties, and specifications. *Renewable and Sustainable Energy Reviews* 16: 143–169.

20. Janek T, Dobrowolski A, Biegalska A, Mirończuk AM. 2017. Characterization of erythrose reductase from *Yarrowia lipolytica* and its influence on erythritol synthesis. *Microbial Cell Factories* 16: 118.

21. Jers C, Kalantari A, Garg A, Mijakovic I. 2019. Production of 3-hydroxypropanoic acid from glycerol by metabolically engineered bacteria. *Frontiers in Bioengineering and Biotechnology* 7: 1–15.

22. Khan A, Bhide A, Gadre R. 2009. Mannitol production from glycerol by resting cells of *Candida magnoliae*. *Bioresource Technology* 100: 4911–4913.

23. Khan MI, Shin JH, Kim JD. 2018. The promising future of microalgae: Current status, challenges, and optimization of a sustainable and renewable industry for biofuels, feed, and other products. *Microbial Cell Factories* 17: 36.

24. Krauter H, Willke T, Vorlop KD. 2012. Production of high amounts of 3-hydroxypropionaldehyde from glycerol by *Lactobacillus reuteri* with strongly increased biocatalyst lifetime and productivity. *New Biotechnology* 29: 211–217.

25. Kubiak P, Leja K, Myszka K, Celińska E, Spychała M, Szymanowska-Powałowska D, Czaczyk K, Grajek W. 2012. Physiological predisposition of various *Clostridium species* to synthetize 1,3-propanediol from glycerol. *Process Biochemistry* 47: 1308–1319.

26. Kumar B, Bhardwaj N, Agrawal K, Chaturvedi V, Verma P. 2020. Current perspective on pretreatment technologies using lignocellulosic biomass: An emerging biorefinery concept. *Fuel Processing Technology* 199: 106244.

27. Lee JW, Kim HU, Choi S, Yi J, Lee SY. 2011. Microbial production of building block chemicals and polymers. *Current Opinion in Biotechnology* 22: 758–767.

28. Levinson WE, Kurtzman CP, Kuo TM. 2007. Characterization of *Yarrowia lipolytica* and related species for citric acid production from glycerol. *Enzyme and Microbial Technology* 41: 292–295.

29. Li C, Yang X, Gao S, Wang H, Lin CSK. 2017. High efficiency succinic acid production from glycerol via in situ fibrous bed bioreactor with an engineered *Yarrowia lipolytica*. *Bioresource Technology* 225: 9–16.

30. Lin DS, Yen HW, Kao WC, Cheng CL, Chen WM, Huang CC, Chang JS. 2015. Bio-butanol production from glycerol with *Clostridium pasteurianum* CH4: The effects of butyrate addition and in situ butanol removal via membrane distillation. *Biotechnology for Biofuels* 8: 1–12.

31. Lindlbauer KA, Marx H, Sauer M. 2017a. 3-Hydroxypropionaldehyde production from crude glycerol by *Lactobacillus diolivorans* with enhanced glycerol uptake. *Biotechnology for Biofuels* 10: 295.

32. Lindlbauer KA, Marx H, Sauer M. 2017b. Effect of carbon pulsing on the redox household of *Lactobacillus diolivorans* in order to enhance 1,3-propanediol production. *New Biotechnology* 34: 32–39.

33. Liu -H-H, Ji X-J, Huang H. 2015. Biotechnological applications of *Yarrowia lipolytica*: Past, present and future. *Biotechnology Advances* 33: 1522–1546.

34. Lüthi-Peng Q, Dileme FB, Puhan Z. 2002. Effect of glucose on glycerol bioconversion by *Lactobacillus reuteri*. *Applied Microbiology and Biotechnology* 59: 289–296.

35. Ma Y, Gao Z, Wang Q, Liu Y. 2018. Biodiesels from microbial oils: Opportunity and challenges. *Bioresource Technology* 263: 631–641.

36. Mata TM, Martins AA, Caetano NS. 2010. Microalgae for biodiesel production and other applications: A review. *Renewable and Sustainable Energy Reviews* 14: 217–232.

37. Monteiro MR, Kugelmeier CL, Pinheiro RS, Batalha MO, da Silva César A. 2018. Glycerol from biodiesel production: Technological paths for sustainability. *Renewable and Sustainable Energy Reviews* 88: 109–122.

38. Morgunov IG, Kamzolova SV, Lunina JN. 2013. The citric acid production from raw glycerol by *Yarrowia lipolytica* yeast and its regulation. *Applied Microbiology and Biotechnology* 97: 7387–7397.

39. Otte B, Grunwaldt E, Mahmoud O, Jennewein S. 2009. Genome shuffling in *Clostridium diolis* DSM 15410 for improved 1,3-propanediol production. *Applied and Environmental Microbiology* 75: 7610–7616.

40. Papanikolaou S, Aggelis G. 2009. Biotechnological valorization of biodiesel derived glycerol waste through production of single cell oil and citric acid by *Yarrowia lipolytica*. *Lipid Technology* 21: 83–87.

41. Papanikolaou S, Galiotou-Panayotou M, Fakas S, Komaitis M, Aggelis G. 2008. Citric acid production by *Yarrowia lipolytica* cultivated on olive-mill wastewater-based media. *Bioresource Technology* 99: 2419–2428.

42. Petrov K, Petrova P. 2009. High production of 2,3-butanediol from glycerol by *Klebsiella pneumoniae* G31. *Applied Microbiology and Biotechnology* 84: 659–665.

43. Petrov K, Petrova P. 2010. Enhanced production of 2,3-butanediol from glycerol by forced pH fluctuations. *Applied Microbiology and Biotechnology* 87: 943–949.

44. Pflügl S, Marx H, Mattanovich D, Sauer M. 2012. 1,3-Propanediol production from glycerol with *Lactobacillus diolivorans*. *Bioresource Technology* 119: 133–140.

45. Pflügl S, Marx H, Mattanovich D, Sauer M. 2013. Genetic engineering of *Lactobacillus diolivorans*. *FEMS Microbiology Letters* 344: 152–158.

46. Pflügl S, Marx H, Mattanovich D, Sauer M. 2014. Heading for an economic industrial upgrading of crude glycerol from biodiesel production to 1,3-propanediol by *Lactobacillus diolivorans*. *Bioresource Technology* 152: 499–504.

47. Pyne ME, Sokolenko S, Liu X, Srirangan K, Bruder MR, et al. 2016. Disruption of the reductive 1,3-propanediol pathway triggers production of 1,2-propanediol for sustained glycerol fermentation by *Clostridium pasteurianum*. *Applied and Environmental Microbiology* 82: 5375–5388.

48. Qiu X, Xu P, Zhao X, Du G, Zhang J, Li J. 2020. Combining genetically-encoded biosensors with high throughput strain screening to maximize erythritol production in *Yarrowia lipolytica*. *Metabolic Engineering* 60: 66–76.

49. Rakicka M, Lazar Z, Dulermo T, Fickers P, Nicaud JM. 2015. Lipid production by the oleaginous yeast *Yarrowia lipolytica* using industrial by-products under different culture conditions. *Biotechnology for Biofuels* 8: 104.

50. Regestein L, Doerr EW, Staaden A, Rehmann L. 2015. Impact of butyric acid on butanol formation by *Clostridium pasteurianum*. *Bioresource Technology* 196: 153–159.

51. Rhie MN, Kim HT, Jo SY, Chu LL, Baritugo K-A, et al. 2019. Recent advances in the metabolic engineering of *Klebsiella pneumoniae*: A potential platform microorganism for biorefineries. *Biotechnology and Bioprocess Engineering* 24: 48–64.

52. Russmayer H, Egermeier M, Kalemasi D, Sauer M. 2019. Spotlight on biodiversity of microbial cell factories for glycerol conversion. *Biotechnology Advances* 37: 107395.

53. Rywińska A, Rymowicz W, Żarowska B, Skrzypiński A. 2010. Comparison of citric acid production from glycerol and glucose by different strains of *Yarrowia lipolytica*. *World Journal of Microbiology and Biotechnology* 26: 1217–1224.

54. Rywińska A, Rymowicz W, Zarowska B, Wojtatowicz M. 2009. Biosynthesis of citric acid from glycerol by acetate mutants of *Yarrowia lipolytica* in fed-batch fermentation. *Food Technology and Biotechnology* 47: 1–6.

55. Rzechonek DA, Dobrowolski A, Rymowicz W, Mirończuk AM. 2018. Recent advances in biological production of erythritol. *Critical Reviews in Biotechnology* 38: 620–633.

56. Samul D, Leja K, Grajek W. 2014. Impurities of crude glycerol and their effect on metabolite production. *Annals of Microbiology* 64: 891–898.

57. Sauer M, Russmayer H, Grabherr R, Peterbauer CK, Marx H. 2017. The efficient clade: Lactic acid bacteria for industrial chemical production. *Trends in Biotechnology* 35: 756–769.

58. Singh D, Sharma D, Soni SL, Sharma S, Kumar Sharma P, Jhalani A. 2020. A review on feedstocks, production processes, and yield for different generations of biodiesel. *Fuel* 262: 116553.

59. Thiru M, Sankh S, Rangaswamy V. 2011. Process for biodiesel production from *Cryptococcus curvatus*. *Bioresource Technology* 102: 10436–10440.

60. Vivek N, Sindhu R, Madhavan A, Anju AJ, Castro E, et al. 2017. Recent advances in the production of value added chemicals and lipids utilizing biodiesel industry generated crude glycerol as a substrate – Metabolic aspects, challenges and possibilities: An overview. *Bioresource Technology* 239: 507–517.

61. Workman M, Holt P, Thykaer J. 2013. Comparing cellular performance of *Yarrowia lipolytica* during growth on glucose and glycerol in submerged cultivations. *AMB Express* 3: 58.

62. Yang G, Tian J, Li J. 2007. Fermentation of 1,3-propanediol by a lactate deficient mutant of *Klebsiella oxytoca* under microaerobic conditions. *Applied Microbiology and Biotechnology* 73: 1017–1024.

63. Yang T-W, Rao Z-M, Zhang X, Xu M-J, Xu Z-H, Yang S-T. 2013. Fermentation of biodiesel-derived glycerol by *Bacillus amyloliquefaciens*: Effects of co-substrates on 2,3-butanediol production. *Applied Microbiology and Biotechnology* 97: 7651–7658.

64. Zhou S, Li L, Perseke M, Huang Y, Wei J, Qin Q. 2015. Isolation and characterization of a *Klebsiella pneumoniae* strain from mangrove sediment for efficient biosynthesis of 1,3-propanediol. *Science Bulletin* 60: 511–521.

CHAPTER 23

The Good Fungus – About the Potential of Fungi for Our Future

Valeria Ellena[a,b] and Matthias Steiger

[a] Austrian Centre of Industrial Biotechnology (ACIB GmbH), Vienna, Austria

[b] Institute of Chemical, Environmental and Bioscience Engineering, TU Wien, Vienna, Austria

23.1 INTRODUCTION

When hearing the term fungus our minds most likely fill with images of edible mushrooms picked in the forest or of undesirable molds growing on old bread and in the corners of the bathroom; but fungi are much more than this (Hyde et al., 2019). The biological kingdom of Fungi consists of a large group of widely diverse organisms which can be found growing almost everywhere on earth. Around 120000 fungal species have been described so far, representing, according to recent estimates, just the 3–8% of the millions of fungal species present on earth (Hawksworth and Lücking, 2017). Due to their capability to decompose organic material and to synthesize new organic molecules from it, fungi play an important role in the recycling of nutrients in the ecosystem and in maintaining a balance between living organisms. In particular, fungi release enzymes in the environment that allow them to digest complex organics, creating simpler molecules that they and other organisms can easily ingest. They are, therefore, essential for life on the planet. From simpler single cell yeasts to filamentous molds to more complex multicellular fruiting body-bearing mushrooms, the kingdom Fungi includes species which are not only beneficial for humankind, but fundamental for its survival – and humans have relied on them for millennia.

23.2 FUNGAL BIOTECHNOLOGY: THE ORIGINS

Evidence of the use of fungi for biotechnological processes dates back thousands of years, to the early Neolithic, well before the term biotechnology existed. In ancient times, humankind started to ferment starch and sugars present in cereal and fruits, to obtain alcoholic beverages. Fermentations to produce alcoholic drinks, such as beer and wine, bread, and other foodstuff, are still carried out today (see section 2). However, the increased knowledge of the biological processes involved allows for controlled and improved results, which are now performed at industrial scale. Today filamentous fungi (molds), like *Aspergillus oryzae* and *Penicillium roqueforti*, are utilized in the production of certain food products, such as soya-based products and cheese. However, an important player of the food-based fermentations is the yeast *Saccharomyces cerevisiae* whose importance is not only connected to the food industry, but also strictly intertwined with the advent of biofuels.

Good Microbes in Medicine, Food Production, Biotechnology, Bioremediation, and Agriculture, First Edition. Edited by Frans J. de Bruijn, Hauke Smidt, Luca S. Cocolin, Michael Sauer, David Dowling, and Linda Thomashow.

23.3 FUNGI FOR MOVING FORWARD – BIOFUELS

Long ago in their history, humans started to burn wood or other plant materials as sources of heat and light. Biofuels – energy sources derived from any type of biomass – are, therefore, not an invention of the last century. Until the nineteenth century the main fuels for heating and lighting were still derived from wood and plant oil, respectively.

Around the same time ethanol fuelled the earliest internal combustion engines and later it was used by Henry Ford to power his cars. During the twentieth century, thanks to their decrease in price, fossil resources became the main source of fuels and supplanted biofuels for energy production. However, three recent developments have sparked new interest in biofuels in the last decades: the increase in the price of petroleum, the environmental concerns around the use of fossil resources and the political instability in many of the areas where crude oil is extracted. Among the different types of biofuels, biodiesel and bioethanol are probably the best known and the most commonly produced worldwide. Bioethanol is obtained by microbial fermentation, mainly by yeast, of the sugars contained in the biomass. Biodiesel can be produced from oils through the chemical process of transesterification in the presence of alcohols (see chapter 22).

Biofuels can belong to different generations, depending on the source of the biomass. First generation biofuels make use of food crops, such as sugarcane and corn. In this type of biomass the sugars are readily available to the yeast for fermentation but food crops are not sustainable in the long-term as they compete with human food and animal feed. Second generation biofuels are derived instead from non-food plant materials, such as crop residues, wood, grasses, food processing residues, and municipal organic waste. The use of this type of biomass as source material for biofuels is preferable for two reasons. First, it avoids using food crops for energy generation. Second, it allows waste material that would otherwise be burnt for the most part to be made use of, releasing harmful compounds into the atmosphere. In non-food crops, sugars have to be released from the mixture of cellulose, hemicellulose, and lignin constituting them, in order to be fermented to ethanol. The conversion of lignocellulosic material into simple and fermentable sugars is a crucial step, where fungi can play a key role (Antoni et al., 2007; Guo et al., 2015). Fungi naturally produce hydrolytic enzymes which are secreted outside the cell allowing them to break down complex organic material, such as cellulose, into simpler molecules. These can be then used by the fungus to grow.

FIGURE 23.1 *Trichoderma reesei* growing on a petri dish.

The filamentous fungus *Trichoderma reesei* (Figure 23.1) has an extraordinary capability to produce cellulases and hemicellulases, enzymes which hydrolyze the plant biomass polysaccharides into fermentable sugars.

T. reesei was discovered only 70 years ago as being responsible for massive rotting of military equipment during World War II (Allen et al., 2009). Since then, its decomposing capability has been exploited for human purposes and major advances in the available genetic engineering technologies have allowed a significant boost in its protein production. It has become a cell factory in the cellulase and hemicellulase industry and a pioneer in the conversion of cellulolytic material to biofuel. However, there is still much undiscovered potential about this fungus and further research is ongoing. The new findings will be essential in the transition from a fossil-based to a bio-based fuel and energy production (Bischof et al., 2016).

23.4 FUNGAL ENZYMES TO THE RESCUE FOR SUSTAINABLE INDUSTRIES

Fungal enzymes as those produced by *T. reesei* not only find application in the biofuel industry, in fact, approximately half of the currently commercially used enzymes is derived from fungi. In the food processing industry, for instance, catalases and lipases are used for cheese making and ripening and pectinases for wine and juice clarification (see chapters 12 and 17). In the feed industry, cellulases and xylanases are employed to increase the feed digestibility and absorption by the animals. Proteases and cellulases are added to detergent agents to improve the appearance of fabrics. In the textile sector, cellulases are used for the improvement and the bio-finishing of cellulose-based fabrics, such as cotton. Lipases and proteases also have potential in wastewater treatment. And the list could go on. Fungal enzymes are definitely important tools for the ongoing transition toward a sustainable future: not only are they biodegradable, but they also increase production efficiency, while reducing environmental costs in terms of energy, water, raw material, and waste. All in all, they have the potential to reduce the industrial carbon dioxide emissions of the industries that utilize them (McKelvey and Murphy, 2017).

While the fungus *T. reesei* is the workhorse for cellulase production, there is another filamentous fungus commonly used for enzyme production: *Aspergillus niger* (Figure 23.2).

A. niger enzymes are widely used in the food and feed industry. Some examples include amylases, used to hydrolyze the starch in bread and beer production (see section 2), and phytases, which improve animal feed by liberating the phosphate from plant material.

FIGURE 23.2 *Aspergillus niger* strains growing on a petri dish.

23.5 FUNGAL ORGANIC ACIDS: JACKS OF ALL TRADES

A. niger was defined as a fungus able to shape biotechnology (Cairns et al., 2018) – and for a good reason. In the last 100 years it has become an important source of organic acids, including the well-known citric acid. Contrary to common conviction, nowadays citric acid is not derived from citrus fruits, but it is most commonly obtained by the fermentation of *A. niger*. In the early twentieth century, the food chemist James Currie discovered that the mold *A. niger* had a natural capability to produce this weak organic acid, when grown on sugar-rich substrates. From then on, citric acid production by *A. niger* became an industrial process with a billion-dollar market. More than 2 million tons of citric acid are produced worldwide every year and the majority of it derives from *A. niger* fermentations. Citric acid finds countless applications in a variety of industries. In the food sector it is added to soft drinks, candies and jams where it acts as acidulant and taste enhancer. In pharmaceutical formulations it acts as an effervescent in the presence of carbonates and bicarbonates and it has anticoagulant activity. Furthermore, it is used in detergents for softening the water, in cosmetic products as antioxidant and for pH adjustment, etc.

A. niger produces other organic acids that are worth mentioning. One is gluconic acid, used in the processing and manufacturing of food, leather, and metal and added as a salt to various detergents. Another is fumaric acid, used as acidulant in food and pharmaceuticals and for the manufacturing of polymeric resins.

Thanks to its biotechnological potential, *A. niger* became a research model organism for industrial fungi and it is without any doubt going to be one of the key organisms in the transition toward a bio-based economy. Over the last decades several genetic engineering techniques have been developed to obtain improved strains and further increase the natural producing capabilities of this fungus. Moreover, current research focuses on the potential of the fungus to produce bioactive secondary metabolites which might have pharmaceutical use.

23.6 FUNGAL METABOLITES – WEAPONS AGAINST DISEASES

In 1928 microbiologist Alexander Fleming was working with bacterial colonies of *Staphylococcus*. He observed that bacterial cultures contaminated with a fungus (later identified as *Penicillium rubens*) had been destroyed while bacterial cultures not contaminated were growing as expected. Fleming identified the cause in the antibacterial substance produced by the fungus, penicillin G: the first world's antibiotic. This antimicrobial compound is probably the best known fungal secondary metabolite. The majority of fungi produce small molecules called secondary metabolites that are not directly involved in growth and development but have biological activity, contributing to the survival of the fungus against competing organisms in the environment. Since Fleming's finding, great efforts have been dedicated to discovering other classes of bioactive secondary metabolites from fungi. Notable examples include the anti-cholesterol drug lovastatin produced by *Aspergillus terreus*, the immunosuppressant cyclosporine A formed by *Tolypocladium inflatum* or the anticancer drug Taxol derived from *Taxomyces andreanae*. Thanks to advances in genome sequencing technologies, it is now known that fungal genomes contain a reservoir of genes dedicated to the production of secondary metabolites. Specific conditions are required to produce these secondary metabolites in the laboratory, but their discovery and production are of utmost urgency, particularly in relation to the worldwide increased antibiotic resistance of bacterial and fungal pathogens. Antibiotic resistance – the capability of a certain pathogen to become resistant to a certain drug due to the abuse of antibiotics by humans and in animal farming – is considered a global problem by the World Health Organization. Since none of the already available antimicrobial drugs is expected to be effective against highly resistant pathogens, the discovery of new antibiotics is of extreme importance and the role that fungi can play here is evident. Not only antibiotics, but also new antimycotics and compounds against a wide range of diseases can be potentially discovered in fungi.

Besides their medical use, other fungal secondary metabolites play a crucial role in agriculture, where they improve crop yields, in the food industry as pigments and in the textile sector as dyes (Boruta, 2018).

23.7 FUNGAL PRODUCTS ON DEMAND

So far, we have described some of the natural synthesis capabilities of fungi which can be improved and employed for human benefit. However, with an increasing understanding of the fungal biology and physiology, certain fungal species can be designed to produce compounds which they do not naturally produce. This is for instance the case of lactic acid or of milk proteins such as casein. Lactic acid is naturally produced by certain types of bacteria but, with the right set of modifications, the same yeast that produces bioethanol can be induced to produce lactic acid. This technology has been developed by companies such as *Syconium* and *Corbion* with the aim of sustainably producing bioplastic (PLA) derived from lactic acid (see chapter 24). Casein is found in animal milk and, consequently, in cheese. Production of casein by fungi, as currently attempted by the start-up *Those vegan cowboys*, would allow the production of animal-free cheese, which is more sustainable and environmentally-friendly than traditional cheese.

23.8 "GREEN" FUNGI FOR A SUSTAINABLE FUTURE

Besides producing molecules and compounds which find sustainable applications, fungi as a whole hold great potential toward a green economy. In nature, fungi have developed symbiotic relationships with plants and various strategies to compete against other organisms in order to survive (see chapter 39). Due to these characteristics, certain fungal species or combinations of fungal species, mainly belonging to the genera *Chaetomium*, *Aspergillus*, *Penicillium*, and *Trichoderma*, are used in agriculture as biofertilizers. By improving the uptake of nutrients, the resistance against diseases and plant growth, they increase plant yields. Other fungal species, such as members of the genera *Alternaria*, *Trichoderma*, and *Beauveria*, are utilized as effective agents of biological control against fungal plant pathogens, insects, nematodes, and even weeds. Fungal spores are released in the environment where they act as natural biofertilizers, pesticides, insecticides, or herbicides. The use of fungal biofertilizers and biocontrol agents is an efficient and eco-friendly alternative to chemically synthesized compounds (see chapter 41). Moreover, biocontrol agents do not induce resistance in pests, contrary to chemically synthesized compounds, providing a long-term solution for sustainable agriculture (Savita, 2019).

Living fungi can also be efficiently used to tackle pollution, one of the biggest problems caused by anthropogenic activity in the last decades. Fungi have been successfully tested for their capability to degrade various types of pollutants, including pesticides and herbicides, heavy metals, and toxic textile dyes released into the environment. This strategy, called mycoremediation, seems to be more effective and much less harmful than conventional techniques, such as the use of chemicals or incineration, to clean up contaminated areas. Another source of environmental pollution is caused by plastic. Plastic production has increased drastically in the last century and has reached more than 380 million tons produced per year. Since only a small percentage of it is recycled or incinerated, accumulation of plastic in the environment has become a serious issue that requires an urgent solution (Geyer et al., 2017). Several studies in the laboratory have shown that more than 100 different fungal species have the capability to degrade various types of plastic, using the same enzymes with which they decompose natural polymers in the environment. Further research on the plastic biodegradation capabilities of fungi is, therefore, a pressing need which might contribute to a solution for future life on earth (see chapters 24 and 32). Another interesting and relatively recent development in fungal biotechnology is the use of fungal mycelium for the manufacturing of composite bio-based materials (Jones et al., 2020). Fungal mycelium is composed of interconnected and branching filaments called hyphae (Figure 23.3), constituted of chitin, beta-glucans, and proteins, which bind organic matter while growing. While chitin provides mechanical rigidity and strength, beta-glucans act as a mucilage, conferring a certain degree of flexibility to the structure.

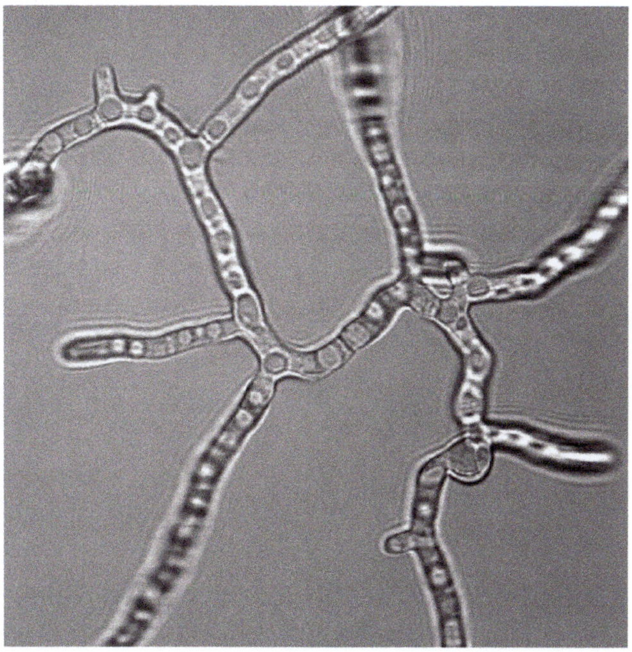

FIGURE 23.3 Hyphae constituting the mycelium of *A. niger* visualized under the microscope.

Materials derived from fungal mycelium are light, due to the porosity of the material, but also very strong. The specific characteristics of the material derived from fungal filaments can be modified by the accurate selection of substrate composition (on which the microorganism grows), type of fungus, and growth conditions. In the last 10 years, a number of start-ups and companies have begun to develop fungal-based material and objects. *Mogu*, for example, developed fungal-based material used to manufacture acoustic panels and floor tiles. *Ecovative* created packaging and foamy materials made of fungi and it even developed mycelium-based meat. *Mycoworks*, whose story started with the use of mycelium as a medium for sculpture by its founder, is developing an alternative to animal leather based on fungi, which is expected to be launched soon to the market. The success of these stories might further expand the applications of mycelium and mycelium composites to other industries, particularly for coatings, due to the hydrophobicity properties of pure mycelium, or for building purposes.

23.9 BIOCOMPUTERS AND LIFE IN SPACE: THE FUTURE OF FUNGAL BIOTECHNOLOGY

In the environment, the intricate networks formed by fungal mycelia are able to integrate and elaborate different types of information (light, chemicals, temperature, gravity, electric fields etc.) and make consequent decisions. These capabilities are being investigated by the consortium *Fungal Architecture*, whose coordinator proposed the development of a computing device made of fungal mycelium, where the information is transmitted through electrical signals and input and output read through the fruiting body, representing the biocomputer interface (Adamatzky, 2018). Such a biocomputer could be used as an environmental sensor to monitor the status of an ecological system. The electrical response of fungal mycelium to different kinds of stimuli also allows us to imagine the development of a self-growing fungal skin that can recognize mechanical and optical information from the environment and smart wearables made of fungi that can sense and report information on the user's body.

All the fungal applications mentioned so far clearly show the great potential of fungi for a sustainable future on earth, free from fossil-based materials and non-biodegradable goods. However, the potential of fungal biotechnology is not limited to this planet, but it holds great promise for space exploration (Cortesão et al., 2020). Filamentous fungi have been often isolated from spacecraft environments such as the International Space Station, showing the capability to survive extreme stressful conditions. Experiments involving the use of living organisms, including fungi, and their biological responses in space have been and are being conducted by different space agency programs. If on the one hand the presence of fungal inhabitants on spacecrafts has to be regarded in terms of health safety and security for the mission (to avoid biodegradation of system components), on the other hand the rational use of fungi in space could open up immense opportunities for long-term space missions. The inherent capabilities of filamentous fungi to produce and decompose a wide range of compounds can be exploited for earth-independent long-duration missions outside low-earth orbit. The idea is that fungi could become allies of astronauts, be designed to recycle their waste and produce medicines, vitamins, enzymes etc. to allow self-sustained life in an isolated environment. Among the various experiments involving fungi, space agencies are also supporting research to investigate the use of mycelium to "grow" liveable habitats for astronauts on the Moon and on Mars.

23.10 CONCLUSIONS

Masters of decomposition and synthesis, fungi have contributed to the good functioning and balancing of the ecosystems since their first appearance on this planet. Although often disregarded, they have become the loyal companions of humans since the dawn of humanity. From their use as a direct food source to their employment as cell factories for the production of a disparate number of compounds, it is difficult to imagine a life on earth without fungi. As knowledge about their biology and genetic engineering tools for their improvement continue to increase, so do their applications. The potential of fungi for the upcycling of waste material into biodegradable products makes them allies in the transition toward a circular economy, ensuring the future of earth and its inhabitants.

ACKNOWLEDGMENTS

The authors acknowledge funding by the COMET center "acib: Next Generation Bioproduction," which is funded by BMK, BMDW, SFG, Standortagentur Tirol, Government of Lower Austria und Vienna Business Agency in the framework of COMET – Competence Centers for Excellent Technologies. The COMET-Funding Program is managed by the Austrian Research Promotion Agency FFG.

REFERENCES

1. Adamatzky A. 2018. Towards fungal computer. *Interface Focus* 8: 20180029.
2. Allen F, Andreotti R, Eveleigh DE, Nystrom J. 2009. Mary Elizabeth Hickox Mandels, 90, bioenergy leader. *Biotechnology for Biofuels* 2: 22.
3. Antoni D, Zverlov VV, Schwarz WH. 2007. Biofuels from microbes. *Applied Microbiology and Biotechnology* 77: 23–35.
4. Bischof RH, Ramoni J, Seiboth B. 2016. Cellulases and beyond: The first 70 years of the enzyme producer. *Trichoderma Reesei. Microbial Cell Factories* 15: 106.
5. Boruta T. 2018. Uncovering the repertoire of fungal secondary metabolites: From Fleming's laboratory to the International Space Station. *Bioengineered* 9: 12–16.
6. Cairns TC, Nai C, Meyer V. 2018. How a fungus shapes biotechnology: 100 years of *Aspergillus niger* research. *Fungal Biology and Biotechnology* 5: 1–14.
7. Cortesão M, Schütze T, Marx R, Moeller R, Meyer V. 2020. Fungal biotechnology in space: Why and how? *Grand Challenges in Fungal Biotechnology*. 501–535. https://doi.org/10.1007/978-3-030-29541-7_18. https://www.springer.com/gp/book/9783030295400?utm_campaign=bookpage_about_buyonpublisherssite&utm_medium=referral&utm_source=springerlink.
8. Geyer R, Jambeck JR, Law KL. 2017. Production, use and fate of all plastics ever made. *Science Advances* 3: e1700782.
9. Guo M, Song W, Buhain J. 2015. Bioenergy and biofuels: History, status, and perspective. *Renewable and Sustainable Energy Reviews* 42: 712–725.
10. Hawksworth DL, Lücking R. 2017. Fungal diversity revisited: 2.2 to 3.8 million species. *Microbiology Spectrum* 5: 79–95.

11. Hyde KD, Xu J, Rapior S, Jeewon R, Lumyong S. 2019. The amazing potential of fungi: 50 ways we can exploit fungi industrially. *Fungal Diversity* 97: 1–136.

12. Jones M, Mautner A, Luenco S, Bismarck A, John S. 2020. Engineered mycelium composite construction materials from fungal biorefineries: A critical review. *Materials and Design* 187: 108397.

13. McKelvey SM, Murphy RA. 2017. Biotechnological use of fungal enzymes. *Biology and Applications* 179: 201–225.

14. Savita SA. 2019. Fungi as biological control agents. *Biofertilizers for Sustainable Agriculture and Environment* 55: 395–411.

CHAPTER 24

Microbes and Plastic – A Sustainable Duo for the Future

Birger Wolter, Henric M.T. Hintzen, Gina Welsing, Till Tiso, and Lars M. Blank
iAMB - Institute of Applied Microbiology, ABBt – Aachen Biology and Biotechnology, RWTH Aachen University, 52074 Aachen, Germany

LIST OF ABBREVIATIONS

AA	Adipic acid
BDO	Butanediol
CO_2	Carbon dioxide
DNA	Deoxyribonucleic acid
EFE	Ethylene forming enzyme
EG	Ethylene glycol
EU	European Union
HAA	Hydroxyalkanoyloxy alkanoate
HDPE	High-density polyethylene
LA	Lactic acid
LCC	Leaf-branch compost cutinase
LDPE	Low-density polyethylene
PBAT	Poly(butylene adipate-co-terephthalate)
PBS	Poly(butylenesuccinate)
PCL	Polycaprolactone
PE	Polyethylene
PET	Polyethylene terephthalate
PHA	Poly(hydroxyalkanoate)
PHV	Poly(hydroxy valerate)
PLA	Poly(lactic acid)
PP	Polypropylene
PS	Polystyrene
PU	Polyurethane
PVC	Polyvinylchloride
RNA	Ribonucleic acid
rPET	Recycled polyethylene terephthalate
TA	Terephthalic acid
TDA	2,4-diaminotoluene

24.1 INTRODUCTION

Plastics are widely used in the global economy. As a material, plastics have revolutionized many sectors due to the application versatility of the different plastic types. Tensile strength, resistance to varying climatic conditions, durability, and low production costs are just some of plastics' beneficial properties. The packaging, building, and automotive industry represent the largest application fields for plastics but also in the electronic, household, and agriculture segments, plastic-based products are indispensable.

Plastics as crafted materials have been used by humankind for thousands of years. There is evidence for the usage of birch tar as glue for spearheads as far back as 50000 years. Among the first industrially produced plastics are rubber (invented in 1839), celluloid (invented in 1869), and viscose (invented in 1892). These first plastics had one common trait: they were produced from

Good Microbes in Medicine, Food Production, Biotechnology, Bioremediation, and Agriculture, First Edition. Edited by Frans J. de Bruijn, Hauke Smidt, Luca S. Cocolin, Michael Sauer, David Dowling, and Linda Thomashow.

bio-based resources. Later, after the establishment of industrial-scale production of crude oil, fossil-based plastics became the most used synthetic materials on earth.

With growing awareness of a sustainable lifestyle, plastics based on renewable resources are moving in focus. Apart from the origin of the raw material (fossil- versus bio-based), plastics are differentiated based on biodegradability (Figure 24.1; see also chapter 32).

Commonly used (conventional) plastics are hardly degraded in nature, while newly developed plastics are often biodegradable by design (Wei et al., 2020) (Figure 24.1). Chemically speaking, plastics are polymers, which are long chains, sometimes branched, and composed of single units – the monomers. Polymers are not just a human invention but also exist abundantly in nature. Materials such as starch, lignin, and cellulose are structural biopolymers, while polynucleotides such as RNA or DNA carry all living organisms' genetic information. The biodegradability of plastics is mainly determined by the chemical bond between the monomers. Conventional plastics often feature carbon–carbon bonds (e.g., PE, PP), which are not susceptible to biodegradation. In contrast, natural polymers are entirely biodegradable (even robust materials such as wood degrade over time). How can the high resistance of synthetic plastics be explained? Natural polymers have formed part of the ecosystems for hundreds of millions of years, thus, offering the microbial communities ample opportunity for the evolution of mechanisms for efficient breakdown of the polymer chains. For synthetic plastics – that have been around for decades only – such mechanisms have not yet evolved.

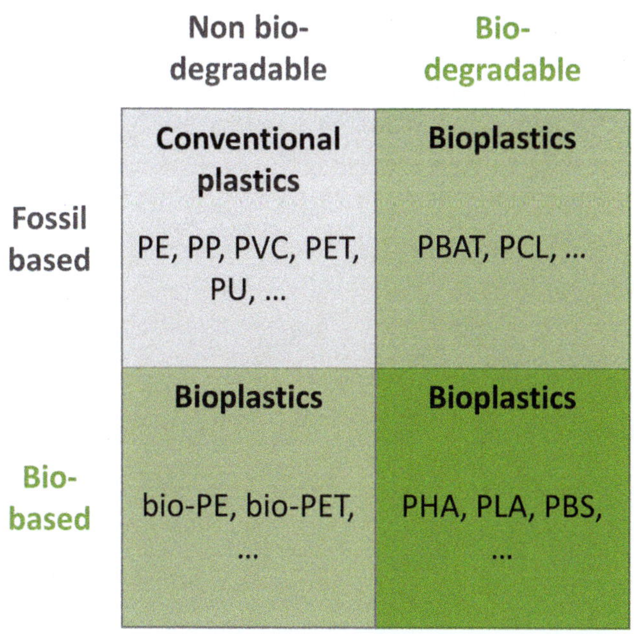

FIGURE 24.1 The four main categories plastics can be divided into. They are discriminated according to fossil- versus bio-based and biodegradable versus non-biodegradable. Both bio-based and bio-degradable plastics are regarded as bioplastics.

24.2 TRAINING MICROBES – GENE TECHNOLOGY AT WORK

At the beginning of the 1970s, researchers started to understand how to engineer DNA rationally and bring this synthetic DNA alive in microbes. A new area, the area of gene technology, was born. The first companies realizing the huge potential in training microbes for a specific application were pharmaceutical producers. A run to produce insulin started, with winners, for example, in Denmark and the US. Tasks that researchers ask microbes to overcome in the plastic crisis are (1) producing enzymes to degrade plastic and (2) using the resulting plastic monomers as a carbon and energy source. In industrial microbiology, sugars dominate, either produced from sugar cane and beet or by plants producing starch such as corn or wheat. Using plastic monomers instead often requires the expansion of the natural metabolism by the monomer degradation pathway. Jay Bailey first introduced such new traits to metabolic networks in 1991 as metabolic engineering (Bailey, 1991). Since then, metabolic engineering has flourished and has been an independent scientific discipline for more than two decades. Metabolic engineering requires information on biochemical reactions and the responsible biocatalysts, the enzymes. Based on this information, gene technology tools are employed to delete or modify genes. Additionally, genes from other organisms can be inserted. Generally, the activity of both native and inserted genes can be up- or downregulated via targeted modifications in the DNA.

The ever-evolving tools of gene technology are used in metabolic engineering for rational cell engineering. Rational engineering in the sense of gene technology requires predictability. It is comparable with a mechanic, who only replaces components whose function is known (Giese et al., 2013).

The synthesis of many molecules is difficult or impossible to achieve via chemical catalysis but can instead be performed via microbial catalysis. Microbes can be enabled to catalyze several chemical conversion steps in a one pot reaction, which would, for instance, require several reaction steps if done via chemical catalysis. Product examples that exceed the million tons yearly production are organic acids like citrate and amino acids like lysine. It is envisaged that most end products of industrial microbiology are produced from renewable carbon sources (i.e., biomass, CO_2, waste) (Blank et al., 2020).

24.3 PLASTICS – PROBLEM OR OPPORTUNITY?

In 2019, worldwide plastic production reached 368 million tons, with more than 50% produced in Asia, of which 99.5% is based on fossil resources (PlasticsEurope, 2020). This massive plastic production raised concerns about low recycling rates and deficient end-of-life solutions for plastics (Bassi et al., 2020). The actual beneficial attributes of plastics for application purposes led to major challenges in their disposal. In particular, packaging material has emerged as a challenge since beverage, food, and industrial packaging are often made for single use only and promptly discarded (Carniel et al., 2020; Geyer et al., 2017). The "take–make–dispose" culture related to single-use plastic seriously harms plastic usage sustainability and requires workable consequences for post-consumer plastics (Blank et al., 2020). It is estimated that 30% of all plastics ever produced are currently in use. But only 12% have been incinerated or recycled (9%), and only 0.9% have been recycled more than once. This means that almost 50% of all plastics ever produced are stored in landfills or, as a result of poor waste management, ubiquitously distributed in the environment (Geyer et al., 2017).

At present, low production costs of fossil-based plastics make the reuse of post-consumer plastic products economically hardly competitive, and products composed of recycled materials struggle to compete with virgin plastics on quality (Nisticò, 2020). Therefore, sustainable recycling strategies are urgently required to increase the recycling rates of consumed plastics and prevent accumulation in terrestrial and marine ecosystems. Instead of branding exhausted plastic as waste, it should be considered a resource - away from linear plastic consumption toward sustainable utilization of plastics within an efficient circular (bio)economy (Figure 24.2).

In this chapter, we will elaborate on why microbes will contribute to a sustainable plastics industry. We will focus on three main reasons why microbes and plastics form a sustainable duo for the future. In the first section, we explain how post-consumer plastics can be used as a feedstock to synthesize valuable products using microbes. The second part will focus on the microbial synthesis of monomers as drop-in educts for sustainable plastic production. In the third part, we will shed light on plastics that microbes can directly produce without the need for chemical polymerization.

24.3.1 Plastic Upcycling

One way to use plastics as novel feedstock (which microbiologists refer to as a substrate) is microbial upcycling. In contrast to recycling approaches, upcycling

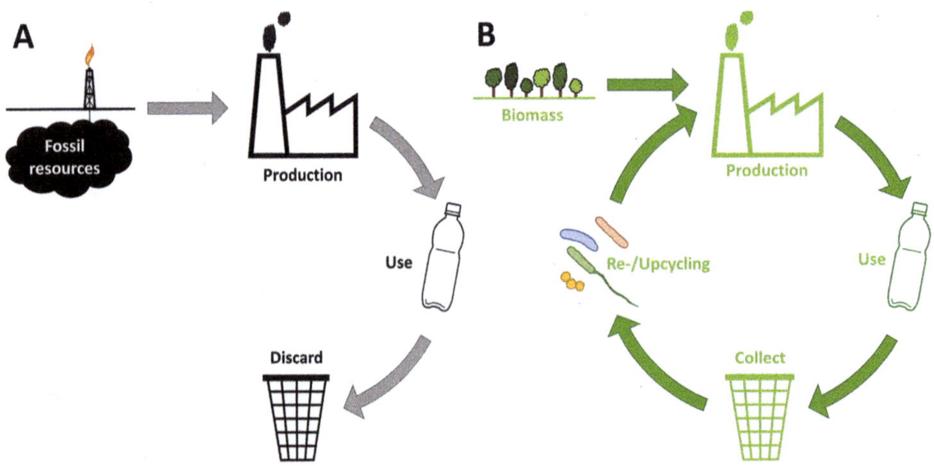

FIGURE 24.2 (A) The state-of-the-art linear plastics industry opposed to (B) the envisioned circular plastics economy in which plastics are reused and the lost fractions are replaced from renewable carbon sources (i.e., biomass, CO_2). The fraction of plastic that cannot be reused is biologically upcycled to bioplastics.

technologies make it possible to increase the value of the resulting products (Vollmer et al., 2020). A major disadvantage of common recycling methods is the decreasing quality and further deterioration of the polymers in every additional recycling cycle. A recycled plastic container rarely becomes a new container after recycling. Instead, it becomes artificial turf or insulation material. Even if some types of plastics can be used a couple of times to manufacture a similar product as previously, so-called downcycling usually occurs after just a few recycling cycles. Downcycling describes the practice of using recycled material for an application of less value than the original purpose of the material (Allwood, 2014). An evolving exception to this is, for instance, bottles composed of recycled PET (rPET). Since 2020 there are beverage bottles on the market consisting of 100% rPET. This "bottle-to-bottle" principle reduces not only plastic waste production but also reduces CO_2 emissions by 44% compared to conventional PET bottles (VILSA, 2019). Biological upcycling strategies are based on the ability of microbes to utilize materials considered as waste. In this manner, upcycling plastic materials is economically feasible and helps minimize plastic waste accumulation (Sohn et al., 2020). Microbial biotechnology can contribute to establish such sustainable upcycling routes on a large scale. For this purpose, research advances in gene technology, and metabolic engineering are helping to identify the microbial ability to degrade plastics and predict degradation pathways (Dvořák et al., 2017).

24.3.2 The Role of Microbes in Plastic Degradation

Microbes are the backbone of all bioconversion processes taking place in every conceivable ecosystem. Bioconversion describes the transformation process of organic materials through living organisms' actions (Kumari et al., 2021). Due to the versatile enzymatic machinery of organisms, they can prosper in extreme surroundings (Singh et al., 2021). Bacteria and fungi belong to the most abundant biological organisms and have recently aroused interest due to their distinct ability to degrade plastic polymers (Pathak and Navneet, 2017; Yoshida et al., 2016). There is already a long list of microbes associated with the natural degradation processes of plastic, which is as diverse as the variety of plastics itself (see also chapter 32). Well-known representatives from this list are *Pseudomonas*, *Streptomyces*, *Rhodococcus*, and *Bacillus* species. Overall, about 17 genera of bacteria have given account to degrade plastics (Kumari et al., 2021; Moharir and Kumar, 2019; Singh et al., 2021).

Before taking a close look at the microbial processes, it is important to clarify the structure of natural and synthetic polymers. As the name already suggests, natural polymers are abundantly found in nature, and among others, play a crucial role in plants to develop biomass. Lignocellulosic plant biomass is composed of cellulose, hemicellulose, and lignin, which are the major building blocks for the strength of plants. Conventional plastics, in turn, are human-made and belong to synthetic polymers. These polymeric chains have unusual bonds seldom occurring in nature. Accordingly, the breakdown of plastic material into its monomers is complicated. At this point, the desired durability of plastic polymers becomes a challenge.

Due to the fast adaptability of microbes, plastics biodegradation happens naturally, but the degradation rates are far too slow compared with the rates of waste generation. Assuming that from now on no new synthetic plastics enter the environment, it would take hundreds of years for the natural biodegradation processes to decompose all the plastic waste already accumulated in nature (Danso et al., 2019; Kumari et al., 2021). The rate of biodegradation depends heavily on the chemical bond entailed in the plastic. Most prominently for polyesters, enzymes exist that can cleave the chemical bonds. These enzymes can be exploited as potent tools for plastic degradation. Biological plastic degradation is an excellent opportunity for the sustainable valorization of plastic wastes, considering plastic as a valuable resource in a circular economy (Blank et al., 2020). However, the true advantage of microbes might not be their ability to degrade natural and synthetic polymers but to synthesize valuable products from the degradation products. A great example of such a microbial conversion process is the synthesis of the ever-increasing bio-based plastic PLA which will be described later (see Section 24.4.2).

24.3.3 Plastic-Eating Microbes

Microorganisms are ubiquitous in nature and their adaptable metabolism and short reproduction cycles make them capable of quickly adjusting to their environment and nutrient availability. At this time, various studies have been conducted to assess the capability of microbes related to plastic degradation.

The recalcitrance of human-made polymers depends on the individual chemical composition of the different polymer types. The chemical structure of polymeric material, as mentioned before, is responsible for its stability, reactivity, and hydrophilicity. High hydrophilicity signifies that a substance readily interacts with water. These aspects chiefly affect the molecular weight, morphology, porosity, and elasticity of the polymers. Obstacles for biodegradability are the facts that most plastic polymers possess high molecular weight, are insoluble in water, and often have crystalline structures (Danso et al., 2019; Singh et al., 2021). Polyethylene (PE), for instance, belongs among polystyrene (PS), polypropylene (PP), and polyvinylchloride (PVC) to the group of highly resistant polymers based on their strong carbon–carbon bonds (Wei and Zimmermann, 2017; Zheng et al., 2005). In contrast, polymers with ester bonds provide a weak point for enzymatic cleavage. However, not all ester bonds containing plastics are biodegradable (Narancic and Connor, 2019).

Polyethylene (PE) is the most produced plastic worldwide and finds application in plastic bags, water bottles, food packaging, and toys (PlasticsEurope, 2020; Wilkes and Aristilde, 2017). In 2010, 15 different bacterial species were found capable of degrading PE. The varying species have been isolated from waste dumpsites in India's Gulf of Mannar region (Balasubramanian et al., 2010). Two of the isolated species received particular attention: *Pseudomonas* and *Arthrobacter*. *Pseudomonas* sp. were found to be the most efficient PE degrading organisms from the dumpsite, followed by *Arthrobacter* sp. More detailed studies of the microorganisms' degradation rate showed that *Arthrobacter* sp. achieved nearly 12% weight loss of commercially available PE and *Pseudomonas* sp. 15% after 30 days of incubation (Balasubramanian et al., 2010).

Polyurethanes (PU) are versatile polymers frequently applied to produce foams, sponges, insulation materials, paint, and coatings (Howard, 2002; Seymour and Kauffman, 1992). With their many different chemical bonds, PUs are quite a recalcitrant group of polymers, therefore not easily amenable for degradation. While some PUs contain polyester bonds, which can be easily cleaved by enzymes (Wei and Zimmermann, 2017), polyether bonds occur more commonly and are less available for enzymatic attack. Magnin et al. (2021) showed how a polyester PU could be recycled by specific enzymatic cleavage of the ester bond, leaving the urethane bond intact. The urethane bond might also be cleaved enzymatically, as shown by a 2018 patent application from the German polymer supplier Covestro (Bittner et al., 2018). Previously, the microorganism *Penicillium* showed the best PU degradation capability of polyester bonds (Magnin et al., 2019). A promising bacterial strain was isolated in 2020 from a dumpsite rich in brittle plastic waste: *Pseudomonas* sp. TDA1 (Espinosa et al., 2020). A significant advantage of the *Pseudomonas* strain is that it not only grows on the different PU monomers, as adipic acid (AA) or 1,4-butanediol (BDO) but also on intermediates and pre-cursor compounds, including the toxic 2,4-diaminotoluene (TDA). TDA is classified carcinogenic and formed during the PU production process, as well as during potential decomposition. Accordingly, it was the first reported organism which can metabolize the hazardous TDA as nutrient (Espinosa et al., 2020). After PU degradation, the different monomers are potential carbon sources for microbes. Introducing bonds that add novel-end-of-life treatment options (such as esters), entail full material recycling as a much-needed approach.

PET is the most widely used synthetic polyester. In 2016 a team from the Kyoto and Keio universities in Japan identified the bacterium *Ideonella sakaiensis* 201-F6 that can degrade PET plastic (Yoshida et al., 2016). The bacterium was isolated from a yard of a PET-bottle recycling factory in Sakai city, Japan. The rate at which *I. sakaiensis* eats PET is improvable. Initial tests on PET films showed that it takes six weeks for *I. sakaiensis* to break down the PET film, which was thinner than PET bottles.

In the course of research on the biodegradability of synthetic plastic polymers also biopolymers from renewable feedstocks are moving into focus. Biopolymers are developed to replace fossil-based polymers and can be defined as polymers that are either bio-based, biodegradable, or both (see Figure 24.1). Thus, the bio-based monomers for these biopolymers originate from renewable sources like plants or even from microbes that can produce bio-monomers (see the following section). Although fossil feedstocks can be replaced by bio-monomers, the challenge regarding the recycling of post-consumer bio-based plastics remains.

Poly(lactic acid) (PLA) unifies the demands of being bio-based and bio-degradable, emerging as one of the most promising candidates as a renewable plastic alternative for the future. PLA has been used in various sectors, especially in packaging for short shelf-life products, including food and beverages (Auras et al., 2017; Rabnawaz et al., 2017; Weber et al., 2002). Indeed, the microbial machinery from the *Comamonas acidovorans* species, the *Rhizopus delemar* species, and the *Amycolatopsis* species can degrade PLA efficiently (Masaki et al.,

2005; Singh et al., 2021). Whereas other reports, e.g., from aquatic systems or soil, show very slow low biodegradability of PLA (Narancic et al., 2018). Nevertheless, due to the good industrial composting possibilities, additional PLA production capacities have been announced. In 2020, more than 390000 tons of PLA have already been produced worldwide (European Bioplastics, 2020).

Other biodegradable polyesters are poly (hydroxyalkanoate) (PHA), poly(butylenesuccinate) (PBS), and poly(butylene adipate-co-terephthalate) (PBAT). For the last two, production capacities of 900000 tons and 450000 tons, respectively, were announced in China only. Some PHAs have good barrier and mechanical properties. However, the breakthrough of PHA is still hampered by its high production costs. If the production costs can be decreased by large-scale PHA production, PHAs could become promising bioplastics for packaging and potentially replace PP-based products (Leong et al., 2017). New plants of about 5000 tons yearly were announced in Japan and the USA.

24.3.4 Enzymes – Molecular Scissors for the Breakdown of Recalcitrant Polymers

Most of the synthetic polymers are remarkably resistant to natural microbial degradation, and only a handful of microorganisms can degrade the polymers as a whole. Microbes identify polymers as carbon and energy source, and before assimilation, they cleave them into the monomeric building bricks. Once the polymers' length has been reduced, the degradation products can be consumed by the microbes (Kumari et al., 2021; Moharir and Kumar, 2019; Wei and Zimmermann, 2017).

The machinery of the already mentioned PET-degrading bacterium *I. sakaiensis* 201-F6 (Yoshida et al., 2016) is the most studied, while faster enzymes have been studied for the last 20 years or so. *I. sakaiensis* produces the two enzymes PETase and MHETas, which catalyze the breakdown of the plastic. The PETase is the key enzyme, which chops the long PET polymers into shorter fractions. Still, additional publications showed that the enzyme is inhibited by some of the released degradation products (Barth et al., 2015). Therefore, the second enzyme, the MHETase, within the overall PET degradation process is likewise important. MHETase degrades primary degradation products and thus preventing the inhibition of the PETase. The combination of the PETase and MHETase enables further conversion into the two PET building blocks: terephthalic acid (TA) and ethylene glycol (EG), as quickly as possible. These two monomers are very valuable for the re-synthesis of PET. However, *I. sakaiensis* is a slow "eater." Therefore, scientists took up the challenge aiming for an effective and affordable PET upcycling solution. In 2020 another landmark related to PET converting enzymes was reached. Via molecular engineering, a covalent bond between the PETase and MHETase was generated, resulting in a physically coupled two-enzyme system six times faster than the native enzymes (Knott et al., 2020). Even though it still takes some time for the bacterium to degrade plastic, scientists are optimistic that the process can be accelerated even further.

At the same time, there were ground-breaking findings of a biotechnological engineered version of another PET-degrading enzyme. Researchers from the Biotechnology Institute in Toulouse and the Start-Up Carbios were able to show how fast such an engineered enzyme can break down amorphous PET into its constituent monomers. The team used biotechnological modification techniques to improve the activity and heat-stability of the native leaf-branch compost cutinase (LCC). Heat-stability is a decisive factor regarding enzymatic activity since degradation processes can be enhanced when performed at higher temperatures. Increased temperatures enable increased accessibility for the enzymes to attack the polymeric bonds. At 65 °C, the modified LCC enzyme showed PET depolymerization rates outperforming all PET degrading enzymes reported so far. Within 10 hours, a minimum of 90% of PET was depolymerized into its monomers. This was a significant improvement compared to the initial 1% after several weeks (Tournier et al., 2020). Further, it could be shown that the biologically recycled PET material can be processed into packaging and bottles since it exhibits the same properties as petrochemical PET. Thereby, it might contribute toward the concept of a circular plastic economy, but cost and scalability, besides many other factors, will have a say in the success.

With all these encouraging findings regarding microbial plastic degradation, it is important to mention that most plastic polymers are today nearly non-biodegradable. Usually, these are plastics that contain saturated carbon–carbon backbone structures, owing to the lack of clearly identified weak points for depolymerization processes. For this reason, further efforts on the identification of metabolic pathways and functionality of microbial enzymes are indispensable. New findings enable potential enhancement via genetic and metabolic engineering to boost the efficient polymer degradation to a required level and beyond. At this point, it is relevant to clarify that none of the biotechnological modified microorganisms containing genetically altered material will ever be applied unregulated in the open environment. Whether on a lab-scale or an industrial scale, microbial polymer degradation processes are and will ever be conducted under contained conditions. By this, any possibility of unregulated release of a genetically modified organism or enzymes is prevented.

24.4 PLASTIC MONOMERS GENERATED BY MICROBES

Microbes have the potential to play a vital role in the production of plastics in the future. As stated before, plastics are polymers consisting of monomers which form long chains. The vast majority of these monomers are obtained from petrochemicals, which entails ecological downsides (e.g., releasing greenhouse gases). One possible solution would be the sustainable production of plastic monomers with microbes feeding on renewable carbon sources (i.e., biomass, CO_2, waste).

The ability to produce plastic monomers can be a natural property of microorganisms, or they can be "trained" to produce monomers of interest. The monomers themselves are either intermediate products or end products in the metabolism of microorganisms. The "training" to produce a specific substance can be performed by gene technology, as explained in the introduction and schematically shown in Figure 24.3. If the monomer's production pathway is known, the necessary missing genes can be implemented from different organisms.

The six most used plastics in the European Union (EU) are PE, PP, PVC, PU, PET, and PS. By now, proof-of-principle studies for the microbial synthesis of monomers of four of these plastics (PE, PU, PET, and PS) exist. Even though there are many demonstrations for the production of plastic monomers with microbes in scientific literature, hardly any process made it to large-scale industrial production. The main reason is that most microbial approaches cannot compete with traditional fossil-based processes in terms of economic efficiency. One example planned to be implemented on an industrial scale is the production of aniline with microbes using CO_2 as a carbon source, with a process developed by Covestro (Covestro, 2017). Aniline is a precursor in the methylene diphenyl diisocyanate (MDI) production, which can be used as a monomer for PU production.

24.4.1 Polyethylene (PE)

The thermoplast polyethylene (PE) consists essentially of a long chain of repeating methylene groups and can be synthesized by polymerization of the monomer

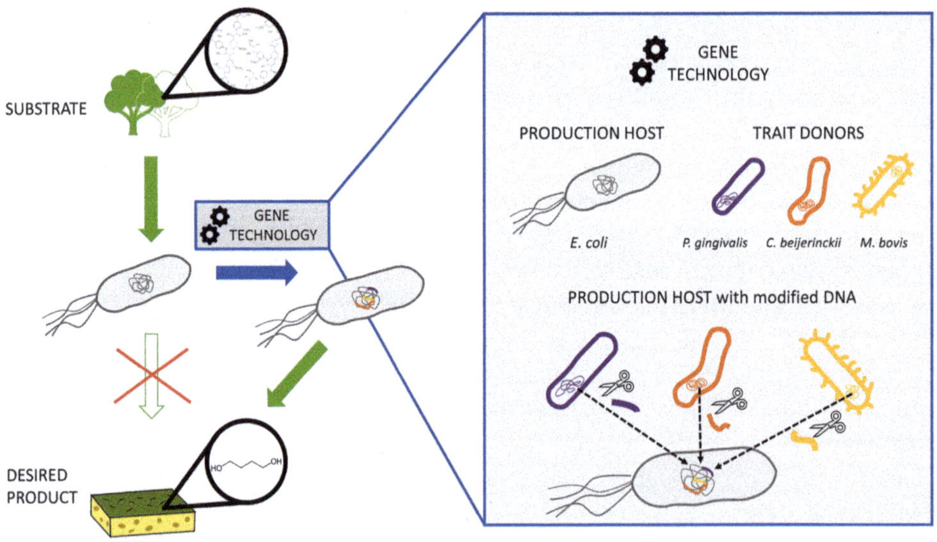

FIGURE 24.3 Schematic representation of metabolic engineering, with the example of modifying *E. coli* for 1,4-butanediol (1,4-BDO) synthesis via gene technology. The production host can use renewable biomass as a carbon and energy source, however it is naturally not able to synthesize 1,4-BDO. With the help of gene technology, genes from other organisms (in this case *P. gingivalis*, *C. beijerinckii*, and *M. bovis*) are inserted into the genome of *E. coli*. These genes code for enzymes, which are not available in *E. coli* and complement the pathway for the synthesis of 1,4-BDO. After the successful insertion of the genes, *E. coli* is able to synthesize 1,4-BDO.

ethylene. The properties of PE depend on the degree of branching, which defines the density of the material. Low-density polyethylene (LDPE) is soft and flexible and is used in wraps, bags, films, squeeze bottles, mulch film, or lids. High-density polyethylene (HDPE) is a stiff, strong, tough material, which is hard to semi-flexible. Typical applications are jugs, bottles, toys, or crates. PEs are the most used plastics worldwide, with a market share of 30% in the EU. Apart from being the monomer of PE, ethylene is one of the most widely used chemical feedstock in the petrochemical industry, with a global demand of 150 million tons per year (Petrochemical, 2015). The most common currently used production process is based on crude petroleum products.

Luckily, microbes can also produce ethylene. Naturally, ethylene occurs in plants as a hormone with several vital functions (He et al., 2021). Still, also microorganisms are known to produce ethylene where it serves, i.e., in non-specific defense signaling (Gottwald et al., 2012) and as a mediator in virulence (Weingart et al., 2001). There are three known biological pathways for the production of ethylene. The majority of the ethylene producers use the so-called α-ketoglutarate pathway, in which the intermediate α-ketoglutarate is converted by the ethylene forming enzyme (EFE) to ethylene in one step (Völksch and Weingart, 1997). So, it is feasible to implement this pathway in a variety of production hosts. Furthermore, it has the highest reported yields of the three pathways. Therefore, the α-ketoglutarate dependent pathway has been widely utilized in several hosts such as *E. coli* (e.g., Fukuda et al. (1992)), *S. cerevisiae* (e.g., Pirkov et al., 2008), *P. putida* (e.g., Wang et al. (2010)), *P. syringae* (e.g., Ishihara et al., 1996), *Trichoderma viride* (e.g., Tao et al., 2008), *T. reesei* (e.g., Chen et al., 2010), and the cyanobacteria *Synechococcus elongatus* (e.g., Carbonell et al., 2019), *Synechococcus* sp. (e.g., Sakai et al., 1997), and *Synechocystis* sp. PCC 6803 (e.g., Guerrero et al., 2012). The EFE has been identified in the *Pseudomonas* and *Penicillium* genus (Fukuda et al., 1986; Nagahama et al., 1991). Most studies used the EFE from *P. syringae* pv. *glyinea*.

Ethylene is an example where one advantageous pathway was identified and implemented into alternative production strains. This is not always the case, as demonstrated in the following sections.

24.4.2 Polyurethane (PU)

Polyurethane (PU) is an alternating copolymer consisting of two types of monomers. The name originates from the characteristic urethane bonds and not from the monomer structure. Polyurethanes are a category of polymers, consisting of diols and polyols as monomers, which leads to a considerable variety of monomers and consequently polymers. They are mainly obtained by reactions between polyols, typically polyesters and polyethers, and polyisocyanates, either aromatic or aliphatic. The infinite variety of PU architectures leads to different thermosetting or thermoplastic polymers. Overall, PUs are often used for long-term applications, including foams for furniture and insulation.

Exemplarily, PU monomers for which industrial or semi-industrial microbiology options are available are 1,4-butanediol (1,4-BDO) and 2,3-butanediol (2,3-BDO), respectively, while adipic acid (AA) is much worked on.

24.4.2.1 2,3-butanediol (2,3-BDO)

2,3-BDO has a global market of 74000 tons per year (Kim et al., 2017). The current major production route is the synthesis from cracked fossil gases, which need very harsh conditions like high temperature and high pressure (Gräfje et al., 2000). A mild alternative production could be the microbial synthesis of 2,3-BDO, which is also naturally produced by various microbes. There, it prevents intracellular acidification (Vivijs et al., 2014), regulates energy supply, as well as energy and carbon storage (Kim et al., 2017; Xiao and Xu, 2007). The genome sequences of some 2,3-BDO producers are known and publicly available, including *Paenibacillus polymyxa* (Xie et al., 2015), *Klebsiella pneumonia* (Tong et al., 2015), *K. oxytoca* (Shin et al., 2012), *Bacillus licheniformis* (Li et al., 2012), *Serratia marcescens* (Li et al., 2014), *Raoultella ornithinolytica* (Shin et al., 2013), and *Enterobacter cloacae* subsp. *dissolvens* (Xu et al., 2012). Among these are also potent 2,3-BDO producers with high yields, *e.g., K. pneumoniae* and *K. oxytoca*. However, these organisms are classified as pathogenic, which makes them difficult to use for large-scale fermentations due to safety reasons (Yang et al., 2017). There has been an effort to implement the 2,3-BDO synthesis pathway in organisms, which are harmless but are not natural producers. Reports on the successful implementation of 2,3-BDO synthesis include organisms like *S. cerevisiae* (e.g., Kim et al., 2014), *C. glutamicum* (e.g., Radoš et al., 2015), *Bacillus amyloliquefaciens* (e.g., Yang et al., 2015), *E. coli* (e.g., Xu et al., 2014), and the cyanobacteria *S. elongatus* (e.g., Oliver et al., 2013). An important aspect regarding sustainability of the whole process is the biomass feedstock (substrate) the microbes use as carbon and energy source. In early stages of development, most researchers use sucrose and starch-based feedstock as proof of principle. However, this leads to the food versus fuel discussion. Consequently, most researchers try to implement production on low-value raw materials like lignocellulose or plastic and thus add value. There are publications about 2,3-BDO production on starch, sugarcane molasses, biodiesel-derived glycerol, industrial waste gases, and lignocellulose as substrate.

24.4.2.2 1,4-butanediol (1,4-BDO)

The global production demand of 1,4-BDO is more than 2 million tons per year (Burgard et al., 2016). Currently, 1,4-BDO is mainly produced via energy-intensive processes from petrochemicals. In nature, there is no microorganism known to produce 1,4-BDO (Barton et al., 2015; Burgard et al., 2016). However, there has been a heterologous (derived from different organisms) biosynthetic pathway constructed and successfully introduced into *E. coli* (illustrated in Figure 24.3) (Barton et al., 2015; Burgard et al., 2016; Yim et al., 2011). It starts with succinyl-CoA or α-ketoglutarate and converts these to 1,4-BDO in a series of six reaction steps. This pathway is an excellent example of the construction of an artificial pathway. Six genes from three different organisms, namely *Porphyromonas gingivalis*, *Clostridium beijerinckii*, and *Mycobacterium bovis*, are inserted into one production host (*E. coli*), where the heterologous enzymes in combination with the homologous (naturally occurring in the organism) enzymes produce the product of interest. Bio-1,4-BDO is commercially available.

24.4.2.3 Adipic Acid (AA)

Adipic acid is a platform chemical and the most important dicarboxylic acid in industry, with a yearly production of 3 million tons (de Jong et al., 2012). Currently, the most important industrial production starts with the petrochemical precursor benzene. This process releases the potent greenhouse gas nitrous oxide (N_2O), which is 300 times more potent than the greenhouse gas CO_2. Studies suggest that a replacement by microbial methods based on renewable biological feedstock could reduce emissions up to 95% (Aryapratama and Janssen, 2017; Skoog et al., 2018).

In nature, there is no adipic acid-producing microbe known. However, various metabolic production routes have been constructed and successfully implanted into microbes. There are two main production schemes, the direct fermentation of adipic acid and the indirect fermentation. In the latter muconic acid or glutaric acid is produced via fermentation and subsequently converted to adipic acid by chemical hydrogenation (Skoog et al., 2018). Organisms in which a constructed adipic acid production pathway has been successfully implemented include *T. fusca* (e.g., Deng and Mao, 2015), *E. coli* (e.g., Kallscheuer et al., 2017), *Candida* spp. (e.g., Beardslee and Picataggio, 2012), *P. putida* (e.g., Vardon et al., 2015), and *S. cerevisiae* (e.g., Chen et al., 2018). Most studies use glucose as a substrate but also production on fatty acids and lignin compounds have been shown. Adipic acid is an example where not one clear advantageous pathway has been identified, but several pathways and production methods are being investigated using different microbes.

24.4.3 Polyethylene Terephthalate (PET)

Polyethylene terephthalate (PET) is composed of ethylene glycol (EG) and terephthalic acid (TA). It is a clear and resistant thermoplastic polymer. It finds applications as containers (bottles, jars, and tubes), semirigid sheets for thermoforming (trays and blisters), or thin oriented films (bags and wrappers).

24.4.3.1 Ethylene Glycol (EG)

The current global demand for EG is 28 million tons annually. It is produced by chemical synthesis from ethylene oxide, with two main production processes. Both rely on chemical processes with high operation costs (Salusjärvi et al., 2019). Several different metabolic pathways for EG production are constructed in engineered microbes since no natural EG producing microbes are known. Host organisms used for the implementation of a metabolic pathway are *E. coli* (e.g., Deng et al., 2015), *Corynebacterium glutamicum* (e.g., Zahoor et al., 2014), *S. cerevisiae* (e.g., Koivistoinen et al., 2013; Salusjärvi et al., 2017), and *K. Lactis* (e.g., Koivistoinen et al., 2013). One exemplary pathway starts with xylose as a substrate, converted to EG in four reaction steps (Zhang et al., 2020). Three enzymes of this pathway are native to *E. coli*, and the first reaction step is catalyzed by an enzyme, which is coded by a gene taken from *Caulobacter crescentus*.

24.4.3.2 Terephthalic Acid (TA)

Terephthalic acid is currently synthesized through a series of oxidation reactions from p-xylene, mainly produced from fossil fuels by a process known as the Amoco process (Ghiaci et al., 2012; Tomás et al., 2013). With a yield of 95 mol-%, this process is truly optimized but still has several drawbacks, apart from being fossil-based, including high energy requirement, high temperature, and high pressure use (Luo and Lee, 2017). There is no process known for the direct microbial production of TA from renewable biomass as a feedstock. However, there have been several processes developed for p-xylene production from renewable biomass. As a result, there is an effort to establish microbial synthesis of TA with p-xylene as a substrate. There has been a publication on the successful construction of such a metabolic pathway in *E. coli* (Luo and Lee, 2017). This is a first step toward the microbial production of TA. However, proof of principle for the direct microbial production from biomass without prior production steps is still missing.

24.4.4 Polystyrene (PS)

Polystyrene (PS) is an aromatic hydrocarbon polymer synthesized by the polymerization of styrene. It can be categorized into solid and foamed PS, whereby typical applications are food packaging or laboratory ware. It is also used as building insulation and for protective packaging as it has good insulating and cushioning properties, respectively. Additionally, it finds application as a blending material with other plastics to produce electronics, appliances, automobile parts, toys, or other equipment. The global demand for styrene monomers in 2012 was more than 26 million tons (Lian et al., 2016). The current production process solely relies on the petrochemical industry, where it is produced via the petro leum-derived ethylbenzene in an energy-intensive process, which requires extensive thermal energy (Lee et al., 2019).

Proof-of-principle studies also exist for the microbial production of styrene. Styrene was naturally synthesized in various microorganisms, including *Penicillium camemberti* and certain plant species (McKenna and Nielsen, 2011; Pagot et al., 2007). However, due to extremely low production efficiencies, there is a need for optimization by genetic engineering. Synthetic styrene pathways have been successfully constructed and inserted in various microorganisms, including *E. coli (e.g.*, Lee et al., 2019), *S. cerevisiae (e.g.*, McKenna et al., 2014), *P. taiwanensis* (Otto, 2020), and *Streptomyces lividans* (e.g., Fujiwara et al., 2016). All publications include the insertion of two key enzymes (a lyase and a carboxylase). The heterologous inserted genes were taken from different organisms, including *Arabidopsis thaliana* (McKenna and Nielsen, 2011), *S. cerevisiae* (McKenna and Nielsen, 2011), and *Streptomyces maritimus* (Lee et al., 2019).

24.5 MICROBE-MADE PLASTICS

Bio-based plastics are currently being discussed as a replacement for fossil-based plastics. One example of plant-based plastic products is bio-degradable bags made of thermoplastic starch. Usually, plant-derived raw materials need to be purified to obtain the desired molecules (e.g., starch), which are modified in a second step to be used as plastic. In contrast, some bacteria can directly produce bio-degradable plastics and natural polymers with plastic-like applications that can be applied like conventional plastics without chemical modification or polymerization.

24.5.1 Using Bacterial "Fat" as Plastic

The biopolymer poly(hydroxyalkanoate) (PHA) is naturally produced by many bacteria and can be used as a bio-based, bio-degradable plastic without subsequent chemical modification. When the bacteria experience limitation of nutrients such as nitrogen or phosphorous but have access to a carbon and energy source, they accumulate PHA within their cells. Analogous to fat storage in humans the process of PHA accumulation is also a way to store carbon and energy. So, this "bacterial fat" is already polymerized within the bacterial cell, and thus, only extraction from the cells is required to obtain PHA plastic.

The plastic properties vary considerably depending on the monomers incorporated into PHA. The monomers are derivatives of fatty acids, namely hydroxy fatty acids. The enzyme PHA synthase links the monomers to each other and, thus, produces the polymer. Depending on the bacterial strain, monomers with a varying number of carbons are produced. Concerning the use of PHA as plastic, the length of the monomers is decisive because it contributes to different properties of the resulting plastics. Moreover, some microorganisms produce PHA types that consist of two or more monomers and are referred to as copolymers. Taken together, these factors result in highly diverse polymer properties. Accordingly, different PHA types can be shaped into a variety of products. Examples of a rather flexible PHA are pure poly(hydroxybutyrate) (PHB) plastics consisting of monomers with four carbon atoms. Such PHAs can be utilized as packaging films or plastic bags, conventionally produced from PVC or LDPE, respectively (Lee et al., 2020). Food containers, which are often produced from PP, could instead be produced from a rather stiff

PHA copolymer with monomers consisting of four and five carbon atoms (Lee et al., 2020). The Swiss company Prodir produces a PHA ballpen (QS40 True Biotic), which has been certified as biodegradable in sea water, sweet water as well as a garden or industrial compost (Pagani Pens SA, 2021).

A variety of bacteria, as well as some yeasts, are capable of PHA production. The PHA produced by each strain not only depends on the strain itself but also the substrate (feedstock) fed. For example, the Italian company Bio-On Srl. uses sugar beets to feed the bacterium *Cupriavidus necator* for the production of PHB and a copolymer of PHB and poly(hydroxy valerate) (PHV) consisting of monomers with five carbon atoms. A Brazilian company utilizes a bacterium of the *Alcaligenes* family to produce the same PHAs feeding the bacteria with sucrose from sugar cane. Bacteria can also be fed with a starch-rich substrate such as the edible root crop cassava (also known as yuca or mandioca), as does a Chinese company to produce PHA. While several bacterial strains naturally produce high amounts of PHA, numerous studies report increased PHA production obtained by metabolic engineering. Another Chinese company, BluePHA, employs such engineered bacteria to produce customized PHA, including pure polymers such as PHV and PHB and a variety of copolymers consisting of monomers with three, four, five, or six carbon atoms (Kourmentza et al., 2017). The first applications for their PHAs are mulch films used in agriculture to suppress weeds or prevent water evaporation.

24.5.2 Lactic Acid Plastic

Poly(lactic acid) (PLA) is synthetic bio-based plastic consisting of lactic acid (LA) monomers. These monomers are produced by microorganisms and can be polymerized to PLA after derivatization of LA to lactide. LA is a small organic acid and was first discovered in sour milk in 1780 by the pharmaceutical chemist Carl Wilhelm Scheele (John et al., 2007). In humans, LA is produced in exerted muscles once the oxygen is consumed (Gupta et al., 2007). Moreover, many fermented foods such as yogurt, sourdough bread, or vegetables (*e.g.*, sauerkraut) contain LA, contributing to the characteristic taste (Holzapfel et al., 2003). During fermentation, so-called lactic acid bacteria convert the sugars in milk, flour, or vegetables (e.g., glucose, fructose, sucrose) to LA (Holzapfel et al., 2003).

24.6 MICROBES WORK MORE PRECISELY THAN CHEMISTS

LA is not only a natural product in humans and microorganisms but also produced industrially. Until the early 1990s, petrochemical resources were the raw material for chemical LA production. Today the chemical route only makes up about 10% of the LA production worldwide (Gupta et al., 2007). The majority of LA (90%) is produced by bacteria (Zhou et al., 2006). Traditionally, the best choice was *Lactobacilli* because they produce few unwanted byproducts. Nowadays, pH-resistant yeast and thermophilic *Bacilli* are used as whole-cell catalysts. The microbial route to LA offers several advantages: Low consumption of fossil resources, reduced CO_2 emission, low production costs, and a nontoxic catalyst, i.e., the microbe (Singhvi et al., 2019). The most significant advantage is, however, the selectivity achieved by bacteria.

To understand this advantage, we need to know that LA can occur in two structural forms, called enantiomers, L- and D-LA. L-LA and D-LA are like mirror images of each other, which cannot be overlaid – just as our hands compare to each other. Chemical LA production yields a racemic, i.e., 50 : 50 mixture of the D- and L-enantiomer. In microbial fermentation – depending on the bacterium – only one of the two enantiomers is produced. For the polymerization of LA to PLA, the purity of the LA (either D or L form) is essential since minor amounts of the other enantiomer have a significant influence on PLA properties (Singhvi et al., 2019). However, the consequences of this effect are not only negative. Polymer scientists use the effect to fine-tune, for example, the stability of PLA at high temperatures (Singhvi et al., 2019).

24.6.1 How Do You Make a Polymer from Lactic Acid?

While LA synthesis is primarily performed by microbes, the polymerization – linking a large number of LA monomers – into PLA is performed chemically. The most common way is the so-called ring-opening polymerization of lactide, a circular molecule consisting of two LA units. Polymerization of lactide requires heavy metals as catalysts, while polycondensation – another polymerization method – is achieved using solvents (Singhvi and Gokhale, 2013; Singhvi et al., 2019). Notably, the PHA synthesis machinery is a possible alternative to produce

PLA. Indeed, the PHA synthase can be engineered to synthesize PLA polymers from cheap raw materials (Singhvi et al., 2019). A one pot production would be of interest; however, a highly efficient production process (high titer, rate, and yield), which can compete with the current chemical route, has to be established.

24.6.2 Engineered Enzymes as Environmentally Friendly Catalysts

To date, no microorganism naturally capable of polymerizing LA to PLA is known. Hence, microbiologists employed gene technology to facilitate bacterial PLA polymerization. Like the LA monomers in PLA, the PHA monomers are also linked by ester bonds (Taguchi et al., 2008). This similarity made it possible to engineer the PHA synthase – that naturally polymerizes hydroxy fatty acid monomers to the biopolymer PHA (see Section 24.5.1) – for the polymerization of D-LA monomers. Japanese scientists engineered the PHA synthase from one bacterium, *Pseudomonas* sp. 61-3, and then expressed it in another bacterium, *Escherichia coli*. This engineered bacterium produced PLA to a concentration of 0.5% of its cell dry weight (Taguchi et al., 2008). Then, Korean scientists additionally engineered the metabolism of the *E. coli* strain bearing the engineered PHA synthase, and the strain synthesized PLA up to 11% of the cell dry weight (Jung et al., 2010).

In future, the transfer of this strategy to bacteria of the genus *Lactobacillus*, might be a promising strategy to achieve high LA synthesis and its microbial polymerization to PLA, since these bacteria naturally produce high amounts of LA (Singhvi et al., 2019).

24.7 MICROBIAL PRODUCTS AS PLASTICIZERS (HAA)

Conventional plastics often require the addition of plasticizers to adapt their properties to a specific application. Generally, plasticizers can be divided into two groups based on how they interact with the plastic monomers. External plasticizers have no chemical bonds to the polymer chains and are only maintained within the polymer due to secondary chemical bonds (e.g., hydrogen bonds, electrostatic or van der Waals interactions). These secondary bonds are less stable. Therefore, external plasticizers may leak from the polymer as it ages or under harsh conditions (high temperature, extreme pH), which can be observed in everyday life. For example, plastic garden furniture turns brittle over the years, and finally, small cracks start showing up since the plasticizer has leaked from the furniture over time. In contrast, internal plasticizers interact with the plastic monomers via primary chemical bonds and are unlikely to leak from the plastic polymer (Bocqué et al., 2016).

Different organic waste types have been studied as raw materials for various plasticizers (Jia et al., 2018). The use of such bio-based plasticizers generally requires chemical modification of the raw material. French scientists directly applied an unmodified bacterial product as a plasticizer in a PU polymer, poly(amide urethane) (Tiso et al., 2020). Hydroxyalkanoyloxy alkanoate (HAA) is a derivative of hydroxy fatty acids synthesized by several bacterial species from the class proteobacteria (Germer et al., 2020). Other bacteria have been genetically engineered to synthesize HAA (Blesken et al., 2020; Germer et al., 2020; Tiso et al., 2020). This versatile molecule type initially aroused interest because of its surfactant character. Also, HAAs have been studied as a starting material for biofuels (Mensah et al., 2020; Meyers et al., 2019). An advantage of HAA in contrast to PHA is that it is secreted by the bacteria that produce it (Tiso et al., 2020). Accordingly, the bacteria do not need to be lysed to obtain the HAAs so that the bacteria can potentially be used several times to produce HAAs. Ultimately, such a "recycling" of microorganisms make the industrial application more economical.

As many as 53 different HAA types are known and distinguished by the kind of hydroxy fatty acids they are made of. The fatty-acid chains can have a length of four to 18 carbon atoms (Germer et al., 2020). As a consequence of the different lengths of HAAs, they have been suggested as internal plasticizers. When HAAs form part of poly(amide urethane), shorter HAA chains will produce a more rigid plastic, while longer chains cause the plastic to be more flexible. Different bacteria produce a mix of HAAs, usually with one predominant chain length (Germer et al., 2020). Scientists are currently working to exploit this natural potential. To date, the HAA tested as a plasticizer contains mainly chains of ten carbon atoms. Soon HAA mixtures with longer carbon chains will be tested to compare their plasticizing effect in poly(amide urethane).

24.8 SUGARS AND MORE FOR PLASTIC-LIKE APPLICATIONS

Microorganisms produce a broad spectrum of sugar polymers, which play a role in various physiological processes. Xanthan, for example, is produced by bacteria of the genus *Xanthomonas* and secreted by the cells. Alginate is another secreted polymer that can be synthesized by bacteria and algae. Other polymers are made from organic acids (polymalate), amino acids (polyglutamate, cyanophycin), or mainly consist of one element (polyphosphate). The suggested applications are as diverse as the polymers, and some of these polymers are also discussed to replace oil-based plastics (Freitas et al., 2011; Zikmanis et al., 2020).

Pullulan consists of repeating units of maltotriose, which itself is composed of three glucose units. The special characteristics of pullulan are determined by the special sugar bonds (α-1,4- and -1,6-glycosidic bonds) (Rehm, 2010). The fungus *Aureobasidium pullulans* produces pullulan for protection against desiccation or predation. A Japanese company has been commercializing pullulan products since the 1970s. Today they offer edible films mainly for food packaging, which dissolve in the mouth. Specialized pullulan films can compete with conventional polystyrene films in terms of strength, transparency, and rigidity (Abhilash and Thomas, 2017; Leathers, 2003). Such films are also oxygen impermeable and can regulate humidity. Hence, special pullulan films exhibit favorable properties for food packaging and coating (Farris et al., 2014; Freitas et al., 2011). Since the market price of pullulan is quite high (US$25 per kg), the number of commercial pullulan products is still limited (Farris et al., 2014; Kraśniewska et al., 2019). In this case, gene technology might be a promising tool to improve pullulan production at microbe level (Kraśniewska et al., 2019).

24.9 CONCLUSION

This chapter has summarized some of the challenges accompanying the staggering accumulation of plastics in every ecosystem of the planet, and possible solutions (microbial)biotechnology has to offer. Starting with novel upcycling concepts by engineering microbes for the transformation of plastic waste and proceeding via the microbial production of monomers for bio-based plastics, the chapter ended with the usage of microbes for the production of novel bio-based plastics. All these aspects of sustainable plastic biotechnology rely heavily on the application of microbes, thus underlining the emerging importance of microbes in the chemical industry.

Although some approaches are already being pursued on an industrial scale, most ideas are still in their infancy. Much more research is needed for the widespread application of microbes in the plastics industry. The lessons learned in lignocellulosic biotechnology may guide plastic biotech as the main aspects are mirrored in the usage of hydrolysates. Lignocellulosic biomass is the most abundant raw material on earth, and a few examples exist that use it as feedstock for bioethanol production. The challenges are similar, as cleaving bonds in this polymer network of lignocellulose is also challenging.

Considering the advances made in biotechnology in recent years, microbes are likely to revolutionize the plastics industry provided that the excellent research carried out can be translated and a fair price of the CO_2 footprint is implemented.

ACKNOWLEDGMENTS

The authors have received funding from the European Union's Horizon 2020 research and innovation program under grant agreement no. 870294 for the project MIX-UP. Further, the German Federal Ministry of Education and Research is acknowledged for funding under the grant agreement No. 031B0899 in the project MIPLACE within the framework of the ERA-CoBioTech program of the EU.

REFERENCES

1. Abhilash M, Thomas D. 2017. Biopolymers for biocomposites and chemical sensor applications. *Biopolymer Composites in Electronics*, pp. 405–435.
2. Allwood JM. 2014. Squaring the circular economy. *Handbook of Recycling*, pp. 445–477.
3. Aryapratama R, Janssen M. 2017. Prospective life cycle assessment of bio-based adipic acid production from forest residues. *Journal of Cleaner Production* 164: 434–443.
4. Auras R, Harte B, Selke S. 2004. An overview of polylactides as packaging materials. *Macromolecular Bioscience* 4(9): 835–864.
5. Bailey JE. 1991. Toward a science of metabolic engineering. *Science* 252(5013): 1668.
6. Balasubramanian V, Natarajan K, Hemambika B, Ramesh N, Sumathi CS, et al. 2010. High-density polyethylene (HDPE)-degrading potential bacteria from marine ecosystem of gulf of mannar, India. *Letters in Applied Microbiology* 51(2): 205–211.
7. Barth M, Oeser T, Wei R, Then J, Schmidt J, et al. 2015. Effect of hydrolysis products on the enzymatic deg-

radation of polyethylene terephthalate nanoparticles by a polyester hydrolase from. *Thermobifida Fusca. Biochemical Engineering Journal* 93: 222–228.

8. Barton NR, Burgard AP, Burk MJ, Crater JS, Osterhout RE, et al. 2015. An integrated biotechnology platform for developing sustainable chemical processes. *Journal of Industrial Microbiology and Biotechnology* 42(3): 349–360.
9. Bassi SA, Boldrin A, Faraca G, Astrup TF. 2020. Extended producer responsibility: How to unlock the environmental and economic potential of plastic packaging waste? *Resources, Conservation and Recycling* 162: 105030.
10. Beardslee T, Picataggio S. 2012. Bio-based adipic acid from renewable oils. *Lipid Technology* 24(10): 223–225.
11. Bittner N, Nefzger H, Behnken G, Jaeger G, Behnken S, et al. 2018. Neue Urethanasen für den enzymatischen Abbau von Polyurethanen. *Covestro Deutschland AG*, Patent Number: 18179032.0.
12. Blank LM, Narancic T, Mampel J, Tiso T, O'Connor K. 2020. Biotechnological upcycling of plastic waste and other non-conventional feedstocks in a circular economy. *Current Opinion in Biotechnology* 62: 212–219.
13. Blesken CC, Bator I, Eberlein C, Heipieper HJ, Tiso T, et al. 2020. Genetic cell-surface modification for optimized foam fractionation. *Frontiers in Bioengineering and Biotechnology* 8: 572892.
14. Bocqué M, Voirin C, Lapinte V, Caillol S, Robin -J-J. 2016. Petro-based and bio-based plasticizers: Chemical structures to plasticizing properties. *Journal of Polymer Science Part A: Polymer Chemistry* 54(1): 11–33.
15. Burgard A, Burk MJ, Osterhout R, Van Dien S, Yim H. 2016. Development of a commercial scale process for production of 1,4-butanediol from sugar. *Current Opinion in Biotechnology* 42: 118–125.
16. Carbonell V, Vuorio E, Aro E-M, Kallio P. 2019. Enhanced stable production of ethylene in photosynthetic cyanobacterium *Synechococcus elongatus* PCC 7942. *World Journal of Microbiology and Biotechnology* 35(5): 77.
17. Carniel A, Gomes ADC, Coelho MAZ, De Castro AM 2020. Process strategies to improve biocatalytic depolymerization of post-consumer PET packages in bioreactors, and investigation on consumables cost reduction. *Bioprocess and Biosystems Engineering*.
18. Chen N, Wang J, Zhao Y, Deng Y. 2018. Metabolic engineering of *Saccharomyces cerevisiae* for efficient production of glucaric acid at high titer. *Microbial Cell Factories* 17(1): 67.
19. Chen X, Liang Y, Hua J, Tao L, Qin W, et al. 2010. Overexpression of bacterial ethylene-forming enzyme gene in *Trichoderma reesei* enhanced the production of ethylene. *International Journal of Biological Sciences* 6(1): 96.
20. Covestro AG 2017. Bio-based aniline - A scientific breakthrough for greater sustainability. Retrieved from https://www.covestro.com/en/sustainability/lighthouse-projects/bio-anilin.
21. Danso D, Chow J, Streit WR. 2019. Plastics: Environmental and biotechnological perspectives on microbial degradation. *Applied and Environmental Microbiology* 85(19).
22. De Jong E, Higson A, Walsh P, Wellisch M. 2012. Bio-based chemicals value added products from biorefineries. *IEA Bioenergy* 34.
23. Deng Y, Mao Y. 2015. Production of adipic acid by the native-occurring pathway in *Thermobifida fusca* B6. *Journal of Applied Microbiology* 119(4): 1057–1063.
24. Deng Y, Mao Y, Zhang X. 2015. Metabolic engineering of *E. coli* for efficient production of glycolic acid from glucose. *Biochemical Engineering Journal* 103: 256–262.
25. Dvořák P, Nikel PI, Damborský J, de Lorenzo V. 2017. Bioremediation 3.0: Engineering pollutant-removing bacteria in the times of systemic biology. *Biotechnology Advances* 35(7): 845–866.
26. Espinosa MJC, Blanco AC, Schmidgall T, Atanasoff-Kardjalicff AK, Kappelmeyer U, et al. 2020. Toward Biorecycling: Isolation of a soil bacterium that grows on a polyurethane oligomer and monomer. *Frontiers Microbiology* 11: 404.
27. European Bioplastics. 2020. Bioplastics market data. Retrieved from https://www.european-bioplastics.org/market.
28. Farris S, Unalan IU, Introzzi L, Fuentes-Alventosa JM, Cozzolino CA. 2014. Pullulan-based films and coatings for food packaging: Present applications, emerging opportunities, and future challenges. *Journal of Applied Polymer Science* 131(13).
29. Freitas F, Alves VD, Reis MAM. 2011. Advances in bacterial exopolysaccharides: From production to biotechnological applications. *Trends in Biotechnology* 29(8): 388–398.
30. Fujiwara R, Noda S, Tanaka T, Kondo A. 2016. Styrene production from a biomass-derived carbon source using a coculture system of phenylalanine ammonia lyase and phenylacrylic acid decarboxylase-expressing *Streptomyces lividans* transformants. *Journal of Bioscience and Bioengineering* 122(6): 730–735.

31. Fukuda H, Fujii T, Ogawa T. 1986. Preparation of a Cell-free Ethylene-forming System from. *Penicillium Digitatum. Agricultural and Biological Chemistry* 50(4): 977–981.

32. Fukuda H, Ogawa T, Ishihara K, Fujii T, Nagahama K, et al. 1992. Molecular cloning in Escherichia coli, expression, and nucleotide sequence of the gene for the ethylene-forming enzyme of *Pseudomonas syringae* pv. *phaseolicola* PK2. *Biochemical and Biophysical Research Communications* 188(2): 826–832.

33. Germer A, Tiso T, Müller C, Behrens B, Vosse C, et al. 2020. Exploiting the natural diversity of RhlA acyltransferases for the synthesis of the rhamnolipid precursor 3-(3-hydroxyalkanoyloxy)alkanoic acid. *Applied and Environmental Microbiology* 86(6): e02317–02319.

34. Geyer R, Jambeck JR, Law KL. 2017. Production, use, and fate of all plastics ever made. *Science Advances* 3(7): e1700782.

35. Ghiaci M, Mostajeran M, Gil A. 2012. Synthesis and characterization of Co–Mn nanoparticles immobilized on a modified bentonite and its application for oxidation of p-xylene to terephthalic acid. *Industrial and Engineering Chemistry Research* 51(49): 15821–15831.

36. Giese B, Koenigstein S, Wigger H, Schmidt JC, Von Gleich A. 2013. Rational engineering principles in synthetic biology: A framework for quantitative analysis and an initial assessment. *Biological Theory* 8(4): 324–333.

37. Gottwald S, Samans B, Lück S, Friedt W. 2012. Jasmonate and ethylene dependent defence gene expression and suppression of fungal virulence factors: Two essential mechanisms of Fusarium head blight resistance in wheat? *BMC Genomics* 13(1): 1–22.

38. Gräfje H, Körnig W, Weitz H-M, Reiß W, Steffan G, et al. 2000. Butanediols, butenediol, and butynediol. In: *Ullmann's Encyclopedia of Industrial Chemistry*, pp. 1–12.

39. Guerrero F, Carbonell V, Cossu M, Correddu D, Jones PR. 2012. Ethylene synthesis and regulated expression of recombinant protein in *Synechocystis* sp. PCC 6803. *PloS One* 7(11): e50470.

40. Gupta B, Revagade N, Hilborn J. 2007. Poly(lactic acid) fiber: An overview. *Progress in Polymer Science* 32(4): 455–482.

41. He Y, Wu B, Xia W, Zhao K-Y, Qin Y, et al. 2021. Metabolic engineering of *Zymomonas moblis* for ethylene production from straw hydrolysate. *Applied Microbiology and Biotechnology* 105(4): 1709–1720.

42. Holzapfel W, Schillinger U, Buckenhüskes H. 2003. Verfärbung von Sauerkraut. In: Farnworth ER (Ed.), *Handbook of Fermented Functional Foods* (Vol. 2). Boca Raton: CRC Press.

43. Howard GT. 2002. Biodegradation of polyurethane: A review. *International Biodeterioration and Biodegradation* 49(4): 245–252.

44. Ishihara K, Matsuoka M, Ogawa T, Fukuda H. 1996. Ethylene production using a broad-host-range plasmid in *Pseudomonas syringae* and *Pseudomonas putida. Journal of Fermentation and Bioengineering* 82(5): 509–511.

45. Jia P, Xia H, Tang K, Zhou Y. 2018. Plasticizers derived from biomass resources: A short review. *Polymers (Basel)* 10(12).

46. John RP, Nampoothiri KM, Pandey A. 2007. Fermentative production of lactic acid from biomass: An overview on process developments and future perspectives. *Applied Microbiology and Biotechnology* 74(3): 524–534.

47. Jung YK, Kim TY, Park SJ, Lee SY. 2010. Metabolic engineering of *Escherichia coli* for the production of polylactic acid and its copolymers. *Biotechnology Bioengineering* 105(1): 161–171.

48. Kallscheuer N, Gätgens J, Lübcke M, Pietruszka J, Bott M, et al. 2017. Improved production of adipate with *Escherichia coli* by reversal of β-oxidation. *Applied Microbiology and Biotechnology* 101(6): 2371–2382.

49. Kim S-J, Kim J-W, Lee Y-G, Park Y-C, Seo J-H. 2017. Metabolic engineering of *Saccharomyces cerevisiae* for 2, 3-butanediol production. *Applied Microbiology and Biotechnology* 101(6): 2241–2250.

50. Kim S-J, Seo S-O, Park Y-C, Jin Y-S, Seo J-H. 2014. Production of 2,3-butanediol from xylose by engineered *Saccharomyces cerevisiae. Journal of Biotechnology* 192: 376–382.

51. Knott BC, Erickson E, Allen MD, Gado JE, Graham R, et al. 2020. Characterization and engineering of a two-enzyme system for plastics depolymerization. *Proceedings of the National Academy of Sciences* 117(41): 25476.

52. Koivistoinen OM, Kuivanen J, Barth D, Turkia H, Pitkänen J-P, et al. 2013. glycolic acid production in the engineered yeasts *saccharomyces cerevisiae* and. *Kluyveromyces Lactis. Microbial Cell Factories* 12(1): 82.

53. Kourmentza C, Placido J, Venetsaneas N, Burniol-Figols A, Varrone C, et al. 2017. Recent advances and challenges towards sustainable polyhydroxyalkanoate (PHA) production. *Bioengineering (Basel)* 4(2).

54. Kraśniewska K, Pobiega K, Gniewosz M. 2019. Pullulan–biopolymer with potential for use as food packaging. *International Journal of Food Engineering* 15(9).

55. Kumari A, Chaudhary DR, Jha B. 2021. Microbial Degradation of Plastics and Its Biotechnological Advancement. *Environmental Biotechnology* 3: 1–30.

56. Leathers TD. 2003. Biotechnological production and applications of pullulan. *Applied Microbiology Biotechnology* 62(5-6): 468–473.

57. Lee CH, Sapuan SM, Ilyas RA, Lee SH, Khalina A 2020. Development and processing of PLA, PHA, and other biopolymers. *Advanced Processing, Properties, and Applications of Starch and Other Bio-Based Polymers*, pp. 47–63.

58. Lee K, Bang HB, Lee YH, Jeong KJ. 2019. Enhanced production of styrene by engineered *Escherichia coli* and *in situ* product recovery (ISPR) with an organic solvent. *Microbial Cell Factories* 18(1): 79.

59. Leong YK, Show PL, Lan JC-W, Loh H-S, Lam HL, et al. 2017. Economic and environmental analysis of PHAs production process. *Clean Technologies and Environmental Policy* 19(7): 1941–1953.

60. Li L, Su F, Wang Y, Zhang L, Liu C, et al. 2012. Genome sequences of two *thermophilic bacillus* licheniformis strains, efficient producers of platform chemical 2,3-butanediol. *Journal of Bacteriology* 194(15): 4133.

61. Li L, Wang Y, Li K, Su F, Ma C, et al. 2014. Genome Sequence of *meso*-2,3-butanediol-producing strain *serratia marcescens* ATCC 14041. *Genome Announcements* 2(3): e00590–00514.

62. Lian J, McKenna R, Rover MR, Nielsen DR, Wen Z, et al. 2016. Production of biorenewable styrene: Utilization of biomass-derived sugars and insights into toxicity. *Journal of Industrial Microbiology and Biotechnology* 43(5): 595–604.

63. Luo ZW, Lee SY. 2017. Biotransformation of p-xylene into terephthalic acid by engineered. *Escherichia Coli. Nature Communications* 8(1): 15689.

64. Magnin A, Entzmann L, Bazin A, Pollet E, Averous L. 2021. Green recycling process for polyurethane foams by a chem-biotech approach. *ChemSusChem*.

65. Magnin A, Hoornaert L, Pollet E, Laurichesse S, Phalip V, et al. 2019. Isolation and characterization of different promising fungi for biological waste management of polyurethanes. *Microbial Biotechnology* 12(3): 544–555.

66. Masaki K, Kamini NR, Ikeda H, Iefuji H. 2005. Cutinase-like enzyme from the yeast *Cryptococcus* sp. strain S-2 hydrolyzes polylactic acid and other biodegradable plastics. *Applied and Environmental Microbiology* 71(11): 7548–7550.

67. McKenna R, Nielsen DR. 2011. Styrene biosynthesis from glucose by engineered. *E. Coli. Metabolic Engineering* 13(5): 544–554.

68. McKenna R, Thompson B, Pugh S, Nielsen DR. 2014. Rational and combinatorial approaches to engineering styrene production by. *Saccharomyces Cerevisiae. Microbial Cell Factories* 13: 123–123.

69. Mensah JB, Hergesell AH, Brosch S, Golchert C, Artz J, et al. 2020. Catalytic deoxygenation of bio-based 3-hydroxydecanoic acid to secondary alcohols and alkanes. *Green Chemistry* 22(11): 3522–3531.

70. Meyers J, Mensah JB, Holzhäuser FJ, Omari A, Blesken CC, et al. 2019. Electrochemical conversion of a bio-derivable hydroxy acid to a drop-in oxygenate diesel fuel. *Energy and Environmental Science* 12(8): 2406–2411.

71. Moharir RV, Kumar S. 2019. Challenges associated with plastic waste disposal and allied microbial routes for its effective degradation: A comprehensive review. *Journal of Cleaner Production* 208: 65–76.

72. Nagahama K, Ogawa T, Fujii T, Tazaki M, Tanase S, et al. 1991. Purification and properties of an ethylene-forming enzyme from *Pseudomonas syringae* pv. *phaseolicola* PK2. *Microbiology* 137(10): 2281–2286.

73. Narancic T, Connor KE. 2019. Plastic waste as a global challenge: Are biodegradable plastics the answer to the plastic waste problem? *Microbiology* 165(2): 129–137.

74. Narancic T, Verstichel S, Reddy Chaganti S, Morales-Gamez L, Kenny ST, et al. 2018. Biodegradable plastic blends create new possibilities for end-of-life management of plastics but they are not a panacea for plastic pollution. *Environmental Science and Technology* 52(18): 10441–10452.

75. Nisticò R. 2020. Polyethylene terephthalate (PET) in the packaging industry. *Polymer Testing* 90: 106707.

76. Oliver JWK, Machado IMP, Yoneda H, Atsumi S. 2013. Cyanobacterial conversion of carbon dioxide to 2,3-butanediol. *Proceedings of the National Academy of Sciences of the United States of America* 110(4): 1249–1254.

77. Otto M 2020. *Microbial catalysis of renewable resources into aromatics via the central metabolic precursor phenylalanine.* (Dissertation). RWTH Aachen University, Apprimus Verlag.

78. Pagani Pens SA 2021. QS40 True Biotic - Created by nature for nature. Retrieved from https://www.prodir.com/en/pens/qs40-true-biotic/overview.

79. Pagot Y, Belin J-M, Husson F, Spinnler H-E. 2007. Metabolism of phenylalanine and biosynthesis of styrene in *Penicillium camemberti*. *Journal of Dairy Research* 74(2): 180–185.

80. Pathak VM, Navneet. 2017. Review on the current status of polymer degradation: A microbial approach. *Bioresources and Bioprocessing* 4(1).

81. Petrochemical. 2015. *The ethylene technology report*. Retrieved from http://www.researchandmarkets.com/research/wdql3r/the_ethylene.

82. Pirkov I, Albers E, Norbeck J, Larsson C. 2008. Ethylene production by metabolic engineering of the yeast. *Saccharomyces Cerevisiae. Metabolic Engineering* 10(5): 276–280.

83. PlasticsEurope. 2020. Plastics - the facts 2020: an analysis of European plastics production, demand and waste data. Retrieved from https://www.plasticseurope.org/en/resources/publications/4312-plastics-facts–2020.

84. Rabnawaz M, Wyman I, Auras R, Cheng S. 2017. A roadmap towards green packaging: The current status and future outlook for polyesters in the packaging industry. *Green Chemistry* 19(20): 4737–4753.

85. Radoš D, Carvalho AL, Wieschalka S, Neves AR, Blombach B, et al. 2015. Engineering *Corynebacterium glutamicum* for the production of 2,3-butanediol. *Microbial Cell Factories* 14(1): 171.

86. Raf T, Bordado JCM, Gomes JFP. 2013. p-xylene oxidation to terephthalic acid: A literature review oriented toward process optimization and development. *Chemical Reviews* 113(10): 7421–7469.

87. Rehm BHA. 2010. Bacterial polymers: Biosynthesis, modifications and applications. *Nature Reviews Microbiology* 8(8): 578–592.

88. Sakai M, Ogawa T, Matsuoka M, Fukuda H. 1997. Photosynthetic conversion of carbon dioxide to ethylene by the recombinant cyanobacterium, *Synechococcus* sp. PCC 7942, which harbors a gene for the ethylene-forming enzyme of *Pseudomonas syringae*. *Journal of Fermentation and Bioengineering* 84(5): 434–443.

89. Salusjärvi L, Havukainen S, Koivistoinen O, Toivari M. 2019. Biotechnological production of glycolic acid and ethylene glycol: Current state and perspectives. *Applied Microbiology and Biotechnology* 103(6): 2525–2535.

90. Salusjärvi L, Toivari M, Vehkomäki M-L, Koivistoinen O, Mojzita D, et al. 2017. Production of ethylene glycol or glycolic acid from D-xylose in *Saccharomyces cerevisiae*. *Applied Microbiology and Biotechnology* 101(22): 8151–8163.

91. Seymour RB, Kauffman GB. 1992. Polyurethanes: A class of modern versatile materials. *Journal of Chemical Education* 69(11): 909.

92. Shin SH, Kim S, Kim JY, Lee S, Um Y, et al. 2012. Complete genome sequence of *klebsiella oxytoca* KCTC 1686, used in production of 2,3-butanediol. *Journal of Bacteriology* 194(9): 2371.

93. Shin SH, Um Y, Beak JH, Kim S, Lee S, et al. 2013. Complete genome sequence of *raoultella ornithinolytica* strain B6, A 2,3-butanediol-producing bacterium isolated from oil-contaminated soil. *Genome Announcements* 1(3): e00395–00313.

94. Singh G, Yadav S, Chowdhary K, Sharma S 2021. Role of microbes in plastic degradation. *Recent Developments in Microbial Technologies*, pp. 281–299.

95. Singhvi MS, Gokhale DV. 2013. Biomass to biodegradable polymer (PLA). *Rsc Advances* 3(33): 13558–13568.

96. Singhvi MS, Zinjarde SS, Gokhale DV. 2019. Polylactic acid: Synthesis and biomedical applications. *Journal of Applied Microbiology* 127(6): 1612–1626.

97. Skoog E, Shin JH, Saez-Jimenez V, Mapelli V, Olsson L. 2018. Biobased adipic acid – The challenge of developing the production host. *Biotechnology Advances* 36(8): 2248–2263.

98. Sohn YJ, Kim HT, Baritugo KA, Jo SY, Song HM, et al. 2020. Recent advances in sustainable plastic upcycling and biopolymers. *Biotechnology Journal* 15(6): e1900489.

99. Taguchi S, Yamada M, Matsumoto K, Tajima K, Satoh Y, et al. 2008. A microbial factory for lactate-based polyesters using a lactate-polymerizing enzyme. *Proceedings of the National Academy of Sciences of the United States of America* 105(45): 17323–17327.

100. Tao L, Dong H-J, Chen X, Chen S-F, Wang T-H. 2008. Expression of ethylene-forming enzyme (EFE) of *Pseudomonas syringae* pv. *glycinea* in. *Trichoderma Viride. Applied Microbiology and Biotechnology* 80(4): 573–578.

101. Tiso T, Narancic T, Wei R, Pollet E, Beagan N, et al. 2020. Bio-upcycling of polyethylene terephthalate. *bioRxiv* 2020(03): 16.993592.

102. Tong Y-J, Ji X-J, Liu L-G, Shen M-Q HH. 2015. Genome sequence of *Klebsiella pneumoniae* CICC10011, a promising strain for high 2,3-butanediol production. *Genome Announcements* 3(4): e00802–00815.

103. Tournier V, Topham CM, Gilles A, David B, Folgoas C, et al. 2020. An engineered PET depolymerase to break down and recycle plastic bottles. *Nature* 580(7802): 216–219.

104. Vardon DR, Franden MA, Johnson CW, Karp EM, Guarnieri MT, et al. 2015. Adipic acid production from lignin. *Energy & Environmental Science* 8(2): 617–628.

105. VILSA. 2019. rPET – 100% recyceltes PET. Retrieved from https://vilsa.de/nachhaltigkeit/rpet/?gclid=CjwKCAjw6fCCBhBNEiwAem5SOyD4vfsr3xV2l7m1skJi1zdP_t8bBVBBos5SDruMbo7pM7sar4DjMxoCS5gQAvD_BwE.

106. Vivijs B, Moons P, Geeraerd AH, Aertsen A, Michiels CW. 2014. 2,3-Butanediol fermentation promotes growth of *Serratia plymuthica* at low pH but not survival of extreme acid challenge. *International Journal of Food Microbiology* 175: 36–44.

107. Völksch B, Weingart H. 1997. Comparison of ethylene-producing *Pseudomonas syringae* strains isolated from kudzu (*Pueraria lobata*) with *Pseudomonas syringae* pv. *phaseolicola* and *Pseudomonas syringae* pv. *glycinea*. *European Journal of Plant Pathology* 103(9): 795–802.

108. Vollmer I, Jenks MJF, Roelands MCP, White RJ, van Harmelen T, et al 2020. Beyond mechanical recycling: Giving new life to plastic waste. *Angewandte Chemie Internationale Edition Englisch.*

109. Wang J-P, Wu L-X, Xu F, Lv J, Jin H-J, et al. 2010. Metabolic engineering for ethylene production by inserting the ethylene-forming enzyme gene (*efe*) at the 16S rDNA sites of *Pseudomonas putida* KT2440. *Bioresource Technology* 101(16): 6404–6409.

110. Weber CJ, Haugaard V, Festersen R, Bertelsen G. 2002. Production and applications of biobased packaging materials for the food industry. *Food Additives and Contaminants* 19(supl): 172–177.

111. Wei R, Tiso T, Bertling J, O'Connor K, Blank LM, et al. 2020. Possibilities and limitations of biotechnological plastic degradation and recycling. *Nature Catalysis.*

112. Wei R, Zimmermann W. 2017. Microbial enzymes for the recycling of recalcitrant petroleum-based plastics: How far are we? *Microbial Biotechnology* 10(6): 1308–1322.

113. Weingart H, Ullrich H, Geider K, Völksch B. 2001. The role of ethylene production in virulence of *Pseudomonas syringae* pvs. *glycinea* and. *Phaseolicola. Phytopathology* 91(5): 511–518.

114. Wilkes RA, Aristilde L. 2017. Degradation and metabolism of synthetic plastics and associated products by *Pseudomonas* sp.: Capabilities and challenges. *Journal of Applied Microbiology* 123(3): 582–593.

115. Xiao Z, Xu P. 2007. Acetoin metabolism in bacteria. *Critical Reviews in Microbiology* 33(2): 127–140.

116. Xie N-Z, Li J-X, Song L-F, Hou J-F, Guo L, et al. 2015. Genome sequence of type strain *Paenibacillus polymyxa* DSM 365, a highly efficient producer of optically active (R,R)-2,3-butanediol. *Journal of Biotechnology* 195: 72–73.

117. Xu Y, Chu H, Gao C, Tao F, Zhou Z, et al. 2014. Systematic metabolic engineering of *Escherichia coli* for high-yield production of fuel bio-chemical 2,3-butanediol. *Metabolic Engineering* 23: 22–33.

118. Xu Y, Wang A, Tao F, Su F, Tang H, et al. 2012. Genome sequence of *enterobacter cloacae* subsp. *dissolvens* SDM, an efficient biomass-utilizing producer of platform chemical 2,3-butanediol. *Journal of Bacteriology* 194(4): 897.

119. Yang T, Rao Z, Zhang X, Xu M, Xu Z, et al. 2015. Economic conversion of spirit-based distillers' grain to 2,3-butanediol by. *Bacillus Amyloliquefaciens. Process Biochemistry* 50(1): 20–23.

120. Yang T, Rao Z, Zhang X, Xu M, Xu Z, et al. 2017. Metabolic engineering strategies for acetoin and 2,3-butanediol production: Advances and prospects. *Critical Reviews in Biotechnology* 37(8): 990–1005.

121. Yim H, Haselbeck R, Niu W, Pujol-Baxley C, Burgard A, et al. 2011. Metabolic engineering of *Escherichia coli* for direct production of 1,4-butanediol. *Nature Chemical Biology* 7(7): 445–452.

122. Yoshida S, Hiraga K, Takehana T, Taniguchi I, Yamaji H, et al. 2016. A bacterium that degrades and assimilates poly (ethylene terephthalate). *Science* 351(6278): 1196–1199.

123. Zahoor A, Otten A, Wendisch VF. 2014. Metabolic engineering of *Corynebacterium glutamicum* for glycolate production. *Journal of Biotechnology* 192: 366–375.

124. Zhang Z, Yang Y, Wang Y, Gu J, Lu X, et al. 2020. Ethylene glycol and glycolic acid production from xylonic acid by. *Enterobacter Cloacae. Microbial Cell Factories* 19(1): 89.

125. Zheng Y, Yanful EK, Bassi AS. 2005. A review of plastic waste biodegradation. *Critical Reviews in Biotechnology* 25(4): 243–250.

126. Zhou S, Shanmugam KT, Yomano LP, Grabar TB, Ingram LO. 2006. Fermentation of 12% (w/v) glucose to 1.2 m lactate by *Escherichia coli* strain SZ194 using mineral salts medium. *Biotechnology Letters* 28(9): 663–670.

127. Zikmanis P, Kolesovs S, Semjonovs P. 2020. Production of biodegradable microbial polymers from whey. *Bioresources and Bioprocessing* 7(1): 36.

CHAPTER 25

Food Waste as a Valuable Carbon Source for Bioconversion – How Microbes do Miracles

Rajat Kumar, Varsha Bohra, Manu Mk, and Jonathan W. C. Wong
Institute of Bioresource and Agriculture and Department of Biology, Hong Kong Baptist University, Kowloon Tong, Hong Kong

25.1 INTRODUCTION

Food is an integral component of life and is essentially required by all living organisms for their survival. The complexity of food consumed varies among life forms at different levels, i.e., from micro-organisms relying on simpler forms of carbohydrates, fats, and proteins to higher eukaryotes like humans feeding on complex food such as vegetables, pulses, meat, and dairy products (Sharma et al., 2020). However, as the world population is rising rapidly, so is the food mismanagement and waste generation by humans. Sources include residential (kitchens), commercial (restaurants), institutional (academic, cafeteria), and industrial (food processing plants), all generating high amounts of food waste. According to the latest UN report, an estimated 17% of food generated globally is wasted each year. This adds up to an astonishing 1.03 billion tons of food (UNEP, 2021). Such a huge volume of food wastage further contributes to an estimated 8–10% of the greenhouse gases released as global emissions. Food waste generation also depletes other resources such as water, fertile land, labor, and energy. This sheds light on the possible impacts of food waste management choices over economy, climate, biodiversity, and health in all continents (Kaza et al., 2018).

According to the Environmental Protection Agency (EPA), food waste can be tracked and classified into two types. The pre-consumption food waste is any fresh produce or associated inedible parts which are not considered optimal and hence removed from the human food supply chain, while post-consumption food waste is the cooked food after our meal either at home or in the restaurants (EPA, 2015). Once discarded, food waste becomes a major fraction of municipal solid waste, that can go up to 30–50% in some countries. Traditionally, this waste is disposed of by landfilling or incineration that will permanently remove it from the world's nutrient pool. At the same time, it may release highly toxic chemicals and gases in the air, which can cause respiratory and neurological diseases. Unlike other types of wastes, food waste contains high amounts of trapped nutrients (starch, protein, fat, cellulose) and moisture, which makes it rot and decompose easily (Palaniveloo et al., 2020). The breakdown also leaches out these nutrients into nearby soil and water, making them unfit for future use.

Numerous national and international programs are now being established to find economically and ecologically sustainable alternatives to manage food waste with the aim to complete the nutrient cycle and maintain a

Good Microbes in Medicine, Food Production, Biotechnology, Bioremediation, and Agriculture, First Edition. Edited by Frans J. de Bruijn, Hauke Smidt, Luca S. Cocolin, Michael Sauer, David Dowling, and Linda Thomashow.

circular economy. The global framework of the United Nations (UN) further emphasizes these aspects in all 17 Sustainable Development Goals (SDGs). Moreover, the food waste problems are more explicitly targeted in Goal 2 (zero hunger, food security, improved nutrition, and sustainable agriculture) and Goal 12 (responsible consumption and production) (Fleetwood, 2020). Both of these aim at maintaining sustainable production of food along with ecosystem protection, improving land quality, and minimizing waste pollution. For successful attainment of these goals, one approach is through following microbiological treatment of food waste for biotransformation into environmentally friendly products (Wang, 2013). Composition of food waste is mainly attributed as 30–60% carbohydrate, 5–10% proteins and 10–40% lipids (w/w) (Pleissner and Lin, 2013). Infused micronutrients such as Ca^{2+}, K^{+}, Na^{+}, Mg^{2+}, Fe^{3+}, Mn^{2+}, Zn^{2+}, P, and S are also found in high quantities (Hafid et al., 2017). This allows food waste to work as an accessible source of carbon, nitrogen, and other essential elements for bioconversion. The initial transformation is mainly carried out by microorganisms including bacteria and fungi which break down the complex molecules of food waste into smaller particles (Luo et al., 2019; Tsui and wong, 2019). The smaller particles are simultaneously utilized by function specific microorganisms under different settings, resulting in recovery of by-products with different applications. These microbial end-products include (i) biofertilizers through composting, (ii) bioenergy through anaerobic fermentation, and (iii) bio-based compounds (biopolymers, biosurfactants, biocatalysts, etc.) through aerobic fermentation (Wang, 2013).

Food waste inherently becomes colonized by diverse microbial populations which secrete a range of hydrolytic enzymes to decompose food waste into high-quality organic fertilizer with elevated humic content, which is preferred for agricultural use as they are free from chemical fertilizers and pesticides. The bacteria and fungi of this microbiome mainly belong to the genus *Bacillus*, *Brevibacillus*, *Pseudomonas*, *Staphylococcus*, *Mycobacterium*, and *Streptomyces*, *Fusarium*, *Aspergillus*, *Thermo*, respectively, to name a few (Franke-Whittle et al., 2014; Palaniveloo et al., 2020; Wu et al., 2018). Potential microbes of these genus are also capable of producing bioenergy (biomethane) when subjected to anaerobic fermentation. Bacteroidetes, firmicutes, clostridium species, and methanogens are key microbes associated with renewable energy production (Wang et al., 2020). Further, fermentative bacteria can be used to derive value added compounds. The diversity for these compounds ranges from active precursors formation (volatile fatty acids, alcohols) (Wang et al., 2020), bioplastic production (polyhydroxyalkanoates) (Nielsen et al., 2017; see also chapter 24), and purified enzymes (amylase, cellulase, lipase, protease) (Pleissner and Lin, 2013; Tišma et al., 2019) to surface active agents (rhamnolipids, sophorolipids) (Kieliszek et al., 2020; Pérez-Armendáriz et al., 2019).

These important microbial communities involve both culturable and non-culturable strains. Furthermore, out of the total microbial population, it is estimated that only 1% of bacteria and 5% of fungi have been explored in laboratories (Du et al., 2011; Torres-León et al., 2018). Therefore, further research and use of microorganisms has promising potential to safely utilize food waste with high efficiency and convert trash to treasure. This can contribute toward decreasing waste accumulation, reducing dependency on fossil fuels, and inducing bioeconomy within societies.

25.2 BIOFERTILIZERS

Biofertilizers are natural compounds having microbial inoculants mixed with biodegradable substances that promote plant growth through seed or soil interactions. This complex living mixture is prepared through the process known as composting. It is defined as an aerobic process mediated by microorganisms that implicates the hydrolytic conversion of food waste into a stabilized and sanitized product called biofertilizer (Figure 25.1) (Selvam et al., 2021). Food waste used during composting consists of fruits, vegetables, grains, eggshells, and dairy. Microorganisms such as bacteria, fungi, and actinomycetes decompose the biodegradable contents present in food waste by utilizing carbon and nitrogen as the energy sources along with oxygen and moisture. The microbial activities result in the production of water, heat, carbon dioxide, and compost enriched with humic substances. During this bioprocess, the temperature in the composting pile rises and helps to destroy pathogenic microorganisms, making the produced compost safer for use (Manu et al., 2017).

Typically, the composting process includes three phases which are classified based on the temperature of the composting pile (Wong et al., 2016) (Figure 25.2). The initial stage is the mesophilic phase where the microorganisms start their activities by utilizing simple compounds such as sugars, amino acids, etc. The microorganisms responsible for this initial

FIGURE 25.1 Food waste degradation by microorganisms under the composting process to generate biofertilizer.

beginning are mostly mesophilic bacteria (*Leuconostoc paramesenteroides*, *Pediococcus acidilactici*, and *Staphylococcus piscifermentans*) and to a lesser extent the fungi (*Aspergillus*, *Chaetomium*, *Humicola*, *Mucor*, *Penicillium*, and *Thermomyces*). As a result of these microbial activities, heat will be produced and the temperature of the composting pile increases rapidly. This temperature elevation above 45 °C starts a second phase of composting, i.e., thermophilic phase. During this phase, the activities of mesophiles will subside due to destruction or inability to multiply at higher temperature. Hence, a new set of microorganisms called thermophilic bacteria (*Bacillus*, *Terrabacter*, *Aneurinibacillus*, and *Brevibacillus*) and fungi (*Scytalidium*, *Rhizomucor*, *Paecilomyces*, *Symbiobacterium*, and *Thermomyces*) become active which are tolerant to elevated temperature up to 55 to 65 °C but a temperature of higher than 70 °C will cause the destruction of beneficial microorganisms. Unlike mesophiles, thermophiles have the ability to degrade complex organic matter such as cellulose, hemicellulose, lignin, and fats. As a result, significant reduction in organic carbon is expected after this stage. Further, the microbial activities will reduce due to the unavailability of degradable substrates which will result in the reduction of temperature toward ambient level and this phase is called the cooling phase. In this phase, some of the mesophiles will become activated again to consume the remaining organic substrates in the composting pile. The dominance in bacterial group shifts from phylum Fermicutes to phylum Bacteoroidetes. During the cooling phase, the polymerization of organic compounds takes place to form complex organic structures called humic substances (Rawat and Johri, 2013; Selvam et al., 2021). The presence of humic substances act as a soil enhancer and helps in plant growth when compost is applied into the soil.

If food waste is discarded or dumped anywhere, the microbial activities will start naturally, however, irresponsible disposal of food waste causes several issues

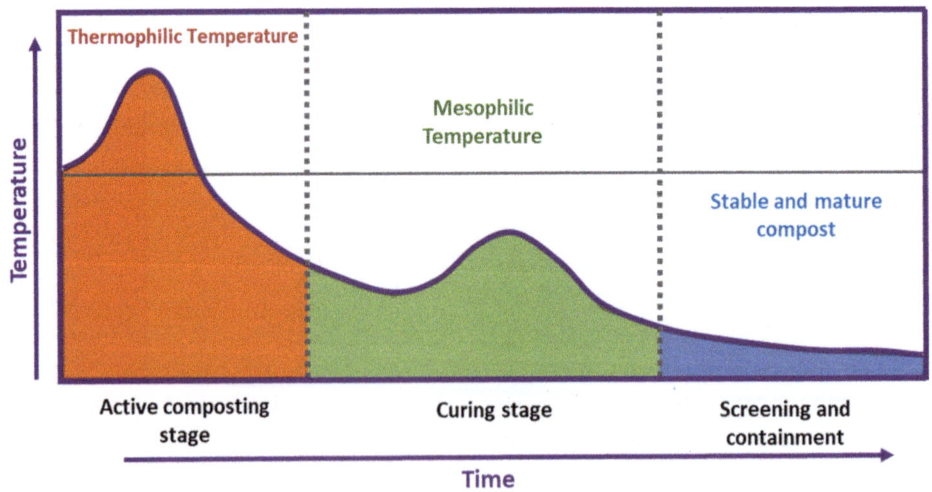

FIGURE 25.2 Typical composting stages observed during composting (Source: Wong et al., 2017).

like odor nuisance, greenhouse gas emission, attracting rodents, and delayed degradation process. Hence, treatment of food waste under controlled and optimized conditions is necessary which allows good quality biofertilizer/compost to be produced with minimum environmental issues.

To improve the biodegradation rate of food waste, significant research on microbial function has also been carried out over the last decade (Rastogi et al., 2020). Effective microbe (EM) cultures have been widely used to improve food waste composting. Commercially available EM cultures generally consist of a mixture of dominant and non-dominant microbial species including cellulolytic and lignocellulolytic microbes. Bacterial species such as *Pseudomonas, Bacillus, Cellulomonas*, and *Thermoactinomycetes* have been found to be capable of producing cellulase enzymes which are accountable for cellulose degradation in food waste. Similarly, fungal species such as *Trichoderma, Aspergillus*, and white-rot fungi secrete extracellular enzymes responsible for cellulose and lignin degradation during composting. Biofertilizers are also enriched with various nitrogen-fixing (*Azotobacter, Rhizobium, Anaebaena*; see chapter 47), phosphate solubilizing (*Pseudomonas* sp.), and phythohormone (auxin) producing bacteria to ensure good quality soil over long periods (Umesha et al., 2018).

Researchers have addressed several varieties of microbes to determine their efficacy in food waste composting, as listed in Table 25.1. Utilizing these in composting can be seen as a robust technology to treat the large quantity of food waste to produce nutrient rich biofertilizer. By composting food waste, a closing loop approach of nutrient recycling could be achieved for improving soil quality with minimum environmental issues.

25.3 BIOENERGY

Food waste, due to its high organic content and better biodegradability, exhibits high potential for the production of bioenergy in the anaerobic digestion (AD) process. AD is one of the major biological-based approaches for organic waste management that has received considerable interests owing to its high valorization value for waste. The end-product of the process is biogas (methane) that can be harvested as renewable energy and a nutrient-rich digestate that can be transformed into biofertilizer for nutrient reclamation (Figure 25.3) (Bong et al., 2018). AD is a multi-stage biochemical process in which the complex organic materials undergo a series of stages involving a completely diverse assemblage of bacteria and methanogenic archaea/ microbial community at every stage (Jang et al., 2015).

Conversion of the complex organic compound to methane is a multistep process which involves the interaction of different groups of organisms which divides the AD process into four steps, i.e., hydrolysis, acidogenesis, acetogenesis, and methanogenesis.

25.3.1 Hydrolysis

This phase includes the breaking of undissolved compounds like cellulose, protein, and fat into monomers like polysaccharides, fatty acids, and peptides by exoenzymes of facultative and obligatory anaerobic bacteria. These exoenzymes include cellulase, protease, and lipase. This step results in end product formation of soluble saccharides, amino acids, glycerol, and long chain fatty acids. The equation for this stage can be represented as:

$$C_6H_{10}O_5 + 2\,H_2O \rightarrow C_6H_{12}O_6 + 2H_2 \quad [1]$$

TABLE 25.1 Summary of studies on microbial additives during food waste composting

Microbial Additive	Effect on Composting	References
Ligno-cellulolytic consortium	Improved compost maturity	Xu et al. (2019)
White-rot fungi (*Phanerochaete chrysosporium, Trametes versicolor, and Fomes fomentarius*)	Accelerated degradation and increased germination index (maturity indicator)	Voberkova et al. (2017)
Thermotolerant cellulolytic, proteolytic, amylolytic, and oil degrading bacterial strains	Increased organic matter degradation, production of mature compost in short duration	Awasthi et al. (2017)
Lactic acid bacteria, yeast, and phototrophic bacteria	Shortening of composting duration to produce non-phytotoxic and matured compost	Manu et al. (2017)
Cellulolytic, thermophilic, actinomycetes	Increased content of humic substances	Zhao et al. (2017)

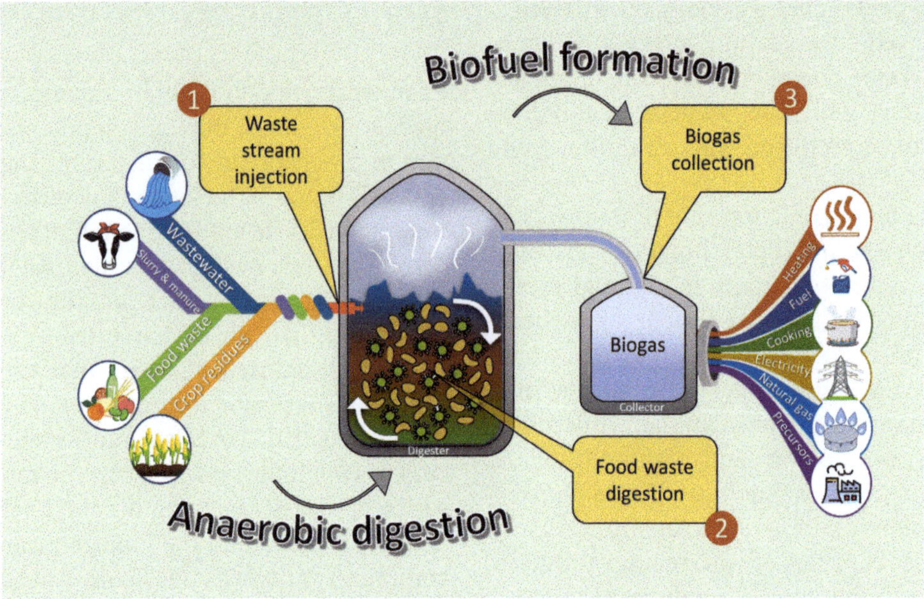

FIGURE 25.3 Food waste degradation through anaerobic digestion and bioenergy formation for various applications.

Hydrolysis is a slow process and acts as the rate-limiting stage of AD in the degradation of recalcitrant compounds like lignocellulose using hydrolase or lyases enzymes to convert a complex polymer into the monomer. *Firmicutes* and *Bacteroides* are the two phyla mostly consisting of hydrolytic bacteria along with genera *Clostridium*, *Bacteroides*, *Acetivibrio*, and *Thermotoga* (Strauber et al., 2012). It has been reported that complex cellulose can be degraded by *Clostridium lochhadii, Butyrivibrio fibrosolvens, Ruminococcus albus, Clostridium stercorarium, Clostridium thermoculum, Bacteroides succinogenes, Clostridiumcelobioporus, Ruminococcus flavefaciens*, and *Micromonospora bispora*. Genera *Enterobacterium* and *Streptococcus* are mainly associated with complex polymer degradation.

25.3.2 Acidogenesis

The monomers produced in hydrolysis are consumed by various obligate and facultative anaerobic bacteria and degraded to short chain organic acids (C1 to C5) (like acetate) in this phase. Hydrolyzed products are converted into further simple molecules of low molecular weight components which include CO_2, ammonia, and organic acid alcohols. The reactions for this step can be represented by the following equations.

$$C_6H_{12}O_6 \rightarrow 2CH_3CH_2OH+2CO_2 \quad [2]$$

$$C_6H_{12}O_6 \rightarrow 2CH_3CHOHCOOH \quad [3]$$

The acid phase bacteria are mostly facultative anaerobes which include genera *Flavobacterium, Bacillus, Clostridium, Pseudomonas*, and *Micrococcus* (Ali Shah et al., 2014). Acidogens belong to major phyla *Proteobacteria, Chloroflexi, Bacteroidetes, Actinobacteria, Thermotogae*, and *Firmicutes* (Yamada et al., 2005).

25.3.3 Acetogenesis

Acetogenic bacteria use products from the acidogenic phase and convert them into H_2, CO_2, and acetate when hydrogen partial pressure is low, and when pressure is higher butyric, propionic, valeric, etc. acids are formed. Homoautotropic acidogenic bacteria produce acetic acid by using hydrogen and carbon dioxide. The equation for this step can be represented by the following equations.

$$CH_3CH_2COOH+2H_2O \rightarrow CH_3COOH+CO_2+3H_2 \quad [4]$$

$$CH_3CH_2CH_2COOH+2H \rightarrow 2CH_3COOH+2H_2 \quad [5]$$

$$CH_3CH_2OH \rightarrow CH_3COOH+2H \quad [6]$$

This step results in the formation of the substrate which is required for methanogenesis.

Syntrophic acetogenes play an important role in this stage to convert long chain fatty acids into acetate. Oxidation of propionate is particularly important since the excess of propionate can cause an inhibitory effect on the AD process. Hydrogen partial pressure below

10^{-4} atm is a necessary requirement to provide favorable conditions for syntrophic acetogenesis (Lowe et al., 1993). Genera *Syntrophobacte, Smithllela, Pelotomaculum*, and *Syntrophomonas* are mainly acetic acid producers, while *Syntrophus* and *Syntrophomonas* are the well-known genera for oxidizing butyrate and long chain fatty acid into acetate (Sousa et al., 2007). *Methanobacterium suboxydans* and *Methanobacterium propionicum* are responsible for the conversion of pentanoic acid to propionic acid and propionic acid to acetic acid, respectively. Study of syntrophic acetogenes has a major limitation as most of them are non-cultivable therefore in-depth study requires other molecular approaches like genomics and metagenomics.

25.3.4 Methanogenesis

In the final stage, the obligatory anaerobes involved in methane generation decompose lower molecular weight compounds (CH_3COOH, H_2, CO_2) to produce methane (CH_4) and CO_2. The group of microbes called acetoclastic methanogens, hydrogenotrophic methanogens, methylotrophic methanogens all belonging to the domain *Archaea* are mainly involved in this step. Acetoclastic methanogens convert acetic acid into methane while hydrogenotrophic methanogens reduce CO_2 and hydrogen into methane and they show proximity with syntrophic acetogenic bacteria. Methylotrophic methanogens produce methane from methanol, methylamines like one carbon containing compounds. The methane production reaction can be shown in the following equations:

$$CH_3COOH \rightarrow CH_4 + CO_2 \qquad [7]$$

$$2CH_3CH_2OH \rightarrow CH_4 + 2CH_3COOH \qquad [8]$$

$$4H_2 + CO_2 \rightarrow CH_4 + 2H_2O \qquad [9]$$

Co-enzyme-M and F_{420} play important role in conversion of CO and formate into methane. Acetate and carbonyl transformation in the process of methane formation is also carried out by co-enzyme-M (Appels et al., 2010).

Methanogens constitute the largest defined group within the archaeal domain. They are immensely diverse at phylogenetic level and consistent with the extremes of anaerobic habitat in which they flourish (Bapteste et al., 2005). Methanogens can grow at a pH ranging from 4.5 to 9.5, and temperatures ranging from 0 °C to 110 °C (Schleper et al., 2005). Most species reduce CO_2 to CH_4 with formate, hydrogen, or carbon monoxide (CO_2 reduction pathway). A few species convert the methyl group of acetate (Acetolactic pathways), or one carbon substrate methanol, methyl sulfide, or methylamines (Methylotrophic pathway) to methane (Liu et al., 2008). It is also found that different methanogens have different biogas production capacities depending on substrate availability (Ferry JG, 2008). *Methanosaeta, Methanosarcina Methanobrevibacter, Methanobacterium*, and *Methanothermobacter* are commonly found in the AD process. *Methanospirillum, Methanobacterium, Methanothermobacter, Methanoculleus*, and *Methanobrevibacter* are common hydrogenotrophic methanogens (Venkiteshwaran et al., 2015). Genera *Methanosarcina and Methanosaeta* are facultative and obligate acetoclastic methanogens, respectively. *Methanosaeta* shows maximum affinity with acetate as substrate and it has filamentous morphology with the ability to form granular aggregation. Some species may follow the route of hydrogenotrophic as well as methylotrophic for generation of methane (Conklin et al., 2006). Along with less affinity toward acetate, *Methanosarcina* shows faster growth rate so it out-competes *Methanosaeta* when acetate concentration prevails in the environment (Hori et al., 2006).

Increased demand for energy generation makes AD an attractive option for food waste reduction along with energy generation. AD plays a crucial role in reducing the greenhouse gas footprint of food waste by onsite capturing of CO_2 and CH_4 while contributing additional carbon credit through renewable energy generation. Additionally, functioning of AD requires lower energy input and footprint as compared to incineration, composting, and landfilling (Ventura et al., 2014).

25.3.5 Bio-products

The general interest of utilizing food waste as a substrate has also driven the use of microorganisms to produce products with medical, environmental, and industrial applications. The types of products vary on several fronts, ranging from chemical aspects to application viewpoints. Production of these microbial-based chemical products takes place via a process known as fermentation (Figure 25.4). The main aim of this process is to achieve maximum growth of cells and ultimately high product yields. Microorganisms generally follow a common pattern of growth which requires control over the substrate concentration and other environmental

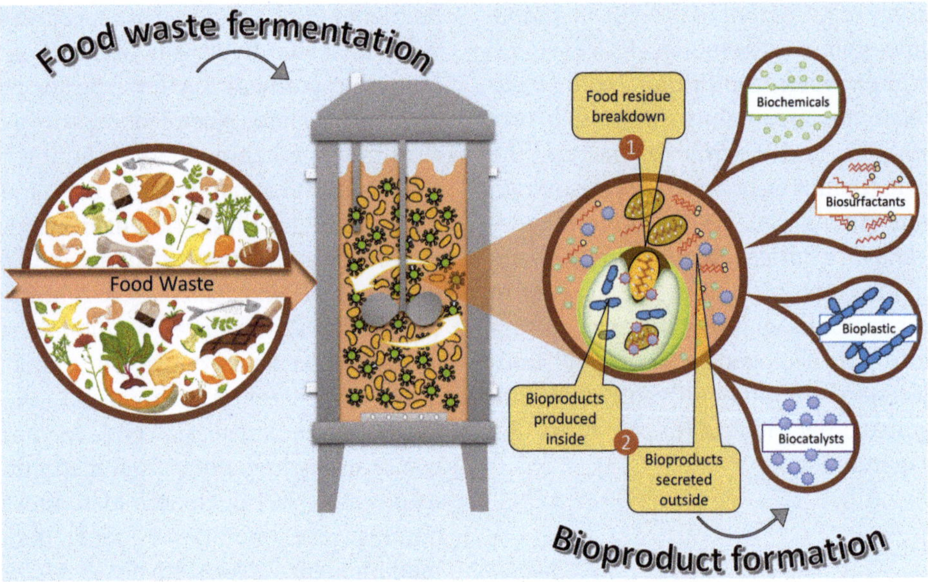

FIGURE 25.4 Microbial mediated bioconversion of food waste to numerous bioproducts.

conditions (oxygen, pH, temperature). During fermentation, optimum conditions are provided where the microbes acclimatize to the food by producing enzymes that they need to survive. This generates simpler forms of carbohydrates, proteins, and lipids which act as building blocks for bioproducts. Lactic acid bacteria (LBA) are one such group which highly specialize in this natural way of breakdown and bioconversion. *Streptococcus thermophilus*, *Lactococcus lactis*, *Leuconostoc* spp., and *Lactobacillus* spp. along with genera *Pediococcus*, *Oenococcus*, and *Weissella* play a pivotal role. Valuable fermented products have also been derived from members of genus *Bacillus*, *B. coagulans*, *B. amyloliquefaciens*, *B. licheniformis*, *B. cereus*, and *B. subtilis* (Sabater et al., 2020) alone or in combination with other bacteria and/or fungi.

25.3.6 Biochemicals

Biochemicals like organic acids (amino acids, lactic acid, succinic acid, and 3-Hydroxypropionic acid) and alcohols (ethanol, butanol, glycerol) have received significant attention as platform chemicals. They are used as precursors to many known petrochemical products, antimicrobial agents, and electrolyte solutions (Du et al., 2011). Most of them are natural products of microbes, or intermediates in major metabolic pathways. Numerous bacteria (*Lactobacillus, Actinobacillus, Aerobacter, Pseudomonus, Micrococcus, Bacillus, Brevibacterium, Corynebacterium, Arthrobacter*), fungi (*Penicillium, Fusarium, Aspergillus*), and yeast (*Saccharomyces, Zygosaccharomyces, Pichia, Torulaspora, Candida, Yarrowia*) are native producers of these chemicals (Ciani et al., 2008; Du et al., 2011; Naraian and Kumari, 2017). The choice of organism influences by-product formation, quality, and applicability. Approximately 85% of organic acid demand accounts for food industry applications as flavor elevators, condensers, humectants, and thickeners. Alcohols on the other hand are applicable in beverages, fuels, lubricants, solvents, and plasticizers (Fillet and Adrio, 2016). Laboratory experiments with bacteria have successfully demonstrated the use of fruit juice residues, potato starch, cassava powder, whey, waste banana, and fish discards for synthesis of biochemicals (Du et al., 2011; Novik et al., 2017).

25.3.7 Bioplastics

Microorganisms store carbon within the cell in the form of polyesters known as polyhydroxyalkanoates (PHAs). These polyesters on isolation, can be molded into polymers which have the ability to replace synthetic plastic in many sectors. They are biodegradable, non-toxic, resistant to ultraviolet light, and insoluble in water (Stavroula et al., 2020). Numerous vegetable residue fermentations have been used for culturing PHA synthesizing microbes. *Cupriavidus necator*, *Haloferax mediterranei*, *Azotobacter chroococcum*, *Burkholderia cepacian*, *Bacillus megaterium*, and *Alcaligenus eutrophus* are a few examples of bacteria which have progressed PHA synthesis (Nielsen et al., 2017; see chapter 24). Biological polymers based on PHAs can be used in various applications, including industrial (food packaging and bottling) (Poltronieri and Kumar, 2017), biomedical (drug carriers, bandages, tissue scaffolds) (Grigore et al., 2019), and agriculture (animal nets, grow bags, mulch films) (Amelia et al., 2019).

25.3.8 Biosurfactants

Biosurfactants are ecofriendly chemicals with surface active properties to disperse water insoluble compounds. They are synthesized on microbial cell surfaces or excreted into the environment. The structure is composed of two parts, a polar (hydrophilic) head and a nonpolar (hydrophobic) tail. Accordingly, biosurfactants are classified as glycolipids (rhamnolipids, sophorolipids), phospholipids, lipopeptides, and lipoproteins (Kapellos, 2017). Major microorganisms that produce biosurfactants are bacterial species of *Bacillus subtilis*, *Klebsiella oxitoca*, *Pseudomonas aeruginosa*, and *Streptococcus thermophilus*, (Khire, 2010) and fungal species of *Candida apicola*, *Candida batistae*, and *Starmerella bombicola* (Claus and Van Bogaert, 2017). These microbes have made biosurfactant production more economical through entrapping nutrition from sources such as waste cooking oil (Pérez-Armendáriz et al., 2019), bakery waste (Patowary et al., 2018), lignocellulosic, and dairy waste (Abdel-Mawgoud and Stephanopoulos, 2018). Furthermore, biological surfactants have found uses in industrial (food ingredients, cosmetics, lubricants, detergents) (Drakontis and Amin, 2020), pharmaceutical (anti-microbials, anti-tumors, anti-adhesives) (Rodrigues et al., 2006), and environmental (bioremediation, enhanced oil recovery, metal extraction) (Mulligan, 2005; Pacwa-Plociniczak et al., 2011) sectors.

25.3.9 Biocatalysts

Biocatalysts are an important category of compounds derived from microbes in the form of enzymes. They catalyze numerous biochemical and metabolic reactions at rates 10 billion times higher than normal chemical reactions. Most of the microorganisms that produce these enzymes require stable nutrient rich conditions. The reactions catalyzed by these enzymes are complex and hence we categorize them into various types i.e., carbohydrases (xylanases, amylases, maltase, invertase, cellulase), proteases, lipases, and catalase (Tamang et al., 2016). Some prominent bacterial producers of enzymes are *Bacilli*, *Geobacillus*, *Thermoactinomyces*, *Pseudomonas*, and *Clostridium*. The diversity of enzyme producers is more prominent among fungal species such as *Polyporus sp.*, *Pleurotus sp.*, *Trichoderma sp.*, *Aspergillus sp*, *Candida sp.*, *Penicillium sp.*, *Rhizopus*, and *Mucor* (Gurung et al., 2013; Nigam, 2013). Fermentation of soya bean, wheat bran, brewer's spent grain, olive pomaces, and other starchy residues provide stable nutrient rich environments to synthesize enzymes (Sabater et al., 2020; Torres-León et al., 2018). Their technological applications are explored in many areas of agriculture, food processing, textile, wood processing, cosmetics, and environmental pollution control (Liu and Kokare, 2017).

Bio-based products provide a cleaner and safer approach against petrochemical based products. Microorganisms used for fermentation are part of Qualified Presumption of Safety and generally recognized as safe (GRAS) groups (Du et al., 2011; Torres-León et al., 2018). Therefore, they can be freely used without fear of causing any harm to humans or the environment. Researchers are now working in various fronts to elevate efficiency and usability of untapped microbes for food waste valorization on a global scale.

25.4 CONCLUSIONS

The increase in population and urbanization has led to elevated food waste generation along with its mismanagement where the uncontrolled accumulation instigates a huge stress. Traditional approaches to dealing with food waste involves landfilling and/or incineration that perpetuates serious health and environmental hazards. This works against the ambitious goals targeted for sustainable development by the UN. In contrast, biological treatment can provide safer alternatives to this global problem, all in alignment with SDGs. The ability to transform waste food trash into treasure greatly relies on the use of microorganisms. Evolution of these tiny life forms has driven their skills to secrete various factors that facilitate bioconversion and use it as a strong resource. The released nutrients are safely inherited within the microbial cell and assembled into bio-products and bioenergy. These further act as green alternatives to conventional petrochemical based products inducing a circular bio-economy. Reliance on such ecofriendly materials could prove to be a promising path toward elevating productivity, maintaining ecosystems, strengthening climate protection, and promoting land quality, which will benefit both people and the planet.

REFERENCES

1. Abdel-Mawgoud AM, Stephanopoulos G. 2018. Simple glycolipids of microbes: Chemistry, biological activity and metabolic engineering. *Synthetic and Systems Biotechnology* 3: 3–19.
2. Amelia TSM, Govindasamy S, Tamothran AM, Vigneswari S, Bhubalan K. 2019. Applications of PHA in Agriculture. In: Kalia VC (Ed.), *Biotechnological Applications of Polyhydroxyalkanoates.* Springer Singapore, Singapore, pp. 347–361.
3. Appels L, Degrève J, Van Der Bruggen B, Van Impe J, Dewil R. 2010. Influence of low temperature thermal pre-treatment on sludge solubilisation, heavy metal release and anaerobic digestion. *Bioresource Technology* 101: 5743–5748.
4. Awasthi MK, Selvam A, Lai KM, Wong JWC. 2017. Critical evaluation of post-consumption food waste composting employing thermophilic bacterial consortium. *Bioresource Technology* 245: 665–672.
5. Bapteste É, Brochier C, Boucher Y. 2005. Higher-level classification of the Archaea: Evolution of methanogenesis and methanogens. *Archaea* 1: 353–363.
6. Bong CPC, Lim LY, Lee CT, Klemeš JJ, Ho CS, et al. 2018. The characterisation and treatment of food waste for improvement of biogas production during anaerobic digestion–A review. *Journal of Cleaner Production* 172: 1545–1558.
7. Ciani M, Comitini F, Mannazzu I. 2008. Fermentation. In: Jørgensen SE, Fath BD (Eds.), *Encyclopedia of Ecology.* Academic Press, Oxford, pp. 1548–1557.
8. Claus S, Van Bogaert INA. 2017. Sophorolipid production by yeasts: A critical review of the literature and suggestions for future research. *Applied Microbiology and Biotechnology* 101: 7811–7821.
9. Conklin A, Stensel HD, Ferguson J. 2006. Growth kinetics and competition between Methanosarcina and Methanosaeta in mesophilic anaerobic digestion. *Water Environment Research* 78: 486–496.
10. Drakontis CE, Amin S. 2020. Biosurfactants: Formulations, properties, and applications. *Current Opinion in Colloid & Interface Science* 48: 77–90.
11. Du J, Shao Z, Zhao H. 2011. Engineering microbial factories for synthesis of value-added products. *Journal of Industrial Microbiology & Biotechnology* 38: 873–890.
12. EPA. 2015. Reducing wasted food & Packaging: A guide for food services and restaurants. United States Environmental Protection Agency, Pacific Southwest, United States. Available at https://www.epa.gov/sites/production/files/2015-08/documents/reducing_wasted_food_pkg_tool.pdf accessed 21 March 2021.
13. Ferry J. 2008. Chapter 13 - Acetate-based methane production. In: Wall J, Harwood C, Demain A (Eds.), *Bioenergy*. ASM Press, Washington, DC, pp. 155–170.
14. Fillet S, Adrio JL. 2016. Microbial production of fatty alcohols. *World Journal Microbiol Biotechnol* 32: 152.
15. Fleetwood J. 2020. Social justice, food loss, and the sustainable development goals in the era of COVID-19. *Sustainability* 12: 5027.
16. Franke-Whittle IH, Confalonieri A, Insam H, Schlegelmilch M, Körner I. 2014. Changes in the microbial communities during co-composting of digestates. *Waste Management* 34: 632–641.
17. Grigore ME, Grigorescu RM, Iancu L, Ion RM, Zaharia C, et al. 2019. Methods of synthesis, properties and biomedical applications of polyhydroxyalkanoates: A review. *Journal of Biomaterials Science, Polymer Edition* 30: 695–712.
18. Gulhane M, Pandit P, Khardenavis A, Singh D, Purohit H. 2017. Study of microbial community plasticity for anaerobic digestion of vegetable waste in Anaerobic Baffled Reactor. *Renewable Energy* 101: 59–66.
19. Gurung N, Ray S, Bose S, Rai V. 2013. A broader view: Microbial enzymes and their relevance in industries, medicine, and beyond. *BioMed Research International* 2013: 329121.
20. Hafid HS, Nor 'Aini AR, Mokhtar MN, Talib AT, et al. 2017. Over production of fermentable sugar for bioethanol production from carbohydrate-rich Malaysian food waste via sequential acid-enzymatic hydrolysis pretreatment. *Waste Management* 67: 95–105.
21. Hori T, Haruta S, Ueno Y, Ishii M, Igarashi Y. 2006. Dynamic transition of a methanogenic population in response to the concentration of volatile fatty acids in a thermophilic anaerobic digester. *Applied and Environmental Microbiology* 72: 1623–1630.
22. Jang HM, Ha JH, Park JM, Kim MS, Sommer SG. 2015. Comprehensive microbial analysis of combined mesophilic anaerobic–thermophilic aerobic process treating high-strength food wastewater. *Water Research* 73: 291–303.
23. Kapellos GE. 2017. Chapter 2 - Microbial strategies for Oil biodegradation. In: Becker SM (Ed.), *Modeling of Microscale Transport in Biological Processes.* Elsevier Academic Press, Cambridge, MA, USA, pp. 19–39.
24. Kaza S, Yao LC, Bhada-Tata P, Van Woerden F. 2018. What A Waste 2.0: a Global Snapshot of Solid Waste

Management to 2050. World Bank, Washington, DC. https://openknowledge.worldbank.org/handle/10986/30317License:CCBY3.0IGO, accessed on 15 March 2021.

25. Khire JM. 2010. Bacterial biosurfactants, and their role in microbial enhanced oil recovery (MEOR). *Advances in Experimental Medicine and Biology* 672: 146–157.

26. Kieliszek M, Piwowarek K, Kot AM, Pobiega K. 2020. The aspects of microbial biomass use in the utilization of selected waste from the agro-food industry. *Open Life Sciences* 15: 787–796.

27. Liu X, Kokare C. 2017. Chapter 11 - Microbial enzymes of use in industry. In: Brahmachari G (Ed.), *Biotechnology of Microbial Enzymes*. Academic Press, Cambridge, MA, USA, pp. 267–298.

28. Liu Y, Whitman WB. 2008. Metabolic, phylogenetic, and ecological diversity of the methanogenic archaea. *Annals of the New York Academy of Sciences* 1125: 171–189.

29. Lowe SE, Jain MK, Zeikus JG. 1993. Biology, ecology, and biotechnological applications of anaerobic bacteria adapted to environmental stresses in temperature, pH, salinity, or substrates. *Microbiology and Molecular Biology Reviews* 57: 451–509.

30. Luo L, Kaur G, Wong JWC. 2019. A mini-review on the metabolic pathways of food waste two-phase anaerobic digestion system. *Waste Management & Research* 37: 333–346.

31. Manu MK, Kumar R, Garg A. 2017. Performance assessment of improved composting system for food waste with varying aeration and use of microbial inoculum. *Bioresource Technology* 234: 167–177.

32. Mulligan CN. 2005. Environmental applications for biosurfactants. *Environmental Pollution* 133: 183–198.

33. Naraian R, Kumari S 2017. Microbial production of organic acids. In: Microbial Functional Foods and Nutraceuticals, pp. 93–121.

34. Nielsen C, Rahman A, Rehman A, Walsh M, Miller C. 2017. Food waste conversion to microbial polyhydroxyalkanoates. *Microbial Biotechnology* 10: 1338–1352.

35. Nigam PS. 2013. Microbial enzymes with special characteristics for biotechnological applications. *Biomolecules* 3: 597–611.

36. Novik G, Meerovskaya O, Savich V 2017. Waste degradation and utilization by lactic acid bacteria: Use of lactic acid bacteria in production of food additives. Bioenergy and Biogas.

37. Pacwa-Plociniczak M, Plaza GA, Piotrowska-Seget Z, Cameotra SS. 2011. Environmental applications of biosurfactants: Recent advances. *International Journal of Molecular Sciences* 12: 633–654.

38. Palaniveloo K, Amran MA, Norhashim NA, Mohamad-Fauzi N, Peng-Hui F, et al. 2020. Food waste composting and microbial community structure profiling. *Processes* 8: 723.

39. Patowary K, Das M, Patowary R, Kalita M, Deka S. 2018. Recycling of bakery waste as an alternative carbon source for rhamnolipid biosurfactant production. *Journal of Surfactants and Detergents* 22: 373–384.

40. Pérez-Armendáriz B, Cal-y-mayor-luna C, El-Kassis EG, Ortega-Martínez LD. 2019. Use of waste canola oil as a low-cost substrate for rhamnolipid production using. *Pseudomonas Aeruginosa. AMB Express* 9: 61.

41. Pleissner D, Lin C. 2013. Valorisation of food waste in biotechnological processes. *Sustainable Chemical Processes* 1: 21.

42. Poltronieri P, Kumar P. 2017. Polyhydroxyalcanoates (PHAs) in Industrial Applications. In: Martínez LMT, Kharissova OV, Kharisov BI (Eds.), *Handbook of Ecomaterials*. Springer International Publishing, Cham Switzerland, pp. 1–30.

43. Rastogi M, Nandal M, Khosla B. 2020. Microbes as vital additives for solid waste composting. *Heliyon* 6: e03343.

44. Rawat S, Johri B. 2013. Role of thermophilic microflora in composting. In: Satyanarayana T, Littlechild J, Kawarabayasi Y (Eds.), *Thermophilic Microbes in Environmental and Industrial Biotechnology*. Springer, Dordrecht, pp. 137–169.

45. Rodrigues L, Banat IM, Teixeira J, Oliveira R. 2006. Biosurfactants: Potential applications in medicine. *Journal of Antimicrobial Chemotherapy* 57: 609–618.

46. Sabater C, Ruiz L, Delgado S, Ruas-Madiedo P, Margolles A. 2020. Valorization of vegetable food waste and by-products through fermentation processes. *Frontiers in Microbiology* 11: 581997.

47. Schleper C, Jurgens G, Jonuscheit M. 2005. Genomic studies of uncultivated archaea. *Nature Reviews Microbiology* 3: 479–488.

48. Selvam A, Wang X, Wong J. 2021. Chapter 5 - Food waste composting: Challenges and possible approaches. In: Wong J, Kaur G, Taherzadeh M, Pandey A, Lasaridi K (Eds.), *Current Developments in Biotechnology and Bioengineering*. Elsevier, Cambridge, MA, USA, pp. 137–162.

49. Sharma P, Gaur VK, Kim SH, Pandey A. 2020. Microbial strategies for bio-transforming food waste into resources. *Bioresource Technology* 299: 122580.

50. Sousa DZ, Smidt H, Alves MM, Stams AJ. 2007. *Syntrophomonas zehnderi* sp. nov., an anaerobe that degrades long-chain fatty acids in co-culture with *Methanobacterium formicicum*. *International Journal of Systematic and Evolutionary Microbiology* 57: 609–615.
51. Stavroula K, Malamis S, Katherine-Joanne H. 2020. Polyhydroxyalkanoates (PHAs) from household food waste: Research over the last decade. *International Journal of Biotechnology & Bioengineering* 6: 26–36.
52. Sträuber H, Schröder M, Kleinsteuber S. 2012. Metabolic and microbial community dynamics during the hydrolytic and acidogenic fermentation in a leach-bed process. *Energy, Sustainability and Society* 2: 1–10.
53. Tamang JP, Shin DH, Jung SJ, Chae SW. 2016. Functional Properties of microorganisms in fermented foods. *Frontiers in Microbiology* 7: 578.
54. Tišma M, Tadić T, Budžaki S, Ostojčić M, Šalić A, et al. 2019. Lipase production by solid-state cultivation of *Thermomyces Lanuginosus* on by-products from cold-pressing oil production. *Processes* 7: 465.
55. Torres-León C, Ramírez-Guzman N, Londoño-Hernandez L, Martinez-Medina GA, Díaz-Herrera R, et al. 2018. food waste and byproducts: An opportunity to minimize malnutrition and hunger in developing countries. *Frontiers in Sustainable Food Systems* 2: 52.
56. Tsui TH, Wong JWC. 2019. A critical review: Emerging bioeconomy and waste-to-energy technologies for sustainable municipal solid waste management. *Waste Disposal & Sustainable Energy* 1: 151–167.
57. Umesha S, Singh PK, Singh RP. 2018. Chapter 6 - Microbial biotechnology and sustainable agriculture. In: Singh RL, Mondal S (Eds.), *Biotechnology for Sustainable Agriculture*. Woodhead Publishing, pp. 185–205.
58. UNEP. 2021. Food waste index report 2021. United Nations Environment Programme, Nairobi, Kenya. Available at https://www.unep.org/resources/report/unep-food-waste-index-report-2021. accessed on 1 March 2021.
59. Venkiteshwaran K, Bocher B, Maki J, Zitomer D. 2015. Relating anaerobic digestion microbial community and process function: Supplementary issue: Water microbiology. *Microbiology Insights* 8: MBI–S33593.
60. Ventura JRS, Lee J, Jahng D. 2014. A comparative study on the alternating mesophilic and thermophilic two-stage anaerobic digestion of food waste. *Journal of Environmental Sciences* 26: 1274–1283.
61. Voberkova S, Vaverkova MD, Buresova A, Adamcova D, Vrsanska M, et al. 2017. Effect of inoculation with white-rot fungi and fungal consortium on the composting efficiency of municipal solid waste. *Waste Management* 61: 157–164.
62. Wang L. 2013. Production of bioenergy and bioproducts from food processing wastes: A review. *Transactions of the ASABE* 56: 217–230.
63. Wang Q, Li H, Feng K, Liu J. 2020. Oriented fermentation of food waste towards high-value products: A review. *Energies* 13: 5638.
64. Wong JWC, Selvam A, Awasthi MK. 2016. Chapter 9 - Composting for organic waste management. In: Wong JWC, Surampalli RY, Zhang TC, Tyagi RD, Selvam A (Eds.), *Sustainable Solid Waste Management*. ASCE publication, pp. 233–272.
65. Wong JWC, Wang X, Selvam A. 2017. Chapter 4 - Improving compost quality by controlling nitrogen loss during composting. In: Wong JWC, Tyagi RD, Pandey A (Eds.), *Current Developments in Biotechnology and Bioengineering*. Elsevier, pp. 59–82.
66. Wu S, Xu S, Chen X, Sun H, Hu M, et al. 2018. Bacterial Communities changes during food waste spoilage. *Scientific Reports* 8: 8220.
67. Xu J, Jiang Z, Li M, Li Q. 2019. A compost-derived thermophilic microbial consortium enhances the humification process and alters the microbial diversity during composting. *Journal of Environmental Management* 243: 240–249.
68. Yamada T, Sekiguchi Y, Imachi H, Kamagata Y, Ohashi A, et al. 2005. Diversity, localization, and physiological properties of filamentous microbes belonging to Chloroflexi subphylum I in mesophilic and thermophilic methanogenic sludge granules. *Applied and Environmental Microbiology* 71: 7493–7503.
69. Zhao Y, Zhao Y, Zhang Z, Wei Y, Wang H, et al. 2017. Effect of thermo-tolerant actinomycetes inoculation on cellulose degradation and the formation of humic substances during composting. *Waste Management* 68: 64–73.

SECTION 4

GOOD MICROBES AND BIOREMEDIATION

Co-Edited by David Dowling and Frans J. de Bruijn

CHAPTER 26

Microbial-based Bioremediation at a Global Scale

The Challenges and the Tools

Victor de Lorenzo, Esteban Martínez-García, and Tomás Aparicio
Environmental Synthetic Biology Laboratory. National Center of Biotechnology CSIC. Campus of the Universidad Autónoma de Madrid, 28049 Spain

26.1 INTRODUCTION

Environmental pollution by chemical emissions resulting from human activities has coexisted with humanity since the onset of the Neolithic era (Grattan et al., 2016). However, the problem began to be noticeable with the arrival of the industrial revolution, later with the development of synthetic chemistry and peaked in 1980–1990, when disasters like the Bhopal leak of methyl isocyanate (Broughton, 2005) and the Exxon Valdez 1989 oil spill in Alaska (Peterson et al., 2003). These incidents exposed to global public opinion the risks associated with large-scale chemical manufacturing. Since then, awareness and data of the impact of pollution on all types of ecosystems has continued to increase. This, in turn, has elicited decades of research efforts for developing technologies to mitigate the problem or – in the best case – eliminate it altogether (Kumar et al., 2011). Microorganisms have been the key actors in this endeavor, as they are subject to natural evolution processes that often result in the solution of a complex biochemical challenge brought about by environmental introduction of new chemical species and misplacement of the existing one (e.g. petroleum components) in niches where they naturally do not belong. Furthermore, the development of genetic engineering and recombinant DNA technology since the late 1970s provided both a detailed understanding of the molecular mechanism underlying pollutant biodegradation processes and a methodology to enhance the catalytic properties of otherwise naturally-occurring strains (Pieper and Reineke, 2000; Sayler and Ripp, 2000). Unfortunately, despite unwarranted expectations and lack of sufficient knowledge on how to make genetically engineered microorganisms work efficaciously and safely in the open, this approach was virtually abandoned years later (Cases and de Lorenzo, 2005) and non-GMO approaches to bioremediation and environmental biocatalysis were refocused on, a trend that has prevailed until recently (de Lorenzo, 2009). As a matter of fact, both public concerns about the release of GMOs and the development of green chemistry and its promise of new industrial processes with little or no pollution kept the motivation for developing GE microorganisms and environmental clean-up agents at bay. However, the last 10 year have witnessed entirely new environmental challenges that received little or no notice when the field started and which have entirely changed the set of

Good Microbes in Medicine, Food Production, Biotechnology, Bioremediation, and Agriculture, First Edition. Edited by Frans J. de Bruijn, Hauke Smidt, Luca S. Cocolin, Michael Sauer, David Dowling, and Linda Thomashow.

priorities about science-based interventions on ecosystems damaged by noxious emission. The *first* is the chemical nature of the pollutants (de Lorenzo, 2017). While 30 years ago the focus was on xenobiotic molecules (e.g., PCBs, nitroaromatics, halo-organics etc.) present at toxic levels in specific sites, now the most worrisome molecules are those of greenhouse gases (CO_2, CH_4, N oxides, fluorinated gases), which are throughout the planet's atmosphere at comparatively low concentrations. While most of these gases are perfectly natural, their increased levels due to industrial discharges, even if marginal, became co-causes of the ongoing climate change and its devastating consequences. A *second* feature of the current environmental circumstances is the unexpected impact of plastic residues in soil and (mostly) in aquatic ecosystems (Cózar et al., 2014; Galloway and Lewis, 2016; see chapters 24 and 32). Some studies predict that the total amount of plastics released into the oceans will soon be higher than the whole of the fish biomass found in them. The impact is not only mechanical, but the abundance of non-degradable microplastics and micropolymers from synthetic textiles cause these materials to reach the trophic chains of many ecosystems with multitude of noxious outcomes. The *third* major source of the global environmental emergency is expansion of drylands and soil desertification (Reynolds et al., 2007). The green cover of the planet seems to be shrinking at an accelerated rate and less and less soil is available to either primary plant life or agricultural exploitation (in many cases, desertification follows agricultural overexploitation; Berdugo et al., 2020). Other emerging issues have been identified as well (de Lorenzo, 2017) such as micropollutants, lignocellulosic residues, air particulates, shortage and poor quality of drinking water, scorched soil, dearth of assimilable N and P, that add to the idea that problems now have a different dimension and engage a set of actors quite unlike those that represented the main concerns of early environmental biotechnology and microbial ecology. New issues ask for new approaches and solutions. But what could they be?

26.2 BIOREMEDIATION BEYOND THE TIPPING POINT

A welcome but somewhat naïve view of how to address contemporary environmental issues entertains a reversible linear or threshold-dependent relation between the cause of the problem and its effects (Figure 26.1). Under this frame, any damage caused by an excessive stress on an ecosystem, whether chemical (i.e., pollutants) or physical (e.g., overheating, desiccation) will go back to normal as soon as the insult ceases. On this basis the emphasis ought to be on prevention (avoiding the build-up of the stress) and remediation (attenuating the intensity of the already existing stress or removal of it completely).

This idea has dominated the whole realm of environmental biotechnology for decades and it has in fact yielded (and keeps on yielding) amazing dividends: metabolic engineering for industrial processes, biodegradable polymers, decarbonation of energy, large-scale recycling, etc. Even extensive bioremediation interventions often work well with science-guided non-GMO bioaugmentation and biostimulation tactics (Adam et al., 2015). This was perfectly feasible for the type of early environmental pollution issues mentioned above. But unfortunately, the same strategies are quite unlikely to work for emerging global-scale challenges. The main reason is that climatic change fostered by greenhouse gases, widespread plastic pollution, desertification, and other new means may be subject to threshold-dependent behavior endowed with a transition or catastrophic fold bifurcation (Figure 26.1 C). In this case, stress increase makes one stable state reach a tipping point before dropping to the next state away from the sustainability zone. This originates irreversible shifts in the form and function of the system (Scheffer et al., 2001). In these cases, simply ceasing the cause the catastrophe will not make the system to return to its original equilibrium, which may in fact be permanently ruined. At the present time, not less than nine such tipping points[1] have been identified where ongoing climate change could push parts of the Earth into irrevocable change. On this basis, it seems clear that reactive strategies that were adequate for scenarios before the tipping point will not be effective beyond it. Instead, proactive counterparts need to be put in place. Which are the components of such strategies and associated technologies?

26.3 THE ENVIRONMENTAL MICROBIOME AS A GLOBAL CATALYST

The starting point for entertaining large-scale interventions aimed not just at containing but also reverting the levels global emissions is identifying possible drivers of the process. One possibility is to consider the global warming problem as a mere physicochemical challenge

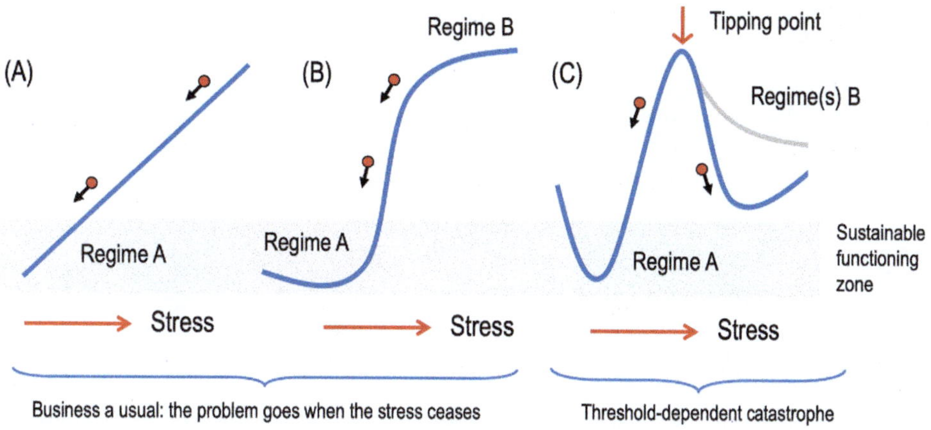

FIGURE 26.1 Stability regimes of natural ecosystems. (A) Reversible linear: the system goes through incremental deterioration following increasing stress, but it is fully reversible along the same trajectory if the insult goes back to initial level. (B) Threshold-dependent, nonlinear behavior: upon exposure to enough pressure, the system moves from one functioning state to virtual collapse – but can return to normal when the burden ceases. (C) Non-linear threshold-dependent with catastrophic fold bifurcation. In this scenario, stress makes the system reach a tipping point followed by irreversibly running into either a complete shift of regime or enter into prolonged instability (gray line). Source: Modified from Scheffer M, et al. 2001.

where the objective is to cool the Earth surface by modifying artificially the planet's albedo (i.e., the reflection of solar radiation out of the total radiation). This is one of the bases of so-called geoengineering, the second foundation of which being large-scale carbon sequestration by physicochemical means (Matter et al., 2016) or other (in principle) promising methods, e.g., reforestation (Bastin et al., 2019) or ocean fertilization (Vaughan and Lenton, 2011; see also chapter 21). While the technical aspects of geoengineering have already been discussed for decades, the idea has generally been received with skepticism if not straight rejection by many environmental stakeholders for a large number of reasons, including fears of unintended consequences and lack of global governance (Corner and Pidgeon, 2010). From a more practical perspective a major limitation of the approach is that whatever geoengineering action we can entertain involves large investments in infrastructures and workforce able to support massive operations if they are to cause a significant impact on global warming. This implies that such installations will be largely bound to be localized in specific sites rather than evenly distributed through the biosphere. Furthermore, geoengineering may help counteracting greenhouse gases and mitigate global warming, but it leaves aside the other large chemical pollution and physical deterioration issues mentioned before. This scenario opens up the opportunity to consider an alternative approach based on capitalizing on the planet-wide omnipresence and catalytic power of the environmental microbiome.

Recent calculations estimate that the global weight of microorganisms/plants on Earth contributes about a quarter of the whole biomass of the biosphere (Bar-On et al., 2018), the bulk of which (80%) being supplied by plants (Figure 26.2). However, 50%–90% of the total organic matter of the vegetal biomass is itself composed of catalytically inert lignocellulose (Johansen, 2016), which increases the ratio of available enzymatic capacity and catalytic diversity of microorganisms (~ 80% of which bacteria) to a much higher figure.

Furthermore, geological records indicate that environmental microorganisms were causative agents of major changes to the composition of the atmosphere at given times in the history of the Earth (Gumsley et al., 2017; Planavsky et al., 2014), thereby accrediting their ability to bring about massive transformations. Finally, both microbes and their DNA propagate very quickly, as extensively spreading pathogens and antibiotic resistance genes repeatedly indicate (Normark and Normark, 2002). It thus seems that a plausible driver of very-large-scale environmental remediation is already in place. The prospect of manipulating it in a beneficial direction is made possible by our growing understanding of the microbiomes which connect all types of biological systems (de Lorenzo, 2018). This adds to the growing opportunities for programming the microbiome with the conceptual and material tools of systems and synthetic biology (DeLisi, 2019; McCarty and Ledesma-Amaro, 2019).

The idea of designing self-propagating microorganisms to cause very-large-scale climate changes is at

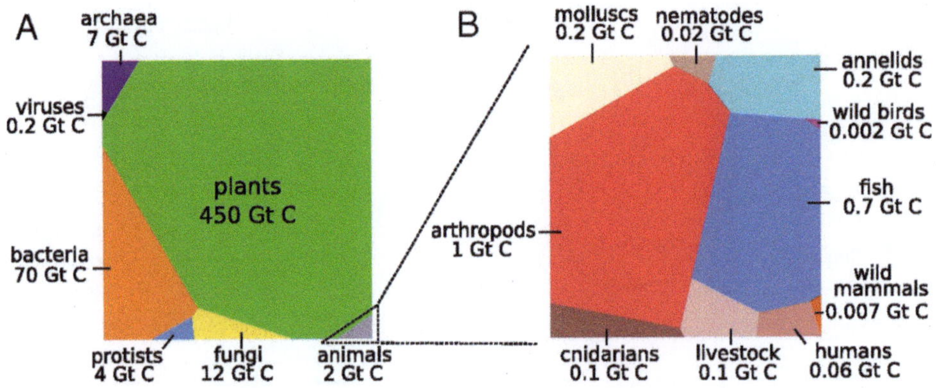

FIGURE 26.2 Breakdown contributors to the whole biosphere biomass. The entire planet hosts ~ 550 gigatons of carbon fixed in the biomass of the different actors in the natural kingdoms. Plants are the dominant contributor (~ 80%) in net weight but, as discussed in the text, about half of it is formed by just one component: lignocellulose. This increases the share of catalytic corpus in favor of microorganisms, the biomass of which is largely composed by enzymes (figure from Bar-On et al., 2018).

the basis of the *terraforming* concept (Rivera-Valentín, 2019) which has traditionally been part of the agenda of astrobiology. The question is whether a similar notion could be applicable not to creation of niches friendly to life on other planets, but for restoring ecological cycles and balances in sites and scenarios well beyond tipping points on Earth (Conde-Pueyo et al., 2020). To advance in this direction we need to leave aside momentarily public perception issues and focus on the specific scientific and technical bottlenecks that need to be overcome to make large-scale bioremediation a reality. Note that the issues at stake are not those of pathway assembly. Since the availability of massive DNA synthesis (Hughes and Ellington, 2017) and a plethora of methods for genome engineering (Wannier et al., 2021), this is no longer an important issue. Technologies for enzyme and pathway improvement/construction/optimization in given hosts belong to the standard toolbox of contemporary biotechnology. Many of these engineered properties have a direct utility in the type of large-scale applications that we entertain here: CO_2 fixation (Antonovsky et al., 2016; Flamholz et al., 2020; Gleizer et al., 2019), plastic biodegradation (Tournier et al., 2020), removal of micropollutants (Chen et al., 2016), elimination of chlorinated compounds (Ang et al., 2005), etc. The challenge is thus not so much generating new beneficial activities (there are plenty of them both naturally occurring and nurtured in the laboratory), but releasing them into the target niche, the dimension of which can go from relatively small to planet-wide. To this end, the emphases switch from engineering the catalytic property to designing social and physical traits of the cognate microbial agents and release strategies to secure the efficacious delivery of such activities in time and space.

26.4 DESIGNING AGENTS FOR SPREADING NEW TRAITS THROUGH THE ENVIRONMENTAL MICROBIOME

By the same token that an active principle can be developed in a pharmaceutical laboratory and then the way of delivery to sick patients has to be figured out, bioremediation agents also require specific formats and ways of supplying to the environment for an effective action. Surprisingly, this aspect (which has been called environmental galenics; Cases and de Lorenzo, 2005) has received comparatively far less attention than the creation of the activity proper. A number of formulations have been proposed over the years to spread bacteria in open niches, for the sake of plant growth (Vidhyasekaran et al., 1997), pest control in crops (Burges and Jones, 1998), soil rehabilitation (Raina et al., 2008), and dispersion of oil spills (John et al., 2016). In these cases, the efforts focused on designing the carrier materials (i.e., the *excipient* in galenic terminology) and/or additives for enhancing the efficacy of available natural or engineered microorganisms. However, improvement of the physical characteristics of the agents proper was hardly addressed. Fortunately, this issue (the interplay between the genetic parts of interest of a microorganism and the rest of the physicochemical properties of the cells) is currently being filled by the concept of *chassis* (de Lorenzo et al., 2020).

In the jargon of contemporary synthetic biology and bioengineering, a chassis is the basic genetic, biochemical, and physical frame of a specific microbial type where genetic implants of interest can be inserted at will and in a fashion relatively independent of the rest of the components (Committee et al., 2020). This notion thus allows separation of engineering efforts toward either activity/pathway design or chassis design, the ultimate agent of interest being the microorganism fortified with both features. Obviously, the starting point for developing chassis for different purposes is an already existing strain of a species that is habituated to thrive in the very niches where new activities are to be delivered. But then, the initial isolate must go through a series of genomic edits to enhance its predictability and reliability as well as securing a safety stamp by regulatory authorities (Committee et al., 2020). In that respect, not every recipient of recombinant DNA qualifies as a chassis, as the strain in question must fulfill a considerable number of requirements to minimize any risks while optimizing its efficacy (de Lorenzo et al., 2020). Furthermore, strains developed for environmental release must also be endowed with a higher colonization capacity to overcome the resistance of existing communities to incorporate new members in their metabolic and genetic networks (He et al., 2014). Fortunately, all these traits can be programmed with the tools of synthetic biology.

26.5 BACTERIAL CHASSIS FOR ENVIRONMENTAL INTERVENTIONS

A number of bacterial platforms for building effective bioremediation agents and have been already proposed depending on the specific scenario. Obviously, terrestrial, marine, aerial ecosystems etc. can be targeted much better with microorganisms that are either indigenous to the same location or used to thrive under the expected operation conditions. A provisional list of possible chassis for different application needs has been published recently (de Lorenzo et al., 2020) and will not be addressed here. The archetypal case of *Pseudomonas putida* can be picked to showcase the roadmap that goes from an environmental isolate with promising, pre-evolved qualities all the way to become a bona fide chassis for operation under harsh environmental settings. This bacterial species is a habitual resident of soil and plant roots in sites with a history of pollution with chemical waste and it is thus habituated to endure different types of physicochemical stresses (Nikel and de Lorenzo, 2018). This property can be traced to a distinct biochemical frame that favors generation of metabolic currency for counteracting redox insults over intermediates for biomass build-up (Nikel et al., 2021). One specific strain (*P. putida* KT2440) has emerged in recent years as a testbed for designing many types of modifications that could be desirable in a chassis for eventual large-scale environmental use. Changes have included genome size reduction (Leprince et al., 2012; Martinez-Garcia et al., 2020; Martínez-García et al., 2014), elimination of parasitic prophages (Martínez-García et al., 2015) and energy-wasting but otherwise useless structures (Martínez-García et al., 2014), and display of adhesins (Espeso et al., 2021; Martinez-Garcia et al., 2020) and other protein structures (Dvorak et al., 2020) on the cell surface. Editing the surfome, i.e., the complement of proteins exposed on the outermost layer of the bacterial envelope is particularly interesting for two reasons. First, as sketched in Figure 26.3 the same surface can be exploited for anchoring adhesins (for instance, nanobodies) (Glass and Riedel-Kruse, 2018; McCarty and Ledesma-Amaro, 2019) that make cells stick to matching interaction partners on other microorganisms (whether bacterial of fungal).

This enables the formation of consortia with new catalytic properties (Che and Men, 2019; Zhou et al., 2015) and in which a division of biodegradative labor can be programmed. But also, the same strategy can be entertained for causing adhesion to abiotic surfaces coated with molecular motifs that can be recognized by the adhesins. (Figure 26.3 C). A separate opportunity of surfome engineering involves exploitation of the cell envelope as a scaffold for assembly of complex multi-enzymatic complexes inspired on cellulosomes (Figure 26.4). Although such structures appear naturally on distinct types of Gram-positive bacteria, the growing availability of display tools for environmental bacteria (Ting et al., 2020) including *Pseudomonas* (Tozakidis et al., 2020) has recently enabled functional expression of cellulosome components on the surface of *P. putida* cells (Dvorak et al., 2020).

This example shows how contemporary tools of genome manipulation facilitate construction of bacteria with complex phenotypes with a potential for post-tipping point interventions (Figure 26.1) that were unthinkable a decade ago. In any case, these developments, which are at this point still in an embryonic state, are bound to multiply the already numerous benefits of adopting environmental bacteria as the basis for sound and effective bioremediation

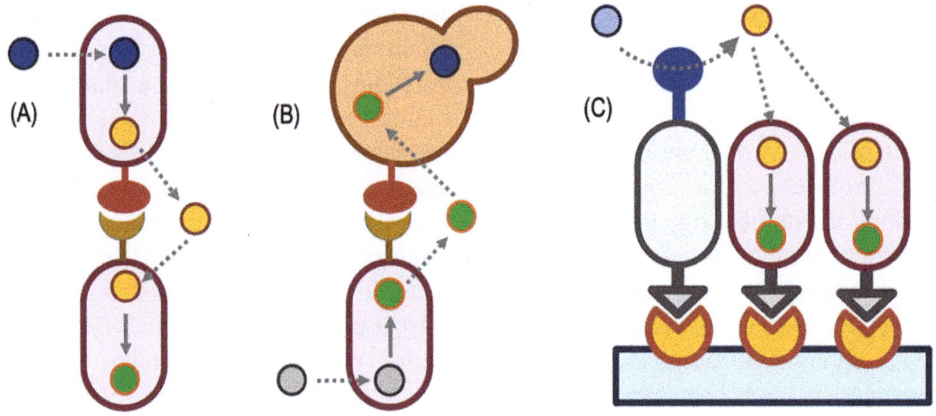

FIGURE 26.3 Engineering physical traits in bacterial chassis for environmental bioremediation. The ability to stick to surfaces of other bacteria (A), to different microbial types like yeast or fungi (B) or abiotic surfaces (C) is a most desirable trait in agents tailored for rehabilitation of polluted niches. This capacity is entered through genetically encoded ectopic display of adhesins (for instance, nanobodies) presented on the cell surface which are specific for molecular targets presented by the interaction partner. In this way, besides its own catalytic activity, the chassis acts as a nucleation factor for the formation of syntropic microbial communities and/or catalytic consortia and biofilms on surfaces of interest.

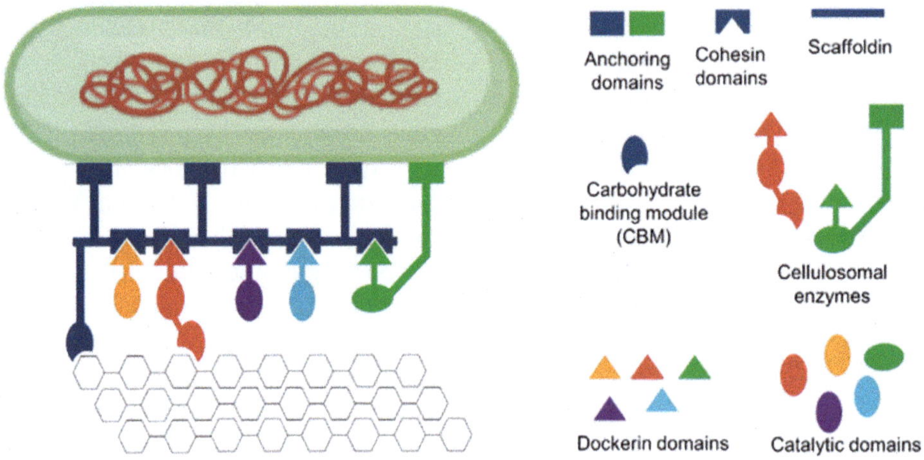

FIGURE 26.4 The bacterial surface as a scaffold for multiprotein complexes. Cellulosomes are highly structured multiprotein clusters found on the envelope of many types of cellulose-degrading bacteria (Artzi et al., 2017). Besides enzymes proper, they adopt 3D molecular architectures of different complexity which are shaped by connecting parts, e.g., cohesins, scaffoldins, and dockerins. The same parts can be genetically moved to other bacteria for display of the same enzymes (or some of them; Dvorak et al. 2020) or fused to heterologous polypeptides for a variety of applications. It is thus possible to combine a catalytic activity located on the cell envelope with the native or engineered metabolic capacity of the carrier.

chassis (Calero and Nikel, 2019; de Lorenzo et al., 2020; Nikel and de Lorenzo, 2018).

Despite the prospects opened up by synthetic biology in the field of environmental management, note that having a good chassis does not ultimately solve the problem of overcoming the resistance to colonization mentioned above. The phenomenon has been extensively studied in the intestinal microbiome (Lawley and Walker, 2013) but operates in all types of ecosystems. This makes establishment of a new member of an otherwise well-balanced ecosystem difficult unless there is a massive selective pressure, and that may not occur. What then could be done to secure establishment of an intruder bearing environmentally beneficial activities in a pre-colonized target niche? There are different mechanisms for such an exclusion of newcomers as well as strategies of invaders to overcome the antagonism of the already settled community members. In particular, the so-called type 6 secretion system (T6SS) is increasingly recognized as a powerful instrument used by bacteria to become established in the host gut (Sana et al., 2016). This system injects toxins in bacteria that enters into contact with the newcomer, thereby leaving an empty niche that can be occupied by the fresh arrival. Once more, this system can be domesticated and repurposed with the tools of synthetic biology for the sake of enabling penetration of engineered bacteria and selective elimination of competitors in otherwise impermeable bacterial communities (Ting et al., 2020). *P. putida* does have endogenous T6SSs (Bernal et al., 2017) that could be exploited to this end.

26.6 INOCULATION OF NEWCOMERS IN EXISTING MICROBIAL NICHES: NO PIECE OF CAKE

The concept of bioremediation largely relies on the beneficial catalytic action of *live* biological agents cells once stimulated to grow in given sites or directly inoculated in target niches. Note however that two additional scenarios can be entertained. One, which has been adopted in some specific cases, involves the use of dead biomass from either naturally occurring or recombinant bacteria that still carry enzymes for inactivation or mitigation of the toxicity of the contaminants at stake. The advantage of this approach (reviewed in Dvořák et al., 2017) is that the catalytic materials are not alive and therefore even in the case they were engineered they do they do not qualify as GMOs. The drawback is, obviously, that the same materials do not self-replicate and therefore scale-up might not be straightforward. This has not been an obstacle for the onset of a whole branch of synthetic biology that bets on the development of cell-free biological materials for environmental sensing and remediation (Karig, 2017). The second scenario that affects direct inoculation of live bacteria as catalytic agents stems from a longstanding observation on microbial ecology: protozoan grazing (Azam et al., 1983; Simek et al., 2001). Even if colonization resistance is defeated and a new member in introduced in a balanced microbial niche, any overgrowth may create a thriving scenario for predators and protist grazers (Cunningham et al., 2009). This often means that microbial biomass excess is removed virtually at the same rate it is produced (Cunningham et al., 2009). Note that bioaugmentation depends on the activity of the released bioremediation agent, which in turn relies on high enough numbers. But given the unavoidable action of grazers, these numbers may never be sufficient for a durable activity. Some possible solutions have been proposed to overcome this bottleneck, for instance slow release of the bacteria from plastic tubes (Mertens et al., 2006) or polymeric beads (Park et al., 2021) that protect from grazing. But again, scale up of the process might be challenging. Fortunately, marine seaweeds and invertebrates often have a wealth of bacteria on their surfaces that produce compounds to deter settling of other organisms because they produce anti-grazing small molecules (Armstrong et al., 2000) that make them unpalatable to possible predators. One could entertain endowing bioremediation agents with biosynthetic routes for such chemicals and therefore providing some protection against eukaryotic scavengers. But in any case, it seems clear that in many instances the laws of microbial ecology do not ease the establishment and action of engineered newcomers in a microbial niche at enough levels to make a difference in their catalytic action on pollutants. Are there other avenues to overcome these limitations, in particular when considering very large bioremediation interventions?

26.7 PROGRAMMING LARGE-SCALE HORIZONTAL GENE TRANSFER

It is a well-established fact that given the right selective pressure DNA propagates very quickly all the way from an individual bacterial cell to the planet-wide environmental microbiome. The evidence for this comes largely from studies on antibiotic resistance genes (Normark and Normark, 2002). These illustrate, e.g., that in the few decades between the first use of penicillin in 1942 and the present time, genes endowing resistance to the first beta-lactams are so globally widespread that the earlier antibiotics are altogether useless. While attention and concern about horizontal gene transfer (HGT) and their mechanisms (Soucy et al., 2015) has logically focused on propagation of traits that have a negative effect on human health (e.g., antibiotic resistances, virulence factors; Sun et al., 2019), the same channels are available for circulating DNA segments among strains, species, and even biological kingdoms (Husnik and McCutcheon, 2018).

On this background, it could be possible to entertain an alternative to inoculation of bioremediating agents based instead on propagation of the key beneficial traits by HGT through the whole membership of a pre-established, natural microbial community. In this case, the challenge would not be so much engineering the chassis as discussed above, but designing a DNA sequence capable of effective self-propagation and bearing the genes encoding the traits of interest under the control of promiscuous expression signals. One decisive step in this direction has been the recent reassembly of the entire conjugal transfer machinery of broad-host range plasmid RP4 in a synthetic transposon that admits functional cargoes (Figure 26.5; Silbert et al., 2021). Any DNA segment or replicon thereof inserted with this genetic device acquires a potential for conjugal transfer to a variety of recipients (Figure 26.6).

It might be possible to merge this scheme to bacterial versions of gene drives (French et al., 2020; Valderrama et al., 2019) for fostering a self-selective pressure and thus bring about the spread of valuable activities without modifying community composition. These are promising possibilities which still require considerable research.

26.8 CONCLUSION

We live in the present time in the midst of a climate emergency with ample evidence of being at the verge of trespassing – or having trespassed already on a good number of key tipping points. This scenario changes the timing, the urgency and the nature of the technologies necessary to counteract environmental deterioration as compared to the picture we had at the onset of bioremediation as a scientific and technical discipline in the mid-1980s. Preventive measures granted, the principal allies for reverting the damage already caused by urban and industrial waste continue to be environmental microorganisms, whether naturally occurring or fortified through genetic engineering and synthetic biology approaches as discussed above. However, realizing the dimension of the challenge means that new avenues for implementing interventions at a very large scale need to

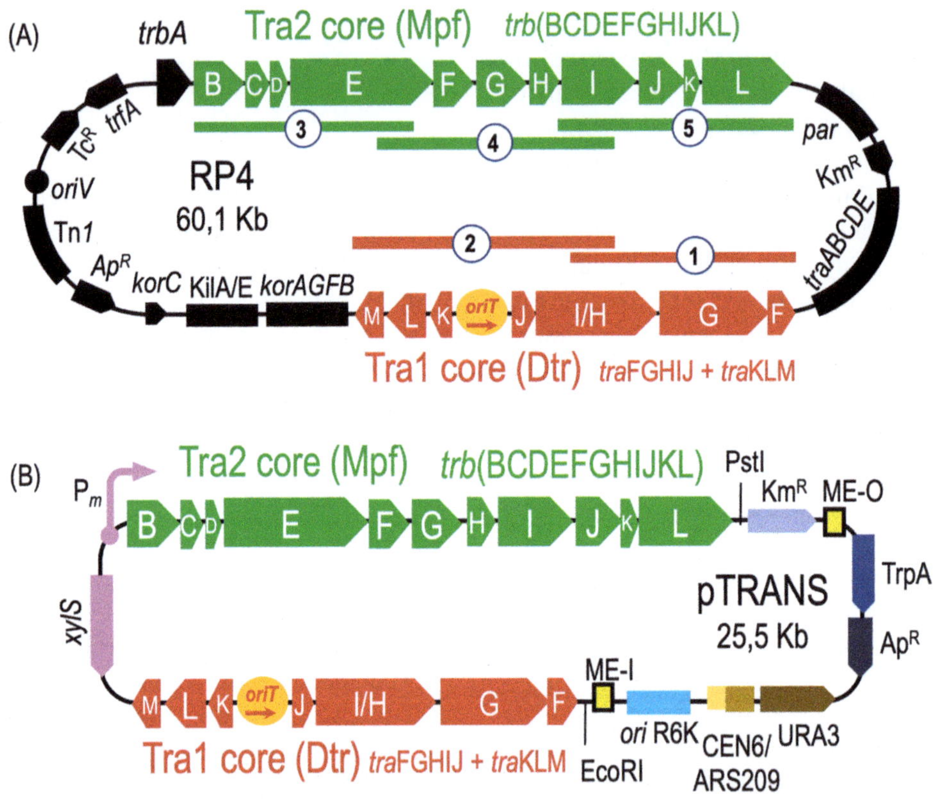

FIGURE 26.5 Refactoring the conjugation machinery of promiscuous plasmids as a tool for promoting HGT. Self-conjugative plasmid RP4 (also called RK2) is one of the most promiscuous mobile elements found in the bacterial world. (A) The non-adjacent segments involved in plasmid delivery to many different types of recipients are highlighted in green and red, with an indication of the genes involved in each case. The five DNA fragments (1–5) used for reconstructing the conjugal machinery in a different context are indicated. (B) Construction of a genetic device for replicon mobilization between different strains and species. The parts of RP4 plasmid involved in conjugation were reassembled in a single DNA segment flanked by the termini of the Tn*5* transposon and placed in a suicide delivery vector. Once inserted into a target replicon, the cognate DNA acquires the ability to enter new hosts and thus propagate through horizontal gene transfer.

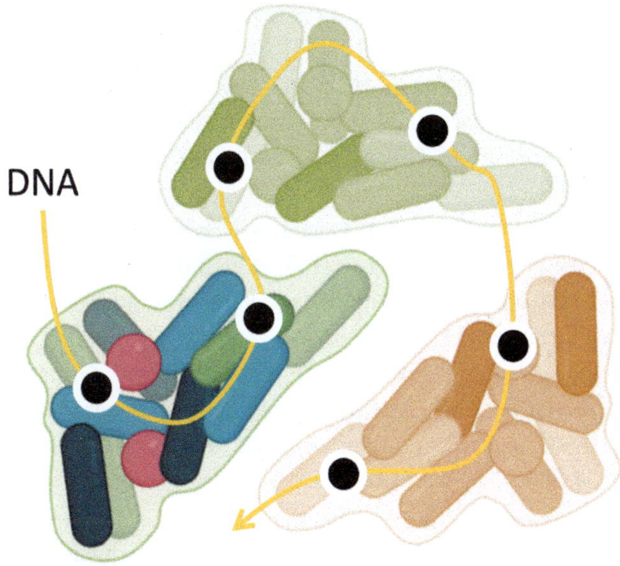

FIGURE 26.6 Engineering propagation of beneficial traits through the environmental microbiome. Insertion of a DNA mobilizing device in a donor strain enables the corresponding replicon to move conjugally into another recipient. Depending on the organization of the device, it may stop in the first transfer event or it may further jump into another replicon of the earlier recipient, thereby mobilizing it in turn into another one and so on – thereby creating a sort of HGT chain reaction. If the mobile element is associated to an environmentally beneficial trait (e.g., biodegradation of an intrinsically toxic molecule), the whole community may acquire the ability to express it, thereby amplifying the corresponding bioremediation effect.

be explored. This involves development of much better thought-over carriers of beneficial activities as well as improved environmental galenic methods for spreading them at an unprecedented extent. In this respect, if a central concern of pre-tipping point bioremediation was containment of engineered microorganisms and their recombinant DNA (Schmidt and de Lorenzo, 2016; Schmidt and Pei, 2015), it is plausible that the challenge of the post-tipping point developments will be exactly the opposite. Namely, how to propagate designer bacterial agents and/or rationally written DNA for maximizing the catalytic abilities of the extant environmental microbiome in a safe and efficacious fashion. Obviously, this cannot happen without developing sound concepts and metrics for assessing risks and offer decision makers and the public realistic cost-benefit choices. It is puzzling that despite the intense controversies on the potential risks associated to deliberately releasing recombinant bacteria to the environment the scientific literature has virtually no quantitative predictive models of large scale spreading of either engineered strains or recombinant DNA. Deployment of massive bioremediation interventions thus ask not only for new genetic tools and better engineering of strains and processes but also reliable propagation prototypes and references that take into account the best available ecological theory (Conde-Pueyo et al., 2020; Sole, 2015) together with involvement of different stakeholders. Whether or not this succeeds will make a difference in the future of our planet.

ACKNOWLEDGMENTS

The work in Authors' Laboratory is funded by the SETH (RTI2018-095584-B-C42) (MINECO/FEDER) and SyCoLiM (ERA-COBIOTECH 2018 – PCI2019-111859-2) Projects of the Spanish Ministry of Science and Innovation, the MADONNA (H2020-FET-OPEN-RIA-2017-1-766975), SynBio4Flav (H2020-NMBP-TR-IND/H2020-NMBP-BIO-2018-814650), and MIX-UP (MIX-UP H2020-BIO-CN-2019-870294) Contracts of the European Union and the InGEMICS-CM (S2017/BMD-3691) Project of the Comunidad de Madrid – European Structural and Investment Funds (FSE, FECER).

NOTE

1 https://www.carbonbrief.org/explainer-nine-tipping-points-that-could-be-triggered-by-climate-change.

REFERENCES

1. Adam GO, Fufeyin PT, Okoro SE, Ehinome I. 2015. Bioremediation, biostimulation and bioaugmention: A review. *International Journal of Environmental Bioremediation and Biodegradation* 3: 28–39.
2. Ang EL, Zha H, Obbard JP. 2005. Recent advances in the bioremediation of persistent organic pollutants via biomolecular engineering. *Enzyme and Microbial Technology* 37: 487–496.
3. Antonovsky N, Gleizer S, Noor E, Zohar Y, Herz E, et al. 2016. Sugar synthesis from CO2 in Escherichia Coli. *Cell* 166: 115–125.
4. Armstrong E, Boyd KG, Burgess JG. 2000. Prevention of marine biofouling using natural compounds from marine organisms. In: *Biotechnology Annual Review*, Elsevier, pp. 221–241.
5. Artzi L, Bayer EA, Moraïs S. 2017. Cellulosomes: Bacterial nanomachines for dismantling plant polysaccharides. *Nature Reviews Microbiology* 15: 83.

6. Azam F, Fenchel T, Field JG, Gray J, Meyer-Reil L, Thingstad F. 1983. The ecological role of water-column microbes in the sea. *Marine Ecology Progress Series* 10: 257–263.
7. Bar-On YM, Phillips R, Milo R. 2018. The biomass distribution on Earth. *Proceedings of the National Academy of Sciences* 115: 6506.
8. Bastin JF, Finegold Y, Garcia C, Mollicone D, Rezende M, Routh D, et al. 2019. The global tree restoration potential. *Science* 365: 76.
9. Berdugo M, Delgado-Baquerizo M, Soliveres S, Hernández-Clemente R, Zhao Y, Gaitán JJ, et al. 2020. Global ecosystem thresholds driven by aridity. *Science* 367: 787–790.
10. Bernal P, Allsopp LP, Filloux A, Llamas MA. 2017. The *Pseudomonas putida* T6SS is a plant warden against phytopathogens. *The ISME Journal* 11: 972–987.
11. Broughton E. 2005. The Bhopal disaster and its aftermath: A review. *Environmental Health* 4: 1–6.
12. Burges HG, Jones KA. 1998. Formulation of bacteria, viruses and protozoa to control insects. In: *Formulation of Microbial Biopesticides*. Kluwer Academic Publishers, Dordrecht, pp. 33–127.
13. Calero P, Nikel PI. 2019. Chasing bacterial chassis for metabolic engineering: A perspective review from classical to non-traditional microorganisms. *Microbial Biotechnology* 12: 98–124.
14. Cases I, de Lorenzo V. 2005. Genetically modified organisms for the environment: Stories of success and failure and what we have learned from them. *International Microbiology* 8: 213–222.
15. Che S, Men Y. 2019. Synthetic microbial consortia for biosynthesis and biodegradation: Promises and challenges. *Journal of Industrial Microbiology and Biotechnology* 46: 1343–1358.
16. Chen Y, Stemple B, Kumar M, Wei N. 2016. Cell surface display fungal laccase as a renewable biocatalyst for degradation of persistent micropollutants bisphenol A and sulfamethoxazole. *Environmental Science and Technology* 50: 8799–8808.
17. Committee ES, More S, Bampidis V, Benford D, Bragard C, Halldorsson T, et al. 2020. Evaluation of existing guidelines for their adequacy for the microbial characterisation and environmental risk assessment of microorganisms obtained through synthetic biology. *EFSA Journal* 18: e06263.
18. Conde-Pueyo N, Vidiella B, Sardanyés J, Berdugo M, Maestre FT, de Lorenzo V, Solé R. 2020. Synthetic biology for terraformation: Lessons from Mars, Earth, and the microbiome. *Life* 10: 14.
19. Corner A, Pidgeon N. 2010. Geoengineering the climate: The social and ethical implications. *Environment: Science and Policy for Sustainable Development* 52: 24–37.
20. Cózar A, Echevarría F, González-Gordillo JI, Irigoien X, Úbeda B, Hernández-León S, et al. 2014. Plastic debris in the open ocean. *Proceedings of the National Academy of Sciences* 111: 10239–10244.
21. Cunningham JJ, Kinner NE, Lewis M. 2009. Protistan predation affects trichloroethene biodegradation in a bedrock aquifer. *Applied and Environmental Microbiology* 75: 7588–7593.
22. de Lorenzo V. 2009. Recombinant bacteria for environmental release: What went wrong and what we have learnt from it. *Clinical Microbiology and Infection* 15: 63–65.
23. de Lorenzo V. 2017. Seven microbial bio-processes to help the planet. *Microbial Biotechnology* 10: 995–998.
24. de Lorenzo V. 2018. Environmental microbiology to the rescue of planet earth. *Environmental Microbiology* 20: 1910–1916.
25. de Lorenzo V, Krasnogor N, Schmidt M. 2020. For the sake of the Bioeconomy: Define what a synthetic Biology Chassis is! *New Biotechnology* 60: 44–51.
26. DeLisi C. 2019. The role of synthetic biology in climate change mitigation. *Biology Direct* 14: 1–5.
27. Dvorak P, Bayer EA, de Lorenzo V. 2020. Surface display of designer protein scaffolds on genome-reduced strains of Pseudomonas Putida. *ACS Synthetic Biology* 9: 2749–2764.
28. Dvořák P, Nikel PI, Damborský J, de Lorenzo V. 2017. Bioremediation 3.0: Engineering pollutant-removing bacteria in the times of systemic biology. *Biotechnology Advances* 35: 845–866.
29. Espeso DR, Martínez-García E, de Lorenzo V. 2021. Quantitative assessment of morphological traits of planktonic bacterial aggregates. *Water Research* 188: 116468.
30. Flamholz DR, Dugan E, Blikstad C, Gleizer S, Ben-Nissan R, Amram S, et al. 2020. Functional reconstitution of a bacterial CO2 concentrating mechanism in Escherichia Coli. *eLife* 9: e59882.
31. French KE, Zhou Z, Terry N. 2020. Horizontal 'gene drives' harness indigenous bacteria for bioremediation. *Scientific Reports* 10: 15091.
32. Galloway TS, Lewis CN. 2016. Marine microplastics spell big problems for future generations. *Proceedings of the National Academy of Sciences* 113: 2331–2333.

33. Glass DS, Riedel-Kruse IH. 2018. A synthetic bacterial cell-cell adhesion toolbox for programming multicellular morphologies and patterns. *Cell* 174(649-658): e616.
34. Gleizer S, Ben-Nissan R, Bar-On YM, Antonovsky N, Noor E, Zohar Y, et al. 2019. Conversion of Escherichia coli to generate all biomass carbon from CO2. *Cell* 179: 1255–1263.
35. Grattan JP, Adams RB, Friedman H, Gilbertson DD, Haylock KI, et al. 2016. The first polluted river? Repeated copper contamination of fluvial sediments associated with Late Neolithic human activity in southern Jordan. *Science of the Total Environment* 573: 247–257.
36. Gumsley AP, Chamberlain KR, Bleeker W, Söderlund U, de Kock MO, et al. 2017. Timing and tempo of the Great Oxidation event. *Proceedings of the National Academy of Sciences* 114: 1811–1816.
37. He X, McLean JS, Guo L, Lux R, Shi W. 2014. The social structure of microbial community involved in colonization resistance. *The ISME Journal* 8: 564–574.
38. Hughes RA, Ellington AD. 2017. Synthetic DNA synthesis and assembly: Putting the synthetic in synthetic biology. *Cold Spring Harbor Perspectives in Biology* 9: a023812.
39. Husnik F, McCutcheon JP. 2018. Functional horizontal gene transfer from bacteria to eukaryotes. *Nature Reviews Microbiology* 16: 67.
40. Johansen KS. 2016. Lytic polysaccharide monooxygenases: The microbial power tool for lignocellulose degradation. *Trends in Plant Science* 21: 926–936.
41. John V, Arnosti C, Field J, Kujawinski E, McCormick A. 2016. The role of dispersants in oil spill remediation: Fundamental concepts, rationale for use, fate, and transport issues. *Oceanography* 29: 108–117.
42. Karig DK. 2017. Cell-free synthetic biology for environmental sensing and remediation. *Current Opinion Biotechnology* 45: 69–75.
43. Kumar A, Bisht B, Joshi V, Dhewa T. 2011. Review on bioremediation of polluted environment:: A management tool. *International Journal of Environmental Sciences* 1: 1079.
44. Lawley TD, Walker AW. 2013. Intestinal colonization resistance. *Immunology* 138: 1–11.
45. Leprince A, de Lorenzo V, Völler P, van Passel MW, Martins dos Santos VA. 2012. Random and cyclical deletion of large DNA segments in the genome of *Pseudomonas putida*. *Environmental Microbiology* 14: 1444–1453.
46. Martinez-Garcia E, Fraile S, Rodriguez-Espeso D, Vecchietti D, Bertoni G, de Lorenzo V. 2020. The naked cell: Emerging properties of a surfome-streamlined *Pseudomonas putida* strain. *ACS Synthetic Biology* 9: 2477–2492.
47. Martínez-García E, Jatsenko T, Kivisaar M, de Lorenzo V. 2015. Freeing *Pseudomonas putida* KT2440 of its proviral load strengthens endurance to environmental stresses. *Environmental Microbiology* 17: 76–90.
48. Martínez-García E, Nikel PI, Aparicio T, de Lorenzo V. 2014. *Pseudomonas* 2.0: Genetic upgrading of *P. putida* KT2440 as an enhanced host for heterologous gene expression. *Microbial Cell Factories* 13: 1–15.
49. Martínez-García E, Nikel PI, Chavarría M, de Lorenzo V. 2014. The metabolic cost of flagellar motion in *Pseudomonas putida* KT2440. *Environmental Microbiology* 16: 291–303.
50. Matter JM, Stute M, Snaebjornsdottir SO, Oelkers EH, Gislason SR, Aradottir ES, et al. 2016. Rapid carbon mineralization for permanent disposal of anthropogenic carbon dioxide emissions. *Science* 352: 1312–1314.
51. McCarty NS, Ledesma-Amaro R. 2019. Synthetic biology tools to engineer microbial communities for biotechnology. *Trends in Biotechnology* 37: 181–197.
52. Mertens B, Boon N, Verstraete W. 2006. Slow-release inoculation allows sustained biodegradation of gamma-hexachlorocyclohexane. *Applied Environmental Microbiology* 72: 622–627.
53. Nikel PI, de Lorenzo V. 2018. *Pseudomonas putida* as a functional chassis for industrial biocatalysis: From native biochemistry to trans-metabolism. *Metabolic Engineering* 50: 142–155.
54. Nikel PI, Fuhrer T, Chavarría M, Sánchez-Pascuala A, Sauer U, de Lorenzo V. 2021. Reconfiguration of metabolic fluxes in *Pseudomonas putida* as a response to sub-lethal oxidative stress. *The ISME Journal* https://doi.org/10.1038/s41396-020-00884-9.
55. Normark BH, Normark S. 2002. Evolution and spread of antibiotic resistance. *Journal of Internal Medicine* 252: 91–106.
56. Park H, Kim H, Kim GY, Lee MY, Kim Y, Kang S. 2021. Enhanced biodegradation of hydrocarbons by *Pseudomonas aeruginosa*-encapsulated alginate/gellan gum microbeads. *Journal of Hazardous Materials* 406: 124752.
57. Peterson CH, Rice SD, Short JW, Esler D, Bodkin JL, et al. 2003. Long-term ecosystem response to the Exxon Valdez oil spill. *Science* 302: 2082–2086.
58. Pieper DH, Reineke W. 2000. Engineering bacteria for bioremediation. *Current Opinion in Biotechnology* 11: 262–270.

59. Planavsky NJ, Asael D, Hofmann A, Reinhard CT, Lalonde SV, Knudsen A, et al. 2014. Evidence for oxygenic photosynthesis half a billion years before the Great Oxidation event. *Nature Geoscience* 7: 283–286.
60. Raina V, Suar M, Singh A, Prakash O, Dadhwal M, Gupta SK, et al. 2008. Enhanced biodegradation of hexachlorocyclohexane Sphi(HCH) in contaminated soils via inoculation with *Sphingobium indicum* B90A. *Biodegradation* 19: 27–40.
61. Reynolds JF, Smith DMS, Lambin EF, Turner B, Mortimore M, Batterbury SP, et al. 2007. Global desertification: Building a science for dryland development. *Science* 316: 847–851.
62. Rivera-Valentín EG. 2019. Reimagining terraforming. *Nature Astronomy* 3: 883–884.
63. Sana TG, Flaugnatti N, Lugo KA, Lam LH, Jacobson A, Baylot V, et al. 2016. *Salmonella Typhimurium* utilizes a T6SS-mediated antibacterial weapon to establish in the host gut. *Proceedings of the National Academy of Sciences* 113: E5044–E5051.
64. Sayler GS, Ripp S. 2000. Field applications of genetically engineered microorganisms for bioremediation processes. *Current Opinion in Biotechnology* 11: 286–289.
65. Scheffer M, Carpenter S, Foley JA, Folke C, Walker B. 2001. Catastrophic shifts in ecosystems. *Nature* 413: 591–596.
66. Schmidt M, de Lorenzo V. 2016. Synthetic bugs on the loose: Containment options for deeply engineered (micro) organisms. *Current Opinion Biotechnology* 38: 90–96.
67. Schmidt M, Pei L. 2015. Improving biocontainment with synthetic biology: Beyond physical containment. In: McGenity T, Timmis KN, Nogales B (Eds.), *Hydrocarbon and Lipid Microbiology Protocols*. Springer-Verlag, Berlin Heidelberg, pp. 185–199.
68. Silbert J, de Lorenzo V, Aparicio T. 2021. Refactoring the conjugation machinery of promiscuous plasmid RP4 into a device for conversion of Gram-negative isolates to Hfr strains. *ACS Synthetic Biology* https://doi.org/10.1021/acssynbio.0c00611.
69. Simek K, Pernthaler J, Weinbauer MG, Horňák K, Dolan JR, Nedoma J, et al. 2001. Changes in bacterial community composition and dynamics and viral mortality rates associated with enhanced flagellate grazing in a mesoeutrophic reservoir. *Applied and Environmental Microbiology* 67: 2723–2733.
70. Sole R. 2015. Bioengineering the biosphere? *Ecological Complexity* 22: 40–49.
71. Soucy SM, Huang J, Gogarten JP. 2015. Horizontal gene transfer: Building the web of life. *Nature Reviews Genetics* 16: 472–482.
72. Sun D, Jeannot K, Xiao Y, Knapp CW. 2019. Horizontal gene transfer mediated bacterial antibiotic resistance. *Frontiers in Microbiology* 10: 1933.
73. Ting SY, Martínez-García E, Huang S, Bertolli SK, Kelly KA, Cutler KJ, et al. 2020. Targeted depletion of bacteria from mixed populations by programmable adhesion with antagonistic competitor cells. *Cell Host and Microbe* 28: 313–321.
74. Tournier V, Topham CM, Gilles A, David B, Folgoas C, Moya-Leclair E, et al. 2020. An engineered PET depolymerase to break down and recycle plastic bottles. *Nature* 580: 216–219.
75. Tozakidis IE, Lüken LM, Üffing A, Meyers A, Jose J. 2020. Improving the autotransporter-based surface display of enzymes in *Pseudomonas putida* KT2440. *Microbial Biotechnology* 13: 176–184.
76. Valderrama JA, Kulkarni SS, Nizet V, Bier E. 2019. A bacterial gene-drive system efficiently edits and inactivates a high copy number antibiotic resistance locus. *Nature Communications* 10: 5726.
77. Vaughan NE, Lenton TM. 2011. A review of climate geoengineering proposals. *Climatic Change* 109: 745–790.
78. Vidhyasekaran P, Rabindran R, Muthamilan M, Nayar K, Rajappan K, Subramanian N, Vasumathi K. 1997. Development of a powder formulation of *Pseudomonas fluorescens* for control of rice blast. *Plant Pathology* 46: 291–297.
79. Wannier TM, Ciaccia PN, Ellington AD, Filsinger GT, Isaacs FJ, Javanmardi K, et al. 2021. Recombineering and MAGE. *Nature Reviews Methods Primers* 1: 1–24.
80. Zhou K, Qiao K, Edgar S, Stephanopoulos G. 2015. Distributing a metabolic pathway among a microbial consortium enhances production of natural products. *Nature Biotechnology* 33: 377–383.

CHAPTER 27

Ecopiling

Beneficial Soil Bacteria, Plants, and Optimized Soil Conditions for Enhanced Remediation of Hydrocarbon Polluted Soil

Robert Conlon, Mutian Wang, Xuemei Liu Germaine, Rajesh Mali, David Dowling, and Kieran J. Germaine
enviroCORE Research Centre, Department of Science and Health, Institute of Technology, Carlow, Ireland

27.1 INTRODUCTION

Petroleum hydrocarbons are the most common environmental contaminant. Hydrocarbon pollution is prominent due to the essential functions petroleum products provide in modern society and as a result the demand on petroleum has led to a large number of environmental pollution events. In 2006 a study carried out by the European Environment Agency (EEA) found that 34% of contaminated sites in Europe were affected by petrochemicals and 54% when polycyclic aromatic hydrocarbons (PAHs) and benzene, toluene, ethylbenzene, and xylenes (BTEX) compounds are included in the evaluation (Pinedo et al., 2013).

Crude oils are mixtures of usually short chained hydrocarbons composed of complex mixtures of paraffinic, alicyclic and aromatic hydrocarbons. They are produced over long geological time periods by the incomplete decomposition of plant and animal organic matter. Typically, crude oil is characterized by 83–87% carbon and 10–14% hydrogen. Small amounts of nitrogen, oxygen, sulphur, and metals are also found in crude oil (Ahmed et al., 2018). Diesel fuels are made from the distillation of crude petroleum separated by fractional distillation. They are composed of approximately 64% aliphatic hydrocarbons, 1–2% olefinic hydrocarbons, and 35% aromatic hydrocarbons (Ahmed et al., 2018). Some of the most difficult compounds to break down found in hydrocarbon contaminated soil are PAHs. PAHs are high molecular weight organic compounds consisting of two or more fused benzene rings (Chunhui et al., 2017; Meneses et al., 2013). They are non-polar compounds and, therefore have poor water solubility, but can be dissolved in organic solvents or lipids and they have high melting points. PAHs that contain six or less rings are considered "small" PAHs and those with more than six are considered "large" PAHs (Hussein et al., 2016; Thamaraiselvan et al., 2015).

The simplest PAHs are naphthalene, anthracene and phenanthrene which are two and three ring structures, respectively. As the number of aromatic rings increase, the water solubility of the compound decreases and resistance to degradation increases. PAHs are classed as persistent organic pollutants (POPs) because they are recalcitrant and can persist in an environment for long periods of time, which may damage the ecosystem or human health (Meneses et al., 2013). They are found in the atmosphere, ice, marine, and freshwater environments. Due to their lipophilic nature they are most commonly found in, and have the greatest capacity to accumulate in, soils and sediments. Sources of PAHs are both natural and man-made, meaning that they are ubiquitous in the

Good Microbes in Medicine, Food Production, Biotechnology, Bioremediation, and Agriculture, First Edition. Edited by Frans J. de Bruijn, Hauke Smidt, Luca S. Cocolin, Michael Sauer, David Dowling, and Linda Thomashow.

environment (Chunhui et al., 2017; Meneses et al., 2013; Thamaraiselvan et al., 2015). They are mostly formed during the incomplete burning of organic materials. PAHs are almost always found as a mixture of two or more different PAHs. Although they can occur naturally, PAHs formed due to human activity account for the majority of those found in the environment (Chunhui et al., 2017; Hussein et al., 2016).

Many of the compounds found in petrochemicals, including PAHs, are known to have toxic, mutagenic, and carcinogenic effects. The US EPA has classified seven PAHs to have a genotoxic effect which may cause cancer. These are benz[a]anthracene, benzo[a]pyrene, benzo[b]fluoranthene, benzo[k]fluoranthene, chrysene, dibenz[ah]anthracene, and indeno[1,2,3-cd] pyrene. The structure of some PAH reactive metabolites allows the compounds to bind to proteins and DNA and interfere with the replication process. The toxic effect of some PAHs is caused by their interference with the function of cell membranes and enzyme systems of an organism (Chunhui et al., 2017; Hussein et al., 2016; Thamaraiselvan et al., 2015).

27.2 REMEDIATION OF HYDROCARBONS

There is an array of ways to remediate hydrocarbon contaminated soil such as thermal treatment (soil incineration), soil washing, chemical oxidation, use of nanotechnology, phytoremediation, bioremediation, and mycoremediation. The use of Fenton reagents is a well-developed technology used for the remediation of recalcitrant organic compounds from both soil and water. The technology has primarily been used to treat pesticides, PAHs, polychlorinated biphenyls (PCBs), and petroleum hydrocarbons. The reaction involves the generation of hydroxide radicals after the addition of H_2O_2 and Fe^{2+} salts (Peluffo et al., 2018). Sometimes the addition of iron ions isn't necessary if there is already an endogenous source of iron. Persulphate can be used instead of hydrogen peroxide and may be preferred as it is more stable and therefore transported further into the soil's subsurface. Activated persulphate radicals react with organic chemicals causing partial or complete mineralization, making it a noteworthy option for remediation. In one study thermal activation of persulfate achieved over 90% degradation of PAHs after seven days of treatment. However, it is important to note that this method had a negative impact on the soil's characteristics, such as drastically decreasing the soil's pH by between 2.8 and 6.0 pH units. Also, even with low concentrations of oxidants used, a significant decrease was seen in the soil's bacterial populations (Peluffo et al., 2018), and can be too expensive if treating very large volumes of soil.

27.3 BIOREMEDIATION

Bioremediation is the exploitation of the metabolic capabilities of bacteria, fungi, algae and plants to degrade or sequester pollutants in the environment. It can be used for many environmental pollutants such as PAHs, PCBs, petroleum hydrocarbons, and pesticides. Compared with techniques such as chemical oxidation and thermal treatment, it is considered to have a low environmental impact and be relatively cost-effective (Das and Chandran, 2011). The main disadvantage of bioremediation is relatively slow degradation rates. The rate at which compounds are degraded will depend on their molecular weight, water solubility, bioavailability, and the condition of the soil itself (pH, redox potential, temperature, presence of other essential nutrients such as nitrogen, phosphorus, and iron). Biodegradation using natural populations of microorganisms is one of the major mechanisms by which hydrocarbon waste can be removed from the environment. Once a microorganism with the suitable metabolic pathways is identified in a polluted environment, nutrients and oxygen can be adjusted to achieve optimal growth and degradation (Das and Chandran, 2011; Liu et al., 2016). The two main approaches to bioremediation are biostimulation and bioaugmentation. Biostimulation is the addition of appropriate nutrients to the existing microbial populations to stimulate the degrading microbes (Das and Chandran, 2011). Contaminants in soil are often degraded slowly over time even without intervention. This is due to natural attenuation, i.e., the presence of natural occurring bacteria with the ability to survive in the contaminated soil and use the contaminants as a food or energy source (Gielnik et al., 2019). This can be an extremely slow process if the population sizes of these indigenous degraders are very low ($>10^2$ cfu g^{-1} soil). In such cases, bioaugmentation can be used, where known contaminant-degrading bacteria are added to the contaminated soil at the polluted site.

Most petroleum hydrocarbons are degraded under aerobic conditions by bacteria and fungi. Several bacteria are known to feed exclusively on hydrocarbons.

When treating a site contaminated by a complex mixture of hydrocarbons, using a single hydrocarbon degrading microbe is not sufficient to effectively degrade the many different chemicals present. Instead, mixed populations with various enzymatic capabilities are used to increase efficiency (Das and Chandran, 2011). The presence of very high concentrations of petroleum organics or heavy metals in site soils can be toxic or inhibit the growth and reproduction of microbes responsible for biodegradation. Conversely, very low concentrations of organic material will result in diminished levels of microbial activity. In general, soil concentrations of total petroleum hydrocarbons (TPH) in the range of 10000 to 50000 ppm, or heavy metals exceeding 2500 ppm, are considered inhibitory and/or toxic to most microorganisms. If TPH concentrations are greater than 10000 ppm, or the concentration of heavy metals is greater than 2500 ppm, then the contaminated soil should be thoroughly mixed with clean soil to dilute the contaminants so that the average concentrations are below toxic levels. The disadvantage of this is that the volume of contaminated soil is increased significantly (US EPA, 2017).

27.4 BIOPILES

Biopiles, also known as biocells, bioheaps, biomounds, and compost piles, are used to reduce concentrations of petroleum constituents in excavated soils through the use of biodegradation. This technology involves heaping contaminated soils into piles (or "cells") and stimulating aerobic microbial activity within the soils through the aeration and/or addition of minerals, nutrients, and moisture. The enhanced microbial activity results in degradation of adsorbed petroleum-product constituents through microbial respiration (Baldan et al., 2015). Biopiles are similar to landfarms in that they are both above-ground, engineered systems that use oxygen, generally from air, to stimulate the growth and reproduction of aerobic bacteria which, in turn, degrade the petroleum constituents adsorbed to soil. While landfarms are aerated by tilling or plowing, biopiles are aerated most often by forcing air by injection or extraction through slotted or perforated piping placed throughout the pile. Biopiles have proved to be very successful, rapid and reliable systems for removing organic pollutants from soil. However, they can be expensive to set up and operate as they require extensive air pumping systems and the vented air may require treatment to remove organic pollutants so as not to contaminate the atmosphere. Biopiling has been used successfully to treat crude oil, diesel, and lubrication oil contaminated soil. Coulon et al. (2010) found that TPH levels had dropped down to 22% of initial levels after 30 weeks. In a study carried out by Wang et al. (2012), 50% degradation of TPH was observed after 220 days in biopiles used to treat oily sludge. Mao et al. (2009) used biopiling with active aeration and nutrient amendment to remediate soil which was grossly contaminated with mineral oil (10000–20000 mg kg^{-1} DM) and after 18 weeks TPH levels were shown to be reduced by 80%. In their study they constructed relatively small biopiles and the aeration process involved mixing the biopile soil every two weeks. This method may not be logistically and economically feasible for very large volumes of soil.

27.5 PHYTOREMEDIATION

Phytoremediation is the use of plants and their associated microorganisms to remove pollutants from either soil or water (see also chapters 28 and 29). It is an environmentally sustainable technology that is solar powered and cost effective. Phytoremediation can be used to treat petroleum hydrocarbons, chlorinated solvents, pesticides, explosives, heavy metals, and radionuclides in both soil and/or water (Truu et al., 2015). The main weaknesses of phytoremediation are the length of time it takes for contaminants to be remediated, unpredictable or inconsistent results and plants may not be equipped to break down hydrocarbons so efficiently. Another problem with phytoremediation is the toxic effect that low molecular weight hydrocarbons, such as naphthalene, has on plants when used to remediate polluted soils. This issue can be overcome by introducing rhizospheric or endophytic bacteria with hydrocarbon degrading capabilities to help protect the plant from the toxic effect (Germaine et al., 2009, 2014).

Factors which effect the efficacy of phytoremediation are plant species, pollutant concentration, and soil characteristics such as pH and nutrient levels. Several studies suggest the success of phytoremediation technologies will rely on understanding plant-microbiome interactions. Therefore, steps required to improve phytoremediation efficiency are creating favorable soil conditions, increasing contaminant bioavailability, plant growth conditions, and exploitation of most useful plant associated microorganisms (Thijs et al., 2016). Currently there is lack of extensive knowledge on what is referred to as the "metaorganism level," i.e., an understanding of the host and microbiome together. Evidence has shown that use of appropriate rhizospheric or endophytic microbes can improve the efficiency of phytoremediation (Thijs et al., 2016; see also chapters 28 and 29). There are several different processes involved in phytoremediation which exploit different mechanisms, e.g., phytoextraction (uptake of heavy metals), phytostabilization (stabilization of compounds in the root zone), and rhizoremediation (degradation of organic compounds in the rhizosphere) (Agnello et al., 2016).

27.6 RHIZOREMEDIATION OF TOTAL PETROLEUM HYDROCARBONS

The rhizosphere is the area around plant roots that is enriched with much higher populations of active microbes than are present in the bulk soil. Rhizoremediation takes advantage of this increased microbial population and activity in order to degrade organic pollutants in the soil. Organic compounds in soil can be degraded by both aerobic and anaerobic means, however, TPHs are not very susceptible to anaerobic bioremediation, especially if the hydrocarbons are recalcitrant. Therefore, aerobic bioremediation is considered the best option for TPH remediation. Rhizoremediation provides a microaerophilic environment around the roots as they interact with the microbes in the soil. This creates conditions suitable for aerobic bacterial growth. Plant exudates provide a substrate for the bacteria to grow in greater numbers as well as adjusting the pH of the soil. There are four ways that biodegradation of TPHs can be improved by exudates, they are: direct degradation, enhanced pollutant bioavailability, structural analogy and co-metabolism, stimulation by energy, and nutrient flow.

Plant root exudates can contain extracellular enzymes such as laccases and peroxidases (Martin et al., 2014) which are capable of degrading certain hydrocarbon compounds. Bioavailability of TPHs can be improved by the secretion of lipophilic chemicals such as rhamnolipid. The increased solubility of the TPHs makes them more susceptible to microbial attack. Bioavailability may also be increased by secreted carboxylates such as citrate and oxalate. Degradation by structural analogy and co-metabolism describes the principle where plant exudates and secondary metabolites are similar in structure to many hydrocarbons. The structural similarities between the exudates and pollutants stimulate the co-metabolic process improving hydrocarbon degradation as a result. Co-metabolism is considered to be the main factor in TPH degradation. Root exudates stimulate the energy and nutrient from within the rhizosphere, this ultimately increases microbial activity and increases TPH degradation. In a year 10–40% of photosynthetically derived products can be secreted into the soil for the purpose of providing food and energy for microbes which live in the rhizosphere (Williams and de Vries, 2019).

27.7 ECOPILING

Microbes are more efficient at degrading some hydrocarbons more than others, e.g., linear alkanes > branched alkanes > small aromatics > cyclic alkanes. High molecular weight hydrocarbons are particularly difficult to break down (Das and Chandran, 2011). Combining bioremediation strategies such as bioaugmentation and biostimulation with phytoremediation appears to be an effective way to remove recalcitrant hydrocarbons from large volumes of contaminated soil. Ecopiling is a modification of traditional passive biopiling in that, instead of enclosing the biopile with black plastic, the pile is planted with suitable phytoremediation plants in order to promote rhizoremediation (Germaine et al., 2014). The ecopile process involves bio-stimulation of indigenous hydrocarbon degraders, bio-augmentation through inoculation with known PAH degrading consortia and phytoremediation, through the effect of root growth and penetration throughout the soil and the resulting stimulation of microbial activity in the rhizosphere. Ecopiles are typically constructed using contaminated soil excavated, treated with nutrients (nitrogen and phosphorus fertilizer) and seeded with hydrocarbon degrading microbial inoculants. The base layer of soil (~0.5 m deep) is placed over a heavy-duty polythene liner and 50 mm perforated drainage pipes placed at approximately 1 m centers, laterally across the pile to allow for passive ventilation (Figure 27.1). The ecopile is then raised in consecutive 0.5 m layers, comprising contaminated soil and drainage piping to a height of 2–3 m. The ecopiles are constructed in a trapezoidal shape with a 2:1 slope from base to top. Finally, each ecopile is capped with uncontaminated topsoil (~5 cm deep) and seeded with suitable plant species such as clover and ryegrass.

Between 2009 and 2012, Germaine et al. (2014) used nine Ecopiles to remediate over 5000 m^3 of TPH

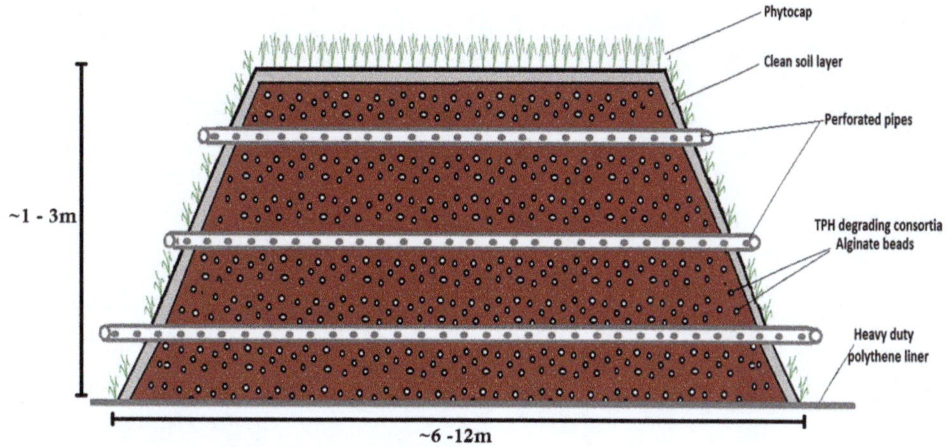

FIGURE 27.1 Structure of an ecopile.

contaminated soil at a former industry site in Carlow, Ireland. After 52 weeks 50% degradation was achieved and after three years the soil had achieved the remediation target level of < 50 mg TPH kg^{-1} soil (Germaine et al., 2014). In another ecopiling project a small-scale ecopile was established on a former creosote treatment facility in Tipperary, Ireland (Liu et al., 2016). The soil consisted of a clayey sediment, heavily contaminated with PAHs (575 ± 63 mg PAHs kg^{-1}). The site was a difficult case due to the long period of time the PAHs had been in the soil meaning they were tightly bound to the soil particles. In addition to this, the soil had a high clay content which has strong adsorption and absorption properties (Liu et al., 2016). Two 1 ton ecopiles were set up containing PAH degrading microorganisms and nutrients (nitrogen and phosphorous, ratio of 25:4). Uncontaminated top soil 5 cm deep was added to provide viable conditions for rye grass and clover to grow during the sensitive germination stage. The ecopiles were set up in August 2013 and monitored over a two-year period (Figure 27.2A).

PAH concentrations were determined at 0, 56, 161, and 730 days. After 56 days 66% and 65% reductions were seen, at 161 days PAHs had reduced by 86% and 81%. At the end of the 730 days PAHs had been reduced by 95% and 94%. Storey et al. (2018) later investigated the bacterial community structure associated with contaminated soil from this site (Figure 27.3). They found that the bacterial diversity of this soil was dominated by two phyla; the Actinobacteria and the Proteobacteria. At the genus level the dominant microbes in the soil were *Micrococci* (9%), *Acidobacter* (9%), *Sphingomonas* (3%), *Bradyrhizobium* (2.6%), and *Lysobacter* (0.9%), with a significant portion of unclassified organisms. They also demonstrate the presence of key genes involved in the degradation of PAHs, and which are important in the natural attenuation of the soil.

Between 2014 and 2016, ecopiles were used to treat two crude oil sludge dump sites in Gudao, Shandong, China (Figure 27.2B–27.2C). The soil was contaminated with up to 40000 ppm of crude oil and consisted of a saline–alkaline soil. The average daily temperature in this region was 25 ^{O}C and provided ideal conditions for degradation. However, during the summer months, air temperatures did reach up to 40 ^{O}C, which required that the ecopiles be irrigated daily to maintain a soil moisture content of 20–30%. After a 20 month period TPH levels were reduced to 100–500ppm. Ecopiles were also utilized by Marchand et al. (2018) to treat hydrocarbon contaminated soil from a 40-year-old municipal landfill site at Moskogen, in southern Sweden. The small scale ecopiles consisted of 3 m^2 soil with a height of about 0.7 m and were planted with either alfalfa (*Medicago sativa* L) either in monoculture or co-cropped with sunflower (*Helianthus annuus* L). They found a 40–50% reduction in hydrocarbons after five months, compared to untreated soils.

In 2019, a large-scale ecopiling project was initiated in order to treat over 6000 m^3 of heavy fuel oil contaminated soils in Carlow, Ireland (Wang et al., 2021). Analysis of the soils before ecopiling showed average TPH levels (C_{10}–C_{40}) of 9294 mg kg^{-1}. The wide range of TPH levels (~2000–26700 mg kg^{-1}) detected in seven soil samples taken at the site demonstrated the heterogeneous dispersal of the hydrocarbons in the soil. TPH analysis showed that the vast majority of the hydrocarbons present were high molecular weight compounds (C_{16}–C_{35} aliphatics and aromatics, but mostly aromatics). Seven

FIGURE 27.2 Ecopiles used to treat (A) creosote contaminated soil in Ireland (2013), (B) and (C) crude oil contaminated soil in China (2014–2016), and (D) heavy fuel oil contaminated soil in Ireland (2019).

large ecopiles were constructed (Figure 27.2D), typically 20 m long, 3–4 m high, and 5–6 m wide. These were augmented with a consortium of total petroleum hydrocarbon (TPH) degrading bacteria that had been isolated from contaminated soil on the same site. This consortium was isolated by incubating 10 g of TPH contaminated soil in 500 ml minimal media (Abraham et al., 2002) at 30 °C, 100 rpm for two weeks and sub-culturing in the same media supplemented with a either diesel oil (D), crude oil (CO), engine oil (EO), an extract from the contaminated soil (SE), a mixture of all four oils (MO), and a sample of combined enrichments used to inoculate the ecopiles (MC). These were subcultured every two weeks for three months. The microbial composition of these consortia can be seen in Figure 27.4. After the third sub-culturing enrichment rapid growth of the consortia were observed with ODs in excess of 1.0 after 3–4 days. After the sixth subculture samples were removed after two weeks of growth, DNA was extracted from each enrichment and the community was analyzed through 16S RNA gene amplicon seuencing. The top 10 most abundant bacterial genera present in the enrichments can be seen in Figure 27.4.

As can be seen in Figure 27.4 similar genera of bacteria were present in all five enrichments even though they were all using different oils as a sole carbon and energy sources. *Extensimonas* and *Pseudomonas* made up the two most abundant genera in nearly all the enrichments (with the exception of the engine oil (EO) enrichment, where *Extensimonas* and *Acinetobacter* were the two most dominant genera present). In addition to these three genera, *Fulivomonas* and *Lysobacter* were present in significant numbers in most of the enrichments. Members of the genus *Pseudomonas* have been

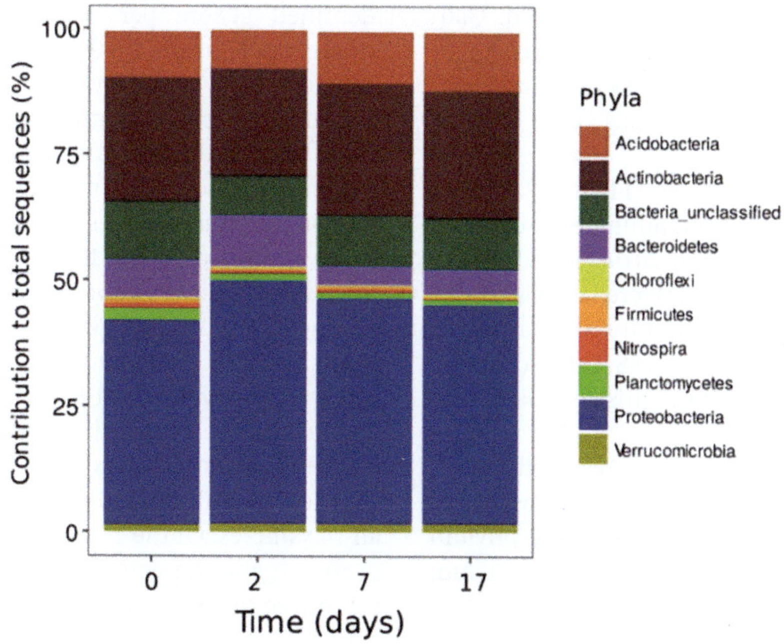

FIGURE 27.3 Bacterial communities present in PAH contaminated soil originating from a creosote contaminated site in Tipperary Ireland. (Storey et al. 2018 / Licenced under CC BY-4.0).

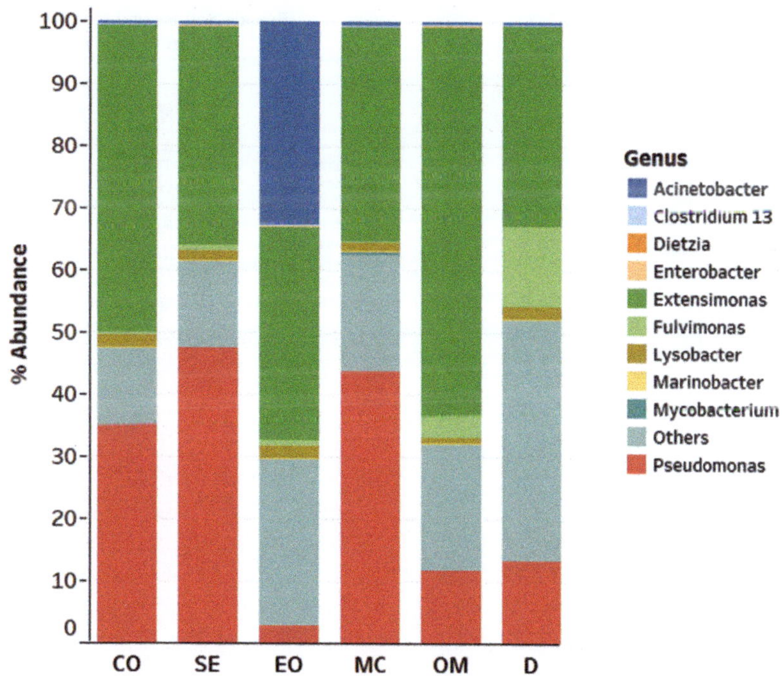

FIGURE 27.4 Bacterial communities present in TPH degrading enrichments used to prepare inoculants for the ecopiles in Carlow Ireland (2019). CO: crude oil, SE: soil extract, EO: engine oil, MC: mix culture; OM: mixed oils, D: diesel. Bars represent the average of two samples per enrichment ($n = 2$).

widely reported to be associated with hydrocarbon degradation (Garrido-Sanz et al., 2019; Sarkar et al., 2020). In contrast, there are very few reports of members of the genus *Extensimonas* linked to hydrocarbon degradation. Gao et al. (2020) did show that *Extensimonas* was consistently present in 50 crude oil well heads in Shandong, China. Although it was not one of the dominant genera observed, it was consistently present at over 0.5% relative abundance in each of the 50 samples analyzed. Gauchotte-Lindsay et al. (2019) also reported *Extensimonas* species that were associated with hydrocarbon degradation (2,4-Di-tert-butylphenol) in soils from former gas works.

A mixture of all five enrichments (Figure 27.4 (MC)) was used to inoculate the TPH contaminated soil during ecopile construction in Carlow in 2019, in addition to fertilizer (N:P). Samples of the soils were taken after two weeks and the microbial community was analyzed through 16S rRNA (bacterial) and 18S rRNA (Fungal) amplicon sequencing. The results of this analysis can be seen in Figure 27.5.

In the ecopile soil, nine phyla represented 99% of the bacterial community. The remaining 1% consisted of more than 36 different phyla. Bacteria in the phylum *Proteobacteria* made up 60–75% of the reads. Phylum *Actinobacter* made up the next largest group (12–24%, average of 17%), while *Firmicutes* made up 3–8% of the community. In total, 734 different bacterial genera were identified in the TPH contaminated soil with 489 ± 42 identified genera per ecopile. These made up (on average) 50% of the total sequencing reads per ecopile sample. At the genus level, the 14 ecopile soil samples had similar bacterial communities (Figure 27.5). Given that all these ecopiles are constructed from the same contaminated soil, this similarity in bacterial community is to be expected. Ten genera made up between 39–55% of the identified genera found in soil samples (average 46%). *Lysobacter*, a Gram-negative genus, was the most represented individual genus across all 14 Ecopile samples (14–39%). *Lysinomonas*, *Luteimonas*, *Mycobacterium*, *Pseudomonas*, and *Extensiomonas* typically made up 1–6% of the bacterial communities. A total of 1850 different bacterial OTUs were identified across the 14 samples, of which, 1115 OTUs (60%) were common to all 14 samples. Of the 10 main genera that were present in the oil enrichments that were used to inoculate the soil, three of these were observed in the soil after two weeks (*Pseudomonas*, and *Extensiomonas and Lysobacter*). These were all present in the soil in significant abundance.

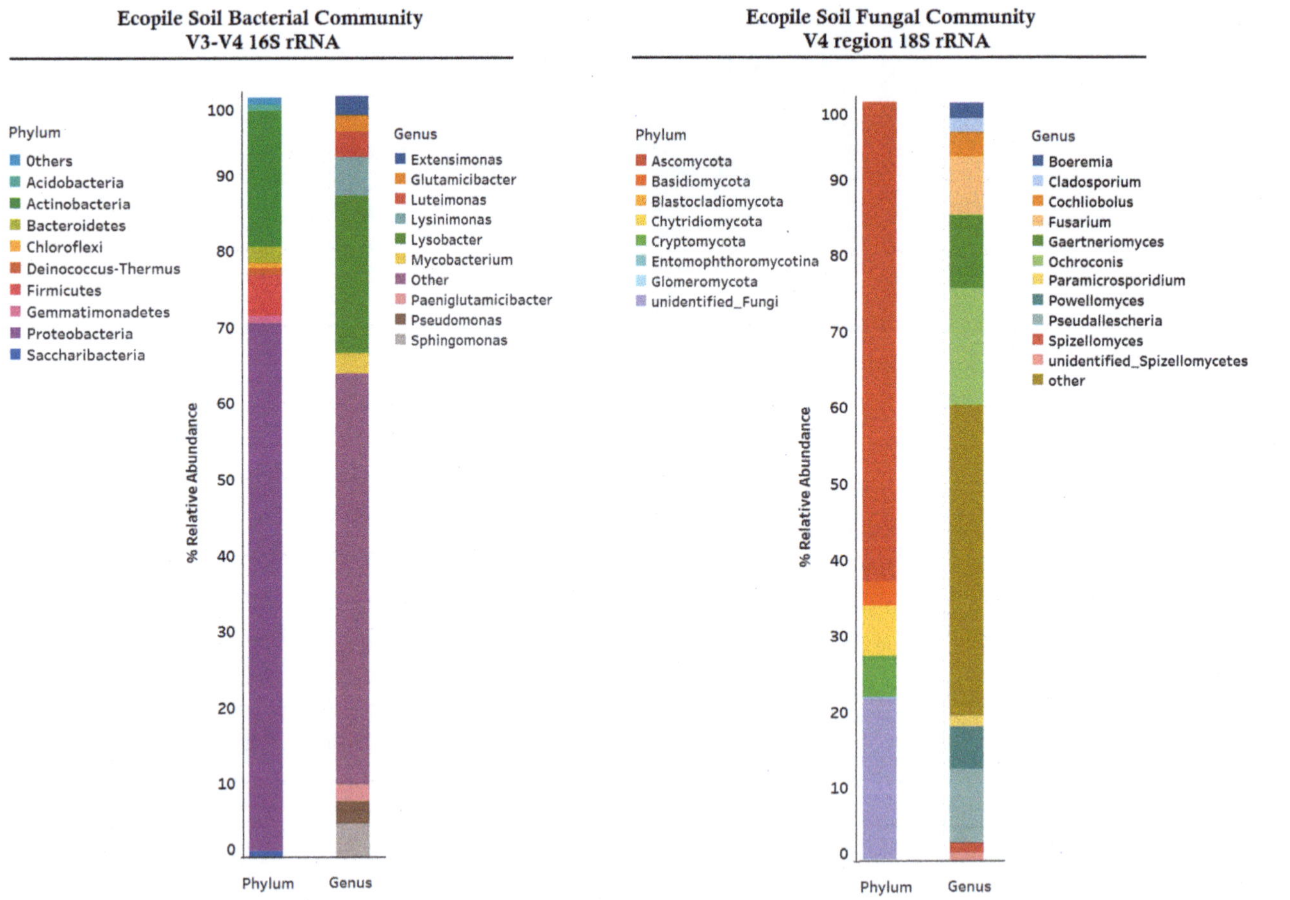

FIGURE 27.5 Top 10–12 most abundant bacterial and fungal phyla and genera present in the ecopile soil in July 2019. The data is the average of the analysis of 14 individual ecopile soil samples ($n = 14$).

Analysis of the fungal microbiome of the ecopile soils based on 18S rDNA sequencing found that, on average, 55% (range 48–64%) of the sequence reads could be assigned to the Kingdom fungi. The remaining sequences were assigned to other Eukaryotic soil organisms such as protists, plants, and animals. A total of seven fungal phyla were identified in the microbiome of the ecopile soil. Unidentified fungal reads accounted for 5–50% of the fungal reads (average 17%). Reads assigned to the phyla *Ascomycota* made up the vast majority (average 70%) of the reads. At the genus level, 145 different fungal genera were identified in the TPH contaminated soil with 66–90 different genera per sample. Figure 27.5 shows the most abundant fungal genera in the ecopile soil samples. In total, 66 of these genera belonged to the phylum *Ascomycota*, 39 to *Basidiomycota*, 17 were unassignable at phyla level, 12 to *Chytridomycota*, 4 to *Glomeromycota*, 3 to *Entomophthoromycota*, and 2 to *Crytomycota*.

After 12 months (July 2019–July 2020) the TPH levels in the ecopiles decreased by an average of 65% (Figure 27.6). The vast majority of this degradation occurred in the summer/autumn months (July–November) while little or no degradation occurred in the winter (November–February). Lower levels of TPH were detected in both the November 2019 and February 2020 samples than were detected in the July 2020 samples. These lower levels in November and February had suggested that TPH degradation rates were up 80%. The high TPH levels detected in July 2020 were mainly associated with two out of the 14 soil samples tested. Both of these originated from the same ecopile (ecopile 7). This ecopile had the highest levels of TPH (average 4600 ppm) in July 2019 and this had dropped to 2300 ppm by July 2020 (representing a 50% TPH degradation rate over the 12-month period). The soil samples taken in July 2020 were taken from the very center of the ecopile using a mechanical digger. TPH biodegradation rates in the center of the ecopiles are expected to be slower due to the lower oxygen levels and the fact that the roots of the ryegrass are unlikely to penetrate this far into the pile. However, a degradation rate of 50% in the center of the biopile was still considered very satisfactory after just 12 months.

27.8 CONCLUSION

Ecopiling is a multi-factorial bio-remediation process involving bio-stimulation, bio-augmentation, and phyto-remediation for the removal of hydrocarbons from

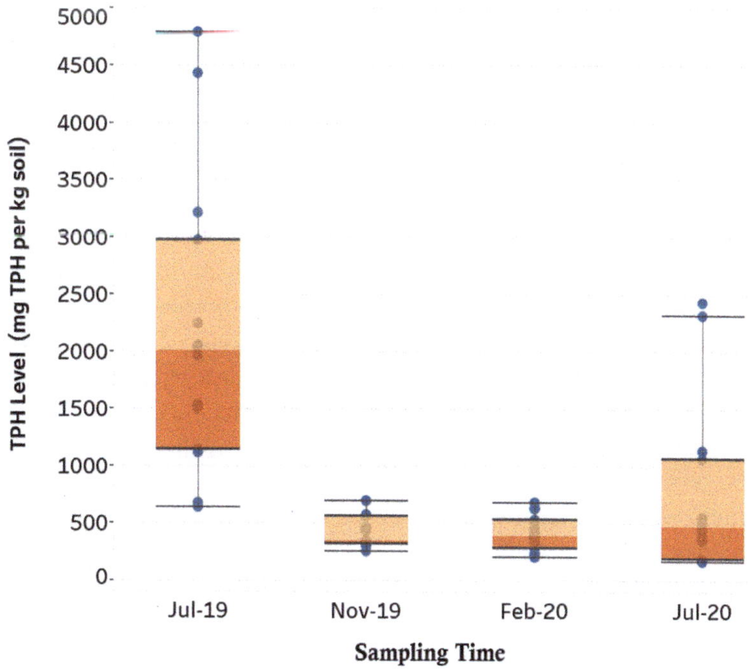

FIGURE 27.6 TPH levels in the ecopile soil samples over the course of 12 months (Jul-2019 to Jul-2020). The large reduction in TPH levels between sampling in Jul-19 and Nov-19 indicates that the ecopiles were more active in this timeframe. The boxplots represent the average TPH concentration across all seven ecopiles ($n = 14$).

contaminated soils. Ecopiling has proved to be a successful technology for the treatment of soil contaminated with PAHs, crude oil, heavy fuel oil, and soils from municipal landfills. The process is most suitable for remediation projects with large volumes of contaminated soil, sufficient space available to construct the ecopiles, and most importantly, have medium to long-term timescales (2–4 year). Many former abandoned industrial and crude oil contaminated sites fit in to this category. One of the key advantages to the process is the reduced cost of the remediation process. Once the ecopiles are constructed, there is little additional cost in terms of labor and maintenance, although the longer process time may incur additional monitoring costs. The success of the ecopiling process is reliant on the activity of many different good microbes that have the ability to metabolize the various hydrocarbon constituents found in the soil. Many of these key microbes such as *Pseudomonas* species have been extensively studied, yet many more, including *Extensiomonas and Lysobacter*, require detailed investigationin order to fully exploit the potential of these beneficial microbes in the ecopiling process.

ACKNOWLEDGMENTS

Many parts of the work presented here was funded under the European Union's Horizon 2020 research and innovation program "GREENER" under grant agreement No. 826312, and the Institute of Technology Carlow Postgraduate Presidents Awards.

REFERENCES

1. Abraham WR, Nogales B, Golyshin PN, Pieper DH, Timmis KN. 2002. Polychlorinated biphenyl-degrading microbial communities in soils and sediments. *Current Opinion in Microbiology* 5: 246–253.
2. Agnello AC, Bagard M, van Hullebusch ED, Esposito G, Huguenot D. 2016. Comparative bioremediation of heavy metals and petroleum hydrocarbons co-contaminated soil by natural attenuation, phytoremediation, bioaugmentation and bioaugmentation-assisted phytoremediation. *Science of the Total Environment* 563-564: 693–703.
3. Ahmed A, Ossai IC, Hassan A, Hamid FS. 2018. A review on environmental contamination of petroleum hydrocarbons and its biodegradation. *International Journal of Environmental Sciences & Natural Resources* 11: 63–69.
4. Baldan E, Basaglia M, Fontana F, Shapleigh JP, Casella S. 2015. Development, assessment and evaluation of a biopile for hydrocarbons soil remediation. *International Biodeterioration and Biodegradation* 98: 66–72.
5. Chunhui W, Wu S, Zhou S, Shi Y, Song J. 2017. Characteristics and source identification of Polycyclic Aromatic Hydrocarbons (PAHs) in Urban Soils: A review. *Pedosphere* 27: 17–26.
6. Coulon F, Awadi AM, Cowie W, Mardlin D, Pollard S, Cunningham C, et al. 2010. When is a soil remediated? Comparison of biopiled and windrowed soils contaminated with bunker-fuel in a full-scale trial. *Environmental Pollution* 158: 3032–3040.
7. Das N. and Chandran P. (2011). Microbial Degradation of Petroleum Hydrocarbon Contaminants: An Overview. *Biotechnology Research International* 2011: 13 pages. Article ID 941810, 2011. https://doi.org/10.4061/2011/941810
8. EPA US. 2017. Biopiles. In: *How to Evaluate Alternative Cleanup Technologies for Underground Storage Tank Sites: A Guide for Corrective Action Plan Reviewers*, Chapter:4.
9. Gao P, Tian H, Li G, Zhao F, Xia W, et al. 2020. *Environmental Conditions and Diffusion-limited Microbial Transfer Drive Specific Microbial Communities Detected at Different Sections in Oil-production Reservoir with Water-flooding*. In Press.
10. Garrido-Sanz D, Redondo-Nieto M, Guirado M, Pindado Jiménez O, Millán R, et al. 2019. Metagenomic insights into the bacterial functions of a diesel-degrading consortium for the rhizoremediation of diesel-polluted soil. *Genes* 14:10(6): 456.
11. Gauchotte-Lindsay C, Aspray TJ, Knapp M, Ijaz UZ. 2019. Systems biology approach to elucidation of contaminant biodegradation in complex samples – Integration of high-resolution analytical and molecular tools. *Faraday Discussions* 218: 481–504.
12. Germaine K, Byrne J, Liu X, Keohane J, Culhane J, et al. 2014. Ecopiling: A combined phytoremediation and passive biopiling system for remediating hydrocarbon impacted soils at field scale. *Frontiers in Plant Science* 5: 1–6.
13. Germaine K, Keogh E, Ryan D, Dowling D. 2009. Bacterial endophyte-mediated naphthalene phytoprotection and phytoremediation. *FEMS Microbiology Letters* 296: 226–234.
14. Gielnik A, Pechaud Y, Huguenot D, Cébron A, Riom JM, et al. 2019. Effect of digestate application on microbial

respiration and bacterial communities' diversity during bioremediation of weathered petroleum hydrocarbons contaminated soils. *Science of the Total Environment* 670: 271–281.

15. Hussein I, Abdel-Shafya M, Mansour S. 2016. A review on polycyclic aromatic hydrocarbons: Source, environmental impact, effect on human health and remediation. *Egyptian Journal of Petroleum* 25 1: 107–123.
16. Liu X, Kiwanuka S, Cleary K, Ryan D, Dowling D, Germaine K. 2016. Use of Ecopiling to remediate PAH-contaminated storm-water lagoon sediment. *Journal of Bioremediation and Biodegradation* 7: 1–8.
17. Mao D, Lookman R, Van De Weghe H, Van Look D, Vanermen G, et al. 2009. Detailed analysis of petroleum hydrocarbon attenuation in biopiles by high performance liquid chromatography followed by comprehensive two-dimensional gas chromatography. *Journal Chromatography. A* 1216: 1524–1527.
18. Marchand C, Menc M, Jani Y, Kaczala F, Notini P, et al. 2018. Pilot scale aided-phytoremediation of a co-contaminated soil. *Science of the Total Environment* 618: 753–764.
19. Martin BC, George SJ, Price CA, Ryan MH, Tibbett M. 2014. The role of root exuded low molecular weight organic anions in facilitating petroleum hydrocarbon degradation: Current knowledge and future directions. *Science of the Total Environment* 472: 642–653.
20. Meneses HE, Bandeira GC. 2013. *Handbook of Polycyclic Aromatic Hydrocarbons: Chemistry, Occurrence, and Health Issues*. Nova Science Publishers.
21. Peluffo M, Rosso JA, Morelli IS, Morab VC. 2018. Strategies for oxidation of PAHs in aged contaminated soil by batch reactors. *Ecotoxicology and Environmental Safety* 151: 76–82.
22. Pinedo J, Ibáñez R, Lijzen JPA, Irabien A. 2013. Assessment of soil pollution based on total petroleum hydrocarbons and individual oil substances. *Journal of Environmental Management* 130C: 72–79.
23. Sarkar J, Saha A, Roy A, et al. 2020. Development of nitrate stimulated hydrocarbon degrading microbial consortia from refinery sludge as potent bioaugmenting agent for enhanced bioremediation of petroleum contaminated waste. *World Journal of Microbiology and Biotechnology* 36: 1–20.
24. Storey S, Ashaari MM, Clipson N, Doyle E, de Menezes AB. 2018. Opportunistic bacteria dominate the soil microbiome response to phenanthrene in a microcosm-based study. *Frontiers in Microbiology* 21: 1–13.
25. Thamaraiselvan R, Peramaiyan R, Natarajan N, Boopathy L, et al. 2015. Exposure to polycyclic aromatic hydrocarbons with special focus on cancer. *Asian Pacific Journal of Tropical Biomedicine* 5: 182–189.
26. Thijs S, Sillen W, Rineau F, Weyens N, Vangronsveld J. 2016. Towards an enhanced understanding of plant-microbiome interactions to improve phytoremediation: Engineering the metaorganism. *Frontiers in Microbiology* 7: 1–15.
27. Truu J, Truu M, Espenberg M, Nõlvak H, Juhanson J. 2015. Phytoremediation and plant-assisted bioremediation in soil and treatment wetlands: A review. *The Open Biotechnology Journal* 9: 85–92.
28. Wang M, Garrido-Sanz D, Sansegundo-Lobato P, Redondo-Nieto M, Conlon R, et al. 2021. Soil microbiome structure and function in Ecopiles used to remediate petroleum contaminated soil. *Frontiers In Environmental Science* 9: 1–18.
29. Wang X, Wang Q, Wang S, Li F, Guo G. 2012. Effect of biostimulation on community level physiological profiles of microorganisms in field-scale biopiles composed of aged oil sludge. *Bioresource Technology* 111: 308–315.
30. Williams A, de Vries FT. 2019. Plant root exudation under drought: Implications for ecosystem functioning. *New Phytologist* 25: 1899–1905.

CHAPTER 28

Plant–Microbe Interactions in Environmental Restoration

Ondrej Uhlik, Jachym Suman, Jakub Papik, Michal Strejcek, and Tomas Macek
University of Chemistry and Technology, Prague, Faculty of Food and Biochemical Technology, Department of Biochemistry and Microbiology, Prague, Czechia

28.1 INTRODUCTION TO PLANT–MICROBE INTERACTIONS

Plants and microorganisms on our planet have evolved side by side. This co-evolution has resulted in numerous plant–microbe interactions, many of which are essential today for the welfare of either partner and proper functioning of the entire ecosystems (Singer et al., 2003). Such interactions (Figure 28.1) include, to name two prominent examples, mycorrhizal mutualism, in which the plant provides their fungal partners with organic carbon and fungi provide the plant with nitrogen, phosphorus and water and which seems to have been crucial for the evolution of terrestrial plants (Delaux et al., 2015; see chapter 39), and mutualistic nitrogen fixation, during which specific (e.g., leguminous) plants attract and then host in their roots (or stems) rhizobia that fix atmospheric nitrogen (Mus et al., 2016; see chapter 37).

It has been shown that the plant selects what microorganisms will thrive in its vicinity (Pascale et al., 2020; Sasse et al., 2018). The major tool the plant uses for this purpose is rhizodeposition, which is the release of compounds from roots into the surrounding soil, either through root exudation or root tissue turnover (Jones et al., 2009; Lambers et al., 2009; Wichern et al., 2008). Root exudation occurs at a significant cost for the plant – in fact, more than 10% of plant-fixed carbon is deposited in the surrounding soil (Jones et al., 2009; Lakshmanan et al., 2014), although this can reach up to 30–40% in young seedlings (Badri and Vivanco, 2009) – but it happens for a reason. The chemicals contained within the root exudates alter the physicochemical conditions of the soil and both the composition and activity of resident microbial communities (Neumann et al., 2009). Molecules that are rhizodeposited include several sugars, amino acids, organic acids, fatty acids, sterols, growth factors, vitamins, enzymes, flavonoids, nucleotides, plant hormones, alcohols, alkyl sulfides, inorganic ions, and gaseous molecules (for a review see Dennis et al. (2010)). Thereby, the rhizosphere, which is the thin layer of soil directly influenced by the roots, is a particularly active spot (hot-spot) for plant–microbe interactions (Chaudhry et al., 2005; Philippot et al., 2013). Consequently, a greater density of microorganisms develops in the rhizosphere than bulk soil (Dennis et al., 2010). Plants can also influence soil microorganisms beyond the rhizosphere via litter deposition onto the top soil (Chen et al., 2006). One of the major changes in the composition of soil microbial communities that can be ascribed to the plant is the enrichment of potentially plant-beneficial microorganisms (Bais et al., 2006; Otieno et al., 2015; Philippot et al., 2013; Tian et al., 2020).

The plant–microbe interactions are beneficial to mankind in multiple ways. Leaving aside the proper

Good Microbes in Medicine, Food Production, Biotechnology, Bioremediation, and Agriculture, First Edition. Edited by Frans J. de Bruijn, Hauke Smidt, Luca S. Cocolin, Michael Sauer, David Dowling, and Linda Thomashow.

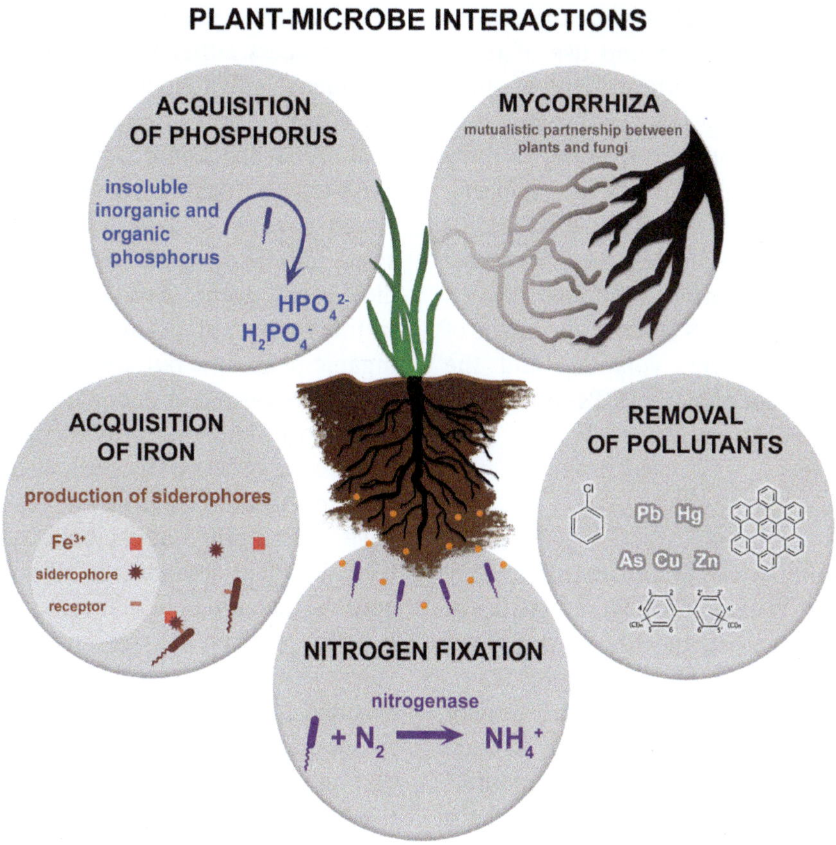

FIGURE 28.1 Examples of plant–microbe interactions.

functioning of the ecosystems, which is a topic way beyond the scope of this chapter, efficiently tailored plant–microbe interactions can increase crop yield, stress resistance, plant defense systems, or expedite environmental restoration (Sasse et al., 2018). The latter process will be discussed in this chapter. However, before proceeding any further, a short introduction to the topic of pollutants and their biodegradation has to be made.

28.2 POLLUTANTS AND THEIR BIODEGRADATION

28.2.1 Persistent Organic Pollutants

Since the industrial revolution, significant quantities of toxic chemicals have been released into the environment, some of which have accumulated at particular sites and exerted negative effects on local biota and ecosystem functioning. These chemicals, which we call pollutants, have entered the environment either intentionally, such as in the case of organochlorine pesticides, or unintentionally, such as in the case of leakage from industrial facilities. Organic pollutants were massively manufactured and used mainly in the last century. A good example of such a group of pollutants is polychlorinated biphenyls (PCBs). PCBs were produced by catalytic chlorination of biphenyl to a specified weight percent of chlorine. Each commercial PCB mixture sold under the names Aroclor, Fenclor, Kanechlor, Delor, or others contained 60 to 90 congeners, i.e., variants differing in the number of chlorine atoms and their configurations on the biphenyl moiety (Macková et al., 2010). Such mixtures of PCBs found use as/in dielectric fluid in capacitors and transformers, hydraulic fluids, lubricants, plasticizers, heat exchange fluids, paint additives, pigment extenders, carbonless copy paper, etc. Global world production of PCBs has been estimated to be over 2 million tons (Holoubek, 2001). Eventually, PCBs were included in the so called *dirty dozen*, i.e., the initial 12 persistent organic pollutants (POPs) which were covered by the international treaty Stockholm Convention on

Persistent Organic Pollutants in 2001 with the aim to eliminate or restrict the production and use of these toxic and recalcitrant chemicals with an adverse effect on human and environmental health (Lallas, 2001). In addition to PCBs, the Stockholm Convention initially covered 11 other pollutants or their groups, including dichlorodiphenyltrichloroethane (DDT), hexachlorobenzene (HCB), and other pesticides, and was later expanded to include other priority pollutants. In fact, the recalcitrant nature of the POPs is the major cause of legacy contamination. Moreover, volatility and the ability of long-range transport of many POPs contributes to global contamination (Choi and Wania, 2011).

28.2.2 Pollutants of Natural Origin

However, not all pollutants are of the anthropogenic origin; some chemicals that exhibit the characteristics of pollutants, such as persistence in the environment, accumulation in the food chain, and subsequent detrimental effects on environmental functions and human health, have an origin in natural materials such as crude oil or natural gas. Typical examples of such compounds are petroleum hydrocarbons, including long-chain alkanes, monoaromatics, or polyaromatic hydrocarbons (PAHs) (Cerniglia, 1992; Shuttleworth and Cerniglia, 1995; see also chapter 27). PAHs are of particular concern due to toxicity, mutagenicity, and carcinogenicity of many of them (Nisbet and LaGoy, 1992). Environmental depositions of PAHs come from multiple sources, including direct aerial fallout, leakage or spills from industrial or household sources, improper disposal of petroleum products, or simply natural oil seeps and surface water run-off (Cerniglia, 1992).

28.2.3 Biodegradation and Bioremediation

Remediation of polluted sites typically involves physical and chemical processes like excavation ensued by landfilling or incineration of the contaminated material; these processes are rather expensive and difficult to execute. An alternative approach to attenuate or even completely eliminate environmental pollution is the employment of the metabolic activity of living organisms and/or their enzymes in a process termed bioremediation or biorestoration (Macková et al., 2006; McGuinness and Dowling, 2009; see chapter 27). Although many organisms, including plants or insects, take part in this process (Singer, 2006; Singer et al., 2004), the key players are microorganisms.

In essence, bioremediation relies on biodegradation, i.e., a process, either a single biological reaction or rather a sequence of reactions, that results in the simplification of a substrate and a decrease or loss of its toxicity (Olson et al., 2003; Shackelford, 2013). From the environmental perspective, ideal biodegradation is mineralization, which is the complete degradation of an organic substrate into its inorganic constituents, most often carbon dioxide and water (Olson et al., 2003; Ramos et al., 1994). In this respect, plants are able to take up many types of organic pollutants but are not dependent on these compounds as carbon or energy sources. For such purposes, the majority of plants are equipped with very potent photosynthetic systems which allow them to be self-sufficient in terms of carbon. This is the reason why plants do not have such a range of catabolic systems as microorganisms do. Plant systems usually transform organic compounds, including pollutants and allelopathic chemicals, in order to decrease their phytotoxicity (Macek et al., 2000). Thus plants themselves can be used for the removal of some pollutants from the contaminated soil, water, sediments, or even air in the process called phytoremediation (Schnoor et al., 1995; see chapter 27); however, the mineralization of the pollutants without the activity of microorganisms is not achieved.

Mineralization consists mostly of a series of reactions, some of which are oxidative while others are reductive, depending on the nature of the molecule, specifically on the oxidation state of carbon in the organic molecule. The lower the oxidation state of carbon, the more susceptible it is for oxidation and vice versa. Therefore, the mineralization of highly chlorinated PCBs and other halogenated pollutants in which the carbon oxidation state can be as high as +4, is likely to be initiated by a reduction in order to lower the oxidation state of carbon.

In essence, the specific way of biodegradation of a pollutants is associated with the structural complexity of the compound and follows the general logic of microbial metabolism. In the majority of chemoorganotrophic microorganisms, a range of organic substrates can be catabolized (Chain et al., 2006; Garrido-Sanz et al., 2020; Harwood and Parales, 1996). The catabolism is initiated by so called peripheral, or upper, pathways which funnel the substrates into a smaller range of intermediates which are then further transformed into central metabolism intermediates by lower degradation pathways, and finally are mineralized (Figure 28.2). In order for the microorganism to use a wide range of organic substrates efficiently, enzymes of the peripheral metabolism have

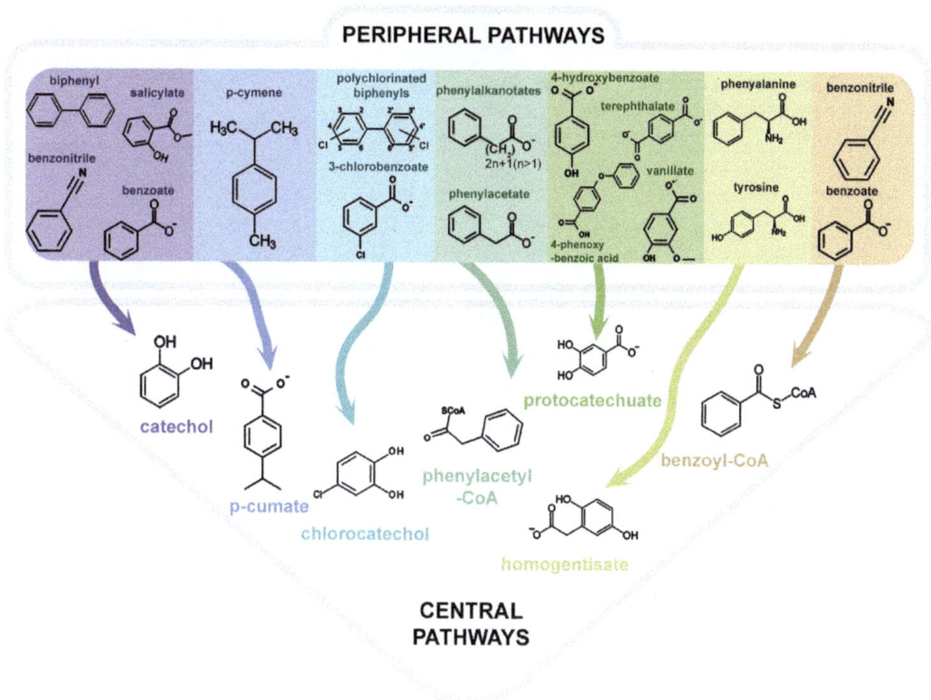

FIGURE 28.2 Typical logic of microbial metabolism. A wide range of compounds are metabolized by enzymes of peripheral pathways into a limited number of degradation intermediates which are then funneled into central pathways.

often broad substrate specificity, whereas the enzymes of central metabolic pathways are mostly strictly specific.

28.3 CATABOLISM VERSUS COMETABOLISM

If the microorganism disposes of proper catabolic enzymes, the particular pollutant can be catabolized, and its degradation yields energy. This is the case of, for instance, biphenyl and 4-chlorobiphenyl or naphthalene (Pothuluri and Cerniglia, 1994; Sylvestre, 1980). More complex compounds can be transformed cometabolically. Cometabolism (Figure 28.3) is a phenomenon resulting from a broad substrate specificity of enzymes of the peripheral pathways due to which these enzymes transform not only their primary substrates but, fortuitously, also other structurally similar compounds, often pollutants (Horvath, 1972).

For instance, PCBs do not commonly serve as a carbon source for bacteria; however, many of the 209 PCB congeners can be degraded via the pathway that is used by bacteria to degrade biphenyl (Macková et al., 2006, 2010). Several strains of the genera *Pseudomonas*, *Alcaligenes*, *Achromobacter*, *Burkholderia*, *Comamonas*, *Sphingomonas*, *Ralstonia*, or *Acinetobacter* among Gram-negative, and *Rhodococcus*, *Corynebacterium*, *Arthrobacter*, and *Bacillus* among Gram-positive bacteria have been isolated that metabolize biphenyl and cometabolize PCB congeners (Furukawa, 2000; Vergani et al., 2019).

Similar to biphenyl-degrading aerobic bacteria, which can often cometabolize multiple PCB congeners (Furukawa, 1994; Pieper and Seeger, 2008; Sylvestre, 1980), naphthalene-degrading bacteria can cometabolize some more complex PAHs; phenol or toluene-degrading bacteria can cometabolize trichloroethylene, etc. (Macková et al., 2006). In soil, the primary substrates for cometabolism that feed the microbes are often provided by plants (Figure 28.3). This represents one of the important examples of plant–microbe interactions with a direct impact on environmental restoration (Musilová et al., 2016; Singer et al., 2003).

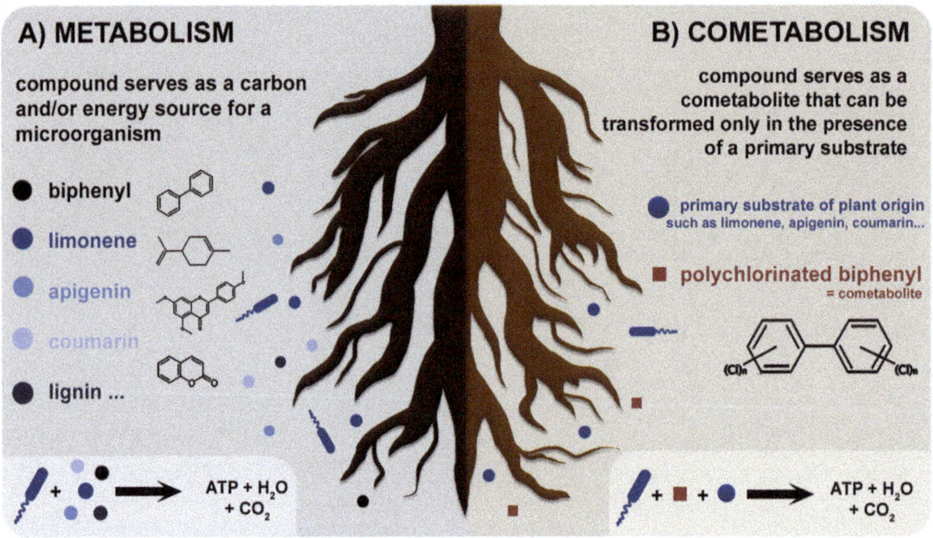

FIGURE 28.3 Difference between catabolism and cometabolism. Secondary plant metabolites (SPMs) often serve as primary substrates of enzymes that degrade pollutants.

28.4 SECONDARY PLANT METABOLITE HYPOTHESIS

Plants are a vast source of different kinds of chemicals. The greatest diversity is attributed to secondary plant metabolites (SPMs), of which over 200000 structures have been described (Yonekura-Sakakibara and Saito, 2009). SPMs, along with other compounds, such as carbohydrates or amino acids, enter the soil through rhizodeposition (Jones et al., 2009; Lambers et al., 2009; Wichern et al., 2008) and appear to be of outstanding importance in terms of controlling soil ecology: not only can they serve as carbon and/or energy sources for bacteria (Donnelly et al., 1994) but they often have antimicrobial activity (Singer et al., 2003) or the ability to disrupt bacterial quorum sensing (Koh et al., 2013) and some SPMs have been predicted to regulate the decomposition of organic matter (Loranger et al., 2002; Northup et al., 1998). Among other roles, SPMs in plants act as allelopathic chemicals as well as chemicals protecting the plant from herbivores (Rausher, 2001). As insects and other herbivores constantly develop resistance to SPMs, plants in turn need to modify their SPM content and composition to provide effective protection. These changes are also likely to affect soil microbial communities, which respond by developing detoxifying enzymes with a broad substrate specificity (Singer et al., 2003). Many SPMs have been reported to structurally resemble anthropogenic pollutants (Singer et al., 2003, 2004), which might explain why anthropogenic pollutants can be cometabolized in the presence of SPMs. This then leads to the hypothesis that enzymes that are also fortuitously involved in the degradation of anthropogenic pollutants originally evolved for the detoxification and/or degradation of SPMs. One such example which supports this hypothesis is enzymes involved in the degradation of PCBs (Figure 28.3), which will be discussed in the following section.

28.4.1 SPM Hypothesis and the Degradation of PCBs

Although biphenyl is not commonly present in the environment, biphenyl-degrading bacteria are ubiquitous in soils and sediments (Focht, 1995). One explanation for this phenomenon might be that these bacteria utilize biphenyl-related structures that are not foreign in the biosphere; for instance lignin, the second most abundant biomolecule on Earth which comprises approximately 25% of terrestrial biomass, contains biphenyl structures as the major branching points (Chang and Jiang, 2020). It is not surprising then, that efficient PCB-degrading bacteria have been isolated from termite guts where a lignin-rich diet is decomposed (Chung et al., 1994). Eventually, Sylvestre and colleagues postulated that the biphenyl degradation pathway in soil bacteria may have evolved to serve different ecological functions

(Pham and Sylvestre, 2013) including the metabolism of SPMs (Pham et al., 2015; Toussaint et al., 2012).

With reference to bioremediation, several microbes have been isolated that are able to cometabolize PCBs with SPMs as primary substrates (Figure 28.3). One of the first reports showed that naturally occurring phenolics, such as apigenin, catechin, myricetin, or coumarin, supported the growth of PCB-degrading bacteria and cometabolism of several di- to pentachlorobiphenyls (Donnelly et al., 1994). Gilbert and Crowley (1997), Hernandez et al. (1997), Singer et al. (2000), and Singer et al. (2001) demonstrated that similar capabilities can be ascribed to terpenes, such as carvone, limonene, or *p*-cymene, and salicylic acid. Additionally, studies were conducted that elucidated the role of SPMs as natural inducers of the expression of biodegradative pathways (Figure 28.4). Park et al. (1999) identified carvone to induce the activity of the ring-cleaving enzyme of the biphenyl-degradation pathway in a Gram-positive degrader of the genus *Arthrobacter*, and later reports appeared summarizing that SPMs act as inducers of hydroxylating dioxygenases which initiate the degradation of biphenyl and PCBs (Kim et al., 2003; Master and Mohn, 2001). Induction of aromatic ring hydroxylase dioxygenase genes was systematically studied by the team of Sylvestre et al. They identified that isoflavone induced the biphenyl degradation pathway in a model PCB-degrader *Pandoraea pnomenusa* B356 (Pham et al., 2012), and that flavonoids induced the biphenyl degradation pathway in the rhizobacterium *Rhodococcus erythropolis* U23A, although they did not support the growth of the bacterium (Pham et al., 2015; Toussaint et al., 2012). Recent results of our team suggest that SPMs of different structures, including phenolic acids, flavonoids, coumarins, and monoterpenes, induce the transcription of the aromatic ring (e.g., biphenyl) hydroxylating dioxygenase genes in pseudomonads (Zubrová et al., 2021) and rhodococci, with the latter being often much more efficiently induced than by biphenyl (Figure 28.4).

28.4.2 SPM Hypothesis and Other Pollutants as Examples

Similar to promoting the degradation of PCBs, SPMs have been reported to promote the degradation of other pollutants. In accordance with that, the rhizosphere has been shown to stimulate the expression of various pollutant-degrading genes, including those of petroleum hydrocarbon or organochloride degradation pathways, a phenomenon which is termed a rhizosphere effect (Pagé et al., 2015). Of particular note is salicylate, a naphthalene degradation intermediate, which is frequently connected with enhanced degradation of the PAHs, some of the most persistent petroleum hydrocarbons (Chen and Aitken, 1999; Powell et al., 2008; Schell, 1985; Tian et al., 2003). More recently, the secondary compound hypothesis was also proposed for chlorinated solvents; certain monoaromatic SPMs were found to stimulate the degradation of *cis*-1,2-dichloroethylene (cDCE) (Fraraccio et al., 2017).

28.5 CONCLUSIONS

The tremendous metabolic capabilities of microorganisms to degrade pollutants initially limited research in bioremediation only to bacteria. Such research started about 70 years ago with the investigation of the fate of

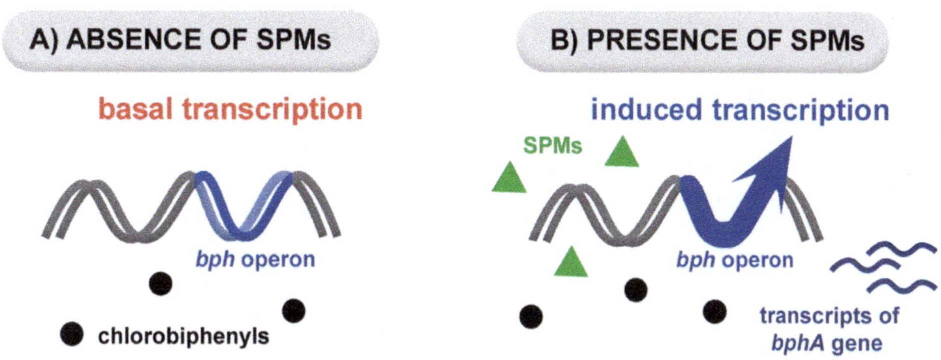

FIGURE 28.4 Capability of secondary plant metabolites (SPMs) to promote microbial biodegradation by inducing the transcription of biodegradative genes (in this case *bphA* gene which encodes a large subunit of biphenyl dioxygenase and which is the first gene of the biphenyl operon).

pesticides in soils. It was another 30 years before the potential of plants was seen as a co-players in the field of bioremediation (Macková et al., 2006). The benefits of plant–microbe interactions for bioremediation were later recognized (Anderson et al., 1993; Schnoor et al., 1995) and widely investigated (Macková et al., 2006; McCutcheon and Schnoor, 2003). In the last decade or two, a fascinating development of molecular methods has rapidly advanced the research on plant–microbe interactions and shed new light on bioremediation and rhizoremediation; these methods span from high-throughput nucleic acid sequencing methods, which enable us to study the extensive biodiversity in the environment, via methods revealing the link between the metabolic process in the environment and its microbial executors to methods focusing on the investigation of microbial and plant metabolites (Badri et al., 2009; Jansson and Hofmockel, 2018; Uhlik et al., 2013). The list of these methods is way beyond the scope of this chapter; however, they have and will continue to increase our understanding of plant–microbe interactions in the future and help us more efficiently harness these interactions for efficient rhizoremediation. Thereby, good plants will continue stimulating good microbes in order to collaborate on a good thing – environmental restoration.

ACKNOWLEDGMENTS

The completion of this work was made possible by the support from the INTER-EXCELLENCE program of the Ministry of Education, Youth and Sports of the Czech Republic under grant no. LTAUSA19013 and Czech Science Foundation under grant no. 20-00291S.

REFERENCES

1. Anderson TA, Guthrie EA, Walton BT. 1993. Bioremediation in the rhizosphere. *Environmental Science & Technology* 27: 2630–2636.
2. Badri DV, Vivanco JM. 2009. Regulation and function of root exudates. *Plant, Cell and Environment* 32: 666–681.
3. Badri DV, Weir TL, van der Lelie D, Vivanco JM. 2009. Rhizosphere chemical dialogues: Plant–microbe interactions. *Current Opinion in Biotechnology* 20: 642–650.
4. Bais HP, Weir TL, Perry LG, Gilroy S, Vivanco JM. 2006. The role of root exudates in Rhizosphere interactions with plants and other organisms. *Annual Review of Plant Biology* 57: 233–266.
5. Cerniglia CE. 1992. Biodegradation of polycyclic aromatic hydrocarbons. *Biodegradation* 3: 351–368.
6. Chain PS, Denef VJ, Konstantinidis KT, Vergez LM, Agullo L, et al. 2006. *Burkholderia xenovorans* LB400 harbors a multi-replicon, 9.73-Mbp genome shaped for versatility. *Proceedings of the National Academy of Sciences of the United States of America* 103: 15280–15287.
7. Chang H-M, Jiang X. 2020. Biphenyl structure and its impact on the macromolecular structure of lignin: A critical review. *Journal of Wood Chemistry and Technology* 40: 81–90.
8. Chaudhry Q, Blom-Zandstra M, Gupta S, Joner EJ. 2005. Utilising the synergy between plants and rhizosphere microorganisms to enhance breakdown of organic pollutants in the environment. *Environmental Science and Pollution Research* 12: 34–48.
9. Chen CR, Condron LM, Xu ZH, Davis MR, Sherlock RR. 2006. Root, rhizosphere and root-free respiration in soils under grassland and forest plants. *European Journal of Soil Science* 57: 58–66.
10. Chen S-H, Aitken MD. 1999. Salicylate stimulates the degradation of high-molecular weight polycyclic aromatic hydrocarbons by *Pseudomonas saccharophila* P15. *Environmental Science And Technology* 33: 435–439.
11. Choi S-D, Wania F. 2011. On the reversibility of environmental contamination with persistent organic pollutants. *Environmental Science and Technology* 45: 8834–8841.
12. Chung S-Y, Maeda M, Song E, Horikoshij K, Kudo T. 1994. A Gram-positive polychlorinated biphenyl-degrading bacterium, *Rhodococcus erythropolis* strain TA421, isolated from a termite ecosystem. *Bioscience, Biotechnology, and Biochemistry* 58: 2111–2113.
13. Delaux P-M, Radhakrishnan GV, Jayaraman D, Cheema J, Malbreil M, et al. 2015. Algal ancestor of land plants was preadapted for symbiosis. *Proceedings of the National Academy of Sciences* 112: 13390–13395.
14. Dennis PG, Miller AJ, Hirsch PR. 2010. Are root exudates more important than other sources of rhizodeposits in structuring rhizosphere bacterial communities? *FEMS Microbiology Ecology* 72: 313–327.
15. Donnelly PK, Hegde RS, Fletcher JS. 1994. Growth of PCB-degrading bacteria on compounds from photosynthetic plants. *Chemosphere* 28: 981–988.

16. Focht DD. 1995. Strategies for the improvement of aerobic metabolism of polychlorinated biphenyls. *Current Opinion in Biotechnology* 6: 341–346.

17. Fraraccio S, Strejček M, Dolinová I, Macek T, Uhlík O. 2017. Secondary compound hypothesis revisited: Selected plant secondary metabolites promote bacterial degradation of *cis*-1,2-dichloroethylene (cDCE). *Scientific Reports* 7: 8406.

18. Furukawa K. 1994. Molecular genetics and evolutionary relationship of PCB-degrading bacteria. *Biodegradation* 5: 289–300.

19. Furukawa K. 2000. Biochemical and genetic bases of microbial degradation of polychlorinated biphenyls (PCBs). Journal of General and Applied Microbiology 46: 283–96.

20. Garrido-Sanz D, Redondo-Nieto M, Martín M, Rivilla R. 2020. Comparative genomics of the *Rhodococcus* genus shows wide distribution of biodegradation traits. *Microorganisms* 8: 774.

21. Gilbert ES, Crowley DE. 1997. Plant compounds that induce polychlorinated biphenyl biodegradation by *Arthrobacter* sp. strain B1B. *Applied and Environmental Microbiology* 63: 1933–1938.

22. Harwood CS, Parales RE. 1996. The *b*-ketoadipate pathway and the biology of self-identity. *Annual Review of Microbiology* 50: 553–590.

23. Hernandez B, Koh S-C, Chial M, Focht D. 1997. Terpene-utilizing isolates and their relevance to enhanced bio transformation of polychlorinated biphenyls in soil. *Biodegradation* 8: 153–158.

24. Holoubek I. 2001. Polychlorinated biphenyl (PCB) contaminated sites worldwide. In: Robertson LW, Hansen LG (Eds.), *PCBs: Recent Advances in Environmental Toxicology and Health Effects*, University Press of Kentucky, Lexington, KY, USA, pp. 17–26.

25. Horvath RS. 1972. Microbial co-metabolism and the degradation of organic compounds in nature. *Bacteriological Reviews* 36: 146–155.

26. Jansson JK, Hofmockel KS. 2018. The soil microbiome—from metagenomics to metaphenomics. *Current Opinion in Microbiology* 43: 162–168.

27. Jones DL, Nguyen C, Finlay RD. 2009. Carbon flow in the rhizosphere: Carbon trading at the soil–root interface. *Plant and Soil* 321: 5–33.

28. Kim B-H, Oh E-T, So J-S, Ahn Y, Koh S-C. 2003. Plant terpene-induced expression of multiple aromatic ring hydroxylation oxygenase genes in *Rhodococcus* sp. strain T104. *The Journal of Microbiology* 41: 349–352.

29. Koh C-L, Sam C-K, Yin W-F, Tan LY, Krishnan T, et al. 2013. Plant-derived natural products as sources of anti-quorum sensing compounds. *Sensors* (Basel, Switzerland) 13: 6217–6228.

30. Lakshmanan V, Selvaraj G, Bais HP. 2014. Functional Soil Microbiome: Belowground Solutions to an aboveground problem. *Plant Physiology* 166: 689–700.

31. Lallas PL. 2001. The Stockholm Convention on persistent organic pollutants. *American Journal of International Law* 95: 692–708.

32. Lambers H, Mougel C, Jaillard B, Hinsinger P. 2009. Plant–microbe-soil interactions in the rhizosphere: An evolutionary perspective. *Plant and Soil* 321: 83–115.

33. Loranger G, Ponge J-F, Imbert D, Lavelle P. 2002. Leaf decomposition in two semi-evergreen tropical forests: Influence of litter quality. *Biology and Fertility of Soils* 35: 247–252.

34. Macek T, Macková M, Káš J. 2000. Exploitation of plants for the removal of organics in environmental remediation. *Biotechnology Advances* 18: 23–34.

35. Macková M, Dowling D, Macek T. 2006. Phytoremediation and rhizoremediation. In: Marcel Hofman, Jozef Anné (Eds.), *Focus on Biotechnology*, 9A. Springer, Dordrecht, Netherlands, pp. 300

36. Macková M, Uhlík O, Lovecká P, Viktorová J, Nováková M, et al. 2010. Bacterial degradation of polychlorinated biphenyls. In: Loy A, Mandl M, Barton LL (Eds.), *Geomicrobiology: Molecular and Environmental Perspective*. Springer, Dordrecht, Netherlands, pp. 347–366.

37. Master ER, Mohn WW. 2001. Induction of *bphA*, encoding biphenyl dioxygenase, in two polychlorinated biphenyl-degrading bacteria, psychrotolerant *Pseudomonas* strain Cam-1 and mesophilic *Burkholderia* strain LB400. *Applied and Environmental Microbiology* 67: 2669–2676.

38. McCutcheon SC, Schnoor JL. 2003. *Phytoremediation: Transformation and Control of Contaminants*, Vol. 121. John Wiley & Sons, Inc, Hoboken, NJ.

39. McGuinness M, Dowling D. 2009. Plant-Associated Bacterial Degradation of Toxic Organic Compounds in Soil. *International Journal of Environmental Research and Public Health* 6: 2226–2247.

40. Mus F, Crook MB, Garcia K, Garcia Costas A, Geddes BA, et al. 2016. Symbiotic nitrogen fixation and the challenges to its extension to nonlegumes. *Applied and Environmental Microbiology* 82: 3698–3710.

41. Musilová L, Rídl J, Polívková M, Macek T, Uhlík O. 2016. Effects of secondary plant metabolites

on microbial populations: Changes in community structure and metabolic activity in contaminated environments. *International Journal of Molecular Sciences* 17: 1205.

42. Neumann G, George TS, Plassard C. 2009. Strategies and methods for studying the rhizosphere—the plant science toolbox. *Plant and Soil* 321: 431–456.

43. Nisbet ICT, LaGoy PK. 1992. Toxic equivalency factors (TEFs) for polycyclic aromatic hydrocarbons (PAHs). *Regulatory Toxicology and Pharmacology* 16: 290–300.

44. Northup RR, Dahlgren RA, McColl JG. 1998. Polyphenols as regulators of plant-litter-soil interactions in Northern California's Pygmy Forest: A positive Feedback? *Biogeochemistry* 42: 189–220.

45. Olson PE, Reardon KF, Pilon-Smits EAH. 2003. Ecology of rhizosphere bioremediation. In: McCutcheon SCS, Jerald L (Eds.), *Phytoremediation: Transformation and Control of Contaminants*, Vol. 121. John Wiley & Sons, Inc, Hoboken, NJ, pp. 317–353.

46. Otieno N, Lally R, Kiwanuka S, Lloyd A, Ryan D, et al. 2015. Plant growth promotion induced by phosphate solubilizing endophytic *Pseudomonas* isolates. *Frontiers in Microbiology* 6.

47. Pagé AP, Yergeau É, Greer CW. 2015. *Salix purpurea* stimulates the expression of specific bacterial xenobiotic degradation genes in a soil contaminated with hydrocarbons. *Plos One* 10: e0132062.

48. Park Y-I, So J-S, Koh S-C. 1999. Induction by carvone of the polychlorinated biphenyl (PCB)-degradative pathway in *Alcaligenes eutrophus* H850 and its molecular monitoring. *Journal of Microbiology and Biotechnology* 9: 804–810.

49. Pascale A, Proietti S, Pantelides IS, Stringlis IA. 2020. Modulation of the Root Microbiome by Plant Molecules: The basis for targeted disease suppression and plant growth promotion. *Frontiers in Plant Science* 10: 1741.

50. Pham TT, Pino Rodriguez NJ, Hijri M, Sylvestre M. 2015. Optimizing polychlorinated biphenyl degradation by flavonoid-induced cells of the rhizobacterium *Rhodococcus erythropolis* U23A. *Plos One* 10: e0126033.

51. Pham TT, Sylvestre M. 2013. Has the bacterial biphenyl catabolic pathway evolved primarily to degrade biphenyl? The diphenylmethane case. *Journal of Bacteriology* 195: 3563–3574.

52. Pham TTM, Tu Y, Sylvestre M. 2012. Remarkable ability of *Pandoraea pnomenusa* B356 biphenyl dioxygenase to metabolize simple flavonoids. *Applied and Environmental Microbiology* 78: 3560–3570.

53. Philippot L, Raaijmakers JM, Lemanceau P, van der Putten WH. 2013. Going back to the roots: The microbial ecology of the rhizosphere. *Nature Reviews Microbiology* 11: 789–799.

54. Pieper DH, Seeger M. 2008. Bacterial metabolism of polychlorinated biphenyls. *Journal of Molecular Microbiology and Biotechnology* 15: 121–138.

55. Pothuluri JV, Cerniglia CE. 1994. Microbial metabolism of polycyclic aromatic hydrocarbons. In: Chaudry GR (Ed.), *Biological Degradation and Bioremediation Toxic Chemicals*. Chapman and Hall, London, UK, pp. 92–124.

56. Powell SN, Singleton DR, Aitken MD. 2008. Effects of enrichment with salicylate on bacterial selection and PAH mineralization in a microbial community from a bioreactor treating contaminated soil. *Environmental Science and Technology* 42: 4099–4105.

57. Ramos JL, Díaz E, Dowling D, Lorenzo V, Molin S, et al. 1994. The behavior of bacteria designed for biodegradation. *Nature Biotechnology* 12: 1349–1356.

58. Rausher MD. 2001. Co-evolution and plant resistance to natural enemies. *Nature* 411: 857–864.

59. Sasse J, Martinoia E, Northen T. 2018. Feed your friends: Do plant exudates shape the root microbiome? *Trends in Plant Science* 23: 25–41.

60. Schell MA. 1985. Transcriptional control of the *nah* and *sal* hydrocarbon-degradation operons by the *nahR* gene product. *Gene* 36: 301–309.

61. Schnoor JL, Licht LA, Mccutcheon SC, Wolfe NL, Carreira LH. 1995. Phytoremediation of organic and nutrient contaminants. *Environmental Science and Technology* 29: A318–A323.

62. Shackelford CD. 2013. *Geoenvironmental Engineering. Reference Module in Earth Systems and Environmental Sciences*. Elsevier.

63. Shuttleworth KL, Cerniglia CE. 1995. Environmental Aspects of PAH Biodegradation. *Applied Biochemistry and Biotechnology* 54: 291–302.

64. Singer AC. 2006. The chemical ecology of pollutant biodegradation: Bioremediation and phytoremediation from mechanistic and ecological Perspectives. In: Macková M, Dowling D, Macek T (Eds.), *Phytoremediation and Rhizoremediation*. Springer, Dordrecht, Netherlands, pp. 5–21.

65. Singer AC, Crowley DE, Thompson IP. 2003. Secondary plant metabolites in phytoremediation and biotransformation. *Trends in Biotechnology* 21: 123–130.

66. Singer AC, Gilbert ES, Luepromchai E, Crowley DE. 2000. Bioremediation of polychlorinated biphenyl-contaminated soil using carvone and surfactant-grown bacteria. *Applied Microbiology and Biotechnology* 54: 838–843.

67. Singer AC, Jury W, Luepromchai E, Yahng CS, Crowley DE. 2001. Contribution of earthworms to PCB bioremediation. *Soil Biology and Biochemistry* 33: 765–776.

68. Singer AC, Thompson IP, Bailey MJ. 2004. The tritrophic trinity: A source of pollutant-degrading enzymes and its implications for phytoremediation. *Current Opinion in Microbiology* 7: 239–244.

69. Sylvestre M. 1980. Isolation method for bacterial isolates capable of growth on *p*-chlorobiphenyl. *Applied and Environmental Microbiology* 39: 1223–1224.

70. Tian L, Ma P, Zhong J-J. 2003. Impact of the presence of salicylate or glucose on enzyme activity and phenanthrene degradation by. *Pseudomonas Mendocina. Process Biochemistry* 38: 1125–1132.

71. Tian T, Reverdy A, She Q, Sun B, Chai Y. 2020. The role of rhizodeposits in shaping rhizomicrobiome. *Environmental Microbiology Reports* 12: 160–172.

72. Toussaint J-P, Pham T, Barriault D, Sylvestre M. 2012. Plant exudates promote PCB degradation by a rhodococcal rhizobacteria. *Applied Microbiology and Biotechnology* 95: 1589–1603.

73. Uhlik O, Leewis M-C, Strejcek M, Musilova L, Mackova M, et al. 2013. Stable isotope probing in the metagenomics era: A bridge towards improved bioremediation. *Biotechnology Advances* 31: 154–165.

74. Vergani L, Mapelli F, Šuman J, Cajthaml T, Uhlík O, Borin S. 2019. Novel PCB-degrading Rhodococcus strains able to promote plant growth for assisted rhizoremediation of historically polluted soils. Plos One 14: e0221253.

75. Wichern F, Eberhardt E, Mayer J, Joergensen RG, Müller T. 2008. Nitrogen rhizodeposition in agricultural crops: Methods, estimates and future prospects. *Soil Biology and Biochemistry* 40: 30–48.

76. Yonekura-Sakakibara K, Saito K. 2009. Functional genomics for plant natural product biosynthesis. *Natural Product Reports* 26: 1466–1487.

77. Zubrová A, Michalíková K, Semerád J, Strejček M, Cajthaml T, Šuman J, Uhlík O. 2021. Biphenyl 2, 3-dioxygenase in Pseudomonas alcaliphila JAB1 is both induced by phenolics and monoterpenes and involved in their transformation. Frontiers in Microbiology 12: 917.

CHAPTER 29

Microbial Endophytes for Clean-up of Pollution

Robert J. Tournay and Sharon L. Doty
School of Environmental and Forest Sciences, College of the Environment, University of Washington, Seattle 98195-2100, WA

29.1 INTRODUCTION

Pollutants can be classified as either inorganic, such as heavy metals or other elements that cannot be degraded, or organic pollutants that are carbon-based and could be potentially degraded. Conventional methods of remediation include excavation to another site for storage, incineration of the contaminated soil to vaporize the pollutants, or capping of the site to limit public exposure to the contaminants. In the case of polluted water, pump-and-treat is a common conventional method. Alternatively, chemicals are pumped down into the contaminated aquifer to react with the pollutants. All of these methods are expensive and can render the site beyond the financial capability of the site owner for remediation, leaving the public exposed to potential harm.

Phytoremediation, the use of plants to remediate pollutants, offers several advantages over conventional remediation, as it is less expensive, requires lower energy and chemical inputs, enjoys strong public acceptance, and provides a range of ecosystem services; e.g., wildlife habitat, soil stabilization, carbon sequestration, and the potential for secondary market products (Gerhardt et al., 2017; McCutcheon and Schnoor, 2003; see also chapters 27 and 28). For example, the phytoremediation of moderately contaminated lands using native woody species (e.g., poplar and willow) could be a valuable source of feedstock for the production of bioethanol, or other biochemicals, from cellulosic biomass (Isebrands and Richardson, 2014). Ecosystem restoration combined with agroforestry could restore economic productivity to agricultural lands while providing a range of forest ecosystem services.

To maximize the potential of phytoremediation, research is focusing on improving phytoremediation success by addressing the problems of phytotoxicity and the relatively slower rate of remediation. One approach is to use genetic engineering of the plants (Dhankher et al., 2011; Pilon-Smits and Freeman, 2006; Rylott et al., 2015). While this approach has been successful, regulatory hurdles have restricted its deployment. A more practical approach to improvement of phytoremediation has been the partnering of specific microorganisms with plants to overcome some of the limitations of phytoremediation. Endophytes, the microorganisms within plants, have been recognized for providing a wide range of direct and indirect benefits to their host plants (Bulgarelli et al., 2013; Santoyo et al., 2016), and recent investigations identifying endophytes capable of detoxifying anthropogenic pollutants suggests a potential role in phytoremediation (Doty, 2008; Rylott, 2014; Thijs et al., 2016; Weyens et al., 2009a, 2009b).

Since the methods of phytoremediation of inorganic and organic pollutants differ substantially, this review will address the two classes separately.

Good Microbes in Medicine, Food Production, Biotechnology, Bioremediation, and Agriculture, First Edition. Edited by Frans J. de Bruijn, Hauke Smidt, Luca S. Cocolin, Michael Sauer, David Dowling, and Linda Thomashow.

29.2 ORGANIC POLLUTANTS

29.2.1 Background on Conventional Remediation and Phytoremediation of Common Organic Pollutants

Through intentional dumping or accidental spills, organic pollutants including solvents, polycyclic aromatic hydrocarbons (PAH's), petroleum, and pesticides are common environmental pollutants. Trichloroethylene (TCE), a degreaser and solvent and one of the most common organic pollutants of soil and groundwater, will serve as our example of conventional remediation strategies. TCE is a known human carcinogen (IARC, 2012), making its remediation a high priority. Common conventional remediation strategies for TCE include methods to increase its volatilization such as air sparging, or pumping, and treating through sorption onto barriers (Kuppusamy et al., 2016). Averaging between USD $700000 to $3000000 per site, not including annual operating costs (ESTCP, 2010), the costs of conventional engineering methods can put clean-up of a site beyond the financial capacity of the land owner. Another conventional method for remediation of organic pollutants is bioaugmentation or bioremediation (Dangi et al., 2019). For bioremediation of TCE, this method usually involves the addition of large quantities of the anaerobic bacterium *Dehalococcoides* that is capable of complete degradation of TCE to harmless ethylene (Vainberg et al., 2009). While this strategy can be quite effective, there are numerous challenges in its implementation. The site must first be made anaerobic, usually done through injection of enough carbon sources to cause oxidation of the substrate to deplete available oxygen. The bioremediation strain of *Dehalococcoides* is sensitive to pH, requiring buffers to be added to the site. A major hurdle is that the metabolic pathway can stall at the much more potent carcinogen, vinyl chloride, or at cis-DCE, rendering the site more dangerous. To counter this problem, a strain certified to contain all the genes required for the full metabolic pathway must be used. Effective bioremediation is expensive due to these site preparations and the need for large quantities of the certified strain.

Phytoremediation is an alternative strategy for remediation of TCE that has proven to be successful at numerous sites (Kang, 2014). Poplar trees (*Populus* sp.) are metabolically capable of degrading TCE to less toxic forms (Shang et al., 2001), and with its extensive root system and ability to take up large volumes of water (Isebrands and Richardson, 2014), it is often the tree of choice for phytoremediation of TCE. And since the tree is basically a solar-driven "pump and treat" system, it is much less expensive than conventional methods. Major drawbacks, however, are that TCE can have phytotoxic effects, impacting the health and growth of the trees, and that the metabolism of TCE is considered to be too slow (Doty et al., 2007). Overexpression of the gene encoding the enzyme in the rate-limiting step in TCE metabolism can greatly enhance degradation of TCE (Doty et al., 2000, 2007). However, restrictions on the use of transgenic plants render this solution unusable at this time.

Conventional methods for remediation of polycyclic aromatic hydrocarbons (PAHs), another important class of pollutants, include incineration of the soil or capping of the site to limit human exposure. Despite the hydrophobicity of PAHs, plants have been demonstrated to take up some of these recalcitrant chemicals (Ballach et al., 2003; Kuhn et al., 2004; Spriggs et al., 2005). However, successful phytoremediation of PAHs has been limited due to their high toxicity, causing growth inhibition, reduced transpiration, chlorosis, and wilting of plants (Germaine et al., 2009). Similarly, phytoremediation of herbicides, pesticides, explosives, and many other organic pollutants has been hampered by the phytotoxic effects on the plants.

29.2.2 Introduction to Endophyte-assisted Phytoremediation of Organic Pollutants

Partnering specific microorganisms and plants can effectively circumvent the problems of direct bioremediation with bacterial strains and of phytoremediation. In this strategy, the plant and bacteria work together, with the plant effectively taking up the pollutant with its wide-spreading roots, and the endophytic bacteria degrading the pollutant within the plant (Doty, 2008; Khan and Doty, 2011; Weyens et al., 2009b), thus reducing the toxicity within the plant and enabling the plant to take up more of the pollutant. With the endophytic bacteria and the plant's own capacity to metabolize the pollutant, the combined effort can effectively degrade the chemicals more fully. Through photosynthesis, the plant can provide the microorganisms with suitable growth substrate since the pollutant itself is generally less than ideal to support bacterial growth. The overall technology is quite simple and adaptable, making it applicable in many settings. For example, the tree species most suited for the specific environment of the contaminated site can be utilized by simply adding the pollutant-degrading endophyte strain to it. Since endophytes generally have a very broad host range including both broad-leaved eudicots and narrow-leaved monocots such as grasses, they can be inoculated onto most plant species of choice for the site. Since the microorganisms do not need to be engineered to be effective, the technology can be readily utilized without regulatory barriers.

The key to the success of the technology is in isolating the most effective endophyte strains for a particular pollutant. To achieve this challenging goal, we can look to what nature has already selected. Plants can harbor dozens of different bacterial species, and this internal microbial community can be altered according to the environmental conditions (Siciliano et al., 2001). Siciliano et al. showed that the genes encoding catabolic pathways increased within the root endophyte population in response to the presence of a given pollutant. Since there was plant specificity in the resulting endophytic populations, it seemed that some plant species have the ability to recruit, or selectively expand, the necessary bacteria to detoxify pollutants (Siciliano and Germida, 1999; Siciliano et al., 2001). Similar research was conducted with endophyte populations of poplar trees growing at a phytoremediation field trial site with BTEX as the major groundwater contaminant (Moore et al., 2006). The following sections highlight successes of using plant–microbe partnerships for enhanced remediation of some classes of organic pollutants.

29.2.3 Endophyte-assisted Phytoremediation of TCE

A novel endophyte was isolated and characterized from a hybrid poplar (*Populus deltoides x nigra clone* DN177) that is resistant to high levels of TCE (Kang et al., 2012). Degradation experiments demonstrated that this unique endophyte, identified as PDN3, dechlorinated 80% of TCE in two days without the addition of any inducing phenolic compound. In 24 hours, PDN3 degraded TCE concurrent with a chloride concentration increase with a molar ratio of TCE removal to chloride generation of 1:3, indicating that this endophyte completely degraded TCE. This was the first report demonstrating that a naturally occurring poplar endophyte can rapidly degrade TCE aerobically without the addition of environmentally-harmful inducing substrates. The endophyte strain was successfully deployed at a TCE-contaminated field site (Doty et al., 2017). While the uninoculated poplar trees showed signs of TCE phytotoxicity, the trees provided with the endophyte demonstrated superior health and growth. In terms of phytoremediation, the PDN3-inoculated trees had more of the TCE metabolic products than the control (uninoculated) trees. This work represented the first case of endophyte-assisted phytoremediation with trees using a natural endophyte strain. In an earlier study, poplar trees were inoculated *in situ* with the modified endophytic strain containing genes encoding toluene ortho-monooxygenase that co-metabolizes TCE (Weyens et al., 2010). Although no increased uptake of TCE was reported, there was a reduction in the phytovolatilization of TCE.

29.2.4 Endophyte-assisted Phytoremediation of BTEX, PAHs, and Petroleum

Another common class of pollutants is BTEX that includes benzene, toluene, ethylbenzene, and xylene. In a ground-breaking study by Barac and colleagues, the concept of engineering endophytes for phytoremediation was proven to be successful (Barac et al., 2004). The catabolic plasmid from a yellow lupine endophyte was conjugatively transferred to a native endophyte, providing the genes for toluene degradation. The effectively-inoculated plants had the ability to significantly reduce the phytovolatilization of toluene, by up to 70%. This research was extended to poplar trees that are more suitable for phytoremediation research (Taghavi et al., 2005). Remarkably, it was found that *in planta* horizontal gene transfer occurred such that the plasmid conferring toluene degradation moved directly into the native poplar endophyte population without the need for prior co-culturing.

Research on using endophytes to improve phytoremediation of polycyclic aromatic hydrocarbons (PAHs) has also been quite successful. Using conjugative transfer of a plasmid conferring PAH degradation into an endophytic *Pseudomonas putida* strain, Germaine and colleagues effectively reduced phytotoxicity of naphthalene (Germaine et al., 2009). The modified endophyte protected the inoculated peas, resulting in significantly higher germination rates in soil containing the pollutant. Inoculated pea plants had higher transpiration rates in hydroponics containing naphthalene, compared to uninoculated controls, indicating that there was reduced phytotoxicity. In terms of potential phytoremediation improvements, the inoculated plants reduced the amount of naphthalene remaining in soils, removing 37% more of the PAH than the uninoculated pea plants.

In another study, endophyte strains from poplar and willow were screened for the ability to utilize the PAHs as sole carbon and energy sources (Khan et al., 2014). The best growing endophyte, termed PD1, was isolated from *Populus deltoides*, and was identified to be a strain of *Pseudomonas putida*. A hydroponics-based experiment was then used to assess the protective effect of PD1 strain against the phytotoxicity of phenanthrene on willow. Results after three weeks of exposure to the toxin indicated that PD1 provided strong protective effects to the plants in the pollutant-saturated solution. There was an increase in root and shoot lengths and an overall increase in the biomass of the inoculated plants compared to the uninoculated plants. In the uninoculated control plants, the roots turned brown and the leaves fell off, rendering these plants ineffective for phytoremediation. The endophyte-inoculated plants were able to remove more of the phenanthrene from solution, and in a soil experiment with willows and grasses, inoculated plants removed up to 40% more of the phenanthrene. Following the success of the lab studies, the endophyte strain was deployed at a PAH-contaminated field site (Landmeyer et al., 2020). The inoculated trees were able to grow larger than the control trees despite the high concentration of the pollutant, indicating that the strain helped reduce the phytotoxic impacts.

Phytoremediation of petroleum can also be enhanced with the help of microorganisms. For example, grasses able to grow well in crude oil-contaminated soil harbor hydrocarbon-degrading endophytes (Fatima et al., 2015). Addition of an endophytic consortium consisting of two of these strains resulted in 78% crude oil remediation in the inoculated grasses compared to up to 66% for the uninoculated plants (Fatima et al., 2016). Importantly, the uninoculated plants had reduced growth in the contaminated soils compared to uncontaminated soils whereas the endophyte-inoculated plants showed increased growth despite the pollutant. While the majority of research papers on bioaugmentation for improved phytoremediation of petroleum focus on grasses, trees could also likely benefit from endophyte inoculations. Poplar trees have been used for phytoremediation of soils contaminated petroleum hydrocarbons, even under environmentally-challenging conditions such as in the boreal forests of Scandinavia (Lopez-Echartea et al., 2020). In a study comparing 20 poplar clones and two willow clones and different sizes, the survival rate in soils heavily contaminated with hydrocarbons ranged from 26–100%, with large (60 cm) poplar stakes of commercial clones performing especially well (Zalesny et al., 2005). Since poplar and willow are propagated through cuttings rather than seeds, they harbor a wealth of well-established endophytes. More research needs to be conducted on the endophytes within these commercial clones that are used primarily for phytoremediation projects.

29.2.5 Endophyte-assisted Phytoremediation of Herbicides and Pesticides

One of the first lab examples of the effectiveness of endophyte-assisted phytoremediation was on the herbicide 2,4-dichlorophenoxyacetic acid, commonly known as 2,4-D. In this ground-breaking study, Germaine and colleagues inoculated pea plants with a genetically tagged poplar endophyte that naturally possessed the ability to degrade 2,4-D (Germaine et al., 2006). Compared to uninoculated plants, the inoculated plants showed a higher capacity for herbicide removal from soil and yet showed no 2,4-D accumulation in their aerial tissues, indicating that the herbicide was completely degraded within the plants by the poplar endophyte.

Chlorpyrifos is a common broad-spectrum organophosphorus pesticide used worldwide. Though it is used against agricultural pests, it also kills beneficial arthropods including pollinator bees (Cox, 1995). Chlorpyrifos has major human health impacts, interfering with the nervous system (US EPA). Phytoremediation can be used to help remove chlorpyrifos in constructed wetlands designed to capture the run-off from agricultural fields (Moore et al., 2002). Riparian buffers which are planting strips to protect streams from agricultural chemicals could be another way to absorb the organophosphorus pesticide. The riparian species, poplar and willow, were tested in hydroponics experiments and were shown to take up and fully degrade chlorpyrifos over several weeks (Lee et al., 2012). Some bacteria are able to degrade the pesticide and may offer a mechanism to increase the speed and effectiveness of phytoremediation of this important pollutant. For examples, the root endophyte *Pseudomonas* strain BF1-3 can hydrolyze the organophosphorus pesticide (Barman et al., 2014). Ryegrass inoculated with a chlorpyrifos-degrading endophyte, *Mesorhizobium* sp. strain HN3, improved plant health and degradation of the pesticide within the plant (Jabeen et al., 2016).

29.2.6 Endophyte-assisted Phytoremediation of More Recalcitrant Organic Pollutants

Recalcitrant pollutants are those that are especially difficult to remediate. An important mechanism by which microorganisms could help improve phytoremediation of recalcitrant organic pollutants is through production of surfactants that release the chemicals bound tightly to soil particles, thus allowing for plant uptake of the pollutant (Feng et al., 2017; Li et al., 2012; Soleimani et al., 2010; Wenzel, 2009). Surfactants can be added directly to contaminated soils; however, they can be costly, not only monetarily but environmentally. If a rain event occurs too soon after surfactants are administered, for example, the newly released pollutant can more readily leach into the aquifer or surrounding aquatic ecosystems. Precise calibration of the amount of surfactant to add to soils to release the pollutant for plant uptake without rendering the site phytotoxic is another challenge. Using biosurfactant-producing microorganisms, on the other hand, could allow for a small but steady amount of the pollutant to be released, improving plant uptake of the chemical at a rate it can metabolize without harm. Indeed, addition of consortia of strains specifically chosen for strong biosurfactant production did successfully improve phytoremediation of hydrophobic organic pollutants (Pathak and Keharia, 2014).

A class of pollutants that is particularly recalcitrant to remediation is the polychlorinated biphenyls (PCBs) which are hydrophobic, chemically stable, and are characterized as persistent organic pollutants (POPs). PCBs bind tightly to soil, rendering plant uptake of these pollutants challenging (Aken et al., 2010). Since metabolism of PCBs is energetically expensive, additional carbon sources are needed to support microbial growth, thus lending itself well to the use of plant-microbe partnerships for effective remediation. Genes required for PCB degradation can be transferred from PCB-degrading bacteria into bacterial strains capable of efficient colonization of plant roots (Aken et al., 2010). Several research groups successfully engineered strains in this way, resulting in improved remediation of this important class of toxic pollutants (110–116) (Brazil et al., 1995; Villacieros et al., 2005).

An emerging area of concern is the prevalence of microplastics in the environment (see also chapters 24 and 32). In agriculture, sources of microplastics include insulating plastic films, controlled-release chemical fertilizer capsules, and biosolids applications (Song et al., 2018; Horton et al., 2017; Song et al., 2020). These microplastics can be absorbed by plants; therefore, phytoremediation of this type of pollutant is a possibility. Plastic-degrading microorganisms could potentially be harnessed and partnered with appropriate plants. Some of the bacterial strains with the ability to degrade plastics such as polyvinyl chloride and polystyrene are *Pseudomonas* species that can associate with plants. Endophyte-assisted phytoremediation of microplastics will be an important new technology to develop.

29.3 INORGANIC POLLUTANTS

29.3.1 Background of Inorganic Pollutants

A legacy of industrial, manufacturing, and agricultural activities has resulted in widespread environmental contamination in both urban and rural areas. Because of their persistence in the environment and toxicity at low levels, the heavy-metals and metalloids (collectively referred to here as HM), e.g., cadmium (Cd), lead (Pb), mercury (Hg), and arsenic (As), are of particular concern. Unlike organic pollutants, which can be entirely or partially detoxified by various microorganisms, HM are non-degradable and can only be transformed into less toxic and/or less bioavailable forms. This means that HM pollution accumulates in the environment, representing a serious, long-term threat to human health and ecosystems.

Generally, metallic chemical elements with a relatively high density and that are toxic at low concentrations are classified under the broader term heavy metals (Burd et al., 2000; Ma et al., 2011). HM are widely distributed in the Earth's crust, although typically at low concentrations. They enter the environment via the natural weathering of volcanic, sedimentary, and metamorphic rocks, and the biochemical activities of microbes, and are then dispersed via the erosive forces of wind and water. While natural hot spots of some HM do occur, these natural processes are often exacerbated as a result of human activities that alter environmental conditions, e.g., exposure of HM bearing ores in mining operations, or the over-extraction of groundwaters for agriculture (Yuan et al., 2014).

Whether through incidental emissions or intentional applications, elevated levels of HM occur in both urban and rural environments as a result of a wide range of activities, such as mining and smelting operations, the consumption of fossil fuels, applications of agrochemicals, manufacture of electronics, the production and consumption of consumer, industrial, and military products, and the improper disposal of solid wastes and wastewaters (Yuan et al., 2014). Public consumption of modern electronic devices and the growing markets for rechargeable batteries will continue to drive the demand for HMs (Wang et al., 2002; Yang and Hoffman, 1984).

Globally, it is estimated there are more than 5 million sites covering 100 million hectares of land contaminated by various HM, and the risk of public exposure is increasing with urbanization (Burd et al., 2000). In China nearly 20% of the agricultural land is believed to be affected by HM, while the European Union reports more than 200000 sites as contaminated with HM (Burd et al., 2000). In the United States, the Environmental Protection Agency lists more than 500000 sites at least moderately contaminated with HM, with an additional 1327 sites listed under the Superfund program (ATSDR, 2020).

Humans are exposed to HM through food, water, air, and soil, which enter the body via inhalation, ingestion, and dermal contact, acting as systemic toxins that are associated with different neurotoxic, carcinogenic, mutagenic, teratogenic, and cardiotoxic effects (Ma et al., 2009; Sheng et al., 2008; Zhang et al., 2011). While all HM are toxic at certain thresholds, some do have a physiological role. For example, the essential HMs copper (Cu), iron (Fe), manganese (Mn), and zinc (Zn) are required for biochemical processes, and are non-toxic below certain thresholds, where the non-essential HMs, such as Cd, Pb, Hg, and As, have no known biological function and are toxic at trace levels for most organisms (Zhang et al., 2011).

29.3.2 Phytoremediation of Inorganic Pollutants

Because they cannot be degraded, the phytoremediation of HMs requires specific strategies that differ from those used for remediation of organic pollutants. While the approach depends on the nature of the pollutant, the site conditions, and type of plant(s) used, the most common strategies can be categorized as: phytoextraction, phytofiltration, and phytostabilization (Salt et al., 1995). Phytoextraction, and phytofiltration in aquatic systems, involves the uptake of contaminants by the roots, followed by the translocation and accumulation in the harvestable tissues, while phytostabilization sequesters HMs in and around the roots by converting the metals to less bioavailable forms, preventing mobilization and leaching (Pilon-Smits and Pilon, 2002; Salt et al., 1995). Phytostabilization, analogous to the conventional capping as described below, contains the HM on site without the need for a physical barrier, while contributing to ecosystem integrity. Conversely, phytoextraction via accumulation in the harvestable plant tissues provides a more efficient method of remediating HM polluted media than conventional excavation methods. Not only do the plant tissues accumulate a higher concentration of HM by mass, they can also be further processed in situ (e.g., via chipping) for transportation or storage, or more interestingly, ex situ (e.g., via incineration) in order to recover the HM for safe storage or recycling. This concept has been demonstrated through phytomining activities to extract precious metals, such as gold and nickel, into plant tissues, which can then be incinerated to recover the metals (Abou-Shanab et al., 2003). Less common is phytovolatilization, in which some HMs, such as Hg, As, and Se, can be taken up by the roots and volatilized, diluting their concentration over a larger area (O'Brien et al., 2012; Santoyo et al., 2016).

Conventional technologies for the remediation of HM use various physical (e.g., excavation or capping) or chemical (e.g., chelating agents or soil washing) methods to either remove (ex situ) or contain (in situ) the pollutants (as reviewed in Sessitsch et al. (2012)). In practice, these methods can be effective and are often selected by regulatory bodies and project managers because they are predictable, broadly applicable, and meet regulatory timelines. However, conventional remediation methods are very expensive and can also be energy intensive, produce secondary pollution, and disrupt habitats (Alqueres et al., 2013). Consequently, the scale of the problem and the high cost of traditional engineering-based remediation methods have limited clean-up efforts to only the most contaminated sites, resulting in increased public exposure to carcinogens and a loss of economic productivity for moderately affected sites.

Phytoremediation, which leverages the natural abilities of some plants to tolerate and accumulate HM, is a low-cost, environmentally sustainable technology to expand the reach of conventional remediation efforts. Phytotoxicity and habitat suitability are the essential considerations in plant selection for the phytoremediation of HM, and early research focused on identifying plant species with high tolerance to the various HM (O'Brien et al., 2012).

Hyperaccumulators are plant species that have the potential to accumulate HMs in their shoots to levels far greater than non-accumulating plants, often at levels several orders of magnitude greater than those of other species growing in the same conditions (Silver and Phung, 2005). More than 400 hyperaccumulators have been identified for various HMs, including plants from some well-known families, such as Asteraceae, Brassicaceae, Fabaceae, and Poaceae (Li et al., 2012). The family Brassicaceae, with genus *Thlaspi*, *Alyssum*, and *Brassica* contain the most known hyperaccumulator species (Zhu et al., 2014). While some hyperaccumulators are specialists, like the arsenic hyperaccumulator, *Pteris vitatta*, which has been shown to accumulate 13800 PPM (mg kg^{-1}) of As in its fronds. Other plants can accumulate multiple HM to high levels, e.g., the pennycress *Thlaspi praecox* Wulf. (Brassicaceae) accumulate high levels of Cd (5950 PPM), Pb (3500 PPM), and Zn (14590 PPM) in their above ground tissues (Zhu et al., 2014). While able to accumulate high levels of HM, hyperaccumulators tend to be slow growing, have low biomass, and narrow habitat ranges, limiting their use in phytoremediation (Rajkumar et al., 2013).

More recently, woody plant species, such as poplars and willows, are being recognized as a good alternative to HM hyperaccumulators. While they accumulate lower concentrations of HM in their above-ground tissue than do hyperaccumulators, woody trees compensate for the decreased concentration with high biomass, deep roots, and broad habitat ranges (Babu et al., 2015; Shin et al., 2012; Wan et al., 2012; Zhu et al., 2014). For example, in a comparison of data between separate studies, Rosenberg et al. (1977) reported that while the hybrid poplar (*P. tremula* × *P. alba*) accumulated a lower concentration of Cd in the shoot tissues than the herbaceous Cd hyperaccumulator *T. caerulescens* (11 mg kg^{-1} versus 40 mg kg^{-1}, dry biomass), the *Populus* trees extracted more total Cd per plant (24.2 mg versus 1.2 mg per plant, dry biomass) and per hectare (estimated at 250 g Cd ha^{-1} versus 125 g Cd ha^{-1}).

Phytoremediation offers a good alternative to conventional remediation technologies, especially in moderately contaminated sites (see also chapters 27 and 28). Improving phytoremediation technology requires reducing phytotoxicity, increasing uptake and translocation of the HM into the harvestable tissues, and improving the rate and efficiency of HM extraction. Accordingly, research has included selective breeding programs, genetic transformation, and harnessing the plant microbiome (Farooq et al., 2016). The first two methods, selective breeding and genetic transformation, are subject to long-generation times of plants and expensive regulatory hurdles, respectively. The plant microbiome, on the other hand, presents a novel and less explored approach to improving phytoremediation outcomes.

29.3.3 Endophytes for Enhanced Remediation of Inorganic Pollutants

Most research on beneficial microbe–plant interactions in phytoremediation has mainly focused on the rhizosphere, as it is the soil–plant interface (Paez-Espino et al., 2009). Microorganisms in the rhizosphere can assist in the phytoextraction of HM by (1) increasing root–soil surface area through increased root biomass via the synthesis of auxin, (2) increasing solubility and changing the speciation of HM through alteration of soil pH or the production of organic ligands, and/or (3) production of organic acids and siderophores that complex cationic ions or desorb anionic ions. However, as the plant–microbiome constituents most intimately associated with the host plant, endophytes are in a unique position to assist in improving HM phytoremediation outcomes.

Endophytes can assist in phytoremediation through reducing phytotoxicity and stimulating host plant growth and development. Studies indicate that endophytes have the capacity to alleviate metal-induced phytotoxicity and enhance biomass production of plants growing under HM stress in laboratory and greenhouse conditions. The endophyte *Rahnella* sp. JN6 isolated from *Polygonum pubescens* growing in multi-HM contaminated soils showed very high Cd, Pb, and Zn tolerance as well as the synthesis of the important plant growth promotors, 1-aminocyclopropane-1-carboxylic acid (ACC) deaminase, indole-3-acetic acid (IAA), and siderophores (Paez-Espino et al., 2009). In subsequent pot experiments, the Cd-hyperaccumulators *Amaranthus hypochondriacus, A. mangostanus*, and *Solanum nigrum* inoculated with *Rahnella* JN27 isolates showed significant increases in biomass and Cd concentrations in both the roots and shoots as compared to un-inoculated controls (Farooq et al., 2016). While the exact mechanisms by which the endophyte promoted increased tolerance were not identified in this study, it is notable that the endophytes were inoculated in novel (i.e., different from the original) hosts and that tolerance and accumulation of Cd were improved in the novel hosts. Broad host range is common in culturable endophytes and represents an important implication for the design of endophyte-assisted phytoremediation systems, one that allows a pairing of the best endophyte(s) and plant(s) for a given pollutant and habitat/site conditions.

Endophytes can also benefit phytoremediation through interactions with the host plant's abiotic stress response. Ethylene, an essential plant hormone, is important for normal plant growth and development, and plays a key role in plant response to a wide range of biotic and abiotic stresses, including metal stress (Farooq et al., 2016). However, an overproduction of ethylene in response to stress can inhibit plant developmental processes, such as root elongation, lateral root growth, and root hair formation, which can be as detrimental to plant health as the actual stressor (Mukherjee et al., 2018; Wang et al., 2002; Xu et al., 2016; Zhu et al., 2014). Many endophytic bacteria produce the enzyme 1-aminocyclopropane-1-carboxylase (ACC) deaminase which hydrolyzes ACC, a precursor in the synthesis of ethylene, into a-ketobutyrate and ammonia, which is used by the microbes as a source of nitrogen (Mukherjee et al., 2018). Several studies have shown that heavy metal-resistant and ACC deaminase-producing bacteria can enhance metal uptake by plants (Mukherjee et al., 2018). Specifically, it was demonstrated that *Brassica napus* inoculated with the high ACC deaminase-producing endophytes, *Ralstonia* sp. J1-22-2, *Pantoea agglomerans* Jp3-3, and *Pseudomonas thivervalensis* Y1-3-9, were found to increase the biomass and above-ground tissue Cu contents when cultivated in Cu-contaminated substrates compared to the uninoculated controls (Flemming and Wingender, 2010).

HM exposure can induce severe oxidative stress in plants by disturbing cellular equilibria between the generation and the neutralization of reactive oxygen species (ROS). ROS, such as superoxide anion (O_2^-) and hydrogen peroxide (H_2O_2), are constitutively produced at low levels during normal metabolic processes and at high levels in response to various stresses, including HM and invasion by pathogens (Pal and Paul, 2008). Successful colonization of host plants by endophytes requires that they evade the host's defense response, and so it is not surprising that genes for the detoxification of reactive oxygen species are prevalent among endophytes (Harrison et al., 2007; More et al., 2014; Sharma et al., 2019) For example, the endophyte *Gluconacetobacter diazotrophicus* PAL5 showed increased expression of genes encoding for ROS-detoxifying enzymes such as superoxide dismutase and glutathione reductase during the colonization of rice roots (Harrison et al., 2007). As with ethylene, the overproduction of ROS species can be detrimental to plant health, and a delicate homeostasis must be maintained (Babu et al., 2013). The expression of ROS-detoxifying genes by endophytes may moderate the host plant response to HM stress, improving phytoremediation outcomes.

HM tolerance is widespread in microorganisms, including endophytes, and most exhibit tolerance to one or more heavy metals (Shin et al., 2012). HM tolerant endophytes belonging to a wide range of taxa have been isolated from both hyperaccumulator and non-hyperaccumulator plants (Li et al., 2012). For example, a total of 41 As-tolerant endophyte strains, representing 12 genera, were isolated from the shoot and root tissues of the arsenic hyperaccumulators, *P. vittata* and *P. multifada* (Zhu et al., 2014). These isolates showed tolerance to both arsenate (AsV) and arsenite (AsIII), as well as the synthesis of key plant growth promoting products, i.e., indole acetic acid (IAA) and siderophores (Zhu et al., 2014) In a survey of non-hyperaccumulators growing in soils moderately contaminated with arsenic, cadmium, and lead, our own research has identified arsenic tolerant endophytes in nearly every plant family assayed, including Ericaceae, Berberidaceae, Rosaceae, Pinaceae, Sapindaceae (Tournay et al., manuscript in preparation).

Endophytic bacteria have several mechanisms for the resistance and detoxification of HM that may benefit the host plant in phytoremediation. These include efflux of metal ions to the exterior of the cell, transformation of metal ions into less toxic forms, the sequestration of metals on the cell surface or in extracellular polymers, and precipitation, adsorption/desorption or biomethylation (Rajkumar et al., 2013). Unlike with organic pollutants which can be metabolically degraded, there is much less evidence that endophytes interact directly with the toxic metals to alleviate HM stress. However, some recent studies suggest that endophytic bacteria may alleviate metal phytotoxicity via mechanisms that reduce *in planta* concentrations, e.g., precipitation, adsorption, intracellular accumulation and sequestration, or via biotransformation of the toxic metal ions to less- or non-toxic forms, e.g., chelation to metal-binding ligands (Babu et al., 2013; Shin et al., 2012; Wan et al., 2012; Zhu et al., 2014).

Arsenate (AsV), the most common arsenic species in oxidized environments, is a chemical and physical phosphate analog and is taken up by plants via phosphate transport pathways (Rosenberg et al., 1977). Once taken up by plants, AsV is reduced to AsIII via arsenate reductase (ARC) in the roots, and then either extruded out of the roots or complexed with phytochelatins (PCs), where the AsIII-PCs complexes are then sequestered in the vacuoles (Farooq et al., 2016). Microbes acquire and detoxify arsenate through a similar pathway, acquisition via phosphate uptake and reduction via arsenate reductase (*arsC*) (Paez-Espino et al., 2009). However, with most microbes, the majority of the AsIII is rapidly expelled from the cell via an arsenite efflux system (*arsB, arsAB*) (Paez-Espino et al., 2009). Interestingly, despite its 100-fold higher toxicity, about 90% of As in plant tissues is found in the form of AsIII, even when plants are exposed to AsV (Farooq et al., 2016). This is likely because AsIII reacts efficiently with the thiol-containing compound, glutathione, and phytochelatins (PCs) (Farooq et al., 2016). The microbial arsenate detoxification operon *ars* is widely distributed in endophytes, and the ability of endophytes to reduce AsV to AsIII is well documented (Mukherjee et al., 2018; Wang et al., 2012; Xu et al., 2016; Zhu et al., 2014). An As-tolerant endophytic consortium of seven endophytes isolated from *Lantana camara* was found to improve the As-phytoremediation efficiency of *Solanum nigrum* (Mukherjee et al., 2018). Notably, it was determined that the endophytic consortium expanded the glutathione pool and increased arsenate reductase activity, and the reduction of As(V) to As(III) was increased both in shoots and roots of consortium-treated plants as compared to the un-inoculated controls; 68% and 51%, respectively (Mukherjee et al., 2018).

Another potential detoxification mechanism in which endophytes might directly interact with HM *in planta* is the formation of biofilms. Biofilms are assemblages of microorganisms embedded in a matrix of extracellular polymeric substances (EPS), and they form in response to a wide range of environmental stresses, e.g., shear, desiccation, UV, predation, toxins, and bactericides (Flemming and Wingender, 2010). Biofilms consist of cells immobilized in a water saturated EPS matrix that is composed primarily of polysaccharides and proteins, with smaller amounts of nucleic acids, lipids, extracellular DNA (eDNA), and other substances (Pal and Paul, 2008). In clinical bacterial strains, biofilms have been associated with increased microbial resistance to both antibiotics and toxic metals, while the biofilms of environmental bacteria isolated from acid-mine drainage have been investigated for their ability to remove metals from industrial waste waters (Harrison et al., 2007; More et al., 2014; Sharma et al., 2019). The complex EPS architecture provides an environment rich in charged functional groups, e.g., carboxylate ($R\text{-}COO^-$), phosphate ($R\text{-}HPO_4$), sulfhydryl (R-SH), aminos ($R\text{-}NH_3^+$), and phenol ($R\text{-}C_6H_4OH$), that may act as potential binding sites for metal ions (Harrison et al., 2007). The HM adsorption capabilities of endophytic biofilms might act as a sequestration or detoxification mechanism by reducing free HM concentrations *in planta*, reducing phytotoxicity and improving phytoremediation outcomes. Transmission electron microscopy (TEM) used in an *in vitro* study of the endophytic bacteria *Pseudomonas koreensis* AGB-1, isolated from roots of *Miscanthus sinensis*, cultured in a soil extract containing 10 mg As, 5 mg Cd, 5 mg Cu, 30 mg Pd, and 20 mg Zn L^{-1} revealed HM accumulations in the extracellular matrix and at bacterial cell surfaces (Babu et al., 2015). Another *in vitro* study utilizing TEM revealed that Pb accumulated in both the extracellular matrix and within the cells of the endophyte *Pseudomonas marginalis* MN3-4 (Shin et al., 2012). The oxyanion arsenate (AsV) may also be sequestered in endophytic biofilms. Our research with the endophyte *Rahnella* sp. PD12R, isolated from *Vaccinium parvifolium* growing on arsenic contaminated soils, produces prodigious amounts of EPS under arsenic stress, and is highly tolerant to both AsV and AsIII, growing at 500 mM and 20 mM respectively (Tournay et al., manuscript in preparation). For both AsV and AsIII, EPS synthesis increased with increasing arsenic concentration in the media, and the arsenic preferentially localized in the EPS compared to the cells (Tournay et al., manuscript in preparation). While these results are promising, to date there have been no reports of endophytic biofilms accumulating HM *in planta.*

Although they are not hyperaccumulators, the secondary ecological and economic benefits of woody tree species, e.g., poplars and willows, has made them attractive candidates in the phytoremediation of moderately HM contaminated soils. Over the past several decades, field trials have been conducted at former agricultural, industrial, and mining sites contaminated with Cd, Zn, Pb, As, Ni, and Cu; mainly in Europe, but also in Asia and North America (as reviewed in Vangronsveld et al., 2009; Zalesny et al., 2019). Despite these efforts, the use of woody tree species in the phytoremediation of HM contaminated soils is still at the developmental stage, with the most significant disadvantage of this technology being the time required to complete the projects (Vangronsveld et al., 2009).

Recently the potential of the plant-microbiome has been recognized for its potential to improve outcomes in the phytoremediation of HM contaminated soils. Most research in this area involves rhizospheric microbes with the goal of increasing the bioavailability and uptake of HM into the host plant, and as with phytoremediation research in general, most of the focus has been on the hyperaccumulators (Burd et al., 2000; Das et al., 2016;

De et al., 2011; Mishra et al., 2017). Much less attention has been given to the potential role of endophytes to improve tolerance, uptake, and translocation of the HM once they have entered the host plant, particularly in woody tree species.

29.4 CONCLUSIONS

Environmental pollution is a serious threat that must be addressed. Conventional methods of remediation have proven effective but their high cost has rendered many contaminated sites untreated, with repercussions to humanity and the environment. Genetic engineering of plants, often with genes from pollutant-degrading bacteria, has been a successful approach for overcoming some of the challenges of phytoremediation. However, current regulations limit their utilization. Phytoremediation partnered with microorganisms adapted to deal with pollutants is now a proven technology that is readily deployable. With widespread adoption, endophyte-assisted phytoremediation has the potential to restore damaged ecosystems and reduce public exposure to carcinogens, ultimately reducing the incidence of cancer and other pollution-caused diseases.

REFERENCES

1. Abou-Shanab RI, Angle JS, Delorme TA, Chaney RL, van Berkum P, et al. 2003. Rhizobacterial effects on nickel extraction from soil and uptake by Alyssum murale. *New Phytologist* 158: 219–224.
2. Aken BV, Correa PA, Schnoor JL. 2010. Phytoremediation of polychlorinated biphenyls: New trends and promises. *Environmental Science and Technology* 44: 2767–2776.
3. Alqueres S, Meneses C, Rouws L, Rothballer M, Baldani I, et al. 2013. The bacterial superoxide dismutase and glutathione reductase are crucial for endophytic colonization of rice roots by Gluconacetobacter diazotrophicus PAL5. *Molecular Plant Microbe Interactions* 26: 937–945.
4. ATSDR's Substance Priority List [ATSDR]. 2020c. Agency for Toxic Substances and Disease Registry, Division of Toxicology and Human Health Services. [accessed 2020, Oct 30]: http://www.atsdr.cdc.gov/spl/index.html#2019spl.
5. Babu AG, Kim JD, Oh BT. 2013. Enhancement of heavy metal phytoremediation by Alnus firma with endophytic Bacillus thuringiensis GDB-1. *Journal of Hazardous Materials* 250-251: 477–483.
6. Babu AG, Shea PJ, Sudhakar D, Jung IB, Oh BT. 2015. Potential use of *Pseudomonas koreensis* AGB-1 in association with *Miscanthus sinensis* to remediate heavy metal loid -contaminated mining site soil. *Journal of Environmental Management* 151: 160–166.
7. Ballach H-J, Kuhn A, Wittig R. 2003. Biodegradation of anthracene in the roots and growth substrate of poplar cuttings. *Environmental Science and Pollution Research* 10: 308–316.
8. Barac T, Taghavi S, Borremans B, Provoost A, Oeyen L, et al. 2004. Engineered endophytic bacteria improve phytoremediation of water-soluble volatile organic pollutants. *Nature Biotechnology* 22: 583–588.
9. Barman DN, Haque MA, Islam SM, Yun HD, Kim MK. 2014. Cloning and expression of ophB gene encoding organophosphorus hydrolase from endophytic *Pseudomonas* sp. BF1-3 degrades organophosphorus pesticide chlorpyrifos. *Ecotoxicology and Environmental Safety* 108: 135–141.
10. Brazil GM, Kenefick L, Callanan M, Haro A, deLorenzo V, et al. 1995. Construction of a rhizosphere pseudomonad with potential to degrade polychlorinated biphenyls and detection of *bph* gene expression in the rhizosphere. *Applied and Environmental Microbiology* 61: 1946–1952.
11. Bulgarelli D, Schlaeppi K, Spaepen S, Ver Loren van TE, Schulze-Lefert P. 2013. Structure and functions of the bacterial microbiota of plants. *Annual Review of Plant Biology* 64: 807–838.
12. Burd GI, Dixon DG, Glick BR. 2000. Plant growth-promoting bacteria that decrease heavy metal toxicity in plants. *Canadian Journal of Microbiology* 46: 237–245.
13. Cox C. 1995. Chlorpyrifos part 3: Ecological effects. *Journal of Pesticide Reform* 15: 13–19.
14. Dangi AK, Sharma B, Hill RT, Shukla P. 2019. Bioremediation through microbes: Systems biology and metabolic engineering approach. *Critical Reviews of Biotechnology* 39: 79–98.
15. Das S, Jean JS, Chou ML, Rathod J, Liu CC. 2016. Arsenite-oxidizing bacteria exhibiting plant growth promoting traits isolated from the rhizosphere of *Oryza sativa* L: Implications for mitigation of arsenic contamination in paddies. *Journal of Hazardous Materials* 302: 10–18.
16. De MS, Rivelli AR, Kuffner M, Sessitsch A, Wenzel WW, et al. 2011. Interactions between accumulation of trace elements and macronutrients in *Salix caprea* after inoculation with rhizosphere microorganisms. *Chemosphere* 84: 1256–1261.

17. Dhankher OP, Pilon-Smits EA, Meagher RB, Doty SL. 2011. Biotechnological approaches for phytoremediation. In: Altman A, Hasegawa PM (Eds.), *Plant Biotechnology and Agriculture: Prospects for the 21st Century*. Elsevier.
18. Doty SL. 2008. Tansley Review: Enhancing phytoremediation through the use of transgenics and endophytes. *New Phytologist* 179: 318–333.
19. Doty SL, Freeman JL, Cohu CM, Burken JG, Firrincieli A, et al. 2017. Enhanced degradation of TCE on a superfund site using endophyte-assisted poplar tree phytoremediation. *Environmental Science and Technology* 51: 10050–10058.
20. Doty SL, James CA, Moore AL, Vajzovic A, Singleton GL, et al. 2007. Enhanced phytoremediation of volatile environmental pollutants with transgenic trees. *Proceedings of the National Academy of Sciences* 104: 16816–16821.
21. Doty SL, Shang QT, Wilson AM, Tangen J, Westergreen A, et al. 2000. Enhanced metabolism of halogenated hydrocarbons in transgenic plants containing mammalian P450 2E1. *Proceedings of the National Academy of Sciences* 97: 6287–6291.
22. ESTCP. 2010 ESTCP Cost and Performance Report: Bioaugmentation for Groundwater Remediation Project ER-0515 Environmental Security and Technology Certification Program United States Department of Defense.
23. Farooq MA, Islam F, Ali B, Najeeb U, Mao B, et al. 2016. Arsenic toxicity in plants: Cellular and molecular mechanisms of its transport and metabolism. *Environmental and Experimental Botany* 132: 42–52.
24. Fatima K, Afzal M, Imran A, Khan QM. 2015. Bacterial rhizosphere and endosphere populations associated with grasses and trees to be used for phytoremediation of crude oil contaminated soil. *Bulletin of Environmental Contamination and Toxicology* 94: 314–320.
25. Fatima K, Imran A, Amin I, Khan QM, Afzal M. 2016. Plant species affect colonization patterns and metabolic activity of associated endophytes during phytoremediation of crude oil-contaminated soil. *Environmental Science and Pollution Research* 23: 6188–6196.
26. Feng NX, Yu J, Zhao HM, Cheng YT, Mo CH, et al. 2017. Efficient phytoremediation of organic contaminants in soils using plant-endophyte partnerships. *Science and the Total Environment* 583: 352–368.
27. Flemming HC, Wingender J. 2010. The biofilm matrix. *Nature Reviews Microbiology* 8: 623–633.
28. Gerhardt KE, Gerwing PD, Greenberg BM. 2017. Opinion: Taking phytoremediation from proven technology to accepted practice. *Plant Science* 256: 170–185.
29. Germaine KJ, Keogh E, Ryan D, Dowling DN. 2009. Bacterial endophyte-mediated naphthalene phytoprotection and phytoremediation. *FEMS Microbiology Letters* 296: 226–234.
30. Germaine KJ, Liu X, Cabellos GG, Hogan JP, Ryan D, et al. 2006. Bacterial endophyte-enhanced phytoremediation of the organochlorine herbicide 2 4-dichlorophenoxyacetic acid. *FEMS Microbiology Ecology* 57: 302–310.
31. Harrison JJ, Ceri H, Turner RJ. 2007. Multimetal resistance and tolerance in microbial biofilms. *Nature Reviews Microbiology* 5: 928–938.
32. Horton, A. A., Walton, A., Spurgeon, D. J., Lahive, E. & Svendsen, C. (2017). Microplastics in freshwater and terrestrial environments: Evaluating the current understanding to identify the knowledge gaps and future research priorities. *Science of the Total Environment* 586, 127-141.
33. IARC. 2012. Monographs on the evaluation of carcinogenic risks to humans: Trichloroethylene tetrachloroethylene and some chlorinated compounds. *International Agency for Research on Cancer IARC World Health Organization Lyon.*
34. Isebrands JG, Richardson J. 2014. *Poplars and Willows: Trees for Society and the Environment*. FAO/CAB, Rome.
35. Jabeen H, Iqbal S, Ahmad F, Afzal M, Firdous S. 2016. Enhanced remediation of chlorpyrifos by ryegrass *Lolium multiflorum* and a chlorpyrifos degrading bacterial endophyte *Mezorhizobium* sp HN3. *International Journal of Phytoremediation* 18: 126–133.
36. Kang JW. 2014. Removing environmental organic pollutants with bioremediation and phytoremediation. *Biotechnology Letters* 36: 1129–1139.
37. Kang JW, Khan Z, Doty SL. 2012. Biodegradation of trichloroethylene TCE by an endophyte of hybrid poplar. *Applied and Environmental Microbiology* 78: 3504–3507.
38. Khan Z, Doty SL. 2011. Endophyte-assisted phytoremediation. *Current Topics in Plant Biology* 12: 97–105.
39. Khan Z, Roman D, Kintz T, delas Alas M, Yap R, et al. 2014. Degradation phytoprotection and phytoremediation of phenanthrene by endophyte Pseudomonas putida PD1. *Environmental Science and Technology* 48: 12221–12228.
40. Kuhn A, Ballach H-J, Wittig R. 2004. Studies in the biodegradation of 5 PAHs phenanthrene pyrene fluoran-

thene chrysene and benzo a pyrene in the presence of rooted poplar cuttings. *Environmental Science and Pollution Research* 11: 22–32.

41. Kuppusamy S, Palanisami T, Megharaj M, Venkateswarlu K, Naidu R. 2016. In situ remediation approaches for the management of contaminated sites: A comprehensive overview. In: P de Voogt (Ed.), *Reviews of Environmental Contamination and Toxicology*, Springer, Switzerland, pp. 1–117.
42. Landmeyer JE, Rock S, Freeman JL, Nagle G, Samolis M, et al. 2020. Phytoremediation of slightly brackish polycyclic aromatic hydrocarbon contaminated groundwater from 250ft below land surface: A pilot scale study using salt-tolerant endophyte-enhanced hybrid poplar trees at a superfund site in the Central Valley of California April November 2019. *Remediation* 2020: 1–17.
43. Lee KY, Strand SE, Doty SL. 2012. Phytoremediation of chlorpyrifos by Populus and Salix. *International Journal of Phytoremediation* 14: 48–61.
44. Li HY, Wei DQ, Shen M, Zhou ZP. 2012. Endophytes and their role in phytoremediation. *Fungal Diversity* 54: 11–18.
45. Lopez-Echartea E, Strejcek M, Mukherjee S, Uhlik O, Yrjala K. 2020. Bacterial succession in oil-contaminated soil under phytoremediation with poplars. *Chemosphere* 243: 125242.
46. Ma Y, Prasad MN, Rajkumar M, Freitas H. 2011. Plant growth promoting rhizobacteria and endophytes accelerate phytoremediation of metalliferous soils. *Biotechnology Advances* 29: 248–258.
47. Ma Y, Rajkumar M, Freitas H. 2009. Inoculation of plant growth promoting bacterium *Achromobacter xylosoxidans* strain Ax10 for the improvement of copper phytoextraction by Brassica juncea. *Journal of Environmental Management* 90: 831–837.
48. McCutcheon SC, Schnoor JL. 2003. *Phytoremediation: Transformation and Control of Contaminants.* John Wiley & Sons Inc, New Jersey.
49. Mishra J, Singh R, Arora NK. 2017. Alleviation of heavy metal stress in plants and remediation of soil by Rhizosphere microorganisms. *Frontiers in Microbiology* 8: 1706.
50. Moore FP, Barac T, Borremans B, Oeyen L, Vangronsveld J, et al. 2006. Endophytic bacterial diversity in poplar trees growing on a BTEX-contaminated site: The characterisation of isolates with potential to enhance phytoremediation. *Systematic and Applied Microbiology* 29: 539–556.
51. Moore MT, Schulz R, Cooper CM, Smith Jr S, Rodgers Jr JH. 2002. Mitigation of chlorpyrifos runoff using constructed wetlands. *Chemosphere* 46: 827–835.
52. More TT, Yadav JS, Yan S, Tyagi RD, Surampalli RY. 2014. Extracellular polymeric substances of bacteria and their potential environmental applications. *Journal of Environmental Management* 144: 1–25.
53. Mukherjee G, Saha C, Naskar N, Mukherjee A, Mukherjee A, et al. 2018. An endophytic bacterial consortium modulates multiple strategies to improve arsenic phytoremediation efficacy in solanum nigrum. *Scientific Reports* 8: 6979.
54. O'Brien JA, Daudi A, Butt VS, Bolwell GP. 2012. Reactive oxygen species and their role in plant defence and cell wall metabolism. *Planta* 236: 765–779.
55. Paez-Espino D, Tamames J, de LV, Canovas D. 2009. Microbial responses to environmental arsenic. *Biometals* 22: 117–130.
56. Pal A, Paul AK. 2008. Microbial extracellular polymeric substances: Central elements in heavy metal bioremediation. *Indian Journal of Microbiology* 48: 49–64.
57. Pathak KV, Keharia H. 2014. Application of extracellular lipopeptide biosurfactant produced by endophytic *Bacillus subtilis K1* isolated from aerial roots of banyan *Ficus benghalensis* in microbially enhanced oil recovery. *MEOR 3 Biotech* 4: 41–48.
58. Pilon-Smits EAH, Freeman JL. 2006. Environmental cleanup using plants: Biotechnological advances and ecological considerations. *Frontiers in Ecology and Environment* 4: 203–210.
59. Pilon-Smits EAH, Pilon M. 2002. Phytoremediation of metals using transgenic plants. *Critical Reviews in Plant Science* 21: 439–456.
60. Rajkumar M, Prasad MN, Swaminathan S, Freitas H. 2013. Climate change driven plant-metal-microbe interactions. *Environment International* 53: 74–86.
61. Rosenberg H, Gerdes RG, Chegwidden K. 1977. Two systems for the uptake of phosphate in *Escherichia coli*. *Journal of Bacteriology* 131: 505–511.
62. Rylott EL. 2014. Endophyte consortia for xenobiotic remediation: The root to success? *Plant and Soil* 385: 389–394.
63. Rylott EL, Johnston EJ, Bruce NC. 2015. Harnessing microbial gene pools to remediate persistent organic pollutants using genetically modified plants–a viable technology? *Journal of Experimental Botany* 66: 6519–6533.
64. Salt DE, Blaylock M, Kumar NPBA, Dushenkov V, Ensley BD, et al. 1995. Phytoremediation: A novel strat-

egy for the removal of toxic metals from the environment using plants. *Bio/Technology* 13: 468–474.

65. Santoyo G, Moreno-Hagelsieb G, Del CO-M, Glick BR. 2016. Plant growth-promoting bacterial endophytes. *Microbiology Research* 183: 92–99.
66. Sessitsch A, Hardoim P, Doring J, Weilharter A, Krause A, et al. 2012. Functional characteristics of an endophyte community colonizing rice roots as revealed by metagenomic analysis. *Molecular Plant Microbe Interactions* 25: 28–36.
67. Shang QT, Doty SL, Wilson AM, Howald WN, Gordon MP. 2001. Trichloroethylene oxidative metabolism in plants: The trichloroethanol pathway. *Phytochemistry* 58: 1055–1065.
68. Sharma D, Misba L, Khan AU. 2019. Antibiotics versus biofilm: An emerging battleground in microbial communities. *Antimicrobial Resistance and Infection Control* 8: 76.
69. Sheng XF, Xia JJ, Jiang CY, He LY, Qian M. 2008. Characterization of heavy metal-resistant endophytic bacteria from rape *Brassica napus* roots and their potential in promoting the growth and lead accumulation of rape. *Environmental Pollution* 156: 1164–1170.
70. Shin MN, Shim J, You Y, Myung H, Bang KS, et al. 2012. Characterization of lead resistant endophytic *Bacillus* sp MN3-4 and its potential for promoting lead accumulation in metal hyperaccumulator Alnus firma. *Journal of Hazardous Materials* 199-200: 314–320.
71. Siciliano SD, Fortin N, Mihoc A, Wisse G, Labelle S, et al. 2001. Selection of specific endophytic bacterial genotypes by plants in response to soil contamination. *Applied and Environmental Microbiology* 6: 2469–2475.
72. Siciliano SD, Germida JJ. 1999. Enhanced phytoremediation of chlorobenzenzoates in rhizosphere soil. *Soil Biology and Biochemistry* 31: 299–305.
73. Silver S, Phung LT. 2005. Genes and enzymes involved in bacterial oxidation and reduction of inorganic arsenic. *Applied and Environmental Microbiology* 71: 599–608.
74. Soleimani M, Afyuni M, Hajabbasi MA, Nourbakhsh F, Sabzalian MR, et al. 2010. Phytoremediation of an aged petroleum contaminated soil using endophyte infected and non-infected grasses. *Chemosphere* 81: 1084–1090.
75. Song, C., Sarpong, C. K., He, J., Shen, F., Zhang, J., Yang, G., Long, L., Tian, D., Zhu, Y. & Deng, S. (2020). Accelerating phytoremediation of degraded agricultural soils utilizing rhizobacteria and endophytes: a review. *Environmental Reviews* 28, 115-127.
76. Song, Y., Xu, M., Li, X., Bian, Y., Wang, F., Yang, X., Gu, C. & Jiang, X. (2018). Long-Term Plastic Greenhouse Cultivation Changes Soil Microbial Community Structures: A Case Study. *Journal of Agricultural and Food Chemistry* 66, 8941-8948.
77. Spriggs TA, Banks K, Schwab P. 2005. Phytoremediation of polycyclic aromatic hydrocarbons in manufactured gas plant-impacted soil. *Journal of Environmental Quality* 34: 1755–1762.
78. Taghavi S, Barac T, Greenberg B, Borremans B, Vangronsveld J, et al. 2005. Horizontal gene transfer to endogenous endophytic bacteria from poplar improves phytoremediation of toluene. *Applied and Environmental Microbiology* 71: 8500–8505.
79. Thijs S, Sillen W, Rineau F, Weyens N, Vangronsveld J. 2016. Towards an enhanced understanding of plant-microbiome interactions to improve phytoremediation: engineering the metaorganism. *Frontiers in Microbiology* 7: 341.
80. Tournay RJ, Firrincieli A, Parikh S, Sivitilli D, Doty SL (manuscript in preparation). Isolation and characterization of arsenic-tolerant endophytes and the potential of endophytic biofilms in arsenic tolerance.
81. Vainberg S, Condee CW, Steffan RJ. 2009. Large-scale production of bacterial consortia for remediation of chlorinated solvent-contaminated groundwater. *Journal of Industrial Microbiology and Biotechnology* 36: 1189–1197.
82. Vangronsveld J, Herzig R, Weyens N, Boulet J, Adriaensen K, et al. 2009. Phytoremediation of contaminated soils and groundwater: Lessons from the field. *Environmental Science Pollution Research* 16: 765–794.
83. Villacieros M, Whelan C, Mackova M, Molgaard J, Sanchez-Contreras M, et al. 2005. Polychlorinated biphenyl rhizoremediation by *Pseudomonas fluorescens* F113 derivatives using a *Sinorhizobium meliloti* nod system to drive *bph* gene expression. *Applied and Environmental Microbiology* 71: 2687–2694.
84. Wan Y, Luo S, Chen J, Xiao X, Chen L, et al. 2012. Effect of endophyte-infection on growth parameters and Cd-induced phytotoxicity of Cd-hyperaccumulator *Solanum nigrum* L. *Chemosphere* 89: 743–750.
85. Wang KL, Li H, Ecker JR. 2002. Ethylene biosynthesis and signaling networks. *Plant Cell* 14(Suppl): S131–S151.
86. Wang X, Rathinasabapathi B, de Oliveira LM, Guilherme LR, Ma LQ. 2012. Bacteria-mediated arsenic oxidation and reduction in the growth media of arsenic hyperaccumulator. *Pteris Vittata. Environmental Science and Technology* 46: 11259–11266.

87. Wenzel WW. 2009. Rhizosphere processes and management in plant-assisted bioremediation phytoremediation of soils. *Plant and Soil* 321: 385–408.

88. Weyens N, Truyens S, Dupae J, Newman L, Taghavi S, et al. 2010. Potential of the TCE-degrading endophyte *Pseudomonas putida* W619-TCE to improve plant growth and reduce TCE phytotoxicity and evapotranspiration in poplar cuttings. *Environmental Pollution* 158: 2915–2919.

89. Weyens N, van der Lelie D, Taghavi S, Newman L, Vangronsveld J. 2009a. Exploiting plant-microbe partnerships to improve biomass production and remediation. *Trends in Biotechnology* 27: 591–598.

90. Weyens N, van der Lelie D, Taghavi S, Vangronsveld J. 2009b. Phytoremediation: Plant-endophyte partnerships take the challenge. *Current Opinion in Biotechnology* 20: 248–254.

91. Xu JY, Han YH, Chen Y, Zhu LJ, Ma LQ. 2016. Arsenic transformation and plant growth promotion characteristics of As-resistant endophytic bacteria from As-hyperaccumulator. *Pteris Vittata. Chemosphere* 144: 1233–1240.

92. Yang SF, Hoffman NE. 1984. Ethylene biosynthesis and its regulation in higher plants. *Annual Review of Plant Physiology* 35: 155–189.

93. Yuan M, He H, Xiao L, Zhong T, Liu H, et al. 2014. Enhancement of Cd phytoextraction by two *Amaranthus* species with endophytic *Rahnella* sp JN27. *Chemosphere* 103: 99–104.

94. Zalesny RS, Bauer EO, Hall RB, Zalesny JA, Kunzman J, et al. 2005. Clonal variation in survival and growth of hybrid poplar and willow in an in situ trial on soils heavily contaminated with petroleum hydrocarbons. *International Journal of Phytoremediation* 7: 177–197.

95. Zalesny RS, Headlee WL, Gopalakrishnan G, Bauer EO, Hall R, et al. 2019. Ecosystem services of poplar at long-term phytoremediation sites in the Midwest and Southeast United States. *Wiley Interdisciplinary Reviews Energy and Environment* 8: 1–26.

96. Zhang YF, He LY, Chen ZJ, Wang QY, Qian M, et al. 2011. Characterization of ACC deaminase-producing endophytic bacteria isolated from copper-tolerant plants and their potential in promoting the growth and copper accumulation of Brassica napus. *Chemosphere* 83: 57–62.

97. Zhu LJ, Guan DX, Luo J, Rathinasabapathi B, Ma LQ. 2014. Characterization of arsenic-resistant endophytic bacteria from hyperaccumulators Pteris vittata and Pteris multifidi. *Chemosphere* 113: 9–16.

CHAPTER 30

Metagenomics of Bacterial Consortia for the Bioremediation of Organic Pollutants

Daniel Garrido-Sanz, Paula Sansegundo-Lobato, Marta Martin, Miguel Redondo-Nieto, and Rafael Rivilla

Department of Biology, Universidad Autónoma de Madrid, 28049, Spain

30.1 INTRODUCTION

Human activities often result in the pollution of the environment. Pollution is defined as "the presence of a chemical or substance out of place and/or present at a higher than normal concentration that has adverse effects on any non-targeted organism" (FAO and ITPS, 2015). Pollution can have naturally occurring origins, such as forest fires or volcanic eruptions. However, the anthropogenic origin is considered the main source of modern environmental pollution (Cachada et al., 2018).

Organic pollutants are compounds derived from carbon species and include several groups that can be broadly divided into aromatic and aliphatic. Aromatic compounds are constituted by one or more aromatic rings and include monoaromatic compounds such as benzene, toluene, ethylbenzene and xylene, and polycyclic aromatic hydrocarbons (PAHs), constituted by two or more fused aromatic rings. Related to aromatics are dioxins, furans, and dioxin-like compounds, including polychlorinated biphenyls (PCBs). On the other hand, aliphatic pollutants are mainly alkanes: saturated hydrocarbons that vary in chain length and ramifications and are the major components of petroleum fuels (National Research Council, 1985). Aliphatic compounds may also be chlorinated. Although alkanes are naturally produced by bacteria, green algae, plants, and animals, release of petroleum hydrocarbons into the environment by accidental spills or by leakages from subterranean storage tanks and pipelines cause a profound impact on living organisms and the ecosystems (Labud et al., 2007; Sikkema et al., 1995).

Different microorganisms, especially bacteria, are known to degrade many of these organic pollutants throughout different and sometimes intricate catabolic pathways, in a process known as biodegradation (Vert et al., 2012). This mechanism usually involves several enzymatic steps in a pathway that ultimately ends with the complete oxidation of the compound, known as mineralization, which also provides the carbon and energy necessary for cell growth. Some organic pollutants, however, are partially degraded due to the absence of a specific enzyme within its pathway. This might result in dead-end bioproducts that can be either less or more toxic than the parent compound, depending on the specific substrate, or be more persistent in the environment by a more limited bioavailability than the parent compound (Cámara et al., 2004; Pepper et al., 2011). On the other hand, cometabolism is defined as "the process by which a contaminant is fortuitously degraded by an enzyme or cofactor produced during the microbial metabolism of another compound," but the obtained energy from the oxidation is not used for cell growth (McCarty, 1988). In addition, the complete mineralization or other pollutant's biotransformation

Good Microbes in Medicine, Food Production, Biotechnology, Bioremediation, and Agriculture, First Edition. Edited by Frans J. de Bruijn, Hauke Smidt, Luca S. Cocolin, Michael Sauer, David Dowling, and Linda Thomashow.

might occur within complex trophic microbial interactions: the products resulting from the metabolism of a specific microorganism might be further degraded/mineralized by a second group of microorganisms (Garrido-Sanz et al., 2018).

The biodegradative potential exhibited by bacteria can be exploited to restore polluted sites, a process known as bioremediation and defined as "the process whereby organic wastes are biologically degraded under controlled conditions to an innocuous state, or to levels below concentration limits established by regulatory authorities" (Mueller et al., 1996; see chapters 27–29). Compared to chemical and physical remediation techniques, bioremediation is often less expensive, can result in the complete elimination of the hazardous pollutant, and is considered a sustainable environmental practice, since it stimulates natural processes (Russell, 1992; US EPA, 2008). However, bioremediation is limited to those chemicals that can be biodegraded and the presence of the microorganisms harboring the appropriate metabolic capabilities.

Multiple bioremediation techniques can be used for the restoration of polluted sites. The choice of technique depends on the characteristics of the pollutant and the site. Among these, bioaugmentation consists of the addition of cultured microorganisms with the ability to degrade the target pollutant to augment the indigenous populations and improve the degradation (Vogel, 1996). This technique can be applied when the autochthonous populations are not capable of degrading pollutants present in complex mixtures, they are in low abundance, the speed of the decontamination is a major factor, and when inoculation may reduce the lag period to start the bioremediation process. For this technique to be successful, the introduced microorganisms must be able to compete with the indigenous populations, including predation by protozoa and viruses. They should also be able to survive the foreign environmental conditions. The factors affecting the bioremediation process are summarized in Figure 30.1. In addition, bioavailability is also a major determinant of the bioremediation process.

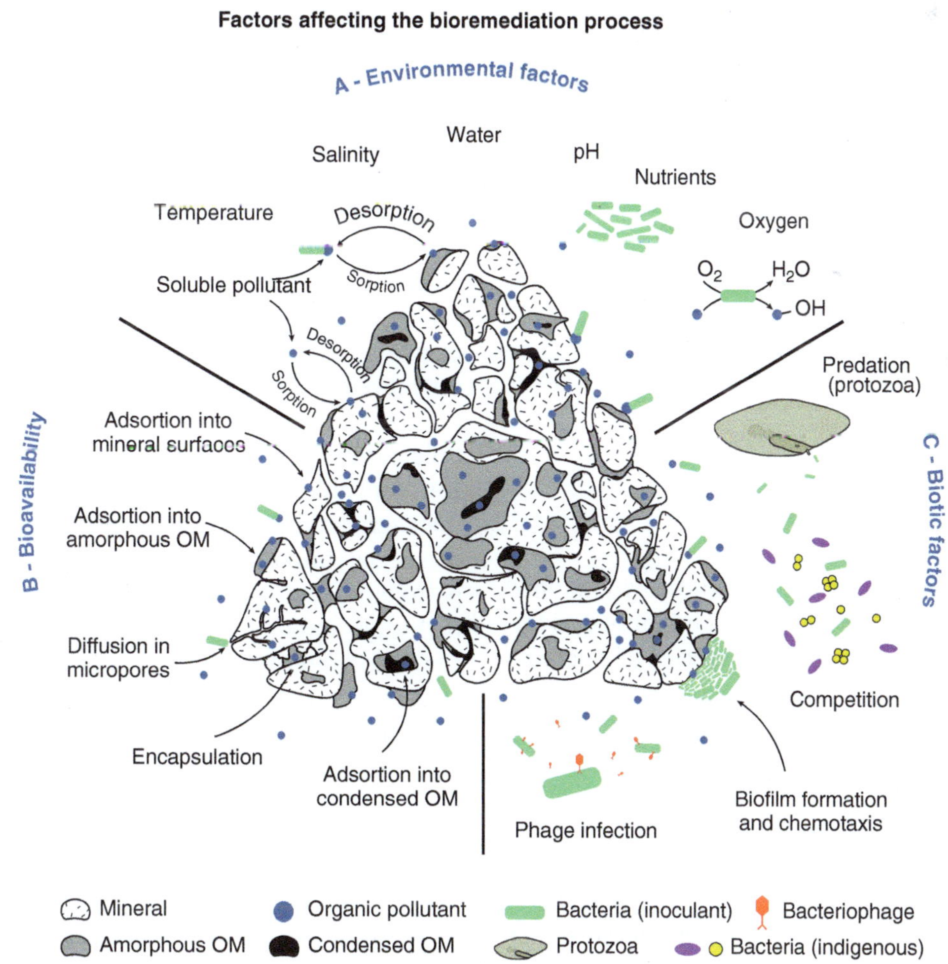

FIGURE 30.1 Schematic representation of the environmental factors (A), bioavailability (B), and biotic factors (C) affecting the bioremediation process in a hypothetical soil environment. OM, organic matter.

Bioavailability can be defined as "the extent to which a contaminant is available for biological conversion which is a function of the biological system, physicochemical properties of the pollutant and environmental factors" (Naidu et al., 2008). The bioavailability of organic pollutants is negatively affected by several factors including water solubility, sorption, and matrix composition (Juhasz et al., 2000; Maier, 2000).

Recent advances in metagenome sequencing have resulted in a profound understanding of the microbial community dynamics in response to pollution and bioremediation treatments (Fuentes et al., 2016; Ribicic et al., 2018), and help to monitor and identify key genes involved in the bioremediation process, which increase our knowledge of the functional degradative potential of microbial communities (Bharagava et al., 2019; Duarte et al., 2017; Techtmann and Hazen, 2016). Metagenomics applied to bacterial consortia isolated by enrichment culture techniques, allows the isolation of that part of the population present in a specific environment with a direct involvement in the degradation of the target pollutant(s) (Garrido-Sanz et al., 2018). Bacterial consortia also show a reduction of the complexity of the initial community, being a key feature to achieve higher sequencing depth and can serve to elucidate the tropic relationships or metabolic interdependence through functional metagenomic analyses (Garrido-Sanz et al., 2018, 2019).

In this chapter, we review the metagenomic approaches that are being currently used to analyze the role of bacterial consortia in the biodegradation of organic pollutants, emphasizing the potential for predicting the degradative potential of specific members of a bacterial consortium.

30.2 BACTERIAL CONSORTIA

The term "consortium" was first introduced by Johannes Reinke in 1872 (Reinke, 1872) and can be defined as the mutualistic association of different species of microorganisms in physical and/or biochemical contact with one another that turns them into a unit (Kull, 2010). In soil microbiology, the term "consortium" is more specifically referred as "an association of microorganisms from different species living in metabolic interdependence" (Popa, 2004) or that fulfill a certain biochemical transformation (Brenner et al., 2007). The relationships between the consortium members are commonly measured on the effect that one species has upon the other rather than in the identification of the specific mechanism of the relationship, which can be indirect or without any communication between organisms (Kull, 2010).

Different interaction mechanisms within the members of microbial consortia have been identified and previously reviewed in Zhang et al. (2018) and summarized in Figure 30.2. These include synergistic division of resources (e.g., cross-feeding), stimulation of microbial growth and biotransformation, enhanced tolerance to inhibitors or toxic compounds, antagonistic interactions that leads to production of beneficial metabolites and enzymes, and assembled biotransformation pathways (Zhang et al., 2018). For bioremediation purposes, however, most studies focus on assembled biotransformation, where the complete mineralization of a pollutant is achieved by different reactions or pathways compartmentalized within specific members of the consortium (Gilbert et al., 2003).

30.2.1 Advantages and Limitations

Historically, the use of single-isolate inoculants or synthetic combinations of different co-cultured bacterial species has been the predominant bioaugmentation-based bioremediation approach. However, at the same time many were reported unsuccessful (Bouchez et al., 2000; Songzhe et al., 2009; Thompson et al., 2005; Wagner-Dobler, 2003). The reasons for these failures are usually the inability to adapt and persist in a foreign environment.

Bacterial consortia, specifically consortia isolated from environmental samples rather than synthetic, present several advantages in bioremediation technologies. Compared to single isolates, consortia can contain multiple pathways that can result in the complete mineralization of the pollutant (Garrido-Sanz et al., 2018) and the degradation of a wide range of compounds, which is of special importance in complex pollutants such as petroleum hydrocarbons. The presence of redundant pathways among different members of the consortium can also improve the consortium robustness against changes in the environment. Compared to environmental samples, bacterial consortia are easier to analyze given that they are usually composed of a limited number of genotypes. Particularly, consortia isolated by means of enrichment culture using the target pollutant(s) as sole carbon and energy source, result in the isolation of that part of the population that is actively involved in the degradation of the pollutant, while most of the genotypes present in the original environment do not

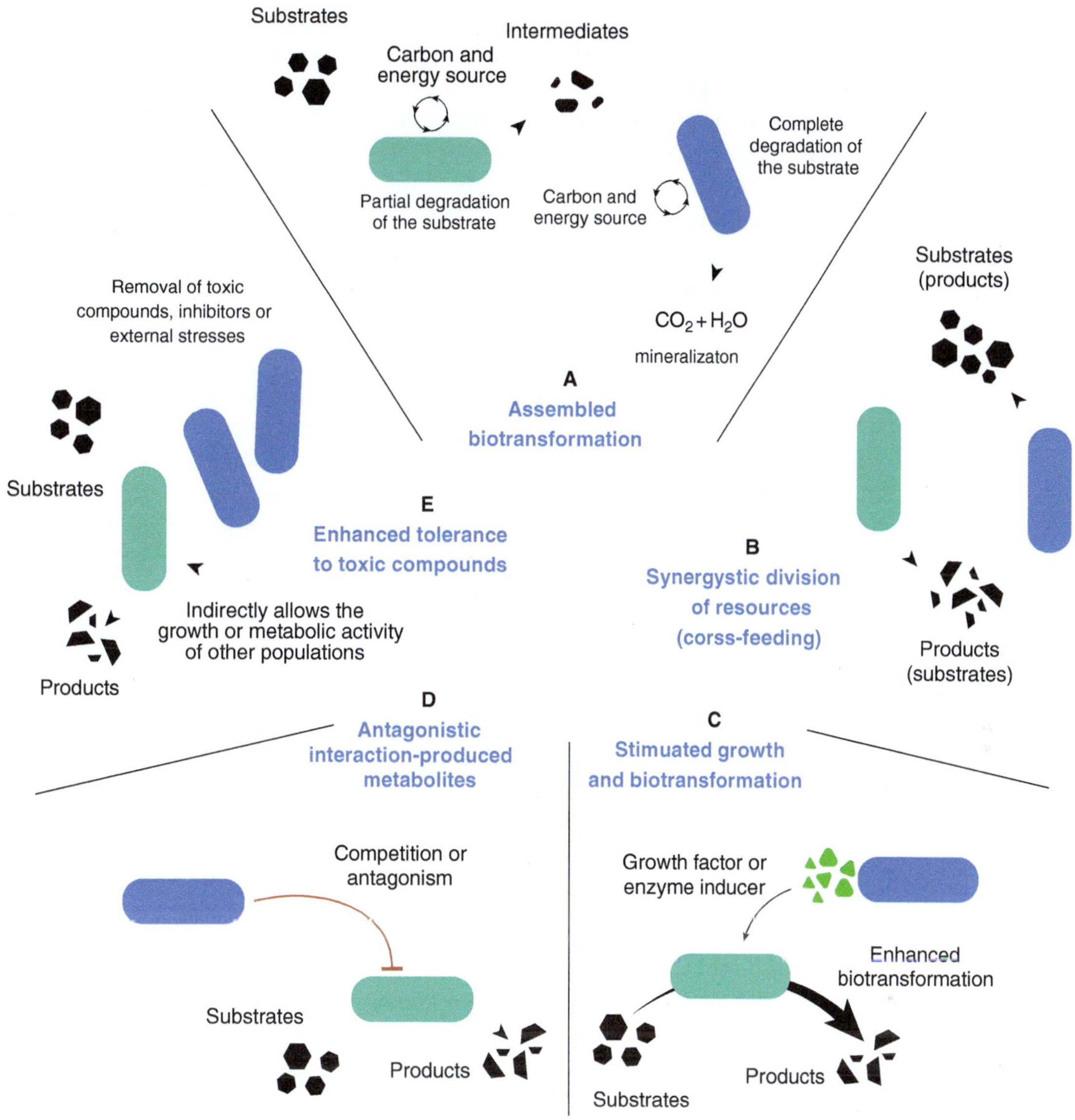

FIGURE 30.2 Schematic representation of interaction mechanisms between consortia members. (A) Assembled biotransformation, (B) synergistic division of resources, (C) stimulated growth and biotransformation, (D) antagonistic interaction-produced metabolites, and (E) enhanced tolerance to toxic compounds. Source: Zhang et al. (2018) under Creative Commons Attribution 4.0 International (CC BY 4.0).

participate in the degradation process. Nonetheless, although the selection of bacterial consortia through enrichment culture will result in a reduction of the original community towards those microorganisms capable of feeding on the target pollutant or involved in its degradation, it is by far the best approach to isolate naturally-occurring bacterial consortia and reducing the complexity present in an environmental sample, while maintaining the ability to biodegrade the pollutant.

30.3 ISOLATION FROM THE ENVIRONMENT: METHODS AND PROBLEMS

Bacterial consortia for bioremediation purposes can be isolated by enrichment culture techniques directly from environmental samples such as soil or water. These techniques consist in the culture of the environmental sample in a mineral salt medium supplemented with the target pollutant(s) as the sole carbon and energy source. Through several subcultures, only that part of the initial population that can grow using the target pollutant or their metabolic derivatives as sole carbon and energy source is selected (Garrido-Sanz et al., 2018). However, it is known that only a small fraction of bacteria is cultivable. It is generally accepted that this inability to culture environmental-derived bacteria is due to the absence of specific nutrients or compounds that are present in the environment but not supplemented within the culture media. Although there have been several attempts to improve the isolation of otherwise unculturable bacteria, including by providing diffusible elements present in the environment (Puspita et al., 2012), the addition of yeast extract in a very low concentration (0.005%) that does not serve as carbon source, can provide some of such elements. Despite all the efforts, it is very likely that part of the environmental diversity that participates in the degradation of a pollutant is lost during the enrichment culture process. In addition, by selecting a fraction of the original environmental population, the equilibrium of the population involved in the biodegradation process might also be altered. Theoretically, factors such as competition for nutrients, production of antibiotics, and other elements that bacteria use to compete with others together with environmental factors, shape the general community structure. However, these other populations could be lost when isolating bacteria through enrichment culture and, therefore, other traits that maintain the original structure are completely ignored. These traits may include the ability of bacteria to effectively form biofilms or chemotaxis toward pollutants (Singh and Olson, 2008; Songzhe et al., 2009).

30.3.1 Metagenomics

One of the main handicaps for the characterization of complex environmental populations is the failure to obtain pure cultures for all bacterial species in the biosphere. For that reason, different approaches have emerged that allow us to obtain an insight into the composition of a bacterial population. In 1998 Handelsman (Handelsman et al., 1998) coined the term "metagenomics" as the set of techniques for analyzing and characterizing the bacterial population within a niche based on its genomic DNA. Although these techniques originally gave a great impulse to the knowledge of bacterial populations ecology, it has not been until recently with the emergence of the next generation sequencing platforms (NGS) and the advances in the computing capacity when a great amount of information has been generated in this field. NGS techniques can sequence billions of DNA sequences in parallel providing statistical robustness to studies that adopt this approach. Depending on the source of DNA there are two kinds of analysis: targeted metagenomics and shotgun metagenomics (Figure 30.3).

30.3.1.1 Targeted Metagenomics

Targeted metagenomics also known as amplicon metagenomics or metabarcoding relies on the amplification of a marker gene by PCR and its subsequent sequencing. In bacteria the most used marker is the 16S rRNA gene that encodes for the small ribosomal subunit RNA molecules of ribosomes. This gene was described as a good distinguishing factor of the populations (Pace et al., 1986) because it presents the following properties:

(a) It is a key element in all prokaryotic organisms. There is no bacterium that does not have at least one copy of this gene.
(b) It is an ancient gene and its structure and sequence is highly conserved.
(c) In its sequence we find highly conserved regions that remain stable between different taxonomic groups and variable regions that allow them to be identified.

The sequence is long enough, around 1500 nucleotides, to give statistical robustness to the dissimilarity positions among two sequences.

For these reasons, this gene has been used as a marker or phylogenetic clock for a long time to differentiate the members of the bacteria kingdom either in pure cultures or in complex communities. As mentioned above, for targeted metagenomics of 16S rRNA or 16S profiling, this gene must be amplified by PCR, but the length of the amplified fragment is constrained by the NGS system. Most of the NGSs systems, including Illumina, are not suitable to sequence the full length of the 16S gene. As a workaround, only the variable regions of the gene are amplified since they are the ones that will provide the distinctive information. Thus, in the literature there are many studies with different regions or pairs of regions that are selected depending on the objectives of the researchers. However, not all of them will be equally effective in amplifying the 16S sequences of all organisms. Therefore, we must be very careful with the region that we want to amplify, since one part of the population can be favored over another. To avoid this bias that

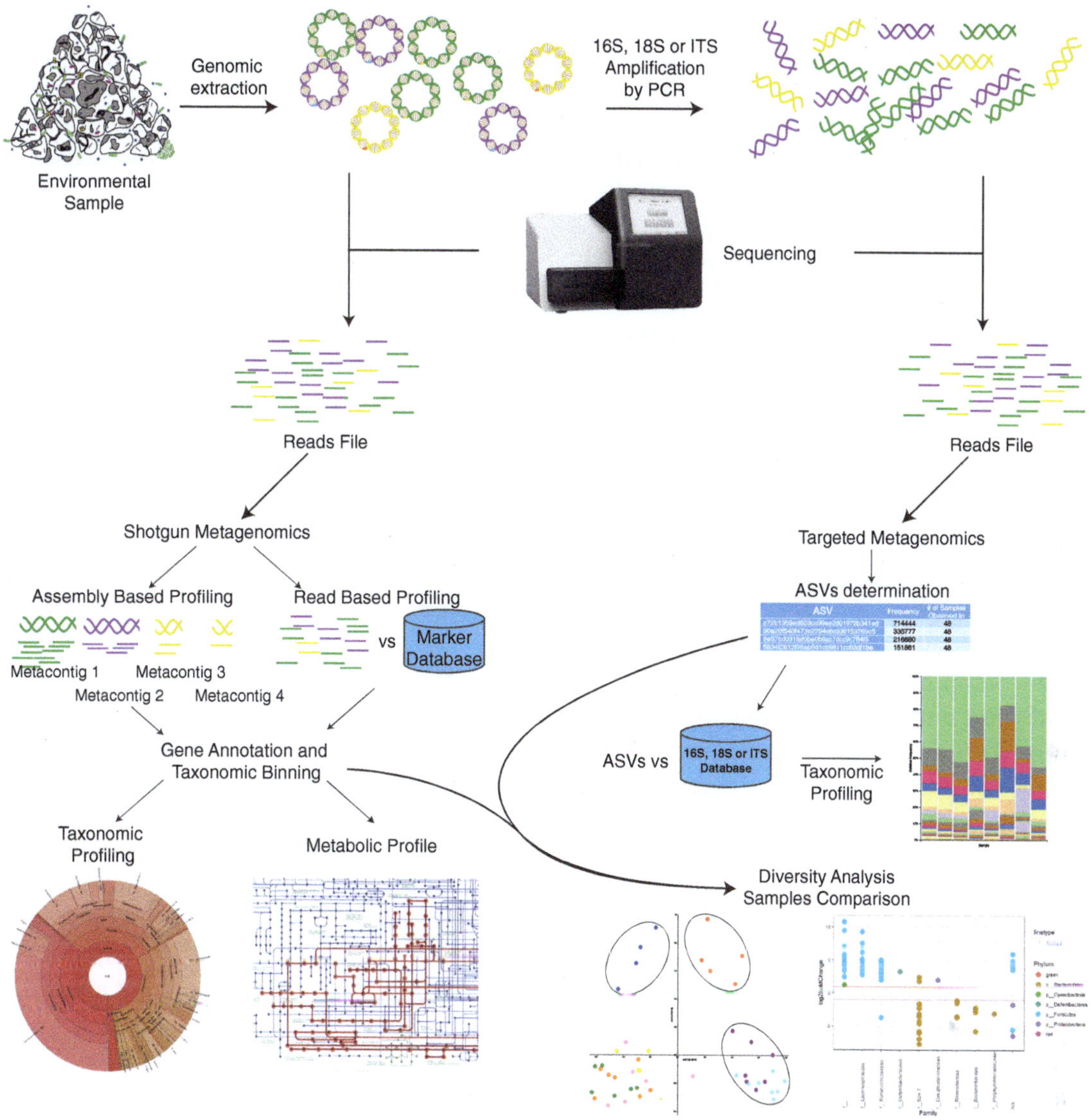

FIGURE 30.3 Schematic representation of current pipelines for shotgun and targeted metagenomics. Sequencing is carried out directly on sample genomic DNA (shotgun metagenomics) and/or a PCR amplification of a marker gene (targeted metagenomics). In the case of targeted metagenomics, distinctive and unique sequences (ASVs) are determined from reads and compared against the corresponding database to get a taxonomic profile. In the case of shotgun metagenomics reads can be assembled to recreate DNA metacontigs of the original population and the genes in those contigs to get the putative metabolic profile (i.e., xenobiotics degradation pathway in red) and the taxonomic profile. Reads can also be compared against mOTUs database to get a taxonomic profile of the sample. Finally, taxonomic profiles including abundance for each taxa can be used to perform diversity studies looking for differences in the populations among samples.

could affect the cross-comparison among studies, several projects have emerged with the aim of standardizing protocols. For instance, The Earth Microbiome Project (Gilbert et al., 2014) has as objective to build an atlas of the microbial taxonomic and functional diversity of soils along the planet. To achieve this goal, not only are all their standard protocols available at their webpage but also the sequencing data coming from their collaborators.

The amplified region, or amplicon, is sequenced and the resulting sequences, called reads when using NGSs, are processed by bioinformatic analysis in integrated pipelines such as QIIME2 (Bolyen et al., 2019) and phyloseq (Paul and Holmes, 2013). Not long ago, reads were grouped in bins of sequences with a homology of at least 97%. These bins are known as OTU (operational taxonomic unit). Each OTU would be represented by a model sequence that results from the consensus of the sequences clustered within the OTU. Concerning diversity meaning, each OTU would represent a distinct unit of the population. Nevertheless, each OTU does not have to correspond to a bacteria species. Two individuals are considered to belong to the same species when their corresponding sequence homology is above 99%, while two individuals with a homology above 97% are considered to belong to the same genus. Recently, with the development of new algorithms and the increase in computing power, the latest studies are choosing to determine not the OTUs but what are known as ASVs (amplicons sequence variants) in which the sequences are differentiated with a resolution of a single nucleotide. Regardless of the approach used, OTUs and ASVs are assigned to a taxonomic category by direct comparison with specific 16S sequence databases such as RDP or SILVA (Quast et al., 2013). Thus, targeted metagenomics will provide us with data regarding the members of a community, including their relative abundance and their putative taxonomic assignation. This data can be further studied to compare diversity within a sample or among samples. Hence, we can determine which biotic or abiotic environmental factors are influencing the populations and which individuals are affected or advantaged by the changes. In the case of hydrocarbon-contaminated soils we could determine which ASVs are present and which are maintained or enriched when the nature of the hydrocarbons is changed, providing a clue as to the type of bacteria that could feed on that contaminant.

Though targeted metagenomic studies are reliable, we must respect certain considerations:

(a) As explained above, relative abundances can differ depending on amplification conditions.
(b) A bacterial species can harbor several copies of 16S rRNA gene. Thus, different ASVs might belong to the same species.

Taxonomic assignation is strongly influenced on the quality of the reference database, and an assignation to the species level should be considered with precaution (Edgar, 2018).

Finally, based on the taxonomic classification a functional metabolic model of the sample could be extrapolated using data from fully sequenced genomes present in the community. This model usually has serious limitations as not all the microorganisms have been sequenced. Besides, in nature, events such as horizontal gene transfer among individuals of different species or gene loss and duplication are very common, therefore the actual metabolic map does not have to agree with the predicted one. To have a better resolution on this matter shotgun metagenomics approaches are used.

30.3.1.2 Shotgun Metagenomics

Although determining populations using amplicons is cost-effective and provides us with some resolution in terms of identifying taxonomic groups, we cannot determine what adaptive and metabolic capacities those groups have. Shotgun metagenomics is based on sequencing the whole genomic DNA obtained from an environmental sample. Though is not a recent methodology, the first publications date from 2004 (Tyson et al., 2004; Venter et al., 2004), it has still to face certain difficulties.

(a) DNA extraction and purification protocols must maintain a balance between obtaining the highest amount of DNA versus the integrity of the DNA. Some of the protocols that use physical methods to separate bacteria from soil and to lysate them can break DNA into small fragments that would hamper bioinformatic analysis.
(b) Microbial populations in a niche are not homogeneous and depending on the sequencing coverage, minority populations might not appear.

Environmental samples can contain hundreds of strains. Bioinformatic analyses must find an optimal solution to cluster the sequenced reads into their corresponding strains avoiding mixtures that could lead to the formation of chimeras.

There are two strategies when processing the samples: read-based profiling and assembly-based profiling.

30.3.1.2.1 Read-based Profiling

This low computing consuming approach is based on the direct comparison of reads against already sequenced genomes. Thus, we can obtain both a functional and a taxonomic profile of the community. But, as we rely on a database, there is a probability to get false positives and data coming from new strains would not be characterized. However, this problem diminishes over time due to the increasing number of genomes sequenced. Taxonomic assignation for each read is usually carried out using LCA algorithm (lowest common ancestor) by which the result of the taxonomic classification is the last common ancestor among the hits of each query against the reference database. Another alternative for achieving a taxonomic profile is to compare reads against a database of phylogenetic marker genes. As opposed to what happens in targeted metagenomics, not a single marker is used but several, usually present in a single copy in the genome. The reference database is constructed from the profiles of the marker genes for each species, each profile constituting what is known as marker gene-based operational taxonomic units (mOTUs) (Milanese et al., 2019). In this sense, by using a database of mOTUs instead of full genomes, results can be obtained more quickly and at a lower computing cost. Other approaches are described and compared in Kibegwa et al. (Kibegwa et al., 2020).

30.3.1.2.2 Assembly-based Profiling

Reads obtained by NGS are pieces of genomic DNA that have been randomly fragmented. There are algorithms that can solve this big puzzle by assembling overlapping reads in a bigger sequence known as contig. Algorithms as de Bruijn graphs are used both for single genome assembly and metagenomic assembly, but metagenome DNA has not a homogeneous coverage as it depends on the diversity of the sample. Species in low abundance will be represented by a low percentage of reads compared to the most abundant species. Therefore, in this kind of approach it is more likely to obtain larger contigs belonging to abundant species and smaller and fragmented contigs for the low abundant species. Furthermore, the more complex a community is, the more difficult its meta-assembly becomes. Specialized programs such as MEGAHIT (Li et al., 2015) and metaSPAdes (Nurk et al., 2017) are commonly used in these tasks, and commonly yield a high number of fragmented contigs together with their relative abundance related to the amount of reads that compose each contig. These contigs are subjected to gene annotation using both direct comparison against known genomes and by modeling software able to predict contig regions that are likely to encode a gene. By direct comparison of the genes within a contig against genome databases, contigs are clustered into bins and each bin is assigned to a taxonomic category so that we can get the taxonomic profile of the community.

Finally, annotation is commonly associated to gene entries in KEGG and COG databases which allows the functional and metabolic pathways of the community to be depicted. However, genomes in databases are not fully annotated and the functional maps are mainly composed of a central metabolism and some other accessory pathways of biotechnological interest. The advances in genome sequencing have followed an exponential curve but for acquiring a better result of functional profiling we depend on other techniques such as (meta)-transcriptomics and proteomics, and mutant techniques generation as CRISPR to complete genome databases.

30.4 CONSORTIUM-BASED BIOREMEDIATION: CASE STUDIES

30.4.1 Petroleum Hydrocarbon Bioremediation

Soil and water pollution by petroleum hydrocarbons, such as diesel fuel, is one of the most serious environmental concerns due to the toxicity and recalcitrant characteristics of their constituents, which include branched aliphatic, PAHs, and substituted aromatic hydrocarbons (see also chapters 27–29). The application of local bacterial consortia isolated from the polluted sites in bioremediation processes might be more convenient than introducing allochthonous strains, since indigenous microorganisms are already specialized and adapted to the polluted site (Bento et al., 2005; Wu et al., 2013). Furthermore, using autochthonous hydrocarbon-degrading microorganisms in rhizoremediation applications has been suggested as an effective technique for the restoration of diesel-polluted soils (Garrido-Sanz et al., 2019).

Garrido-Sanz et al. (2019) isolated and characterized an indigenous bacterial consortium from a ship fuel-polluted rhizospheric soil that can grow aerobically using diesel and aliphatic and aromatic constituents of diesel as the sole carbon and energy source. The bacterial composition of the consortium and the identification of key genes involved in the pollutant degradation process were analyzed by metagenomic approaches. Both sequencing of 16S rRNA amplicons and whole metagenome shotgun sequencing and annotation, showed that the consortium is dominated by *Pseudomonas*, which was not unexpected since this is one of the most abundant genera in hydrocarbon-polluted soils (Yergeau et al., 2012). Other relevant taxa within the consortium are the *Aquabacterium* genus and the Sphingomonadaceae family, which, along with *Pseudomonas*, could be responsible for the initial oxidation of different chain-length alkanes, as they harbor most of the identified genes encoding alkane hydroxylases. Moreover, the initial step in PAH biodegradation is also primarily attributed to bacteria of the Sphingomonadaceae family. Metagenomic analysis of the consortium growing in diesel showed the presence of 47 genes encoding alkyl hydroxylases, the first enzyme in alkane biodegradation pathways and 83 genes encoding hydroxylating dioxygenases, the enzyme which initiates the biodegradation of aromatic and PAH compounds. This abundance of genes implicated in bioremediation, together with their apparent redundancy, probably contributes to the robustness of the system. Bioremediation feasibility of the original diesel-polluted soil was evaluated through rhizoremediation (integrated plant-bacteria systems) microcosms assays inoculated with the consortium and using *Medicago sativa* plants. The results showed a clear TPHs reduction in both aliphatic and aromatic fractions after four months, indicating that the bacterial consortium isolated in the study could serve as an inoculum for effective rhizoremediation of diesel-polluted soils. Interestingly, the combined use of plants and the consortium showed a synergistic effect, with a degradation activity higher than that of the sum of the individual treatments.

Guerra et al. (2018) isolated two oil-degrading consortia from drill cuttings residues through an enrichment culture procedure using petroleum as the sole carbon and energy source. A comparative metagenomic approach was used for the identification of taxa and genes involved in the hydrocarbon degradation pathways in both original drill cuttings samples and isolated consortia. The predominant phyla in the consortia were *Firmicutes* and *Brevibacillus*, *Paenibacillus*, and *Bacillus* being the most abundant genera, whose ability to degrade hydrocarbons has been reported (Prince et al., 2010). Some biodegradation pathways were significantly more abundant in the consortia than in the drill cuttings, such as those related to chloroalkane, chlorobenzene, chlorocyclohexane, and toluene degradation. Moreover, genes encoding enzymes specifically involved in aliphatic and aromatic hydrocarbon degradation were identified within the consortia metagenomes, such as alkane monooxygenase, cytochrome P450, and methane monooxygenase regarding aliphatic compounds, and xylene monooxygenase, benzoate dioxygenase, toluene monooxygenase, catechol dioxygenase, and phenol hydroxylase regarding aromatic compounds. Furthermore, the consortia were evaluated for alkane degradation in the drill cuttings through microcosms assays, and the results showed an effective removal of the alkane fractions.

30.5 AROMATICS BIOREMEDIATION

Polychlorinated biphenyls (PCBs) are a group of synthetic persistent organic chemicals whose physical and chemical properties led to their widespread manufacture (National Research Council, 1979). Although their production and commercialization have been forbidden for decades due to their exposure-related health effects in humans, they are still significantly present in the environment since they are recalcitrant toxics. One of the principal strategies to mitigate the persistence of PCBs is through aerobic bioremediation, in which the enzymes of the biphenyl upper degradation pathway are usually implicated (Furukawa and Fujihara, 2008). Therefore, in biodegradation research, biphenyl is commonly used as a mineralizable PCB analog (Garrido-Sanz et al., 2018; Vergani et al., 2017).

Garrido-Sanz et al. (2018) reported the isolation and metagenomic characterization of a bacterial consortium from a rhizospheric soil near a gas station that is able to grow with biphenyl, and other aromatic compounds such as benzoic acid and protocatechuate acid, as the sole carbon and energy source. Both the relative abundance based on 16S rRNA amplicons and CDSs showed that the biphenyl-degrading consortium is clearly dominated by *Pseudomonas*, but other relevant genera that are present are *Bordetella*, *Achromobacter*, *Stenotrophomonas*, *Rhodococcus*, and *Cupriavidus*. Furthermore, the metagenomic analysis allowed the assignation of specific reactions and pathways to specific populations. Metagenomic analysis allowed the modeling of all degradation pathways from biphenyl to tricarboxylic acid (TCA) cycle intermediates (Figure 30.4). *Rhodococcus* was the only genus responsible for initiating the degradation of biphenyl into benzoate, since all biphenyl dioxygenase copies identified in the metagenome were only assigned to this genus. Moreover, several gene clusters (*bph*, *etb*, and *nar* gene clusters), whose involvement in both biphenyl and PCBs degradation has been reported (Iwasaki et al., 2007; Kimura et al., 2006; Taguchi et al., 2007), were found in the metagenome and the degradation pathways were completely reconstructed in the *Rhodococcus* present in the consortium. On the other hand, all the enzymes involved in the further degradation of benzoate to TCA intermediates through different aerobic pathways were also screened and found in the metagenome of the biphenyl-degrading consortium, mainly *Pseudomonas* and *Bordetella* being the genera responsible of these lower degradative pathways. PCB

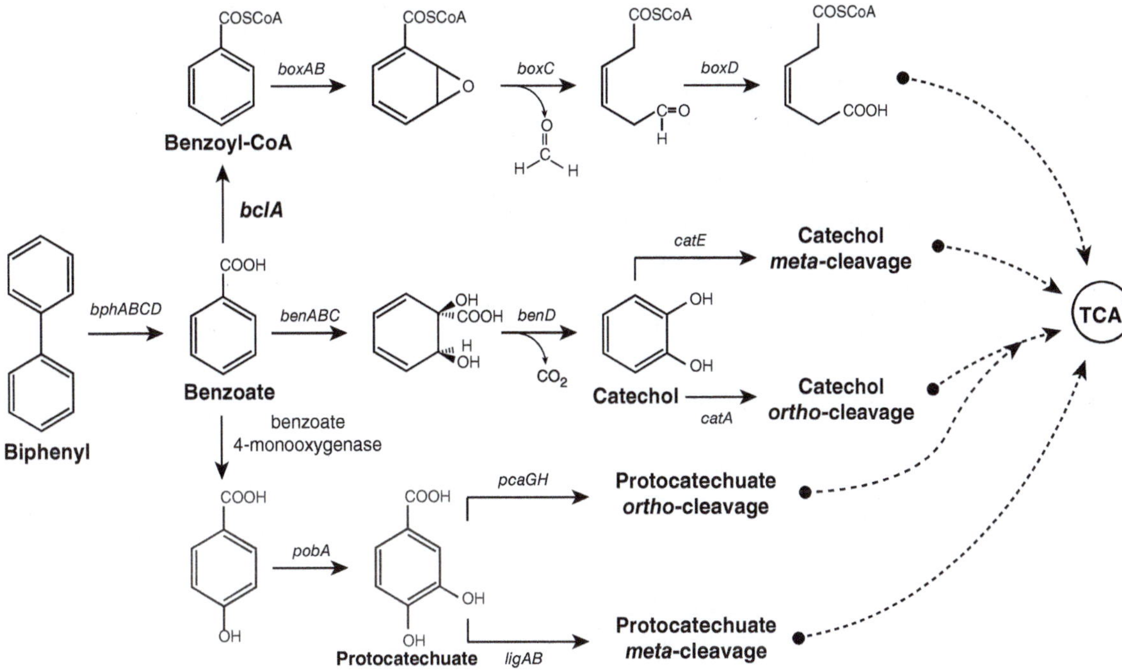

FIGURE 30.4 Schematic representation of the aerobic biodegradation pathway of biphenyl and its metabolic products (benzoate, catechol, and protocatechuate) into intermediates of the tricarboxylic acid (TCA) cycle. The different pathways were inferred from the metagenome of a biphenyl degrading consortium. Adapted from Garrido-Sanz et al. (2018).

bioremediation feasibility using this consortium has not been evaluated through microcosms assays yet, but the isolation of *Rhodococcus* sp. WAY2 from the consortium and its ability to co-metabolize 23 PCBs congeners has been reported by Garrido-Sanz et al. (2020).

Polycyclic aromatic hydrocarbons (PAHs) are constituted by multiple fused aromatic rings, and they are considered by the US EPA (United States Environmental Protection Agency) as prior pollutants due to their persistence in the environment and potential deleterious effects on human health, since many PAHs are known carcinogens and/or mutagens (see chapters 27–29). Wu et al. (2013) isolated several PAH-degrading bacterial consortia from three different long-term petroleum-contaminated locations through a successive enrichment culture procedure using anthracene, phenanthrene, and pyrene as the sole carbon and energy sources. All the consortia were further used as inoculums in microcosms assays to evaluate the PAH degradation feasibility in one of the original polluted soils sampled in the study. The results showed that all PAH-degrading consortia were capable of enhancing the degradation, although it is worth mentioning that the highest degradation rates were obtained for the consortia originally isolated from the soil used in the microcosmos assays. This suggests that degrading strains isolated from the same soil to which they were further reintroduced are more likely to survive and metabolize PAHs during bioremediation than microorganisms isolated from somewhere else. Moreover, specific strains from the consortia were isolated and identified through DNA extraction and 16S rRNA gene sequencing. The isolated strains mainly belong to *Stenotrophomonas*, *Bacterium*, *Klebsiella*, *Bacillus*, *Citrobacter*, and *Raoultella* genera. Future whole-metagenome sequencing of the consortium and its metagenomic analysis would be convenient to gain insights into the presence of key genes involved in the already observed PAH-degradation process.

30.6 PROSPECTS AND CONCLUSIONS

In this chapter, we have shown that bioremediation is benefitting from advances in DNA sequencing and metagenomic analysis. A metagenomic study of a polluted area can predict the metabolic potential of the microbial community present in the site and identify the populations that may play a relevant role in biodegradation. As it has been shown above, enrichment cultures can be used to isolate and characterize inoculants for bioaugmentation. Natural consortia can also be used to isolate specific strains of interest that could be the starting point for synthetic or semi-synthetic consortia presenting improvements over the natural counterparts. Microorganisms other than the degrading populations might also be of interest in these consortia. For instance, there might be populations able to produce surfactants that help to emulsify the organic pollutants, making it more available to the degrading microbia. Other populations can participate in building biofilms that enhance the biodegradation process or can feed on biodegradation by-products that might otherwise accumulate (see chapter 29). Metagenomics are useful to answer two relevant questions in bioremediation: which microorganisms are present and what are these microorganisms doing? Sequencing depth is normally enough to get a clear picture of microbial composition and activity. However, to get an in-depth knowledge of the bioremediation process, the help of other omic analysis will be relevant. Metagenomic studies can be complemented by metatranscriptomic analysis, i.e., the analysis of genes transcribed by a community. Metatranscriptomics is usually performed by RNA-Seq, a method that involves RNA extraction, reverse transcription of this RNA, and DNA sequencing by NGS. Although more complex than metagenome analysis, it gives similar resolution to metagenomes. Comparing the metatranscriptome with the metagenome, provides information about which genes are expressed at higher (or lower) levels, revealing the activity of the different populations. Metaproteomics, the analysis of proteins produced by a microbial community, might also contribute to a clearer picture of the bioremediation process. However, the power of metaproteomics is lower, and only a small part of the proteins can be extracted and analyzed. While typical metagenomic or metatranscriptomic analysis yields hundreds of thousands of genes identified, metaproteomics yields hundreds to a few thousands of proteins identified. The improvement of metaproteomics will in the future allow the identification of the proteins that are carrying the enzymatic reactions in bioremediation. Finally, meta metabolomics, the identification of the metabolites produced by the community, can allow the identification of metabolic intermediates in the bioremediation process. A combination of omics will allow the determination of carbon fluxes during the biodegradation process, both at the polluted sites and within the consortia, and to model bioremediation.

ACKNOWLEDGMENTS

Research in the authors' laboratory is currently funded by GREENER-H2020 (EU), grant number 826312 and MICINN/FEDER EU, grant number RTI2018-0933991-B-I00.

REFERENCES

1. Bento FM, Camargo FA, Okeke BC, Frankenberger WT. 2005. Comparative bioremediation of soils contaminated with diesel oil by natural attenuation, biostimulation and bioaugmentation. *Bioresource Technology* 96: 1049–1055.
2. Bharagava RN, Purchase D, Saxena G, Mulla SI. 2019. Applications of metagenomics in microbial bioremediation of pollutants: From genomics to environmental cleanup. In: S Das, HR Dash (Eds.), *Microbial Diversity in the Genomic Era*. Academic Press, pp. 459–477.
3. Bolyen E, Rideout JR, Dillon MR, Bokulich NA, Abnet CC, et al. 2019. Reproducible, interactive, scalable and extensible microbiome data science using QIIME 2. *Nature Biotechnology* 37: 852–857.
4. Bouchez T, Patureau D, Dabert P, Juretschko S, Dore J, et al. 2000. Ecological study of a bioaugmentation failure. *Environmental Microbiology* 2: 179-190.
5. Brenner K, Karig DK, Weiss R, Arnold FH. 2007. Engineered bidirectional communication mediates a consensus in a microbial biofilm consortium. *Proceedings of the National Academy of Sciences* 104: 17300–17304.
6. Cachada A, Rocha-Santos T, Duarte AC. 2018. Soil and pollution: An introduction to the main issues. In: AC Duarte, A Cachada, T. Rocha-Santos (Eds.), *Soil Pollution*. Elsevier, pp. 1–28.
7. Cámara B, Herrera C, Gonzalez M, Couve E, Hofer B, et al. 2004. From PCBs to highly toxic metabolites by the biphenyl pathway. *Environmental Microbiology* 6: 842–850.
8. Duarte M, Nielsen A, Camarinha-Silva A, Vilchez-Vargas R, Bruls T, et al. 2017. Functional soil metagenomics: Elucidation of polycyclic aromatic hydrocarbon degradation potential following 12 years of in situ bioremediation. *Environmental Microbiology* 19: 2992–3011.
9. Edgar R. 2018. Taxonomy annotation and guide tree errors in 16S rRNA databases. *PeerJ* 6: e5030.
10. FAO and ITPS. 2015. Status of the world's soil resources (SWSR)– Main report. *Food and Agriculture Organization of the United Nations and Intergovernmental Technical Panel on Soils*, Rome, Italy, p. 650.

11. Fuentes S, Barra B, Caporaso JG, Seeger M. 2016. From rare to dominant: A fine-tuned soil bacterial bloom during petroleum hydrocarbon bioremediation. *Applied and Environmental Microbiology* 82: 888–896.
12. Furukawa K, Fujihara H. 2008. Microbial degradation of polychlorinated biphenyls: Biochemical and molecular features. *Journal of Bioscience and Bioengineering* 105: 433–449.
13. Garrido-Sanz D, Manzano J, Martín M, Redondo-Nieto M, Rivilla R. 2018. Metagenomic analysis of a biphenyl-degrading soil bacterial consortium reveals the metabolic roles of specific populations. *Frontiers in Microbiology* 9: 232.
14. Garrido-Sanz D, Redondo-Nieto M, Guirado M, Pindado Jiménez O, Millán R, et al. 2019. Metagenomic insights into the bacterial functions of a diesel-degrading consortium for the rhizoremediation of diesel-polluted soil. *Genes* 10: 456.
15. Garrido-Sanz D, Sansegundo-Lobato P, Redondo-Nieto M, Suman J, Cajthaml T, et al. 2020. Analysis of the biodegradative and adaptive potential of the novel polychlorinated biphenyl degrader *Rhodococcus* sp. WAY2 revealed by its complete genome sequence. *Microbial Genomics* 6: e000363.
16. Gilbert ES, Walker AW, Keasling JD. 2003. A constructed microbial consortium for biodegradation of the organophosphorus insecticide parathion. *Applied Microbiology and Biotechnology* 61: 77–81.
17. Gilbert JA, Jansson JK, Knight R. 2014. The Earth Microbiome Project: Successes and aspirations. *BMC Biology* 12: 69.
18. Guerra AB, Oliveira JS, Silva-Portela RC, Araújo W, Carlos AC, et al. 2018. Metagenome enrichment approach used for selection of oil-degrading bacteria consortia for drill cutting residue bioremediation. *Environmental Pollution* 235: 869–880.
19. Handelsman J, Rondon MR, Brady SF, Clardy J, Goodman RM. 1998. Molecular biological access to the chemistry of unknown soil microbes: A new frontier for natural products. *Chemistry & Biology* 5: R245–R249.
20. Iwasaki T, Takeda H, Miyauchi K, Yamada T, Masai E, et al. 2007. Characterization of two biphenyl dioxygenases for biphenyl/PCB degradation in a PCB degrader, *Rhodococcus* sp. strain RHA1. *Bioscience, Biotechnology, and Biochemistry* 71: 993–1002.
21. Juhasz A, Megharaj M, Naidu R. 2000. Bioavailability: The major challenge (constraint) to bioremediation of organically contaminated soils. In: D Wise, DJ Trantolo, EJ Cichon, HI Inyang, U. Stottmeister (Eds.), *Environmental Science and Pollution Control Series*, pp. 217–242.
22. Kibegwa FM, Bett RC, Gachuiri CK, Stomeo F, Mujibi FD. 2020. A comparison of two DNA metagenomic bioinformatic pipelines while evaluating the microbial diversity in feces of Tanzanian small holder dairy cattle. *BioMed Research International* 2020: 2348560.
23. Kimura N, Kitagawa W, Mori T, Nakashima N, Tamura T, et al. 2006. Genetic and biochemical characterization of the dioxygenase involved in lateral dioxygenation of dibenzofuran from *Rhodococcus opacus* strain SAO101. *Applied Microbiology and Biotechnology* 73: 474–484.
24. Kull K. 2010. Ecosystems are made of semiosic bonds: Consortia, umwelten, biophony and ecological codes. *Biosemiotics* 3: 347–357.
25. Labud V, Garcia C, Hernandez T. 2007. Effect of hydrocarbon pollution on the microbial properties of a sandy and a clay soil. *Chemosphere* 66: 1863–1871.
26. Li DH, Liu CM, Luoi RB, Sadakane K, Lam TW. 2015. MEGAHIT: An ultra-fast single-node solution for large and complex metagenomics assembly via succinct de Bruijn graph. *Bioinformatics* 31: 1674–1676.
27. Maier RM. 2000. Bioavailability and its importance to bioremediation. In: JJ Valdes (Ed.), *Bioremediation*. Springer, pp. 59–78.
28. McCarty PL. 1988. Bioengineering issues related to in situ remediation of contaminated soils and groundwater. In: GS Omenn (Ed.), *Environmental Biotechnology*. Springer, pp. 143–162.
29. Milanese A, Mende DR, Paoli L, Salazar G, Ruscheweyh HJ, et al. 2019. Microbial abundance, activity and population genomic profiling with mOTUs2. *Nature Communications* 10: 1–11.
30. Mueller JG, Cerniglia CE, Pritchard PH. 1996. Bioremediation of environments contaminated by polycyclic aromatic hydrocarbons. *Biotechnology Research Series* 6: 125–194.
31. Naidu R, Semple KT, Megharaj M, Juhasz A, Bolan N, et al. 2008. Bioavailability: Definition, assessment and implications for risk assessment. *Developments in Soil Science* 32: 39–51.
32. National Research Council. 1979. *Polychlorinated Biphenyls: A Report*. National Academies.
33. National Research Council. 1985. *Oil in the Sea: Inputs, Fates, and Effects*, Vol. 1. National Academies Press.
34. Nurk S, Meleshko D, Korobeynikov A, Pevzner PA. 2017. metaSPAdes: A new versatile metagenomic assembler. *Genome Research* 27: 824–834.

35. Pace NR, Stahl DA, Lane DJ, Olsen GJ. 1986. The analysis of natural microbial populations by ribosomal RNA sequences. In: KC Marshall (Ed.), *Advances in Microbial Ecology, Springer*, pp. 1–55.

36. Paul JM, Holmes S. 2013. phyloseq: An R package for reproducible interactive analysis and graphics of microbiome census data. *PLoS ONE* 8(4): e61217.

37. Pepper IL, Gerba CP, Brusseau ML. 2011. *Environmental and Pollution Science. Elsevier.*

38. Popa R. 2004. *Between Necessity and Probability: Searching for the Definition and Origin of Life. Springer Science & Business Media.*

39. Prince RC, Gramain A, Mcgenity TJ. 2010. Prokaryotic hydrocarbon degraders. In: TJ McGenity (Ed.), *Handbook of Hydrocarbon and Lipid Microbiology*, pp. 1672–1692.

40. Puspita ID, Kamagata Y, Tanaka M, Asano K, Nakatsu CH. 2012. Are uncultivated bacteria really uncultivable? *Microbes and Environments* 27(4): 356–366. ME12092.

41. Quast C, Pruesse E, Yilmaz P, Gerken J, Schweer T, et al. 2013. The SILVA ribosomal RNA gene database project: Improved data processing and web-based tools. *Nucleic Acids Research* 41: D590–D596.

42. Reinke J. 1872. Ueber die anatomischen Verhältnisse einiger Arten von Gunnera L. Nachrichten von der. *Königl. Gesellschaft der Wissenschaften und der Georg-Augusts-Universität zu Göttingen* 9: 100–108.

43. Ribicic D, Netzer R, Hazen TC, Techtmann SM, Drabløs F, et al. 2018. Microbial community and metagenome dynamics during biodegradation of dispersed oil reveals potential key-players in cold Norwegian seawater. *Marine Pollution Bulletin* 129: 370–378.

44. Russell DL. 1992. *Remediation Manual for Petroleum Contaminated Sites*. CRC Press.

45. Sikkema J, De Bont JA, Poolman B. 1995. Mechanisms of membrane toxicity of hydrocarbons. *Microbiological Reviews* 59: 201–222.

46. Singh R, Olson MS. 2008. Application of bacterial swimming and chemotaxis for enhanced bioremediation. In: V Shah (Ed.), *Emerging Environmental Technologies*. Springer, pp. 149–172.

47. Songzhe F, Hongxia F, Shuangjiang L, Ying L, Zhipei L. 2009. A bioaugmentation failure caused by phage infection and weak biofilm formation ability. *Journal of Environmental Sciences* 21: 1153–1161.

48. Taguchi K, Motoyama M, Iida T, Kudo T. 2007. Polychlorinated biphenyl/biphenyl degrading gene clusters in *Rhodococcus* sp. K37, HA99, and TA431 are different from well-known *bph* gene clusters of *Rhodococci*. *Bioscience, Biotechnology, and Biochemistry* 71: 1136–1144.

49. Techtmann SM, Hazen TC. 2016. Metagenomic applications in environmental monitoring and bioremediation. *Journal of Industrial Microbiology & Biotechnology* 43: 1345–1354.

50. Thompson IP, Van Der Gast CJ, Ciric L, Singer AC. 2005. Bioaugmentation for bioremediation: The challenge of strain selection. *Environmental Microbiology* 7: 909–915.

51. Tyson GW, Chapman J, Hugenholtz P, Allen EE, Ram RJ, et al. 2004. Community structure and metabolism through reconstruction of microbial genomes from the environment. *Nature* 428: 37–43.

52. US EPA. 2008. Green remediation: Incorporating sustainable environmental practices into remediation of contaminated sites. *Technology Primer.*

53. Venter JC, Remington K, Heidelberg JF, Halpern AL, Rusch D, et al. 2004. Environmental genome shotgun sequencing of the Sargasso Sea. *Science* 304: 66–74.

54. Vergani L, Mapelli F, Marasco R, Crotti E, Fusi M, et al. 2017. Bacteria associated to plants naturally selected in a historical PCB polluted soil show potential to sustain natural attenuation. *Frontiers in Microbiology* 8: 1385.

55. Vert M, Doi Y, Hellwich KH, Hess M, Hodge P, et al. 2012. Terminology for biorelated polymers and applications (IUPAC Recommendations 2012). *Pure and Applied Chemistry* 84: 377–410.

56. Vogel TM. 1996. Bioaugmentation as a soil bioremediation approach. *Current Opinion in Biotechnology* 7: 311–316.

57. Wagner-Dobler I. 2003. *Microbial Inoculants: Snake Oil or Panacea. Bioremediation: A Critical Review.* Horizon Scientific Press, Norfolk, pp. 259–289.

58. Wu M, Chen L, Tian Y, Ding Y, Dick WA. 2013. Degradation of polycyclic aromatic hydrocarbons by microbial consortia enriched from three soils using two different culture media. *Environmental Pollution* 178: 152–158.

59. Yergeau E, Sanschagrin S, Beaumier D, Greer CW. 2012. Metagenomic analysis of the bioremediation of diesel-contaminated Canadian high arctic soils. *PloS One* 7: e30058.

60. Zhang S, Merino N, Okamoto A, Gedalanga P. 2018. Interkingdom microbial consortia mechanisms to guide biotechnological applications. *Microbial Biotechnology* 11: 833–847.

CHAPTER 31

Soil Microbial Fuel Cells for Energy Harvesting and Bioremediation of Soil Contaminated with Organic Pollutants

Bongkyu Kim, Jakub Dziegielowski, and Mirella Di Lorenzo

Department of Chemical Engineering, University of Bath, BA2 7AY, Bath, UK
Centre for Biosensors, Bioelectronics and Biodevices, University of Bath, BA2 7AY, Bath, UK

31.1 INTRODUCTION TO SOIL MICROBIAL FUEL CELLS

Intensive industrial and agricultural activities, uncontrolled waste disposal and environmental accidents, such as oil spills, put our ecosystems under increasing threats (Gordon et al., 2018; Koshlaf et al., 2019). Land severely suffers from this pollution, which can reach levels that exceed the natural self-purification capacity of soils, posing risks not only to natural ecosystems but also to human health (Rodríguez-Eugenio et al., 2018). There is, therefore, the need for appropriate techniques to effectively remediate contaminated soils (Chen et al., 2015). Traditional chemical-based remediation processes have several drawbacks, including the introduction of additional chemicals into the environment, limited purification capacity, and high operating costs (Yao et al., 2012). Bioremediation strategies, such as bioaugmentation, biostimulation, and phytoremediation, represent, therefore, a particularly attractive alternative to chemical treatments (Cundy et al., 2013; see chapters 27–29). In this context, soil microbial fuel cell (SMFC) has great potential as an environmentally friendly bioremediation technology, with significant advantages over other methods, such as cost-effectiveness and possibility to concurrently produce electricity for autonomous operations (Wu et al., 2018).

In biological fuel cells, like SMFCs, the chemical energy of organic matter is converted into electric energy thanks to the action of specific types of microorganisms, defined as electrigens or electroactive (Wang and Ren, 2013). Although the concept of biological fuel cells and energy from microorganisms was introduced in 1911 (Potter, 1911), it was only in the early 2000s that greater interest in this technology was raised, followed by increasing enhancements (Cai et al., 2020; Kim et al., 2002). Particularly, researchers have focused on the development of microbial fuel cells for wastewater treatment and energy generation. The interest in sediment, plant, and soil microbial fuel cells is more recent (Doherty et al., 2015; Zabihallahpoor et al., 2015). These types of biological fuel cells have been developed for applications such as powering remote sensors (Donovan et al., 2013), bioremediation (Wu et al., 2018), and energy harvesting (Dziegielowski et al., 2020). In particular, soil microbial fuel cells have shown the ability to remove organic pollutants from contaminated soil, such as phenol (Huang et al., 2011), diesel (Lu et al., 2014a), polycyclic aromatic hydrocarbon (Yu et al., 2017), hexachlorobenzene (Cao et al., 2017), and atrazine (Wang et al., 2018).

Good Microbes in Medicine, Food Production, Biotechnology, Bioremediation, and Agriculture, First Edition. Edited by Frans J. de Bruijn, Hauke Smidt, Luca S. Cocolin, Michael Sauer, David Dowling, and Linda Thomashow.

This chapter provides an overview on operating principles of SMFCs and factors affecting performance, as well as challenges and future perspectives in their use for soil bioremediation.

31.2 WORKING PRINCIPLE OF SMFC

Compared to other types of biological fuel cells, SMFCs can be characterized by a very simple design and minimum maintenance requirements. In these devices, the soil provides the source of microorganisms and organic matter required for the process (Dunaj et al., 2012; Huang et al., 2011). As such, in contrast to other types of biological fuel cells, for the case of SMFCs there is no need to inoculate electroactive microorganisms and supply a fuel to the system. Consequently, the technology can be easily installed in the field (Li et al., 2017).

Figure 31.1 depicts a "flat plate" SMFC, considered to be one of the simplest and most studied designs. In such a configuration, the anode is buried in the soil and the cathode is exposed to air (Dziegielowski et al., 2020). The air permeability gradient within the soil allows for a natural stratification with more and more anaerobic regions through the soil depth, which eliminates the need for a membrane (Huan et al., 2014). At the anode, a biofilm of electroactive microorganisms generates electrons and protons from organic matter via metabolic pathways (Yu et al., 2020). These electrons are transferred to the electrode and flow through an external circuit, thus generating current, to reach the cathode, where they combine with the protons to reduce oxygen in air to water (Chen et al., 2012; Huan et al., 2014).

In the flat plate geometry, the soil provides support to the fuel cell structure and acts as electrode separator, a source of electroactive microbes and a source of fuel. The design is therefore drastically simplified, capital and operating costs are also very low, which are clear advantages over other types of bioelectrochemical systems (BESs). Several design variants have been explored, as shown in Figure 31.2.

A plant microbial fuel cell (PMFC) incorporates plants as the source of organics in the soil, to sustain long-term operations (Figure 31.2B and C). A variety of plant species, including paddy rice, salt marsh, freshwater species, grass species, and wetland plants, have been successfully tested in PMFCs (Kabutey et al., 2019), and a marked increase in the output power density demonstrated (Liu et al., 2013). Sediment microbial fuel cells are distinguished by their operation under water, as shown in Figure 31.2D. In this system, the anode is embedded in the sediment and the cathode is suspended in water. Considering the vast availability of sediment in the world's ocean floor, sediment microbial fuel cells represent an exciting opportunity for renewable energy (Straume et al., 2019). Initial prototypes implemented in marine sediments achieved power densities of up to 20 mW m^{-2} (Erable et al., 2013), which has been further increased to 100 mW m^{-2} with design optimization (Dumas et al., 2008). However, these systems

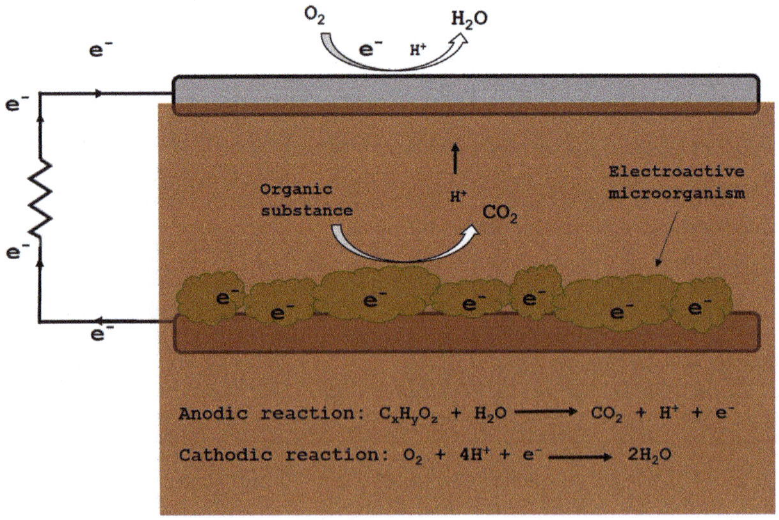

FIGURE 31.1 Schematic and working principle of soil microbial fuel cells.

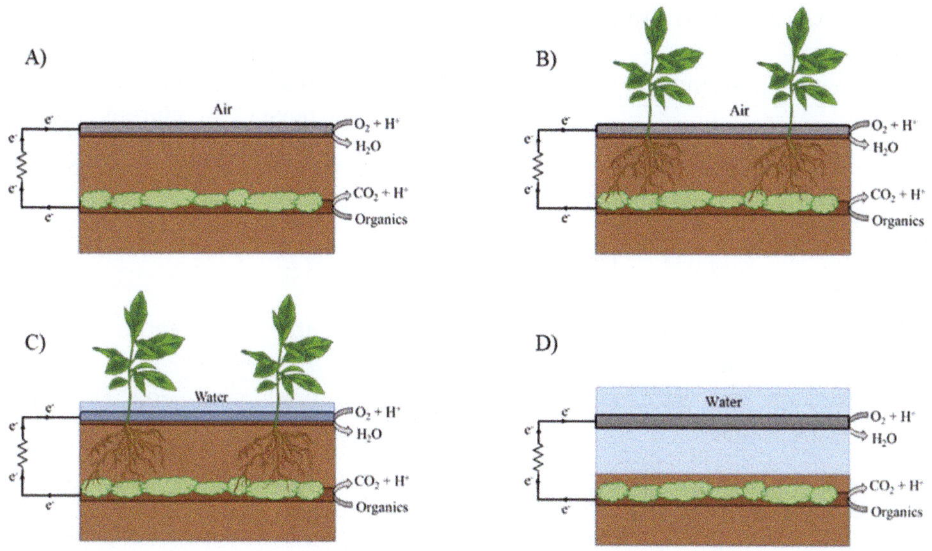

FIGURE 31.2 Design variants with soil-based anolyte: (A) soil microbial fuel cell; (B) plant microbial fuel cell; (C) wetland plant microbial fuel cell; (D) sediment microbial fuel cell.

often struggle with high ohmic losses, caused by significant electrode spacing, oxygen limitations, fouling, and corrosion, thus high performance has only been sustained in a laboratory setting for no more than 45 days (Dumas et al., 2008).

31.3 KEY FACTORS INFLUENCING THE PERFORMANCE

The electrochemical performance of SMFCs depends on many factors such as: electrode material (Yu et al., 2020); design configuration (Lu et al., 2014b; Yu et al., 2017; Zhang et al., 2015); and soil properties including porosity (Dunaj et al., 2012), temperature (Gong et al., 2021), pH (Lu et al., 2014b), conductivity (Wang et al., 2012), organic matter content and nature (Dunaj et al., 2012), and moisture condition (Wang et al., 2012). On this basis, the systems have been subject to various enhancement measures, resulting in power density yields increasing 100 folds, reaching 0.679 Wm^{-2} (Wetser et al., 2015). Nonetheless, the reported high performances were only recorded in a lab setting, sustained for a short duration, and any additional power requirement (e.g., for aeration, pumping) was neglected. Table 31.1 summarizes the performance of solid-based microbial fuel cells in field tests.

31.3.1 Electrode Material

The material and structure of the electrode used for the SMFC system, influence not only the electron transfer efficiency and electrochemical reaction rates but also the biofilm attachment. Biocompatibility, resistance to corrosion, high electrical conductivity, chemical stability, high specific and electroactive surface area, and low cost, are some of the most desired properties for the selection of appropriate electrode material (Sonawane et al., 2017). For this reason, carbon-based electrodes, such as graphite or carbon felts, foams, rods, and granules, are the preferred choices (Zhang et al., 2019).

To further enhance material properties, various electrode modification techniques have been introduced. Anodic modifications are common, and involve the addition of important functional groups, via a chemical treatment, to the surface of the electrode, to facilitate bacterial adhesion, and accelerate electron transport (Zhu et al., 2011). Smith et al. showed the success of such a strategy, with electrochemically treated graphite felt anodes, increasing current densities by 39% compared to untreated electrodes (Smith et al., 2015). Similarly, Dziegielowski et al. enhanced SMFC performance by 78%, by treating the anodes with a two-step acid-thermal process (Dziegielowski et al., 2020).

TABLE 31.1 Field test performance of different types of soil microbial fuel cell variants

System	Reactor design	Anode	Cathode	Catalyst	Membrane	Max power (mWm^{-2})	Max current (mAm^{-2})	Period (day)	Reference
Soil MFC	Flat plate	Graphite felt	Graphite felt	None	None	40*	180*	140	(Dziegielowski et al., 2020)
Plant MFC	Tubular	Graphite felt	Graphite felt	None	Ultrafiltration membrane	22♦	-	160	(Wetser et al., 2017)
	Flat plate	Graphite felt	Graphite felt	Bio-catalyst	Bipolar membrane	10*	28*	151	(Wetser et al., 2015)
	Multi-electrode	Graphite granules	Graphite felt	None	Cation exchange membrane	88♦	200♦	221	(Helder et al., 2013)
Wetland MFC	Flat plate	Graphite disk	Carbon cloth	Activated carbon	None	18♦	105♦	90	(Lu et al., 2015)
	Multi-electrode	Graphite felt	Graphite felt	Platinum	None	140♦	-	93	(Ueoka et al., 2016)
	Flat plate	Graphite granules	Carbon felt	Biocatalyst	None	-	55♦	60	(Habibul et al., 2016)
Sediment MFC	Flat plate	Graphite felt	Graphite felt	Platinum	None	19♦	132♦	55	(Kouzuma et al., 2013)
	Flat plate	Graphite plate	Graphite plates	None	None	54*	-	50	(Donovan et al., 2013)

♦ Value presented in the paper.
* Calculated value based on the information in the paper.

At the cathode, the oxygen reduction reaction (ORR) rate is usually enhanced with the use of a catalyst (Rahimnejad et al., 2015). The most widely studied is platinum, yet its high cost and instability in neutral and alkaline media makes it impractical for SMFCs (Kodali et al., 2016). In SMFCs, biocathodes represent, therefore, the most practical option for field operations (Wetser et al., 2015).

31.3.2 Reactor Design

Unlike conventional microbial fuel cell designs consisting of two chambers, SMFCs are limited to a single chamber configuration (Flimban et al., 2019). For this reason, the flat-plate geometry (Figure 31.3A) is the most studied set-up. Although cheap, effective, and stable, such a design brings undesirable limitations to the scale-up process, due to the necessity of cathode exposure to air, resulting in significant coverage of land (Dziegielowski et al., 2020).

To address the issue and reduce the footprint of the technology, tubular designs have been explored. In these designs, the anode is usually wrapped outside a membrane and the cathode placed inside and exposed to air, via a gas diffusion layer, as shown in Figure 31.3B. This configuration allows for the complete submersion of the reactor in the soil and the scaling up can be achieved by simply piling together multiple SMFCs (Casula et al., 2020). Nonetheless, oxygen crossover from cathode to anode can significantly lower the performance (Wetser et al., 2017). The use of a ceramic separator has also been explored for this type of geometry (see Figure 31.3C), but the limited permeability of the ceramic material is too restrictive for effective ion transfer from anode to cathode, significantly increasing the internal resistance, resulting in power densities of only around 1.67 mWm^{-2} (Casula et al., 2020).

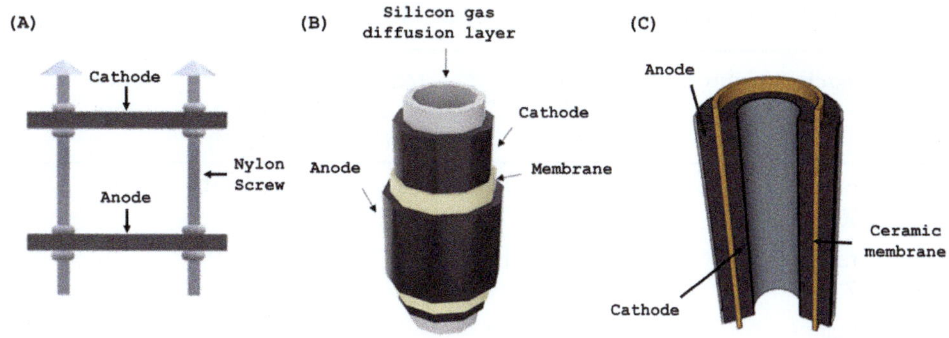

FIGURE 31.3 SMFC reactor designs: (A) flat-plate design; (B) tubular design for underground operation; (C) tubular design with ceramic membrane.

31.4 SOIL PROPERTIES

The properties of the soil markedly influence the performance of SMFCs (Dunaj et al., 2012). The porosity and water diffusivity of the soil affect the transport of species, including oxygen, water, and organics through the soil (Casula et al., 2020; Singh et al., 2020). Generally, a more porous soil leads to faster mass transfer, facilitating substrate consumption. It has been shown that the addition of large-grained sand into soil, as a means of increasing porosity, leads to a decrease in internal resistance by 46% and an increase in substrate uptake (Doherty et al., 2015).

The water content of soil is another key factor in SMFCs, as water is an indispensable substance for the metabolism of microbes, as well as a medium for the exchange of substance by microorganisms. In addition, a higher moisture content promotes the dissolution of hydrophilic compounds, making them easier to consume by the microorganisms, which in turn enhances electron generation (Zhang et al., 2019). Finally, water in soil facilitates the transport of ions, and, consequently, soil moisture greatly influences performance (Dziegielowski et al., 2020).

31.5 SMFCs TO BIOREMEDIATE CONTAMINATED SOILS

In addition to other bioremediation strategies, SMFCs can provide an effective means to effectively treat soils contaminated by petroleum hydrocarbons, polycyclic hydrocarbons (PAHs), pesticides, and chlorophenols. Table 31.2 provides an overview of the use of soil microbial fuel cell technology for the bioremediation of contaminated soils.

31.5.1 Petroleum Hydrocarbon

One of the first attempts to use SMFCs for the bioremediation of petroleum contaminated soil is reported by Lu et al. (2014b). A tubular configuration was used (Figure 31.4A), and a removal efficiency of up to 89% of total petroleum hydrocarbon (TPH) was demonstrated. The study also proved that the SMFC enhances the treatment efficacy showing maximum 241% increase of degradation compared to natural bioremediation, and an estimated maximum distance of influence was estimated of 90 cm (Lu et al., 2014b). The same group also reported a 63.5–78.7% increase in the bioremediation of polyaromatic hydrocarbons (PAH) in soil with respect to natural degradation (Lu et al., 2014a). In this study, *Comamonas testosteroni*, *Pseudomonas putida*, and *Ochrobactrum anthropi*, which are uncommon electroactive microorganisms, were identified on the anodic biofilm, while in the soil sample, hydrocarbon oxidizing bacteria were dominant (Lu et al., 2014a).

The influence of electrode spacing on the bioremediation efficacy was investigated with an SMFC consisting of an air-cathode, made by coating a stainless-steel mesh with activated carbon, placed at the bottom of the contaminated soil in which three anodes were inserted at an increasing distance from the cathode (Figure 31.4B). The result was that the removal efficiency of the SMFCs was influenced by the anode distance from the cathode, with the highest removal efficiency occurred at the shortest electrode distance. In this case removal efficiencies of 18%, 36% and 29% were shown, respectively,

TABLE 31.2 SMFC studies for bioremediation of contaminated soil by various pollutants

Target pollutant	Reactor design	Removal efficiency (%)	Max power (mWm^{-2})	Max current (mAm^{-2})	Period (Day)	Reference
Phenol	Tubular (250 ml)	90.1	29.45	145♦	10	(Huang et al., 2011)
PHC	U-tube (2.7 L)*	15.2 (TPHs)	0.85	5.48*	25	(Wang et al., 2012)
PHC (Diesel)	Tubular (3 L)*	78.7 (TPHs)	17.7	86	64	(Lu et al., 2014a)
PHC (Diesel)	Tubular (50 L)	89.7 (TPHs)	8.8	70.4	120	(Lu et al., 2014b)
PHC	Multiple electrode (324 ml)*	12 (TPHs)	37	102	180	(Li et al., 2014)
PHC	Multiple electrode (324 ml)*	22 (TPHs)	0.0938*	95.2*	135	(Doherty et al., 2015)
PHC	Multiple electrode (324 ml)	12.5 (TPHs)	7.52*	26.38*	135	(Zhang et al., 2015)
PHC	Carbon fiber mixed (720 ml)*	30 (TPHs)	17.3	> 180♦	144	(Li et al., 2016)
Polycyclic aromatic hydrocarbon	Flat plate (1.27 L)*	54.2 (Anthracene)	12.1	> 100♦	175	(Yu et al., 2017)
Hexachlorobenzene	Flat plate (150 ml)	71.15	77.5	> 450♦	56	(Cao et al., 2015)
Hexachlorobenzene	Flat plate (150 ml)	71.14	70.8	> 420♦	56	(Cao et al., 2017)
Atrazine	Flat plate	91.69	-	-	63	(Wang et al., 2017)
Atrazine	Flat plate (150 ml)	95	70.57	-	63	(Wang et al., 2018)

* Calculated value based on the information in the paper
♦ Maximum current confirmed from polarization curve.

on TPH, PAH (phenanthrene, fluoranthene, pyrene, and chrysene) and total n-alkanes (Li et al., 2014).

The influence of electrode spacing on PAH bioremediation efficacy was also investigated by Yu et al. (2017) with a flat-plate geometry SMFC, confirming better treatments at shorter electrode spacing. Under minimum electrode spacing, removal efficiencies of 161%, 146%, 131%, respectively, for anthracene, phenanthrene, and pyrene were observed.

To enhance the mass transfer rate of the pollutants in soil, the use of sand to be added to the contaminated soil has been suggested (Doherty et al., 2015) to increase the soil permeability. This strategy has demonstrated to lead to greater current and power outputs (95.2 mA m^{-2} and 93.8 uW m^{-2} with the addition of sand versus 68 mA m^{-2} and 47.6 uW m^{-2} without adding the sand), as well as a better removal of TPH (22% with sand compared to 12% without sand).

The use of carbon fibers mixed with the contaminated soil has also been suggested (Figure 31.4C) (Li et al., 2016). Carbon fibers in soil can enhance the soil conductivity and reduce the internal resistance, thus improving the overall electrochemical performance. The use of carbon fibers in soil has in fact led to a current output generated by the SMFC 16 times higher (120.5 mA m^{-2} compared to 7.6 mA m^{-2} without carbon fibers) and a power output approximately 20 times greater (17.3 mW m^{-2} compared to 0.8 mW m^{-2} without fibers). The use of the carbon fibers also enhanced the efficacy of TPH removal, 30% in the presence of carbon fibers versus 15% in the absence, and N-alkanes removal, 55% versus 42%.

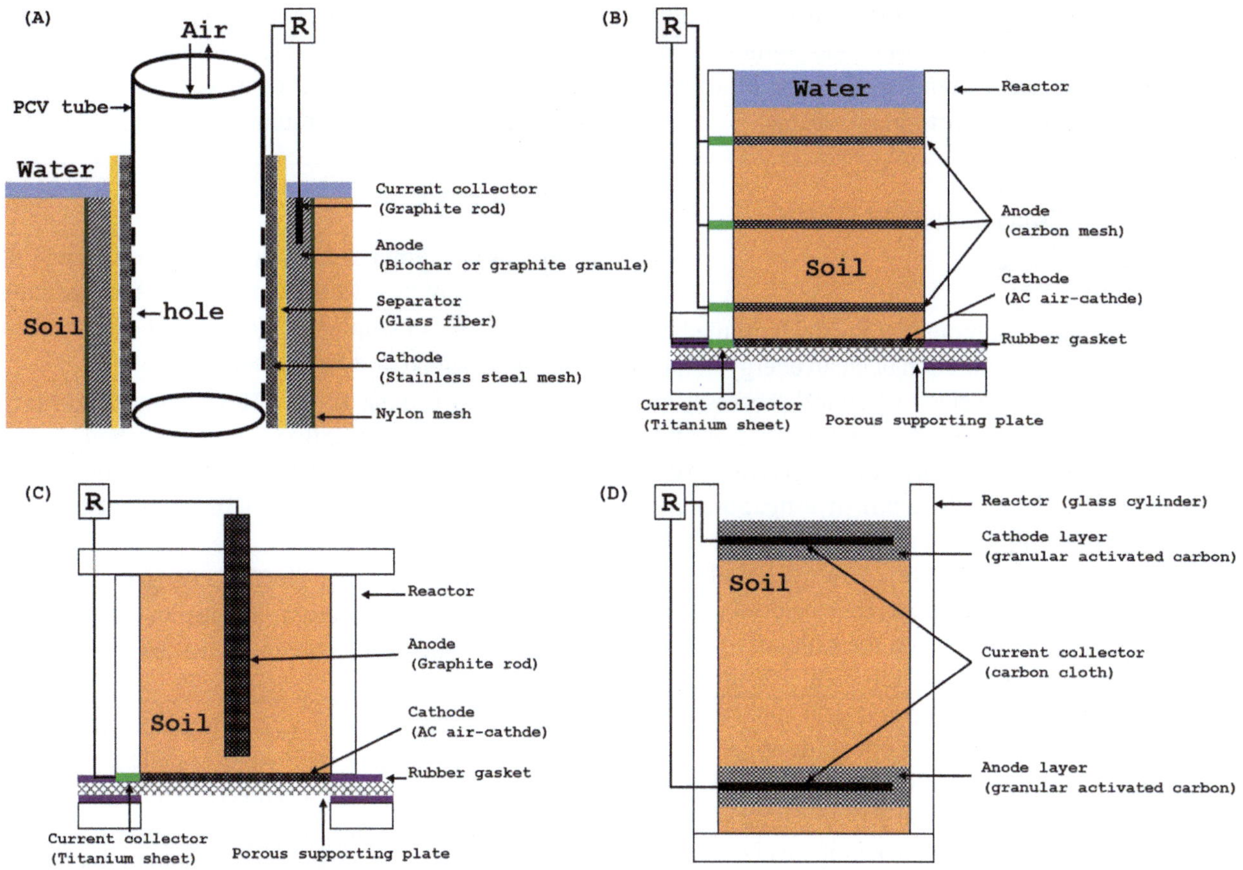

FIGURE 31.4 Schematic of configuration of soil microbial fuel cell constructed for case studies. Adapted from (A) (Lu et al., 2014b); (B) (Li et al., 2014); (C) (Li et al., 2016); (D) (Cao et al., 2015).

31.5.2 Pesticides

The bioremediation of hexachlorobenzene (HCB), a refractory organic pesticide, by SMFC technology has been reported (Figure 31.4D) (Cao et al., 2015). A removal efficiency of HCB of about 71% compared to 38% in a control system was observed. This study also showed a decrease on the treatment efficacy with the increase in the concentration of HCB. To confirm the effect of current production on the bioremediation efficacy, the SMFCs were operated under several values of external resistance, 2000 Ω, 1000 Ω, 510 Ω, and 10 Ω, demonstrating a greater degree of removal under the lower values of external resistance. The use of an anionic surfactant, such as sodium dodecyl sulfate (SDS), has also been suggested to enhance HCB removal by SMFCs (Cao et al., 2017). A removal efficiency of 88.63% was in fact observed when SDS was added to the soil, compared to a 67.82% efficacy in the absence of SDS (Cao et al., 2017).

Atrazine removal by SMFC technology has also been demonstrated (Wang et al., 2017). In this study, the effect of the electrode distance and the influence of initial concentration on removal efficacy were investigated. The result was that the highest removal efficiency of atrazine (91.69%) was observed in the SMFC with the shortest electrode distance and lowest atrazine concentration (Wang et al., 2017). Electroactive microorganisms, *Deltaproteobactera* and *Geobacter*, were observed as dominant species near the anode, while, surprisingly, it observed that atrazine-degrading *Pseudomonas* species were present only in very low proportions (Wang et al., 2017). In a follow-up research by the same team, the influence of the applied external resistance on the efficacy degradation of atrazine, HCB and mixture of both by SMFCs was investigated (Wang et al., 2018). The removal efficiency and degradation rate in all cases increased with decreasing external resistance. Under the lowest external resistance (20 Ω), the removal efficiency of atrazine and HCB was 95% and

78%, respectively (Wang et al., 2018). Another study demonstrated that 83% of atrazine was removed by SMFC at a five time faster than that the case of a non-sterilized soil (Domínguez-Garay et al., 2016).

31.6 CONCLUSIONS AND FUTURE PERSPECTIVE

SMFC technology has great potential for soil bioremediation associated with the generation of clean energy. It has been demonstrated that the electrochemical processes in SMFCs enhance the pollutant degradation in soil and shorten the treatment period. The technology is, however, still in its infancy and more in-depth investigations are required for practical implementations. Lab-scale studies of sample pollutants must progress toward field-based research with complex soil matrices (Dziegielowski et al., 2020), along with suitable strategies for scale-up must be developed (Casula et al., 2020; Lu et al., 2014b). In addition, as a measure for scale-up and field applicability, there must be an approach for developing the appropriate SMFC structure. Strategies already considered for liquid microbial fuel cell systems could be explored first, such as modularization (An et al., 2014), stacking (Walter et al., 2020), and multiple electrode arrangements (Kim and Chang, 2018). Despite design challenges that still need to be overcome, tubular configurations appear to be the most suitable for practical applications (Casula et al., 2020; Huang et al., 2011). Mathematical models can also assist in the development of highly performing SMFCs stacks and help predict the radius of bioremediation action of each SMFC unit so that effective SMFCs arrays for the treatment of large contaminated sites can be developed (Casula et al., 2020; Lu et al., 2014b).

Efforts must be made to secure efficiently produced electric energy and to provide an appropriate place for use. Along with removal efficacy enhancement, the current generated must also be enhanced with the use of a customized power management system (Dziegielowski et al., 2020). The current generated could be used to power low-power electrical devices, such as sensors that monitor soil/environment condition over time (Donovan et al., 2011; Rossi et al., 2017), LED light (Walter et al., 2020), small devices such as a cell phone (Prasad and Tripathi, 2020), and a small water treatment system (Dziegielowski et al., 2020). Follow-on research necessarily must provide an in-depth investigation on the efficacy of the SMFC technology in cooperation with other soil bioremediation strategies, such as eco-piling (Germaine et al., 2015; see chapter 27), phytoremediation (Sarwar et al., 2017), bioaugmentation (Cycoń et al., 2017; see chapter 27–29), and biostimulation (Adams et al., 2015).

REFERENCES

1. Adams GO, Fufeyin PT, Okoro SE, Ehinomen I. 2015. Bioremediation, biostimulation and bioaugmention: A review. *International Journal of Environmental Bioremediation and Biodegradation* 3: 28–39.
2. An J, Kim B, Jang JK, Lee H-S, Chang IS. 2014. New architecture for modulization of membraneless and single-chambered microbial fuel cell using a bipolar plate-electrode assembly (BEA). *Biosensors and Bioelectronics* 59: 28–34.
3. Cai T, Meng L, Chen G, Xi Y, Jiang N, et al. 2020. Application of advanced anodes in microbial fuel cells for power generation: A review. *Chemosphere* 248: 125985.
4. Cao X, Song H-L, C-y Y, X-n L. 2015. Simultaneous degradation of toxic refractory organic pesticide and bioelectricity generation using a soil microbial fuel cell. *Bioresource Technology* 189: 87–93.
5. Cao X, Yu C, Wang H, Zhou F, Li X. 2017. Simultaneous degradation of refractory organic pesticide and bioelectricity generation in a soil microbial fuel cell with different conditions. *Environmental Technology* 38: 1043–1050.
6. Casula E, Kim B, Chesson H, Di Lorenzo M, Mascia M. 2020. Modelling the influence of soil properties on performance and bioremediation ability of a pile of soil microbial fuel cells. *Electrochimica Acta* 368: 137568.
7. Chen M, Xu P, Zeng G, Yang C, Huang D, et al. 2015. Bioremediation of soils contaminated with polycyclic aromatic hydrocarbons, petroleum, pesticides, chlorophenols and heavy metals by composting: Applications, microbes and future research needs. *Biotechnology Advances* 33: 745–755.
8. Chen Z, Huang Y-C, Liang J-H, Zhao F, Zhu Y-G. 2012. A novel sediment microbial fuel cell with a biocathode in the rice rhizosphere. *Bioresource Technology* 108: 55–59.
9. Cundy A, Bardos R, Church A, Puschenreiter M, Friesl-Hanl W, et al. 2013. Developing principles of sustainability and stakeholder engagement for "gentle" remediation approaches: The European context. *Journal of Environmental Management* 129: 283–291.

10. Cycoń M, Mrozik A, Piotrowska-Seget Z. 2017. Bioaugmentation as a strategy for the remediation of pesticide-polluted soil: A review. *Chemosphere* 172: 52–71.
11. Doherty L, Zhao Y, Zhao X, Hu Y, Hao X, et al. 2015. A review of a recently emerged technology: Constructed wetland–microbial fuel cells. *Water Research* 85: 38–45.
12. Domínguez-Garay A, Boltes K, Esteve-Núñez A. 2016. Cleaning-up atrazine-polluted soil by using microbial electroremediating cells. *Chemosphere* 161: 365–371.
13. Donovan C, Dewan A, Heo D, Lewandowski Z, Beyenal H. 2013. Sediment microbial fuel cell powering a submersible ultrasonic receiver: New approach to remote monitoring. *Journal of Power Sources* 233: 79–85.
14. Donovan C, Dewan A, Peng H, Heo D, Beyenal H. 2011. Power management system for a 2.5 W remote sensor powered by a sediment microbial fuel cell. *Journal of Power Sources* 196: 1171–1177.
15. Dumas C, Basseguy R, Bergel A. 2008. DSA to grow electrochemically active biofilms of. *Geobacter Sulfurreducens. Electrochimica Acta* 53: 3200–3209.
16. Dunaj SJ, Vallino JJ, Hines ME, Gay M, Kobyljanec C, et al. 2012. Relationships between soil organic matter, nutrients, bacterial community structure, and the performance of microbial fuel cells. *Environmental Science and Technology* 46: 1914–1922.
17. Dziegielowski J, Metcalfe B, Villegas-Guzman P, Martínez-Huitle CA, Gorayeb A, et al. 2020. Development of a functional stack of soil microbial fuel cells to power a water treatment reactor: From the lab to field trials in North East Brazil. *Applied Energy* 278: 115680.
18. Erable B, Lacroix R, Etcheverry L, Féron D, Délia M-L, et al. 2013. Marine floating microbial fuel cell involving aerobic biofilm on stainless steel cathodes. *Bioresource Technology* 142: 510–516.
19. Flimban SG, Ismail IM, Kim T, Oh S-E. 2019. Overview of recent advancements in the microbial fuel cell from fundamentals to applications: Design, major elements, and scalability. *Energies* 12: 3390.
20. Germaine KJ, Byrne J, Liu X, Keohane J, Culhane J, et al. 2015. Ecopiling: A combined phytoremediation and passive biopiling system for remediating hydrocarbon impacted soils at field scale. *Frontiers in Plant Science* 5: 756.
21. Gong L, Amirdehi MA, Miled A, Greener J. 2021. Practical increases in power output from soil-based microbial fuel cells under dynamic temperature variations. *Sustainable Energy & Fuels* 5: 671–677.
22. Gordon G, Stavi I, Shavit U, Rosenzweig R. 2018. Oil spill effects on soil hydrophobicity and related properties in a hyper-arid region. *Geoderma* 312: 114–120.
23. Habibul N, Hu Y, Wang Y-K, Chen W, Yu H-Q, et al. 2016. Bioelectrochemical chromium (VI) removal in plant-microbial fuel cells. *Environmental Science and Technology* 50: 3882–3889.
24. Helder M, Strik DP, Timmers RA, Raes SM, Hamelers HV, et al. 2013. Resilience of roof-top plant-microbial fuel cells during Dutch winter. *Biomass and Bioenergy* 51: 1–7.
25. Huan D, Yi-Cheng W, Zhang F, Huang Z-C, Zheng C, et al. 2014. Factors affecting the performance of single-chamber soil microbial fuel cells for power generation. *Pedosphere* 24: 330–338.
26. Huang D-Y, Zhou S-G, Chen Q, Zhao B, Yuan Y, et al. 2011. Enhanced anaerobic degradation of organic pollutants in a soil microbial fuel cell. *Chemical Engineering Journal* 172: 647–653.
27. Kabutey FT, Zhao Q, Wei L, Ding J, Antwi P, et al. 2019. An overview of plant microbial fuel cells (PMFCs): Configurations and applications. *Renewable and Sustainable Energy Reviews* 110: 402–414.
28. Kim B, Chang IS. 2018. Elimination of voltage reversal in multiple membrane electrode assembly installed microbial fuel cells (mMEA-MFCs) stacking system by resistor control. *Bioresource Technology* 262: 338–341.
29. Kim HJ, Park HS, Hyun MS, Chang IS, Kim M, et al. 2002. A mediator-less microbial fuel cell using a metal reducing bacterium, Shewanella putrefaciens. *Enzyme and Microbial Technology* 30: 145–152.
30. Kodali M, Gokhale R, Santoro C, Serov A, Artyushkova K, et al. 2016. High performance platinum group metal-free cathode catalysts for microbial fuel cell (MFC). *Journal of the Electrochemical Society* 164: H3041.
31. Koshlaf E, Shahsavari E, Haleyur N, Osborn AM, Ball AS. 2019. Effect of biostimulation on the distribution and composition of the microbial community of a polycyclic aromatic hydrocarbon-contaminated landfill soil during bioremediation. *Geoderma* 338: 216–225.
32. Kouzuma A, Kasai T, Nakagawa G, Yamamuro A, Abe T, et al. 2013. Comparative metagenomics of anode-associated microbiomes developed in rice paddy-field microbial fuel cells. *PLoS One* 8: e77443.

33. Li X, Wang X, Weng L, Zhou Q, Li Y. 2017. Microbial fuel cells for organic-contaminated soil remedial applications: A review. *Energy Technology* 5: 1156–1164.

34. Li X, Wang X, Zhang Y, Cheng L, Liu J, et al. 2014. Extended petroleum hydrocarbon bioremediation in saline soil using Pt-free multianodes microbial fuel cells. *Rsc Advances* 4: 59803–59808.

35. Li X, Wang X, Zhao Q, Wan L, Li Y, et al. 2016. Carbon fiber enhanced bioelectricity generation in soil microbial fuel cells. *Biosensors and Bioelectronics* 85: 135–141.

36. Liu S, Song H, Li X, Yang F. 2013. Power generation enhancement by utilizing plant photosynthate in microbial fuel cell coupled constructed wetland system. *International Journal of Photoenergy*

37. Lu L, Huggins T, Jin S, Zuo Y, Ren ZJ. 2014a. Microbial metabolism and community structure in response to bioelectrochemically enhanced remediation of petroleum hydrocarbon-contaminated soil. *Environmental Science and Technology* 48: 4021–4029.

38. Lu L, Xing D, Ren ZJ. 2015. Microbial community structure accompanied with electricity production in a constructed wetland plant microbial fuel cell. *Bioresource Technology* 195: 115–121.

39. Lu L, Yazdi H, Jin S, Zuo Y, Fallgren PH, et al. 2014b. Enhanced bioremediation of hydrocarbon-contaminated soil using pilot-scale bioelectrochemical systems. *Journal of Hazardous Materials* 274: 8–15.

40. Potter MC. 1911. Electrical effects accompanying the decomposition of organic compounds. *Proceedings of the Royal Society of London. Series b, containing papers of a biological character* 84: 260–276.

41. Prasad J, Tripathi RK. 2020. Scale-up and control the voltage of sediment microbial fuel cell for charging a cell phone. *Biosensors and Bioelectronics* 172: 112767.

42. Rahimnejad M, Adhami A, Darvari S, Zirepour A, Oh S-E. 2015. Microbial fuel cell as new technology for bioelectricity generation: A review. *Alexandria Engineering Journal* 54: 745–756.

43. Rodríguez-Eugenio N, McLaughlin M, Pennock D. 2018. *Soil Pollution: A Hidden Reality*. FAO.

44. Rossi M, Tosato P, Gemma L, Torquati L, Catania C, et al. 2017. Long range wireless sensing powered by plant-microbial fuel cell. *Paper read at Design, Automation and Test in Europe Conference and Exhibition (DATE), 2017.*

45. Sarwar N, Imran M, Shaheen MR, Ishaque W, Kamran MA, et al. 2017. Phytoremediation strategies for soils contaminated with heavy metals: Modifications and future perspectives. *Chemosphere* 171: 710–721.

46. Singh VK, Kumar D, Kashyap P, Singh PK, Kumar A, et al. 2020. Modelling of soil permeability using different data driven algorithms based on physical properties of soil. *Journal of Hydrology* 580: 124223.

47. Smith RE, Davies TJ, Baynes NDB, Nichols RJ. 2015. The electrochemical characterisation of graphite felts. *Journal of Electroanalytical Chemistry* 747: 29–38.

48. Sonawane JM, Yadav A, Ghosh PC, Adeloju SB. 2017. Recent advances in the development and utilization of modern anode materials for high performance microbial fuel cells. *Biosensors and Bioelectronics* 90: 558–576.

49. Straume EO, Gaina C, Medvedev S, Hochmuth K, Gohl K, et al. 2019. GlobSed: Updated total sediment thickness in the world's oceans. *Geochemistry, Geophysics, Geosystems* 20: 1756–1772.

50. Ueoka N, Sese N, Sue M, Kouzuma A, Watanabe K. 2016. Sizes of anode and cathode affect electricity generation in rice paddy-field microbial fuel cells. *Journal of Sustainable Bioenergy Systems* 6: 10.

51. Walter XA, You J, Winfield J, Bajarunas U, Greenman J, et al. 2020. From the lab to the field: Self-stratifying microbial fuel cells stacks directly powering lights. *Applied Energy* 277: 115514.

52. Wang H, Cao X, Li L, Fang Z, Li X. 2018. Augmenting atrazine and hexachlorobenzene degradation under different soil redox conditions in a bioelectrochemistry system and an analysis of the relevant microorganisms. *Ecotoxicology and Environmental Safety* 147: 735–741.

53. Wang H, Li L, Cao X, Long X, Li X. 2017. Enhanced degradation of atrazine by soil microbial fuel cells and analysis of bacterial community structure. *Water, Air, and Soil Pollution* 228: 308.

54. Wang H, Ren ZJ. 2013. A comprehensive review of microbial electrochemical systems as a platform technology. *Biotechnology Advances* 31: 1796–1807.

55. Wang X, Cai Z, Zhou Q, Zhang Z, Chen C. 2012. Bioelectrochemical stimulation of petroleum hydrocarbon degradation in saline soil using U-tube microbial fuel cells. *Biotechnology and Bioengineering* 109: 426–433.

56. Wetser K, Dieleman K, Buisman C, Strik D. 2017. Electricity from wetlands: Tubular plant microbial fuels with silicone gas-diffusion biocathodes. *Applied Energy* 185: 642–649.

57. Wetser K, Sudirjo E, Buisman CJ, Strik DP. 2015. Electricity generation by a plant microbial fuel cell with an integrated oxygen reducing biocathode. *Applied Energy* 137: 151–157.

58. Wu Y, Jing X, Gao C, Huang Q, Cai P. 2018. Recent advances in microbial electrochemical system for soil bioremediation. *Chemosphere* 211: 156–163.

59. Yao Z, Li J, Xie H, Yu C. 2012. Review on remediation technologies of soil contaminated by heavy metals. *Procedia Environmental Sciences* 16: 722–729.

60. Yu B, Feng L, He Y, Yang L, Xun Y. 2020. Effects of anode materials on the performance and anode microbial community of soil microbial fuel cell. *Journal of Hazardous Materials* 401: 123394.

61. Yu B, Tian J, Feng L. 2017. Remediation of PAH polluted soils using a soil microbial fuel cell: Influence of electrode interval and role of microbial community. *Journal of Hazardous Materials* 336: 110–118.

62. Zabihallahpoor A, Rahimnejad M, Talebnia F. 2015. Sediment microbial fuel cells as a new source of renewable and sustainable energy: Present status and future prospects. *RSC Advances* 5: 94171–94183.

63. Zhang X, Li X, Zhao X, Li Y. 2019. Factors affecting the efficiency of a bioelectrochemical system: A review. *RSC Advances* 9: 19748–19761.

64. Zhang Y, Wang X, Li X, Cheng L, Wan L, et al. 2015. Horizontal arrangement of anodes of microbial fuel cells enhances remediation of petroleum hydrocarbon-contaminated soil. *Environmental Science and Pollution Research* 22: 2335–2341.

65. Zhu N, Chen X, Zhang T, Wu P, Li P, et al. 2011. Improved performance of membrane free single-chamber air-cathode microbial fuel cells with nitric acid and ethylenediamine surface modified activated carbon fiber felt anodes. *Bioresource Technology* 102: 422–426.

CHAPTER 32

Biotechnology for the Management of Plastics and Microplastics

Loriane Murphy and John Cleary
enviroCORE, Department of Science & Health, Institute of Technology Carlow, Kilkenny Road, Carlow, R93 V960, Ireland

32.1 INTRODUCTION

Large-scale production and use of plastics began around 1950, and since then they have become increasingly ubiquitous materials due to their combination of useful properties, including cost-effectiveness, low density, ease of processing, water-resistance, and durability. Using various polymer types and processing techniques, plastics can be produced with a wide range of mechanical properties to suit different requirements. They have consequently been used in a multitude of products and applications, including packaging, textiles, consumer products, electrical and electronic devices, transport, industrial equipment, and construction. They are now produced on enormous scales; global production of polymer resins and fibers grew from 213 million metric tons (Mt) in 2000 to 381 Mt in 2015 (Geyer et al., 2017). Their production continues to increase annually, and this trend is projected to continue in coming decades. Most of the plastics produced in high volumes are chemically synthetized from petroleum. The six most used plastics, low-density polyethylene (LDPE), high-density polyethylene (HDPE), polypropylene (PP), polystyrene (PS), polyvinylchloride (PVC), and polyethylene terephthalate (PET), account for almost 70% of global production (Geyer et al., 2017).

Due to their widespread use, careless disposal, and inefficient waste management systems, plastics pollution has become a global environmental issue, and in recent years this has been highlighted in the scientific literature and popular media alike (see also chapter 24). In this context, the durability that has contributed to the success of plastics has become a serious liability. Depending on the chemical and physical properties of the specific polymer, plastic items are expected to persist in the environment for timescales of years to hundreds or even thousands of years (Barnes et al., 2009). So widespread are plastics in the environment that their presence has been suggested as a geological indicator of the proposed Anthropocene era (Zalasiewicz et al., 2016).

While the problem of plastic waste in general has been recognized (if not effectively managed) for several decades, the issue of microplastics pollution has only received significant attention within the last 20 years. Microplastics are typically defined as polymeric materials with size <5 mm They are often categorized as primary microplastics and secondary microplastics. Primary microplastics are produced for direct use in products including cosmetic and personal hygiene products, or as intermediates in the polymer processing industry. Secondary microplastics can be formed by mechanical wear on larger plastic items during their normal use, for example as motor vehicle components,

Good Microbes in Medicine, Food Production, Biotechnology, Bioremediation, and Agriculture, First Edition. Edited by Frans J. de Bruijn, Hauke Smidt, Luca S. Cocolin, Michael Sauer, David Dowling, and Linda Thomashow.

in shipping and fisheries, clothing and textiles, or in industrial, construction, and agricultural materials and equipment. Secondary microplastics are also formed by chemical, physical, and biological degradation of any plastic item which is exposed to the environment for significant periods of time. There have been some recent improvements such as the withdrawal of polymeric microbeads from consumer products, implementation of plastic bag taxes, banning of some single use plastic items, and improvements in the management of plastic waste by recycling. Nevertheless, the ongoing increase in the use of plastics, and the concomitant abundance of plastic litter in the environment, ensures that secondary microplastics will continue to be produced for hundreds of years into the future.

Although the size and unobtrusiveness of microplastics allowed them to go largely unnoticed in the environment for many years, there is now significant concern in relation to their effects on organisms, ecosystems, and human health. In this chapter we will describe the environmental effects of microplastics, discuss some of the key biotechnological approaches to the management of plastic waste in general, and examine how effectively these approaches can be applied to the problem of microplastics pollution.

32.2 MICROPLASTICS AND THEIR ENVIRONMENTAL EFFECTS

Microplastics in environmental samples are frequently categorized in terms of their chemical identity, color, and morphology (bead/sphere, pellet, film, fiber, foam, fragment). These properties can give some indication of the origin of the particle. The amount of plastic present can be quantified by mass or by number of particles and varies widely depending on the location and nature of the sample. In aquatic environments, for example, surface, sub-surface, and sediment samples may have different types and abundances of plastics, due to density differences between the various polymers. However, growth of biofilms on microplastic particles can increase their overall density, allowing lighter plastics to sink to deeper water and sediments.

Furthermore, while the upper size limit of 5 mm is widely accepted, there is significant variation in the lower limit; in practice this is determined by the pore size, mesh size, or sieve opening used in a study, which in turn may be influenced by environmental variables such as water turbidity, sample processing requirements, and detection methods available to the researchers. Comparability of studies has also been limited by the variety of sampling and analytical methods used, and by a lack of relevant standards, although this is beginning to be addressed (ASTM, 2020a, b). Nevertheless, numerous studies have confirmed that microplastics are near-ubiquitous in the global environment, being found in the water and sediments of rivers and lakes (Cedro and Cleary, 2015; Corcoran et al., 2015; Horton et al., 2017a, b; Lechner et al., 2014), wetlands (Gbogbo et al., 2020; Helcoski et al., 2020; Townsend et al., 2019), estuaries (Gallagher et al., 2016; Li et al., 2020; Yonkos et al., 2014), and the marine environment (Coyle et al., 2020; Law et al., 2014; Thompson, 2015; Yang, 2021). Microplastics are also found in wastewater (Sun et al., 2019), soil (Wang J et al., 2019), air (Gasperi et al., 2018), drinking water (Oßmann, 2021; Weber, 2020), and various foods and beverages (Mortensen et al., 2021).

Microplastics can be easily ingested, intentionally or otherwise, by aquatic organisms, hence entering ecosystem food webs. Some species have been found to preferentially ingest microplastics over food items, for example, Rotjan et al. (2019) observed that the coral *Astrangia poculata* ingested more polyethylene microbeads than brine shrimp eggs of similar size and shape (Figure 32.1) when both were available in equal abundance. Once ingested, microplastics may accumulate in the gastrointestinal tract of organisms, causing obstructions or false satiety, but smaller particles can also pass biological barriers, penetrating tissue and accumulating in organs (Kashiwada, 2006; Von Moos et al., 2012). In addition to the physical effects of microplastics ingestion, there is significant concern in relation to various toxic substances contained in plastics which may be leached into organisms after ingestion. These include residual monomers and catalysts from the polymerization process, as well as additives such as flame retardants, plasticizers, antioxidants, UV stabilizers, and colorants. Due to their hydrophobicity, microplastics have been shown to adsorb and accumulate harmful organic compounds (Hartmann et al., 2017; Koelmans et al., 2016) and metals (Munier and Bendell, 2018; Wang, F. et al., 2019). Moreover, microplastics have also been shown to support the growth of microorganisms and have the potential to act as vectors for pathogens (Kirstein et al., 2016; Murphy et al., 2020; Silva et al., 2019).

The effects of microplastics in aquatic environments, and particularly the marine setting, have been extensively studied. Ingestion and adverse effects have been observed in relation to numerous organisms, as

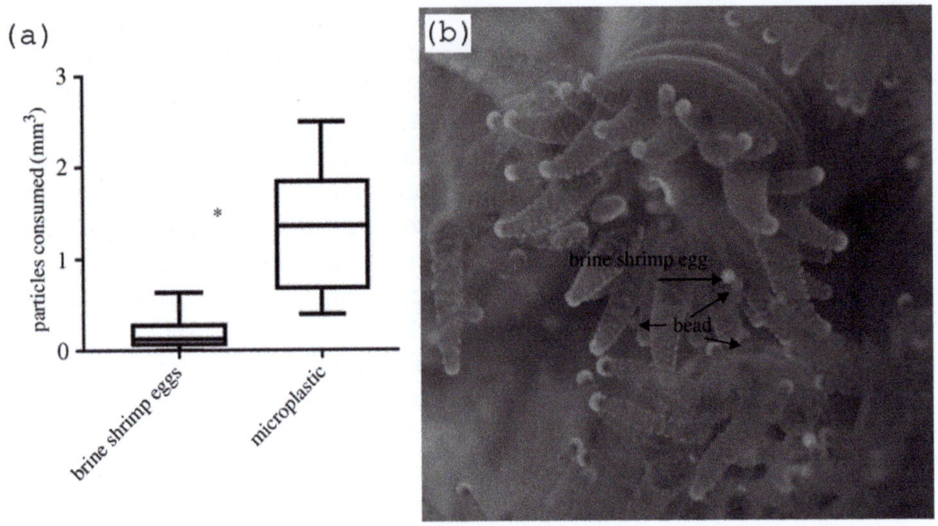

FIGURE 32.1 Comparison of brine shrimp egg (BSE) consumption versus microplastic bead (MP) consumption. (a) Ingestion of MPs and BSEs in a 50/50 choice assay. (b) Size and shape similarity of BSE and MP beads. Error bars in (a) represent ±1 s.e.m. Adapted from Rotjan et al. (2019).

reviewed by Gallo et al. (2018), Ganesh Kumar et al. (2020), and Khalid et al. (2020). Affected organisms include rotifers, copepods, diatoms, krill, gastropods, shrimp, sea urchins, mussels, fish, and sea birds. There is less information on the effects of microplastics on freshwater organisms, but similar effects can reasonably be predicted. Ma et al. (2016) observed harmful effects on *Daphnia magna* following exposure to microplastics with adsorbed phenanthrene. Studies on larval and adult zebrafish showed that microplastics and nanoplastics caused locomotive changes, intestinal damage, and change in metabolic profiles (Chen et al., 2017; Lei et al., 2018; Lu et al., 2016; Sleight et al., 2017).

Effects have also been observed in terrestrial systems. Cytotoxic effects were observed in filamentous fungi (*Aspergillus oryzae* and *Aspergillus nidulans*) exposed to nanoplastics in soil (Nomura et al., 2016). Earthworms have been commonly used as test species due to their potential to ingest, accumulate, and transport microplastics in soil. Huerta Lwanga et al. (2017a) found that *Lumbricus terrestris* exposed to polyethylene microplastics had reduced growth rate and the microplastics were concentrated in the earthworm casts. The same group (Huerta Lwanga et al., 2017b) found evidence for transfer of micro- and macro-plastic debris along a terrestrial food chain, with microplastics abundance increasing from soil to earthworm casts to chicken feces. Exposure to microplastics altered the diversity of the gut microbiome in springtails (*Folsomia candida*) while growth and reproduction were significantly inhibited (Zhu et al., 2018).

It is also posited that microplastics in soil can affect plants. The presence of microplastics can affect soil physical and chemical parameters, which may directly affect the performance of plants (Qi et al., 2020; Rillig et al., 2019). But since rhizosphere microorganisms, such as *Azotobacter* and mycorrhizal fungi, are important to plant growth, microplastic-induced changes to the structure of soil microbial communities could also indirectly affect plant performance. Toxicological responses to microplastics have been reported in plants including wheat, leeks, broad beans, celery, corn, cucumber, and cress (Pignatelli et al., 2020; Zhou et al., 2020). Plant responses vary with the type and dose of microplastics and their interaction with contaminants such as cadmium (Wang et al., 2020).

To date there is no direct evidence of negative impacts on human health due to microplastics. A 2020 report on the detection of monomers and polymer additives in human tissues (American Chemical Society, 2020) was widely misreported as the detection of microplastics in such tissues. Nevertheless, the probability of ingestion and inhalation of microplastics is obvious, given their ubiquity. In 2018, Schwabl and co-workers (Schwabl et al., 2019) detected microplastics in each of eight samples of human feces analyzed (from volunteers in Japan, Russia, the Netherlands, United Kingdom, Italy, Poland, Finland, and Austria). A median of 20 microplastics per 10 g of feces were identified. Nine types of plastic were detected, with polypropylene and polyethylene terephthalate being most abundant. Prata et al. (2019a) have examined the potential human health effects of microplastics exposure and identified several pathways of potential toxicity, including oxidative stress and cytotoxicity, disruption of energy homeostasis and metabolism, translocation of microplastics to circulatory system and distant tissues, disruption of immune function, neurotoxicity, and microplastics as vectors of microorganisms and toxic chemicals.

Significant further research is required to clarify the harmfulness of microplastics and elucidate the mechanisms by which harm occurs. However, the range of established and potential negative impacts on organisms, ecosystems, and health, is sufficient to emphasize the importance of reducing the vast quantities of plastic waste entering the environment, and of remediating the plastics and microplastics which are already there.

32.3 BIOTECHNOLOGICAL APPROACHES TO MANAGEMENT OF PLASTIC WASTE

The sheer volume of plastic production, use, and disposal, combined with the current inefficiency of global plastic waste management systems (Jambeck et al., 2015; Lebreton et al., 2017), represents a daunting challenge. It has been projected that overall plastics consumption will grow to about 1 billion tonnes by the end of the current century (OECD, 2013). Ravenstijn estimated that globally, in 2010, approximately one million mobile phones, ten million plastic cups, one billion plastic bottles, and ten billion plastic bags were disposed of every day (Ravenstijn, 2010).

Despite improvements in recycling technology, the plastics economy is largely linear; some plastics are unsuitable for recycling, many plastic items are improperly disposed of by users, either as litter or into incorrect waste streams, and potentially recyclable plastic waste is sent to landfill or discharged to rivers and oceans on enormous scales (Jambeck et al., 2015; Schmidt et al., 2017). In addition to the environmental effects of plastics and microplastics, this results in two further problems; bulky, low-density, and durable plastic items take up dwindling landfill space, while a constant cycle of greenhouse gas emissions is maintained – more plastics are required to replace the ones thrown away, requiring more oil and energy inputs.

No single solution will solve this global problem. A combination of political/regulatory, social, and technological improvements will be required. Regulatory measures such as plastic bag charges, banning of microplastic beads in consumer hygiene and cosmetic products, and reducing single-use plastics have been implemented in many countries, but there is much to do in terms of implementing and enforcing existing legislation, as well as in improving source control and waste management. Social strategies can also contribute; these are based on educating and engaging citizens to change their behaviors – for example promoting recycling and use of biodegradable materials while reducing demand and consumption of disposable plastics.

Figure 32.2 illustrates the more desirable alternative to the linear economy; a circular economy in which the loop is closed by recycling and/or biodegradation of polymeric materials. In current recycling technology, plastics are shredded and reformed into pellets for further processing, whereas biodegradation-based recycling ideally disassembles the waste plastic to monomers which can be used to form new virgin plastics. It is unlikely that complete efficiency and a complete closing of the circle can be achieved; the objective of circular economy is to minimize loss of plastics to landfill or energy recovery, by maximizing the efficiency of recycling and reuse. This in turn minimizes the need for use of new raw materials.

A major impediment to a circular economy for plastics is the fact that most of the current high-production plastics are not readily biodegradable. In this context, there are two major ways in which biotechnology can contribute. The first is the development of bioplastics, specifically biodegradable plastics. Though often conflated, the distinction is an important one; bioplastics are manufactured from biological materials but many bioplastics are not biodegradable, and conversely, some biodegradable plastics are produced from fossil resources. The second major contribution of biotechnology is in relation to the use of organisms or biomolecules capable of degrading plastics which are not typically biodegradable.

32.3.1 Biodegradable Bioplastics for Circular Economy

Research into biodegradation of plastics has gathered momentum over the last two decades (see also chapter 24). Jenkins et al. (2020) list multiple degrading organisms, and the relevant enzymes, for a selection of

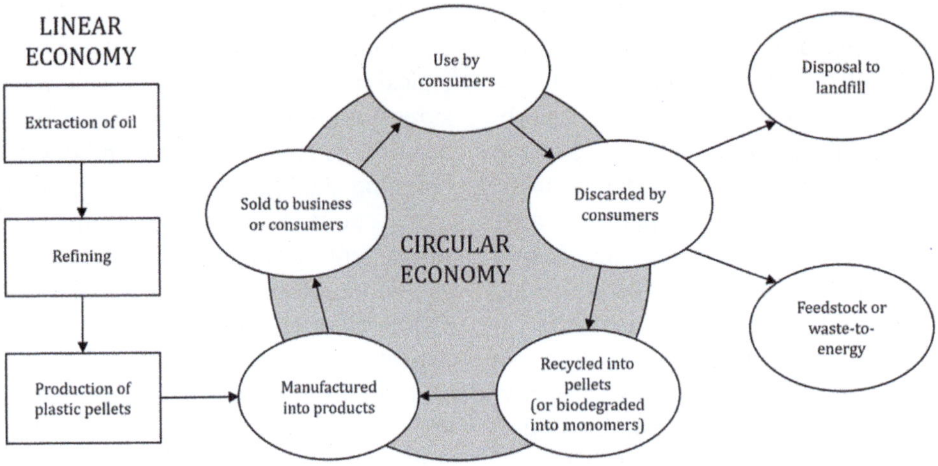

FIGURE 32.2 Life cycle of plastics in linear and circular economies. Adapted from Prata et al. (2019b).

polymers including polybutylene succinate (PBS), polycaprolactone (PCL), polyurethane (PU), and polyethylene terephthalate (PET). The actual degradability of existing biodegradable polymers is, however, often overestimated, perhaps due to a lack of clear distinction between the meanings of "compostable" (can be converted into H_2O and CO_2 within a certain time scale and under specifically defined conditions, as specified in various standards, leaving behind no harmful residues) and "biodegradable" (can be more than 90% broken down by enzymes produced by living organisms). The latter definition, for instance, allows for significant residual material, perhaps in the form of microplastics or larger fragments.

Degradation of PET by the bacterium *Ideonella sakaiensis* gained significant popular attention when it was reported by Yoshida et al. (2016a). PET is used extensively, notably in water and soft drinks bottles, and its consequent accumulation in the environment is significant. By screening natural microbial communities exposed to PET in the environment, Yoshida and coworkers isolated *Ideonella sakaiensis* 201-F6, capable of using PET as its major energy and carbon source. The organism produces two enzymes capable of hydrolyzing PET and the reaction intermediate mono(2-hydroxyethyl) terephthalic acid, resulting in the degradation of PET into two environmentally benign monomers, terephthalic acid and ethylene glycol (Figure 32.3). Thin films of PET were almost completely degraded in six weeks (Yoshida et al., 2016a).

PET has become a prominent example of the potential of biotechnology to close the production-use-recycling loop. The polyester hydrolase TfH from *Thermobifida fusca* can be used to produce low crystalline PET films (Mueller, 2006). These low crystalline films can be completely depolymerized by microbial polyester hydrolases. The resulting monomers can subsequently be used for the synthesis of virgin PET, or in the synthesis of other added-value products. It should be noted however, that bottle-grade PET, which has higher crystallinity, had much lower enzymatic degradability (Yoshida et al., 2016a).

Austin et al. (2018) engineered the PETase enzyme produced by *Ideonella sakaiensis* 201-F6, resulting in improved PET degradation. Apart from suggesting that additional protein engineering to increase PETase performance may be possible, this highlights the need for further investigation of structure/activity relationships for improved biodegradation of synthetic polymers.

Other plastics that are simultaneously bio-based and biodegradable, including polylactic acid (PLA), polybutylene succinate (PBS), and polyhydroxyalkanoates (PHA), also offer opportunities in terms of circular economy.

PLA is a polyester widely used for packaging. PLA is chemically synthesized from lactic acid, a fermentation product from plant resources, and is thereby defined as a bioplastic. Temperatures above the glass transition temperature of PLA (approx. 60 °C) are usually required for the onset of PLA hydrolysis, making its degradation difficult at ambient temperatures. However, several enzymes have been identified as degrading PLA at lower temperatures, including proteases from *Amycolatopsis*, lipases from *Paenibacillus amylolyicus*, and fungal

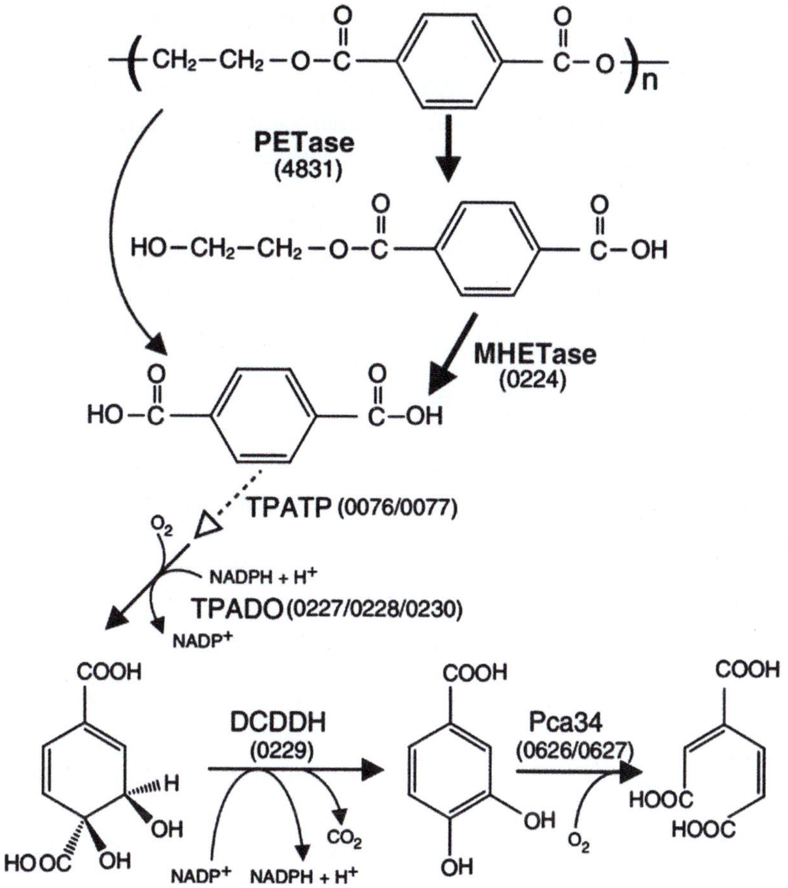

FIGURE 32.3 Predicted pathway for PET degradation pathway by Ideonella sakaiensis 201-F6. Extracellular PETase hydrolyzes PET to produce MHET and TPA. MHETase, a predicted lipoprotein, hydrolyzes MHET to TPA and EG. TPA is incorporated through the TPA transporter (TPATP) and catabolized by TPA 1,2-dioxygenase (TPADO), followed by 1,2-dihydroxy-3,5-cyclohexadiene-1,4-dicarboxylate dehydrogenase (DCDDH). The resultant PCA is ring-cleaved by PCA 3,4-dioxygenase (Pca34). Adapted from Yoshida et al. (2016a).

protease and cutinase from *Tritirachiium album* and *Aspergillus oryzae*, respectively (Kawai, 2010).

PBS was formerly produced only from fossil resources, but can now be obtained using a fully bio-based process by polycondensation reactions of succinic acid and 1,4-butanediol, each produced from renewable feedstocks (Babu et al., 2013; Bechthold et al., 2008). PBS is biodegradable by microbial lipases and cutinases produced by *Actinomycetes* (Jarerat et al., 2002) and *Fusarium solani* (Abe et al., 2010). *Fusarium solani* strain WF-6 degraded 2.8% of the PBS during a 14-day experiment in a sterile soil environment. Degradability was enhanced by co-culturing strain WF-6 with newly isolated bacterial strain *Stenotrophomonas maltophilia* YB-6, which itself did not show PBS-degrading activity (Abe et al., 2010).

PHAs are biopolymers produced in more than 300 microorganisms with diverse monomer compositions, although the most commonly reported polymers are polyhydroxybutyrate (PHB) and polyhydroxyvalerate (PHV) (Dias et al., 2006). A variety of PHA homopolymers, random copolymers, and block copolymers can be generated by engineering of three synthesis pathways including the acetoacetyl-CoA pathway, in situ fatty acid synthesis, and b-oxidation cycles, as well as by PHA synthase specificity. PHA structures and properties can therefore be tailored via genome editing or process engineering (Chen et al., 2015).

Due to its biodegradability, biocompatibility, and physical properties similar to petroleum-derived thermoplastics such as polyethylene and polypropylene, PHB is seen as one of the most promising bioplastics. It is used for storage of carbon and intracellular energy in many microorganisms, including *Bacillus*, *Pseudomonas*, and *Alcaligenis euterophus*. It can be biodegraded by depolymerases in fungi and bacteria (Chen et al., 2015; Yeo et al., 2018). Although challenges remain in terms of production cost, thermal stability, and complete biodegradability, PHB could emerge as the most environmentally sustainable biopolymer (Isola et al., 2017).

32.3.2 Biodegradation of Synthetic Plastics

Despite some significant progress, considerable research, development, and process optimization will be required before biodegradable bioplastics can replace petroleum-based plastics on large scales. Since these synthetic plastics are and will continue to be produced in huge volumes for decades to come, it is imperative that we also close the loop on these materials. To that end, numerous researchers have investigated the potential of a wide variety of organisms and enzymes to degrade plastics such as polyethylene (PE), polypropylene (PP), polystyrene (PS), polyvinyl chloride (PVC), polyethylene terephthalate (PET), polyurethane (PU), polycarbonate (PC), polyamide (PA), and others.

Many plastic-degrading microorganisms have evolved their ability from enzymes which degrade plant biomass. Cutinases which hydrolyze cutin, a plant aliphatic polyester, are also able to hydrolyze ester bonds in polymers such as PET and PU, while lignin-modifying enzymes are involved in the degradation of low density polyethylene (LDPE) (Jenkins et al., 2020). In fact, due to its leading position in production volumes, polyethylene, in low-density and high-density forms (LDPE and HDPE) has attracted a large share of research attention. Pathak and Navneet (2017) listed 46 organisms which have been identified as biodegraders of PE. Surely with this array of polyethylene-munching microorganisms available, PE's days as a recalcitrant waste material are numbered? Unfortunately, it is not so straightforward; biodegradability of polymers is influenced by factors including chemical structure, molecular weight, morphology, and hydrophobicity. High molecular weight and crystalline morphology generally hinder biodegradation. PE has a semi-crystalline structure, high hydrophobicity, and is highly resistant to biodegradation. Biodegradation of PE typically proceeds slowly and may require specific physicochemical treatments to facilitate the process. Such treatments include UV irradiation, photo-oxidation, thermal treatment, and oxidation with nitric acid (Pathak and Navneet, 2017).

Hadad et al. (2005) described the biodegradation of polyethylene by *Brevibacillus borstelensis*. An 11% weight reduction was observed after 30 days, for PE films of low density and previously treated with UV radiation.

Gyung Yoon et al. (2012) studied the biodegradation of polyethylene by a soil bacterium and alkane hydroxylase gene (*alkB*) cloned recombinant cells. A mesophilic bacterium identified as *Pseudomonas* sp., isolated from crude-oil contaminated beach soil was able to biodegrade low molecular weight polyethylene. Up to 28.6% of the carbon was mineralized into CO_2 after 80 days at 37 °C, however the biodegradability decreased with increase in molecular weight of the polymer.

Park and Kim (2019) investigated the degradation of micro-sized polyethylene by mesophilic mixed bacterial culture isolates obtained from a municipal landfill sediment. *Bacillus* sp. and *Paenibacillus* sp. were the most dominant when exposed to the PE. The dry weight of particles was reduced by 14.7% after 60 days. Yuan et al. (2020) listed 12 examples of bacterial strains capable of degrading microplastics (including PE in five cases) under laboratory conditions and in the environment. Sources of the microorganisms included mangrove ecosystem sediment, waxworms, soils, compost, and digester sludge. Weight losses of the plastics, where reported, typically ranged from 3–11% over 28–40 days.

To achieve higher rates of biodegradation, the use of microbial consortia has been proposed. Sowmya et al. (2015) achieved 27% weight loss in 10 mm polyethylene discs after three months using a fungal consortium of *Curvularia lunata*, *Alternaria alternate*, *Penicillium simplicissimum*, and *Fusarium* sp. while the use of the same species individually resulted in only 0.7–7.7% degradation.

In recent years biodegradation of PE by invertebrates or their gut isolates have been investigated. Yang et al. (2014) isolated LDPE-degrading bacteria, *Enterobacter asburiae* YT1 and *Bacillus* sp. YP1 from the gut of waxworms (larvae of *Plodia interpunctella*, the Indian Mealmoth) (Figure 32.4). Bacterial isolates identified from the gut of *P. interpunctella* were able to degrade approximately 6–11% of PE films and released 12 water-soluble products. Bombelli et al. (2017) showed that caterpillars of the wax moth *Galleria mellonella* could consume PE films in the form of shopping bags. Leaving 100 worms in contact with a bag for 12 hours resulted in a 92 mg loss in mass. When worm homogenate was left in contact with PE films, a mass loss of 13% was observed over 14 hours, and chemical changes were observed between treated and untreated

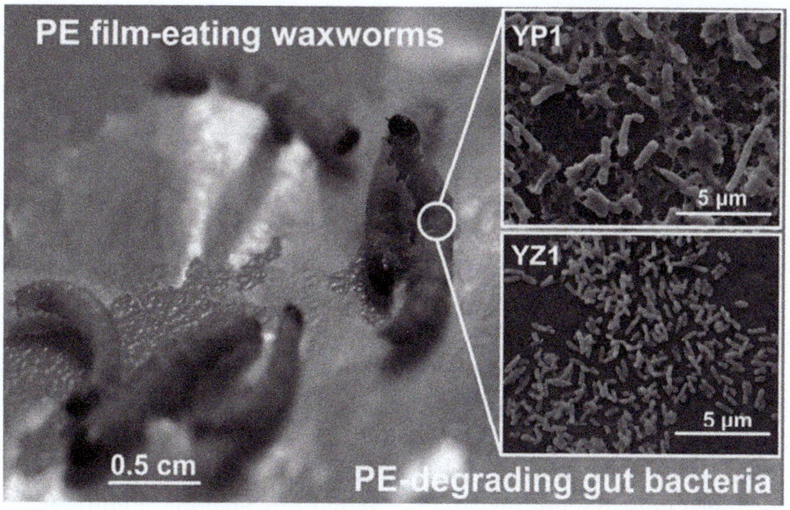

FIGURE 32.4 Waxworms and their gut bacteria engaged in the biodegradation of polyethylene. Reproduced from Yang et al. (2014).

samples. Other invertebrates have also been identified as efficient PE biodegraders, including larvae of Yellow Mealworms (*Tenebrio molitor*) (Billen et al., 2020), Dark Mealworms (*Tenebrio obscurus*) (Brandon et al., 2018), Lesser Waxworms (*Achroia Grisella*) (Kundungal et al., 2019), and land snails (*Achatina fulita*) (Song et al., 2020).

While the rates of biodegradation achieved using invertebrates is significantly higher than typically observed using microorganisms, there are practical complications which may limit their application in waste management. These include the costs and practical issues of sustaining insect cultures to produce the PE-eating larvae, and the generation of microplastics due to incomplete degradation.

It is beyond the scope of this chapter to review the full array of microorganisms and enzymes used to biodegrade the various synthetic polymers (see also chapter 24). Mohanan et al. (2020) have recently and comprehensively reviewed the area of microbial and enzymatic degradation of synthetic plastics. However, the preceding case study on the biodegradation of polyethylene serves to highlight the potential, as well as some of the current limitations, of biodegradation as an approach to the management of plastic waste.

Biodegradation is a complex process, occurring via a series of steps, namely:

Biofilm formation is the initial attachment of microorganisms to the plastic surface, typically via exopolysaccharides.

Biodeterioration begins with superficial degradation of the plastic surface, generally via hydrolysis or oxidation reactions influenced by abiotic parameters such as presence of water, light, and thermal energy. Microbial communities produce extracellular enzymes which attack the polymer structure.

Biofragmentation destabilizes the long carbon chains of polymers by the action of oxygenase enzymes by hydrolases such as proteases and lipases. The polymer is broken down into smaller units; oligomers, dimers, and monomers.

Bioassimilation occurs when monomers are assimilated by the microorganisms as sources of energy and carbon, leading to increase in microbial biomass.

Mineralization refers to complete degradation of the polymer to final products such as CO_2, CH_4, N_2, and H_2O.

Biodegradation has typically been quantified by one of several variables, including substrate weight loss, changes in the mechanical properties or chemical structure of the polymer, and emission of carbon dioxide. Direct comparison of the biodegradative potential of different organisms or enzymes is often complicated by wide variations in the properties of the plastics studied, the conditions used, and the way in which biodegradation is assessed. Occasionally this had led to disputes in the literature, such as a claim that authors "exaggerated degradation efficiency" leading to "rather misleading conclusions" by using a material with un-representatively low crystallinity (Yang et al., 2016; Yoshida et al., 2016b).

It is valid to point out that many studies of biodegradation have been carried out under optimized conditions which are perhaps not representative of "real-world" applications in waste management. It is understandable that researchers wish to achieve quantifiable results on reasonable timescales; this does not represent a problem provided that the resulting rates or extents of biodegradation are presented realistically. It may well be that achieving efficient biodegradation processes for many polymers will require the use of elevated temperatures, UV treatment, or the addition of acid or other chemicals. If so, it is important that these factors be taken into account when assessing the overall potential of a particular biodegradative route, rather than treating them as a technical afterthought for later consideration. Danso et al. (2019) have also noted that since the commercial plastics typically used in biodegradation studies often contain significant proportions of chemical additives, more attention could be paid to biochemical mechanisms of degradation, in order to verify that the polymer itself is being degraded as opposed to the additives. In this case the representativeness of the material studied is correct, but still has the potential to give misleading (or at least misinterpretable results) over a limited timescale. As a simplistic illustration, if 5% of the material mass is lost after a certain time, but that loss comes mainly or exclusively from the additive content, the biodegradative potential of the microorganism or enzyme can be greatly overestimated. Paco et al. (2018) have commented that many studies on biodegradation should be read as proof-of-concept.

To date biodegradation of synthetic polymers has not been implemented on a commercial scale due to the slow degradation rates achievable. However, many promising results have been obtained in what is still a relatively young research area. To maximize the undoubted potential of polymer biodegradation, a number of future research needs have been identified (Ganesh Kumar et al., 2020; Jenkins et al., 2020) and are discussed below.

Protein engineering of enzymes. The potential of this approach has been demonstrated by a number of studies. Wei et al. (2016) engineered polyester hydrolase enzymes TfCut2 from *Thermobifida fusca* KW3 and LC-cutinase (LCC) from a compost metagenome by exchanging selected amino acid residues of TfCut2 involved in substrate binding with those present in LCC. Enzyme variants with increased PET hydrolytic activity at 65°C were obtained. Variant G62A and double variant G62A/I213S caused a weight loss of PET films of more than 42% after 50 hours of hydrolysis, corresponding to a 2.7-fold increase compared to the wild type enzyme. Ma et al. (2018) created an enhanced artificial PETase enzyme through rational protein engineering of key hydrophobic sites. Degradation rate of a PET film reached 22.5 mg $\mu mol^{-1} \cdot L^{-1}$ PETase per day. Islam et al. (2019) reported the fusion of anchor peptide Tachystatin A2 to the bacterial cutinase Tcur1278, accelerating the degradation of polyester-polyurethane nanoparticles by a factor of 6.6 compared to wild-type Tcur1278.

Strain engineering. A recent approach is to engineer organisms with enhanced biodegradative potential. Huang et al. (2018) inactivated twin arginine translocation complexes and observed 3.8-fold enhanced PETase activity in *Bacillus subtilis*. Moog et al. (2019) used the photosynthetic microalga *Phaeodactylum tricornutum* as a chassis to generate a microbial cell factory for production of an engineered PETase into the surrounding culture medium. The PETase was active against industrially shredded PET in a saltwater-based environment at mesophilic temperatures (21 °C).

Metagenomic analysis is a powerful approach to studying the diversity of enzymes and microorganisms, potentially allowing identification of novel and highly specific enzymes with activity against synthetic polymers (see also chapter 30). Pinnell and Turner (2019) reported the shotgun metagenomic sequencing of biofilms on plastic and bioplastic microcosms at the sediment-water interface of a coastal lagoon. Findings indicated that bioplastic (polyhydroxyalkanoate, PHA), but not PET, promoted a rapid and significant shift in benthic microbial diversity, selecting for microbes that participate in bioplastic degradation and sulfate reduction, leading the authors to caution against possible unintended effects on benthic biogeochemical activities if plastic pollution is replaced by bioplastic pollution.

In silico genome mining. Almeida et al. (2019) employed an in silico-based screening approach on the genus *Streptomyces* in a search for potential PETase homologs. From a total of 52 genomes analyzed, three PETase-like enzymes were identified, all of which were derived from marine-sponge associated *Streptomyces* isolates. A candidate PETase-like gene (SM14est) was identified in *Streptomyces* sp. SM14. Molecular docking experiments indicated that the SM14est enzyme possessed the capacity to bind plastics as substrates. Polyesterase activity was confirmed using a polycaprolactone (PCL) plate clearing assay following heterologous expression of SM14est in *Escherichia coli*. The study highlights the potential for high-throughput screening to identify novel genes and enzymes with biodegradative activity, as well as to inform greater understanding of biodegradation mechanisms.

Combining physicochemical and biological treatment. As alluded to above, thermal energy, UV radiation, and chemical treatments can be used to facilitate biodegradation. Plastics can also be made more biodegradable by increasing their hydrophilicity, shortening the polymer chain, reducing the crystallinity, or increasing the ratio of more labile bonds such as esters or amides to carbon–carbon bonds. Blending synthetic plastics with natural, more easily degraded polymers such as starch or cellulose, can result in an increase of polymer biodegradability (Arutchelvi et al., 2008). Increased attention on design for degradability and developing plastics to be degraded under specific conditions or by specific substances could yield significant results.

Designing Plastic Degrading Consortia. Microorganisms typically live in diverse and complex communities, exemplified by the biofilms which form on environmental surfaces including plastics. If synergistic consortia of biodegrading microorganisms can be designed, biodegradation rates can potentially be greatly increased relative to single organism approaches, and more stable communities of biodegraders might also result.

32.3.3 Biotechnology for Microplastics Management and Remediation

To state that it is better to prevent microplastics from entering the environment than to attempt to remediate them is to state the obvious. Yet it is perhaps worth emphasizing that technologies and practices which can avoid or limit the occurrence of microplastics, both primary and secondary, should be our first priority. In this context, we might consider the effectiveness of wastewater treatment plants (WTPs) in retaining or eliminating microplastics. WTPs are significant point sources of microplastic contamination into the aquatic environment, deriving from sources including fibers and fragments of artificial textiles in washing machine run-off, and particles from tire wear and other plastic fragments on urban surfaces. Although existing WTPs have not generally been designed with microplastics in mind, they have been found to be quite effective in removing them from treated water. Lares et al. (2018) found that a membrane bioreactor (MBR) process had a slightly better removal efficiency of microplastics (99.4%) compared to a conventional activated sludge (CAS) based process (98.3%). Two papers by Talvitie and co-workers (2017a, b) compared rapid gravity sand filter (RSF), dissolved air flotation (DAF), and membrane bioreactor (MBR) used as tertiary treatment. Wastewater containing microplastics was treated and all three technologies showed removal rates of microplastics from primary and secondary effluents higher than 95%. An analysis by Iyare et al. (2020) examined reported rates of microplastics removal in 21 studies from 16 countries and found average removal rates of 72%, 88%, and 94% at primary, secondary, and tertiary treatment stages, respectively.

The high removal rate of microplastics from wastewater, of course, raises questions as to their fate. Since the residence time in wastewater treatment systems is likely too short for biodegradation to be very significant, it seems that microplastics end up in the sludge component. Mahon et al. (2016) characterized microplastics in sludge samples from 7 WTPs in Ireland. Abundances ranged from 4196 to 15385 particles kg^{-1} (dry weight), but lower abundances were found in sludge after anaerobic digestion compared to sludges after thermal drying or lime stabilization, suggesting that some biodegradation may take place in anaerobic digestors. Nevertheless, where sludge is land-spread rather than used for energy recovery, a significant microplastics burden will be transferred to agricultural soils, some of which will in turn enter aquatic systems. There is certainly scope for biotechnological optimization of biological wastewater and sludge treatment processes to reduce microplastics emissions from WTPs.

There is no inherent difference between the biodegradation of microplastics compared to that of larger plastic waste; hence many studies on the potential biodegradation of environmental microplastics use or cite some of the same approaches, organisms, and enzymes as we have discussed in previous sections. Due to their small size and high surface-to-volume ratio, microplastics might be considered as ideal candidates for biodegradation, yet there are a number of barriers to successful adoption of this approach. Unlike the carefully controlled conditions of laboratory studies on plastics biodegradation, or even the more robust but still controllable environment of a waste processing facility, microplastics are widely dispersed in an uncontrolled environment and we do not have effective means to collect or accumulate them. In terms of the abiotic factors that can promote biodegradation, environmental microplastics will generally have encountered helpful abiotic factors such as UV light, oxygen, and water, although depending on their environmental fate, those factors may or may not be present during the biodegradation process. Temperature may be lower than optimal; pH will typically be moderate but again not necessarily optimal. Mineral nutrients such as N, P, or Fe may be limiting, and the microplastics themselves may have adsorbed toxic or inhibitory substances such as metals or organic pollutants. Moreover, biodegrading organisms will have to compete with complex communities of well-established and -adapted wild organisms, albeit existing microfilms may have conditioned and perhaps roughened the microplastic surface.

All in all, it would seem that advanced biotechnological approaches such as metagenomic analyses of microbial communities associated with environmental plastics and microplastics, or from locations subject to contamination by plastics or other petroleum derivatives, followed by protein or strain engineering, offer the best opportunities for identification of efficient and environmentally robust biodegraders.

There are further caveats; as previously discussed, Pinnell and Turner (2019) observed a shift in benthic microbial diversity, due to the presence of a bioplastic, prompting their reminder of the possibility of unintended consequences if biodegradable materials become prevalent environmental contaminants. Introduction of efficient biodegrading organisms, bioengineered or otherwise, also has potential to affect existing communities and ecosystems. Some authors (Shen et al., 2019; Paço et al., 2018) point out another possible negative impact in the form of shortened lifetime for useful plastic items exposed to efficient biodegradation. Rai et al. (2020) caution that "newly developed materials should only be applied after their environmental fate, degradation time, and ecotoxic responses are thoroughly understood."

Nevertheless, considering the existing and future burden of microplastics pollution, it is appropriate to examine some options. Huerta Lwanga et al. (2018) used a bacterial consortium isolated form the gut microbiome of the earthworm *Lumbricus terrestris*. 60% of the LDPE microplastic initially present decayed after 21 days in the presence of bacteria, while the size distribution of the microplastics shifted to smaller sizes. Volatile compounds such as octadecane, eicosane, docosane, and tricosane were detected in the soil treatment that included both microplastics and bacteria, but not in the controls.

Paco et al. (2017) used the marine fungus *Zalerion maritimum* to biodegrade PE microplastics in simulated seawater. A slight increase in biomass of *Z. maritimum* corresponded to a loss in the mass of the plastic particles, which decreased by 43% after 14 days. Infrared analysis (ATR-FTIR) was used to monitor changes in polymer chemistry, with increases in the hydroxyl and carbonyl regions noted, compatible with oxidation of the polymer.

In addition to the bacterial biodegraders listed by Yuan et al. (2020), the same authors discuss fungal biodegradation of microplastics, as well as the use of bacterial consortia and biofilms. Bacterial consortia reviewed gave reported weight losses typically ranging from 7% to 22%, notably higher than the typical rates of degradation by single strains. Nevertheless, this review highlights the need for (i) better understanding of the processes involved, (ii) further optimization of MP degradation conditions, (iii) attention to be paid to anaerobic microorganisms, and (iv) greater exploitation of molecular techniques.

32.4 CONCLUSIONS

Biodegradation of plastics in general is at a stage of great opportunity while still facing major barriers to its successful large-scale adoption. There is much that can be achieved by moving toward a circular economy for plastics and biotechnology can make two major contributions; by developing biodegradable bioplastics which can be fully recycled and by improving the capability to biodegrade synthetic petroleum-based plastics. There are challenges with respect to each of these. Examples of microorganisms and enzymes that are shown to biodegrade even the most recalcitrant plastics abound, yet their implementation in waste management is hampered by slow rates of degradation and the need for additional treatments and controlled conditions. Various biodegradable polymers have been developed and some have seen significant commercial uptake, yet if not managed properly, these materials will take much longer to fully biodegrade in the environment than under industrial conditions, and the degradation process still generates large quantities of potentially harmful microplastics. In some applications, achieving the required balance between material biodegradability and product lifetime is difficult, and in others, the use of toxic or inhibitory additives in plastics may also complicate the success of biodegradative technologies.

Biodegradation of microplastics is similarly challenged but in different ways; here the main barrier is the dispersed nature of microplastics contamination in the environment, and the resulting inability to provide controlled and predictable conditions to facilitate the operation of biodegradative organisms and biomolecules.

There is great potential in the exploitation of advanced biotechnological approaches which could improve our ability to tailor the properties of biodegradable bioplastics on one hand, and optimize the performance of biodegradative technologies on the other. Significant progress has been made and must continue (see also chapter 24).

However, the ability to achieve a circular economy for plastics based on biotechnology has not yet arrived, nor has the ability to efficiently bioremediate microplastics in the environment. There is a more immediate way to reverse the ever-growing problem of environmental contamination by plastic waste, followed by its inevitable fragmentation into innumerable microplastics. Lebreton et al. (2017) examined plastic emissions from rivers to the world's oceans. The top 20 polluting rivers accounted for more than two-thirds (67%) of the global annual input; 15 of these rivers are in Asia with others in South America and Africa. China is the country with the highest mass of mismanaged waste (Jambeck et al., 2015); however, large amounts of this plastic are imported from other countries. It is technologically more feasible, albeit economically and politically more demanding, to tackle the global issue of plastic waste by improving and expanding current waste management technologies and practices and by implementing them more evenly on a global level.

ACKNOWLEDGMENTS

We acknowledge the support that Loriane Murphy has received from the Irish Research Council under the Government of Ireland Postgraduate Scholarship programme, project ID: GOIPG/2019/2786.

REFERENCES

1. Abe M, Kobayashi K, Honma N, Nakasaki K. 2010. Microbial degradation of poly(butylene succinate) by *Fusarium solani* in soil environments. *Polymer Degradation and Stability* 95: 138–143.
2. Almeida EL, Carrillo Rincón AF, Jackson SA, Dobson ADW. 2019. In silico screening and heterologous expression of a polyethylene terephthalate hydrolase (PETase)-like enzyme (SM14est) with polycaprolactone (pcl)-degrading activity, from the marine sponge-derived strain *Streptomyces* sp. SM14. *Frontiers in Microbiology* 10: 2187.
3. American Chemical Society. 2020. Methods for microplastics, nanoplastics and plastic monomer detection and reporting in human tissues. News release, August 17, 2020. Available online at: https://www.acs.org/content/acs/en/pressroom/newsreleases/2020/august/micro-and-nanoplastics-detectable-in-human-tissues.html
4. Arutchelvi J, Sudhakar M, Arkatkar A, Doble M, Bhaduri S, et al. 2008. Biodegradation of polyethylene and polypropylene. *Indian Journal of Biotechnology* 7: 9–22.
5. ASTM. (2020a) D8332-20: Standard practice for collection of water samples with high, medium, or low suspended solids for identification and quantification of microplastic particles and fibers. ASTM International 2020. Available online at: https://www.astm.org/Standards/D8332.htm Accessed 27 July 2021.
6. ASTM. (2020b) D8333-20: Standard practice for preparation of water samples with high, medium, or low suspended solids for identification and quantification of microplastic particles and fibers using Raman spectroscopy, IR spectroscopy, or pyrolysis-GC/MS. ASTM International 2020. Available online at: https://www.astm.org/Standards/D8333.htm Accessed 27 July 2021.
7. Austin HP, Allen MD, Donohoe BS, Rorrer NA, Kearns FL, et al. 2018. Characterization and engineering of a plastic-degrading aromatic polyesterase. *Proceedings of the National Academy of Sciences* 115: E4350–E4357.
8. Babu RP, O'Connor K, Seeram R. 2013. Current progress on bio-based polymers and their future trends. *Progress in Biomaterials* 2: 8.
9. Barnes DKA, Galgani F, Thompson RC, Barlaz M. 2009. Accumulation and fragmentation of plastic debris in global environments. *Philosophical Transactions of the Royal Society B: Biological Sciences* 364: 1985–1998.
10. Bechthold I, Bretz K, Kabasci S, Kopitzky R, Springer A. 2008. Succinic acid: A new platform chemical for biobased polymers from renewable resources. *Chemical Engineering and Technology* 31: 647–654.
11. Bombelli P, Howe CJ, Bertocchini F. 2017. Polyethylene bio-degradation by caterpillars of the wax moth *Galleria mellonella*. *Current Biology* 27: R292–R293.
12. Billen P, Khalifa L, Van Gerven F, Tavernier S, Spatari S. 2020. Technological application potential of polyethylene and polystyrene biodegradation by macro-organisms such as mealworms and wax moth larvae. *Science of the Total Environment* 735: 139521.
13. Brandon AM, Gao SH, Tian R, Ning D, Yang SS, Zhou J, et al. 2018. Biodegradation of polyethylene and plastic mixtures in mealworms (larvae of *Tenebrio molitor*) and effects on the gut microbiome. *Environmental Science & Technology* 52: 6526–6533.

14. Cedro A, Cleary J. 2015. Microplastics in Irish freshwaters: A preliminary study. *CEST 2015: The 14th International Conference on Environmental Science and Technology*, 3–5 September 2015, Rhodes, Greece.
15. Chen GQ, Hajnal I, Wu H, Lv L, Ye J. 2015. Engineering biosynthesis mechanisms for diversifying polyhydroxyalkanoates. *Trends in Biotechnology* 33: 565–574.
16. Chen Q, Gundlach M, Yang S, Jiang J, Velki M, et al. 2017. Quantitative investigation of the mechanisms of microplastics and nanoplastics toward zebrafish larvae locomotor activity. *Science of the Total Environment* 584-585 Supplement C: 1022–1031.
17. Corcoran PL, Norris T, Ceccanese T, Walzak MJ, Helm PA, et al. 2015. Hidden plastics of Lake Ontario, Canada and their potential preservation in the sediment record. *Environmental Pollution* 204: 17–25.
18. Coyle R, Hardiman G, Driscoll KO. 2020. Microplastics in the marine environment: A review of their sources, distribution processes and uptake into ecosystems. *Case Studies in Chemical and Environmental Engineering* 2: 100010.
19. Danso D, Chow J, Streit WR. 2019. Plastics: Microbial degradation, environmental and biotechnological perspectives. *Applied and Environmental Microbiology* 85: e01095–19.
20. Dias JML, Lemos PC, Serafim LS, Oliveira C, Eiroa M, et al. 2006. Recent advances in polyhydroxyalkanoate production by mixed aerobic cultures: From the substrate to the final product. *Macromolecular Bioscience* 6: 885–906.
21. Gallagher A, Rees A, Rowe R, Stevens J, Wright P. 2016. Microplastics in the Solent estuarine complex, UK: An initial assessment. *Marine Pollution Bulletin* 102: 243–249.
22. Gallo F, Fossi C, Weber R, Santillo D, Sousa J, et al. 2018. Marine litter plastics and microplastics and their toxic chemicals components: The need for urgent preventive measures. *Environmental Sciences Europe* 30: 13.
23. Ganesh Kumar A, Anjana K, Hinduja M, Sujitha K, Dharani G. 2020. Review on plastic wastes in marine environment – Biodegradation and biotechnological solutions. *Marine Pollution Bulletin* 150: 110733.
24. Gasperi J, Wright SL, Dris R, Collard F, Mandin C, et al. 2018. Microplastics in air: Are we breathing it in? *Current Opinion in Environmental Science and Health* 1: 1–5.
25. Gbogbo F, Takyi JB, Billah MK, Ewool J. 2020. Analysis of microplastics in wetland samples from coastal Ghana using the Rose Bengal stain. *Environmental Monitoring and Assessment* 192: 208.
26. Geyer R, Jambeck JR, Law KL. 2017. Production, use, and fate of all plastics ever made. *Science Advances* 3: e1700782.
27. Gyung Yoon M, Jeong Jeon H, Nam Kim M. 2012. Biodegradation of polyethylene by a soil bacterium and *alkB* cloned recombinant cell. *Journal of Bioremediation and Biodegradation* 03: 145.
28. Hadad D, Geresh S, Sivan A. 2005. Biodegradation of polyethylene by the thermophilic bacterium *Brevibacillus borstelensis*. *Journal of Applied Microbiology* 98: 1093–1100.
29. Hartmann NB, Rist S, Bodin J, Jensen LH, Schmidt SN, et al. 2017. Microplastics as vectors for environmental contaminants: Exploring sorption, desorption, and transfer to biota. *Integrated Environmental Assessment and Management* 13: 488–493.
30. Helcoski R, Yonkos LT, Sanchez A, Baldwin AH. 2020. Wetland soil microplastics are negatively related to vegetation cover and stem density. *Environmental Pollution* 256: 113391.
31. Horton AA, Svendsen C, Williams RJ, Spurgeon DJ, Lahive E. 2017a. Large microplastic particles in sediments of tributaries of the River Thames, UK – Abundance, sources and methods for effective quantification. *Marine Pollution Bulletin* 114: 218–226.
32. Horton AA, Walton A, Spurgeon DJ, Lahive E, Svendsen C. 2017b. Microplastics in freshwater and terrestrial environments: Evaluating the current understanding to identify the knowledge gaps and future research priorities. *Science of the Total Environment* 586: 127–141.
33. Huang X, Cao L, Qin Z, Li S, Kong W, et al. 2018. Tat-independent secretion of polyethylene terephthalate hydrolase PETase in *Bacillus subtilis* 168 mediated by its native signal peptide. *Journal of Agricultural and Food Chemistry* 66: 13217–13227.
34. Huerta Lwanga E, Gertsen H, Gooren H, Peters P, Salánki T, et al. 2017a. Incorporation of microplastics from litter into burrows of. *Lumbricus Terrestris*. *Environmental Pollution* 220: 523–531.
35. Huerta Lwanga E, Mendoza Vega J, Ku Quej V, Chi J, de los Sanchez A, et al. 2017b. Field evidence for transfer of plastic debris along a terrestrial food chain. *Scientific Reports* 7: 14071.
36. Huerta Lwanga E, Thapa B, Yang X, Gertsen H, Salanki T, et al. 2018. Decay of low-density polyethylene by bacteria extracted from earthworm's guts: A potential for soil restoration. *Science of the Total Environment* 624: 753–757.
37. Islam S, Apitius L, Jakob F, Schwaneberg U. 2019. Targeting microplastic particles in the void of diluted suspensions. *Environment International* 123: 428–435.

38. Isola C, Sieverding HL, Raghunathan R, Sibi MP, Webster DC, et al. 2017. Life cycle assessment of photodegradable polymeric material derived from renewable bioresources. *Journal of Cleaner Production* 142: 2935–2944.
39. Iyare PU, Ouki SK, Bond T. 2020. Microplastics removal in wastewater treatment plants: A critical review. *Environmental Science: Water Research and Technology* 6: 2664–2675.
40. Jambeck JR, Geyer R, Wilcox C, Siegler TR, Perryman M, et al. 2015. Plastic waste inputs from land into the ocean. *Science* 347: 768–771.
41. Jarerat A, Pranamuda H, Tokiwa Y. 2002. Poly(L-lactide)-degrading activity in various actinomycetes. *Macromolecular Bioscience* 2: 420–428.
42. Jenkins S, Martínez I Quer A, Fonseca C, Varrone C. 2020. Microbial degradation of plastics: New plastic degraders, mixed cultures and engineering strategies. In: Jamil N, Kumar P, Batool R (Eds.), *Soil Microenvironment for Bioremediation and Polymer Production*. Scrivener Publishing LLC, 215–238.
43. Kashiwada S. 2006. Distribution of nanoparticles in the see-through Medaka (*Oryzias latipes*). *Environmental Health Perspectives* 114: 1697–1702.
44. Kawai F. 2010. Polylactic acid (PLA)-degrading microorganisms and PLA depolymerases, green polymer chemistry: Biocatalysis and biomaterials. In: Cheng HN, Gross RA (Eds.), *ACS Symposium Series: Green Polymer Chemistry: Biocatalysis and Biomaterials Volume 1043, Polylactic Acid (PLA)-degrading Microorganisms and PLA Depolymerases*. American Chemical Society, 405–414.
45. Khalid N, Aqeel M, Noman A, Hashem M, Mostafa YS. 2020. Linking effects of microplastics to ecological impacts in marine environments. *Chemosphere* 264: 128541.
46. Kirstein IV, Kirmizi S, Wichels A, Garin-Fernandez A, Erler R, et al. 2016. Dangerous hichhikers? Evidence for potentially pathogenic *Vibrio* spp. on microplastic particles. *Marine Environmental Research* 120: 1–8.
47. Koelmans AA, Bakir A, Burton GA, Janssen CR. 2016. Microplastic as a vector for chemicals in the aquatic environment: Critical review and model-supported reinterpretation of empirical studies. *Environmental Science & Technology* 50: 3315–3326.
48. Kundungal H, Gangarapu M, Sarangapani S, Patchaiyappan A, Devipriya SP. 2019. Efficient biodegradation of polyethylene (HDPE) waste by the plastic-eating lesser waxworm (*Achroia grisella*). *Environmental Science and Pollution Research* 26: 18509–18519.
49. Lares M, Ncibi MC, Sillanpää M, Sillanpää M. 2018. Occurrence, identification and removal of microplastic particles and fibers in conventional activated sludge process and advanced MBR technology. *Water Research* 133: 236–246.
50. Law KL, Morét-Ferguson SE, Goodwin DS, Zettler E, DeForce E, et al. 2014. Distribution of surface plastic debris in the Eastern Pacific Ocean from an 11-year data set. *Environmental Science and Technology* 48: 4732–4738.
51. Lebreton LC, Van der Zwet J, Damsteeg JW, Slat B, Andrady A, et al. 2017. River plastic emissions to the world's oceans. *Nature Communications* 8: 15611.
52. Lechner A, Keckeis H, Lumesberger-Loisl F, Zens B, Krusch R, et al. 2014. The Danube so colourful: A potpourri of plastic litter outnumbers fish larvae in Europe's second largest river. *Environmental Pollution* 188: 177–181.
53. Lei L, Wu S, Lu S, Liu M, Song Y, et al. 2018. Microplastic particles cause intestinal damage and other adverse effects in zebrafish. *Danio Rerio and Nematode Caenorhabditis Elegans. Science of the Total Environment* 619–620: 1–8.
54. Li Y, Lu Z, Zheng H, Wang J, Chen C. 2020. Microplastics in surface water and sediments of Chongming Island in the Yangtze Estuary, China. *Environmental Sciences Europe* 32: 15.
55. Lu Y, Zhang Y, Deng Y, Jiang W, Zhao Y, et al. 2016. Uptake and accumulation of polystyrene microplastics in zebrafish (*Danio rerio*) and toxic effects in liver. *Environmental Science and Technology* 50: 4054–4060.
56. Ma Y, Huang A, Cao S, Sun F, Wang L, et al. 2016. Effects of nanoplastics and microplastics on toxicity, bioaccumulation, and environmental fate of phenanthrene in fresh water. *Environmental Pollution* 219: 166–173.
57. Ma Y, Yao M, Li B, Ding M, He B, Chen S, et al. 2018. Enhanced poly(ethylene terephthalate) hydrolase activity by protein engineering. *Engineering* 4: 888–893.
58. Mahon AM, O'Connell B, Healy MG, O'Connor I, Officer R, et al. 2016. Microplastics in sewage sludge: Effects of treatment. *Environmental Science & Technology* 51: 810–818.
59. Mohanan N, Montazer Z, Levin DB. 2020. Microbial and enzymatic degradation of synthetic plastics. *Frontiers in Microbiology* 11: 2837.
60. Moog D, Schmitt J, Senger J, Zarzycki J, Rexer K-H, et al. 2019. Using a marine microalga as a chassis for polyethylene terephthalate (PET) degradation. *Microbial Cell Factories* 18: 171.

61. Mortensen NP, Fennell TR, Johnson LM. 2021. Unintended human ingestion of nanoplastics and small microplastics through drinking water, beverages, and food sources. *NanoImpact* 21: 100302.
62. Mueller RJ. 2006. Biological degradation of synthetic polyesters — Enzymes as potential catalysts for polyester recycling. *Process Biochemistry* 41: 2124–2128.
63. Munier B, Bendell LI. 2018. Macro and micro plastics sorb and desorb metals and act as a point source of trace metals to coastal ecosystems. *PLoS ONE* 13: e0191759.
64. Murphy L, Germaine K, Dowling DN, Kakouli-Duarte T, Cleary J. 2020. Association of potential human pathogens with microplastics in freshwater systems. In: Cocca M et al. (Eds.), *Proceedings of the 2nd International Conference on Microplastic Pollution in the Mediterranean Sea. ICMPMS 2019*. Springer Water, pp. 112–120.
65. Nomura T, Tani S, Yamamoto M, Nakagawa T, Toyoda S, Fujisawa E. et al. 2016. Cytotoxicity and colloidal behavior of polystyrene latex nanoparticles toward filamentous fungi in isotonic solutions. *Chemosphere* 149: 84–90.
66. OECD. 2013. Biotechnology for the environment in the future: Science, technology and policy. In: *OECD Science, Technology and Industry Policy Papers*, Vol. 3. OECD Publishing, Paris.
67. Oßmann BE. 2021. Microplastics in drinking water? Present state of knowledge and open questions. *Current Opinion in Food Science* 41: 44–51.
68. Paço A, Duarte K, Pinto da Costa J, Santos PSM, Pereira R, et al. 2017. Biodegradation of polyethylene microplastics by the marine fungus. *Zalerion Maritimum. Science of the Total Environment* 586: 10–15.
69. Paço A, Jacinto J, Pinto da Costa J, Santos PSM, Vitorino R, et al. 2019. Biotechnological tools for the effective management of plastics in the environment. *Critical Reviews in Environmental Science and Technology* 49: 410–441.
70. Park SY, Kim CG. 2019. Biodegradation of micro-polyethylene particles by bacterial colonization of a mixed microbial consortium isolated from a landfill site. *Chemosphere* 222: 527–533.
71. Pathak VM, Navneet. 2017. Review on the current status of polymer degradation: A microbial approach. *Bioresources and Bioprocessing* 4: 15.
72. Pignattelli S, Broccoli A, Renzi M. 2020. Physiological responses of garden cress (*L. sativum*) to different types of microplastics. *Science of the Total Environment* 727: 138609.
73. Pinnell LJ, Turner JW. 2019. Shotgun metagenomics reveals the benthic microbial community response to plastic and bioplastic in a coastal marine environment. *Frontiers in Microbiology* 10: 1252.
74. Prata JC, da costa JP, Lopes I, Duarte AC, Rocha-Santos T. 2019a. Environmental exposure to microplastics: An overview on possible human health effects. *Science of the Total Environment* 702: 134455.
75. Prata JC, Silva ALP, da costa JP, Mouneyrac C, Walker TR, et al. 2019b. Solutions and integrated strategies for the control and mitigation of plastic and microplastic pollution. *International Journal of Environmental Research and Public Health* 16: 2411.
76. Qi Y, Ossowicki A, Yang X, Huerta Lwanga E, Dini-Andreote F, et al. 2020. Effects of plastic mulch film residues on wheat rhizosphere and soil properties. *Journal of Hazardous Materials* 387: 121711.
77. Rai PK, Lee J, Brown RJC, Kim K-H. 2020. Micro- and nanoplastic pollution: Behavior, microbial ecology, and remediation technologies. *Journal of Cleaner Production* 291: 125240.
78. Ravenstijn J. 2010. Bioplastics in the consumer electronics industry. *Industrial Biotechnology* 6: 252–263.
79. Rillig MC, Lehmann A, de Souza Machado AA, Yang G. 2019. Microplastic effects on plants. *New Phytologist* 223: 1066–1070.
80. Rotjan RD, Sharp KH, Gauthier AE, Yelton R, Lopez EMB, et al. 2019. Patterns, dynamics and consequences of microplastic ingestion by the temperate coral. *Astrangia Poculata. Proceedings of the Royal Society B: Biological Sciences* 286: 20190726.
81. Schmidt C, Krauth T, Wagner S. 2017. Export of plastic debris by rivers into the sea. *Environmental Science & Technology* 51: 12246–12253.
82. Schwabl P, Köppel S, Königshofer P, Bucsics T, Trauner M, et al. 2019. Detection of various microplastics in human stool. *Annals of Internal Medicine* 171: 453.
83. Shen M, Zeng G, Zhang Y, Wen X, Song B, et al. 2019. Can biotechnology strategies effectively manage environmental (micro)plastics? *Science of the Total Environment* 697: 134200.
84. Silva MM, Maldonado GC, Castro RO, de Sá Felizardo J, Cardoso RP, et al. 2019. Dispersal of potentially pathogenic bacteria by plastic debris in Guanabara Bay, RJ, Brazil. *Marine Pollution Bulletin* 141: 561–568.
85. Sleight VA, Bakir A, Thompson RC, Henry TB. 2017. Assessment of microplastic-sorbed contaminant bioavailability through analysis of biomarker gene expression in larval zebrafish. *Marine Pollution Bulletin* 116: 291–297.
86. Song Y, Qiu R, Hu J, Li X, Zhang X, et al. 2020. Biodegradation and disintegration of expanded poly-

styrene by land snails. *Achatina Fulica. Science of the Total Environment* 746: 141289.

87. Sowmya H, Ramalingappa B, Nayanashree G, Thippeswamy B, Krishnappa M. 2015. Polyethylene degradation by fungal consortium. *International Journal of Environmental Research* 9: 823–830.

88. Sun J, Dai X, Wang Q, van Loosdrecht MCM, Ni B-J. 2019. Microplastics in wastewater treatment plants: Detection, occurrence and removal. *Water Research* 152: 21–37.

89. Talvitie J, Mikola A, Koistinen A, Setälä O. 2017a. Solutions to microplastic pollution – Removal of microplastics from wastewater effluent with advanced wastewater treatment technologies. *Water Research* 123: 401–407.

90. Talvitie J, Mikola A, Setälä O, Heinonen M, Koistinen A. 2017b. How well is microlitter purified from wastewater? A detailed study on the stepwise removal of microlitter in a tertiary level wastewater treatment plant. *Water Research* 109: 164–172.

91. Thompson RC. 2015. Microplastics in the marine environment: Sources, consequences and solutions. In: Bergmann M, Gutow L, Klages M (Eds.), *Marine Anthropogenic Litter*. Springer, pp. 185–200.

92. Townsend KR, Lu HC, Sharley D, Pettigrove V. 2019. Associations between microplastic pollution and land use in urban wetland sediments. *Environmental Science and Pollution Research* 26: 22551–22561.

93. Von Moos N, Burkhardt-Holm P, Kohler A. 2012. Uptake and effects of microplastics on cells and tissue of the blue mussel *Mytilus edulis L.* after an experimental exposure. *Environmental Science and Technology* 46: 11327–11335.

94. Wang F, Yang W, Cheng P, Zhang S, Zhang S, et al. 2019. Adsorption characteristics of cadmium onto microplastics from aqueous solutions. *Chemosphere* 235: 1073–1080.

95. Wang F, Zhang X, Zhan S, Zhang S, Sun Y. 2020. Interactions of microplastics and cadmium on plant growth and arbuscular mycorrhizal fungal communities in an agricultural soil. *Chemosphere* 254: 126791.

96. Wang J, Liu X, Li Y, Powell T, Wang X, et al. 2019. Microplastics as contaminants in the soil environment: A mini-review. *Science of the Total Environment* 691: 848–857.

97. Weber F, Kerpen J, Wolff S, Langer R, Eschweiler V. 2020. Investigation of microplastics contamination in drinking water of a German city. *Science of the Total Environment* 755: 143421.

98. Wei R, Oeser T, Schmidt J, Meier R, Barth M, et al. 2016. Engineered bacterial polyester hydrolases efficiently degrade polyethylene terephthalate due to relieved product inhibition. *Biotechnology and Bioengineering* 113: 1658–1665.

99. Yang H, Chen G, Wang J. 2021. Microplastics in the marine environment: Sources, fates, impacts and microbial degradation. *Toxics* 9: 41.

100. Yang J, Yang Y, Wu WM, Zhao J, Jiang L. 2014. Evidence of polyethylene biodegradation by bacterial strains from the guts of plastic-eating waxworms. *Environmental Science and Technology* 48: 13776–13784.

101. Yang Y, Yang J, Jiang L. 2016. Comment on "A bacterium that degrades and assimilates poly(ethylene terephthalate)." *Science* 353: 759.

102. Yeo JCC, Muiruri JK, Thitsartarn W, Li Z, He C. 2018. Recent advances in the development of biodegradable PHB-based toughening materials: Approaches, advantages and applications. *Materials Science and Engineering: C* 92: 1092–1116.

103. Yonkos LT, Friedel EA, Perez-Reyes AC, Ghosal S, Arthur CD. 2014. Microplastics in four estuarine rivers in the Chesapeake Bay, U.S.A. *Environmental Science & Technology* 48: 14195–14202.

104. Yoshida S, Hiraga K, Takehana T, Taniguchi I, Yamaji H, et al. 2016a. A bacterium that degrades and assimilates poly(ethylene terephthalate). *Science* 351: 1196–1199.

105. Yoshida S, Hiraga K, Takehana T, Taniguchi I, Yamaji H, et al. 2016b. Response to Comment on "A bacterium that degrades and assimilates poly(ethylene terephthalate)." *Science* 353: 759.

106. Yuan J, Ma J, Sun Y, Zhou T, Zhao Y, et al. 2020. Microbial degradation and other environmental aspects of microplastics/plastics. *Science of the Total Environment* 715: 136968.

107. Zalasiewicz J, Waters CN, Ivar do Sul JA, Corcoran PL, Barnosky AD, et al. 2016. The geological cycle of plastics and their use as a stratigraphic indicator of the Anthropocene. *Anthropocene* 13: 4–17.

108. Zhou Y, Wang J, Zou M, Jia Z, Zhou S, et al. 2020. Microplastics in soils: A review of methods, occurrence, fate, transport, ecological and environmental risks. *Science of the Total Environment* 748: 141368.

109. Zhu D, Chen QL, An XL, Yang XR, Christie P, et al. 2018. Exposure of soil collembolans to microplastics perturbs their gut microbiota and alters their isotopic composition. *Soil Biology and Biochemistry* 116: 302–310.

CHAPTER 33

Bio-electrochemical Systems for Monitoring and Enhancement of Groundwater Bioremediation

Rory Doherty, Altaf AlBaho, and Lily Roney

School of Natural & Built Environment Queen's University Belfast, Belfast BT7 1NN

33.1 INTRODUCTION

It has been suggested that biotechnologists are on the cusp of a renaissance in environmental engineering of biological systems as their understanding of microbial ecology and engineering principles is rapidly developing (Curtis et al., 2003). To do this, biotechnologists must embrace and adopt seemingly disparate approaches often from differing disciplines. Often there is no need to develop new systems from first principles rather the biotechnologist can learn and incorporate good engineering design and tools into better and more efficient biotechnological systems. In 2010, the global turnover for the monitoring and remediation of contaminated land (both soil and groundwater) has been estimated around €34 billion. In addition, the current EU market size has been established in a range of between €2 billion and 10 billion including the assessment and equipment costs which is considered as 6–31% share of global market. Geographically, the extent of contaminated land in the UK was represented around 63750 ha. Although Germany, the Netherlands, and the UK are the three major leaders in EU states markets, the US is categorized as a global market leader in the remediation industry of contaminated land (EPEC, 2011). The concept of bioremediation has a long history dating back to the 1940s with microbiologists like Zobell (1946) investigating the ability of microorganisms to metabolize petroleum hydrocarbons. Bioremediation is a form of environmental biotechnology developed to prevent, recover, or manage pollutants utilizing microorganisms' metabolic capabilities to transform harmful substances (see also other chapters of section 4). The Exxon Valdez oil spill of 1989 in Prince William Sound, Alaska, is a widely studied environmental disaster (EPA, 1989). The oil tanker Exxon Valdez after striking a reef released over 11 million gallons of crude oil onto the coastline. This spill was known as the "most expensive oil spill in history" exceeding $7 billion in clean-up as well as fines and penalties (Lyon and Weiss, 2010). The success of clean-up efforts here were fundamental in boosting interest in the potential applicability of bioremediation technologies. As bioremediation has become an accepted technology for remediation engineers there is a need to improve and optimize systems. The ability to monitor biological processes in real time offers the potential for cost savings and improved project management. The ability to speed up and optimize remediation mechanisms thereby increasing profit margins for remediation contractors is a key incentive to adopt new and improving technologies such as bio-electrochemical systems.

Good Microbes in Medicine, Food Production, Biotechnology, Bioremediation, and Agriculture, First Edition. Edited by Frans J. de Bruijn, Hauke Smidt, Luca S. Cocolin, Michael Sauer, David Dowling, and Linda Thomashow.

33.2 LAND/GROUNDWATER CONTAMINATION AND REMEDIATION DESIGN

Land contamination is used to describe a broad range of soil and groundwater conditions where potentially toxic substances can cause harm to human health or the environment. Elevated levels of potentially toxic substances can occur at sites for many reasons, for example naturally from the soil or bedrock (geogenic sources), or man-made (anthropogenic sources) (McIlwaine et al., 2017). Former industrial and urban areas may have left a legacy of contamination due to specific spillages and releases as point sources, or from widespread airborne pollution as diffuse sources. A risk-based approach is the preferred method to manage potentially toxic substances. This involves identifying pollutant linkages that consist of three components, which together form a hazard.

1. *A source* – a potentially toxic substance that could cause harm.
2. *A receptor* – the body that could be adversely affected by the source (e.g., human health, ecology, water resources, or property).
3. *A pathway* – the route whereby a source becomes in contact with the receptor.

For example, spilled petroleum hydrocarbons (*source*) can migrate to the water table where they dissolve in groundwater and move in a plume (*pathway*) toward a drinking water abstraction well (*receptor*).

The risk is the probability of the hazard occurring and the size of the consequences. There can be multiple pollutant linkages when dealing with land contamination. One source could potentially move in a variety of pathways to multiple receptors. Often there can be many sources. The development and prioritization of these pollutant linkages builds into a conceptual model of our understanding of a land contamination site and is the basis of the risk assessment process. Where the risk is quantified and exceeds a threshold there is often the need for remediation or clean up. If there is no pollutant linkage, then no risk exists. Each part can exist independently and not be a risk. This concept forms the basis of all land and groundwater remediation design, which uses a series of tools and technologies to break the pollutant linkages and successfully manage the risk. This is achieved by:

(a) Source management: Removing, reducing, or changing the properties of the source (e.g., bioremediation).
(b) Pathway management: Disrupting or preventing the pathway (e.g., using a physical barrier).
(c) Receptor management: Changing receptor behavior (e.g., restricting groundwater abstraction).

There is no treatment panacea for land and groundwater contamination issues. The site-specific conditions will determine the approach and methodologies required to manage various and compound issues found. In general, there are ways of grouping remediation technologies as outlined below.

33.2.1 In Situ Versus Ex Situ Methods

In situ remediation methods happen within the subsurface. This can be through the addition of chemicals, nutrients, biological agents, stabilization agents, heat, electrical current, etc. via boreholes, reactive chambers, augers, trenching tools, and a variety of civil engineering plants. A drawback of this approach is that conditions in the subsurface can be very heterogeneous and there is no guarantee that all subsurface contamination will be managed successfully. Ex situ methods remove the contamination to surface via excavation or pumping methods where it can manage through a "treatment train" of applicable technologies. While these methods give a greater surety that the contamination is being successfully managed the material management costs can quickly become prohibitive.

33.2.2 Active Versus Passive Methods

Remediation techniques are often termed as passive systems when no further energy is needed once installed. These systems rely on harnessing natural physical, geochemical, and biological processes such as dilution, sorption, and biodegradation. This requires a detailed understanding and design to successfully manage contaminated sites. Examples of passive methods include Permeable Reactive Barriers (Davis et al., 2010) and Monitored Natural Attenuation (Prommer et al., 2006) and often require detailed monitoring over the life of the project. An active system typically requires equipment and consumables with a continuous energy input. Often the plant required is at the surface with contaminated materials moved between the various configurations depending on the setup. There is a greater control over the total amount and final quality of contaminated material treated but this often comes at a significant energy and consumable cost (Bayer and Finkel, 2006).

33.3 SUSTAINABLE REMEDIATION

In order to balance social, environmental, and economic approaches the concept of sustainable remediation has been developed (Nathanail et al., 2017) to promote best practice when managing land and groundwater issues. Here criteria are aligned to the UN sustainable development goals including emissions to air, soil, and water; social factors such as health, safety, and community involvement can be balanced alongside economic drivers such as direct and indirect costs (CL:AIRE, 2020). This allows the design of remediation technologies to be evaluated and improved upon and for new technologies to be designed and optimized using this approach.

33.4 VERIFICATION OF REMEDIATION

One aspect of the design of new remediation technologies is the methods by which the verification of remediation will take place. Often it is incorrectly assumed that this is only some form of chemical analysis that relates to the final state of the project. The lines of evidence for verification should also be based around how the pollutant linkage is broken and should be monitored from the start of the remediation process (Bone, 2010). If biodegradation due to monitored natural attenuation is proposed as a groundwater management method (pathway management) then evidence of enhanced microbial activity at the plume fringe should be provided from the start of the project.

One promising innovative technology for the management of contaminated groundwater is bio-electrochemical systems (BESs).

33.5 BIO-ELECTROCHEMICAL SYSTEMS

Bio-electrochemical systems are engineered devices that transfer electrons from an anode to a cathode using microbial activity. These are not new systems with the earliest reported BES using from *E. coli* cultures in 1911 (Potter, 1911). The microbes can act as a mechanism that sources electrons through degradation or oxidation of compounds, or they can be the transfer mechanism of the electrons to electrodes. The bacteria that can transfer electrons are known as electro-active, and the mode of electron transport is called extracellular electron transport (EET) (Nakamura et al., 2013). There is a variety of methods by which microbes can do this, from indirect transfer through mediators (Velasquez-Orta et al., 2010), cytochrome transfer directly through the cell (Lovley, 2008), to pili that are hair-like appendages on the surface of the cell (Reguera, 2018), to "cables" of filamentous bacteria (Müller et al., 2016). Generally, the process of EET involves the transfer of metabolically generated electrons by microbes capable of externally moving electrons to an electrode or across a redox boundary. The general laboratory set of up of a BES consists of two chambers, an anode and a cathode, separated by a membrane known as an ion exchange membrane (Figure 33.1). Within the anode electrons are generated by the anaerobic biodegradation or oxidation of compounds. In our case these are often organic compounds. These electrons are passed through microbial communities onto electrodes that are usually made from graphite. Graphite is a good electron conductor and supplies an environment for dense microbial growth. The electrons are transferred from the anode electrode via an electronic conductor such as a wire to a cathode electrode where they are consumed or reduced. The consumption of electrons can be a biological process or a simple chemical one.

BESs have appeared in various forms depending upon the desired result of the technology. These systems are versatile in their construction and application including microbial fuel cells (MFC; see chapter 31), microbial electrolysis cells (MEC), microbial desalination (MDC), and microbial electrosynthesis systems (MES) (Bajracharya et al., 2016; Cecconet et al., 2020). In addition, integration of BESs with other technologies to produce hybrid systems often occurs to increase efficiency.

33.5.1 BES Hydrocarbon Remediation

Bioremediation of soils and groundwater contaminated by hydrocarbons is a well-established technique (Atlas, 1991). Here, microbes utilize hydrocarbons as an energy source either using electron acceptors such as oxygen (oxidation) or electron donors such as sulfate or nitrate (reduction) often breaking them down to little more than water and carbon dioxide. Aerobic bioremediation is often preferred by remediation contractors as a faster method of contaminant breakdown (Kao et al., 2008), but this is an inefficient method due to energy costs

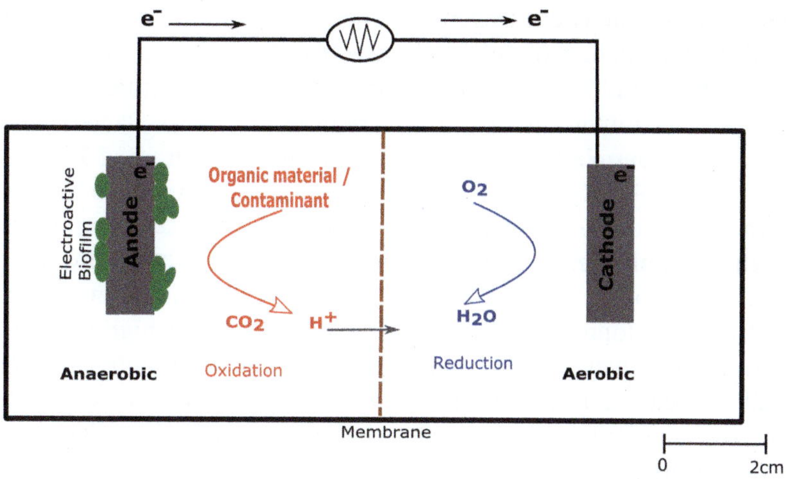

FIGURE 33.1 Simple bio-electrochemical system laboratory setup where organic contaminants are oxidized by microbes in an anaerobic environment releasing electrons which are passed to the anode by biomass on the electrode surface. The anode chamber is separated from the cathode by a membrane that allows protons to pass through. The electrons travel from the anode to cathode by an external wire.

associated with aeration and the relatively low solubility of oxygen in water (Huang et al., 2011). A study by Morris et al. (2009) utilized a BES to enhance the anaerobic biodegradation of diesel range organics (DRO) over a period of three weeks. The BES was designed as a double-cell anode and cathode. The working electrode material was a stainless-steel scrubber, placed inside an anode chamber connected to the cathode via a stainless-steel wire. The aerated cathode material had conductive silver epoxy also covered in silicone. The anode chamber held a mixture of phosphate buffer and diesel contaminated groundwater sourced from a refinery in Wyoming, USA, as inoculum. The experiment was successful with an 80% decrease of DRO concentrations in three separate BES. The microbial characterization found a > 98% similarity to microbes with denitrifying capabilities the majority classified as Gamma-Proteobacteria, including, *Pseudomonas sp.* (99%), *Citrobacter sp.* (98%), and *Stenotrophomonas sp.* (99%). Links were also found between microbes and the ability to utilize multiple electron acceptors such as, chlorinated inorganics, sulphate, and arsenate. The term "total petroleum hydrocarbons" or "TPHs" encompasses a broad range of several hundred hydrocarbon compounds all originating from crude oil, which in turn are utilized to produce petroleum products (Kuppusamy et al., 2020). It is not practical to measure each chemical included in crude oil and other petroleum products, due to the vast number of more than 17000 hydrocarbon compounds found (Daghio et al., 2017). A study by Lu et al. (2014) utilized BESs of a tubular configuration which could be scaled up in future and positioned down boreholes and groundwater monitoring wells. The petroleum hydrocarbon remediation efficiency of two different anode materials, biochar and carbon cloth, was compared. After 64 days at a 1 cm distance, the TPH removal was greater from the active BES with the Biochar anode reaching 78.7%, compared to the 73.1% at the carbon cloth anode. TPH degradation after 64 days at 5 cm distance for the active biochar anode and carbon cloth anode was 56.8% and 51.5%, respectively. TPH depletion rates were reduced as the distance from the anode was increased, the electrodes influence decreases with distance; this is attributed to reduced bacterial activity as well as contaminant mass transfer. Despite this, degradation rates within an active BES were still significantly higher 31–37% in comparison to controls used. The bacterial community *Proteobacteria (including Alpha-, Beta-, Gamma-*, and *Delta-)* dominated all sampling accounting for 28.7–94.0% of readings during the phylogenetic analysis; all expressing electrochemical activity. The abundance of *Proteobacteria* reacted positively to hydrocarbon contamination increasing from 57% to 69% post-remediation. A unique consortium of bacterial communities is actively working aiding the biodegradation of petroleum hydrocarbons. At the genus level *Betaproteobacteria* were detected on both the anode as well as in the soil, including genera such as *Achromobacter*, *Bordetella*, *Comamonas* and *Pusillimonas*. *Enterobacter*, *Pseudomonas*, *Pseudoxanthomonas*, and *Stenotrophomonas* were identified genera of *Gamma Proteobacteria*. Interestingly, within the genera of *Enterobacter* and *Pseudomonas* both hydrocarbon-degrading bacteria as well as electrogens were found. The successful biodegradation of volatile organic compounds, such as those of BTEX (Benzene, Toluene, Ethylbenzene, Xylene) is important due to their significance making up 15% of petroleum contaminants (Nelson-Smith, 1972). Of the BTEX compounds, Toluene is regarded as the most degradable and efficient removal was observed utilizing a graphite electrode BES within marine sediments (Daghio et al., 2016a). Sulphate-reducing was observed alongside toluene degradation, sulfate reducing bacteria playing a key role in this contaminant's breakdown belonging to the *Desulfobulbaceae* and *Desulfobacteraceae* families (Daghio et al., 2016b). Both families have organisms capable of electroactivity, for example *Desulfobulbaceae* are known as "cable bacteria" or long filamentous bacteria that can transfer electrons across redox zones over centimeter distances (Cupples, 2011). Benzene is regarded as the most volatile and hazardous of the BTEX compounds. Contaminated groundwater treated using a BES experienced 100% Benzene removal, 80% at the anode compartment and the remaining 20% removed in the effluent passing through the cathode (Wei et al., 2015). BTEX compounds typically contaminate groundwater sources due to their lack of soil adsorption and motility (Sarkar et al., 2013). Most aromatic hydrocarbon compounds have similar structures and so a microbial community that successfully degrades for example Benzene, may also have an affinity toward similar compounds. Many iron and sulfate-reducing microbial communities have been shown to degrade BTEX compounds; these studies carried out using soils having indigenous microbes capable of degrading BTEX. A direct link can be found between the presence of petroleum pollutants at high concentrations and bacteria belonging to a particular phylum. Reported hydrocarbon-degrading bacteria include: *Actinobacteria, Proteobacteria, and Firmicutes* (Thomas et al., 2020). Specific genera from these phyla undergo enrichment in the presence of hydrocarbons such as, *Gordonia* and *Rhodococcus* subgeneras of *Actinobacteria, Sphingomonas, Acinetobacter, and Pseudomonas* subgeneras of *Proteobacteria, and Bacillus and Exiguobacterium* subgeneras of *Firmicutes.* Petroleum-contaminated aquifers tend to have anaerobes such as, Beta-proteobacteria, Gamma-proteobacteria, as well as Cytophaga–Flexibacter–Bacteroidetes group (Feris et al., 2004). Geobacter is an example of a Delta-proteobacteria which is found abundantly in sedimentary environments and can grow under anaerobic conditions solely utilizing aromatic compounds as its source of energy (Shi et al., 1999). The species Geobacter is also electroactive containing EET abilities, shuttling electrons from the substrate to the electrode within the BES (Yang et al., 2020). Bacteria like the latter capable of anaerobically degrading BTEX compounds are attractive as these compounds are typically found in anoxic environments. According to Sarkar et al. (2013) a preference was shown by all pseudomonads toward BTEX, this inclination toward these volatile compounds by specific genera highlights their potential introductory role in simple hydrocarbon breakdown. Weelink et al. (2010) supports this concept of pseudomonads and their role in BTEX degradation as well as a few members of Beta, Delta, or Epsilon Proteobacterial genera. Complex groundwater contamination is often found at Gasworks sites where mixtures of light aromatic and aliphatic hydrocarbons are often found alongside heavy metals, ammonia, and cyanide compounds. These hydrocarbons are often difficult to manage using standard biodegradation approaches. (Kirmizakis et al., 2019) found that a BES made from graphite combined with granular activated carbon was able to deal with 99% of a highly contaminated groundwater compared with only a 33% reduction using a graphite BES and glass beads. The GAC BES was also able to outperform the GAC in removing lighter aromatic fractions that made up 81% of the groundwater contamination. These light aromatic compounds, such as naphthalenes and methyl-naphthalenes, are "contaminants of concern" that the regulators focus on during validation of remediation due to their toxicity and mobility. They also found that Beta-proteobacteria in particular *Psuedomondaceae* were key to all the samples and reflected the dominant degrading communities within the gasworks groundwater. *Rhodocyclaceae* and *Comamonadaceae* were also found to increase with the BES chambers suggesting a possible electrogenic capability. In general, the BES technology while not perfect as a stand-alone technology does offer the possibility of greater technical efficiency when used in conjunction with other remediation approaches.

33.5.1.1 BES Design and Remediation Technology

BES are a narrow sub-section of the bioremediation "toolbox" and in themselves may be problematic to scale up and use as a novel remediation technology. In the lab the components of BES are expensive (membranes) and delicate (carbon brushes, felt, carbon paper, electrodes).

For the BES to be a cost-effective and reliable solution, BES developers need to embrace the principles of sustainable remediation as the technology is scaled up. Rather than "reinvent the wheel" to solve these problems an alternative approach should be to develop conceptual biogeochemical models of accepted and reliable remediation technologies and modify them to become BES friendly. This will allow the methods to progress more rapidly through technology readiness levels (TRLs) from lab scale prototypes (TRL 4) to field demonstration (TRL 7) to full commercial applications (TRL 9) (Holtmann and Harnisch, 2019). By "piggy backing" on accepted remediation technologies the BES can improve upon and quickly find a technological niche that becomes widely accepted by industry. Stand-alone technologies are rarely used in remediation design and are quickly combined into an efficient "treatment train" that optimizes performance and increases the applicability of the methods used.

33.5.1.2 Scaling up BES Construction and Design

Electrode design parameters for BES should provide (1) suitable substrate for microbial growth, (2) be able to conduct/transfer electrons, (3) should have as large a surface area as possible for efficient degradation, and (4) they should be self-supporting (will not be damaged under their own weight or the weight of functioning technology/contaminated material).

In addition, if we are to apply sustainable remediation and eco design principles (Knight and Jenkins, 2009) they should also be (1) economically viable, (2) recycled or at least not be made using hazardous chemicals or energy intensive processes, (3) not cause harm to human health or the environment, and (4) they should be designed for ease of recovery, reuse, remanufacture/recycling. If we consider the above criteria for our BES materials selection, there are materials already in use by the remediation community that could be adapted for BES as well as materials that can be reused, recovered, or recycled for BES applications. Exchange membranes are used to separate the anode and cathode chambers, in the laboratory these are often used from bespoke materials such as Nafion and Ultrex which are difficult to manufacture, scale up, and are expensive. A good exchange membrane requires conductivity to allow for the efficient transfer of protons to the cathode, while preventing the transfer of the remaining material such as the substrate. Nafion is a popular choice; however, the materials excessive costs at \$200 per m^2 Kim et al. (2014) to \$1400 by Oliot et al. (2016); hinders attempts to scale up and commercialize BESs. Nafion membranes have also been demonstrated to allow oxygen into areas of the system which require anaerobic conditions (Min et al., 2005). Alternative and cheaper membranes include j-cloths, biodegradable bags, ceramics, and natural rubber. The use of waste materials like latex rubber gloves (Winfield et al., 2013) has been tested against conventional cation exchange membranes to test applicability for larger scale systems. The latex gloves feature a "lattice-type structure" allowing for proton transfer without a delay period (Winfield et al., 2014). This material exhibits a porous network enabling the protons to flow through with ease. The power output results revealed that the conventional BES produced higher power overall; however, the BES utilizing natural rubber gloves produced results with greater stability and reliability over a longer time. It was also observed that deterioration of the latex gloves within the BES aided degradation rates, due to the generation of pores allowing proton exchange improving system performance. An issue with utilizing certain membranes in which deterioration excessively occurs is the possibility of short-circuiting the BES. However, the system in the study by Winfield et al. (2014) remained unbroken and workable up to day 493, when the experiment concluded, and the BES was dismantled. This study shows the applicability of biodegradable waste materials as ion exchange membranes, both environmentally and economically superior to conventional membranes. A further step to increase the economic sustainability of scaling up these systems when considering the principles of sustainable remediation, is the option of no membrane (Cecconet et al., 2020). It has been noted that the membrane could be a "log jam" when scaling up BESs (Logan et al., 2006). Alternative approaches have considered focusing on mechanisms at the cathode to work around this (Rabaey and Keller, 2008), it has been noted that at large scale water table fluctuation (movement of a cathode) can turn electron transport across redox boundaries on and off (Heenan et al., 2017). A membrane-less BES also called a "bioelectric well," was utilized in a study by Palma et al. (2019), with a granular graphite filled glass cylinder acting as the anode separated from the stainless steel mesh (the cathode) via a plastic mesh. The bioelectric well was successful in the overall degradation of BTEX-contaminated groundwater. The dominant bacterial community found on the graphite bioanode was Geobacter (>34%), indicating the microbe's capability to degrade BTEX compounds, as well as electroactive abilities. Within a BES the electrode utilized has a significant impact upon the efficiency and performance of the system, particularly encouraging biofilm growth as well as electrochemical activity. Current options around BES electrodes in larger, field scale applications are carbon based granular activated carbon (GAC) or graphite granules (GGs) (Huggins et al. 2016). GGs tend to be the cheaper option of the two but are still relatively expensive for large scale systems, both averaging between 500 to 2500 US\$ ton^{-1}. These materials are based on feedstock that require substantial amounts of energy during extraction as well as production, resulting in the discharge of harmful gases into the atmosphere. For example, during its life cycle GAC which is typically produced using coal or petroleum, both non-renewable sources, requires a secondary form of activation (thermal or chemical). Granular activated carbon (GAC) has been used as a sorption media by the remediation industry for decades. It is used in a variety of active and passive systems such as pump and treat, permeable reactive barriers PRB, and wastewater treatment systems. Recently it has been noted that activated carbon can act as an important electrical conduit in direct interspecies electron transport (Lovley, 2017) as well as promoting degradation in oil contaminated sediments in anaerobic conditions (Bonaglia et al., 2020). It has been known for some time that biodegradation and sorption can occur together in activated carbon systems (Xiaojian et al., 1991). GAC has significant quantities of graphite due to its high temperature formation which gives it good electrical properties. GAC however is often manufactured from fossil fuels; GAC made from coconut husks or biochar electrodes (Huggins et al., 2014) would be a better choice environmentally and economically when considering sustainable remediation principles.

An alternative choice for electrode material is the use of biochar; this option is viewed as more environmentally friendly, significantly reducing the associated carbon footprint as well as costs. Biochar is a by-product from the thermal decomposition of bio-waste, which is often locally sourced and of little commercial value significantly reducing costs of feedstock. Biochar's can be produced from various materials, feedstock sources include, residues produced from forestry and agriculture (e.g., sawdust and animal manures) as well as construction and municipal waste (Lehmann and Joseph, 2012). The price of biochar electrodes averages between 51 and 381 US\$ ton^{-1}. Here we propose a series of industry standard remediation approaches where a BES could be coupled with an accepted remediation technology to enhance biodegradation and act as a real time biosensor to monitor microbial activity.

33.5.1.3 Pump and Treat BES (Ex Situ/Active Remediation Technology)

GAC is often used in pump and treat systems to sorb hydrocarbons and other recalcitrant organic compounds. The systems are often simple where contaminated groundwater is pumped to surface and through a series of drums holding GAC that sorbs the contaminants. In order to prevent breakthrough of contaminants, once the GAC has reached its maximum sorptive capacity, the drums are connected in series with the output of the first drum monitored. This approach could be easily modified so that the drums in series become the chambers of the BES (Figure 33.2a). A non-conductive drum filled with GAC/Biochar would operate primarily as a sorption agent but with the addition of cheap ion exchange membrane material at the top would allow the drum operate as an air cathode BES.

Other options would be to connect the drums in series with a wire and allow them to act as anode and cathode. This approach uses the sorptive capacity of GAC to capture the contamination and then to promote anaerobic biodegradation of the contaminants as a BES. This provides the added benefit of the GAC being able to sorb contaminants at higher levels than a standard contaminant degrading BES would operate under (Kirmizakis et al., 2019). The conductive GAC allows such a BES to work as a real-time biosensor that can act as a proxy when contaminant loading affects microbial activity. Such a system offers the remediation contractor cost savings in operation by having an additional mechanism to promote degradation and extend the operating life of the GAC before replacement. Such savings are attractive to practitioners who need to maintain competitive commercial margins.

33.5.1.4 Biopile BES (Ex Situ/Active/Semi-Passive Remediation Technology)

Biopiles are contaminated soils that are excavated, screened with addition of nutrient and bulking materials then placed into mounds, often with leachate collection systems (see chapter 27). They can be covered or enclosed within large sheds. They are static; the soil is not turned but can have forced aeration systems if required (Jørgensen et al., 2000). Biopiles can take weeks to months

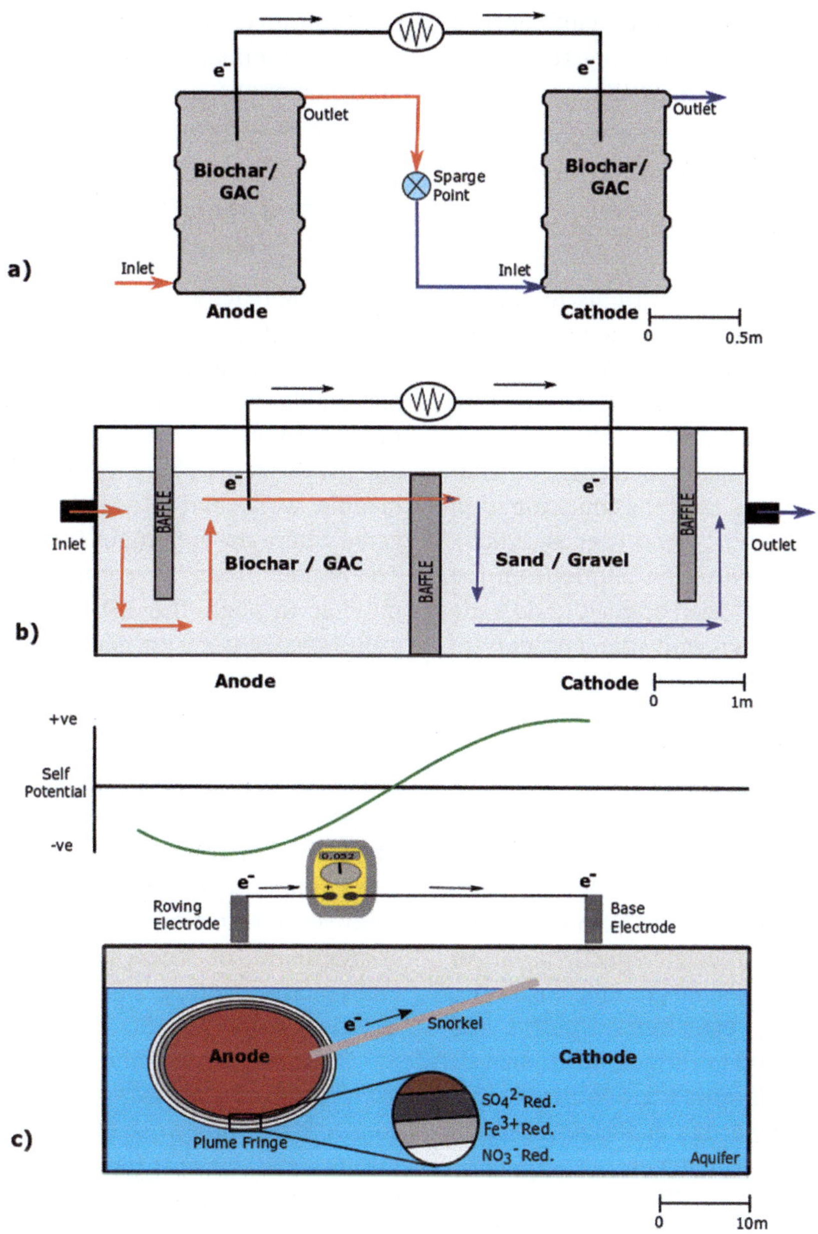

FIGURE 33.2 Standard remediation designs that can be converted into a BES (a) pump and treat using biochar, (b) a permeable reactive barrier (PRB) using biochar for the anode and sparged sand and gravel for the cathode, and (c) enhanced monitored natural attenuation using a snorkel electrode with self potential signal measured at surface.

for the remediation to be completed, due to their static nature a drawback can be dead zones where biodegradation is not successful. Biochar is used as a soil amendment and has been applied to ex situ bioremediation (Qin et al., 2013). Normally the effective zone of influence of a BES in soils is usually limited to a few centimetres around an electrode. If Biochar is used as a soil amendment within a Biopile for example added at the screening stage where it would be homogenized in the soil, then there is the opportunity for longer distance bioelectric remediation. Effective distances of contaminant removal up to 16 cm from the electrode were in a soil BES that used Biochar as a soil electron conductor (Cai et al., 2020). The ability to effectively monitor the zone of influence from the electrode would allow greater control over an integrated Biopile-BES and allow them to be effectively scaled up. Such methods to prove the zone of influence in soils have already been demonstrated using geoelectric monitoring approaches (Fachin et al., 2012).

33.5.1.5 Plume Biostimulation (In Situ Active/Semi-Passive Remediation Technology)

Biostimulation is the addition of nutrients/microbes/amendments directly into a contaminated plume through injection into boreholes. A "closed loop" system can be generated where the amendments are injected into the center of a plume and recovered down gradient from a pumping well where required further amendments are added and reinjected until the contamination has been reduced sufficiently (Iwamoto et al., 2000). This promotes the degradation of the plume from the inside out. Biostimulation has been applied to degrade organic contaminants (BTEX) with the addition of ammonium acetate to promote fermentative methanogenic conditions (Ramos et al., 2013), or to change the properties of contaminants through biomediated precipitation of metals such as Uranium (Yabusaki et al., 2007) by the addition of acetate. The Rifle site in Colorado, USA, was a uranium ore processing facility where tailings had contaminated a shallow groundwater aquifer on the flood plain of the Colorado River. To monitor the effectiveness of acetate injections in near real time, a series of down borehole graphite electrodes (anodes) in and around the contaminant plume and injection zone were connected to surface electrodes (air cathodes) over a vertical of distance 6 m (Williams et al., 2010). BES sensors outside the area of acetate injection showed a negligible response in current density while those within injection areas were able to track delivery of acetate that correlated with uranium removal from groundwater. The use of down borehole electrodes that produce electrical signals that can act as a proxy for microbial activity is an alternative application for a BES. Here the BES is not a tool that manages the source or pathway of contamination but is a biosensor that provides verification that the proposed sub-surface microbial activity during remediation is taking place. Traditional sampling approaches for sub-surface microbial activity and chemical transformation are expensive requiring significant field and analytical experience. The Biosensor BES has the potential to be a key tool in the scaling up, monitoring, and verification of BES and other biodegradation remediation technologies. Plume biostimulation can be combined with monitored natural attenuation (below) to become enhanced monitored natural attenuation where the process of plume management is further enhanced to more efficiently manage remediation timeframes (Lv et al., 2019).

33.5.1.6 PRB BES (In Situ/Semi-Passive Remediation Technology)

Permeable reactive barriers are in ground systems that effectively capture contaminant plumes. They have a variety of designs and formulations from subsurface permeable "walls" or zones of higher hydraulic conductivity than that the natural subsurface, to funnel and gate systems where the plume is captured by impermeable walls and directed toward more permeable gates where the treatment is focused (Gavaskar, 1999). The systems can be passive where the plumes are moving through the PRB with the natural flow of groundwater with no need for more energy input or addition of materials. Some systems are semi-passive where the groundwater plume passes through a series of sequential cells or chambers where differing processes such as sparging, biodegradation, and sorption by GAC can take place (Gibert et al., 2007). In these cases, the reactive cells are designed to manage complex contaminant plumes, such as those from gasworks, where each reactive cell is designed to manage different contaminants of concern. Sequential barriers with several reactive chambers or cells are good candidates for BES design and retrofit.

PRB gates or chambers can vary in size and shape from large chambers several meters deep that are connected by pipework (Ferguson et al., 2008) to flow through systems that have a series of interceptors and baffles to extend the flow regime and retention time (Davis et al., 2010). These chambers and partitions within the PRB are designed to work within different biogeochemical environments which would easily allow retrofits such GAC/Biochar anodes and cathodes to work within. In Figure 33.2b we show how a PRB reactive cell could be converted into a BES. The groundwater inflows into a baffled chamber filled with GAC/ Biochar that as before acts as an electrode and a sorption media. The central baffle divides the chamber into anaerobic anode where there is the standard anaerobic oxidation of organic contaminants and production of electrons. The anode of the PRB chamber is connected by an external wire to the cathode section. The groundwater in the anode section overflows the central baffle becoming aerated as it enters the cathode section. Depending on the design, flow through, and retention time of the chamber this section could be also be sparged through the base to ensure it is properly aerated.

The applicability of BES to PRBs has also recently been proposed (Cecconet et al., 2020) where a series of vertical tubular electrodes and pumps circulate groundwater within boreholes vertically as a bioelectric well (Palma et al., 2018). This configuration requires a good understanding of the zone of influence of the pumped tube BES beyond the borehole construction to ensure that borehole spacing leaves no "dead zones" that would allow contaminant breakthrough. It uses a series of groundwater circulation wells to act as the "permeable barrier" which requires detailed understanding and monitoring of localized flow regimes (Johnson and Simon, 2007).

33.5.1.7 Monitored Natural Attenuation BES – (In Situ/Passive Remediation Technology)

Monitored natural attenuation or intrinsic remediation usually refers to a combination of processes that are biological, physical, and chemical that act upon subsurface contamination so that its mass and spatial distribution is held in check, a "steady state" plume (Maier and Grathwohl, 2006) or preferably, is reducing. These processes can include dilution and dispersion (transport) in groundwater, sorption onto the aquifer matrix, volatilization into soil gas, and biodegradation of the plume. These processes must be demonstrated to be occurring without human intervention but must also be closely monitored to supply evidence that the mechanisms of plume control and reduction are constantly occurring and can be predicted/modeled over time. Our understanding of the concepts and mechanisms around plume spread and reduction is changing and becoming more focused (Prommer et al., 2006). There is a move away from the concept of longitudinal redox zonation or thermodynamic ladder (Bethke et al., 2011) within a groundwater plume where there are wide zones of differing electron acceptors such as sulphate reduction, iron reduction, and nitrate reduction. This is replaced with the idea of a plume fringe where microbial activity and distribution of electron acceptors is focused into a small area at the outer shell of the plume (Meckenstock et al., 2015). This would allow steady state plumes to have sharp chemical gradients close to uncontaminated groundwater with a wide degree of diversity, intense microbial activity, and biodegradation at the plume fringe. Such steady state contaminant plumes anaerobically oxidizing organic matter at the plume fringe act like the biological anode in a BES with the surrounding uncontaminated groundwater acting as a cathode (Doherty et al., 2010; Revil et al., 2010). This still requires an electronic conductor that would act as the external wire in a lab BES. Various mechanisms for the electronic conductor have been proposed at various sites from conductive metallic wastes, to bioprecipitation of magnetite along smear zones at the plume fringe (Heenan et al., 2017). One way of further stimulating the process of degradation is the efficient transfer of electrons from plume anode to clean aquifer cathode. This could be achieved by the emplacement of simple, passive "snorkel" electrodes that directly bridge the gap between anode and cathode (Figure 33.2c). Snorkel electrodes are a simple short-circuited BES design that does not supply current and are made from graphite rods or similar materials (Erable et al., 2011). The "oil spill snorkel" has further provided proof of concept that snorkels are a simple and effective tool that can work to enhance monitored natural attenuation with a variety of natural consortia of electroactive microbes (Matturro et al., 2017; Viggi et al., 2015).

33.5.1.8 Monitoring the Microbe Geo-electric?

A need for real-time monitoring of groundwater systems so that researchers and remediation practitioners can evaluate and react to remediation strategies is needed, but neglected, by the research community (Zhang et al., 2017). As BES systems become larger there need to be technologies that can be used to effectively monitor their performance non-intrusively, economically and in near real time. The BES itself has proven to be an effective biosensor by directly monitoring the electric current it produces; however, this relates directly to the performance and design of the electrodes not of the whole remediation system. Like the industry accepted remediation technologies there are already a variety of standard techniques and an emerging scientific discipline that measures electrical current sources in the subsurface. The discipline is "bio-geophysics" (Atekwana and Slater, 2009) which seeks to not only measure physical and chemical properties of the subsurface but also microbial activity and its interaction with the subsurface. It has emerged as environmental geophysics that monitored contaminant plumes started to consider the biogeochemical mechanisms that can contribute to geologic current sources. Self-potential is one of the oldest and simplest tools in the biogeophysics toolbox. It is the passive measurement of potential difference at or beneath the surface using non-polarizing electrodes connected to a voltmeter. Sixty years ago, the method was demonstrated to locate and map large graphite deposits (Sato and Mooney, 1960), the authors realized that the graphite deposits must act as non-reacting conductors or massive electrodes that transfer electrons across redox zones. When applied to contaminated environments and groundwater plumes this model later became known as the geobattery model (Bigalke, 1997) and biogeobattery model (Revil et al., 2010; Risgaard-Petersen et al., 2014) as the chemistry and microbiology of electron production mechanisms became better understood. A remarkably simple application of the SP method to a BES system would be observing that a snorkel electrode is working correctly. Snorkels are effectively a short-circuited BES with no way of measuring electron transfer (Hoareau et al., 2019). The SP method would measure the electric field generated by the electron flux from the anode to the cathode through the electrode. If this electrode was emplaced horizontally or at a low incline through a steady state plume (Figure 33.2c) or contaminated sediment (which would also increase the available area of degradation) then the SP electrodes at surface would measure the electric field produced by the snorkel that would result in a dipole (Doherty et al., 2015). A single traverse across a plume would produce a simple positive and negative response indicating the spatial extent of the electric field and subsurface microbial activity, a grid of surface monitoring points would supply more detailed spatial resolution. SP already has been used to define the zone of influence of in situ electrodes due to the degradation of organic matter in an analogue geobattery.

SP signals are non-unique and can arise due to, and are often used to measure electrokinetic (groundwater flow), electrochemical (groundwater plume chemistry), or microbially mediated redox potentials (Jouniaux et al., 2009). A well-designed remediation system with known parameters (e.g., porosity and flow) would be able to account for and differentiate between these signals attributing them to mechanisms of transport or degradation. Other standard geophysical techniques can be used to monitor processes that would be expected to occur in BES systems. Bulk resistivity (the inverse of conductivity) is a method which measures the resistivity of the subsurface which has been used to monitor brownfield sites, contaminant plumes, and gas emissions over time (Mendonça et al., 2015). When resistivity is normalized against other geoelectric methods such as induced polarization that monitors the decay of an applied current in the subsurface, this removes the fluid effects and allows the identification of new precipitates that clog sediment pore throats and changes in cation exchange properties of mineral surfaces. These methods were used to identify the spatial extent of a cathode of a geobattery around a gasworks groundwater plume (Doherty et al., 2010) and have been used to show changes in biogeochemical subsurface environments following restoration works (McAnallen et al., 2018). The biogeophysics discipline offers a wide range of techniques that will be of use to biotechnologists who need to monitor the effectiveness of bio-electrochemical systems beyond the lab and into the field

33.6 CONCLUSION

In conclusion, BES are a highly innovative system that can help biotechnologists design systems for the management of polluted environments. Like other applications of the BES to help solve technological problems they need to be integrated or hybridized with other approaches (MES etc.). The design of BES to deal with contamination must include the principles of sustainable remediation and eco design when considering the materials that are to be used in the system. To be accepted and taken up by industry they must piggyback on accepted technologies and improve on industry standard remediation efficiencies. New disciplines such as biogeophysics can supply the tools that will help biotechnologists scale up and monitor in real time large scale engineered solutions for BES remediation.

REFERENCES

1. Atekwana EA, Slater LD. 2009. Biogeophysics: A new frontier in earth science research. *Reviews of Geophysics* 47: 4004.
2. Atlas RM. 1991. Microbial hydrocarbon degradation—bioremediation of oil spills. *Journal of Chemical Technology & Biotechnology* 52: 149–156.
3. Bajracharya S, Sharma M, Mohanakrishna G, Dominguez Benneton X, Strik DPBTB, et al. 2016. An overview on emerging bioelectrochemical systems (BESs): Technology for sustainable electricity, waste remediation, resource recovery, chemical production and beyond. *Renewable Energy*. 98: 153–70

4. Bayer P, Finkel M. 2006. Life cycle assessment of active and passive groundwater remediation technologies. *Journal of Contaminant Hydrology* 83: 171–199.
5. Bethke CM, Sanford RA, Kirk MF, Jin Q, Flynn TM. 2011. The thermodynamic ladder in geomicrobiology. *American Journal of Science* 311: 183–210.
6. Bigalke J. 1997. The geobattery model: A contribution to large scale electrochemistry. *Electrochimica Acta* 42: 3443–3452.
7. Bonaglia S, Broman E, Brindefalk B, Hedlund E, Hjorth T, et al. 2020. Activated carbon stimulates microbial diversity and PAH biodegradation under anaerobic conditions in oil-polluted sediments. *Chemosphere* 248: 126023.
8. Bone B 2010. Verification of remediation of land contamination: SC030114/R1. UK Environment Agency, Bristol. ISBN: 9781849111751.
9. Cai X, Yuan Y, Yu L, Zhang B, Li J, et al. 2020. Biochar enhances bioelectrochemical remediation of pentachlorophenol-contaminated soils via long-distance electron transfer. *Journal of Hazardous Materials* 391: 122213.
10. Cecconet D, Sabba F, Devecseri M, Callegari A, Capodaglio AG. 2020. In situ groundwater remediation with bioelectrochemical systems: A critical review and future perspectives. *Environment International* 137: 105550.
11. CL:AIRE. 2020. Supplementary report 2 of the SuRF-UK framework: Selection of indicators/criteria for use in sustainability assessment for achieving sustainable remediation. *Contaminated Land: Applications in Real* Environments *(CL:AIRE)*, Haddenham. ISBN: 9781905046348.
12. Cupples AM. 2011. The use of nucleic acid based stable isotope probing to identify the microorganisms responsible for anaerobic benzene and toluene biodegradation. *Journal of Microbiological Methods* 85: 83–91.
13. Curtis TP, Head IM, Graham DW. 2003. Theoretical ecology for engineering biology. *Environmental Science & Technology* 37: 64A–70A.
14. Daghio M, Aulenta F, Vaiopoulou E, Franzetti A, Arends JBA, et al. 2017. Electrobioremediation of oil spills. *Water Research* 114: 351–370.
15. Daghio M, Vaiopoulou E, Patil SA, Suárez-Suárez A, Head IM, et al. 2016a. Anodes stimulate anaerobic toluene degradation via sulfur cycling in marine sediments. *Applied and Environmental Microbiology* 82: 297–307.
16. Daghio M, Vaiopoulou E, Patil SA, Suárez-Suárez A, Head IM, et al. 2016b. Anodes stimulate anaerobic toluene degradation via sulfur cycling in marine sediments. *Applied and Environmental Microbiology* 82: 297–307.
17. Davis CA, Slater LD, Kulessa B, Ferguson AS, Atekwana EA, et al. 2010. Self-potential signatures associated with an injection experiment at an in situ biological permeable reactive barrier. *Near Surface Geophysics* 8: 541–551.
18. Doherty R, Kulessa B, Ferguson AS, Larkin MJ, Kulakov LA, et al. 2010. A microbial fuel cell in contaminated ground delineated by electrical self-potential and normalized induced polarization data. *Journal of Geophysical Research* 115: G00G08.
19. Doherty R, McPolin B, Kulessa B, Frau A, Kulakova A, et al. 2015. Microbial ecology and geoelectric responses across a groundwater plume. *Interpretation* 3: SAB9–SAB21.
20. EPA. 1989. Bioremediation of Exxon Valdez oil spill. *US Environmental Protection Agency*, Washington.
21. EPEC. 2011. Recommendation to the European Commission. *European Policy Evaluation Consortium* 1–123.
22. Erable B, Etcheverry L, Bergel A. 2011. From microbial fuel cell (MFC) to microbial electrochemical snorkel (MES): Maximizing chemical oxygen demand (COD) removal from wastewater. *Biofouling* 27: 319–326.
23. Fachin SJ, Abreu EL, Mendonça CA, Revil A, Novaes GC, et al. 2012. Self-potential signals from an analog biogeobattery model. *Geophysics* 77.
24. Ferguson AS, Gibert O, Kalin RM 2008. Technology demonstration project report : TDP17 a biological sequential reactive barrier (SEREBAR): Performance at a former manufactured gas plant (FMGP). *CL:AIRE*. ISBN: 9781905046140.
25. Feris KP, Hristova K, Gebreyesus B, Mackay D, Scow KM. 2004. A shallow BTEX and MTBE contaminated aquifer supports a diverse microbial community. *Microbial Ecology* 48: 589–600.
26. Gavaskar AR. 1999. Design and construction techniques for permeable reactive barriers. *Journal of Hazardous Materials* 68: 41–71.
27. Gibert O, Ferguson AS, Kalin RM, Doherty R, Dickson KW, et al. 2007. Performance of a sequential reactive barrier for bioremediation of coal tar contaminated groundwater. *Environmental Science & Technology* 41: 6795–6801.
28. Heenan JW, Ntarlagiannis D, Slater LD, Beaver CL, Rossbach S, et al. 2017. Field-scale observations of a transient geobattery resulting from natural attenuation of a crude oil spill. *Journal of Geophysical Research: Biogeosciences* 122: 918–929.

29. Hoareau M, Erable B, Bergel A. 2019. Microbial electrochemical snorkels (MESs): A budding technology for multiple applications. A mini review. *Electrochemistry Communications* 104: 106473.

30. Holtmann D, Harnisch F. 2019. Electrification of biotechnology: Quo vadis? In: Harnisch F, Holtmann D (Eds.), *Bioelectrosynthesis*. Advances in Biochemical Engineering/Biotechnology, Vol. 167. Springer, Cham, pp. 395–411. ISBN: 9783030032999.

31. Huang DY, Zhou SG, Chen Q, Zhao B, Yuan Y, et al. 2011. Enhanced anaerobic degradation of organic pollutants in a soil microbial fuel cell. *Chemical Engineering Journal* 172: 647–653.

32. Huggins T, Wang H, Kearns J, Jenkins P, Ren ZJ. 2014. Biochar as a sustainable electrode material for electricity production in microbial fuel cells. *Bioresource Technology* 157: 114–119.

33. Iwamoto T, Tani K, Nakamura K, Suzuki Y, Kitagawa M, et al. 2000. Monitoring impact of in situ biostimulation treatment on groundwater bacterial community by DGGE. *FEMS Microbiology Ecology* 32: 129–141.

34. Johnson RL, Simon MA. 2007. Evaluation of groundwater flow patterns around a dual-screened groundwater circulation well. *Journal of Contaminant Hydrology* 93: 188–202.

35. Jørgensen KS, Puustinen J, Suortti AM. 2000. Bioremediation of petroleum hydrocarbon-contaminated soil by composting in biopiles. *Environmental Pollution* 107: 245–254.

36. Jouniaux L, Maineult A, Naudet V, Pessel M, Sailhac P. 2009. Review of self-potential methods in hydrogeophysics. *Comptes Rendus Geoscience* 341: 928–936.

37. Kao CM, Chen CY, Chen SC, Chien HY, Chen YL. 2008. Application of in situ biosparging to remediate a petroleum-hydrocarbon spill site: Field and microbial evaluation. *Chemosphere* 70: 1492–1499.

38. Kim Y, Shin SH, Chang IS, Moon SH. 2014. Characterization of uncharged and sulfonated porous poly(vinylidene fluoride) membranes and their performance in microbial fuel cells. *Journal of Membrane Science* 463: 205–214.

39. Kirmizakis P, Doherty R, Mendonça CA, Costeira R, Allen CCR, et al. 2019. Enhancement of gasworks groundwater remediation by coupling a bio-electrochemical and activated carbon system. *Environmental Science and Pollution Research* 26: 9981–9991.

40. Knight P, Jenkins JO. 2009. Adopting and applying eco-design techniques: A practitioners perspective. *Journal of Cleaner Production* 17: 549–558.

41. Kuppusamy S, Maddela NR, Megharaj M, Venkateswarlu K 2020. Bioavailability of total petroleum hydrocarbons. In: *Total Petroleum Hydrocarbons*. Springer International Publishing, Cham.

42. Lehmann J, Joseph S. 2012. *Biochar for Environmental Management: Science and Technology*. Routledge. ISBN: 9781849770552.

43. Logan BE, Hamelers B, Rozendal R, Schröder U, Keller J, et al. 2006. *Microbial Fuel Cells: Methodology and Technology. Environmental Science & Technology.*

44. Lovley DR. 2008. The microbe electric: Conversion of organic matter to electricity. *Current Opinion in Biotechnology* 19: 564–571.

45. Lovley DR. 2017. Syntrophy goes electric: Direct interspecies electron transfer. *Annual Review of Microbiology* 71: 643–664.

46. Lu H, Wang Y, Wang H. 2019. Determination of major pollutant and biogeochemical processes in an oil-contaminated aquifer using human health risk assessment and multivariate statistical analysis. *Human and Ecological Risk Assessment: An International Journal* 25: 505–526.

47. Lu L, Yazdi H, Jin S, Zuo Y, Fallgren PH, et al. 2014. Enhanced bioremediation of hydrocarbon-contaminated soil using pilot-scale bioelectrochemical systems. *Journal of Hazardous Materials* 274: 8–15.

48. Lyon S, Weiss DJ. 2010. Oil spills by the numbers - center for American progress [WWW Document]. *Center for American Progress*. https://www.americanprogress.org/issues/green/news/2010/04/30/7620/oil-spills-by-the-numbers (accessed 14 August 2020).

49. Maier U, Grathwohl P. 2006. Numerical experiments and field results on the size of steady state plumes. *Journal of Contaminant Hydrology* 85: 33–52.

50. Matturro B, Cruz Viggi C, Aulenta F, Rossetti S. 2017. Cable bacteria and the bioelectrochemical snorkel: The natural and engineered facets playing a role in hydrocarbons degradation in marine sediments. *Frontiers in Microbiology* 8: 952.

51. McAnallen L, Doherty R, Donohue S, Kirmizakis P, Mendonça C. 2018. Combined use of geophysical and geochemical methods to assess areas of active, degrading and restored blanket bog. *Science of the Total Environment* 621: 762–771.

52. McIlwaine R, Doherty R, Cox SF, Cave M. 2017. The relationship between historical development and potentially toxic element concentrations in urban soils. *Environmental Pollution* 220: 1036–1049.

53. Meckenstock RU, Elsner M, Griebler C, Lueders T, Stumpp C, et al. 2015. Biodegradation: Updating the

concepts of control for microbial cleanup in contaminated aquifers. *Environmental Science & Technology* 49: 7073–7081.

54. Mendonça CA, Doherty R, Amaral ND, McPolin B, Larkin MJ, et al. 2015. Resistivity and induced polarization monitoring of biogas combined with microbial ecology at a brownfield site. *Interpretation* 3: SAB43–SAB56.
55. Min B, Cheng S, Logan BE. 2005. Electricity generation using membrane and salt bridge microbial fuel cells. *Water Research* 39: 1675–1686.
56. Morris J, Jin S, Crimi B, Pruden A. 2009. Microbial fuel cell in enhancing anaerobic biodegradation of diesel. *Chemical Engineering Journal* 146: 161–67
57. Müller H, Bosch J, Griebler C, Damgaard LR, Nielsen LP, et al. 2016. Long-distance electron transfer by cable bacteria in aquifer sediments. *ISME Journal* 10: 2010–2019.
58. Nakamura R, Kai F, Okamoto A, Hashimoto K. 2013. Mechanisms of long-distance extracellular electron transfer of metal-reducing bacteria mediated by nanocolloidal semiconductive iron oxides. *Journal of Materials Chemistry A* 1: 5148–5157.
59. Nathanail CP, Bakker LMM, Bardos P, Furukawa Y, Nardella A, et al. 2017. Towards an international standard: The ISO/DIS 18504 standard on sustainable remediation. *Remediation* 28: 9–15.
60. Nelson-Smith A. 1972. *Oil Pollution and Marine Ecology. Springer*, US. ISBN: 9780306307232.
61. Oliot M, Galier S, Roux De Balmann H, Bergel A. 2016. Ion transport in microbial fuel cells: Key roles, theory and critical review. *Applied Energy* 183: 1682–1704.
62. Palma E, Daghio M, Franzetti A, Petrangeli Papini M, Aulenta F. 2018. The bioelectric well: A novel approach for in situ treatment of hydrocarbon-contaminated groundwater. *Microbial Biotechnology* 11: 112–118.
63. Palma E, Espinoza Tofalos A, Daghio M, Franzetti A, Tsiota P, et al. 2019. Bioelectrochemical treatment of groundwater containing BTEX in a continuous-flow system: Substrate interactions, microbial community analysis, and impact of sulfate as a co-contaminant. *Nature Biotechnology* 53: 41–48.
64. Potter MC. 1911. Electrical effects accompanying the decomposition of organic compounds. *Proceedings of the Royal Society of London. Series B, Containing Papers of a Biological Character* 84: 260–276.
65. Prommer H, Tuxen N, Bjerg PL. 2006. Fringe-controlled natural attenuation of phenoxy acids in a landfill plume: Integration of field-scale processes by reactive transport modeling. *Environmental Science & Technology* 40: 4732–4738.
66. Qin G, Gong D, Fan MY. 2013. Bioremediation of petroleum-contaminated soil by biostimulation amended with biochar. *International Biodeterioration and Biodegradation* 85: 150–155.
67. Rabaey K, Keller J. 2008. Microbial fuel cell cathodes: From bottleneck to prime opportunity? *Water Science and Technology* 57: 655–659.
68. Ramos DT, Mlb DS, Chiaranda HS, Alvarez PJJ, Corseuil HX. 2013. Biostimulation of anaerobic BTEX biodegradation under fermentative methanogenic conditions at source-zone groundwater contaminated with a biodiesel blend (B20). *Biodegradation* 24: 333–341.
69. Reguera G. 2018. Harnessing the power of microbial nanowires. *Microbial Biotechnology* 11: 979–994.
70. Revil A, Mendonça CA, Atekwana EA, Kulessa B, Hubbard SS, et al. 2010. Understanding biogeobatteries: Where geophysics meets microbiology. *Journal of Geophysical Research: Biogeosciences* 115: 1–22.
71. Risgaard-Petersen N, Damgaard LR, Revil A, Nielsen LP. 2014. Mapping electron sources and sinks in a marine biogeobattery. *Journal of Geophysical Research: Biogeosciences* 119: 1475–1486.
72. Sarkar P, Rai AR, Ghosh S. 2013. Degradation of aromatic petroleum hydrocarbons (BTEX) by a solvent tolerant bacterial consortium. *Journal of Urban and Environmental Engineering* 7: 274–279.
73. Sato M, Mooney HM. 1960. The electrochemical mechanism of sulphide self-potentials. *Geophysics* XXV: 226–249.
74. Shi Y, Zwolinski MD, Schreiber ME, Bahr JM, Sewell GW, et al. 1999. Molecular analysis of microbial community structures in pristine and contaminated aquifers: Field and laboratory microcosm experiments downloaded from. *Applied and Environmental Microbiology* 65: 2143–2150.
75. Thomas GE, Cameron TC, Campo P, Clark DR, Coulon F, et al. 2020. Bacterial community legacy effects following the Agia Zoni II oil-spill, Greece. *Frontiers in Microbiology* 11: 1706.
76. Velasquez-Orta SB, Head IM, Curtis TP, Scott K, Lloyd JR, et al. 2010. The effect of flavin electron shuttles in microbial fuel cells current production. *Applied Microbiology and Biotechnology* 85: 1373–1381.
77. Viggi CC, Presta E, Bellagamba M, Kaciulis S, Balijepalli SK, et al. 2015. The "oil-spill snorkel": An innovative bioelectrochemical approach to accelerate hydrocar-

bons biodegradation in marine sediments. *Frontiers in Microbiology* 6: 881.

78. Weelink SAB, Mha VE, Stams AJM. 2010. Degradation of BTEX by anaerobic bacteria: Physiology and application. *Reviews in Environmental Science and Biotechnology* 9: 359–385.
79. Wei M, Harnisch F, Vogt C, Ahlheim J, Neue TR, et al. 2015. Harvesting electricity from benzene and ammonium-contaminated groundwater using a microbial fuel cell with an aerated cathode. *RSC Advances* 5: 5321–5330.
80. Williams KH, Nevin KP, Franks A, Englert A, Long PE, et al. 2010. Electrode-based approach for monitoring in situ microbial activity during subsurface bioremediation. *Environmental Science & Technology* 44: 47–54.
81. Winfield J, Chambers LD, Rossiter J, Greenman J, Ieropoulos I. 2014. Towards disposable microbial fuel cells: Natural rubber glove membranes. *International Journal of Hydrogen Energy* 39: 21803–21810.
82. Winfield J, Ieropoulos I, Rossiter J, Greenman J, Patton D. 2013. Biodegradation and proton exchange using natural rubber in microbial fuel cells. *Biodegradation* 24: 733–739.
83. Xiaojian Z, Zhansheng W, Xiasheng G. 1991. Simple combination of biodegradation and carbon adsorption-the mechanism of the biological activated carbon process. *Water Research* 25: 165–172.
84. Yabusaki SB, Fang Y, Long PE, Resch CT, Peacock AD, et al. 2007. Uranium removal from groundwater via in situ biostimulation: Field-scale modeling of transport and biological processes. *Journal of Contaminant Hydrology* 93: 216–235.
85. Yang K, Ji M, Liang B, Zhao Y, Zhai S, et al. 2020. Bioelectrochemical degradation of monoaromatic compounds: Current advances and challenges. *Journal of Hazardous Materials* 398: 122892.
86. Zhang S, Mao G, Crittenden J, Liu X, Du H. 2017. Groundwater remediation from the past to the future: A bibliometric analysis. *Water Research* 119: 114–125.
87. Zobell CE. 1946. Action of microorganisims on hydrocarbons. *Bacteriological Reviews* 10: 1–49.

SECTION 5

GOOD MICROBES AND AGRICULTURE

Co-Edited by Linda Thomashow and Frans J. de Bruijn

CHAPTER 34

Beneficial Microbes for Agriculture

From Discovery to Applications

Gabriele Berg[1,2,3], Peter Kusstatscher[1], Birgit Wassermann[1,2], Tomislav Cernava[1,2], and Ahmed Abdelfattah[3]

[1]Institute of Environmental Biotechnology, Graz University of Technology, 8010 Graz, Austria
[2]Austrian Center for Industrial Biotechnology, Graz, Austria
[3]Leibniz Institute for Agricultural Engineering and Bioeconomy (ATB), 14469 Potsdam, Germany

34.1 INTRODUCTION

The fundamental discoveries of the nineteenth century by Hermann Hellriegel in 1888 about the symbiotic association between nitrogen fixing bacteria and legume roots (Hellriegel and Wilfarth, 1888; see chapter 37), followed by those of Lorenz Hiltner, who coined the term "rhizosphere" to describe the area in close proximity to plant roots (Hiltner, 1902), have opened a new venue for scientists to discover, and apply beneficial microbes to improve crop health and growth. The objective of numerous studies in the following century was to identify these beneficial microbes, multiply them, and bring them back to the field (Berg, 2009). However, several obstacles have hindered progress toward achieving these objectives, mainly due to methodological limitations during that era. The main problem was that the majority of microorganisms were not discoverable by cultivation-dependent techniques; which was the only available method until the 1970s. In addition, it was impossible to understand the diverse and complex interactions of microorganisms, including pathogens, and this limitation often resulted in inconsistent results when laboratory results were transferred to the field. Since the end of the last century, however, the research field of environmental microbiology has gradually been revolutionized by molecular techniques and next-generation sequencing technologies. Microbiome research was established at the beginning of the twenty-first century and nowadays provides an interdisciplinary platform for many fields including agriculture, food science, biotechnology, and plant pathology (Berg et al., 2020). Early indications about the importance of microbiomes to their hosts have led to the development of novel concepts such as the *hologenome theory of evolution* to describe the role of microorganisms in the evolution of plants and animals (Zilber-Rosenberg and Rosenberg, 2008). Such a holistic view has been crucial to our current understanding of plant–microbial interactions and has paved the way for new discoveries and applications, which are described in this chapter.

The use of beneficial microbes, instead of chemicals for plant nutrition and protection, provides an environmentally friendly alternative based on restoring the natural balance in systems affected by human activities. Intensive agriculture, domestication, and excessive use of chemical fertilizers and pesticides have resulted in significant biodiversity loss, pollution, soil degradation, deforestation, dead zones, ozone depletion, climate change, multi-resistant pathogens, and disease outbreaks. Moreover, the misuse of antibiotics and pesticides along the food supply chain has resulted in

Good Microbes in Medicine, Food Production, Biotechnology, Bioremediation, and Agriculture, First Edition. Edited by Frans J. de Bruijn, Hauke Smidt, Luca S. Cocolin, Michael Sauer, David Dowling, and Linda Thomashow.

antimicrobial resistance (AMR) which is responsible for the death of 700,000 people every year worldwide (WHO). Due to all of these human activities, our epoch is called the Anthropocene (Lewis and Maslin 2015). Within the planetary boundaries concept, it can be assumed that four of nine boundaries have already been crossed as a result of various human activities. These include climate change, loss of biosphere integrity, land-system change, and altered biogeochemical cycles (Rockström et al., 2009). Understanding plant microbiomes, their interplay and beneficial functions can play a key role in finding solutions to these pressing issues.

The overall findings in microbiome research have shown that the microbiota are an integral part of plants and are strongly interconnected with the surrounding environment. In turn, this has placed the term "plant health" in a new perspective – one that is embedded in environmental and planetary health concepts (Flandroy et al., 2018; Horton et al., 2014).

34.2 BENEFICIAL MICROBES CAN BE PART OF MICROBIOME MANAGEMENT CONCEPTS

In the past, the general idea for plant protection was to identify, isolate, and apply specific, beneficial microbes for disease protection (Wilson and Pusey, 1985). Nowadays, we are moving toward more complex approaches for the development of sustainable solutions for agriculture. The increasing availability of microbiome data driven by advances in -omics technologies has led to substantial increases in our understanding of the microbiome's potential to enhance the productivity and sustainability of diverse systems. This also changed the possibilities and visions of applied microbiome research, which is to improve the health of humans, animals, plants, and whole ecosystems.

In general, microbiomes can be managed either directly, by applying: (i) microbiome transplants, (ii) microbiota-active metabolites, (iii) microbes with beneficial properties that modulate the microbiome, or (iv) indirectly, by changing environmental conditions in order to also shift the structure and function of microbiomes, e.g., to restore a healthy state from dysbiosis. Beneficial microbes can be part of microbiome management concepts, but by microbiome modulation, it is also possible to support indigenous beneficial microbes. Moreover, by applying combined strategies, synergistic effects can be reached. An example was shown by Tao et al. (2020), who used organic fertilizer and *Bacillus amyloliquefaciens* W19; together they stimulate indigenous beneficial *Pseudomonas* populations.

In an agricultural context, this provides the basis for tailored treatments that can be used for next-generation precision farming in a sustainable manner. This concept considers that not all individual hosts and their associated indigenous microbiomes will respond in the same way to microbial treatments. Instead, future developments will rely on a fundamental understanding of particular host–microbe, environment–microbe, and microbe–microbe interactions that mediate microbiome assembly and functional capacities across diverse settings. In their conceptual framework for microbiome managemnt, Stegen et al. (2018) suggest greater crosstalk across different research areas, e.g., leveraging specific ecological concepts from environmental microbiome science to guide optimization of strategies to manipulate human microbiomes toward improved health (Stegen et al., 2018).

34.3 BENEFICIAL MICROBES ARE EMBEDDED IN THE PLANT MICROBIOME: FACTS AND PROBLEMS

Plants are naturally colonized by a high variety of microorganisms and are able to form complex and dynamic interactions with their associated microbes. Such microbes associated with the plant are termed as the plant's microbiota, which nowadays also are seen as the plant's second genome and contribute a broad range of beneficial traits (Turner et al., 2013). Although the composition of the plant microbiota changes from plant to plant and even over time for single plants, bacteria from the phyla *Proteobacteria, Bacteroidetes, Firmicutes, Actinobacteria*, and fungal phyla such as *Ascomycota* and *Basidiomycota* are dominant members of the plant microbiota. In addition, several genera such as *Pseudomonas, Agrobacterium, Methylobacterium, Sphingomonas, Erwinia, Aureobasidium, Cladosporium, Alternaria, Coniothyrium, Resinicium*, and *Fusarium* have commonly been reported to be part of the core community of different plant species (Abdelfattah et al., 2021; Trivedi et al., 2020; Whitehead et al., 2021).

The plant and its associated microbiota can also be regarded as a meta-organism, the so-called "holobiont" (Vandenkoornhuyse et al., 2015). The plant-associated microbiota contribute with multiple functions to the holobiont, including (i) seed germination and growth, (ii) nutrient supply, (iii) resistance against biotic and abiotic stress factors, and (iv) production of bioactive metabolites (Berg et al., 2016; Turner et al., 2013). Recently, new functions of the plant microbiota were identified; the plant-associated microbiota are able to alter aboveground herbivory (Hol et al., 2010), attract predators and parasites of herbivores (Pieterse et al., 2013), influence the flowering time (Wagner et al., 2014), and participate in cold acclimation of plants (Etemadi et al., 2018). On the other hand, the composition and structure of the microbiota are affected by several factors including genotype, seasonality, agricultural management, environmental factors, and soil properties, and in continuous interactions with the below- and above-ground environments (Abdelfattah et al., 2019, 2020, 2016; Berg and Smalla, 2009; Faticov et al., 2021; Liu et al., 2018; Wassermann et al., 2019a).

Within the plant, microbial communities are embedded in complex networks, where beneficial microbes have diverse functions for microbial fitness, population dynamics, and functional capacities (Barberán et al., 2012; Trivedi et al., 2020). The interactions can either be between microorganisms of the same species or between different species, genera, families, and domains of life. Moreover, the interactive patterns within these webs may be positive, negative, or neutral (Faust and Raes, 2012). Antagonistic interactions are classified as negative – microbial antagonists are able to suppress pathogens. They are naturally occurring in the plant microbiome; their proportion can account for up to 80% in natural plant microbiomes (Berg, 2009). Healthy plant microbiomes are characterized by a high microbial diversity and high evenness within the plant-associated microbial communities. Interestingly, all healthy plant microbiomes naturally also contain potential pathogens (Wassermann et al., 2019a). Manzotti et al., (2020) showed that after isolation and re-inoculation, they could exert their pathogenicity on the same host plant. This is an interesting observation underlying the conclusion that the balance and evenness within the plant microbiome is crucial for plant health. Each healthy plant microbiome is characterized by key stone species within the core microbiome and a dense network of positive interactions among the constituents (Berg et al., 2020). Novel tools enable assembly of the genomes of plant-associated microorganisms, also known as metagenome-assembled genomes (MAGs), from metagenomic data, which allows investigators to study them in detail without need for cultivation (Wicaksono et al., 2020). Concluding, each plant microbiome contains a certain, specific proportion of beneficial microbes, which can be used for plant germination, growth, health, and diverse functions beyond. Due to its immense potential to improve future agricultural applications, the plant microbiome was recently identified as a key to the next green revolution (National Academies of Sciences, Engineering and Medicine, 2019).

However, currently, only a very small number of beneficial microorganisms is used for agricultural applications. The first reason is that the plant microbiome of only a few plant species is currently studied. Most plant microbiome studies focus on crops, which represent only a very small proportion of the approximately half a million species within the plant kingdom. The impact of plant domestication on the associated microbiota is well-studied (Pérez-Jaramillo et al., 2018), showing the tendency to loss of microbial diversity. More studies on native/endemic and tropical vegetation are required to avoid a substantial part of the plant microbiota becoming extinct along with their hosts, before its discovery. In addition, the overwhelming majority of plant-associated microbes are not yet cultivable, with only <10% currently thought to be cultured. Interestingly, 85 out of the currently established 118 phyla have not had a single species cultivated to date (Overmann et al., 2019). For example, so far, there is no plant-associated archaeon in culture, although their proportion in the rhizosphere can be up to 30% (Taffner et al., 2019). A third point is that only a few strains have currently reached the market for potential application (Droby et al., 2016; see chapter 42). While in research studies, the potential of several beneficial genera is well studied, e.g., *Pseudomonas* (Haas and Défago, 2005; Weller et al., 2002), *Burkholderia* (Compant et al., 2008), *Bacillus* (Santoyo et al., 2012), or *Paenibacillus* (Rybakova et al., 2016), products are mainly *Bacillus*-based. Similarly, fungal species such as *Candida oleophila, C. sake, Cryptococcus albidus, Metschnikowia fructicola*, and *Aureobasidium pullulans* have been tested and used commercially, although may have shown inconsistent performance (Droby et al., 2016; Zhimo et al., 2021). This is mainly due to the low number of plant species already studied, the low proportion of cultivable strains, and the low numbers of strains already exploited. These problems, however, can be solved by (i) more studies on the native plant microbiota and concepts to discover beneficial plant-associated microbes, (ii) efforts to enhance the number of cultivable strains, and (iii) new methods for fermentation and formulation. Therefore, in the following

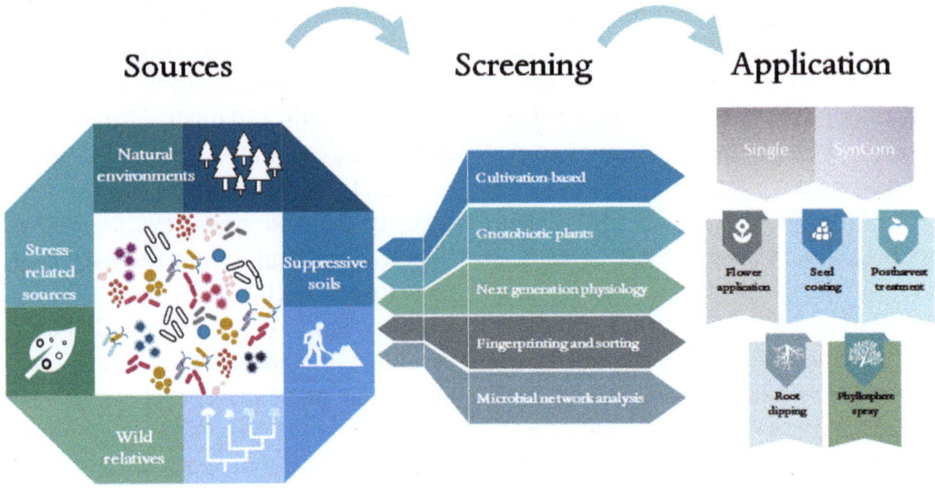

FIGURE 34.1 Strategies for the discovery and application of plant-associated and beneficial microorganisms.

sections, we will focus on concepts to discover novel beneficial plant-associated microbes by exploiting plant species and bio-resources, cultivation, and screening strategies and application (Figure 34.1).

34.4 CONCEPTS FOR DISCOVERING AND CAPTURING NOVEL BENEFICIAL MICROBES

34.4.1 Exploiting Novel Bio-resources for Discovering Beneficial Microbes

34.4.1.1 Wild Relatives – Exploiting The "Back to the Root" Concept

It has been reported in recent years that domestication and intensive agriculture have resulted in loss of biodiversity in general and in microbial diversity in particular (Pérez-Jaramillo et al., 2018). Thus, in the search for new microbes that can have a beneficial effect on plants, it has been only natural to focus on wild or non-domesticated relatives. This was especially promising considering that host-associated microbiomes were found to have developed by coevolution, implying that the application of microbes isolated from wild relatives would restore diversity rather than adding new community members (Abdelfattah et al., 2021). Breeding changed the composition of the microbiota by up to 50% (Cardinale et al., 2012). Pérez-Jaramillo and colleagues (2016) proposed the "back to the roots" approach for crop plants, which presents an interesting opportunity for unraveling native microbiomes of wild relatives and ancient heirloom breeds of crop cultivars to preserve beneficial microbes for agriculture. In addition, harnessing seed microbiomes of wild relatives of crop plants provides additional promising biological resources, potentially enabling the preservation of matching symbioses between plants and their specific microbiota (Berg and Raaijmakers, 2018).

However, understanding the impact of breeding on the plant microbiome also has revealed correlations that can be of interest to discover new beneficial plant–microbe relationships. Breeding often has been directed toward increasing the yield and resistance against pathogens and pests. Interestingly, both yield and resistance can be mediated directly by plant-associated microorganisms. Indirectly, the microbiome has also been shaped in this direction as well. There are several recent studies showing the correlation between cultivars, resistance toward specific pathogens and microbiota, e.g., in the pathosystems pumpkin-*Erwinia* (Adam et al., 2018), beans-*Fusarium* (Mendes et al., 2019), or sugar beet-*Rhizoctonia* (Wolfgang et al., 2020). *Rhizoctonia*-tolerant cultivars of sugar beet, mainly mediated by the "Fort Collins Resistance" (USDA-ARS Sugarbeet Research Unit), resulted in an enrichment of bioactive *Pseudomonas* strains in the rhizosphere. Isolated strains of *Pseudomonas* are also able to induce that resistance in the pathosystem, which shows the potential to discover new biocontrol agents (Wolfgang et al., 2020).

34.4.1.2 Suppressive Soils – Exploiting "Soil Immune Response"

Disease-suppressive soils have been documented since the last century, and identified as soils that have the ability to prevent or reduce the establishment and development of soil-borne diseases (Weller et al., 2002; see chapter 36). Suppressive soils owe their disease-reducing capacity to two different types of suppression: general and specific suppression. General suppression is universal to all soils and describes the vast and nonspecific power of the total soil microbial community to outcompete invading pathogens. The functional upgrade to this is specific suppression, which is triggered by individual microbes or microbial consortia that are highly effective in suppressing specific soil-borne pathogens (Schlatter et al., 2017). Specific suppression is likely induced by the interactions within the pathosystem that lead to a release of metabolites, which in turn enrich the disease-suppressive microorganisms (Chapelle et al., 2016). Although this mechanism requires time to adapt, the suppressive microbes are able to remember the pathogen at a later point of invasion and can react efficiently. For most specifically disease-suppressive soils, the microbes and mechanisms involved in pathogen control are unique (Larkin et al., 1996; Lemanceau and Alabouvette, 1993; Olatinwo et al., 2006; Weller et al., 2002). Responsive microbiota have been shown to produce specific bioactive molecules that mediate this suppression (Cha et al., 2016; Mendes et al., 2011). Additionally, the involvement of specific antifungal volatiles has been described (Carrión et al. 2018; Schmidt et al. 2015).

Intensive agricultural management, including soil pasteurization, fumigation, and chemical pesticides, can destroy specific suppression, resulting in conductive soils where only general suppression is active (Raaijmakers and Mazzola, 2016). However, 90 years ago, the transplantation of soil to geographically distant locations have already been reported to induce specific suppression (Henry, 1932). In spite of that, the conditions within suppressive soils are highly complex and thus, general solutions to suppress soil-borne diseases in agricultural settings are still missing (Raaijmakers and Mazzola, 2016). Novel chemical, molecular, and next-generation sequencing techniques, including community structure and network analyses, now offer opportunities to understand the hidden interactions and mechanisms and to make suppressive soils a promising option for broader application than currently imagined (see chapter 36).

34.4.2 Natural Systems – Harvesting the Best from Nature

Natural environments and other unaffected ecosystems are another resource for discovering beneficial microorganisms (Berg et al., 2013). A targeted selection strategy to obtain cultivable microorganisms from promising bioresources using bait plants was developed by Zachow and colleagues (2013), who used plants to recruit and select specific microorganisms with beneficial features from natural environments. They used alpine mosses and plants as highly diverse bioresources, and seedlings of maize, oilseed rape, sorghum, and sugar beet as bait plants. A crop-specific selection from native bioresources was found enriched in bait plant rhizosphere communities, and their potential role as stress protecting agents (SPA) was successfully evaluated following the cultivation of captured bacteria. Another interesting bioresources are lichens, which are able to colonize harsh environments due to the efficient microbial symbiosis consisting of a fungal host, photosynthetic algae and cyanobacteria, and thousands of associated specific bacteria. The latter play key roles in pathogen defense, nutrient supply, growth promotion, and detoxification of metabolites (Grube et al., 2015). They are an interesting source for agricultural applications as well, and this was already exploited: a wide range of native lichen bacteria was shown to efficiently antagonize plant pathogenic fungi (Cernava et al., 2015). Similar functions were described for *Sphagnum* plants grown in undomesticated, highly evolved, old bog ecosystems where no substantial fungal disease had so far been described. Almost 50% of *Sphagnum*-associated bacterial isolates possessed antagonistic activity toward several fungal plant pathogens (Opelt et al., 2007). Such bog ecosystems even have the potential to address important human and planetary health issues such as the expanding antibiotic resistance crisis, as has recently been pointed out (Obermeier et al., 2019). By analyzing isolates, a fosmid library, and the metagenome, a highly diversified resistome targeting a wide array of natural and synthetic drug classes was assigned to the *Sphagnum* plant. These cumulative findings provide clear evidence that natural ecosystems can be exploited as a source of beneficial organisms for plant protection, even among members of other species including phylogenetically distant organisms. Moreover, investigating natural systems can help to resolve the antibiotic resistance background of undisturbed environments, thereby providing a better understanding of ecosystem plasticity as well as of clinically relevant resistance mechanisms.

34.4.3 Stressed Microbiomes – Exploiting the "Pathobiome" Concept

Microbiomes are dynamic systems (Berg et al., 2020). While the beneficial interplay of the host and its microbiome is essential for maintaining the health of the holobiont, diseases are often correlated with microbial dysbiosis. In the context of dysbiosis, the "pathobiome" concept, which integrates pathogenic agents within biotic environments, was established and applied to multiple pathosystems (Vayssier-Taussat et al., 2014). In the field and after harvest, distinctive microbial mixtures can be observed in combination with healthy and diseased plants, as has been shown for several fruits and vegetables including olive, sugar beet, kiwi, mango, and onion (Abdelfattah et al., 2015; Diskin et al., 2017; Kusstatscher et al., 2019a; Wu et al., 2019; Yurgel et al., 2018). High pathogen pressure induces not only plant defense mechanisms but also enriches plant beneficial and antagonistic microorganisms (Garbeva et al., 2004; Weller et al., 2002). This interaction can be used to increase the probability of isolating beneficial microbes with specific antagonistic activities. The approach originally was used to recover biocontrol antagonists against postharvest rots in apples (Wilson et al., 1993). More recently, Kusstatscher and colleagues employed it to obtain beneficial bacteria from sugar beets that were selected from fields with high pathogen pressure (Kusstatscher et al., 2019a). They found that up to 50% of isolated bacterial strains were antagonistic to the plant pathogen *Fusarium oxysporum*. Similarly, Zachow et al. found a high proportion (28%) of antagonistic strains in close proximity to sclerotia (dormant resting bodies) of the fungal plant pathogen *Rhizoctonia solani* (Zachow et al., 2011). Another study has shown that sclerotia themselves harbor rich bacterial communities and can be implemented as a promising source from which to isolate antagonistic bacteria (Mülner et al., 2019). The identification of key features in microbial communities of healthy and diseased plants furthermore provides novel strategies to prevent diseases because of the possibility of implementing early countermeasures (Kusstatscher et al., 2019b). In addition, comparative analyses of microbial communities in healthy and diseased plants can provide information about the etiology of the disease, especially those that have complex or multiple causal agents (Abdelfattah et al., 2017). Abiotic stresses such as drought, frost, salinity, high temperature, and soil contaminants of pesticide residues, heavy metals, and/or radioactive material were found to shift the microbial communities associated with plants (Chen et al., 2021). Among the microbes characterizing this shift, some were recruited by the plant or enriched in the soil to mitigate the negative effects on the host (Xu and Coleman-Derr, 2019). Screening those environments for microbes might uncover useful reservoirs of beneficial microorganisms that can be employed to combat biotic and abiotic stresses.

34.4.4 New Strategies for the Isolation and Cultivation of Beneficial Microbes

34.4.4.1 Cultivation and Isolation of Beneficial Microbes

Since the discovery of microorganisms, culture media have remained a key tool for microbial isolation and identification. Commonly, screening for beneficial microorganisms from environmental samples such as soil or plant tissues includes the implementation of selective or semi-selective culture media. The resulting pure cultures can then be further identified and characterized with a variety of biochemical, physiological, and molecular methods. However, culture-dependent techniques are generally limited to the microorganisms that can be grown under laboratory conditions, and such isolates are estimated to represent only a small fraction of the total microbial diversity that can be described by advanced molecular techniques such as metagenomics. A broader range of culture conditions could increase the probability that previously uncultivated microbes might be grown under suitable growth requirements, hence increasing the number isolated microorganisms. In this regard, culturomics, a new name for advanced culture techniques originally developed by Lagier and colleagues (2012), which implements diversified and multiple combinations of various growth media, culture conditions, atmospheres, and incubation times as a high throughput strategy of cultivation, can be used to fill the gap between culture-dependent and culture-independent methods. In addition, developing and using new culture media that resemble the native habitat of microorganisms, and/or enriching media with substrates extracted from native habitats has been shown to support the growth of bacteria that previously could not be grown in culture (Bonnet et al., 2020). For instance, the use of plant-based culture media, the addition of plant extract additives, signaling compounds, and/or co-culturing improved the recovery of microorganisms (Mourad et al., 2018; Nour et al., 2012; Sarhan et al., 2019). Regardless of the method used, screenings in the laboratory can target the potential of microbes to fix nitrogen, to solubilize nutrients, to antagonize pathogens, as well as to provide specific secondary metabolites such as phytohormones that affect their hosts (Hardoim et al., 2015). The microorganisms identified in this way can then be screened either *in vitro*, *in situ*, or *in vivo* for those functions within the targeted system. Plant inoculation with the selected microbes is preferably done first under controlled conditions to compensate for biotic and abiotic variability in open systems.

34.4.5 Linking Cultivation-dependent and -independent Methods for Better Cultivation

Microscopy techniques such as confocal laser scanning microscopy (CLSM) combined with fluorescence *in situ* hybridization (FISH-CLSM) has been shown to be an indispensable tool in microbiome research. The most common use of FISH-CLSM is to visualize microbial colonization patterns and to track microbial inoculants (Figure 34.2). Zachow and colleagues (2010) implemented visualizing techniques to monitor different biocontrol strains *in situ* in order to identify their different colonization niches and the applicability of potential

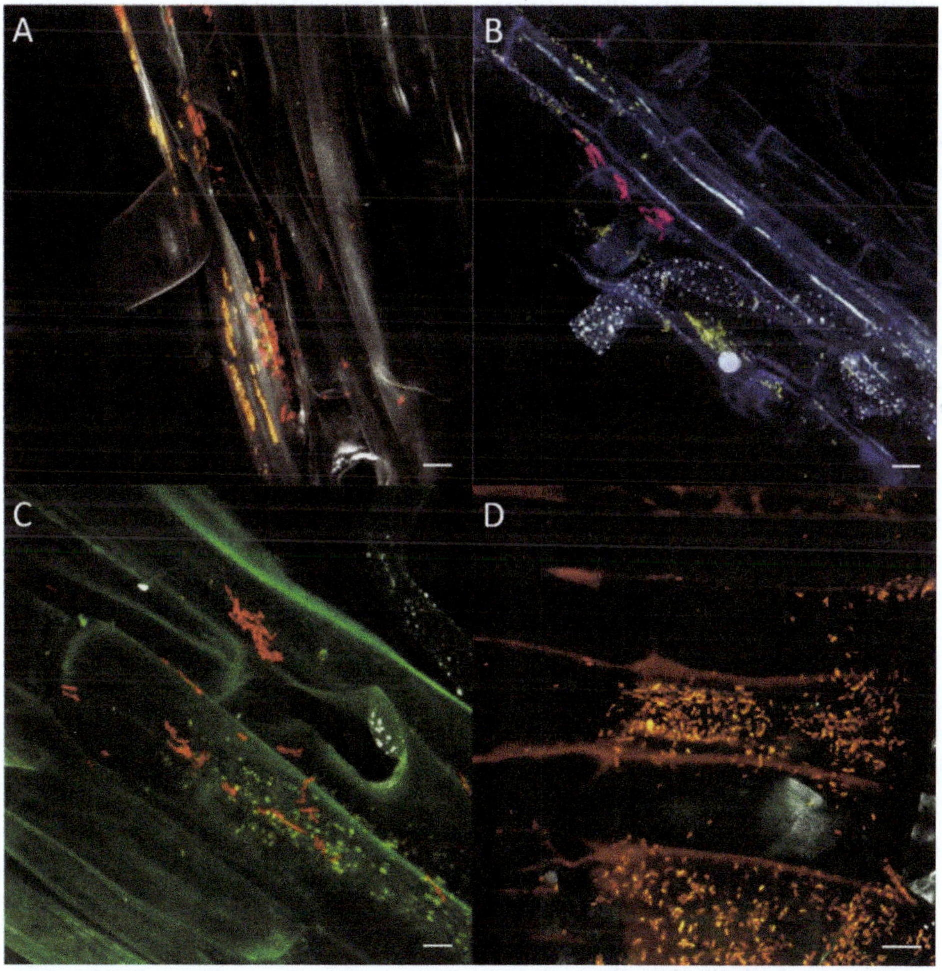

FIGURE 34.2 CLSM images visualizing the colonization of oilseed rape roots by FISH-stained bacterial communities. A, B: Endorhizosphere colonization by native seed-borne bacteria stained with FISH probes specific for Gammaproteobacteria (fluorescing yellow), Firmicutes (pink), and Eubacteria (red). C, D: Visualization of root-colonizing capacity by biocontrol strains (C: Bacillus, fluorescing red; D: Pseudomonas, fluorescing yellow) applied as seed treatments. Bar indicates 10 μm.

combinations based on colonization efficiencies. FISH-CLSM can also be used in combination with various next-generation sequencing techniques to confirm specific results (Cardinale et al., 2015).

Microbial communities that are found in natural habitats are embedded in complex interactive networks (Proulx et al., 2005). The complexity of such systems represents a great challenge to understanding microbial functions and interactions. Here, microbial network analyses can play a fundamental role in disentangling interactive systems and identifying core and hub taxa. For instance, ecological niches are often found to be dominated by only a few specific groups of microorganisms. While those groups might have major functional roles in their habitat, they often require the presence of other taxa to maintain the stability of the community. They are commonly called keystone taxa or hub microorganisms and frequently are present in low abundance, yet their absence can result in a dramatic shift in microbiome structure and function or in dysbiosis (Banerjee et al., 2018; Toju et al., 2018). Information from network analysis can be used to discover potential inter-microorganism associations and to identify core species (Cernava et al., 2015). For example, comparative network analysis of healthy and diseased plants can identify those microorganisms which have an important role in the healthy microbiome, e.g., core hubs; and those with negative correlation, with phytopathogens. The data inferred from such networks can be verified by constructing synthetic communities (SynComs) consisting of those microorganisms. Downstream assessment of SyncComs can be evaluated by Raman spectroscopy, in microfluidics devices, or in plants. Further exploitation of network analysis can result in mathematical models to predict ecosystem reactions to external factors (Barberán et al., 2012; Faust and Raes, 2012); all of which can help establish the foundation for microbiome management. However, whereas all theoretical and microscopic images can help to identify beneficial microbes, they must be confirmed in *in planta* assays.

34.4.6 Next Generation Physiology, Fingerprinting, and Cell Sorting

Microbiomes in their natural habitat harbor diverse microorganisms that are metabolically interdependent and influenced by a wide range of physiochemical and environmental factors. Cultivable microorganisms can have different physiological characteristics *ex situ* than within their indigenous communities. Current techniques which aim to capture microbes in their environments such as metagenomics, metatranscriptomics, or metabolomics are destructive methods, i.e., once imaged, the sample can no longer be monitored or studied. Here, next-generation physiology, defined as "any combination of techniques that analyses the phenotype of an individual cell in a microbiome in a non-destructive way, which enables the physical separation of this cell based solely on its phenotype for subsequent, downstream applications" (Hatzenpichler et al., 2020), represents a progressive step toward a greater understanding of microbial ecophysiology in native habitats with higher precision. For instance, Raman microspectroscopy can be used in combination with or without, phenotype probing and a manual or automated cell sorting method to study heterogeneous microbial populations *in situ*. Phenotype probing includes the use of stable isotopes such as ^{13}C, ^{15}N, D_2O, or substrate analog probe such as fluorophores or clickable molecules (Hiramatsu et al., 2019; Lee et al., 2019; Song et al., 2016). Promising single cell or community separation technologies are now starting to be commercially available, including optical tweezers, microdissection, microdroplets, and electrode-based sorting. Label-free Raman flow cytometry was developed recently to overcome the drawbacks associated with fluorescent labeling, especially cell toxicity and interference (Hiramatsu et al., 2019; Song et al., 2016). High throughput screening in plants, using single-cell chemical phenotyping with Raman spectroscopy and an inoculation system like layered microfluidics (Li et al., 2016), the RootChip (Grossmann et al., 2011), or microcosmos (Abdelfattah et al., 2021) was previously proposed. Despite the promising advantages, these technologies are yet to be used for the discovery and evaluation of plant beneficial microorganisms.

34.5 CONCEPTS FOR APPLICATION OF BENEFICIAL MICROBES

34.5.1 Designing Consortia – A Promising Alternative to Single Microbes?

The introduction of beneficial microbial strains into agricultural systems was previously shown to have a tremendous impact. Nevertheless, a mixed inoculum of diverse isolates rather than a single strain could provide higher efficacy and open the way toward improved microbial products. Complex inoculants showed stronger plant disease inhibition (Hu et al., 2016; Yasmin et al., 2016), and growth promotion (Baas et al., 2016; Timm et al., 2016) compared to single strains. Microbial consortia using so called "helper strains" provide a valuable strategy to increase the efficacy of plant beneficial microbes. Loján and colleagues (2017) showed that a plant growth-promoting microbe also increases hyphal branching and germ tube elongation of a beneficial arbuscular mycorrhizal fungus (AMF; see chapter 39) and therefore assists the development of the symbiotic relationship (Loján et al., 2017). Co-inoculation of *Pseudomonas* and *Stenotrophomonas* strains both biocontrol strains producing volatile organic compounds (VOCs) showed increased plant growth promotion due to the stimulatory VOCs they synthesized (Rojas-Solís et al., 2018). In addition to laboratory screenings, the ongoing microbiome research era provides the basis to develop more complex communities for future applications (Massart et al., 2015). While single strain applications are still the easiest and most widely used strategies, novel findings, and current research trends indicate an increasing potential for the development of multi-strain applications in the future.

34.6 TARGETED APPLICATIONS ALONG THE FOOD SUPPLY CHAIN

Microorganisms that are utilized as individual or multi-strain inoculants can be applied to plants along the food supply chain, including pre- and post-harvest. For example, a keystone species identified by one of the previously described strategies, can be applied to plant seeds as a seed coating formulation, or later, to the seedling, by root dipping or treatments introduced near the roots of seedlings. The seed was recently identified as a perfect target for next-generation microbials in order to substitute missing microbial taxa and functions (Berg and Raaijmakers, 2018). For example, seed applications, with *Bacillus amyloliquefaciens*, have been shown to protect plants against soil-borne pathogens under greenhouse conditions (Chen et al., 2013) and root and rhizosphere treatment with *Phaseolus vulgaris*, a nodule-inhabiting endophyte, was found to increase plant biomass and the number of nodules in common beans (Chihaoui et al., 2015). Flowers, being the linking point of the mother plant to the next generation, are a crucial step for seed microbiome assembly and a potential target in biological applications. For instance, *Aureobasidium pullulans* (BoniProtect, Bio-Ferm, Tulln, Austria), is used against several diseases including fire blight, one of the most devastating diseases in pome fruit (Lima et al., 2015). Furthermore, it was recently discovered that introducing *Paraburkholderia phytofirmans* PsJN to plant flowers results in its inclusion in progeny seed microbiomes (Mitter et al., 2017). Therefore, beneficial traits can be introduced at that stage to increase seed viability, shelf life, and the health of a plant's offspring.

The post-harvest microbiota are closely linked with our food microbiota, which can also be managed for desired functional properties of food products regarding safety and preservation issues, and organoleptic or health properties (Kusstatscher et al., 2020). Several biocontrol agents are being currently used and are commercially available against postharvest diseases in fruits and vegetables. Those include, Biosave™ (*Pseudomonas syringae* Van Hall), Aspire™ (*Candida oleophila*), Yieldplus™ (*Cryptococcus albidus*), Shemer, Bayer, Leverkusen, Germany, (*Metschnikowia fructicola*), Candifruit™ (*Candida sake*), Nexy™, (*Candida oleophila*), and BoniProtect™ (*Aureobasidium pullulans*). Additionally, the use of fungal and bacterial consortia was recently shown to protect fruits such as apple and grapefruit from postharvest rots (Wassermann et al., 2019b; Zhimo et al., 2020). This represents a wide-open field for microbiome-based applications that will benefit from the emerging amount of data on food ecosystems.

34.7 THE VISION AND FUTURE CHALLENGES

Since the beginning of agriculture, humans have relied on their environment for crop health and productivity. During the last decades large inputs of chemical products have temporarily increased crop yields; however, they also have had devastating effects on global soil biodiversity leading to long-term soil health deficiencies (Wall et al., 2015). Current trends now recognize and are heading toward a reduction of chemical inputs and more sustainable agricultural practices that are partially based on microbial products. Going "back to the roots," as suggested by Pérez-Jaramillo and colleagues (2016), has set a new trend, restoring the former biodiversity and utilizing lost microbial traits to solve current agricultural issues. The use of beneficial inoculants and their functional molecules to replace chemical products could provide the key for maintaining a healthy agricultural system and global health. A successful expansion of applications based on beneficial microbes is connected to several challenges. After identification of a new and potentially beneficial strain, excessive analyses must be performed to confirm previously observed results and guarantee efficacy under a wide range of conditions. Human and plant safety is a key element to confirm before extensive field trials can be undertaken. Moreover, application techniques as well as formulation are important to provide successful and reliable products for farmers. Although technologies have been developed to prolong shelf life and increase the stability of the microbial inoculum (Vassilev et al., 2020), current applications focus on a minority of microbial spore producers only. Toward commercialization, the registration of effective and safely applicable microbial strains is key to success (see chapter 42). Nevertheless, this step is the most laborious and time-consuming due to regulations between countries, and many initially highly promising microbes never make it to market (De Silva et al., 2019; Kamilova et al., 2015). The field of microbiome research provides an extended toolbox to facilitate registration processes in the future and will thus play an increasingly important role.

34.8 CONCLUDING REMARKS

Microbiome research has clearly broadened our view of agriculture and led to a reassessment of current agricultural practices. Intensive practices that were established during the Anthropocene have led to a progressive loss of microbial biodiversity which has already resulted in long-term effects on plant and planetary health. The implementation of nature-based microbial products in agriculture is one of the first steps to restoring microbial diversity and functionality. Excessive research of the past several decades has provided numerous ways to harness beneficial traits from natural systems, and ongoing technological advances will increase our knowledge of complex interactive microbial networks. We are now beginning to rethink agricultural practices and restore global sustainability. The innovative, alternative, and sustainable solutions in the management of diseases and improvement of a plant's tolerance to biotic and abiotic stress should result in reduced dependence on chemical pesticide and fertilizer application. This on one hand should help limit the exposure of soil to chemicals and the contamination of ground and underground water and air. On the other hand, it will be automatically followed by beneficial effects on the population density of natural enemies and pollinators in support of more sustainable production of agricultural products.

ACKNOWLEDGMENTS

The authors received funding from the European Union's Horizon 2020 research and innovation program under grant agreement No. 817946 (Excalibur) [G.B.] and by European Union through Marie Skłodowska-Curie Actions – Individual Fellowships (MSCA-IF) grant agreement 844114 [A.A.]. We have received support from the ERA-NET SusCrop project "potato-METAbiome" funded by the Federal Ministry Republic of Austria for Agriculture, Regions and Tourism (project No. 101392) and from The COMET center acib: Next Generation Bioproduction is funded by BMK, BMDW, SFG, Standortagentur Tirol, Government of Lower Austria und Vienna Business Agency in the framework of COMET – Competence Centers for Excellent Technologies. The COMET-Funding Program is managed by the Austrian Research Promotion Agency FFG, both to [G.B.].

REFERENCES

1. Abdelfattah A, Cacciola SO, Mosca S, Zappia R, Schena L. 2017. Analysis of the fungal diversity in citrus leaves with greasy spot disease symptoms. *Microbial Ecology* 73: 739–749.
2. Abdelfattah A, Freilich S, Bartuv R, Zhimo VY, Kumar A, Biasi A, Salim S, Feygenberg O, Burchard E, Dardick C, et al. 2021. Global Analysis of the Apple Fruit Microbiome: Are All Apples the Same? *Environmental Microbiology*. https://doi.org/10.1111/1462-2920.15469
3. Abdelfattah A, Li Destri Nicosia MG, Cacciola SO, Droby S, Schena L. 2015. Metabarcoding analysis of fungal diversity in the phyllosphere and carposphere of olive (*Olea europaea*). *PLOS ONE* 10: e0131069.
4. Abdelfattah A, Sanzani SM, Wisniewski M, Berg G, Cacciola SO, Schena L. 2019. Revealing cues for fungal interplay in the plant–air interface in Vineyards. *Frontiers in Plant Science* 10.
5. Abdelfattah A, Tack AJ, Wasserman B, Liu J, Berg G, Norelli J, Droby S, Wisniewski M. 2021. Evidence for host–microbiome co-evolution in apple. *New Phytologist*. https://nph.onlinelibrary.wiley.com/doi/full/10.1111/nph.17820

6. Abdelfattah A, Whitehead SR, Macarisin D, Liu J, Burchard E, et al. 2020. Effect of washing, waxing and low-temperature storage on the postharvest microbiome of apple. *Microorganisms* 2020(8): 944.
7. Abdelfattah A, Wisniewski M, Droby S, Schena L. 2016. Spatial and compositional variation in the fungal communities of organic and conventionally grown apple fruit at the consumer point-of-purchase. *Horticulture Research* 3: 16047.
8. Abdelfattah A, Wisniewski M, Schena L, Tack A. 2021. Experimental evidence of microbial inheritance in plants and transmission routes from seed to phyllosphere and root. *Environmental Microbiology*. https://doi.org/10.1111/1462-2920.15392.
9. Adam E, Bernhart M, Müller H, Winkler J, Berg G. 2018. The Cucurbita pepo seed microbiome: Genotype-specific composition and implications for breeding. *Plant and Soil* 422: 35–49.
10. Baas P, Bell C, Mancini LM, Lee MN, Conant RT, Wallenstein MD. 2016. Phosphorus mobilizing consortium Mammoth P™ enhances plant growth. *Peer Journal* 4: e2121.
11. Banerjee S, Schlaeppi K, Van Der Heijden MG. 2018. Keystone taxa as drivers of microbiome structure and functioning. *Nature Reviews Microbiology* 16: 567–576.
12. Barberán A, Bates ST, Casamayor EO, Fierer N. 2012. Using network analysis to explore co-occurrence patterns in soil microbial communities. *The ISME Journal* 6: 343–351.
13. Berg G. 2009. Plant–microbe interactions promoting plant growth and health: Perspectives for controlled use of microorganisms in agriculture. *Applied Microbiology and Biotechnology* 84: 11–18.
14. Berg G, Raaijmakers JM. 2018. Saving seed microbiomes. *The ISME Journal* 12: 1167–1170.
15. Berg G, Rybakova D, Fischer D, et al. 2020. Microbiome definition re-visited: Old concepts and new challenges. *Microbiome* 8: 1–22.
16. Berg G, Rybakova D, Grube M, Köberl M. 2016. The plant microbiome explored: Implications for experimental botany. *Journal of Experimental Botany* 67: 995–1002.
17. Berg G, Smalla K. 2009. Plant species and soil type cooperatively shape the structure and function of microbial communities in the rhizosphere. *FEMS Microbiology Ecology* 68: 1–13.
18. Berg G, Zachow C, Müller H, Philipps J, Tilcher R 2013. Next-generation bio-products sowing the seeds of success for sustainable agriculture. In: Multidisciplinary Digital Publishing Institute.
19. Bonnet M, Lagier JC, Raoult D, Khelaifia S. 2020. Bacterial culture through selective and non-selective conditions: The evolution of culture media in clinical microbiology. *New Microbes and New Infections* 34: 100622.
20. Cardinale M, Grube M, Castro JV, Jr, Müller H, Berg G. 2012. Bacterial taxa associated with the lung lichen Lobaria pulmonaria are differentially shaped by geography and habitat. *FEMS Microbiology Letters* 329: 111–115.
21. Cardinale M, Grube M, Erlacher A, Quehenberger J, Berg G. 2015. Bacterial networks and co-occurrence relationships in the lettuce root microbiota. *Environmental Microbiology* 17: 239–252.
22. Carrión VJ, Cordovez V, Tyc O, Etalo DW, de Bruijn I, de Jager VCL, Medema MH, Eberl L, Raaijmakers JM. 2018. Involvement of Burkholderiaceae and sulfurous volatiles in disease-suppressive soils. *The ISME Journal* 12(9): 2307–2321.
23. Cernava T, Müller H, Aschenbrenner IA, Grube M, Berg G. 2015. Analyzing the antagonistic potential of the lichen microbiome against pathogens by bridging metagenomic with culture studies. *Frontiers in Microbiology* 6.
24. Cha J-Y, Han S, Hong H-J, et al. 2016. Microbial and biochemical basis of a Fusarium wilt-suppressive soil. *The ISME Journal* 10: 119–129.
25. Chapelle E, Mendes R, Bakker PH, Raaijmakers JM. 2016. Fungal invasion of the rhizosphere microbiome. *The ISME Journal* 10: 265–268.
26. Chen F, Wang M, Zheng Y, et al. 2013. The effect of biocontrol bacteria on rhizosphere bacterial communities analyzed by plating and PCR-DGGE. *Current Microbiology* 67: 177–182.
27. Chen X, Wicaksono WA, Berg G, Cernava T. 2021. Bacterial communities in the plant phyllosphere harbour distinct responders to a broad-spectrum pesticide. *Science of the Total Environment* 751: 141799.
28. Chihaoui S-A, Trabelsi D, Jdey A, Mhadhbi H, Mhamdi R. 2015. Inoculation of *Phaseolus vulgaris* with the nodule-endophyte *Agrobacterium* sp. 10C2 affects richness and structure of rhizosphere bacterial communities and enhances nodulation and growth. *Archives of Microbiology* 197: 805–813.

29. Compant S, Nowak J, Coenye T, Clément C, Ait Barka E. 2008. Diversity and occurrence of Burkholderia spp. in the natural environment. *FEMS Microbiology Reviews* 32: 607–626.

30. De Silva N, Brooks S, Lumyong S, Hyde K. 2019. Use of endophytes as biocontrol agents. *Fungal Biology Reviews* 33: 133–148.

31. Diskin S, Feygenberg O, Maurer D, Droby S, Prusky D, Alkan N. 2017. Microbiome alterations are correlated with occurrence of postharvest stem-end rot in mango fruit. *Phytobiomes* 1: 117–127.

32. Droby S, Wisniewski M, Teixidó N, Spadaro D, Jijakli MH. 2016. The science, development, and commercialization of postharvest biocontrol products. *Postharvest Biology and Technology* 122: 22–29.

33. Etemadi M, Zuther E, Müller H, Hincha DK, Berg G. 2018. Ecotype-dependent response of bacterial communities associated with Arabidopsis to cold acclimation. *Phytobiomes* 2: 3–13.

34. Faticov M, Abdelfattah A, Roslin T, Vacher C, Hambäck P, Blanchet FG, Lindahl BD, Tack AJM. 2021. Climate warming dominates over plant genotype in shaping the seasonal trajectory of foliar fungal communities on oak. *New Phytologist* 231(5): 1770–1783.

35. Faust K, Raes J. 2012. Microbial interactions: From networks to models. *Nature Reviews Microbiology* 10: 538–550.

36. Flandroy L, Poutahidis T, Berg G, et al. 2018. The impact of human activities and lifestyles on the interlinked microbiota and health of humans and of ecosystems. *Science of the Total Environment* 627: 1018–1038.

37. Garbeva PV, Van Veen J, Van Elsas J. 2004. Microbial diversity in soil: Selection of microbial populations by plant and soil type and implications for disease suppressiveness. *Annual Reviews Phytopathology* 42: 243–270.

38. Grossmann G, Guo W-J, Ehrhardt DW, et al. 2011. The RootChip: An integrated microfluidic chip for plant science. *The Plant Cell* 23: 4234–4240.

39. Grube M, Cernava T, Soh J, et al. 2015. Exploring functional contexts of symbiotic sustain within lichen-associated bacteria by comparative omics. *The ISME Journal* 9: 412–424.

40. Haas D, Défago G. 2005. Biological control of soil-borne pathogens by fluorescent pseudomonads. *Nature Reviews Microbiology* 3: 307–319.

41. Hardoim PR, Van Overbeek LS, Berg G, et al. 2015. The hidden world within plants: Ecological and evolutionary considerations for defining functioning of microbial endophytes. *Microbiology and Molecular Biology Reviews* 79: 293–320.

42. Hatzenpichler R, Krukenberg V, Spietz RL, Jay ZJ. 2020. Next-generation physiology approaches to study microbiome function at single cell level. *Nature Reviews Microbiology* 18: 241–256.

43. Hellriegel H, Wilfarth H, 1888. Untersuchungen uber die Stickstoffnahrung der Gramineen und Leguminosen.

44. Henry AW. 1932. Influence of soil temperature and soil sterilization on the reaction of wheat seedlings to *Ophiobolus graminis* Sacc. *Canadian Journal of Research* 7: 198–203.

45. Hiltner L. 1902. *Die Keimungsverhältnisse Der Leguminosensamen Und Ihre Beeinflussung Durch Organismenwirkung* (Parey J). Springer.

46. Hiramatsu K, Ideguchi T, Yonamine Y, et al. 2019. High-throughput label-free molecular fingerprinting flow cytometry. *Science Advances* 5: eaau0241.

47. Hol WG, De Boer W, Termorshuizen AJ, et al. 2010. Reduction of rare soil microbes modifies plant–herbivore interactions. *Ecology Letters* 13: 292–301.

48. Horton R, Beaglehole R, Bonita R, Raeburn J, Mckee M, Wall S. 2014. From public to planetary health: A manifesto. *The Lancet* 383: 847.

49. Hu J, Wei Z, Friman V-P, et al. 2016. Probiotic diversity enhances rhizosphere microbiome function and plant disease suppression. *MBio* 7: e01790–01716.

50. Kamilova F, Okon Y, De Weert S, Hora K. 2015. Commercialization of microbes: Manufacturing, inoculation, best practice for objective field testing, and registration. In: Lugtenberg B (Ed.), *Principles of Plant–microbe Interactions*. Springer, 319–327.

51. Kusstatscher P, Cernava T, Abdelfattah A, Gokul J, Korsten L, Berg G. 2020. Microbiome approaches provide the key to biologically control postharvest pathogens and storability of fruits and vegetables. *FEMS Microbiology Ecology* 96: fiaa119.

52. Kusstatscher P, Cernava T, Harms K, et al. 2019a. Disease incidence in sugar beet fields is correlated with microbial diversity and distinct biological markers. *Phytobiomes Journal* 3: 22–30.

53. Kusstatscher P, Zachow C, Harms K, et al. 2019b. Microbiome-driven identification of microbial indicators for postharvest diseases of sugar beets. *Microbiome* 7: 112.

54. Lagier JC, Armougom F, Million M, Hugon P, Pagnier I, Robert C, Bittar F, Fournous G, Gimenez G, Maraninchi M, et al. 2012. Microbial culturomics: paradigm shift in the human gut microbiome study. *Clinical Microbiology and Infection* 18(12): 1185–1193.

55. Larkin RP, Hopkins DL, Martin FN. 1996. Recovered from a disease-suppressive soil. *Pathology* 86: 812–819.

56. Lee KS, Palatinszky M, Pereira FC, et al. 2019. An automated Raman-based platform for the sorting of live cells by functional properties. *Nature Microbiology* 4: 1035–1048.

57. Lemanceau P, Alabouvette C. 1993. Suppression of fusarium wilts by fluorescent pseudomonads: Mechanisms and applications. *Biocontrol Science and Technology* 3: 219–234.

58. Lewis, S and Maslin, M. 2015. Defining the Anthropocene. *Nature* 519: 171–180. https://doi.org/10.1038/nature14258

59. Li R, Lv X, Zhang X, Saeed O, Deng Y. 2016. Microfluidics for cell-cell interactions: A review. *Frontiers of Chemical Science and Engineering* 10: 90–98.

60. Lima G, Sanzani SM, De Curtis F, Ippolito A. 2015. Biological control of postharvest diseases. In: Wills RBH, Golding JB (Eds.), *Advances in Postharvest Fruit and Vegetables Technology*, 65–81.

61. Liu J, Abdelfattah A, Norelli J, et al. 2018. Apple endophytic microbiota of different rootstock/scion combinations suggests a genotype-specific influence. *Microbiome* 6: 18.

62. Loján P, Demortier M, Velivelli SL, et al. 2017. Impact of plant growth-promoting rhizobacteria on root colonization potential and life cycle of *Rhizophagus irregularis* following co-entrapment into alginate beads. *Journal of Applied Microbiology* 122: 429–440.

63. Massart S, Martinez-Medina M, Jijakli MH. 2015. Biological control in the microbiome era: Challenges and opportunities. *Biological Control* 89: 98–108.

64. Manzotti A, Bergna A, Burow M, Jørgensen HJL, Cernava T, Berg G, Collinge DB, Jensen B. 2020. Insights into the community structure and lifestyle of the fungal root endophytes of tomato by combining amplicon sequencing and isolation approaches with phytohormone profiling. *FEMS Microbiology Ecology* 96 (5): May 2020, fiaa052. https://doi.org/10.1093/femsec/fiaa052

65. Mendes LW, Gonçalves De Chaves MG, Da Fonseca MDC, Mendes R, et al. 2019. Resistance breeding of common bean shapes the physiology of the rhizosphere microbiome. *Frontiers in Microbiology* 10: 2252.

66. Mendes R, Kruijt M, De Bruijn I, et al. 2011. Deciphering the rhizosphere microbiome for disease-suppressive bacteria. *Science* 332: 1097–1100.

67. Mitter B, Pfaffenbichler N, Flavell R, et al. 2017. A new approach to modify plant microbiomes and traits by introducing beneficial bacteria at flowering into progeny seeds. *Frontiers in Microbiology* 8: 11.

68. Mourad EF, Sarhan MS, Daanaa H-SA, et al. 2018. Plant materials are sustainable substrates supporting new technologies of plant-only-based culture media for in vitro culturing of the plant microbiota. *Microbes and Environments* 33: 40–49.

69. Mülner P, Bergna A, Wagner P, et al. 2019. Microbiota associated with sclerotia of soilborne fungal pathogens – A novel source of biocontrol agents producing bioactive volatiles. *Phytobiomes Journal* 3: 125–136.

70. National Academies of Sciences, Engineering and Medicine. 2019. *Science Breakthroughs to Advance Food and Agricultural Research by 2030*. National Academies Press, Washington, DC.

71. Nour EH, Hamza MA, Fayez M, Monib M, Ruppel S, Hegazi NA. 2012. The crude plant juices of desert plants as appropriate culture media for the cultivation of rhizospheric microorganisms. *Journal of Advanced Research* 3: 35–43.

72. Obermeier -M-M, Taffner J, Bergna A, et al. 2019. Unravelling native plant resistomes–The Sphagnum microbiome harbours versatile and novel antimicrobial resistance genes. *bioRxiv*, 695973.

73. Olatinwo R, Yin B, Becker JO, Borneman J. 2006. Suppression of the plant-parasitic nematode *Heterodera schachtii* by the fungus *Dactylella oviparasitica*. *Phytopathology* 96: 111–114.

74. Opelt K, Berg C, Berg G. 2007. The bryophyte genus *Sphagnum* is a reservoir for powerful and extraordinary antagonists and potentially facultative human pathogens. *FEMS Microbiology Ecology* 61: 38–53.

75. Overmann J, Huang S, Nübel U, Hahnke RL, Tindall BJ. 2019. Relevance of phenotypic information for the taxonomy of not-yet-cultured microorganisms. *Systematic and Applied Microbiology* 42: 22–29.

76. Pérez-Jaramillo JE, Carrión VJ, De Hollander M, Raaijmakers JM. 2018. The wild side of plant microbiomes. *Microbiome* 6: 143.

77. Pérez-Jaramillo, J.E., Mendes, R. Raaijmakers, J.M. 2016. Impact of plant domestication on rhizosphere

microbiome assembly and functions. *Plant Mol Biol* 90: 635–644. https://doi.org/10.1007/s11103-015-0337-7

78. Pieterse CM, Poelman E, Van Wees S, Dicke M. 2013. Induced plant responses to microbes and insects. *Frontiers in Plant Science* 4: 475.
79. Proulx SR, Promislow DE, Phillips PC. 2005. Network thinking in ecology and evolution. *Trends in Ecology & Evolution* 20: 345–353.
80. Raaijmakers JM, Mazzola M. 2016. Soil immune responses. *Science* 352: 1392–1393.
81. Rockström J, Steffen W, Noone K, et al. 2009. Planetary boundaries: Exploring the safe operating space for humanity. *Ecology and Society* 14.
82. Rojas-Solís D, Zetter-Salmón E, Contreras-Pérez M, Del Carmen Rocha-Granados M, et al. 2018. *Pseudomonas stutzeri* E25 and *Stenotrophomonas maltophilia* CR71 endophytes produce antifungal volatile organic compounds and exhibit additive plant growth-promoting effects. *Biocatalysis and Agricultural Biotechnology* 13: 46–52.
83. Rybakova D, Cernava T, Köberl M, Liebminger S, Etemadi M, Berg G. 2016. Endophytes-assisted biocontrol: Novel insights in ecology and the mode of action of *Paenibacillus*. *Plant and Soil* 405: 125–140.
84. Santoyo G, Mdc O-M, Govindappa M. 2012. Mechanisms of biocontrol and plant growth-promoting activity in soil bacterial species of *Bacillus* and *Pseudomonas*: A review. *Biocontrol Science and Technology* 22: 855–872.
85. Sarhan MS, Hamza MA, Youssef HH, et al. 2019. Culturomics of the plant prokaryotic microbiome and the dawn of plant-based culture media–A review. *Journal of Advanced Research* 19: 15–27.
86. Schlatter D, Kinkel L, Thomashow L, Weller D, Paulitz T. 2017. Disease Suppressive Soils: New Insights from the Soil Microbiome. *Phytopathology* 107: 1284–1297.
87. Schmidt R, Cordovez V, de Boer W, Raaijmakers J, Garbeva P. 2015. Volatile affairs in microbial interactions. *The ISME Journal* 9(11): 2329–2335.
88. Song Y, Yin H, Huang WE. 2016. Raman activated cell sorting. *Current Opinion in Chemical Biology* 33: 1–8.
89. Stegen JC, Bottos EM, Jansson JK. 2018. A unified conceptual framework for prediction and control of microbiomes. *Current Opinion in Microbiology* 44: 20–27.
90. Taffner J, Cernava T, Erlacher A, Berg G. 2019. Novel insights into plant-associated archaea and their functioning in arugula (Eruca sativa Mill.). *Journal of Advanced Research* 19: 39–48.
91. Tao C, Li R, Xiong W, Shen Z, Liu S, Wang B, Ruan Y, Geisen S, Shen Q, Kowalchuk GA. 2020. Bio-organic fertilizers stimulate indigenous soil Pseudomonas populations to enhance plant disease suppression. *Microbiome* 8(1): 137.
92. Timm CM, Pelletier DA, Jawdy SS, et al. 2016. Two poplar-associated bacterial isolates induce additive favorable responses in a constructed plant–microbiome system. *Frontiers in Plant Science* 7: 497.
93. Toju H, Peay KG, Yamamichi M, et al. 2018. Core microbiomes for sustainable agroecosystems. *Nature Plants* 4: 247–257.
94. Trivedi P, Leach JE, Tringe SG, Sa T, Singh BK. 2020. Plant–microbiome interactions: From community assembly to plant health. *Nature Reviews Microbiology* 18: 607–621.
95. Turner TR, James EK, Poole PS. 2013. The plant microbiome. *Genome Biology* 14: 1–10.
96. Vandenkoornhuyse P, Quaiser A, Duhamel M, Le Van A, Dufresne A. 2015. The importance of the microbiome of the plant holobiont. *New Phytologist* 206: 1196–1206.
97. Vassilev N, Vassileva M, Martos V, et al. 2020. Formulation of microbial inoculants by encapsulation in natural polysaccharides: Focus on beneficial properties of carrier additives and derivatives. *Frontiers in Plant Science* 11: https://doi.org/10.3389/fpls.2020.00270
98. Vayssier-Taussat M, Albina E, Citti C, et al. 2014. Shifting the paradigm from pathogens to pathobiome: New concepts in the light of meta-omics. *Frontiers in Cellular and Infection Microbiology* 4: 29.
99. Wagner MR, Lundberg DS, Coleman-Derr D, Tringe SG, Dangl JL, Mitchell-Olds T. 2014. Natural soil microbes alter flowering phenology and the intensity of selection on flowering time in a wild Arabidopsis relative. *Ecology Letters* 17: 717–726.
100. Wall DH, Nielsen UN, Six J. 2015. Soil biodiversity and human health. *Nature* 528: 69–76.
101. Wassermann B, Cernava T, Müller H, Berg C, Berg G. 2019a. Seeds of native alpine plants host unique microbial communities embedded in cross-kingdom networks. *Microbiome* 7: 108.
102. Wassermann B, Kusstatscher P, Berg G. 2019b. Microbiome Response to hot water treatment and potential synergy with biological control on stored apples. *Frontiers in Microbiology* 10: https://doi.org/10.3389/fmicb.2019.02502

103. Weller DM, Raaijmakers JM, Gardener BBM, Thomashow LS. 2002. Microbial populations responsible for specific soil suppressiveness to plant pathogens. *Annual Review of Phytopathology* 40: 309–348.

104. Whitehead SR, Wisniewski M, Droby S, Abdelfattah A, Freilich S, Mazzola M. 2021. The Apple Microbiome: Structure, Function, and Manipulation for Improved Plant Health. In: Korban SS (Ed.), *The Apple Genome. Compendium of Plant Genomes*. Springer, Cham. https://doi.org/10.1007/978-3-030-74682-7_16

105. Wicaksono WA, Cernava T, Berg C, Berg G. 2020. Reconstruction of bacterial metagenome-assembled genome sequences from alpine bog vegetation. *Microbiology Resource Announcements* 9.

106. Wilson CL, Pusey P. 1985. Potential for biological control of postharvest plant diseases. *Plant Disease* 69: 375–378.

107. Wilson CL, Wisniewski ME, Droby S, Chalutz E. 1993. A selection strategy for microbial antagonists to control postharvest diseases of fruits and vegetables. *Scientia Horticulturae* 53: 183–189.

108. Wolfgang A, Zachow C, Müller H, et al. 2020. Understanding the impact of cultivar, seed origin, and substrate on bacterial diversity of the sugar beet rhizosphere and suppression of soil-borne pathogens. *Frontiers in Plant Science* 11: 1450.

109. Wu W, Lei J, Hussain M, Cao S, Du B, Wang R. 2019. Structure and function of the fruit microbiome in healthy and diseased kiwifruit. *Pakistan Journal of Agricultural Sciences* 56: 577–585.

110. Xu L, Coleman-Derr D. 2019. Causes and consequences of a conserved bacterial root microbiome response to drought stress. *Current Opinion in Microbiology* 49: 1–6.

111. Yasmin S, Zaka A, Imran A, et al. 2016. Plant growth promotion and suppression of bacterial leaf blight in rice by inoculated bacteria. *PloS One* 11: e0160688.

112. Yurgel SN, Abbey L, Loomer N, Gillis-Madden R, Mammoliti M. 2018. Microbial communities associated with storage onion. *Phytobiomes* 2: 35–41.

113. Zachow C, Grosch R, Berg G. 2011. Impact of biotic and a-biotic parameters on structure and function of microbial communities living on sclerotia of the soil-borne pathogenic fungus *Rhizoctonia solani*. *Applied Soil Ecology* 48: 193–200.

114. Zachow C, Fatehi J, Cardinale M, Tilcher R, Berg G. 2010. Strain-specific colonization pattern of Rhizoctonia antagonists in the root system of sugar beet. *FEMS Microbiology Ecology* 74(1): 124–135.

115. Zachow C, Müller H, Tilcher R, Donat C, Berg G. 2013. Catch the best: novel screening strategy to select stress protecting agents for crop plants. *Agronomy* 3(4): 794–815.

116. Zhimo VY, Biasi A, Kumar A, et al. 2020. Yeasts and bacterial consortia from kefir grains are effective biocontrol agents of postharvest diseases of fruits. *Microorganisms* 8: 428.

117. Zilber-Rosenberg I, Rosenberg E. 2008. Role of microorganisms in the evolution of animals and plants: The hologenome theory of evolution. *FEMS Microbiology Reviews* 32: 723–735.

CHAPTER 35

Biological Control of Soilborne Plant Diseases

Linda Thomashow and David M. Weller

USDA, Agricultural Research Service Wheat Health, Genetics and Quality Research Unit 365 Johnson Hall, Washington State University Pullman, Washington, USA

35.1 INTRODUCTION

Biological control, the practice of managing the population of one organism to reduce the population of another, unwanted organism, is thought to have been applied to agricultural systems as early as 2000 years ago to control plant insect pests (van Lenteren and Godfray 2005). More recently, basic and applied efforts to understand the ecology and evolution of plants and their associated microbial populations in order to increase agricultural productivity and control plant diseases are being reported with increasing frequency (Cordovez et al., 2019; Finkel et al., 2017). Benefitting especially from advances over the past decade in high-throughput genomic sequencing and bioanalytical techniques, these studies have led to recognition that the plant microbiome, like that of the human gut, is an important functional driver in its relationship with the host. Together, these microbiomes and their hosts are now seen as co-evolved holobionts (Vandenkoornhuyse et al., 2015), tightly integrated functional entities in which the microbiota expand the host's capabilities by facilitating nutrient acquisition, stress tolerance, and defense against pests and pathogens. These fundamental advances, along with ecological and environmental concerns and public demand for a safe, healthy food supply for a growing human population, have led to a resurgence of interest in biological control, and especially in the control of soilborne plant pathogens, as an important player in future, more sustainable agricultural practices.

35.2 BIOLOGICAL CONTROL OF SOILBORNE PATHOGENS, NO LONGER A COTTAGE INDUSTRY

As of 2017, over 450 entries were listed by the Organic Materials Review Institute as microbial inoculants or microbial products for use as crop fertilizers or management tools. (Finkel et al., 2017). Now applied mainly as augmentative agents released seasonally, in large numbers, over 30 million hectares worldwide to control insects and plant diseases, biological control products have reduced residual pesticide levels and gained in popularity partly because they are healthier than synthetic pesticides for users, farming communities, and plants (van Lenteren et al., 2018). Industry improvements including reduced production costs, better quality control, distribution and release methods, and more complete guidance for farmers have further stimulated acceptance among growers, as have the increased cost of conventional pesticides, the insurgence of pest resistance, and demonstrated evidence of effective performance by biologicals. Consumers increasingly seek organically-raised

Good Microbes in Medicine, Food Production, Biotechnology, Bioremediation, and Agriculture, First Edition. Edited by Frans J. de Bruijn, Hauke Smidt, Luca S. Cocolin, Michael Sauer, David Dowling, and Linda Thomashow.

foods, and public concern for policies favoring eco-sustainable agricultural practices protective of natural biodiversity worldwide, are further encouraging more sustainable methods of pest management.

Many of the reasons for the improved performance of plant growth promoting and biological control bacteria (PGPB) are due to better understanding of the physiology of the agents themselves. For example, stock cultures of commercial PGPB are now maintained and preserved such that they retain their full biological activity because it is recognized that faster-growing mutants that have lost important environmental fitness or antagonistic traits often accumulate when strains from nature are kept in the laboratory or formulated inappropriately. Similarly, target pathogens are more accurately diagnosed so that appropriate control treatments with known activity spectra can be applied when the target pathogen is susceptible to treatment. Finally, the control agent itself is introduced at a concentration sufficient to assure that it is present at an effective dose over the window of opportunity for control of the pathogen, overcoming potential failures associated with poor establishment in the infection site.

Historically, the search for PGPB was carried out by individuals or small research groups. The activity of candidate isolates often was assessed with soil isolates in the Petri dish rather than under environmental conditions that undoubtedly impact the isolate's activity in the field, completely avoiding assessment of its capacity for growth promotion or competitiveness in the environment. In contrast, current efforts are part of enormous screening operations, often conducted in the field and by industry, that yield thousands of candidate strains from among microbial populations recovered from target crop plants. Such strain collections, biased toward culturable isolates rather than the much more diverse populations that include nonculturable taxa from the soil, are then screened genetically and phenotypically to identify and characterize traits of potential interest. Further and more rigorous screening then must evaluate plant colonization and persistence in the infection court in competition with the pathogen and the indigenous microbiota, as well as the ability to persist over the time needed for protection. Only strains with desirable attributes are then advanced to a thorough evaluation of safety, development of scaled-up production and formulation methods, and registration. Collectively, the cost of these procedures can be well in excess of one million US dollars (see chapter 42).

Despite large-scale screening projects and fundamental progress toward understanding the mechanisms involved in management of plant pathogens, spore-forming *Bacillus* species that were among the first to be commercialized, are heat-tolerant, easy to formulate, and remain shelf-stable until applied, usually early in the growing season, still dominate lists of PGPB (van Lenteren et al., 2018). One species of another sporeformer, *Pasteuria nishizawae*, has been approved for control of nematodes. Fewer Gram-negative species such as *Pseudomonas*, once thought to be intolerant of formulation, also are now marketed to control soil- and seed-borne fungal pathogens of turfgrass, vegetable, and cereal crops in the USA and the European Union. Some of these pseudomonads produce phenazine antibiotics active against fungal pathogens and will be discussed in greater detail below. *Pantoea agglomerans*, another Gram-negative species, is sprayed in orchards to control fire blight of apples and pears, and a few *Streptomyces* species, which are well known for their capacity to synthesize antibiotics and form spores, also have been commercialized.

35.3 THE HOLOBIONT: FUNCTIONAL COORDINATION OF THE MICROBIOME AND ITS PLANT HOST

The rhizosphere, the thin layer of soil adjacent to and influenced by terrestrial plant roots (Figure 35.1), is host to a complex assemblage of commensal, mutualistic, or pathogenic fungi, bacteria, and archaea. These microorganisms are derived initially from seed-associated and resident soil populations that differ in response to the chemical and physical properties of the surrounding environment. Nurtured by root exudates, mucilage, sloughed cells, and other rhizodeposits synthesized from the carbon fixed photosynthetically in the foliar tissues and translocated to the roots (Hassan et al., 2019), these pioneer populations, as well as immigrant bacteria that must evade surveillance by the host plant's own immune system (see below), carry out dynamic interactions resulting in colonization of the rhizosphere, the root surface or rhizoplane, and the root interior (the endosphere), in processes that are still actively under intensive investigation.

35.4 ROOT COLONIZATION: BREACHING THE BARRIER

Colonization is a dynamic process governed by microbial determinants as well as those of the host plant, including hormone-regulated local and systemic defense

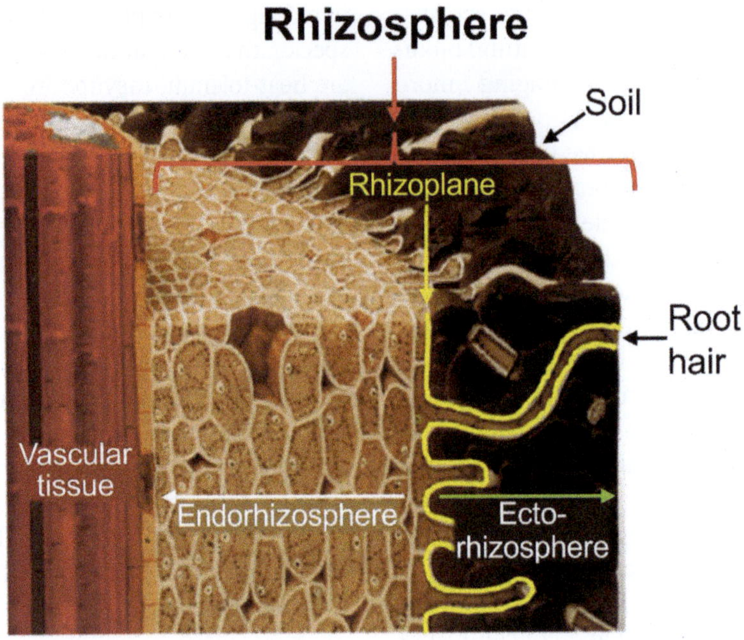

FIGURE 35.1 The rhizosphere is the zone of soil immediately adjacent to and influenced by nutrients including exudates, mucilage, and cells sloughed from the root surface. Some microbes recruited from the soil can traverse the root surface (rhizoplane) and colonize inner layers of root tissue (endorhizosphere), where they can assimilate host plant nutrients and secrete metabolites influencing root health. Figure courtesy of Wikimedia.org/Wikipedia/Commons/2/27/Rhizosphere_composition.jpg.

mechanisms; physical attributes such as growth stage, root age and morphology; and above- and below-ground physiological status as influenced by biotic and abiotic stresses that also can act directly upon soil microorganisms. Colonizers do not occupy the entire root surface but instead, establish patchy populations where nutrients are available near metabolites produced by border cells at the root cap, at exudation sites where root hairs emerge, in grooves between epidermal root cells, and where soil particles have abraded the root surface. Varying degrees of spreading over the root surface from established colonies also can occur via biofilms, hydrated matrices that enhance bacterial survival in environments challenged for water and nutrient availability. Additionally, cells released from biofilms can serve as inoculum for further colonization by transport through water-filled pores in the soil or by swimming to other colonization sites. Many *Bacillus* spp. form copious biofilms, and electron microscopy also has revealed abundant biofilms formed by phenazine antibiotic producing cells of *Pseudomonas synxantha* on the roots of wheat grown under semi-arid conditions over thousands of acres on the Columbia Plateau in the Pacific Northwest (Figure 35.2); LeTourneau et al., 2018). Numerous studies indicate that biofilm formation on roots has the capacity to provide greater drought tolerance to plants limited for adequate water. Exudates from drought-stressed plants also differ in amount and composition compared to those from irrigated plants, as does the composition and function of the

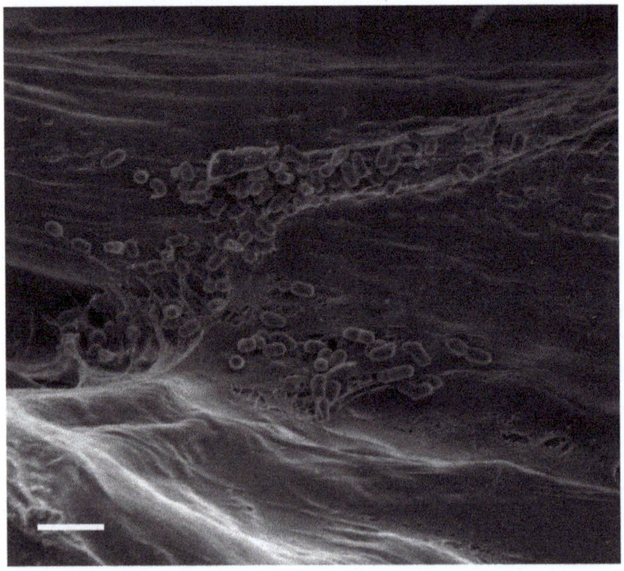

FIGURE 35.2 Cells of *Pseudomonas synxantha* 2-79 encased in a biofilm on roots of wheat. Bar indicates 2μ. Figure courtesy of Dr. Melissa Letourneau.

associated microbiota, which undergo continual successions of bacterial populations, effectively consuming available nutrients to form a dynamic physical and metabolic barrier that must be breached by introduced root-invading beneficial microbes including rhizobia (see chapter 37) and mycorrhizae (see chapter 39).

35.5 HOST IMMUNITY: INDUCED SYSTEMIC RESISTANCE LINKS ROOTS AND FOLIAR TISSUES

Triggered by recognition of microbe-associated molecular patterns (MAMPs) or other molecular elicitors that are conserved among a wide range of neutral, beneficial, and pathogenic microorganisms, the plant immune system has an active role in controlling the rhizosphere microbial composition. By binding to MAMPS, which include such shared molecules as flagellin, elongation factor Tu, lipopolysaccharide, and chitin, plant pattern recognition receptors (PRRs) activate a multilayered immune signaling cascade that within minutes can eliminate potential pathogens through ion fluxes and bursts of highly toxic reactive oxygen species. Subsequent downstream events can lead to the deposition of the physical barrier callose, antimicrobial compound accumulation, and systemic defense hormone regulation. The plant hormones salicylic acid (SA) and jasmonic acid (JA), acting in concert with other hormones like auxin, ethylene, abscisic acid, cytokinins, brassinosteroids and gibberellin, are the central modulators that coordinate this complex plant response network that ultimately must distinguish among and limit the establishment of pathogens while permitting colonization by beneficial microbes like PGPB, rhizobia, and mycorrhizae.

Given their long history of co-evolution, it is not surprising that plant-colonizers and successful pathogens have developed diverse strategies to evade or short-circuit the host immune system (Texeira et al., 2019; Yu et al., 2019). Beneficials and pathogens alike have evolved variant MAMPs that are no longer recognized by the plant. Others degrade or obscure MAMPs to prevent immune recognition, efficiently eliminating the transient reactive burst of toxic oxygen that occurs initially upon detection. Many microorganisms produce molecules that can interfere with or block the plant's immune signaling pathway. Rhizobia, in particular, both suppress MAMP-induced immune responses and produce symbiosis-related molecules recognized by the host that initiate symbiosis signaling (see chapter 37). Beneficial *Pseudomonas* species can suppress root immunity by producing organic acids that lower the environmental pH and alter the conformation of cell surface molecules, thus interfering with MAMP recognition. Some *Bacillus* species reduce biofilm formation on roots to prevent recognition and defense activation. Numerous examples have documented the modulation of the root microbiome composition upon disruption or interference with SA- or JA-dependent signaling, providing further evidence that the immune system controls access and determines which microbes can colonize the roots. In addition, activation of the immune system in above ground tissues by insects or pathogens has resulted in alterations of the rhizosphere microbiome composition in several plant species, indicating that the influence of the immune system extends throughout the entire plant.

Stimulation of the plant immune system by unrelated rhizosphere mutualistic and commensal bacteria also can induce systemic immunity (ISR) in plants, an important indirect mechanism of defense against pathogens (Figure 35.3). ISR results in enhanced sensitization that "primes" a rapid and intense systemic resistance reaction that occurs upon subsequent microbial or insect challenge, even in above-ground plant parts. The phenomenon is now known to be widespread, but host plant genotype-specific, among diverse monocots and dicots colonized by PGPB. It is, however, distinct from systemic acquired resistance (SAR), which can occur after a localized disease resistance response in plants that are genetically resistant to certain pathogens. The two forms of immune response function independently of one another and operate by entirely different molecular mechanisms: whereas ISR is regulated by a signaling pathway involving the plant hormones jasmonic acid and ethylene, signaling in SAR is mediated by salicylic acid and is accompanied by the accumulation of a suite of so-called pathogenesis-related (PR) proteins.

35.6 DIRECT ANTAGONISM OF PATHOGENS: MICROBIAL WARFARE IN THE RHIZOSPHERE

Despite the plethora of microorganisms influenced by the plant, there are comparatively few investigations of interbacterial interactions other than those mediated by direct antagonism of root pathogens by PGPB.

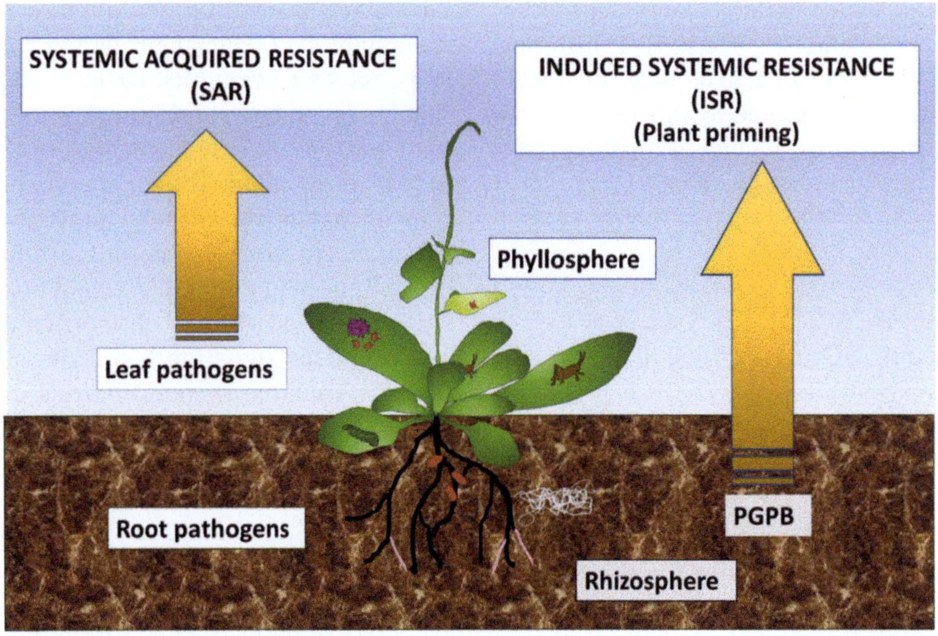

FIGURE 35.3 Systemic acquired resistance (SAR) is a plant defense response induced by direct wounding or attack by insects or pathogens. The signal is translocated by salicylic acid. In contrast, induced systemic resistance is an indirect mechanism of defense induced by commensal or mutualistic bacteria. The signal is translocated by jasmonic acid and ethylene. Figure courtesy of Cody Willmore.

Given the complexity of the root colonization process and the variety of biotic and abiotic factors that influence it, it is not surprising that introduced PGPB become unevenly distributed on root systems, varying in population size from root to root by up to several orders of magnitude. Some roots are thus likely to be completely unprotected, except perhaps by mechanisms provided by the plant itself. Even when mixtures of bacteria have been introduced to overcome these limitations, any resulting benefits have largely been additive, rather than by introduced bacteria working together. In one exception to this conclusion, however, a consortium of three unrelated bacterial species, *Stenotrophomonas, Xanthomonas*, and *Microbacterium*, was recovered from the rhizosphere of the model plant *Arabidopsis thaliana* after activation of the plant's own defenses by inoculation of the leaves with the downy mildew pathogen *Hyaloperonospora arabidopsidis* (Berendsen et al., 2018). Together, the three strains interacted synergistically in biofilm formation, induced systemic resistance against the downy mildew, and promoted the growth of *Arabidopsis*. Individually, however, they neither formed biofilms nor affected the plant significantly, indicating that upon pathogen infection, plants can adjust their root microbiomes to recruit beneficial consortia that function together as PGPB.

35.7 CHEMICAL WARFARE IN THE RHIZOSPHERE

In contrast to the mechanism of direct competition, there are numerous examples implicating control of plant pathogens by antibiosis, a form of antagonism mediated by small organic compounds (antibiotics) that, at low concentration, are deleterious to the growth or metabolic activities of other microorganisms. Hundreds of such antibiotics have been described in the literature and new ones are discovered weekly, often by analysis of genomic DNA sequences of soil-dwelling bacteria and fungi. Few antibiotics have directly been shown to be present in natural microbial habitats due to degradation or the paucity of nutrients available to support the energetically expensive process of synthesizing them, and many of them lack potency, are toxic, or are otherwise unsuitable for pharmacological use. Some are used to promote animal growth in livestock production and a few may still be applied as agricultural pesticides, although both practices are discouraged to reduce accumulation in the environment of pharmaceuticals that contribute to the selection of resistant human pathogens.

Direct evidence of antibiotic activity in nature was circumstantial or indirect as recently as 40 years ago, and although unidentified "secondary metabolites" were suspected of being active in soils, it was not until pyrrolnitrin (now modified as the commercial fungicide Fludioxinil) was isolated from *Pseudomonas fluorescens* Pf-5 (now known as *P. protegens*), that the first biochemical documentation of their involvement was obtained. This foundational study, and others like it involving other *Pseudomonas* PGPB, is the root of an environmentally friendly and sustainable means of plant protection that has grown increasingly more important with appreciation of the need to feed a growing human population. Combined with the availability of 'omics technology and new sensitive bioanalytical tools, this work has yielded insights into the mechanisms and metabolites that underpin pest and pathogen control by a few model strains and enabled recognition of new strains for commercial development.

35.8 ANTIBIOTICS AND THE SUSTAINABILITY OF WHEAT IN THE PACIFIC NORTHWEST

Phenazine antibiotics produced by some species of *Pseudomonas* and *Streptomyces* are members of a large family of compounds that mediate electron transport from phenazine-producing cells to the surrounding environment. These compounds can facilitate respiration by cells buried in dense biofilms under otherwise anaerobic conditions and they also have broad antibiotic activity, interfering with the metabolism of target plant pathogens. The nonpathogenic strain *Pseudomonas chlororaphis* O6, which produces phenazine-1-carboxylic acid (PCA) acid and its derivative compounds, has been commercialized for biological control (van Lenteren et al., 2018), and root colonization by *P. chlororaphis* M71 recently was shown to activate tomato for enhanced tolerance to water stress. (Brilli et al., 2019). PCA itself has gained pesticide certification and is marketed in China as Shenqinmycin for control of fungal pathogens of melon, pepper and rice (Zhao et al., 2018). PCA produced in the field by a natural strain of *Pseudomonas synxantha* has been recovered directly from the roots of wheat grown in the low precipitation zone (150–300 mm annual rainfall) of the Columbia Plateau in the Pacific Northwest of the USA (Mavrodi et al., 2018), which also harbors *Streptomyces* species capable of phenazine production (Dar et al., 2020). Soil moisture is a critical determinant of the dynamics and activity of these bacterial communities, and genes for synthesis of PCA were expressed even under very dry conditions (Le Tourneau et al., 2018). The bacteria formed robust biofilms on roots in the dry soil, whereas they produced no biofilm on roots in moist soils under conditions of low stress. These results, and the fact that PCA-nonproducing mutants generated no biofilms, indicate clearly that the antibiotic promotes biofilm development as an adaptation of the bacteria to drought stress, and that the biofilm forms a protective water-retentive matrix around the root. Whether wheat grown under these conditions is primed for drought stress tolerance or if the biofilm matrix itself influences the retention of soil particles and organic matter on the highly erodible Columbia Plateau remain to be determined.

In contrast to PCA synthesis on dryland wheat, another antibiotic, 2,4-diacetylphloroglucinol (DAPG), is important in the higher rainfall areas of the Columbia Plateau and eastward to the foothills of the Rocky Mountains (Mavrodi et al., 2012). DAPG producers have been recovered from the roots of diverse plant species worldwide, with the biosynthesis genes broadly conserved and embedded in the genomes of at least six species of *Pseudomonas*. The antibiotic is active against a wide variety of organisms including viruses, bacteria, fungi, and nematodes. Like PCA, DAPG is capable of inducing stress resistance, at least in some plants, and at higher concentrations it can even be phytotoxic. In *Arabidopsis*, tolerance to DAPG is shaped by a variety of biological processes, cellular components, and molecular functions. Membrane activity plays a key role among these, as do enzymes active in the detoxification of reactive oxygen species (Chae et al., 2020). The best example of biological control by DAPG may be the suppression of take-all disease of wheat caused by fungal root pathogen *Gaeumannomyces graminis* var. *tritici* known as take-all decline (TAD), which is described in detail in Chapter 36 of this volume.

Perhaps in no other place in the world are the roles of DAPG and PCA so intimately intertwined with agricultural productivity as they are in the Inland Pacific Northwest, the largest continuous cropping system in the western US. This region is unique in the number of microclimates in which wheat is grown either in a wheat-fallow rotation in the drier areas, or in rotation with barley, canola, and pulse crops in areas of higher annual precipitation or irrigation. Just as management practices and soil moisture drive the incidence and severity of root diseases throughout the region, so also do they determine which bacteria respond to the need for defense. PCA producers are active against diseases such as Rhizoctonia root rot and Fusarium crown rot, and their colonization frequency on the roots of wheat increases as soil moisture declines. In contrast, DAPG producers remain below the detection limit in soils where PCA producers are present, but are enriched on wheat in soils where precipitation exceeds 380 mm annually. DAPG and PCA producers do not act in concert to protect wheat roots, but are selectively enriched at different ends of the soil moisture spectrum. Results from over 135 years of farming in the area illustrate and emphasize the key roles of crop management and the environment in determining not only the major soilborne pathogens in these agroecosystems, but more importantly, the members of the microbiome poised to control the outcome of these plant-pathogen interactions.

35.9 DUAL CONTROL OF PATHOGENS AND INSECTS

Pseudomonas protegens strains Pf-5 and CHA0 are remarkable not only for the rich diversity of plant growth-promoting and broad-spectrum antimicrobial compounds they produce, but also for their toxicity to agriculturally important insect pests. Both strains are highly competitive root colonizers active against damping-off diseases, root rots, and wilts caused by major plant pathogens on various crops, and both have been studied in considerable detail. The genomic DNA sequence of strain Pf-5, the first DNA sequence from a biological control strain, revealed the expected genes for production of HCN, iron-sequestering siderophores and the antibiotics pyrrolnitrin, pyoluteorin, and DAPG, as well as numerous unexpected "orphan" gene clusters that together occupy a large portion of the Pf-5 genome. Notable among these are clusters coding for synthesis of the cyclic lipopeptide orfamide A, derivatives of the macrolide antibiotic rhizoxin, and the potent insecticidal toxin Fit.

Two different systems have been developed to investigate insect infection by ingestion of strains Pf-5 and CHA0 (Loper et al., 2016; Vacheron et al., 2019). For Pf-5, mid-second instar larvae of the fruit fly *Drosophila melanogaster* fed Pf-5, even at sublethal doses, were delayed in the onset of metamorphosis. The bacteria became established in the gut cavity or internal tissues of the fly larvae and induced expression of the gene encoding the antimicrobial peptide diptericin. Larvae that survived pupation had morphological defects in the eye, leg, and wing discs consistent with impaired ability to obtain or use nutrients needed for growth and development.

The events subsequent to ingestion of the closely related strain CHA0 by herbivorous lepidopteran larvae have revealed the involvement not only of toxins like Fit, cyclic lipopeptides and rhizoxins, but also of lytic enzymes including chitinase and phospholipase (Figure 35.4).

Once ingested, the bacteria resist antimicrobial cecropins central to insect antimicrobial defenses and colonize the gut, breach the peritrophic membrane, and invade and proliferate in the hemocoel, causing septicemia and eventually killing the insect. Expression in the gut of a type six secretion system by CHA0 (Vacheron et al., 2019) disrupts the commensal gut microbiota of the insect, promoting host colonization. Fit, an insect toxin produced in the hemocoel, has a major role in the toxicity of CHA0 to lepidopterans, but it does not contribute to the activity of Pf-5 against *Drosophila melanogaster*, where rhizoxin exhibits the greatest toxicity. The cyclic lipoprotein orfamide A contributes significantly to oral toxicity in both the *Drosophila* and lepidopteran systems, and chitinase produced by strain Pf-5 potentially enhances transit of the bacteria or their metabolites through the peritrophic membrane of *Drosophila*.

Whereas bacteria colonizing fruits and foliage can readily be dispersed by insects and birds, it is less apparent how root-colonizing PGPB with seemingly limited capacity for transmission might be disseminated in nature. However, such bacteria may have greater ecological versatility than previously thought. Following ingestion of strain CHA0 by larvae of a root-pest, the cabbage root fly *Delia radicum*, the bacteria persisted throughout the insect's pupal and adult stages and were transferred to new plant hosts by emerging flies that laid eggs to initiate another cycle of root infestation (Flury et al., 2019). In contrast to its lethality in lepidopterans, CHA0 had little effect on *Delia* development or its ability to damage plant roots. Similar results were obtained in a more limited study with the black vine weevil *Ohiorhynchus sulcatus*, indicating that at least with some insects, CHA0 (and potentially, other rhizobacteria) can maintain a commensal rather than a pathogenic lifestyle. Given that the toxicity of CHA0 and Pf-5 is limited to very few lineages of *Pseudomonas*, the role of insects as vectors for the dispersal of rhizosphere bacteria is worthy of further investigation.

FIGURE 35.4 Infection of the cabbage butterfly *Pieris brassicae* by *Pseudomonas protegens* CHA0.1–3:Larvae were fed on an artificial diet inoculated with a low dose of *P. protegens* CHA0. More than 90% of the larvae survived but the bacteria persisted throughout development, interfering with pupation and emergence of the adult and impairing reproduction and the number of progeny in the next generation. Pictures 1–3 show untreated animals and 4–7, the animals of 4th instar larvae fed with *P. protegens* CHA0. The bacteria were detected in living and dead larvae, pupae and adult insects. Figure courtesy of Pilar Vesga: (Flury et al., 2019) https://doi.org/10.1038/s41396-018-0317-4.

35.10 CLASSIC BIOCONTROL AGENTS: BACILLUS

Their broad antifungal activity and generation of shelf-stable spores made strains of *Bacillus* species an early choice for commercial development as biological control agents, but an understanding of the rhizosphere ecology of these bacteria is only now beginning to emerge. Upward of 12% of the genomes of biological control strains of *B. subtilis* and *B. velezensis*, the model species for plant-associated bacilli, encode a diverse assortment of antimicrobials including lytic enzymes, volatiles, lantibiotics, polyketides, and cyclic lipopeptides (CLPs) (Andríc et al., 2020). Synthesis of cyclic lipopeptides, especially, can increase upon interaction with fungal plant pathogens, but there are only a few examples of interbacterial interactions affecting their production. Bacilli *in vitro* exhibit variations in colony morphology, motility, biofilm formation, and sporulation indicating that they can sense and protect themselves from bacterial competitors, but curiously, despite the arsenal of weapons they possess, they do not seem to influence the composition of the microbiome when introduced into soil.

The three structurally diverse families of CLPs synthesized by *Bacillus* spp., the surfactins, iturins, and fengycins, trigger ISR in the host plant and play specific, complementary roles in the biological activity and ecological fitness of the producing strains. Surfactins typically are the most abundant of the three; they facilitate gliding motility and have an essential role in biofilm formation and rhizosphere colonization, but have limited activity against plant pathogens. Conversely, iturins and fengycins are broadly active against a wide range of plant pathogenic fungi. Within each CLP family, the peptide chain is invariant in length but amino acids at specific positions can vary, and the variants also can differ in cyclization and in the length, branching, and isometry of the fatty acid portion of the molecule, providing for enormous structural diversity. Strains can produce representatives of multiple CLP families and multiple analogues of a single compound, accounting for differences in biological control activity even among individual isolates within a species.

Although CLPs are the major antifungal weapons in the *Bacillus* arsenal, it has been proposed that concentrations of CLPs produced in situ may not be sufficient to mediate direct control of fungal pathogens. Instead, ISR may be mainly responsible for antifungal activity. However, it seems probable that ISR and CLPs, along with a plethora of other metabolites produced by the bacteria, probably work together to inhibit both fungal pathogens and bacterial competitors, and that these interactions differ on a strain-by-strain basis. Along with CLPs, most species of the *B. subtilis* group produce such oligopeptides as the highly conserved siderophore bacillibactin, which competes with phytopathogens for iron, and the antibiotics bacilycin, chlorotetaine, bacitracins, and rhizocticins, which target bacterial cell wall synthesis. Three classes of antibacterial polyketide antibiotics, the difficidins, macrolactins, and bacillaenes, mainly inhibit bacterial protein synthesis, and a group of ribosomally synthesized compounds produced by specific species or strains are active against Gram-positive bacteria. These include subtilin, ericin, sersacidin, amylolycin, and amylocyclicin. It is regrettable that this rich array of compounds remains relatively uninvestigated.

35.11 PERCEPTION AND RESPONSE OF BACILLUS TO FUNGAL AND BACTERIAL COMPETITORS

Numerous reports (Andrić et al., 2020) document the stimulation of CLP biosynthesis by *Bacillus* spp. exposed to fungal phytopathogens, with the particular strain of *Bacillus* and the species of the interacting fungus determining the specific CLPs produced. Thus, *B. subtilis* 98S produced iturins and fengycins when confronted with *Pythium aphanidermatum* or *Fusarium oxysporum* but not *Botrytis cinerea*, and *B. velezensis* SQR9 overproduced the iturin bacillomycin D but not fengycins. In contrast, *B. velezensis* FZB42 increased production of fengycins but not iturins when confronted with *Rhizomucor variabilis*. Collectively, these results suggest that *Bacillus* strains sense specific soluble signals presented by fungal competitors and respond by modulation of CLP synthesis. The identity of such hypothetical signals, the full range of responses they trigger in *Bacillus*, and whether they act at inhibitory or sub-inhibitory levels, is yet to be determined.

Compared to their antagonistic interactions with phytopathogenic fungi, interactions of *Bacillus* spp. and other soil and rhizosphere bacteria have resulted mainly in alterations to their physiological development (Andrić et al., 2020). One exception of significance is the synthesis by *B. subtilis* of the polyketide bacillaene as a protective response to competition with *Streptomyces* soil isolates. Biofilm production and enhanced sliding motility also have been observed, and sub-inhibitory concentrations of chloramphenicol produced by *S. venezuelae* were identified as triggering sliding motility in *B. subtilis* to avoid a competitive interaction. Many potential bacterial competitors promote sporulation in *Bacillus* upon interaction *in vitro*, and presumably also in the soil. One exception to this stimulation of developmental processes is the antibiotic DAPG produced by *Pseudomonas* spp., which acts at subinhibitory concentrations to alter colony morphology and block biofilm formation and sporulation of *B. subtilis*.

35.12 STREPTOMYCES: "PLANTS' BEST FRIENDS"?

Roughly a third of known bioactive metabolites produced by microorganisms have been isolated from *Streptomyces* and more than 60% of all antibiotics have come from these bacteria. It is, therefore, surprising that even though they form desiccation-tolerant spores in culture and should be comparatively easy to formulate, *Streptomyces* have received comparatively little attention as biological control agents until recently, this may be partly because they grow more slowly than *Pseudomonas* and *Bacillus*, especially in the rhizosphere.

S. globisporus S4-7, a member of the *S. griseus* group, is one of the best-characterized biological control agents to date and was isolated from the roots of strawberry grown in a soil naturally suppressive of a devastating wilt disease caused by the fungal pathogen *Fusarium oxysporum f.* sp. *Fragariae* (Cha et al., 2016). Selected on the basis of bacterial community DNA analyses and phenotypic characteristics, strain S4-7 inhibited *F. oxysporum in vitro* and protected strawberry from wilt disease. Further genetic dissection of the antifungal arsenal of S4-7 revealed gene clusters coding for 30 predicted secondary metabolites including four lantipeptides, two bacteriocins, two siderophores, two butyrolactones, three nonribosomal peptide synthetases (NRPSs), six terpenes, two ectoines, three polyketide synthetases, four NRPSs linked to polyketide synthetases, one melanine cluster, and a lassopeptide. Of these, the ribosomally synthesized and posttranslationally modified LanM lantipeptide named grisin contributed substantially to inhibition of the wilt pathogen and is the first class II lantipeptide from *Streptomyces* reported to have a biological function. The *lanM* gene itself has proven useful as a molecular marker for the presence in soils of populations of *Streptomyces* capable of mediating soil suppressiveness of the wilt pathogen (Kim et al., 2019a).

35.13 ENDOSPHERE COLONIZATION AND A TRIPARTITE SYSTEM BENEFITTING POLLINATORS

Given their intimate association within the host, bacterial endophytes inhabiting the internal tissues of plants and producing phytohormones, antibiotics and bioactive volatiles are ideally situated to participate in plant defense against pathogens as well as in plant development and growth (Bakker et al., 2010, Pinski et al., 2019). *Streptomyces* make up a substantial part of the endophytic microbial community, and the endophyte *S. globisporus* SP6c4, which is 99% identical to strain S4-7, enters strawberry roots and moves upward in the plant endosphere in a process actively controlled by the host (Kim et al., 2019b). Isolated originally from pollen, strain SP6c4 protects strawberry foliar tissues including flowers from *Botrytis cinerea*. it also can move downward, back to the roots, and surprisingly, it is transferred among flowers by pollinators, which it protects from common entomopathogens (Figure 35.5).

Moreover, by "hitching a ride" on insects, the bacteria translocate to other plants that may gain protection as the *Streptomyces* move systemically throughout them. Another endophytic strain of *Streptomyces* from *Arabidopsis thaliana* colonizes the root tissue not only intercellularly, but also intracellularly, with no apparent damage to the host. Moreover, exposure of a collection of *Streptomyces* endophytes from *Arabidopsis* to the phytohormones jasmonic acid, salicylic acid, and especially indoleacetic acid, resulted in a marked increase in production of antibiotics inhibitory to *B. subtilis* and *Escherichia coli* (van der Meij et al., 2018). Given the frequency with which they reside in association with plants, the breadth, diversity and potential impacts of the bioactive

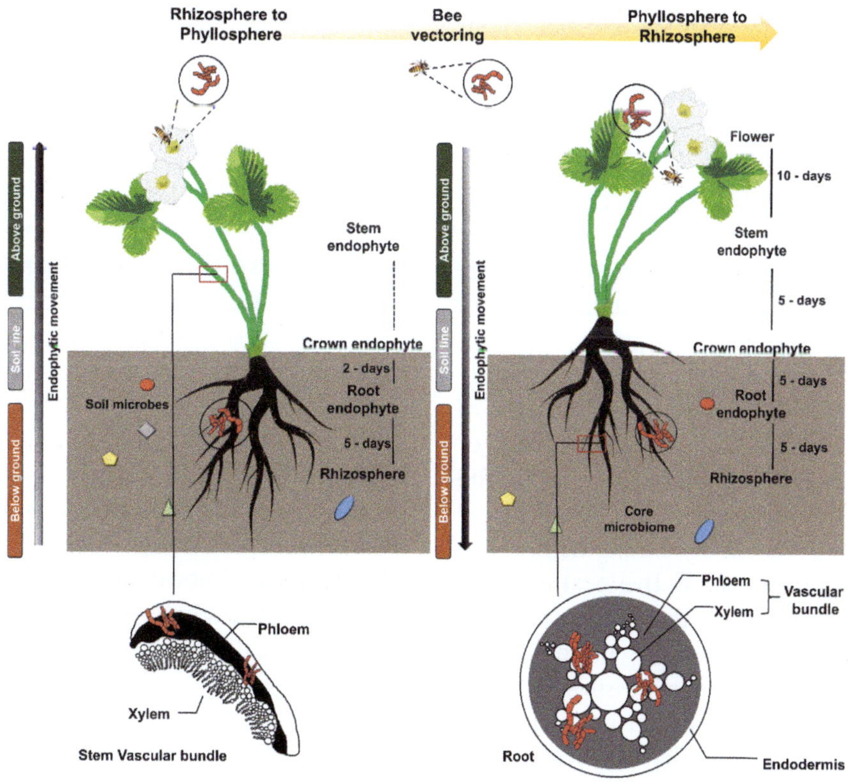

FIGURE 35.5 The rhizosphere bacterium *Streptomyces globisporus* SP6c4 colonizes the roots of strawberry plants, moves through the root and stem endosphere to the flowers, and colonizes the tissues of a pollinating bee. The bee transfers the bacterium to flowers of other, nearby plants, where it travels through the plant vascular system to the roots and rhizosphere soil. Figure courtesy of Youn-Sig Kwak.

metabolites these bacteria produce, and their responsiveness to the host metabolome, it is not surprising that within the microbiome, *Streptomyces* may yet emerge as the "plant's best friend" (Viaene et al., 2016).

35.14 BIOLOGICAL CONTROL: SAFETY AND ECOSUSTAINABILITY

With a projected annual growth rate of about 15% over the next five years, the global biological control market clearly has begun to respond to consumer demand for healthier and more sustainable alternatives to agricultural dependence on synthetic chemical fertilizers and pesticides. Still, bottlenecks on the side of industry must be overcome to sustain or even boost further growth: costs of product development must decline, shelf life must be improved, products must be made more compatible with existing agricultural practices and equipment, and farmers must become more comfortable with the technology.

Consumers also need to be assured that microbials pose minimal risks to human health and the environment because of the existence of rigid governmental regulations limiting their release and the increasingly sophisticated tools applied to evaluate and document their safety. Thus, plant and animal pathogens are now eliminated from consideration as potential PGPB based on cheap and readily available genomic DNA sequence analysis, rather than on inferences based on phenotypic properties in culture. Further, it is recognized that plants and their associated microbiomes have co-evolved for millennia, providing strong biological buffers against the persistence of introduced agents. Indeed, millions of regulatory dollars and hundreds of studies have shown that concentrations of introduced agents consistently return to low or undetectable levels within months of application or even faster, with no permanent or unwanted impact on the environment. Analysis of the mechanisms involved in biological control has revealed that many microbials function by strengthening the resistance of the plant or by making the local environment unsuitable for pathogens; those that produce inhibitory metabolites synthesize miniscule quantities in discrete, environmentally supportive sites, and the metabolites themselves have short life spans in the environment because they are rapidly degraded by indigenous microbial populations. Any metabolite that poses a potential risk to humans or the environment is investigated in depth. In addition, the buildup of resistance by pathogens, as typically occurs from overuse of chemical pesticides, has not been observed in response to the use of microbials, probably because the metabolites they produce inhibit multiple targets in the pathogen.

35.15 CLOSING THOUGHTS

Efforts over the past 30 years have provided glimpses into the highly regulated tripartite relationships among plants, their microbial partners, and pathogens. The biochemical basis of antagonism between PGPB and pathogens has been revealed in considerable detail, but the molecular signals driving these antagonistic interactions are still virtually unknown. Most notably, it has become apparent that the host plant has a much more active role in these interactions than previously imagined; its capacity to perceive and respond systemically to microbes and their metabolites overshadows virtually all relationships within the holobiont, imposing a previously unforeseen level of control over the entire microbiome. Exudates from the roots provide not only nutrients, but also signals including flavonoids, growth regulators, steroids, terpenoids, alkaloids, coumarins, and vitamins, many of which can influence microbiome composition, gene expression, and the production of secondary metabolites. Conversely, the associated microbes can induce alterations in the plant metabolome, affecting the quality and quantity of metabolites including bioactive phytochemicals and high value natural plant products. A much deeper understanding of these underlying physiological and molecular complexities must be developed if we are to fully reap the benefits of the insights we have gained for the benefit of agriculture. The foundations laid over the past three decades also must be broadened to include the overarching effects of the soil environment: its composition and quality, the availability of water and nutrients, and the impact of cultural practices on plants and their associated microbes. The knowledge gained from investigation of model systems, the availability of 'omics tools, and the combined efforts of scientists, information technology experts, and farmers must now be applied to improve the performance of agronomically relevant crops.

ACKNOWLEDGMENTS

Linda Thomashow acknowledges the USDA Wheat Health, Genetics and Quality Research Unit for support during the preparation of this chapter. She acknowledges wikimedia.org/wikipedia/commons/2/27/Rhizosphere_composition.jpg, Dmitri Mavrodi, Melissa LeTourneau, Cody Willmore, Pilar Vesga (Flury et al., 2019 https://doi.org/10.1038/s41396-018-0317-4), and Youn-Sig Kwak for Figures 35.1, 35.2, 35.3, 35.4, and 35.5, respectively.

REFERENCES

1. Andrić S, Meyer T Ongena M. 2020. *Bacillus* Responses to Plant-associated Fungal and Bacterial Communities. *Frontiers in Microbiology* https://doi.org/10:3389/fmicb.2020.01350.
2. Bakker MG, Glover JD, Mai JG Kinkel LL. 2010. Plant Community Effects on the Diversity and Pathogen Suppressive Activity of Soil Streptomycetes. *Applied Soil Ecology* 46: 35–42.
3. Berendsen RL, Vismans G, Yu K, Song Y, De Jonge R, et al. 2018. Disease-induced Assemblage of a Plant-beneficial Consortium. *The ISME Journal* 12: 1496–1507.
4. Brilli F, Pollastri S, Raio A, Baraldi R, Neri L, et al. 2019. Root Colonization by *Pseudomonas Chlororaphis* Primes Tomato (*Lycopersicum Esculentum*) Plants for Enhanced Tolerance to Water Stress. *Journal of Plant Physiology* 232: 82–93.
5. Cha J-Y, Han S, Hong H-J, Cho H, Kim D, et al. 2016. Microbial and biochemical basis of a Fusarium Wilt-suppressive soil. *The ISME Journal* 10: 119–129.
6. Chae D-H, Kim D-R, Cho G, Moon S, Kwak Y-S. 2020. Genome-wide Investigation of 2,4-diacetylphloroglucinol Protection Genes. *Arabidopsis Thaliana. Molecular Plant-Microbe Interactions* 33: 1072–1079.
7. Cordovez V, Dini-Andreote F, Carrión VJ, Raaijmakers JM. 2019. Ecology and evolution of plant microbiomes. *Annual Review of Microbiology* 73: 69–88.
8. Dar D, Thomashow LS, Weller DM, Newman DK. 2020. Global landscape of phenazine biosynthesis and biodegradation reveals species-specific colonization patterns in agricultural soils and crop microbiomes. *eLife* 9: e59726. https://doi.org/10.7554/eLife.59726.
9. Finkel OM, Castrillo G, Herrera Paredes S, GonzálezS, Dangl JL. 2017. Understanding and exploiting plant beneficial microbes. *Current Opinion in Plant Biology* 38: 155–163.
10. Flury P, Vesga P, Dominguez-Ferreras A, Tinguely C, Ullrich CI, et al. 2019. Persistence of Root-colonizing *Pseudomonas Protegens* in herbivorous insects throughout different developmental stages and dispersal to new host plants. *The ISME Journal* 13: 860–872.
11. Hassan MK, McInroy JA, Kloepper JW. 2019. The interactions of Rhizodeposits with plant Growth-promoting Rhizobacteria in the Rhizosphere: A review. *Agriculture* 9: 142. https://doi.org/10:3390.agriculture9070142.
12. Kim D-R, Cho G, Jeon C-W, Weller DM, Thomashow LS, et al. 2019b. A Mutualistic Interaction between *Streptomyces* Bacteria, Strawberry plants and pollinating bees. *Nature Communications* (2019) 10: 4802. https://doi.org/10.1038/s41467-019012785-3.
13. Kim D-R, Jeon C-W, Shin J-H, Weller DM, Thomashow L, et al. 2019a. Function and Distribution of a Lantipeptide in Strawberry Fusarium Wilt-disease-suppressive soils. *Molecular Plant Microbe Interactions* 32: 306–312.
14. LeTourneau MK, Marshall MJ, Cliff JB, Bonsall RF, Dohnalkova AC, et al. 2018. Phenazine-1-carboxylic acid and soil moisture influence Biofilm development and turnover of Rhizobacterial Biomass on wheat root surfaces. *Environmental. Microbiology* 20: 2178–2194.
15. Loper JE, Henkels MD, Rangel LI, Olcott MH, Walker FL, et al. 2016. Rhizoxin analogs, Orfamide A and Chitinase production contribute to the Toxicity of *Pseudomonas Protegens* Strain Pf-5 *to Drosophila Melanogaster. Environmental Microbiology* 18: 3509–3521.
16. Mavrodi D, Mavrodi O, Elbourne L Tetu S, Bonsall R, et al. 2018. Long-term irrigation affects the dynamics and activity of the Wheat Rhizosphere microbiome. *Frontiers in Plant Science* 9: 345. https://10.3389/fpls.2018.00345.
17. Mavrodi OV, Mavrodi DV, Parejko JA, Thomashow LS, Weller DM. 2012. Irrigation differentially impacts Populations of indigenous Antibiotic-producing *Pseudomonas* spp. In the Rhizosphere of wheat. *Applied and Environmental Microbiology* 78: 3214–3220.
18. Pinski A, Betekhtin A, Hupert-Kocurek K, Mur LAJ, Hasterok R. 2019. Defining the genetic basis of Plant-endophytic bacteria interactions. *International Journal of Molecular Sciences* 20: 1947. https://doi.org/10.3390/IJMS20081947.
19. Texeira PJPL, Colaianni NR, Fitzpatrick CR, Dangl JL. 2019. Beyond pathogens: Microbiota interactions with the plant immune system. *Current Opinion in Microbiology* 49: 7–17.
20. Vacheron J, Péchy-Tarr M, Brochat S, Heiman CM, Stojiljkovic M, et al. 2019. T6SS contributes to gut Microbiome invasion and killing of an Herbivorous Pest insect by Plant-beneficial. *Pseudomonas Protegens. The ISME Journal* 13: 1318–1329.

21. van der Meij A, Willemse J, Schneijderberg MA, Geurts R, Raaijmakers JM, et al. 2018. Inter- and Intracellular Colonization of *Arabidopsis* roots by Endophytic Actinobacteria and the impact of plant hormones on their Antimicrobial activity. *Antonie Van Leeuwenhoek* 111: 679–690.

22. van Lenteren JC, Bolckmans K, Köhl J, Ravensberg WJ, Urbaneja A. 2018. Biological Control using invertebrates and microorganisms: Plenty of new opportunities. *BioControl* 63: 39–59.

23. van Lenteren JC, Godfray HCJ 2005. European science in the Enlightenment and the discovery of the Insect Parasitoid Life Cycle in the Netherlands and Great Britain. *Biological Control* 32: 12–24.

24. Vandenkoornhuyse P, Quaiser A, Dunhamel M, Le Van A, Dufresne A. 2015. The importance of the microbiome of the plant Holobiont. *New Phytologist* 206: 1196–1206.

25. Viaene T, Langendries S, Beirinckx S, Maes M, Goormachtig S. 2016. *Streptomyces* as a plant's Best Friend? *FEMS Microbiology Ecology* 92: fiw119 https://doi.org/10.1093/femsec/fiw119.

26. Yu K, Pieterse CMJ, Bakker PAHM Berendsen RL. 2019. Beneficial microbes going underground of root immunity. *Plant, Cell & Environment* 42: 2860–2870. https://doi.org/10.1111/pce.13632.

27. Zhao X, Chen Z, Yu L, Hu D, Song B. 2018. Investigating the Antifungal activity and mechanism of a microbial pesticide Shenqinmycin against *Phoma* sp. *Pesticide Biochemistry and Physiology* 147: 46–50.

CHAPTER 36

Classification, Discovery, and Microbial Basis of Disease-Suppressive Soils

David M. Weller[1], Melissa LeTourneau[1], and Mingming Yang[2]

[1]U. S. Department of Agriculture, Agricultural Research Service, Wheat Health, Genetics and Quality Research Unit, Pullman, WA 99164-6430, USA

[2]Department of Agronomy, Northwest A&F University, Yangling 712100, P. R. China

36.1 MICROBE-BASED PLANT DEFENSE OF ROOTS

Crop plants generally lack resistance to many common soilborne fungal pathogens such as *Rhizoctonia*, *Pythium*, *Gaeumannomyces*, *Fusarium*, *Verticillium*, and *Sclerotinia* and parasitic nematodes such as *Pratylenchus* spp. As a result, plant roots rely on rhizodeposition to recruit, enrich, and support microorganisms as their first line of defense against soilborne pathogens (Schlatter et al., 2017; Weller et al., 2002; 2007). Microbial defenders in the soil and root microbiomes answer the "SOS" issued by attacked roots through competition with pathogens for nutrients and infection courts, antibiosis, parasitism, and/or predation of the pathogen, signal interference, and/or by inducing systemic resistance in the plant host (Lugtenberg and Kamilova, 2009; Pieterse et al., 2014). In addition, some microbial defenders can improve plant growth directly through mobilization of nutrients, production of growth hormones and reduction of the effects of abiotic stresses caused by extreme environmental conditions such as drought and salinity (Vaishnav et al., 2015). For example, plants over-produce ethylene during abiotic stress responses which is detrimental, but microbes that produce the ethylene-degrading enzyme 1-aminocyclopropane-1-carboxylate deaminase (ACC deaminase) can prevent the buildup of ethylene, thus mitigating damage to the plant (Glick et al., 2007). Activities of microbial defenders that directly improve growth also facilitate the plant's recovery from pathogen attack through mechanisms of enhanced root production to outgrow the pathogen.

36.2 DEFINITIONS AND EXAMPLES OF DISEASE-SUPPRESSIVE SOILS

Disease-suppressive soils are the premier examples of natural microbe-based defense of plant roots. In the classic books on biological control of plant pathogens, Baker and Cook (1974) and Cook and Baker (1983) defined the characteristics of suppressive soils that we continue to use today. More recently, Schlatter et al. (2017) described them as "soils in which, because of their microbial makeup and activity, a pathogen does not establish or persist, establishes but causes little or no disease, or establishes and causes disease at first but then the disease declines with successive cropping of a susceptible host even though the pathogen may still persist in the soil." Suppressive soils become apparent when the incidence or severity of a disease is less in a field or soil than expected compared to the disease in nearby fields or soil. Historically, soil suppressiveness has been described as

Good Microbes in Medicine, Food Production, Biotechnology, Bioremediation, and Agriculture, First Edition. Edited by Frans J. de Bruijn, Hauke Smidt, Luca S. Cocolin, Michael Sauer, David Dowling, and Linda Thomashow.

resulting from "general" or "specific" suppression or a combination of both. Suppressive soils have been described for diseases caused by a wide variety of soilborne pathogens including fungi such as *Gaeumannomyces tritici*, *Fusarium oxysporum*, *Thielaviopsis basicola*, and *Rhizoctonia solani*; oomycota such as *Aphanomyces euteiches*, *Phytophthora cinnamomic*, *Phytophthora infestans*, *Pythium splendens*, *and Pythium ultimum*; nematodes such as *Heterodera avenae*, *H. schachtii*, *Meloidogyne* spp., and *Criconemella xenoplax*; bacteria such as *Streptomyces scabies* and *Ralstonia solanacearum*; and other eukaryotes such as *Plasmodiophora brassicae* (Expósito et al., 2017; Weller et al., 2002).

36.3 GENERAL SUPPRESSION

This type of suppression is the capacity of soils, regardless of location or type, to inhibit a pathogen(s) to a limited extent because of the collective activity among organisms in the soil and the root microbiome competing with the pathogen(s) (Cook, 2014; Schlatter et al., 2017; Weller et al., 2002). General suppression is therefore an inherent characteristic of soils resulting from the thousands of microbial species typically found in a gram of soil. The well-described phenomenon of "soil fungistasis," which describes the inability of fungal spores to germinate in soil under conditions seemingly favorable for germination, is an important component of general suppression (Bakker et al., 2020). General suppression is effective against a wide variety of soilborne pathogens, but it cannot be transferred between fields or soils. Moreover, it is reduced by steaming or fumigating the soil, which lowers the population size or diversity of the soil microbial community; and it is eliminated by autoclaving or gamma irradiation. Conversely, general suppression can be increased by any practice that enhances the population size, diversity, or activity of the soil microbiome (Weller et al., 2002). Soils that show only a basal level of pathogen inhibition are designated as conducive (non-suppressive) when compared to soils with specific suppression or soils where general suppression has been greatly enhanced by management practices.

Members of the soil microbiome often live under conditions of nutrient starvation or water stress, and soilborne pathogens survive such conditions as dormant structures or hyphae embedded in plant residue. These organisms reproduce rapidly when optimal water, food, and environmental conditions return to the soil. Seed exudates and root rhizodeposits such as exudates, secretions, lysates, and mucilages fuel the renewed growth and reproduction of soil organisms, as does infection of roots and crowns by pathogens. As much as 40% of the plant's fixed carbon is released by rhizodeposition (Badri and Vivanco, 2009; Curl and Truelove, 1986; Lynch, 1990; Whipps, 1990), but the composition of rhizodeposits varies greatly according to the plant host species or cultivar, its physiological state and environmental conditions (De-la-pena et al., 2008; Lynch, 1990; Nguyen, 2003; Phillips et al., 2004). General suppression can therefore be envisioned as a microbial "furnace" that is "burning" nutrients and thus reducing their availability to pathogens. The more rapid the burn rate by the soil microbiome, the greater the level of general suppression that can reduce pathogen growth and root infection (Cook, 2014). Although general suppression has limited potential for controlling soilborne diseases, its efficacy can be significantly elevated by various organic inputs such as composts, green manures, and cover crops. These inputs promote a more extensive and more diverse microbial community with elevated metabolism, essentially increasing the "microbial burn" (Cook, 2014; Mazzola and Freilich, 2017; Stone et al., 2004; Vida et al., 2016). An excellent example of the enhancement of general suppression was observed in US Washington State apple orchards where the application of brassica seed meal to the soil resulted in significantly improved control of apple replant disease as compared to a treatment of soil fumigation or non-treated soil (Mazzola et al., 2015).

Soilborne pathogens are formidable foes for general suppression to control because agricultural soils can contain high populations of pathogen survival structures such as oospores, chlamydospores, sporangia, and sclerotia that can remain viable for years or decades. These propagules are easily "awakened" from a dormant state by seed and root exudates, especially amino acids, sugars and volatile compounds, and they rapidly germinate in preparation for root infection. For example, in fields of the Inland Pacific Northwest, USA, soils typically contain 350–400 propagules of *Pythium* spp. in the top 10–15 cm of the soil where seeds are sown. *Pythium* propagules begin to break dormancy within minutes of exposure to seed exudates, germinate within 1.5 hours, and can infect wheat seeds in as little as 24 hours (Cook and Veseth, 1991; Whipps and Lumsden, 1991). Dicotyledonous crops that contain more seed exudates than wheat and other monocotyledonous crops can be infected even faster. For example, colonization of cucumber seeds by *Pythium* occurs as early as four hours after planting, with maximum colonization by eight hours (Nelson, 1987). *Pythium* spp. can infect seeds and seedlings at temperatures as low as 5–10 °C, thus evading and out-competing antagonists that require higher temperatures for growth and antibiotic or lytic enzyme production. Many soilborne pathogens like *Pythium, Rhizoctonia*, and *Verticillium* have a broad host range, thus rendering ineffective the practice of crop rotation to reduce the inoculum potential of such pathogens, and addition of organic material that enhances general suppression can also increase the inoculum potential of most soilborne pathogens.

36.4 SPECIFIC DISEASE SUPPRESSION

Historically, specific suppression has been researchers' primary consideration with regard to suppressive soils. Baker and Cook (1974) and Cook and Baker (1983) summarized earlier literature on suppressive soils and listed the characteristics of specific suppression that are still accepted today. Specific suppression is a highly effective disease-control mechanism, but it usually targets only a single disease and pathogen or species of the same genus of pathogen. It is similarly transferrable in the field and under controlled conditions by adding a small amount (1–10% w/w) of a suppressive soil to a conducive soil. It is eliminated by pasteurization for 30 minutes at a temperature of at least 55–60 °C or by fumigation, for example with methyl bromide (Weller et al., 2002). Specific suppression results from the activity of a population of an individual microbial species or a consortium of specific microbes, and therefore the level of disease control obtained by the suppressive soil can be duplicated by adding the suppressive microbe or microbes into a conducive soil. Transferability and specificity against a single target pathogen and disease are two traits that most distinguish specific from general suppression. The effectiveness of specific suppression likely results from the ability of the responsible microbe(s) to precisely target the pathogen in key sites such as infection courts on the roots or pathogen propagules prior to or after germination. Such targeted attacks on the pathogen allow organisms involved in specific suppression to achieve disease control with population sizes of as small as 10^5 CFU g^{-1} root or soil (Raaijmakers and Weller, 2001).

Long before the microbial bases of specific disease suppression were identified, specific suppressive soils were classified by the origin of the suppressiveness, as either "long standing" or "induced." Long standing refers to suppression of unknown origin that is naturally associated with the soil and persists in the absence of the host plant (Weller et al., 2002). Two classic examples of long-standing suppression are found in soils of the US Salinas Valley, CA (Scher and Baker, 1980) and the Chateaurenard region of France (Alabouvette, 1986) that are suppressive to Fusarium wilts (caused by forma specialis of *F. oxysporum*) of a variety of different crops. The suppressiveness is specific to Fusarium wilts and is not effective against diseases caused by non-vascular *Fusarium* species such as *F. solani* or other soilborne pathogens. The timing of the development of suppressiveness in these soils is not known. In contrast, "induced" suppressive soils are initiated and sustained by practices such as crop monoculture, continuously growing crops susceptible to the target disease or addition of pathogen inoculum into the soil (Hornby, 1983, 1998; Raaijmakers and Mazzola, 2016; Schlatter et al., 2017; Weller et al., 2002). Induced suppressive soils occur worldwide and the best examples include take-all decline (Chng et al., 2015; De Souza et al., 2003; Schlatter et al., 2017), Rhizoctonia root rot decline (Roget, 1995; Schillinger and Paulitz, 2014), and potato scab decline (Lorang et al., 1989; Meng et al., 2012; Menzies, 1959).

36.5 MICROBIAL BASIS OF SPECIFIC SUPPRESSIVE SOILS

The microbial basis of most long-term and induced suppressive soils remains poorly understood (Weller et al., 2007); however, the microbial mechanisms of disease suppressiveness currently are a major topic of investigation in the fields of plant pathology, soil microbiology, and microbial ecology. This is due to shifts in consumer demand, government regulations, and environmental sustainability concerns surrounding the use of chemical pesticides. In fact, for the control of some devastating soilborne diseases, the induction of natural suppression or the application of microbes responsible for the suppressiveness may be the only economically viable approach for controlling these diseases. Fusarium wilt disease of banana, caused by *Fusarium oxysporum* f. sp. *cubense* race 4, which is threatening worldwide banana production, is one example where there is no effective sustainable conventional management strategy (Fu et al., 2020; Shen et al., 2019).

Historically, specific suppression has been thought to result primarily from the activity of individual species or select groups of microbes within the soil and/or rhizosphere microbiome (Schlatter et al., 2017; Weller et al., 2002). One example is the natural suppression of crown gall, caused by *Agrobacterium tumefaciens*. Almost five decades ago, New and Kerr (1972) reported that the incidence of crown gall on almond in South Australia correlated with the ratio of pathogenic to closely-related non-pathogenic agrobacteria, implicating non-pathogenic *A. radiobacter* in the natural suppression of the disease. This finding resulted in the development of one of the most successful biocontrol agents, the non-pathogenic *Agrobacterium* strain K84 that controls crown gall. Similarly, effects of naturally occurring non-pathogenic *Agrobacterium* strains suppressing crown gall has been observed in other orchards (Lamovšek et al., 2014).

Take-all disease, caused by *Gaeumannomyces tritici* (*Gaeumannomyces graminis* var. *tritici*) (Hernandez-Restrepo et al., 2016), is one of the most important root diseases of wheat (Cook, 2003). Take-all decline (TAD), a classic example of induced suppression, "is defined as the spontaneous reduction in the incidence and severity of take-all and increase in yield occurring with continuous monoculture of wheat or barley and following a severe attack of the disease" (Weller, 2015). As is typical for specific suppression, TAD is transferable by adding TAD soil to conducive soils and is eliminated by soil pasteurization or fumigation. Growing crops not affected by take-all reduces or eliminates suppression. However, fields that have transitioned into TAD retain a "memory" of suppressiveness such that even when TAD soils are cropped for several years to non-host crops, TAD suppressiveness is restored after only a couple cycles of wheat or barley, bypassing the need for years of wheat or barley monoculture to regain suppression (Weller, 2015). Other types of specific suppressive soils also show similar "memory" characteristics (Weller et al., 2002).

Cook (2003) estimated that in the US Pacific Northwest (PNW, states of Washington, Oregon, and Idaho) roughly 0.8 million ha of wheat suffer little damage from take-all, probably due to TAD. In the PNW, TAD results from the build-up of populations of 2,4-diacetytlphloglucinol (DAPG)-producing members of the *Pseudomonas fluorescens* complex to a density above 10^5 CFU g^{-1} root, the threshold required for take-all control (Raaijmakers and Weller, 1998, 2001; Weller et al., 2007). *P. brassicacearum* comprises at least 60% of the DAPG producers on wheat roots in PNW TAD soils. DAPG-producing pseudomonads and *G. tritici* are natural inhabitants of native grasses and sage brush that originally covered most wheat-growing regions of the PNW. When commercial wheat production began in the PNW in the 1860s, DAPG producers and *G. tritici* migrated from native plants, beginning a 160-year association with the roots of wheat. Suppressiveness of PNW TAD soils is lost when DAPG-producers are eliminated, and conducive soils become suppressive when DAPG producers are introduced by mixing in 1–10% w/w TAD soil (Raaijmakers and Weller, 1998). DAPG is isolated from the rhizosphere of wheat grown in PNW TAD but not in conducive soils (Raaijmakers et al., 1999). Similarly, take-all is suppressed when a DAPG-producing strain, but not a DAPG-deficient mutant of that strain, is added to a conducive soil (Weller et al., 2007).

All PNW isolates of *G. tritici* tested to date are inhibited by DAPG (Kwak et al., 2009), but the level of disease suppression is positively related to the sensitivity of the isolate to the antibiotic (Kwak et al., 2009; Mazzola et al., 1995). Yang et al. (2018) also demonstrated that the wheat cultivar influences the expression of take-all suppression in a TAD soil (Yang et al., 2018). This phenomenon is most likely due to differences in the amount of antibiotic produced in the rhizosphere of a cultivar rather than differences in population sizes of the DAPG producers on the roots (Kwak et al., 2012). Although DAPG producers are known to be the primary drivers of TAD in the PNW, further research will clarify our understanding of the influence of wheat genotype and soil and rhizosphere microbes on the development of TAD (Bakker et al., 2020; Chng et al., 2015; Sanguin et al., 2009).

One of the most interesting recent discoveries of induced suppression occurred in a commercial 15-year strawberry monoculture field in the Republic of Korea, which showed suppression of Fusarium wilt caused by *Fusarium oxysporum* f. sp. *fragariae*. Cha et al. (2016) applied traditional methods along with modern community analysis and bioanalytical techniques to dissect the basis of the suppressive soil by comparing it to soil from a conducive field (3-years strawberry monoculture) located only a few meters away. Actinobacteria were shown to be key players in the defense of strawberry roots against *F. oxysporum* in the suppressive soil. One representative strain isolated from protected strawberry, *Streptomyces* sp. S4-7 (reclassified as *S. globisporus*; Kim et al., 2019), produces a heat-stable antifungal thiopeptide antibiotic inhibitory to *F. oxysporum* and suppresses wilt disease when added to the conducive soil; however, antibiotic- deficient mutants of S4-7 lost the ability to suppress Fusarium wilt. Strain S4-7 aggressively colonizes the rhizosphere of strawberry, but subsequent studies showed that other *S. globisporus* strains in the soil, genetically identical to S4-7, enter the endosphere, move systemically upward in the strawberry plant and establish a tripartite mutualism with the plant and bees that pollinate the flowers. *S. globisporus* strains colonize the flowers, protecting them against gray mold caused by *Botrytis cinerea*; pollinator bees transport *Streptomyces globisporus* among flowers of other plants, and as a "fare for the ride" to other plants, *Streptomyces globisporus* protects the bees against entomopathogens (Kim et al., 2019).

The application of next generation sequencing and other "omics" technologies to analyze soil and root microbiomes, coupled with classical methods of characterizing suppressive soils, has revealed that the microbial basis of some suppressive soils is much more complex than previously thought and involves consortia of microbes (Expósito et al., 2017). The evolution in our understanding of the complexity of the microbial basis of suppressiveness is especially well exemplified in studies of the Chateaurenard Fusarium wilt suppressive soil, which originally was ascribed to the activity of non-pathogenic *F. oxysporum* (Alabouvette, 1986, 1999). Mazurier et al. (2009) later demonstrated that suppression also involved fluorescent *Pseudomonas* spp. producing phenazine antibiotics. Mechanisms that have been associated with these and other long-standing Fusarium wilt-suppressive soils include competition for carbon and induction of resistance by non-pathogenic *F. oxysporum* along with antibiosis and competition for iron by antibiotic- and siderophore-producing pseudomonads (Alabouvette, 1986; Lemanceau and Alabouvette, 1991; Scher and Baker, 1980; Weller et al., 2002). More recently, Siegel-Hertz et al. (2018) conducted a comparative microbiome analysis of the Chateaurenard wilt suppressive soil and a conducive soil and showed that operational taxonomic units (OTUs) found to be more abundant in the suppressive soil were affiliated with the bacterial genera *Adhaeribacter, Massilia, Microvirga, Rhizobium, Rhizobacter, Arthrobacter, Amycolatopsis, Rubrobacter, Paenibacillus, Stenotrophomonas,* and *Geobacter*. OTUs detected exclusively in the suppressive soil were affiliated with the fungal genera *Acremonium, Chaetomium, Cladosporium, Clonostachys, Fusarium, Ceratobasidium, Mortierella, Penicillium, Scytalidium,* and *Verticillium*. However, which members of these bacterial and fungal genera play a key role in the suppressiveness will require much more research, especially where the aim is to identify individual microbes or groups of microbes that can duplicate suppression when introduced into a conducive soil.

The microbial basis of *Rhizoctonia solani* suppressive soils appears to be similarly complex. For example, in the Netherlands, a sugar beet field developed induced suppressiveness to *R. solani* after several years of severe Rhizoctonia disease. Under greenhouse conditions, the soil from the field retained its suppressiveness, and PhytoChip-based metagenomics of the rhizosphere of sugar beet seedlings grown in suppressive and conducive soil along with culture-based methods revealed that Pseudomonadaceae, Burkholderiaceae, Xanthomonadales, Lactobacillaceae, and Actinobacteria were more abundant in the suppressive soil than in the conducive soil (Mendes et al., 2011). Metagenomic analysis (Carrión et al., 2019) later showed that in the suppressive soil, sugar beet plants establish a second line of defense against *R. solani* consisting of a consortium of root-endophytic Chitinophagaceae and Flavobacteriaceae with enhanced enzymatic activities associated with fungal cell-wall degradation and secondary metabolite biosynthesis encoded by non-ribosomal peptide synthetases (NRPSs) and polyketide synthases (PKSs) (Carrión et al., 2019).

In both Australia and in the US PNW, Rhizoctonia root rot and bare patch disease caused by *R. solani* AG-8 is the most important soilborne disease of wheat and barley grown with reduced tillage. Symptoms of this disease in the field often appear as patches of severely stunted or dead plants that can vary in size from only a few centimeters to many meters across. Rhizoctonia decline occurs in fields sown continuously to crops that are susceptible to *R. solani* AG-8 (i.e., wheat, barley), and just like TAD, the decline is characterized by several years of severe disease followed by a spontaneous reduction in disease incidence and severity (Roget, 1995; Schillinger and Paulitz, 2014). In the PNW, Schillinger and Paulitz (2014) monitored the number and size of Rhizoctonia bare patches in a 15-year no-till cropping systems field study of continuous spring wheat and a spring wheat-spring barley rotation. Bare patches first appeared in the third year of the study, peaked in size and number in the sixth year, declined in subsequent years, and totally disappeared from the plots of spring wheat and the spring wheat-spring barley rotation in the twelfth and subsequent years. Yin et al. (2013) used pyrosequencing to analyze bacterial communities from bulk and rhizosphere soil of spring wheat from inside and outside diseased and recovered patches and found considerable differences in microbial communities. For example, the rhizosphere of healthy wheat outside the patches had a higher frequency of bacteria in the class *Acidobacteria* and the genus *Gemmatimonas* than diseased wheat from inside the patches. *Dyella* and *Acidobacteria* subgroup Gp7 occurred at higher frequencies in recovered patches. More recently, Yin et al. (2021) induced suppression to Rhizoctonia root rot in a conducive soil by growing multiple cycles of spring wheat after addition of *R. solani* AG-8 inoculum. Analysis of the wheat rhizosphere community of the suppressive soil revealed that it was distinct from that of wheat grown in soil with no pathogen. Successive monoculture with the pathogen increased the abundance of *Chitinophaga*, *Pseudomonas*, *Chryseobacterium*, and *Flavobacterium*, as well as the growth-promoting nitrogen-fixing bacteria *Pedobacter, Variovorax*, and *Rhizobium*.

The application of "omics" approaches to suppressive soils has facilitated a major leap forward in our understanding the microbial basis of suppression. These techniques have streamlined the identification of microbes that alone or as part of consortia are potentially involved in suppressiveness and revealed new putative biocontrol agents as candidates for commercial development.

36.6 CONCLUDING REMARKS

In three decades, there may be 9–11$^+$ billion people on Earth to feed using no more farmland and water than is currently available for agricultural production. This is an extremely daunting challenge for farmers, especially for those in developing countries, because all parts of the production system chain, from farm to processing to transport, are challenged by consumer and government regulations to produce and deliver food more sustainably and with less carbon emissions and pollution. Thus, farmers must grow more crops using less fertilizer and synthetic chemical pesticides and less tillage to reduce erosion, yet these practices are the very ones that helped to propel the Green Revolution 60 years ago. In future agricultural systems, soilborne pathogens causing root rots, crown rots, seed, and seedling damping-offs and wilts will be even more formidable barriers to the expansion of global crop production due to the restrictions on soil fumigants and chemical seed treatments and the reduction in tillage and expansion of no-till practices. Beginning in the 1950s, production of many high-value crops such as fruits and vegetables became dependent on the use of broad-spectrum gaseous soil fumigants like methyl bromide for disease, insect, and weed control. Even as methyl bromide has been phased out over the past two decades, governments and consumers are finding replacement chemicals and fumigants increasingly unacceptable in agriculture.

Soilborne pathogens and other pests have always been a challenge to control because plants often lack resistance and chemical seed treatments are usually only effective during the seedling phase of the disease. Many scientists think that roots and the rhizosphere and associated microbiomes holds the key to the second "Green Revolution" (Gewin, 2010). Innovative new crop varieties and management practices will be developed that facilitate the recruitment and utilization of beneficial microbes from the soil microbiome for growth promotion, defense against pathogens and pests, and abiotic stresses (Schlatter et al., 2020). Disease-suppressive soils are not only sustainable and practical controls for soilborne diseases, but also are model systems for elucidating how roots signal, enrich, and sustain elite microbial defenders who answer the plant's SOS when they are attacked or suffering from environmental stresses.

REFERENCES

1. Alabouvette C. 1986. Fusarium wilts suppressive soils from the Chateaurenard region: Review of a 10-year study. *Agronomie* 6: 273–284.
2. Alabouvette C. 1999. Fusarium wilt suppressive soils: An example of disease-suppressive soils. *Australasian Plant Pathology* 28: 57–64.
3. Badri DV, Vivanco JM. 2009. Regulation and function of root exudates. *Plant, Cell and Environment* 32: 666–681.
4. Baker KF, Cook RJ. 1974. *Biological Control of Plant Pathogens*. Freeman, San Francisco, p. 433.
5. Bakker PAHM, Berendsen RL, Van Pelt JA, Vismans G, et al. 2020. The soil-borne identity and microbiome-assisted agriculture: Looking back to the future. *Molecular Plant* 13: 1394–1401. https://doi.org/10.1016/j.molp.2020.09.017.
6. Carrión V J, Perez-Jaramillo J, Cordovez V, Tracanna V, de Hollander M, et al. 2019. Pathogen-induced activation of disease-suppressive functions in the endophytic root microbiome. *Science* 366: 606–612.
7. Cha J-Y, Han S, Hong H-J, Cho H, Kim D, et al. 2016. Microbial and biochemical basis of a Fusarium wilt-suppressive soil. *The ISME Journal* 10: 119–129.
8. Chng S, Cromey M, Dodd S, Stewart A, Butler RC, Jaspers MV. 2015. Take-all decline in New Zealand

wheat soils and the microorganisms associated with the potential mechanisms of disease suppression. *Plant Soil* 397: 239–259. https://doi.org/10.1007/s11104-015-2620-4.

9. Cook RJ. 2003. Take-all of wheat. *Physiological and Molecular Plant Pathology* 62: 73–86.
10. Cook RJ. 2014. Plant health management: Pathogen suppressive soil. In: Van Alfen NK (Ed.) *Encyclopedia of Agriculture and Food Systems*. Elsevier, Amsterdam, Netherlands, pp. 441–455.
11. Cook RJ, Baker KF. 1983. *The Nature and Practice of Biological Control of Plant Pathogens*. American Phytopathology Society, St. Paul, MN, p. 539.
12. Cook RJ, Veseth RJ. 1991. *Wheat Health Management*. APS Press, Eagan, MN, USA, p. 152.
13. Curl EA, Truelove B. 1986. *The Rhizosphere*. Springer-Verlag, Berlin, p. 288.
14. De-la-pena C, Lei Z, Watson BS, Sumner LW, Vivanco JM. 2008. Root-microbe communication through protein secretion. *Journal of Biological Chemistry* 283: 25247–25255.
15. De Souza JT, Weller DM, Raaijmakers JM. 2003. Frequency, diversity, and activity of 2,4-diacetylphloroglucinol-producing fluorescent *Pseudomonas* spp. in Dutch take-all decline soils. *Phytopathology* 93: 54–63. https://doi.org/10.1094/PHYTO.2003.93.1.54.
16. Expósito RG, de Bruijn I, Postma J, Raaijmakers JM. 2017. Current insights into the role of rhizosphere bacteria in disease suppressive soils. *Frontiers in Microbiology* 8: 2529. https://doi.org/10.3389/fmicb.2017.02529.
17. Fu L, Xiong W, Dini-Andreote F, Wang B, Tao C, Ruan Y, et al. 2020. Changes in bulk soil affect the disease-suppressive rhizosphere microbiome against Fusarium wilt disease. *Frontiers of Agricultural Science and Engineering* 7: https://doi.org/10.15302/J-FASE-2020328.
18. Gewin V. 2010. An underground revolution. *Nature* 466: 552–553.
19. Glick BR, Todorovic B, Czarny J, Cheng Z, Duan J, McConkey B. 2007. Promotion of plant growth by bacterial ACC deaminase. *Critical Reviews in Plant Science* 26: 227–242.
20. Hernandez-Restrepo M, Groenewal J, Elliott M, Canning G, McMillan V, Crous PW. 2016. Take-all or nothing. *Studies in Mycology* 83: 19–48.
21. Hornby D. 1983. Suppressive soils. *Annual Review of Phytopathology* 21: 65–85.
22. Hornby D. 1998. *Take-all of Cereals: A Regional Perspective*. CAB International, Wallingford, U.K, p. 384.
23. Kim D-R, Cho G, Jeon C-W, Weller DM, Thomashow LS, et al. 2019. A mutualistic interaction between *Streptomyces* bacteria, strawberry plants and pollinating bees. *Nature Communications* 10: 4802. https://doi.org/10.1038/s41467-019012785-3.
24. Kwak Y-S, Bakker PAHM, Glandorf DCM, et al. 2009. Diversity, virulence and 2,4-diacetylphloroglucinol sensitivity of *Gaeumannomyces graminis* var. *tritici* isolates from Washington State. *Phytopathology* 99: 472–479.
25. Kwak Y-S, Bonsall RF, Okubara PA, Paulitz TC, Thomashow LS, Weller DM. 2012. Factors impacting the activity of 2,4-diacetylphloroglucinol-producing *Pseudomonas fluorescens* against take-all of wheat. *Soil Biology and Biochemistry* 54: 48–56.
26. Lamovšek J, Stare BG, Urek G. 2014. Isolation of non-pathogenic *Agrobacterium* spp. biovar 1 from agricultural soils in Slovenia. *Phytopathologica Mediterranea* 53: 130–139.
27. Lemanceau P, Alabouvette C. 1991. Biological control of fusarium diseases by fluorescent *Pseudomonas* and non-pathogenic Fusarium. *Crop Protection* 10: 279–286. https://doi.org/10.1016/0261-2194(91)90006-D.
28. Lorang JM, Anderson NA, Lauer FI, Wildung DK. 1989. Disease decline in a Minnesota potato scab plot. *American Potato Journal* 66: 531.
29. Lugtenberg B, Kamilova F. 2009. Plant-growth-promoting rhizobacteria. *Annual. Review of Microbiology* 63: 541–556. https://doi.org/10.1146/annurev.micro.62.081307.162918.
30. Lynch JM. 1990. Microbial metabolites. In: Lynch JM. (Ed.), *The Rhizosphere*. John Wiley & Sons, Chichester, UK, pp. 177–206.
31. Mazurier S, Corberand T, Lemanceau P, Raaijmakers JM. 2009. Phenazine antibiotics produced by fluorescent pseudomonads contribute to natural soil suppressiveness to Fusarium wilt. *The ISME Journal* 3: 977–991.
32. Mazzola M, Freilich S. 2017. Prospects for biological soil-borne disease control: Applications of indigenous versus synthetic microbiomes. *Phytopathology* 107: 256–263. http://dx.doi.org/10.1094/PHYTO-09-16-0330-RVW.
33. Mazzola M, Fujimoto DK, Thomashow LS, Cook RJ. 1995. Variation in sensitivity of *Gaeumannomyces graminis* to antibiotics produced by fluorescent *Pseudomonas* spp. and effect on biological control of take-all of wheat. *Applied and Environmental Microbiology* 61: 2254–2559.
34. Mazzola M, Hewavitharana SS, Strauss SL. 2015. Brassica seed meal soil amendments transform the

rhizosphere microbiome and improve apple production through resistance to pathogen reinfestation. *Phytopathology* 105: 460–469.

35. Mendes R, Kruijt M, de Bruijn I, Dekker E, van der Voort M, Schneider JHM, et al. 2011. Deciphering the rhizosphere microbiome for disease-suppressive bacteria. *Science* 332: 1097–1100.

36. Meng QX, Yin JF, Rosenzweig N, Douches D, Hao JJ. 2012. Culture-based assessment of microbial communities in soil suppressive to potato common scab. *Plant Disease* 96: 712–717.

37. Menzies JD. 1959. Occurrence and transfer of a biological factor in soil that suppresses potato scab. *Phytopathology* 49: 648–652.

38. Nelson EB. 1987. Rapid germination of sporangia of *Pythium* species in response to volatiles from germinating seeds. *Phytopathology* 77: 1108–1112.

39. New PB, Kerr A. 1972. Biological control of crown gall: Field measurements and glasshouse experiments. *Journal of Applied Bacteriology* 35: 279–287.

40. Nguyen C. 2003. Rhizodeposition of organic C by plants: Mechanisms and controls. *Agronomie* 23: 375–396.

41. Phillips DA, Fox TC, King MD, Bhuvaneswari TV, Teuber LR. 2004. Microbial products trigger amino acid exudation from plant roots. *Plant Physiology* 136: 2887–2894.

42. Pieterse CM, Zamioudis C, Berendsen RL, Weller DM, Van Wees SCM, Bakker PAHM. 2014. Induced systemic resistance by beneficial microbes. *Annual Review of Phytopathology* 52: 347–375.

43. Raaijmakers JM, Bonsall RF, Weller DM. 1999. Effect of population density of *Pseudomonas fluorescens* on production of 2,4-diacetylphloroglucinol in the rhizosphere of wheat. *Phytopathology* 89: 470–475.

44. Raaijmakers JM, Mazzola M. 2016. Soil immune responses: Soil microbiomes may be harnessed for plant health. *Science* 352: 1392–1393.

45. Raaijmakers JM, Weller DM. 1998. Natural plant protection by 2,4-diacetylphloroglucinol-producing *Pseudomonas* spp. in take-all decline soils. *Molecular Plant-Microbe Interactions* 11: 144–152.

46. Raaijmakers JM, Weller DM. 2001. Exploiting genotypic diversity of 2,4-diacetylphloroglucinol-producing *Pseudomonas* spp.: Characterization of superior root-colonizing *P. fluorescens* strain Q8r1-96. *Applied and Environmental Microbiology* 67: 2545–2554.

47. Roget DK. 1995. Decline in root rot (*Rhizoctonia solani* AG-8) in wheat in a tillage and rotation experiment at Avon, South Australia. *Australian Journal of Experimental Agriculture* 35: 1009–1013.

48. Sanguin H, Sarniquet A, Gazengel K, Moënne-Loccoz Y, Grundmann GL. 2009. Rhizosphere bacterial communities associated with disease suppressive stages of take-all decline in wheat monoculture. *New Phytologist* 184: 694–707.

49. Scher FM, Baker R. 1980. Mechanism of biological control in a Fusarium-suppressive soil. *Phytopathology* 70: 412–417.

50. Schillinger WF, Paulitz TC. 2014. Natural suppression of Rhizoctonia bare patch in a long-term no-till cropping systems experiment. *Plant Disease* 98: 389–394.

51. Schlatter D, Kinkel L, Thomashow L, Weller D, Paulitz T. 2017. Disease suppressive soils: New insights from the soil microbiome. *Phytopathology* 107: 1284–1297.

52. Schlatter DC, Yin C, Hulbert S, Paulitz TC. 2020. Core rhizosphere microbiomes of dryland wheat are influenced by location and land use history. *Applied and Environmental Microbiology* 86: e02135–19. https://doi.org/10.1128/AEM.02135-19.

53. Shen Z, Xue C, Penton CR, Thomashow LS, Zhang N, Wang B, et al. 2019. Suppression of banana Panama disease induced by soil microbiome reconstruction through an integrated agricultural strategy. *Soil Biology and Biochemistry* 128: 164–174.

54. Siegel-Hertz K, Edel-Hermann V, Chapelle E, Terrat S, Raaijmakers JM, Steinberg C. 2018. Comparative microbiome analysis of a Fusarium wilt suppressive soil and a Fusarium wilt conducive soil from the Châteaurenard region. *Frontiers in Microbiology* 9: 568. https://doi.org/10.3389/fmicb.2018.00568.

55. Stone AG, Scheverell SJ, Darby HM. 2004. Suppression of soil-borne diseases in field agricultural systems: Organic matter management, cover cropping, and other cultural practices. In: Magdoff F, Well RR (Eds.), *Soil Organic Matter in Sustainable Agriculture.* CRC Press, Boca Ratón, FL, pp. 131–177.

56. Vaishnav A, Kumari S, Jain S, Varma A, Choudhary DK. 2015. Putative bacterial volatile-mediated growth in soybean (*Glycine max* L. Merrill) and expression of induced proteins under salt stress. *Journal of Applied Microbiology* 199: 539–551.

57. Vida C, Bonilla N, de Vicente A, Cazoria FM. 2016. Microbial profiling of suppressiveness-induced agricultural soil amended with composted almond shells. *Frontiers in Microbiology* 7: 4.

58. Weller DM. 2015. Take-All decline and beneficial pseudomonads. In: Lugtenberg B (Ed.), *Principles of Plant-Microbe Interactions - Microbes for Sustainable Agriculture.* Springer, Heidelbeerg, New York, Dordrecht, London. pp. 363–370. p. 448.

59. Weller DM, Landa BB, Mavrodi OV, Schroeder KL, De La Fuente L, Blouin Bankhead S, et al. 2007. Role of 2,4-diacetylphloroglucinol-producing fluorescent *Pseudomonas* spp. in the defense of plant roots. *Plant Biology* 9: 4–20.

60. Weller DM, Raaijmakers JM, McSpadden Gardener BB, Thomashow LS. 2002. Microbial populations responsible for specific suppressiveness to plant pathogens. *Annual Review of Phytopathology* 40: 309–348.

61. Whipps JM. 1990. Carbon economy. In: Lynch JM (Ed.), *The Rhizosphere.* The Rhizosphere. John Wiley and Sons, Chichester, UK, pp. 59–97.

62. Whipps JM, Lumsden RD. 1991. Biological control of *Pythium* species. *Biocontrol Science and Technology* 1: 75–90.

63. Yang M, Mavrodi DV, Thomashow LS, Weller DM. 2018. Differential response of wheat cultivars to *Pseudomonas brassicacearum* and take-all decline soil. *Phytopathology* 12: 1363–1372.

64. Yin C, Vargas JMC, Schlatter DC, Hagerty CH, Hulbert SH, Paulitz TC. 2021. Rhizosphere community selection reveals bacteria associated with reduced root disease. *Microbiome* 9: 86. https://doi.org/10.1186/s40168-020-00997-5.

65. Yin CT, Hulbert SH, Schroeder KL, Mavrodi O, Mavrodi D, et al. 2013. Role of bacterial communities in the natural suppression of *Rhizoctonia solani* bare patch disease of wheat (*Triticum aestivum* L.). *Applied and Environmental Microbiology* 79: 7428–7438. https://doi.org/10.1128/AEM.01610-13.

CHAPTER 37

Biological Nitrogen Fixation

Frans J. de Bruijn[1] and Mariangela Hungria[2]
[1]INRAE/CNRS Laboratoire des Interactions Plantes-Microorganismes et Environment, 24 Chemin de Borde Rouge-Auzeville CS 52627, 31326 Castanet-Tolosan Cedex, France
[2]EMBRAPA Soja, Cx. Postal 231, 86001-970, Londrina, Paraná, Brazil

37.1 INTRODUCTION

Biological nitrogen fixation is a process by which nitrogen (N_2) in the atmosphere is converted into ammonia (NH_3) (de Bruijn, 2015a; Postgate, 1998). Atmospheric nitrogen or elemental nitrogen (N_2) is relatively inert: it does not easily react with other chemicals to form new compounds. Fixation processes free up the N-atoms from their diatomic form (N_2) to be used in other ways. Nitrogen fixation, natural and synthetic, is essential for all forms of life because N is required to biosynthesize basic building blocks of plants, animals, and other life forms, e.g., nucleotides for DNA and RNA and amino acids for proteins. Therefore, nitrogen fixation is essential for agriculture and the manufacture of fertilizer (Free Wikipedia Encyclopedia; de Bruijn, 2015a; Postgate, 1998).

Nitrogen fixation also refers to other biological conversions of nitrogen, such as its conversion to nitrogen dioxide. Microorganisms that fix nitrogen are bacteria called diazotrophs (de Bruijn, 2015a; Postgate, 1998). Some higher plants, and some animals (termites), have formed associations (symbioses) with diazotrophs.

The first nitrogen-fixing microbe to be discovered was *Clostridium pasteurianum* obtained by S. Winogradsky in 1893. The second free-living diazotroph, *Azotobacter chroococum* was first reported by the Dutch microbiologist M.W. Beijerinck in 1901. Symbiotic nitrogen fixation was discovered by the German agronomists H. Hellriegel and H. Willfarth in 1886–8 (de Bruijn, 2015a; Postgate, 1998).

Biological nitrogen fixation (BNF) occurs when atmospheric nitrogen is converted to ammonia by an enzyme called nitrogenase (Newton, 2015; Postgate, 1998). The simplified reaction for BNF is

$$N_2 + 8e^- + 8H^+ + 16\,MgATP \rightarrow 2\,NH_3 + H_2 + 16\,MgADP + 16\,Pi$$

The process is coupled to the hydrolysis of 16 equivalents of ATP and is accompanied by the co-formation of one molecule of H_2. In free-living diazotrophs, the nitrogenase-generated ammonium is assimilated into glutamate through the glutamine synthetase/glutamate synthase pathway (de Bruijn, 2015a; Postgate, 1998). Enzymes responsible for the nitrogenase action are very susceptible to destruction by oxygen (O_2), called "the oxygen paradox" (de Bruijn, 2015c). Many bacteria cease production of the enzyme in the presence of O_2, others fix N_2 only in anaerobic conditions, while yet others fix by employing other mechanisms, such as binding the free O_2 to the protein leghemoglobin (Free Wikipedia Encyclopedia; Postgate, 1998; de Bruijn, 2015a,b,c).

37.2 FREE-LIVING DIAZOTROPHS

Obligate anaerobes cannot tolerate oxygen even if they are not fixing N_2. They live in habitats low in O_2, such as soils and decaying vegetable matter, for example

Good Microbes in Medicine, Food Production, Biotechnology, Bioremediation, and Agriculture, First Edition. Edited by Frans J. de Bruijn, Hauke Smidt, Luca S. Cocolin, Michael Sauer, David Dowling, and Linda Thomashow.

Clostridium. Sulphate-reducing bacteria are important in ocean sediments (e.g. *Desulfovibrio*), and some Archean methanogens fix N_2 in muds and animal intestines, as well as in the deep ocean.

Facultative anaerobes can grow either with or without O_2, but they only fix N_2 anaerobically. Often, they respire O_2 as rapidly as it is supplied, keeping the amount of free O_2 low. Examples include *Klebsiella pneumoniae*, *Bacillus polymyxa*, *Bacillus macerans*, and *Escherichia intermedia* (see de Bruijn, 2015b,c).

Aerobes require O_2 to grow, yet their nitrogenase is still debilitated if exposed to O_2. *Azotobacter vinelandii* is the most studied of these organisms. It uses very high respiration rates, and protective compounds, to prevent O_2 damage. Many other species also reduce the O_2 levels in this way, but with lower respiration rates and lower O_2 tolerance (Free Wikipedia Encyclopedia; Postgate, 1998; de Bruijn, 2015a,b,c).

Oxygenic photosynthetic bacteria generate O_2 as a by-product of photosynthesis, yet some are able to fix N_2 as well. These are colonial bacteria that have specialized cells (heterocysts) that lack the O_2 generating steps of photosynthesis. Examples are *Anabeana cylindrica* and Nostoc commune. Other cyanobacteria lack heterocysts and can fix N_2 only in low light and O_2 levels (e.g., *Plectonema*).

Anoxygenic photosynthetic bacteria do not generate O_2 during photosynthesis, having only a single photosystem which cannot split water. Nitrogenase is expressed under N limitation. Normally, the expression is regulated via negative feedback from the produced ammonium ion; however, in the absence of N_2, the product is not formed, and the by-product H_2 continues unabated (biohydrogen). Example species: *Rhodobacter sphaeroides*, *Rhodopseudomonas palustris*, *Rhodobacter capsulatus* (Free Wikipedia Encyclopedia; de Bruijn, 2015a,b,c; Postgate, 1998).

37.3 SYMBIOTIC NITROGEN-FIXING BACTERIA

Rhizobia are Gram-negative bacteria with the ability to establish N_2-fixing symbioses on legume roots and on the stems of some aquatic legumes. Rhizobia invade their host plants via root-hairs or crack entry infection to enter the root nodule, which the host is constructing for the potentially beneficial microsymbionts (de Bruijn, 2015a,c; Sprent, 2001). The symbiosis between leguminous plants and alpha- or betaproteobacterial species (rhizobia) is the best known and also agronomically most important. During this interaction the rhizobia differentiate into a form called "bacteroids" and are contained in intracellular compartments called "symbiosomes" within a specific plant organ called the "nodule" (see Figure 37.1) where they fix N_2. Within the nodule, nitrogen compounds are synthesized and exported to the host plant, contributing to its nutrition and growth (see Section 37.5). Similarly, soil bacteria called *Frankia* induce the formation of nitrogen-fixing nodules on actinorhizal plants (de Bruijn, 2015a,c; Postgate, 1998).

The Legume family – Fabaceae – includes taxa such as clovers, soybeans, beans, alfalfa, lupines, kudzu, and peanuts. Not all legumes are able to establish symbiosis with rhizobia. The ability is restricted to some subfamilies, with the Papilionoideae comprising the majority of the nodulated legumes.

In addition to enhancing plant growth, when the plant dies, the fixed nitrogen is released, making it available to other plants and also contributes to fertilizing the soil. Therefore, in many traditional and organic farming practices, fields are rotated through various types of crops, which may include clover, alfalfa, pigeon pea, leucena, often referred to as "green manure" (Free Wikipedia Encyclopedia; de Bruijn, 2015a).

Legumes also include major food and feed crop species, such as soybean, pea, clover, chickpea, alfalfa, and mungbean, representing the second largest group of food and feed crops grown globally. They are cultivated on 12–15% of available arable land and are responsible for more than 25% of the world's primary crop production, with 247 million tons of grain legumes produced annually (de Bruijn, 2015a; Ferguson et al., 2010). In

FIGURE 37.1 Nodulation at the root crown, associated with inoculant applied to seeds in an area "traditionally" cropped with soybean in Brazil.

addition to food and feed crops, legumes such as soybeans and *Pongamia pinnata* have gathered a great deal of attention as future sustainable biofuel sources because of their high seed oil content (Ferguson et al., 2010).

Although by far the majority of plants able to form N_2-fixing root nodules are in the legume family Fabaceae, there are a few exceptions: *Parasponia*, is a tropical Celtidaceae also able to interact with rhizobia. Actinorhizal plants such as alder (Alnus) and bayberry, can also harbor N_2-fixing nodules, thanks to a symbiotic association with *Frankia* actinobacteria and are widespread in nature (Franche et al., 2009). These plants belong to 25 genera distributed among eight plant families. The ability to fix N_2 is far from universally present in these families. For instance, of 122 genera in the Rosaceae, only four genera are capable of fixing N_2. All these families belong to the orders Cucurbitales, Fagales, and Rosales, which together with the Fabales form a clade of eurosids. In this clade, Fabales were the first lineage to branch off; thus, the ability to fix N_2 may be plesiomorphic and subsequently lost in most descendants of the original nitrogen-fixing plant; however, it may be that the basic genetic and physiological requirements were present in an incipient state in the last common ancestors of all these plants, but only evolved to full function in some of them (Free Wikipedia Encyclopedia; de Bruijn, 2015a,b,c). The symbioses of members of the genera Azolla, Cycas, and Gunnera are less known (Lindström et al., 2015; Osborne and Bergman, 2009).

Associative N_2 fixers form a group of mutualist systems in which there is some interdependence between the partners though both can grow satisfactorily apart (Postgate, 1998; de Bruijn, 2015a,b,c). They involve grasses (Gramineae) primarily. The prospect of finding bacteria with associative relationships with rice, other cereals, or graminaceous crops has provoked considerable interest in those diazotrophic bacteria which inhabit the rhizosphere of such plants (de Bruijn, 2015a,c; Postgate, 1998). In recent years several types have been reported, for example species of the genera *Herbaspirillum*, *Gluconacetobacter*, *Azoarcus*, and *Paraburkolderia* (Postgate, 1998; de Bruijn, 2015a,b,c).

37.4 EVOLUTION AND TAXONOMY OF NITROGEN-FIXING ORGANISMS

Mineral nitrogen (NH_4^+, NO_3^-) fertilization of crop plants in agriculture forms the basis for modern agriculture, often at the expense of the utilization of biological nitrogen fixation (BNF) (Stoddard et al., 2009). However, whereas it is estimated that one kilogram of industrially reduced nitrogen requires 1.5 kilograms of oil, BNF is not dependent on fossil fuels but uses energy derived from the sun. In agriculture, BNF by legumes in symbiosis with N_2-fixing bacteria can, under optimal growth conditions, provide all the nitrogen necessary for plant growth (Lindström et al., 2015). In the case of symbiotic BNF with plants or photosynthetic microorganisms, the solar energy is captured by the photosynthetic system and stored in carbon compounds which are transported to the site of N_2 fixation, in the case of legumes the nodules (Lindström et al., 2015; Udvardi and Poole, 2013). These in turn use their metabolic machinery for the production of ATP and reducing power needed for BNF. In non-symbiotic organisms the energy is provided by corresponding energy-yielding mechanisms. It is estimated that the reduction of one mole of N_2 requires 16–24 moles of ATP (Franche et al., 2009). Thus, BNF is always an investment for plants and bacteria, both in the formation of nodules and in the process of fixing N_2. In their review, (Lindström and Mousavi 2010) summarize key features of BNF and its importance for the input of N to the biosphere and also give some recent figures of amounts of N fixed in diverse systems. The nitrogenase enzyme complex is the center for the BNF process. From an evolutionary point of view, the origin of nitrogenase is still unclear. A set of genes named *nif* (for example *nifH*, *nifD*, and *nifK* encoding the core subunits of nitrogenase) is present in most N_2-fixing organisms (for a review see Franche et al., 2009; Newton, 2015). The genes *nifHDK* and 16S rRNA can be used as genetic markers when studying the occurrence N_2-fixing organisms in diverse taxa (Figure 37.2) (Franche et al., 2009).

An important component of nitrogenase enzyme complexes is the metal cofactor. The genes *nifHDK* encode a nitrogenase with an iron–molybdenum (Fe–Mo) cofactor. However, some nitrogen fixers use vanadium (V, genes *vnfHDK*) instead of Mo in the cofactor and others just iron (genes *anfHDK*) (Schüddekopf et al., 1993). It is a wide-spread perception that V- and Fe-nitrogenases are ancient forms of Mo-nitrogenase. However, (Boyd et al., 2011) by examining multiple nitrogenase proteins phylogenetically and structurally, came to a different conclusion and proposed that Mo-nitrogenase first emerged within the hydrogenotrophic methanogenic archaea already before the "great oxidation event" (GOE), which occurred 2.45–2.22 billion years ago (Anbar et al., 2007). The alternative V- and Fe-nitrogenases then evolved in response to Mo limitation, and Mo-nitrogenase became prevalent when

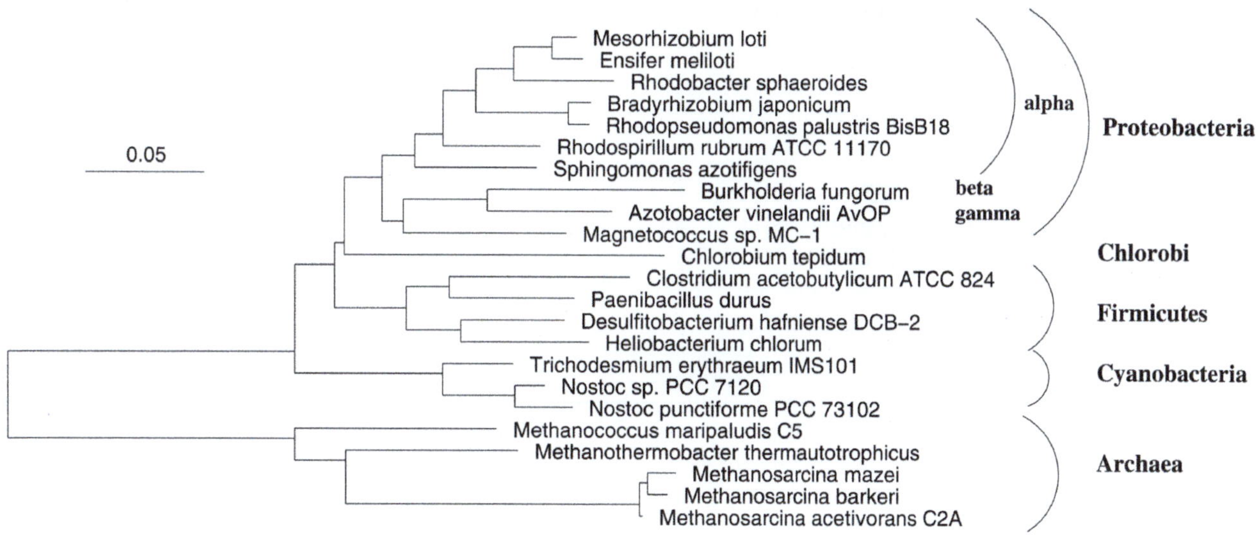

FIGURE 37.2 Phylogenetic 16S rRNA gene tree with prokaryotes carrying *nif* genes (by courtesy of German Jurgens). Reprinted from (Franche et al. 2009) with the kind permission of Springer Verlag.

bioavailable Mo increased in the biosphere at the time of the GOE. According to the analyses by (Fani et al., 2000), *nifDK* and *nifEN* (involved in the biosynthesis of the Fe-Mo-cofactor) constitute a paralogous gene family which arose via gene duplications, perhaps even before the appearance of the last common ancestor (LUCA). They further proposed that the *nifDKEN* genes were present before the divergence of Bacteria and Archaea.

This suggestion was supported by results compiled in a mini-review by (Zehr et al. 2003), who most comprehensively describe how widespread nitrogenase (represented by *nifH*) is across diverse ecosystems and prokaryotic taxa. (Hartmann and Barnum 2010) inferred phylogenies of 60 *nifK* and *nifDK* sequences across diverse prokaryotic phyla (represented by 16S rRNA gene sequence) and proposed that both horizontal and vertical gene transfer could explain the observed patterns though within phyla vertical transmission seemed to be prevalent.

The evolution of symbiotic nitrogen fixation is, in comparison with the evolution of *nif* genes, a recent phenomenon. The symbiosis between leguminous plants and alpha- or betaproteobacterial species (rhizobia) is the best known and also agronomically most important. Rhizobia invade their host plants via root-hairs or crack entry infection to enter the root nodule, which the host is constructing for the potentially beneficial microsymbionts (de Bruijn, 2015a,c; Sprent, 2001). Recently, similarities between arbuscular mycorrhizal and rhizobial infections have led to the proposal that N_2-fixing root symbioses have the same evolutionary origin as mycorrhizal ones. These aspects have recently been reviewed both in terms of signaling factors (Gough and Cullimore, 2011) and a central signaling pathway in the plant (Singh and Parniske, 2012). Figure 37.3 shows a range of evolutionary events regarding symbioses. Mycorrhizal symbioses are thought to predate the rhizobial ones, which implies that the rhizobia adopted pathways in the plants already evolved in response to mycorrhizal infections (Lindström et al., 2015).

37.5 NODULATION AND NITROGEN FIXATION

The symbiosis requires partner recognition and specificity, which are partially mediated by plant-exuded molecules, mainly flavonoids (Hassan and Mathesius, 2015). Nodulation also involves the development of root nodules from root cortical and pericycle cells, which is controlled in part through polar auxin transport and auxin accumulation. Flavonoids control auxin transport and breakdown during nodule development. Thus, flavonoids have roles in the rhizosphere as signals to rhizobia, and in addition act inside the root to control nodule development. Flavonoids act in several ways during the symbiosis, such as chemically attracting rhizobia, inducing nodulation (*nod, nol, noe)* gene synthesis in rhizobia, acting as determinants of host specificity, as regulators of root development or by altering plant defense responses (Cooper, 2004; Hassan and Mathesius, 2015). The regulation of *nod* genes in rhizobia by root-secreted flavonoids has been well studied. The nodulation genes encode enzymes that synthesize

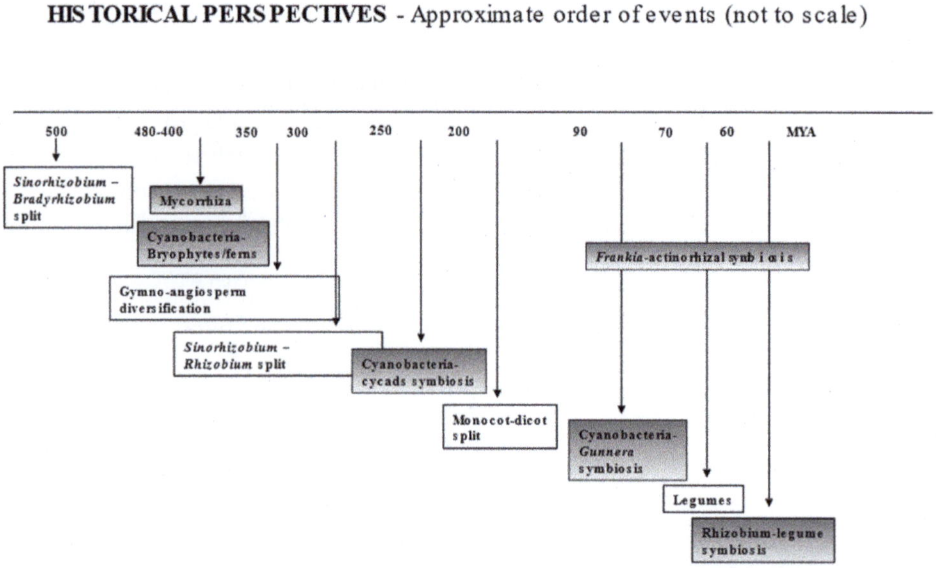

FIGURE 37.3 Approximate order of historical events important for the evolution of nitrogen-fixing symbioses, compiled from the sources (Turner and Young 2000), (Hungria et al., 2002), (Lavin et al. 2005), (Osborne and Bergman 2009), (Sprent 2008), and (Wang et al. 2010) and references in them. Reprinted from (Lindström and Mousavi, 2010) with the kind permission of John Wiley & Sons Ltd.

species-specific lipochitooligosaccharide molecules known as Nod factors. The Nod factors are required for formation of nodules and the invasion of rhizobia in the host plant (Hassan and Mathesius, 2015; Spaink, 1995). They give a (high) degree of specificity to the interaction depending on their structure (side groups to the sugar backbone) and are received by specialized plant receptors.

Nodulation is a multi-stage developmental process, and plant hormones are involved at every stage, from influencing the initial Nod factor production by rhizobia to the formation of nodules (Breakspear et al., 2015). During nodulation, rhizobia typically attach near the tip of a growing root hair which curls around the attached rhizobia to entrap them, creating the so-called infection pocket. The infection thread initiates from this infection pocket as a narrow tube around 1 μM in diameter (Breakspear et al., 2015). The infection thread then grows through the root hair cell as it becomes colonized by the rhizobia. The infection thread subsequentially grows into the subtended outer cortical cells and finally into the cells of the growing nodule primordium, where it branches and eventually releases bacteria enclosed by host membrane into nodule cells where they fix nitrogen after differentiation into bacteroids (Breakspear et al., 2015). As mentioned before, in some symbioses rhizobia invade their host plant via crack entry to form the nodules.

37.6 INOCULUM PRODUCTION

In several important legume crops the contribution of BNF under field conditions is maximized with the use of inoculants carrying elite rhizobial strains. For this rhizobia must be grown in (large) fermenters and the inoculants produced are usually coated on legume seeds before sowing. There are many factors and phases involved in the production of high quality legume inoculants which are the result of many years of research and development input from researchers, quality assurance/quality control (QA/QC) refinements and industry processes. Failure to observe the standards required to produce a highly effective inoculant and poor application technique renders the practice of inoculation suboptimal. Inoculant quality is most commonly defined in terms of effectiveness of the rhizobial strain, numerical standards for rhizobia and determinants of shelf life. Numerical standards have been established and changed through decades and also vary with the plant species and the country. For example, in Brazil, Argentina, Uruguay, and Paraguay (Mercosur Market) the minimum concentration for rhizobial inoculants must be 10^9 rhizobia per g or mL of product. From these numbers the recommendation evolved to minimal concentrations applied to the seeds, e.g., 1.2 million cells per soybean seed in Brazil (Santos et al., 2019). For less common formulations, such as granules and freeze dried products, Australia has assigned a value of 10^{11} rhizobia ha^{-1} to inoculants applied directly to the soil. Products that are easy to use, such as pre-inoculated pasture legume seeds, are broadly adopted; however, quality can be poor and this has implications for the contribution of BNF over vast areas of pasture and rangelands (Deaker et al., 2015). The generation of highly efficient inocula exceeds the scope of this review, but has a very important impact on the efficacy of their use (Deaker et al., 2015; Santos et al., 2019).

37.7 APPLICATION OF SYMBIOTIC NITROGEN FIXATION IN AGRICULTURE

Since the symbiotic nitrogen fixing symbiosis is the most utilized and studied, an example of its agricultural application is presented. It involves the interactions between nodulating and N_2-fixation proficient bradyrizobia and soybean with an emphasis on Brazilian agriculture.

Soybean is the most important legume cropped worldwide, with important roles in human and animal nutrition, besides broad industrial applications in terms of oil production for the generation of biofuel. The United States and South American countries account to most of the world grain production, but cropped areas in "low-income food-deficit countries" are increasing, highlighting an important role as a protein source for impoverished populations (Hungria and Mendes, 2015; Vanlauwe et al., 2019). The legume has been rated as the highest contribution of biological nitrogen fixation (BNF) among the grain producers, with reports of rates of up to 450 kg N ha^{-1} (Ormeño-Carrillo et al., 2013). South American countries represent an excellent example of success of BNF with soybean, reaching high yields with no N-fertilizer; furthermore, yield average increases of 8% associated with annual reinoculation are consistent (Hungria and Mendes, 2015). Low contributions of BNF are usually associated with countries with breeding and cropping-managements with high inputs of N-fertilizers. Interestingly, only a few *Bradyrhizobium* strains have been used in inoculants for the soybean crop worldwide. These strains have a high capacity of BNF, but there are possibilities of searching or developing new effective strains, also including important traits as higher competitiveness and tolerance to pesticides (Hungria and Mendes, 2015). Besides the economic importance, saving billions of dollars with the replacement of N-fertilizers, the environmental impacts of BNF by soybean should be emphasized, with lower emission of GHG and pollution of water reservoirs (Hungria and Mendes, 2015). Despite considerable differences in yield among countries, e.g., averages of 2925 and 2666 kg ha^{-1} in North and South America, respectively, of 1202 in Africa and as low as 990 kg ha^{-1} in countries such as Nigeria (USDA, 2020), one may conclude that reasonable yields may be achieved even with low-input management practices (Hungria and Mendes, 2015). One major example is the improvement of BNF performance with soybean in Africa with the application of phosphorus (Vanlauwe et al., 2019). Considering the high demand of N by soybean crops, and that the great majority of the soybean in South America, Africa, and in some Asian countries is cropped on soils of very low N content and with no application of N-fertilizer, we may conclude that, in a variety of environments, biological nitrogen fixation makes significant contributions to soybean crops (Hungria and Mendes, 2015).

In an exercise to quantify the global contribution of soybean BNF, (Herridge et al. 2008) came up with a figure of 16.4 Tg annually, representing 77% of the N fixed by all crop legumes. However, the increasing shift production from USA, with high inputs of N-fertilizers to the soybean-maize cropping system, to the low N-fertilizer input agriculture practiced in South America and Africa, should soon result in significantly greater contributions from BNF (Hungria and Mendes, 2015).

The few reports of low contributions of BNF by soybean require evaluation of limiting factors related to such cases. We hypothesize that the majority of reports of low BNF contributions result from breeding and cropping-managements with high inputs of N-fertilizers. Besides soybean, important contributions of BNF are found in other legumes, such as the fodder crops bitter lupin and alfalfa, with up to 527 and 470 kg ha^{-1} of N, respectively (Ormeño-Carrillo et al., 2013). However, for all symbioses, attention should be paid to global environmental changes, with emphasis on increasing periods of drought that may represent serious limitations to BNF, highlighting the need to identify genotypes with high BNF and high tolerance of moisture deficiency (Chen et al., 2007; Purcell, 2009; Sinclair et al., 2007). Other factors may be critical to the success of the symbiosis, such as incompatibility with seed treatment with pesticides (e.g., Campo et al., 2009; Hungria and Mendes, 2015). The need for selecting strains with higher capacity of fixing nitrogen has long been recognized (Baldwin and Fred, 1929; Fred et al., 1932); however, since the late 1920s, Fred and colleagues commented that native soil strains were not always effective, requiring a screening process to increase productivity by inoculation (Fred et al., 1932).

The most important soybean breeding programs have been in the USA, Brazil, and Argentina, and as these countries do not represent centers of genetic domestication of the host plant, when their soils have not been previously cropped with the legume and in the absence of inoculants, populations of compatible rhizobia are small or absent (e.g., Brutti et al., 1998; Elmore, 1984; Ferreira and Hungria, 2002; Vargas et al., 1994). In such cases, soybean cropped for the first time in soils of low N content show significant responses to inoculation with selected strains (Hungria et al., 2006; Hungria and Mendes, 2015; McLoughlin et al., 1990), as exemplified in Figure 37.4.

37.8 ECONOMIC, SOCIAL, AND ENVIRONMENTAL IMPLICATIONS

Economic returns of BNF with soybean in South America are outstanding, with the Brazilian case used here as an example. Considering that for each 1000 kg of soybean produced, the plant requires 80 kg of N, the crop needs 240 kg N ha^{-1} to achieve a yield of 3000 kg ha^{-1}. Generally, Brazilian soils are poor in N and usually supply only 15 to 30 kg N/crop, in which case the plants need an additional 210–225 kg N ha^{-1}. Considering an N-utilization efficiency of 50%, the N application need is doubled, to ≈ 420 kg N (Hungria and Mendes, 2015). Therefore, considering the crop season of 2019/2020 in Brazil, the price of the N-fertilizer at the local market in 2020, the area cropped of 36.95 Mha, and the national mean yield of 3,373 kg ha^{-1}, we have that the cost of applying N-fertilizer would cost ≈ US$ 14.4 billion nationally. A similar picture is found in other countries of South America; clearly, the continent could not produce half of the world's yield of soybean without BNF (Hungria and Mendes, 2015).

FIGURE 37.4 Soybean grown in the first year areas and soils poor in N. Contrasting effect of green plants inoculated and fixing nitrogen and yellowish non-inoculated plants. Experiment performed in the Planaltina, central Brazil, in the 1980s (Photo M.A.T. Vargas, Embrapa).

The economy related to soybean agribusiness affects thousands of farmers and attendant industry workers, with far-reaching social implications. Similarly, soybean and BNF profoundly affect small-holding farmers and the economies of several African countries as exemplified in the project N2Africa (http://www.n2africa.org.br; Vanlauwe et al., 2019). There is no doubt that, in these countries, there is potential for BNF to play increasingly important roles in ensuring food security and improving life quality.

The environmental effects of using N-fertilizers should not be forgotten. Pollution of aquifers, rivers, reservoirs, and lakes are associated with leaching and runoff of nitrate. In addition, the emission of GHG related to N-fertilizers is an important issue. Brazil again is a good example, as the government launched a program called "ABC" (Agriculture of Low Carbon Consumption) and among the targets is the adoption of agricultural practices involving BNF, with the commitment of an area expansion of 5.5 million hectares with a mitigation potential of 10 million tons of CO_2. Considering the synthesis, transport and application of N-fertilizers in Brazil, (Hungria et al. 2013) used a rate of 4.5 kg of e-CO_2 kg^{-1} (CO_2 equivalent) of N-fertilizer, and adding to that an estimate of 6.2 kg of e-CO_2 kg^{-1} due to direct and indirect emissions of N_2O, we have a figure of 10.7 kg of e-CO_2 kg^{-1} of fertilizer. Applying these numbers to Brazil we would conclude that BNF mitigated the emission of about 165 million tons of e-CO_2 in the last crop season.

37.9 CONCLUSIONS

Biological nitrogen-fixation by beneficial diazotropic microbes is an inexpensive and environmentally sound alternative to chemical fertilizers. It has been particularly and successfully applied in agriculture to legumes, but as pointed out by de Bruijn [2015d] is a promising technology for cereals as well.

ACKNOWLEDGMENTS

Frans J. de Bruijn acknowledges the LIPM, INRAE, CNRS, and the Labex Tulip for supporting his Editorial Activities. M. Hungria has the leadership of the Brazilian National Institute of Science and Technology-Plant-Growth Promoting Microorganisms for Agricultural Sustainability and Environmental Responsibility (CNPq 465133/2014, Fundacao A1raucaria-STI 043/2019, CAPES).

REFERENCES

1. Anbar AD, Duan Y, Lyons TW, Arnold GL, Kendall B, et al. 2007. A whiff of oxygen before the great oxidation event? *Science* 317: 1903–1906.
2. Baldwin IL, Fred EB. 1929. Strain variation in the root nodule bacteria of clover. *Rhizobium Trifolii. Journal of Bacteriology.* 17: 17–18.
3. Boyd ES, Hamilton TL, Peters JW. 2011. An alternative path for the evolution of nitrogen fixation. *Frontiers Microbiology* 205: 1–6.
4. Breakspear A, Liu C, Cousins DR, Roy S, Guan D, Murray JD. 2015. The role of hormones in rhizobial infection. In: de Bruijn FJ (Ed.), *Biological Nitrogen Fixation.* Wiley-Blackwell, Hoboken, NJ, USA, pp. 557–566.
5. Brutti L, Rivero E, Basurco JCP, Nicolás M, Iriarte L, Abbiati N, et al. 1998. Persistence of *Bradyrhizobium japonicum* in arable soils of Argentina. *Applied Soil Ecology* 10: 87–94.
6. Campo RJ, Araujo RS, Hungria M. 2009. Nitrogen fixation with the soybean crop in Brazil: Compatibility between seed treatment with fungicides and bradyrhizobial inoculants. *Symbiosis* 48: 154–163.
7. Chen P, Sneller CH, Purcell LC, Sinclair TR, King CA, Ishibashi T. 2007. Registration of soybean germplasm lines R01-416F and R01-581F for improved yield and nitrogen fixation under drought stress. *Journal Plant Regulators* 1: 166–167.
8. Cooper JE. 2004. Multiple responses of rhizobia to flavonoids during legume root infection. *Advances in Botanical Research Incorporating Advances in Plant Pathology* 41: 1–62.
9. de Bruijn FJ. (Ed.). 2015a. *Biological Nitrogen Fixation.* Wiley-Blackwell, Hoboken, NJ, USA, pp. 1–1196.
10. de Bruijn FJ. (Ed.). 2015b. Introduction. In: *Biological Nitrogen Fixation.* Wiley-Blackwell, Hoboken, NJ, USA, pp. 1–4.
11. de Bruijn FJ. 2015c. Biological nitrogen fixation. In: Lugtenberg B (Ed.), *Principles of Plant-Microbe Interactions.* Springer International Publishing, Switzerland, pp. 215–224.
12. de Bruijn FJ. 2015d. The quest for biological nitrogen fixation in cereals: A perspective and prospective. In: de Bruijn (Ed.), *Biological Nitrogen Fixation.* Wiley Blackwell, Hoboken, NJ, USA, pp. 1089–1101.
13. Deaker R, Hartley E, Gemell G. 2015. Making the most of high quality legume inoculants. In: de Bruijn FJ (Ed.), *Biological Nitrogen Fixation.* Wiley-Blackwell, Hoboken, NJ, USA, pp. 991–1002.
14. Elmore RW 1984. Soybean inoculation - When is it necessary? Historical materials from University of Nebraska-Lincoln Extension. Paper 743.G84-737. Available at http://digitalcommons.unl.edu/extensionhist/743.
15. Fani R, Gallo R, Liò P. 2000. Molecular evolution of nitrogen fixation: The evolutionary history of the *nifD, nifK, nifE* and *nifN* genes. *Molecular Biology and Evolution* 51: 1-1.
16. Ferguson BJ, Indrasumunar A, Hayashi S, Lin M-H, Lin Y-H, Reid DE, Gresshoff PM. 2010. Molecular analysis of legume nodule development and autoregulation. *Journal of Integrative Plant Biology* 52:: 61–76.
17. Ferreira MC, Hungria M. 2002. Recovery of soybean inoculant strains from uncropped soils in Brazil. *Field Cropping Research* 79: 139–152.
18. Franche C, Lindström K, Elmerich C. 2009. Nitrogen-fixing bacteria associated with leguminous and non-leguminous plants. *Plant Soil* 321: 35–59.
19. Fred EB, Baldwin IL, McCoy E. 1932. *Root Nodule Bacteria of Leguminous Plants.* The University of Wisconsin Press, Madison, p. 343.
20. Gough C, Cullimore J. 2011. Lipo-chitooligosaccharide signaling in endosymbiotic plant-microbe interactions. *Molecular Plant Microbe Interactions* 24(8): 867–878.
21. Hartmann LS, Barnum SR. 2010. Inferring the evolutionary history of Mo-dependent nitrogen fixation from phylogenetic studies of *nifK* and *nifDK. Journal of Molecular Evolution* 71: 70–85.
22. Hassan S, Mathesius U. 2015. Flavonoids play multiple roles in symbiotic root-rhizosphere interactions. In: de Bruijn FJ (Ed.), *Biological Nitrogen Fixation.* Wiley-Blackwell, Hoboken, NJ, USA, pp. 501–509.
23. Herridge DF, Peoples MB, Boddey RM. 2008. Global inputs of biological nitrogen fixation in agricultural systems. *Plant Soil* 311: 1–18.
24. Hungria M, Campo RJ, Mendes IC, Graham PH. 2006. Contribution of biological nitrogen fixation to the

N nutrition of grain crops in the tropics: The success of soybean [*Glycine max* L. Merr.] in South America. In: Singh RP, Shankar N, Jaiwal PK (Eds.), *Nitrogen Nutrition and Sustainable Plant Productivity*. Studium Press, LLC, Houston, Texas, pp. 43–93.

25. Hungria M, Mendes IC. 2015. Nitrogen fixation with soybean: The perfect symbiosis. In: de Bruijn FJ (Ed.), *Biological Nitrogen Fixation*. Wiley-Blackwell, Hoboken, NJ, USA, pp. 1009–1023.

26. Hungria M, Mendes IC, Mercante FM. 2013. A fixação biológica do nitrogênio como tecnologia de baixa emissão de carbono: Avaliação nas culturas do feijoeiro e da soja. Embrapa Soja, Londrina, 22 p. (Documentos, 337)Kistner C, Parniske M. 2002. Evolution of signal transduction in intracellular symbiosis. *Trends Plant Science* 7: 511–518.

27. Lavin M, Herendeen PS, Wojciechowski MF. 2005. Evolutionary rates analysis of *Leguminosae* implicates a rapid diversification of lineages during the Tertiary. *Systematic Biology* 54: 575–594.

28. Lindström K, Aserse AA, Abdollah-Mousavi S.. 2015. Evolution and Taxonomy of nitrogen-fixing organisms with emphasis on rhizobia. In: de Bruijn FJ (Ed.), *Biological Nitrogen Fixation*. Wiley-Blackwell Publishers, Hoboken, NJ, USA, pp. 21–46.

29. Lindström K, Mousavi SA. 2010. *Rhizobium* and other N-fixing symbioses. In: *Encyclopedia of Life Sciences (ELS)*. John Wiley & Sons, Ltd, Chichester. DOI; 10.1002/9780470015902. a0021157.

30. McLoughlin TJ, Hearn S, Alt SG. 1990. Competition for nodule occupancy of introduced *Bradyrhizobium japonicum* strains in a Wisconsin soil with a low indigenous bradyrhizobia population. *Canadian Journal Of Microbiology* 36: 839–845.

31. Newton WE. 2015. Recent advances in understanding nitrogenases and how they work. In: de Bruijn FJ (Ed.), *Biological Nitrogen Fixation*. Wiley-Blackwell Publishers, Hoboken, NJ, USA, pp. 7–20.

32. Ormeño-Carrillo E, Hungria M, Martínez-Romero E. 2013. Dinitrogen-fixing prokaryotes. In: Rosenberg E, DeLong EF, Lory S, Stackebrandt E, Thompson F (Eds.), *The Prokaryotes*. Springer-Verlag, Berlin and Heidelberg, pp. 427–451.

33. Osborne B, Bergman B. 2009. Why does *Gunnera* do it and other angiosperms don't? An evolutionary perspective on *Gunnera-Nostoc* symbiosis. In: Pawlowski K (Ed.), *Prokaryotic Symbionts in Plants*. Springer, Germany, pp. 207–224.

34. Postgate JU. 1998. *Nitrogen Fixation 3rd*, Cambridge University Press, Cambridge, UK.

35. Purcell LC. 2009. Physiological responses of N_2 fixation to drought and selecting genotypes for improved N_2 fixation. In: Emerich DW, Krishnan HB (Eds.), *Nitrogen Fixation in Crop Production*. American Society of Agronomy, Madison, USA, pp. 211–238.

36. Santos MS, Nogueira MA, Hungria M. 2019. Microbial inoculants: Reviewing the past, discussing the present and previewing an outstanding future for the use of beneficial bacteria in agriculture. *AMB Express* 9: 205.

37. Schüddekopf K, Hennecke S, Liese U, Kutsche M, Klipp W. 1993. Characterization of *anf* genes specific for the alternative nitrogenase and identification of *nif* genes required for both nitrogenases in. *Rhodobacter Capsulatus. Molecular Microbiology* 8: 673–684.

38. Sinclair TR, Purcell LC, King A, Sneller CH, Cheng P, Vadez V. 2007. Drought tolerance and yield increase of soybean resulting from improved symbiotic N_2 fixation. *Field Cropping Research* 101: 68–71.

39. Singh S, Parniske M. 2012. Activation of calcium- and calmodulin-dependent protein kinase (CCaMK), the central regulator of plant root endosymbiosis. *Current Opinion Plant Biology* 15: 444–453.

40. Spaink HP. 1995. The molecular basis of infection and nodulation by rhizobia: The ins and outs of sympathogenesis. *Annual Reviews Phytopathology* 33: 345–368.

41. Sprent JI. 2001. *Nodulation in Legumes*, Royal Botanic Gardens, Kew, UK.

42. Sprent JI. 2008. 60Ma of legume nodulation. What's new? What's changing? *Journal Experimental Botany* 59: 1081–1084.

43. Stoddard FL, Hovinen S, Kontturi M, Lindström K, Nyänen A. 2009. Legumes in Finnish agriculture: History, present status and future prospects. *Agriculture Food Science* 18: 191–205.

44. Turner SL, Young JPW. 2000. The glutamine synthetases of rhizobia: Phylogenetics and evolutionary implications. *Molecular Biology Evolution* 17: 309–319.

45. Udvardi M, Poole PS. 2013. Transport and metabolism in rhizobia-legume symbioses. *Annual Reviews Plant Biology* 64: 781–805.

46. USDA (United States Department of Agriculture). 2020. World Agricultural Production (Circular Series WAP 3-20 March 2020) Available at https://apps.fas.usda.gov.psdonline/circulars/production.pdf.Retrieved on March 1, 2020.

47. Vanlauwe B, Hungria M, Kananpiu P, Giller KE. 2019. The role of legumes in the sustainable intensification of smallholder agriculture: Lessons learned

and challenges for the future. *Agriculture Ecosystems Environment* 284: 106583.

48. Vargas MAT, Mendes IC, Suhet AR, Peres JRR. 1994. Inoculation of soybean in Cerrado soils with established populations of. *Bradyhrizobium Japonicum. Reviews Microbiology* 25: 245–250.

49. Wang B, Yeunn LH, Xue J-Y, Liu Y, Ané J-M, Qiu Y-L. 2010. Presence of three mycorrhizal genes in the common ancestor of land plants suggests a key role of mycorrhizas in the colonization of land by plants. *New Phytologist* 186: 514–525.

50. Zehr JP, Jenkins BD, Short SM, Steward GF. 2003. Nitrogenase gene diversity and microbial community structure: A cross-system comparison. *Environmental Microbiology* 5: 539–554.

CHAPTER 38

A Primer on the Extraordinary Efficacy and Safety of Bacterial Insecticides Based on *Bacillus Thuringiensis*

Brian Federici

Department of Entomology and Institute for Integrative Genome Biology University of California, Riverside Riverside, California 92521

38.1 INTRODUCTION

After World War II, broad spectrum synthetic chemical insecticides were used throughout the world for controlling insect pests and vectors of disease, with many millions of pounds still used annually. Despite benefits, their detrimental effects on non-target vertebrate and invertebrate populations, along with the evolution in insecticide resistance, led to the development of organic and IPM programs. This favored development of bacterial insecticides based primarily on *Bacillus thuringiensis* (Bt) due to its relatively high specificity for target populations and lack of significant effects on most non-target species including humans. Nowadays, most products used in agriculture and forestry are based on Bt subsp. *kurstaki* (Btk), but Bt subsp. aizawai (Bta), which has a slightly different Cry protein complement, is also commonly used, especially against lepidopteran species of the genus *Spodoptera*, which includes important pests such as the beet armyworm (*S. exigua*) and fall armyworm (*S. frugiperda*). With respect to controlling nuisance and vector mosquitoes and black flies as well as certain other related pests such as mushroom flies, Bt subsp. *israelensis* (Bti) is the dominant species used in commercial products. The insecticidal proteins of Bti are also Cry proteins, but differ from those of Btk and Bta (Federici and Siegel, 2007; Glare and O'Callaghan, 2000; Nester et al., 2002). In addition, Bti contains a cytolytic (Cyt) protein, important because it synergizes with its Cry proteins and delays the evolution of resistance in mosquito and black fly populations.

The most significant technical advance, however, for reducing the use of chemical insecticides over the last two decades of the twentieth century, was the development of insecticidal commodity crops genetically engineered with Bt Cry proteins that target specific pest species (Charles, 2001; Federici, 2005; Gould et al., 2002). These crops, commonly referred to as genetically modified organisms (GMOs), were first released in 1996, and were quickly adopted by farmers in more than 25 countries, with the combined annual acreage currently being more than 400 million acres, averaging about 60% of the major commodity crops worldwide. The principal Bt crops are corn engineered to control caterpillar and beetle pests, cotton to control different caterpillar and beetle pests, and soybeans and rice to control caterpillar pests.

The above Bt developments have been the subject of numerous reviews since 2000, several mentioned below, but the focus of this chapter is to briefly review the key

Good Microbes in Medicine, Food Production, Biotechnology, Bioremediation, and Agriculture, First Edition. Edited by Frans J. de Bruijn, Hauke Smidt, Luca S. Cocolin, Michael Sauer, David Dowling, and Linda Thomashow.

properties of Bt and then summarize studies of the past several years using examples that deal with human safety and environmental effects of Bt insecticides and GMO crops. The use of Bt products have benefited humanity by greatly reducing the use of chemical insecticides and their effects on non-target organisms, but not without controversy.

38.2 SUMMARY OF BT BIOLOGY AND ITS MODE OF ACTION

Bt is a common Gram-positive, spore-forming aerobic bacterium that can be readily cultured on simple microbiological media from a variety of environmental sources including soil, water, plant surfaces, grain dust, dead insects, and insect feces. When nutrients and environmental conditions are sufficient for growth, the spore germinates, producing a vegetative cell that grows and reproduces by binary fission. Cells continue to multiply until one or more nutrients becomes insufficient for continued vegetative growth. Cells then sporulate, producing a spore and a parasporal body (PB) that separate from each other when cells lyse at the end of sporulation. The PB is composed of crystals of one or more insecticidal Cry or Cry and Cyt proteins (Figure 38.1), the crystalline form having evolved for environmental stability. These proteins are actually protoxins that must be activated by proteolytic cleavage under alkaline conditions, which is typical of caterpillar and larval mosquito midguts (stomachs) (Adang et al., 2014; Hechel, 2020). Once activated, Cry proteins must bind to protein receptors on midgut microvilli, where they then form cationic pores. This causes an influx of cations followed by water,

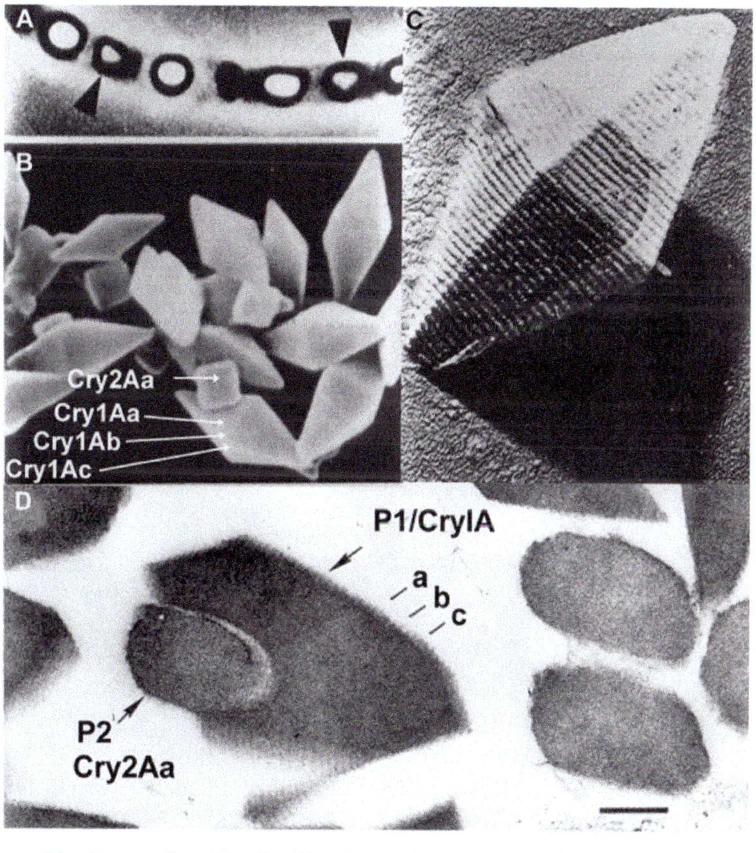

FIGURE 38.1 Typical parasporal bodies produced by *Bacillus thuringiensis*. A. Sporulated cells of *B. t.* subspecies *kurstaki* (HD-1) prior to lysis. The arrowheads point to the parasproral body adjacent to the spore. B and C. Bipyramidal crystals that contain Cry proteins. In Btk the bipyramidal crystal contains Cry1Aa, Cry1Ab, and Cry1Ac, whereas in *B. t.* subspecies *aizawai* (Bta) the crystal is composed of Cry1Aa, Cry1Ac, Cry1C, and CryD. Note that in B there is also a cuboidal crystal, which contains Cry2Aa. This crystal is usually embedded in the bipyramidal crystal, as shown in D. P1 and P2 were the designations assigned to the two crystals prior to the era of Bt molecular biology, early during which it was determined that the HD-1 isolate of Btk contained three Cry proteins, which accounted for its broad target spectrum, and why it was the first isolate commercialized. These Cry proteins were the first engineered to construct Bt cotton, maize, and soybeans.

ultimately lysing midgut cells and leading to larval death, typically with 24 hours (Figure 38.2). Cyt proteins also must be proteolytically activated, but do not require protein receptors; rather, they are highly hydrophobic and bind directly to the lipid membrane of microvilli, causing lesions much larger than cationic channels formed by Cry proteins (Federici et al., 2010).

Genes encoding more than 300 Cry proteins and 20 Cyt proteins have been cloned and sequenced. An important aspect of most Bt subspecies is that the *cry* and *cyt* genes are typically encoded on plasmids, not integrated into the Bt chromosome. Thus, if heated during vegetative growth to 40 °C, Bt is cured of the plasmids coding for its insecticidal protein genes, and thus does not produce PBs. Without the PB, such isolates are not insecticidal, and moreover share several characteristics consistent with other bacilli, including *B. cereus*, which can cause food poisoning.

In addition, Bt can also produce other types of insecticidal proteins during vegetative growth, referred to as vegetative insecticidal proteins (VIPs, not discussed here).

Below is a summary of the steps from ingestion to midgut binding that account for the specificity and safety of Bt insecticides and Bt crops.

(i) Cry and Cyt crystals must be ingested. This is one reason sucking invertebrates like aphids and spiders are not sensitive to proteins in Bt insecticides and Bt crops.

(ii) After ingestion, Bt endotoxin crystals must be activated. Effective activation requires that crystals dissolve under alkaline digestive juices, pH 8–10, and be cleaved at both the N-terminus and the C-terminus. This pH range is typical of the midgut lumen of target insects including lepidopteran and mosquito larvae. Most, but not all, non-target invertebrates have neutral or only slightly acidic or basic midguts. Importantly, under the highly acidic stomach conditions of most vertebrates, especially humans (pH 1–2), Cry and Cyt crystals may dissolve, but once in solution the proteins are rapidly degraded to non-toxic peptides by gastric juices, typically within two minutes.

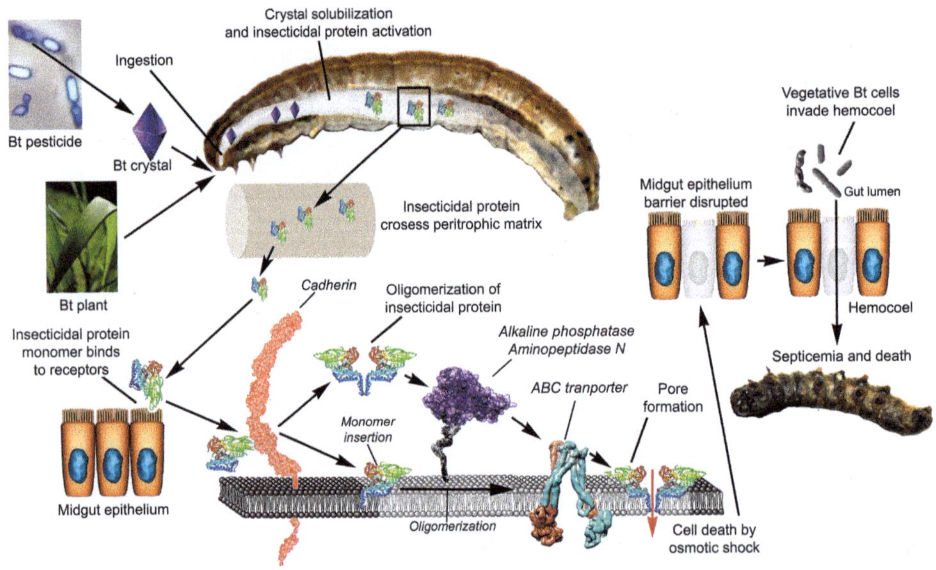

FIGURE 38.2 Schematic illustration of the mode of action of a typical Cry protein from ingestion to death of host midgut cells. Ingestion is followed by solubilization and then activation under alkaline conditions (pH 8–11) in the larval midgut lumen. Activated toxin monomers bind to microvillar "receptors" and then to cadherin protein subunits, after which the monomers oligomerize. The oligomers then bind to midgut proteins such as alkaline phosphatases, aminopeptidases, or ABC transporters, and then insert into the microvillar membrane forming cation channels (often referred to as pores). Other data suggest that monomers can insert directly in the microvillar membrane, and then oligomerize forming a cation channel. The cation channels lead to water inflow into cells followed by cell lysis. It is thought that binding can also activate cell death pathways. After significant cell gut damage, the pH drops to eight or below, spores germinate, invade, and colonize larvae, resulting in death 24–48 hours after ingestion. Illustration courtesy of Professor Juan Luis Jurat-Fuentes, University of Tennessee, Knoxville, Tennessee, USA.

(iii) Once activated, the toxin must bind to protein or lipid "receptors" on the midgut microvillar membrane. Most chewing insects that ingest insecticidal crystals do not have appropriate receptors or binding conditions, and thus are not killed.
(iv) After binding to a midgut receptor, the toxin must enter the cell membrane and form cation-selective channels.

In Bt crops, Cry proteins have been engineered for optimal activity, and generally whether full length or truncated, do not form typical Cry crystals, basically circumventing dissolution in the second phase of activation. Nevertheless, whether produced in plant cells as a full length or truncated protoxin, Cry proteins must still be properly activated, which happens in plant cells or after ingestion.

A very important part of the Bt life cycle is that spores are ingested with Cry and Cyt crystals, and after the activated proteins begin destroying midgut cells, the alkaline gut juices leak into the hemocoel, raising the pH, which paralyzes the larvae through actions on exposed nerves. Simultaneously, the hemolymph leaks into the midgut lumen, reducing the alkaline pH to eight or lower. Lowering the pH to at least eight is required for Bt spores to germinate. As Cry and Cyt proteins are not toxic to humans, any putative food poisoning can only take place after the enterotoxins coded for by the Bt chromosome (or *B. cereus* chromosome) are synthesized. Thus, because the active ingredients of bacterial insecticides based on Btk, Bta, or Bti consist only of spores and Cry and Cyt proteins, they have two inherent, i.e., built-in, safety measures for humans, (1) they contain no enterotoxins capable of causing food poisoning, and (2) the highly acidic pH and proteolytic enzymes of the human stomach rapidly degrade Bt insecticidal proteins and prevent spore germination. This is because the supernatant that may contain these toxins is discarded, or if present, is degraded during the formulation process (spore/parasporal body concentration by centrifugation and spray-drying).

In caterpillars and mosquito larvae, vegetative cells germinated from spores invade and colonize the larvae, eventually completing the life cycle by sporulating and producing new parasporal bodies.

Bt crops are not engineered with spores, so they have no Bt enterotoxins. The primary targets for their engineered Cry proteins are the first instar larvae that emerge from the eggs, i.e., the most sensitive stage. While the Cry proteins can kill these larvae, the process is likely accelerated by other bacteria on the plant surface that enter the midgut when the larvae begin feeding.

38.3 SUMMARY OF EARLIER STUDIES ON BT SAFETY

During the 1990s several concerns emerged about Bt safety due to its close relationship to *B. cereus*, the latter which by then was known to produce protein toxins during vegetative growth that could cause vomiting (emetic toxins) or diarrhea (enterotoxins). Additionally, the development of genetically engineered insect-resistant crops based on Cry proteins became a controversial new technology that triggered new concerns about the safety of Bt and its Cry proteins to humans and non-target organisms. These concerns resulted in a variety of studies that confirmed the safety of Bt insecticides to humans and most non-target organisms, and showed that these novel crops based on Cry proteins were remarkably safe for vertebrates and non-target organisms as well (Bishop et al., 1999; Federici and Siegel, 2007; McClintock et al., 1995; Mehere et al., 2002; Siegel, 2001). It was determined, for example, that Bt strains used in commercial products were capable of producing enterotoxins during vegetative growth. However, no evidence was found that these were present in commercial products at levels that could cause illness, or that these caused outbreaks of gastrointestinal disease as a result of proper or even improper use of Bt products. Moreover, detailed epidemiological studies carried out in the late 1990s in Canada and New Zealand found no significant confirmed or lasting health impacts on residential suburban human populations treated aerially with commercial Btk formulations to control invasive or natural lepidopteran pests (The New Zealand Experience, 1997; Auckland Regional Public Health Service, 2003; Pearce et al., 2002; Petrie et al., 2003; Teschke et al., 2001; Valadares de Amorim et al., 2001). In addition, an unusual level of public concern arose, fanned by the media, over the use of genetic engineering techniques to produce food crops (Charles, 2001; Gould et al., 2002). This had the positive effect of resulting in numerous extensive studies to determine the effects of Bt crops on non-target organisms in the laboratory, along with field studies under commercial growing conditions.

Aside from the safety to humans of Bt products used to control invasive forest pests as in the cases listed above in Canada and New Zealand, Bt products, especially those based on Btk, are used routinely to control caterpillar forests pests in many countries. A concern here is the effect of Bt on non-target lepidopterans, especially endangered species. The latter populations were reduced, but did recover (Boulton, 2004; Boulton et al., 2007; Liebhold et al., 2021; Nunez-Mir et al., 2021; Owens and Johns, 2021). With respect to songbirds that feed on caterpillars and adults, no significant reductions in populations were observed associated with decreases in gypsy moth populations with the exception of the spotted towhee, *Pipilo maculatus* (Sopuck et al., 2002), which apparently eventually returned to normal levels.

It should be realized that studies of direct effects of the Bts used in insecticides on humans are rare because strains and formulations are tested on surrogate vertebrates, primarily rats and rabbits. Data obtained by various health agencies and published summaries of these data demonstrate the Btk and Bti strains used in commercial formulations are not infective or toxic to these test vertebrate animals (Table 38.1). Similar results were obtained repeatedly despite the very large amounts of test materials used in these studies, which are 100 to more than 1000-fold the amount of material used to control insect pests. For example, no health effects were found when rats, over a period of three weeks, were fed massive doses of Bt strains, 1×10^{12} spore/crystal mixtures per rat (Bishop et al., 1999). Thus, most assessments of the safety of Bt to humans are based on a lack of any confirmed food poisonings resulting from the use of Bt insecticides on vegetable crops during their first 50 years of use.

38.4 KEY SIMILARITIES AND DIFFERENCES BETWEEN B. THURINGIENSIS AND B. CEREUS

The close relationship between these two species raised concerns by some investigators in northern Europe that Bt rather than *B. cereus* may be the cause of occasional outbreaks of food poisoning in humans. Food poisoning caused by *B. cereus* is due to two types of toxins, emetic toxins, which cause vomiting, and enterotoxins, which cause gastrointestinal discomfort that often leads to diarrhea (Granum and Lund, 1997; Gharib et al. 2020; Stenfors Arnesen et al., 2008; Jessberger et al., 2020). Several studies carried out over the past three decades have demonstrated that commercial strains of Bt do not produce emetic toxins, but do contain genes for enterotoxins capable of being synthesized during vegetative growth. The *B. cereus* emetic toxin is the cyclic peptide cereulide. Cereulide is denatured by digestive enzymes, so the onset of symptoms, nausea and vomiting, is normally observed soon (0.5 to 5 hours) after ingestion of contaminated food. This indicates that cereulide must be present at an elevated concentration (10^5–10^8 cells g^{-1}) at the time of ingestion to produce emetic toxicity.

The enterotoxins of *B. cereus* are HBL (Hemolysin BL), a hemolytic toxin considered to be its primary virulence factor causing diarrhea; NHE, a non-hemolytic toxin causing mild diarrhea; and the rarer enterotoxins T and cytotoxin K (CytK).

TABLE 38.1 Food sources and key pathogens that cause food poisoning*

Food	Primary pathogens
Poultry	*Campylobacter species Salmonella*
Vegetables and leafy greens	*Escherichia coli, Salmonella, Listeria*
Fish and shellfish	*Gambierdiscus toxicus alga*
Rice	*Bacillus cereus*
Delicatessen meats	*Staphylococcus aureus Listeria*
Unpasteurized dairy	Brucella, Campylobacter
	E. coli, Listeria, Salmonella
Eggs	*Salmonella*
Fruit	*Listeria*
Sprouts	*Salmonella, E. coli, Listeria*

Source: www.healthline.com/nitrition/foods-that-cause-food-poisoning.
*Note that *Bacillus* cereus is not listed as a key pathogen of vegetables and leafy greens, but it is for rice.

38.5 ENTEROTOXINS OF B. THURINGIENSIS

The phenotypic similarities of *B. thuringiensis* and *B. cereus*, and the significant overlap of their genomic characteristics, suggested that under conditions appropriate for spore germination and vegetative growth, *B. thuringiensis* could produce enterotoxins similar to those of *B. cereus*. By screening soil isolates of Bt using commercial test kits for enterotoxin production, it was shown that 83% of new isolates tested positive for enterotoxin production. In another study, Bt strains were screened for enterotoxin genes using PCR, and for potential enterotoxicity by testing culture fluid for cytotoxicity. Six strains of Btk were analyzed, and five were determined to contain genes coding for enterotoxins T, the HBL complex, and the NHE complex. Other studies of commercial Btk strains indicated that they were capable of producing these enterotoxins during vegetative growth, and more recent studies have shown that during vegetative growth, different strains of Bt from soils, vegetable crops, and even associated with rare cases of food poisoning can produce the above and certain other enterotoxins (Johler et al., 2018; McIntyre et al., 2008). Thus, it has been accepted for decades that Bt strains, including those used in commercial insecticides, encode enterotoxin genes. However, the associations with food poisonings are correlations, not proof that any illness was caused by Bt, especially because *B. cereus* was present in a large majority of the same sources.

38.6 SUMMARY OF EARLIER STUDIES ON BT SAFETY

Overall, these studies show that under appropriate conditions most Bt strains can produce enterotoxins during vegetative growth. Nevertheless, despite the large quantities used in agriculture and forestry, there is no known case where any commercial Bt product has been confirmed as the cause of a food poisoning event. In fact, earlier studies showing Bt-like strains, likely natural food contaminants from grains and nuts, at concentrations of vegetative cells of 10^7/ml or gram, were not correlated with any known food poisoning. Because commercial Bt insecticides are used commonly on vegetable and fruit crops, especially organic crops, a possibility exists that under conditions favorable for spore germination and vegetative growth, enterotoxins could be produced, and perhaps in quantities capable of causing symptoms of food poisoning in humans. Nevertheless, to keep food poisoning in perspective, most cases are attributed to handling and eating foods such as poultry, beef and other meats, fish and shellfish, eggs, pasta, and rice. When vegetables are the source of food poisoning, most events are attributed to strains of *Escherichia coli* and *Salmonella enterica*, not *B. cereus* or *B. thuringiensis*. Overall, as shown in Table 38.1, the most common causes of poisoning are *E. coli, S. enterica*, numerous species of *Campylobacter, Clostridium perfringens, Staphylococcus aureus*, and viruses such as Norovirus (e.g., the Norwalk virus). Thus, even if such events due to Bt occur, data from the last 40 years indicate such events are mild and extremely rare, far rarer than food poisoning by the above bacteria and viruses. Also, it helps to avoid food poisoning by pathogens by using recommended food handling precautions such as normal washing and cooking procedures.

38.7 SAFETY OF BT BACTERIAL INSECTICIDES TO HUMANS

While the safety of Bt insecticides to humans has been recognized for decades, recently several reports by the European Food Safety Association (EFSA) have suggested that products based on Bta and Btk are the cause of a limited number of food poisonings in Europe. In a relatively recent opinion, EFSA (2016) reported that a case of food poisoning in Germany implicated Bt as a possible cause of this event. Briefly, in July 2012, a German family of five ate a meal of cheese noodles, and three members also ate a salad (EFSA, 2016). The latter three became ill with diarrhea. Analysis of food samples showed the salad presumably contained 3×10^4 CFU g^{-1} of Bt, whereas the cheese noodles contained 6.0×10^3 CFU g^{-1} of *B. cereus*. The Bt had characteristics of *B. t. aizawai* (Bta), the strain used in a commercial product applied to the salad crop eaten. Repeat sampling of salad from the supermarket where it was purchased found Bta concentrations of 4×10^4 colony forming units (CFU) g^{-1} and 1.5×10^5 CFU g^{-1}. These levels could not reliably implicate Bt because there were two possible etiological agents involved, Bt and *B. cereus*. We do not know the infective doses of Bt, or if it is infective for humans. However, estimates of infective doses for *B. cereus* are commonly 10^5–10^7 CFU g^{-1}(Frederiksen et al., 2006; Rosenquist et al., 2005). Moreover, even 10^3–10^4 CFU g^{-1}, which constitutes a low dose of *B. cereus*, can cause an infection in some individuals based on our knowledge of this organism and epidemiology in general. The fact that the three individuals who became ill also happened to eat the salad could therefore be a coincidence.

In addition to Bt spores and Cry proteins, commercial products contain remnants of fermentation media and protective efficacy enhancers such as UV protectants and ingredients to increase adherence to crops. These can be washed off, but not totally. Thus, it is not surprising that CFUs in the range of 10^3–10^6 g^{-1} are found on fresh vegetables in supermarkets. If consuming such rates of Bt spores caused diarrhea, we would expect people who consume organic crops to routinely report such problems. Moreover, tests at the above doses showed that Bt strains were not infectious for mammals (Bishop et al., 1999, 1999; McClintock et al., 1995; Mehere et al., 2002; Siegel, 2001).

Despite an abundance of data in the earlier literature on the safety of Bt to humans and other mammals, and rebuttals to data claiming Bt was a probable cause of food poisonings, the EFSA continues to question the safety of commercial products based on Btk and Bta to humans and non-target organisms at current application rates (EFSA, 2020a,b,c, 2021). Despite numerous studies in the literature, an example of typical EFSA concerns are the following (EFSA, 2021), dealing with Btk strain EG2348: "Satisfactory information was not provided leading to issues not being finalised for the potential effects, infectivity and pathogenicity to non-target terrestrial organisms (bees and non-target arthropods) for representative uses in open field and walk-in tunnels and for the potential effects, infectivity and pathogenicity to soil organisms and to non-target aquatic organisms (algae and aquatic plants) for all representative uses. Satisfactory information was not provided leading to issues not being finalised for the potential effects of toxins/secondary metabolites such as crystal proteins to nontarget terrestrial organisms (birds, wild mammals, bees, non-target arthropods and soil organisms) for representative uses in open field and walk-in tunnels and to non-target aquatic organisms (fish, freshwater invertebrates, algae and aquatic plants) for all representative uses.

In addition to these recent EFSA concerns, a recent paper questions again the safety of Bt to humans based on recent cases of food poisoning in France (Bonis et al., 2021). In this "outbreak," 19 out of 49 people tested positive for Btk and no other bacterium. The techniques used to identify Btk were acceptable. However, the study's design was flawed in that it did not test for viruses, e.g., the Norwalk virus, and moreover, appropriate controls were not used. For example, because Btk strains are common in nature, the contamination may not have been from a commercial product. Even more importantly, it could have been from a commercial product, but other people eating the same foods may not have been infected because they followed appropriate food washing and handling procedures. If, in fact, Bt strains in commercial products do cause outbreaks of food poisoning, then a way to determine this would be to sample people who routinely eat uncooked organic vegetable crops sprayed with commercial Bt products, determine the level of Bt on these crops prior to and after preparation for consumption, and then monitor these individuals for a few days to determine whether they develop intestinal discomfort followed by diarrhea, also testing them for other food-poisoning pathogens.

An interesting aspect of the Bonis et al. (2021) paper is that the authors cite many of the papers and opinions dealing with correlations between Bt and food poisoning, and the ESFA (2016) opinion on the family food poisoning in Germany, but not papers that criticize the methodology used to make these correlations, for example, Raymond and Federici (2017, 2018).

38.8 SAFETY AND NON-TARGET EFFECTS OF BTI INSECTICIDES

Bti was discovered serendipitously in 1976 when Goldberg and Margalit (1977) were searching for mosquito-parasitic *Coelomomyces* fungi where mosquito larvae where breeding near a Bedouin camp in the Negev Desert of Israel. Within a few years it was known that this new isolate of Bt, officially named *B. t.* subsp. *israelensis* (Bti) had a narrow host range, primarily restricted to flies of the suborder Nematocera (order Diptera), which contains mosquitoes, black flies, craneflies, and others such as various midges. Bti was shown to have little or no activity against flies of the other two dipteran suborders, Brachycera (horse, deer, and robber flies) or Cyclorrhapha (house and fruit flies). Like Btk and Bta, Bti produced a parasporal body during sporulation containing three major Cry proteins, Cry11Aa, Cry4Aa, and Cry11Aa, along with one major Cyt protein, Cyt1Aa. Bti's Cry proteins also target the midgut microvilli and require alkaline conditions, pH 8–11, for activation. The protein has no contact activity, and thus has to be eaten by larvae to be effective.

Bti's high environmental safety compared to chemical insecticides, and its efficacy, led to its rapid development and use around the world. One of Bti's early successes in the 1980s was its alternating use with temephos (Abate), a chemical insecticide, in reducing larval black fly populations that transmit the filarial worm in western Africa that causes Onchocerciasis, commonly known as River Blindness. Bti's use in the World Health Organization's Onchocerciasis Control Program (OCP), combined with new drugs that killed the microfilarial worms, reduced the prevalence of River Blindness by over 90% (WHO, 1995). In the OCP, Bti was applied during the dry season to streams when the water flow was low compared to the rainy season, when temephos was used. Bti rescued the OCP because black fly populations began to show resistance to Abate. Alternating the temephos with Bti significantly delayed the evolution of resistance, and simultaneously protected most of the non-target invertebrate populations.

The environmental safety and success of Bti was also shown in many control programs that target nuisance and vector mosquitoes, especially those based on chemical insecticides. Thus, Bti's use continues to expand worldwide today, being a major component of mosquito and vector control programs throughout the US, Europe, Australia, China, and many countries in South America and Africa. A remarkable trait of Bti, especially compared to chemical insecticides, is that virtually no significant resistance has evolved under field conditions, or even under heavy selection in the laboratory. Various studies have shown that this is due to its Cyt1Aa protein, which creates large lesions in the membrane, allowing Cry proteins to function (Federici, 2005; Federici et al., 2010; Park et al., 2005).

Bti is not directly of concern to human health, meaning food poisoning; however, it is not without some controversy. In most cases this is because its use for mosquito control can reduce populations of related nematocerous flies (e.g., midge larvae) and other non-target invertebrates, for example, predaceous insects such as aquatic beetles, larval and adult dragonflies, and damselfies that feed on mosquito larvae and adults, and thus eliminate segments of the diet of vertebrates such as fish and birds. An important aspect of the consequences of reducing mosquito populations is the effect on feeding preferences of birds and fish, which are often not included in target impact studies.

There are many examples of these reductions of higher levels of the food chain, but the extent varies widely depending on specific ecological habitats, some reporting no level of reductions, others, moderate significance. For example, in a study of the effect of mosquito control in the protected Carmargue area of southern France, it was found that reduced mosquito and other fly populations due to Bti treatments over a period of three years correlated with a reduction of their invertebrate predators such as spiders and dragonflies, which in turn reduced clutch size and fledgling survival from 3.2 to 2.3 per nest in populations of the house martin, *Delichon urbicum* (Poulin et al., 2010). In a subsequent study in the Carmargue, it was found that the effect on the house martin populations was correlated to significant reductions in the odonata (dragonfly and damselfly) populations, in which both species diversity and numbers were reduced by about 50% over a period of five years (Jakob and Poulin (2016). Alternatively, in Germany's Rhine Valley, where Bti has been used yearly since 1980 for mosquito control, virtually eliminating the use of chemical insecticides, Timmerman and Becker (2017) studied the impact of Bti treatments on the adult chironomid populations. They found little variation in these between the Bti-treated and non-treated areas. They monitored feeding by the house martin in these areas and found that midges were eaten, but not preferred. Thus, even if there were reductions in midges, the house martin, the same species studied in the Carmargue, would not have been affected significantly by Bti treatments.

38.9 SAFETY AND NON-TARGET EFFECTS OF BT CROPS

Ever since molecular biology advanced to cloning genes and introducing them from one species to another, the possibility of creating transgenic crops was apparent. This became a probability in 1981 when the first Bt genes coding for Cry proteins were identified (Nester et al., 2002, for a brief review; Charles, 2001, for the early history of how agricultural biotechnology led to an industry worth today about $100 billion). Bt crops remain a controversial issue, one example being that they are not permitted to be grown in most European countries, or in countries such as Japan and New Zealand. This resistance to GMO crops has led to increased scrutiny of their safety, not so much to human safety at this point, but to the environmental consequences on non-target populations. This subject is too complicated to review here; however, I briefly summarize below some of the key studies that both validate Bt crop safety as well as the concerns about its effects on non-target insects.

Most of these studies have been long-term, taking place over periods from two to six years, and many are ongoing today. None have shown significant impacts on human health or, compared to chemical insecticides, marked negative impacts on the various non-target populations studied (Romeis et al., 2004, 2006; Shelton et al., 2009a,b; Carstens et al., 2011; Sanders et al., 1998; Lang et al., 2019). Moreover, all showed significant reductions in the use of chemical insecticides (Naranjo et al., 2005a,b).

Perhaps a reasonable summary at this point about Bt crops comes from the MS thesis of De Bock (2019), "...the use of a *B. thuringiensis*-containing plant protection product poses a low risk for public health that does not outweigh its plant protection capacity, contributing to food security. It can therefore be used as an interesting and environmentally friendly alternative for the future."

38.10 SUMMARY AND CONCLUSIONS

The first Bt insecticides in the 1960s were only active against lepidopteran pests. However, because these insect larvae cause significant economic losses, especially in high value vegetable and fruit crops, Bt products accelerated the development of organic crops and IPM programs. Moreover, they were considered a green technology owing to their specificity and lack of major non-target effects. Today, various Bt insecticides are used worldwide and remain the cornerstone in crop protection programs where caterpillars are the main pests, and in environmentally compatible mosquito and black fly control programs. Their development greatly reduced the use of chemical insecticides with no proven negative effects on human health or on most target organisms. Thus, Bt after 50 years of use remains a major part of green technologies. Once it was determined that the active ingredients of Bt were insecticidal proteins, GMO crops were developed using the science and technology of molecular biology to integrate Cry proteins into major commodity crops, again reducing the use of chemical insecticides for pest control. Even though initially controversial, GMO crops have been widely adopted and today represent more than 60% of all major crops grown worldwide. They have been and will continue to be accepted by most farmers. Controversies remain; Bt GMOs reduced chemical insecticide usage; the stacking of herbicides with Bt genes in new varieties of GMO corn, cotton, and soybeans has increased use of chemical herbicides. Moreover, recent controversies have been raised about minor non-target effects such as those that may reduce bird and fish populations at the top of food chains. At present, the benefits still outweigh the risks compared to the use of chemical insecticides.

Finally, to keep things in perspective with respect to Bt insecticides, as noted by Raymond and Federici (2018), "We believe that requiring more and more data (or more and more restrictions) to address ever more imaginative risk hypotheses for products that are not pathogenic to humans and that pass all safety tests with flying colors is not helpful for farmers and consumers with a stake in environmentally friendly food production. Bt products, aside from their use in integrated pest management programs, are some of the few insecticides permitted in organic agriculture. People who eat organic vegetables sprayed with Bt products routinely consume viable Bt spores when products are used just prior to harvest. Bt products have been used in organic agriculture for more than 30 years, without any evidence of causing diarrheal disease. Moreover, if restrictions are placed on the use of Bt in agriculture and forestry, it is likely there will be a return to using broad spectrum synthetic chemical insecticides, which we are sure EFSA will agree, have higher risks for non-target organisms, including humans. The same applies to GMO crops."

REFERENCES

1. Adang M, Crickmore N, Jurat-Fuentes JL. 2014. Diversity of *Bacillus thuringiensis* crystal toxins and mechanism of action. *Advances in Insect Physiology* 47: 39–87.
2. Auckland Regional Public Health Service. 2003. Human health considerations in the use of btk-based insecticide foray 48B for asian gypsy moth in hamilton. summary report prepared for the ministry of health, ministry of agriculture and forestry, and waikato DHB public health unit. p. 45.
3. Bishop AH, Johnnson C, Perani M. 1999. The safety of *Bacillus thuringiensis* to mammals investigated by oral and subcutaneous dosage. *World Journal of Microbiology and Biotechnology* 15: 375–380.
4. Bonis M, Felten A, Pairaud S, Dijoux MV, et al. 2021. Comparative phenotypic, genotypic and genomic analysis of bacillus thuringiensis associated with foodborne outbreaks in France. https://doi.org/10.1371/journal.pone.0246885.
5. Boulton TJ. 2004. Responses of non-target lepidoptera to Foray 48B *Bacillus thuringiensis* var. kurstaki on Vancouver Island, British Columbia. *Environmental Toxicology and Chemistry* 23: 1297–1304.

6. Boulton TJ, Otvos IS, Halwas KL, Rohls DA. 2007. Recovery of nontarget lepidoptera on Vancouver Island, Canada: One and four years after a gypsy moth eradication program. *Environmental Toxicology and Chemistry* 4: 738–748.

7. Carstens K, Anderson J, Bachman, De Schrijver A, Dively G, et al. 2011. Genetically modified crops and aquatic ecosystems: Considerations for environmental risk assessment and non-target organism testing. *Transgenic Research* https://doi.org/10.1007/s11248-011-9569-8.

8. Charles D. 2001. *Lords of the Harvest - Biotech Big Money, and the Future of Food.* Basic Books, A member of the Perseus Books Group, New York, NY, p. 348.

9. De Bock T. 2019. Food safety evaluation of the use of *Bacillus thuringiensis* as biological control agent in the primary production of butterhead lettuce. In: *MS Thesis in Bioscience Engineering: Food Science and Nutrition.* Ghent University, Ghent, Belgium.

10. EFSA. 2016. Risks for public health related to the presence of *Bacillus cereus* and other *Bacillus* spp. including *Bacillus thuringiensis* in foodstuffs. *EFSA Journal* 14: 1–93.

11. EFSA. 2020a. Peer review of the pesticide risk assessment of the active substance *Bacillus thuringiensis* ssp. *Kurstaki Strain SA-12* https://doi.org/10.2903/j.efsa.2020.6262.

12. EFSA. 2020b. Peer review of the pesticide risk assessment of the active substance *Bacillus thuringiensis* ssp. *Aizawai GC-91* https://doi.org/10.2903/j.efsa.2020.6293.

13. EFSA. 2020c. Peer review of the pesticide risk assessment of the active substance *Bacillus thuringiensis* ssp. *Aizawai Strain ABTS-1857* https://doi.org/10.2903/j.efsa.2020.6294.

14. EFSA. 2021. Peer review of the pesticide risk assessment of the active substance *Bacillus thuringiensis* subsp. *Kurstaki Strain EG2348* https://doi.org/10.2903/j.efsa.2021.6495.

15. Federici BA. 2005. Insecticidal bacteria: An overwhelming success for invertebrate pathology. *Journal of Invertebrate Pathology* 89: 30–38.

16. Federici BA, Park H-W DK, Bideshi DK. 2010. Overview of the basic biology of *Bacillus thuringiensis* with emphasis on genetic engineering of bacterial larvicides for mosquito control. *The Open Toxicology Journal* 3: 154–171.

17. Federici BA, Siegel JP. 2007. Chapter 3, pp. 46 – 101. In: Assessment of the safety of *Bacillus thuringiensis* and *Bt* crops used for insect control. In: Hammond BG (Ed.), *Food Safety of Proteins in Agricultural Biotechnology*. CRC Press, Taylor and Francis, London, UK, pp. 46-101.

18. Frederiksen K, Rosenquist K, Jorgensen K, Wilcks A. 2006. Occurrence of natural *Bacillus thuringiensis* contaminants and residues of *Bacillus thuringiensis*-based insecticides on fresh fruits and vegetables. *Applied and Environmental Microbiology* 72: 3435–3440.

19. Gharib AA, El-Hamid MIA, El-Aziz NL, Yonan EY, Allam MO. 2020. *Bacillus cereus*: Pathogeneicity, viability and adaptation. *Advances in Animal and Veterinary Sciences* 8: 34–40.

20. Glare TR, O'Callaghan M. 2000. *Bacillus Thuringiensis: Biology, Ecology and Safety*. J Wiley & Sons, Chichester, UK. P. 350.

21. Goldberg LJ, Margalit J. 1977. A bacterial spore demonstrating rapid larvicidal activity against *Anopheles sergentii, Uranotaenia unguiculata, Culex univitattus, Aedes aegypti*, and *Culex pipiens*. *Mosquito News* 37: 355–358.

22. Gould FL, Andow DA, Blossy B, Chapela I, Ellstrand N, et al. 2002. *Environmental Effects of Transgenic Plants: The Scope and Adequacy of Regulation*. National Academy Press, Washington, DC, p. 320.

23. Granum PE, Lund SJ. 1997. *Bacillus cereus* and its food poisoning toxins. *FEMS Microbiology Letters* 157: 223–228.

24. Hechel DG. 2020. How do toxins from *Bacillus thuringiensis* kill insects? An evolutionary perspective. *Archives of Insect. Biochemistry and Physiology* doi:// 104: e21673.

25. Jakob C, Poulin B. 2016. Indirect effects of mosquito control using Bti on dragonflies and damselflies (Odonata) in the Carmargue. *Insect Conservation and Diversity* 9: 161–169.

26. Jessberger N, Dietrich R, Granum PE, Martlbauer E. 2020. The bacillus cereus food infection as multifactorial process. https://doi.org/10.3390/toxins12110701.

27. Johler S, Kalbhenn EM, Heini N, Brodmann P, Gautsch S, Bagcioglu et al. 2018. Enterotoxin production of *Bacillus thuringiensis* isolates from biopesticides, foods, and outbreaks. *Frontiers in Microbiology* 9: 1915. https://doi.org/10.3389/fmicb.2018.01915.

28. Lang A, Lee M, Dolek M, Berchtold J, Otto M. 2019. Laboratory tests with lepidoptera to assess non-target effects of Bt maize pollen: Analysis of current stud-

ies and recommendations for a standardized design. *Environmental Sciences Europe* https://org/10.1186/s12302-019-0220-2.

29. Liebhold AM, Leonard D, Marra JL, Pfister SE. 2021. Area-wide management of invading gypsy moth (*Lymantria dispar*) populations in the USA. In: Hendrichs J, Pereiria R, Vreysen MJB (Eds.), *Area-Wide Integrated Pest Management: Development and Field Applications*. CRC Press, Boca Raton, Florida, USA, pp. 551–560.

30. McClintock JT, Schaffer CR, Sjoblad RD. 1995. A comparative review of the mammalian toxicity of *Bacillus thuringiensis*-based pesticides. *Pest Management Science* 45: 95–105.

31. McIntyre L, Bernard K, Beniac D, Isaac-Renton JL, Naseby DC. 2008. Identification of *Bacillus cereus* group species associated with food poisoning outbreaks in British Columbia, Canada. *Applied and Environmental Microbiology* 74: 7451–7453.

32. Mehere SM, Bodhankar SL, Anukumar A, Dhuley JN, Khodape DJ, Naik SR. 2002. Toxicity studies of microbial insecticide *Bacillus thuringiensis* var. *kenyae* in rats, rabbits and fish. *International Journal of Toxicology* 21: 99–105.

33. Naranjo S. 2005a. Long-term assessment of the effects of transgenic Bt cotton on the abundance of nontarget arthropod natural enemies. *Environmental Entomology* 34: 1193–2010.

34. Naranjo S. 2005b. Long-term assessment of the effects of transgenic Bt cotton on the function of the natural enemy community. *Environmental Entomology* 34: 1211–1223.

35. Nester EW, Thomashow LS, Metz M, Gordon M. 2002. 100 years of *Bacillus thuringiensis*: A critical Scientific Assessment. In: *A Report from the American Academy of Microbiology*. DC, Washington, p. 18.

36. The New Zealand Experience. 1997. Public health protective service health report on aerial spray of foray 48B to control the white-spotted tussock moth.

37. Nunez-Mir G, Walter JA, Johnson DM. 2021. Assessing risk of *Lymantria dispar* L. invasion management to monarch butterflies (Danaus plexippus). In: *Conservation Science and Practice*. https://doi.org/10.1111/csp2.359.

38. Owens E, Johns R. 2021. *Healthy Forest Partnership Report on Water Monitoring within 2020 Research Areas for Foray 76B Bacillus Thuringiensis Var Kurstaki (Btk) and Tebufenozide*. Report Atlantic Forestry Centre, Canadian Forest Service, Fredericton, New Brunswick, Canada, pp. 1–8.

39. Park H-W, Bideshi HK, Wirth MC, Johnson JJ, Walton W, Federici BA. 2005. Recombinant larvicidal bacteria with markedly improved efficacy against *Culex* vectors of West Nile Virus. *American Journal of Tropical Medicine and Hygiene* 72: 732–738.

40. Pearce M, Habbick B, Williams J, Eastman M, Newman M. 2002. The effects of aerial spraying with *Bacillus thuringiensis kurstaki* on children with asthma. *Canadian Journal of Public Health* 92: 21 – 25. *Environ. Health Rev.* 46(19): 2002.

41. Petrie K, Thomas M, Broadbent E. 2003. Symptom complaints following aerial spraying with biological insecticide Foray 48B. *New Zealand Medical Journal* 116: 1070.

42. Poulin B, Lefebvre G, Paz L. 2010. Red flag for green spray: Adverse trophic effects of Bti on breeding birds. *Journal of Applied Ecology* 47: 885–889.

43. Raymond B, Federici B. 2017. In defense of *Bacillus thuringiensis*, the safest and most successful microbial insecticide available to humanity – A response to EFSA. *FEMS Microbial Ecology* https://doi.org/10.1093/femsec/fix084.

44. Raymond B, Federici BA. 2018. An appeal for a more evidence based approach to biopesticides safety in the EU. *FEMS Microbiology Ecology* 94: 1–5.

45. Romeis J, Dutton A, Bigler F. 2004. *Bacillus thuringeinsis* toxin (Cry1Ab) has no direct effect on larvae of the green lacewing *Chrysoperla carnea* (Neuroptera: Chrysopidae). *Journal of Insect Physiology* 50: 175–183. 2004.

46. Romeis J, Meissle M, Bigler F. 2006. Transgenic crops expressing *Bacillus thuringeinsis* toxins and biological control. *Nature Biotechnology* 1: 63–71.

47. Rosenquist H, Smidt L, Anderson SR, Jenson GB, Wilcks A. 2005. Occurrence and significance of *Bacillus cereus* and *Bacillus thuringiensis* in ready-to-eat food. *FEMS Microbiology Letters* 250: 129–136.

48. Sanders PR, Lee TC, Groth ME, Astwood JD, Fuchs RL. 1998. Safety assessment of insect-protected corn. pp. 241-256. In: Thomas JA (Ed.), *Biotechnology and Safety Assessment*, 2nd ed. Taylor & Francis, London.

49. Shelton AM, Naranjo SE, Romeis J, Hellmich RL, Wolt JD, et al. 2009a. Appropriate analytical methods are necessary to assess nontarget effects of insecticidal proteins in GM crops through meta-analysis (response to Andow et al., 2009). *Environmental Entomology* 38: 1533–1538.

50. Shelton S, Naranjo SE, Romeis J, Hellmich JRL, Wolt JD, et al. 2009b. Setting the record straight: A rebuttal to an erroneous analysis on transgenic insecticidal crops and natural enemies. *Transgenic Research* 18: 317–322.

51. Siegel JP. 2001. The mammalian safety of *Bacillus thuringiensis* based insecticides. *Journal of Invertebrate Pathology* 77: 13–21.

52. Sopuck L, Ovaska K, Whittington B. 2002. Responses of songbirds to aerial spraying of the microbial insecticide *Bacillus thuringiensis* var. *kurstaki* (FORAAY 48B) on Vancouver Island, British Columbia Canada. *Environmental Toxicology and Chemistry* 8: 1664–1672.

53. Stenfors Arnesen LP, Fagerlund A, Granum PE. 2008. From soil to gut: *Bacillus cereus* and its food poisoning toxins. *FEMS Microbiology Reviews* 32: 579–606.

54. Teschke K, Chow Y, Bartlett K, Ross A, van Netten C. 2001. Spatial and temporal distribution of airborne *Bacillus thuringiensis* var. *kurstaki* during an aerial spray program for gypsy moth eradication. *Environmental Health Perspectives* 109: 47–54.

55. Timmerman U, Becker N. 2017. Impact of routine *Bacillus thuringiensis israelensis* (BTI) treatment on the availability of flying insects as prey for aerial feeding predators. *Bulletin of Entomological Research* 107: 705–714.

56. Valadares de Amorim G, Whittome B, Shore B, Levin DB. 2001. Identification of *Bacillus thuringiensis* subsp. *kurstaki* strain HD1-like bacteria from environmental and human samples after aerial spraying of Victoria, British Columbia, Canada, with Foray 48B. *Applied and Environmental Microbiology* 67: 1035–1043. 2001.

57. World Health Organization. 1995. Onchocerciasis and its control: Report of a WHO expert committee on onchocerciasis control. Vol. 852, pp. 1–103.

CHAPTER 39

Life of Microbes Inside the Plant: Beneficial Fungal Endophytes and Mycorrhizal Fungi

Luisa Lanfranco and Valentina Fiorilli
Department of Life Sciences and Systems Biology, University of Turin, 10125, Torino, Italy

39.1 THE PLANT MICROBIOTA

Plants are not alone: in many ecosystems, plants are greatly outnumbered by microorganisms such as bacteria and fungi that may live in close proximity or even in strict association with plant hosts (Müller et al., 2016; Vandenkoornhuyse et al., 2015). Complex microbial communities thrive on the outer surfaces as well as in internal tissues of plants; this microbial component, referred as the plant microbiota (all microorganisms) or plant microbiome (all microbial genomes), may influence plant growth and health by modulating a wide range of biological processes (Trivedi et al., 2020). Pioneering reports on the plant microbiota focused on the characterization of bacterial assemblages in the model species *Arabidopsis thaliana* (Bulgarelli et al., 2012), but more recent papers also consider fungi so that the term plant *mycobiota* is also currently used (Bonfante et al., 2019). High-throughput sequencing techniques applied to environmental samples have revealed that fungi proliferate in many different environments (soil, air, water) but, remarkably, the greatest numbers of fungi are found in the soil or associated with plants, at an estimated fungal/plant species ratio of 17/1 (Nilsson et al., 2019).

Living together, which is described as a symbiosis (Martin and Schwab, 2012), implies different possible types of interactions ranging from parasitic or pathogenic, where one partner gains benefit at the expense of the other, to mutualistic, where reciprocal benefits occur. The knowledge of the nature of these interactions and on how these fungal symbionts can influence plant physiology and metabolism is of fundamental relevance because it is recognized nowadays that plants likely host a still overlooked reservoir of beneficial fungi that may contribute to supporting their survival and growth under biotic and abiotic stresses. Some of these plant–fungal associations may function as key elements in the development of sustainable programs of crop production and environmental protection which, in the context of global climate change, represent emerging challenges to modern societies (see chapter 23).

In this chapter we focus on fungal endophytes with a recognized beneficial impact on plants and on a special group of endophytes, the mycorrhizal fungi. Mycorrhizal fungi are root endophytes which form specialized plant–fungal interfaces for resource exchange. In addition, they are phylogenetically distinct from most groups of endophytes (Porras-Alfaro and Bayman, 2011; Smith and Read, 2008; and see also chapter 29).

We will describe here the interaction at the cellular level, emphasize the functional significance of the associations and the impact of the fungal symbionts on the host plant, and illustrate potential applications.

Good Microbes in Medicine, Food Production, Biotechnology, Bioremediation, and Agriculture, First Edition. Edited by Frans J. de Bruijn, Hauke Smidt, Luca S. Cocolin, Michael Sauer, David Dowling, and Linda Thomashow.

39.2 FUNGAL ENDOPHYTES OF PLANTS

Notwithstanding the knowledge of plant-associated microbes has been significantly advanced by molecular high-throughput sequencing techniques (Nilsson et al., 2019), the study of plant microbiota has very ancient roots. Anton de Bary, in 1886, coined the term *endophytes* to define microorganisms, normally bacteria and fungi, that live within plant tissues (Martin and Schwab, 2012; Wilson, 1995). The term endophyte therefore refers to the location, i.e. inside plant tissues, irrespective of being inside the cells or among the cells, and does not specify the nature of the relationship established with the host plant. Indeed, the majority of endophytes do not affect the plant at all; some are pathogenic and others may even be beneficial. The outcome of the association may depend on the partners involved, and the functional roles can also vary during the life cycle of the microbe or in response to plant or environmental factors (Porras-Alfaro and Bayman, 2011).

It has been shown that each plant species has a highly variable and, in some cases, impressively rich community of fungal endophytes – up to 90 species have been described within a single leaf of a tropical tree (Bayman, 2006). Plant organs of a single plant also often differ greatly in the dominant taxa of endophytes present, but we are far from having a global picture of their functional significance.

Considering this heterogeneity, we will here give examples of some of the most well characterized fungal endophytes known to have a positive impact on the host plant. They include the shoot grass endophytes of the genus *Epichloë spp.* (Ascomycota, Sordariomycetes, Clavicipitaceae), the well-known biocontrol agents belonging to *Trichoderma* spp. (Ascomycota, Sordariomycetes, Hypocreaceae; see also Chapter 41), the generalist root endophyte *Serendipita indica* (Basidiomycota, Sebacinales), and the recently described root/systemic endophyte *Colletotrichum tofieldiae* (Ascomycota, Sordariomycetes, Glomerellaceae) identified in natural samples of the model species *Arabidopsis thaliana.*

39.2.1 Epichloë spp.

The genus *Epichloë*, belonging to the family Clavicipitaceae within Ascomycota, forms symbiotic relationships with many grasses of the Pooideae subfamily (Schardl et al., 2004). Phylogenetic analysis places *Epichloë* species within a group of endophytic and plant pathogenic fungi, with a common ancestor probably derived from an animal pathogen (Spatafora et al., 2007).

Epichloë endophytes grow asymptomatically in the intercellular spaces of the aerial tissues of the host plant. Notably, the expansion of hyphae occurs in synchrony with the growth of the newly developing leaf, while the root is essentially uncolonized (Herd et al., 1997; Tan et al., 2001). These endophytes, in most cases, are exclusively seed-transmitted from previously infected hosts (Schardl et al., 2004), a mode of transmission that represents another signature with ecological relevance of the intimate plant-fungus association.

The nature of this symbiosis has been described as "defensive mutualism" (Bastias et al., 2017): *Epichloë* endophytes are indeed able to produce several alkaloids, bioactive secondary metabolites that are detrimental for herbivores (Schardl et al., 2004). Genomic studies and chemical analyses have been instrumental in deciphering the alkaloid biosynthetic pathways (Schardl et al., 2013a; Young et al., 2015); it is now well established that alkaloid profiles depend not only on the specific fungal strain but also on the host plant phenology and on environmental factors (Ball et al., 1995; Justus et al., 1997; Saikkonen et al., 2013). Moreover, the quali- and quantitative alkaloid profile (i.e., chemical structure and amount of specific compounds) is a key determinant in the defense efficiency against a given attacker.

In the late 1970s, *Epichloë* endophytes were considered undesirable, as they were found to lead to serious health problems in livestock (Bacon et al., 1977; Fletcher and Harvey, 1981). Later, however, since the yield of endophyte-free grasses was lower compared to endophyte-infected grasses, a major effort was undertaken to select livestock-friendly endophytes that still produced alkaloids with insecticidal and nematocidal activities (Cagnano et al., 2019) and artificially infected grass cultivars were developed for the market (Johnson et al., 2013).

In order to gain a deeper understanding of the biological system, the *Epichloë festucae* – ryegrass *(Lolium perenne)* model was characterized at the cellular and molecular levels. The benefits provided by *E. festucae* were found to extend far beyond protection against herbivores (the only issue that has been mechanistically explained). Improvement in pathogen resistance (Tian et al., 2008), drought tolerance, and enhanced growth (Hahn et al., 2008), and better ability to compete with other plant species (Quigley, 2000) have also been reported.

To gain insight into the molecular changes induced by endophyte infection, a comparison of the gene expression profiles of perennial ryegrass infected or not with *E.*

festucae was carried out (Dupont et al., 2015). Endophyte infection triggered a strong reprogramming of host genes, in particular favoring secondary metabolism at a cost to primary metabolism. This is similar to what was found in plant pathogenic interactions but in stark contrast to what has been observed in response to fungal colonization by mycorrhizal fungi where only a modest host gene modulation occurs.

A transcriptomic analysis of the *E. festucae–L. perenne* interaction also was carried out with a focus on the developmental stages of different host tissues (Schmid et al., 2017). Notably, fungal genes annotated as encoding small secreted proteins (SSP), which can function to control colonization of host tissue in pathogenic and mutualistic symbioses (Kloppholz et al., 2011; Plett et al., 2011; Toruño et al., 2016), were highly expressed in hyphae within emerging leaves. By contrast, genes involved in synthesizing anti-herbivore compounds were highly expressed in the fungal mycelium of mature plant tissues, in agreement with the alkaloid profiles. But *E. festucae*-infected plants can have a second line of defense: endophyte colonization induces a systemic effect in that plant. Young as well as mature plant tissues display the activation of genes mainly involved in hormone biosynthesis and signaling and in stress and pathogen resistance (Dupont et al., 2015; Schmid et al., 2017). Considering, in particular, defense against chewing insects, it has been proposed that *Epichloe* endophytes can use not only alkaloids but also can activate a level of intrinsic immunity in the plant by promoting innate defense responses (Bastias et al., 2017). These data suggest that *E. festucae* is able to trigger an activated status (Martinez-Medina et al., 2016), in analogy to a mechanism that has been proposed for mycorrhizal fungi (Pozo and Azcón-Aguilar, 2007). This activated status could be the basis of the increased resilience of the host plants to a broad variety of stresses (Figure 39.1).

39.2.2 *Trichoderma* spp.

Filamentous fungi belonging to the genus *Trichoderma* thrive in the rhizosphere (the soil adjacent to and influenced by plant roots), and are able to colonize the root system at early seedling developmental stages (Harman

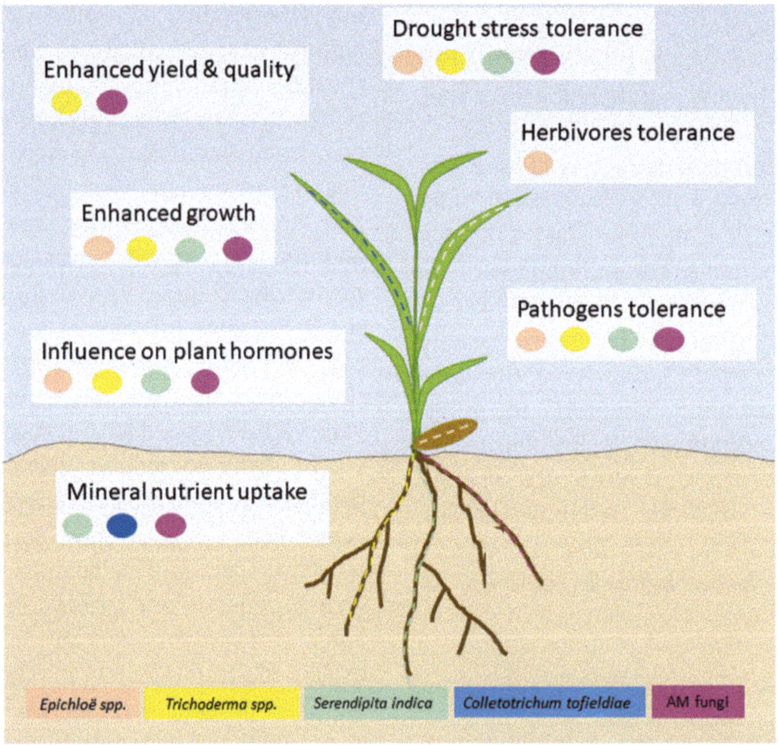

FIGURE 39.1 Beneficial effects of plant fungal endophytes and mycorrhizal fungi. The plant-associated endophytes *Epichloë* (light pink), *Trichoderma* (yellow), *Serendipita indica* (turquoise), *Colletotrichum tofieldiae* (blue), and the AM fungi (purple) colonize different plant organs (dashed lines). They can provide several benefits to the host plant directly, via nutrient acquisition, or indirectly, through the modulation of plant hormones and the induction of plant defenses to respond to pathogen attack. Each microorganism is represented by dot with corresponding color code.

et al., 2004). As early as 90 years ago, *Trichoderma* species were recognized as pathogens of other soil fungi (Weindling, 1932). Since then, the knowledge of *Trichoderma* biology has evolved continuously, leading to the successful development and marketing of *Trichoderma*-based products to be used in agriculture as biocontrol agents to control plant pathogenic fungi (Harman et al., 2004; see also Chapter 41).

Studies focused mainly on *Trichoderma harzianum* have demonstrated that parasitism by fungi, even beneficial ones, relies on the production of many hydrolytic enzymes, which act in synergy with fungitoxic metabolites (Lorito et al., 2010). However, these may not be the primary mechanisms of biocontrol, since root colonization also stimulates plant defense mechanisms initiated by extensive physiological and metabolic changes mediated mainly by jasmonic acid and another plant hormone, ethylene (Contreras-Cornejo et al., 2015; Martínez-Medina et al., 2017). A strain of *T. harzianum* also has been found to produce a nitrogen heterocyclic compound called harzianic acid, which has various biological properties including plant growth promotion, antimicrobial activity and the ability to chelate soil iron, facilitating its solubilization and, as a consequence, its availability to plants (Vinale et al., 2013).Thus, the multiple benefits provided to the host plant encompass not only protection from pathogens, but also improvement of plant growth and vigor and enhanced defense responses against various stresses (Figure 39.1). Collectively, these reasons explain why the new generations of *T. harzianum* formulations are therefore described as plant protectants and biostimulants of crop yield and quality (see chapter 41).

39.2.3 *Serendipita indica*

Another plant endophyte that has been well characterized in recent decades is *Serendipita indica* (syn. *Piriformospora indica;* Varma et al., 1999), a filamentous fungus belonging to Sebacinales (Basidiomycota). It is worth noting that the order Sebacinales include endophytes that associate with roots (i.e. *S. indica, S. vermifera*, and *S. herbamans*), mycorrhizal fungi forming symbioses with orchids, as well as saprotrophs, but not pathogenic fungi. The root associated Sebacinales likely evolved from saprotrophic ancestors (Garnica et al., 2016) and are described as ubiquitously distributed generalists. In particular, *S. indica* colonizes the intercellular spaces of plant roots and the intracellular regions of epidermal and cortical cells, but never the central cylinder of the root. In early colonization stages the fungus is surrounded by a plant-derived plasma membrane, indicating that the colonized cells are viable. Later, it switches to a host cell death-associated phase, where thin fungal hyphae are found in dead or dying cells, although massive necrotic areas are never observed (Deshmukh et al., 2006; Fesel and Zuccaro, 2016).

Benefits for the host plant include increased biomass, possibly as a consequence of enhanced assimilation of mineral nutrients such as phosphate (Yadav et al., 2010), and a higher tolerance against biotic and abiotic stresses (Sherameti et al., 2008; Figure 39.1). The *S. indica-Arabidopsis thaliana* interaction has emerged as a model to study the systemic effect leading to an increase tolerance to leaf pathogens. In that specific case, the protection against the powdery mildew *Golovinomyces orontii* was demonstrated to be mediated by the jasmonate hormonal signaling pathways (Stein et al., 2008).

39.2.4 *Colletotrichum tofieldiae*

Another intriguing endophyte, the recently characterized *Colletotrichum tofieldiae*, was originally isolated from natural populations of asymptomatic *A. thaliana* plants in Spain (García et al., 2013). It was also detected in trees and orchids across the Eurasian continent (Damm et al., 2009; Tao et al., 2013) suggesting a very broad host range. The fungus colonizes the root and occasionally spreads into the aerial parts of the plant through the central cylinder, reminiscent of a pathogenic association; indeed *C. tofieldiae* is closely related to pathogenic *Colletotrichum* species. *C. tofieldiae* was shown to promote plant growth as a consequence of improved mineral nutrition, thanks to endophyte-mediated transfer of phosphorus. Remarkably, phosphate transfer only occurs under low phosphorus conditions (Hiruma et al., 2016), which is similar to what is observed in the arbuscular mycorrhizal symbiosis (Figure 39.1). This result has opened a window on how non-mycorrhizal plants cope with low nutrient availability, since *A. thaliana*, like most Brassicaceae, has lost the ability to form mycorrhizal symbioses during evolution.

The comparison between the genomes of *C. tofieldiae* and its pathogenic relative, *C. incanum*, and between the transcriptomes of both fungi and the host plant *Arabidopsis*, was instrumental in highlighting the genomic traits related to the transition from pathogenicity to mutualism (Hacquard et al., 2016). The *C. tofieldiae* genome has a narrowed repertoire of genes encoding secreted proteins, chitin-binding and secondary metabolism-related proteins compared to that of *C. incanum*. Moreover, *C. tofieldiae*-colonized plants show limited activation of pathogenicity-related genes compared to *C. incanum*-infected plants. It is worth noting that *A. thaliana* roots induced defense genes when colonized by *C. tofieldiae* under conditions of high phosphorus concentrations (Hacquard et al., 2016). This finding pointed for the first time to an interdependency between plant immunity and plant nutrition. A similar activation of plant defense-related genes mirrored by alterations in the root-inhabiting fungal community was observed in maize plants grown under high phosphorus concentrations (Yu et al., 2018). Lowering plant defenses under low phosphorus availability therefore appears to be a strategy to favor the recruitment of beneficial soil microbes that can help to overcome the nutritional stress.

Fungal endophytes are thus ubiquitous and taxonomically very diverse organisms. Phylogenetic analyses and data from whole genome sequencing, although still limited to only a few taxa, have given important insights into the evolutionary origin of endophytism. As mentioned before, some of endophytes, such as *Epichloë* and *C. tofieldiae*, are clearly derived from fungi with a pathogenic life-style, whereas the root-associated Sebacinales (i.e., *S. indica*) are most probably derived from a saprotrophic ancestor (Fesel and Zuccaro, 2016; Lahrmann et al., 2015). The genomic signature of this origin is a reduced repertoire in *S. indica* of genes involved in secondary metabolite production compared to that found in the genomes of *Epichloë spp.* and *C. tofieldiae*.

Despite these differences, colonization of the roots of *A. thaliana* by *S. indica* or *C. tofieldiae* share common features: both fungi initially penetrate the epidermis by undifferentiated hyphae and spread in the cortical tissue intercellularly and intracellularly; at later stages of both interactions cell death, mainly limited to the epidermal layer, is observed. Moreover, in both cases, to restrict fungal growth inside the root and keep a well-balanced endophytic interaction, an intact plant immune system is required (Fesel and Zuccaro, 2016). Further comparative analysis of additional interactions will help to define commonalities.

A further level of complexity in the plant–fungal endophyte interaction is the presence within the microbial symbiont of endobacteria (Bonfante et al., 2019) and viruses (Bao and Roossinck, 2013). Notably, these bacteria can influence several aspects of the fungal biology including the outcome of plant–fungal symbioses. One example from the prokaryotic world is *Rhizobium radiobacter* strain F4, which was detected inside *S. indica*; the endobacterium improves the fitness of *S. indica* and contributes to the beneficial effects exerted by the fungal endophyte on the host plant (Guo et al., 2017).

Fungal viruses, also called mycoviruses, also are emerging as an abundant component of the microbiota of fungal endophytes (Bao and Roossinck, 2013) that may have an impact on fungal fitness and, as a consequence, on the plant host, in a tripartite interaction. A historic case is the virus that makes the fungal pathogen *Cryphonectria parasitica* less virulent for the host plant, which led to the identification of a powerful biocontrol agent (Choi and Nuss, 1992). Another example refers to thermotolerance: a mycovirus, living inside the fungal endophyte *Curvularia protuberate* was shown to confer heat tolerance to the fungus and to the host plant *Dichanthelium lanuginosum*, allowing their growth in the geothermal soils of Yellowstone National Park (Márquez et al., 2007). The precise molecular mechanisms at the basis of the phenomenon are not known but osmolytes and reactive oxygen species were differentially accumulated in plants colonized with the virus-infected endophyte.

From these few examples it is clear that plants can host a taxonomically broad spectrum of fungal endophytes; this heterogeneity inevitably mirrors functional diversity potentially leading to a range of significance for the host plant growth and health. Digging into this world is indispensable to expanding our knowledge on endophytism and its many beneficial applications to agriculture.

39.3 MYCORRHIZAL FUNGI: ANCIENT ALLIES OF PLANTS

Under the word "mycorrhizas" – literally fungus–root interaction – we include several types of interactions which have originated over 400 million years of co-evolution between plants and fungi (Genre et al., 2020). About 85% of vascular and non-vascular plants develop mycorrhizas (Brundrett and Tedersoo, 2018). This number and their scattered geographical distribution are a clear indication of their evolutionary and ecological success (Davison et al., 2015). Four main mycorrhizal types are recognized – ectomycorrhiza, arbuscular, orchid and ericoid mycorrhizas – that feature specific morphological traits and different combinations of partners (Figure 39.2).

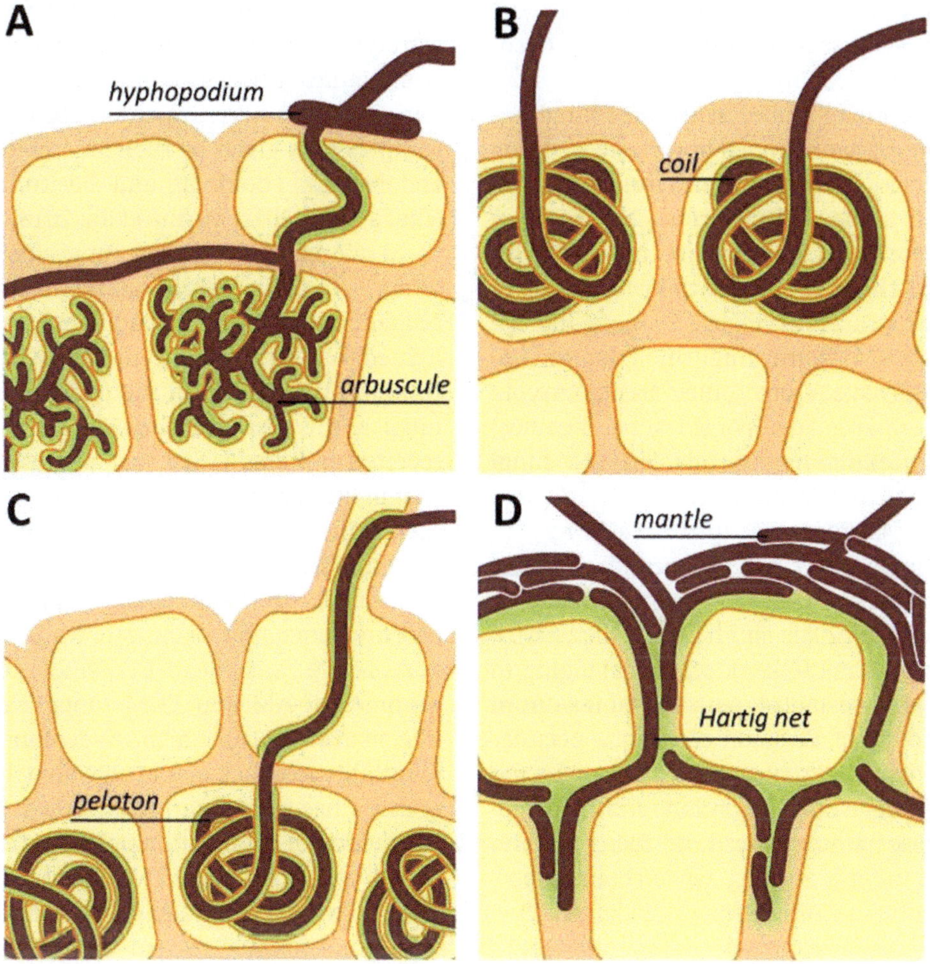

FIGURE 39.2 Scheme of the main cellular features of ectomycorrhizas, arbuscular, orchid, and ericoid mycorrhizas (from Genre et al., 2020). In arbuscular mycorrhizas, fungal hyphae contact epidermal cells through a flattened structure, the hyphopodium. Penetrating hyphae then develop via inter- and intracellular structures to reach the inner cortical cells, where they form highly branched tree-like structures called arbuscules. Fungal hyphae are always separated from the plant cell cytoplasm by an extracellular interface (green) surrounded by a membrane that develops from an extension of the host plasma membrane (orange). This interface is thought to be the main site of nutrient exchange. In ericoid mycorrhizas, hyphae penetrate the thick surface cell walls of ericaceae roots and produce dense hyphal coils inside each epidermal cell that also are surrounded by an interface compartment (green) and an extension of the plant membrane. In orchid mycorrhizas, the best described colonization process is in protocorms, the post-embryonic developmental phase after seed germination. Hyphae penetrate epidermal hair cells and reach the cortical tissue, where they form large hyphal coils, called "pelotons," within a membrane-delimited symbiotic interface. In ectomycorrhizas, fungal hyphae form a tissue known as the "sheathing mantle," which engulfs the whole root tip. Inner hyphae further develop between epidermal cells, reaching different depths into the cortical tissue depending on the host plant. These intraradical hyphae, called the Hartig net, never penetrate the cell lumen, and form an intercellular interface (green).

Ectomycorrhizas (ECM) are established between soil-borfungi (predominantly belonging to Basidiomycota and Ascomycota) and shrubs and forest trees including woody species with commercial value, such as poplar, birch, oak, pine, and spruce (Wiensczyk et al., 2002). In ECM the young root tips are enveloped by fungal hyphae that generate a pseudoparenchymatous tissue known as the "sheathing mantle." The mantle's inner hyphae further develop between epidermal and cortical root cells, leading to the so-called "Hartig net," but intraradical hyphae never penetrate the cell lumen (Figure 39.2). Together with the extraradical mycelium, these fungal structures are directly involved in the mobilization, absorption, and translocation of mineral nutrients and water from soil to roots. ECM has a massive potential in forest management, including reforestation, phytostabilization, and arboriculture activities. The promotion of abiotic stress tolerance provided by ECM to the host plants may also have a positive impact on the likelihood of plant adaptation to climate changes (Bennet and Classen, 2020).

Furthermore, a growing interest in ECM fungi lies in the production of worldwide-appreciated fruiting bodies such as boletes, chanterelles, and truffles (Danell et al., 1997; Mello et al., 2012; Murat et al., 2015).

The other mycorrhizal types are all endomycorrhizas, since they are characterized by the occurrence of fungal intracellular structures (Figure 39.2). In the ericoid mycorrhizas (ERM), some fungi belonging to Leotiomycetes (Ascomycota) or some Basidiomycota penetrate the epidermal cell walls of Ericaceae (heather) roots and form intracellular hyphal coils. The ERM fungi help the host species to colonize low mineral and acid soil (i.e., heathland), since they play a key role in releasing soil nutrients through the efficient decomposition of a wide variety of complex and recalcitrant organic substrates (Perotto et al., 2018; Smith and Read, 2008).

In orchid mycorrhizas (ORM), plants belonging to the Orchidaceae family establish with fungal taxa from several similar groups of Basidiomycota in a peculiar symbiotic interaction (Dearnaley et al., 2012). In the early stages of their development, orchids show a mycoheterotrophic phase, since instead of fixing carbon themselves via photosynthesis, they rely on their mycorrhizal partners to obtain carbohydrates, likely extracted by the fungus from soil organic matter (Merckx, 2012). ORM fungi colonize the functional cells of the host by forming dense hyphal coils called pelotons (Peterson et al., 2004; Figure 39.2). In the adult phase several orchids switch exclusively to photosynthesis; however, at maturity some of them continue to obtain carbon from mycorrhizal fungi or utilize both strategies (Merckx, 2012). Beside ensuring the orchids' livelihood, ORM fungi also play a pivotal role in orchid propagation and conservation (Batty et al., 2002).

The arbuscular mycorrhizal (AM) symbiosis is established between a small group of soil fungi that belong to the Glomeromycotina subphylum (Spatafora et al., 2016) and an extremely wide range of land plants, including the majority of staple crops. Among mycorrhizal types, the AM symbiosis is therefore the dominant association (72%), followed by ORM (10%), ECM (3%), and ERM (1%) (Brundrett and Tedersoo, 2018).

AM fungi are characterized by peculiar features such as multinucleate hyphae and spores and the apparent absence of a sexual cycle (Corradi and Brachmann, 2017; Kamel et al., 2016; Lanfranco et al., 2016). They are also obligate biotrophs, that is, uncultivable in the absence of host plants, with very large genomes extremely rich in transposable genetic elements (Chen et al., 2018b; Kobayashi et al., 2018; Morin et al., 2019; Sun et al., 2019; Venice et al., 2019). A further level of complexity is given by their microbiota, which include endobacteria (Bonfante and Desirò, 2017) and viruses (Ikeda et al., 2012; Turina et al., 2018).

Although mycorrhizal fungi are highly diverse in terms of their evolutionary history, the genome sequencing data (see the database MycoCosm) has highlighted some symbiotic signatures (Genre et al., 2020). A common trait found in the genome of all mycorrhizal fungi is the presence of several genes encoding small secreted proteins (SSP) that may serve to manipulate host plant metabolism and enhance colonization. A few such proteins have been characterized and are indeed shown to interfere with plant defense responses (Kloppholz et al., 2011; Plett et al., 2011). A limited repertoire of genes encoding plant cell wall-degrading carbohydrate-active enzymes (CAZymes) characterizes the genomes of AM and ECM fungi (Kohler et al., 2015; Martin et al., 2016), which are evolutionarily oldest. The low ability to attack plant cell wall polysaccharidic components has been interpreted as a strategy to limit the production of degradation products that are well known to activate plant defense responses. These common features may have originated by processes of convergent evolution, possibly related to adaptation to a lifestyle dependent upon living host tissues as food sources (Genre et al., 2020).

The current knowledge on mycorrhizal symbioses, especially with regard to cellular and molecular aspects, mainly refers to the AM symbiosis because it has been investigated more due to its wide host range and geographical distribution. For this reason, in the following paragraphs we will more deeply consider some aspects of this mycorrhizal association.

39.3.1 The AM Symbiosis: From the Origin to the Present

The origin and the evolution of mycorrhizal symbioses have been deciphered in the light of paleontological and, more recently, phylogenomic data (Genre et al., 2020; Strullu-Derrien et al., 2018). These data identify the AM symbiosis as the most ancient mycorrhizal type. The coevolution of land plants with AM fungi is witnessed by fossil records of early land plants from the Rhynie chert dated to be 407 million years old, where arbuscule-like structures were identified in the rhizomes of extinct plants such as *Aglaophyton majus* (Remy et al., 1994; Strullu-Derrien et al., 2018). To support the hypothesis that the evolutionary history of land plants and AM fungi are inextricably linked, phylogenetic analyses indicates that plant genes required for AM formation are present in the genomes of the closest algal relatives to land plants (Delaux et al., 2015). The occurrence of AM symbiosis in liverworts, that are the earliest diverging group among living plants, clearly shows that this mutualistic association predates the development of true roots (Brundrett, 2002; Field et al., 2012) and suggests that AM fungi assisted the early land plants to cope with nutrient acquisition from a new environment, the harsh and primeval soil (Pirozynski and Malloch, 1975).

Nutrient transfer appears to be a dominant function, as AM fungi provide access to phosphorus, which is poorly mobile in soil, and also, to a lesser extent, nitrogen, sulfur, and other essential nutrients (Wang et al., 2017). As a consequence, under limiting nutrient availability, mycorrhizal plants often display a growth effect compared to non-mycorrhizal plants. But the AM symbiosis is multifunctional, as the benefits at the level of the entire plant extend beyond improved mineral nutrition. AM fungi offer plant protection against pathogens (Cameron et al.; Fiorilli et al., 2018; Miozzi et al., 2020; Pozo and Azcón Aguilar, 2007), positively influence plant water potential (Porcel and Ruiz-Lozano, 2004), and can enhance drought tolerance (Bahadur et al., 2019; Balestrini et al., 2020; Chitarra et al., 2016; Fracasso et al., 2020). These benefits are not only a consequence of a better nutritional status, but involve several metabolic and physiological changes at local and systemic levels (Fiorilli et al., 2018; Fracasso et al., 2020; Miozzi et al., 2020). The AM symbiosis acts, therefore, as a natural mechanism against biotic and abiotic stresses.

In addition, AM fungi confer several ecosystem benefits in natural and agricultural environments (van der Heijden et al., 2015). They are the main drivers of carbon sequestration in the rooting zone (Johnson et al., 2016), and, thanks to their hyphal network, they improve soil quality by promoting particle aggregation, thereby limiting erosion by wind and water, and they decrease the risks of ground water contamination (Chen et al., 2018b).

AM fungi have a major role in controlling the composition of plant communities (van der Heijden et al., 2015). Moreover, each fungal mycelium can colonize several host plants of the same or different species, leading to the formation of the so-called common mycorrhizal networks (CMNs). A strongly interconnected plant community can potentially gain stability, because weaker individuals might profit from the mineral nutrient supply of the CMN at the expense of older plants, a phenomenon known as facilitation. These multiple benefits of AM fungi translate into significant ecological services not only in natural contexts, but also in agricultural sites. In the current context of growing concerns about environmental issues, AM fungi are proposed as crucial actors of ecosystem functioning and in the so-called Microbial Revolution, i.e. the route to increase, through the exploitation of the plant microbiota, sustainable crop productivity (Baez-Rogelio et al., 2017).

39.3.2 How to Establish a Mutualistic AM Relationship

The establishment of a functional AM requires a fine coordination at the cellular and metabolic levels of both partners; a number of signaling molecules and genes play a role in this inter-kingdom communication (Lanfranco et al., 2018; MacLean et al., 2017; Pimprikar and Gutjahr, 2018). Under conditions of limiting inorganic phosphorus, host plant roots actively exude strigolactones, a class of compounds (Waters et al., 2017) that are perceived by AM fungi and promote spore germination, hyphal branching, nuclear divisions, and respiratory metabolism (Besserer et al., 2008; Salvioli et al., 2016). In turn, the fungal partners release structural carbohydrate-derived molecules, including lipo-chito-oligosaccharides (Maillet et al., 2011) and short chitin tetra- and pentamers (Genre et al., 2013), which are perceived by the root surface epidermal cells, where they trigger the activation of the so-called common symbiosis signaling pathway – the same pathway involved in the nitrogen fixing legume symbiosis (MacLean et al., 2017; Oldroyd, 2013), which includes several evolutionarily conserved genes needed for fungal accommodation (see chapter 37). The CSSP induces the expression of a range of symbiosis-supporting genes (Hohnjec et al., 2015; Schmitz and Harrison, 2014) among which are transcription factors that coordinate the expression of a set of downstream genes involved in specific plant symbiotic functions such as phosphate uptake and lipid synthesis and transport (Hartmann et al., 2019; Pimprikar and Gutjahr, 2018). Notably, a recent comparative survey of hundreds of plant genomes and transcriptomes showed that a core set of CSSP members and downstream genes are conserved in all plant lineages hosting intracellular symbionts including AM, ORM and ERM, as well as nitrogen-fixing symbioses, but these genes have been lost in plant lineages hosting extracellular symbionts (including ECM) or those that are not symbiotic (Radhakrishnan et al., 2020).

At a cellular level, fungal hyphae contact the surface cells of host roots through swollen adhesion structures called hyphopodia. Penetrating hyphae develop inside the root cortex, taking an intercellular and/or intracellular route. Once the inner root cortex is reached, hyphae penetrate single cells, where they give rise to highly branched structures called arbuscules (Lanfranco et al., 2018; Figure 39.2). Intracellular hyphae, including large and fine branches of arbuscules, are surrounded by the perifungal membrane, a plant derived specialized membrane, which functions to host the fungal structures and separate them from the plant cell contents (Bonfante, 2018). Arbuscules are believed to be the main sites for nutrient transfer between partners. Different studies based on functional genomics analyses have provided evidence of these nutritional exchanges between partners (Javot et al., 2007; Keymer and Gutjahr, 2018; Rich et al., 2017; Roth and Paszkowski, 2017; Wang et al., 2017; Willmann et al., 2013). Mycorrhiza-specific phosphate transporters, located at the membrane surrounding arbuscules and responsible of the uptake of inorganic phosphorus ions released by the fungus, have been pinpointed in different host species (Harrison et al., 2002; Paszkowski et al., 2002), providing a mechanistic basis for the well-known benefit of phosphorus transfer. Only recently, a nitrate transporter from rice (OsNPF4.5), exclusively expressed in the cells containing arbuscules, was shown to play a key role in mycorrhizal nitrate acquisition; the up-regulation of related genes in maize and sorghum suggests that this symbiotic route of nitrogen uptake might be conserved in cereal species (Wang et al., 2020).

On the fungal side, a six-carbon sugar transporter possibly involved in the uptake of compounds from the host has been characterized (Helber et al., 2011). However, one of the most important discoveries in the biology of AM fungi was obtained from genomic analyses that showed that they lack genes encoding for the synthesis of fats (Wewer et al., 2014). Their dependence on a living host for these molecules was then supported by several other studies showing that AM fungi rely on lipids received from the host plant (Jiang et al., 2017; Keymer et al., 2017; Luginbuehl et al., 2017). This finding has shed light on the obligate nutritional nature of AM fungi and has suggested new strategies for large scale production of AM inocula (Kameoka et al., 2019).

There is now great interest in deciphering whether AM fungi prefer carbohydrates or fats (Rich et al., 2017) derived from the host and to clarify the forces controlling the exchange of nutrients in this mutualistic association. Toby Kiers has proposed the "biological market" theory in which one partner preferentially allocates resources in direct proportion to the relative amount of benefits it receives from the other partner (Kiers et al., 2011; Werner and Kiers, 2015). However, this simple mechanism does not always apply and the issue deserves further investigation (Kiers et al., 2016; Walder and van der Heijden, 2015).

39.3.3 Genetic Variation among the Partners Affects the Outcome of the Symbiosis

With the increasing availability of genomic resources, natural or induced (via domestication) genetic variation has become a promising field of investigation that will be instrumental in further exploration of the genetic determinants controlling the establishment and functioning of mycorrhizal symbioses. It is well known that the outcome of the AM symbiosis can be strongly influenced by the specific genetic makeup of the plant and the fungal partners (Lanfranco et al., 2018). Despite the low morphological variation, AM fungi show a high level of genetic variability and high functional diversity, even considering isolates belonging to the same species (Koch et al., 2017). On the other hand, responsiveness to the AM symbiosis can differ among plant cultivars of the same species and, in addition, it is affected by soil nutritional status (Chu et al., 2013; Sawers et al., 2010), pointing out a complex genotype-environment interaction. Sawers and colleagues (2017) investigated AM-responsiveness in several American maize lines and identified a correlation between the performance of the symbiosis (in terms of shoot phosphorus content and dry weight) with the amount of associated hyphae outside the root. The plant molecular determinants at the basis of the fungal performance are still unknown; it has been hypothesized that it could be related to the amount of carbohydrates and lipids released by the plant.

Plant genetic variation also determines the root colonization level. In this context, Lehnert et al. (2017) monitored root colonization on 94 bread wheat genotypes colonized by a mixed inoculum of three AM fungal species. Genomic regions containing genes related to defense and cell wall metabolism were reported to be associated with the low colonization level detected in some genotypes (Lehnert et al., 2017).

It has also been suggested that domestication, which was likely carried out under highly fertilized conditions, may have decreased the ability of plants to be colonized by, and to respond positively to, AM fungi (Lehmann et al., 2012). To address this issue, Stefani and colleagues (2020) recently surveyed the AM fungal communities associated with different durum wheat varieties released at different times in the history of durum wheat breeding. Using deep 18S rDNA sequencing under field conditions, low variation in the AM fungal diversity associated with the different cultivars was found. It has been hypothesized that the genetic variation among plant genotypes seems to be too narrow to select for specific plant–AM fungal associations; thus, in this case breeding has only marginally altered the symbiotic signaling systems and the core of genes involved in mycorrhization (Stefani et al., 2020).

In another interesting work, Martín-Robles et al. (2018) compared the AM-responsiveness in different crops and in their wild progenitors. They showed that growth benefits exerted by the AM symbiosis were dependent on inorganic phosphrus availability; while wild progenitors positively responded to the AM symbiosis irrespective of phosphorous availability, in domesticated plants the growth effect became negligible when phosphorus availability increased. These data suggest that domestication selected for AM independence at high phosphorus concentrations, which might support the possibility to increase yield in the absence of the fungus-associated carbon drain.

We envisage that further investigations of natural variations on both the fungus and the plant side will have an impact on fundamental as well as on applied research.

39.3.4 AM Translational Research: Pitfalls and Successes

Due to the several benefits that mycorrhizal fungi can provide, ranging from the level of the individual host plant to the entire ecosystem, these fungi are of high relevance for applied purposes. The impressively wide host range, which includes major crops, implies a large potential for AM fungi, especially in agroecosystems (Berruti et al., 2016; Chen et al., 2018b). However, results of the benefits provided by AMF to host plants obtained in glasshouses may not be directly relevant in the field since, under these conditions, several complex variables contribute to shape the outcome of the symbiosis. A very important issue to consider is how benefit is defined; one parameter, often used under controlled conditions, is the mycorrhizal growth response, MGR, the difference in growth between colonized and non-colonized plants at a defined level of resource supply (Janos, 2007; Johnson et al., 2015). However, under field conditions, crop yield may represent a more interesting parameter. As commented by Ryan and Graham (2002) in a field meta-analysis study, a benefit in terms of increased yield is not often evident, especially when cereals are considered. But to measure the activity of AM on crop productivity in field conditions is not easy, mainly due to the need to have non-mycorrhizal (control) plants for comparison. Despite some inconsistencies, there are many crops for which AM fungi have been clearly shown to increase yield, including cassava (Ceballos et al., 2013; Rodriguez and Sanders, 2015), potato (Hijri, 2016), and chickpea (Rocha et al., 2019). Moreover, increased temporal productivity stability (Yang et al., 2014), enhanced yield quality (Bona et al., 2017; Lehmann et al., 2014; Torres et al., 2018), and higher tolerance to biotic stresses (Fiorilli et al., 2018) have also been observed (Figure 39.1). Remarkably, even in the absence of a yield increase, the reduction in the amount of water and fertilizer required to achieve the same yield, will influence profitability and reduce the environmental impact. As advocated by Rillig et al. (2019) all these aspects have to be considered: following the concept of agroecology, based on the enhancement of ecosystem services to improve sustainable agricultural production (Bender et al., 2016; DeClerck et al., 2016), the development of production systems, where all the potential benefits provided by mycorrhizal symbionts as well as by other beneficial soil microbes is maximized, needs to be pursued in the future.

39.4 CONCLUSIONS AND PERSPECTIVES

Plants represent a preferential ecological niche for many fungi which may provide several benefits to the host (Figure 39.1). Even though the mycorrhizal symbiosis – at least the AM – has received great attention, the world of plant endophytes is still overlooked. Based on the current data on the description of novel fungi and the discovery of new metabolites, endophytes seem to be an extremely valuable resource for bioprospecting.

There is also a clear need to integrate endophytic and mycorrhizal fungal research, as important insights on how the host plant can recognize the symbiont, manage its accommodation, and the establishment of a beneficial interaction can be identified (Hilbert et al., 2020).

In many mycorrhizal studies, endophytes are ignored or reported as contaminants, but we can envisage complex interactions with functional significance (Lace et al., 2015). Studying the plant microbiota as a whole system will be the next challenge to gain a more comprehensive knowledge of the functions of the fungal symbionts living inside a plant (Thomashow et al., 2019; van der Heijden and Hartmann, 2016).

Mycorrhizal fungi and endophytes enable plant hosts to survive in hostile environments and can play a key role in improving plant responses to climate change. The development of tailored microbial inocula for specific crops and environmental conditions appears to be a promising contribution to solve urgent societal needs, an improvement of environment-friendly agricultural crop production systems, and an increase of global food security.

ACKNOWLEDGMENTS

Research in LL and VF laboratory is supported by 60% Projects (University of Torino), TOMRES from the European Union's Horizon 2020 research and innovation program under grant agreement no. 727929 and by Competitive Research Grant (CRG2017) from King Abdullah University of Science and Technology.

REFERENCES

1. Bacon CW, Porter JK, Robbins JD, Luttrell ES. 1977. *Epichloe typhina* from toxic tall fescue grasses. *Applied and Environmental Microbiology* 34: 576–581.
2. Baez-Rogelio A, Morales-García YE, Quintero-Hernández V, Muñoz-Rojas J. 2017. Next generation of microbial inoculants for agriculture and bioremediation. *Microbial Biotechnology* 10: 19–21. https://doi.org/10.1111/1751-7915.12448.
3. Bahadur A, Batool A, Nasir F, Jiang S, et al. 2019. Mechanistic insights into arbuscular mycorrhizal fungi-mediated drought stress tolerance in plants. *International Journal of Molecular Sciences* 20: 4199.
4. Balestrini R, Brunetti C, Chitarra W, Nerva L. 2020. Photosynthetic traits and nitrogen uptake in crops: Which is the role of arbuscular mycorrhizal fungi? *Plants* 9: 1105.
5. Ball OJ, Prestidge RA, Sprosen JM. 1995. Interrelationships between *Acremonium lolii*, peramine, and lolitrem B in perennial ryegrass. *Applied Environmental Microbiology* 61: 1527–1533.
6. Bao X, Roossinck MJ. 2013. Multiplexed interactions: Viruses of endophytic fungi. *Advances in Virus Research* 86: 37–58. https://doi.org/10.1016/B978-0-12-394315-6.00002-7.
7. Bastias DA, Martínez-Ghersa MA, Ballaré CL, Gundel PE. 2017. *Epichloë* fungal endophytes and plant defenses: Not just alkaloids. *Trends in Plant Science* 22(11): 940–948.
8. Batty AL, Dixon KW, Brundrett MC, Sivasithamparam K. 2002. Orchid Conservation and Mycorrhizal Associations. In: Sivasithamparama K, Dixon KW, Barrett RL (Eds.), *Microorganisms in Plant Conservation and Biodiversity*, Springer, Dordrecht.
9. Bayman P. 2006. Diversity, scale and variation of endophytic fungi in leaves of tropical plants.In: Bailey MJ, Lilley AK, Timms-Wilson TM (Eds.), *Microbial Ecology of Aerial Plant Surfaces*, Oxfordshire, UK, CABI, pp. 37–50.
10. Bender SF, Wagg C, van der Heijden MG. 2016. An underground revolution: Biodiversity and soil ecological engineering for agricultural sustainability. *Trends in Ecology and Evolution* 31: 440–452.
11. Bennet AE, Classen AT. 2020. Climate change influences mycorrhizal fungal-plant interactions, but conclusions are limited by geographical study bias. *Ecology* 101(4): e02978.
12. Berruti A, Lumini E, Balestrini R, Bianciotto V. 2016. Arbuscular mycorrhizal fungi as natural biofertilizers: Let's benefit from past successes. *Frontiers in Microbiology* 6: 1559. https://doi.org/10.3389/fmicb.2015.01559.
13. Besserer A, Becard G, Roux C, Jauneau A, et al. 2008. GR24, a synthetic analogue of strigolactones, stimulates mitosis and growth of the arbuscular mycorrhizal fungus *Gigaspora rosea* by boosting its energetic metabolism. *Plant Physiology* 148: 402–413.
14. Bona E, Cantamessa S, Massa N, Manassero P, et al. 2017. Arbuscular mycorrhizal fungi and plant growth-promoting pseudomonads improve yield, quality and nutritional value of tomato: A field study. *Mycorrhiza* 27: 1–11.
15. Bonfante P, Desirò A. 2017. Who lives in a fungus? The diversity, origins and functions of fungal endobacteria living in Mucoromycota. *The ISME Journal* 11(8): 1727–1735. doi: 10.1038/ismej.2017.21.
16. Bonfante P. 2018. The future has roots in the past: The ideas and scientists that shaped mycorrhizal research. *New Phytologist* 220: 982–995.

17. Bonfante P, Francesco Venice F, Lanfranco L. 2019. The mycobiota: Fungi take their place between plants and bacteria. *Current Opinion in Microbiology* 49: 18–25.

18. Brundrett MC. 2002. Coevolution of roots and mycorrhizas of land plants. *New Phytologist* 154: 275–304.

19. Brundrett MC, Tedersoo L. 2018. Evolutionary history of mycorrhizal symbioses and global host plant diversity. *New Phytologist* 220: 1108–1115. https://doi.org/10.1111/nph.14976.

20. Bulgarelli D, Rott M, Schlaeppi K, Loren V, van Themaat E, et al. 2012. Revealing structure and assembly cues for *Arabidopsis* root-inhabiting bacterial microbiota. *Nature* 488(7409): 91–95.

21. Cagnano G, Roulund N, Jensen CS, Forte FP, Asp T, et al. 2019. Large scale screening of *Epichloë* endophytes infecting *Schedonorus pratensis* and other forage grasses reveals a relation between microsatellite-based haplotypes and loline alkaloid levels. *Frontiers in Plant Science* 10: 765. https://doi.org/10.3389/fpls.2019.00765.

22. Ceballos I, Ruiz M, Fernández C, Peña R, et al. 2013. The in vitro mass-produced model mycorrhizal fungus, *Rhizophagus irregularis*, significantly increases yields of the globally important food security crop cassava. *PloS One* 8(8): e70633. https://doi.org/10.1371/journal.pone.0070633.

23. Chen M, Arato M, Borghi L, Nouri E, et al. 2018b. Beneficial services of arbuscular mycorrhizal fungi – from ecology to application. *Frontiers in Plant Science* 9: 1270. https://doi.org/10.3389/fpls.2018.01270.

24. Chitarra W, Pagliarani C, Maserti B, Lumini E, et al. 2016. Insights on the impact of arbuscular mycorrhizal symbiosis on tomato tolerance to water stress. *Plant Physiology* 171: 1009–1023. https://doi.org/10.1104/pp.16.00307.

25. Choi GH, Nuss DL. 1992. Hypovirulence of chestnut blight fungus conferred by an infectious viral cDNA. *Science* 257: 800–803.

26. Chu Q, Wang X, Yang Y, Chen F, et al. 2013. Mycorrhizal responsiveness of maize (*Zea mays* L.) genotypes as related to releasing date and available P content in soil. *Mycorrhiza* 23: 497–505.

27. Contreras-Cornejo HA, López-Bucio JS, Méndez-Bravo A, Macías-Rodríguez L, Ramos-Vega M, et al. 2015. Mitogen-activated protein kinase 6 and ethylene and auxin signaling pathways are involved in *Arabidopsis* root-system architecture alterations by. *Trichoderma Atroviride. Molecular Plant-Microbe Interactions* 28: 701–710. https://doi.org/10.1094/MPMI-01-15-0005-R.

28. Corradi N, Brachmann A. 2017. Fungal mating in the most widespread plant symbionts? *Trends in Plant Science* 22: 175–183.

29. Damm U, Woudenberg JHC, Cannon PF, Crous PW, 2009. *Colletotrichum* species with curved conidia from herbaceous hosts. *Fungal Diversity* 39: 45.

30. Danell E, Camacho FJ. 1997. Successful cultivation of the golden chanterelle. *Nature* 385: 303.

31. Davison J, Moora M, Öpik M, Adholeya A, et al. 2015. Fungal symbiont. Global assessment of arbuscular mycorrhizal fungus diversity reveals very low endemism. *Science* 349(6251): 970–973. https://doi.org/10.1126/science.aab1161.

32. Dearnaley JDW, Martos F, Selosse M. 2012. Orchid mycorrhizas: Molecular ecology, physiology, evolution and conservation aspects. In: Esser K (Ed.), *The Mycota Volume IX – Fungal Associations*, 2nd edition. Berlin,Springer Verlag, pp. 207–230. *(13)*.

33. DeClerck FAJ, Jones SK, Attwood S, Bossio D, et al. 2016. Agricultural ecosystems and their services: The vanguard of sustainability? *Current Opinion in Environmental Sustainability* 23: 92–99.

34. Delaux PM, Radhakrishnan GV, Jayaraman D, Cheema J, et al. 2015. Algal ancestor of land plants was preadapted for symbiosis. *Proceedings of the National Academy of Sciences of the United States of America.*112: 13390–13395.

35. Deshmukh S, Hückelhoven R, Schäfer P, Imani J, Sharma M, et al. 2006. The root endophytic fungus *Piriformospora indica* requires host cell death for proliferation during mutualistic symbiosis with barley. *Proceedings of the National Academy of Sciences of the United States of America USA* 103: 18450–18457.

36. Dupont PY, Eaton CJ, Wargent JJ, Fechtner S, et al. 2015. Fungal endophyte infection of ryegrass reprograms host metabolism and alters development. *New Phytologist* 208: 1227–1240.

37. Fesel PH, Zuccaro A. 2016. Dissecting endophytic lifestyle along the parasitism/mutualism continuum in. *Arabidopsis. Current Opinion in Microbiology* 32: 103–112.

38. Field KJ, Cameron DD, Leake JR, Tille S. 2012. Contrasting arbuscular mycorrhizal responses of vascular and non-vascular plants to a simulated Palaeozoic CO_2 decline. *Nature Communications* 3: 835.

39. Fiorilli V, Vannini C, Ortolani F, Garcia-Seco D, et al. 2018. Omics approaches revealed how arbuscular mycorrhizal symbiosis enhances yield and resistance to leaf pathogen in wheat. *Scientific Reports* 8(9625): https://doi.org/10.1038/s41598-018-27622-8.

40. Fletcher LR, Harvey IC. 1981. An association of a *Lolium* endophyte with ryegrass staggers. *New Zealand Veterinary Journal* 29: 185–186. https://doi.org/10.1080/00480169.1981. 34839.
41. Fracasso A, Telò L, Lanfranco L, Bonfante P, et al. 2020. Physiological beneficial effect of *Rhizophagus intraradices* inoculation on tomato plant yield under water deficit conditions. *Agronomy* 10: 71.
42. Fraceto LF, Maruyama CR, Guilger M, Mishra S, Chetan K, et al. 2018. Trichoderma harzianum-based novel formulations: Potential applications for management of Next-Gen agricultural challenges. *Journal of Chemical Technology and Biotechnology* 93: 2056–2063.
43. García E, Alonso Á, Platas G, Sacristá S. 2013. The endophytic mycobiota of *Arabidopsis Thaliana*. *Fungal Diversity* 60:71-89.
44. Garnica S, Riess K, Schön ME, Oberwinkler F, et al. 2016. Divergence times and phylogenetic patterns of Sebacinales, a highly diverse and widespread fungal lineage. *Plos One* 11: e0149531.
45. Genre A, Chabaud M, Balzergue C, Puech-Pagès et al. 2013. Short-chain chitin oligomers from arbuscular mycorrhizal fungi trigger nuclear Ca^{2+} spiking in *Medicago truncatula* roots and their production is enhanced by strigolactone. *New Phytologist* 198: 190–202.
46. Genre A, Lanfranco L, Perotto S, Bonfante P. 2020. Unique and common traits in mycorrhizal symbioses. *Nature Reviews Microbiology* 18(11): 649–660.
47. Guo H, Glaeser SP, Alabid I, Imani J, et al. 2017. The abundance of endofungal bacterium *Rhizobium radiobacter* (syn. *Agrobacterium tumefaciens*) increases in its fungal host *Piriformospora indica* during the tripartite sebacinalean symbiosis with higher plants. *Frontiers Microbiology* 8: 629.
48. Hacquard S, Kracher B, Hiruma K, Münch PC, et al. 2016. Survival trade-offs in plant roots during colonization by closely related beneficial and pathogenic fungi. *Nature Communication* 7: 11362.
49. Hahn H, McManus MT, Warnstorff K, Monahan BJ, et al. 2008. *Neotyphodium* fungal endophytes confer physiological protection to perennial ryegrass (*Lolium perenne* L.) subjected to a water deficit. *Environmental and Experimental Botany* 63: 183–199.
50. Harman GE, Howell CR, Viterbo A, Chet I, Lorito M. 2004. *Trichoderma* species – opportunistic, avirulent plant symbionts. *Nature Review Microbiology* 2: 43–56. https://doi.org/10.1038/nrmicro797.
51. Harrison MJ, Dewbre GR, Liu J. 2002. A phosphate transporter from *Medicago truncatula* involved in the acquisition of phosphate released by arbuscular mycorrhizal fungi. *The Plant Cell* 14: 2413–2429.
52. Hartmann RM, Schaepe S, Nübel D, Petersen AC, et al. 2019. Insights into the complex role of GRAS transcription factors in the arbuscular mycorrhiza symbiosis. *Scientific Reports* 9: 3360. https://doi.org/10.1038/s41598-019-40214-4.
53. Helber N, Wippel K, Sauer N, Schaarschmidt S, et al. 2011. A versatile monosaccharide transporter that operates in the arbuscular mycorrhizal fungus *Glomus sp.* is crucial for the symbiotic relationship with plants. *The Plant Cell* 23: 3812–3823.
54. Herd S, Christensen MJ, Saunders K, Scott DB, Schmid J. 1997. Quantitative assessment of *in planta* distribution of metabolic activity and gene expression of an endophytic fungus. *Microbiology* 143: 267–275.
55. Hijri M. 2016. Analysis of a large dataset of mycorrhiza inoculation field trials on potato shows highly significant increases in yield. *Mycorrhiza* 26: 209–214.
56. Hilbert M, Novero M, Rovenich H, Mari S, Grimm C, et al. 2020. MLO differentially regulates barley root colonization by beneficial endophytic and mycorrhizal Fungi. *Frontiers in Plant Science* 16: 1678.
57. Hiruma K, Gerlach N, Sacristán S, Nakano RT, et al. 2016. Root endophyte *Colletotrichum tofieldiae* confers plant fitness benefits that are phosphate status dependent. *Cell* 165: 464–474.
58. Hohnjec N, Czaja-Hasse LF, Hogekamp C, Küster H. 2015. Pre-announcement of symbiotic guests: Transcriptional reprogramming by mycorrhizal lipochitooligosaccharides shows a strict co-dependency on the GRAS transcription factors NSP1 and RAM1. *BMC Genomics* 16: 994.
59. Ikeda Y, Shimura H, Kitahara R, Masuta C, Ezawa T. 2012. A novel virus-like double-stranded RNA in an obligate biotroph arbuscular mycorrhizal fungus: A hidden player in mycorrhizal symbiosis. *Molecular Plant-Microbe Interactions* 25: 1005–1012.
60. Janos DP. 2007. Plant responsiveness to mycorrhizas differs from dependence upon mycorrhizas. *Mycorrhiza* 17: 75–91.
61. Javot H, Penmetsa RV, Terzaghi N, Cook DR, et al. 2007. A *Medicago truncatula* phosphate transporter indispensable for the arbuscular mycorrhizal symbiosis. *Proceedings of the National Academy of Sciences of the United States of America* 104: 1720–1725.

62. Jiang Y, Wang W, Xie Q, Liu N, et al. 2017. Plants transfer lipids to sustain colonization by mutualistic mycorrhizal and parasitic fungi. *Science* 356: 1172–1175. https://doi.org/10.1126/science.aam9970.

63. Johnson DS, Warren RS, Deegan LA, Mozdzer TJ. 2016. Saltmarsh plant responses to eutrophication. *Ecological Applications: A Publication of the Ecological Society of America* 26: 2647–2659. https://doi.org/10.1002/eap.1402.

64. Johnson LJ, Acm DB, Briggs LR, Caradus JR, et al. 2013. The exploitation of epichloae endophytes for agricultural benefit. *Fungal Diversity* 60: 171–188. https://doi.org/10.1007/s13225-013-.

65. Johnson NC, Wilson GWT, Wilson JA, Miller RM, et al. 2015. Mycorrhizal phenotypes and the law of the minimum. *New Phytologist* 205: 1473–1484. https://doi.org/10.1111/nph.13172.

66. Justus M, Witte L, Hartmann T. 1997. Levels and tissue distribution of loline alkaloids in endophyte-infected. *Festuca Pratensis. Phytochemistry* 44: 51–57.

67. Kamel L, Keller-Pearson M, Roux C, Anè J-M. 2016. Biology and evolution of arbuscular mycorrhizal symbiosis in the light of genomics. *New Phytologist* 213: 531–536.

68. Kameoka H, Tsutsui I, Saito K, Kikuchi Y, et al. 2019. Stimulation of asymbiotic sporulation in arbuscular mycorrhizal fungi by fatty acids. *Nature Microbiology* 4(1654-1660): https://doi.org/10.1038/s41564-019-0485-7.

69. Keymer A, Gutjahr C. 2018. Cross-kingdom lipid transfer in arbuscular mycorrhiza symbiosis and beyond. *Current Opinion in Plant Biology* 44: 137–144.

70. Keymer A, Pimprikar P, Wewer V, Huber C, Brands M, et al. 2017. Lipid transfer from plants to arbuscular mycorrhiza fungi. *Elife* 6: e29107. doi: 10.7554/eLife.29107.

71. Kiers ET, Duhamel M, Beesetty Y, Mensah JA, et al. 2011. Reciprocal rewards stabilize cooperation in the mycorrhizal symbiosis. *Science* 333: 880–882.

72. Kiers ET, West SA, Wyatt GA, Gardner A, et al. 2016. Misconceptions on the application of biological market theory to the mycorrhizal symbiosis. *Nature Plants* 2(16063): https://doi.org/10.1038/nplants.2016.63.

73. Kloppholz S, Kuhn H, Requena N. 2011. A secreted fungal effector of *Glomus intraradices* promotes symbiotic biotrophy. *Current Biology* 21: 1204–1209.

74. Kobayashi Y, Maeda T, Yamaguchi K, Kameoka H, et al. 2018. The genome of *Rhizophagus clarus* HR1 reveals a common genetic basis for auxotrophy among arbuscular mycorrhizal fungi. *BMC Genomics* 19: 465. https://doi.org/10.1186/s12864-018-4853-0.

75. Koch AM, Antunes PM, Maherali H, Hart MM, Klironomos JN. 2017. Evolutionary asymmetry in the arbuscular mycorrhizal symbiosis: Conservatism in fungal morphology does not predict host plant growth. *New Phytologist* 214: 1330–1337.

76. Kohler A, Kuo A, Nagy LG, Morin E, Barry KW, et al. 2015. Convergent losses of decay mechanisms and rapid turnover of symbiosis genes in mycorrhizal mutualists. *Nature Genetics* 47: 410–415.

77. Lace B, Genre A, Woo S, Faccio A, et al. 2015. Gate crashing arbuscular mycorrhizas: In vivo imaging shows the extensive colonization of both symbionts by *Trichoderma atroviride*. *Environmental Microbiology Reports* 64–77. https://doi.org/10.1111/1758-2229.12221.

78. Lahrmann U, Strehmel N, Langen G, Frerigmann H, et al. 2015. Mutualistic root endophytism is not associated with the reduction of saprotrophic traits and requires a non compromised plant innate immunity. *New Phytologist* 207: 841–857.

79. Lanfranco L, Bonfante P, Genre A. 2016. The mutualistic interaction between plants and arbuscular mycorrhizal fungi. *Microbiology Spectrum* 4: FUNK-0012-2016.

80. Lanfranco L, Fiorilli V, Gutjahr C. 2018. Partner communication and role of nutrients in the arbuscular mycorrhizal symbiosis. *New Phytologist* 220: 1031–1046. https://doi.org/10.1111/nph.15230.

81. Lehmann A, Barto EK, Powell JR, Rillig MC. 2012. Mycorrhizal responsiveness trends in annual crop plants and their wild relatives – A meta-analysis on studies from 1981 to 2010. *Plant and Soil* 355: 231–250.

82. Lehmann A, Veresoglou SD, Leifheit EF, Rillig MC. 2014. Arbuscular mycorrhizal influence on zinc nutrition in crop plants – A meta-analysis. *Soil Biology and Biochemistry* 69: 123–131.

83. Lehnert H, Serfling A, Enders M, Friedt W, et al. 2017. Genetics of mycorrhizal symbiosis in winter wheat (*Triticum aestivum*). *New Phytologist* 215: 779–791.

84. Lorito M, Woo SL, Harman GE, Monte E. 2010. Translational research on *Trichoderma*: From 'omics to the field. *Annual Review of Phytopathology* 48: 395–417.

85. Luginbuehl LH, Menard GN, Kurup S, Van Erp H, et al. 2017. Fatty acids in arbuscular mycorrhizal fungi are synthesized by the host plant. *Science* 356: 1175–1178.

86. MacLean AM, Bravo A, Harrison MJ. 2017. Plant signalling and metabolic pathways enabling arbuscular mycorrhizal symbiosis. *The Plant Cell* 29: 2319–2335.

87. Maillet F, Poinsot V, Andre O, Puech-Pages V, et al. 2011. Fungal lipochitooligosaccharide symbiotic signals in arbuscular mycorrhiza. *Nature* 469: 58–63.

88. Márquez LM, Redman RS, Rodriguez RJ, Roossinck MJ. 2007. A virus in a fungus in a plant: Three-way symbiosis required for thermal tolerance. *Science* 315: 513–515. https://doi.org/10.1126/science.1136237.

89. Martin B, Schwab E. 2012. Current usage of symbiosis and associated terminology. *International Journal of Biology* 5: 32–45.

90. Martin F, Kohler A, Murat C, Veneault-Fourrey C, Hibbett DS. 2016. Unearthing the roots of ectomycorrhizal symbioses. *Nature Review Microbiology* 14: 760–773.

91. Martinez-Medina A, Flors V, Heil M, Mauch-Mani B, et al. 2016. Recognizing Plant Defense Priming. *Trends in Plant Science* 21(10): 818–822. https://doi.org/10.1016/j.tplants.2016.07.009.

92. Martínez-Medina A, Van Wees SCM, Pieterse CMJ. 2017. Airborne signals by *Trichoderma* fungi stimulate iron uptake responses in roots resulting in priming of jasmonic acid-dependent defences in shoots of *Arabidopsis thaliana* and. *Solanum Lycopersicum. Plant Cell Environment* 40: 2691–2705. https://doi.org/10.1111/pce.13016.

93. Martín-Robles N, Morente-López J, Freschet G, Poorter H, Roumet C, et al. 2018. Root traits of herbaceous crops: Pre-adaptation to cultivation or evolution under domestication? *Functional Ecology* 33(2): 273–285. 10.1111/1365-2435.13231ff.ffhal-0230965.

94. Mello A. 2012. in Edible ectomycorrhizal mushrooms. In: Zambonelli A, Bonito GM (Eds.), *Soil Biology*, Vol. 34. Springer, pp. 73–81.

95. Merckx VSFT. 2012. *Mycoheterotrophy: The Biology of Plants Living on Fungi*. Springer, New York, Dec 6, 2012 Science.

96. Miozzi L, Vaira AM, Brilli F, Casarin V, et al. 2020. Arbuscular mycorrhizal symbiosis primes tolerance to *Cucumber Mosaic Virus* in tomato. *Viruses* 12:: 675. https://doi.org/10.3390/v12060675.

97. Morin E, Miyauchi S, San Clemente H, Chen ECH, et al. 2019. Comparative genomics of *Rhizophagus irregularis, R. cerebriforme, R. diaphanus* and *Gigaspora rosea* highlights specific genetic features in Glomeromycotina. *New Phytologist* 222: 1584–1598.

98. Müller DB, Vogel C, Bai Y, Vorholt JA. 2016. The plant microbiota: Systems-level insights and perspectives. *Annual Review of Genetics* 50: 211–234.

99. Murat C. 2015. Forty years of inoculating seedlings with truffle fungi: Past and future perspectives. *Mycorrhiza* 25: 77–81.

100. Nilsson RH, Anslan S, Bahram M, Wurzbacher C, Baldrian P, et al. 2019. Mycobiome diversity: High-throughput sequencing and identification of fungi. *Nature Review Microbiology* 17: 95–109.

101. Oldroyd GE. 2013. Speak, friend, and enter: Signalling systems that promote beneficial symbiotic associations in plants. *Nature Reviews Microbiology* 11: 252–263.

102. Paszkowski U, Boller T. 2002. The growth defect of *lrt1*, a maize mutant lacking lateral roots, can be complemented by symbiotic fungi or high phosphate nutrition. *Planta* 214: 584–590.

103. Paszkowski U, Scott Kroken S, Roux C, Briggs SP. 2002. Rice phosphate transporters include an evolutionarily divergent gene specifically activated in arbuscular mycorrhizal symbiosis. *Proceeding of the National Academy of Sciences U S A* 99(20): 13324–13329.

104. Perotto S, Daghino S, Martino E. 2018. Ericoid mycorrhizal fungi and their genomes: Another side to the mycorrhizal symbiosis? *New Phytologist* 220: 1141–1147.

105. Peterson RL, Massicotte HB, Melville LH. *Mycorrhizas: Anatomy and Cell Biology* CABI Publishing, 2004.

106. Pimprikar P, Gutjahr C. 2018. Transcriptional regulation of arbuscular mycorrhiza development. *Plant and Cell Physiology* 59: 673–679.

107. Pirozynski KA, Malloch DW. 1975. The origin of land plants: A matter of mycotrophism. *BioSystems* 6: 153–164.

108. Plett JM, Kemppainen M, Kale SD, Kohler A, et al. 2011. A secreted effector protein of *Laccaria bicolor* is required for symbiosis development. *Current Biology* 21: 1197–1203.

109. Porcel R, Ruiz-Lozano JM. 2004. Arbuscular mycorrhizal influence on leaf water potential, solute accumulation, and oxidative stress in soybean plants subjected to drought stress. *Journal of Experimental Botany* 55: 1743–1750. https://doi.org/10.1093/jxb/erh188.

110. Porras-Alfaro A, Bayman P. 2011. Hidden fungi, emergent properties: Endophytes and microbiomes. *Annual Review Phytophatology* 49: 291–315.

111. Pozo MJ, Azcón-Aguilar C. 2007. Unraveling mycorrhiza-induced resistance. *Current Opinion in Plant Biology* 10: 393–398.

112. Quigley PE. 2000. Effects of *Neotyphodium lolii* infection and sowing rate of perennial ryegrass (*Lolium perenne*) on the dynamics of ryegrass/ subterranean clover (*Trifolium subterraneum*) swards. *Australian Journal Agricultural Research* 51: 47–56.

113. Radhakrishnan GV, Keller J, Rich MK, et al. 2020. An ancestral signalling pathway is conserved in intracellular symbioses-forming plant lineages. *Nature Plants* 6: 280–289. https://doi.org/10.1038/s41477-020-0613-7.

114. Remy W, Taylor TN, Hass H, Kerp H 1994. Four hundred-million-year-old vesicular arbuscular mycorrhizae. *Proceedings of the National Academy of Sciences of the United States of America* 91: 11841–11843.

115. Rich MK, Nouri E, Courty P-E, Reinhardt D. 2017. Diet of arbuscular mycorrhizal fungi: Bread and butter? *Trends in Plant Science* 22: 652–660.

116. Rillig MC, Aguilar-Trigueros CA, Camenzind T, Cavagnaro TR, et al. 2019. Why farmers should manage the arbuscular mycorrhizal symbiosis. *New Phytologist* 222: 1171–1175.

117. Rocha I, Duarte I, Ma Y, Souza-Alonso P, et al. 2019. Seed coating with arbuscular mycorrhizal fungi for improved field production of chickpea. *Agronomy* 9: 471.

118. Rodriguez A, Sanders IR. 2015. The role of community and population ecology in applying mycorrhizal fungi for improved food security. *The ISME Journal* 9: 1053–1061. https://doi.org/10.1038/ismej.2014.207.

119. Roth R, Paszkowski U. 2017. Plant carbon nourishment of arbuscular mycorrhizal fungi. *Current Opinion in Plant Biology* 39: 50–56.

120. Ryan MH, Graham JH. 2002. Is there a role for arbuscular mycorrhizal fungi in production agriculture? *Plant and Soil* 244: 263–271.

121. Saikkonen K, Pedro G, Halander M. 2013. Chemical ecology mediated by fungal endophytes in grasses. *Journal of Chemical Ecology* 39: 962–968.

122. Salvioli A, Ghignone S, Novero M, Navazio L. 2016. Symbiosis with an endobacterium increases the fitness of a mycorrhizal fungus, raising its bioenergetics potential. *The ISME Journal* 10: 130–144.

123. Schardl CL, Leuchtmann A, Spiering MJ. 2004. Symbioses of grasses with seedborne fungal endophytes. *The Annual Review of Plant Biology* 55: 315–340.

124. Schardl CL, Young CA, Hesse U, Amyotte SG, et al. 2013a. Plant-symbiotic fungi as chemical engineers: Multi-genome analysis of the clavicipitaceae reveals dynamics of alkaloid loci. *PLoS Genetic* 9: e1003323.

125. Schmid J, Day R, Zhang N, Dupont PY, et al. 2017. Host tissue environment directs activities of an *Epichlöe* endophyte, while it induces systemic hormone and defense responses in its native perennial ryegrass host. *Molecular Plant Microbe Interaction* 30: 138–149. https://doi.org/10.1094/MPMI-10-16-0215-R.

126. Schmitz AM, Harrison MJ. 2014. Signaling events during initiation of arbuscular mycorrhizal symbiosis. *Journal of Integrative Plant Biology* 56: 250–261. https://doi.org/10.1111/jipb.12155.

127. Sherameti I, Tripathi S, Varma A, Oelmüller R. 2008. The root-colonizing endophyte *Pirifomospora indica* confers drought tolerance in *Arabidopsis* by stimulating the expression of drought stress–related genes in leaves. *Molecular Plant Microbe Interaction* 21: 799–807.

128. Smith SE, Read D. 2008. *Mycorrhizal Symbiosis*, 3rd edition. Academic Press.

129. Spatafora JW, Chang Y, Benny GL, Lazarus K, et al. 2016. A phylum-level phylogenetic classification of zygomycete fungi based on genome scale data. *Mycologia* 108: 1028–1046.

130. Spatafora JW, Sung GH, Sung JM, Hywel-Jones NL, et al. 2007. Phylogenetic evidence for an animal pathogen origin of ergot and the grass endophytes. *Molecular Ecology* 16: 1701–1711.

131. Stefani F, Dupont S, Laterrière M, Knox R, et al. 2020. Similar arbuscular mycorrhizal fungal communities in 31 durum wheat cultivars (*Triticum turgidum* L. var. durum) under field conditions in Eastern Canada. *Frontiers in Plant Science* 11: 1206. https://doi.org/10.3389/fpls.2020.01206.

132. Stein E, Molitor A, Kogel KH, Waller F. 2008. Systemic resistance in Arabidopsis conferred by the mycorrhizal fungus *Piriformospora indica* requires jasmonic acid signaling and the cytoplasmic function of NPR1. *Plant Cell Physiology* 49: 1747–1751.

133. Strullu-Derrien C, Selosse MA, Kenrick P, Martin FM. 2018. The origin and evolution of mycorrhizal symbioses: From palaeomycology to phylogenomics. *New Phytologist* 220: 1012–1030.

134. Sun X, Chen W, Ivanov S, MacLean AM, et al. 2019. Genome and evolution of the arbuscular mycorrhizal fungus *Diversispora epigaea* (formerly *Glomus versiforme*) and its bacterial endosymbionts. *New Phytologist* 221: 1556–1573.

135. Tan YY, Spiering MJ, Scott V, Lane GA, et al. 2001. *In planta* regulation of extension of an endophytic fun-

gus and maintenance of high metabolic rates in its mycelium in the absence of apical extension. *Applied and Environmental Microbiology* 67: 5377–5383.

136. Tao G, Liu ZY, Liu F, Gao YH, et al. 2013. Endophytic *Colletotrichum* species from *Bletilla ochracea* (Orchidaceae), with descriptions of seven new species. *Fungal Diversity* 61: 139–164.

137. Thomashow LS, LeTourneau MK, Kwak Y-S, Weller DM. 2019. The soil-borne legacy in the age of the holobiont. *Microbial Biotechnology* 12: 51–54.

138. Tian P, Nan Z, Li C, Spangenberg G. 2008. Effect of the endophyte *Neotyphodium lolii* on susceptibility and host physiological response of perennial ryegrass to fungal pathogens. *European Journal of Plant Pathology* 122: 593–602.

139. Torres N, Antolin MC, Goicoechea N. 2018. Arbuscular mycorrhizal symbiosis as a promising resource for improving berry quality in grapevines under changing environments. *Frontiers in Plant Science* 9: 18.

140. Toruño TY, Stergiopoulos I, Coaker G. 2016. Plant-pathogen effectors: Cellular probes interfering with plant defenses in spatial and temporal manners. *Annual Review of Phytopathology* 54: 419–441. https://doi.org/10.1146/annurev-phyto-080615-100204.

141. Trivedi P, Leach JE, Tringe SG, Sa T, Singh BK. 2020. Plant-microbiome interactions: From community assembly to plant health. *Nature Reviews Microbiology* https://doi.org/10.1038/s41579-020-0412-1.

142. Turina M, Ghignone S, Astolfi N, Silvestri A, Bonfante P, Lanfranco L. 2018. The virome of the arbuscular mycorrhizal fungus *Gigaspora margarita* reveals the first report of DNA fragments corresponding to replicating nonretroviral RNA viruses in fungi. *Environmental Microbiology* 20: 2012–2025.

143. van der Heijden MG, Hartmann M. 2016. Networking in the plant microbiome. *PLoS Biology* 14: e1002378.

144. van der Heijden MGA, Martin FM, Selosse MA, Sanders IR. 2015. Mycorrhizal ecology and evolution: The past, the present, and the future. *New Phytologist* 205: 1406–1423. https://doi.org/10.1111/nph.13288.

145. Vandenkoornhuyse P, Quaiser A, Duhamel M, Le Van A, Dufresne A. 2015. The importance of the microbiome of the plant holobiont. *New Phytologist* 206: 1196–1206.

146. Varma A, Savita V, Sudha, Sahay N, Butehorn B, Franken P. 1999. *Piriformospora indica*, a cultivable plant-growth-promoting root endophyte. *Applied and Environmental Microbiology* 65: 2741–2744.

147. Venice F, Ghignone S, Salvioli di Fossalunga A, Amselem J, et al. 2019. At the nexus of three kingdoms: The genome of the mycorrhizal fungus *Gigaspora margarita* provides insights into plant, endobacterial and fungal interactions. *Environmental Microbiology* 22: 122–141.

148. Vinale F, Nigro M, Sivasithamparam K, Flematti G, Ghisalberti EL, et al. 2013. Harzianic acid: A novel siderophore from *Trichoderma harzianum*. *FEMS Microbiology Letters* 347(2): 123–129.

149. Walder F, van der Heijden MGA. 2015. Regulation of resource exchange in the arbuscular mycorrhizal symbiosis. *Nature Plants* 1: 15159.

150. Wang S, Chen A, Xie K, Yang X, Luo Z, et al. 2020. Functional analysis of the OsNPF4.5 nitrate transporter reveals a conserved mycorrhizal pathway of nitrogen acquisition in plants. *Proceeding of the National Academy of Sciences United States of America* 117(28): 16649–16659. doi: 10.1073/pnas.2000926117.

151. Wang W, Shi J, Xie Q, Jiang Y, Yu N, et al. 2017. Nutrient exchange and regulation in arbuscular mycorrhizal symbiosis. *Molecular Plant* 10: 1147–1158.

152. Waters MT, Gutjahr C, Bennett T, Nelson DC. 2017. Strigolactone signaling and evolution. *Annual Review of Plant Biology* 68: 291–322.

153. Weindling R. 1932. *Trichoderma lignorum* as a parasite of other soil fungi. *Phytopathology* 22: 837–845.

154. Werner GD, Kiers ET. 2015. Partner selection in the mycorrhizal mutualism. *New Phytologist* 205:1437-1442. https://doi.org/10.1111/nph.13113.

155. Wewer V, Brands M, Dörmann P. 2014. Fatty acid synthesis and lipid metabolism in the obligate biotrophic fungus *Rhizophagus irregularis* during mycorrhization of. *Lotus Japonicus. The Plant Journal* 79: 398–412.

156. Wiensczyk A, Gamiet S, Durall D, Jones M, et al. 2002. Ectomycorrhizae and forestry in British Columbia: A summary of current research and conservation strategies. Br Columbia. *Journal Ecosystem Managment* 2: 1–19.

157. Willmann M, Gerlach N, Buer B, Polatajko A, et al. 2013. Mycorrhizal phosphate uptake pathway in maize: Vital for growth and cob development on nutrient poor agricultural and greenhouse soils. *Frontiers in Plant Science* 4: 533.

158. Wilson D. 1995. Endophyte: The evolution of a term, and clarification of its use and definition. *Oikos* 73: 274–276.

159. Yadav V, Kumar M, Deep DK, Kumar H, Sharma R, et al. 2010. A phosphate transporter from the root endophytic fungus *Piriformospora indica* plays a role in phosphate transport to the host plant. *Journal of Biological Chemistry* 285: 26532–26544.

160. Yang GW, Liu N, Lu WJ, Wang S, et al. 2014. The interaction between arbuscular mycorrhizal fungi and soil phosphorus availability influences plant community productivity and ecosystem stability. *Journal of Ecology* 102: 1072–1082.

161. Young CA, Schardl CL, Panaccione DG, Florea S, et al. 2015. Genetics, genomics and evolution of ergot alkaloid diversity. *Toxins* 7: 1273–1302. https://doi.org/10.3390/toxins7041273.

162. Yu P, Wang C, Baldauf JA, Tai H, et al. 2018. Root type and soil phosphate determine the taxonomic landscape of colonizing fungi and the transcriptome of field-grown maize roots. *The New Phytologist* 217: 1240–1253. https://doi.org/10.1111/nph.14893.

CHAPTER 40

Aromatherapy

Improving Plant Health through Microbial Volatiles

Ana Shein Lee Diaz and Paolina Garbeva

Department of Microbial Ecology, Netherlands Institute of Ecology (NIOO-KNAW), Droevendaalsesteeg 10, 6708 PB, Wageningen, The Netherlands

40.1 BACKGROUND

During their entire lifecycles, plants are involved in intimate interactions with microbes that are essential for plant growth and health. Secondary metabolites, produced by plants and microbes alike, are important for plant–microbe interactions and communication. Secondary metabolites are organic compounds which are not directly involved in primary metabolic functions such as growth, development, or reproduction. Both plants and microorganisms can produce a wide variety of secondary metabolites including volatile and non-volatile compounds. Compared to non-volatile compounds which tend to accumulate around the cells that produce them, volatiles can diffuse easily in both gas and water phases and have been shown to play an important role in mediating chemical interactions. Volatile organic compounds (VOCs) are small molecules (<300 Da) belonging to different chemical classes that can evaporate and diffuse easily (Schmidt et al., 2015). These physiochemical properties make volatiles ideal candidate metabolites for long-distance plant–microbe interactions. Microbial volatiles are the first microbial signals or chemical weapons to reach a target organism, and thus are considered early alert signals. Volatile blends can include alkenes, alcohols, aldehydes, acids, ketones, benzenoids, pyrazines, terpenes, and S-derivates.

The ability of microorganisms to release chemically diverse volatile compounds has been overlooked for a long time. However, over the last few years, scientists have realized that the blend of volatiles released by microorganisms can be very complex and includes many unknown compounds (Effmert et al., 2012; Lemfack et al., 2014). The reason why soil microorganisms produce so many volatiles is not clear. Emission of volatiles by microorganisms is known to greatly vary depending on their phylogenetic origin, their physiological state and the prevailing nutrient conditions. Furthermore, emission of volatiles can be specifically triggered by the presence of an interacting partner. This suggests that microbial volatiles are not mere by-products of primary metabolism, but that they are released in nature specifically when needed.

Over the last few years, evidence has been accumulating that supports the idea that microbial volatiles can have a positive or negative effect on the growth of other organisms such as plants, fungi, and nematodes. This chapter will focus on volatile mediated plant–microbe interactions that are important for plant growth and health (Figure 40.1).

Good Microbes in Medicine, Food Production, Biotechnology, Bioremediation, and Agriculture, First Edition. Edited by Frans J. de Bruijn, Hauke Smidt, Luca S. Cocolin, Michael Sauer, David Dowling, and Linda Thomashow.

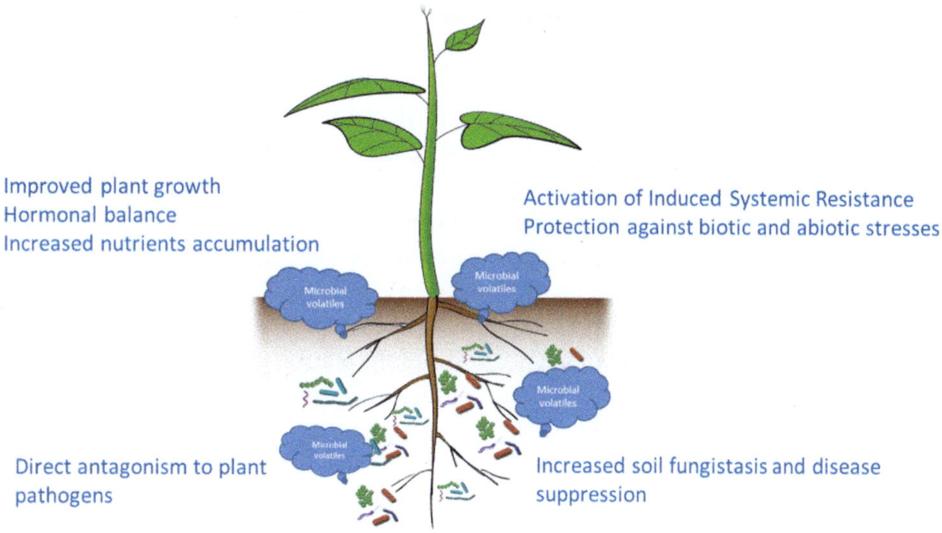

FIGURE 40.1 Beneficial effects of microbial volatiles on the plant.

40.2 MICROBIAL VOLATILES IMPROVING PLANT GROWTH

In 2003, Ryu et al. revealed that exposing the model plant *Arabidopsis thaliana* to volatiles emitted by different *Bacillus* strains led to drastic plant growth promotion. This study was followed by many reports of growth modulation of *A. thaliana* and a few other plant species by volatile blends emitted by phylogenetically diverse bacteria and fungi. It is now well-established that microbial volatiles can increase root biomass and lateral root formation as well as leaf surface (Ditengou et al., 2015; Ryu et al., 2003, 2004). For example, the rhizosphere bacterium *Pseudomonas fluorescens* SS101 was found to promote plant growth both *in vitro* and in soil. The volatile compounds responsible for this effect were identified as 13-tetradecadien-1-ol, 2-butanone and 2-methyl-n-1-tridecene (Park et al., 2015).

Growth promotion of plants exposed to microbial blends is partially due to the carbon dioxide generated as the result of bacterial respiration. Hence, a good experimental setup with appropriate controls is required to avoid artifacts or misinterpretations of results (Piechulla et al., 2017). In addition to carbon dioxide, many microbial volatile compounds including 2,3- butanediol, indole, and dimethyl disulfide promote plant growth when applied as pure compounds, suggesting a substantial contribution of diverse organic molecules to the overall stimulation of plant growth by volatiles. The rhizospheric *Streptomyces* and *Microbacterium* species were able to enhance the shoot and root biomass of *Arabidopsis* plants via volatiles and induced upregulation of plant genes involved in sulfur and nitrogen metabolism in shoot and root tissues, respectively (Cordovez, 2016; Cordovez et al., 2018). The profiling of microbial volatiles revealed abundant sulfur-containing compounds including dimethyl disulfide and dimethyl trisulfide.

The mechanisms underlying growth promotion by microbial volatiles are still poorly understood, but a few studies suggest that they are diverse. For example, hormonal manipulation through interference with auxin regulation was shown to underlie increases in indole-induced root and shoot biomass triggered by Gram-negative bacterial indole producers. Exposure to *Bacillus amyloliquefaciens* GB03 volatiles also triggered auxin-dependent growth promotion.

Other phytohormones such as cytokinins, ethylene, and abscisic acid are involved in a plant's response to growth-promoting bacterial volatiles. Beyond hormonal balance, microbial volatiles can influence the uptake of specific nutrients such as iron and sulfur by plants by increasing either their availability or their uptake. In the case of iron, bacterial volatiles emitted by *B. amyloliquefaciens* GB03 acted through three different mechanisms on *A. thaliana*: by triggering enhanced rhizosphere acidification leading to higher iron availability; by inducing upregulation of an extracellular ferric reductase enabling Fe^{+}(III) to Fe^{+}(II) reduction necessary for exterior iron uptake; and by inducing overexpression of the major plant iron transporter. Both the acidification and higher expression of the reductase were also observed in *Medicago sativa* upon exposure to the volatiles of its nitrogen-fixing symbiont *Sinorhizobium meliloti*.

Using a pot-in-jar system, Martín-Sánchez and coauthors tested the ability of microbial volatiles to diffuse freely in soil and affect the nutritional status of maize from a distance (Martín-Sánchez et al., 2020). The results showed that volatiles emitted belowground by the rhizosphere bacteria *Serratia plymuthica* and the fungal pathogen *Fusarium culmorum*, in monoculture or interactions, differentially impacted the content of nutrients in maize plants such as Fe, Zn, Cu, and Mo.

Increased plant growth is also associated with enhancing the availability of reduced sulfur by S-containing volatiles such as dimethyl disulfide (a volatile compound commonly emitted by soil bacteria) (Meldau et al., 2013).

40.3 MICROBIAL VOLATILES INDUCING SYSTEMIC RESISTANCE IN PLANTS

The plant immune system has evolved to recognize molecules associated with pathogenic microbes, such as flagellin and other microbial components (Jones and Dangl, 2006). The recognition of these elicitor molecules, known as microbe/pathogen-associated molecular patterns (MAMPs and PAMPs, respectively), triggers a defense response in the plant. However, MAMPs, which occur in beneficial microbes, can also be recognized through a process where the plant activates its defenses without triggering the hypersensitive responses typical of pathogen recognition (Van Wees et al., 2008). This specific beneficial-microbe recognition process is known as microbial induced systemic resistance (ISR), and leads to an enhanced plant immune status (Pieterse et al., 2014). The discovery of beneficial microbes as helpers in plant defense through ISR opened the possibility of searching for more microbial elicitors that could help plants cope with stresses. In 2004, a cornerstone study by Ryu and collaborators, showed that ISR could also be triggered by microbial volatiles; two *Bacillus* species could induce a defensive response in Arabidopsis plants to help fight against the pathogen *Erwinia carotovora* via volatile seed priming (Ryu et al., 2004). The responsible compounds, 2,3-butanediol and acetoin, seemed to regulate the plant's defenses through the ethylene pathway. This study was the starting point for a new area of research on microbial volatile-mediated ISR as well as for the search into how microbial volatile organic compounds (mVOCs) could have an impact on a plant's tolerance to stress.

Since then, many studies have reported the activation of ISR through beneficial-microbe volatiles (Thomas et al., 2020). Such volatiles are mainly produced by beneficial bacteria belonging to the genera *Bacillus* spp. and *Pseudomonas* spp., and beneficial fungi such as *Trichoderma* spp (see chapter 41). The specificity of ISR-inducing mVOCs is not yet known. For example, 2,3-butanediol has been reported to have a broad spectrum, impacting different plant species (*A. thaliana*, *N. benthamiana*, and *C. sativus*) and protecting the host plants against different pathogens (Song et al., 2019; Song and Ryu, 2013; Wu et al., 2018). Most studies have focused on ISR-inducing mVOCs against bacterial and fungal pathogens; however, some studies against insect pests (Song and Ryu, 2013) and viruses (Choi et al., 2014) have also been reported. Still, ISR-inducing volatiles produced to fight against other biotic stresses such as oomycetes, insect pests, and viruses have been less explored. Another aspect regarding the specificity of volatiles relies on the emitting microbe: volatiles from other microbes besides beneficial microbes might also have a role in ISR activation. A recent study has shown that sorghum plants might distinguish between volatiles emitted by beneficial, commensal, and pathogenic microbes and activate defense responses accordingly (Hernández-Calderón et al., 2018). The specificity or general character of ISR-inducing volatiles is still under study, since to date no volatile receptor has been found in plants (Bailly, 2020) and the molecular mechanism for volatile-mediated ISR activation has not yet been deciphered.

ISR-inducing volatiles produced by microbes present a broad chemical diversity including such compounds as alcohols, sulfur compounds, phenols, esters, and long-chain hydrocarbons. Volatile ISR-inducing assays mainly rely on chemically synthesized volatiles previously detected by chromatography in microbial volatile blends. A volatile's effect on ISR can vary depending on the method used (often in *in vitro* plant cultures), the volatile dose and the ratio among certain compounds. Lee and collaborators found that *Arabidopsis* plants responded strongly to volatiles from *Paenibacillus polymyxa* as compared to volatiles from the known ISR-inducing strain *Bacillus subtilis* GB03 due to the presence of a long chain volatile (tridecane) (Lee et al., 2012). However, tridecane introduced as pure compound reduced plant growth and the effect was observed mainly at short distances, indicating that the unique volatile blend from *P. polymyxa* and the volatile concentration are key for the systemic resistance activation. ISR-inducing volatiles also have been reported to be successful in more complex systems such as greenhouses or hydroponic cultures (Huang et al., 2012; Naznin et al., 2014; Song et al., 2019) and even in a field trial where isomer-dependent ISR activity was observed (Kong et al., 2018). Further research will shed light on the context-dependency that regulates the volatile-mediated ISR activation.

The ISR signaling pathway often involves the activation of local transcription factors that lead to changes in jasmonic acid (JA), salicylic acid (SA), and ethylene (ET) phytohormone signaling (Broekgaarden et al., 2015; van Wees et al., 2000). In direct plant–microbe interactions, the plant receptors bind the MAMPs elicitors such as bacteria flagellin, fungal chitin, lipopolysaccharides, antibiotics, and other molecules that trigger the local activation of defense genes (Stringlis et al., 2018). This signal cascades systemically throughout the plant tissues, inducing the broad-spectrum ISR-enhanced defensive status. Although volatile-mediated ISR signaling is not yet fully described, it is known that it impacts certain genes belonging to the jasmonic acid and ethylene pathways (Kant et al., 2009; Thomas et al., 2020). Interestingly, ISR-inducing volatiles seem to alter different phytohormonal pathways since these volatiles have been shown to be related to ET in a JA-independent way (Ryu et al., 2004) or also by activating the two antagonistic JA and SA pathways (Naznin et al., 2014). Therefore, microbial volatiles impact the plant's physiology by altering the gene expression of defense-related genes such as PAL, the production of defense compounds, and volatile production in aboveground green tissues. For example, it has been observed that volatiles can induce the expression of genes related to C6 volatile production above ground (Kishimoto et al., 2005). However, in some instances, whether the volatile exerts a direct effect through pathogen inhibition or triggers ISR is under debate, since not all studies provide clear evidence of activated genes and phytohormonal pathways upon VOC perception (Garbeva and Weisskopf, 2020; Sharifi and Ryu, 2016).

Thus, ISR-inducing microbial volatiles are an interesting alternative to chemical pesticides and might play an important role in future agricultural systems. Some pure compounds such as dimethyl-disulfide (DMDS) have been successfully tested in field trials showing ISR-activation in tomato against fungal pathogens (Tyagi et al., 2020). However, in view of an applied approach, many fundamental and applied aspects still require further research. In addition to understanding of the molecular mechanisms behind the volatile-mediated ISR induction and the phytohormonal associated changes, important aspects such as volatile application mode, number and concentration of doses, selection of host genotype, and consideration of multi-species interactions, among others, will contribute to advancements in the field.

One of the bottlenecks of ISR-inducing volatile and microbes' application in agriculture is the inherent context-dependency of the ISR process. ISR is a complex plant physiological response to an elicitor and for its activation it requires a specific context suitable for the interaction (Walters et al., 2013), making ISR therefore, highly unpredictable. One factor is the plant genotype; it determines the plant's defense potential as well as the compatibility with ISR microbial strains (Berendsen et al., 2012; Tucci et al., 2011) and maybe with other elicitors such as VOCs . Volatile-mediated ISR activation has been mainly studied in model plants, and more research on commercial crops is necessary to understand the mechanisms of volatile-ISR responses in major crop families such as Solanaceae, legumes, and monocotyledons.

40.4 MICROBIAL VOLATILES DIRECTLY INHIBITING PLANT PATHOGENS

Many microbial volatiles act not only as signals but also as antimicrobial compounds that can directly inhibit fungal and bacterial plant pathogens. The germination and growth of fungi is often negatively affected by direct or indirect contact with soils, a phenomenon named soil fungistasis (Garbeva et al., 2011) and microbial volatiles can play important roles there. Chuankun et al. (2004) revealed a strong correlation between inhibition and germination of fungal spores that were in direct contact with soils, as well as the germination of spores exposed to microbial volatiles emitted from those soils. The emission of suppressive volatiles was stopped after soil sterilization. In a recent study by van Agtmaal et al. (2018), two pathogenic fungi (*Fusarium oxysporum* and *Rhizoctonia solani*) and a plant pathogenic oomycete (*Pythium intermedium*) growing on nutrient-poor agar were exposed to microbial volatiles emitted from soil samples obtained from 50 agricultural fields. Most soil samples emitted growth-suppressing volatiles, but the extent of growth inhibition per soil differed strongly for the three pathogens.

Next to soil fungistasis, microbial volatiles can play an important role in disease suppressiveness. Recent screening of a large collection of field soils for suppressiveness to *Fusarium culmorum* revealed volatile-mediated soil suppressiveness in several, but not all suppressive soils (Ossowicki et al., 2020). This indicates that even though volatile emission is one of the factors that may contribute to disease suppressiveness to *F. culmorum*, it is not a general one. Volatile compounds emitted by the soil bacteria *Pseudomonas chlororaphis* subsp. *aureofaciens* (3-methyl butanol, phenylethyl alcohol and 2-methyl butanol) can inhibit conidial germination and mycelial growth of the fungal pathogen causing black rot of sweet potato (Zhang et al., 2019). The volatiles emitted by another pseudomonad, *P. fluorescens* WR-1, inhibited the replication of the bacterial pathogen *Ralstonia solanacearum* that causes tomato wilt (Raza et al., 2016). The volatile blend emitted by the tomato isolate *Pseudomonas donghuensis* P482 has strong antifungal (*Rhizoctonia solani*, *Fusarium culmorum*) and antioomycete *Pythium ultimum*) activity, but not antibacterial activity (*Agrobacterium* sp. and *Pseudomonas fluorescens*). Many studies have reported powerful antifungal and anti-oomycete activities of bacterial volatiles, with high variation in fungal sensitivity. For example, Garbeva et al. (2014) observed that the plant pathogens *Rhizoctonia solani*, *Fusarium culmorum*, and *Pythium ultimum* were the most sensitive, while the saprotrophic fungi *Chaetomium* sp., *Mucor hiemalis* and *Trichoderma harzianum* were the most resistant. This is in line with other reports on the difference in volatile sensitivity between pathogenic and saprotrophic fungi. More examples of microbial volatiles with antagonistic activity against plant pathogens are reviewed by Bitas et al. (2013), de Boer et al. (2019); and Garbeva and Weisskopf (2020).

40.5 APPLICATION OF MICROBIAL VOLATILE COMPOUNDS

Microbial volatiles can improve plant growth and health and are therefore considered as a new and important source of safe and environmentally friendly gaseous agrochemicals. One advantage of volatile compounds is that they have a short half-life and evaporate rapidly. Hence, they are therefore unlikely to have a long-term effect and leave residues on the fruit and vegetable products. Most of our knowledge on the emission of volatiles by plant beneficial microbes originates from laboratory studies on single strains. However, few recent studies have shown that microbial volatiles are able to protect plants against diseases *in planta* (De Vrieze et al., 2015; Guyer et al., 2015) or even in greenhouse and field conditions (Choi et al., 2014; Song and Ryu, 2013), and they raise hope that bacterial volatiles might be a suitable source of new plant protection products of reduced toxicity for the environment and human health.

The most advanced studies in regard to microbial volatile application in field sites have tested single pure compound application. For example, recently, the optimal concentration of the volatile compound 2,3-butanediol for disease control against *Cucumber mosaic virus* and *Tobacco mosaic virus* was tested in the greenhouse and in the field (Kong et al., 2018). In the field trial, in particular, the authors show an isomer-dependent effect since the treatment with 2 R,3 R-butanediol and 2 R,3S-butanediol significantly reduced the incidence of naturally occurring viruses. Furthermore, one single volatile compound can show various beneficial functions, such as dimethyl disulfide, which improves plant growth by enhancing the availability of reduced sulfur, protects plants against pathogens by directly inhibiting them and inducing systemic resistance in plants. Dimethyl disulfide was recently patented as a novel soil fumigant (under the name PALADIN®) which targets nematodes and soilborne pathogens. Another example is indole, a ubiquitous microbial volatile that can act both as bioprotectant and a biostimulant by promoting root growth and priming defenses via jasmonic acid signaling (Chung et al., 2016). However, research on the application of other microbial volatiles in agriculture is still in its infancy, as most studies to date were conducted under controlled laboratory conditions where volatile dispersion is controlled in sealed environments and precise temperature and humidity. The studies assessing the effects of volatile application under open conditions have been summarized in recent reviews from Song et al. (2019) and Garbeva and Weisskopf (2020).

Most studies have focused on the effect of single compounds in one concentration, but microbial volatiles are released as a blend with specific concentrations and ratios, and nearly all volatile compounds reveal dose-dependent effects. Testing individual compounds can lead to opposite effects, which may not explain the function of a full volatile bouquet. Plant–microbial volatile interactions are considered to be unique, as plants might react differently to microbial volatiles emitted by beneficial or pathogenic microbes (Hernández-Calderón et al., 2018). In other instances, however, plants seem not to react differently to pathogen and beneficial microbes' volatiles (Moisan et al., 2019), bringing into question the specificity of volatiles and the plants' mechanisms for volatile recognition. In addition, plant defense is closely related to plant growth, and the nutritional status of the plant might also influence the defense responses upon volatile perception. A recent study shows how some bacterial volatiles can actually induce hypersensitive stress responses in plants if they are already undergoing nutrient deprivation (Morcillo et al., 2020). Furthermore, there is a lack of studies evaluating other unexpected effects of the bioactive volatiles on non-target plant beneficial microorganisms. For example, it was observed that microbial volatiles from bacteria can have an impact on native fungal soil communities (Yuan et al., 2017).

Concerning the implementation of microbial volatiles, our knowledge on the potential use of those compounds in large-scale agriculture and horticulture is still limited. In agriculture systems, volatiles have to be applied under open-field conditions, which are very different and variable.

The plant beneficial effects mediated by microbial volatiles opens many new avenues for the application of either volatile-emitting strains or the pure compounds themselves. Possible application techniques of pure volatiles are: (1) drenching/spraying application (as used for non-volatile plant protection products), (2) an application by air diffusion (e.g., as used for mating-disruption pheromones), or (3) encapsulation into a specific matrix that allows slow release (Figure 40.2). Encapsulation technology is gaining interest in agriculture and especially in application of volatiles, as it increases their stability and enables controlled release (Sharifi and Ryu, 2020). Most of the encapsulation methods produce macro- or nano-capsules that are suitable for spray application.

Inoculation of volatile-producing strains is another common approach. However, when considering that inoculated strains will integrate into complex microbial communities, the fact that interspecific interactions can strongly influence volatile emission should be kept in mind (Schmidt et al., 2017; Tyc et al., 2015). A major obstacle for the success of inoculated strains is that their establishment under field conditions is hampered by abiotic and biotic factors (de Boer, 2017). With respect to the latter, introduced biocontrol strains have to compete for nutrients with indigenous microbes that are better adapted to the local environmental conditions (Chaparro et al., 2012; Compant et al., 2019).

Given the superior competitive abilities of indigenous microbes, it may be more profitable to make use of these microbes and develop strategies to stimulate activities that suppress pathogens. This can be indicated as

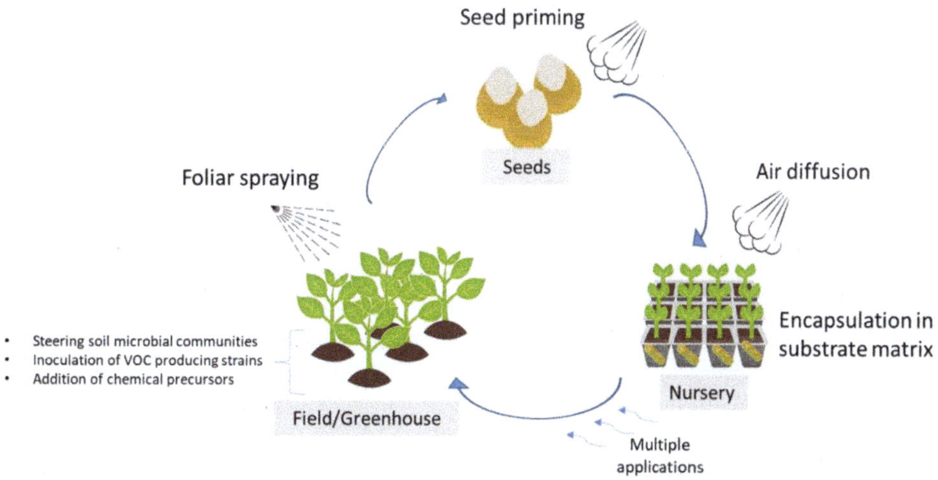

FIGURE 40.2 Application modes of microbial volatiles according to plant developmental stages. From top to right: seeds can be exposed to microbial volatiles (as pure compounds) though seed priming. Upon germination, certain crops are grown in nurseries during early life-stages. The infrastructure might allow application of volatiles through air diffusion or spraying and also by mixing encapsulated volatile compounds in the substrate mixtures. Due to their short half-life, volatiles can be applied in multiple instances for prevention or maintenance of the desired effect. In open field the application can be either foliar (spraying) but also by belowground application; (1) steering soil microbial communities for production of desired compounds, (2) inoculation of volatile-producing microbial strains, and (3) addition of metabolic precursors.

"steering of indigenous soil microbial communities" and is an approach that is receiving increasing attention. Composition and activity of soil microbial communities is strongly influenced by the availability and composition of organic energy sources. In arable soils, these can be resources supplied by plants (root exudates, plant remainders) or by farmers (manure, compost). The composition of these organic sources can also affect the production of microbial volatiles. Similarly, from *in vitro* studies it is known that medium composition has a major impact on the blend of volatiles produced by microorganisms (Garbeva et al., 2014; Lazazzara et al., 2017). Therefore, selection of appropriate organic materials appears to be essential to realize enhanced production of plant beneficial volatiles by soil microbial communities.

Another still-unexplored approach is seed coating with volatile-emitting strains or the pure compounds, which may increase the effectiveness of seedling protection. Yet more research is needed to find the best way and time (e.g., plant growth stage) for applying volatiles or volatile-emitting strains. Overall, the application of microbial volatiles or volatile-producing strains is expected to maintain a healthy balance in soil and plant microbial communities, which is by contrast strongly impacted by synthetic fungicides and pesticides. Microbial volatiles present a new source of bioactive natural compounds with high crop tolerance, high efficacy, and reduced environmental impact.

40.6 CONCLUSION

Effective, sustainable alternatives for the use of chemical fungicides and pesticides in horticulture and agriculture are urgently needed. In recent years, evidence has accumulated that microbial volatiles have a strong biological effect on plants and their pathogens, turning microbial volatiles from overlooked metabolites to potential bioactive compounds for crop health. Despite the multiple reports of strong plant beneficial activity of microbial volatiles, many fundamental and applied questions still remain unanswered. A better understanding of the synthesis of plant protective microbial volatiles, the factors stimulating their emission, the mechanisms of action and possible application in greenhouse and field conditions will ultimately lead to the development of new, environmentally friendly volatile-mediated crop protection strategies.

REFERENCES

1. Bailly A. 2020. How plants might recognize rhizospheric bacterial volatiles. In: Ryu C-M, Weisskopf L, Piechulla B (Eds.), *Bacterial Volatile Compounds as Mediators of Airborne Interactions*, Springer Singapore, Singapore, pp. 139–165.
2. Berendsen RL, Pieterse CMJ, Bakker PAHM. 2012. The rhizosphere microbiome and plant health. *Trends in Plant Science* 17: 478–486.
3. Bitas V, Kim H-S, Bennett JW, Kang S. 2013. Sniffing on microbes: Diverse roles of microbial volatile organic compounds in plant health. *Molecular Plant–microbe Interactions* 26: 835–843.
4. Broekgaarden C, Caarls L, Vos IA, Pieterse CMJ, Van Wees SCM. 2015. Ethylene: Traffic controller on hormonal crossroads to defense. *Plant Physiology* 169: 2371–2379.
5. Chaparro JM, Sheflin AM, Manter DK, Vivanco JM. 2012. Manipulating the soil microbiome to increase soil health and plant fertility. *Biology and Fertility of Soils* 48: 489–499.
6. Choi HK, Song GC, Yi H-S, Ryu C-M. 2014. Field evaluation of the bacterial volatile derivative 3-pentanol in priming for induced resistance in pepper. *Journal of Chemical Ecology* 40: 882–892.
7. Chuankun X, Minghe M, Leming Z, Keqin Z. 2004. Soil volatile fungistasis and volatile fungistatic compounds. *Soil Biology and Biochemistry* 36: 1997–2004.
8. Chung J, Song GC, Ryu C-M. 2016. Sweet scents from good bacteria: Case studies on bacterial volatile compounds for plant growth and immunity. *Plant Molecular Biology* 90: 677–687.
9. Compant S, Samad A, Faist H, Sessitsch A. 2019. A review on the plant microbiome: Ecology, functions, and emerging trends in microbial application. *Journal of Advanced Research* 19: 29–37.
10. Cordovez V. 2016. *Volatile-mediated Interactions in the Rhizosphere. Wageningen University.*
11. Cordovez V, Schop S, Hordijk K, Dupre de Boulois H, Coppens F, et al. 2018. Priming of plant growth promotion by volatiles of root-associated microbacterium spp. *Applied and Environmental Microbiology* 84: https://doi.org/10.1128/AEM.01865-18.
12. de Boer W. 2017. Upscaling of fungal-bacterial interactions: From the lab to the field. *Current Opinion in Microbiology* 37: 35–41.

13. de Boer W, Li X, Meisner A, Garbeva P. 2019. Pathogen suppression by microbial volatile organic compounds in soils. *FEMS Microbiology Ecology* 95: https://doi.org/10.1093/femsec/fiz105.

14. de Vrieze M, Pandey P, Bucheli TD et al. 2015. Volatile organic compounds from native potato-associated *Pseudomonas* as potential anti-oomycete agents. *Frontiers in Microbiology* 6: 1295.

15. Ditengou FA, Müller A, Rosenkranz M, et al. 2015. Volatile signalling by sesquiterpenes from ectomycorrhizal fungi reprogrammes root architecture. *Nature Communications* 6: 6279.

16. Effmert U, Kalderás J, Warnke R, Piechulla B. 2012. Volatile mediated interactions between bacteria and fungi in the soil. *Journal of Chemical Ecology* 38: 665–703.

17. Garbeva P, Hol WHG, Termorshuizen AJ, Kowalchuk GA, de Boer W. 2011. Fungistasis and general soil biostasis – A new synthesis. *Soil Biology and Biochemistry* 43: 469–477.

18. Garbeva P, Hordijk C, Gerards S, de Boer W. 2014. Volatiles produced by the mycophagous soil bacterium Collimonas. *FEMS Microbiology Ecology* 87: 639–649.

19. Garbeva P, Weisskopf L. 2020. Airborne medicine: Bacterial volatiles and their influence on plant health. *The New Phytologist* 226: 32–43.

20. Guyer A, De Vrieze M, Bönisch D, et al. 2015. The anti-phytophthora effect of selected potato-associated *Pseudomonas* strains: From the laboratory to the field. *Frontiers in Microbiology* 6: 1309.

21. Hernández-Calderón E, Aviles-Garcia ME, Castulo-Rubio DY, et al. 2018. Volatile compounds from beneficial or pathogenic bacteria differentially regulate root exudation, transcription of iron transporters, and defense signaling pathways in Sorghum bicolor. *Plant Molecular Biology* 96: 291–304.

22. Huang C-J, Tsay J-F, Chang S-Y, Yang H-P, Wu W-S, Chen C-Y. 2012. Dimethyl disulfide is an induced systemic resistance elicitor produced by *Bacillus cereus* C1L. *Pest Management Science* 68: 1306–1310.

23. Jones JDG, Dangl JL. 2006. The plant immune system. *Nature* 444: 323–329.

24. Kant MR, Bleeker PM, Wijk MV, Schuurink RC, Haring MA. 2009. Plant volatiles in defence. In: *Advances in Botanical Research*, Elsevier, pp. 613–666.

25. Kishimoto K, Matsui K, Ozawa R, Takabayashi J. 2005. Volatile C6-aldehydes and allo-ocimene activate defense genes and induce resistance against *Botrytis cinerea* in. *Arabidopsis Thaliana. Plant & Cell Physiology* 46: 1093–1102.

26. Kong HG, Shin TS, Kim TH, Ryu C-M. 2018. Stereoisomers of the bacterial volatile compound 2,3-butanediol differently elicit systemic defense responses of pepper against multiple viruses in the field. *Frontiers in Plant Science* 9: 90.

27. Lazazzara V, Perazzolli M, Pertot I, Biasioli F, Puopolo G, Cappellin L. 2017. Growth media affect the volatilome and antimicrobial activity against *Phytophthora infestans* in four *Lysobacter* type strains. *Microbiological Research* 201: 52–62.

28. Lee B, Farag MA, Park HB, Kloepper JW, Lee SH, Ryu C-M. 2012. Induced resistance by a long-chain bacterial volatile: Elicitation of plant systemic defense by a C13 volatile produced by. Paenibacillus Polymyxa. *Plos One* 7: e48744.

29. Lemfack MC, Nickel J, Dunkel M, Preissner R, Piechulla B. 2014. mVOC: A database of microbial volatiles. *Nucleic Acids Research* 42(Database issue): D744–748.

30. Martín-Sánchez L, Ariotti C, Garbeva P, Vigani G. 2020. Investigating the effect of belowground microbial volatiles on plant nutrient status: Perspective and limitations. *Journal of Plant Interactions* 15: 188–195.

31. Meldau DG, Meldau S, Hoang LH, Underberg S, Wünsche H, Baldwin IT. 2013. Dimethyl disulfide produced by the naturally associated bacterium *Bacillus* sp. B55 promotes *Nicotiana attenuata* growth by enhancing sulfur nutrition. *The Plant Cell* 25: 2731–2747.

32. Moisan K, Cordovez V, van de Zande EM, Raaijmakers JM, Dicke M, Lucas-Barbosa D. 2019. Volatiles of pathogenic and non-pathogenic soil-borne fungi affect plant development and resistance to insects. *Oecologia* 190: 589–604.

33. Morcillo RJ, Singh SK, He D, et al. 2020. Rhizobacterium-derived diacetyl modulates plant immunity in a phosphate-dependent manner. *The EMBO Journal* 39: e102602.

34. Naznin HA, Kiyohara D, Kimura M, Miyazawa M, Shimizu M, Hyakumachi M. 2014. Systemic resistance induced by volatile organic compounds emitted by plant growth-promoting fungi in Arabidopsis thaliana. *PLoS One* 9: e86882.

35. Ossowicki A, Tracanna V, Petrus MLC, et al. 2020. Microbial and volatile profiling of soils suppressive to *Fusarium culmorum* of wheat. *Proceedings. Biological Sciences/the Royal Society* 287(1921): 20192527.

36. Park Y-S, Dutta S, Ann M, Raaijmakers JM, Park K. 2015. Promotion of plant growth by *Pseudomonas fluorescens* strain SS101 via novel volatile organic compounds. *Biochemical and Biophysical Research Communications* 461: 361–365.

37. Piechulla B, Lemfack MC, Kai M. 2017. Effects of discrete bioactive microbial volatiles on plants and fungi. *Plant, Cell & Environment* 40: 2042–2067.
38. Pieterse CMJ, Zamioudis C, Berendsen RL, Weller DM, Van Wees SCM, Bakker PAHM. 2014. Induced systemic resistance by beneficial microbes. *Annual Review of Phytopathology* 52: 347–375.
39. Raza W, Wang J, Wu Y, et al. 2016. Effects of volatile organic compounds produced by *Bacillus amyloliquefaciens* on the growth and virulence traits of tomato bacterial wilt pathogen Ralstonia solanacearum. *Applied Microbiology and Biotechnology* 100: 7639–7650.
40. Ryu C-M, Farag MA, Hu C-H, et al. 2003. Bacterial volatiles promote growth in Arabidopsis. *Proceedings of the National Academy of Sciences of the United States of America* 100: 4927–4932.
41. Ryu C-M, Farag MA, Hu C-H, Reddy MS, Kloepper JW, Paré PW. 2004. Bacterial volatiles induce systemic resistance in Arabidopsis. *Plant Physiology* 134: 1017–1026.
42. Schmidt R, Cordovez V, de Boer W, Raaijmakers J, Garbeva P. 2015. Volatile affairs in microbial interactions. *The ISME Journal* 9: 2329–2335.
43. Schmidt R, de Jager V, Zühlke D, et al. 2017. Fungal volatile compounds induce production of the secondary metabolite Sodorifen in *Serratia plymuthica* PRI-2C. *Scientific Reports* 7: 862.
44. Sharifi R, Ryu C-M. 2016. Are bacterial volatile compounds poisonous odors to a fungal pathogen *Botrytis cinerea*, alarm signals to Arabidopsis seedlings for eliciting induced resistance, or both? *Frontiers in Microbiology* 7: 196.
45. Sharifi R, Ryu C-M. 2020. Formulation and agricultural application of bacterial volatile compounds. In: Ryu C-M, Weisskopf L, Piechulla B (Eds.), *Bacterial Volatile Compounds as Mediators of Airborne Interactions*, Springer Singapore, Singapore, pp. 317–336.
46. Song GC, Riu M, Ryu C-M. 2019. Beyond the two compartments Petri-dish: Optimising growth promotion and induced resistance in cucumber exposed to gaseous bacterial volatiles in a miniature greenhouse system. *Plant Methods* 15: 9.
47. Song GC, Ryu C-M. 2013. Two volatile organic compounds trigger plant self-defense against a bacterial pathogen and a sucking insect in cucumber under open field conditions. *International Journal of Molecular Sciences* 14: 9803–9819.
48. Stringlis IA, Proietti S, Hickman R, Van Verk MC, Zamioudis C, Pieterse CMJ. 2018. Root transcriptional dynamics induced by beneficial rhizobacteria and microbial immune elicitors reveal signatures of adaptation to mutualists. *The Plant Journal for Cell and Molecular Biology* 93: 166–180.
49. Thomas G, Withall D, Birkett M. 2020. Harnessing microbial volatiles to replace pesticides and fertilizers. *Microbial Biotechnology* 13: 1366–1376.
50. Tucci M, Ruocco M, De Masi L, De Palma M, Lorito M. 2011. The beneficial effect of *Trichoderma* spp. on tomato is modulated by the plant genotype. *Molecular Plant Pathology* 12: 341–354.
51. Tyagi S, Lee K-J, Shukla P, Chae J-C. 2020. Dimethyl disulfide exerts antifungal activity against *Sclerotinia minor* by damaging its membrane and induces systemic resistance in host plants. *Scientific Reports* 10: 6547.
52. Tyc O, Zweers H, de Boer W, Garbeva P. 2015. Volatiles in inter-specific bacterial interactions. *Frontiers in Microbiology* 6: 412.
53. van Agtmaal M, Straathof AL, Termorshuizen A, Lievens B, Hoffland E, de Boer W. 2018. Volatile-mediated suppression of plant pathogens is related to soil properties and microbial community composition. *Soil Biology and Biochemistry* 117: 164–174.
54. van Wees SC, de Swart EA, van Pelt JA, van Loon LC, Pieterse CM. 2000. Enhancement of induced disease resistance by simultaneous activation of salicylate- and jasmonate-dependent defense pathways in Arabidopsis thaliana. *Proceedings of the National Academy of Sciences of the United States of America* 97: 8711–8716.
55. Van Wees SCM, Van der Ent S, Pieterse CMJ. 2008. Plant immune responses triggered by beneficial microbes. *Current Opinion in Plant Biology* 11: 443–448.
56. Walters DR, Ratsep J, Havis ND. 2013. Controlling crop diseases using induced resistance: Challenges for the future. *Journal of Experimental Botany* 64: 1263–1280.
57. Wu L, Li X, Ma L, Borriss R, Wu Z, Gao X. 2018. Acetoin and 2,3-butanediol from *Bacillus amyloliquefaciens* induce stomatal closure in *Arabidopsis thaliana* and Nicotiana benthamiana. *Journal of Experimental Botany* 69: 5625–5635.
58. Yuan J, Zhao M, Li R, et al. 2017. Microbial volatile compounds alter the soil microbial community. *Environmental Science and Pollution Research International* 24: 22485–22493.
59. Zhang X, Niu M, Teixeira da Silva JA, et al. 2019. Identification and functional characterization of three new terpene synthase genes involved in chemical defense and abiotic stresses in Santalum album. *BMC Plant Biology* 19: 115.

CHAPTER 41

Trichoderma for Biocontrol and Biostimulation – A Green Fungus Revolution in Agriculture

Sheridan Lois Woo[1,3,4,5] and Matteo Lorito[2,3,4,5]

[1]Department of Pharmacy, University of Naples Federico II, 80131 Naples, Italy
[2]Department of Agricultural Sciences, University of Naples Federico II, Portici, 80055 Naples, Italy
[3]Institute for Sustainable Plant Protection, National Research Council (CNR-ISPP), Portici, 80055 Naples, Italy
[4]BAT Center-Interuniversity Center for Studies on Bioinspired Agro-Environmental Technology, University of Naples Federico II, Portici, 80055 Naples, Italy
[5]Task Force on Microbiome Studies, University of Naples Federico II, 80131 Naples, Italy

41.1 MODERN AGRICULTURE WITH OLD PROBLEMS

The challenges to modern agriculture have not changed over time. On a worldwide perspective the main concerns remain food security and nutrition, providing a livelihood for farmers, and protecting the environment (FAO; OECD). In order to improve crop production, new technologies are required to enhance the existing agricultural systems by expanding available land for cultivation (i.e., use of previously marginal lands), increasing the intensity/frequency of cropping, and augment crop yields (FAO). The climate changes occurring with global warming, along with the environmental problems associated to intensive agricultural farming practices, cause a loss in biodiversity, modifications in the geographic distribution of organism (pests and pathogens), the contamination of soil, air, and water from the inputs (chemical fertilizers and phytosanitary products) all presenting negative impacts not only to the agroecosystem, but also to human health.

The year 2020 was declared as the International Year of Plant Health (IYPH) by the United Nations General Assembly, which endorsed the slogan "Protecting Plants, Protecting Life," presenting the concept by the Food and Agriculture Organization of the United Nations (FAO) that "plants are the source of the air we breathe and most of the food we eat." It is estimated by the FAO that 40% losses are incurred to agrifood crops by plant pests (e.g., insects) and disease pathogen infestations. Plant health must be protected by using preventive measures, in order to reduce the interventions required when full-blown epidemics occur, thus decreasing the economic costs, and diminishing the risks to human and environmental health. The key to obtaining such a sustainable agroecosystem is to adopt an approach of cultivating crops with minimal chemical inputs (http://www.fao.org/plant-health-2020/about/en).

To this end, many agricultural policies implemented by diverse governing bodies throughout the world are moving toward a more ecological perspective of cultivation. For example, in the European Community, Directive 2009/128/EC has established a framework for achieving sustainable use of pesticides by reducing risks and impacts on human health and the environment by prompting the use of integrated pest

Good Microbes in Medicine, Food Production, Biotechnology, Bioremediation, and Agriculture, First Edition. Edited by Frans J. de Bruijn, Hauke Smidt, Luca S. Cocolin, Michael Sauer, David Dowling, and Linda Thomashow.

management. This method of crop protection employs management practices to decrease the use of chemical phytosanitary products for pest and pathogen control, by incorporating alternative non-chemical strategies such as agronomic and biological control techniques.

In plant pathology, biological control can be defined as the use of specific microorganisms that can interfere with plant pathogens; a method of plant protection that is nature-friendly; an ecological approach to overcome the problems caused by standard chemical methods of disease control (Chet et al., 2006). The concept of biological control or biocontrol, as an alternative technique to the use of chemical products for plant disease control became popular with the publications of Baker and Cook (1974), followed by Cook and Baker (1983). Subsequently, works by Papavizas (1985), then Chet (1987), endorsed the concept of biocontrol, and explicitly highlighted the antagonistic characteristics of a fungus belonging to the genus *Trichoderma* against soilborne phytopathogens, thus promoting its potential as a biological control or biocontrol agent (BCA) for plant diseases (see also chapter 35).

41.2 TRICHODERMA WHO'S WHO

Trichoderma (teleomorph Hypocrea) are filamentous fungi belonging to the division Ascomycota, Order Hypocreales and Family Hypocreaceae. This genus was first named by Persoon in 1794, but the fungus became of interest to agriculture only after the subsequent observations and descriptions by Weindling (1932, 1934). The common name for *Trichoderma* is "green mold," because of the clearly visible, yellow green to dark green asexual conidia that it prolifically produces. These reproductive fungal structures of conservation serve as an important basis for the development of the biological products for agriculture.

Trichoderma fungi are ecological opportunists that can be found naturally in diverse environments as saprophytes, mycoparasites, or cosmopolitan microorganisms, and more significantly for crops and agriculture, they are associated to the rhizosphere, the soil zone surrounding the plant root system. Due to the release of root exudates e.g., proteins, sugars, and other bioactive compounds in the rhizosphere, the chemical-physical activity of the plant influences the presence, growth, and interactions, both positive and negative, of the various microorganisms found in the plant-soil community or microbiota. *Trichoderma*, as a component of the plant microbiota, plays an important biological, chemical and ecological role in natural cycles, as well as the interactions that have valuable impacts on crop growth and agricultural production.

Recently, a consolidated analysis of *Trichoderma* taxonomic diversity indicated that the classification of the genus was represented by more than 460 species names, of which 375 were considered as valid species and barcoded (Cai and Druzhinina, 2021). The species of agricultural interest are mainly mycoparasites and plant growth promoters (PGP) that do not belong to a single closely united taxonomic group but are principally distributed among the Sections of Trichoderma and Longibrachiatum, and the Clades Harzianum and Virens (Druzhinina, personal communication). Many *Trichoderma* are associated to plants in the zones where they grow, but are normally avirulent, not disease agents; although some species are reported as pathogens: green mold to (edible) mushrooms, to immune-compromised patients where in these cases, they are disease agents due to their opportunistic ability as secondary colonizers.

Weindling (1934), shortly after his initial description of *Trichoderma*, noted that *T. lignorum* was a parasite of the plant root pathogen *Rhizoctonia solani* and other soil fungi. This was the first of many observations to be followed in the subsequent years, that attested to the antagonistic abilities of this fungus to various plant pathogens, such as *Armillaria mellea* (Bliss, 1951), *Sclerotium rolfsii* (Wells et al., 1972), *Fusarium* (Zhang et al., 1996), *Botrytis cinerea* (Elad and Kapat, 1999), *Pythium*, *Phytophthora*, and *Rhizopus* species (Howell, 2002). The parasitic characteristic of numerous *Trichoderma* against a variety of important plant disease agents, soon characterized this fungus as a biological control agent (BCA), providing the first opportunities for its applications in agriculture as a beneficial microbe (see also chapter 35).

41.3 IN THE BEGINNING – TRICHODERMA, A BCA OF PLANT PATHOGENS

The selection of potential *Trichoderma* BCAs can be based on observations of parasitized phytopathogens or in vitro experiments that demonstrate growth inhibition of the targeted microorganisms in plate confrontation cultures (Figure 41.1). However, the final confirmation of the "true" antagonistic capability of the chosen strain is determined in assays with the plant growing in soil, by combining the BCA, pathogen in the real field conditions. *Trichoderma* employs the "classical" biological control mechanisms against disease agents: direct parasitism of the plant pathogen, secretion of hydrolytic

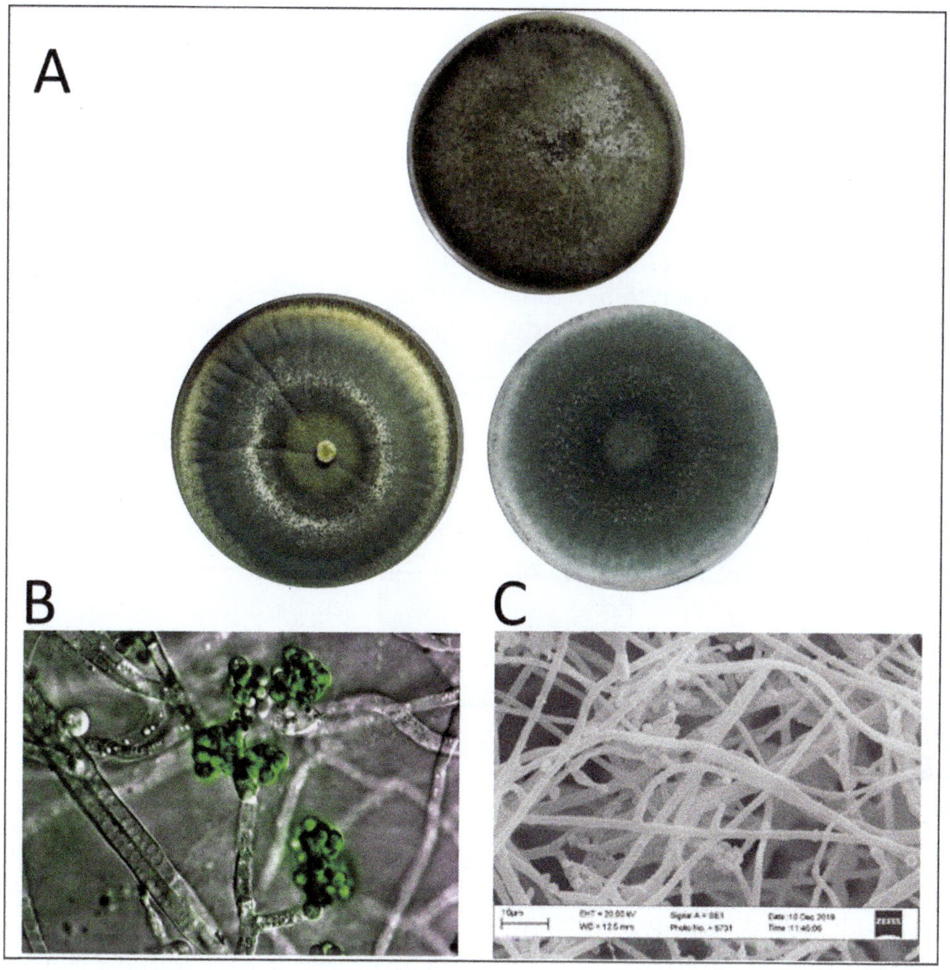

FIGURE 41.1 (A) Plate cultures of *Trichoderma* demonstrating the green sporulation (above), and (B) the fungal hyphae and conidiophores as seen under the optical (below left) and (C) scanning electron (right) microscopes.

enzymes that degrade the target microorganism, production of bioactive compounds or secondary metabolites with antibiosis activity, and competition for niches in the rhizosphere or nutritional food sources (Benitez et al., 2004; Harman et al., 2004; Hermosa et al., 2012, 2013; Howell, 2003; Lorito et al., 2010; Vinale et al., 2008).

Briefly, parasitism is a nutritional rapport established between two organisms, whereby the parasite benefits by directly feeding on a host that suffers as a consequence of this action. This process may damage the host, with or without causing death, thus in the case of biocontrol by *Trichoderma*, the antagonist inhibits the pathogen and the development of disease symptoms on the plant. The mode of action is regulated by the secretion of lytic enzymes, such as proteases, chitinases and glucanases, which degrade the cell wall constituents of the target disease agent (Harman et al., 2004; Lorito et al., 2010; Vinale et al., 2008). The antagonism by *Trichoderma* spp. against fungal plant pathogens clearly defines this fungus as a mycoparasite.

Competition concerns a biological interaction between organisms or species that coexist in the same ecological niche, for the use of limited resources – space and essential nutrients, such as amino acids, oxygen, carbohydrates, and micronutrients. The quantitative and qualitative reduction of nutrients by the *Trichoderma* in the rhizosphere decreases the chances of success of plant infection by the pathogen (Harman et al., 2004; Vinale et al., 2008).

Antibiosis involves a process in which fungi and bacteria produce secondary metabolites capable of inhibiting growth and development, even causing death, of other microbes in the vicinity. *Trichoderma* spp. are capable of producing a wide range of volatile organic compounds (VOCs; see chapter 35) and non-volatiles that affect the interactions with disease causing agents, other microorganisms in the rhizosphere, as well as the plant (Vinale et al., 2008; Vinale and Sivasithamparam, 2020).

All of these biological control processes may be activated singularly or in synchrony; having a solitary, additive, or synergistic functions which produce

FIGURE 41.2 Evaluation of *Trichoderma* activity as a biocontrol agent (BCA). *In vitro* plate confrontation assays against (A) foliar and post-harvest pathogen *Botrytis*, and (B) soilborne root rot pathogen *Rhizoctonia; in vivo* bioassays of (C) *Botrytis* on leaves of tomato, without (*Trichoderma* upper) and with BCA (lower); (D) soil pathogen *Sclerotium rolfsii* on peppers – wilted plants with no BCA; and (E) *Botrytis* in post-harvest on apples (Trt = water, Trt2 = BCA, Trt4 = *Botrytis*, Trt5 = *Botrytis* + BCA.

inhibitory effects on the plant pathogen. However, it should be noted that the results obtained on disease control are variable, depending upon the *Trichoderma* species or isolate utilized. Therefore, the first discovery of *Trichoderma*'s plant beneficial effects was largely attributed to its direct antagonism of the invading detrimental organisms, by having a negative impact on the pathogen structures and inhibiting their biological activities, thus permitting the host plant to overcome the primary disease damage incurred and consequently recuperating new growth and development (Figure 41.2).

41.4 INDUCED RESISTANCE TO BIOTIC FACTORS FOR CROP PROTECTION

Over time it became apparent that there were additional beneficial processes attributable to *Trichoderma* in the interactions with the plant that positively influenced crop protection to pathogens and pests. In particular, *Trichoderma* were demonstrated to have indirect protective effects to pathogen and pest attack, able to activate plant defense mechanisms, such as induced systemic resistance (ISR) or systemic acquired resistance (SAR). These two processes involve the plant's ability to recognize and respond to biotic or abiotic stress agents and autonomously defend itself by resisting the threats of attack or damage (see chapter 35). In brief, the plant resistance response to pathogen invasion can be described by the zigzag model presented in Lorito et al. (2010), involving the recognition of molecules released by a disease agent which activate a signaling cascade of defense mechanisms, resulting in the production of compounds that provide an effective level of resistance to protect the plant from pathogen attack and disease development. Over time, the plant response diminishes in confrontation to the pathogen, also due to counteractive effectors produced by the attacker, that lower the defense capacity below the threshold for effective resistance, thus allowing disease symptoms develop.

When *Trichoderma* is involved in the antagonist–plant–pathogen interaction, the BCA once recognized by the plant, stimulates an earlier elicitation in the plant defense response to counteract the pathogen effectors, thus stabilizing the amplitude of resistance above the effective threshold (no disease). Furthermore, as time passes, the presence of *Trichoderma* prevents a reduction in plant resistance, maintaining a level of effective defense that provides plant protection, thus minimizing disease (Lorito et al., 2010). Importantly, the regulation of ISR with *Trichoderma* establishes a priming effect or state of alert that renders the plant more sensitive and ready to defend itself to eventual contacts with any challenge organism. Upon attack, the primed plant is able to activate a more rapid and more intense response to defend from the invader (Alfiky and Weisskopf, 2021; Conrath 2006; Harman et al., 2004; Hermosa et al., 2012; Lorito et al., 2010; Pieterse et al., 2009; Shoresh et al., 2010; Van der Ent et al., 2009). Moreover, this interaction stimulates the cascade of important gene signaling pathways, the hormone-mediated salicylic acid (SA) and jasmonic acid (JA), that activate the SA-dependent defense of SAR or the JA/ET-dependent defense of ISR respectively, and the production of immunity compounds (such as antifungal or oxidative enzymes) noteworthy for contrasting diverse biotic stress factors, such as pathogens and pests.

Interestingly, the protective effect of *Trichoderma* is not limited only to disease causing agents, but also to other damaging pests. Recently, it was found that larvicidal activity of mosquito were obtained by direct treatments of *T. asperellum* spores, that acted as an effective entomopathogen, and its methanolic extracts that caused a loss of structural integrity of the larval membrane causing cell dehydration (Podder and Ghosh, 2019). The first field trials of *T. longibrachiatum* spores as a biopesticide against *Leucinodes orbonalis*, a pest of eggplant, provided good control (Ghosh and Pal, 2016). It was also noted that other *Trichoderma* compounds, such as commercial chitinases were efficient in attacking the peritrophic matrix in the control lepidopteran pests (Berini et al., 2016). Moreover, other mechanisms of insect control were also indicated when *Trichoderma* was applied to tomato. The BCA conferred ISR to the treated plants that protected them to infestations by insects having different feeding habits, the noctuid moth *Spodoptera littoralis* and the aphid *Macrosiphum euphorbiae* (Coppola 2019a; 2019b). Direct negative effects were noted on the survival and development of the insects feeding on the *Trichoderma* treated plants. Remarkably, indirect effects were also observed, whereby the application of *Trichoderma* to tomatoes activated the release of VOCs (see chapter 40) that influenced insect behavior, increasing the attractivity of the parasitoid wasp (*Aphidius ervi*) to the plants infested with the aphid prey. Additionally, as noted above with the ISR defense response to pathogens, the BCA treated tomato was similarly primed, able to counteract the subsequent infestation by the aphids by incremented resistance to insect attack.

Furthermore, crop protection was also noted in the effective biocontrol activity of *Trichoderma* against various plant-parasitic nematodes, including *Meloidogyne javanica* (Medeiros et al., 2017; Sharon et al., 2001) and *M. incognita* (Martínez-Medina et al., 2017). The mode of action that beneficial fungi, including *Trichoderma*, mycorrhizae, and other endophytes, used successfully in the biocontrol of the nematodes also involved the activation of resistance in the plant (Poveda, 2020). The systemic resistance induced in tomato to *M. incognita* was complex and involved both SA-mediated (SAR) and JA-related (ISR) defense responses that depended upon the stage of nematode development (e.g., eggs, juveniles, reproduction), associated to the phases of invasion and parasitism of the plant roots (Martínez-Medina et al., 2017). Interestingly, it was noted that *T. atroviride* was able to activate ISR against *M. javanica*, as well as influence the heritability of this resistance phenomenon in the subsequent generations of tomato treated with *Trichoderma*, suggesting an epigenetic effect (de Medeiros et al., 2017).

41.5 INDUCED RESISTANCE TO ABIOTIC STRESS – CULTIVATION IN ADVERSE CONDITIONS

The crop protection properties of *Trichoderma* to confer resistance of the plant to biotic stress was also noted when the plants were subjected to abiotic stress. Seed treatments with *Trichoderma* were found to improve corn seed germination and seedling vigor to oxidative, cold temperature and allelopathic stress (Bjorkman et al., 1998). Similarly, *Trichoderma* treatments of tomato seed alleviated the negative effects of osmotic and salinity stress, low temperatures, and loss of seed quality, resulting in uniform seed germination and strong seedling vigor; an effect attributable to plant physiological protection against oxidative damage

(Mastouri et al., 2010). Furthermore, *Trichoderma* were able to confer tolerance to plants subjected to stress by salinity, drought, heat, cold, heavy metals in a variety of crops including cereals, mustards, tomato, cacao, sunflower, maize (Hidangmayum and Dwivedi, 2018, plus references within; Sood et al., 2020). The mechanisms that produced the plant benefits by the fungal BCA for protection to abiotic stress included induced systemic resistance, phytohormonal effects (i.e., cytokinins, abscisic acid-ABA), increased antioxidants (e.g., SOD-superoxide dismutase, CAT-catalase, and POD-peroxidase) (Mastouri et al., 2010; Sood et al., 2020). However, studies highlighted that plants responded differently to a combination of multiple abiotic stresses than to individual stress (Atkinson and Urwin, 2012). For example, on grapevines exposed to heat and drought stress it was demonstrated that *Trichoderma* T39-induced resistance efficacy was not affected by the single stress factor; however, when the plant was subjected to both stresses together, decreasing trends in ISR were observed (Roatti et al., 2013). This observation regarding the possibility to increase plant resistance to abiotic factors by *Trichoderma* applications is a very important consideration in the perspective of changing global climatic conditions, where warmer and drier seasons are predicted in several regions that can negatively affect plant growth (Boulanger and Messerlin, 2010).

41.6 BIOSTIMULATION – PLANT GROWTH PROMOTION AND INCREASED PRODUCTION

Other well noted positive effects that *Trichoderma* may have on plants important for agriculture, are those associated to plant growth promotion (PGP), the biostimulation activity that translates to improved crop productivity and yields (see also chapter 34). Plant enhancement noted with *Trichoderma* treatments can be observed with increases in: seed germination (percentage, rate, and uniformity) and seedling emergence; expansion of the root system architecture (volume, extension, secondary roots, and root hair development); above-ground vegetation growth (height, leaf number/area, branches, tillers, biomass); flowers and fruits (number, dimension, production); plus the yield parameters (quantity, fresh/dry weight, maturity, color, quality. etc.). These increased growth parameters have been observed both with treatments of the living fungus BCA as well as with the *Trichoderma* produced secondary metabolites (SMs), in diverse crops, for example: maize, soybeans, various horticultural crops (lettuce, tomato, pepper, radishes, cucumber), cereals, and fruits, strawberry, apples, grapes (Alfiky and Weisskopf, 2021; Harman et al., 2004; Harman and Uphoff, 2019; Hermosa et al., 2012; Hidangmayum and Dwivedi, 2018 (and References within); Sood et al., 2020 (and references within); (Lombardi et al., 2020a, 2020b; Pascale et al., 2017).

Numerous plant bioactive compounds produced by different *Trichoderma* strains, produce positive effects on plant growth due to their hormonal-like properties that are similar to those produced by the plant indole-3-acetic acid (IAA) or auxins (Vinale et al., 2008, Vinale and Sivasithamparam, 2020). For example, it was noted that the fungal secondary metabolites harzianolide and 6-pentyl-a-pyrone had elongation effects on treated etiolated pea stems, and treatments with a hydrophobin stimulated lateral root and root hair formation in germinating seedlings (Vinale et al., 2008). Furthermore, some SMs such as harzianic acid functioned as siderophores (Vinale et al., 2013), the low molecular weight compounds able to chelate or bind to iron ($Fe3+$) in the soil environment, contributing to increased bioavailability of $Fe2+$ forms in the rhizosphere for plant uptake and assimilation of this important, but not readily available microelement, thus improving iron nutrition and growth in various plant species.

A review by Hermosa et al. (2012) clearly documented that the growth-promoting activity of *T. atroviride* on tomato was attributable both to the fungus–plant interaction in the rhizosphere as well as the effect of the fungal produced compounds on the key components of biosynthetic pathways influencing plant growth. It was noted that *Trichoderma* production of IAA hormone and the enzyme 1-aminocyclopropane-1-carboxylic acid deaminase (ACCD) were the principal fungal compounds that modulated the PGP process. The principal influential point of the fungus–plant cross-talk involves the regulation of 1-aminocyclopropane-1-carboxylic acid (ACC), a precursor to ethylene (ET), by the presence of the *Trichoderma* IAA and/or ACCD. Decreases in ACC diminishes ET biosynthesis that subsequently effects ABA biosynthesis, impacting gibberellin signaling. The increased gibberellin activity in turn affects degradation of growth-repressing DELLA proteins, and this deregulation consequently stimulates plant growth/development (Alfiky and Weisskopf, 2021; Hermosa et al., 2012; Viterbo et al., 2010).

What was interesting in this same model is that the *Trichoderma* regulation of these biosynthetic pathways also influenced the plant defense response to biotic or abiotic stress. In this case, the *Trichoderma* IAA influence on the plant ACC production will cause an increase in ET and ABA biosynthesis, decreasing gibberellin signaling and degradation of growth-repressing DELLA proteins, that reduces plant growth. Instead, the effect is on the pathway of JA signaling and JA/ET dependent ISR defense, or the pathway of SA signaling and the SA-dependent SAR defense response, both processes involved in indirect crop protection. Therefore, there is a balance occurring between the physiological processes of plant growth and plant defense, a trade-off between the processes of one or the other that can be modulated by *Trichoderma*.

In summary, the interaction of *Trichoderma* with the plant: (1) is complex, and it involves a dynamic molecular cross-talk between the fungus and the plant; (2) is a continuous trade-off occurring between the plant immunity response by ISR and/or SAR and PGP; and (3) the IAA and ACCD molecules released by *Trichoderma* will determine the processes activated in the plant that result in stimulated growth or triggering resistance (Figure 41.3).

41.7 PLANT PHYSIOLOGICAL BENEFITS – IMPROVED QUALITY

Other benefits to the plant by *Trichoderma* depend upon the fungal species/strains, the different compounds produced and their varying bioactive properties. The production of *Trichoderma* siderophores, phosphate-solubilizing enzymes, phytohormones such as cytokinin-like (Zeatin) and gibberellin-like (GA3 or GA4) molecules can enhance crop production (Hidangmayum and Dwivedi, 2018, and references within). Coumaric, glucuronic, and citric acids can aid in the release of phosphorus ions which are not readily bioavailable to plants in most soils (Harman et al., 2004). For example, it was found that *T. harzianum* strain 22 was able to solubilize and sequester minerals in the hyphae when cultured in liquid substrate (Altomare et al., 1999). This implies that during plant colonization by the fungus, hyphal lysing could release minerals in the rhizosphere that increases the availability of P, Fe, and Zn to the plant. Similarly, with *T. asperelloides* strain T-203, it was noted that Fe and P bioavailability was improved in the rhizosphere by 30% and 90%, respectively; and similarly, another species, *T. asperellum* also enhanced P and Fe availability to cucumber, resulting in significant dry weight, shoot length, and leaf area increase (Yedidia et al., 1999).

FIGURE 41.3 Plant growth promotion or biostimulant effects with *Trichoderma* treatments (always on the right) compared to water control (from upper left to right): (A) lettuce – more uniform development; (B) more strawberry fruits from treated plants; (C) wall rocket increased growth; (D) *Brachypodium* increased tillering effect; (E) tomato fruit formation, and (F) harvested yield.

The availability of other important minerals influenced by *Trichoderma* include Cu, Na, and Zn, plus micronutrients such as Mn, that play a role not only in nutrition influencing plant growth and vigor but are also involved in metabolism of defense processes affecting plant resistance to biotic or abiotic stress (Altomare et al., 1999). The biological activity of *Trichoderma* in plant nutrient availability was observed with cucumber whereby *T. asperellum* T-6 applications increased $Fe2^+$ and siderophore content in soil plus the activity of $Fe2^+$ and $Fe3^+$ chelate reductase (Harman et al., 2004). Furthermore, Colla et al. (2014) reported that *T. atroviride* strain MUCL45632 could produce two types of siderophores (hydroxamate and catechol), as well as auxin-like compounds that stimulated growth of lettuce and zucchini. These studies highlight that *Trichoderma* application in soil assists the plant in reduction of $Fe3^+$ to $Fe2^+$, which consequently boosts its solubilization and uptake.

Physiological effects with *Trichoderma* can produce important advantages to plant vigor, resistance to biotic/abiotic stress, and quality. The colonization of maize rhizosphere by *T. virens* induced higher photosynthetic rates and systemic increases in the uptake of CO_2 in leaves (Vargas et al., 2009); whereas increased nitrogen use efficiency was observed with *T. harzianum* T22 in maize (Harman et al., 2004; Mastouri et al., 2010), and *T. virens* (GV41) in lettuce (Fiorentino et al., 2018). In wall rocket treated with a protein hydrolysate biostimulant in combination with *Trichoderma*, produced an increase in N, PO4, Mg, and Na, whereas the single treatments enhanced lipophilic and hydrophilic antioxidant activity, as well as ascorbic acid and chlorophyll b content (Caruso et al., 2020). In tomato, *T. harzianum* T22 treatments increased crop productivity by 40%, plus improved lycopene, asparagine, and GABA that effect properties of the fruits. Interestingly, either T22 or a *Trichoderma* secondary metabolite 6PP significantly increased lycopene, a principal lipophilic antioxidant in tomato fruits, notable for prolonging shelf-life quality. Furthermore, lycopene has also been documented as having important human health properties such as cardiovascular protection or prevention of oxidative stress-mediated carcinogenesis (Carillo et al., 2020). Similarly, in strawberries, it was observed that applications with a *Trichoderma* fungal inoculum or their secondary metabolites to the plant increased growth and physiological processes that was reflected in the produced fruit (Lombardi et al., 2020a; 2020b). Ripened strawberries demonstrated an accumulation of metabolites and proteins involved in important physiological processes of growth and defense, producing beneficial compounds such as anthocyanins and other antioxidants that may positively influence fruit quality plus consumer health effects. Additionally, it was demonstrated by Marra et al. (2019) that the applications to soybean of different *Trichoderma* strains or metabolites, singly or in combination, affected the fatty acid profiles, increasing the monounsaturated lipids such as oleic acid that increases the health value of the oil products. Therefore, it can be noted that *Trichoderma* applications to crops can be considered both as a biofertilizer aiding in nutrient availability, and a probiotic to improve the plant properties that can provide health benefits to the consumer.

41.8 TRICHODERMA PRODUCTS IN AGRICULTURE

The real proof of *Trichoderma*'s success in agriculture can be testified by the more than 250 biological products distributed commercially worldwide, in which the fungus is the active ingredient or a component in the bioformulation (Woo et al., 2014). The initial introduction of *Trichoderma* to the agricultural market occurred in the late 1980s, when it was registered as a plant protection product (PPP) in the USA, specifically as a microbial BCA, a biofungicide for controlling various pathogens. However, these mBCA biofungicides, targeted specifically for crop protection, account for only about 10–15% of the total *Trichoderma* containing products in circulation. In fact, *Trichoderma* formulations are also marketed globally as plant biostimulants, biofertilizers, soil amendments-integrators, and inducers of natural plant resistance to biotic and abiotic stress. Most of these products, are not registered, but distributed with disclaimers indicating improved crop production, increased seed germination, vegetative growth, and yield, attesting to the renowned capacity of *Trichoderma* to have advantages other than plant disease control, which are accepted as a positive "side-effects" associated with their use in the field (Harman et al., 2010; Woo et al., 2014). Furthermore, *Trichoderma* is able to stimulate plant defense responses with crop protection effects to diverse biotic, i.e., pathogens and pests, or abiotic, drought, and salinity stress. The fungus–plant interaction also improves the nutrient ability (bioavailability) of macro- and micronutrients in the rhizosphere, increased uptake and assimilation by the plant, thus acting as a biofertilizer (Altomare et al., 1999; Harman, 2000; Woo, 2014). Other positive properties attributed to *Trichoderma* include decomposition and biodegradation of organic materials, bioremediation or biosanification of contaminated sites.

Trichoderma formulations can be used as applications during all phases of production: at pre-planting to seed or propagation material; at time of seeding or transplant by watering as a drench, a root dip or soil incorporation; during phases of plant growth and development with foliar spray, watering or irrigation; during harvest or post-harvest, for product transformation, conservation, and at time of pruning. The biological products are available for use in various farming systems including the greenhouse, nursery, protected and open field, orchards, plus hydroponics; on many diverse cultures – cereals, horticultural, small fruits and ornamental crops, woody plants, to name a few (Woo et al., 2014).

41.8.1 Agricultural Sustainability – Reducing the Disparity

Sustainable agriculture is a widely used term with broad and variable definitions. FAO presents a concept of sustainable food and agriculture with considerations to food security: availability, access, utilization, and stability; and a need for an equilibrium among social, economic, and environmental principles. "Sustainability" in the future faces many challenges to resolve. In 2050, it is estimated that world populations will increase by 22.6% to 9.3 billion (United Nations, 2013), requiring enhanced agricultural food production of 60% globally and 100% in developing countries to meet consumer demands. The availability of additional new land for cultivation is scarce, and much of the territory that is already in use is subject to degradation (salinization, erosion, contamination), drought, and desertification. Other difficulties include limited water and nutrient resources, climate change (i.e., increasing temperatures, reduced precipitation), losses in biodiversity, shifts in pathogen and pest dynamics that create damage to crops not previously known. Agricultural production in large scale industrial farming systems of developed countries has almost reached the maximum threshold of yields that can be offered by the existing technologies. Furthermore, this method with the implementation of high chemical inputs has resulted in negative trade-offs to human and environmental health, a situation signifying that alternatives to existing products must be found.

Certainly, there is not an easy solution, but the answer cannot involve a high monetary cost by the farmers, particularly in developing countries. However, assistance needs to be obtained through diverse government or non-government organizations (NGOs), providers from private foundations, national, and international aid organizations, agricultural research institutes or centers, etc., or even project calls that address agricultural development in low income countries or marginal geographical zones, i.e., Africa, Asia, Latin America. An overall change in management, the development of long-term strategies with government policies to confront the diverse risks, to compensate the uncertainties in the agricultural sector to a minimum is required. This may involve changes to existing "traditional" farming practices, by regulating the inputs that help increase productivity, efficiency and sustainability of the agri-food system, focusing on components including the environment, biodiversity, and natural resources.

One solution could be to increase the possibility for subsistence farmers in developing countries to improve their farming systems, helping local smallholder producers to become autonomous and auto-sufficient, thus providing food security, social and economic prospects for the sustainability of the resident community. An important aspect begins with the use of crop varieties that are adapted to the region of cultivation, and this requires a holistic understanding of the climatic conditions (rain and drought seasons, precipitation, temperatures, etc.), agro-ecosystem components (soil properties, topography, pathogens, pests etc.), as well as the selection of a culture appropriate for the existing farming practices and suitable to the diet preferences of the local consumers.

41.8.2 A Promising Role for *Trichoderma* in Agricultural Development

Trichoderma can play a small, but significant role in this conversion process. This plant beneficial microbe can be inserted in the transformation from conventional agriculture with chemical inputs, to an integrated pest management system with reduced pesticide and fertilizer use, then eventually toward organic farming with the elimination of chemical products. *Trichoderma* can serve as biopesticides and enhancers of plant resistance to biotic and abiotic stress, thus acting as crop protectors. Their activity as biofertilizers can aid in nutrient availability and uptake, and as biostimulators of plant growth to increase yield productivity, plus improve quality and nutritional value of the produce. This latter point is a positive aspect when considering that many non-Western cultures depend upon legume crops as the principal protein food source in their diets, in the absence of animal products, and that functional foods can provide important health benefits.

Global climate change is another concern due to negative effects on cultivation with adverse conditions such as temperature and precipitation variations above and below the norms for optimal crop production. The ability of some *Trichoderma* species to induce systemic resistance in plants, may offer increased tolerance to unfavorable abiotic factors such as high temperature, drought, salinity, pollutants, providing a possibility for increased crop tolerance and cultivation in adverse conditions. In addition, this may open new opportunities for farmers to cultivate in zones that were until now considered unsuitable for agriculture. Land availability is an aspect that will become progressively more important for future requirements where global demands are to increase crop production to ensure food security, thus there will be the necessity to extend the farming to marginal lands.

How can scientific knowledge about *Trichoderma* be usefully employed in the development of sustainable agriculture worldwide, particular in farming systems with limited economic and natural resources? Some reflections are provided based on diverse experiences of the authors in different developing regions in Central America, Honduras, Costa Rica, and in Africa, Kenya.

The use of appropriate *Trichoderma* species in the agricultural system of the geographic region is important for successful applications and results. Basic research is required to obtain or test an isolate that is adapted to the local field and climate conditions, for example an autochthonous fungus with the beneficial biocontrol and/or biostimulant characteristics would be ideal. The differential interaction of the crop genotype with different *Trichoderma* strains has been noted (Bazghaleh et al., 2020; Tucci et al., 2011), and this may modify the cross-talk between the microbe and the hot that determines the positive effects to the plant, such as the defense response or PGP, as well as the ability of the fungus to colonize the plant. Therefore, to confront this issue of fungus–plant specificity, it could be important to obtain the microbe from the rhizosphere of the desired crop of which it is intended is to apply this isolate. This would permit a targeted selection of the optimal plant-microbe combination, a specific crop and its "chosen" biological agent or a "designer" beneficial microbe (Figure 41.4).

A mixture of microbes with combined diverse adaptations and plant beneficial effects may supply advantages that are additive or synergistic to help overcome limitations of the microbial strain. Diverse farming systems of crops, locations, agro-ecosystems, field conditions, etc. may require a multi-microorganism integrated approach. A microbial consortium, for example, *Trichoderma* and nitrogen-fixing bacteria such as *Azotobacter* or *Rhizobia* on legume crops (see chapter 37) could

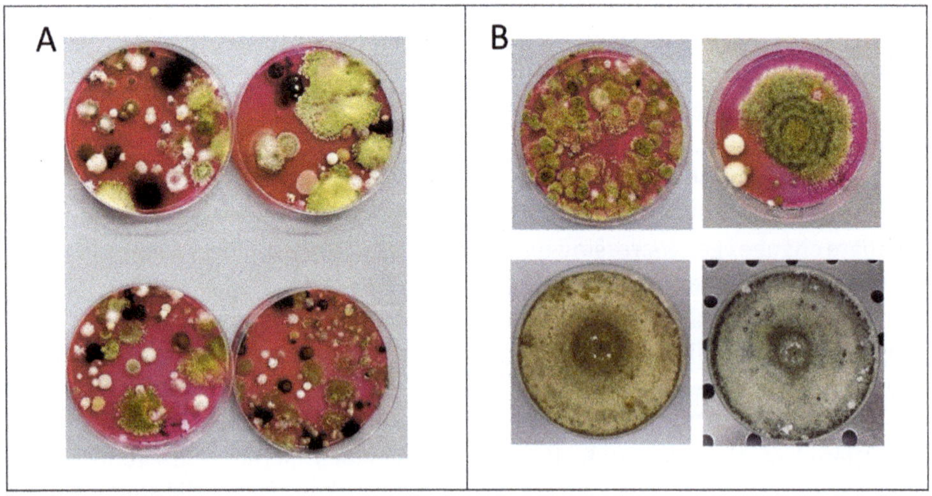

FIGURE 41.4 (A) Isolation of fungi from soil rhizosphere samples on a selective substrate and counts of colony forming units (left), (B) selection, and purification *Trichoderma* isolates (right).

improve productivity by combining various biostimulant effects – hormones, siderophores, changes in root system architecture, affecting nutrient availability, absorption, uptake and assimilation, plus crop protection to pathogens and pests. The use of living microorganisms provides advantages not only in the growing season of application, but also in the subsequent season, providing further disease control, plant nutrients when the treated crop is used as a green manure incorporation to the soil, or inserted in a crop rotation system (legume–cereal). Since these microbes continue to live in the soil-rhizosphere zone, the beneficial effects may persist overtime with population establishment without the need for repeated applications, participating in important biological processes in the agro-ecosystem.

Other than the field aspects, it is also important that adequate conditions are guaranteed for the conservation of prepared fungal inoculum to ensure viability, standardization of the biological activity and efficacy in the field application. This is a problem in many developing countries, hot desert, or tropical regions, where there is not the possibility for refrigeration or suitable storage requirements at the farms. Another limiting factor for the inoculum preparation is the availability of water for liquid formulations. For example, in many African communities, such as that observed in Kitui county in Kenya, a water source is only accessible at a single centralized supply point located in the principal municipality, and distribution to the farms/homes occurs manually, transported by animals or humans for kilometers in rationed volumes, which means that use of this valuable resource is not plausible for agriculture. This implies that the microbial product must be based on highly durable fungal conservation structures, i.e., spores for *Trichoderma*. Moreover, both the inoculum and application procedures need to have a reduced water requirement for their preparation, but in order to function, the fungus necessitates water for spore germination and colonization. With the goal to overcome the water limiting factor, many farmers plan seeding of the crops in coincidence with the arrival of the seasonal rains, thus applications of an associated fungal treatment alongside of the seed also provides the possibility take advantage of this precipitation cycle.

It is possible to develop *Trichoderma* products that are economical and adapted to the farming system described above in developing countries by using an in-house small scale production model similar to that described by Harman et al. (2010) for local grower use. The technology could involve personnel from local governments, NGOs, research institutions, universities, etc. to assist in the process development, by providing biological and technical material, know-how, and the coordination of the local authorities, farmers and their families in the diverse activities. The inoculum should be prepared shortly prior to the anticipated seeding time to avoid potential problems of conservation, contamination, and formulation. The starting point could be the selection of an indigenous *Trichoderma* isolate from the region. In Costa Rica, rice was used as an economical substrate in the local production of the *Trichoderma* inoculum (Figure 41.5). The humid sterilized substrate was produced in breathable plastic bags, then the fungal culture was grown in this semi-solid state to obtain spores for the field applications. In developing countries, the use of a potential food source in the process to produce the fungal inoculum is not an option; however, alternative substrates could be used, such as chopped or ground compost, mulch, manure, or other re-cycled organic materials, that could be wetted with low volumes of water. The difficulty could be to maintain sterility during the production process, but if the concentration of the *Trichoderma* inoculum is high, the presence of other microbes in the substrate may not be a problem, unless they have fungicidal or fungistatic effects. At the time of seeding in the field, a "spoonful" of the inocula could be added to the hole alongside the seed being planted.

Another possible treatment, for example if a limited quantity of a *Trichoderma* spore formulation is available, such as a commercial product, could be to perform a seed-coating treatment with a spore inoculum, a process that was tested in Kenya (Figure 41.6). A spore suspension was prepared in low volumes of water, with or without an adhesive sticker, in concentrations from 10^7 to 10^8 spores per ml, the seeds were added to the slurry, stirred constantly to cover the seed surface uniformly with the *Trichoderma* inoculum, left to air dry, then seeded in the traditional manner. Both the substrate inoculum, seed-coating production and application techniques would allow for an efficient use of water, permit the possibility of *Trichoderma* to be present with the seed, the germinating seedlings, growing plants and in the zone of the developing root system, taking advantage of the seasonality of available water in many geographical climatic regions. Such treatment methods could provide promising advantages to producers in small scale subsistence farming systems, whereby these procedures could be adapted and integrated into the organizational activities of the local community for providing potential social and economic gains.

FIGURE 41.5 Production of *Trichoderma* inoculum on rice or other solid organic substrate: (A) rice sterilized in plastic bags with breathable strips; (B) colonization of fungus on the humid rice after five days; (C) sporulation on rice about 10 days after inoculation; (D) incubation of BCA inoculated rice bags on shelves; (E) ground *Trichoderma*-rice inoculum ready for field application; (F) pure *Trichoderma* spores collected from washed rice inoculum.

41.9 CONCLUSIONS

The first green revolution in agriculture occurred in the 20-year period from the late 1960s to 1980s generating notable increases in crop production due to selective breeding and development of improved genetic varieties with higher productivity, that resulted in a doubling of rice and wheat yields in Asia (Pingali, 2012). This revolution contributed to food distribution to many of the world's populations and dramatically reduced hunger, malnutrition, and poverty. However, the continued need to increase yields subsequently resulted in the introduction of chemicals, synthetic fertilizers, and pesticides, to intensify agricultural production in the existing farming systems on the same available land. The results from the overuse of these chemical products have produced negative impacts to the environment and health. New revolutions to agriculture include the use of novel biotechnological applications for plant breeding improvement (CRISPR/Cas9, GMO crops), and digital technologies to develop systems for efficient management practices (precision farming). But many of these high-tech prospectives for modernization and advancement will only provide possibilities to producers in developed countries, and these opportunities for farmers in low-income countries will not exist. With a constant growth in the human population, the economic disparity gap will continue to widen, but the challenges will remain for the need to increase agricultural production, land availability, reduce hunger and poverty. The new second green revolution needs to provide innovative strategies for sustainable agriculture, also to nations with poorer economies. The use of the green fungus *Trichoderma* in the farming system may contribute to the development of an equilibrated agroecosystem by decreasing harmful chemical inputs, reducing risks to the environment and human health, providing economic and social gains to farmers, but more importantly, aid in offering food security to the populations in need.

FIGURE 41.6 Field applications of *Trichoderma* in Kenya: (A) preparation of field before seasonal rains, (B) selection of adapted crop cultures – green grams and sorghum; (C) seed coating treatment showing a spore suspension in a vial; (D) untreated cleaned green gram seed; (E) mixing of seed with *Trichoderma* in a container; (F) air drying, the *Trichoderma*-coated seeds ready for seeding; (G) crop production after seasonal rains (no irrigation).

ACKNOWLEDGMENTS

This research was partially supported by the following projects: European Union Horizon 2020 Research and Innovation Program, ECOSTACK (grant agreement no. 773554); and MUIR- PRIN 2017 [grant number PROSPECT 2017JLN833]; Consultancy agreement with CEFA (Project Contract No. KENYA-AID: 10306/CEFA/KEN); Consultancy agreement with IZSM (Project Contract No. D01 6309 of 14/12/2016).

REFERENCES

1. Atkinson NJ, Urwin PE. 2012. The interaction of plant biotic and abiotic stresses: from genes to the field. *J Exp Bot*. Jun 63(10): 3523–43. doi: 10.1093/jxb/ers100
2. Alfiky A, Weisskopf L. 2021 Deciphering *Trichoderma*–Plant–Pathogen Interactions for Better Development of Biocontrol Applications. *J. Fungi* 7: 61. https://doi.org/10.3390/jof701006
3. Altomare C, Norvell WA, Harman GE. 1999. Solubilization of phosphates and micronutrients

by the plant-growth-promoting and biocontrol fungus *Trichoderma harzianum* Rifai. *Applied and Environmental Microbiology* 65: 2926–2933.

4. Baker KF, Cook RJ. 1974. *Biological Control of Plant Pathogens*. WH Freeman and Company, San Francisco, 433 pp.
5. Bazghaleh N, Prashar P, Woo S, Vandenberg A. 2020. Effects of Lentil Genotype on the Colonization of Beneficial *Trichoderma* Species and Biocontrol of *Aphanomyces* Root Rot. *Microorganisms* 8: 1290. https://doi.org/10.3390/microorganisms8091290
6. Benitez T, Rincon AM, Limon MC, Codon AC. 2004. Biocontrol mechanisms of *Trichoderma* strains. *International Microbiology* 7: 249–260.
7. Berini F, Caccia S, Franzetti E, Congiu T, Marinelli F, et al. 2016. Effects of *Trichoderma viride* chitinases on the peritrophic matrix of Lepidoptera. *Pest Management Science* 72: 980–989. https://doi.org/10.1002/ps.4078.
8. Björkman T, Blanchard L, Harman G. 1998. Growth enhancement of shrunken-2 (sh2) sweet corn by *Trichoderma harzianum* 1295-22: Effect of environmental stress. *Journal of the American Society of Horticultural Science* 123: 35–40.
9. Bliss DE. 1951. The destruction of *Armillaria mellea* in citrus soils. *Phytopathology* 41: 665–683.
10. Boulanger Pierre H, and Messerlin Patrick A. 2010. 2020 European Agriculture: Challenges and Policies. Washington DC/Brussels: The German Marshall Fund of the United States, Economic Policy Paper Series 10. http://www.gmfus.org/publications/index.cfm.
11. Cai F, Druzhinina IS. 2021. In honor of John Bissett: Authoritative guidelines on molecular identification of *Trichoderma*. *Fungal Diversity* https://doi.org/10.1007/s13225-020-00464-4.
12. Carillo P, Woo SL, Comite E, El-Nakhel C, Rouphael Y, et al. 2020. Application of *Trichoderma harzianum*, 6-pentyl-α-pyrone and plant biopolymer formulations modulate plant metabolism and fruit quality of plum tomatoes. *Plants* 9: 771. https://doi.org/10.3390/plants9060771.
13. Caruso G, El-Nakhel C, Rouphael Y, Comite E, Lombardi N, et al. 2020. *Diplotaxis tenuifolia* (L.) DC yield and quality as influenced by cropping season, protein hydrolysates, and *Trichoderma* applications. *Plants* 9: 697. https://doi.org/10.3390/plants9060697.
14. Chet I. 1987. *Trichoderma*-application, mode of action, and potential as a biocontrol agent of soilborne pathogenic fungi. In: Chet I (Ed.), *Innovative Approaches to Plant Disease Control*. John Wiley & Sons, New York, pp. 137–160.
15. Chet I, Viterbo A, Brotman Y, Lousky T. 2006. *Enhancement of Plant Disease Resistance by the Biocontrol Agent Trichoderma*. Life Sciences Open Day. Weizmann Institute of Science.
16. Colla G, Rouphael Y, Di Mattia E, El-Nakhel C, Cardarelli M. 2014. Co-inoculation of *Glomus intraradices* and *Trichoderma atroviride* acts as a biostimulant to promote growth, yield and nutrient uptake of vegetable crops. *Journal of the Science of Food and Agriculture* 95: 1706–1715. https://doi.org/10.1002/jsfa.6875.
17. Cook RJ, Baker KF. 1983. *The Nature and Practice of Biological Control of Plant Pathogens*. American Phytopathological Society.
18. Coppola M, Cascone P, Di Lelio I, Woo SL, Lorito M, et al. 2019a. *Trichoderma atroviride* P1 colonization of tomato plants enhances both direct and indirect defense barriers against insects. *Frontiers in Physiology* 10: 813. https://doi.org/10.3389/fphys.2019.00813.
19. Coppola M, Diretto G, Digilio MC, Woo SL, Giuliano G, et al. 2019b. Transcriptome and metabolome reprogramming in tomato plants by *Trichoderma harzianum* strain T22 primes and enhances defense responses against aphids. *Frontiers in Physiology* 10: 745. https://doi.org/10.3389/fphys.2019.00745.
20. Conrath U. 2006. Systemic acquired resistance. *Plant Signal Behav* Jul 1(4): 179–84. doi: 10.4161/psb.1.4.3221.
21. Elad Y, Kapat A. 1999. The role of *Trichoderma harzianum* protease in the biocontrol of *Botrytis cinerea*. *European Journal of Plant Pathology* 105(1): 77–189.
22. FAO (Food and Agriculture Organization). http://www.fao.org/3/y3557e/y3557e03.htm (consulted 05 November 2020).
23. FAO. http://www.fao.org/plant-health-2020/about/en (consulted 05 November 2020).
24. Fiorentino N, Ventorino V, Woo SL, Pepe O, De Rosa A, et al. 2018. *Trichoderma*-Based biostimulants modulate rhizosphere microbial populations and improve N uptake efficiency, yield, and nutritional quality of leafy vegetables. *Frontiers in Plant Science* 9: 743. https://doi.org/10.3389/fpls.2018.00743.
25. Ghosh SK, Pal S. 2016. Entomopathogenic potential of *Trichoderma longibrachiatum* and its comparative evaluation with malathion against the insect pest. *Leucinodes Orbonalis*. *Environmental Monitoring and Assessment* 188: 37. https://doi.org/10.1007/s10661-015-5053-x.
26. Harman GE. 2000. Myths and Dogmas of Biocontrol Changes in Perceptions Derived from Research on

Trichoderma harzianum T-22. *Plant Dis* Apr 84(4): 377–393. doi: 10.1094/PDIS.2000.84.4.377

27. Harman GE, Howell CR, Viterbo A, Chet I, Lorito M. 2004. *Trichoderma* species - opportunistic, avirulent plant symbionts. *Nature Reviews Microbiology* 2: 43–56.
28. Harman GE, Obregon MA, Samuels GJ, Lorito M. 2010. Changing models for commercialization and implementation of biocontrol in the developing and the developed world. *Plant Disease* 94: 928–939. https://doi.org/10.1094/ PDIS-94-8-0928.
29. Harman GE, Uphoff N. 2019. Symbiotic root-endophytic soil microbes improve crop productivity and provide environmental benefits. *Scientifica* 9106395: 25. https://doi.org/10.1155/2019/9106395.
30. Hermosa R, Belén RM, Cardoza RE, Nicolás C, Monte E, Gutiérrez S. 2013. The contribution of *Trichoderma* to balancing the costs of plant growth and defense. *International Microbiology* 16: 69–80.
31. Hermosa R, Viterbo A, Chet I, Monte E. 2012. Plant-beneficial effects of *Trichoderma* and of its genes. *Microbiology* 158: 17–25.
32. Hidangmayum A, Dwivedi P, Katiyar D. et al. 2019. Application of chitosan on plant responses with special reference to abiotic stress. *Physiol Mol Biol Plants* 25: 313–326. https://doi.org/10.1007/s12298-018-0633-1
33. Howell CR. 2002. Cotton seedling preemergence damping-off incited by *Rhizopus oryzae* and *Pythium* spp and its biological control with *Trichoderma* spp. *Phytopathology* 92: 177–180.
34. Howell CR. 2003. Mechanisms employed by *Trichoderma* species in the biological control of plant diseases: The history and evolution of current concepts. *Plant Disease* 87: 4–10.
35. Howell CR. 2006. Understanding the mechanisms employed by *Trichoderma virens* to effect biological control of cotton diseases. *Phytopathology* 96: 178–180.
36. Lombardi N, Caira S, Troise AD, Scaloni A, Vitaglione P, et al. 2020a. *Trichoderma* applications on strawberry plants modulate the physiological processes positively affecting fruit production and quality. *Frontiers in Microbiology* 11(1364): https://doi.org/10.3389/fmicb.2020.01364.
37. Lombardi N, Salzano AM, Troise AD, Scaloni A, Vitaglione P, et al. 2020b. Effect of *Trichoderma* bioactive metabolite treatments on the production, quality, and protein profile of strawberry fruits. *Journal of Agricultural and Food Chemistry* 68(27): 7246–7258. https://doi.org/10.1021/acs.jafc.0c01438.
38. Lorito M, Woo SL. 2015. *Trichoderma*: A multipurpose tool for IPM, pp 345-353. In: Lugtenberg B (Ed.), *Principles of Plant-Microbe Interactions,- Microbes for Sustainable Agriculture.* Springer, 448 pp. https://doi.org/10.1007/978-3-319-08575-3_36.
39. Lorito M, Woo SL, Ge H, Monte E. 2010. Translational research on *Trichoderma*: From 'omics to the field. *Annual Review of Phytopathology* 48: 395–417. https://doi.org/10.1146/annurev-phyto-073009-114314.
40. Marra R, Lombardi N, d'Errico G, Tisi J, Scala G, et al. 2019. Application of *Trichoderma* strains and metabolites enhances soybean productivity and nutrient content. *Journal of Agricultural and Food Chemistry* 67: 1814–1822. https://doi.org/10.1021/acs.jafc.8b06503.
41. Martínez-Medina A, Fernandez I, Lok GB, Pozo MJ, Pieterse CM, Van Wees SC. 2017. Shifting from priming of salicylic acid-to jasmonic acid-regulated defences by *Trichoderma* protects tomato against the root knot nematode. *Meloidogyne Incognita. New Phytologist* 213: 1363–1377. https://doi.org/10.1111/ nph.14251.
42. Mastouri F, Björkman T, Harman GE. 2010. Seed treatment with *Trichoderma harzianum* alleviates biotic, abiotic, and physiological stresses in germinating seeds and seedlings. *Phytopathology* 100: 1213–1221.
43. Medeiros, H., Araújo Filho, J., Freitas, L. et al. 2017. Tomato progeny inherit resistance to the nematode *Meloidogyne javanica* linked to plant growth induced by the biocontrol fungus *Trichoderma atroviride. Sci Rep* 7: 40216 https://doi.org/10.1038/srep40216
44. OECD. https://www.oecd.org/agriculture/understanding-the-global-food-system/how-we-feed-the-world-today/ (consulted 05 November 2020).
45. Papavizas GC. 1985. *Trichoderma* and *Gliocladium*: Biology, ecology, and potential for biocontrol. *Annual Review of Phytopathology* 23: 23–54. https://doi.org/10.1146/annurev.py.23.090185.000323.
46. Pascale A, Vinale F, Manganiello G, Nigro M, Lanzuise S, Ruocco M, et al. (2017). *Trichoderma* and its secondary metabolites improve yield and quality of grapes. *Crop Prot.* 92: 176–181. doi: 10.1016/j.cropro.2016.11.010
47. Pieterse CMJ, Leon-Reyes A, Van der Ent S. & Van Wees SCM. 2009. Networking by small-molecule hormones in plant immunity. *Nat Chem Biol* 5: 308–316.
48. Pingali PL 2012. Green revolution: Impacts, limits, and the path ahead. *Proceedings of the National Academy of Science* 109: 12302–12308. https://doi.org/10.1073/pnas.0912953109.

49. Pocurull M, Fullana AM, Ferro M, Valero P, Escudero N, et al. 2020. Commercial formulates of *Trichoderma* induce systemic plant resistance to *Meloidogyne incognita* in tomato and the effect is additive to that of the Mi-1.2 resistance gene. *Frontiers in Microbiology* 10: 3042. https://doi.org/10.3389/fmicb.2019.03042.
50. Podder D, Ghosh SK. 2019. A new application of *Trichoderma asperellum* as an anopheline larvicide for eco friendly management in medical science. *Science Reports* 9: 1108. https://doi.org/10.1038/s41598-018-37108-2.
51. Poveda J, Abril-Urias P, Escobar C. 2020. Biological control of plant-parasitic nematodes by filamentous fungi inducers of resistance: *Trichoderma*, mycorrhizal and endophytic fungi. *Frontiers in Microbiology* 11: 992. https://doi.org/10.3389/fmicb.2020.00992.
52. Roatti B, Perazzolli M, Gessler C, Pertot I 2013. Abiotic stresses affect *Trichoderma harzianum* T39-induced resistancetodownymildewingrapevine. *Phytopathology* 103: 1227–1234. 10.1094/PHYTO-02-13-0040-R
53. Sharon E, Bar-Eyal M, Chet I, Herra-Estrella A, Kleifeld O, Spiegel Y. 2001. Biological control of the root-knot nematode *Meloidogyne javanica* by *Trichoderma harzianum*. *Phytopathology* 91: 687–693.
54. Shoresh M, Mastouri F, Harman G. 2010. Induced systemic resistance and plant responses to fungal biocontrol agents. *Annual Review of Phytopathology* 48: 21–43.
55. Sood M, Kapoor D, Kumar V, Sheteiwy MS, Ramakrishnan M, et al. 2020. *Trichoderma*: The "secrets" of a multitalented biocontrol agent. *Plants* 9: 762. https://doi.org/10.3390/plants9060762.
56. Tucci M, Ruocco M, De Masi L, De Palma M, Lorito M. 2011. The beneficial effect of *Trichoderma* spp. on tomato is modulated by the plant genotype. *Molecular Plant Pathology* 12: 341–354. https://doi.org/10.1111/J.1364-3703.2010.00674.X.
57. United Nations. 2013. World Population Prospects: The 2010 Revision and World Urbanization Prospects: The 2011 Revision. Available at http://esa.un.org/unpd/wup/unup/index_panel1.html.
58. Vargas WA, Mandawe JC, Kenerley CM. 2009. Plant-derived sucrose is a key element in the symbiotic association between *Trichoderma virens* and maize plants. *Plant Physiology* 151: 792–808.
59. Van der Ent S, Van Hulten M, Pozo MJ, Czechowski T, Udvardi MK, Pieterse CMJ and Ton J. 2009. Priming of plant innate immunity by rhizobacteria and β-aminobutyric acid: differences and similarities in regulation. *New Phyt.* 183: 419–431 doi: 10.1111/j.1469-8137.2009.02851.
60. Vinale F, Sivasithamparam K. 2020. Beneficial effects of *Trichoderma* secondary metabolites on crops. *Phytotherapy Research* 2020: 1–8. https://doi.org/10.1002/ptr.6728.
61. Vinale F, Nigro M, Sivasithamparam K, Flematti G, Ghisalberti EL, Ruocco M, Varlese R, Marra R, Lanzuise S, Eid A, Woo SL, Lorito M. 2013. Harzianic acid: a novel siderophore from *Trichoderma harzianum*. *FEMS Microbiology Letters* 347: 103–106.
62. Vinale F, Sivasithamparam K, Ghisalberti E, Marra R, Woo SL, Lorito M. 2008. *Trichoderma*-plant-pathogen interactions. *Soil Biology and Biochemistry* 40: 1–10.
63. Viterbo A, Landau U, Kim S, Chernin L, Chet I. 2010. Characterization of ACC deaminase from the biocontrol and plant growth-promoting agent *Trichoderma asperellum* T203. *FEMS Microbiology Letters* 305: 42–48.
64. Weindling R. 1932. *Trichoderma lignorum* as a parasite of other soil fungi. *Phytopathology* 22: 837–845.
65. Weindling R. 1934. Studies on a lethal principle effective in the parasitic action of *Trichoderma lignorum* on *Rhizoctonia solani* and other soil fungi. *Phytopathology* 24: 1153–1179.
66. Wells HD, Bell DK, Jaworski CA. 1972. Efficacy of *Trichoderma harzianum* as a biocontrol for *Sclerotium rolfsii*. *Phytopathology* 62: 442–447.
67. Woo SL, Ruocco M, Vinale F, Nigro M, Marra R, et al. 2014. *Trichoderma*-based products and their widespread use in agriculture. *The Open Mycology Journal*, Special Issue on Fungal-Fungal Interactions: 71–126. https://doi.org/10.2174/1874437001408010071.
68. Yedidia I, Benhamou N, Chet I. 1999. Induction of defense responses in cucumber plants (*Cucumis sativus* L.) by the biocontrol agent. *Trichoderma harzianum*. *Applied and Environmental Microbiology* 65: 1061–1070.
69. Zhang J, Howell CR, Starr JL. 1996. Suppression of Fusarium colonization of cotton roots and fusarium wilt by seed treatments with *Gliocladium virens and Bacillus subtilis*. Biocontrol Science and Technology 6: 175–187.

CHAPTER 42

Companies and Organizations Active in Agriculture and Horticulture

Ben Lugtenberg

Institute of Biology, Leiden University, Sylviusweg 72, 2333 BE Leiden, The Netherlands

42.1 INTRODUCTION

Good microbes can promote plant growth either directly or indirectly (Lugtenberg and Kamilova, 2009; Lugtenberg, 2015; Lorito et al., 2010). Direct plant growth promotion can be achieved by several mechanisms. (i) Biofertilization: the microbe delivers a specific nutrient to the plant, such as nitrogen, iron, or phosphorus. Some microbes, such as the fungi *Mycorrhiza* (see chapter 39) and *Trichoderma* (see chapter 41), improve plant growth and robustness by using more than one mechanism. (ii) Root growth stimulation: the microbe produces the hormone auxin. (iii) Plant stress control: the microbe produces the enzyme ACC (1-aminocyclopropane-1 carboxylate) deaminase, which reduces the level of the stress hormone ethylene (see chapter 34).

Indirect plant growth promotion occurs by protection of the plant against pathogens, most often fungi but also bacteria and viruses. It can be based on the following mechanisms. (i) Antibiosis: the beneficial microbe produces an antibiotic which stops growth of the pathogen. (ii) Competition for nutrients and niches: the beneficial microbe outcompetes the pathogen for both nutrient acquisition and niche occupation (colonization). (iii) Induced systemic resistance: the beneficial microbe activates and sensitizes the entire plant for defense against diverse pathogens. (iv) Colonization of hyphae of pathogenic fungi, thereby disabling their activity (see chapter 35).

In contrast to resistance against chemicals, resistance against microbial disease control agents hardly plays a role because a beneficial microbe often uses multiple mechanisms of action. This makes resistance a rare phenomenon: in case resistance occurs against one mechanism, the beneficial microbe can still act through the other mechanism(s).

Many beneficial microbes have been marketed. The role of companies in the complicated, time-consuming, and expensive process from harvesting seeds to food on the plate will be discussed in this chapter.

Seed companies such as ENZA Zaden (https://www.enzazaden.com/) harvest the seeds. In order to protect them against pathogens and give them a good start after germination, seeds are often coated with beneficial microbes which will colonize the germinating seedlings and outcompete the pathogens. Therefore the seeds, certainly those for high value crops, often go to specialized firms, e.g., Incotec (https://www.incotec.com/en-gb) to make them ready for sowing. There they are cleaned of harmful microbes, for example by heat treatment. They are coated with a film which can contain plant protection products and plant growth stimulators, including microbes. Moreover, seeds are "primed" so that they germinate fast and homogeneously. They are also made uniform in size, given a sufficient weight, and their surface is smoothed, all in order to become compatible with sowing equipment.

Microbes are produced by specialized companies. Before large scale production starts, it is important to

Good Microbes in Medicine, Food Production, Biotechnology, Bioremediation, and Agriculture, First Edition. Edited by Frans J. de Bruijn, Hauke Smidt, Luca S. Cocolin, Michael Sauer, David Dowling, and Linda Thomashow.

know the following information. Is there a market need? Has the strain been patented? Is the strain safe for humans and the environment so that it can be registered as a product? Can it be grown easily? What are the target crops? Indoor or outdoor, or post-harvest control? Who are the competitors? What are the market size and the distribution channels? What should the price be in order to make the product profitable?

The industrial manufacturing process includes scaling up growth, formulation, and packaging. The goals of formulation include the delivery of the microbes in good condition to their targets and the enhancement of their activity. The microbes can be applied as a powder, as granules, or as a liquid. Formulations may contain additives such as carriers, nutrients, stickers, protectants, and emulsifiers. Endospores are easier to keep alive than vegetative cells. This explains why a spore-former such as *Bacillus* is more often used under practical conditions than *Pseudomonas* despite the fact that *Pseudomonas* spp. can be excellent biocontrol strains under laboratory conditions. Microbial exo-metabolites that are directly or indirectly involved in beneficial effects of strains can also be included in the formulations (Kamilova et al., 2015).

Registration of a product is a very complex process. Therefore, most microbe-producing companies involve specialized consultants who help the producer to assess the relevant data and facilitate the interaction with the authorities by writing the relevant dossiers (see Section 42.6.).

42.2 EXAMPLES OF IMPORTANT MICROBES

42.2.1 Arbuscular Mycorrhizas

Mycorrhizal fungi live in symbiosis with more than 80% of the land plants (Lanfranco and Fiorilli, in this volume). They provide the plant with water and nutrients. In return, they receive carbon from the plant. Mycorrhizal fungi can be considered as an extension of the root system. Since they are thinner than roots, they can reach narrower sites and because they extend far beyond the root system, they can provide the plant with nutrients from a much larger area than the root itself (Lugtenberg and Kamilova, 2009; see chapter 39).

Companies selling mycorrhizal inoculants include Agrifutur (http://www.agrifutur.com), MycAgro Lab (http://www.mycagrolab.com/31.html), Mycosat (https://www.micosat.it), Premieretech (http://www.mykepro.com/about-myke-pro.aspx), The Energy and Resources Institute (TERI) (https://www.teriin.org/), and Mycorrhizal Applications (https://mycorrhizae.com/). The latter website is very informative on how the symbiosis works.

42.2.2 Bacillus

Bacillus spp. can be used as biocontrol agents and plant growth promoters. The model strain for these purposes is *Bacillus velezensis* FZB42 (Fan et al., 2018). It has many mechanisms of action against plant-pathogenic viruses, bacteria, fungi, and nematodes. It contains 13 gene clusters encoding secondary metabolites with putative antibiotic action. Formulations based on spores of *Bacillus* spp., especially of *B. subtilis*, are widely used against pathogens.

ABiTEP GmbH (www.abitep.de), now a member of the Andermatt Biocontrol Group, is a producer of microbial products for use in agriculture and gardening. The *Bacillus* products for plant growth promotion and biocontrol have an excellent scientific basis. The same applies for the production, downstream processing and formulation (spray drying, freeze drying, liquid formulation). The products of ABiTEP GmbH are different *Bacillus* strains used as soil enhancers or as biostimulants (FZB24®, RhizoVital®42, RhizoVital®P45, and RhizoVital®C5) or mixtures of *Bacillus* strains for nutrient mobilization and organic waste degradation (ABIFERT® and ABiSAN®). *Bacillus*-based products are also sold by other companies such as LLC Bisolbi-Inter, Bayer, which sells SERENADE® since it acquired Agraquest and Total Bio, whichsells a product for composting based on *B. subtilus*, *B. amyloliquefasciens*, and *B. pumilus*.

42.2.3 Bacillus thuringiensis

Bacillus thuringiensis (*Bt*) is widely used as a biocontrol agent to control insect pests (De Maagd, 2015). It produces several toxins, most of which are localized in parasporal crystals. The product is a dry mixture of spores and crystals. *Bt* products are not only used in high margin crops such as vegetables and fruits, but also against forestry pests. Several decades ago, genes encoding *Bt* toxins began to be introduced into crop plants. Companies selling *Bt* products include the Indian firm UPL, Certis, Becker, and Valent BioSciences (http://wrir4.ucdavis.edu/events/2017_SLR_Meeting/Presentations/Insect%20Management/1%20Eldridge%20R%20Insecticides%20Bt%20Valent

%20BioSciences.pdf). This website explains nicely how the product works. More information about Bt is presented in Chapter 38 by Federici in this volume.

42.2.4 Nitrogen-fixing Bacteria

Plants require relatively large amounts of the element nitrogen (N) for growth. Chemical N-fertilizers are often used to support plant growth. Leguminous plants, such as alfalfa, pea, and soybean, can use nitrogen from the air (N_2) through an interaction with microbes. The latter can be free-living, such as *Azospirillum brasiliense*, or can live inside the plant, such as *Bradyrhizobium japonicum*. For more details about nitrogen fixation by bacteria, see Chapter 37 by de Bruijn and Hungria in this volume.

Soybean is one of the most important legume crops in the world. Approximately 80% of the world's soybean is produced in the Americas. Due to a symbiosis with *Bradyrhizobium*, so-called nodules are formed in which atmospheric N_2 is reduced to NH_4 which can be used by the plant as a source of nitrogen. This process reduces the need for synthetic N-fertilizers (Chang et al., 2015).

Many companies sell nitrogen-fixing bacteria as inoculants. Agrifutur was already mentioned. Agrocete sells inoculants based on *Bradyrhizobium japonicum* and *B. tropici*. They also sell *Azospirillum brasiliense* (https://www.agrocete.com.br/pt). Total Bio sells a product based on *B. japonicum* (https://www.totalbio.com.br/). Rizobacter sells nitrogen fixing bacteria with special attention given to resistance against high temperature and drought (www.rizobacter.com.ar).

(Brady)rhizobia produce Nod-factors (de Bruijn and Hungria, this book, Chapter 37) and arbuscular mycorrhizae produce Myc factors (Lanfranco and Fiorilli, this book, Chapter 39), both of which are essential for plant infection by these symbiotic microorganisms. These molecules have similar structures and are lipo-chitooligosaccharides (LCOs). Studies by Stewart Smith and colleagues have shown that LCOs, combined with the host specific *(Brady)rhizobium* bacterium and applied to legume seeds, lead to improved plant growth parameters and to greater crop yields. Field trials have shown that this effect can also be achieved by LCOs alone and can be extended to non-legumes such as corn (Smith et al., 2015). Commercial products for enhanced crop yield were launched for soybeans, peanuts, peas/lentils, alfalfa, and for non-legumes such as corn. The product line based on LCOs was recently extended by the product BioRise 2 for both soybeans and corn. It contains not only an LCO (for enhancing root growth) but also the fungus *Penicillium bilaiae* (https://www.agprofessional.com/article/acceleron-introduces-biorise-2-seed-treatment). It is claimed that *Penicillium bilaiae* releases phosphate bound in the soil, resulting in an 85% enhanced mycorrhizal colonization and that this in turn results in an increased root volume and an increase in the uptake of water and nutrients via the roots.

Already since the beginning of genetic modification of plants, it was predicted that cereals would be constructed in which the nitrogen-fixing ability of microbes would be expressed. To my knowledge all such attempts have failed so far.

42.2.5 Phosphate-solubilizing Bacteria

(Barea and Richardson, 2015) Plants can take up free phosphate (P_i) only as HPO_4^{2-} and $H_2PO_4^-$. Free P_i is only 1% of the total phosphate in soil. Free P_i diffuses very slowly, resulting in a P_i -depletion zone around the root. This problem is usually overcome by adding chemical P-fertilizer (P_2O_5) from phosphate mines. This has two disadvantages. Firstly, only 10% of it is taken up by the plant. Part of the remainder pollutes groundwater and oceans. Secondly, the amounts in phosphate mines are limited: the estimated reserve is for only 80 to 100 years. Fortunately, there are microbes present in soil which can release P_i from organic and inorganic P-sources present in soil. These sources are mainly apatite (inorganic) and phytate (organic). Inorganic P is solubilized by producing organic acids, whereas organic P is enzymatically degraded by acid and alkaline phosphatases and phytases.Phosphate-solubilizing microbes include strains of *Bacillus*, *(Brady)rhizobium*, and *Pseudomonas*, and of *Mycorrhizae* and *Trichoderma*. Rizobacter (www.rizobacter.com.ar) sells *Pseudomonas fluorescens* as a P-solubilizer for fertilization of crops, such as wheat, corn, and sunflower.

42.2.6 Rhizobium rhizogenes

Agrobacterium tumefaciens causes crown gall disease, which can result in severe losses in fruit and wine growing orchards. A strain of an avirulent *A. radiobacter*, now called *Rhizobium rhizogenes*, controls the disease by producing agrocin84, which selectively kills certain classes of tumorigenic *Agrobacterium* strains (Kerr and Htay, 1974). The product is commercially sold as "NOGALL" by companies such as Evergreen (https://www.evergreengrowers.com/nogall-group-nogall.html) and Bio-Care Technology (http://bio-caretechnology.com/nogall-biological-control-treatment/).

42.2.7 Trichoderma

Trichoderma is one of the most versatile beneficial microbes applied in agriculture (Lorito and Woo, 2015; Woo and Lorito, this book, Chapter 41). Many of its strains promote plant growth and control diseases. The latter property can be based on a combination of antibiosis, root colonization, competition for nutrients, mycoparasitism, and induced systemic resistance. *Trichoderma* cannot only kill fungi but also nematodes. *Trichoderma*-based products are distributed in over 100 countries. India alone has over 300 products on the market. The total market for *Trichoderma*-based products is estimated to exceed one billion dollars worldwide.

BioWorks (https://www.bioworksinc.com/) sells RootShield PLUS +, which is based on two *Trichoderma* strains. Koppert Biological Systems (https://www.koppert.com/) is one of the major producers. Rizobacter (www.rizobacter.com.ar) sells *Trichoderma* fungi effective against the important pathogenic fungus *Fusarium graminearum*.

42.2.8 Pseudomonas

Products based on *Pseudomonas* strains and their metabolites have been marketed. They have been described in Chapter 35 of this volume by Linda Thomashow.

42.3 GREAT DISCOVERIES

A potentially valuable product is often patented to protect its application. Discoveries at universities are published, often in specialized journals. The journal *Biocontrol* (https://www.springer.com/journal/10526) is an initiative of the IOBC (International Organization for Biocontrol). The journal *Molecular Plant-Microbe Interactions* (MPMI) publishes primary research and reviews on the molecular biology and genetics of pathological, symbiotic and associative interactions between microbes and plants. It is an activity of the IS-MPMI (International Society of Molecular Plant-Microbe Interactions) (www.ismpmi.org). This organization also sponsors a bi-annual congress. The magazine *2BMonthly Newsletter* (http://www.2bmonthly.com) is a newsletter on biocontrol and biostimulants and is focussed on industrial interests.

In the following, three examples of great discoveries which I consider as role model examples for students, are discussed.

42.3.1 "The Indiana Jones of Fungus Hunters"

That is how Gary A. Strobel was described by *Forbes Magazine*. For me, he is the king of bioprospecting. Based on information from native people, he collected medicinal plants from rain forests (Strobel and Daisy, 2003). He isolated the plant's fungal endophytic microbes and identified their bioactive beneficial compounds. Examples are antibiotics (including volatile ones), antiviral compounds, immunosuppressive compounds, antioxidants, and agents against cancer, diabetes, and insects. He owns over 50 patents. He has licenced more than 20 strains to industry. With his son Scott, a professor at Yale University, he took a group of students to the Amazonian rain forest (Strobel and Strobel, 2007). Recently, Gary Strobel edited a special issue of the *Journal of Fungi* entitled "Fungal Endophytes in Plants" (Strobel, 2017).

42.3.2 From Volatiles in the Rain Forest to Rocket Fuel and Plant Protection in Desert Soil

Examples of some striking discoveries on volatiles from novel fungal endophytes are the following:

Muscodor albus produces a mixture of more than 28 volatile compounds. It kills a broad range of plant- and human-pathogenic fungi and bacteria. A patent was issued for its use for the treatment of human waste (Strobel et al., 2001) and was commercialized as a portable environmental toilet. Marrone BioInnovations took a licence on *Muscador albus* for the sterilization of soils (Mycofumigation), as an alternative for methylbromide. The latter compound degrades the ozone layer and therefore has been banned.

Muscodor vitigenus appeared to produce naphthalene as its sole volatile. We have already known for a long time that naphthalene is an insect repellent, the active ingredient in common mothballs.

Gliocladium sp. appeared to be the producer of the volatile anti-oomycete and anti-fungal antibiotic [8] annulene. It is unstable, flammable, and explosive. In WW II this compound was used as a rocket fuel (Stinson et al., 2003).

Taxol is the first billion dollar anti-cancer drug. It is a natural product from the yew tree. It inhibits mitosis by inhibiting depolymerization of microtubules. Taxol is also produced by the fungal endophyte *Taxomyces andreanae* (Stierle et al., 1993).

David Ezra, a former postdoc of Gary Strobel, and Orna Liarzi isolated the endophytic fungus (*Daldinia* cf.

concentrica) from an olive tree in Israel. It produces a biologically active mixture of secondary volatile metabolites effective against post-harvest fungal pathogens (Liarzi et al., 2016). These volatiles are also active against soil-borne fungal pathogens and nematodes. Ezra and Liarzi, together with Luxembourg Industries (https://luxembourg.co.il/) and Copia Agro (http://copia-agro.com), are developing a soil disinfection pre-planting product (SoilFender) for the control of fungal pathogens, nematodes, weeds, and pests. It is based on *trans-2-octenal*, the most active metabolite of the mixture from the endophyte (Liarzi et al., 2020).

42.3.3 Using Fungi to Adapt Plants to Climate Change

Thermotolerance generated by plant-fungal symbiosis (Redman et al., 2002) is an academic discovery which resulted in extremely valuable applications. Panic grass in Yellowstone National Park grows at 65°C. It appeared to harbour the fungus *Curvularia protuberata*. When the plant was cured of the fungus, neither plant nor fungus alone could grow above 38°C. After the plant was reinoculated with the fungus, growth at 65°C was restored. So, for thermotolerant growth, plant and fungus need each other (Márquez et al., 2007). Or, seen from the plant's point of view, the plant needs the fungus to adapt to high temperatures.

Inspired by these findings, Regina Redman and Rusty Rodriguez asked whether fungi are also involved in adaptation of plants to other stress conditions. Indeed, similar grasses grown under stress conditions such as high salinity, aridity, or cold harboured similar, so-called class 2 fungal endophytes. They designated the phenomenon "habitat-adapted symbiosis" (Rodriguez et al., 2008, 2009).

They then hypothesized that endophytic fungi can be used to adapt crop plants to climate change. Indeed, they did so, not only for rice (Redman et al., 2011) and other cereals, but also for pulses, vegetables, and industrial crops. The technology looks simple. A suspension of fungal spores is sprayed on seeds. After growth, the plant contains the fungus between its cells. The microbial products are named *BioEnsure*®. Based on these discoveries, Rodriguez and Redman established two companies. The first, Adaptive Symbiotic Technologies (AST) (http://www.adaptivesymbiotictechnologies.com/), produces the microbial product *BioEnsure*®, which enables crops to resist drought, salinity, and temperature stress. Average yield increases of 2–6% under low stress conditions and 10–50% under high stress conditions have been obtained. Successful field experiments were performed with corn, rice, soybean, wheat, and cotton. Safety tests showed that the fungus does not grow into the seed or fruit and does not establish in soil.

The second, Symbiogenics (http://www.symbiogenics.org/) is a non-profit corporation. The objectives are to perform basic research and to educate high school science teachers. Results of this amazing success story were presented by Rusty Rodriguez in the TEDxRainier Talk "Unlocking the power of symbiosis in a warming world" (https://www.youtube.com/watch?v = jJslxcgo-Gg). To my big surprise, most plant breeders I met are not aware of this incredibly efficient method of breeding plants for plant adaptation to other habitats. And it does not require genetic modification!

42.3.4 Agrobacterium Tumefaciens, the Natural Genetic Engineer

The plant pathogen *A. tumefaciens* causes crown gall disease. It contains a large, transmissible plasmid, called the Ti (tumor-inducing) plasmid. During infection, the bacterium delivers a segment of its Ti plasmid, the T-DNA, into plant cells. The bacterium has been converted to a vector for the genetic modification of plants. To this end, the T-DNA was modified to remove the tumor-inducing genes. Strains with such a disarmed "helper" Ti plasmid can deliver any DNA of interest into plants (Hoekema et al., 1983). Cloning vectors with such artificial T-DNAs are called binary vectors. Because of its ease and low cost, *Agrobacterium*-mediated DNA delivery has become the most popular vector system in plant research. Since patenting, it has been used widely to create genetically modified crops. Later, *Agrobacterium* was also shown to transform yeast and fungi in the laboratory. This has led to the wide application of *Agrobacterium*-mediated transformation in (especially plant-pathogenic) fungi, which are difficult to transform otherwise (Bundock et al., 1995).

42.4 EFFICIENT TRANSLATION OF FUNDAMENTAL RESEARCH TO INDUSTRY

Cooperation between academia and companies is not always smooth. The two groups also often hold separate conferences. A pleasant exception is the biannual symposium series "miCROPe" organized by Angelica Sessitsch in Vienna (https://www.micrope.org/). Good examples of a smooth translation of fundamental science into applications are the following:

42.4.1 Nanjing Agricultural University

At Nanjing Agricultural University, the group "Organic Fertilizer and Soil Microbiome" in the College of Resources and Environmental Sciences, under direction of Qirong Shen, investigates how beneficial microorganisms, added in bio-organic fertilizers to soils, colonize the rhizosphere and roots to promote plant growth and help host plants to become antagonistic to soil-borne pathogens. They apply their results directly on their own premises, for example, for the production of composts. The technologies they developed have been applied in over 1000 companies. The group also has several field stations where they test the biocontrol effects of their bio-organic fertilizers on crops such as banana, watermelon, tomato, and cucumber.

42.4.2 Grasslanz Technology Ltd

Ryegrass can be toxic for grazers. The New Zealand research institute AgResearch Limitedhas managed to eliminate toxicity. They found that toxicity is caused by the common fungal endophyte *Epichloë festucae* var. *lolii*, which produces three major secondary metabolites. (i) Lolitrem B, a neurotoxin. (ii) Ergovaline, a toxin causing vasoconstriction. (iii) Peramine, an insect feeding deterrent. The first two compounds cause diseases in animals grazing the grass: lolitrem B causes ryegrass staggers in sheep and the vasorestriction caused by ergovaline results in heat stress, causing fescue toxicosis in cattle. Peramine is beneficial for grass because it protects it against a major insect pest, Argentine stem weevil. AgResearch Limited has isolated *Epichloë* strains lacking the mammalian toxins but producing the insect deterrent, thereby ensuring plant persistence (Johnson et al., 2013; Lugtenberg et al., 2016). Similar discoveries were made by AgResearch in the US, where cattle grazing on tall fescue pastures harboring *Epichloë* endophytes contracted fescue toxicosis.

The fundamental research results of AgResearch Limited were the basis for establishing Grasslanz Technology Ltd. (www.grasslanz.com). This company invests into applied research and development to deliver effective forage and microbial technologies to end-user farmers. Their technologies are licensed to appropriate companies for manufacture and sale. Grasslanz has invested in the development of *Epichloë* endophyte technologies for improving resistances to biotic and abiotic stresses of temperate grasses such as ryegrass and tall fescue, resulting in the branded products AR1 and AR37 for use with ryegrass, and MaxQ II and MaxP, for use with tall fescue. They are widely used in New Zealand, Australia, USA, and South America.

42.4.3 LLC Bisolbi-Inter

LLC Bisolbi-Inter (http://www.bisolbi.ru) is a spin-off company of the All-Russia Research Institute for Agricultural Microbiology (ARRIAM), established by Igor Tikhonovich. It conducts research on agricultural microbiology, environmental biotechnology, and bioremediation of technologically polluted lands. The Bisolbi brand unites a group of biotechnological companies involved in research and development for agricultural production. Bisolbi-Inter produces biopesticides based on *Bacillus amiloliquefaciens* and *B. subtilis*, microbiological fertilizers based on *B. subtilis*, *B. pumilus* and *B. megaterium* and growth stimulators based on *Azotobacter chroococcum*, *Pseudomonas fluorescens*, and *Paenibacillus* sp. It has registered products in Russia, Asia, Europe, and Africa. LLC Bisolbi-Inter also supports conversion of agricultural farms from conventional to organic production.

42.5 EXAMPLES OF COMPANIES PRODUCING MICROBES

Agrifutur(http://www.agrifutur.com/en) has its headquarters in Italy and focuses on products targeted on growth promotion of legumes and cereals. For legumes, they sell individual rhizobia and combinations of rhizobia with mycorrhizal fungi. Products for cereals are based on combinations of mycorrhizal fungi, *Trichoderma harzianum*, and various bacteria. They also perform collaborative research on pest control with the beneficial fungi *Beauvaria*, *Fusarium*, and *Metarhizum* and the bacterium *Pantoea agglomerans*.

Indigo Agriculture https://www.bpia.org/member-company/indigo-agriculture) is an agriculture company that uses discoveries in the plant microbiome to enhance crop's resistance to pests, diseases, and environmental stresses. Indigo was created by the founders of Seres Therapeutics, a company focused on human health through microbiome enhancement.

Koppert Biological Systems (https://www.koppert.com/) produces many different microbes for the cultivation of food and ornamental crops. It is a family company with third generation Kopperts on the board. Back in

1967, the grandfather, Jan Koppert, was a dedicated cucumber grower. However, he became allergic to chemicals. Therefore, he introduced natural enemies, beneficial insects, and mites, for crop protection. This became the core business of the company, next to bumblebees for pollination. Presently, microbial products including bacteria and fungi are being produced to combat diseases and pests, strengthen crops, and improve the uptake of nutrients. Koppert has more than 25 subsidiaries all over the world and its products are being applied in more than 100 countries. Specialized consultants visit growers, inspect their crops, and offer specific advice. A very successful microbial product, Trianum, is a biological fungicide based on the fungus *Trichoderma harzianum* strain T-22, which combats various soil-borne diseases and promotes plant growth (see chapter 41).

Mycorrhizal Applications (http://mycorrhizae.com/) is a wholly owned subsidiary of Valent BioSciences Corporation. They produce and market species of mycorrhizal fungi that accelerate plant vigor by greatly increasing the surface absorbing area of roots, producing a healthier and more efficient root system. These specialized beneficial fungi colonize plant roots to create an extensive mycelial network within the surrounding soils, dramatically increasing access to soil volume for maximized nutrient and water uptake to optimize crop yield (see chapter 39).

42.6 EXAMPLES OF COMPANIES FOR REGISTRATION AND OTHER FORMS OF CONSULTATION

42.6.1 Requirements for Registration

For field crops, the EU (European Union) is divided into three evaluation zones: the Northern, Central, and Southern Zones. For greenhouse uses, seed treatment and post-harvest treatment, the EU is considered as a single zone.

The authorization process of plant protection products consists of evaluation of dossiers (i) of the active microorganism and (ii) of the formulated product. Evaluation is first at the EU level followed by the national level.

To receive approval of the active microorganism the following steps must be successful. (i) The dossier must be supported by a Rapporteur Member State (RMS). (ii) The European Food and Safety Authority (EFSA) and the other member states provide their assessment about risk for humans and environment. (iii) The EFSA releases its scientific report. (iv) The Standing Committee for Food Chain and Animal Health, consisting of national risk managers, votes about approval. (v) The EC (European Commission) adopts this decision and publishes a Regulation on approval. Altogether, this process takes 2.5 to 3.5 years.

For evaluation of a single formulated product, the Zonal RMS evaluates the dossier in which the emphasis is on efficacy. Their authorization can be used for mutual recognition by the member states in the same zone. The zonal authorization process takes 12–18 months. Authorization based on mutual recognition takes a minimum of 120 days. Application fees vary substantially among countries.

Although application fees for registration of biopesticides in general are lower than for chemicals, the total fee is typically between 80000 and 200000 Euro ($97000–$122000). There are no harmonized fees.

In the USA, the registration process for biopesticides includes (i) federal registration in the biopesticide branch of the US-Environmental Protection Agency (EPA) and (ii) registration at the state level. After submission, the biopesticide branch of US-EPA can review dossiers of the active microorganism as well as of up to four of its formulated products. The evaluation focuses on safety of the microorganism and its formulated products to humans and the environment. The legal timeframe for the evaluation is 18 months. The application fee is $30000. However, it is possible to get an exemption from the EPA application fee via Inter-Regional Research Project N4, supporting "minor use pesticides."

Registration at the state level includes registration of the product either by the Department of Pesticide Regulation or by the Department of Agriculture of individual states. Data requirements vary tremendously. For instance, California requires the submission of a complete dossier including local efficacy trial reports, whereas in Ohio, filling in a simple administrative application form is sufficient. Application fees are in the range of a few hundred to a few thousand dollars. Timeframes vary from a few days to several months.

Genetically modified (GM) microbes cannot be registered in the European Union for political reasons based on the attitude of the public and influenced by non-scientific organizations. The attitude of the US is more rational: it is based on the assumption that regulation should focus on the nature of the products, rather than the process in which they were produced. See https://www.loc.gov/law/help/restrictions-on-gmos/usa.php#Release. Nevertheless no GM microbes have been registered in the USA.

42.6.2 Companies Specialized in Registration

APIS (applied insect science) (https://appliedinsect-science.com/) specializes in regulatory testing for microbial pesticides and provides a dedicated contract research facility for biopesticides. It has a good laboratory practice (GLP) laboratory providing guidelines and studies for active microbes and formulated products. They test samples for microbial contaminants and perform "phys-chem tests" and studies on storage stability. Moreover, they study aquatic and mammalian toxicology and terrestrial ecotoxicology. The latter include tests on honeybees and bumblebees, as well as on non-target arthropods and earthworms.

Linge Agroconsultancy (www.lingeagroconsultancy.nl) is a small consulting company for plant protection product registration. They are specialized in the registration of active substances, biological and chemical pesticides, fertilizers, biostimulants, adjuvants, and biocides in Europe.

Knoell (www.knoell.com) is one of the pioneers in the compilation of registration dossiers. It has subsidiaries in Europe, USA, Brazil, and several countries in Asia. Knoell not only provides project management, study management, and monitoring, but also prepares and defends dossiers for biological and chemical active substances and plant protection products. Biopesticide products include microorganisms. Knoell's regulatory team helps to identify missing information in the form of data gap analysis, and provides scientific and regulatory support for all dossier sections. These include: identity of a microorganism, biological properties, physical-chemical properties of the product, analytical methods, toxicology, residues, environmental fate and behavior, ecotoxicology, and efficacy. The company helps to define the required data package and advises the most appropriate regulatory approaches for client products. Recently, Knoell expanded their services to include regulatory advice to companies operating in the area of biostimulants and other fertilizing products, drawing on their experience in registration of (biological) plant protection products and industrial chemicals.

42.6.3 Small General Consultation Firms

Consultants providing general information are often independent retired professionals with wide experience in the complex business of agriculture and horticulture.

Helmut Junge was the co-founder of ABiTEPGmbH, a company specialized in the production of *Bacillus* plant protection products. After his retirement, he started the consultancy firm "Dr. Junge Bioberatung" (www.bioberatung.org) with the goal of advising farmers, gardeners, and companies in their efforts to develop or use organic products. His strength is the evaluation of complex cultivation methods and the development of integrated strategies for environmentally friendly crop production.

In 2015, Dr. Rainer Borriss, professor emeritus of Bacterial Genetics at Humboldt University Berlin (https://www.researchgate.net/profile/Rainer_Borriss), assumed responsibility of the firm Nord Reet UG (https://www.nordreet.de/), which then started the business "Bacillus consulting." The company is specialized in the isolation, metabolic, and genetic characterization of plant-associated bacilli with plant growth promoting activity and the potential to be developed as biofertilizers, bio-stimulants, and biocontrol agents. Based on rich experience in research and application, they advise companies to develop their own bioformulations for use in sustainable agriculture. Contact: info@nordreet.de.

42.7 ACQUISITIONS, MERGERS, AND ALLIANCES

Many chemical firms have recognized the demand for biological products and have become active in the biological crop protection business. For example, BASF acquired Becker Underwood, Bayer acquired Agraquest, Novozymes and Monsanto established The BioAg Alliance (which was later terminated), and Syngenta acquired Pasteuria. Recently, the biological firm Andermatt bought another biological firm, ABiTEP.

Also patents and commercial products easily change from one owner to another. For example, the finding by Stewart Smith that LCOs promote plant growthwas initiated in The Nitragin Co., Milwaukee, USA. Ownership by international companies subsequenty moved to Allied Signal (USA), then Lipha (France), followed by Merck (Germany) and Novozymes (Denmark). Novozymes formed an alliance with Monsanto which was later terminated. The latter firm was recently purchased by Bayer. Presently, Acceleron, part of the Bayer group, sells products based on LCOs: http://www.acceleronsas.com/About/Pages/default.aspx).

42.8 ORGANIZATIONS IN BIOCONTROL SCIENCE

IBMA (https://www.ibma-global.org) (International Biocontrol Manufacturers Association) is the voice of the biocontrol technologies industry.

BPIA (https://www.bpia.org/) (Biological Products Industry Alliance) promotes the development of safe and effective biological products including biopesticides, biostimulants, and biofertilizers as beneficial tools for commercial agriculture, forestry, golf courses, home gardens, horticulture, ornamentals and public health. The website is informative about various categories of microbial products for agriculture and horticulture.

ANPII (http://www.anpii.org.br/) is an association of microbe-producing companies, mainly based in South America, which include Agrocete, Lallamand, Novozymes, Rizobacter, and Stoller.

IFOAM – Organics International (https://directory.ifoam.bio/) is an organization of companies involved in organic agriculture.

ACKNOWLEDGMENTS

This chapter is dedicated to José Miguel Barea who passed away in 2018. He was a world leader of arbuscular mycorrhiza research and of microbial phosphate solubilization. I thank Paola Bonfante, Rainer Borriss, John Caradus, Vladimir Chebotar, Ruud de Maagd, David Ezra, Paul Hooykaas, Helmut Junge, Mariangela Hungria, Faina Kamilova, Orna Liarzi, Choong-Min Ryu, Stewart Smith, Gary Stacey, Linda Thomashow, and David Weller for advice.

REFERENCES

1. Barea JM, Richardson AE. 2015. Phosphate mobilization by soil microorganisms. In: Lugtenberg B (Ed.), *2015. Principles of Plant-microbe Interactions*. Springer, ISBN 978-3-319-08574-6, 225–234.
2. Bundock PA, den Dulk-Ras A, Beijersbergen A, Hooykaas PJJ. 1995. Trans-kingdom T-DNA transfer from *Agrobacterium tumefaciens* to *Saccharomyces Cerevisiae*. *The EMBO Journal* 14: 3206–3214.
3. Chang W-S, Lee H-I, Hungria M. 2015. Soybean production in the Americas. In: Lugtenberg B (Ed.), *Principles of Plant-microbe Interactions*. Springer, ISBN 978-3-319-08574-6, pp. 393–400.
4. De Maagd RA. 2015. *Bacillus thuringiensis*-based products for insect pest control. In: Lugtenberg B (Ed.), *Principles of Plant-microbe Interactions*. Springer, Cham, Switzerland, ISBN 978-3-319-08574-6, pp. 185–192.
5. Fan B, Wang C, Song X, Ding X, Wu L, et al. 2018. *Bacillus velezensis* FZB42: The Gram-positive model strain for plant growth promotion and biocontrol. *Frontiers in Microbiology* 9: article 2491.
6. Hoekema A, Hirsch P, Hooykaas PJJ, Schilperoort RA. 1983. A binary plant vector strategy based on separation of *vir* and T region of the *Agrobacterium tumefaciens* Ti plasmid. *Nature* 303: 179–180.
7. Johnson LJ, Bonth ACM, Briggs LR, Caradus JR, Finch SC, et al. 2013. The exploitation of Epichloae endophytes for agricultural benefit. *Fungal Diversity* 60: 171–188.
8. Kamilova F, Okon Y, de Weert S, Katja Hora K. 2015. Commercialization of microbes: Manufacturing, inoculation, best practice for objective field testing, and registration. In: Lugtenberg B (Ed.), *Principles of Plant-microbe Interactions*. Springer, Cham, Switzerland, ISBN 978-3-319-08574-6, pp. 319–327.
9. Kerr A, Htay K. 1974. Biological control of crown gall through bacteriocin production. *Physiological Plant Pathology* 4: 37–44.
10. Liarzi O, Bar E, Lewinsohn E, Ezra D. 2016. Use of the endophytic fungus *Daldinia cf. concentrica* and its volatiles as biocontrol agents. *PLoS ONE* 11(12): e0168242. https://doi.org/10.1371/journal.pone.0168242.
11. Liarzi O, Benichis M, Gamliel A, Ezra D. 2020. Synthetic volatile mixture, of a fungal origin, controls *Sclerotium rolfsii*, both in vitro and in soil. *Pest Management Science* 76: 2068–2071. https://doi.org/10.1002/ps.5744.
12. Lorito M, Woo S. 2015. *Trichoderma*: A multi purpose tool for integrated pest management. Ch 36. In: Lugtenberg B (Ed.), *Principles of Plant-microbe Interactions*. Springer, ISBN 978-3-319-08574-6, pp. 345–354.
13. Lorito M, Woo SL, Harman GE, Monte E. 2010. Translational research on *Trichoderma*: From 'omics to the field. *Annual Review of Phytopathology* 48: 395–417.
14. Lugtenberg B. (Ed.). 2015. *Principles of Plant-microbe Interactions – Microbes for Sustainable Agriculture*. Springer, ISBN 978-3-319-08574-6; 448.
15. Lugtenberg B, Kamilova F. 2009. Plant-growth-promoting rhizobacteria. *Annual Review of Microbiology* 63: 541–556.

16. Lugtenberg BJJ, Caradus JR, Johnson LJ. 2016. Fungal endophytes for sustainable crop production. *FEMS Microbiology Ecology* 92: 1–17.
17. Márquez LM, Redman RS, Rodriguez RJ, Roossinck MJ. 2007. A virus in a fungus in a plant: Three-way symbiosis required for thermal tolerance. *Science* 315: 513–515.
18. Redman RS, Kim YO, Woodward CJDA, Greer C, Espino L, et al. 2011. Increased fitness of rice plants to abiotic stress via habitat adapted symbiosis: A strategy for mitigating impacts of climate change. *PLoS ONE* 6: e14823.
19. Redman RS, Sheehan KB, Stout RG, Rodriguez RJ, Henson JM. 2002. Thermotolerance generated by plant/fungal symbiosis. *Science* 298: 1581.
20. Rodriguez RJ, Henson J, Van Volkenburgh E, Hoy M, Wright L, et al. 2008. Stress tolerance in plants via habitat-adapted symbiosis. *The ISME Journal* 2: 404–416.
21. Rodriguez RJ, White Jr JF, Arnold AE, Redman RS. 2009. Fungal endophytes: Diversity and functional roles. *New Phytologist* 182: 314–330.
22. Smith S, Habib M, Diaz-Zorita M. 2015. LCO applications provide improved responses with legumes and non-legumes. In: de Bruijn FJ (Ed.), *Biological Nitrogen Fixation*. pp. 1077–1085.
23. Stierle A, Strobel GA, Stierle D. 1993. Taxol and taxane production by *Taxomyces Andreanae*. *Science* 260: 214–216.
24. Stinson M, Ezra D, Hess WM, Sears J, Strobel GA. 2003. An endophytic *Gliocladium sp.* of *Eucryphia cordifolia* producing selective volatile antimicrobial compounds. *Plant Science* 165: 913–922.
25. Strobel G, Daisy B. 2003. Bioprospecting for microbial endophytes and their natural products. *Microbiology Molecular Biology* 67: 491–502.
26. Strobel GA. (Ed.). 2017. *Journal of Fungi. Special Issue: Fungal Endophytes in Plants. MDPI*. Switzerland, 231.
27. Strobel GA, Dirkse E, Sears J, Markworth C. 2001. Volatile antimicrobials from *Muscodor albus*, a novel endophytic fungus. *Microbiology* 147: 2943–2950.
28. Strobel SA, Strobel GA. 2007. Plant endophytes as a platform for discovery-based undergraduate science education. *Nature Chemical Biology* 3: 356–359.

Index

Note: Page references in *italics* refer to figures and tables.

Good Microbes in Medicine, Food Production, Biotechnology, Bioremediation, and Agriculture, First Edition. Edited by Frans J. de Bruijn, Hauke Smidt, Luca S. Cocolin, Michael Sauer, David Dowling, and Linda Thomashow.

C